AF680593

BAHAMAS

IN

WORLD WAR II

BAHAMAS

IN

WORLD WAR II

A Military Chronology

1939-1945

ERIC WIBERG

Island Books

by the same author

about Bahamas & Bermuda:

U-Boats in the Bahamas
U-Boats off Bermuda
Mailboats of the Bahamas
Swan Sinks
Drifting to the Duchess (book)
Drifting (script, with Paolo Pilladi)
Lyford Cay International School History (2020)
The Lyford Cay Foundation History (2021)
Åke Wiberg, Bahamas Wibergs (with Mats Larsson)

to October, 2020:

U-Boats in New England
Boston Harbor
Tanker Disasters
Published Writing
Round the World in the Wrong Season
Napoleon's Battles (with Felix Wiberg)
Juvenilia
Scars, Sea Stories
Yacht Voyages
Travel Diaries

in production:

Tankers East of Suez
Surviving St. George's
First Fifty

private:

Diaries, 1984-2019
Albums, Seven Volumes

Published by Island Books, Boston, MA, USA
© Eric Troels Wiberg, 2020

ISBN (hardback) 978-0-9983759-3-9 , 0-9983759-3-4
ISBN (ebook) 978-0-9983759-4-6 , 0-9983759-4-2
Library of Congress Control Number, PCN: 2017951460

All rights reserved. No part of this publication may be reproduced in any manner without the prior written permission of the publisher, except in the case of brief quotations embodied in articles or reviews. This publication owes its compilation to the hard work of hundreds of other researchers, editors, designers, cartographers, and writers, who are acknowledged in the bibliography and notes.

For information, or to contact the author please email eric@ericwiberg.com
Book cover, layout, editing by Abdul Rehman Qureshi writingpanacea@gmail.com

Printed in the United States of America

for our parents,

Jane McDermid Wiberg & Hon. Sten Anders Wiberg, Esq.

They moved to Nassau in 1965: she from New York, he from Stockholm.

Their four children could not ask for more supportive, encouraging mentors.

Forewords

Air Commodore G. R. Pitchfork MBE, BA, FRAeS, RAF (Rtd.)

In the introduction to this book Eric Wiberg begins with the quote: "Nothing happened in the Bahamas during World War II." It is a quote that I too have come across and I suspect that it is a common thought amongst many. After reading this book, that illusion will disappear forever.

During my years of RAF service, and the years that I have been studying and researching RAF history, most of my attention has been directed to the campaigns and actions in Europe and the Mediterranean. The Battle of Britain, the role of Bomber Command and the crucial involvement of the USAAF have provided opportunities for deep research.

Shamefully, the gallant exploits of the men of Coastal Command, and their role in the Battle of the Atlantic and the defeat of the U-boat scourge, receives less attention. Perhaps it is a case of "out of sight – out of mind" since maritime patrols involve a vast amount of time on patrol over the far away oceans and almost always with very little action. Hence, the routine of uneventful, but extremely necessary, patrols appears less exciting and therefore has received little attention. However, to the historian these activities were crucial to the survival of the western world that simply could not have carried the war to the enemy without the essential supplies that crossed the Atlantic.

In researching these operations I became aware of the role of No. 111 Operational Training Unit (OTU) based at Nassau in the Bahamas. This led me to a closer look at their activities and in doing so, I also became aware of the vital role of RAF Ferry Command, the essential life-line for the movement of the aircraft from factories in North America to the squadrons based in Great Britain and the Middle East. This often-overlooked contribution to the Allied victory, and aviation history, relied heavily on facilities made available in the Bahamas. The very presence of these two organizations inevitably led to the establishment of other vital support facilities and thus the build-up of facilities and personnel in the region.

These activities outlined above were familiar to me, although not in the detail that the author has provided, but there is a great deal more in this book that is new to me and adds greatly to my knowledge of the scale of operations in this more distant land.

The degree and extent of Eric Wiberg's research is astonishing and it is difficult to imagine that he has left any stones unturned. There are aspects of actions in World War Two that will be completely unknown to the great majority of people. In recording these events, the author has not only created a very interesting read, but he has produced an invaluable research document for historians and for future generations and I commend him for producing this book.

Graham Pitchfork
Winchcombe, England

The Bahamas are ranking high among the beautiful places on earth. Since decades, each year thousands of tourists enjoy the amenities of the archipelago. But from February 1942 on, the peaceful islands became a World War II hotspot, when the flow of unarmed merchant ships in Bahamian waters was suddenly and deliberately attacked by U-boats of the evil empire of Nazi Germany and its Italian Ally, after their governments had declared a state of war with the United States of America two months before. In the following months the heinous submarine attacks killed hundreds of civilian merchant sailors and sent 130 ships with nearly 700,000 gross tons to the bottom of the ocean.

For more than two years the German and Italian U-Boat commanders wrought havoc among American and Allied shipping before they were eventually driven out of these waters by Allied naval and air forces, which ultimately sank no less than three of these underwater marauders. In the meantime, Britain and the United States worked hard to turn the Bahamas into an important link in the chain of military and supply convoys to transport vital raw materials and an ever-growing stream of troops, weapons and equipment from America's arsenal of democracy to the frontlines in Europe and elsewhere on the world. Wartime activities and the presence of many thousands of military serviceman and women between 1942 and 1945 catapulted the once quiet and sleepy islands into modernity.

Eric Wiberg, long-time Bahamas resident and well-known for his previous publications on Axis U-Boat operations in American waters and other maritime topics, diligently collected a mass of information on many aspects of the wartime effort in and around the Bahamas during World War II. This global conflict spread rapidly into the daily life of the islanders thousands of miles away from the frontlines in Europe. Apart from actual combat actions, the war effort resulted in numerous training, reconnaissance and building activities in the area.

The reverberations of this episode in history can still be felt or seen today. Researchers interested in the history of the Bahamas or naval operations during World War II in the Caribbean Sea will greatly appreciate the various data from many archives, neatly assembled in easy-to-read tabular form. Anyone who has ever spent time for research inside of an archive building knows fairly well how much effort and motivation it takes to compile the encyclopaedic information in this book. Because information is the foundation of understanding, this book will serve this purpose magnificently.

Dr. Axel Niestlé
Author & Historian
Dubendorf, Germany

Eric Wiberg and I both grew up in the Bahamas, but unlike him, I was old enough to live through World War II. We always felt safe and secure, thousands of miles from the fighting, or so I thought.

Recently, I met a 96-year-old gentleman who assured me that he had been in the Bahamas in 1943, as a crew-member of an Italian submarine. He explained that Italy had switched sides, and they came as co-combatants. I told Eric about this visit, and he was able to locate the sub in question as having been in Guantanamo Bay, at the appropriate time, and discuss it with the Italian naval officer in person.

Eric and I have stayed in touch, and he has kept me current on his research and findings. I am constantly impressed by what he finds and envy his ability and dedication to research.

This book opens a new chapter on World War II.

Captain Paul C. Aranha
Pilot, Author, Historian
Nassau, N.P., Bahamas

Acknowledgements

With a project of such scope, it is best to thank hundreds of contributors by city, region, and country. I have been fortunate to visit archives in the US, Bahamas, Germany, Bermuda, the UK, and Australia, but to really crack mysteries, I find it is more efficient to recruit the help, often voluntary, of local researchers with vastly more local knowledge.

Australia:

The Australian War Memorial holds valuable resources of Bahamian history, including a very rare complete set of *The Liberator*, which was a unique weekly magazine printed in Nassau with *The Nassau Guardian* and serving the thousands of personnel in several RAF bases in the Bahamas. I was also able to purchase numerous excellent photos and paintings from their collection. It is an extraordinary resource for Royal Australian Air Force personnel who served in the Bahamas. Specifically, Jane Robertson, Acting Online Reference Manager, Research Centre, and Ms. Kellidie Saunders at the Australian War Memorial, which our family visited from Nassau in 1989, were very helpful.

Special thanks to B24 historian Robert Livingstone of Queensland, Australia, whose father served at No. 111(C)OTU in Nassau in World War II, starting with B24's then moving to B25s. Bob has been extraordinarily generous with his expertise, time, photographs, and in helping make the aircraft silhouettes as accurate as possible. Bob is the author of *Under the Southern Cross* and *Strike Back*, maintains b-24.weebly.com, advised on the Hollywood movie *Unbroken*, and is published in *Warbirds Worldwide, Air Forces Monthly, Flightpath, Aeroplane, Australian Flying*, and *The Aviation Historian*.

Bahamas:

The Bahamas Historical Society membership, from President Andrea P. Major, Jim and Anne Lawlor, Prof. Stephen Aranha and Capt. Paul C. Aranha, Dr. Gail Saunders, Ms. Delaney and team at the National Archives, Ms. Willamae Johnson, Librarian for the LJM Maritime Academy, and Ms. Tanya Hunt-Major at Christ Church Cathedral. At the University of the Bahamas (formerly College of the Bahamas), I thank the following professors, primarily of history: Tracey Thompson and professors Curry, Rolle, Pintard-Newry and Ulentin. Stuart Cove of eponymous diving firm, was helpful in confirming the location of RAF bomber wrecks off Simms Point (Lyford Cay), and Clifton Bay, even relating that he refurbished an item. Many individuals have provided help: Stephen Holowesko hung a sign recognizing the assistance given by the Red Cross at his distillery and brewery named Pirate Republic on Woodes Rogers Way.

The Antiquities, Monuments, and Museums Corporation hosted me for several discussions over years, individuals at the Lands and Surveys Department are always helpful and enthusiastic, at the Abaco National Trust Forest, ranger Marcus Davies and his wife Siren were exemplary hosts, as was everyone in the settlement of Crossing Rocks and Sandy Point during several trips there. Orjan Lindroth in Crossing Harbor and Marcus Mitchell with Darby Island, Exuma, enabled me to see important sights from the air. *The Nassau Tribune* and *Nassau Guardian* were very helpful with archival materials. Mari Anderson, Fritz Damler, Rev. Newton Williamson, and Rev. Capt. Collie's daughter, also a Reverend, were extremely helpful. In Nassau Pericles Maillis, Craig Symonette, Ronald G. Lightbourn, Amanda Coulson, Jane Lloyd, and innumerable other people including my parents for their enthusiasm for history and support of my interests, and for attending all my lectures, were inspirational and helpful. I cannot name them all. Whether it was visiting sites in Abaco, Grand Bahama, hosted by Peter Johnston and family at Little Harbor, or being shown around San Salvador, where Greeks fled from Italians shelling their ship, or Hope Town, or any of myriad corners of the Bahamas where World War II landed, including lovely Andersons at Acklins, everyone has been helpful and engaged.

Finally there are so many areas of expertise in the Bahamas: Dr. Gail Saunders and Mr. Murray on forts and landmarks, Wayne Neely on historic hurricanes, Anne and Jim Lawlor on Harbour Island, and Williams, Darius author *Rail & Locomotive History of the Bahamas*, Marcus Mitchell and Ray d'Arville on marine salvage, others on archaeology, or diving, or flying: one could write a book thanking them for their books! I thank the book sellers, at the Linen Shop, who place my book in the window, Logos, Buy the Book, University of the Bahamas, Lyford Cay Marina, at the airport, on Bay Street, the hotels and resorts, and any I missed, I think you very much. Neil Sealey at Media Enterprises and his missed colleague Larry Smith If I have omitted anyone, over a decade after beginning, I do apologize.

Barbados:

The family which owned the schooner *Vivian P. Smith*, Canadian-built and sunk east of the Turks & Caicos with a cargo of salt with hapless donkeys on deck, were very helpful and provided a lovely photo of her sister ship as well. Specifically, Erica Luke, a member of the multi-generational shipping firm Eric Hassell & Son Ltd. of Bridgetown, and a prominent member of the Caribbean Shipping Association.

Bermuda:

I thank Elena Strong, President of the Bermuda Heritage Museum, Dr. Edward Harris, Martin Buckley at the Bookmart at Brown & Co. in Hamilton, Tom Singfield and co-author Ewan Partridge, historians, Jonathan Land Evans, author of a book on Bermuda in World War II, and Ms. Ellen Hollis in the National Library. Several other Bermudians have helped in innumerable other ways, however that was not for the main focus of this book, but a previous one, including Bernews.com.

Brazil:

I thank Capt. Plinio R. Calenzo and his spouse, Akza Janifer Zugliato Calenzo of the *Instituto Brasileiro de Rebocagem* and *Rebocar É Preciso* in Rio de Janeiro, Brazil, for translating my 2016 book, *U-Boats of the Bahamas and Turks & Caicos*, the research for which formed the basis of this compendium.

Canada:

Capt. Jerry and Mrs. Charla Mason in Victoria, British Columbia, is a retired US Navy aviator in anti-submarine missions, and has again proved that a book of this scope cannot be accurately written without the support that the two of them so generously supply. Thanks are not enough for their stellar effort translating and sharing German U-boat patrol, KTB, or daily war logs, for and with folks like me. It gives the project depth and colour. In Montreal, Platon Alexiades continued to supply translation and other helpful input regarding Italian submarines from the BetaSom flotilla which formed its vanguard, and Roman Hruby made the first book on Bahamas a reality by volunteering to edit it.

Cayman Islands:

The extended family of the captain of the small steamer *Mariana*, lost off Mayaguana to U-Boat attack in calm conditions yet with the loss of all persons despite the Germans survivors getting clear of the ship, was much appreciated, despite the tragedy. I thank the relatives of Capt. Ivan Elroy Hurlestone specifically for sharing background, stories and images.

Cuba:

Maximino Gomez Alvarez of AIDNHC, the Association for the Investigation and Dissemination of the Naval History of Cuba, who has supported this project with innumerable detailed studies of bases in Cuba in World War II, Cuban servicemen and merchant mariners killed in the region at that time, and the names of individual sailors killed by U-boat attack.

Germany:

Kai Steenbuck, Mrs. Annemarie Bredow, and Peter Monte at the Deutsches U-Boot-Museum, (*Stiftung Traditionsarchiv Unterseeboote E.V.), Archiv für Internationale Unter-wasserfahrt* in Cuxhaven-Altenbruch, where continuously helpful and supportive, including spending time with me in Hamburg. Dr. Axel Niestlé continued to be an excellent, thorough and informative source of intelligence and data on the loss of U-84 under Uphoff to a Bermuda-based bomber under Lt. Thomas R. Evert, USNR. Dr. Niestlé is the leading authority on the final fates of all German U-boats in World War II, and his work supporting an international team lead to the award of a Distinguished Flying Cross at the U.S. Senate Building in Washington, DC in November, 2019. I am extremely indebted to Dr. Niestlé , author of the definitive text *German U-Boat Losses During World War II: Details of Destruction*, for volunteering to write the *Foreword*.

Iceland:

Guðmundur Helgason, founder of Uboat.net provides an invaluable service, and has for a long time. Without this site, which is now one of the few remaining like it, daily positions, commander and flotilla details and boat and officer histories and the names of allies killed in action would not be so easily accessible, and might even become forgotten.

India:

Aditi Rae performed stellar work extracting and typing every individual's name for the names project, as well as typing in hundreds of pages from both the Gulf Sea Frontier as well as RAF 111 O.T.U. base diary into readable form, and merging them. She has been a stalwart support for over half a decade, and I cannot thank her enough. Sanchit Goel, founder of Onomatopo boutique design firm, is also in New Delhi. Since about 2014, he and his team have produced at least 16 book covers for Island Books, covering a wide gamut. As reluctant as most authors would be to admit it, she and Abdul in Pakistan largely "wrote," or typed out, the essential core Operational Record Books and recitation of names and ranks which became the bedrock of these texts, including complex interpretations of the Gulf Sea Frontier ORBs, which require considerable technical understanding

Italy:

Cristiano D'Adamo of the history website Regia Marina.net dedicated to the Royal Italian Navy, always provides excellent accurate information. Also, since it has the .eu ending but exists as a voluntary database globally, Tony Allen and the many volunteers who maintain records and images of nearly 200,000 wrecks at wrecksite.eu deserve thanks for the instant and accurate information on ship losses.

Jamaica:

Oftentimes a researcher will take a risk by buying a book which seems a good fit, and that the contents will be topical to their research, only to be disappointed within a few minutes of the book's arrival; typically evinced by the fact the word "Bahamas" and the book index never made each other's acquaintance. A surprise exception has been an online archive of Kingston's *The Daily Gleaner*. This newspaper is the oldest Jamaican daily. Since it was founded in 1834, it is also the oldest continually published, English language newspaper in the Western Hemisphere.

Though I did not know of it before this project, *The Daily Gleaner* has proved itself of invaluable merit and accuracy in corroborating obscure Caribbean vessels like *Content S.*, the *Foundation Aranmore* (which ran her over off Cuba), the schooners *Ival* and *Vivian P. Smith*, and several others. Their records of ship movements in and out of Jamaica, which was central in the Caribbean, between Panama and the Windward Passage, provided a pleasant trove using sites like newsapaperarchive.net.

New Zealand:

The New Zealand Air Force replied to an inquiry sharing whatever they had on their soldiers who served in the Bahamas in World War II. Sydney Waters' book on the

New Zealand Shipping Company's struggles in World War II include the SS *Matakana*, wrecked in the Plana Cays before U-boats arrived, and the SS *Opawa* lost northeast of Bermuda.

Pakistan:

Abdul Rehman Qureshi of Writing Panacea has been absolutely a rock of reliability and outstanding production, discipline and organization. Without he and Aditi in New Delhi, I could not conclude a project like this in less than two decades. Abdul is always level headed, results-focussed, on or under budget and he always delivers on every aspect, from typing in a military and nautical lingo that is new to him, to organizing interior and cover design and layout, and formatting a large and complex composition of 50 or more moving parts. Abdul has patiently and always accurately and professionally cooperated closely with me to produce over 15 books with well over 5,000 pages and a million words in recent years. His capacity to see large complex projects like this through continues to greatly impress me, and I rely upon him heavily.

Switzerland:

Rainer Kolbicz, responsible for Allied loss data for Uboat.net, does a very thorough job.

Trinidad & Tobago:

Gaylord T. M. Kelshall's seminal work *U-Boat War in the Caribbean*, from 1988 has inspired me then and to this day. It is an extraordinary work, especially when one considers that Kelshall had no internet. His assistant and loyal friend, Jerome Lee at their museum in Chaguaramas, Trinidad, has always helpfully relayed messages to and from Mr. Kelshall, as well as sharing with me a tremendous database on West Indians who volunteered to serve Britain and her allies in both world wars, much of which is online.

Turks & Caicos Islands:

Dr. Michael P. Pateman, Director of the Turks & Caicos National Museum Foundation and manager of the Turks and Caicos Museum, has always quickly replied to inquiries and published several articles relating to this research in their prestigious quarterly newsletter, *The Astrolabe*. There were no bases, two ships lost, and one large RAF aircraft lost in TCI.

The United Kingdom:

Tony Cooper and volunteers at the convoyweb.org.uk, were extraordinarily helpful at providing cheerfully prompt and highly detailed and accurate details of all convoys from around the world that interacted with those in the Bahamas area in World War II. The story of how Arnold Hague, John K. Burgess, Dominique Lemaire, Don Kindell and many others saved hundreds of thousands of pages of documents from destruction, devoting a lifetime to contributing in ways which were not allowed to be remunerated, is truly exceptional. It is typical that when I visited him Tony drove hours to the airport and hosted me. He and colleague, data-guru Mike Holdoway extend their gratitude to an original Convoy Project colleague, Rodney Agutter, as well as the professionals at the Guildhall Library in London for their inestimable assistance and support over the years.

In London, Simon Fowler continues to show his mettle as an extraordinary researcher in The National Archives in Kew outside London and also adroitly finding other resources. After about a decade of working together, Simon is able to warn me of likely dead-ends before I start down them. Institutions including the Imperial War Museum and the Royal Air Force Museum have been extraordinarily helpful, as has the Commonwealth War Graves Commission, which maintains the peaceful and very helpful Tower Hill Memorial commemorating 36,087 allied civilian merchant seamen lost at sea during both World Wars. It is along the Thames at Trinity Square, near St. Katharine's Docks in London.

Derek Waller, author, former officer in the RAF, and an expert on post-war U-Boat dispositions, introduced me to the RAF historical community, in particular Wing Commander Colin Cummings of the RAF Historical Society. Louise Clark and Sonia Taylor at the Royal Air Force Museum in Cosford were helpful, as was Val Hurlstone in their London museum. Robin Grant ,who visited Nassau from Scotland and with wife Sandra and Capt. Paul Aranha laid flowers at the RAF Cemetery on 1st May 2018 to commemorate his father Robert Grant who served with OTC111 during WW2 , Jack Burgess BEM , OTC111 and Scottish Saltire Aircrew Association along with all other RAF personnel who served during WW2. Helping the Bahamas Historical Society to keep alive the RAF history.

In 2020, Robin Grant was kind enough to introduce me to an acclaimed authority on RAF aviation history - Air Commodore Graham Pitchfork, MBE, BA FRAeS. After studying this book, the Air Commodore graciously agreed to pen the Foreword, providing a level of aviation expertise and an RAF historical perspective which the author is entirely unqualified to provide.

Air Commodore Pitchfork served in the Royal Air Force for 36-years. He graduated from the RAF College Cranwell before serving globally as a navigator, instructor and senior air staff officer. He is the author of 14 books and contributes to numerous magazines and historical journals and has delivered over 300 presentations. He also writes the aviation entries for the *Oxford Dictionary of National Biography*. The Air Commodore serves on the committee of the RAF Historical Society and is the Archivist of the Aircrew Association. He is a Fellow of the Royal Aeronautical Society since 1991, and has received numerous awards, including the 2012 Guild of Air Pilots and Air Navigators Award for Aviation Journalism.

It is an honour and privilege to have Air Commodore Pitchfork lend his tremendous experience and insight to this project.

The United States:

Mike Constandy of Westmoreland Research was instrumental in gathering all the many hundreds of pages of the Gulf Sea Frontier (GSF) diaries for this project, going back years, as well as the Eastern Sea Frontier and all the individual survivor statements from Allied ships in the region. Dr. Robert M. Browning, Jr., former USCG Historian, provided insight and help regarding disposition of US-flagged ships lost. Diver Mike Barnett shared his expertise on Allied ships lost on Florida's east coast. The USCG and US Navy Historical Branch have been very helpful and supportive, particularly by sharing the GSF or Gulf Sea Frontier daily diaries, which are available at the National Archives and Records Administration (NARA), Washington, DC.

Since we were introduced by a U.K. Admiralty Hydrographer in the Republic of Panama in around 2012, Robert Eller Prat, who has decades of experience as a cartographer with the prestigious National Geographic organization has produced perhaps close to 50 custom charts of individual battles, survival stories, and macro-level overviews of entire zones of conflict a million square miles in breadth. For this book Bob gets credits for all charts and a few graphs as well. His work product is self-evident and thus gets its own entire section. Thank you for your efficiency, friendliness and fairness over the years, Bob.

Many others in the USA, from the airfield historian for what was Morrison and is now West Palm Beach Airport, to local historians along the Florida coast and in Georgia and Virginia as well have provided insight, detail, and perspective. A few times a month I receive inquiries from the 850 or so blog articles posted online, many of them informative and proving the adage that if you want to learn about a topic, publish a book on it, and folks will quickly correct what you got wrong! Since I cannot name everyone who helped, all I can try to do is produce a readable book.

Uruguay:

I thank the National Navy of Uruguay for publishing my research on the SS *Maldonado*.

Over 11 years and at least 15 countries, there are bound to be more individuals who helped years ago with other related books, whether *U-Boats in the Bahamas*, *U-Boats off Bermuda*, *Drifting to the Duchess,* or *Swan Sinks*, who I have omitted to name here, to whom I here apologize. I greatly appreciate every moment, every email, every encouragement provided by family as well.

Please see a detailed *Bibliography* and online resources at the end for further recognition.

Guide to Using this Book

To serve those with more than a general interest, this book is designed to be a detailed cross-reference that will help answer the questions that many family members of participants, as well as ship and aviation history enthusiasts, will have. Cross-referencing the contents of this book is likely to be easiest for those equipped with the name/s of either a person's, a ship, aircraft, U-boat or *Sommergibili* they served with. The name or number of the military unit they served with, which includes chapters of the Red Cross and Imperial Order of the Daughters of Empire. Most importantly perhaps, are the dates they were believed to have served in the Bahamas theatre.

Armed with some or all of that information, here are specific groups one can search to find more information on in the pages to follow. Please try to remember that the book itself is an index. In this case, the "index" insomuch as there is one, is really the 50-item Table of Contents at the beginning. If we put the data fields in appendices, there might be more appendix than main text, so all relevant information is in the main text to make cross-referencing easier.

While women both served and died aboard allied merchant ships in this theatre at this time, the mortality among the women is low in terms of percentage of the overall number. In cases when masculine phraseology ("he," "the men," etc.) is employed, it is not always accidental, since the data drives those terms. In other words, in scenarios where no women were present, as with aboard military aircraft and ships, the language employed is not always gender-inclusive, as to make it so might be inaccurate or misleading.

Merchant Mariners:

Fortunately, the first, last, even middle names as well as ranks and final ship served upon have been carefully tabulated by many sources, including the US Navy, many historians, volunteers, merchant marine academies and government agencies. As a result, many names are obtainable at sites like uboat.net (free, no password required), wrecksite.eu (ditto), and national resources like a volunteer site for Norwegian sailors of World War II named warsailors.org and a variety other sources from Dutch, Canadian, British, US, Australian and many more national archives.

As a result, over 95% of the nearly 2,000 Allied merchant mariners from all over the world who perished in the waters in and around the Bahamas are obtainable and shared here. Simply go to the *Killed in Action* section. All names are listed alphabetically, with next names, rank, and vessel. It was not necessary to include a date, as each ship in the *Merchant Ship Losses* has its own date of sinking. We can never know precisely when many of these poor souls perished; if in the actual sinking and attack, or weeks later on a raft or lifeboat, so to be consistent the date of death is given, in this study at least, as the date of the sinking.

Survivors of sunken ships are not listed. While every effort is made to name and honour the dead from all causes and sides, it would be overly burdensome and inaccurate to attempt to list every merchant mariner who traversed a million square wartime miles over more than half a decade.

Once the reader has a person's name and/or that of a ship, they may then obtain the dates, ports, departure ports and cargo details in the *Convoys Section*, at least in the event the ship convoyed. Then one can read the *Daily Diary* to find out which submarine sank the ship, the U-boat or Italian submarine, whether other allied vessels were involved in rescuing other men, and so forth. That also enables one to seek the ship name on one of several charts in the *Charts Section*, and see the comparative death rate in relation to other ships.

Furthermore, there is extensive information in the *Convoys Section* that enables readers to look up the ship names of all vessels known to have convoyed through Bahamas and see a list of all convoys that ships that a particular person sailed in, as well as to and from what ports, and when. In fact, some ships have up to twenty convoys which can then be cross-referenced by date in the daily diary section. The details of the short regular convoys between Miami and Nassau are within the *Daily Diary* section, and are basically predictable; ships like the Kelly family's *Ena K.*, *Betty K.*, and *Jean Brillant* and others took turns going back and forth like clockwork, literally thousands of times. Details of convoy escorts, delays, VIP passengers, and reports of them being overdue are all noted.

In summary; with a bit of doggedness one can track:

- the name of a merchant marine sailor through all convoys their ship sailed on,
- basic details of the attack, such as general location in text, chart and spreadsheet
- the name of the U-boat or Italian submarine which sank it,
- the fate of other sailors on the merchant ship
- and the fates of the ships which rescued survivors

All with a bit of persistence. Much of the information is in these pages. As a note, there were female crew members on allied merchant ships, though not many; they mostly served in the galley and stewards' department. Women appear more often aboard passenger ships, staying in paid cabins. They did die, in some instances, such as Canadian *Lady Ships*, and the SS *San Jacinto*, sunk on its way from New York to San Juan with the loss of seven women, a full half of the number lost overall.

Finally, one can read the progress of the U-boat or Italian submarine which sank the Allied ship, and track its movements, and the fate of the U-boat and its commanders and crew.

Aviators:

The Allied aviators about whom this study shares the most are those who sadly perished in the line of duty for their country. Most of the aviators killed were young pilots from Europe, the Americas, and the Antipodes who lost their lives in training accidents within 100 nautical miles of either Oakes Field or Windsor Field on New Providence Island, outside of Nassau. However, some, confusingly memorialized by their RAF units in the Nassau cathedral and cemetery, were attached to the transport wing (No. 113 Transport Wing and others, including staging posts and terminal staging posts), were lost as far away as Belem, Brazil and Khartoum, Sudan. Those buried or memorialized at the RAF (aka Western) Cemetery in Nassau, and memorialized in Christ Church Cathedral include men from the Czech Republic, Australia, New Zealand, Canada, USA, and beyond.

The aviators who were non-RAF, such as US Navy pilots and crew on patrols, missions, transport duties, and other, that flew over (or may not even have flown over) the Bahamas area are nebulous. Because they are presumed to have downed, ditched planes which exploded or broke apart and died "somewhere" between airfields either end of the Bahamas in an area before full Radar and other coverage, their fates remain unknown and are more challenging to pin-point. They may be listed just by plane and in a rough area of ASR, which stands for aircraft search and rescue.

As regards RAF personnel, not all died in airplanes; at least two were killed by propellers on ground in front of a hangar at night, and one was drowned in a sailing accident. A lorry accident led to at least one amputation. Those accidents are recorded in the *Daily Diary* portion and would necessitate having a date to find, or, or course, reading the entire document. Although women of several nationalities certainly served in medical, administrative, and other roles on RAF bases in Bahamas in the war, as well as very ably leading I.O.D.E. and Red Cross efforts in Nassau and beyond, no females are known to have died in the line of duty on shore in this region and time-frame.

In summary, to assist readers searching for an RAF or USN serviceman or woman by name, it is recommended that the following sections be consulted:

- *Killed in Action Section,*
- *Buried in the RAF Cemetery Section,*
- *Names Section*/s for both No. 113 Transport Wing and No. 111 O.T.U.,
- *Daily Diary*, preferably armed with start date, by details of an airplane, such as tail number or type. If one has service records for an individual, and then perhaps some dates they served in the Bahamas to narrow it down, it may be possible to match a specific aircraft with that individual airman. They are all identified by type and identifier number, even some by manufacturer number (US Navy).
- RAF, USN, or other, military unit. Knowing this may provide just general answers as to where and for how long individual units served in the Bahamas. As with other sections, the names of all survivors are not recorded; only those mentioned in the service diaries, which include courts martial.

Regarding the *Daily Diary* item above, to date there has been no precise and comprehensive list of the up to 10,000 persons who arrived, trained, departed at and locally enlisted in and supported not only the RAF bases in Nassau and staging posts beyond, but the US Navy base on Exuma and the RAF outpost on San Salvador. Therefore, a researcher will do best to begin the search with a name.

Every single word of both RAF base diaries has been thoroughly typed up and then combed through by several editors, all looking for any names of personnel over a period of many years. In some cases, each name was transcribed at times up to 25 times, since individuals might have shared the same name, moved from base to base, and were promoted. All of this is to help ensure loved ones' names that were recorded and will be perpetuated. If a reader is unable to verify the precise service performed by the individual they are researching, it is probably because they are not among the thousands of dead in the theatre.

U-Boat Crew:

Commanders of German U-boats are listed with details of birth and fate and ranks, in the section on *U-boat Patrols*. Non-commanders are listed in the overall KIA, or *Killed in Action Section*. This section includes the names of roughly 150 Germans KIA aboard U-84, U-176, and U-157 in this region in WWII. It has, using German records, as accurate and complete a listing by last, first, middle names and ranks. One can then cross-reference the name with a U-boat, go to the *U-boat Section*, and determine the precise days that U-boat was in the Bahamas patrolling, and read those exact days in the daily diary to determine what, if anything noteworthy, happened to and by that submarine that day, including aircraft and warship sightings, bombings, and sinkings.

Since one book cannot be all things to all readers, websites including Uboat.net and Uboatarchive.net and others, some of which have closed down in the time it has taken to write this book, have names or pointers to the names of German crewmen. The *Deutsches' U-Boot Museum* in Cuxhaven and German naval and U-Boat memorials and organization might be another source. German mariners are treated the same as allied; KIA who perished in the region are named, those that survived to at least make it outside are area, are not. Commanders are named in all instances, because of the nature of U-boat warfare.

Civilians:

Nurses, the Red Cross, and essentially anyone recorded at No. 113 Transport Wing or No. 111 Operational Training Unit (OTU) are in the *Names Section* at end. Per the *Aviators Section* explanation above, no precise and comprehensive list of the service men and women who served the military in the Bahamas in World War II is known to this author or researchers in the UK and US. Every word of both RAF base diaries (No. 113 Transport Wing and No. 111 O.T.U.) has been scoured for any names of personnel, and each name transcribed.

There are references to dignitaries in the Bahamas in the Red Cross and church and other organizations such as the I.O.D.E. (Imperial Order of the Daughters of Empire), and to Bahamian (British) aviators and others who served on the base. Many civilians were involved in the construction, security, evacuation, rescue efforts, and other roles. But the resultant list is patchy at best, and far from thorough.

Though not named, we must recognize the importance of such groups as

- labour force who constructed the airport,
- constabulary who policed the servicemen in and around the town,
- police downtown who put down labour unrest later etched in the national psyche,
- workers who dug, largely by hand, a shallow ditch across Hog Island,
- thousands of Bahamian mariners who operated hundreds of merchant vessels, under steam, diesel and power, connecting the Bahamas, with US and Caribbean,
- the craftsmen in shipyards around the colony, but particularly Symonette Shipyards, Ltd. on Hog Island, owned and run by Sir Roland Symonette of The Current, Eleuthera, which produced two mine-sweepers on order of the Royal Navy and repaired hundreds of allied vessels, and others in Harbour Island, Abaco, Ragged Island, Andros, and throughout the islands, and
- Bahamian laborers who went to the US under The Contract during the war.

Because this book is a military history, and many of the extraordinary accomplishments of the men and women of the groups above have been deservedly recognized in other texts, their names are not added here, unless they enlisted in the Bahamas contingent of the RAF and/or visited the bases with sufficient impact to have been cited in the diary. The names here are for those killed in the line of military duty, or in the merchant marine directly or indirectly supporting their nations', and the allies constant demand for both raw and finished products, civilian and military.

While any name, unit number, or plane or vessel or U-boat number has been recorded here, one resource cannot provide every name without perhaps being exclusively that: a list of names, which this book is not. In the event that readers are unable to match their expectations and data with the data provided herein, take heart inasmuch that by reading this story, details of what that person and their comrades accomplished on a daily basis ought still to be informative.

Sample Paragraph from Diary Section, Interpreted & Explained

Sample 1: U-Boat Prevailing:

6th March 1942

R.Smg. *Enrico Tazzoli* NE of Bahamas sinks the Norwegian SS *Tønsbergfjord* a third of way between Bermuda and NE Bahamas. One died, and remaining 32 were rescued. Captain's boat with 18 saved by Dutch SS *Telamon* on 13th March and landed in Jeremie, Haiti on 16th March, while others were picked up by Norwegian Tanker *Velma* on 18th. Same submarine sank the Dutch SS *Astrea*. R.Smg. *Enrico Tazzoli* begins single week-long, six-vessel sinking rampage in the region E of Bahamas. BetaSom sinks SS *Astrea,* SS *Tønsbergfjord,* SS *Montevideo,* and SS *Cygnet* only 4 nm off Dixon's Point, San Salvador, resulting in 30 survivors being landed there. She also sinks SS *Daytonian* and tanker *Athelqueen* 50 or so nm SE of Abaco, resulting in 4 sailors drowning on both. The submarine made contact with the *Athelqueen* and damaged her bow, necessitating a return to France. *U-128* shells *coup de grace* day after *O. A. Knudsen* sinking, then east for France. Survivors from *O. A. Knudsen* set course for Hole in the Wall, Abaco. Leaving Olaus Johansen, they continue to Nassau aboard *Content S.* on the 8th. Master of SS *Nueva Andalusia* reports burning object, probably ship, NE of Eleuthera; it was *O. A. Knudsen.* R.Smg. *Finzi* sank tanker *Melpomene* E of Turks and Caicos, N of Puerto Rico. U-126 N of Windward Passage, south through Caicos Passage

Italian submarine sinks Norwegian ship; both names in italics. R.Smg. = Italian for Regia Sommergibili; Royal Submarine

Fate of allied sailors.

Further attacks by Italians, same day.

Summary of Italian submarine's activities.

Projection of Italian attacks, patrol.

German U-boat's mop-up after 2-day battle against a Norwegian ship.

Outcome for Norwegian crew, man.

Other Italian and German submarine positions that day, whether approaching leaving. N = North, S = South, and so on

Sample 2: Aircraft Prevailing:

3rd January 1944

Local flying training continued throughout the day from Oakes and Windsor Fields. Liberators FL-964 LD, BZ-758LS, BZ-744 airborne from Windsor Field. They landed at Boca Chica and were briefed regarding exercises, then returned after completion. Liberator FL-994 LA airborne from Windsor landed at Miami returned from Miami and landed at base. B25's FV-946 FF, FW-123 FJ, FW-149 FN, FV951 FC, FW-123 FK airborne for air exercise Number 6 and landed after completion. B25's FV-946 FF, FW-123 FJ, FW-149 FN, FR-381 DC, FV-949 FD airborne for night navigation exercise, landed after completion. B-25's FV-945 FD, FV-946 FF, FW-176 FS, FW-147 FO, FW-145 FM, FW-154 FR airborne and landed after exercise completion. Liberators FL-967 LJ airborne and landed at Boca Chica briefed regarding exercise, completed exercise, then landed back at base. Tanker *Oklahoma* is NORTH in Straits passing Palm Beach. Tanker *Evans Creek* is N passing Matanilla Shoal, NW Bahamas. USS PC-1212 and USS PC-1195 are N in Straits. Tanker *Eano* and USS YT-338 are south in Straits. Gunboat *Cuba* and tanker *Opequon* are north in Straits. USS *Knave* (AM-256) are west in Straits. SS *Seatrain New Orleans* is southwest in Straits.

Oakes Field = Main Field, is older and smaller, taken over by No. 113 RAF Transport Wing from March, 1943. Satellite Field = Main Field, larger, for No. 111 RAF O.T.U. from 1942, further west of Nassau.

Liberator = large B24 bomber, and FL964 is the "tail number," often with "FL" painted on sides. Boca Chica is the Naval Air Station at Key West, Florida.

From 111 O.T.U. ORB, Operational Record Books: Air exercise Number 6 = training manoeuvre, as are Kingsley and Queensley, arbitrarily named for various tasks from landing to scouting to night bombing.

Merchant marine ship movements are largely from the US military's Gulf Sea Frontier GSF Daily War Diary with charts and legend. Straits = Straits of Florida. Matanilla Shoal = northwest of Grand Bahama.

USS PC are US Navy patrol craft, and USS YT is a USN yard tug, or workhorse. Seeing (AM-256) simply indicates the specific vessel number in military codes; a bit like a license plate, since there may be more than one type or name of ship.

Abbreviations

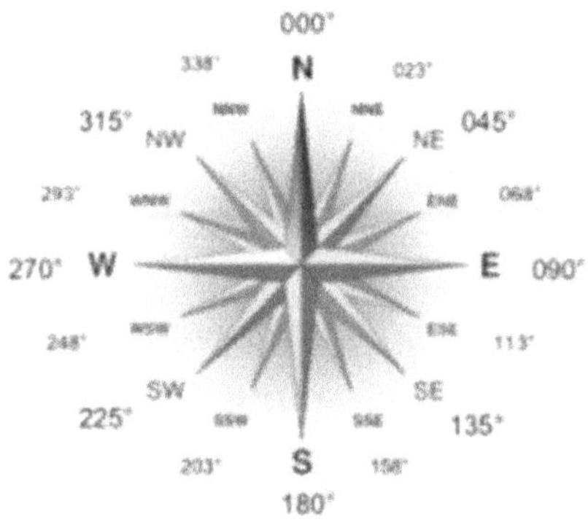

Boxing a Compass; Directions

NNE: north-northeast
NE: northeast
ENE: east-northeast
ESE: east-southeast
SE: southeast
SSE: south-southeast
SSW: south-southwest
SW: southwest
WSW: west-southwest
WNW: west-northwest
NW: northwest
NNW: north-northwest
SB: southbound
NB: northbound
WB: westbound
EB: eastbound
SEB: southeast bound
SWB: southwest bound
NEB: northeast bound
NWB: northwest bound

Geography:
WP: Windward Passage
OBC: Old Bahama Channel
SOF: Straits of Florida (Straits)
NEP: Northeast Providence Channel
NWP: Northwest Providence Channel
NP: New Providence Island, on which Oakes Field and Windsor Field lay.

Relevant geographic features:
Clifton Point, forming Clifton Bay
Clifton Bay, Clifton canal/creek, ASR
Clifton Pier fuel terminal
Old Fort Bay, east of Simms Point/Lyford
Delaport Point, west of Cable Beach
Simms Point – former cay, Lyford Cay
South Ocean, south of Windsor Field
Coral Harbour, southwest of Windsor Field
Lyford Cay; mostly swamp, marina in Clifton Bay held the RAF ASR boats
Montagu Foreshore, east of Nassau, hotel,
South West Bay
Cable Beach
Goodman's Bay, east of Cable Beach
Maura's Creek
Eastern Foreshore, southeast from Montagu
Nassau Harbour, Hog Island, Athol Island

Islets or Cays:
North Cay (later Balmoral Island)
Goulding Cay (off Lyford Cay)
Hog Island (later Paradise Island)
Athol Island
Crystal Cay
Long Cay
Rose Island
Salt Cay (aka Blue Lagoon, SEALS trained)

Communities (partial only):
Yamacraw
Fox Hill
Grants Town
Nassau
Adelaide
Gambier
Delaport
Camperdown

Forts:
Fort Charlotte
Fort Fincastle
Fort Montagu
Fort Nassau (British Colonial Hotel)

Lakes:
Lake Killarney (between air fields)
Laker Cunningham (north and between)

Military Ranks; Equivalent, RAF, USN:

Italian Commanders:

Capitano di Corvetta: Corvette Captain
Capitano di Fregata: Frigatte Captain
Tenente di Vascello: Ship of the Line Lieutenant

German Commanders:

Fregattenkapitän: Commander
Kapitän zur See: Captain
Kapitänleutnant: Lieutenant Captain
Korvettenkapitän: Lieutenant Commander
Oberleutnant zur See: Lieutenant zur See (Junior Grade)

RAF, Definitions of Abbreviations:
2nd Lt.: Second Lieutenant
A/Cdr.: Air Commodore
A/CPL: Acting Corporal
A/F/L F/O: Acting Flight Lieutenant
A/F/S: Acting Flight Sergeant
A/S/L, F/Lt.: Acting Squadron Leader
A/Snr/Sister: Senior Nurse
A/V/M: Air Vice Marshal
AC1: Aircraftman First Class
AC1 (CPL): Corporal
AC1 (LAC): Leading Aircraftman
AC1 (SGT): Sergeant
AC2: Aircraftman Second Class
AC2 (CPL): Corporal
AC2 (LAC): Leading Aircraftman
AC2 (SGT): Sergeant
Advocate S/L: Squadron Leader (Lawyer)
Brig.: Brigadier
Capt.: Captain
Captain F/L: Flight Lieutenant
CoE: Church of England (i.e. Christ Church Cathedral)
Col.: Colonel
Comdr.: Commander
CPL: Corporal
CPO: Chief Petty Officer
Dir. Admin. and Finance: Director of Administration and Finance
Barrister-at-Law, Esquire: lawyer, attorney
F/L: Flight Lieutenant
F/L A/S/L: Acting Squadron Leader

F/L (A/W/C): Acting Wing Commander
F/Lt.: Flight Lieutenant
F/O: Flying Officer
F/O: A/F/L: Acting Flight Lieutenant

F/S, F/Sgt.: Flight Sergeant
G/C, G/Capt.: Group Captain
G/C A/A/Cdr.: Acting Air Commodore
Genl.: General
Gun. Sgt.: Gunnery Sergeant
LAC: Leading Aircraftman
LAC CPL: Corporal
LAC F/S: Flight Sergeant
LAC SGT: Sergeant
Lt.: Lieutenant
Lt. Cdr.: Lieutenant Commander
Lt. Col.: Lieutenant Colonel

Lt. Genl. Lieutenant General
Maj.: Major
Maj. Genl.: Major General
Marine Officer, F/O: Flying Officer
Matron: Senior Nurse
N/Sister: Senior Nurse
OC RAF Ferry Command: Air Officer Commanding RAF Ferry Command, Air Chief Marshal
P/O: Pilot Officer
Pres. W/C, AFC Judge: Wing Commander President of a Court Martial
Rear Admiral
Rev. S/L, S/L (Revd.): Squadron Leader (Religious)
Reverend: Religious Minister
S/C: Squadron Commander
S/L, S/Ldr.: Squadron Leader
S/L A/G/C: Acting Group Captain
S/L A/W/C: Acting Wing Commander
S/L W/C: Wing Commander
S/O: Squadron Officer
S/Sister: Senior Nurse
Sctry. to Sctry. of State for Air: Secretary to the Secretary of State for Air
SMO W/C: Senior Medical Officer, Wing Commander
SGT: Sergeant
Sister: Senior Nurse
W/C Dr.: Wing Commander, Doctor (Medical)
W/Cdr., W/C: Wing Commander
W/O: Warrant Officer
Wing Cmdr., 2nd in Cmnd. RAF Nassau: Wing Commander, Second in Command, RAF Nassau
A/S/L: Acting Squadron Leader

RAF, Allied Ranks:
Air Chief Marshal
Air Commodore
Air Marshal
Air Ministry
Air Ministry, Works Directorate
Air Vice Marshal
Auditor, Air Ministry
Chief Technical Officer
Group Captain
Chief Operating Officer, Caribbean Wing, US Ferry Command
Czechoslovak Minister of National Defence
Deputy Head of BAC
RAF Delegation, Washington
Governor of Bermuda
Governor of the Bahamas
Station Chaplain (Religious)
Sub-Lieutenant
Superintendent of Prisons
Under Secretary of State for Air
Vice E/O: Vice Equipment Officer

Military, General:
RAF: Royal Air Force
MR: maximum range
ME: maximum endurance
ASR: aircraft search and rescue (for plane, from sea and air)
SR: search and rescue
A/C: aircraft
NAS; Naval Air Station
Sqdn, Sqdrn: squadron of aircraft
B24: USAAF Bomber Aircraft
B25: USAAF Bomber Aircraft
RNAS: Royal Naval Air Station
OTU: Operational Training Unit (RAF)
TSP: Transport Staging Post (RAF) ?
NSC: Nassau Sailing Club
ETA: estimated time of arrival
NM: nautical miles (1.18 land or statute miles)
M/V, MV: motor vessel
SS: steam ship
MT: motor transport
PMRAF: Princess Mary's Royal Air Force Nursing Service
PV: passenger vessel
A/S: anti-submarine

ASW: anti-submarine warfare, air or sea
W/T: wireless (or radio) telegraphy
U/S: unserviceable
USCG: US Coast Guard
OC: Officer in Charge
SNO: Senior Naval Officer
ComGulfSeaFron: Commander, Gulf Sea Frontier, Senior US Naval officer
ComEasSeaFron: Commander, Eastern Sea Frontier, Senior US Naval officer
CNO: Chief of Naval Operations, the most senior US Naval officer
MTB: Motor Torpedo Boat
MM: Royal Navy Minesweeper type
MB: motorboat
RAFTC: Royal Air Force Transport Command
WAAF: Women's Auxiliary Air Force

Civilian, Miscellaneous:
IODE: Imperial Order of the Daughters of Empire

Introduction

"Nothing happened in the Bahamas during World War II."

Hearing this extraordinary assertion from an erudite colleague in a bookstore in Nassau over a decade ago, as I began amassing over 50 three-ring folders of research for what became *U-Boats in the Bahamas*, I decided that, however mundane, thankless and costly the process, I would have to literally set the record straight. So I wrote this book to share some of the millions of things that did happen in the then-British colony of the Bahamas from 1939 into 1946. What better way to do so than to use the actual War Diaries of both belligerents: Germans and Italians in submarines, and for the mainly British and US allies, logs detailing merchant ships, bomber planes, many warships, but mostly destroyers and sub-chasers, convoys and airships?

We must put the impact of World War II on the Bahamas in perspective. At one extreme is denial; saying that nothing of consequence happened. In the middle is the approach most often taken; to focus on several events which helped define a nascent nation, namely, and in no particular order, The Project, which brought work to thousands of Bahamian laborers, The Riot, which sowed the bloody seeds of a labour movement which would later help propel the black majority to power after the war, the Duke and Duchess of Windsor, who brought international media attention, high-end visitors and 'glamour' to the colony, The Contract, which brought exposure and income to poorer, mostly black, Bahamians, what I will call The Trial, a sensation following the brutal tar-and-feather, slash-and-burn murder of the colony's most influential man, Sir Harry Oakes, while his family waited for him at their home in Bar Harbor, Maine, and finally The airports; fittingly named Oakes Field and Windsor Field, along with others built in Exuma, on San Salvador and the Turks & Caicos, which permitted the Bahamas to conveniently capitalize on a post-war tourism boom which has supported it for eight decades since.

One could over-state it, as in the assertion on Wikipedia that "During World War II, the Allies cantered their flight training and antisubmarine operations for the Caribbean in the Bahamas." The word "cantered" may be strong, but for Canadian and US factories, Nassau provided an efficient means of delivery for thousands of aircraft bound for Africa and beyond. Yes, there were also major bases in Trinidad, Panama and beyond, but the Bahamas were critical for providing airways safe from winter weather and the more experienced *Luftwaffe* pilots, and in which 5,000 or so men became certified in bombing techniques which flattened Europe and sank U-boats. Here is an assertion which would seem extraordinary were it not written in 1955, nine years after the departure of the RAF, by A. Deans Peggs, Ph.D., Head Master of Government High School, Nassau, in his *"Short History of the Bahamas"*:

> *With the outbreak of war in 1939, memories of the last war caused Bahamians to fear the worst. With the entry of the United States into the war [1941], the tourist traffic virtually ceased. The colony was saved from depression, however, by the construction of a Royal Air Force training station and Ferry Command base. ...Consequently, what might have been a severe depression as in the First World War became a period of great prosperity, with the Colony's revenue exceeding even the peak of the bootlegging years.*

Here are some facts which will surprise even many of us from the Bahamas:

$62 million (in USD equivalent to 2020 rates) was spent locally to build the bases under the Lend-Lease Agreement of 1941. This is per the National Archives; in 1943 GBP 629,978 on labour by contractors, and GBP 250,000 on local merchants ($44.46 M + $17.65 M = $62.1 M).

Over 25% of the population of New Providence during the war were RAF and other Allied personnel. The New Providence population in 1943 was 29,391 (per the *Van Voast-American Museum of Natural History Bahama Islands Expedition*), and at least 8,000 RAF airmen and nurses stayed on New Providence. With 3,000 permanent staff on one base alone, 5,000 trainees, 2,000 transit flights through a different base, where aviators all stayed at the Royal Victoria (3,000 of them named herein), almost certainly the number tops 10,000. That is more than a third of all people living on New Providence, the island surrounding Nassau, the capital.

The first airport built, Oakes Field (aka Main Field) was over a mile long, at 6,000' x 150' wide, with three runways, the others being 5,000' x 150'. Windsor Field (aka Satellite Field) was 1.33 miles long: 7,000' x 150', with two other runways of 5,000' x 150'. Both were capable of operating at night as well as in various wind conditions, using different strips, with numerous hangars and tie-down arrangements. During hurricanes they evacuated to Palm Beach or Cuba.

The US Naval Air Station on Great Exuma's runway was the translucent Atlantic Ocean, fitted with underwater runway lights, to mark the right place to land.

The Red Cross and War Materials Committee provided entertainment, hospitality and distraction for the aviators. They also fulfilled the unenviable tasks of burying the many who died, and the IODE and others have kept their graves honourably for the decades after the war ended. The Red Cross in 1943 sent 2,140 boxes with 22,835 hand-collected, hand-packed items to the UK.

Much is made of the Gallant Five and the hundreds of Bahamians who volunteered and served in Allied forces in the UK, Canada, and the US, many with distinction. Yet in Nassau, Rosemary Kelly was the first Bahamian to volunteer for the military, as did Ann Wanklyn, Peggy Moseley Millar, Rosalie Knowles, Jean Sweeting, Winks Brownrigg, Joan Winders, Mary Brown, and Peggy Hilton, all of whom served in the RAF Women's Auxiliary Air Force, or WAAF.

Not only did Commandant Terese Stratton lead the Red Cross branch in Nassau, reporting to the Duchess of Windsor, but her daughter Joan served in the WAAF and her son left Queens College to fly against Germany in the RAFVR. Sgt. Peter David Stratton was shot down and killed over the North Sea off Noordwijk, Netherlands on 24 June, 1943. While hundreds of men left the colony for the war, many more remained behind, performing key security functions such as guarding the sprawling bases and protecting Clifton Pier's vital fuel depot. To reach the terminal, deep-water petroleum tankers threaded past the dozens of German submarines which transited just north of Nassau using the Northeast and Northwest Providence Channels.

As with any story which was denied to the public for over 30 years and kept in national archives thousands of miles from Nassau, a degree of fantasy inevitably supplanted reality: the fiction of U-boat commanders or crew landing in the Bahamas pushed out the hard facts that 112 submarines indeed invaded, but never landed. Ditto rumours of Fifth Columns operating out of Darby Cay, Exumas, Little Harbour, Abaco, even Hog Island, opposite Nassau. One rumour, finally dismissed along with all the others in the 2011 *Bahamas Handbook*, was that a German submarine was supplied with fuel and other items at a submarine "pen," or just offshore Hog Island. Besides the fact that this grossly insults the patriotism of Bahamians by suggesting they participated in treason, it is also an affront to the British and Bahamian servicemen protecting the population at the time, since both U-boats and the RAF arrived within months of each other.

Even though the operation would have required the illegal accrual, storage, shuttling and transfer of diesel to U-boats on the high seas requiring some 100 persons and many months of preparation, no witness ever reported hundreds of tons of rationed, high grade diesel being loaded into a submarine. Nor, for the record, did it happen at any time during the war, anywhere in the Bahamas. Perhaps people were arrested on suspicion of signalling to the enemy, however that same enemy is emphatic about never having received any help locally. That is because the Germans did not require help to sink some 130 ships; they had each other, the Italians, and resupply ships and submarines of their own. And they made more economical use of the diesel and electric engines – German engines – than the Allies ever gave them credit for. To double their range they filled half their bathrooms with food, filled bunks with supplies, slept two to a bunk, travelled on the surface to save fuel, made fresh water using condensers, enabling fuel to be put in water tanks, and steamed slower.

The Swedish owner of both Electrolux and Hog Island named Axel Wenner-Gren and the Duke of Windsor seemed made for each other; The Duke and Duchess were enamoured of people with money and large yachts; Simpson had witnessed many air crashes as the wife of a US Air Force officer, and preferred to travel by sea. Wenner-Gren literally owned the Bank of the Bahamas, as well as *Southern Cross*, the largest yacht in the world at the time. Their palatial homes overlooked each other across narrow Nassau Harbour. They both sought out wealth and status in others. The down-side to their affinity for each other was that the Duke was caught on camera being chummy towards Adolf Hitler in Germany (something which mortified Churchill and most Allies). What's worse, he appears to have really meant it when he professed admiration for the Nazis, and Wenner-Gren had tried to appease the British by befriending Hermann Göring, with whom he had a Swedish connection. Wenner-Gren also thought he could help to make peace, but he was dismissed precisely because he actually had *no sway* over the Germans, even though no treasonous acts were proven.

Interestingly, another leading politician who lost his job through appeasement had spent six years in Bahamas: This was Neville Chamberlain, the British Prime Minister from 1937 to 1940, who infamously uttered "peace in our time" after meeting Hitler. At the age of 21 Neville and his brother were sent to Andros, Bahamas (there is an island of similar name in Greece), because the Governor to the Bahamas, Sir Ambrose Shea, convinced their spendthrift father, Joseph, Secretary of State for the Colonies, that they could make a fortune in sisal (aka agave) there. Early in 1891 the duo leased 26,000 acres (40.6 square miles) in North Andros, improved the infrastructure, reduced alcohol consumption, set up a banking system, and ultimately lost $8 million USD in today's currencies. Neville returned to the UK after nearly seven years in the Bahamas early in 1897, successfully built ship's cabins, and rose to political heights.

Wenner-Gren returned to the Bahamas after the war and invested in North Andros, starting in the early 1950s with the Andros Bahamas Development Company and the development of Andros Town. Within a decade the area was chosen as the site for the US Navy's Atlantic Undersea Test and Evaluation Center (AUTEC), abutting Fresh Creek, in 1959.

On the other side, U-boat commander Herbert A. Werner wrote to me flatly dismissing the notion of German sailors landing in the Bahamas, even though during the war two companies of soldiers were sent to protect the Duke of Windsor.

The Duke of Windsor, is often credited with nobly seeing the war through as Governor, even though the Duke and Duchess left the Bahamas in May 1945 immediately after the war in Europe was over. This gives a misleading impression that the Duke somehow served his term and finished the job, but neither is historically correct. While technically they remained in the colony for just over a day in May, the Duke quit just under two months before VE-Day, stopped work in April, and left the colony months before the war ended. There was scramble to fill his place, and it was only achieved by taking an able administrator from Bermuda, just barely in time (two weeks) before the war ended, and while the RAF activities in No. 113 Transport Wing were still in full swing.

The fact is that, just as he abdicated as King Edward VIII after 326 days on 10 December, 1936, so too did HRH the Duke of Windsor quit his role as Governor of the Bahamas before World War II ended, in fact even before Victory in Europe (VE) Day – which of course no one could be certain of until it happened. The Duke resigned on the March 16, 1945, roughly two months before Germany fell. The Duke's replacement, Sir William Lindsay Murphy, was did not assume the Governorship until July 28, 1945, having vacated his role as Colonial Secretary of Bermuda to become Governor of the

Bahamas, following postings to Sri Lanka and before that in Rhodesia. The Windsors embarked on 2 May aboard the Canadian ferry *Jean Brillant*, and the RAF assigned the B24 Liberator with tail number NT, to escort ferry to Miami. One wonders what the soldiers, aviators and sailors throughout the islands thought of their most senior leader retreating with the world's largest war - which killed 3% of all humans on the planet - still raging.

It is however fair to say that the Windsors lent glamour to the local social scene, especially among the well-heeled tourists who continued to flock to the islands. And there is no doubt that the war years proved beneficial to the economy. Many wealthy Europeans took shelter in Nassau. Employment opportunities increased with the expansion of Oakes Airfield, and in the construction of a new air base at Windsor Field. Nassau became an air force town. Thousands of uniformed men from the Royal Air Force, the Royal Canadian Air Force, and the United States Army Air Force mingled downtown. Some attended churches and found friends in hospitable families. After the war, servicemen spread great reports of the Bahamas' beauty, and their return visits with family members boosted the expansion of tourism.

The fact is that without the RAF and the US Navy and air forces in the Bahamas World War II these are the likely outcomes:

- No building and employment boom
- No surge in the capital's population beyond 25%
- No Riot
- No Contract
- No Windsors
- No Airports to start a post-war tourism boom with after the war
- No RAF (later Prospect Ridge) Hospital
- No Belmont School
- No wrecks, graves on 3+ islands, boat bases, airplane dive-sites, archaeological finds
- No exposure to the greater world for Bahamian participants in the military and labour
- No tens of thousands of young Allied servicemen, women to enthuse about Bahamas

Whether the Oakes saga might have played out, the fact that it happened in 1943 made it inexorably linked to the war years, ditto any intrigue surrounding Wenner-Gren. Forgetting all the local politics and intrigue surrounding a handful of controversial, powerful men, the theatre of war surrounding the Bahamas was a significant battlefield, with some 2,000 killed on both sides, over 140 vessels lost to war-related causes as well as hundreds of aircraft; the Bahamas alone lost close to 200 aviators, none to enemy action except Isadore Stressel of the airship *K-74* off Cay Sal Bank in July of 1943. Tens of thousands of merchant and allied naval ships transited the region, manned by millions of personnel. 700,000 tons of that shipping was sent to the bottom. At least 30 allied military units, most of them RAF, others US, and manned by servicemen from the Empire and the world are listed below:

1. Bahamas Defence Company (BDC), North Caribbean Force, aka the Bahamas Battalion. Oct. 1942-1945. Under Major Rutherford, Capt. Lightbourn, Adjutant in Aug. 1943, and Lt. Roberts in Feb., 1944. May 1944 Lt. Wenzel Granger, Battalion served Italy, Egypt.
2. US Army Engineering Department, subcontracted Pleasantville, Inc. of USA to build both Oakes Field in Grants Town and Windsor Field in the Pine Barrens out west.
3. RAF No. 1318 (Communications) Flight, formed in Nassau, 1944, disbanded, June, 1946.
4. RAF Ferry Training Flight, under No. 111 OTU, formed in Nassau1942/3, disbanded Aug. 3, 1942, became No. 3 Ferry Crew Pool, part of the Service Aircrew Pool, Dorval.
5. RAF No. 7 Ferry Crew Pool, attached to RAF No. 113 Transport Wing, Oakes Field
6. RAF No. 7 Ferry Unit, No. 113 Transport Wing, Oakes Field, formed July 5th, 1944 in Nassau, from parts of No. 3 Ferry Crew Pool, Reykjavik and Dorval (itself from Nassau to Aug. 3, 1943), from Jan. 1945 to disbandment in Nassau on 1 Oct. 1945
7. RAF No. 7 Aircraft Preparation Unit, formed Nassau 5 July, 1944 from the Transatlantic Aircraft Preparation & Dispatch Unit (AP&DU), within No. 45 Group, and disbanded on 1 Oct. 1945.
8. RAF No. 88 Terminal Staging Post (TSP), Attached to No. 113 Transport Wing, Oakes Field. There are also references in the diaries to No. 80 SP Amphibious in Bermuda, No. 82 SP in New Jersey, No. 83 SP in Gander, Canada, No. 89 SP in Jamaica, No. 90 in Ascension, No.95 SP in Trinidad at Piarco Field, No. 96 SP at Borinquen Field, Puerto Rico, No. 97 SP in Miami, No. 98 SP in Natal and No. 99 SP in Belem, Brazil.
9. RAF No. 111 (Coastal) Operational Training Unit (OTU), at RAF Windsor (Satellite) Field, Nassau. Formed 20 Aug. 1942, at Oakes (Main) Field, to train general reconnaissance crews on US-built aircraft, using 35 Mitchells, 29 Liberators, and 3 amphibious aircraft. Worked with the US Navy and the US Army Air Force against Axis submarines. On 25 Jul. 1945, planes flew to RAF Lossiemouth in the UK, completed by 1 Sept. 1945. "One of the largest in the Empire."
10. RAF No. 18 Group, supported No. 111 OTU with 25 Liberator VIIIs and 7 Wellington XIII/XIV bombers, also left Oakes Field on 12 Sept. 1945 for Lossiemouth, UK.
11. RAF No. 113 (HQ, South Atlantic) Wing (Transport Command), RAF Oakes Field, (HQ No. 45 (Atlantic Transport) Group Dorval, Canada. Formed at Nassau on 15 Feb. 1943, "to control ferrying operations over the South Atlantic." On 11 Mar.1943 named No. 113 (Transport) Wing, on 3 July, 1945 became No. 88 Terminal Staging Post

12. RAF No. 45 (Transport) Wing disbanded Nassau 15 June,1946, and Dorval 29 Sep., 1946
13. RAF No. 250 Air-Sea Rescue Unit, Harbour I. Marine Base, Dunmore Town, Eleuthera.
14. RAF No. 250 Air-Sea Rescue Unit, Lyford Cay Marine Base, Clifton Bay, W. New Prov.
15. RAF No. 250 Air-Sea Rescue Unit, Montagu Bay Marine Base, north-eastern Nassau
16. Princess Mary's Royal Air Force Nursing Service, 6 nurses at a time at the RAF Hospital
17. RAF Out Island Commission, Jan., 1944 to recruit Bahamians to the RAF Bahamas Defence Company, Bahamas Air Service Squadron
18. RAF San Salvador Signals Detachment, amphibious, providing navigational aid to transport aircraft on long-distance flights. Cockburn Town, San Salvador
19. RAF Station Hospital, Skyline Drive, Prospect Ridge, Lake Cunningham
20. RAF Station Military Band (a subset of No. 111 OTU)
21. RAF Station Choir (a subset of No. 111 OTU)
22. RAF Rugby Team (a subset of No. 111 OTU)
23. RAF Editorial Board, *The Liberator* magazine, (a subset of No. 111 OTU)
24. Windsor Field, aka Satellite Field, built after Oakes/Main Field, west New Providence
25. Oakes Field, aka Main Field, built first, barely southeast of Nassau, New Providence
26. US Navy Naval Air Station (NAS) Great Exuma, amphibious aircraft base, George Town, southern Exuma Islands, Lieut. Commander D. C. Higgins, commanding officer
27. US Navy OSS Maritime Unit, Group I. Underwater Training, Underwater Demolition Team UDT 10, and others, trained by Navy SEALS, UK's Office of Strategic Services, and others. Salt Cay, northeast of Hog Island and Athol Island, Nassau, for 1 year, 1944.
28. US Coast Guard, US Navy rescue station, Abraham's Bay, Mayaguana. "Rescue station, barracks, canteen, support facilities, generators, water-makers, radios, radio towers, a good timber dock, where they tied up PT-type boats." (Eugene Pyfrom, *A Great Adventure*, Nassau, 1980s, p.3, *Bahama Site for Defence is Isolated*, Somerset, PA, US, Dec. 4, 1940)
29. WAAF, Women's Auxiliary Air Force; recruited in Bahamas, trained in Jamaica
30. West Indian Auxiliary Territorial Service (ATS), 1943-1945), a female branch of the British Army. Vernell Albury, Earl Grant, Sybil Lightbourn, Shirley Wright, and Zoe Cumberbatch went to Jamaica to train and serve, from secretarial to manning AA-guns.
31. Bahamas Air Service Squadron, Jan. 1944 with 100 "coloured troops" at Oakes Field, New Providence, Flt. Lt. Cullum
32. Volunteer Defence Force, 1939-1945, Nassau & Family Islands, protected installations.
33. Bahamas Batmen Establishment, Oakes Field. Authorized by the RAF Delegation

in Washington in July, 1944

34. Garrison of Pictou Highlanders (Company), Nassau (Canadians, protecting the Governor)
35. Garrison of Veterans Guard of Canada, Nassau (mission to protect Government House)
36. Queen's Own Cameron Highlanders, reopened, occupied the Fort Montagu Hotel, 1942
37. In Turks & Caicos, US forces built an airstrip on South Caicos in 1944 for a base, and by January maintained a 104-foot rescue boat operated by ATC for the USAAF.
38. Walker Cay, aka Advance Base Lano, North Bahamas: from 1 Sept.1942, activated under USN Ensign E. S. Williams, A-V(N), USNR. Thereafter, USN 7th Naval District COM7. Included a 120-foot concrete barge (YOG), a 50-foot long tank landing lighter, two 33-foot boats equipped to rearm aircraft, two 30-foot bowser boats, used for re-fuelling aircraft, and two 26-foot whaleboats as shuttles.
39. Cat Cay, Bimini, kept a 42-foot rescue boat operated by ATC (Air Transport Command, USAAF) for ASR (aircraft search and rescue) under command of US Army.
40. NAS Great Exuma had a 63-foot rescue boat operated by ATC, for the USAAF.
41. Mayaguana maintained a 104-foot rescue boat operated by ATC for the USAAF.
42. Naval Air Facility (NAF) Great Exuma, under command of USN NAS Guantanamo, was replaced by NAAF, Naval Auxiliary Air Facility Great Exuma on Sept. 25, 1944
43. NAAF Great Exuma, under USN, Guantanamo, replaced NAS Great Exuma Sept., 1944
44. Elbow Cay & Pelican Cay, Abaco were carefully scouted as seaplane bases in Sept. 1942

Note: One cannot include, nor ignore, all those who served. Some added names of Bahamians who served are Basil Johnson, Rudolph Pyfrom, Lt. Augustus Roberts, Capt. Reney, Major d'Arcy Rutherford, Lt. David Smith, Lt. William Pemberton, Kendal Isaacs, Fane Solomon, Benson McDermott, and W/O Norman F. Aranha, a gunner on RAF Liberators; any omissions are regretted. Bahamas inhaled Europeans and exhaled Bahamians to Europe.

The US Navy built its amphibious aircraft base (Naval Air Station Exuma) as well as an airstrip at Georgetown, Exuma, as part of the Destroyers for Bases Agreement of 1940. They also had a consular official, naval attaché (US Consul John W. Dye and USN Lt. (jg) Aaron Bilgore), and air liaison in Nassau for most of the war, with aircraft and pilots ebbing and flowing as needed; USN planes would land on search and rescue missions and stay for weeks. Similarly, British Staging Posts and Terminal Staging Posts sprung up on other islands, such as Trinidad, and within existing bases, such as No. 88 TSP at No. 113 Transport Wing, Nassau. Smaller bases at San Salvador (RAF) and Mayaguana (US Navy) suggest there may have been other small outposts which are not included. These also ebbed and flowed with the exigencies of war, with occasional glimpses caught as personnel were recorded transferring between bases.

Looking at the wartime contributions of the RAF alone, many Bahamians and students of World War II generally are unaware (in part due to these records only becoming available in the 1970s), of the scale and magnitude of their operations. Here are highlights. The RAF:

- Built two large airfields in Nassau and smaller bases on other islands
- Maintained a permanent staff of 3,000 over nearly 3 years; with churn over 7,000
- Hired thousands of locals over the years to build bases, join contingents, etc.
- Ferried over 2,000 aircraft – most of them large bombers requiring 11 men each
- Trained over 5,000 airmen in aviation warfare, bombing, anti-submarine, etc.
- Built infrastructure including roads, hangars, hospital, chapel, cemetery.
- Developed and trained an RAF Choir, football teams, a rugby team, and so on
- Recreated at the Nassau Sailing Club, the RAF Seafarer's Club, contributed to ZNS Radio, published their own magazine *The Liberator*, frequented the IODE café in the Masonic Hall, the Bahamian Club, the Canadian YMCA, Emerald Beach Club at Cable Beach, Hobby Horse race track, the Royal Victoria Hotel, the Fort Montagu Beach Hotel, the British Colonial, Windsor, Lucerne, Prince George, and Rozelda hotels, went deep-sea fishing on Tony Pulitzer's yacht *Opitsa*, and conducted parades by marching to town and back for church memorial services.
- Flew to the rescue of numerous civilian and commercial vessels including ferries, mailboats and overdue yachts.
- Rescued fellow airmen, or retrieved their remains and identified evidence.
- Buried over 80 comrades in Nassau and commemorated over 100 others missing.
- The main aircraft they flew, trained and ferried were land-based bombers, including Mitchells, Baltimores, Mosquitoes and Liberators, as well as many others including fighters and amphibious craft.

In her memoir *The Windsors I Knew*, based on her experiences as the Duchess' Private Secretary from 1940-1944, Jean D. Hardcastle-Taylor relates how the influx of over 2,500 labourers from Out Islands into the communities of Grants Town, Bain Town, Fort Hill, Kemp Road, Fox Hill, McCullough Corner, Mackey Street, and Johnson Road (Turnquest, p.72), as well as the thousands of RAF, US military and civilian construction staff onto the bases, created "the water problem facing Nassau: The sudden increase in our population brought about by the arrival of evacuees from England [20] and troop and Air Force activities here had given the island a water problem" (p.36, notes). When Hardcastle-Taylor identified and proffered the services of a highly experienced Scottish engineer with experience in catchments, aquifers and water farming, the introduction fizzled over discussions of whether they were Bahamian, white, or black, and why white policemen could not be brought in to police white areas, as in Bermuda, and so forth.

On a more substantive level, Hardcastle-Taylor gives the back-story of the United Services Canteen in the iconic estate known as the Bahamian Club, which in recent decades has forlornly peered over the brush at The Fish Fry at Arawak Cay, flanked by Fort Charlotte, the Bahamas Humane Society and Ardastra Gardens.

> *One of the Duchess of Windsor's greatest successes in the Bahamas was her United Services Canteen in Nassau, a place where the forces could spend their leisurely hours. The canteen was located in the famous and fashionable Bahamian Club, owned by Mr. Frederick Sigrist [Sopwith's partner and British aviation pioneer], who lived partly in ...the Bahamas. ...The Duchess planned the arrangements for the canteen directly with the RAF staff, and spent many hours superintending the alterations and decorations (The Windsors I Knew, p.70)*

She described how renovations were made by the American airport construction firm Pleasantville Constructors, and that the canteen was run by Nassuvian Miss Bessie Armory, and was separate from the RAF Officer's Club and those of the US military, and the IODE Canteen in the Masonic Lodge on Bay Street. The grand opening took place on December 24, 1942 and throughout its existence, by all accounts it was a joyous and lively place. They building even had its own Military Police presence. (*The Windsors I Knew*, p.70-71)

Since the pivotal military tenant, change-agent, and employer in the Bahamas was the Operational (Coastal) Training Unit No. 111 (aka 111 OTU), a short synopsis of its history is appropriate. The unit was formed at Oakes Field (Main Field) on August 20th, 1942, and "its main mission was to train crews for maritime patrol service (for Coastal Command, as well as ACSEA), on aircraft supplied by the USA according to the Lend-Lease Act." (text by S.C.L., RAAF, father to aviation historian Robert Livingston, of Brisbane, who kindly shared it in 2020). As soon as training started in November of 1942, the newer, larger Windsor Field (Satellite Field) opened, which hosted "C" Squadron's B-24 Liberators, with Squadron A and Squadron B, using twin-engine B-25 Mitchell aircraft staying behind at Oakes, at least until No. 113 Transport Wing arrived from Canada the following year. The first Liberators for training arrived at Nassau on 28 February, 1943, eventually totalling 29 during 1943, or more than one fortnightly.

Equipment is technical. Each plane was named, starting with a base identifier of the letters L, M, or N, and then the letter after that being specific to the aircraft, as in LA, LB, and LC up to LX3. Fifteen of the new Liberators were described as GR fitted with Mk. V, and ASG-3 radar sets, and a further 8 were "Dumbo" fitted, with ASG-1 radar. Yet 6 more were type III, and 8 more were fitted with Mk. VI and allocated after April of 1944, with 16 more GR M. VIII's in 1945; many were fitted with Leigh Lights behind their engines. The two-letter combinations assigned to No. 111 OTU aircraft were H3 and X3 from 1942 to 1945, with 3G used later. As you will see specific aircraft numbers in training and the annals of wrecks and deaths will include, in order of delivery:

A Squadron: B-25 Mitchell I FK-179 AR, from August, 1942 to July, 1945
C Squadron: B-25 Mitchell II FR-378 CL, from August, 1942 to July, 1945
F Squadron: B-25 Mitchell II FW-150 FP, from August, 1942 to July, 1945
J Squadron: B-25 Mitchell III KJ-671 JZ, from August, 1942 to July, 1945
L Squadron: B-24 Liberator III FL-993 LB & Liberator V BZ806 LU, Aug., 1942-July, 1945
M Squadron: B-24 Liberator V BZ-810 MA, BZ-811 MB, from Aug., 1942 to July, 1945
N Squadron: B-24 Liberator VI EV-896 NA & KH-262 NH, from Aug., 1942 to July, 1945
X3 Squad.: B-24 Liberator VIII KK-333 X3-C & Wellington XIV PG-183 X3-CC, '44-Jul.'45

Other aircraft identified specifically include Mk. VI EV880 NC, Liberator GR Mk. V's BZ808 LY, BZ722 LK, BZ813 MD, BZ812 MC, and BZ744 LL. Sadly both BZ746 LM and Liberator B MK III FL994 LA collided mid-air while approaching Windsor Field on 23 February, 1945. A Halifax III, VI RG386 D, and possibly other types may have been utilized. A Grumman Goose on loan from the US Navy was active in the Bahamas which may have been Goose I FP742 and Goose I FP740.

Although it is possible that one-off anomalies of aircraft types used Nassau as a layover or waystation, and those were recorded in the ORBs or base diaries, the core aircraft for the training unit were B-24 Liberators and B-25 Mitchells. The countries of different services used different names for the same aircraft, just as the Japanese World War II aircraft Mitsubishi A6M was simply called *Zero* by its adversaries. A Baltimore to the British was a Martin A-30 to the US, a Catalina was a PBY, a Dakota was also a C-47 Skytrain, a Mariner was also known as a PBM, and a Liberator was a Consolidated B-24; and this is a simplification of how complex it could be.

These more than 20 aircraft were used to train the following "classes" at Nassau:

- No. 202 Advanced Flying School
- No. 205 Advanced Flying School
- No. 206 Advanced Flying School
- No. 215 Advanced Flying School
- RAF Central Flying School Detachment
- No. 102 Flying Refresher School
- No. 1 Flying Training School
- No. 3 Flying Training School
- No. 6 Flying Training School
- No. 7 Flying Training School
- No. 22 Flying Training School
- No. 9 Advanced Flying Training School
- No. 22 Service Flying Training School

A lesser-known fact is that, as soon as they graduated from their courses, the 5,000 students then delivered aircraft across the South Atlantic towards the Middle East, and were expected to join combat units. After VE-Day, many of the Liberators and Mitchells went back to USA and the balance of RAF aviators and aircraft returned to the UK, via Bermuda and Canada, starting on 23 July, 1945. They were all delivered to RAF Lossiemouth in Scotland by 29 August, 1945, with the arrival of the last of 18 aircraft. By the 1st of September, No. 111 OTU had handed over Oakes Field, and on the 12th Oakes Field and some aircraft for transport and conversion purposes were handed over to RAF No. 113 Transport Wing, which remained through much of 1946. On the 21st of May, 1946 No. 111 OTU was officially disbanded in Lossiemouth, Scotland.

No. 113 Transport Wing was created as part of No. 45 (Air Transport) Group with its HQ at Dorval airfield near Montreal, and shared the base with No. 88 RAF TSP (Terminal Staging Post). This unit's origins lay with the creation of AFTERO, the Atlantic Ferry Organization in July 1940. Its purpose was to transfer aircraft from the US factories to the UK airfields before the US entered the war in December, 1941. No. 113 Wing arrived in the Bahamas after No. 111 OTU and stayed later. They arrived first on 22 February, 1943, opened in March of 1943, and closed completely on 31 May, 1946. Their purpose first and foremost was the safe delivery of over 2,000 aircraft from North America to theatres of war in North Africa, Europe, and the Middle East via the Caribbean, South America, Ascension Island and Africa. Without Nassau a lynchpin providing access to the vital South Atlantic route would not have got, to use a pun, off the ground, and a vastly more perilous, though shorter route via the North Atlantic, would have constrained the delivery of aircraft at critical times to Tunisia and as far as Rangoon in Burma.

Aircraft used by No. 113 and its affiliated Communications Squadron included:

- Liberator II AL-504
- Ventura V JS-955
- Hudson III, IIIA, VI FH-315
- Spartan 7W KD-10
- Dakota III FL-528
- Catalina IIIA FP-530
- Norseman FR-406
- Baltimore IIIA, IV FA-114
- Marauder III FB-455

So, who were the people covered herein, who operated all of the military and civilian equipment? Roughly 6,000 German and Italian submariners attacked; they and some submarines repeatedly. Roughly 10,000 British and Allied servicemen and women are estimated to have rotated through the several RAF bases and posts in the Bahamas, with thousands more from the US Navy in Exuma and Mayaguana, and the USAAF in Nassau.

The figures for the Allied merchant marine that were actually attacked were 5,844 sailors, gunners and passengers on the ships, of whom 1,323 were killed in or after the attacks, and 4,613 survived, of whom 321 made it to shore in the Bahamas or the adjacent Turks and Caicos Islands. The names, ranks, vessels and, where possible, dates of all 1,915 personnel killed in the line of duty are listed herein; 1,323 Allied merchant mariners, 183 RAF aviators, 254 US Navy personnel and 155 German sailors. Also 2,399 people listed in the diaries of No. 111 OTU and 369 in those of No. 113 Transport Wing are included, giving a total of 4,683 named personnel.

Regarding convoys, 200 of them specifically to and from the Bahamas, calling mostly at Nassau or a smattering in Exuma, are listed. In the channels through and around the Bahamas but not to the colony, 2,756 convoys are listed with dates of arrival and departure and from and to where. If each of those convoys included an average of five merchant vessels and two naval escorts, that equate to 19,292 allied vessels in convoys. The fact is that many convoys had more ships, and some had none, and were cancelled. They were to act like escalators: a reliable and convenient way for ships to assemble, board and move regularly up the coast, from one place to another: into ports laden, back empty.

If each allied ship carried an average of 40 persons, then some 1.4 million allied merchant mariners transited the area during the war. Added to these are British, US, and other naval and even army armed guards to fire weapons against U-boats and try to defend the merchant ships from attacks by German raider ships, Italian and German submarines, and the *Luftwaffe*, where they operated.

The fact is that determining precisely how many ships entered a million-square-mile body of ocean during wartime is impossible: it would require studying of the records of every port from 80 years ago, starting with knowing which ports to search. It would require not just train-spotting, but equivocates better to setting up a position next to a major highway and then trying to count the traffic going both ways, over a period of over five years, by vehicle type. Eighty years ago. So the number of ships sailing inbound independently from all over the world, continuously, to all US and Canadian east coast ports, is almost certainly much higher, this author suggests that at least 1.4 million allied merchant sailors traversed the Bahamas and surrounding waters, and with shore-based military personnel from all sides added, the figure is sure to exceed 2 million. How could all that activity be done over so many years in a dozen countries by less than 600,000 individuals, in and out of uniform?

While it is impossible to pinpoint exactly how many Allied service personnel entered the region in transit and on anti-submarine and convoy escort missions from over 140 air, naval, rescue, supply, refuel, airship and other bases in and around the Bahamas, it is safe to say that there were hundreds of thousands involved from base construction to manning, administration, maintenance and direct military operations, which resulted in the destruction of three German U-boats, which simply means *under-sea-boat* in German. So, including the rim of nations which supplied airfields and naval bases to the west, south, southeast and northeast (Florida, Cuba, Haiti and the Dominican Republic, Puerto Rico, other Caribbean islands down to Trinidad, and Bermuda), and all servicemen on the water and the air, over 2 million persons participated in this theatre and contributed to an ultimate Allied victory. While 25% of mariners on ships struck by U-boats, on average, perished, 2,000 overall deaths out of some 2 million is a very low death rate, though of course that would matter not to families of those killed.

Axis submarines seeking to sever the Achilles heel of Allied supply lines, destroyed or attacked 690,928 gross registered tons of Allied shipping in this region alone, including many of the most valuable tankers and ore carriers. Over a third of all Allied losses globally were in the Americas in early 1942. Literally millions of tons of shipping passed through the region; particularly as certain ship types are well suited to ply much of the same routes most of their working lives, such as the bauxite route for aluminium and tanker trades into specific ports where there are refineries and tank farms. Some ports have facilities to receive gypsum, dolomite, aragonite, cement, chrome ore, refined or crude petroleum, acids, frozen meats, or livestock on the hoof, and others do not. So many ships, like the submarines, returned multiple times.

Unfortunately, even a study which tries to be as comprehensive as this, cannot include every contributor. Ironically, since it was a colony and the invaders and many of the defenders were imported from Europe, and most of the Bahamian servicemen were sent to Europe and abroad, the vast majority of the participants named and detailed in this study are non-Bahamian Europeans. However, back then, Bahamians were also British subjects, so it was literally a different world, and the world war was all-consuming.

Since it is important to recognize all participants, if only once or twice, here are groups from, or based in the Bahamas, whose hard-won contributions deserve mention:

- Bahamian contingents raised and sent overseas such as
  - Caribbean Regiment and any other Allied branch (not a focus here)
  - RAF trainees at No. 111 OTU
  - RAF "batmen," or officer valets, at No. 111 OTU
- Labour for the construction of Oakes Field (aka Main Field)
- Labour for the construction of Windsor Field (aka Satellite Field)
- Labour to construct bases in Exuma & Mayaguana (USN), San Salvador (RAF)
- The many grassroots IODE volunteers who set up canteens and worked with the Red Cross, the Masonic Lodge/s, and on drives to procure all manner of useful donations for the war effort, from bandages to scrap metal to recreational rooms.
- The Red Cross, run in the Bahamas by the Duchess of Windsor, who inherited the role from Lady Dundas, the wife of the previous Governor, founder of the Branch.
- Civil Defence Forces, some of whom patrolled bases, Montagu & Clifton Pier
- Civil Servants sent by the British colonial government on postings that included teachers, engineers, technicians, doctors and many other administrative roles.
- Finally, there were unusual participants who nevertheless required the Government's attention, including French escapees from the penal colony

on Devil's Island off French Guyana. After the Germans took over much of France, the prison complex was largely abandoned, leaving thousands of desperate criminals on a diaspora for freedom anywhere they would be accepted; ideally via the US back to France by way of the UK to join General de Gaulle.

- The military police and local, as well as British, police sent to maintain order, who found themselves amidst two of the country's defining moments; deadly labour riots relating to wage differences between the US and British for work on the air fields, and the murder of Sir Harry Oakes, still unsolved 77 years later.
- Turks & Caicos Islands; locals manned an anti-aircraft battery on Grand Turk. In 1942 Cable & Wireless connected with the Dominican Republic and Jamaica, an RAF bomber crashed in the islands, & 2 ships arriving and 1 departing were sunk.

What does this study miss, or fail to capture? In the interest of managing expectations and forging equanimity with the reader, the following are not covered well herein:

- Bahamian politics, social, national events, rather than international ones
- Non-military activities, including cultural, as well as inter-island transport
- US Navy, US Army Air Force, US Civil Air Patrol (CAP), USN Airships, or Blimps, US Army, and even US fleet deployments, including aircraft carriers, cruisers, battleships and destroyers, are covered only inasmuch as they were used in tandem with, rescued, convoyed with, and interacted in any way with, forces actively protecting the Bahamas, like the RAF and USN fliers.
- Florida, Cuba, Puerto Rico and the dozen other locations manning over 130 bases large and small. The Bahamas are the central focus of this work, so only a fraction of the many activities in other areas are covered. That said, the extensive data on 3,898 ships and 2,956 convoys, plus the interweaving Gulf Sea Frontier Diaries, and RAF ORBs succeed in providing much of the interplay between the areas.

Bermuda, with very detailed aviation and naval records, and exclusively occupying a large swathe of ocean with frequent contact with the Bahamas, is an exception, in that its World War II activities to its south are well covered.

Although at the outset of this book the goal was to be about events not people, and data not personalities, because the diaries for both RAF bases include VIPs and people of global impact, from royalty to entrepreneurs, Peruvian pilots to French penal colony escapees, and a young man daydreaming out of a bus window who lost his arm, as well as nurses and technicians, and an accountant sent to Harbour Island to investigate "a claim for compensation submitted by the owner of The Little Boarding House, in which personnel of Air Sea Rescue were billeted during 1943 to 1945." The building still stands today, its owners well known.

The editorial net was opened a bit to include 2,768 people welcomed aboard or working at the RAF bases, in the hopes that locals and the families of servicemen alike will make a connection. Not every discovery will be enjoyed; as new aircraft such as the Mosquito were being rolled out, trainees had to test them, with they themselves flying them for the first time. One poor fellow is recorded having damaged two of them on back-to-back days; however he emerged uninjured from both episodes.

Despite its many failings, the pages ahead do contain a bit of local colour; brushes with Bahamian ships, shipyards, shipwrights, and merchants, like the entrepreneurial family who leased out their van to the base when wheels were rare. Many of Nassau's hotels are cited, even in passing, as are spots like Whale Cay, where Marion aka Joe Carstairs held court, and remote Bahamian spots like Cay Sal Bank, Cay Lobos opposite Cuba, Mira Por Vos Passage and Hogsty Reef, Mayaguana and Ragged Island, where a battle with a U-boat took place in which three Allied ships where hit.

During the war, for example, The Little Boarding House was Harbour Island's only hotel, owned by Ms. Marion Johnson and Ms. Hattie Thompson, who managed the blue shingled building on the corner of Murray Street (named after John Murray, 4th Earl of Dunmore, Governor of the Bahamas from 1787 to 1796, for whom the town is named), and Bay Street. Ms. Johnson was then headmistress of the Harbour Island primary school.

One quickly learns that not all official correspondence needed to be tedious. Take these two back-to-back entries, post-war, in the No. 113 Transport Wing diary: "Major J.U. Moorhead, representing the U.S. War Department, visited this Unit with instructions that all Post Engineers activities cease forthwith and withdraw U.S. Corps of Engineers detachment by 31st March, 1946. [*Next entry*]: Vital services of US Post Engineers Department taken on to RAF civilian payroll." Officers read about a crash of an aircraft in Yundum, outside Banjul, the capital of The Gambia, which had taken off from Nassau. And during the war we learn that "one German prisoner under armed escort passed through in transit from Jamaica to Canada. He was accommodated at Fort Charlotte Barracks and proceeded to Dorval next day." It seems an age since the Governor Sir Woodes Rogers recorded in the early 1700's that enemy prisoners were housed under guard at Fort Charlotte, overlooking Nassau Harbour; efforts are still underway to locate Roger's 1732 grave in Nassau.

During the war cooperative behaviour, which would seem unthinkable only a few months after it in the Cold War, flourished. Under Batista, Cuba was a major supporter of the US and Allied effort generally, and the US provided vessels and aircraft for Cuba's many air bases and cooperation. Once Castro took over, it was as though Cuba had not sunk U-176 or helped the war effort at all. During the war the US provided its allies with much equipment, including for the Arctic Convoys and the Soviet Air Force and Navy. So, occasional flotillas or military convoys of Russian assault boats, whether under command of US or Soviet forces was not clear, transited the Bahamas and escorted convoys.

In other instances, a little tugboat being carried on a larger freighter popped free when the freighter was sunk, only to be sunk by a befuddled German U-boat crew who found the *Letitia Porter* floating with no evidence of ever been crewed. Many more stories, of lifeboat survivors of the *Potlatch* missing the entire Antilles Islands, being rejected by a sailing boat, finding iguanas but no people, donkeys, but having no ability to kill and eat them, and of Norwegian ships' captains refusing aid in favour of sailing to shore themselves, occur throughout.

One local character based in Nassau was nicknamed *The Last Schoonerman*; Capt. Lou Kenedy lost his schooner *Wawaloam* to a U-boat between Barbados and Nova Scotia, but insisted the U-boat skipper tow him to shore, and when rescued, insisted that the Irish ship deliver his crew to a Canadian warship to expedite their arriving at a destination of their choice. Among other odd side-effects of war, rather than let the Germans know they had invented radar, the RAF put out fake news stories that they were instead feeding their pilots extra doses of Vitamin A, enabling them to see better and thus find U-boats from planes without radar. This drove up public demand for Vitamin A, and since shark liver oil is a source, the shark fishery in Green Turtle Cay, Abaco skyrocketed and was sustained throughout the war, according to the Albert Lowe Museum. With economies disrupted, inventive fishermen went out in channels to fish out valuable bales of rubber off Provo and Harbour Island, flour, whatever they could find that had drifted from sunken ships. Salvage masters even received and rewarded some of this.

Following the war, US and British officials increased commercial ties with countries like Uruguay, Peru and Argentina, hosting Mr. Dodero of Dodero Airlines, Argentina; perhaps, as the Allies sold Liberty Ships as cheap surplus after the war, some aircraft were also being sold commercially – or militarily. The evacuation of thousands of serviceman and aircraft during several hurricanes was done with admirable military precision; as was the final pull out of troops later. Locals were employed in various functions including transport and baggage details. Extra defences were in place on the anniversary of the labour riots. Parades continued right through until Victory Japan, or VJ Day and into the final departure in the second half of 1946.

Every year, special ceremonies were held to honour the RAF Ensign in Christ Church, the fallen dead on Remembrance Day at the Cenotaph, as well as those comrades who lie in the RAF Cemetery, still as lovely as ever thanks to the efforts of IODE volunteers. When a cook drowned in an off-duty sailing accident, he was buried "with full military honours." Local troops were assigned to guard a vital fuel depot and defend Clifton Pier following a fire and then to fill in for a reduction in available guards from overseas. One night, a truck clipped the tail of a large transport plane, the driver did not confess, and a senior official was held up for days. Servicemen recreated and rested at the Royal Victoria, Montagu Beach, British Colonial, and the Jungle Club.

The extraordinary case of British evacuee children who spent almost the entire war in Nassau illustrates how difficult it can be to separate pure military from civilian activities. I learned of these students by speaking with Alan Heald, a radio operator aboard the British tanker *Athelqueen*, which was torpedoed off Abaco in early March 1942. Heald and almost all his colleagues survived being sunk. Their ship was then accidentally rammed by the colourful Italian commander Count Carlo Fecia di Cossato in the R.Smg. *Enrico Tazzoli*. Sadly, several British sailors drowned trying to swim from the outer reef to Hope Town, Abaco, and the remaining survivors were taken to Nassau by the yacht-turned-mailboat *Content S.*, once the pride of the New York Yacht Club, then owned by local businessman Richard Sawyer (Heald is still convinced it was the Royal Yacht).

Many British sailors laid over in Nassau to recuperate and await another ship back to the UK via the US and Halifax. During that time they visited the churches, theatres, clubs, canteens, beaches, forts, stores, and other attractions. Heald, who was taught how to plait palm fronds in Abaco and danced on a balcony with the daughter of an ice-factory owning family named Farrington, told me of the pleasure he took of interacting with the friendly teachers and volunteers who would cheering up a small group of evacuee students. They were well looked after, as Sir Harry Oakes provided his four-storied, 18th-century Clerihew House on Bay Street for their campus, and the Duke and Duchess of Windsor visited them, and included them in a Christmas feast at Government House their first year, in 1940.

So, how did the students find their way to Nassau? The headmaster, Max de Wharton Burr, befriended Harry Oakes in the 1920's and was invited to send children to Nassau, whence the American-Canadian gold magnate had decamped, having purchased a lake believed to have been drained of gold; Oakes emptied the lake and became one of the richest men in the empire on the gold he found under it. The voyages from the UK were short-lived and fraught with very real danger. Between 12 and 24 August, 1940 a party of 16 students (the only three girls were sisters), from Belmont School, a British preparatory school in West Sussex, and four adult escorts and teachers, successfully crossed the North Atlantic from Liverpool. From the decks of the British SS *Orduna*, the children witnessed some of the six ships (of 50 in the convoy) being sunk by U-boats, despite an escort of four destroyers. It was a traumatic voyage, at least until the party arrived safely in Bermuda.

Despite the Lend-Lease Act, in late 1940, the Panama Treaty established a neutrality zone essentially connecting Halifax, Bermuda, Antigua and Trinidad, beyond which German and Italian submarines did not penetrate until early 1942. The near-miss on the way out and the tragedy of the British SS *City of Benares*, sunk by Heinrich Bleichrodt in U-48 a few weeks later, which killed 77 of 90 evacuee children and 260 out of 407 total people on board, caused Winston Churchill to cancel the Children's Overseas Reception Board (CORB) evacuation plan as a result of outcry by parents as well as the public. So, being unable to sail back home, the Belmont School pupils became long-term inhabitants of the Bahamas for most of the war, and part of the community. The students performed in the Christ Church Choir, practised First Aid and performed water sports and drama, among many activities. ,

The teaching staff were drawn from the island, and included Kenneth Brown, an announcer on Radio Nassau; Father Holmes, a local priest later to become a bishop; Baroness Anna Trolle, a Swede, and Mrs. Marcelle Goldsmith, mother of Teddy and Jimmie (later Sir James), who joined the school as pupils. As the school history relates, "Belmont became very popular with the local residents, the numbers rising to 36 boys and 16 girls, ...and close to 100 by the end of the year. These included the ADC to the Governor's children, Sir Harry Oakes's two sons and a daughter, and Norman Solomon, whose uncle, Sir Kenneth Solomon, was Speaker of the Bahamian House of Assembly."

The children of prominent families, British and Bahamian, attended, including Peter and Diane Christie, Hubert Hauck, Roy Solomon (who went on to the armed services), David Donald, and Gale Kelly.

As a postscript, what became known as Belmont Bahamas repatriated its original students and instructors by ship, air and train. The trip from Nassau to Miami, New York, New Orleans to Lisbon on the neutral SS *Magallanes*, and via British Short Sunderland Flying Boat to Foynes in Ireland and Northolt, England, concluded safely at Lichfield, Staffordshire, England, on March 2, 1944. Their arrival and survival, was chronicled in both the *Daily Gleaner* in Jamaica, and the *London Evening News*. Belmont Bahamas student James Goldsmith dropped out of Eton, became an industrialist, then a Member of the European Parliament for France in the EU. He and his family are larger-than-life; his son ran for Mayor of London, losing to Sadiq Khan, and daughter Jemima Marcelle married Imran Khan, the world-renown cricketer who would later be elected leader of Pakistan.

Alan Heald, the radio operator who survived the torpedoing of the *Athelqueen* and the loss of shipmates on a reef at night, would not discuss the balance of his voyage with this author, or indeed with his own family. The nearest one can surmise is that before he arrived home, two other ships were shot out from him. As a result of these undisclosed traumas, he was categorized DMS, meaning Distressed Merchant Sailor, deemed unfit for sea, and provided with work in a factory assembling and repairing radio sets. He retired to Preston, near Liverpool and never shared the details of the other traumas.

To illustrate how the RAF structured its various units and sub-units, like the marching band, the chaplaincy, the medics, transport, marine and many other units including mail, mortuary, accident investigation, legal for courts martial and to publish *The Liberator* weekly magazine, this passage from the No. 111 OTU Operational Record Book from February 1, 1944, should be instructive:

> Introduction of Bahamas Air Service Squadron: The unit was formed upon the Station for the recruitment of coloured troops. The unit was to have been called the "Native Detachment." This name was changed to the Bahamas Air Service Squadron for the purpose of recruiting. The recruiting appeal was augmented by a short talk by the Officer Commanding No. 111 OTU, over the local Radio Station. The number [of recruits] who reported on the 10th of January 1944 was 104. There were only 2 absentees. The Officer Commanding the Squadron, Flight Lieutenant Cullum and the Adjutant Flying Officer Evans, instructed the recruits on their obligations upon joining the RAF.
>
> The men were quartered in a self-contained unit. NCO's [non-commissioned officers] from both the RAF and the Bahamas Defence Company have been seconded to the unit for the purpose of instructing the recruits in foot and arms drill. As yet [Feb. 1, 1944] only foot drill has been given to the recruits. An Out Island Commission composed of Officer Commanding, the Adjutant, and a medical officer from the Station Hospital made a tour of the islands for the purpose of recruiting men for the Squadron. The commission set sail on the 28 Jan. 1944.

On 29 Jan. 1944 the Squadron was represented by a flight [group] of men at the Dedication Parade for the RAF Ensign at Nassau Cathedral. After the parade to the Cathedral the Flight took part in the march past HRH the Governor, who took a salute, supported by the Officer Commanding troops in the Bahamas, Captain R. N. Waite.

It is fascinating that the RAF was both flat-footed (intending to name a new unit "Native Detachment,") and also media-astute: they recruited via radio programs and sent fairly high-ranking RAF officers in uniform aboard sleek modern aircraft to remote airfields to recruit impressionable young men who were perhaps looking for a way to reach the larger world: they offered a patriotic and respectable way for them to leave home in wartime. At the same time, it is clear that the recruits were in a "self-contained" unit; in other words, they were segregated. To provide further context, British and Allied servicemen were prohibited under penalty of court-martial from visiting Nassau's over-the-hill communities, particularly night clubs and so forth; again, the enforcement of a *de-facto* colour bar.

Court-martial cases in the record books were focused mainly on adjudicating young men for going AWOL, stealing equipment or trucks, driving drunkenly or recklessly, being caught in forbidden night spots at forbidden hours, breaking and entering, theft of property and participating in fights. Punishments usually consisted of confinement for a period of weeks or months. They were dealt with quickly, and often quite severely. Other committees were set up to investigate crashes, damage, wrecks, accidents and airborne fatalities.

Often VIP officers were flown in from Canada, Trinidad, Bermuda, the US and beyond to officiate and inspect everything from technical aspects to engineering, and some foreign dignitaries were hosted on the way from South American countries to Washington DC, and the smaller bases had their soccer teams flown in for friendly matches. On special days like the King's Birthday or VE and VJ days, the bases were closed to all but essential personnel and extra fleets of busses were provided for trips to town and the beach. On Christmas Eve the officers broke protocol by serving the enlisted men dinner; an especially popular occasion.

The focus of this study is to faithfully record what actually happened, when, by whom caused, and upon whom the effect was felt. Readers seeking overall strategic views and sweeping opinions, or analyses of tactics and strategy or post-war assessment, will not find it in these pages. The goal of this book is very simple: to bring the often hard-to-access original raw data concerning day-to-day events of a tucked-away-seeming, yet geographical lynch-pin of a colony to life in a synthesized and readable way. It is as much written by the participants themselves, in their action and words, as by any modern writer. Indeed whole sections of the No. 111 OTU are typed verbatim from the base diaries or Operational Record Books (ORBs).

While the scope is ambitious, the goal is simply to bring a period of a small colony's history to light, and to place it alongside other critical epochs in the Bahamas over its 500-year-history. The Bahamas found itself from early 1942 onwards straddling three major conduits of oil and supplies from the Western Hemisphere to the UK: the Windward Passage from the Venezuelan oil fields and Guyanese ore deposits from which metals were made. Ships also utilized the Caicos Passage, Turks Island Passage, Mayaguana Passage, Crooked Island Passage, Old Bahama Channel, Santaren Channel, Mira Por Vos Passage, Nicholas Channel and others to vulnerably thread their way through the Bahamas with shallow reefs on both sides.

The western flank of the Bahamas created a narrow passage which tankers heading north from the oil patch in the US Gulf, particularly Galveston and Houston, were forced to use the Straits of Florida. The passage allows laden tankers and dry-bulk ships from the Panama Canal to ride the Gulf Stream northwards around Maravilla Bank and Walker's Cay above Grand Bahama. Or ships might opt to turn east, north of Bimini, and take the Northwest Providence Channel from Bimini to the Berry Islands south of Grand Bahama, and the Northeast Providence Channel. That way, going close north of Nassau, they passed between southern Abaco and North Eleuthera at Spanish Wells and Harbour Island. Both Allied merchant ships and German U-boats accessed the Straits of Florida and the wide North Atlantic using these routes, later heavily convoyed.

It is hoped that this *Daily Diary Section*, with its dozens of supporting charts and data, will be a starting point for further research into exactly who undertook those actions and what greater outcome resulted therefrom. The Bahamas archipelago are a strategic choke point, straddling a divining rod-shaped trading route: to the left, or west, is the Gulf of Mexico, and to the right the ore and oil route from South America. After they join above the Bahamas, ships headed up the stem towards Halifax assemble into convoys to head across the Atlantic to the UK, as demonstrated in the chart below:

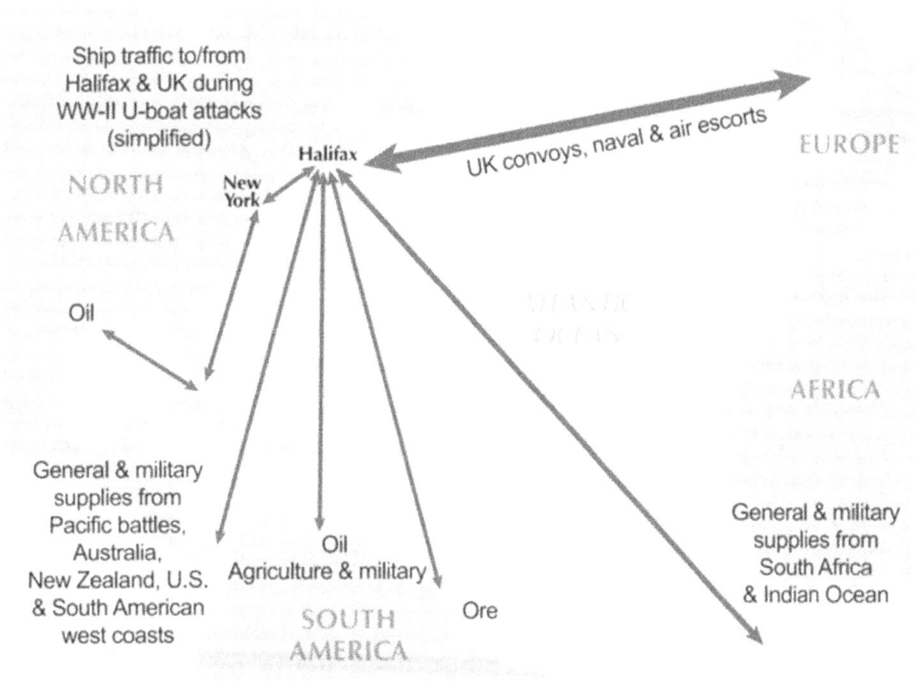

Then the empty ships must reverse course and go back again. This study covers every relevant day from leading up to formal initiation of war in September 1939 to May 1945, into 1946, even, to the closure of most bases. It includes details of:

- all attacks on Allied merchant ships in the area
- all U-Boat patrols by both 5 Italian submarines & 107 German missions
- all convoy transits, including convoy and ship names, dates and ports
- air traffic from RAF air bases and overlapping USN operations in the same area
- transcription from USN Gulf Sea Frontier diaries of daily maps and entries detailing US air squadrons, convoys Miami-Nassau, all of the Straits of Florida.

The progeny of this book has been manifold, but on a visit to the Bahamas National Archives, I was given a short pamphlet covering resources for both world wars. It was helpful but not enough, and I felt that we all deserve to know more about this pivotal. Never, before or since, has the Bahamas played such an important, sustained part in world events. Starting in 2009, I stumbled upon two massive treasure troves of data in Washington DC and London: the Gulf Sea Frontier Diaries in the US National Archives and Records Administration (NARA) and the complete daily war diaries (ORB's or Operational Record Books) of the RAF No. 111 OTU and RAF No. 113 (Transport) Wing, both based in Nassau in the war. I was determined that students, researchers, visitors and World War II devotees around the globe will no longer have to hunt overseas in order to learn what happened in and around the Bahamas from 1939 to 1945. I gave copies of both to the Bahamas National Archives. But that would not result in them reaching many persons, except those already in the know.

Amazingly, huge data clusters were only discovered in the months leading up to publication: the entire base diary for No. 113 Transport Wing RAF was obtained in London and typed from scratch by a dedicated team overseas. Books on aviation history focussing on Bermuda, on USN aircraft losses, Banana River and Morrison Field aviation histories, on Puerto Rican and Cuban military history barely arrived in time. Then the COVID-19 shut-downs began, and I decided that 11 years of work has been enough: that this would not continue to be a never-ending project, and if there were missed threads, like getting all the original *The Liberator* magazines from Canberra or London or Nassau, that could be done by other enthusiasts, the principle being that *a bird in the hand.....*

So now, with the publication of this hybrid chronology, encyclopaedia, time-line and database, the veil is pierced. I am more a messenger and project manager than an author. This is not the *final* word on the topic. Rather I hope it will be a *first* word for others to grasp and follow up. May it be a modest foundation of a building which most persons using it don't see; for students and others to hang their aspirations and accomplishments on going forward. Since this book is data-driven, it may seem to be top-heavy on fact. Despite this every effort has been made to merge information and condense it into readable, palatable bits. In the pages ahead readers are able to study a chart, then easily flip a few pages to read the actual data underlying the symbols on that chart, and from there to take in photos and scale-models of silhouettes showing aircraft, naval and merchant vessels to clarify.

In order to manage the experience, my production team and I have composed the following, all clearly listed in the *Table of Contents*, to guide readers through the process.

- Table of Contents
- How to Use this Book Section
- Definitions of Abbreviations
- Charts, Graphs, and Diagrams
- Notes on Sources Section
- Bibliography, Websites, Interviews Section
- Data lists, from convoys to KIA, patrols, sightings of and by U-boats
- Names of individuals in RAF and Bahamian units
- Names of all servicemen killed in all branches on all sides

There is no index, because this book *is* itself an index. If you are looking for long and florid narrative, descriptive sentences, there aren't many. If, as the reader, you revel in finding mistakes and errors, you will probably find hundreds of them, particularly as the book is being rushed to press at the outset of a global pandemic. However, rest assured that if you are willing to stick with this book, you will also find thousands, if not hundreds of thousands of details, facts and minutiae which this author hopes will motivate and inspire authors and researchers going forward. Nothing could please me more than to have a new author utilize this one, and for she or him to create a better one in a style that they prefer.

Readers of this book will probably have very specific targets, like the bomber pilots and submariners detailed herein. The audience will probably be a global, polyglot, dedicated and small cabal of individuals looking for very specific solutions to riddles: ships, individuals, aircraft, units, dates, names, or all of the above. The importance of this book is to present, almost on a basic, baseline level, that *very much happened* in the Bahamas during World War II. This book should illustrate through contemporary accounts by participants in virtually real time, that the residents of the colony should be proud of their Islands' daily battle to fend off an often-superior enemy, and to hold up a bulwark of freedom and democracy at the footstep of our ally, the United States.

Many Allied pilots sallied forth in planes from New Providence and other bases, from Florida to Bermuda to Trinidad and Puerto Rico or Cuba, and never returned. Thousands of Allied merchant sailors perished in or near our waters, as did 155 German officers and men in three submarines, the sailor from a US Navy airship, and hundreds of their colleagues killed in a hurricane (including the destroyer USS *Warrington*), or in the friendly fire of a minefield in Key West (USS *Sturtevant*). This is not only their story – it is our story. In the face of considerable death, there was also new life borne of these interactions: Jack Lowe, a Bahamian serviceman from Abaco who moved to Nassau to contribute to the war effort, writes of the understandable outcomes of Nassau having, as he put it in his memoir

> *...become an Air Force Town... some found friends in hospitable families. Jack Joudrey, stationed in Nassau for a year with the Royal Canadian Air Force, lived at the Fort Montagu Hotel. He met our sister Eulah at the Paris Ship downtown. Often Jack came into the shop, attracted to the beauty and fragrance of Eulah. Jack Joudrey left Nassau, survived the horrors of air battle in Europe, and was later honoured with war medals. Eulah and Jack corresponded until their marriage in 1946. They lived in his birthplace, Nova Scotia. A number of other Bahamian women married servicemen at the same time. Some relocated in England, the United States, and Canada. Our cousin Joyce Russell married a Canadian, and they lived in Toronto (My Life, p.95-96).*

Another RAF officer, Harold Woodman, married his Bahamian wife LaGloria Maria Pyfrom, and a number his colleagues returned to run Bahamas Airways: Wing Commander Eric H. Coleman, formerly officer in charge of the base became the first Director of Civil Aviation in May of 1946, and Flight Lieutenant Wilfred "Tubby" Welch joined BAL in 1951, per Paul C. Aranha's book on Bahamas Airways. A young American personal secretary to the Duchess of Windsor, Jean Wallace Drewes, met and married Brian Hardcastle-Taylor and married at Government House on December 31, 1943.

Before we take off on this story together, a personal vignette. In the late 1970s my brothers John and James and I explored, as boys are wont to do, many square miles of forests, swamps and lakes around our home on Skyline Drive. The remains of the RAF Hospital started half a mile southeast of our home, between Sanford Drive, on which the Duke and Duchess of Windsor lived, at Sigrist House on Prospect Ridge. The RAF Hospital was intentionally placed on a ridge so that patients would have fresh breezes, and between the Oakes Field and Windsor Field, where the 3,000 or more residents were likely to become ill or be injured.

When the RAF Station Hospital was completed on 21 March, 1943, its leader, Dr. R. H. Pratt, had fallen ill, and so Dr. C. R. McLaughlin assumed command. Sadly, this was not uncommon, with the SMO Dr. Bell being hospitalized there and the senior base commanders (OinC, for Officer in Charge), including the first, John Alexander became seriously ill and was replaced, underscoring the need for good care. The Senior Medical Officers (SMO) for the hospital included Dr. Rassin, Dr. Lockhart-Mummery, Dr. Pearce, Dr. Lyons, Dr. Primrose, Group Surgeon Dr. Luth and Dr. Hughes. The Chief Medical Officer for the colony was Dr. John Merrill Cruikshank. They were all ably supported by half-a-dozen matrons and nurses from the Princess Mary's Royal Air Force Nursing Service, or PM RAFNS, founded in 1918 by Dame Joanna Cruickshank, DBE, RRC, British Army and RAF Matron-in-Chief.

Of these doctors, Dr. Rassin returned to the Bahamas and created Doctor's Hospital on Collins Avenue and Shirley Street, Dr. Cruikshank oversaw the stabilization of two starving survivors from the *Anglo-Saxon*, a ship shelled by the German armed raider *Widder*. The men, Tapscott and Widdicombe, arrived on Eleuthera in October, 1940 after 70 days and 2,300 miles in a lifeboat which is now in the Imperial War Museum in London. Cruikshank became Chief Medical Officer of Fiji and British possessions in the Pacific. Dr. Lyons went to the aid of the survivors of the *O. A. Knudsen* in Cross Harbour, Abaco, in March, 1942 aboard the Carl and Richard Sawyer-owned mailboat *Content S.* The medical community as a whole tended to 255 allied survivors of German and Italian submarines in Nassau in 1942.

What remains of the RAF Hospital? Between 1947 and 1953 the Bahamian government leased and reactivated it while the Bahamas General Hospital was being constructed. During those 6 or more years it was known as Prospect Ridge Hospital, and many patients were treated with a new cure for tuberculosis. Also known as Prospect Hospital, it accommodated 160 patients in general wards in five pavilions, with maternity, children's tuberculosis and a private ward (Munnings, p.83).As children we found dozens of concrete or wooden and stucco buildings intact, in tidy rows, many with rooves still on them. There was a main avenue with perhaps a dozen yellow-coloured buildings perhaps 20-feet-square on each side. These were presumably accommodation either for staff or patients, or both. Different out-buildings, for food preparation, waste disposal, guards, and so forth, including a dispensary with many shelves, could be found. There was little glass; the windows were louvered to allow air and sunlight in. My two main recollections were of thousands of discarded and empty vials of morphine and fascinating medicinal bottles, which had survived where wood had rotted and metal rusted.

The site is bounded by Stanford Drive to the north, Skyline Heights Drive to the west, Ficus Drive and Prospect Lane to the south, and Prospect Ridge Road to the east. More precisely, the line between the back of Bahamas Red Cross and the imposing old gate at Tiger Findlayson's old Star Acres estate covers the focal area at the highest point. The highest probability of finding a ruin is 100 feet before where Prospect Ridge Road meets the JFK Drive roundabout. In 2009 and from satellite in 2020 there is a small square structure still standing. Resources on this topic include *Westward: The Walk of a Bahamian Doctor,* by Dr. Harold Munnings, Jr., which covers the colourful and moving history of modern medicine in the Bahamas. *Artist on His Island, A Study in Self Reliance,* by sculptor Randolph W. Johnston, published in 1975, makes a specific reference to traveling from Little Harbour, Abaco, to the hospital on Prospect Ridge in the 1950s. Also, when reading accounts of travel by mailboat, references are made by Bahamians to the hospital, with the son of a Premier born there.

For many months after the war, the RAF and USN continued to help Bahamian vessels and persons in distress, just as the locals helped find and rescue survivors of many air crashes. In February, 1946 a local recipient of the Distinguished Flying Cross for leading bombers in combat over France, Squadron Leader Lester Brown was sent to the Pine Ridge Lumber Camp, in Grand Bahama to save lives. A railway engine boiler had exploded, killing four men, and injuring four others, one of whom had lost a limb. Assisted by aviators Nay and Miles, and joined by SMO Dr. Lockhart-Mummery, they raced in a large amphibious Grumman Goose on loan from the US. As the base diary reads: "Plasma was administered to one of the injured, who had lost an arm as a result of the explosion, but he died on the airfield."

Among the first Bahamians to volunteer, Brown was one of the Gallant Five, including Fane Solomon, John Maura, John Maillis and Ivor Thompson. The month before the Grumman had been sent by the Colonial Secretary to race to bring back a "serious ambulance case" in Cherokee Sound, Abaco. The 2010 census for the community shows 169 persons live in Cherokee Sound, and the dock is so shallow that vessels have to anchor off and shuttle in. At the time even that was cut off from decent roads. Earlier in the war, Captain Granville Bethel of that port made lemonade from lemons when he salvaged the lifeboat from the Norwegian tanker *O. A. Knudsen* and converted it to a whale-bottomed miniature mailboat to help his neighbours in equally isolated Crossing Rocks nearby. The name Captain Bethel christened his unusual craft with? *Beluga*. A beach to the south, at Cross Harbour, is known as Olaus Johansen Beach since his grave was found there in 2015. Another RAF flight was sent to Mangrove Cay, Andros, for an urgent medical evacuation.

Regarding RAF aircraft crashes, with the exception of New Providence, northeast Andros is where most of the planes that were found ended up. Most of them of course simply went missing at sea with, if fortunate, a self-inflating dinghy to mark the spot, but rarely any bodies. Thousands of hours were flown by aviators seeking lost and missing comrades; a lonely, frustrating duty often turning up patches of seaweed thought to be oil, and wooden wreckage actually from boats, but at times tail numbers and in one case off Booby Rocks, wreckage visible in less than 20 feet of water.

One does not have to look far for evidence of massive engineering structures begun during World War II: the National Stadium sits on what was Oakes Field, and Atlantis is astride the canal which Axel Wenner-Gren funded. Windsor Field is now The Linden Pindling International Airport, the Naval Air Station in Georgetown became Exuma's main airport, Morrison Field is now Palm Beach International Airport, and so on. While the Montagu Beach Hotel, Royal Victoria's main buildings, Emerald Beach Club, Prince George Hotel, Lucerne, Rozelda Hotel, Hobby Horse, and others are gone, as is Symonette's Shipyard, many other structures such as the Windsor Hotel and the footprint of Clerihew House and many other edifices can still be found. *Reminiscing: Memories of old Nassau*, in two volumes, by Valeria Moseley Moss and Ronald G. Lightbourn describe what has been lost and preserved.

What lies beneath? Skin-diving and SCUBA enthusiasts will be interested to learn from obscure and declassified records just how many RAF and other planes crashed; this study found at least 75 accidents, involving over 300 personnel. The planes came down at places like Soldier Road, within a mile or two miles of the nation's largest airport, or 400 yards from Balmoral Island, then North Cay, in eight feet of water (my brothers and I dove that one in the 1970s and 1980s), and another large bomber off Simms Point, Goulding Cay, near Lyford Cay where rescue boats where kept; two bodies were recovered. Another plane landed on rocks near Bannerman Town, Eleuthera, and another was visible in the sand off Acklins, as was one off San Salvador. One plane was salvaged from a lake on New Providence, however it is thought to have been scrapped after the war. Custom-made charts shows dozens of air wrecks on New Providence, some along major roads, though most, like most ruins, burned and were scattered by the accidents and pillaging. Still, there must be much for the archaeologist with a metal detector in these pages.

Importantly many, presumably hundreds, of nurses, ranked as Sisters, Matrons and the like, rotated through No. 111 OTU. These are virtually the only women, aside from heads of the Red Cross and IODE and spouses and family of VIPs such as Governors, HRH Princess Alice, traveling with her husband, the Earl of Athlone. The most senior woman VIP who arrived as a guest of the RAF was Squadron Officer Lady Bowhill, wife of Air Chief Marshal Bowhill, who adroitly predicted the position of the *Bismark*, allowing British aircraft to find, disable and sink it. There were many interesting VIP's transiting both RAF bases in Bahamas, on their way to and from many air bases along the entire eastern seaboard, particularly Morrison Field, Key West, Norfolk, Virginia, Elizabeth City New Jersey, and beyond.

In Canada, flights often included Dorval near Montreal as well as Gander, and Halifax. Island ports of call included Bermuda, Borinquen on Puerto Rico, Guantanamo on Cuba, Jamaica, St. Vincent, Piarco Field on Trinidad, Ascension in the South Atlantic, and emergency landing strips in between, of which many were built during the war, creating tourism access following it. On mainland South America ,planes called and fuelled at Georgetown, British Guyana (Demerara) and Belem, Brazil. The landing fields in Africa are too numerous to list here, but the primary ones were Accra, Ghana, in West Africa, Khartoum in Sudan, north-eastern Africa, and fields in South Africa.

It was not uncommon for the base diaries to record sanguine notes about pilots setting off for Cairo, Egypt, or making an emergency landing near Ascension and being rescued, or landing on a desert strip in West Africa. And there are tantalizing details like local pilot, Squadron Leader D. Lester Brown earning a medal for service in Burma. There are many references to the Middle East, and one can surmise that the aircraft passing through Nassau made it as far as India and Myanmar (Burma) and probably supported the critical supply lines over almost impassable mountains known as *The Hump*, as the India-China Ferry by air was known, crossing the southern Himalayas. So the Bahamas can lay claim to having helped push the Allies over *The Hump* in World War II. On top of this, "World War II changed aviation in The Bahamas. Many Bahamians who volunteered for service with allied air forces pursued careers in civil aviation on their return," (Cleare, *History of Tourism in the Bahamas*). Just some names of those who contributed after the war include former RAF Ferry Command pilot Collyn Rees, Lambert Albury, Bobby Hall, Jack Graham (ex-RAF), and Sid Larkin.

Continuing the vignette with another, towards the coast, friends including Colin Green and Paul Bowers were fond of looking for turtles and exploring the woods, lakes and ponds behind the Westward Villas Subdivision at Cable Beach. Colin lived on Yorkshire Street, and Paul on Hampshire Street. Once, while exploring the road which connects them, in the lot on the southeast corner of Eton Avenue and Hampshire Street, we discovered a dummy aircraft bomb which looked like a cartoon caricature, tear-drop shape, with the round fins at the back. It was small, metal, about two feet long, and painted a durable white, as it showed no rust streaks. It was light, and looked merely decorative, with no detonation function.

We assumed at the time, and I believe to this day, that in the 1940s an RAF pilot jettisoned that very bomblet into a then-uninhabited forest for practice, or for a lark, and that we had the fortune to find it. In these pages we learn that the RAF participated in a show to impress Bahamians, whereby the police indicated an area for an RAF plane to bomb with dud bombs, like the one we found, being dropped at the parade ground at Fort Charlotte. As they hit, 'fake' explosions with dust and a loud noise, were set off from the ground, as though the bomb blew up. Of course it would have been less popular for us to have walked around the community with a bomb than the usual spears that we used for fishing. We left it where it was. Perhaps some treasure hunter will rediscover it, and marvel at what a story it must have told, as we did.

On a local scale, this is the story of how that bomb ended up in our neighbourhood. It populates as quickly as it folded; from mid-1942 to mid-1946; four years. It is unlikely that the Bahamas will play so central a role in global military events for some time. As a point-person on a global team, I am also still that wide-eyed child who wanted to know how the biggest war ever left so much in our little backyard. Many of us may have resignedly thought that ours was also just the backyard of world history. It was not.

During World War II, the Bahamas was closer to the hub of the air and sea war than to the rim. The Bahamas *truly mattered* to the outcome of World War II. If it failed? Not enough planes for Europe in time, not enough oil for Britain's many war machines, or ores to make metals with, possibly even insufficient food for Allied soldiers and civilians. As a youngster, I needed to know why there was a plane on the reef, an aerial bomb in the back yard, and vials of morphine across the street. Now that I found out; please allow me to share what I've learned. I hope that you enjoy, and share in some of that wonderment here. It may make you exclaim:

"A lot happened in the Bahamas in World War II!"

E. T. W.
Boston, US
October, 2020

Table of Contents

Charts, Graphs, & Visual Guides

All original charts created on commission by the author, between 2009 and 2020 by cartographer Robert Eller Pratt of Graphic Design and Illustration, Milton, Delaware, US.

Chart 1: Convoy Traffic Through the Bahamas

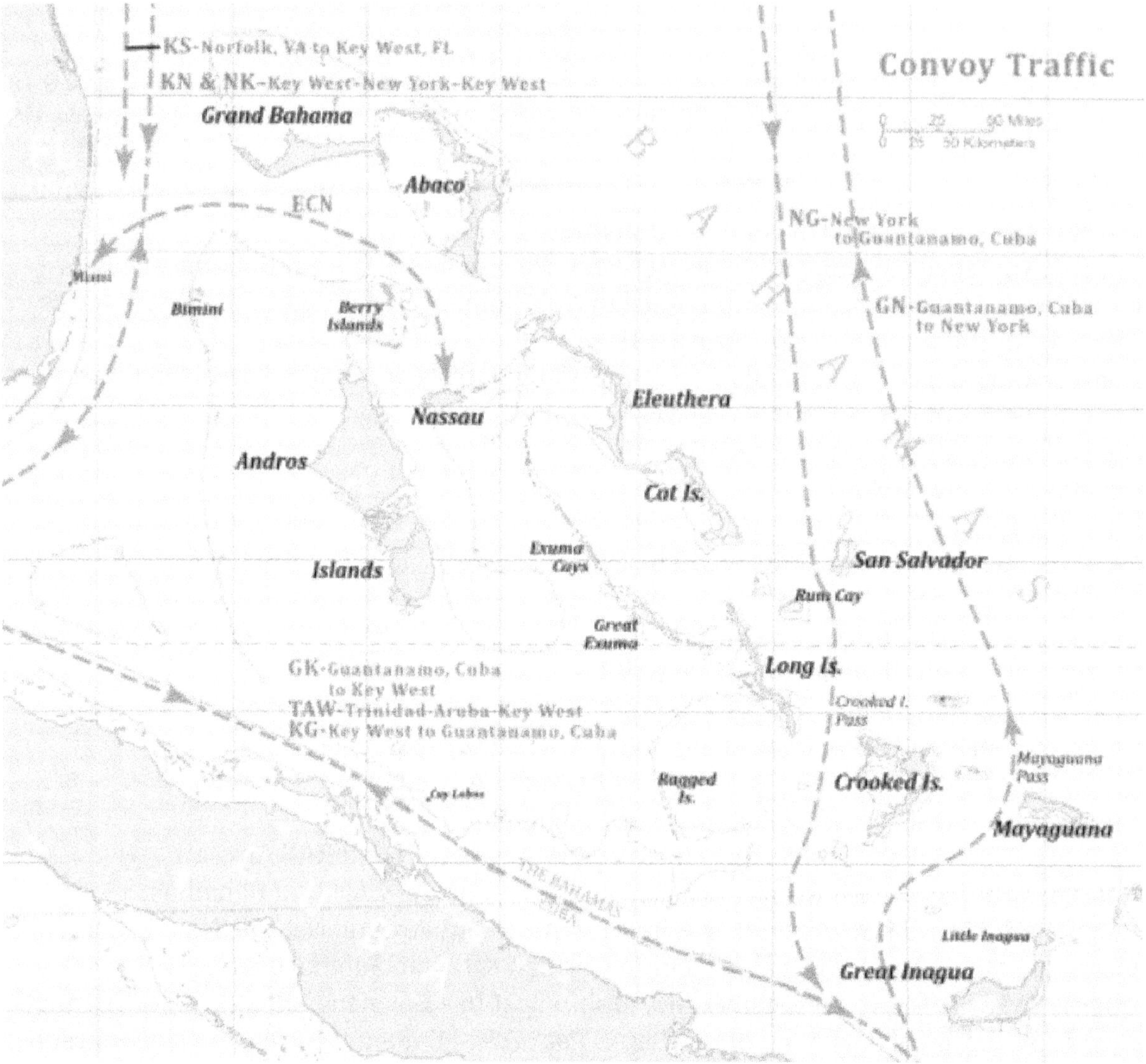

Chart 2: Distances in Nautical Miles Between Nassau and Principal Bahamas Islands

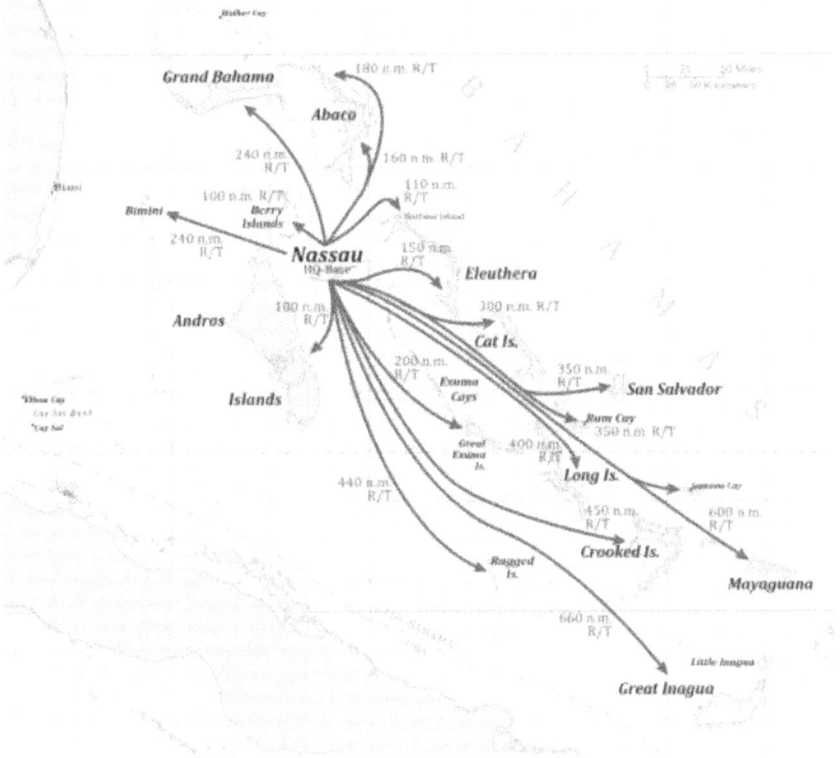

Chart 3: All Allied Merchant Ship Losses to Axis Submarines & Combined Death Rates

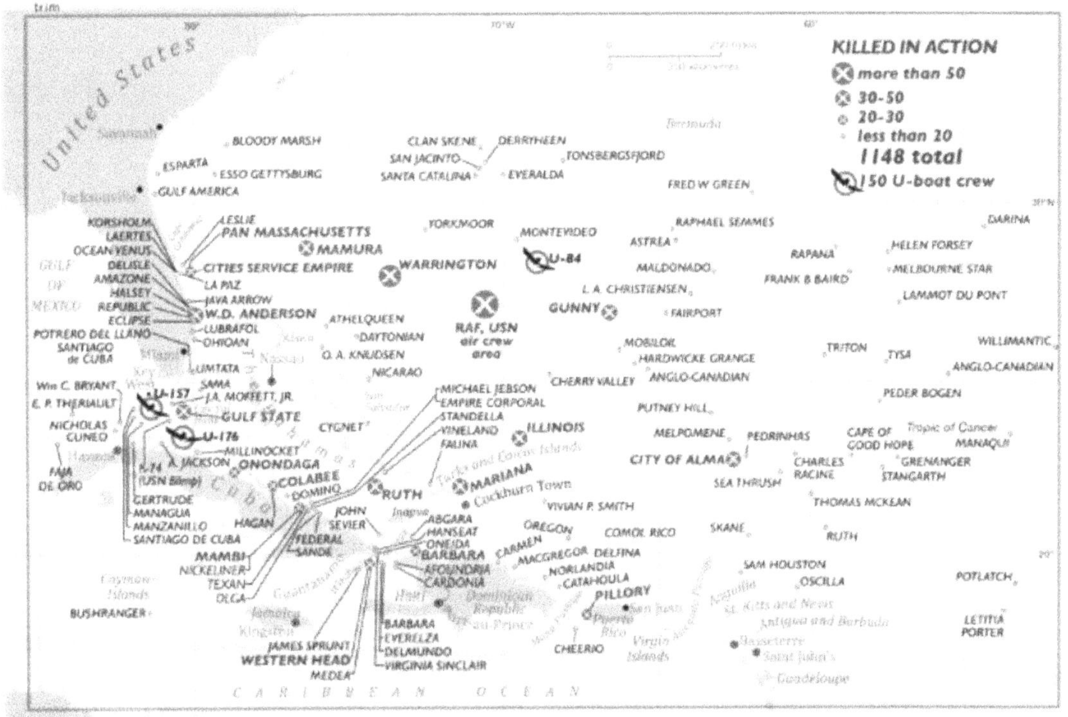

Chart 4: All Italian Submarine Patrols, 1942

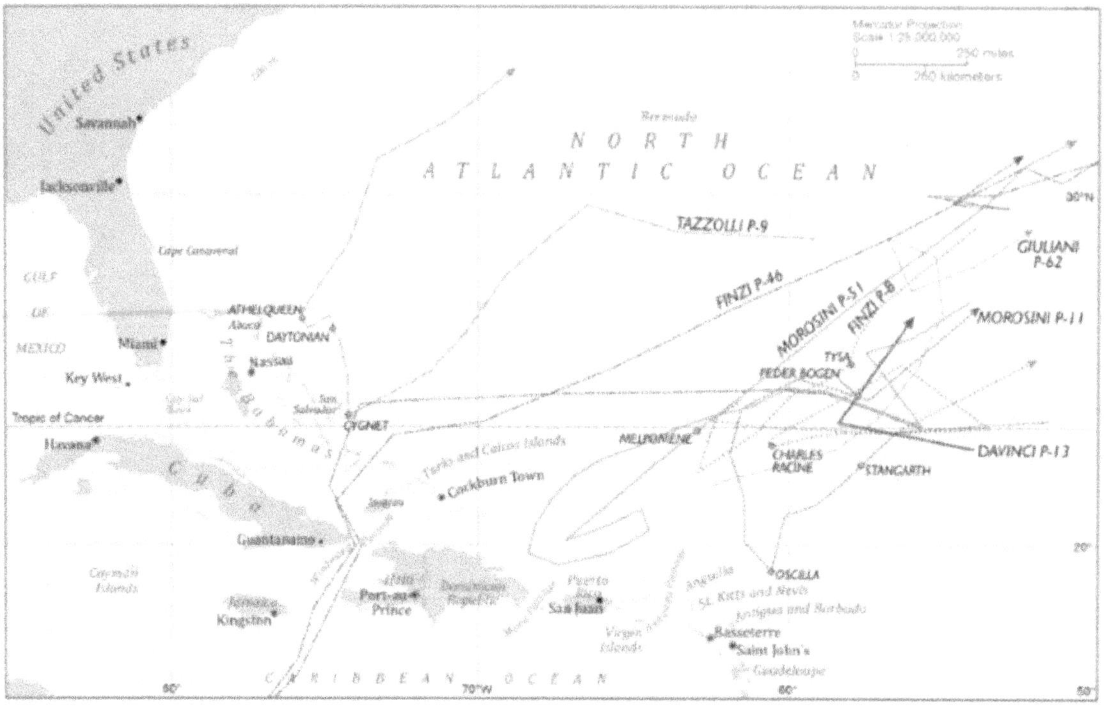

Chart 5: German U-Boat Patrols 1 to 19 (Missing Patrols are Italian Patrols in Chart 4

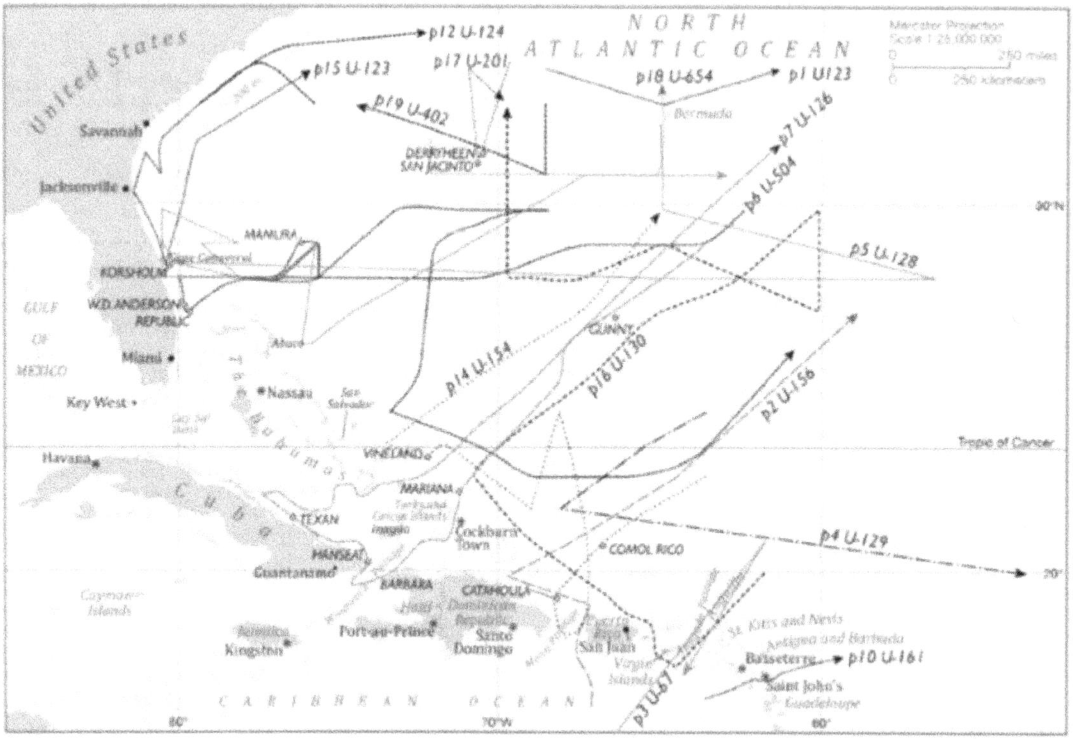

Chart 6: German U-Boat Patrols 20 to 29

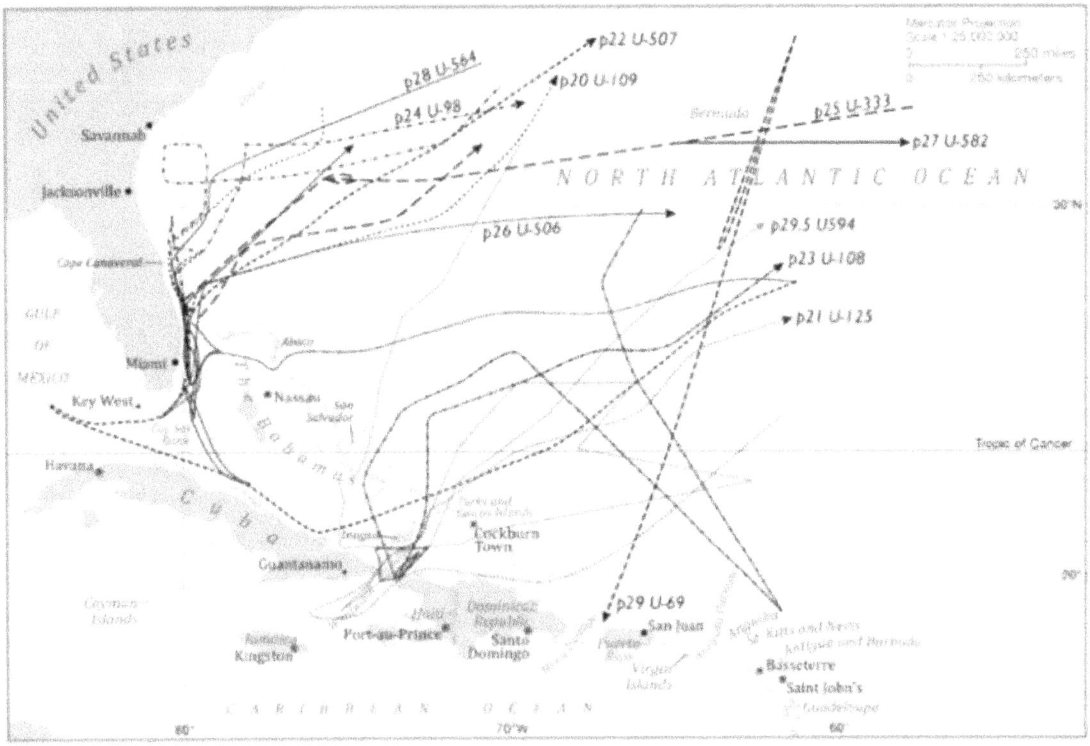

Chart 7: German U-Boat Patrols 30 to 39

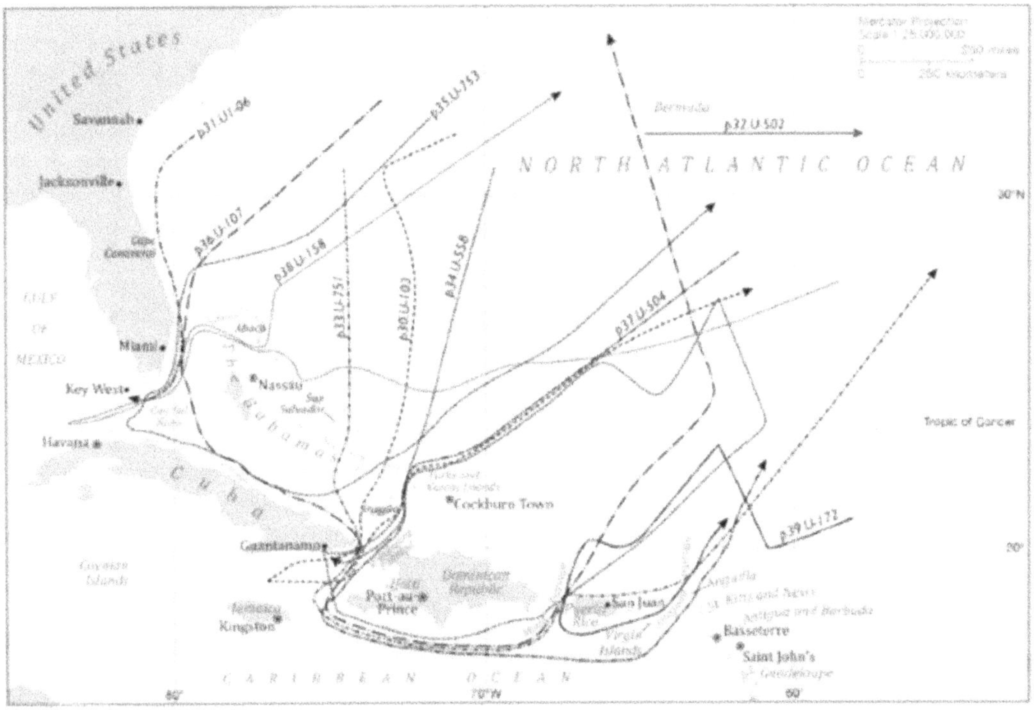

Chart 8: German U-Boat Patrols 40 to 49

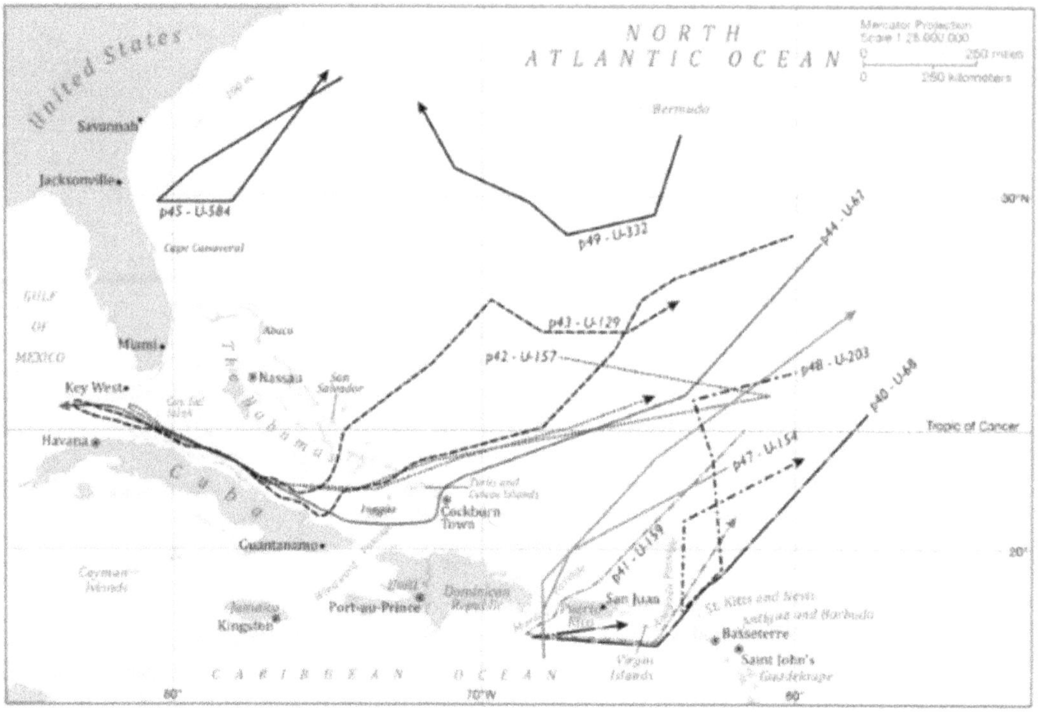

Chart 9: German U-Boat Patrols 50 to 59

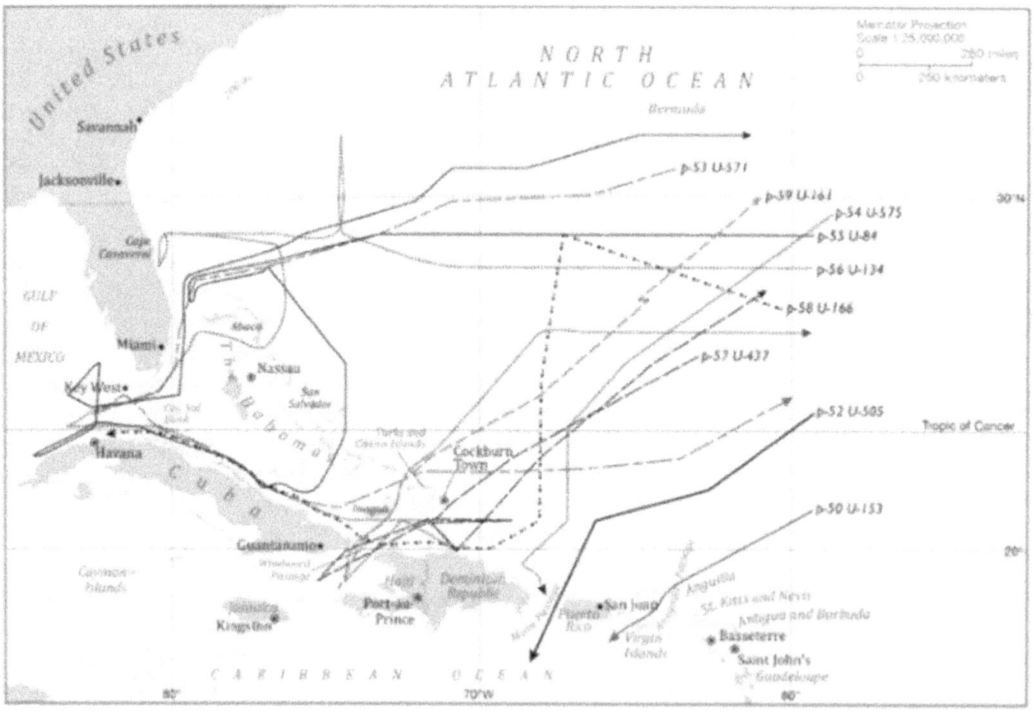

Chart 10: German U-Boat Patrols 60 to 69

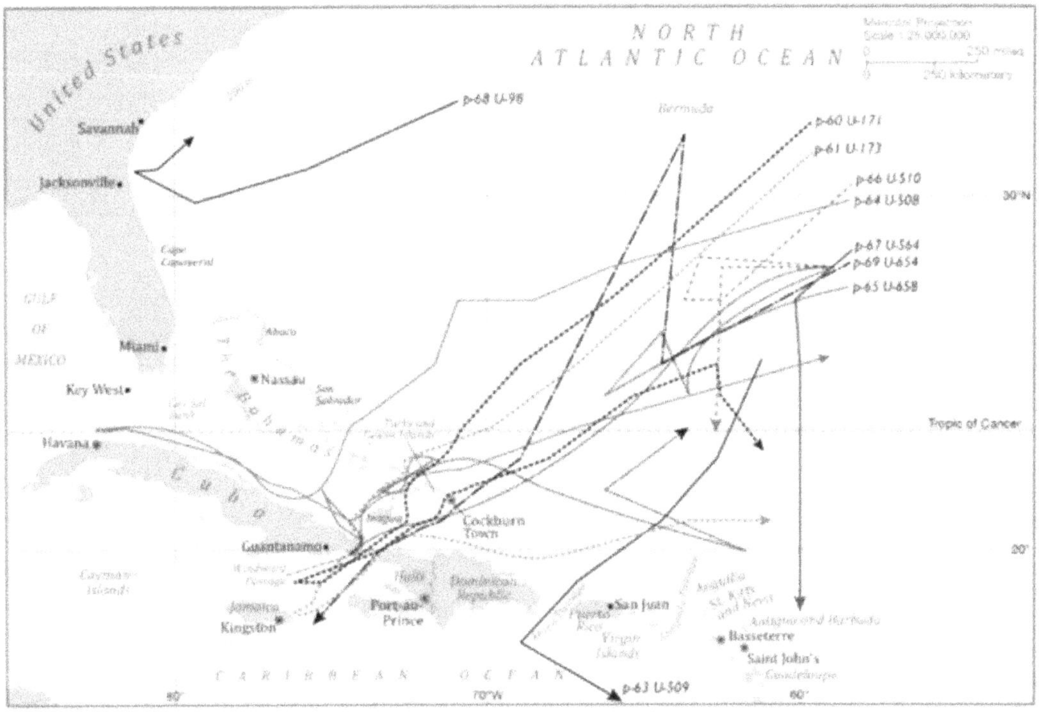

Chart 11: German U-Boat Patrols 70 to 79

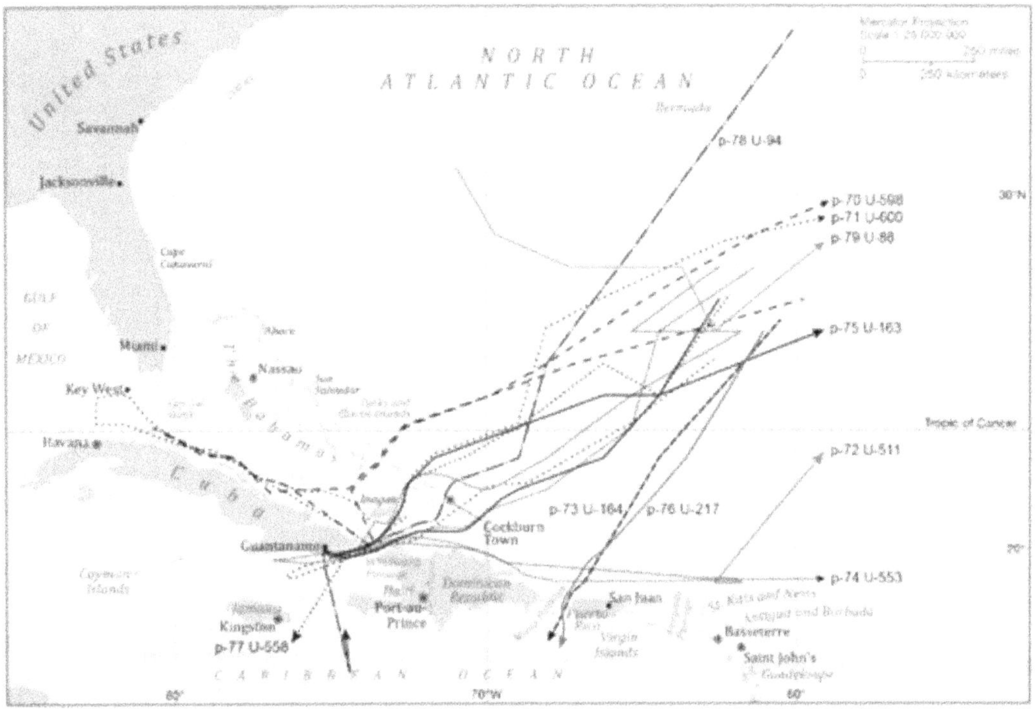

Chart 12: German U-Boat Patrols 80 to 89

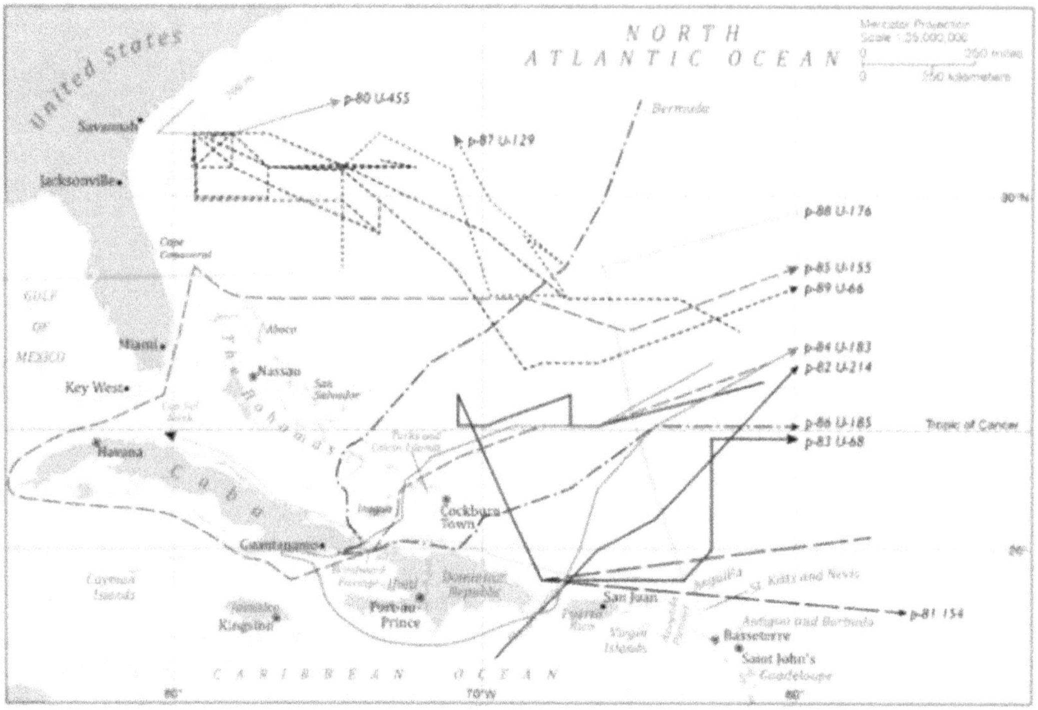

Chart 13: German U-Boat Patrols 90 to 99

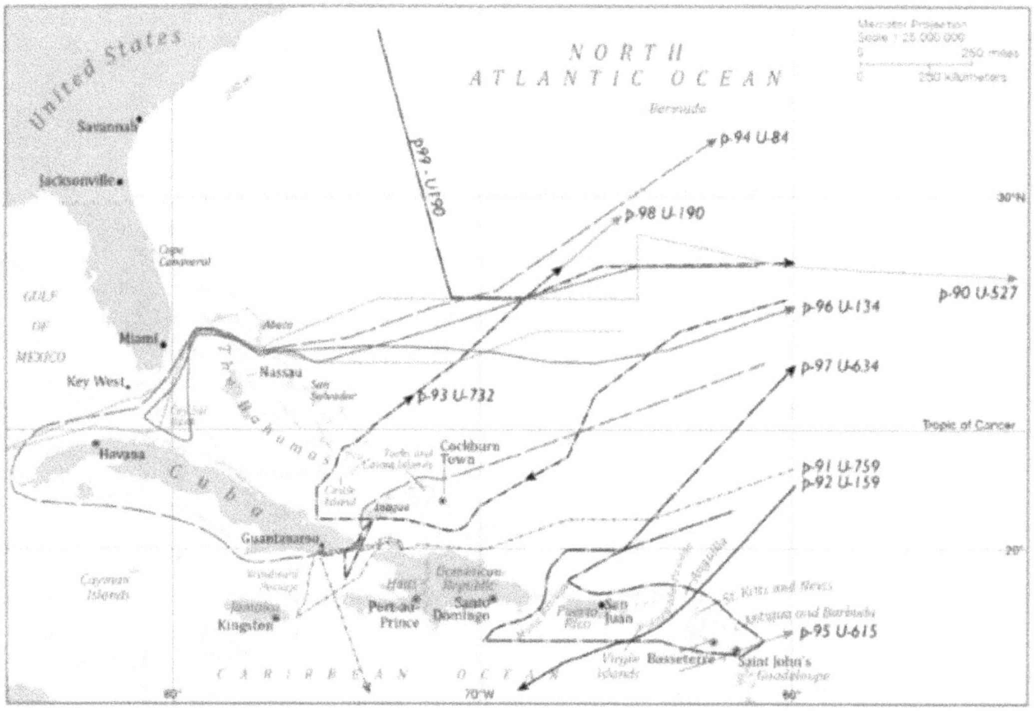

Chart 14: German U-Boat Patrols 100 to 111

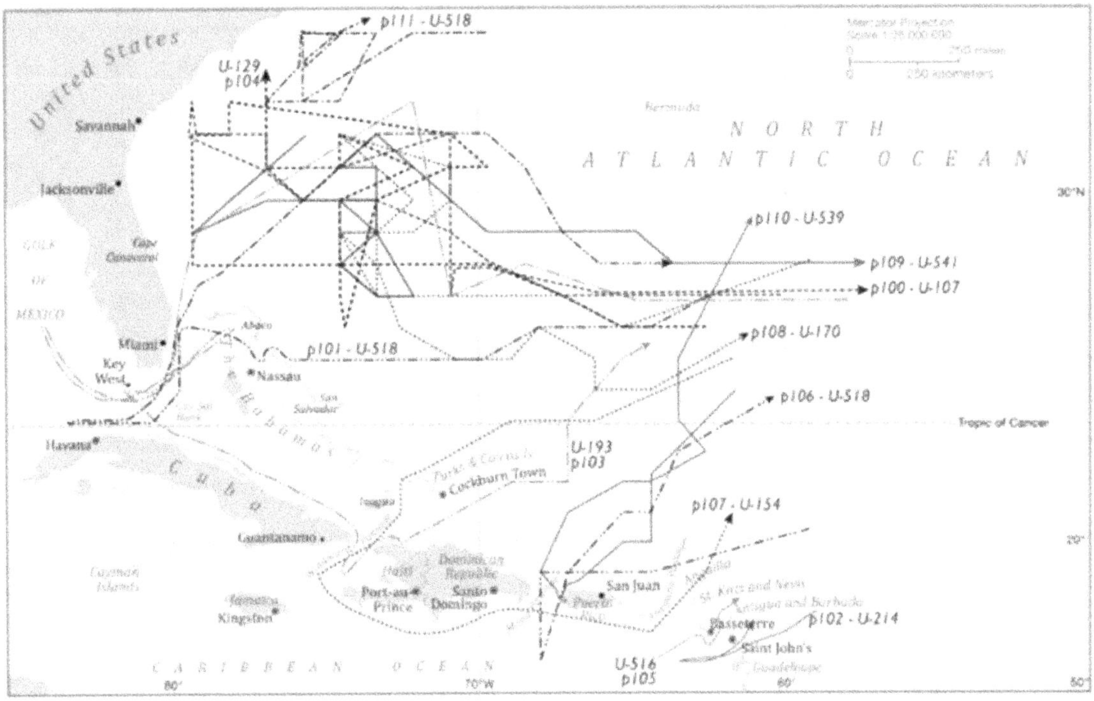

Chart 15: German U-Boat and Allied Military Vessel Losses

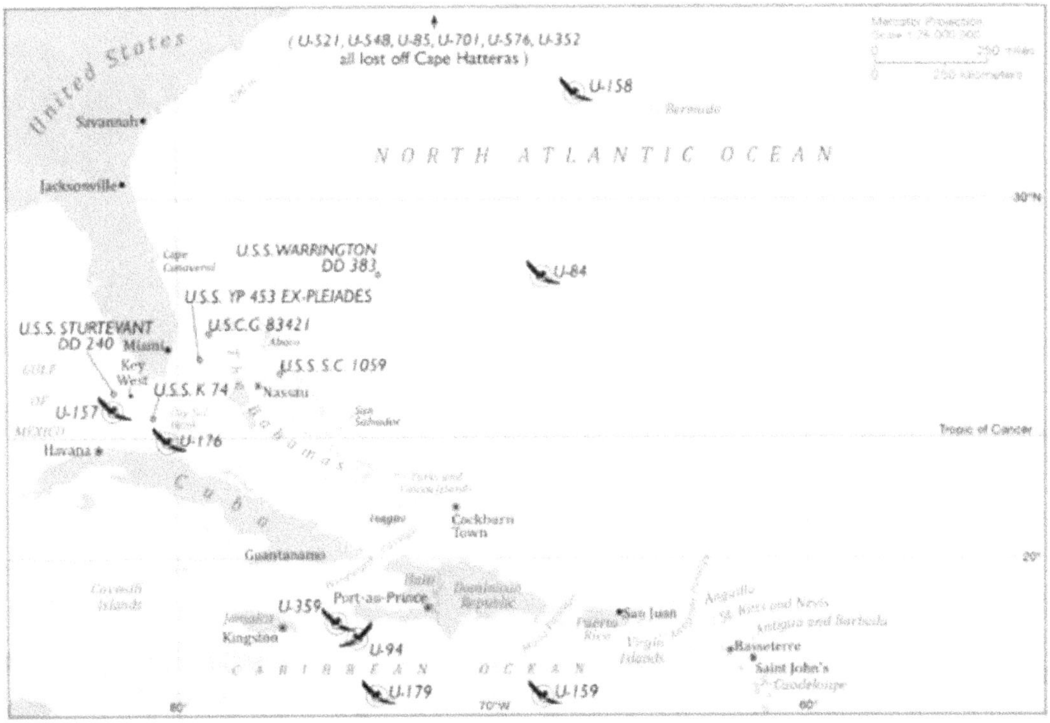

Chart 16: All 143 German U-Boat Patrols Around Bermuda

Chart 17: Allied – RAF, USN, Cuban – Bases in Region

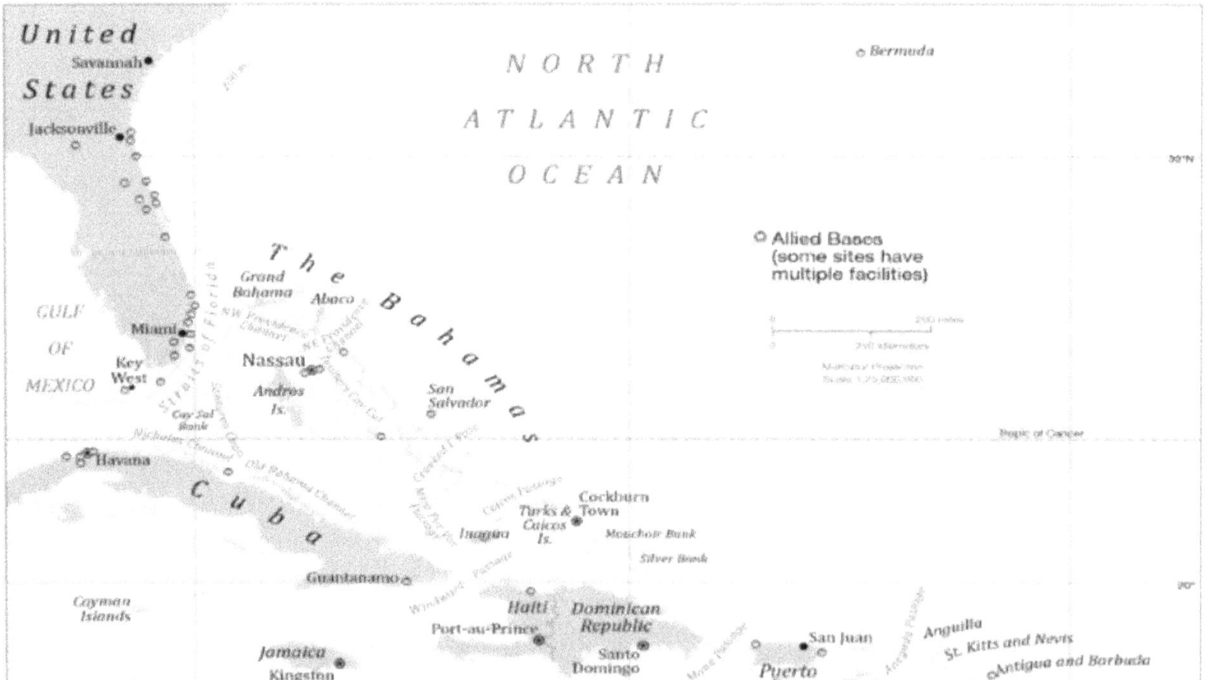

Chart 18: German and Italian Submarine Sightings of USN and RAF Aircraft

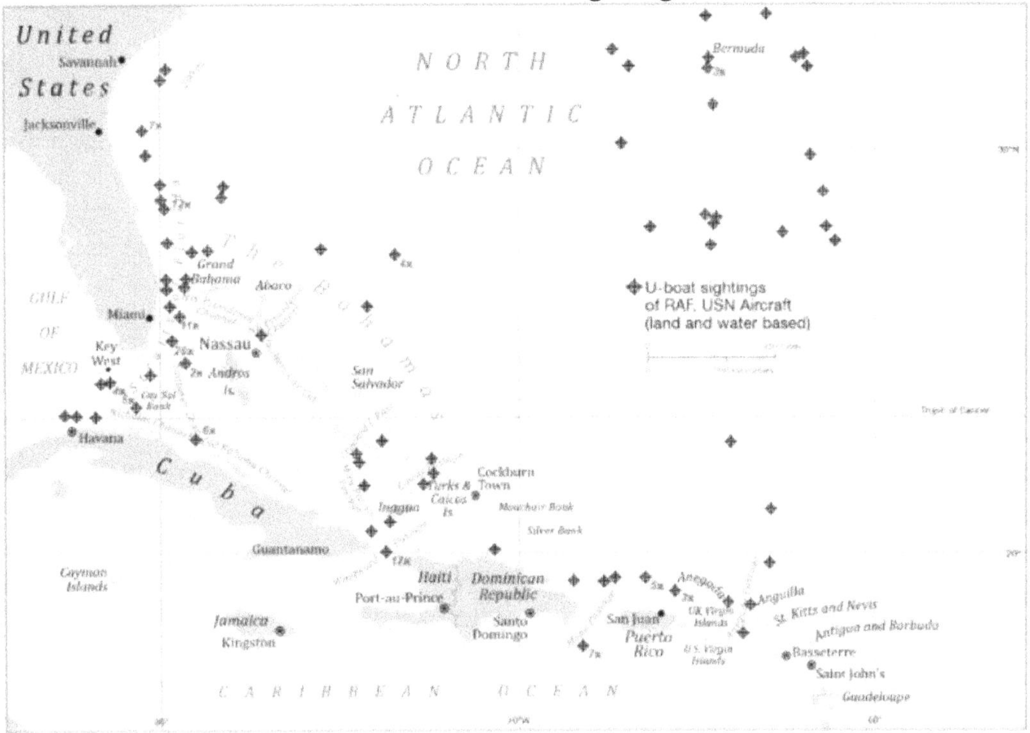

Chart 19: RAF and USN Aircraft Crashes and Accidents in & Around Bahamas

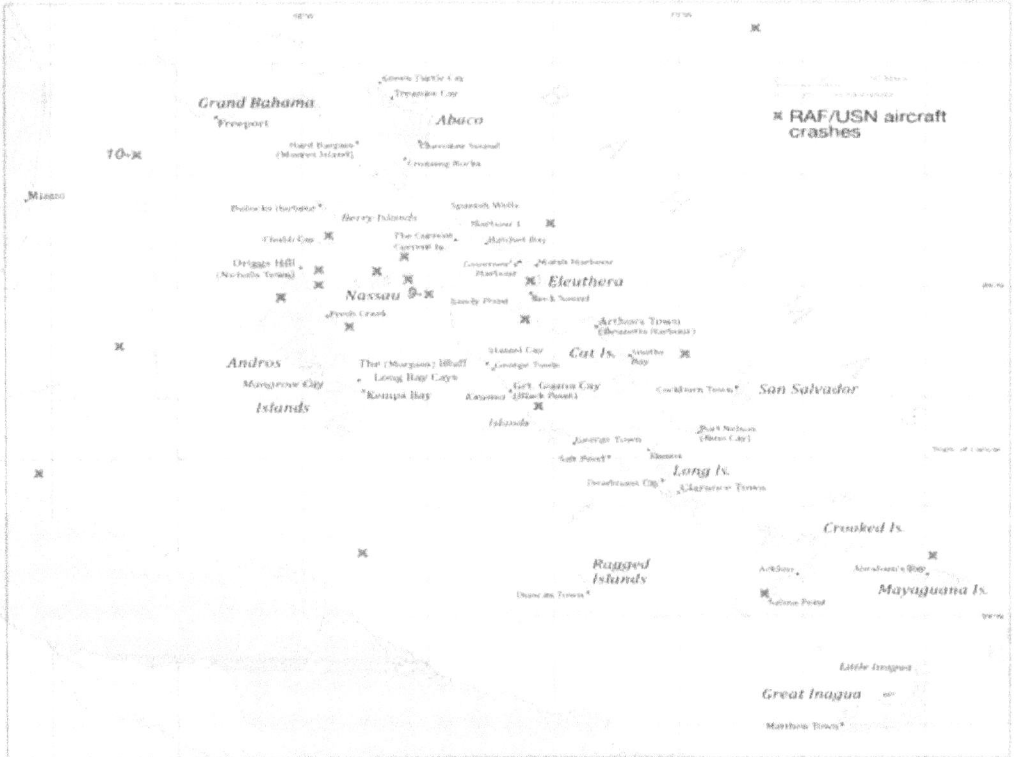

Chart 20: RAF and USN Aircraft Crashes and Accidents Around Nassau, New Providence

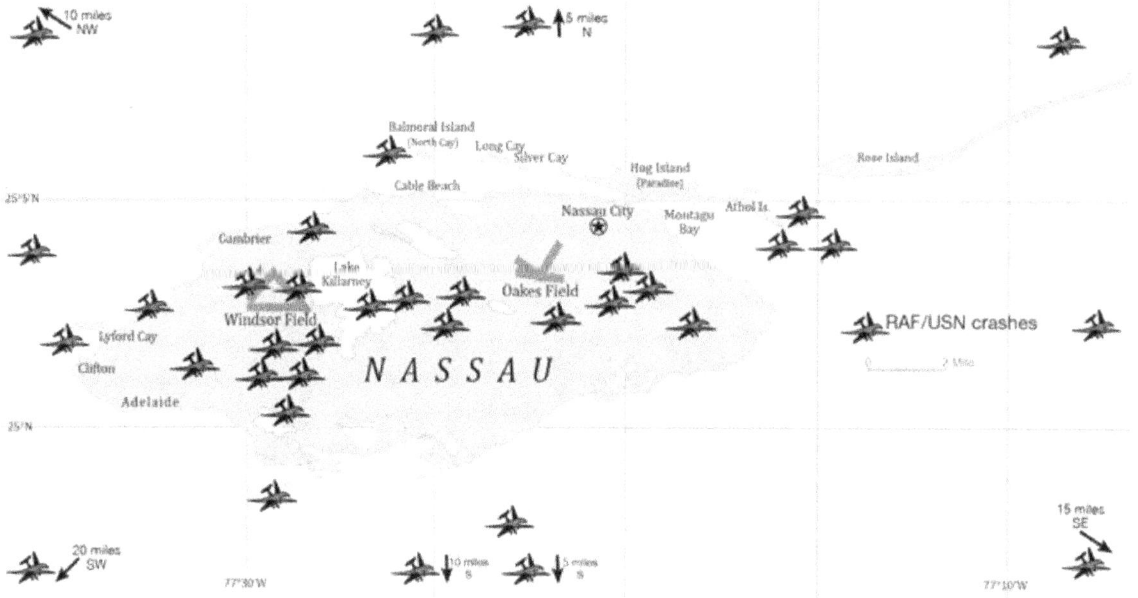

Chart 21: Allied Air and Naval Attacks in German & Italian Submarines

Chart 22: Loss of USN *K-74*, Only USN Airship Lost to a German U-Boat, *U-134*, Channels

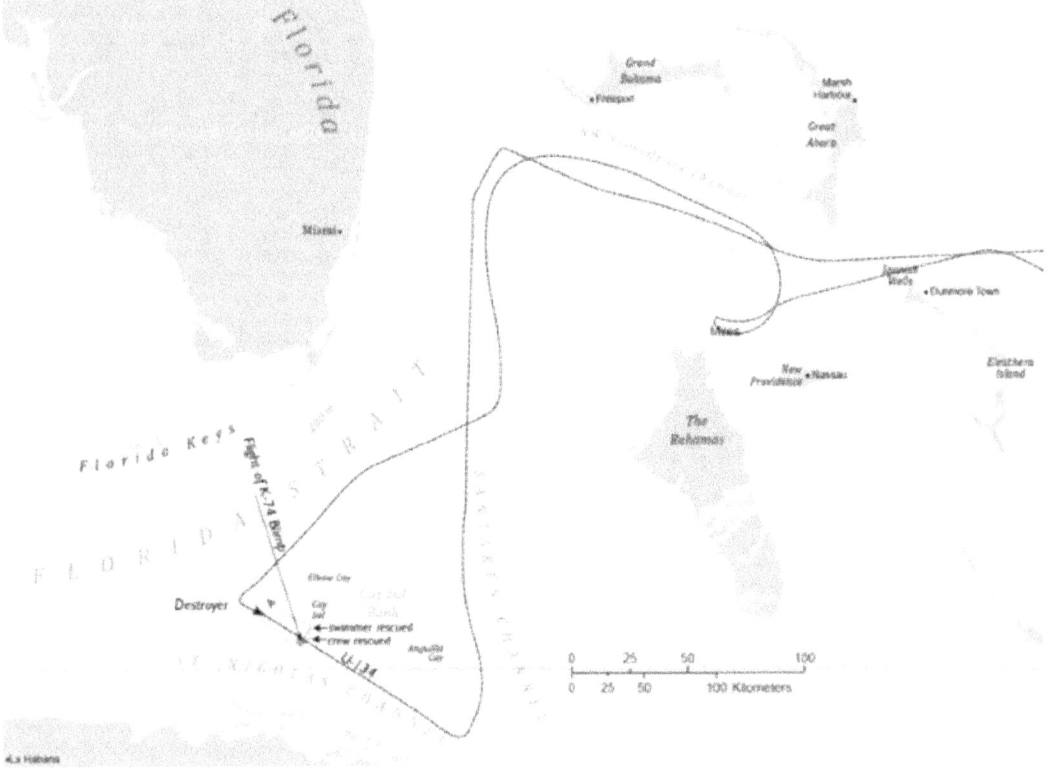

Chart 23: Rescue of Survivors of US-flag SS *Potlatch* in Bahamas by Marion "Joe" Carstairs

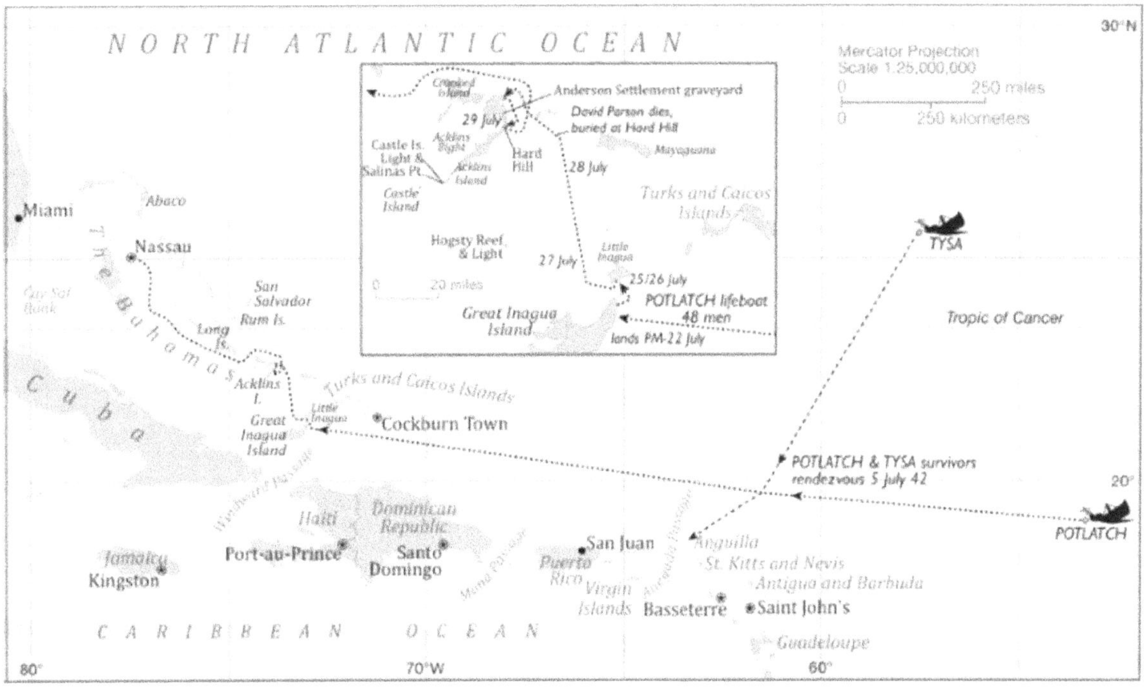

Chart 24: Channels, Passages, and Cuts Through Bahamas and Greater Antilles

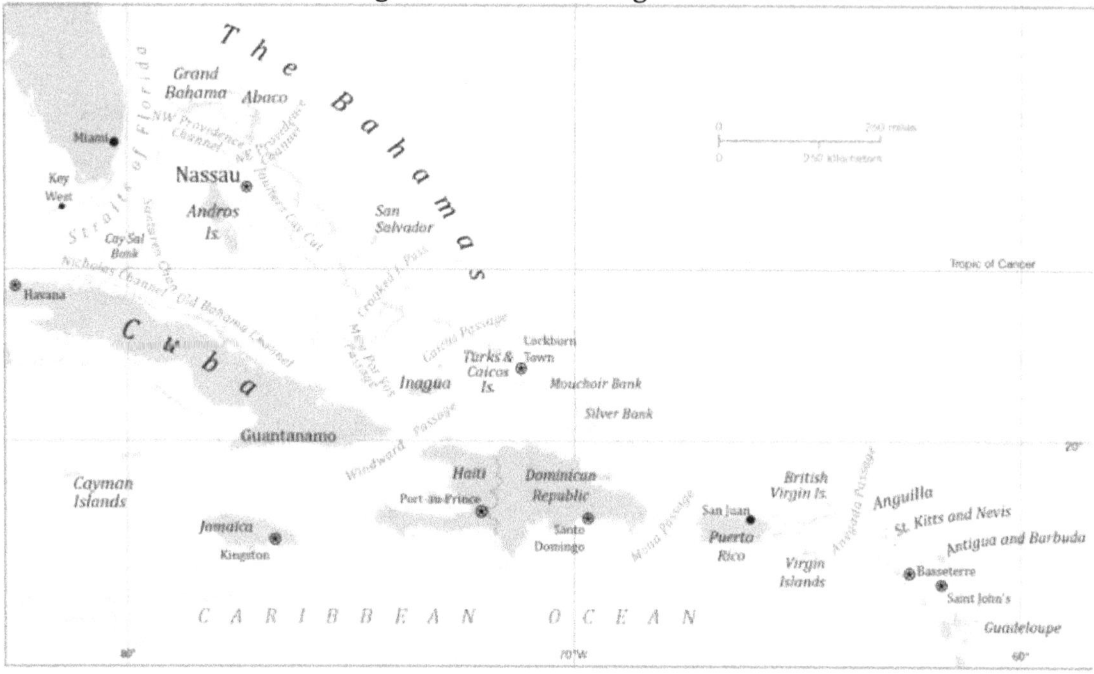

Chart 25: Names of Prominent Islands, Communities, Ports and Landmarks in Bahamas

Graph 1: Allied and Axis Killed in Action by Service

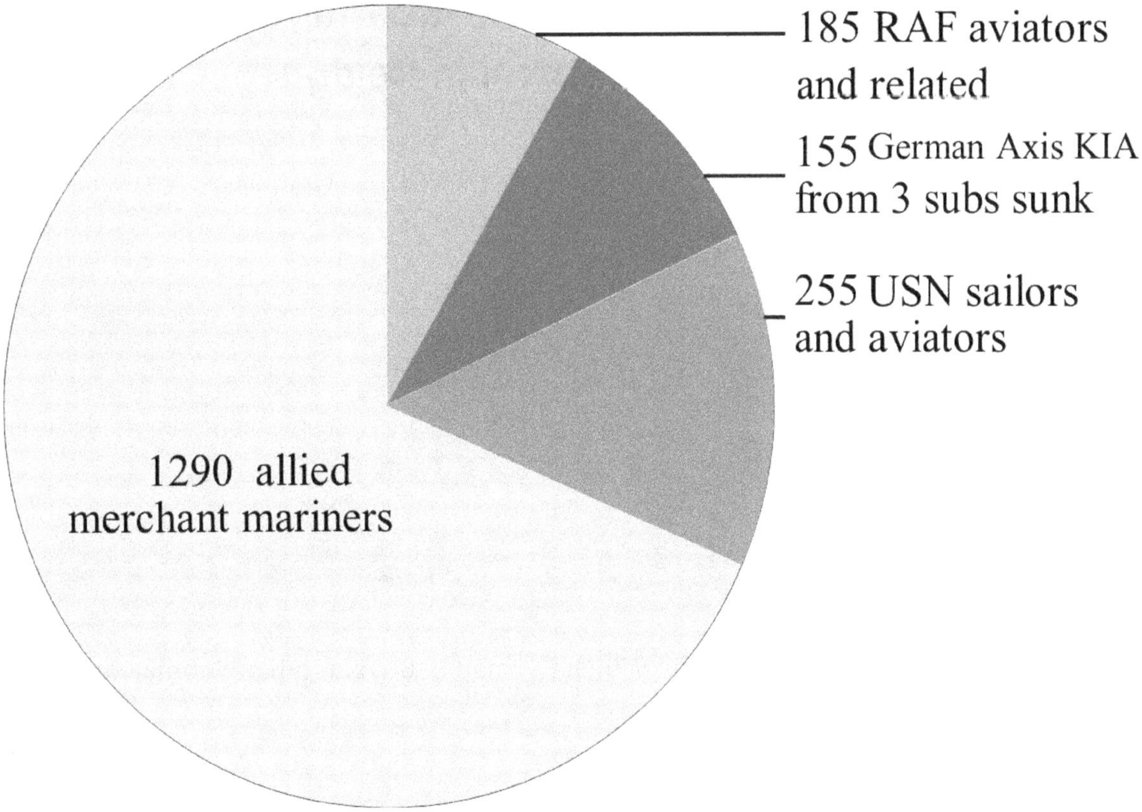

Diagram 1: Allied Merchant Ships Sunk by Axis Submarines, by Date

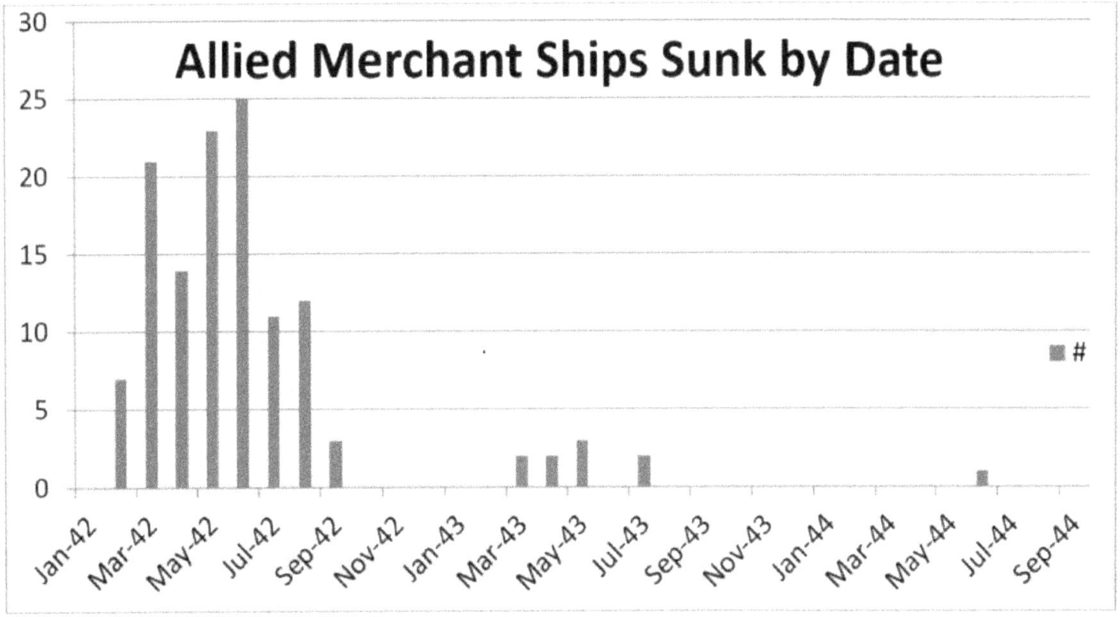

Diagram 2: Patrol Days by Start of U-Boat or Italian Submarine Patrol in Region

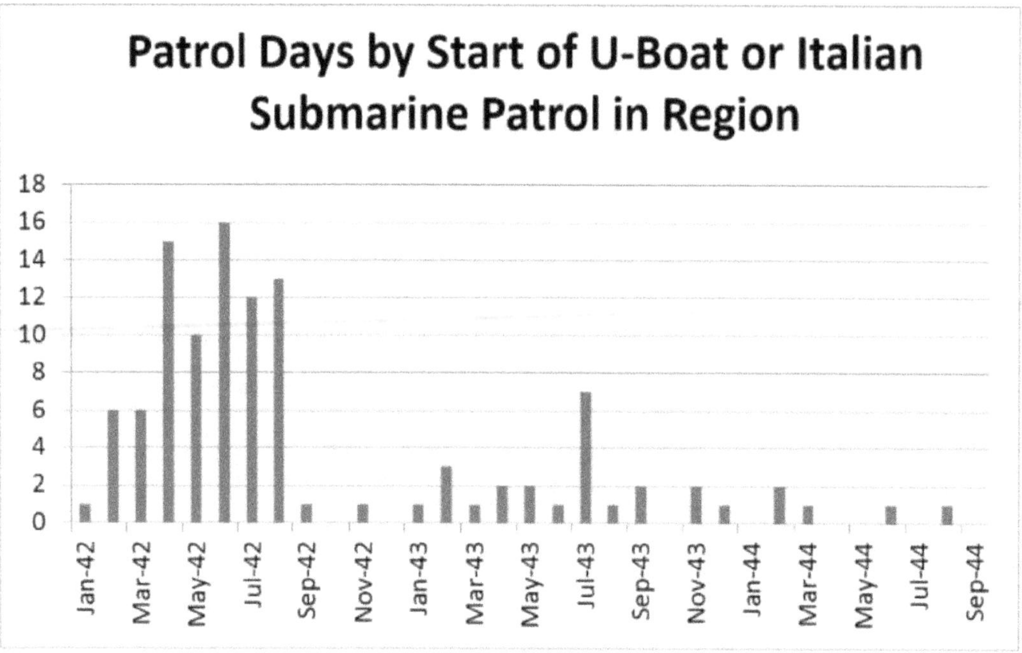

Diagram 3: Allied Ships Sunk by Which German U-Boat & Italian Submarines

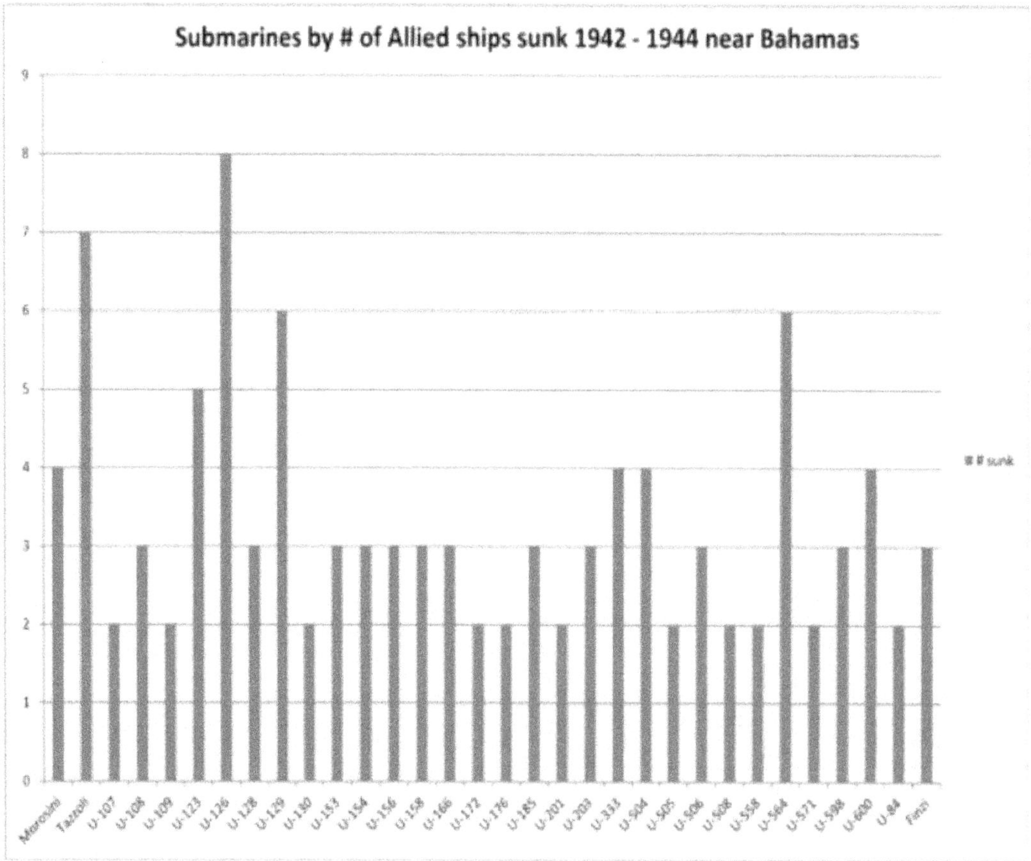

Allied Military Bases in Region (150)

Antigua:

1. Antigua Naval Air Station (NAS), Crabbs, Parham Harbour, British West Indies
2. Antigua US Army Airfield, Crabbs, Parham Harbour

Bahamas:

1. Bahamas Air Service Squadron, Jan. 1944 with 100 "coloured troops" at Oakes Field, New Providence, Fl. Lt. Cullum
2. Bahamas Batman Establishment, Oakes Field. Authorized by RAF Delegation Washington, July, 1944
3. Bahamas Defence Company (BDC, under Major Rutherford, Capt. Lightbourn, Adjutant in Aug. 1943, and Lt. Roberts in Feb., 1944)
4. Garrison of Pictou Highlanders (Company), Nassau (to protect Government House)
5. Garrison of Veterans Guard of Canada, Nassau (mission to protect Government House)
6. Main Field (Oakes Field), western New Providence, behind the city of Nassau, to the south
7. RAF 7 Ferry Crew Pool, Attached to RAF No. 113 Transport Wing, Oakes Field (aka Main Field), New Providence
8. RAF 7 Ferry Unit, Attached to RAF No. 113 Transport Wing, Oakes Field, New Providence
9. RAF 88 Terminal Staging Post (TSP), Attached to RAF No. 113 Transport Wing, Oakes Field
10. RAF No. 111 (Coastal) Operational Training Unit, RAF, Windsor (Satellite) Field, Nassau, NP
11. RAF No. 113 (South Atlantic) Wing (Transport Command), RAF Oakes Field, with HQ at No. 45 (Atlantic Transport) Group Dorval, Canada, outside Montreal
12. RAF No. 250 Air-Sea Rescue Unit, Harbour Island ASR marine base, Dunmore Town, Eleuthera
13. RAF No. 250 Air-Sea Rescue Unit, Lyford Cay ASR marine base, Clifton Bay, western New Providence
14. RAF No. 250 Air-Sea Rescue Unit, Montagu Bay ASR marine base, north-eastern Nassau, NP
15. RAF Out Island Commission, Jan., 1944 to recruit Bahamians to RAF, Bahamas Defence Company, Bahamas Air Service Squadron
16. RAF San Salvador Signals Detachment, amphibious, providing navigational aid to

transport aircraft on long-distance flights. Cockburn Town, San Salvador
17. RAF Station Hospital, Skyline Drive, Prospect Ridge, Lake Cunningham area, between fields
18. RAF Station Military Band (subset of No. 111 O.T.U.)
19. Satellite Field (later Windsor Field, after Oakes Field, aka Main Field), central-western New Providence
20. NAS Great Exuma, marine base, 63-foot rescue boat operated by ATC, for the USAAF
21. Mayaguana, marine base for 104-foot rescue boat operated by ATC for the USAAF
22. Naval NAS Great Exuma, original base, downgraded to NAAF on Sept. 25, 1944
23. Naval Air Facility (NAF) Great Exuma, under command of USN NAS Guantanamo
24. NAAF Great Exuma, under command of USN, Guantanamo, from Sept. 1944
25. Walker Cay, aka Advance Base Lano, marine resupply base, North Bahamas: from 1 Sept.1942
26. Cat Cay, Bimini, marine base for 42-foot rescue boat operated by ATC (Air Transport Command, USAAF) for ASR (aircraft search and rescue) under command of US Army

Bermuda:

1. NAS Bermuda, St. David's Parish, Bermuda
2. NOB Bermuda, St. David's Parish, Bermuda
3. Kingsley Field, Bermuda

Cuba:

1. Caimanera US Naval Base, Guantanamo, Oriente Province Cuba (southeast)
2. Columbia Military Airport, Havana
3. Comandante Manuel Pérez Alfonso, Aviation Camp (training), Rancho Boyeros, Havana
4. Casablanca Naval Base and Arsenal, Havana
5. Guantanamo Bay US Naval Base, Guantanamo Bay, Oriente Province
6. Havana Bay Subchaser fleet for amphibious aircraft, Pan American Air Lines, Havana
7. Prestón Aerodrome (anti-submarine patrol), Oriente Province, southeast
8. San Julian (Caribbean Command), Pinar del Río
9. Squadron V/S 62, Cayo Frances San Pascual; Las Villas Province
10. Agustín Parla Aeronaval Base, Mariel (Pinar del Río)
11. Batista Field (Caribbean Command), San Antonio de los Baños

Haiti:

1. Cap-Haïtien Haitian Coast Guard base, Cap-Haïtien, north coast

Puerto Rico:

1. 10th Bombardment Squadron, B-18 Bolo, Borinquen
2. 12th Bombardment Squadron, B-18 Bolo, Borinquen

3. 20th Troop Carrier Squadron (Panama Air Depot), Douglas C-47 Skytrain, Borinquen
4. 35th Bombardment Squadron, B-18 Bolo, Borinquen
5. Borinquen Army Airfield, Punta Borinquen
6. Roosevelt Roads Naval Air Station, Ceiba, off Vieques
7. Roosevelt Roads Naval BaseCeiba, off Vieques, Puerto Rico
8. 417th Bombardment Squadron, Douglas B-18 Bolo, Borinquen
9. 44th Bombardment Squadron (40th Bombardment Group), B-18 Bolo, Borinquen
10. 4th Tactical Reconnaissance Squadron (72d Reconnaissance Group), Borinquen

United States (Primarily Florida; see also Puerto Rico):

1. 8th Pursuit Wing, III Interceptor Command, Morrison Field, West Palm Beach
2. 26th Antisubmarine Command, Boca Chica NAS, Key West NAS
3. 26th Antisubmarine Wing (Squadrons), Marathon, NAS, Florida Keys
4. 26th Antisubmarine Wing (Squadrons), Flagler Beach Airport, north of Daytona
5. 26th Antisubmarine Wing (Squadrons), Lantana Airport, West Palm Beach
6. 26th Antisubmarine Wing (Squadrons), Meacham Naval Outlying Field, Key West
7. 26th Antisubmarine Wing (Squadrons), Morrison Field, West Palm Beach
8. 427th Army Air Force Base Unit, Morrison Field, West Palm Beach
9. Air Bombers School, Jacksonville NAS, Jacksonville
10. Air Bomber's Training Unit, Banana River NAS, Jacksonville
11. Air Bombers Training Unit, Jacksonville NAS, Jacksonville
12. Air Operational Base, Vero Beach NAS, Vero Beach, between Canaveral, Palm Beach
13. Air Operational Training Base, Daytona NAS, Daytona
14. Air Operational Training Base, Palatka, Florida, west of St. Augustine
15. Airship Headquarters Squadron Detachment 21, Air Station (Sea Duty), Richmond
16. Anti-Submarine Development Detachment, Atlantic Surface Div. (Sea Duty), Lauderdale
17. Anti-Submarine Warfare Training Unit, Naval, Auxiliary Air Station, Key West NOB
18. Army Air Forces Antisubmarine Command, 26th Antisubmarine Wing (Squadrons), Jax.
19. Army Air Forces Antisubmarine Command, Flagler Beach Airport Flagler Beach
20. Army Air Forces Antisubmarine Command, Lantana Airport, west of Palm Beach
21. Army Air Forces Antisubmarine Command, Marathon NAS, Marathon Island, Keys
22. Army Air Forces Antisubmarine Command, Meacham Naval Outlying Field, Key West
23. Army Air Forces Antisubmarine Command, Morrison Field, West Palm Beach
24. Banana River Naval Air Station (NAS), Jacksonville
25. Boca Chica Anti-Submarine Warfare Training Unit, NAAS Boca Chica, Key West NAS
26. Chief of Naval Air Operational Training, Jacksonville
27. Civil Air Patrol (CAP) base, Flagler Beach Airport, north of Daytona
28. Coast Guard Air Station, Miami NAS

29. Coast Guard Patrol Base, Marathon, Florida Keys
30. Coast Guard Training Station Palm Beach
31. Convoy Control Center, Key West NOB, Florida Keys
32. Daytona Naval Air Station (NAS), Daytona
33. Deland Naval Air Station (NAS), Deland, southwest of Daytona
34. Destroyer Escort Training Unit, Miami NAS
35. Dinner Key US Naval Air Facility, Miami
36. District Intelligence Officer, Miami NAS
37. Fleet Air Detachment, Deland NAS, Deland, southwest of Daytona
38. Fleet Sonar School, Submarine Training, Mine Warfare, Key West NOB
39. Fort Lauderdale Naval Air Station (NAS), Port Everglades
40. Gulf Sea Frontier, Anchorage Station, Key West NOB
41. HQ, Caribbean Division, Air Transport Command, 1100th AAF Base Unit, West Palm
42. Homestead Army Air Base (US Army), Homestead, south of Miami
43. HQ, Caribbean Wing, Air Transport Command (ATC), West Palm Beach
44. HQ, 25th Bombardment Group, Borinquen
45. HQ, 26th Antisubmarine Wing, Miami Army Air Field, Miami
46. HQ, 45th Bombardment Group (I Bomber Command), Miami Army Air Field, Miami
47. Jacksonville 1 Bomber Command, Jacksonville
48. Jacksonville Army Airfield, Jacksonville
49. Jacksonville Naval Air Station (NAS), Jacksonville
50. Key West NAS, Florida Keys, USA
51. Key West Naval Air Station (NAS), Key West, Florida Keys
52. Key West Naval Operating Base (NOB), Key West, Florida Keys
53. Key West Repair Base, Naval Station, Key West, Florida Keys
54. Lake City Naval Air Station (NAS), west of Jacksonville
55. Liaison Officer with the Soviet Navy Command in the United States, Miami NAS
56. Melbourne Naval Air Station (NAS), Melbourne, between Daytona and Palm Beach
57. Miami Army Air Field, Army Air Forces Antisubmarine Command, Miami
58. Miami NAS, Miami
59. Miami Naval Air Station (NAS), Miami
60. Morison FieldWest Palm Beach
61. Morrison Field Air Transport Command Base, West Palm Beach
62. Naval Air Operational Training Command Boat Facility, New Smyrna, S of Daytona
63. Naval Air Operational Training Command Boat Facility, Port Everglades
64. Naval Air Operational Training Command Boat Facility, Fort Pierce
65. Naval Air Operational Training Command Boat Facility, Mayport, near Jacksonville
66. Naval Auxiliary Air Station, Jacksonville Municipal #1, Jacksonville
67. Naval Auxiliary Air Station, Mayport, east of Jacksonville
68. Naval Aviation Operational Training Base, Sanford, between Orlando and Daytona
69. Naval Aviation Operational Training Base, Fort Lauderdale NAS, Fort Lauderdale

70. Naval Aviation Operational Training Bases, Deland NAS, Deland, SW of Daytona
71. Naval Aviation Operational Training Bases, Jacksonville NAS, Jacksonville
72. Naval Frontier Base, Key West NOB Key West, Florida Keys
73. Naval Radio Direction Finder and Radio Station, Jupiter, north of Palm Beach
74. Naval Training Center, Miami NAS Miami
75. Navy V-12 Unit, University of Miami (School), Miami
76. Office of District Intelligence Officer, Jacksonville NAS, Jacksonville
77. Operational Training Unit VPB2 #3, Banana River NAS, Jacksonville
78. Pan American Airways Training School, Miami NAS, Miami
79. Regional Air Traffic Coordinator, Caribbean, stationed at Miami NAS, Miami
80. Richmond Naval Air Station (NAS), Richmond, southwest of Miami
81. Sanford Naval Air Station (NAS), Sanford, Florida, between Orlando and Daytona
82. Section Base, Inshore Patrol, Fort Lauderdale NAS Fort Lauderdale, Port Everglades
83. Submarine Base, Key West NOB, Key West, Florida Keys
84. Submarine Chaser Training Center (Sub-Chaser School), Pier Two, Miami NAS
85. US Coast Guard Beach Patrol Station No. 3, Marineland, Saint Augustine
86. US Coast Guard Patrol Base, Key West NOB, Key West, Florida Keys
87. US Coast Guard Patrol Base, Port Everglades, Fort Lauderdale
88. US Coast Guard Training Station, Saint Augustine
89. US Naval Auxiliary Air Station, Green Cove Springs, Jacksonville
90. US Navy, Morrison Field, West Palm Beach
91. Vero Beach Naval Air Station (NAS) Vero Beach, between Canaveral and Palm Beach
92. Zone and Branch Intelligence Offices, Key West NOB, Florida Keys
93. Zone Intelligence and Naval Liaison Officers, Saint Augustine

Major Channels, Passages, Cuts Around, Through The Bahamas, Turks & Caicos, & Northern Greater Antilles

Passages Between the Greater Antilles:
Anegada Passage: between Anegada and Anguilla, northeast exit from Caribbean
Mona Passage: east of the Dominican Republic, west of Puerto Rico
Mouchoir Bank & Silver Bank: adjacent shallow discs of reefs, north of Dominican Republic
Windward Passage: southeast of Cuba, west of Haiti

Bahamas' Outer Boundaries:
Nicholas Channel: north coast Cuba, southern coast of Cay Sal Bank, Bahamas
Old Bahama Channel: north coast of Cuba, Havana to Windward Passage, southern Bahamas
Santaren Channel: north and east coasts of Cay Sal Bank, southwestern Bahamas
Straits of Florida: Key West to North of Jacksonville, entire west coast of Bahamas

Through Northern Bahamas:
Northeast Providence Channel: northeast of Nassau, south of Great Abaco, north of Eleuthera
Northwest Providence Channel: north of Great Isaac Light & Bimini, south of Grand Bahama

Through Southern Bahamas & Turks & Caicos:
Caicos Passage: west of Turks & Caicos and east of Mayaguana, south-eastern Bahamas
Crooked Island Passage: west of Crooked Island, south of Long Island, southeast Bahamas
Mira Por Vos Passage: west of Great Inagua, east of Ragged Islands, south-eastern Bahamas

Local Cuts, Passages Between Bahama Islands:
Booby Rocks: between New Providence, Nassau and The Current, North Eleuthera
Devil's Backbone, Eleuthera: a narrow passage from Spanish Wells to Harbour Island
Fleeming Channel, Six Shilling Cay, southwest of The Current, North Eleuthera
Joulters Cay Cut: southwest of Chubb Cay, northeast of Morgan's Bluff, Andros
Ship Channel Cut: north of Ship Channel Cay, Exuma Islands, southwest of Cape Eleuthera

Representative Naval Vessels, Merchant Ships and Aircraft

Representative Naval Vessels

Representative Merchant Ships in the Bahamas Theater in World War II

Ena K., 87`

Jean Brillant, 159`

Vivian P. Smith, 150`

Stede Bonnet, 119`

Content S., 110`

Lady Drake, 437`

Liberty ship, 441`

O. A. Knudsen, 508`

Representative Military Aircraft in the Bahamas Theater in World War II

Douglas A-20 Havoc (Boston)

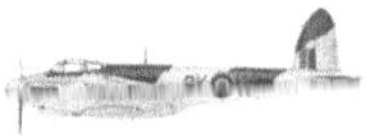

de Havilland Mosquito

B-25 Mitchell

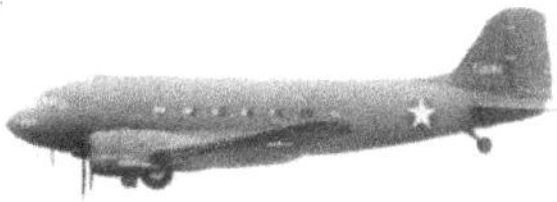

Douglas C-47

Douglas SBD

Martin B-26 Marauder

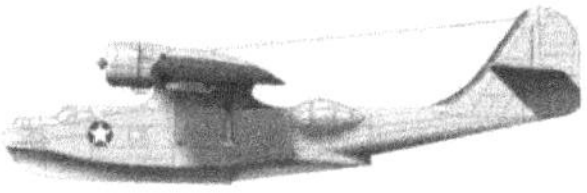

Consolidated PBY Catalina

Grumman G-21

Consolidated B-24 Liberator

Martin PBM Mariner

Martin Baltimore

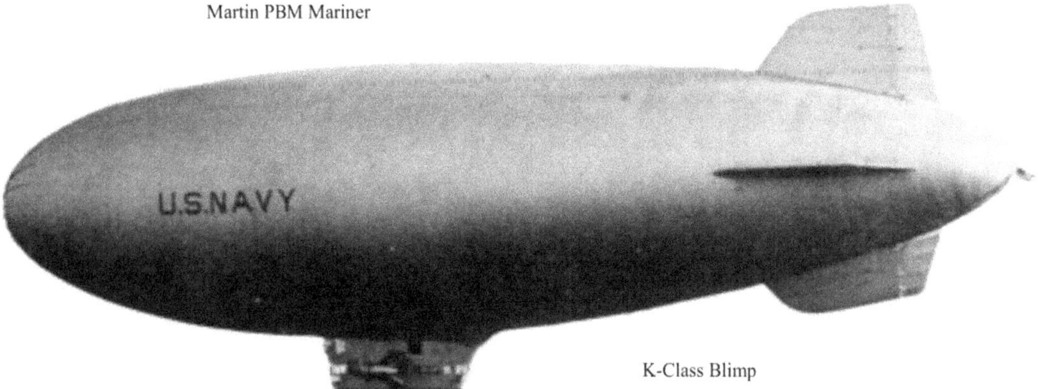

K-Class Blimp

Bahamas WWII Military Chronology: Daily Diary

Ship Types, Abbreviations, United States Navy in World War II

USS AG – Mics. (Auxiliary, general)
USS AM – Minesweeper
USS AO – Tanker (auxiliary, or fleet, oiler)
USS AOG –Tanker (auxiliary oil, gasoline)
USS AOR – Tanker (replenishment oiler)
USS APD – High speed transport
USS ARB – Battle damage repair ship
USS ARG – Repair ship
USS ATA – Tug (auxiliary ocean)
USS ATF – Tug (fleet ocean)
USS ATR – Tug (rescue)
USS AVP - Seaplane tender (USS *Sandpiper* converted minesweeper, *AM-51* to *AVP-9*)
USS CA – Cruiser (heavy)
USS CL – Cruiser (light)
USS CV – Aircraft carrier (also seaplane tender)
USS CVA – Attack aircraft carrier
USS CVE – Escort aircraft carrier (includes HMS *Arbiter*, D-31)
USS CG, or USCG – Coast Guard vessel (i.e., cutter)
USS DD - Destroyer
USS DE – Destroyer escort
USS DM – Minelayer (fast)
USS DMS – Destroyer minesweeper
USS FS – Corvette
USS K - K-class blimp, or non-rigid airship, for ASR (many based in Florida)
USS LCI – Landing craft, infantry
USS LCT(L) – Landing craft tank (large)
USS LST – Landing ship, tank
USS MB – Motor boat
USS ML – Marine launch
USS MLC – Motorized landing craft
USS MM - Minelayer

USS MMA – Minelayer (auxiliary)
USS MMC – Minelayer (coastal)
USS MSC – Minesweeper (coastal)
USS MSO – Minesweeper (ocean)
USS MV – Merchant vessel
USS PB – Patrol boat
USS PC – Patrol craft (was submarine chaser)
USS PCE – Patrol craft, escort
USS PF - Frigate
USS PT – Motor torpedo boat
USS PTC – Patrol torpedo boat, subchaser
USS SC - Submarine Chaser
USS SS – Submarine (attack)
USS W (prefix) – US Coast Guard
USS YAG – Seaplane tender (AV- aircraft carriers were also seaplane tenders)
USS YFB – Ferry boat (launch)
USS YFD – Floating dry dock
USS YHB – House boat (accommodation)
USS YMS – Minesweeper
USS YMT – Motor tug
USS YO – Fuel oil barge (oiler, tank barge)
USS YOG – Gasoline tank barge (oiler, barge)
USS YOGN - Gasoline tank barge (non-self-propelled)
USS YON – Fuel oil barge (non-self-propelled)
USS YP - Patrol craft (training)
USS YRB – Repair and berthing barge
USS YT – Tug (harbour)
USS YTB – Tug (large harbour)
USS YTR – Tug (rescue)
USS ZR – Rigid airship (navy blimp, dirigible)

Abbreviations

A-A, AA – Anti-aircraft gunnery (was also practiced on cays, rocks, floating targets)
A/C - Aircraft
ADC – *Aide-de-Camp*, in this case to HRH the Duke of Windsor
ASDIC – Submarine detection (Anti-submarine Detection Investigation Committee)
ASR – Air sea rescue (in Bahamas ASR launch boats are designated HMS P-…)
ASW - Anti-submarine warfare or exercises
ATC – Air traffic control
BAL - Bahamas Airways (1936 to 1970), commercial airline, at times government subsidized, operated seaplane, amphibious land-based, Bellanca aircraft, tail numbers included *RD-4* in World War II. See P. Aranha's eponymous history
Bahamas Ferry – freight and passengers Nassau to Miami, owned by Kelly, Sawyer, et al
Bahamas Mailboat – inter-island freight, passenger boats, sail, wood, various owners
Bombs – 500 lb SAP, 62-inches long, Mks II, III, IV, 490 lbs., No. 30 tail fuze, dark green, white bands, 11.5" diameter, round tail, duds commonly used, RAF Bahamas
Bounce – Hard landing, often precipitating a stall and crash, fire, fatality (also a drill)
B/T – communication system using telegraphy, drill for radio blind, or silent [with U/T]
Bu.No. – Builder's number; US airplane manufacturer's number for aircraft (a serial #)
Cdr. - Commander
CinC – Commander in Charge
CLA Search – Creeping line ahead search pattern in tandem with other aircraft
CO - Commanding Officer
Comgulfseafron – Commander, Gulf Sea Frontier (GSF; US military, east coast of Florida)
CORT – Convoy operational readiness training (as in **Convoy CORT Div.-67** on 6.11.44)
D/F – Direction finding
DFC – Distinguished Flying Cross (issued by US and UK)
DFCO - Duty Flying Control Officer
D/I – Dangerously ill (e.g. air crash survivors)
DR – Dead reckoning, navigating without many instruments on past tracks, other inputs
DTC – Descent through clouds (exercise)
EC2 – Liberty ship, common non-tanker WWII freighter design from the US
E/T - Form of communication, transmission, telegraphy to/from aircraft (see M/T)
ETA – Estimated time of arrival
EV – Escort vessel (or E/V, used with MV or M/V, for merchant vessel)
FFS - Free French Navy ship; an Allied vessel (as opposed to Vichy French)
FLEX – Fleet exercise
Gulf – US Gulf, aka Gulf of Mexico, broadly the GSF or Gulf Sea Frontier, Comgulfseafron
HDML – Harbour defence motor launch (HMS, British Royal Navy)
HE – His Excellency, or Her Excellency
HM – Her Majesty, or His Majesty (King or Queen)
HMAS – Her Majesty's Australian Ship
HMCS – Her Majesty's Canadian Ship

HMS – Her Majesty's Ship (British Royal Navy)
HMS P- ASR launch, or motor craft (as in HMS *P-190)*
HMM – Minesweeper variant (HMS, British Royal Navy)
HMYMS – Minesweepers (HMS, British Royal Navy)
HNoMS – Royal Norwegian Navy ship (equivalent to USS, or HMS)
HRH – His Royal Highness, or Her Royal Highness (in this context, the Duke of Windsor)
Hurricane Warnings – in Red, Blue, White to degree of severity (worst to least)
I/C – In charge
Kcl – Kilo cycle, expresses frequency of electromagnetic waves; kilo = 1,000, see Khz
Khz - Kilohertz, frequency of electromagnetic waves; kilo = 1,000, see Kcl.; re: radios
KIA – Killed in action
KMNS – Royal Dutch Navy Ship (equivalent to USS or HMS)
Knots – Nautical miles per hour (a nm being 1.18 land miles, a knot is faster than mph)
Launch – small motor boat, as in ASR, or air sea rescue marine craft, nimble, sturdy
Leaflet-dropping - Governor's Harbour, Port Howe, Cat Island, Clarence Town, Long Island, Georgetown, Exuma, Great Sale Cay, Grand Bahama, Hope Town, Abaco, Nicholl's Town, Andros (perhaps others; see September 7 and 8, 1943)
Leigh Light – Bright light on wings of bombers to stun U-boat crew on surface at night
Liberty Ship, or EC2; common non-tanker WWII freighter design from the US
Lt. - Lieutenant
LORAN – Long range navigation system relying on shore towers
MA – Critical point in an accident investigation
M/F, D/F, or MFDF – Medium frequency direction finding, or finder; navigational aid
MIA – Missing in action
Mosquito - British de Havilland DH.98 (twin engine, multi-role, made of wooden)
M/T – Form of communication, transmission, telegraphy to/from aircraft (see E/T)
MT – Mechanized, motor transport; provides road transport for air fields, officers, crew
MV- Merchant vessel (also M/V, used with EV, E/V or escort vessel)
NCO – Non-commissioned officer
N/E – Not operable (as in radio not transmitting)
NM – Nautical mile, or 1.18 statute, or land miles, use by vessels and aircraft
OinC – Officer in Charge, or Officer in Command
Padre – Religious minister, Reverend
PFL – Practice forced landing
PLE – Plane or pilot's limit of endurance
PSR – Point of safe return
RADAR – surface and air recognition sending, receiving radio waves
RAAF – Royal Australian Air Force
RAF – Royal Air Force
RATC – Remote air traffic control
RCAF – Royal Canadian Air Force
RDF – Range direction finding, a system of triangulating position from fixed objects
RN – Royal Navy (Her Majesty's Ship, HMS)
R/N – Radio operator, navigator

RNZAF – Royal New Zealand Air Force
R.Smg – *Regia Sommergibili*; submarine of the Royal Italian Navy (*Regia Marina*); ocean-going large Axis attack submarine of BetaSom flotilla, a German joint-venture
R/T – Radio telegraphy, or radio (see W/T)
SAP - Semi Armour Piercing, as in *British 500 pound SAP Mk. II & III Bombs* used by RAF
S/B – Communications method (possibly side-band)
S/E – Listening or sound detection device gauging emissions (also SE)
SEO – Senior executive officer (often conflated with SOE, below)
SOE – Special operations executive (elite RAF bomber command role, conflated with SEO)
SOE – Safety Operating Envelope (like PLE, plane, pilot limit of endurance; a boundary)
SONAR – Sub-sea recognition using system of pings sent, received (sound propagation)
SONO Buoy – a detection float dropped to the sea that is equipped to detect underwater sounds and transmit them by radio. Useful then as now to detect U-boats.
SOP – Senior officer present
SOS – Save our Souls (*Mayday*, from French for help me; *m'aidez*)
SS – Steam ship; general carrying dry cargo, not a tanker (M/V, MV is motor vessel)
Stall – Loss of control of aircraft, lack of propulsion, often resulting in crash. Causes include bounce landing, machinery or systems failure, weather, pilot error, fuel loss, fire, enemy fire. Challenging to recover from. RAF Bahamas crashes often left insufficient evidence to conclusively determine cause. Fast, unseen, in deep water.
S/T – Steam tanker
RATC - Remote air-traffic control
RU – Russian ship, in the Soviet, or USSR Navy (equivalent to USS or HMS)
U/S – Un-serviceable (as in radio), or Under Stands, Under Stood
USCGC – United States Coast Guard Cutter
USN – United States Navy
USS – United States Ship
USSR – Union of Soviet Socialist Republics (aka Russia)
U/T – Communications system utilizing telegraphy (used with B/T)
V/S – Vertical speed
W/T – Wireless telegraphy, or radio (see R/T, W/P, U/T, B/T, S/E)
W/P – Wireless device for communication (possibly conflated with W/T)
YMCA – Young Men's Christian Association; canteen or social hall without alcohol

Codewords for Locations, Exercises

Adder [reptile series] (patrol exercise)
Alligator [reptile series] (patrol exercise)
Blues (search pattern, see Reds]
Borinquen (US Navy air field in northwest Puerto Rico)
Boca Chica (Key West, Florida, the US Naval Air Station)
Bounce drill; both an exercise and an often fatal landing
Caicos Island, Turks Island, Grand Turks Island, Providenciales: all part of Turks & Caicos
Cay Sal Bank; 2,000-square-mile atoll in Bimini District, Bahamas between Cuba, Florida Keys, Andros. One of the world's largest atolls: resource rich, strategic, forgotten
Cayo Francés; Cuba, opposite Cay Sal Bank, turning point in Old Bahama Channel
Cayo Lobos Light, Old Bahama Channel (position)
Clifford Park, Nassau (practice range) duds bombing demonstration, 10 November, 1944
Conch Sound, northeast Andros, Nicholl's Town, Grumman Goose GA, landed 7 July, 1945
Cobra [reptile series] (patrol exercise)
Cockerel (patrol exercise)
Crocodile [reptile series] (patrol exercise)
Devils Point, Cat Island (destination)
Esco (patrol exercise)
Fleeming Channel, Rose and Current islands (USS *SC-1059* aground, 11 Dec. 1944)
Fox Hill Beach (target position; site of crash by B-25 "FL" on 28 October, 1944)
Frog [reptile series] (patrol exercise)
Great Sale Cay, between Grand Bahama, Little Abaco, and Walker Cay (destination)
Great Stirrup Cay Light, Berry Islands, on Northwest Providence Channel (destination)
Guantanamo, Cuba (US Navy's Naval Operating Base and Naval Air Station)
Harvey's Cay, Central Exuma Cays [aka Harvey Cay] (ASR search boundary)
High Point Cay, South Andros (ASR search boundary)
Hop very short training flight, or quick transit
Joulter Cay, northwest of Morgan's Bluff, north Andros (ASR activity 1 May, 1945)
Kingsley [regal series] (patrol exercise)
Liberty ship (position of vessel)
Lizard [reptile series] (patrol exercise)
Loop short training flight around and back
Lyford Cay ASR Base, was an inlet in Clifton Bay, between Clifton Pier and Goulding Cay, at the Wishing Well Canal, Clifton Bay Drive. Entrance at modern Jaw's Beach. This is the original marina, by H. G. Christie, before they sold it to Lyford Cay Club
Mamba [reptile series] (patrol exercise)
Man Island Light, north of Harbour Island (destination)
Marsh Harbour, Abaco; Grumman Goose GA landed, 10 March, 1945
Moore's Island, Abaco; Grumman Goose GA landed Hard Bargain, 10 March, 1945
Morgan's Bluff, North Andros (ASR search boundary)
Palisadoes, Jamaica; often from Great Exuma, via Guantanamo

Paw Paw Rocks, north of Grand Bahama, B-24's searched, US Privateer, 11 March, 1945
Pelican, seaplane tender USS *Pelican* (AVP-6), a dummy target (position)
Pelican Cay, Abaco (destination)
Platter (patrol exercise)
Ploughshare, dummy target (towed behind a target ship)
Python [reptile series] (patrol exercise)
Queen [regal series] (position, destination)
Queensley [regal series] (patrol exercise)
Reds (search pattern, see Blues)
Riding Rock, San Salvador (destination)
Royal Island, west of Spanish Wells, where seaplanes sometimes moored
Rum Cay (bombing practice, off of)
Sail Rock, Exumas, north of Ship Channel Cay (bombing target)
Salamander [reptile series] (patrol exercise)
Samaná Cay, north of Acklins, Crooked, Mayaguana, east of Long Island (destination)
San Salvador, packages parachuted, include on dropped at "the hotel," 10 May, 1944
Sand Key, one of Florida Keys in USA near Key West, used in training (not Bahamas)
Sandpiper, seaplane tender USS *Sandpiper* (AVP-9) (destination)
Serpent [reptile series] (patrol exercise)
Silver Cay (practice range), on 3 November, 1944 used in demonstration of dud bombs
South Point, Hole-in-the-Wall, Abaco (USN arbitrary name, as in North Point Eleuthera)
Southwest Point [now Freeport], Grand Bahama (ASR search boundary)
Spanish Wells (gunnery exercises)
Tampa (patrol exercise)
Turks Island, Grand Turks Island, Providenciales, Caicos Island: all part of Turks & Caicos
Viper [reptile series] (patrol)
Walker Cay, north of Grand Bahama and Abaco (destination)

Part I

Prequel: September 1, 1933 to December 27, 1939

1st September 1933
Future German U-Boat ace Werner Henke departed Germany for Bahamas on the Hapag passenger liner *Resolute,* which returned to New York on 6th September. He signed on as an Ordinary Seaman on 13th July 1933, while waiting evaluation of his application to the Navy, which was issued. [After distinguished wartime service, he was captured in U-515, imprisoned, and committed suicide by rushing the perimeter wire in captivity outside Washington, DC on 15th June 1944].

15th January 1939
Construction on the first 3,000-foot runway of what became Oakes Field began for Her Majesty's Government, and with Bahamas Airways (BAL) anticipating scheduled service.

1st May 1939
The heavy cruiser USS *Wichita* (CA-45) conducted her shakedown cruise at the Virgin Islands, Cuba, and the Bahamas. Having departed Houston, she then headed north to her builder's yard for post-shakedown repairs.

10th July 1939
The destroyer USS *Borie* (DD-215) was to the north of St. Thomas, Virgin Islands when she reported sighting a submarine which submerged on her approach. This had to have been a French submarine, believed to have been operating from Vichy French Caribbean islands including Guadeloupe and Martinique.

15th August 1939
The German tanker *Altmark* obtained fuel oil in Port Arthur Texas and then transited the Straits of Florida and possibly the Northwest Providence Channel and Northeast Providence Channel to return to Europe.

30th August 1939
The Canadian National west Indies Steamship Company passenger liner *Lady Rodney,* carrying up to 340 passengers, left Montreal on a voyage to the Bahamas, Jamaica, and Cuba via Bermuda. At the outbreak of war, Captain Edward LeBlanc ordered his crew to paint over the ship's white colours with US Navy grey as a means of disguise.

1st September 1939
Three days before the outbreak of war, the German tanker *Altmark* used the same Texas fuel oil to supply the German battleship *Admiral Graf Spee* southwest of the Canary Islands. At the same time the passenger liner *Lancastria* was on a cruise in the Bahamas under Captain Sharp.

3rd September 1939

The United Kingdom declared war on Germany. The British SS *Athenia* was immediately sunk by *U-30*. The *Lady Rodney* arrived in Bermuda, where her master was informed that wartime grey paint was unnecessary. Therefore, he had the crew paint the hull white again.

4th September 1939

The Swedish motor yacht *Southern Cross*, owned by the owner of Electrolux, Axel Wenner-Gren and his American wife, who were attempting to flee Europe for the Bahamas before war broke out, stopped to pick up 376 survivors of *Athenia*. Of these, 236 were later transferred to the US SS *City of Flint*, which landed them at Halifax, Canada. The balance of 140 souls were landed in Miami after the *Southern Cross* diverted from Nassau, its intended destination, where the Wenner-Grens owned Hog (later Paradise) Island. The Admiralty in London requisitioned the passenger liner *Lancastria* as a troop ship and immediately sent her from Nassau to New York. Her hull was painted grey, and a 4-inch gun was mounted on the taffrail. On 17th June 1940 the RMS *Lancastria* was sunk by air attack off France while desperately overloaded with evacuating troops as part of Operation Ariel. Though little-known, the tragedy claimed at least 3,000 lives and as many as 5,800, making it the single largest loss of life aboard an individual British ship.

5th September 1939

US citizens were henceforth required by the Neutrality Act to sail to Bahamas only on US-flagged, and not foreign, or Canadian, ships. Until the attack on Pearl Harbour in December 1941 over two years later, when US ships were requisitioned by their government, British vessels were hard-pressed to compete (or survive) on the busy Miami-Nassau route. When war broke out, the Canadian passenger liner *New Land* was unable to take the Miami-Nassau route as US citizens were forbidden from travelling in a "belligerent ship." The use of unrestricted U-boat warfare off the British Isles, followed by Gunther Prein's audacious penetration of the Royal Navy base in Scapa Flow (in which he sank HMS *Royal Oak*) put critical British supply lines in peril during the period before the fall of France and the Battle of Britain. The Germans raced to keep ahead of both UK and Russian war production by starving the UK of foodstuffs, fuels, heating oils, and other material. At first, they used some 50 U-boats, however with the lifting of restrictions as a result of war, they put pre-existing construction concepts, tested in Netherlands and beyond, into full production in Germany.

20th September 1939

Yacht *Southern Cross* landed 140 survivors of torpedoed SS *Athenia* in Miami *en route* to Nassau. Wenner-Grens owned Hog Island, and commissioned a canal to mollify rioting workers and please the Duke of Windsor, out of his own pockets. He was nevertheless banished.

26th September 1939

The German heavy cruiser, or pocket battleship, *Deutschland* (A-59), acting as a commerce raider rather than attacking other naval ships, departed Germany to disrupt enemy merchant shipping by all possible means. She soon managed to evade a British blockade and broke out into the Atlantic.

5th October 1939

The Allied SS *Stonegate*, on passage from Antofagasta, Chile to Alexandria, Egypt, was sunk by gunfire from the *Deutschland* after her crew had been taken off, 600 nm E-SE of Bermuda, after it managed to send a distress call by radio.

9th October 1939

Deutschland intercepted another ship, the clearly-marked neutral US-flagged SS *City of Flint*, 1,200 nm east of New York. The ship had rescued survivors of the *Athenia* a month before, was stopped and searched, whereon the Germans discovered that her cargo of barrels of lubricating oil (considered a war supply) was bound for England. The ship was seized, the crew of US nationals locked up, a prize crew of Germans was put aboard, and the *City of Flint* was dispatched on circuitous route to Tromsø, Norway, also, like the US, neutral at the time.

14th October 1939

The German pocket battleship *Admiral Scheer*, captained by Theodor Kranke, leaves Germany to disrupt Allied merchant and naval operations and to better the effort by fellow pocket battleship *Deutschland*, which had only limited success. The *Deutschland* sinks Norwegian neutral SS *Lorenz W. Hansen* in the Central Atlantic, east of Bermuda.

15th October 1939

Allied **Convoy KJ-3** departs Kingston, Jamaica escorted by the British light cruiser HMS *Orion*, which was relieved by the heavy cruiser HMS *Berwick* on 8th October and later the Australian light cruiser HMAS *Perth*. Both were relieved on the 15th by the cruiser HMS *Effingham*, following which HMAS *Perth* proceeded towards Bermuda. Though damaged by a hurricane, she arrived safely.

24th October 1939

The German tanker *Emmy Friedrich*, attempting to run a blockade from the Bahamas to Germany, was scuttled by her own crew to avoid capture under the guns of HMCS *Caradoc.*

31st October 1939

Debris from the German tanker *Emmy Friedrich* washed ashore on Fernandina Beach, Florida.

8th December 1939

USS *Wichita* (CA 45) arrived at Guantanamo, whereon her commanding officer, Captain Thomson, assumed command of the newly formed Caribbean Patrol units which were based at Guantanamo and San Juan.

16th December 1939

The German passenger ship *Columbus* (the third largest in the German merchant marine), was attempting to obey an order to return to Germany after disembarking passengers in Havana, and seeking neutrality shelter in Mexico. Several allied ships and USN destroyers shadowed *Columbus*, one by one during her journey. The US vessels were asserting and protecting the nation's neutrality, which would remain in place for a further two years.

17th December 1939

The German pocket battleship *Admiral Graf Spee* was ordered to raid Allied commerce and naval vessels in the north and south Atlantic, and ultimately the Indian Ocean. It sank nine vessels, and at one point loaded prisoners and prizes onto the German tanker *Altmark*.

18th December 1939

USN Destroyers USS *Upshur* (DD-144) and USS *Greer* (DD-145) relieve USS *Ellis* and USS *Cole* in shadowing the German ship *Columbus* up the US east coast. Later, after it was scuttled, USN *Tuscaloosa* (CA-37) disembarked 576 German survivors of the *Columbus* at Ellis Island, New York, the only facility large enough at the time to process them as "distressed mariners," rather than Prisoners of War. Many of them returned to Germany in time to fight the US.

19th December 1939

Destroyer USS *Truxton* trails German *Aruaca* before its final dash to Port Everglades. She was overtaken by the HMS *Orion.* The German SS *Aruaca* reaches sanctuary in Port Everglades in the presence of the destroyer USS *Philip* (DD-76).

21st December 1939

USS *Wichita* (CA-45) leaves Guantanamo, for San Juan, arriving two days later.

22nd December 1939

USS *Philip* relieves USS *Twiggs* (DD-127) during neutrality patrol duty and monitoring the Royal Fleet Auxiliary (RFA) tanker HMS *Patella* off east coast of Florida; while *en route* to Port Everglades, USS *Twiggs* observes British light cruiser HMS *Orion* off Port Everglades, and anchors to keep an eye on it.

23rd December 1939

USS *Wichita* (CA-45) arrives in San Juan from Guantanamo.

24th December 1939

Oakes Field Airport officially opened on Nassau, New Providence. The island's main civilian airport until Nassau International Airport (Windsor Field), opened in 1957.

27th December 1939

USS *Wichita* (CA-45) leaves San Juan, visits St. Thomas, Virgin Islands, briefly on the 28th and 29th of December, before returning to San Juan, where she remained until 2nd January 1940.

Part II

1940-1942:
January 3, 1940 to December 21, 1942

3rd January 1940
USS *Wichita* (CA-45) arrives back at Guantanamo, having exercised locally from 8th to 24th January. She departed Cuban waters as flagship of the newly constituted Antilles Detachment, which included USS *Vincennes* and **Destroyer Squadron (DesRon) 10**. Two days later the force separated, with USS *Wichita* and **Destroyer Division (DesDiv) 82** visiting Willemstad, at Curacao, in the Netherlands west Indies from 26th to 30th January, before getting underway to rendezvous with USS *Vincennes* and her group of destroyers on 31st January, *en route* back to Puerto Rican waters.
8th January 1940
Oakes Field, now sold to Sir Harry Oakes, investor in BAL, is opened.
15th January 1940
RN light cruiser HMS *Dunedin* went to the Americas and west Indies Station, and was based at Bermuda. Five weeks later she intercepted the German merchant ship *Heidelberg*, which was scuttled by HMS *Dunedin* sailors in the north Atlantic.
24th January 1940
USS *Wichita* leaves Guantanamo then returned on the 3rd of February. USS *Vincennes* departed Cuban waters as flagship of the newly constituted Antilles Detachment, including **Destroyer Squadron (DesRon) 10**.
10th February 1940
USS *Wichita* conducted exercises between Guantanamo and Culebra, Puerto Rico through late February. She arrived at Norfolk on 4th March, then to Philadelphia.
1st March 1940
USS *Wichita* leaves Bahamas area for Virginia.
2nd March 1940
HMS *Dunedin* sank the German SS *Heidelberg*.
8th March 1940
Canadian destroyer HMCS *Assiniboine* stops the German SS *Hannover* in the Mona Passage, between Dominican Republic and Puerto Rico, the SS *Hannover* was set on fire by its German crew, who abandoned her. HMS *Dunedin* saves SS *Hannover* from destruction, and tows it to port. The Trujillo government of the Dominican Republic drops question of violation of territorial waters after there were conflicting representations by British and German diplomats regarding SS *Hannover*'s exact position. Two other German steamers; SS *Mimi Horn* and SS *Seattle,* escape capture during SS *Hannover*'s capture and make a break

across the north Atlantic from the Bahamas towards Germany.

28th March 1940

German SS *Mimi Horn* was scuttled to avoid capture in the Denmark Strait.

9th April 1940

German SS *Seattle* lost during early phases of the invasion of Norway on 8th to 9th April, 1940.

25th April 1940

Naval Air Stations (NAS) in St. Lucia, British west Indies, and British Guiana as well as Naval Auxiliary Air Facility (NAAF), Antigua, British west Indies, are established. [Later both an NAS and NAAF were established near Georgetown, Great Exuma, Bahamas].

1st May 1940

The SS *Makatana*, a refrigerated cargo steamer carrying dairy and meat products, 10,000 bales of wool, pelts, and tallow and manned by eighty-two crew was bound from Panama to the UK when she ran aground on Flat Cay Reef on the east end of Plana Cay in the southern Bahamas, due to non-war causes. The crew abandoned ship to go ashore. After the grounding the ship became a total loss, with her master reporting that his vessel was badly damaged. Several days later the status by Long Island Commissioner and wreck master was given as vessel settled. Accident due to routing as a consequence of war, then eight months in. SS *Panama* diverted from its course and picked up seventy-nine persons from open boats, two with broken legs, then proceeded with all survivors except master, chief engineer, and chief mate, to New York.

3rd May 1940

Salvage tug USS *Warbler* enlisted by Lloyd's Agents on "No Cure No Pay" *Lloyds' Open Form* basis to salvage any of ship, machinery, engines, or cargo of *Makatana*. Left Miami on day of casualty (1st May), arrived 3rd May, stayed until at least 10th May, along with a purpose-chartered small steamer and lighter, salvaged some 200 bales of dry wool, possibly some other wool, tallow and pelts, steel billets, all refrigerated cargo and the ship and machinery lost.

4th May 1940

SS *Panama* arrives in New York to disembark seventy-nine survivors of SS *Makatana* from Plana Cay.

8th May 1940

Destroyer USS *Laub*, though the US had yet to enter the war and Bahamas was technically British colonial territory, stood by with salvage tug USS *Warbler* and other salvage equipment to assist the SS *Makatana* operation Plana Cay.

10th May 1940

Floating platform, barge and small steamer to push and pull it to from the wreck site and hold the barge in place, chartered by salvors, arrives Plana Cays to support salvage from SS *Makatana*.

2nd July 1940

U-29 sank SS *Santa Margarita* under British charter with a Yugoslavian crew, southwest of England. Those survivors rescued were discharged in boats off Anguilla, cast adrift by the German raider KS *Widder* from SS *King John* on 13th July, finally landed on Anguilla 18th July.

13th July 1940

Twenty-one survivors of *Santa Margarita* were picked up by British merchant ship *King John*; three were lost when ship sunk by German raider KS *Widder* 240 nm north-northeast of Anguilla.

17th July 1940

Lifeboat of SS *King John* arrives Anguilla, had taken only Captain, Chief Engineer and wounded on board, left remainder 41 in the boats, made for the Lesser Antilles, proved remarkably difficult to dispose of, finally sunk by German raider KS *Widder*, 240 nm north-northeast of Anguilla.

18th July 1940

Last of SS *Santa Margarita* survivors reach land after being cast adrift with *King John* crew by *Widder*. Ruckteschell correct in assumption that men in lifeboats could make landfall, as they reached Anguilla, survivors made full report.

17th August 1940

The Duke and Duchess of Windsor (he was HRH King Edward VIII between 20 January to 11 December, 1936 before abdicating in order to marry Wallis Simpson) arrived in Nassau aboard the Canadian passenger ship *Lady Somers*, from Bermuda. The Windsors had been whisked from neutral Lisbon on orders from Churchill. Despite initial discomfort, the Windsors kept busy, she with running the Red Cross and both with official duties. Despite attempts to restrain their travel and spending, they travelled often to the US and Canada and enjoyed prevailing on wealthy contacts like the Alfred P. Sloans (General Motors) and Wenner-Grens (Electrolux) to use their yachts (*Rene* and *Southern Cross*) and homes (Marion Joe Carstairs at Whale Cay). [After being the only royal known to serve in a civilian post, the Duke resigned the governorship before the end of the war, and left the colony by 1 May, 1945. Others in their circle also fared poorly: the Wenner-Grens were black-listed by the US and banned from using the Panama Canal, the Frenchman who loaned them his castle to be married in, Charles Eugene Bedaux, was later arrested by the FBI for trading with the Germans in north Africa and committed suicide in their custody on 13th January, 1943 in Miami. Their predecessors, the Dundas' were sent to Kampala, Uganda.]

21st August 1940

German Raider KS *Widder* sinks British SS *Anglo Saxon*. Out of the 41-man crew only two survivors drift ten weeks to James Cistern, Eleuthera, Bahamas and became subject of *All Brave Sailors*, by J. Revell Carr, and *Two Survived* by Guy Jones.

2nd September 1940

Anglo-American treaty (Lend-Lease) signed, granted US access to air bases in Bahamas (Exuma, Nassau, Mayaguana, San Salvador, Eleuthera, Inagua) in exchange for 50 older US destroyers.

30th October 1940

Two survivors, Tapscott and Widdecombe, from the SS *Anglo Saxon* lifeboat were discovered near James' Cistern on Eleuthera, northeast Bahamas after months and thousands of miles adrift.

31st October 1940

A Bahamas Airways' Bellanca aircraft rescued the two merchant mariner survivors of German raider KS *Widder*'s attack on the SS *Anglo Saxon*, and took them to Bahamas General Hospital in Nassau.

24th November 1940

SS *Port Hobard* sunk by German pocket battleship *Admiral Scheer* well southeast of Bermuda; the attack leads to change in Admiralty policy; subsequently large convoys were usually escorted by battleship or battle cruisers.

25th November 1940

The New York Times, referring to *Anglo Saxon* vs. German raider KS *Widder* survivors, writes "Raider shells ship east of Bahamas, suggests Nazis use two surface craft. Reports from two British steamships in widely separated parts of the Atlantic Ocean indicated yesterday that two German surface raiders might be roaming the Atlantic."

1st December 1940

SS *Tribesman* sunk by German pocket battleship *Admiral Scheer*, south and east of Bermuda.

8th December 1940

The only regular passenger boat between Miami and Nassau, passenger ferry SS *Munargo* suddenly cancelled forthcoming sailings.

9th December 1940

HRH Duke of Windsor, Governor, and Duchess leave Nassau for Miami as a guest of Axel Wenner-Gren in his motor yacht *Southern Cross*, as *Munarg*o is out of service.

10th December 1940

Windsors arrive Miami, on the *Southern Cross*, exactly four years after the Duke's abdication as King Edward VIII. They are received by a large welcoming crowd in Miami including the governor and four mayors. Prime Minister Churchill did not condone these trips.

13th December 1940

USS *Trippe* assigned to escort USS *Tuscaloosa*, carrying US President Franklin D. Roosevelt (FDR) from Washington, DC via Charleston, Eleuthera, to Jamaica on a post-election-1940 victory voyage. HRH the Duke of Windsor was invited on board. Against the wishes of the Colonial Office in London, Windsor and Drury (his *Aide-De-Camp*) depart Miami in an amphibious airplane from the Miami Coast Guard Air Station. They arrive off Eleuthera and land alongside the USS *Tuscaloosa*, USS *Trippe* and USS *Mayrant* on FDR's insistence for a luncheon and talks. Harry Hopkins and the British Governor of Jamaica were present. A Loening aircraft "buzzed" Oakes Field on its final departure, "to see whether any dust would be raised."

15th December 1940

Windsors as guests of Wenner-Gren on *Southern Cross* tour a Grand Bahama crayfish cannery.

17th December 1940

Windsors return to Nassau in *Southern Cross.* German *Admiral Scheer* sinks SS *Duquesa* southeast of Bermuda.

10th January 1941

Aboard the power yacht *Rene*, the Windsors were guests of American industrialist Alfred P. Sloan (General Motors) on Cat Cay, in Bimini. Nearby Whale Cay, Berry Islands is owned by British heiress, and speed boat racer, Marion Joe Carstairs, owner of *Vergemere IV* [later to rescue *Potlatch* survivors].

11th January 1941

Private Aircraft flew Windsors from Cat Cay, with Sloan, to watch horse races in Florida, earning the chagrin of London's Colonial Office. USS *Wichita* (CA-45) departed Hampton Roads Virginia on 7th January and reached Guantanamo four days later. For next two and half months she participated in fleet manoeuvres in the Bahamas region and took part in practice amphibious landings at Puerto Rico. She called at Portland Bight, Jamaica; Culebra; Guayanilla, Fajardo Roads, and Mayaguez, Puerto Rico. She later arrived at the New York Navy Yard on 23rd March.

14th January 1941

Swede Axel Wenner-Gren placed on the US State Departments "Proclaimed List of Blocked Nationals." *Southern Cross* is prohibited from transiting the Panama Canal or utilizing Allied territory (US and Bahamas included). No specific charges were provided or proven, but related to his efforts at his failed peace-making with Nazis pre-war. It was determined he held no sway with Germans.

16th January 1941

Bahamian Philip Farrington left Nassau to join the RAF, following Leonard Thompson and Rosemary Kelly [he joined Bahamas Airways on May 1, 1946 back home in Nassau].

15th February 1941

USS *Wichita* from 11th January to next-two-and-one-half months, participated in fleet manoeuvres in the Bahamas region and practice amphibious landings at Puerto Rico.

15th March 1941

Destroyer USS *Benson* (DD-421) escorts Presidential yacht *Potomac* (AG-25) from Washington, DC to Bahamas (Mayaguana, Acklins, Crooked Island Eleuthera) for a fishing holiday by FDR.

19th March 1941

USS *Wichita* leaves Puerto Rico for New York and arrives back on 23rd March.

27th March 1941

Alfred Sloan-owned yacht *Rene* takes the Windsors on a tour of the Out Islands of Bahamas with Sergeant Holder and John Hughes, Senior Out Island Commissioner.

3rd April 1941

Yacht *Rene* runs aground in rough weather off Hole-in-the-Wall Light, rescued in early hours.

6th April 1941

USS *Wichita* sailed for Bermuda and reached her destination two days later. She then operated with USS *Tuscaloosa* in the north Atlantic.

15th April 1941

In company with USS *Tuscaloosa*, USS *Wichita* operated from Bermuda into the north Atlantic.

18th April 1941

The Windsors depart Nassau in their power yacht *Gemini*; a gift. They spend weekend in Palm Beach then depart Nassau and stay at the Everglades Club. The US is not at war yet.

21st April 1941

The Windsors return from weekend in Palm Beach Florida on their boat *Gemini*.

3rd June 1941

First scheduled service into Oakes Field, which boasts 4,000-foot, 3,000-foot, and 3,200-foot long runways. A 21-seat Douglas DC-3A of Pan American Airways initiated service.

15th July 1941

Yacht *Southern Cross* with Wenner-Grens on board, leaves Nassau on tour of south America, including Peru, to discuss investments, archaeology; they were invited by the Peruvian President.

23rd September 1941

Bahamian motor vessel *Arawak*, 115', 210-tons, built of wood by John Cleare in Hatchet Bay, Eleuthera, was carrying sawmill equipment with 15 persons under Capt. Roland Roberts Jacksonville to Abaco when she caught fire. US Coast Guard planes saw flames, and a rescue boat met *Arawak*'s lifeboat 4 nm east of Ponce de Leon Inlet, Florida at dawn. All ten crew, most from Green Turtle Cay, Abaco, and five passengers, mostly from Florida, are safe.

23rd November 1941

Collier USS *Proteus,* sister ship to USS *Cyclops* and *Nereus*, went missing after leaving St. Thomas US Virgin Islands, for Portland, Maine, probably due to design and construction flaws compounded by harsh weather. There are no U-boats operated along the route at this time, it being prior to Germany's declaration of war against the US.

2nd December 1941

Tanker *Astral, en route* Aruba to Portugal with 73,000 barrels of kerosene and gasoline, sunk by *U-43*. *U-575* had left her for lack of weaponry, thinking she was neutral US vessel. Crew of 37 all died, ship exploded and sank in minutes. Exact sinking position not known. This is the third US merchantman to be sunk by U-boats prior to the US' entry into war five days later.

7th December 1941

Imperial Japanese Navy attacks US naval bases at Pearl Harbor, Hawaii. US declares war on Japan, diverting resources to the Pacific, and leaving their Atlantic flank exposed to U-boats, particularly after lending 50 destroyers to Britain's Royal Navy. Bahamian Leonard Thompson becomes a pilot in the Royal Canadian Air Force. Passenger liner *New Northland* is replaced by US Army transport ships Miami to Nassau for 1940 and 1941 seasons. US-flagged ships were used only for war duties after Pearl Harbour.

10th December 1941

Germany declares war on the US, eliminating the Panama Treaty and opening unrestricted warfare on US and Allied warships and merchant ships in the Americas and globally. Construction begins on the naval air station on Great Exuma, using local labour, US Navy, US Marines, yet not Seabees. The bases' purpose under the Guantanamo Pact is to support a squadron of seaplanes to patrol the passages through Bahamas, which it did non-stop for three years, to June, 1945. The core site is 324 leased acres on the southeast tip of Great Exuma, and occupied a further 459 acres on high ground. Facilities include a 50-foot timber seaplane ramp, concrete parking area 400 feet by 300 feet, and a 180-foot barge pier, sited on a partly dredged-in beach adjacent to the take-off area, 3 miles long and protected by Stocking Island, one mile offshore, opposite the station. The 79 initial buildings were fabricated and assembled under the San Juan agreement, then 28 Quonset huts we raised, with quarters for 80 officers and 180 enlisted men, a ten-bed medical dispensary, an administration building, a storehouses, chapel, bakery, a power house, and several industrial buildings. There were magazines embedded throughout the tract. Limestone rock was picked up by local labour at a set price per yard, and broken into usable size by women laborers using small hammers. Dredging made a seaplane area at least 6 feet, improved the small-vessel anchorage in Stocking Harbor, Exuma and deepened the approach channel to the anchorage. Some 750,000 cubic yards of spoil fortified a beach area. Seven drilled wells provided fresh water which was filtered, chlorinated, and stored in a 100,000-gallon underground concrete reservoir. The Public Works Department, using local labour, performed construction and maintenance after completion in summer, 1942. The USN collier *Nereus*, sister ship to both *Proteus* and *Cyclops* (lost in March 1918 between US Virgin Islands and Norfolk), goes missing. Its voyage from the USVI to Portland Maine began on first day of war. Loss is likely through a radical coal-loading design which directly or otherwise led to the demise of all three sister ships very suddenly without a trace in rough area, and not due to enemy action, espionage, or fifth column activities.

13th December 1941

After war is declared on them by Hitler, the US impounds German vessels and converts the ex-German merchantman *Aruaca* in Florida. She had ducked into Port Everglades two years earlier on 19th December 1939 to avoid capture. She is impounded and eventually became USN store ship USS *Saturn* (AF-40).

21st December 1941

Royal Navy escort carrier HMS *Audacity* D-10 (ex-German *Hannover*), sails. She was captured by HMCS *Assiniboine* and HMS *Dunedin* in Mona Passage on 8th March, 1940, and converted to an Allied escort carrier.

Part III

Operation *Neuland* Envelopes The Islands: January 1, 1942 to December 30, 1942

1st January 1942

GrossAdmiral Karl Dönitz [WWI U-boat veteran who commanded U-boats of WWII, then the German Navy, then Germany] orders first fleet of U-boats to prepare to proceed to US east Coast for Operation *Pauckenshlag*, or *Drumbeat*. Germans attack the exposed underbelly of the US, particularly Cape Hatteras, which becomes known as Torpedo Junction. Some 600 Allied ships are sunk in the first few months of this campaign, and over 5,000 Allied sailors killed in US waters during the war [twice as twice as many as killed in attacks on Pearl Harbor[. US Marines deploy to assist construction of Naval Air Station Great Exuma. On their one night off weekly, the US sailors enjoy nightly outdoor movies, a PX shop selling candy and beer, raw-lumber recreation rooms, two pool tables, a juke box, and ping pong.

4th January 1942

SS *Chilore* reports a submarine 10 nm south of Sand Key, Tampa, Florida. Distress ended at 9.10 am per ESF [USN Eastern Sea Frontier] Enemy Action Diary. It is a false report.

5th January 1942

US-flag sail training ship *Morning Star* begins five-month cruise of the Bahamas, having set out from Annapolis, Maryland, with "regular prep school students who act as crew, study, and practice celestial navigation" on 1st October, 1941. She arrived in Miami on 30th December 1941 under Captain and Mrs. William M. Pond of Cambridge and Boston, once at Harvard. [The couple owned several large sail training ships, and grounded several times, in one instance rescued by Joe Carstairs of Whale Cay].

15th January 1942

US-flagged ships only are used for war duties, including the shuttle from Nassau to Miami after Pearl Harbor. In 1942 the Canadian firm Clarke Steamship Company thus had to withdraw its ships from the route, leaving it to smaller vessels like the ferry *Ena K.*, owned and operated by the Kelly's of Nassau. The colonial government introduced travel restrictions within Bahamas: anyone arriving in Nassau from other islands was required to obtain a permit from the island's commissioner, giving reason for travel, dates, and length of stay. Also, boats from outside the capital were to report to government officials before disembarking passengers there. Bahamian-built cargo and mail vessels, including the motor yachts *Vergemere IV*, and *Samaná*, are also put to use.

19th January 1942

Canadian passenger liner *Lady Hawkins*, which often called Bahamas, is sunk by *U-66*, 150 nm east of Cape Hatteras with the loss of 251 lives; among the highest casualties in the region.

24th January 1942

SS *Lady Hawkins*' survivors drift for five days, then are rescued by passenger SS *Coamo*, landed in Puerto Rico on 28th January 1942. [SS *Coamo* is sunk on 1st December, 1942 by *U-604*; its 186 dead made it the largest US Merchant Marine loss on a single vessel in the war]. German submarine *U-123* under Hardegen is bound southeast between the Carolinas and Bermuda before turning east to France.

2nd February 1942

Seventh Naval District [7ND] with headquarters at Key West, Florida, is re-established, known as the Gibraltar of the West. Naval Air Stations are established in St. Lucia, British west Indies, and British Guyana; along with a Naval Auxiliary Air Facility in Antigua.

10th February 1942

SS *Mary* in collision with SS *Palimol*, proceeds to San Juan with damage to forepeak. *U-67* enters area briefly via Anegada Passage.

11th February 1942

U-129 enters area northeast of Bahamas, south. *U-67* is still in Anegada Passage, southeast of Bermuda, merchant ship SS *Pan Halcyon* thanks Bermuda aircraft for deterring a U-boat.

12th February 1942

U-129 is bound north of Virgin Islands, east of Anegada.

13th February 1942

U-129 was north of Mona Passage, east of Turks and Caicos.

14th February 1942

SS *Atenas* sights submarine in Florida Straits between Key Largo and Little Bahama Bank. *U-128* enters area by circling Bermuda and heading south, then southeast and west towards Jacksonville. *U-504* enters area bound west for Canaveral, passing north of Grand Bahama. *U-129* exits area north of Anegada.

15th February 1942

U-128 is bound northeast of Bahamas, west to Canaveral

17th February 1942

U-504 is bound northeast of Bahamas, south towards Florida, Grand Bahama, and the Straits. *U-134*, Northwest Providence Channel: Lighthouse Great Stirrup Cay, Berry Islands, in sight. Lighthouse Great Isaac in sight. Lighthouse Bimini in sight. Lighthouse Cat Cay in sight, all burning as in peacetime. Crash dive for aircraft, followed by depth charge. Glass damage, gyrocompass out of service. Damage repaired before re- surfacing.

18th February 1942

Passenger liner SS *Queen Mary* departs New York for the Middle east on her first service as a US troop carrier, often transiting channels in the region. *U-504* is northeast of Abaco, heading east to France.

19th February 1942

Commercial aircraft Pan American Clipper sights submarine between Bermuda and Anegada. US-flagged tanker *Pan Massachusetts* is sunk by *U-128*, 20 nm off Canaveral. USCG Cutter *CG-Forward* (WAGL-160) took lifeboat in tow, some bodies recovered, all survivors brought to Jacksonville. *U-504* is northeast of Grand Bahama, heading east to sink SS *Stangarth* then southeast to Puerto Rico, then France. East of Jacksonville *U-504*';s men find an empty lifeboat floating near a burning tanker blasted into two parts, yet cannot go closer for fear the tall column of smoke will entice an enemy.

20th February 1942

USN Transport USS *William P. Biddle* (AP-15) arrives at Guantanamo and disembarks the 9th Defence Battalion. SS *William C. Atwater* at 4.22 am reports suspicious craft following it, suspected to be a surfaced submarine, 5 nm off Palm Beach. Passenger liner SS *Queen Mary* arrives and departs Key West, changes captains. *U-128* is in Straits between Cape Canaveral and Jacksonville. *U-108*, north of Puerto Rico came across three lifeboats from SS *Norland*, Norwegian ship sunk by U-boat; they are 5 nm away, and have set sails.

21st February 1942

US Admiral Leahy receives instruction to see Vichy French Admiral Darlan immediately regarding *U-156*'s receiving assistance for a badly injured U-boat sailor at Martinique. Vichy France needs to assure US government that no Axis ships or planes will be allowed to enter French ports or territory in the western Hemisphere, else action will be taken.

22nd February 1942

U-128 sinks tanker *Cities Service Empire* 25 nm north of Bethel Shoal, Vero Beach, Florida. Thirty-four survivors on rafts picked up by destroyer USS *Biddle* (DD-151), taken to Fort Pierce. USCG Cutter *Vigilant* (WPC-154) rescues two people as tanker explodes. Master, three armed guards, ten crew killed. Eastern Airlines aircraft sights submarine between two steam ships 7 nm to 10 nm off Fort Pierce. The SS *W. D. Anderson* sights periscope of a submarine 15 nm off Miami. *U- 504* transits over Grand Bahama and northern Bahamas to the Straits. A steam ship is seen lying on her side adrift 7 nm north-northeast of Jupiter Light and is identified as SS *Baldbutte*. At 10.30 pm a U-boat is reported seen charging batteries 17 nm northeast of Jupiter Light. Merchant ship with radio callsign *KHLQ* reported a surfaced U-boat 7 nm east of Mona Island, between Puerto Rico and the Dominican Republic [this is not corroborated by any known U-boat position]. SS *Republic* is torpedoed by *U-504* 3.5 nm northeast of Jupiter Island Light. The ship drifted to Hobe Sound where it sank the following day. Five were killed, with 29 survivors rowing to shore, where they were taken in by residents. They survivors were then taken by truck to Palm Beach. Seven were rescued by the tanker *Cities Service Missouri* and landed in Port Everglades. *U-67* exits area south via Anegada Passage. *U-128*, off Canaveral: crash dives to avoid a flying boat. U-boat then remains submerged until twilight to remain undetected while approaching the coast.

23rd February 1942

US-flagged tanker SS *W. D. Anderson* is sunk by *U-504* 12 nm northeast of Jupiter Point Lighthouse. The only survivor of its 36-man crew swims ashore, is picked up by small fishing boat, then transferred to US Coast Guard vessel *USCG-Trouper* and taken to nearby Stuart. USCG Cutter *USCGC-Vigilant* reports submarine following it on surface in Straits, 5 nm off Jupiter, near the grounded hulk of SS *Republic*, attacked the day before. U-boat sighted on surface by Pan American pilot, 25 nm east of Carysfort Reef, Florida Keys. Passenger liner *Queen Mary* saw *U-161* and *U-129*, taking the same route, via Anegada Passage. *U-128* is off Canaveral, north to Jacksonville in the Straits, west of Grand Bahama. *U-123* is in the Straits, reports an escort of a 4-funneled destroyer ahead of, and screening, protecting three to four large tankers. *U-185* in the Crooked Island Passage reports: near Bird Rick Lighthouse they pass a sailboat then are surprised by land-based aircraft and crash dive near Castle Island Light.

24th February 1942

USN yacht USS *Emerald* (PY-C1) is attacked a suspected U-boat off Canaveral. *U-128* under Heyse is in Straits north of Canaveral up to 3 March. *U-504* under Poske is bound east north of Grand Bahama, heading towards France. *U-84* is in the Old Bahama Channel near Cayo Lobos. Off the entrance to Old Bahama Channel is a sea patrol by destroyer which according to the U-84 commander, is "quite inattentive."

25th February 1942

U-156 is entering Mona Passage north to sink SS *Macgregor* and SS *Oregon* north of Puerto Rico and the Dominican Republic. *U-128* is in the Straits north of Canaveral up to 3 March.

26th February 1942

Disabled U-boat is reported off Fort Pierce, in Straits. In fact. was a disabled Allied vessel, anchored. Navy sent USS *Canfield* out. *U-504* torpedoed SS *Mamura* 230 nm off east coast of Florida, 65 nm north, to the northeast of Abaco's Elbow Cay Light. All 49 crew, 35 of them Chinese, perished. USCG *Cutter CG-Millinocket* is aground at Paredon Grande Key Lighthouse, Old Bahama Channel, north coast of Cuba, south of Andros. U-boat believed sighted 4 nm off Palm Beach, in Straits, plane investigating. May have been *U-128* under Heyse, which was in Straits north of Canaveral up to 3 March. *U-108*, east of San Salvador: destroyer ahead of a large passenger steamer with goal post masts of *President Grant* type. The destroyer ahead steers in large legs ahead. Ran on a collision course for enemy, but before the shot he turned once more and just passed us. In the meantime, another type of detonation, like a torpedo detonation was heard. Failures from being depth charged: battery automatic switch fails, gyro-compass, light bulbs. Listening gear temporarily. Closures of the ballast and reserve fuel oil tank burst open. Submerged. I suspect that a torpedo of the fan was a surface runner, and the destroyer threw the depth charges at the shooting location. A torpedo sound was heard perfectly. A hit at the time of the depth charge run is possible. By my impression before the shot the boat was not noticed ahead of time. Now, destroyer still drives search sources, however throws no more depth charges. Sound bearings are no longer heard. Later, two metallic impacts were heard. Destroyer sounds were louder. After the fall of the shot, went to depth, then a series of well-placed depth charges. Initially three, then two more. *U-753*, southeast of Miami: chased by fighter aircraft, harassed by continuous aircraft of Lockheed Hudson, like Cessna-150, shoulder-winged single-engined (land-based), high-winged single-engined, single float, single-engined aircraft. Biplane, 2-engined high-winged flying boat sport and training aircraft. *U-509*, north of Puerto Rico: sights the conning tower of a fellow U-boat on port beam, in sight for a short time. Probably an opposite sailer; going other direction [this is possibly U-156 which sinks SS *MacGregor* north of Puerto Rico the following day].

27th February 1942

U-156 attacks Aruba, Netherlands Antilles, and five ships are sunk, with two others damaged. *U-183* north in Windward Passage to Old Bahama Channel. SS *MacGregor*, dispersed from **Convoy ON-60**, is sunk by gunfire from *U-156* 25 nm northwest of Cape Viejos, Puerto Rico. One crew lost. Master, twenty-three crew, and six gunners picked up by a Dominican Republic Coast Guard cutter and landed at

Puerto Plata. Royal Italian Navy submarine *Regia Sommergibili*, or R.Smg. *Finzi* enters area southeast of Bermuda, heading in straight line for Mona Passage. *U-128* is in Straits north of Canaveral up to 3 March. *U-129* is bound north of Bahamas: vessel comes high out of horizon, abruptly, so that U-boat can no longer remain unseen on the surface. Crash dives for attack on a submarine of American U-cruiser of the USS *Narwhal* type. Zig zagging. Sound room reports only humming sounds, in that case he was running on E-motors. Weyher identification guide confirms the type is diesel-electric. *U-129* fired a torpedo fan from tubes 1, 3, and 4, yet all missed. It is believed to be a US Navy submarine, making about 13 knots.

28th February 1942

U-156 sinks US-flagged tanker SS *Oregon* 150 nm northeast of Mona Passage. Eight officers and 28 crew abandon ship in one lifeboat and one raft, master, two officers, three crew died during attack. Twenty-six survivors made landfall near Puerto Plata, Dominican Republic, on 4th March. Four survivors on raft are picked up by tanker SS *Gulfpenn* 5.5 days after attack. *U-126* enters area *en route* to Windward Passage. *U-504* is bound northeast of Abaco, east bound. *U-128* is still in Straits north of Canaveral up to 3 March. *U-155* in the Windward Passage, heard 30 depth charges quite distant, and saw two-engined land-based aircraft

1st March 1942

U-156 is bound north of Puerto Rico, east of Turks and Caicos, bound east across the Atlantic for France. *U-126* is bound northeast of Bahamas, bound south. *U-128* is still in Straits, north of Canaveral up to 3 March.

2nd March 1942

Norwegian SS *Gunny* torpedoed by *U-126* 400 nm south of Bermuda and 490 nm northeast of Grand Turk Light. Injured chief engineer dies on 4th day. Other crew are picked up after a week by the Swedish motor merchant *Temnaren* and landed in Progresso, Mexico, where the Norwegian Vice Consul came on board and took them to Merida. *U-156* exits area northeast of Bahamas, east for France. *U-128* leaves Florida for Abaco, then east for France. *U-504* is northeast of Bahamas, east. R.Smg. *Finzi* is heading south from entry point to the south of Bermuda to refuel from R. Smg. *Morosini* which is north-northeast of Anegada.

3rd March 1942

R.Smg. *Enrico Tazzoli* is *en route* west to Bahamas. *U-126* is bound northeast of Bahamas bound west. *U-128* is bound north bound out of Straits for northwest and Northeast Providence Channels and Abaco.

4th March 1942

U-126 east of Crooked Island one day from entering Crooked Island Passage south for Windward Passage and Old Bahama Channel. *U-504* is north of Grand Bahama. *U-128* south, northeast of Abaco *en route* to sinking *O. A. Knudsen*. R.Smg. *Finzi* south to east, towards Anegada. *U-67* is near Anegada, and sighted land bases and float plane aircraft camouflaged white and gray at Sombrero Light.

5th March 1942

SS *Cardonia* is in transit of Old Bahama Channel, sunk off Haiti on 7th March. Survivors taken to Guantanamo. SS *Barbara en route* from Baltimore to San Juan, 60-crew plus passengers, passed east of Bahamas and through Crooked Island Passage before being sunk off *Isle Tortue*, Haiti. *U-128* transited Northeast Providence Channel, then sinks *O. A.*

Knudsen 50 nm east of Abaco. Crew abandon ship in starboard lifeboat and motorboat, boats taken in tow by schooner *E. P. Theriault*, which landed them at Cornwall lumber camp at Cross Harbour, Abaco on 10th March. An able seaman, Olaus Edvard Gamst Johanesen, died of wounds in hospital and is buried on Abaco. [His grave was re-discovered in 2015 and he was memorialized by fellow Norwegians]. *U-128* heads east and is attacked by USN aircraft south of Bermuda. R.Smg. *Enrico Tazzoli* enters area south of Bermuda. R.Smg. *Finzi* still east of Bahamas south towards Anegada. Neutral Portuguese freighter SS *Nyassa* declared to be leaving Bermuda 15 February to Havana, clearly marked.

6th March 1942

R.Smg. *Enrico Tazzoli* is northeast of Bahamas sinks the Norwegian SS *Tønsbergfjord* a third of way between Bermuda and northeast Bahamas. One died, and remaining 32 were rescued. Captain's boat with 18 saved by Dutch SS *Telamon* on 13th March and landed in Jeremie, Haiti on 16th March, while others were picked up by Norwegian tanker *Velma* on 18th. Same submarine sank the Dutch SS *Astrea* on 6th March. R.Smg. *Enrico Tazzoli* begins single week-long, six-vessel sinking rampage in the region east of Bahamas. Betasom sinks SS *Astrea*, SS *Tønsbergfjord*, SS *Montevideo*, and SS *Cygnet* only 4 nm off Dixon's Point, San Salvador, resulting in some 35 survivors being landed there. She also sinks SS *Daytonian* and *tanker Athelqueen* only 50 or so nm southeast of Abaco, resulting in three sailors being buried there. This single patrol stands out as most relevant and effective (from the Axis point of view) of the war. It ended because the submarine made contact with the *Athelqueen* and damaged her bow, necessitating a return to France. *U-128* shells *coup de grace* day after *O. A. Knudsen* sinking, then east for France. Survivors from *O. A. Knudsen* set course for The Hole in the Wall, Abaco. Master of SS *Nueva Andalusia* reports burning object, probably ship, northeast of Eleuthera; it was *O. A. Knudsen*. R.Smg. *Finzi* sank tanker *Melpomene east* of Turks and Caicos, north of Puerto Rico. *U-126* north of Windward Passage, south through Caicos Passage. *U-128*, northeast of Bahamas: signal received from HQ about respecting Prize Rules for Mexican and South American vessels.

7th March 1942

U-126 sinks SS *Barbara* approximately 9 nm north-northwest of West Tortuga Island, Dominican Republic. SS *Cardonia* witnesses sinking of *Barbara* by *U-126* only 7 nm away. She tries to escape by increasing the speed and altering course towards the coast, evades two torpedoes but ultimately sinks. Ten officers, twenty-six crew, and two passengers on board abandon ship in one lifeboat and three rafts. Overall, 26 died, 59 survived. USN Naval Air Station gets message from Key West stating that a ship (*O. A. Knudsen*) was torpedoed near Abaco. *U-128* is bound east for France. R.Smg. *Finzi* torpedoed and sank SS *Skåne*, 4,528 tons, Swedish neutral, north of Antigua. Lifeboat of SS *O. A. Knudsen* sight a local schooner off Hole in the Wall Abaco. SS *Tønsbergfjord* lifeboats see SS *Velma*, SS *Arthur W. Sewall* and SS *Telamon* for multiple survivor rescue episode over the next three to six days. Tanker *Phoebus* transits through the Bahamas, most likely the Caicos Passage from New York to Caripito approaches the stricken *Hanseat*, hit by *U-126* on 6th March off Cape Maysí, northeast Cuba, master warned away from assisting. R.Smg. *Giuseppe* enters area.

8th March 1942

Propaganda piece by Charles Hurd *Navy plotting habits of enemy submarines, final defeat of U-Boats expected when lanes and tasks are charted*, published in *New York Times*. Survivors of *O. A. Knudsen* leave Olaus Johansen, continue to Nassau aboard *Content S*, sent by Colonial government with medical assistance. In Nassau the Norwegians wait a ferry to Miami, leaving injured shipmates in Nassau. R.Smg. *Enrico Tazzoli* sinks SS *Montevideo east* of Bahamas, halfway northeast of Abaco southwest of Bermuda. Fourteen crew were lost, and 35 were rescued 390 nm northeast of Elbow Cay Light. R.Smg. *Enrico Tazzoli* is bound northeast of Bahamas. *U-126* transits Windward Passage north to Old Bahama Channel. *U-128* is bound north, to the northeast of Bahamas when south of Bermuda two USN VP-74 PBM Mariners attack it southwest of Bermuda. German commander Heyse reports Consolidated flying boat out of sun. Two depth charges damaged gyrocompass repeater, light bulbs, depth meter. The submarine survives and continues to France, the planes return to Bermuda. R.Smg. *Finzi* rendezvous with R. Smg. *Morosini* for 50 tons fuel.

9th March 1942

Tanker *Hanseat* hit by two torpedoes from *U-126*, 10 nm north-northeast of Cape Maysí, Cuba in Old Bahama Channel. Danish crew abandons ship in all four lifeboats. *U-126* could have chased and sunk SS *Phoebus* in vicinity, but did not. Tanker *Charles Racine* torpedoed by R.Smg. *Giuseppe Finzi* at northeast of Anguilla. Four lifeboats launched, three of the boats holding 34 men stayed together but the 4th could not be seen in the dark. Survivors picked up by USS *Moffett* and an Argentinian steam ship. Remaining SS *Gunny* survivors picked up by the Swedish MS *Temnaren*. *U-128* exits area northeast of Bahamas, east for France. *U-504* is bound northeast of Bahamas, east of Savannah, south. R.Smg. *Enrico Tazzoli* is bound northeast of Bahamas west. R.Smg. *Finzi* sinks SS *Charles Racine* 300 nm northeast of Anguilla; all survive. *U-128*, off Canaveral: crash dive for destroyer 2 nm away, followed by ASDIC-type locating sounds, or pings. At periscope depth, torpedoed tanker has sunk. Destroyer and freighter on westerly course. Both periscopes smeared by oil; petroleum smoke or outflowing petroleum, so U-boat has no clear picture. They did not pursue destroyer and freighter due to smeared periscopes and inexact position.

10th March 1942

U-504 is bound northeast of Bahamas, heading south. *U-126* is off Cape Maysí, Cuba, heading north. Panamanian tanker *Leda* reports sighting a submarine. Count Carlo Fecia di Cossato in R.Smg. *Enrico Tazzoli* sinks SS *Cygnet*, a Goulandris-owned Greek ship, 4.5 nm from San Salvador's Dixon Point. All 30 survivors were landed on San Salvador and were taken to Nassau by the mailboat SS *Monarch of Nassau*. It was closest any allied vessel was attacked to the shore of any Bahamian Island during the war. R.Smg. *Finzi* refuels from R.Smg. *Morosini*. *U-161* fires two torpedoes near Port Castries, St. Lucia. First sank passenger liner *Lady Nelson* in shallow water and the second sinks SS *Umtata, en route* from Durban to New York. Survivor from *O. A. Knudsen* able seaman Olaus Johannesen who was left behind in Abaco died on 10th March and was buried at Alexandra, Cross Harbor. Survivors in lifeboat of SS *Charles Racine* begin a nearly weeklong raft and lifeboat voyage then other long voyages of repatriation by Argentinian steam ship that rescued them on 15th March. *U-504* is in the Straits near Jupiter Inlet: at periscope depth, one single-engined Martin or Douglas aircraft is seen.

11th March 1942

SS *Afoundria* sends Submarine Sighted Surfaced [SSS] message and chased by a submarine 15 nm south of Cape Pepe, Isle of Pines, Cuba. SS *Texan* is sunk by *U-126* in Old Bahama Channel. Ten officers and 37 men launch two lifeboats, three officers and six men perish but 10 men righted one lifeboat. Cuban fishing vessel *Yoyo* rescues SS *Texan* survivors 40 nm east of Nuevitas, Cuba. *U-504* is bound northeast of Abaco heading south towards San Salvador. R.Smg. *Finzi* rendezvous with R.Smg. *Morosini* well east of Bahamas north towards Europe. R.Smg. *Enrico Tazzoli* transits from San Salvador, the *Cygnet*, to Abaco to attack the *Daytonian* on the 13th March. Crown Lands Officer William Norman Aranha declares a right-of-way for a "a triangular parcel of land situate about 960 feet south of west Bay Street in the vicinity of the Second Mile Post in the Island of New Providence [bordered by] Mrs. Mary Lucille Robinson Reeves, Granville Ferguson, Clarence Clover Mason, originally granted Eliza Ann Forbes.... land being taken on behalf of His Majesty the King..." for roads relating to the new airfields.

12th March 1942

U-504 is bound east of Long Island, Bahamas, heading southeast then bound east to France. R.Smg. *Morosini* and R.Smg. *Finzi* rendezvous with R.Smg. *Da Vinci,* then R.Smg. *Morosini* pursued a tanker. SS *Colabee*, 10 nm off Cape Guajaba Cuba is torpedoed by *U-126* which kills master and one other man.

13th March 1942

USS *Truxton* is directed by plane to find five men 4 nm east of a wrecked ship, 12 nm southeast of Lobos Key, off the northeast of Cuba. Probably survivors of either SS *Olga* or SS *Colabee*, or both. USS *Truxton* was unable to find them. SS *Tønsbergfjord* survivors in Captain's boat, 18 on board, located by Dutch SS *Telamon* in the morning, occupants landed in Jeremie, Haiti. Emergency SSS from a ship with call letters SWKC torpedoed 5 nm off Green Cay, Bahamas (Old Bahama Channel), may be the torpedoed ship reported by SS *Helen*. [Later reports state that Abaconians claim to have a part from *U-324*, however *U-324* never left Norway or Europe]. R.Smg. *Enrico Tazzoli* sinks British SS *Daytonian* east of Abaco. One crew perished, while the remaining 58 were rescued by Dutch SS *Rotterdam*, which landed at Nassau and later were brought to Miami. *U-504* is bound south, east of San Salvador. *U-126 is* in Old Bahama Channel north headed for Santaren Channel then Straits and northwest and Northeast Providence Channels. R.Smg. *Morosini is* well east of Bahamas, north of Anegada. R.Smg. *Finzi* exits area northeast of Bahamas east for Bordeaux, France. Swedish MS *Temnaren* lands final group of SS *Gunny* survivors in Progresso, Mexico, where the Norwegian Consular Corp take care of them.

14th March 1942

SS *Atenas* reports submarine sighting between Key Largo and Bahamas in Straits. USS *Algorab* sighted a green rocket 300 nm northeast of Hope Town, Abaco, relaying that a fast cargo vessel heading south turned away sharply. [No Axis submarines were in vicinity of or fit the description of vessels sighted]. *U-504* is bound east of Turks and Caicos, northeast of the Windward Passage, bound south. R.Smg. *Enrico Tazzoli* is bound north, east of Eleuthera. R.Smg. *Da Vinci* exits area.

15th March 1942

U-161 enters area very briefly, transiting channel south of Anegada Passage east, then north for France. *U-126* is in Straits, is bound northeast of Bahamas, bound east for France. *U-504* is bound north of Mona Passage east of Turks and Caicos. SS *Colabee* drifts ashore and grounds on a shoal. She was pulled off the shoal by SS *Oriente* and anchored on 16th March. Cuban Navy towed her into Nuevitas. After sinking SS *Stangarth,* R.Smg. *Morosini* relocated southwest of the original position, intercepting on the night of the 15th March the Dutch motor tanker SS *Oscilla* which it torpedoes and sinks east of Bahamas, with the result that 51 are later saved and four die. Count Carlo Fecia di Cossato on R.Smg. *Enrico Tazzoli* located and sank British tanker *Athelqueen* east of Hope Town, Abaco. Three crew lost, 46 rescued and three died when the lifeboat spilled over on the reef, forcing some men to swim for it, their bodies never being found. During the attack, R.Smg. *Enrico Tazzoli* collides with the hard-to-sink *Athelqueen.* She is damaged and commander abandons the most successful Italian mission to the region. Survivors of SS *Charles Racine* end a nearly weeklong raft and lifeboat voyage then begin other long voyages of repatriation to Trinidad then New York on an Argentinian steamer.

16th March 1942

R.Smg. *Enrico Tazzoli,* exits region bound north to France. R.Smg. *Morosini* is bound east of Bahamas north of Anegada. *U-126* transits northwest and Northeast Providence Channels, then is bound east for France. SS *Stangarth*, on her maiden voyage, was hit by one torpedo from *U-504,* exploded and sank immediately 300 nm north of San Juan. The master, 39 crew and six gunners were lost. SS *Tønsbergfjord* survivors landed by SS *Telamon* in Port Jeremie, Haiti. *U-161* transits Bahamas from 15 March to 17 March; is *en route* to sink SS *Fairport.*

17th March 1942

U-126 exits area northeast of Bahamas, bound east for France. R.Smg. *Morosini* is bound east of Bahamas bound north of Anegada.

18th March 1942

U-504 exits area bound northeast for France. Fifty-eight survivors of British SS *Daytonian* land in Nassau. SS *Tønsbergfjord* survivors set sail southwest for Florida, with final survivors picked up by the Norwegian tanker SS *Velma* this day.

19th March 1942

Bahamian government sends the fast mailboat *Content S.*, converted by Richard Sawyer into a mailboat, from Nassau to Hope Town to rescue British survivors of the tanker *Athelqueen.* Due to the grounding of a lifeboat on the reef line, three of the sailors who tried to swim to shore at night perished. [Many years later residents with the support of the Wyannie Malone Museum erect a limestone memorial to the dead]. The 46 others were repatriated to England through Nassau.

20th March 1942

U-124 enters area very briefly from Hatteras to the Savannah, Georgia area.

21st March 1942

The New York Times reports "U-Boat sinks ship as plane passes, raider gets Greek steamer off coast five minutes after patrol flies over. Survivors reported that a tragedy occurred when one of the three lifeboats, within sight of land, struck a reef, three of the men who had survived the torpedoing were drowned attempting to swim to shore, only a mile away. The Associated Press said that 171 had survived and six had died in the four sinkings; tanker *Athelqueen* accounts for three deaths and tanker *O. A. Knudsen* for the four." *The Nassau Tribune* reported that "Miamians welcomed the survivors of a torpedoed Dutch ship. One of three United Nations vessels sunk off eastern United States shores. A boat out of Whale Harbour brought the three men ashore."

23rd March 1942

U-124 exits the area north to France via Hatteras. Tanker *Peder Bogen* is sunk by R.Smg. *Morosini east* of Bermuda and Anegada. *U-600* transits Windward Passage.

24th March 1942

SS *Hanseat* survivors arrive in New York on an Allied ship, after leaving Cape Maysí, Cuba.

25th March 1942

R.Smg. *Calvi* intercepted and sank the British SS *Tredinnick* over 1,500 nm northeast of Anegada, however none of the 46 crew were rescued.

27th March 1942

U-123 sinks *Atik* aka *Carolyn* after heated engagement, off Hatteras, fatalities on both sides. After receiving *Atik*'s distress signals USS *Noa* (DD-343), USS *Sagamore* (AT-20), her sister ship USS *Asterion* (AK-100) and several aircraft conducted fruitless search for survivors next few days. Only signs found were debris and five empty rafts sighted by aircraft at between Hatteras and Bermuda on 30th March. All hands lost either in explosion or during gale that hit area shortly after. This ended the US Navy's efforts at decoy ships.

31st March 1942

R.Smg. *Calvi* sinks US tanker *T. C. McCabe*, the first US ship sunk by an Italian submarine in the area. SS *Mangore* headed from Cuba to Florida. Armed Guards sighted surfaced submarine, craft then disappears. Distress traffic ends for both SS *Mangore* and SS *Edward L. Shea*.

1st April 1942

Small convoy transits Straits north from Texas via Key West to New York including SS *Byron Benson* and SS *Gulf of Mexico*, with two escorts, USS *Hamilton* (DD-141), *USCG-Dionne*, and British armed trawler HMS *City of Norwich*.

3rd April 1942

Submarine sighted 250 nm east of Jacksonville. *U-154* enters area bound west to sink *Comol Rico* next day. Twenty-six survivors from the tanker SS *Oregon* in lifeboat made landfall near Puerto Plata, Dominican Republic.

4th April 1942

Tanker *Comol Rico*, carrying molasses, is hit by *U-154* 225 nm north of San Juan. Eight officers, 28 men, and six armed guards left the ship in one lifeboat and three rafts. Second torpedo killed one officer and two men on watch below. USS *Sturtevant* (DD-240) picked survivors and on 9th April 39 survivors were landed at San Juan. *U-94* enters Windward Passage and is sunk southwest of Haiti.

5th April 1942

U-123 enters area south in Straits from Hatteras to Canaveral, 100 nm off coast approaching Savannah. Tanker SS *Catahoula* is sunk by *U-154*, northeast of the Dominican Republic, nine lost and 38 survived who were picked up by USS *Sturtevant* (DD-240). *U-154* exited Windward Passage after sinking SS *Vineland*.

6th April 1942

U-754 sinks Norwegian tanker SS *Kollskegg* 350 nm northwest of Bermuda, four dead and 38 survived. Survivors landed by SS *Bushranger* in Nassau. *U-123* between Grand Bahama and Canaveral in Straits.

7th April 1942

U-123 off Savannah then Jacksonville, then south in Straits from Hatteras. South of Bermuda, three Bermuda aircraft under Carlson, Soverell, and Gannon attacked *U-571* with four Mk 17 bombs, inflicting minor damage. They later also attacked *U-754* unsuccessfully.

8th April 1942

U-123 under Hardegen attacks *Esso Baton Rouge* 13 nm off Brunswick, Georgia. The ship is salvaged and repaired.

9th April 1942

U-123 between Grand Bahama and Canaveral in Straits. USS *Sturtevant* lands 39 survivors of tanker *Comol Rico* in San Juan, having been directed to them by a USN patrol plane on 7th April. *U-504*, off Canaveral: one destroyer, four smokestacks in water too shallow for U-boat commander. Destroyer sounds shift quickly out, then suddenly were not heard. Silent running on electronic, or e-motors.

10th April 1942

U-123 sinks the tanker *GulfAmerica* off Jacksonville. Out of 41-men, 17 people perish, two of the seven-man Armed Guard die. USS *YP-32* aids in rescue of survivors. *U-504* is off Canaveral and reports fast running propellers heard all around, typical destroyer propellers, during the U-boat attack. The first destroyer runs in exactly over the boat. However, throws four depth charges too late at medium range. On the second approach, three depth charges, a bit closer. In the boat all is calm, only after 14-hours of submerged cruise has the air been very consumed, and filled with carbon dioxide.

11th April 1942

SS *Grenanger,* carrying general cargo and coffee, *en route* to New York from Barnaby Island, near Vancouver, Canada, is hit by one of two torpedoes from *U-130.* Survivors are picked up by British SS *Almenara*, towards Sombrero Island, off Anegada. Minesweeper USS *Courier* continues towards St. Thomas. *U-123* is off Jacksonville. *U-128* is off Matanilla Shoals: crash dives for Consolidated or Martin flying boat, approaching. Then another crash dive, for flying boat running on both electronic electric e-motors, rather than emit exhaust, make more noise and suck oxygen from interior.

12th April 1942

SS *Grenanger*'s motorboat takes is lifeboat in tow and heads southwest towards the West Indies. Next day, both boats sail independently. The motor is boat picked up by SS Almenara which later finds another lifeboat. Together they set course for Sombrero Island, arriving on 18th April. *U-123* is off Canaveral, *U-504* is also off Canaveral, and crash dives for aircraft directly ahead; at periscope depth, the aircraft is in sight; a Consolidated flying boat.

13th April 1942

SS *Korsholm, is* shelled by *U-123*, catches fire and sinks 70 nm Off Canaveral. Master, eight crew lost. SS *Leslie, en route* Cuba to New York hit by last torpedo from *U-123*, 3 nm southeast of the Hertzel Shoals Gas Buoy. Most of the nine officers, 22 crew, and one consular passenger abandon ship in two lifeboats and land north of the Canaveral Lighthouse next day. Another survivor picked up by US-flagged tanker *Esso Bayonne* and landed at Key West. [SS *Leslie* was raised in 1954 and broken up].

14th April 1942

U-130 enters area south to Anegada Passage. *U-123* is bound northeast of Abaco, east for France. *U-504*, north of Abaco: pursued by ASDIC pings from surfaced warships, followed by barrage of depth charges, temporarily moved off to observe route between Bermuda and Florida. *U-154*, Mona Passage: crash dive for aircraft, then more aircraft sighted.

15th April 1942

U-123 exits area north to France. *U-154* re-enters area via Mona Passage north. *U-128*, off Canaveral: faint propeller sounds astern, nothing seen, proceeded accordingly and from the shift, the sounds was just offshore. Then a small tugboat, far away. South of Bermuda PBM-1, Bu.No. [builder's number] 1252, flown by Koch, a USN aviator, is on a night sweep, when it spots a surfaced U-boat that dives.

16th April 1942

U-154 is bound north of Puerto Rico, east of Turks and Caicos.

17th April 1942

U-154 is 200 nm east of Long Island, Bahamas, north of Turks and Caicos Islands.

18th April 1942

U-154 is bound north of Puerto Rico and east of Turks and Caicos. An emergency SSS message is received in the Eastern Sea Frontier HQ. A lifeboat from SS *Grenanger* lands at Sombrero Island, north of Anguilla. The men are subsequently rescued by aircraft and USS *Courier* (AMC-72) and landed at St. Thomas, USVI.

19th April 1942

U-154 is bound north of Puerto Rico, east of Turks and Caicos.

20th April 1942

U-154 sinks the Canadian SS *Vineland* north of Turks and Caicos and east of Mayaguana. One crew is lost, master and 33 survivors are questioned by Germans. Lifeboats were picked up later by fishermen from the Turks and Caicos, and the Canadian sailors were repatriated back to Halifax via Haiti and Aruba. *U-125* enters the region via Windward Passage, then exits via the Mona Passage

21st April 1942

U-154 is bound east of Exumas, south of San Salvador, and very close to Plana Cays and Acklins before then turns northeast, bound for France.

22nd April 1942

U-201 sinks SS *Derryheen* 400 nm northeast of Abaco. Master, gunners, crew and a passenger are rescued by SS *Lobos* and taken to Havana. SS *Seatrain Texas* reports three suspicious vessels on surface signalling each other, 25 nm north of Havana. Passenger liner *San Jacinto* is torpedoed by *U-201* 375 nm southeast of Cape Hatteras and north of Grand Bahama. Survivors in six lifeboats and several rafts are rescued by USS *Rowan* (DD-405), and are landed at Norfolk.

23rd April 1942

U-154 is bound east of Exuma, south of San Salvador.

24th April 1942

U-201 exits the Bahamas area, as does *U-154*, which is bound east to France. *U-654* enters the area south, between Florida and Bermuda.

25th April 1942

U-507 enters the region from Bermuda, heading for Old Bahama Channel. *U-402* enters westward, bound for Jacksonville. *U-654* is bound east of Florida. *U-504* is bound north of Walker Cay, Abaco, in a severe storm. It dives to proceed submerged, because hurricane-like squalls arise, reducing its speed to just one knot.

26th April 1942

U-109 enters the area 200 nm east of Savannah, bound south. *U-125* enters the area south bound to Windward Passage, then to off Guantanamo. *U-402* is bound northeast of Abaco. *U-654* continues to patrol between Bermuda and Florida. *U-507* enters the region *en route* to Old Bahama Channel from the Bermuda area. *U-504* is bound north of Abaco, experiences three detonations, thought to be aircraft bombs. As the sea is mirror flat, it is possible that the U-boat is seen by an aircraft at periscope depth.

27th April 1942

U-506 enters the region west from south of Bermuda. *U-109* is bound northeast of Abaco, bound south. *U-108* enters from the Bermuda region after sinking a ship and proceeds towards Haiti. *U-125* is bound north of Puerto Rico, bound south to Windward Passage, having made it to the southeast of the Turks and Caicos. *U-402* is bound northeast of Abaco. *U-654* continues between Bermuda and Florida. *U-507* is bound south towards Old Bahama Channel. Torpedo-damaged SS *Colabee* is towed to Tampa and returned to service in September, 1942. USS *Sturtevant* (DD-240) is sunk by mines laid by the US Navy off Marquesas Key, Florida, because no one informed the Royal Navy or US Navy of the minefield. *U-504*, north of Abaco, launches a failed attack on a steamer and sees the ship as a U-boat trap, or Q-ship, used by the British mostly. The missed torpedo shot cannot be explained, as according to the sound bearings the torpedoes ran correctly. The three detonations may also have been fright-bombs to force the boat deep. The steamer sounds and propeller sounds were perfectly detected.

28th April 1942

U-582 enters southwest of Bermuda on a short one-day incursion south. *U-98* enters west of Bermuda bound south to Jacksonville. *U-333* under Cremer enters west of Bermuda, bound south in Straits and Canaveral. *U-402* exits the region bound north to Savannah. *U-506* is bound west from south of Bermuda. *U-125* is still bound south for the Windward Passage, whilst *U-109* is bound west , bound to Grand Bahama and Abaco. *U-507* reaches Old Bahama Channel to patrol between Cuba and the Bahamas, and *U-654* holds position between Bermuda and Florida, hoping for tonnage avoiding the bloodbaths of the coast. *U-108* is bound east of Bahamas bound south to Windward Passage. *U-504* is north of Abaco when a sailing vessel comes in sight, so course altered to look at the four-masted gaff-rigged schooner of about 450 GRT [tons], heading south. It is deemed too small and too little draft for a torpedo, and the operation of artillery is not possible in the sea state.

29th April 1942

U-109 is bound northeast of Abaco bound south in Straits. *U-582* returns north of Bermuda, departing after a brief sortie, and *U-654* also exits southwest of Bermuda, bound for France. *U-506* is bound west from south of Bermuda. *U-125* is bound north of the Dominican Republic on its way to the Windward Passage. *U-333* continues towards Canaveral, and *U-98* is bound northeast of Abaco, west to Savannah. *U-507* continued up Old Bahama Channel off Cuba and *U-108* sinks US-flagged tanker *Mobiloil* 350 nm northwest of Turks and Caicos. USS *PC-490* manages to retrieve the survivors and land them on 4th May at San Juan. The master had ignored orders to await a convoy off Norfolk and sailed alone, for which he was convicted. *U-128* is bound south of Bermuda, where it reports possible aircraft and warship: searchlight seen 11 nm on horizon, then sub-sea sounds, becoming broader.

30th April 1942

U-564 heads towards the Straits bound south, whilst *U-125* continues south towards the Windward Passage. *U-506* is bound west from south of Bermuda. *U-109* is bound north of Abaco and Grand Bahama, bound south. *U-333* is approaching Canaveral. *U-98* is bound north of Grand Bahama, bound west to Jacksonville. *U-507* is still in Old Bahama Channel, and *U-108* motored east towards Anegada following the sinking of the tanker *Mobiloil*. The US-flagged tanker *Federal* is attacked and sunk by a U-boat 60 nm from Ragged Island, and 6 nm from Cuba's Gibara port. The US Army transport USS *USAT-Yarmouth* circled survivors several times, yet it steamed away, due to the danger of being torpedoed. Some survivors on rafts were picked up by fishing craft from Gibara 2.5 hours after the attack. Fourteen survivors in the boat followed rescue craft to shore. The Bahamian trading schooner *Ival* also puts out from Gibara, on hearing *Federal* being attacked. They are guided to site by smoke. In Nassau a bridge and tea event sponsored by the Nassau Yacht Club and Country Garden Club raised funds for the Survivors Fund. They US Navy's Eastern Sea Frontier reported a U-boat attacking in vicinity Guantanamo. *U-504*, northeast of Bahamas sights a white lifeboat with sails rigged as a cutter. *U-123*, off Jacksonville reports a crash-dive for a large Clipper aircraft civilian airliner, yet assumes the US will not use flying boats to patrol the coastal waters. Commander Hardegen lays the boat on the bottom off St. Augustine, and observes that aircraft are following the coast with position lights set.

1st May 1942

U-108 dog-legs back from Anegada towards Haiti to resume attacks, whilst *U-564* is bound east of Savannah and bound south, and *U-125* exits the region bound towards Guantanamo via the Windward Passage. *U-333* is motoring towards Canaveral, and *U-507* remains in Old Bahama Channel. *U-98* is off Savannah. *U-106* transited the Straits then sank SS *Empire Drake* between Hatteras and Bermuda. The Haitian government re-institutes its rubber cultivation to make up for wartime shortages, by reviving old plantations. USN aircraft on reconnaissance rescue 52 survivors from the US-flag tanker *Mobiloil*. *U-123*, off Canaveral reports 2-engined aircraft without position lights passed directly over at low altitude. One could only see it after hearing it. Too late to crash-dive, probably it was a coincidence, and he did not see us because he flew along. Sighted an empty lifeboat on port side abeam east of Canaveral. [Hardegen was in the *Luftwaffe* before U-boats]. *U-130*, northeast of Anegada: crash dived to avoid aircraft near burning tanker *Esso Boston*.

2nd May 1942

U-564 is off Jacksonville, *U-333* is still south in Straits bound to Canaveral, and *U-506* is in the Northeast Providence Channel. *U-507* continues in the Old Bahama Channel, whilst *U-98* is off Jacksonville. *U-109* is having a destructive patrol: it damages the SS *La Paz* off Canaveral, and various ships, including fruit carrier SS *Worden* went to her aid. *La Paz* is partially salvaged. The German commander misinterprets a radio message when Nicaraguan SS *Worden* goes to *La Paz*'s aid, and reports that *Worden* itself was the victim. *U-108* is bound north of Turks and Caicos and south to Windward Passage via the Bahamas. *U-506*, off Palm Beach, where it reports many aircraft, including seaplanes, constant crash dives, including for land-based aircraft, slow and fast, suspects it was seen. *U-130* is off the Sombrero Passage and Anegada Passage, where it reports several 2-engine land-based aircraft flying out of the clouds. *U-108* is close along the south coast of Bermuda and records an aircraft in sight.

3rd May 1942

U-69 enters the region southeast of Bermuda, making a direct bead for the Mona Passage, *U-333* on course southwest bound to the Straits, whilst *U-507* holds its position in Old Bahama Channel. *U-98* is in Straits off Jacksonville. *U-109* is leaving the carnage it wrought off Canaveral, and *U-563* is lining up attacks on SS *Eclipse* and SS *Ocean Venus*, while *U-506* is still in the Straits. *U-108* is 100 nm northeast of San Salvador and bound south to Crooked Island Passage. SS *Laertes* is torpedoed by *U-109* off Canaveral. The US Navy Eastern Sea Frontier logs that an Allied SS sent an SSS from 180 nm 165° (south- southeast) from Bermuda. *U-506* is in Northeast Providence Channel, then Northwest Providence Channel, bound north in Straits. The Nicaraguan SS *Sama*, a converted Finnish schooner, is sunk in the straits 80 nm northwest of Andros by *U-506.* SS *Ocean Venus* is torpedoed and sunk by *U-564* 12 nm off Canaveral. The same submarine sinks SS *Eclipse* and other ships. *U-126* sinks SS *Mariana* quickly with one torpedo northeast Turks Island, east of Mayaguana. Although survivors were seen to enter boats in fairly smooth weather, foul conditions must have sprung up as none of the eight officers and 28 crew under a Cayman captain survived, despite being close to and upwind of Mayaguana. *U-108*, south of Bermuda, reports that aircraft has moved out of sight, enabling submariners to see southern Bermuda and the lights of Hamilton. *U-109*, northwest of Jacksonville; double shot on destroyer of the *Fletcher* type. Detected by surface runner. Depth charge pursuit. No ASDIC counter-attack. *U-654* of Bermuda sighted a patrol boat or pilot vessel near Hamilton; it was scouting.

4th May 1942

U-506 exits area via Key West. *U-69* heads south-southwest for Mona Passage in a straight line. *U-98* is bound east of Savannah. *U-109* is in the Straits between Grand Bahama and Canaveral. *U-130* re-enters area via Mona Passage north after patrols in Bahamas region since 15th April. SS *Eclipse,* dispersed from **Convoy ON-87**, is torpedoed by *U-564* off Port Everglades. She sinks but later salvaged and towed into Port Everglades, is repaired and returned to service in December 1942. The US-flagged tanker SS *Java Arrow* is torpedoed off Port Everglades and towed there. Civil Air patrol also locates the steam ships SS *Delisle,* SS *Amazone,* and SS *Halsey,* victims of German attack. Thirteen steam ships USN *PC-490* delivers survivors of US-flagged tanker *Mobiloil* sunk on 29th April by *U-108* to San Juan. *U-507,* Old Bahama Channel: crash dive for flying boat type Consolidated; it circles, apparently as an escort. Crash dive again, because the aircraft comes towards me. Nothing more is seen of the steamer and aircraft. The air escort was probably on account of my appearance. Continued transit in the direction of Old Bahama Channel. Crash dive for 2-engined land-based aircraft. *U-654,* Bermuda: slow large seaplane, six miles from Hamilton. *U-506,* east of Bahamas: crash dive for 3-engined land-based aircraft, U-boat was not sighted.

5th May 1942

U-130 is between Dominican Republic and Turks and Caicos *en route* north to skirt Acklins Island and patrol exit channels from Windward Passage. *U-594* is bound north*east* of Abaco, south. SS *Delisle* is sunk by *U-564,* 15 nm off Jupiter Inlet, Florida. The crew of eight officers, 24 men, and four stowaways abandons ship; two perish. Ship repaired. One lifeboat of SS *Delfina* landed at Monte Christi, Dominican Republic and other boat towed to San Juan by *PC-67.* SS *Stanbank* is sunk by *U-103,* with nine crew killed. Master, 34 crew, and four gunners are picked up after 10 days by SS *Rhexenor,* landed at Bermuda. Newspapers report on survivors of a "medium sized US merchant- man" sunk off Cocoa Beach (Canaveral area), Florida. The *Los Angeles Times* claims that crew were "unnecessarily murdered" while lowering a boat. US-flagged SS *Afoundria* is torpedoed and sunk by *U-108,* 8 nm north of Le Male Light, Haiti. No casualties among 38 crew and eight passengers. Net-tender USS *Mulberry* (YN-22) rescues all survivors from SS *Afoundria,* including women, and lands them at Guantanamo. *U-507,* Old Bahama Channel: crash dived for Consolidated flying boat that approached U-boat from astern. At periscope depth, aircraft seen circling astern, then turns suddenly and comes towards U-boat, decreasing altitude substantially, so again forced to go deep. Later, propeller sounds, shifting ahead and disappearing: apparently a follower. Did not attack because, from the experience of the previous day, the Consolidated probably circled above him in escort. Instead, I will get out of the geographically limited Old Bahama Channel into my assigned operations area, where I expect more worthwhile targets. Three shadows in sight, shifting slowly out, maybe a patrol vessel line. *U-98,* off Jacksonville: spurts of flame 15 nm away with subsequent fiery glow observed. On same bearing 7 nm away, a red sail that soon passes out of sight. *U-109* is off Canaveral, where it crash dives in response to a land-based aircraft; the U-boat was not noticed, and proceeded submerged. *U-564* is off Canaveral, where it encounters a *K-47*-class British, or Canadian corvette, in sight, and dived for the attack. The British corvette zig zags shortly before firing. Commander considers a shot uncertain, rationalizing that he can make better use of the torpedo later. Further pursuit useless, due to target going too fast. Surfaced, then crash dive for aircraft. Three aircrafts sighted, briefly in view. Again, crash dive for three aircraft and yet again for another two aircraft. General pattern: diving due to heavy day air patrol. Still the same day two more crash dives for aircraft, and then a large tanker with airplane escort.

6th May 1942

U-130 is bound east of Acklins and Crooked Island Passage south to Windward Passage exits. *U-594* is bound northeast of Bahamas, south towards Crooked Island Passage. *U-564* is off east coast of Florida. *U-109* is off Canaveral. Tanker *Java Arrow* torpedoed twice by *U-333,* 8 nm off Vero Beach, Florida. Two officers on watch below killed. Seven other officers, 32 crew, and six armed-guards are picked up by USS *PC-483* and a USCG craft, and landed at Miami and Fort Pierce. Tug *Ontario* and tug *Bafshe* tow the stricken tanker, escorted by US Coast Guard vessels, to Port Everglades. The Latvian-flagged SS *Abgara* is sunk by *U-108* 15 nm southeast of Great Inagua. Crew abandons ship and 34 survive. SS *Amazone* is sunk by *U-333* under Cremer off Miami. Fourteen crew lost; survivors picked up by USS *PC-484* and landed at Miami. Eight officers and 24 men abandon ship. *U-333* comes alongside lifeboats and Cremer offers assistance, which was declined. USS *PC-451* goes to survivors' aid, but breaks off efforts to conduct an unsuccessful pursuit of *U-333*. Some survivors saved by two fishing vessels, which take lifeboats in tow and land at Gilbert Bar Lifeboat Station. *U-507*, Old Bahama Channel: crash dive for aircraft, probably the Consolidated again. Directly ahead a steamer which forced to pass soon after, as determined by sound bearings. Surfaced, then crash dive for inbound 3-engined land-based aircraft (possibly Consolidated) which came directly from behind. U-boat is just about out of compressed air and battery power and can no longer afford repeated crash dives therefore will remain submerged until darkness. Crash dive for aircraft type Consolidated, that comes directly towards boat, so it seems to have noticed something previously. Surfaced. Crash dive for aircraft that comes towards *U-507* out of the clouds from astern. Seems to be watching continuously. Changed to the south side of the Nicholas Channel to lose the aircraft. *U-333*, off Jacksonville: sighted aircraft heading northeast. *U-507*, Nicholas Channel: two masts with flags and pennants in sight 10 nm away, initially identified as a destroyer group because the positions of the masts relative to one another continually changed. *U-506*, off Miami: two depth charge bombs, crash dived for seaplane, then for six fighter aircraft in two formations, then two more land-based aircraft.

7th May 1942

U-333 is bound northwest of Grand Bahama. *U-130* is bound east of Bahamas, north. *U-109* is in Straits between Grand Bahama and Canaveral. *U-98*, east of St. Augustine: shadow in sight; a destroyer of Samson class; port lookout should have seen him. As a result, he is already very close and I am positioned between him and the moon and show a wide silhouette. If I turn away, he will see me. Therefore, I turn towards him and strive to shoot as fast as possible. I keep the speed and fire a 4-fan of torpedoes. The two left-most torpedoes jump two times in succession, quite high out of the water. The shooting track can be seen easily. After the fan is fired, the destroyer turns hard towards, out-manoeuvres the fan and runs towards me. Crash dive, followed by two depth charges that lie so-so. All the depth meter gauges, except for the Papenburg, and several light bulbs, fail. During the crash dive, the boat hits the ocean bottom about 7° bow down. Because it is too heavy, it cannot be freed again, and instead scoots along the bottom. In the mean-time the destroyer has come to high speed and is positioned off my starboard side. Because I cannot get free of the bottom and get away without making a lot of noise, I stop the motors and secure everything except for the gyro compass and the converter for the listening gear. I send the seamanship personnel to their bunks and the technical watch sits ready at their stations. It is utterly quiet in the boat. The destroyer is easily heard, alternately stopping and proceeding at listening speed. He moves slowly around my stern to my port side. Then he comes back around to the front. ASDIC sounds are not heard. One can hear in the boat when he gets underway. Now he runs at high speed, comes from starboard ahead and over runs the boat about at the level of the control room. In so doing, he throws six depth charges, excellently placed. With this about one ton of water enters through the conning tower hatch. We later determine a set screw is broken at the dog. All of the lights go out, compressed air discharges from the drains of the high- and low-pressure air distributors. One could see how the boat was compressed and shaken back and forth, all hull valves and both exhaust gas flap valves were spring about one thread. In the E-motor room, there is heavy arching with flames at the port main switchboard. Switches did not spring out. The entire order system fails. Both vents of main ballast tank 3 and main ballast and reserve fuel oil tank 4 sprung open. The articulated lever of vent 3 port has a shorn toggle bolt. I get reports from all the rooms that the boat remains tight and no major leaks are present. The people behaved perfectly. For most, these were the first depth charges, at least at such close range. Those not needed lay down on my orders again, the others try as quietly as possible to get an understanding of the failures in their spaces. After the listening device, which was turned off while running off, is turned back on, the destroyer heard again abeam to port astern as he moves around at high speed there. I now have the gyro compass converter secured and the converter for the listening gear operated only now and then. It is now dead quiet in the boat. In my view not a sound can escape. *U-98*, east of Savannah: aircraft with set lights, so U-boat stopped for a moment. Then five other attacks or aircraft sightings. *U-564*, off Miami: aircraft seeking the U-boat, tanker, one escort and four aircraft 3 nm distant. *U-333*, Straits: crash dive for biplane float plane, then fast land-based aircraft on horizon, did not dive. Crash dive for another biplane, float plane. *U-108*, Windward Passage: large land-based aircraft, flies a circle very low over the steamer. Markings: Red star on white circle. In the centre a white ball.

8th May 1942

U-103 enters area west of Bermuda south for Windward Passage. *U-69* exits region south via Mona Passage. *U-594* is bound south, east of Cat Island to Crooked Island Passage then Old Bahama Channel. SS *Ohioan* is torpedoed by *U-564* 10 nm off Boynton Beach, Florida. Eight officers and 29 crew tried to launch two lifeboats, but lost an officer and 14 crew. Survivors climbed on six rafts and were rescued by the US Coast Guard, which lands them in West Palm Beach, where four are hospitalized. *U-564*, off Cat Cay Bimini: allied naval escort vessel detected U-boat by sight or sound, illuminated it by searchlight. Artillery bombardment by escort vessel. Escaped on the surface, submerged patrolling Cat Cay, Bimini.

9th May 1942

U-108 is in southeast Bahamas. *U-103* is bound northeast of Abaco south to Crooked Island Passage. *U-98* is off Jacksonville. Tanker *Lubrafol is* torpedoed by *U-564*, 3.5 nm off Hillsboro Inlet, Florida. Thirteen were killed and seven injured. Her lifeboats are seen by two US Coast Guard boats which tow them clear of the burning oil to a point of safety. Thirty-one survivors are landed at Boynton Beach, and seven bodies were recovered. *U-109*, off Canaveral: three light detonations, suspected aircraft bombs. Suddenly the green and red side lights of an aircraft are in sight ahead. Already approaching very close, so that an undetected crash dive is no longer possible. Stopped the diesels, so that speed comes off the boat and the shining wake disappears. All superfluous people off the conning tower. Machine gun is ready to shoot and manned by gun crew. Boat is ready to crash dive and with Bulkheads closed! ordered. Aircraft directly over-flies the boat at low altitude, recognized as Martin PBM one to two flying boat. *U-109* as not sighted. The pilot seems not to be aware [of the boat], turning from astern in a steep left turn two times around the boat, then approaching the boat directly again ahead, this time a little lower. The drone of the engines can be heard in the control room. But again, nothing happens. The flying boat turns around the boat again and approaches again from ahead at low altitude. Now I must disappear for good. Also, in this approach no bombs or machine gun fire. When the flying boat goes astern I go with a crash dive and moved off submerged. *U-333*, Straits: shadow to starboard: a destroyer, patrolling the route from buoy to buoy. *U-506*, south of Keys: crash dive for aircraft with set lights.

10th May 1942

U-103 is 300 nm east northeast of Abaco, south to the Windward Passage. *U-98* is bound north in Straits then east to France. *U-564* is bound south in Straits. SS *Clan Skene,* torpedoed twice by *U-333*, sinks 300 nm southeast of Cape Hatteras. Nine crew lost, master and seventy-two survivors picked up by USS *McKean* (APD-5), and landed at San Juan. *U-333*, badly damaged by depth charges of an escort ship, is limping back to France. *U-109*, off Vero Beach; suddenly the green and red side lights of an aircraft are in sight ahead. Already approaching very close, so that an undetected crash dive is no longer possible. Stopped the diesels, so that speed comes off the boat and the shining wake disappears. All superfluous people off the conning tower. Machine gun is ready to shoot and manned by gun crew. Boat is ready to crash dive and with Bulkheads closed! ordered. Aircraft directly over-flies the boat and was recognized as a Martin PBM 1-2 flying boat, and U-boat was not sighted. It seems not to be aware [of the boat], turning from astern in a steep left turn two times around the boat, then approaches the boat directly again from starboard ahead of the boat. This time a little lower. The drone of the engines can be heard in the control room. But again, nothing happens. The flying boat turns around the boat again and approaches again from ahead at low altitude. Now I must disappear for good. Also, in this approach no bombs or machine gun fire. When the flying boat gets astern, I go with a crash dive. Moved off submerged. Bethel Shoal Several bright, weak detonations heard are probably aircraft bombs. *U-333*, St. Lucie Inlet: two destroyers had seen me in the mirror flat sea. Depth charges. While going deeper from periscope depth, the 1st depth charges detonated. With these both diving planes failed. Before the diving planes were shifted to hand drive, and the bilges, which now filled continuously due to the already described problem, were pumped, they forced the stern down. In a few seconds the boat developed a tendency to break through the water surface, so that broaching could only be prevented by flooding the negative buoyancy tanks. The boat sank to the bottom like a stone. Two destroyers located [used ASDIC] the boat and overran it very accurately. One heard the depth charges falling and they all lay damn close. At long range the locator sent out an impulse every few seconds. On getting closer the boat was held by an irregular strength impulse which sounded as if the main ballast and fuel oil tanks were being fired at by a light machine gun with short bursts of fire. The better part of the lighting failed immediately from the depth charges, it rained down in the usual manner from the ceiling. All depth gauges and rudder indicators failed. The diesel air intake mast leaked heavily, so that the drain had to be closed. Broaching was only prevented by flooding the negative buoyancy tanks. Again, we lay on the bottom and were attacked anew with depth charges. As the depth charges disappeared, a third destroyer arrived, destroyers patrolled back and forth behind the boat. *U-108*, south of Inagua: sighted an aircraft which crossed the U-boat's course to Haiti. Should have seen us, if he were alert.

11th May 1942

U-103 is bound east of Bahamas, south for Crooked Island Passage. *U-98* is bound north of Grand Bahama, east for France. *U-109* is bound northeast of Grand Bahama, east for France. *U-594* is off Long Island Bahamas, in Crooked Island Passage. She is bound south towards Old Bahama Channel then entry to Windward Passage. *U-108*, Windward Passage: sighted stocky land-based aircraft from Cuba and low-flying amphibious plane headed to Haiti.

12th May 1942

U-108 is between Cuba and Bahamas, west of Inagua. *U-564* is in the northern end of Old Bahama Channel, but turns around and uses Santaren Channel east of Cay Sal to go back north in Straits and out of region. *U-594* is off Cape Maysí Cuba in Old Bahama Channel east towards Windward Passage and Inagua. A mine drifts into Port Everglades and causes evacuation of a five-block section of the beach, then it is removed by truck. *U-564*, off Bimini: small escort vessel, then a crash dive for aircraft. Masts indicate a small convoy of five steamers and one destroyer heading south. Attempt to manoeuvre ahead on the surface. Starboard diesel inoperative, valve disk is broken, therefore proceeding on one engine. U-boat then detected by destroyer, and fired on, result: three depth charges dropped and a crash dive. Two fillings of main ballast and reserve fuel oil tank 4 are blown. Afterwards peace. U-boat immediately went deep to avoid aircraft. After not quite five minutes the U-boat sailors heard the sinking of the steamer which they had torpedoes, in about two minutes. *U-108*, north of Windward Passage: one life boat in sight, sails set, on westerly course, and second life boat must have been hit and destroyed. *U-109*, off Vero Beach: aircraft abeam, so crash dived. Then heard detonations: four aircraft bombs, then eight aircraft bombs.

13th May 1942

U-502 enters area south to Anegada Passage. *U-106* enters region south. *U-502* proceeds south in Straits towards Key West. *U-108* is bound east for France. *U-564* is in Old Bahama Channel off Cuba. *U-103* is in Windward Passage transiting south out of region. *U-594 is* off Inagua, north toward Mayaguana Passage. *U-98* is bound northeast of Abaco, east for France. *U-109* is well northeast of Grand Bahama, east for France. Master and seventy-two survivors of SS *Clan Skene* are landed at San Juan by USS *McKean*. Schooner *Ival* returns to Nassau to tell of its rescue of torpedoed ship SS *Federal* survivors off Gibara, Cuba. *U-69*, north of San Juan: crash dived to avoid Consolidated flying boat on reconnaissance, balanced the boat, then surfaced, land in sight ahead, again crash dived for a 2-engined land-based aircraft flying out of the sun.

14th May 1942

U-751 enters area southwest of Bermuda, south for Windward Passage. *U-502* exits area south via Anegada Passage. *U-103* exits area south through Windward Passage. *U-109* exits area north west of Bermuda then east for France. *U-594* is passing north through Mayaguana Passage, then east. *U-130* is bound north, northeast of Bahamas. *U-108* is bound east, passing San Salvador, east for France. *U-564* sinks the Mexican tanker *Portero del Llano* off the Florida coast which, along with sinking of *Faja de Oro*, leads to declaration of war by Mexico against Germany. Survivors are picked up by USS *PC-536* and taken to Miami. US-flagged sail-training ship SY *Morning Star* concludes a 5-month cruise of Bahamas, and returns to Miami's City Yacht Basin, for a month lay over, during which most of the students volunteer for the USN and sign off. The *Nassau Daily Tribune* reports on Bahamian schooner *Ival*'s rescue of 25 seamen "*Ival* with Cuban naval officers on board, put out in a 40-mile breeze and, guided by the smoke, picked up 25 men in the choppy sea." Captain Bowleg and crew of the *Ival* received high praise. *U-109*, off Canaveral: aircraft with set lights turned towards land, yet come in sight again and flew back. A lighted aircraft abeam, then two more aircraft abeam to starboard. and an aircraft with set lights ahead. That aircraft extinguished lights and passed out of sight. Yet another aircraft flew low over the water out of the moon track towards the boat, causing a crash dive. *U-103*, northeast of Bahamas: aircraft cause a crash dive. *U-109*, off Canaveral: aircraft sighted abeam to starboard.

15th May 1942

U-558 enters area southwest of Bermuda south, heading towards Windward Passage. *U-106* in northern Straits, south towards Key West. *U-564* in Straits north. *U-751* is bound northeast of Bahamas south. *U-594* is northeast of Mayaguana, north of Dominican Republic, east. *U-130* is between Bermuda and Bahamas to the far northeast. Civilian tug *Ontario* and tug *Bafshe* tow SS *Java Arrow* to Port Everglades, Florida. SS *Stanbank*'s seven crew and two gunners were lost. Master, 34 crew, and four gunners picked up by SS *Rhexenor* and landed at Bermuda. *U-333* inadvertently surfaces under the small power yacht *Jay-Tee*, submerges, resurfaces and proceeds. Civilian Air Patrol (CAP) aircraft are authorized to carry bombs along the US coast as a response to predations by U-Boats there. Bahamian laborers from Harbour Island, Eleuthera and other settlements have their departure to the US to work on farms (known as "the contract,") delayed due to the U-Boat threat. Eventually the boat voyage is replaced by flights and the plan, envisioned by the Duke of Windsor and implemented with US support, went forward and proved a boon to Bahamians financially and culturally. At its height in July, 1944 there were 5,762 Bahamians on the contract, working in the US. *U-564*, Straits: depth charge detonations heard far away, and propeller noises in the GHG sound-detection device. At periscope depth see two escort vessels of USCG *Argo* class cutters. One lies stopped, the other moves back and forth at high speed. U-boat moves away submerged. I want to go to periscope depth for another sweep, when the stationary escort vessel moves towards us and passes overhead. To depth to withstand the five depth charges dropped. Heavy concussions in the boat. Failures: much glass and fuse damage, gyro-compass repeater failed, the rolling jaw for the onboard casing of the attack periscope torn from the pressure hull. Rubber-seated valves sprang open, main ballast and reserve fuel oil tanks 4 battered, so that no further pursuit is possible. *U-98*, off Canaveral: attacked by four aircraft and patrol boat. Buzzed by 4-engine flying boat. Pinned down by patrol boat, which did not bomb them. The US Navy's Naval Air

Station Exuma was commissioned, five months after work began.

16th May 1942

U-753 enters area from northeast of Grand Bahama west and south. *U-130* is bound east of Bahamas, northeast of Abaco, north and west Bahamas and Bermuda. *U-558* is bound east of Bahamas, south to Windward Passage. *U-108* is well east of Bahamas east for France. SS *Nicarao* is torpedoed by *U-751* north of San Salvador and east of Rock Sound, Eleuthera. Eight officers, 27 crew and four armed guards abandon ship in two lifeboats, yet both overturn. Survivors jump overboard to swim to three rafts. One officer and seven crew drown. Thirty-one survivors are picked up by tanker *Esso Augusta* and landed at Norfolk.

17th May 1942

U-753 is rounding Grand Bahama to port, south in Straits. *U-751* is bound east of San Salvador, approaching Crooked Island Passage south to Windward Passage after sinking SS *Nicarao*. Dutch SS *Fauna* is sunk by *U-558* 25 nm east of Mayaguana and 25 nm north of Providenciales in the east Caicos Passage. Two killed and 27 survivors are landed in Providenciales the same day. *U-564*, Straits: due to heavy air patrol the U-boat was approached several times by aircraft. Dark night, windy, strongly cloudy. Crash dive for aircraft which dropped one depth bomb, no damage. Then again, crash dive for aircraft, which dropped four well-placed depth charges. Water intake from tube II, as the torpedo had just been moved, and the rear door of the torpedo tube. *U-108* is in the Windward Passage: four aircraft, one a flying boat, came out of thunderstorm, so the boat crash dives.

18th May 1942

U-108 exits area after busy patrol and numerous sinkings, bound towards Azores and France. *U-130* exits area bound towards Hatteras then east for base. *U-98* exits the area bound east to France. *U-594* is bound northeast of Eleuthera and north of Puerto Rico. *U-753* is in Straits south bound towards Key West. *U-564* is at the north end of Straits, bound north. *U-751* is bound northwest of Crooked Island entering Crooked Island Passage. *U-507* re-enters area eastbound via Key West. *U-558* sinks the Dutch SS *Fauna* east of Mayaguana, north of Turks and Caicos. *U-103* is bound southwest of Inagua, reporting crash dives to avoid four aircraft, including bi-plane. *U-98*, off Canaveral: saw aircraft, 4-stack destroyer, fired from three torpedo tubes from roughly one mile, chased off by two bi-plane aircraft. *U-108* is in Windward Passage, reporting fast land-based aircraft over Cap du Môle, Haiti.

19th May 1942

U-594 is bound east of Acklins, north of the Dominican Republic. *U-753* exits the area bound west past Key West. *U-751* exits area. *U-558* is bound northwest of Haiti, bound south to Windward Passage. *U-108* is bound east of Cat Island, reporting a destroyer escorting large steamer southbound, zigzagging. *U-106* is bound south of Jacksonville, reporting that it is forced to submerge by three to four aircraft, from bi-plane to patrol planes. *U-98* is bound north of Canaveral, reporting one aircraft with position lights set. *U-753* is bound south of Key West, reporting constant aircraft harassment in tracking convoy; it dives for aircraft, was sighted. Three aircraft drop scare bombs. The U-boat sent radio message; a convoy of 24 steamers, with sea and air escort, low speed. Then, dived for aircraft twice. Aircraft recognized in the periscope, U-boat dived, aircraft another two times. At the moment cannot endure a listening pursuit with the chattering net guard. Must get out of here as soon as possible. Aircraft seen, slowly passes out of sight. Several aircraft of various types relieved each other at the diving location (near and far circles). Training of US aircraft increases. Aircraft type Vought Sikorsky OSL-V1. Surfaced, then dived due to the same aircraft. Aircraft still in sight. Aircraft circles at the old diving location until was replaced by an aircraft of the same type. Surveillance is carried out with diligence and zeal. Today 12 crash dives!

20th May 1942

U-558 exits the area in Windward Passage south off Guantanamo. *U-106* exits the area west past Key West. Motor fishing yacht FV *Pilar* sets out on six-week "Anti-Submarine patrol" after *U-176* sank SS *Mambi.* New *de facto* German patrol area is off Cuba's central north coast, in Old Bahama Channel. Work begins on RAF Base No.111 OTU to be called "the Project" for security reasons. *U-106*, off Canaveral: aircraft alarm due to airborne reconnaissance along the Florida coast. Then aircraft with set lights in sight. Seven nm from St. Lucie Shoal sighted a patrol vessel, then an illuminated aircraft which was flying away. *U-564* is in the Straits, where it crash dives for two different aircraft. Then crash dived for an airship, followed by another crash dive for aircraft: total four dives. *U-564*, Straits: U-boat attacked a 4-stack destroyer with two torpedoes, then crash dived for aircraft with searchlight. The U-boat commander suspects coordination between land stations or an escort aircraft, taking bearings on *U-564*'s radio transmission to enable switching searchlight on at approximate range from the boat and passing aft of the stern. Submerged attack on destroyer, yet both torpedoes missed! Probably torpedo failures, as the values were exact, and the range very short. Escort aircraft are often seen over the convoy; however, they cannot see us due to the large distance. Good visibility and choppy sea makes contact keeping possible. When aircraft approach within 10 nm we do not dive, but only slow to let the long white wake disappear and show the narrowest silhouette. Crash dive for aircraft, obviously a training flight from flight school Miami.

21st May 1942

U-107 is approaching entry to area. *U-564* is bound north of Grand Bahama, east for France. *U-507* is bound northeast of Bahamas. *U-125* re-enters area via Windward Passage, then east for France. *U-106* sinks Mexican tanker *Faja de Oro* off the Florida coast, causing the subsequent Mexican declaration of war against Germany on 1st June 1942.

22nd May 1942

U-107 enters area *en route* from Hatteras south in Straits and Old Bahama Channel. *U-504* enters area south for Windward Passage. *U-564* exits area north then east to France. *U-507* exits area west of Bermuda east to France. *U-594* is off Matthew Town, Inagua south to Windward Passage. *U-125* is bound southwest of Inagua, north from Windward Passage to France. *U-753 is* entering area via Straits from Key West.

23rd May 1942

U-125 is in Crooked Island Passage, north from Windward Passage to France. *U-107* in Straits south to Windward Passage from Hatteras. *U-594* is in Windward Passage south, exiting area.

24th May 1942

U-158 enters area after sinking SS *Frank B. Baird* southeast of Bermuda two days before and SS *Darina* before that. *U-107* in Straits south to Windward Passage. *U-125* is bound northeast of Crooked Island, east to France. *U-506* rounds Key West north up Straits and past Grand Bahama. *U-504* is bound south towards Windward Passage.

25th May 1942

U-107 is in Santaren Channel leaving Straits south to Windward Passage. *U-506* is in Straits north past Grand Bahama. *U-125* is bound northeast of Turks and Caicos, south to Windward Passage. *U-158* is bound east of Eleuthera, to transit Northeast Providence Channel and Northwest Providence Channel.

26th May 1942

U-158 is 300 nm east of Eleuthera, bound west to region. *U-107* is bound east of Savannah and northeast of Grand Bahama east to France. *U-504* is in Crooked Island Passage south to Windward Passage. *U-506* is bound northeast of Bahamas, east for France. *U-753*, west of Grand Bahama: seven aircraft sighted during manoeuvres in the Straits, several crash dives.

27th May 1942

U-504 exits area in Windward Passage, south. *U-125* exits area north for France after 2nd transit of Windward Passage. *U-107* in Old Bahama Channel south towards Windward Passage. *U-158* is bound northeast of Eleuthera west to Northeast Providence Channel. *U-506* northeast of Bahamas east for France. *U-504*, on a patrol that sank six ships, enters Windward Passage and exits through Windward Islands.

28th May 1942

U-158 is 90 nm northeast of Harbour Island, Eleuthera approaching Northeast Providence Channel and Straits. *U-751* re-enters area to transit Anegada Passage northeast of Bahamas and east for France. *U-506* shells and sinks SS *Yorkmoor* northeast of Bahamas, then is bound south in Straits. Master, 40 crew, and five gunners are picked up by a US Coast Guard cutter and landed at Morehead City. Duke and Duchess of Windsor depart Nassau for a one-month tour of the US, passing the Straits to Miami in their small gifted motor yacht *Gemini*.

29th May 1942

U-751 exits area heading northeast toward France. *U-558* re-enters area via Mona Passage then heads northeast to France. *U-158* transiting to transit Northeast Providence Channel and Northwest Providence Channel, bound west for Straits. SS *Western Head*, registered to Nassau, is sunk in Windward Passage by *U-107*. Twenty-four crew lost, master and five crew picked up by a USN patrol ship and landed at Guantanamo.

30th May 1942

U-506 exits area northeast of Bahamas, east for France. *U-158* is in Straits south from Northwest Providence Channel towards Key West. *U-558* is bound north of San Juan, east to France.

31st May 1942

U-158 exits area west, south of Key West. *U-558* is bound north of Virgin Islands, east for France. *U-103* re-enters via Windward Passage transiting north. *U-753* re-enters area east north of Cuba, south of Key West, east for Old Bahama Channel then France.

1st June 1942

U-558 exits area northeast of Bahamas and east to France. *U-753* is in Old Bahama Channel, east for Caicos Passage. *U-103* is transiting Crooked Island Passage north then east for France. *U-504* exits area via Mona Passage. US Marines complete a half-year deployment to build NAS Exuma.

2nd June 1942

U-172 enters area west to sink SS *City of Alma* and SS *Delfina*. *U-753* in Old Bahama

Channel east for Caicos Passage. *U-103* is bound northeast of Turks and Caicos east to France. *U-594* is bound south of Inagua in the Windward Passage, north towards Inagua. *U-159* sinks SS *Illinois* 400 nm northwest of Puerto Rico. Eight officers and 30 crew picked up by US-flagged tanker *Esso Montpelier* and taken to Port of Spain, Trinidad. SS *Triton* is shelled by *U-558* midway between Anegada and Bermuda. Five crew are killed, and survivors are given the best course to Bermuda or Puerto Rico. Eight survivors are wounded and one killed. Master, 28 crew, and one passenger picked up by US-flagged SS *MormackPort* and landed at New York. R.Smg. *Torelli* is bound northeast of Bahamas when attacked from air. It returned to and was wrecked in Spain. US Navy aircraft carry HRH the Duke of Windsor to Nassau from Washington DC to help quell labour riots around wages for large construction projects like OTU No. 111 and disparities between US and UK wages. US President FDR sends message "75 US Marines under Captain Goodwin, disguised as Military Police and armed with sub-machine guns and tear gas left for Nassau... to preserve peace and order on United States construction projects and protect US property." In what is known as the Nassau Riot or Burma Road Riot, two were killed, 40 injured, and 100 arrested. SS *Domino*, drifting, is attacked by *U-753*, east in Old Bahama Channel near Nuevitas Bay, north of Cuba. Both vessel crew were surprised by the other, and SS *Domino* manned its guns and fought off *U-753*, which simply continued past it. Both craft documented the encounter. An oil slick was seen by the master of SS *Veracruz*, however neither craft was badly damaged. R.Smg. *Enrico Tazzoli* sinks Dutch SS *Astrea* well east of Bahamas.

3rd June 1942

U-594 is east of Inagua, south of Turks and Caicos Islands, north. *U-68* enters area north through Anegada Passage. *U-753* is in Old Bahama Channel east for the Caicos Passage. *U-103* is bound northeast of Turks and Caicos heading northeast for France. *U-172* is bound northeast of Turks and Caicos, south for Mona Passage. US-flagged SS *City of Alma* is sunk by *U-159*, 400 nm northeast of San Juan. Twenty-nine men die, including the master. Two officers and eight men survive on a lifeboat to be picked up by USS *YP-67* and taken to San Juan. SS *Oregon* is torpedoed and sunk off Santo Domingo, Dominican Republic; 26 survivors landed there, but captain and nine members missing. R.Smg. *Enrico Tazzoli* sank SS *Astrea* east of the Bahamas. R.Smg. *Giuseppe Finzi* sank *Melpomene* east of the Bahamas.

4th June 1942

U-594 east of Mayaguana, north of Turks and Caicos heading northeast for France. *U-68* exits region south through Anegada Passage. *U-753* is in Caicos Passage having transited Old Bahama Channel east. *U-159* is bound north of Puerto Rico, south to Mona Passage. *U-172* is bound northeast of Puerto Rico, south to Mona Passage. *U-103* is bound northeast of Turks and Caicos heading northeast for France. *U-106*, off Jupiter Inlet: illuminated aircraft continually sighted, apparently transport aircraft or flying school, because mainly take-offs and landings observed. A few depth charges. Aircraft are insight continually, however are not dangerous. Aircraft with lights sighted over the coast. Crash dived at dawn, patrol vessel ahead. Raised the boat from the bottom. While surfacing, a gunboat which was not heard previously, is seen in the periscope, near the U-boat and dimly lit.

5th June 1942

U-594 is bound east of Cat Island, north of the Dominican Republic, north *en route* to France. *U-753* is bound east of Long Island heading northeast for France. *U-68* is entering the region north through Anegada Passage. *U-103* exits the area east southeast of Bermuda and back to France. *U-172* sinks SS *Delfina* 200 nm east of Turks and Caicos, 100 nm north of San Juan. Eight officers and 23 men abandon ship in chaos; 12 are picked up by USS *PC-67*. Another 15 land at Montecristi, Dominican Republic. Three die. *U-751*, Windward Passage: submarine chaser, one mast and fast for Puerto Banes, then saw a high-winged aircraft.

6th June 1942

U-157 enters region west for Caicos Passage. *U-68* exits region north through Anegada Passage. *U-172* exits region via Mona Passage. *U-753* is bound northeast of Bahamas, east for France. *U-594* is bound northeast of Bahamas heading northeast to France. *U-564*, Straits: lead aircraft of the very high-flying formation changes course before diving on us. A badly placed depth bomb proves that it had seen us. Surfaced. Pursuit continued. Smoke clouds 20 nm off. Crash dived to avoid three aircraft. Surfaced. Crash dive for destroyer. *U-558*, Windward Passage: four flying boats.

7th June 1942

U-753 is bound northeast of Bahamas, east for France. *U-594* is bound northeast of Bahamas, heading northeast to France. *U-157* is bound east of Turks and Caicos, south Crooked Island Passage.

8th June 1942

U-753 exits area northeast of Bahamas, east for France. *U-594* is bound southeast of Bermuda 150 nm, heading northeast towards France. US aircraft based at Cayo Francés, Cuba, drop bombs and depth charges on suspected U-Boat off Cayo Fragoso, near Cayo Confites, where Hemingway in his FV *Pilar* are patrolling against U-boats.

9th June 1942

U-129 enters area south to Old Bahama Channel. *U-157* enters area from north of Anegada and heads west along north coast of Greater Antilles.

10th June 1942

U-67 is bound south in Caicos Passage. *U-157* is off Inagua in Caicos Passage, south to enter Old Bahama Channel. Tanker *L. A. Christensen* is sunk by *U-129,* 500 nm east of Abaco. All 31 men abandon ship in three boats. Survivors are picked up by SS *Bill* and landed at Bermuda.

11th June 1942

U-129 is bound northeast of Bahamas, south. *U-67* clears Caicos Passage south to Old Bahama Channel. *U-157* torpedoes and sinks US-flagged tanker *Hagan* 5 nm north of Cuba, between Cape Roman and Cayo Verde. Thirty-eight of the eight officers, 27 crew, and nine armed guards board two lifeboats; two officers and four crew are lost. Survivors are transferred to Nuevitas by pilot boat and a Cuban Navy patrol vessel.

12th June 1942

U-157 is in Old Bahama Channel, sending his position, which may have doomed them, as they are attacked the next day. *U-67* is bound northwest of Inagua, south to Old Bahama Channel. SS *Hardwicke Grange* is torpedoed twice by *U-129* north of Puerto Rico. Three crew are lost. Master and 19 survivors are landed at Monte Cristi, Dominican Republic, after 13 days in a lifeboat. Another 23 survivors are picked up by the British tanker *Athelprince*, and landed at Nuevitas, Cuba. Sixteen survivors are landed at Môle St. Nicolas, Haiti. Sixteen officers and crew are rescued and landed at Jamaica. The captain is awarded the Order of the British Empire for bravery at sea. Tanker *L. A. Christiansen* survivors make it ashore in Bermuda.

13th June 1942

USCGC *Thetis* (WPC-115) sinks *U-157* 80 nm west of Cay Sal Bank, north of Cuba and east of Florida Keys, with 52 Killed. *U-129* is bound northeast of Bahamas, bound west. *U-753*, southeast of Miami reports two plane bombs, seven depth charges from submarine chaser. Another sub chaser, hunting group, more aircraft, more crash dives to avoid two land-bases aircraft. Forced away from the coasts and reefs, unable to radio, more aircraft, went around Key West. Straits of Florida east of Key West, USCG cutter *CG-Thetis, (WCG-115)*, focused attack by pilot N. C. McCormick vs. *U-157*.

14th June 1942

U-67 is bound north up Old Bahama Channel towards Key West. *U-129* is bound south, to the northeast of Turks and Caicos.

15th June 1942

U-129 is bound south, entering Crooked Island Passage south for Old Bahama Channel. *U-584* enters area southeast of Savannah, bound west to land saboteurs at Punta Vedra, Florida. *U-564*, Straits reports an attack on a convoy frustrated by aircraft: *U-564* can run only a short time at maximum speed, since they must slow for aircraft frequently. When they are forced to dive, on resurfacing the commander finds that the bearing to the convoy is unchanged after 90 minutes, perhaps due to convoy zig zag routines.

16th June 1942

SS *Managua*, carrying potash from Charleston to Havana is sunk by one torpedo from *U-67*, 40 nm west northwest of Elbow Cay, Cay Sal Bank in Straits; all 25 survive. *U-106* is in the Straits off Florida Keys, and reports progress monitored by US Navy Patrol Craft boats. Surveillance shows Alligator Reef is fully illuminated U-boat is forced to submerge by night due to aircraft. Only patrols seen. *U-504*, Windward Passage, reports multiple crash dives for land-based aircraft and flying boats. *U-172* is bound northeast of Puerto Rico: multi-engine aircraft attacked inbound forcing crash dive.

17th June 1942

SS *Millinocket* is sunk by *U-129* 30 nm southwest of Anguilla Cays, 40 nm southeast of Cay Sal in Nicholas Channel. Some men are blown overboard with only one lifeboat and two rafts launched, some jump into water. Twenty-two crew and six armed guards on board are lost. Survivors are picked up by a Cuban motor boat and two Cuban fishing boats and landed at La Isabela then taken to Sagua La Grande. *U-202* inserts four German agents ashore on Ponte Vedra Beach, south of Jacksonville. Fishermen discover the landing, and after Dasch, the leader of a different cell, turn them in, FBI agents capture Germans, who are part of *Operation Pastorius*. *U-172*, Mona Passage: crash dive! Aircraft, then again, crash dive for aircraft, and two more aircraft, very far, so remained on surface. Finally, another crash dive for more aircraft. *U-158*, Caicos Passage: reported Caicos uneventful, Windward Passage strong air cover, bombed twice from air at night.

18th June 1942

U-129 is bound south of Cay Sal in Old Bahama Channel, north.

19th June 1942

U-129 exits area west south of Key West and north of Havana. *U-584* leaves Punta Vedra heading southeast after landing saboteurs. *U-107* re-enters the area after sinking SV *Cheerio* in Mona Passage 8 nm southeast of Mona Island. Survivors were found by USCG *CG-459* and landed at Mayaguez, Puerto Rico. *U-172*, Mona Passage: screen of one or two destroyers escorting a convoy. Because it is about 45 minutes before twilight, *U-172* turned towards them to deliver a fan of torpedoes, however the ships were all suddenly lost from sight in tropical rain.

20th June 1942

U-504 re-enters area via Anegada Passage.

22nd June 1942

U-107 is bound east of Eleuthera, north of Puerto Rico halfway to Bermuda. *U-504* exits area east, bound to France.

24th June 1942

U-159 re-enters area via Anegada passage in transit, bound east to France.

25th June 1942

U-154 enters area bound south for Mona Passage. *U-203* enters area bound south. *U-159* exits area bound east for France.

26th June 1942

U-154 is bound east of Turks and Caicos, south to Mona Passage. *U-158* re-enters area off Key West, north for Bermuda area. *U-203* sinks SS *Pedrinhas*. British SS *Putney Hill* is hit by one torpedo from *U-203* 450 nm east northeast of Puerto Rico. Thirty-seven survivors are picked up after 10 days by HMS *Saxifrage* (K-04). Brazilian SS *Pedrinhas* is sunk by *U-203*, 300 nm northeast of Puerto Rico.

27th June 1942

U-332 enters the area northeast of Bahamas west. *U-154* exits area south via Mona Passage. *U-158* is in the Straits, bound north to the north of Grand Bahama. SS *Potlatch* torpedoed by *U-153* 650 nm east of the Virgin Islands. Seven officers, 32 crew, and 16 escape ship. U-boat under Reichmann (son of a Vicar), provided cigarettes. One officer, five crew lost with ship, and two later died in the month aboard a small lifeboat. Survivors land on Great Inagua, then to uninhabited Little Inagua and then to Acklins Island, before Marion Joe Carstairs, in her yacht *Vergemere IV* rescues them and took them to Nassau. *U-154*, in Mona Passage: three aircraft, then crash dive for aircraft, then more aircraft, U-boat showed its stern, continued on. For next aircraft, boat remained on the surface.

28th June 1942

U-158 exits Straits steaming northeast towards Bermuda next day hits SS *Everalda*. R.Smg. *Morosini* sinks Dutch motor ship *Tysa* 500 nm northeast of Anegada. SS *Sea Thrush* is sunk by *U-505*, 425 nm northeast of San Juan. Sixty-six survivors are picked up by USS *Surprise* (PG-63) and landed in San Juan. Last boat with 16 survivors is spotted by a patrol aircraft 8 nm off St. Thomas, Virgin Islands and brought to the island on 3rd July. Motor tanker SS *Sam Houston* is sunk by *U-203*, 160 nm northeast of the Virgin Islands. An officer and two men below are killed. Eight officers, 29 crew, and nine armed guards take to three lifeboats. Survivors were picked up by USS *Courier* (AMc 72) and landed at St. Thomas. SS *Raphael Semmes* is torpedoed by *U-332* is 875 nm off Canaveral. Nine officers, 26 crew, and two passengers jump overboard. Eighteen survivors cling to wreckage and two rafts. On 16th July the US-flagged SS *Explorer* rescues them and lands them at Jersey City. Master, six officers, 11 crew, and one passenger are lost.

29th June 1942

U-203 is in Sombrero Passage off Anguilla. *U-332* is bound northeast of Bahamas. R.Smg. *Morosini* is well east of Bahamas, south to Anegada. SS *Thomas McKean* is sunk by *U-505* 350 nm northeast of Puerto Rico. Two lifeboats with 29 survivors made landfall at St. Thomas, and one boat with 12 survivors landed at Antigua on 12th July. On 14th July last boat with 14 survivors and one dead arrived at Micheson, Dominican Republic. The master was lost during repatriation as a passenger on SS *Onondaga,* when it was sunk by *U-129* on 23rd July.

30th June 1942

U-505 enters area north of Anegada, west to Mona Passage. *U-571* enters region south in Straits. *U-575* enters the area east of Bahamas west. *U-203* is in Sombrero Passage off Anguilla. *U-332* is bound northeast of Bahamas, enters briefly. R.Smg. *Morosini* sinks the Dutch SS *Tysa* 500 nm northeast of Anegada. All 43 crew are rescued. Survivors of SS *Sea Thrush* in three boats are picked up by USS *Surprise* (PG-63) and landed in San Juan. Four crew died from burns in lifeboats from SS *Sam Houston* before survivors are picked up by USS *Courier* (AMC 72) landed at St. Thomas, where another crewman died of burns in a hospital. *U-158* under Rostin is sunk by US Mariner aircraft USN *VP-74* under Schreder, 465 nm northeast of Abaco Light, 140 m west of Bermuda. Fifty-three Germans killed, also Spanish Captain Bernardo Corescardama from SS *Everalda. U-67,* northwest of Inagua: spotted numerous aircraft in succession; one land-based, then large flying boat, a 2-engined flying boat, all causing crash dives. Then another fling boat when surfaced, another, and a crash dive. Apparently the area off the Bahamas Channel is closely patrolled. Decide to remain submerged until the arrival of darkness, the bird is sticking to me.

1st July 1942

U-437 enters area northeast of Bahamas, bound west. *U-134* enters the area west in Straits, then bound to Key West. *U-153* enters area south, bound to Anegada. R.Smg. *Morosini* and R.Smg. *Finzi* refuel 500 nm northeast of Anegada. *U-505* is bound north of Puerto Rico bound west to Mona Passage. *U-84* enters Windward Passage via Crooked Island Passage. *U-134* is bound south in Straits. *U-129* in the Old Bahama Channel off Cuba reports a crash dive for Lockheed Hudson-type aircraft, then another crash dive for a Consolidated type aircraft.

2nd July 1942

U-332 exits area north bound to Carolina coast. *U-203* exits area bound south to Anegada Passages. *U-84* is bound northeast of Abaco bound west. *U-134* is bound south in Straits. *U-172* re-enters area via Mona Passage, bound east to Anegada Passage. SS *Mary* sends distress signal after collision with SS *Palimol* 400 nm north of San Juan. *U-129,* in Old Bahama Channel off Cay Lobos reports it twice crash dived for aircraft, and again, off Nuevitas: dived for aircraft twice.

3rd July 1942

U-437 is bound east of San Salvador, heading southwest towards Windward Passage. *U-153* exits area bound south through Anegada Passage. *U-505* exits area via Mona Passage bound south. *U-172* is in transit northeast to France from Anegada Passage. Sixteen survivors in the last lifeboat of *Sea Thrush* are spotted by USN aircraft 8 nm north of St. Thomas. It was brought to the island by a rescue craft. Submarine is sighted in middle of Windward Passage. Message intercepted from a U-boat indicates she had spotted US Navy task group proceeding south. ComSeven (commander Seventh Naval District) warns all ships east and north of Sand Key Light, in vicinity of northeast Bahama Banks, to zig-zag and run completely darkened during daylight and bright moonlight, and that enemy U-boats are expected in Straits. *U-171* is *en route* south in Old Bahama Channel via Cuba bound to Windward Passage.

4th July 1942

U-172 exits area northeast of Bahamas, bound east for France. *U-84* is bound north of Abaco and Grand Bahama bound south. *U-437* is bound north of Dominican Republic, bound west to Windward Passage. USS *Corry* (DD-463) rescues four survivors of SS *Ruth*, sunk by *U-153* on 29th June. *U-575* enters area south via Mona Passage. SS *Norlandia* is torpedoed by *U-575* 25 nm northeast of Cape Samaná, San Domingo and ultimately sinks off Haiti. Twenty-one survivors of eight officers and 22 crew abandon ship in two lifeboats. One boat with 14 survivors landed at Samaná, Dominican Republic, the other arrived later. HMS *Saxifrage* (K-04) rescues SS *Putney Hill*'s master, 29 crew, and five gunners northeast of the Mona Passage. They were sunk by *U-203* on 26th June.

5th July 1942

R.Smg. *Finzi* is bound east of Bahamas. R.Smg. *Morosini* east of Bahamas, north of Puerto Rico. *U-437* exits area via Windward Passage south. Seven survivors (of 21) from US-flagged SS *Norlandia* reach Samaná, Dominican Republic. *U-67* in the Old Bahama Channel reports an aircraft flies past, behind the stern, disappears, for approximately a minute an exhaust-pipe flame visible behind a cloud astern. Aircraft flies over the U-boat with throttled motors, then we crash dive. Another aircraft from the bow ahead, illuminates his path with landing lights. Crash dive for silhouette like Dutch patrol vessel of *Rigel* class, but smaller and more compact, also, the after mast is thinner and lower, foremast well free from the superstructure. Armament not determined. Sighted aircraft, high altitude, 23 nm distant, descending, necessitating a crash dive. *U-129*, Old Bahama Channel: crash dive for two aircraft with position lights. Convoy of five ships distinguished in line abreast, large intervals, with no protective naval screen. Then several flying boats on the horizon, which fly off again. *U-67*, south of the Keys, reports aircraft, then two flying boats, one aircraft, another flying boat, then the U-boat dived for aircraft. At periscope depth they sighted a flying boat, then a new flying boat, which disappeared behind a cloud bank. *U-129* is bound south of Key West, reporting a crash dive for flying boat that flies close escort over the convoy.

6th July 1942

U-84 turns east around Abaco to patrol east of Bahamas to the Crooked Island Passage and Old Bahama Channel. R.Smg. *Finzi* is bound east of Bahamas *en route* southwest. HMS *Saxifrage* (K-04) lands 35 survivors of SS *Putney Hill*'s at San Juan; they were sunk 26 June by *U-203* and rescued on 4th July. *U-171* is *en route* west to Cuba after re-entry via Windward Passage. *U-505* is bound northeast of Anegada, reporting several reddish flashes seen to starboard, vessel in sight, one mast, one smokestack, probably a gun boat or coastal patrol vessel, suspect searching for survivors of the steamers which U-505 sank. *U-332* is bound northeast of Bahamas, where it finds life rafts from Latvian SS *Everalda*, sunk by *U-158* under Rostin on 29 June.

7th July 1942

U-166 enters area west to south of Bermuda then south in straight line to Dominican Republic, then bound up Old Bahama Channel. *U-134* is in Straits. SS *Umtata* is sunk by *U-571* 42 nm west of Bimini in the Straits, 10 nm east of Key Biscayne, Miami. She is under tow by US-flagged tug *Edmund J. Moran*. Master and 91 crew picked up by the tug and transferred to USCG Cutter *USCG-Thetis* (WPC-115) and landed at Miami. Destroyer USS *Corry* (DD-463) lands the four survivors of SS *Ruth* at Trinidad. *U-571* is bound north along northwest coast of Cuba bound out of the Old Bahama Channel. *U-509* is 300 nm north of Anegada, reporting that it witnessed another submarine conning tower; this may be U-166, which is bound south of Bermuda west-bound. *U-332*, west of Bermuda: mistook drifting wreckage for warship mastheads.

8th July 1942

U-84 is bound northeast of Abaco, bound south. R.Smg. *Finzi* is bound northeast of Crooked Island Passage, *en route* south to Windward Passage. *U-134* is bound northeast of Grand Bahama. SS *J. A. Moffett Jr.* is torpedoed twice by *U-571,* 3 nm southwest of Tennessee Reef, Florida Keys, west of Cay Sal Bank. Eight officers and 29 crew board two lifeboats and three rafts. Ship grounds on reef. Remaining crew abandon ship, but the arm of the master is caught in the falls, and thus pinned, he bleeds to death. Survivors are picked up by USCG Auxiliary craft *Mary Jean* and *Southbound,* and landed at Craig, Florida. USCG Cutter *USCG-Nike* (WPC-112) saves one man and lands him also at Craig. *U-171* is in Windward Passage, bombed three times by a flying boat. *U-84* is bound northeast of Grand Bahama, reporting a crash dive for aircraft. *U-571* is bound west of Grand Bahama reporting aircraft sighted, five depth-charges heard, patrol vessel. There is an attack on an unknown U-boat reported 107 nm east of Bermuda by a USN Kingfisher of VS-32 inshore patrol squadron.

9th July 1942

U-84 is 200 nm east of Eleuthera, bound south. R.Smg. *Finzi* sights Dixon Hill Light, San Salvador bound south to Windward Passage. *U-166* is bound west to south of Bermuda then south in straight line to Dominican Republic, then up Old Bahama Channel. *U-134* is bound northeast of Abaco, where during approach, a ship is recognized as the Swedish neutral *Venezia*. They do not shoot. Then an incoming radio message: *To Schendel. Sink*. During the second attempt; at the moment of the shot the steamer changed speed and target angle. With his relatively short length and high speed, he manoeuvres and turns so quickly that one cannot get at him. I suspect that he uses a listening device, or his reaction to every shot would be inexplicable. Artillery use does not appear promising to me, because in the darkness I would have to come very close by surprise in order to enable the Gunnery Officer, and surprise is no longer possible with this steamer. Therefore, aborted the pursuit to head for the ordered operations area.

10th July 1942

U-84 is bound east of Crooked Island, bound south to Crooked Island Passage, then Windward Passage and Old Bahama Channel. *U-134* is off Canaveral. R.Smg. *Finzi* passes west of Mira por Vos, in southeast Bahamas, and east of Cape Maysí Cuba, bound south to Windward Passage.*U-84* is off Jacksonville, where at dawn three patrol vessels are in sight. Very small silhouettes, of PC-boat type, which were avoided on the surface. Crash dived for aircraft with searchlights that flew exactly towards the U-boat, despite moonless dark night, and no marine phosphorescence. Again, crash dived for aircraft, in the same circumstances that indicate the planes have Radar or similar. The boat is almost at the same location, because submerged there is a strong counter-current and over the bottom, and so the boat went backwards. Crash dived once more for aircraft. U-boat was stuck in position, and were caught just after surfacing. Then, in the Straits off Miami, a crash dive, due to aircraft approaching. Searchlight lit up about 350' ahead of the boat. Crash dive for aircraft, series of aircraft bombs. Light damage I remained submerged, assume we were in a Radar location and would be caught on the surface again. The battery is nearly empty, and I have still not gained any ground. Therefore, I must come about and attempt to charge by day off the Strait. Surfaced. Ran out of the Strait and charged. There are many aircraft in the air, however, in the clear visibility, I can avoid diving by dodging or turning with them. Outside the Strait the air cover stops rather abruptly. Only the main routes are heavily monitored, otherwise no air reconnaissance despite proximity of coast. *U-571*, off Port Everglades, reports two aircraft; one later extinguishes its lights and flies past off the stern and searches with a searchlight. Crash dived, before he could approach. Surfaced. Aircraft, then two flat shadows pass on the sea to port, which show silhouettes similar to German *Schnell* or *Räumboote* class, only without a mast. Then, off Miami: harassed by frequent scanning of the water surface, with searchlights, aircraft nearly always fly with set position lights, causing crash dives.

11th July 1942

U-161 enters area in Windward Passage north, heading northeast of Bahamas. R.Smg. *Finzi* exits area via Windward Passage. *U-84* is in Old Bahama Channel bound north for Straits. Dominican sailing schooner SV *Carmen* is sunk by *U-166* off Cabarete, northeast Dominican Republic. *U-84* is bound north of Grand Bahama; they could not gain ground in the Florida Straits with the increasing moon, so commander decides to head for the operations area through the Crooked Island Passage and Old Bahama Channel. If these are also heavily monitored, at least I am not going to held up by the current. *U-571*, off Port Everglades, reports one to two aircraft seen, some with set lanterns, some search with searchlights, others throw flares. Crash dive to bottom, patrol vessels drop depth charges distant. *U-134* is off Canaveral reports a crash dive for amphibious bi-plane while crews were doing torpedo maintenance.

12th July 1942

U-84 in Old Bahama Channel bound north. *U-166* is bound northwest of Haiti for Old Bahama Channel. *U-571* is bound south of Key West reports destroyer of *Blue* class and two steamers of medium tonnage in sight. Because the earliest that I would be able to attack this convoy would be east of Key West, we continued westerly transit. With the expected heavy air surveillance by day, I would probably not be able to get in attack position. After attack, a crash dive for aircraft, and submerged for an hour. At periscope depth, the steamer lies with a heavy list, still burning, and later disappears entirely. Two aircraft circle above it.

13th July 1942

U-171 enters area from Bermuda region bound south to Windward Passage. *U-173* enters area south to Crooked Island Passage, then is bound to the Windward Passage. *U-161* is bound north of Turks and Caicos, east of Bahamas, bound north. *U-134* is off Canaveral, bound north. US-flagged SS *Andrew Jackson* is sunk by *U-84* 20 nm off Cardenas Light, Cuba, between

Cay Sal Bank and Marathon Cay, Florida. Three men on watch are killed, eight officers, 30 crew, and 11 armed-guards abandon ship. Survivors land at Vavendaro, Cuba in three lifeboats. US-flagged SS *Oneida*, straggling from **Convoy NG 359**, is sunk by *U-166* 2 nm north of Cape Maysí, Cuba. Eight officers and 21 crew from SS *Oneida* abandon ship on two rafts, two officers and four crew were lost, survivors make landfall 5 nm northwest of Cape Maysí several hours later taken in the Cuban schooner SV *Zoila* to Baracoa, Cuba. *U-134*, off Canaveral, reports no traffic, slight air cover, mostly with float planes. Then, from Canaveral southwards, heavy air surveillance. Aircraft fly at night with set lights and sometimes search the ocean surface with searchlights. *U-166* is bound south of Bermuda, reporting it took six shots at a convoy with two steamers, two destroyers, and all were misses.

14th July 1942

U-84 is in Nicholas Channel. *U-171* is bound northeast of Bahamas, bound south towards Windward Passage. *U-166* is bound west in Old Bahama Channel. Fourteen survivors with one dead shipmate arrive via lifeboat at Micheson, Dominican Republic from SS *Thomas McKean* sunk by *U-505* on 29 June.

15th July 1942

U-166 is in Old Bahama Channel near Cay Sal Bank, bound west. *U-134* is bound northeast of Abaco. *U-576* is bound north in Straits. *U-759* is bound south in the Windward Passage. *U-84* bound from Crooked Island Passage to the Old Bahama Channel reports a crash dive for aircraft, re-surfacing, then another crash dive for aircraft.

16th July 1942

U-134 is bound southeast of Hole in Wall Light, Abaco. *U-171* is bound east of San Salvador and north of Turks and Caicos, bound south to Windward Passage. R.Smg. *Giuliani* enters area from east, is attacked by a Boeing aircraft north of Anegada. *U-166* exits the area bound west past Key West. *U-571* re-enters area southwest of Key West bound east to Old Bahama Channel and Caicos Passage. *U-173* is bound north of Mayaguana, bound south to Windward Passage. *U-161* attacks **Convoy AS-4** 500 nm north of St. Thomas, sinks SS *Fairport*. [All 123 survivors are picked up by USS *Kearny* (DD-432) and landed in New York on 21st July]. Small US-flagged fishing vessel FV *Gertrude* is ordered to stop by *U-166* 30 nm northeast of Havana and west of Cay Sal Bank. All three men abandon ship in motor boat. *U-166* sinks trawler. Motorboat runs out of fuel, drifts, is then spotted by a Civil Air Patrol aircraft 3 nm south of Alligator Reef Lighthouse, Florida. A boat out of Whale Harbour brings them men ashore. *U-571* is bound south of Key West, reporting that one large flying boat with set lights passed the U-boat's wake, flew in a loop, extinguished lights, and circled the boat twice. In the meantime, the boat stopped. Machine was gun ready and the plane targeted. When he flies close by the stern the second time, he uses his searchlight. Crash dive before he can approach again. That he has seen me in the dark night without marine phosphorescence and with the faint wake, I believe is improbable, and so I suspect Radar. A *Blue* type destroyer is directly ahead zig zagging very frequently, emerging out of the twilight. Nearly at the same time as via the detection device, the propeller sounds are heard in the control room with the naked ear. Despite the pumping and high speed of the U-boat during the crash diving, the destroyer did not detect it, which confirmed the bad listening conditions. A clash between the USN vs. R.Smg. *Giuliani* takes place 262 nm southeast of Bermuda [not likely U-161, which was closer to Windward Passage then]. The aircraft was VP-52, a PBY-5 BuNo.2367 (52-P-6) out of NOB Bermuda. Also, a Catalina ferry crew reported sighting a U-boat. *U-134*, off Palm Beach: recognized the sub-chaser type *PC-451* in tandem with sea planes, it ran above the U-boat then headed for Palm Beach. *U-161* under Achilles is in the Windward Passage, where it crash dived six times for aircraft with crossing course, 3 knot adverse current in passage.

17th July 1942

R.Smg. *Giuliani* attacks an unknown ship, is attacked by planes, then moves to northeast of Puerto Rico [see entry for PBY-5 attacking a sub near Bermuda]. *U-84* is in Nicholas Channel. *U-161* is bound east of Bahamas, bound north. *U-171* is bound east of Inagua, in Caicos Passage bound south to Windward Passage from off Bermuda. *U-571* is in Old Bahama Channel bound east for Caicos Passage. *U-173* is off Inagua, then in Windward Passage bound south. *U-437* re-enters area via Windward Passage then is bound east for France. R.Smg. *Finzi* re-enters area via Windward Passage bound north for Crooked Island Passage. *U-67* re-enters via Key West bound to Old Bahama Channel. *U-134* is in Straits off Palm Beach, reports crash dive for aircraft heard close by over the boat, not seen. Other aircraft, more crash dives. Radio message from HQ: *U-boats in the American theatres. After heavy merchant ship losses in June from early July in some areas, the enemy had stopped traffic needlessly. Capture this moment, be tough, make use of every opportunity.*

18th July 1942

U-161 exits area northeast of Bahamas bound east to France. *U-173* exits area bound south in Windward Passage. *U-171* exits area bound south in Windward Passage. *U-67* in Old Bahama Channel is bound east towards Caicos Passage and France. *U-134* is off Bimini to transit Northeast Providence Channel and Northwest Providence Channel. R.Smg. *Giuliani* is bound west from Anegada to Puerto Rico to refuel R.Smg. *Finzi*. Motor yacht, recreational fishing vessel *Pilar*, owned by Ernest Hemingway, returns from frustrating six-week "Anti-Submarine patrol" on same day that *U-134* shoots down USN Blimp *K-74* off Cay Sal Bank to the north. He also missed *U-527* that had been in the area. *U-527* "fired four torpedoes at a fast-moving, zig-zagging merchantman." *U-84* in the Nicholas Channel reports flashing lights between warships were answered by a not-as yet-detected aircraft, and also flashing signals from land. Then crash dived for aircraft. *U-161*, northeast of Eleuthera, discovered drifting rafts.

19th July 1942

U-571 enters Caicos Passage bound north. *U-134* exits area via Key West, bound west. *U-67* is bound north of Inagua bound east out of region through Caicos Passage and Crooked Island Passage. R.Smg. *Finzi* transits Crooked Island Passage bound east. R.Smg. *Giuliani* is bound west from north of Anegada to a point northeast of Puerto Rico. R.Smg. *Morosini* attacks a small military ship, possibly a gun boat or an armed merchant ship, without success. *U-161* is bound northeast of Eleuthera, reporting it sighted a capsized, powered lifeboat, like the cutters carried aboard liners, with two masts with sails. R.Smg. *Giuliani*, northeast of Anegada, is attacked by Boeing plane inbound.

20th July 1942

R.Smg. *Morosini* is bound east of Bahamas, north of Puerto Rico. R.Smg. *Finzi* is bound east of Crooked Island. They sighted a tanker, but could not fire. *U-571* exits Caicos Passage, enters Old Bahama Channel bound east. *U-67* clears the Caicos Passage bound east to France. Armed Merchant Cruiser USS *Asterion* departing New York sailed north of the Bahamas bound to the Windward Passage.

21st July 1942

U-437 is bound east of Bahamas, south to coast of Dominican Republic near Samaná. *U-509* enters area north of Anegada Passage for the Mona Passage. R.Smg. *Giuliani* is bound north of Puerto Rico. *U-571* is bound northeast of Caicos, east of Bahamas heading northeast for France. R.Smg. *Finzi* is bound northeast of Turks and Caicos heading east northeast for France and a rendezvous with R.Smg. *Giuliani.* The US-flagged SS *William Cullen Bryant*, in **Convoy TAW 4J,** is torpedoed by *U-84* between Havana and Key West. All hands including 40-man crew, 12-man Armed Guard, and two USN signalmen abandon ship, yet manage to re-board later. Salvage tug USS *Willett* (ARS-12) and civilian tug *Moran* tow SS *William Cullen Bryant* to Key West, and subsequently she returns to service. Thirty-four survivors, including 10 injured in one lifeboat and two rafts, of the passenger ship *Baja California* are rescued by the Cuban fishing schooner FV *San Ignacio* and taken to Havana Naval Station. *U-161*, northeast of Eleuthera, reports a destroyer, of *Gleaves* class protecting a convoy in four columns. They went immediately on attack course, no depth charge counter attack follows; rather depth charges are thrown irregularly in all directions. There is a destroyer at the shooting location, steaming back and forth. Nothing detected in the sound gear. Went to depth, because of the mirror flat sea. Now and then at periscope depth. Another destroyer. Shortly thereafter in the sound locator quickly stronger growing sounds in the same direction. Believe U-boat was sighted by a destroyer. At the same moment, from astern, an approaching destroyer, at high speed, followed by four well-placed depth charges from a destroyer passing overhead. Damage includes the cracking of all water level glass, deep depth gauges except for the conning tower, remote gyrocompasses, diving plane motors and minor failures. Overall, a nine-hour sound pursuit by three destroyers, detection by sounding charges, depth charges, our thrust bearing leaked. Return to service uncertain. Otherwise fresh water maker out of service, as well as trim pump connection.

22nd July 1942

U-84 is the Straits between Andros and Florida Keys, bound north to France via Grand Bahama. *U-67* exits area bound for France. *U-509* is bound northeast of Puerto Rico bound south for Mona Passage. *U-571* is bound northeast of Puerto Rico, well east of Bahamas bound east for France. *U-129* re-enters area east via Key West bound south to Old Bahama Channel then Crooked Island Passage. Survivors of SS *Anglo Canadian* and British tanker *Athelknight* are in transit on the Cuban SS *Libertad* from the British Virgin Islands to Norfolk or New York. *U-134* is off Bimini, where three times they were forced to crash dive for aircraft. A flying boat was seen approaching in the periscope. Aircraft fly with set lights. Surveillance of the Straits seems to be only by flying boats and float aircraft. R.Smg. *Giuliani*, northeast of Puerto Rico, reports an attack on an unknown ship, then being counter-attacked by planes.

23rd July 1942

R.Smg. *Morosini* is bound east of Bahamas, north of Puerto Rico. *U-437* is bound east of Turks and Caicos and Bahamas, bound east for France. *U-571* exits area bound east for France. R.Smg. *Giuliani* has a rendezvous with R.Smg. *Finzi* to receive water and fuel to the northeast of Puerto Rico. *U-129* sinks SS *Onondaga* in Old Bahama Channel, 5 nm north of Cayo Guillermo, Cuba. Twenty lost, 14 survivors are picked up by Cuban fishing boat FV *Laventina* to be landed in Cuba. *U-129* then heads south towards Crooked Island Passage. Torpedo-damaged SS *William Cullen Bryant* arrives in Key West anchorage. Cuban SS *Libertad* arrives with 20 survivors from British tanker *Athelknight* and SS *Anglo-Canadian*. They embark at St. Kitts, BWI, for transhipment onto a US Naval Transport, which conveys them to a port in the US. *U-161*, southeast of Bermuda, reports three destroyers sighted escorting eight to ten steamers.

24th July 1942

U-437 is bound east of Bahamas bound east to France. R.Smg. *Giuliani* is bound northeast of Puerto Rico heading southwest. *U-84* is bound north of Grand Bahama, bound east to France. Cuban FV *Laventina* lands 14 survivors from SS *Onandaga* at Punta San Juan, Cuba. *U-134*, off Key West: crash dives for aircraft

25th July 1942

U-437 exits area east bound for France. *U-129* is in Old Bahama Channel south towards Crooked Island Passage. *U-84* is bound northeast of Grand Bahama bound east for France. R.Smg. *Finzi* is bound northeast of Puerto Rico, heading northeast refuelling. *U-509* enters Mona Passage, then west up Old Bahama Channel bound to Key West. SS *Veragua* at New York with survivors of SS *Putney Hill* sinking on 25 June north of Puerto Rico.

26th July 1942

U-508 enters area south for Crooked Island Passage and Old Bahama Channel. R.Smg. *Giuliani* exits area east of Anegada. *U-129* is in Crooked Island Passage, bound north to France. R.Smg. *Morosini* is bound east of Bahamas, north of Puerto Rico in loops from 1 July to 24 July. *U-134* re-enters area from south of Key West, bound east towards Old Bahama Channel and Crooked Island Passage. R.Smg. *Morosini*, southeast of Bermuda: U-boat attacked a small military ship, possibly a gun boat or an armed merchant ship, which followed the submarine for a little while, without attacking. The attack was unsuccessful and unconfirmed. *U-84* is off Key West, where it crash dived for aircraft, one of which came too close, then five aircraft were on all sides. One of them circles constantly over smoke clouds. Commander placed three additional men on the bridge to monitor all the aircraft. One flying boat flies towards the U-boat, and depth charge detonations are heard. For the entire night, star shells and light rockets seen, yet do not illuminate the boat. Very active near Key West. Crash dive four times for aircraft. Then, off Port Everglades crash dive for aircraft.

27th July 1942

U-134 is in Old Bahama Channel bound south. *U-508* is bound northeast of Abaco bound south to Crooked Island Passage. *U-129* is bound northeast of Turks and Caicos bound east to France. Fourth cruise of USS *Asterion* north of the Bahamas and Windward Passage area; then bound to New York, arriving on 18th August. *U-84*, off Canaveral: crash dives for aircraft. *U-509*, Mona Passage: crash dives for aircraft. *U-*

134 is in Old Bahama Channel, Cayo Coco; crash dives to avoid flying boat on parallel course, then another crash dive for aircraft. Crash dive for flying boat and Lighthouse Cayo Caimán abeam, 6 nm on south side of Old Bahama Channel. In poor visibility and hazy horizon, a convoy comes in sight, very close to the south edge of the Great Bahama Bank, bound on a westerly course. *U-134* attempts to manoeuvre ahead, yet is forced to submerge by the air escort of 4-engined land-based aircraft. From periscope depth, they see that the convoy consists of eight large freighters, screened forward by two destroyers, and air escort consists of four 4-engined land-based aircraft that fly remote sweeps on all sides.

28th July 1942

R.Smg. *Morosini* on a few more days of patrol with only 85 tons of fuel left, has A rendezvous with R.Smg. *Finzi* to obtain 25 tons of diesel fuel. [She continues operations until on 31st July she reached minimal fuel reserve, and headed for Bordeaux, France]. *U-129* is bound north towards France. *U-134* is bound south of Cay Sal, in Old Bahama Channel bound south towards Crooked Island Passage. Hemingway takes two intelligence officers to Matanzas Cuba aboard his fishing yacht *Pilar* to see whether a cigar factory is doubling as a U-boat supply dump. *U-508* is bound northeast of Bahamas; a SSS signal is received from SS *Mormac Star*. *U-134*, Nicholas Channel: twice crash dived for aircraft. Sixty-eight nm south of Bermuda, VP-52 PBY, 52-P-2, USN said it engaged *U-84*, however this is unlikely, as they are 400 nm apart. Then, 303 nm south of Bermuda, VP-52 PBY5s, 52-P-9 Bu.No.2371, USN engage *U-510*. Later, 230 nm south of Bermuda, 52-P-2 of VP-52 PBY, claim to have engaged a U-boat which may have been *U-508*, since *U-658* and *U-129* were too far away from that location at the time.

29th July 1942

U-658 enters region south to Turks and Caicos, bound for the Windward Passage. R.Smg. *Morosini* exits area east for Bordeaux, [then was sunk on August 7th]. *U-84* exits the region bound east to France. *U-129* is bound northeast of Turks and Caicos bound east to France. *U-508* is bound northeast of Bahamas, where it discovered unoccupied lifeboat of SS *Fairport*, sunk by Achilles in *U-161*. Since the lifeboat is fully equipped, crew of U-boat took food and wage and service books and master roll-cards.

30th July 1942

U-510 enters area from south of Bermuda, bound south. R.Smg. *Finzi* exits area to northeast. *U-508* is bound northeast of Abaco bound south to Crooked Island Passage and Old Bahama Channel. *U-129* is bound northeast of Turks and Caicos bound east to France. *U-134* is off Gibara Cuba, bound south in Old Bahama Channel to Windward Passage. *U-658* is bound east of Eleuthera, bound south to Windward Passage. Women's Auxiliary Force, Royal Navy WAVES (Women Accepted for Volunteer Emergency Service) is established. *U-84* is bound northeast of Bahamas, where sights four destroyers in moonlight, with a large shadow seen behind them.

31st July 1942

U-508 is approaching Crooked Island Passage east of San Salvador. *U-129* is exiting area north to France. *U-134* is bound south in Old Bahama Channel. *U-154* re-enters area via Mona Passage. *U-510* is well northeast of Bahamas south. Motor yacht *Vergemere IV* arrives in Nassau with 47 survivors of SS *Potlatch*, sunk by *U-153* on 27th June. They sailed 980 nm to Inagua, then Acklins, and arrived in Nassau on 1st August.

1st August 1942

U-154 is bound east of Turks and Caicos bound east for France. *U-508* clears Crooked Island Passage bound to Inagua and Cuba. *U-134* is bound south of Turks and Caicos bound east to the Dominican Republic. *U-658* is bound east of Eleuthera bound south to Windward Passage. *U-510* is bound northeast of Bahamas south. USS *Asterion* is cruising north of Bahamas as a decoy in Straits. Final 47 survivors of SS *Potlatch*, retrieved from Acklins by merchant ship *Vergemere IV*, reach Nassau. *U-134* in the Old Bahama Channel off Nuevitas Cuba crash dives to avoid aircraft, then maintains listening speed submerged. *U-84* is close to and south of Bermuda, where it reports Consolidated flying boat which dropped three aircraft bombs, then approached again with three more bombs, causing three more crash dives. A masthead was sighted, from a destroyer, which was avoided on the surface. Then, 315 nm south-southeast of Bermuda, aircraft of US Navy Squadron VP-52, the PBY-5 P-9 (BuNo.2371) engaged *U-510*.

2nd August 1942

U-508 is in Straits north to Key West. *U-154* is bound north of Puerto Rico bound east to France. *U-134* is off central north Hispaniola. SS *Maldonado* is torpedoed by *U-510* and sinks 1,000 nm east of Abaco. [Ship is from neutral Uruguay and all 49 on board survive, however it caused diplomatic furore].

3rd August 1942

U-564 enters area off Anegada. *U-134* is of northwest tip of Haiti near Windward Passage. *U-510* is well east of Bahamas. *U-173* re-enters area via Windward Passage, then north coast of Hispaniola bound east to Anegada, then France. San Juan radio picks up submarine sighted message (SSS) from SS *Daltram*, position not ascertained. Another U-boat is positively sighted in Windward Passage; this was probably *U-173*, possibly *U-134*. *U-134* is in Old Bahama Channel off Cuba, where it crash dive for approaching land-based aircraft, which dropped four depth charges. Rudder out of service. Glass damage, all repaired quickly. Aircraft may have used Radar, as aircraft approached directly out of a continuous cloud deck.

4th August 1942

U-154 exits area bound north to France. *U-508* exits area bound west past Key West. *U-134* is in the Windward Passage. *U-510* is well east of Bahamas. *U-564* is bound south, to the east of Bahamas. *U-173* exits area bound south via Mona Passage. *New York Times* reports that four more ships are lost to U-Boats: small British and Norwegian ships, a US tug, and a fishing boat; possibly the *Gertrude*. *U-134* is in Old Bahama Channel off Nicarao Cuba, reporting it twice crash dived for aircraft. Since both aircraft approached the boat directly at night in a relatively large sea area, Radar locating can be accepted with certainty. Especially since otherwise searching with spotlight without solid evidence would not occur. Then, north of Puerto Plata, *U-134* crash dived for aircraft while doing torpedo maintenance. Then crash dived for aircraft twice more that day. At 6 pm the first 27 men out of 47 from US *SS Potlatch* (11 ship's crew, 16 gunners) depart Oakes Field for Miami aboard a US Army transport plane.

5th August 1942

U-98 returns to area for second patrol, bound south to lay mines at Jacksonville. *U-134* is in Crooked Island Passage bound east to France. *U-173* is bound north of Puerto Rico bound east to France. *U-564* has rendezvous with *U-154* is northeast of Bahamas to take on torpedoes. Swedish neutral SS *Scania* is stopped by *U-564* in the Sargasso Sea, 700 nm east of Eleuthera. Crew and ship are released. At 5 pm, final 20 officers, crew, gunners of SS *Potlatch* depart Nassau for Miami via US Army aircraft, arriving 6.10 pm and entraining for New York under Capt. Lapoint and Lt. Lybrand.

6th August 1942

U-134 is bound north along Crooked Island and Acklins Island, north of Inagua, then bound east to France. *U-654* enters the area bound south for Windward Passage. *U-510* exits area heading southeast towards Anegada. *U-173* exits area northeast of Anegada bound north, to the east to France. *U-98* is bound northeast of Abaco, bound west to Jacksonville. *U-564* is well east of Bahamas, bound south.

7th August 1942

U-658 is bound east of San Salvador, bound south for Turks and Caicos. *U-598* enters area bound south to Crooked Island Passage. *U-134* is bound northeast of Turks and Caicos, bound north to France.

8th August 1942

U-658 is bound east of Turks and Caicos, bound south for Windward Passage. *U-600* enters area bound south for Turks and Caicos and Windward Passage. *U-564* exits area bound south to Anegada. *U-654* is bound northeast of Turks and Caicos bound south. *U-598* is bound east of Turks and Caicos bound west to Crooked Island Passage. *U-98* is 400 nm southeast of Bermuda, bound to Jacksonville to lay mines,. While looking for survivors of neutral SS *Maldonado,* USN aircraft 52-P-9 initiates two attacks against two U-boats, possibly *U-463* and, or *U-505.*

9th August 1942

U-654 is bound northeast of Turks and Caicos, east of Bahamas bound south to Windward Passage. *U-134* is bound east of Bahamas. *U-98* completes mining St. Johns River, Jacksonville with 12 TMB mines, in less than an hour. *U-98* sights aircraft with a white light. On land all the lights of the entrance to Jacksonville are easily distinguished. The air base at the southern mole is illuminated, and the glow of Jacksonville itself is plainly seen. *St. Johns Lightship* is abeam, one nm. Went to electronic motors so as not to be detected in the calm weather, and remained north of the Lightship, so that the silhouette of the boat would not be seen in the bright glow from Jacksonville. No patrol vessels detected up to now. During deployment of minefield, very heavy aircraft traffic, with side lights or white lights at the airfield at the St. Johns south mole: 10-15 aircraft observed in one hour. [One mine sinks USS *Bold* on 10 August]. *U-508* is in Old Bahama Channel at Cay Lobos, where it crash dives before three individual vessels. Action stations. All tubes ready except III ready, since number III outer tube door is not in service. On closing [approaching], commander observes one battleship of *California* class and three destroyers in very close protective screen. One destroyer is ahead, two, close on either side. The formation proceeds at high speed. I want to go straight to attack course for a bow shot on the battleship, but the starboard destroyer a better target. The battleship turns as well and keeps bow right, reducing the U-boat's target angle. I believe I am seen with certainty, and so dive to depth. This measure was incorrect, since nothing happened, and while I went deep at high speed, the formation rushed past close by on the starboard side with a great noise. Range was probably too short for me to be able to shoot with the stern tube. *U-564* is bound southeast of Bermuda, initiating a fuelling, supply, and morale-boosting rendezvous with a U-tanker, or supply submarine, U-463 under Wolfbauer. Other U-boats there are nicknamed after their commanders, as in U-Witt, U-Neitzel, U-Forster, and U-Kölle.

10th August 1942

U-134 exits area bound east for France. *U 654* is bound southeast of Turks and Caicos using Caicos Channel and bound south to Windward Passage. *U-600* under Zurmühlen sinks Barbadian [Bajan] schooner *Vivian P. Smith* using 139 rounds of shelling 40 nm east of Grand Turk Light, east of Turks and Caicos and north of Hispaniola. All live donkeys on deck perished, but 11 human survivors sailed a short distance west to the Turks and Caicos, which they recently left. *U-163* enters area northeast of Bahamas, bound south to Windward Passage. *U-508* is off Havana, and reports a crash dive for Douglas O-46A, then a crash dive for Douglas, and another crash dive for a Curtiss SOC-1, which came in sight again when submerged, using the periscope.

11th August 1942

U-658 is at the entrance to the Windward Passage. *U-511* enters area northeast of Bahamas, bound south for Crooked Island, Caicos Passage and Windward Passage. *U-98* hugs shore bound north from Jacksonville in Straits then exits area for Savannah. *U-598* is in Crooked Island Passage bound south to Old Bahama Channel. *U-600* is bound north of Dominican Republic. *U-654* is bound south of Inagua, northwest of Haiti and bound south into Windward Passage. *U-163* is bound northeast of Bahamas, bound south to Windward Passage.

12th August 1942

U-164 enters area northeast of Bahamas, bound south for Mona Passage. *U-654* exits area bound south via Windward Passage. *U-598* is bound southwest of Inagua seeking convoys coming out of Windward Passage. *U-600* sights Matthew Town Light, Inagua, then is bound south for Windward Passage. Allied **Convoy WAT-13,** *en route* through Windward Passage, bound south, is attacked by *U-658.* Dutch SS *Medea* sinks and 24 survivors are picked up by USN escort then landed at Guantanamo; one crew died of his wounds. *U-508* is between Key West and Havana and attacks **Special Convoy 12** including *Empresa Naviera de Cuba*, *Havana*, *Manzanillo* and *Santiago De Cuba.* Both Cuban steam ships sink west of Cay Sal Bank. Twenty-three were on board *Manzanillo*; 10 killed, 19 survived. *U-163* is bound northeast of Bahamas, bound south to Windward Passage. *U-508* is off Key West, where it crash dives for aircraft with a white light that towed a red light far astern. Observed perfectly. Perhaps a target tug as searchlights play on land. Other aircraft with position lanterns are seen. *U-134* is bound south of Inagua, north of Windward Passage, reporting a *Gleaves*-class destroyer with smokestack and bridge over the horizon, which runs off at high speed in the direction of Guantanamo. Three hundred and fifty nm south of Bermuda USN aircraft of VP-52, 52-P-9 drops two depth charges against *U-163*; at the same time its compatriots *U-463* and *U-505* are also in the vicinity.

13th August 1942

U-553 is bound east of Bahamas, bound south for Windward Passage. *U-217* enters area northeast of Bahamas bound south for Mona Pass. *U-163* is bound northeast of Bahamas, bound south to Windward Passage. *U-598* is following a convoy out of Windward Passage into the Old Bahama

Channel. *U-511* is bound east of Bahamas, bound south to Crooked Islands Passage. SS *Michael Jebsen* is sunk in the Battle of Ragged Island, at 17 nm southwest of Cay Santo Domingo, 35 nm southwest of Ragged Islands, Bahamas, and northwest of Barlovento Point, Cuba, in 7,000 feet of water. *U-600* fires three torpedoes at **Convoy TAW-12** 18 nm south of Cape Maysí, Cuba, but target is missed. The third torpedo hits SS *Everelza*; she explodes and all are killed. *U-600* fires another torpedo which strikes the ship of the convoy commodore, SS *Delmundo.* She sank in five minutes. Crew of nine officers, 32 men, nine armed guards, and eight passengers (including two women) abandon ship in one lifeboat and two rafts. One officer, three men and three passengers die. Survivors are picked up by HMS *Churchill* and taken to Guantanamo. [Master died later in hospital from his wounds]. After attack, convoy is disorganized, as there are no arrangements for someone else to take command. SS *Empire Corporal* in **Convoy TAW-12** is sunk by *U-598* along with SS *Michael Jebsen* and the damaged SS *Standella* 17 nm southwest of Cay Santo Domingo, 35 nm southwest of Ragged Islands. *U-658* then sinks SS *Medea* in **Convoy WAT-13**. Twenty-four survivors are picked up by a US Navy escort and landed at Guantanamo, where one crew dies of wounds. Norwegian SS *Reinholt* was in Old Bahama Channel the morning of 13th August. She left Port of Spain Trinidad and heard SOS from torpedoed ship, likely the SS *Delmundo.* Then, 330 nm south-southeast of Bermuda USN aircraft 52-P-10 attack *U-163.*

14th August 1942

U-217 is bound east of central Bahamas heading southwest for Mona Pass. *U-600* is in Old Bahama Channel north along coast of Cuba. *U-658* is bound south of Windward Passage. *U-163* is bound northeast of Bahamas heading southwest for Windward Passage via Turks and Caicos. *U-511* is bound east of Bahamas, bound south. *U-98* is bound south of Jacksonville, reporting a destroyer on south-easterly course, then two small vessels with flat silhouettes.

15th August 1942

U-658 at mouth of Windward Passage. *U-553* is bound east of Bahamas, south for Windward Passage. *U-164* exits area via Mona. *U-600* is in Old Bahama Channel north along coast of Cuba. *U-217* is bound north of Puerto Rico and east of central Bahamas heading southwest for Mona Pass. *U-598* is in Old Bahama Channel bound north after attacking three ships off Ragged Island. USS *PC-469* arrived in Miami, moored at the Submarine Chaser Training Center, entered a shake down training routine, sailing each day. USN **Air Patrol Squadron Patrol Wing 11** is established at San Juan for operations in Bahamas region. *U-508* off Key West reports it sights one aircraft each with position lanterns, two destroyers, and another. They then crash dive for aircraft with two white lights.

16th August 1942

U-163 is bound east of Turks and Caicos and north of Haiti, heading southwest for Windward Passage. *U-658* exits area to go into Windward Passage and sink three ships off Jamaica. *U-217* exits area via Mona after a four-day transit from Bermuda area. *U-553* is in Caicos Passage bound south for Windward Passage. *U-511* is bound north of Dominican Republic bound west for Caicos Passage. *U-600* is in Old Bahama Channel at Santaren Channel entering Nicholas Channel. *U-598* is bound south of Cay Sal, in Nicholas Channel. *U-508* in the Straits reports five depth charges heard far off. U-boat heard distinct propeller and sub-locating sounds astern on and off. The *Bolde* device may have helped them to get away. In direction of the shooting location an aircraft is seen searching with searchlight. Surfaced, they moved off to execute pressing repairs to the diesel motor. The commander feels it is certain that a strong U-boat hunt will start in his old area. The crew can make good use of a day on the surface. Later, an aircraft flew by them, ahead. Crash dived for aircraft, then surfaced to discover that the aircraft was a heavily flickering star!

17th August 1942

U-553 exits area bound south to Windward Passage. *U-511* is rounding Turks and Caicos bound for Crooked Island Passage. *U-600* is bound south of Cay Sal bound west to Key West. *U-598* is bound south of Cay Sal in Nicholas Channel, bound south of Key West , where it then turns around 180° to go south into Old Bahama Channel. Passenger ship SS *William Cullen Bryant* arrives in New York after a safe transit of Cay Sal Bank under tow without being sunk twice [as happened to SS *Umtata*]. *U-600* is in the Windward Passage, reporting a destroyer and aircraft, *Eagle*-class USN patrol type, and more aircraft. In Old Bahama Channel near Lobos, they crash dive for aircraft, then another crash dive for flying boat.

18th August 1942

U-163 attacks a convoy in Old Bahama Channel bound out of Windward Passage. *U-511* is in Crooked Island Passage, south of Inagua heading south for Windward Passage. *U-598* is in Old Bahama Channel, southwest of Cay Sal, in Santaren Channel. *U-600* is bound south of Key West. *U-658* re-enters area from Windward Passage bound to the Crooked Island Passage. *U-508* sinks SS *Manzanillo* and SS *Santiago De Cuba* in Old Bahama Channel. *U-600*, in Old Bahama Channel near Cay Lobos, reports a crash dive for flying boat.

19th August 1942

U-558 enters area southeast of Bermuda bound south for Windward Passage. *U-511* exits area south via Windward Passage. *U-658* is bound southwest of Inagua rounding north to Crooked Island Passage. *U-508* re-enters area east from Key West bound to Old Bahama Channel via Nicholas Channel. *U-553* is in the Windward Passage and reports a flying boat in sight, so it crash dives. Another flying boat approaching, then ship funnel smoke clouds, above which a flying boat. They crash dive for a flying boat, then another crash dive for flying boat approaching at low altitude. Then up to five aircraft circle continuously at the level of the convoy until darkness. *U-600* in the Old Bahama Channel crash dives for aircraft, then sights a glow to the northwest, yet no aircraft noted.

20th August 1942

U-558 is bound east of Bahamas south to Windward Passage. *U-600* exits area from Nicholas Channel near Havana, bound west. *U-658* is in Crooked Island Passage. *U-508* is in Old Bahama Channel, bound south and east for France. *U-509* enters area via Mona Passage, bound north. First party for RAF No. 111 O.T.U. arrives in Miami on the RMS *Queen Mary*, under G/C Waite.

21st August 1942

U-598 is off northwest Haiti, then bound north to Crooked Island Passage. Duke of Kent Memorial Service, for the Duke of Windsor's brother, at Christ Church Cathedral, Nassau. G/C Waite met with Duke of Windsor and surveyed the site, station. **Task Group TG-38.6**, submarine hunting group, dispatched USS *Charles F. Hughes* to search Old Bahama Channel and Santaren Channel, USS *Lansdale* the Caicos Passage, and USS *Ingraham* to search the Crooked Island Passage for U-boats. USS *Roe* and USS *H. P. Jones* passed Cay Lobos Light in the Old Bahama Channel.

22nd August 1942

U-94 enters area northeast of Bahamas bound south to Windward Passage [t will be sunk there on 28 August]. *U-163* re-enters area from Windward Passage, bound north for Crooked Island Passage.

23rd August 1942

U-163 is in Crooked Island Passage, then bound east for France. *U-598* is bound northeast of Acklins and east for France. *U-658* is bound northeast of Turks and Caicos, east towards Anegada. *U-558 is* in Caicos Passage, south to Windward Passage. *U-94* is well east of Bahamas, west to Windward Passage. *U-600* is along the north coast of Cuba in the Old Bahama Channel, and reports several aircraft very far, which it avoided, then flying boats, also avoided. Crash dived for aircraft, and re-surfaced. More aircraft far away, avoided, followed by sudden crash dive for aircraft, which was very close as it came out of the sun.

24th August 1942

U-163 is bound east of Crooked Island, bound east to France. *U-558* exits area south via Windward Passage. *U-658* is bound north of the Dominican Republic, bound east. *U-598* is bound northeast of Acklins Island, bound east for France. *U-508* is bound south, to the southeast of Inagua, south of Turks and Caicos, bound east for France. South of Bermuda USAAF [Army Air Force] B-17 Kindley Bermuda to Langley VA engages an unknown U-boat. *U-600*, is bound north coast Cuba in Old Bahama Channel, it crash dives for aircraft and remains submerged. Made an all-around sweep using periscope: one aircraft, two destroyers moving back and forth, a destroyer shows bow left, then goes to opposite course, the second destroyer is out of sight. Later sweep: one zig zagging destroyer distant 4 nm then out of sight. Then another destroyer and a flying boat.

25th August 1942

U-94 is bound east of Mayaguana, bound south to Crooked Island Passage. *U-508* is bound northeast of Turks and Caicos Islands, bound east for France. *U-658* is bound north of Mona Passage, bound east to Anegada. *U-558* is bound south in Straits.

26th August 1942

U-94 enters the Crooked Island Passage bound south to Windward Passage. *U-658* is bound northeast of Virgin Islands and bound north of Anegada. *U-553* re-enters area north via Mona Passage, then is bound east to France. North of Mona Passage, at Punta Cana, USAAF B-17 on anti-submarine mission, attacks *U-553*, then sights a periscope, possibly that of *U-598*, 300 nm south of Bermuda.

27th August 1942

U-86 enters area to 400 nm bound northeast of Abaco. *U-163* exits area bound east to France. *U-508* exits area bound east to France. *U-658* is bound northeast of Mona Passage and bound north. *U-598* is bound east of Bahamas and bound east for France. *U-553* is bound north of Puerto Rico and bound east to France. Schooner SV *Sande* is in Windward Passage after leaving a convoy in Old Bahama Channel.

28th August 1942

U-598 exits area east for France. Both *U-658* and *U-553* exit area north of Anegada, then east to France. *U-86* is bound northeast of Abaco. *U-600* re-enters area in Windward Passage, north to Crooked

Island Passage. *U-600*, Straits: five aircraft, three of them causing crash dives: crash dive for aircraft, another crash dive for aircraft. Off Florida Keys, where beacons for 28-Foot-Shoal, Rebecca Shoal, and Loggerhead Key burn as in peacetime. One aircraft far off, avoided. Crash dive for aircraft. From periscope, all around sweep reveals another aircraft. *U-511* is in Windward Passage, where it crash dives for land-based Douglas bombers.

29th August 1942

U-86 is 400 nm northeast of Abaco. *U-600* is bound north of Windward Passage, bound north, entering Crooked Island Passage near Inagua.

30th August 1942

U-600 is exiting Crooked Island Passage, then bound east to France. *U-558* in the Windward Passage reports a multi-engine land-based aircraft, two bombs being dropped, a 4-stack destroyer, and flying boat.

31st August 1942

U-600 is bound northeast of Acklins Island and bound east for France. Merchant schooner *Sande* sinks north of Cuba *en route* Puerto Rico to Cuba and Honduras. Crew's subsequent claim to have seen a U-boat at the site of the sinking was false, and possibly an insurance fraud.

1st September 1942

U-600 is bound northeast of San Salvador, bound east for France. Damaged passenger ship SS *William Cullen Bryant* is bound north in Straits to Jacksonville. [It was torpedoed near Cay Sal Bank, by *U-84* on 29th July]. Walker Cay becomes Advance Base Lano, is activated this day under USN Ensign E. S. Williams, A-V(N), USNR.

2nd September 1942

USS *PC-469* leads first convoy from Key West to Guantanamo, arriving on the on 4th. Three hundred and ten nm south of Bermuda, USN Aircraft BuNo.2352 engages *U-600*.

3rd September 1942

U-600 exits area, bound east for France. SS *William Cullen Bryant* arrives in Jacksonville *en route* to New York for repairs. Report of SS *Queen* being torpedoed and sunk off Lake Worth and Delray Beach is false. Then, 280 nm east southeast of Bermuda, two Bermuda-based PBY-5's attack *U-600*, which dives. And 360 nm south-southeast of Bermuda, 52-P-10 attacks a U-boat with three depth charges, possibly *U-600*. Furthermore, 700 nm southeast of Bermuda, USN aircraft VP-52 52-P10, drop three depth charges on *U-505*, or possibly U-463. Two USN planes from Nassau patrolled north to Advance Base Lano (Walker Cay), observing Pelican Cay and Elbow Cay, Abaco as prospective seaplane and submarine chaser bases. [RAF later designated a practice target location known as Pelican, which was the seaplane tender USS *Pelican* (AVP-6)]. Civilian manager L. D. Apgar is owner of yacht *LaFitte*.

7th September 1942

U-511 re-enters area via Anegada Passage bound east to France. US and Cuba conclude an agreement for naval and military cooperation.

11th September 1942

Convoy KG-1 starts first convoy in the US since WWI from Key West at Guantanamo. USS *PC-469* is one of the escorts.

12th September 1942

U-171 re-enters the area north via Windward Passage.

13th September 1942

U-171 is bound northwest of Haiti. Tanker *Patrick J. Hurley* is sunk northeast of Anegada by *U-512*. Twelve crew, the captain, and four USN gunners are killed.

14th September 1942

U-171 is bound east of Grand Turk, bound east to France. USS *PC-469* arrives in Guantanamo after uneventful first convoy. **Convoy Charleston 15** is bound south in Straits, army plane escorting to Canaveral. SS *Dalwarnic* with SS *Wanigan* are off Canaveral, bound north, with an army plane escorting.

15th September 1942

U-171 is bound north of Puerto Rico and east of Turks and Caicos after exiting Windward Passage. SS *Fernbank* arrives Newport News from Port Everglades after diverting there and Nassau to repair leaking hull. **Convoy NK-502** is bound south in Straits with army plane escorting. Lend-Lease tanker SS *Bunker Hill* is in Straits, northeast to Belfast Ireland with army plane escorting. **Convoy Charleston 15** is bound south to Key West with army plane escorting. The Nuffield Trust funds a Bahamian canteen, run by Duchess of Windsor, United Nations [UN] Services, to also be provisioned by USN, and defended against civil riot by Squadron Leader Vare.

15th September 1942 – roughly

Nassau Harbour was curfewed and closed from 6.30 pm to 7.45 am: essentially it shut down at night to all civilian amphibious aircraft, boating and shipping traffic.

16th September 1942

USN **Air Patrol Squadron Patrol Wing 12** established at Key West, for operations in GSF [Gulf Sea Frontier]. **Convoy SS *Chicopee* and SS *Joseph Hooker*** are bound south in Straits with army plane escorting. **Convoy NK-502** from Norfolk to Key West is bound south in Straits with army plane escorting. **Convoy KN-201** with 13 merchant ships, two navy ships; SS *Thistle*, SS *Mumbles*, USS *Bambo* and HMS *Ram*, is bound north in Straits with army plane escorting. **Convoy PC-551** is bound north in Straits with army plane escorting. **Convoy Miami 12**, with one ship is leaving Miami, east across Straits to Nassau. **Convoy KG-602** is bound south in Old Bahama Channel.

17th September 1942

US Naval Air Station (NAS) De Land, Florida, is established. **Convoy KW-23** is bound north in Straits with army plane escorting. USS *PC-562* is bound north in Straits. SS *Joseph Hooker* with USS *SC-648* is bound south in Straits with army plane escorting. **Convoy Miami 12** is bound south in Straits.

18th September 1942

Tug *Kelvin Moran* and tow with USS *YP-395* is bound north in Straits with army plane escorting. **Convoy SC-676** is bound south in Straits. Lord Knowles, Governor of Bermuda, visited RAF No. 111 OTU with Air Chief Marshall Sir Frederick Bowhill.

19th September 1942

U-171 exits area east to France. **Convoy SS *Cape Alava*** with USS *SC-507* is bound south in Straits with army plane escorting. **Convoy GK-704** is bound north in Old Bahama Channel with army plane escorting.

20th September 1942

U-455 enters area southeast of Savannah. An unnamed USN patrol craft claims an attack on a U-boat. A U-boat contact is reported from a ship that witnessed two K-guns fired. One army plane is moving towards submarine contact. USS *PC-484*, USS *SC-525*, USS *B-188* are all at point of contact. **Convoy SS *Portia*** with USS *YP-395* is bound south in Straits with army plane escorting. SS *Cape Alava* with USS *PC-551* and USS *PC-451* with two army planes escorting, is bound west to Key West.

21st September 1942

U-455 is bound east of Savannah. **Convoy SS *Portia*** with USS *YP-395* is bound south in Straits with army plane escorting. USS

YMS-34 is bound north from Straits. **Convoy USS *Mistletoe*** with USCG *CG-145* is bound north in Straits.

22nd September 1942

Convoy Charleston 37 is bound south in Straits with army plane escorting. **Convoy SS *Norfolk*** and USS *YMS-34* is bound north in Straits from Canaveral. **Convoy NK-503** is bound south in Straits to Palm Beach with army plane escorting.

23rd September 1942

Convoy KN-202 is bound north in Straits with army plane escorting. **Convoy Charlseton-37** is bound south in Straits with army plane escorting. USS *YP-248* is in Straits near SS *Portia*, off Miami. SS *Chedabucto* is bound south in Straits. USS *PC-482* is bound south in Straits Miami to Key West. **Convoy KG-604** is bound east in Old Bahama Channel along Cuba with an army plane escorting. *U-558* in Windward Passage, reports 18 depth charges from USN destroyer. As a result, forward periscope stuffing box jammed, gyro-compass repeater at the chart table, radio direction finder gyro-compass repeater and true bearing on the attack periscope, main rudder position indicator receiver on the bridge failed.

24th September 1942

U-455 exits area northeast of Bahamas, east to France. **Convoy SS *William Tilghman*** with USS *SC-674* and USS *SC-677* are bound south in Straits with army plane escorting. USS *YMS-34* is bound south in Straits from Mayport, near Jacksonville, to Miami. **Convoy SS *Chicopee*** is bound north in Straits with army plane escorting. **Convoy SS *Portia*** with SS *Queen Wilhemina* and USS *YMS-68* is bound south in Straits with army plane escorting.

25th September 1942

USS *Asterion* begins 6th cruise from New York, now as decoy ship USS *AK-63*. **Convoy YO-135** and USS *YP-453* are bound south in Straits. USS *SC-647* is bound south in Straits. **Convoy SS *William Tilghman*** with USS *SC-674* and USS *SC-677* is bound south in Straits with army plane escorting. **Convoy KW-29** is bound south in Old Bahama Channel along Cuba. South of Bermuda a USAAF B-17 bomber reported that it damaged a U-boat.

26th September 1942

Convoy YMS-107 is bound north in Straits from Key West to Miami. **Convoy USS *Icarus*** (YAG-16) with USS *YP-513* is bound east across Straits from Miami to Great Exuma with army plane escorting. USS *SC-645* is bound north in Straits from Key West to Miami. **Convoy USS *Mistletoe*** with USCG *CG-145* is bound north in Straits from Key West to Miami. **Convoy GK-706** is bound north in Old Bahama Channel along Cuba with army plane escorting.

27th September 1942

HMS *Ilex* is bound south in Straits.

28th September 1942

Convoy and USS *Icarus* (YAG-16) is bound south in Straits, escorted by USS *YP-513*, from Nassau to Great Exuma. SS *Seakay* is bound north in Straits from Houston to Seawarren, New Jersey with army plane escorting. *USCGR-7011, USCGR-46022*, and *USCG- 45021* are bound north off northern Cuba in a submarine hunt.

29th September 1942

Military crash boats C-Boats USS *C-9470-1*, USS *C-9470-2*, and SS *Bacoi* are bound north in Straits. USS *Icarus* (YAG-16) is bound south in Straits.

30th September 1942

Convoy NK-504 is bound south in Straits, with army plane escorting. SS *Argyle* is bound southeast *en route* to Great Exuma. **Convoy KN-203** is bound northeast in Straits.

1st October 1942

Convoy USS *PC-469* is heading for Trinidad, its new homeport. SS *La Paz* is off Canaveral. **Convoy KN-203** is bound north, passed Palm Beach. *USCG-7011, USCG-46022*, and *USCG-45021* are on submarine hunt moving south towards Cay Sal Bank. **Convoy NK-504** is bound west in Straits towards Key West with US Navy plane escorting. SS *Argyle* is bound south Miami to Great Exuma passing Eleuthera from east.

2nd October 1942

USS *PC-609* is in Straits north returning to Miami after search of US Coast Guard cutters at Cay Sal Bank. **Convoy NG-309** is bound south, to the east of San Salvador.

3rd October 1942

USS *YO-142* is bound south in Straits from Miami to Key West. **Convoy KW-32** is bound north in Straits, from Key West up Straits. USS *YAG-16* is bound northwest, Great Exuma to Port Everglades, passing Eleuthera.

4th October 1942

USS *YMS-108* is bound north *en route* from Charleston to Key West. SS *Robert Gray* is bound south *en route* Charleston to Smith Shoals with US Navy plane escorting. **Convoy KW-32** is bound north *en route* Key West to Florida east coast ports with US Navy plane escorting. USS *YAG-16* is bound west in Northwest Providence Channel *en route* from Great Exuma to Port Everglades. USS *PC-583* is bound south in Straits, Miami to Key West. Tug USS *Warbler* is bound west in Straits with US Navy plane escorting, returning from Key West after aiding SS *Robert Gray*. SS *Mambi* is *en route* from Port Everglades in Straits, west towards Key West, with US Navy plane escorting. USS *SC-658* is bound north in Straits *en route* Key West to Miami.

5th October 1942

Crew of salvage tug owned by Merritt, Chapman and Scott pulled SS *Umtata* off the reef and towed her to Key West [declared a total loss, *Umtata* was later towed to Galveston then scrapped in January 1943]. USS *YR-26* towed by tug *Umpqua* is escorted USS *SC-670* south from Mayport to Smith Shoals with US Navy plane escorting. **Convoy KW-32** is bound north from Key West in Straits with US Navy plane escorting. Tug USS *Wellfleet*, escorted by USS *YP-31* is bound south in straits from Mayport to Canaveral with US Navy plane escorting. **Convoy SS *Salamonie*** is bound south in Straits from Norfolk to Galveston with US Navy plane escorting. USS *PC-608* is bound south in Old Bahama Channel from Key West to Guantanamo.

6th October 1942

Convoy NK-505 is bound south with two navy planes escorting. **Convoy Jacksonville-10** is bound south in Straits from Jacksonville to Port Everglades and then Key West, US Navy plane escorting. USS *Mistletoe* (WAGL-237), escorted by USCG *CG-145*, is bound east from Port Pierce to Walker Cay, north of Abaco. SS *Lake Capens*, towed by tug USS *Hercules*, escorted by USS *SC-656*, is bound southwest from Port Everglades to Havana with US Navy plane escorting.

7th October 1942

SS *Fred Kellogg*, escorted by USS *SC-515*, is bound south in Straits from Mayport to Smith Shoal, passing Canaveral with US Navy plane escorting. **Convoy NK-505** is bound south in Straits, passing Palm Beach with two navy planes escorting. USS *SC-664* is bound west in southern Straits, Miami to Key West. **Convoy Jacksonville-10** is bound south in Straits past Port Everglades to Key West, with US Navy plane escorting. **Convoy KN-204** is bound north in Straits with US Navy plane escorting. Neutral Portuguese SS *Serpa Pinto* is bound north in Straits, passing Cay Sal Bank, Veracruz Mexico to Baltimore.

8th October 1942

Convoy KN-204 is bound north with two navy plane escorting. SS *Oriente* is bound south, escorted by USS *YP-395*, from west Palm to Miami. SS *Bacoi* is bound north, escorted by USS *SC-665*, from Key West to Port Everglades with US Navy plane escorting. SS *Fred Kellogg* is bound west escorted USS *SC-515* from Mayport to Smith Shoal in Straits with US Navy plane escorting.

9th October 1942

USS *SC-655* and USS *SC-1014* are bound south in Straits from Charleston to Miami. Tug USS *Umpqua* is bound north in Straits from Key West to Mayport. USS *Ticonderoga* (CV-14) is bound south in Straits from Cape Henry to Baton Rouge with US Navy plane escorting. SS *Kaye Marie* is bound south in Straits, escorted by USS *SC-632* and USS *SC-643*, Miami to San Juan. USS *PC-584* is bound southeast in Old Bahama Channel from Key West to Guantanamo. USS *PC-583* is bound southeast from Havana to Guantanamo. USS *SC-515* is bound north in Straits from Key West to Miami. SS *Coloradan* sunk by *U-159 en route* from Iran to New York via Trinidad is actually sunk 200 nm southwest of Cape Town, not in this region. *U-159* was later sunk off Haiti.

10th October 1942

USS *Mistletoe* (WAGL-237) and SS *Wanigan* is bound north from Fort Pierce to St. Augustine, Florida. USS *Ticonderoga* (CV-14) is bound south from Cape Henry to Baton Rouge with US Navy plane escorting. USS *SC-674* is bound west from Miami to Key West in Straits. Tanker *Esso Hartford* is bound southwest from New York to Baytown, Texas in Straits, with US Navy plane escorting.

11th October 1942

USS *FC-619* is bound south in Straits from Charleston to Miami. USS *USAT-James Parker* is bound south from Charleston to Smith Shoals with US Navy plane escorting.

12th October 1942

Convoy KW-40 is bound north in Straits passing Palm Beach with US Navy plane escorting. Tug USS *Hercules* and tow southwest from Port Everglades. USS *Icarus* (YAG-16) and USCG Cutter *CG-156* are bound southwest in Straits. Tug *Edmond J. Moran* and two dredges are bound northwest in Straits near Nicholas Channel.

13th October 1942

Convoy KW-40 is bound north in Straits to Jacksonville with US Navy plane escorting. **Convoy SS *Salmonie*** is bound north in Straits passing Palm Beach with US Navy plane escorting. **Convoy SS *Icarus*** and four escorts is bound south in Nicholas Channel. **Convoy GK-709** is bound north in Straits to Cay Sal Bank, with US Navy plane escorting.

14th October 1942

USN minesweepers USS *YMS-9*, USS *YMS-10*, USS *YMS-49* and USS *YMS-50* are bound south. **Convoy PE-35** is bound south in Straits passing Port Everglades with US Navy plane escorting. USS *YP-391* and USS *YP-407*, and USS *YMS-61* are bound southeast to exit the Old Bahama Channel. USS *SC-524* is bound northeast from Old Bahama Channel. USS *Icarus* (YAG-16) is bound northwest towards Nicholas Channel, with US Navy plane escorting.

15th October 1942

SS *J. A. Moffett* [damaged by *U-517* on 8th July 1942 on Tennessee Reef], was towed to Galveston and scrapped in January 1943 [later salvaged as *J. A. Moffett Jr*, she is written off as a total loss]. USS *Mistletoe* (WAGL-237) and SS *Wanigan* are bound south towards St. Augustine. USS *YMS-9*, USS *YMS-10*, USS *YMS-49*, and USS *YMS-50* are bound south in Straits towards Cuba. Columbian destroyers *Antioquia* and *Caldas* are bound south in Straits. USCG Cutter *GC-Legare* (WSC-144) is bound southeast in Straits towards Old Bahama Channel.

16th October 1942

USS *Mistletoe* (WAGL-237) and SS *Wanigan* are bound west in Straits towards St. Augustine.

17th October 1942

USS *Mistletoe* (WAGL-237) and SS *Wanigan* are bound south in Straits. SS *American Seaman* is bound south in Straits towards Canaveral, with US Navy plane escorting. USCG *CG-499* is bound north from Port Everglades. **Convoy KN-205** is bound northeast in Straits with US Navy plane escorting. USS *YAG-16* is bound north in Straits.

18th October 1942

Convoy KN-205 is bound north in Straits. **Convoy NK-506** is bound south in Straits towards Palm Beach with two navy planes escorting. Bahamas mailboat *Lady Cordeaux* is bound southwest from Port Everglades. **Convoy KW-41** is bound north in Straits to Port Everglades with US Navy plane escorting. Crash-boat USS *C-1999* is bound southwest in Straits. **Convoy GN-12** is bound north in Straits from Nicholas Channel with a navy blimp and navy plane escorting.

19th October 1942

Convoy Charleston-32 is bound south from Charleston to Smith Shoals, with US Navy plane escorting. **Convoy GN** is bound north with a navy blimp and navy plane escorting. **Convoy NK-506** is bound south in Straits to, two navy planes escorting. **Convoy Smith Shoal-25** is bound northeast in Straits to Port Everglades with US Navy plane escorting.

20th October 1942

Convoy Charleston-32 and SS *Kentuckian* is Charleston to Smith Shoals, bound south in Straits with army plane escorting. USS *USAT-Cuba* is bound east in Straits, Miami to Nassau. USS *SC-517* is bound north in Old Bahama Channel, Guantanamo to Miami. USS *YMS-26*, USS *YMS-27*, USS *YMS-27*, and USS *SC-516* are bound north in Straits, Guantanamo to Norfolk. USS *PC-482*, USS *PC-575*, and USCG *Unalga* with **Convoy GN-13** are bound north in Straits bound to Norfolk.

21st October 1942

SS *William C. Atwater* is *en route* from Port Everglades to Jacksonville, north with an army plane escorting. **Convoy USS *PE-41*** is *en route* from Port Everglades to Norfolk bound north, army plane escorting. Tug USS *Warbler* and USS *YP-542* are salvaging SS *J. A. Moffett Jr.* in Straits. HMS *Ocean Transport* is bound south in Straits, bound Port Everglades to Key West.

22nd October 1942

USS *SC-516*, USS *SC-517*, USS *YMS-26,* USS *YMS-24*, and USS *YMS-27* are bound north in Straits, bound Miami to Norfolk. **Convoy Jacksonville-14** is bound south in Straits, from Mayport to Smith Shoals with US Navy plane escorting. SS *Oriente* is bound south in Straits, Lake Worth to Key West with army plane escorting. Tug USS *Warbler* and USS *YP-542* are salvaging SS *J. A. Moffett, Jr.* off Tennessee Reef.

23rd October 1942

Neutral Portuguese SS *Nyassa* is bound north in Straits, Veracruz to Baltimore. **Convoy *PC-482***, Miami to Norfolk, is bound north in Straits. USS *PC-619* is bound north in Straits, Key West to New York. USS *PC-475* is Norfolk from

Guantanamo, bound north. SS *Moysalen* escorted by USS *SC-634*, Port Everglades to Havana, is bound south in Straits with army plane escorting.

25th October 1942

Tanker *Gulfland* is bound south in Straits, Jacksonville to Smith Shoal, army plane escorting. SS *Wanks* is Miami to Nicarao, Cuba bound east. **Convoy KN** is bound north in Straits with army plane escorting. SS *Imperial Chilean Cristobal* is bound southwest in Straits bound to Havana with army plane escorting.

26th October 1942

Convoy NK-507 is bound south in Straits with army plane escorting. **Convoy KW-49** is bound north in Straits to Palm Beach, with army plane escorting. SS *Wanks* escorted by USCG *CG-Wanigan* is bound east in Straits, Miami to Great Exuma.

27th October 1942

USS *SC-637* is Charleston to Miami, south in Straits. SS *Norfolk* with four escorts, Charleston to Smith Shoals, is bound south in Straits with an army plane escorting. **Convoy NK-507** is bound south in Straits, with army plane escorting. SS *Argyle* is bound southwest in Straits with one escort, from Miami to Nicarao, Cuba.

28th October 1942

USS *SC-658* is Key West to Charleston, bound north in Straits. USS *PYC-31* is in Straits from Miami to Tampa with army plane escorting. SS *Mambi* is bound south in Straits with one escort, Port Everglades to Key West.

29th October 1942

Tanker *Esso Harrisburg* is bound north in Straits to New York with army plane escorting. British rescue tug HMS *HMRT Masterful* is bound north in Straits, Key West to Charleston. USCG *CG-Wanigan* is bound west in Straits, Great Exuma to Miami.

30th October 1942

USS *PC-586* is bound south in Straits from Miami to Key West. **Convoy Havana North Coast Cuba-14** is bound north in Straits, Havana to Port Everglades. USS *Icarus* (YAG-16) and USCG Cutter *CG*-156 are bound south in Straits to Cayo Francés. USCG *CG-Wanigan* is bound south in Straits from Great Exuma to Miami.

31st October 1942

USS *USAT-Cuba* is bound north in Straits with army plane escorting, Miami to Charleston. The house boat USS *YHB-14* and USS *YP-218*, are bound south in Straits from Miami to Cayo Francés. USS *Icarus* (YAG-16) and USCG Cutter *CG*-156 are bound south in Straits, Miami to Cayo Francés.

1st November 1942

Convoy Charleston-35 is bound south in Straits from Charleston to Key West, passing Jacksonville with US Navy plane escorting. SS *Guardian* with one escort is bound north in Straits from Palm Beach to Key West, south towards Port Everglades, with one navy plane escorting. **Convoy KW-52** is bound north in Straits from Key West to Palm Beach, north towards Port Everglades with US Navy plane escorting.

2nd November 1942

USS *YMS-109* is bound north in Straits towards Jacksonville. **Convoy Jacksonville-16** is bound south in Straits passing St. Augustine. **Convoy Charleston-35** is bound south in Straits, passing Canaveral. SS *Portland* and **Convoy KW-54** are bound north in Straits. USS *Dahlgren* is bound southwest in Straits. **Convoy KN-207** is bound north in Straits with US Navy plane escorting.

3rd November 1942

SOS from SS *Texan,* which is torpedoed in Old Bahama Channel. SS *Polartank* is bound south towards Canaveral with US

Navy plane escorting. **Convoy NK-508** is bound south towards Palm Beach with US Navy plane escorting. **Convoy Port Everglades-48** is bound north in Straits passing Port Everglades, with Cuban army plane escorting. **Convoy Port Everglades-49** is bound south towards Miami, with US Navy plane escorting. **Convoy KW-55** is bound northeast in Straits with US Navy plane escorting. **Convoy Port Everglades-47** is bound south in Straits to Key West. **Convoy Charleston-35** is bound south in Straits to Key West. **Convoy San Juan-15** is bound north passing Cay Sal Bank towards Key West with US Navy plane escorting. **Convoy KN-53** is bound south, passing Nicholas Channel. **Convoy Charleston-36** is bound south towards Cat Island in the Atlantic.

4th November 1942

Convoy Port Everglades-48 is bound north in Straits near Canaveral with US Navy plane escorting. Tanker *British Diligence* is bound south in Straits towards Canaveral with US Navy plane escorting. SS *Empire Norseman* is bound south towards Palm Beach with US Navy plane escorting. **Convoy Jacksonville-16** is bound east across Straits from Port Everglades to Bimini. SS *Caddo* is bound south in Straits passing Port Everglades with US Navy plane escorting. **Convoy NK-508** is bound south in Straits, two navy planes escorting. Tanker *Polartank* is bound west in Straits with US Navy plane escorting. **Convoy KG-611** is bound south across the Straits, Cay Sal Bank towards Cayo Francés, with US Navy plane escorting.

5th November 1942

U-154 enters area northeast of Anegeda. SS *Empire Norseman* is bound south in Straits passing Canaveral with US Navy plane escorting. SS *Solarium* is bound south in Straits, passing Palm Beach with US Navy plane escorting. USS *YP-403* and barge are anchored near Palm Beach. SS *Empire Coral* is bound south in Straits passing Palm Beach with US Navy plane escorting. USS *YAG-19* east Port Everglades towards Bimini. SS *Samaná* west from Bimini towards Port Everglades. **Convoy Miami-37** is bound northeast in Straits towards Miami. Tanker *British Diligence* is bound southwest passing Miami with US Navy plane escorting. SS *Wralia* south in Straits passing off Miami. SS *Gaspe County* is bound southeast passing Cay Sal Bank.

6th November 1942

U-154 is bound northeast of Dominican Republic in brief incursion. RAF bases declare the day unfit for flying no operational flying owing to hurricane warning which is later warning cancelled. SS *Empire Flint* and SS *Empire Norseman* are bound south in Straits towards Palm Beach. **Submarine Chaser Training Fleet** is bound north in Straits towards Palm Beach. Aircraft carrier USS *Ticonderoga* (CV-14) is bound southwest in Straits passing Fort Everglades. USS *PE-50* is bound west in Straits towards Key West. **Convoy GK-712** is bound north, passing Great Bahama Banks. USS *YAG-19* is bound south in Northeast Providence Channel towards Nassau

7th November 1942

U-154 exits area east from north of Puerto Rico and Dominican Republic. Tanker *Esso Augusta* returns to service [it was damaged by *U-701* on 15th June 1942]. **Submarine Chaser Training Fleet** is bound east from Port Everglades. SS *Markay* is bound northeast in straits. SS *Empire Norseman* is bound west in Straits. SS *Norindies* and tug USS *Montcalm* or on Cay Sal Bank, the shallow triangle between the Keys, Cuba, and Andros. Crown Lands Officer William N. Aranha declares: "*...acting under the* *direction of His Royal Highness the Governor and Commander in Chief of the Bahama Islands I have caused ...to mark the route of a public road which will join Satellite Field situate south of the Ninth (9th) Mile Post on Bay Street in the western District of the Island of New Providence with the junction of Lake Road with Gladstone Road and also for a piece of land to increase the width of the existing Gladstone Road or Lake Road and the Farrington Road to fifty (50) feet – the said pieces of land passing through the undermentioned tracts of land in the western District: The Caves Company Limited, 285 acres, granted 1937, John Petty, 698, 1790), Ebenezer Love, (440, 1791), Rebecca and Mary Baldwin, (231, 1793), Peter Dean, (280, 1788), G. A. Bosfield and S. A. Bosfield, (65, 1890), Robert Johnstone, (153, 1788), William Moss, (312, 1789), John Brown, (550, 1789), Robert Cunningham, (100, 1809), Hon. Stephen Delancy, (300, 1793), and thereof have caused possession to be taken...*" [by order of government, for the expanding airfield].

8th November 1942

SS *Regent Panther* is bound south in Straits passing Palm Beach, US Navy plane escorting. Bahamas ferry *Ena K.* and USS *YP-272* are bound east from Miami with US Navy plane escorting. USS *YAG-19* is bound north from Nassau in Northeast Providence Channel. **Convoy Guantanamo-17** is bound north in Straits, passing Nicholas Channel with US Navy plane escorting.

9th November 1942

A hurricane passes over the Bahamas, hits Long Island, but does not cause much damage. SS *Portland* is bound north in Straits towards Canaveral with US Navy plane escorting. SS *Galoria* is bound south towards Palm Beach. US Coast Guard ships *GC-603* and *GC-604* are bound south in Straits passing Palm Beach. Tanker *Gulf of Maracaibo* is bound north in Straits passing Matanilla Shoal, northwest Bahamas with US Navy plane escorting. **Convoy NG-319** is bound south, to the east of Cat Island in the Atlantic. Seventh Naval District [7ND] supply vessel USS *Niagara* is bound north in Old Bahama Channel from Cayo Francés towards Nicholas Channel.

10th November 1942

SS *Richard Spaight* is bound south in Straits, passing Canaveral with US Navy plane escorting. **Convoy NK-509** is bound south in Straits to Canaveral with US Navy plane escorting. Seventh Naval District supply vessel USS *Niagara* is bound northwest passing Cay Sal Bank in Straits.

11th November 1942

Convoy KW-60 is bound north in Straits to Palm Beach. **Convoy KN-208** is bound north in Straits, west of Bimini. Tank barge USS *YO-142* and USS *YP-272* are bound west in Northwest Providence Channel. **Convoy-37** is bound east in Straits, passing north of Bimini. **Convoy-1** is bound west in Straits, bound Bimini towards Miami. **Convoy V-509** is bound south in Straits with US Navy blimp near Bimini. SS *Richard Spaight* is bound south in Straits. **Convoy Port Everglades-54** is bound south in Straits, bound to Key West.

12th November 1942

SS *San Antonio* is bound south in Straits towards Canaveral, with US Navy plane escorting. SS *Argyle* is bound south in Straits near Port Everglades. SS *Bunker Hill* is bound southwest in Straits passing Cay Sal Bank. **Convoy KG-612** is bound south in Nicholas Channel.

13th November 1942

Tug *Mike Moran* is bound south in Straits passing St. Augustine with US Navy plane escorting. USS *YP-453* is bound north in Straits passing Palm Beach. SS *Nathaniel*

Alexander is bound south in Straits towards Port Everglades with US Navy plane escorting. SS *San Antonio* is bound south in Straits with US Navy plane escorting. Oil barge USS *YO-142* and USS *YP-272* are bound northeast in Straits towards Northwest Providence Channel. **Convoy NG** is bound south, to the east of Cat Island Bahamas, in the Atlantic.

14th November 1942

Army tug USS *LT-145* is bound south in Straits towards Canaveral. Tug *Mike Moran* is bound south in Straits towards Palm Beach, with US Navy plane escorting. **Convoy KW-62** is bound north in Straits with US Navy plane escorting. Dredge USS *San Pablo* is bound west in Straits towards Key West. **Convoy GK-713** is bound west in Straits ,west of Great Bahama Bank, with US Navy plane escorting.

15th November 1942

Tug *Kevin Moran* is bound north in Straits passing Palm Beach with US Navy plane escorting. USS *PC-626* is bound south towards Palm Beach. USS *Icarus* (YAG-16) is bound west in Northwest Providence Channel. SS *Chedabucto* is bound south in Straits. USS *Partridge* (AM-16) is bound northwest in Straits passing Nicholas Channel.

16th November 1942

At No.111 OTU RAF base outside Nassau, a large number of officers attend instructional courses. Tanker *Caddo* transits Bahamas via Straits and northwest and northeast passages *en route* from Baytown Texas to Iceland. SS *Fernbank* left Guantanamo in **Convoy GN-20** to New York, but had to divert to Nassau because its hull was leaking. Tanker *Gulfland*, escorted by USS *SC-695* is bound south in Straits passing St. Augustine with US Navy plane escorting. **Convoy KW-62** is bound north in Straits towards Canaveral with US Navy plane escorting. SS *Appledore* escorted by USS *PC-611* is bound north in Straits passing Port Everglades. USS *Icarus* (*YAG-16*) and USCG *CG-74308* waiting at Nassau for orders.

17th November 1942

Tanker *Gulfland* escorted by USS *SC-695* Mayport to Smith Shoals is bound south in Straits passing Palm Beach. SS *Varida* escorted by USS *AM-89*, USS *AM-90*, and USS *AM-91* are Key West to Norfolk north in Straits. **Convoy NG-321** is bound south, east of Eleuthera. SS *Caddo* is bound north passing Matanilla Shoal, northwest Bahamas with US Navy plane escorting, to Iceland.

18th November 1942

SS *Varida* escorted by USS *AM-89*, USS *AM-90*, and USS *AM-91*, from Key West to Norfolk is bound north in Straits to Canaveral with US Navy plane escorting. USS *YMS-36*, Miami to Mayport north in Straits passing Palm Beach. USS *YO-142* and USS *YP-272* are Port Everglades to Nassau, east in Straits passing Fort Everglades towards Bimini, with US Navy plane escorting. Tanker *Markay* is bound southwest in Straits, from New York to Texas City in US Gulf.

19th November 1942

Convoy KN-209 is bound south in Straits from Mayport to Miami passing Palm Beach. Tanker *Virginia* is Norfolk to Port Arthur Texas, south in Straits passing Palm Beach with US Navy plane escorting. SS *Guardian* and USS *YP-426* are Palm Beach to Havana, south in Straits passing Port Everglades with US Navy plane escorting. Tanker *Bunker Hill*, Port Arthur to New York, is bound northeast, passing out of the Straits. Tug *Edward J. Moran* and USS *PY-13* are Guantanamo to Key West, east in Straits passing Nicholas Channel. Two US navy tugs, two USN *YO*'s, two USN *YC*'s, and two US Navy Submarine Chasers are bound southeast in Old Bahama Channel passing Cay Sal Bank, Key West to Guantanamo.

20th November 1942

Convoy NK-510 is bound south in Straits passing Palm Beach with US Navy plane escorting. USS *SC-646* is bound north in Straits. Tug *Henry W. Card* is bound southwest in Straits, passing Nicholas Channel.

21st November 1942

USS *PE-61* (possibly leading small convoy) is bound north in Straits, Port Everglades to Mayport passing Canaveral with US Navy plane escorting. SS *Nidareid* and USS *PC-584*, from Port Everglades to Nicarao, Cuba are bound south of Straits. SS *Fernbank* from **Convoy GN-20** is bound east in Northeast Providence Channel to Nassau, for repairs. USS *PE-60* (possibly leading small convoy) is bound south in Straits from Port Everglades to Havana with US Navy plane escorting. SS *Samaná* and USS *YP-452* are bound Miami to Cayo Francés, south in Straits towards Cay Sal Bank.

22nd November 1942

USS *YMS-36* is bound south in Straits towards Canaveral. USS *PC-589* is bound north in Straits towards Palm Beach. Tug USS *Hercules* is bound north in Straits. **Convoy GK-714** is bound north in Straits, exiting Nicholas Channel with US Navy plane escorting.

23rd November 1942

Tug USS *Meteor* is bound south in Straits towards Port Everglades. **Convoy KW-74** is bound north in Straits towards Port Everglades, with US Navy plane escorting.

24th November 1942

Convoy Charleston-47 is bound south in Straits passing Canaveral, with US Navy plane escorting. Tug USS *Hercules* towing a US Navy barge are bound southwest in Straits. Tug USS *Humrick* and tow are bound northeast in Straits. **Convoy NG-323** is bound south, to the east of Eleuthera, Bahamas. SS *Atlantida* is bound south in Straits near Cay Sal Bank.

25th November 1942

At No.111 OTU RAF, first flight of B-25's; FK-164, duration of flight 40 minutes, is followed by subsequent local flying. USS *YP-118* and USCG *CG-369* are bound south in Straits passing Canaveral with US Navy plane escorting. SS *Monmouth* is bound northeast with US Navy plane escorting. **Convoy Charleston-47** is bound south in Straits, passing Miami with US Navy plane escorting. SS *Argyle* with USS *AM-96* are bound north in Straits.

26th November 1942

US Navy operation areas in the GSF [Gulf Sea Frontier, which dissects Bahamas] are surveyed from the air from Miami. Local flying. Tanker *Esso Washington* is bound northeast in Straits with US Navy plane escorting. USS *YAG-16* and USCG *CG-83317* are bound west in Northwest Providence Channel. Tanker *Markay* is bound north in Straits, passing Bimini with US Navy plane escorting. **Convoy Port Everglades** is bound south in Straits, with US Navy plane escorting. **Convoy KN-210** is bound north in Straits past Port Everglades with US Navy plane escorting. SS *Samaná* and USS *YP-452* are bound north in Straits.

27th November 1942

SS *Ben Smith* is bound south in Straits passing Daytona, with US Navy plane escorting. **Convoy NK-511** is bound south in Straits passing Canaveral with US Navy plane escorting. **Convoy KN-210** is bound north in Straits with US Navy plane escorting. The straggler from **Convoy KN-210** SS *Lake Ormoc,* is bound north in Straits. USS *Semmes* (AG-24) is bound north in Straits. USS *Siren* (PY-13) is bound north in Straits. USS *YO-142* and USS *YP-408* are bound west towards Walker Cay, north Abaco. USS *YAG-16* and USCG *CG-83317* are bound southwest out of Northwest Providence Channel. SS *Guardian* is bound north in Straits departing from Miami. USS *YP-427* is bound southwest in Straits towards Miami. **Convoy KW-77** is bound west out of Nicholas Channel.

28th November 1942

Four Baltimore planes *en route* for Middle East experience handing affected due to adverse weather conditions. SS *Pompoon* is bound south towards Canaveral with US Navy plane escorting. Convoy straggler SS *Lake Ormoc* is repaired and north in Straits passing Canaveral. USS *Ticonderoga* (CV-14) is bound north in Straits towards Palm Beach. **Convoy NK-511** is bound south in Straits passing Miami with USN Blimp *K-18* and navy plane escorting. SS *Elizabeth Kellogg* and SS *Spencer Kellogg* are bound east in Straits from Miami. SS *Ben Smith* is bound south in Straits with US Navy plane escorting.

29th November 1942

Local flying. SS *Pompoon* and USS *YP-272* are bound south in Straits passing Palm Beach. USS *YO-142* and USS *YP-408* are bound south in Straits towards Port Everglades. USS *SC-638* is bound north in Straits passing Port Everglades. SS *Samaná* and USS *YP-535* are bound east in Straits from Miami. **Convoy NG-324** is bound south, to the east of Eleuthera.

30th November 1942

USS *SC-638*, tug USS *Meteor*, USS *YP-460* and USS *YP-482* are bound north in Straits towards Canaveral. **Convoy Charleston-51** is bound south in Straits passing Palm Beach with US Navy plane escorting. SS *Pompoon* is bound west in Straits. SS *Bunker Hill* is bound south in Straits. Tug USS *Chippewa* is bound north in Nicholas Channel. **Convoy GK-715** is bound north in Nicholas Channel with US Navy plane escorting.

1st December 1942

Four Baltimore and one B-25 planes take off. One Baltimore returns to aerodrome due to undercarriage trouble. USS *PC-559* and USS *PC-561* are bound north in Straits passing Canaveral. Tanker *Esso Augusta* is bound north in Straits passing Port Everglades. Tug *Zwartezee* is bound north in Straits. **Convoy Port Everglades-14** is bound south in Straits. SS *Samaná* is bound west , north of Nassau in Northwest Providence Channel to Bimini. USS *YAG-19* is bound north in Straits passing Andros. A B-25 and four Baltimore aircraft take off. NAS Great Exuma: USS *YAG-22* arrived. Three bombers and two unidentified planes sighted.

2nd December 1942

Baltimore plane that failed to take off yesterday did so today. Second B-25 arrived to station accompanied by a Boston aircraft which later left base. Party led by Commanding Officer, Group Captain and accompanied by Station Signal Officer and Flying Instructor left for San Salvador inspect Air Sea Rescue Base another party left to survey Harbour Island and Eleuthera. **Convoy-28** is bound north in Straits passing Canaveral. **Convoy-D** is bound south in Straits. USS *PE-71* is bound south in Straits. USS *YP-433* is bound north in Straits. One army and one navy plane near two blimps are bound west of Andros. USS *YAG-19* is bound southwest in Northwest Providence Channel to Straits. A group flew to San Salvador to survey an aircraft search and rescue base, another to Harbour Island, for five days. NAS Great Exuma: The mailboat *Alice Mabel* arrived at Georgetown from Long Island. Two Army bombers, four Army planes, one Army transport plane and one unidentified planes are sighted.

3rd December 1942

MV *Guardian* is bound south in Straits passing Port Everglades. USS *PC-625* is bound south in Straits from Miami. **Convoy NG-325** is bound south, to the east of Eleuthera. NAS Great Exuma: Two Army bombers, one Army transport plane and four unidentified planes sighted.

4th December 1942

SS *Goldheels* is bound south in Straits passing Canaveral with army plane escorting. USS *YMS-69* and USS *AM-97* are bound north in Straits towards Canaveral. USS *YMS-36* is bound south in Straits towards Palm Beach. Tanker *Esso Washington* is bound south from Matanilla Shoal, northwest Bahamas. **Convoy KN-211** is bound north in Straits near Miami with army plane and a blimp escorting. NAS Great Exuma: Four unidentified planes sighted at different times.

5th December 1942

SS *David Stone* is bound south in Straits, passing St. Augustine with US Navy plane escorting. SS *Goldheels* is bound southwest in Straits passing Miami with US Navy plane escorting. USS *YAG-19* is bound southwest in Straits. NAS Great Exuma: Sailing sloop *Joseph B.* arrived and left George Town, Exuma.

6th December 1942

Party that had left for San Salvador on 2nd returned to base. **Convoy NK-512** is bound south in Straits passing Canaveral with army plane escorting with a blimp. SS *David Stone* is bound south with army plane escorting passing Port Everglades. NAS Great Exuma: A sailing schooner arrived. Three Army bombers and one unidentified plane sighted.

7th December 1942

Party that left to survey Harbour Island and Eleuthera on 2nd returned to base. **Convoy NK-512** is bound south towards Port Everglades army plane and a blimp. Seventh Naval District vessel *Lazybones* is bound southwest in Straits. Crash boats USS *C-9479* and USS *C-18368* are bound southwest in Straits towards Key West. NAS Great Exuma: Two bombers, two Army planes and one unidentified plane sighted.

8th December 1942

Convoy Motor Torpedo Boat Squadron-13, SS *Marjory*, and SS *Bunker Hill* are bound north in Straits passing Port Everglades. **Convoy GK-716** is bound north in Old Bahama Channel. NAS Great Exuma: Two unidentified planes and two Army transport planes sighted.

9th December 1942

SS *Fernbank* is bound north in Straits, Port Everglades for New York via Newport News Virginia. Motor yacht *Pilar*, Hemingway and crew, at sea off Havana at Bahia Honda, 16 nm out, practicing shooting at U-boats. Tanker *Pan Georgia* is bound south in Straits towards St. Augustine. SS *Baldbutte* is bound south in Straits passing Daytona with US Navy plane escorting. SS *Marjory* is bound north in Straits passing Daytona. Tanker *Sabine Sun* and USS *Barney* (DD-149) are bound south in Straits passing Palm Beach. Tanker *Gulfland* is bound south in Straits towards Port Everglades. Tug USS *Henry W. Card* is bound southwest in Straits passing Nicholas Channel. NAS Great Exuma: Seven planes are sighted, including three Army bombers and a transport plane.

10th December 1942

Army tug USS *LT-113* and tow are bound south in Straits towards St. Augustine. **Convoy KN-212** is bound north in Straits with a navy blimp towards Canaveral with US Navy plane escorting. USS *SC-1023* and USS *SC-738* are bound north in Straits towards Palm Beach. **Convoy KN-82** is bound north in Straits passing Miami. SS *Baldbutte* is bound south in Straits passing Miami. Tanker *Pan Georgia* is bound west in Straits. **Convoy Guantanamo-31** is bound northwest in Straits passing Nicholas Channel with US Navy plane escorting. NAS Great Exuma: A sailing schooner arrives and a twin-engine Army plane is sighted.

11th December 1942

Convoy KW-62 and troop transport USS *USAT-Mexico* are bound north in Straits passing Canaveral with US Navy plane escorting. Tanker *Esso Washington* is bound southwest towards Matanilla Shoal, northwest Bahamas with US Navy plane escorting. Army tug USS *LT-113* is bound south in Straits, passing Banana River. Tanker *Gargoyle* is bound north in Straits passing Banana River. Tanker *R. P. Smith* and tanker *Pan Virginia* are bound north in Straits passing Miami. NAS Great Exuma: Three unidentified planes are sighted overhead.

12th December 1942

Tanker *Pan Virginia* is bound north in Straits towards Daytona with US Navy plane escorting. Army tug USS *LT-113* and barge are bound south in Straits passing Banana River. USCG *CG-Spruce*, USS *YP-426*, USS *YP 427*, USS *YO-142*, and USS *YP-408* are bound southwest in Straits. NAS Great Exuma: Five planes sighted, including a bomber and PBY-2.

13th December 1942

Convoy SS *Chemung* is bound southwest in Straits, east of Bahamas with US Navy plane escorting. USS *PC-494* is bound north in Straits passing Canaveral. **Convoy NK-513** is bound south towards Palm Beach, with US Navy plane escorting. USS *YP-431* is bound south in Straits towards Palm Beach. USS *PC-491* is bound north in Straits towards Palm Beach. USS *SC-514*, USS *SC-640*, and USS *SC-650* are bound west in Straits. USS *PC-549* and USS *PC-9449* are bound southwest in Straits, passing Nicholas Channel. NAS Great Exuma: Mailboat *Alice Mabel* arrives and leaves Georgetown. Two planes and a power boat are sighted.

14th December 1942

Aircraft FK-163 arrives at the RAF station. SS *City of Philadelphia* is bound south in Straits towards Canaveral. SS *Mindi* is bound south in Straits towards Canaveral, with US Navy plane escorting. USS *IX-78* and USS *YP-272* are bound south in Straits towards Palm Beach. USS *PC-491* is bound north in Straits, passing Palm Beach. Tanker *Texas Sun* and USS *SC-738* are bound south towards in Straits Miami. Tanker *Esso Augusta* is bound south in Straits passing Miami. SS *Markay* and **Convoy KW-83** are bound north in Straits passing Port Everglades with US Navy planes escorting. Tanker *Sinclair Opaline* is bound south in Straits. **Convoy NK-513** is bound south in Straits with a navy blimp escorting. USCG *CG-Spruce* and USS *PC-625* are stationed at Cay Sal Bank. NAS Great Exuma: One PBM Mariner left base, and an unidentified plane is sighted overhead.

15th December 1942

Aircraft FK-177 and FK-182 arrive at the RAF Station. USS *SC-750* is bound south in Straits, passing St. Augustine. **Convoy KW-83** is bound north in Straits towards Canaveral with US Navy plane escorting. A plane crashes off Melbourne, Florida, in Straits. SS *City of Philadelphia* is bound south in Straits towards Palm Beach. USS south *Dakota* (BB) and two destroyers are bound north in Straits, passing Palm Beach, with two navy planes escorting. The disabled tug USS *Ionian* and SS *Virginia* are bound south, in the northern Straits. NAS Great Exuma: Mailboat *Alice Mabel* arrives from Long Island, Bahamas, and one plane is sighted.

16th December 1942

USS *IX-78* and USS *YP-272* are bound south in Straits towards Palm Beach. USS *Mistletoe* (WAGL-237) is bound north in Straits towards Palm Beach. NAS Great Exuma: Mailboat *Alice Mabel* left Georgetown, USS *YAG-10* arrived. A PBY from Guantanamo arrived then returned there. A twin-engine plane sighted.

17th December 1942

A Group Captain, Air Commodore, and Flight Lt. arrived at RAF Nassau by air from Washington via Miami. First night flying carried out. Plane FK-180 arrived. Irish SS *Irish Oak* is bound north passing Matanilla Shoal, northwest Bahamas. USS *YP-445*, USS *YP-437*, and USS *YP-443* are bound south in Straits towards Palm Beach. USS *YAG-16* is bound east from Northwest Providence Channel towards Northeast Providence Channel. SS *Wanks* and USS *SC-1041* are bound north in Straits. USS *SC-646* is bound southwest bound in Straits. USS *YAG-20* is bound northwest from Exumas towards Nassau. First night-flying sorties undertaken by RAF. NAS Great Exuma: Five unidentified planes sighted at various times.

18th December 1942

Air Commodore arrives from Montreal by air via Miami. USS *YP-428* and USS YP-*433* are bound south in Straits towards St. Augustine. USS *SC-488* is bound south in Straits, passing Canaveral. SS *Wanks*, SS *Watertown*, and USS *SC-1041* are bound north in Straits, towards Canaveral. USS *SC-766* is bound south in Straits towards Palm Beach. **Convoy KN-213** is bound north in Straits towards Canaveral navy plane escorting. Three Fairlee British Motor Launches (HMS *HMML's*) are bound south in Straits, passing Palm Beach. US Navy tanker, or oiler USS *Benedict* is bound south in Straits, passing Miami. Tanker *Esso Washington* is bound north in Straits, passing Miami. NAS Great Exuma: Two unidentified planes sighted overhead.

19th December 1942

Aircraft FK-173 arrived at RAF station Nassau. USS *YMS-51*, USS *YMS-52*, USS *YMS-53*, and USS *YMS-11* are bound south in Straits passing Canaveral. USS *YP-433* is bound south in Straits towards Port Everglades. USS *YAG-16* is bound west from Bimini towards Port Everglades. Troop transport USS *USAT-Frank Leisenring* and USS *SC-621* are bound east in Straits, from Miami, with a US Navy plane escorting. USS *YAG-20* is bound northwest from Nassau. RAF Unit officers hosted a cocktail party at the Royal Victoria Hotel, downtown Nassau, with the RAF station band playing all evening. NAS Great Exuma: Schooner *Olive Shaw* [may be mailboats *Ollie Ford* or *Olivette*] at anchor. Two unidentified planes overhead.

20th December 1942

Further three P/O (pilot officers) and one S/Ldr. (squadron leader) arrived from Miami by sea. USS *Mistletoe* (WAGL-237) is bound north in Straits passing Canaveral. USS *YMS-37* is bound south in Straits passing Canaveral. USS *YP-428* is bound south in Straits towards Palm Beach. USS *YAG-20* is bound southwest from Northwest Providence Channel towards Miami. USS *PE-73* is bound southeast in Straits. USCG Cutter *CG-Crawford* is bound north from Cayo Francés.

21st December 1942

Squadron Leader arrived from Miami by air. **Convoy NK-514** is bound south in

Straits passing Canaveral, with US Navy plane escorting. Tanker *Esso Hartford* is bound north in Straits towards Canaveral. SS *Chicopee* is bound southwest in Straits towards Palm Beach. SS *Chemung* is bound north in Straits passing a US Navy blimp, bound towards Palm Beach. Spanish-flagged SS *Aldecoa* is bound west in Straits. NAS Great Exuma: Three unidentified planes are sighted overhead.

22nd December 1942

Air Marshall arrives from Washington via Miami by air. Air Commodore and S/Ldr leaves by air for Miami. USS *YO-56* is bound south in Straits passing Canaveral. SS *Virginia* is bound north in Straits towards Canaveral. **Convoy NK-514** is bound south in Straits passing Port Everglades with US Navy blimp USS *K-18*. SS *Frank Leisenring* is bound west in Straits from Bimini towards Miami with US Navy plane escorting. SS *Markay* is bound south in Straits. USS *SC-656* is bound northeast in Straits. **Convoy KG-617** is bound southwest in Nicholas Channel towards Cayo Francés. USS *SC-1040* is bound south in Straits towards Cayo Francés. USS *YP-423* is bound west from Nassau towards Andros. Troop transport USS *USAT-Mexico* is bound south, east of Cat Island. **Convoy GN-29** is bound north, in the Atlantic. NAS Great Exuma: Five planes sighted, two twin-engine Army plane, three unidentified planes. A PBM landed in Great Exuma's bay.

23rd December 1942

Air Marshall left by air for Washington via Miami, and party from the delegation at Washington returned. Air Matron and Nursing Sister Cross arrived from Miami by air. SS *Chiloil* is bound south in Straits towards Palm Beach. US Coast Guard Cutter *CG-Crawford* is bound north in Straits towards Palm Beach. SS *Seatrain New Orleans* is bound northwest in Straits towards Port Everglades. SS *Rio Grande* of Argentina west in Straits. Tanker *Sinclair Opaline* is bound northeast in the Atlantic with US Navy plane escorting. NAS Great Exuma: Five unidentified planes are sighted, though one plane was a 4-engine US Army bomber.

24th December 1942

Three flight officers (F/O) and P/O Clark arrived from Miami by air. USS *AM-92* and Argentinian SS *Rio Diamente* are bound south in Straits towards Palm Beach. SS *Chiloil* is bound west in Straits towards Key West. USS *YMS-11,* USS *YMS-51,* USS *YMS-52*, and USS *YMS-53* are bound west in Straits towards Key West. Opening of the United Services Canteen in Nassau, by the Duchess of Windsor, where the forces could unwind. The canteen is located in the Bahamian Club, owned by Mr. Frederick Sigrist, on west Bay Street, was not affiliated with the RAF, US military, or IODE, though it did have a guard of Military Police. [In 2020 the building still looms out of the brush overlooking Arawak Cay, with Ardastra Gardens behind it, derelict for decades but still standing]. NAS Great Exuma: Six planes are sighted including two 4-engine Army bombers and a 2-engine Army bomber.

25th December 1942

One F/O arrives by air from Miami. Station Commander and Officers served the airmen at Christmas Dinner in Airmen's Mess. Aircraft FK-172 and FK-175 arrived at the RAF station. A US Coast Guard Cutter is bound north of Matanilla Shoal, northwest Bahamas. SS *Pen Sun* is bound north in Straits from Palm Beach. Tanker *Esso Augusta* is bound north in Straits passing Port Everglades. USS *USAT-Frank Leisenring* is bound west in Northwest Providence Channel towards Port

Everglades. USS *YP-421* is bound south from Port Everglades. Minelayers USS *ML-470,* USS *ML-484,* USS *ML-485,* and USS *ML-487* are bound south from Port Everglades. Tanker *Goldheels* is bound north in Straits towards Miami. USS *AM-92* is bound south in Straits from Miami. Tanker *Esso Wilmington* is bound southwest in Straits. **Convoy KN-214** is bound northeast in Straits with US Navy plane escorting. USS *PC-565* is bound north in Straits passing Nicholas Cay to Key West. Officers and unit Commander serve airmen a Christmas dinner; a tradition much in favour with enlisted soldiers.

26th December 1942

Tanker *Delaware Sun* is bound south in Straits, soon to pass Daytona. Tanker *Goldheels* and USS *YP-426* are bound north in Straits passing Daytona. **Convoy KN-214** is bound north in Straits passing Canaveral with US Navy plane escorting. SS *American Trader* is bound south in Straits passing Canaveral. Tanker *Gulfpride* is bound north passing Matanilla Shoal, northwest Bahamas. Tanker *Esso Washington* is bound south in Straits passing Palm Beach. SS *Norwalk* is bound north in Straits passing Port Everglades [it was later lost in a collision in the region]. USS *PC-591* is bound north in Straits towards Miami. Tug USS *YT-330* towing gasoline barge USS *YOG-8,* is bound west in Straits towards Key West.

27th December 1942

Aircraft FR-382 arriveds at the RAF station. USS *SC-740*, USS *Aquila* (AK-47), USS *PC-1191*, USS *SC-714,* USS *SC-739*, and USS *SC-1030* are bound south in Straits towards Canaveral. USS *Bristol* and USS *Edison* are bound north in Straits towards Palm Beach. Tanker SS *Delaware Sun* and USS *YP-414* are bound west in Straits towards Key West. Barge USS *YAG-16* is bound south in Straits. USS *USAT-Frank Leisenring* is bound east passing Northwest Providence Channel towards Nassau. NAS Great Exuma: Three 2-engine planes are sighted, one of them was Douglas cargo plane.

28th December 1942

Two Flight Lieutenants, one P/OD, and one F/O arrived at RAF from Miami by sea. **Convoy NK-515** is bound south in Straits towards Canaveral with US Navy plane escorting. USS *Aquila* (AK-47) is bound south in Straits with US Navy plane escorting. SS *Dixiano* is bound north in Straits. SS *Iroquois* is bound west , passing Nicholas Channel. NAS Great Exuma: Three planes are sighted.

29th December 1942

Aircraft FR-381 arrived at the Station. Tug USS *Wellfleet* towing floating drydock is escorted by USS *SC-1024*, USS *YP-420*, USS *YP-426*, and *USCG-369,* which are bound south in Straits passing Daytona with US Navy plane escorting. **Convoy NK-515** and SS *Salvador* are bound south in Straits towards Palm Beach. Tanker *Cherry Valley* is bound southwest in Straits. SS *Iroquois* is bound north in Straits. NAS Great Exuma: One unidentified plane sighted overhead.

30th December 1942

The Governor landed from a B-25 in company with two other aircraft, to open Satellite Field. RAF Base No.111 OTU (Operational Training Unit) is officially opened at Oakes Field by the Governor, HRH Duke of Windsor. The B-25 Bomber was built in US for transport to meet allied war needs in Africa, Europe, and Asia. RAF represented by Senior Leader Vare, Squadron Leader Crawford, and Wing Commander McBratney. Tanker *Gulfland* and tanker *Gulfmaid* are bound south in Straits passing Canaveral. SS *Iroquois* is bound north in Straits passing Canaveral.

Convoy NK-515 is bound southwest in Straits passing Fort Everglades, with US Navy plane escorting. SS *Virginia* is bound south in Straits passing Bimini. Oil barge USS *YO-36* is bound south in Straits towards Old Bahama Channel and Cayo Francés. **Convoy KG-618** is bound southeast in Straits passing Nicholas Channel towards Old Bahama Channel and Cayo Francés with US Navy plane escorting. NAS Great Exuma: Mailboat *Alice Mabel* arrives and leaves for Long Island. Sailing sloop *Joseph B.* arrives and leaves.

31st December 1942

Chief of Operations, Caribbean Wing, American Ferry Command, Morrison Field arrived by Catalina Amphibian at Satellite returned to Miami same day. Number of pupils arrived for course at Operational Training Wing. Flying transferred from Main to Satellite Field today owing to unserviceability. USS *SC-751* is bound south in Straits towards St. Augustine. SS *Markay* is bound northeast in Straits. Tanker *Gulfland* and tug USS *Wellfleet* are bound south towards Port Everglades with US Navy plane escorting. Tanker *Gulfmaid* is bound west in Straits. NAS Great Exuma: Mailboat *Olive Shaw* and a sailing schooner arrived. *Olive Shaw* left.

Part IV

No. 111 O.T.U. to the Rescue: January 1, 1943 to December 30, 1943

1st January 1943

Aircraft (hereinafter just the tail numbers) FR-393 and FR-380 arrive on the station. SS *E. M. Dow* found herself adversely affected by adding deck houses, guns, and other military equipment. Ship's crew had knowledge of this as well as of the recent foundering of the SS *E. M. Wilcox*. Officers and men stuck to their posts as long as possible, commanding officer is believed to have used good judgment in his decision to abandon an unseaworthy vessel, thereby probably saving the lives of the members of his crew. USS *SC-765* and tanker *L. L. Abshire* are bound south in Straits passing Canaveral. SS *Chicopee* is bound north in Straits passing Canaveral, US Navy plane escorting. USS *USAT-Frank Leisenring* is bound south in Northwest Providence Channel towards Nassau. SS *Montana* is bound west in Straits. NAS Great Exuma: HMS *Lotus* anchored near the base. One unidentified plane is sighted.

2nd January 1943

During WWII Blue Lagoon Island, northeast of Paradise Island, Nassau, was requisitioned for a year by the Allies for use as a secret training base for three teams of British and American underwater demolition squads. Squads would swim several nm around the island. Conversion flying on B-25 aircraft started today. Tug USS *Wellfleet* is bound north in Straits towards Canaveral. Tanker *Esso Wilmington* is bound north in Straits towards Canaveral. **Convoy Motor Torpedo Boat Squadron-14, Group A** is bound south in Straits towards Palm Beach. SS *Walter Jennings* and SS *Magdala* are bound southwest in Straits towards Palm Beach. Tanker *Gulfland* is bound south in Straits passing Miami. Tanker *L. L. Abshire* is bound south in Straits. SS *Northern Sun* is bound northeast in Straits towards Miami. Tanker *Esso Washington*, SS *Ormandale*, tanker *Petrofuel*, and SS *Brandywine* are bound north in Straits. **Convoy KN-215** is bound north in Straits, US Navy blimp escorting. First conversion flying courses initiated. NAS Great Exuma: One unidentified plane sighted.

3rd January 1943

Courses in conversion carried out throughout the day. Tanker *Empire Dickens* is bound north in Straits towards Palm Beach with US Navy plane escorting. USS *YAG-20* is bound southwest past Northwest Providence Channel towards Port Everglades. Tanker *Cities Service Fuel* is bound north in Straits, passing Miami navy plane joining from east. USS *YAG-21* is bound southwest from Nassau. NAS Great Exuma: Three unidentified planes sighted.

4th January 1943

Temperature unfit for training by pupils. During the day and afternoon flights are carried out by instructors. USS *PC-1192* is bound south in Straits passing Daytona. USS *YMS-69* is bound north in Straits passing Canaveral. USS *Thrush* is bound southwest in the Atlantic. Tanker *John Worthington* is bound north in Straits passing Bimini. USS *YO-142* is bound west in Northwest Providence Channel towards Port Everglades. USS *AM-92* is bound north in Straits passing Miami. Tug *Thomas E. Moran* and USS *SC-768* are bound southwest in Straits passing Nicholas Channel towards Old Bahama Channel and Cayo Francés. NAS Great Exuma: An Army bomber and one unidentified plane are sighted. Gasoline barge USS *YAG-21* docked near the base.

5th January 1943

Flying training instruction for pupils. No.111 OTU is started, night flying carried out by the instructors. SS *Brandywine* and USS *SC-701* are bound south in Straits towards St. Augustine. **Convoy NK-516** is bound south in Straits with US Navy plane escorting. USS *YP-441* is bound south in Straits passing Canaveral. SS *Mangore* is bound south towards Palm Beach, US Navy plane joining from west. Tanker *American Trader* is bound north passing Matanilla Shoal, northwest Bahamas. Tanker *Chiloil* is bound north in Straits passing Matanilla Shoal, northwest Bahamas navy plane joining from east. Tanker *Bostonian* is bound north passing Matanilla Shoal, northwest Bahamas, US Navy plane joining from east. SS *Brasil* is bound north in Straits passing Port Everglades. Tanker *Delaware Sun* is bound north in Straits. SS *Harrison Smith* is bound north, US Navy plane joining from Cay Sal Bank. USS *Thrush* (AVP-3) is bound southeast, to the east of Bahamas in the Atlantic. NAS Great Exuma: Four unidentified planes are sighted.

6th January 1943

Tanker *Trimountain*, and tanker *Gulfpride* are bound south in Straits towards Daytona. Tanker *Pan Georgia* is bound southeast in Straits towards Daytona. Tanker *Spencer Kellogg* is bound south in Straits towards Canaveral, US Navy plane joining from west. USS *SC-713* is bound north in Straits towards Canaveral. USS *SC-977* is bound south passing Banana River. USS *SC-511* is bound north in Straits passing Palm Beach. Tanker *Virginia* is bound north in Straits. Tanker *Elizabeth Kellogg* is bound south in Straits towards Bimini. **Convoy NK-516** is bound southwest in Straits towards Port Everglades. Tanker *Delaware Sun* is bound northwest near Northwest Providence Channel, US Navy blimp escorting. Irish SS *Irish Elm* is bound west in Straits from Cay Sal Bank. NAS Great Exuma: an Army bomber sighted.

7th January 1943

USS *SC-696* is bound south in Straits from Canaveral. Tanker. SS *Cape Spencer* and USS *YMS-112* are bound north in Straits, passing Palm Beach. Tanker *Cherry Valley* is bound north emerging from Straits. Tanker *Gulfpride* is bound west in Straits. Tanker *Trimountain* and tanker *Pan Georgia* are bound west in Straits. **Convoy NK-516** is bound west in Straits with US Navy blimp. **Convoy KG-619** is bound southeast in Straits towards Nicholas Channel, with US Navy plane escorting. NAS Great Exuma: an Army bomber is sighted.

8th January 1943

USCG *GC-83372* and USCG *GC-83375* are bound south in Straits passing Palm Beach. Irish SS *Irish Elm* and tanker *Spencer P. Kellogg* are bound west in Straits. **Convoy NG-375** is bound southeast of Bahamas. NAS Great Exuma: USCG *GC-Lotus* and USS *YAG-21* stood out of [departed] harbour. An aircraft rescue boat departed to Nassau. A Naval Transport seaplane arrived from Miami and departed to Guantanamo.

9th January 1943

Tanker *Hagood*, tanker *Gulf Caribbean* and SS *Norfolk* are bound south in Straits passing Canaveral. Tug USS *Samaná* is towing USS *Lazybones* north in Straits. SS *Oklahoma* is bound west in Straits. Irish SS *Irish Elm* is bound west in Straits. NAS Great Exuma: An aircraft rescue boat returned from Nassau. One unidentified plane sighted.

10th January 1943

Two B-25 aircraft, FK-180 and FK-171, are airborne from Nassau, then landed at Jamaica. Then FK-171 left Jamaica same day and landed at Nassau that night. SS *Norwalk* is rammed by the Norwegian SS *Nidareid* 50 nm north of Isabela de Sagua, Cuba. SS *Norwalk*'s complement is 40 crew and 14 naval armed-guard; with third engineer lost. Tanker *Umtata* has salvage crew of Merritt, Chapman, and Scott pull it from reef and tow her to Key West. [It was declared a total loss and later towed to Galveston and scrapped in January 1943]. SS *Christy Payne*, SS *George Barnes*, SS *Sparrows Point*, and USS *SC-1029* are bound south in Straits passing Canaveral. USS *Opal* (PYC-8) and tanker *Goldheels* are bound south in Straits, passing Palm Beach. USS *USAT-Frank Leisenring* is bound west in Straits from Bimini towards port Everglades. Tanker *Esso Wilmington* is bound south in Straits from Bimini. USS *YAG-19* is bound north in Straits from Miami. Tanker *Hagood* is bound south in Straits. *R. P. Smith* is bound north in Straits. **Convoy HK-132** is bound north in Straits, with US Navy blimp. **Convoy GK-720** is bound northwest from Nicholas Channel towards Key West. NAS Great Exuma: One unidentified plane sighted.

11th January 1943

USS *Apache* (AK-67) is bound north in Straits passing Palm Beach. **Convoy KN-216** is bound north in Straits near Matanilla Shoal with US Navy plane escorting. SS *Swivel* is bound north in Straits towards Palm Beach. SS *Christy Payne* is bound south towards Port Everglades. Tanker *L. L. Abshire* is bound north in Straits from Bimini. Tanker *Esso Washington* is bound south in Straits from Bimini. SS *George W. Barnes* is bound southwest in Straits. SS *Flight Command* is bound north in Straits. SS *Vigil* and USS *SC-1119* are bound south towards Nicholas Channel. NAS Great Exuma: One unidentified plane sighted. Patrol Seaplane arrived from and then returned to Guantanamo.

12th January 1943

B-25 FK-180 is airborne from Jamaica and landed at RAF Nassau. Four new aircraft arrived. FK-883 and FK-379 landed from Homestead Air base, Florida, FK-374, and FK-394 from Dorval, Florida. SS *Ormondale* is bound south in Straits towards Canaveral. Tanker *Gulf Caribbean* is bound north in Straits. HMS *Phoebe* and escorts are bound north in Straits passing Palm Beach with US Navy plane escorting. SS *C. A. Canfield* and SS *Flight Command* are bound north in Straits towards Palm Beach. SS *Arizona* is bound south in Straits. **Convoy NG** is bound south, to the east of Bahamas, in the Atlantic. NAS Great Exuma: One unidentified plane is sighted.

13th January 1943

Training flights are carried out by pupils during day. Ventura 895 is airborne at night for Miami. SS *Flight Command* is bound north in Straits towards St. Augustine. USS *YMS-36*, USS *SC-692*, and SS *Garnet Huling* are bound south in Straits towards Canaveral. SS *W. H. Ferguson* is bound south in Straits towards Palm Beach. USS *SC-731*, SS *S. B. Hunt 75*, and SS *Virginia* are also south in Straits towards Palm Beach. SS *Magdala 75* is bound south in Straits passing Palm Beach. SS *Tustem*, and SS *Drupa* are bound south in Straits towards Port Everglades. SS *Pueblo* is bound north in Straits towards Port Everglades. SS *Cistula* is bound south in Straits. SS *Ormondale* is bound west in Straits. SS *Empire Might* and USS *SC-698* are bound north in Straits. **Convoy KW-95** is bound west in Nicholas Channel. USS *USAT-Frank Liesenberg* and USS *YP-427* are bound south towards Nassau. NAS Great Exuma: One unidentified plane and an Army bomber sighted.

14th January 1943

Training flights carried out by pupils during day, night flying exercises completed by the instructors. Liberator FL-914 landed from Trinidad due to be airborne for Dorval, detained due to faulty oil feed pipe. No.111 OTU passed its first pupil pilot who qualified solo. NAS Great Exuma: One unidentified plane and an Army bomber sighted.

15th January 1943

Training flights carried out by pupils of No.111 OTU from Main Field. Instructors continued training from Satellite Field. Liberator 23694 landed from Morrison Air Field [in west Palm Beach]. Tanker *Turret Capes* is bound north in Straits. Tanker *Chiloil* is bound south in Straits towards Palm Beach. Tanker *Oklahoma* is bound north in Straits towards Palm Beach. Tanker *E. G. Suebert,* tanker *Pan North Carolina*, and tanker *Virginia Sinclair* are bound south in Straits. Tanker *James Hawson* is bound west in Straits. **Convoy NK-517** is bound west in Straits with US Navy plane escorting. USS *Albemarle* (AV-5) is bound north in the Atlantic, east of Bahamas. NAS Great Exuma: Naval Transport seaplane arrived from Miami and departed to Guantanamo. Army bomber sighted.

16th January 1943

Training flights carried out during day. Liberator FL-914 airborne and landed at La Guardia Airport New York and aircraft B-25 41/11702 landed from Morrison Field. Tanker *British Prestige* is in collision with another vessel while in **Convoy GN-34**, from Guantanamo to New York, accident happened on approach to New York City. USS *SC-715* south towards Canaveral. Tanker *E. L. Doheny,* USS *SC-985*, and SS *Pocahontas* are bound south in Straits towards Palm Beach. SS *Elizabeth Kellogg* is bound south in Straits passing Palm Beach. Tanker *Gulf Caribbean* is bound north in Straits passing Port Everglades. Tanker *Eagle* is bound north in Straits towards Port Everglades. SS *Seatrain Nola* is bound north in Straits towards Miami. Tanker *Esso Baton Rouge* is bound north in Straits. USS *SC-981* and USS *SC-637* are bound northwest in Straits. **Convoy NG-336** is bound south in the Atlantic, east of Bahamas.

17th January 1943

Aircraft B-17 23831 is airborne for Puerto Rico. USS *PYC-41* south towards Canaveral. SS *Algonquin* is bound south in Straits passing Canaveral. Tanker *Pan Georgia* and SS *Spencer Kellogg* are bound north towards Matanilla Shoal, northwest Bahamas. SS *Cockerel* and SS *Wanks* are bound north out of Straits. USS *YAG-19* is bound northwest in Straits. **Convoy GK-721** is bound west towards Nicholas Channel. NAS Great Exuma: Four transport planes sighted.

18th January 1943

Aircraft C-43 25227 landed from Morrison Field. Tanker *Delaware Sun* and SS *American Trader* are bound south in Straits passing Canaveral. Argentinian SS *Rio Colorado* is bound north in Straits towards Canaveral. Tanker *Esso Norfolk* is bound northeast in Straits passing Canaveral. USS *PYC-41* is bound south in Straits towards Miami. SS *Sparrows Point* and SS *Mississippi* are bound northeast in Straits from Miami. USS *PC-466* and SS *Polonaise* are bound south in Straits from Miami. Tanker *Cities Service Missouri* is bound northeast in Straits from Miami. **Convoy KN-217** is bound northeast in Straits from Miami with US Navy plane escorting. SS *Acavus* is bound southwest in Straits. SS *Castana* is bound northeast in Straits. NAS Great Exuma: Naval Transport seaplane arrived from Guantanamo and departed to Miami. USS *YAG-20* arrived. A 4-engine transport plane sighted.

19th January 1943

A B-25 is airborne to search for aircraft A-20 reported down at sea, landed at Satellite and carried out fruitless search. Liberator 23694 airborne for Puerto Rico. *U-214* enters area north bound via Mona Passage. USS *SC-1025* and USS *SC-741* are bound south in Straits towards Canaveral. USS *SC-997* is bound south in Straits passing Canaveral. SS *Castana* is bound northeast in Straits near Canaveral. SS *Pollycastle* and tanker *Delaware Sun* are bound southwest in Straits towards Palm Beach. SS *Walter Jennings* is bound southwest in Straits towards Port Everglades. USS *PYC-41* is bound south in Straits passing Port Everglades. USS *YP-437* is bound northwest in Straits towards Miami. SS *American Trader* is bound west in Straits. Tanker *Esso Hartford* is bound west in Straits near Cay Sal Bank. NAS Great Exuma: Three Army bombers sighted.

20th January 1943

Towing aircraft DC-3 and its glider landed on the Main Field airborne again in the evening training flights continued from Main Field by pupils and from Satellite by instructors. USS *AMC-67* is bound south towards Daytona. USS *ARS-31*, USS *AM-93*, and SS *Colina* are bound south in Straits towards Daytona. USS *SC-714* is bound north towards Daytona. USS *SC-997* is aground near Daytona. USS *SC-1016* is bound south in Straits towards Canaveral. USS *YP-426* is bound south in Straits towards Palm Beach. SS *Walter Jennings* is bound southwest in Straits near Bimini. Tanker *Goldheels* is bound west in Straits. USS *SC-996* and SS *G. W. Barnes* are bound north in Straits. USS *YO-142* is bound southeast near Cay Sal Bank towards Cayo Francés and Old Bahama Channel. USS *YAG-20* is bound northwest off Exuma towards Nassau. **Convoy NG-337** is bound south in the Atlantic, east of Eleuthera. *U-214* is bound north of San Juan, bound east to France. NAS Great Exuma: USCG *CG-YAG-20* arrives, and an Army bomber is sighted.

21st January 1943

Aircraft FR-372 landed from Dorval, Montreal. Report received from Bahamas mailboat MV *Paddy Halfready* is overdue at Nassau had left from Cornwall, Abaco. Aircraft B-25 FL-184 brought to readiness, report received that boat had arrived at Nassau, signal to return to base was sent to aircraft landed back at Main Field later found that motorboat was located 30 nm west of Mayaguana. **Convoy NK-518** is bound south in Straits towards Canaveral. SS *Christy Payne* and SS *G. W. Barnes* are bound north Off Canaveral. Tug USS *Willett* and tanker *Esso Washington* are bound north towards Palm Beach. Tanker *18 de Marzo* is bound north in Straits towards Palm Beach. SS *Swivel* is bound south in Straits passing Palm Beach. USS *YAG-20* is bound west in Northwest Providence Channel to Port Everglades. SS *Thorshovdi* is bound south in Straits passing Port Everglades. SS *William Boyce Thompson* is bound north in Straits off Port Everglades. SS *Colina* and USS *AM-93* are bound southwest in Straits. SS *W. H. Ferguson* is bound north in Straits. USS *YP-437* is bound southeast, to the east of Cat Island in the Atlantic. *U-214* is bound northeast of Puerto Rico, east of Bahamas south to exit via Mona Passage. MV *Paddy Halfready* missing, drifted from Cornwall Abaco to 30 nm west of Mayaguana, a distance of over 350 nautical miles, nearly the length of the colony.

22nd January 1943

Squadron leader awarded Air Force Cross. Training flights continued from Main Field and Satellite Field. USS *SC-683* is bound south towards Port Everglades. **Convoy NK-518** is bound south in Straits near Port Everglades with US Navy blimp. USS *YAG-19* is bound northwest in Straits towards Miami. SS *Arizona* and SS *Tustem* are bound north out of Straits. USS *YO-142* is bound northwest in Straits. *U-214* exits area north of Anegada Passage before southeast then northeast to Europe. NAS Great Exuma: Naval Transport seaplane arrived from Miami and departed to Guantanamo.

23rd January 1943

Liberator landed from Dorval airborne subsequently for Trinidad. Tug *Edmund P. Moran* is bound north in Straits passing Canaveral. USS *YP-429* is bound south in Straits towards Palm Beach. Tanker *Virginia* is bound north in Straits passing Palm Beach. Tanker *S. B. Hunt* is bound north in Straits near Port Everglades. Tanker *Penn Sun* is bound south in Straits near Port Everglades. SS *Goldheels* and tanker *Arizona* are bound north in Straits. USCG *CG-Spruce* is bound northwest in Straits towards Miami. USS *USAT-Frank Liesenring* is bound south in Northwest Providence Channel towards Nassau. Tanker *Esso Philadelphia* and **Convoy SS *Salamonie*** are bound southwest in Straits towards Palm Beach.

24th January 1943

Tug USS *Willett* is bound south in Straits towards Canaveral. SS *Goldheels* is bound north in Straits towards Canaveral. Tanker *Pan Virginia* and SS *Empire Opal* are bound south in Straits passing Palm Beach. Tanker *Ormondale* and SS *Cape Saint Elias* are bound north in Straits from near Bimini. Spanish SS *Aldecoa* is bound northeast in Straits towards Bimini. **Convoy Motor Torpedo Boat Squadron-14, Group B** is bound north in Straits. SS *Oklahoma* is bound west in Straits. NAS Great Exuma: USS *Greer* arrived and two unidentified planes sighted. Naval Transport seaplane arrives from Guantanamo and departs to Miami.

25th January 1943

B-25 FR-382 is airborne for Mobile landed later. Two Catalina FP-316 and FP-318 are airborne *en route* for Bermuda. New Control tower commenced operation located on northwest side of the runways. USS *PC-608* and tanker *L. L. Abshire* are bound south in Straits passing Canaveral. Tanker *Ormondale* and tanker *Esso Wilmington* are bound north in Straits passing Canaveral. **Convoy Motor Torpedo Boat Squadron-14, Group B** is bound north towards Canaveral. USS *YMS-38* is bound south in Straits passing Palm Beach. Tug USS *Willett* is bound south in

Straits passing Port Everglades. Tanker *Pan Virginia*, USS *YT-194*, tanker *Gulf Caribbean* and SS *Empire Opal* are bound west in Straits. NAS Great Exuma: Two 2-engine Army bombers and an unidentified plane sighted.

26th January 1943

Catalina West-430 is waterborne. Six Albacores, BF-6161, BF-770, BF-615, BF-657, and X-9185, landed from Morrison Field *en route* for RNAS [Royal Navy Air Station] Trinidad. AO's other naval vessels are bound south in Straits towards Palm Beach with US Navy plane escorting. USS *USAT-Liesenring* and tanker *Alpha* are bound north in Straits passing Palm Beach. Tanker *Watertown* and tanker *Bostonian* are bound south in Straits passing Palm Beach. **Task Unit TU-909.10** is bound south in Straits passing Port Everglades. USS *SC-526* is bound north in Straits. Tanker *L. L. Abshire* is bound west in Straits. **Convoy KN-218** is bound east in Straits with US Navy blimp. **Convoy GK-122** is bound north in Straits towards Key West with US Navy plane escorting. NAS Great Exuma: Four unidentified planes sighted.

27th January 1943

Two Catalinas, FP-316 and FP318, are airborne for Bermuda. Six Albacores airborne for Gamaquey, Cuba, *en route* for Trinidad. Night flying carried out from Satellite Field. USS *AM-123* and six USS Sub Chaser's are bound south in Straits passing Canaveral. **Convoy KN-218** is bound north in Straits passing Canaveral with US Navy blimp. SS *Elizabeth Kellogg* is bound north, to the west of Matanilla Shoal, northwest Bahamas. SS *E. L. Doheny* is bound north in Straits out of Northwest Providence Channel. Tanker *Bostonian* is bound southwest in Straits off Miami. USS *Icarus* (YAG-16) is bound south in Straits. NAS Great Exuma: Two Army bombers and an unidentified plane are sighted.

28th January 1943

Catalina west-430 airborne and landed at Bermuda B-25 FL-697 landed from Borinquen. SS *Rimutaka* is bound northeast near Matanilla Shoal, northwest Bahamas. SS *Colina* is bound northeast north of Bimini. SS *Joaquin Miller* and SS *James Hawson* are bound northeast in Straits passing Miami. SS *American Trader* is bound northeast in Straits south of Miami. NAS Great Exuma: An Army bomber and an unidentified plane sighted.

29th January 1943

Liberator AL-590 airborne and landed at Dorval. B-25 FR-382 returned from Mobile landed at Satellite Field. Night flying carried out day flight done from both Main Field and Satellite Field. USS *Broadbill* (AM-58) is bound south in Straits towards Palm Beach. HMS *Hunter* and *USCG-7405* are bound north in Straits, passing Port Everglades with US Navy plane escorting. SS *Zwarteezee* is bound west in Straits. USS *YT-194* towing USS *YF-363* is bound southeast towards Cayo Francés and Old Bahama Channel. NAS Great Exuma: An Army bomber and an unidentified plane sighted. Naval Transport seaplane arrived from Miami and departed to Guantanamo.

30th January 1943

Training of operational crew continued during day and local night flying practice carried out. SS *Wayfarer* and tanker *Pacific* are bound north in Straits towards Palm Beach. Tanker *W. H. Ferguson* is bound south in Straits passing Palm Beach. Crash boat USS *C-1030* is bound north in Straits. Tanker *Salvator* is bound west in Straits. USS *YAG-19*, USS *YP-160*, USS *AM-93*, SS *Ogna*, SS *Cordillera,* and four merchant ships are bound north in Straits. SS *William Ellery* and SS *Empire Keats* are bound north in Old Bahama Channel from Cayo Francés. NAS Great Exuma: Two unidentified planes are sighted.

31st January 1943

B-25 FR-376 arrived from Homestead base, Florida. Tanker *W. B. Thompson* and tanker *Christy Payne* are bound southeast nearing Canaveral. **Convoy SS *Merrimac*** is bound north in Straits. Tanker *Delaware Sun*, seven independent steam ships, and tanker *Liebre* are bound south in Straits. Tanker *Penn Sun* is bound north in Straits. Tanker *Birkenhead* is bound north in Straits passing Palm Beach. Ships for New York are bound north passing Matanilla Shoal, northwest Bahamas. **Convoy NK-519** is bound south in Straits towards Port Everglades with US Navy blimp. SS *Empire Keats,* USS *YAG-16*, and SS *Park Haven* are bound north in Straits near Port Everglades. **Convoy KN-219** is bound north in Straits navy plane escorting. Tanker *Esso Philadelphia* and SS *Egba* are bound north in Straits towards Bimini. Tanker *Gulfmaid* is bound west in Straits. Tanker *Cities Service Ohio* is bound east in Straits. **Convoy KG-622** is bound south in Nicholas Channel with US Navy blimp. NAS Great Exuma: An unidentified plane sighted. Naval Transport seaplane arrived from Guantanamo and departed to Miami.

1st February 1943

B-25 FR-372 is airborne and landed at Havana. Catalina FP-532 landed at Main Field from Dorval, Montreal. Night flying at Satellite pupil flight training during daylight hours. *U-68* is in Windward Passage: damage caused delay, so moving to north of Bahamas for repairs. Destroyer USS *Barney*, tanker *Cities Service Ohio*, and SS *Egba* are bound north in Straits passing Canaveral. USS *YP-444* is bound south in Straits passing Canaveral. USS *YAG-20* is bound southwest in Straits towards Port Everglades. Tanker *Swivel* is bound north in Straits. Tanker *Oklahoma* is bound north in Straits from near Bimini. Irish SS *Irish Elm* is bound northeast in Straits. Tanker *Liebre* is bound west in Straits. NAS Great Exuma: One unidentified plane sighted.

2nd February 1943

B-25 FR-372 returned from Havana. Flight training exercise by day and night carried out. Tanker *Esso Hartford* is bound southwest in Straits. Irish SS *Irish Elm* is bound north, to the north of Matanilla Shoal. USS *SC-1016* is bound north in Straits passing Palm Beach. USS *Icarus* (YAG-16) is bound west in Northwest Providence Channel. SS *Fred R. Kellogg* is bound north in Straits passing Bimini. Tanker *R. P. Smith* is bound north in Straits passing Miami. Tanker *Petrofuel* and US Navy fast torpedo boats USS *PT-199,* USS *PT-200*, and USS *PT-93* are bound north in Straits. NAS Great Exuma: One unidentified plane is sighted.

3rd February 1943

Flight training exercise by day and night carried out. Report received that civil aircraft flying between Miami and Nassau had sighted a drifting boat with three survivors US Navy has requested an air search, B-25 FR-383 airborne for fruitless search and landed back. USS *YDG-1* south towards Daytona. **Convoy Motor Torpedo Boat Squadron-15** is bound north towards Canaveral. **Convoy NK-520** is bound south in Straits passing Canaveral navy plane escorting. Tanker *Virginia* is bound west in Straits. Civilian plane reported seeing a drifting boat with three occupants, B-25 B-8383 searched. NAS Great Exuma: Two 2-engined Army planes were sighted.

4th February 1943

A Ventura B-34 of USAAC. landed from Boorator Field. USS *SC-767* is bound southeast nearing Canaveral. Tanker *18 de Marzo* and SS *Tampico* are bound south in Straits passing Palm Beach. SS *Polycastle* and SS *Atascosa* are bound north in Straits passing Palm Beach. SS *Santa Cecelia* is bound southwest in Straits near Bimini. SS *Alcoa Patriot* is bound north in Straits near Port Everglades. USS *YDG-1* and three US Coast Guard vessels are bound west in Straits. **Convoy NK-520** is bound west in Straits escorted by a navy blimp. NAS Great Exuma: Two 2-engined Army planes sighted.

5th February 1943

Liberator 11660 FL-993 landed. Liberator FL-992 landed from Homestead Air Base, not fitted with special equipment. Liberator FL-942 and FL-964 landed from Homestead, Florida fitted with special equipment. SS *G. W. Barnes* is bound south in Straits nearing Daytona. SS *Arizona* south in Straits towards Palm Beach. **Convoy KW-106** and SS *Polycastle* are bound north in Straits towards Palm Beach. Tanker *Pan Virginia*, USS *YDG-1* and three Coast Guard vessels are bound south in Straits passing Port Everglades. SS *Watertown* is bound north in Straits nearing Port Everglades. USS *Turaco* (AMC-55) is bound north in Straits towards Miami. SS *Spencer Kellogg* and tanker *Esso Rochester* are bound south in Straits. SS *Hawaiian Planter* is bound north in Straits. USS *Icarus* (YAG-16) is bound south in Northwest Providence Channel towards Miami. NAS Great Exuma: Five unidentified planes sighted.

6th February 1943

Hudson FK-400 and B-25 A3-107 landed from Morison Field, in West Palm Beach. Liberator FL-992 also landed from Morrison Field. USS *YN-29* is bound south in Straits passing Canaveral. **Convoy KN-220** is bound north in Straits with US Navy blimp. SS *Cape Cod* is bound south in Straits towards Palm Beach. SS *G. W. Barnes* is bound south in Straits towards Port Everglades. Tug *Swatrzeezee* is bound south in Straits passing Miami. SS *Columbia Star* is bound north in Straits passing Miami. SS *Spencer Kellogg* is bound southwest in Straits. SS *Arizona* is bound west in Straits. NAS Great Exuma: Two unidentified planes sighted.

7th February 1943

Flight training throughout the day. B-25 FR-372 airborne and landed at Havana. SS *America Trader* is bound south in Straits. British Minesweepers HMS *YMS-5*, HMS *YMS-6*, HMS *YMS-7*, and HMS *YMS-8* are bound south in Straits passing Canaveral. British rescue tug HMS *HMRT-Friskey* and USCG *GC-83389* are bound south in Straits towards Palm Beach. SS *W. M. Eustis* and SS *New Zealand Star* are bound north out of Straits. USS *SC-741* is bound west in Straits. NAS Great Exuma: Two unidentified planes sighted. A 3-masted sailing schooner anchored off Stocking Island. Mailboat *Alice Mabel* enters harbour and then leaves for Long Island.

8th February 1943

Flight training throughout the day and night. B-25 FR-373 lands from Havana. Tug USS *Montcalm* is bound south in Straits passing Canaveral. USS *SC-711* is bound southwest in Straits towards Palm Beach. USS *PC-589* is bound southwest in Straits towards Port Everglades. SS *Garnet Hulings* is bound southwest in Straits. SS *Larchmont* is bound northeast in Straits passing Bimini. Tanker *Gulftide* is bound southwest in Straits. SS *American Trader* is bound south in Straits towards Cay Sal Bank. SS *Orart* is bound north in Straits. NAS Great Exuma: One unidentified plane sighted. The schooner entered the harbour.

Mailboat *Alice Mabel* anchors in Georgetown.

9th February 1943

Flight training throughout the day. USS *Arapaho* (AG-68) is bound south in Straits towards Canaveral. SS *Larnleetarn* is bound north in Straits passing Canaveral. USS *Maricopa* (ATR) is the north. Ten USN USS *LCI*'s (landing craft, infantry), USS *Apache*, SS *Empire Airman*, tug *Thomas E. Moran*, and USS *YP-428* are bound south in Straits passing Palm Beach. Tanker *Gulftide* is bound west in Straits. USS *Icarus* (YAG-16) is bound south in Straits. SS *Benjamin Franklin*, SS *W. H. Ferguson*, and SS *Eli Whitney* are bound north in Straits. Tug USS *Montcalm* is bound south in Northeast Providence Channel towards Eleuthera. NAS Great Exuma: Tug USS *YAG-20* anchored.

10th February 1943

Flight training throughout the day. B-25 FR-393 is airborne to search for lost Havocs of USAAC off Andros; they located survivors and landed Catalina of USN standing by SS *Thornaby*, bag dropped to survivors. An ASR (air sea rescue) launch from Lyford Cay was dispatched to Andros, recalled before destination reached. SS *Goldheels* is bound south in Straits. Tanker *Gulfland* is bound north in Straits towards Canaveral. SS *E. L. Doheny* south in Straits passing Canaveral. SS *Virginia* and tanker *Esso Hartford* are bound north in Straits. SS *Benjamin Franklin* and SS *Eli Whitney* are bound north in Straits towards Canaveral. **Convoy NK-521** is bound south in Straits towards Palm Beach with US Navy plane escorting. **Convoy KN-221** is bound north in Straits passing Matanilla Shoal, with US Navy blimp. SS *Port Alma* is bound north in Straits towards Palm Beach. SS *Toteco* is bound southwest in Straits towards Port Everglades. SS *Garnet Hulings* is bound south in Straits passing Port Everglades. SS *Alpha* and *PC-617* are bound southwest in Straits. SS *Christy Payne* is bound north out of Straits. SS *Trimountain* north out of Straits. USS *Arapaho* (AT-68) and USS *SC-1017* are bound west in Straits. Army tug USS *Kidwell* is bound northwest in Old Bahama Channel past Cayo Francés. NAS Great Exuma: Three masted schooner *Island Belle* arrived and then left harbour.

11th February 1943

Flight training throughout the day and night. USS *SC-733* is bound south in Straits towards St. Augustine. Tanker *Delaware Sun* and USS *SC-1026* are bound south in Straits towards Canaveral. Tanker *Toteco*, tanker *Goldheels*, and tanker *Chilbar* are bound south in Straits towards Palm Beach. Tanker *Tydolgas*, tanker *Salvator*, and tanker *Liebre* are bound north in Straits towards Palm Beach. USS *YO-142* and USS *YP-433* are bound east towards Walker Cay, north of Abaco. Tug USS *Samaná* (YAG-28) is bound south towards Nassau. British Royal Navy minesweepers HMS *YMS-5*, HMS *YMS-6*, HMS *YMS-7*, and HMS *YMS-8* are bound southwest in Straits. SS *Winona* and SS *City of Savannah* are bound northwest in Old Bahama Channel near Cayo Francés. SS *Apache* and 20 USN USS *LCI*'s (landing craft, infantry) are bound west in Straits. Tanker *Clio* is bound northwest in Straits passing Nicholas Channel. NAS Great Exuma: Two Army transports and tug USS *YAG-20* are sighted.

12th February 1943

Flight training throughout the day. Liberator FL-968 and FL-962 landed from Homestead, Florida latter unserviceable after landing. USS *YMS 237* and USS *YMS 238* are bound south in Straits towards Palm Beach. Tanker *Gulfpride*, SS *Kotagede* are bound northwest in Straits passing Miami. Tanker *Cities Service Kansas* is bound northwest passing Bimini. Tanker *Delaware Sun* and tanker *Goldheels* are bound southwest in Straits. SS *Royal Star* is bound northeast in Straits. USS *Icarus* (YAG-16) is bound north in Straits. British Royal Navy Motor Torpedo Boats HMS *MTB-333* and HMS *MTB-343* are bound north in Straits. SS *Fort Lajoie* is bound north in Straits. USS *YN-29*, USS *YF-294*, USS *YP-442* and USS *YP-444* are bound south in Straits towards Old Bahama Channel. NAS Great Exuma: One unidentified plane sighted. *Island Belle* anchored in the bay.

13th February 1943

Heavy rainfall early morning and intermittently until early afternoon, so flying was curtailed during heavy rainfall, after which training was carried out during the afternoon. Tanker *Esso Aruba* returns to service after it was damaged by *U-511.* NAS Great Exuma: *Island Belle* stood out to sea. Three Army bombers are sighted.

14th February 1943

Flying continued through the day. Liberator FL-908 is airborne for Borinquen. NAS Great Exuma: Four unidentified planes and two Army transports are sighted.

15th February 1943

Flying continued through the day night flying exercises were carried out. Tanker *Virginia* is bound south in Straits towards Canaveral. Tanker *Polonaise* south in Straits passing Canaveral. Tanker *Pacific* and tug USS *Okisko* are bound south in Straits passing Palm Beach. HMS *Talker* is bound north in Straits passing Bimini, with US Navy plane escorting. **Convoy KN-222** is bound north in Straits with US Navy blimp. Tanker *Manuel Rionda* is bound north in Straits. RAF Base No.113: Transfer of RAF FC Detachment from Nashville, Tennessee to open up HQ, south Atlantic Wing, Nassau commenced. NAS Great Exuma: Two unidentified planes and 14 B-25 bombers were sighted.

16th February 1943

B-25 FR-382 is airborne for Havana returned to Main Field later. B-25's FR-383 and FR-375 carried out navigational exercise east side of Bahama Islands, from Sail Rock, Exumas, north of Ship Channel Cay, Exuma Cays, to Rum Cay; bombs dropped on sea marker off Sail Rock, Exumas first navigational exercise carried out by pupils of this unit. USS *SC-1017* is bound northwest in Straits passing St. Augustine. SS *William Mulholland* is bound north in Straits, passing Canaveral. Tanker *Britamsea* is bound south in Straits passing Canaveral. SS *Arizona* is bound north in Straits near Canaveral. SS *G. W. Barnes* is bound northeast in Straits near Canaveral. SS *Virginia* is bound southwest in Straits towards Palm Beach. Navy tug USS *Okisko* is bound south in Straits, passing Palm Beach. SS *Polonaise* is bound southwest in Straits towards Port Everglades. Tanker *Petrofuel* is bound west north of Nassau towards Northwest Providence Channel. SS *Pacific* is bound south in Straits. SS *Manuel Rionda* is bound southwest in Northeast Providence Channel towards Nassau.

17th February 1943

Flying training done through the day circuits and landings done by two aircraft at night. B-25 FR-383 airborne for navigational exercise course set for Sail Rock, Exumas endeavoured to release 500-pound bomb. They released bomb on reaching destination. Two attempts to attack target carried out but both went unsuccessful, request sent to base to continue flight and to jettison the bombs was carried out, gunnery practice carried out on navigation completion, they returned to base at night. Same exercise done by B-25 FR-375 as aircraft dropped a T-bomb successfully at Sail Rock, Exumas, made dummy runs over target. A convoy was spotted 6 nm from port bow, aircraft approached within 5 nm, convoy was 28 ships with an escort aircraft, set course for base, dropped sea marker and two 100 pound practice bombs, air firing exercises completed when it landed at base. It was noticed that starboard wing was punctured, cause could be bomb fragments. B-25 FR-376 airborne for night flight to Miami, returned late that night. Tanker *Gulfport* is bound south in Straits passing Canaveral. Tanker *Virginia* is bound southwest in Straits. USS *AMC-94*, USS *AMC-95*, and USS *AMC-96* are bound south in Straits passing Palm Beach. USS *SC-978* is bound west in Straits from Bimini. Tanker *Spencer Kellogg* and SS *W. A. Graham* are bound north in Straits. NAS Great Exuma: One Army transport is sighted.

18th February 1943

Flying training done through the day, night flying exercises carried out from Main Field. B-25 FR-382 is airborne for Sail Rock, Exumas dropped bomb 50 yards to starboard of target, navigation exercise carried out between Sail Rock, Exumas and Rum Cay followed by gunnery practice, then it landed at base at night. B-25 FR-374 is airborne from Main Field on a visual reconnaissance at Harbour Island, reconnaissance made of Harbour Island off to Exuma, and a visual reconnaissance made. Landed at base, homing exercise carried out prior to landing. B-25 FR-3823 airborne from Main Field on escort duty for tanker *Hercules* from Nassau to Great Stirrup Cay Light, Berry Islands, circled tanker and left patrol area, then landed back at base. **Convoy NK-523** is bound south in Straits near Daytona with US Navy plane escorting. SS *Taybank*, SS *Tulsa*, SS *W. C. Endicott*, and SS *W. A. Graham* are bound north in Straits passing Canaveral. USS *SC-977* is bound south in Straits near Palm Beach. SS *Zaanland* is bound south in Straits near Palm Beach. SS *Schofield* is bound north in Straits passing Palm Beach. SS *Zebulon Pike* is bound north in Straits, approaching Palm Beach. Tanker *Gulftide* is bound north, near Matanilla Shoal. SS *Britamer* is bound north in Straits near Matanilla Shoal, northwest Bahamas. SS *American Trader* is bound north in Straits near Bimini. SS *John Steele* is bound north, out of Straits. SS *Port Auckland* is bound north out of Straits. SS *Antar* is bound north in Straits. SS *Steel Inventor* is bound north in Straits. Tanker *Gulfport* is bound west in Straits. Tanker *Esso Hartford* is bound west near Cay Sal Bank. SS *Ganymede* is bound northwest in Old Bahama Channel near Cayo Francés. Tanker *Hercules* provided with air escort from Nassau to Great Stirrup Cay Light, Berry Islands, on Northwest Providence Channel by B-25 FR-383 T. NAS Great Exuma: Two unidentified planes and one Army bomber sighted.

19th February 1943

B-25 FR-374 airborne from Main Field set course for Sail Rock, Exumas gunnery and bombing exercises carried out both practice bombs dropped on wave crests 300 rounds from front guns and 400 rounds from rear guns fired landed back at base after carrying out a descent through cloud exercise. B-25 FR-173 K is airborne for visual reconnaissance of Exuma, Eleuthera, and Harbour Island, after satisfactory exercise landed back at base after carrying out a descent through cloud exercise. B-25 FR-381 airborne for escort duty to SS *Manuel Rionda*, escorted until

Abaco Point, then ship left steaming on constant course. SS *George Washington* is bound southwest in Straits. USS *PC-559*, USS *PYC-42*, and USS *SC-1026*, are bound south towards Canaveral. SS *Port Auckland*, SS *Antar*, and SS *Fort Anne* are bound north in Straits towards Canaveral. SS *Steel Inventor* is bound north in Straits passed Palm Beach. **Convoy NK-523** is bound south in Straits with US Navy blimp off Palm Beach. SS *Ganymede*, SS *Moses Rodgers*, US Army tug USS *ST-40* and SS *John Steele* are bound north in Straits near Palm Beach. SS *Empire Galahad* is bound north in Straits, passing Port Everglades. SS *Ship 108,* SS *Empire Dunstan,* and tanker *Bostonian* are bound north in Straits. SS *Gavealoide* is bound north in Old Bahama Channel, passing Cayo Francés. NAS Great Exuma: A Sikorsky landed and departed. Schooner *Island Belle* proceeds to sea via West Passage. One B-25 is sighted.

20th February 1943

B-25 FR-378 airborne for navigation exercise set course for Sail Rock, Exumas dropped 500-pound SAP bomb on foreshore. Attempt made to send signal to base found its radio was unserviceable, returned to base after gunnery exercise. B-25 FR-374 is airborne for navigation Exercise Number 2, set course for Sail Rock, Exumas and carried practice for Anti-Submarine attacks against sea marker depth charges hung up aircraft landed with depth charges still in the racks. B-25 FR-376 CK is airborne for visual reconnaissance of Exuma, Eleuthera, and Harbour Island, landed back at base after carrying out a descent through cloud exercise. SS *Empire Dunstan*, and SS *Bostonian* is bound northeast in Straits. SS *Cacalilao* is bound southwest in Straits towards Palm Beach. USS *SC-693* and US Army tug USS *LT-115* and south in Straits towards Port Everglades. SS *China Mail* is bound northeast near Northwest Providence Channel. SS *George Chamberlain* and SS *Black Point* [later the final ship ever sunk by a German U-boat], and USCG *CG-Vigilant* are bound northwest towards Cay Sal Bank. **Convoy NK-523** is bound southwest in Straits with US Navy blimp. Ferry *Ena K.* and SS *Tekoa* are bound east in Straits. NAS Great Exuma: One B-25 sighted. A three masted schooner sighted east of bay.

21st February 1943

Thirteen new crew arrived for Number 3 course at No.111 OTU. B-25 FR-381 airborne to escort Ferry *Jean Brillant*, failed to meet, recall signal sent as information received that ship had arrived at Nassau. B-25 FR-383 T is airborne on navigation Exercise Number 7, 500-pound SAP Bomb dropped on target at Sail Rock, Exumas oil patch sighted at [coded position] sea marker and two practice bombs dropped there, streak was 60,100 feet in width and irregular curves running northwest to southeast not possible to determine which was front and if the streak was moving. Report received that U-boat had been sighted off San Salvador diverted to search the area, signal not received by aircraft area not searched continued oil patch investigation convoy sighted and aircraft challenged when 3 nm to 4 nm away Aldis lamp not serviceable so replied with recognition cartridge, landed back at base late evening. B-25 FR-378 airborne submarine attack carried out by dropping depth charges and a practice bomb on a sea marker. Diversion signal received to proceed to last reported position of submarine to carry out square search no sighting carried out creeping line ahead then set course for base, landed back at base after carrying out a descent through cloud exercise nothing to report regarding submarine. Oil barge USS *YO-143* is bound south in Straits towards Daytona. Tug USS *Montcalm* and nine APC's are bound south in Straits towards Canaveral. Ferry *Ena K.* and SS *Tekoa*, are bound north in Straits passing Canaveral. SS *Cape Romano* and SS *Kaipara* are bound north in Straits towards Canaveral. Tanker *Edward L. Doheny* and tanker *Alpha* are bound north in Straits towards Canaveral. USS *PC-620* and tanker SS *Garnet Hulings* and SS *William Richardson* are bound north in Straits towards Palm Beach. **Convoy KN-223** is bound north in Straits near Bimini with US Navy plane escorting. SS *China Mail* is bound northeast in Straits, and tanker *Tampico* are bound north in Straits. Tug USS *Willett* and USS *SC-978* are bound west in Straits. Tug USS *Icarus* (YAG-16) is bound south in Straits. No.113: Temporary office in small Control Building loaned by No.111 OTU This was entirely inadequate, and inconvenient being 300 yards from the nearest telephone and too small to accommodate everyone. No RAF FC Transport had arrived, and No.111 OTU could not supply any, and local supplies were found to consist mainly of private cars. We have one old Ford Station Van, which was hired for £5 per week. B-25 FR-374 and B-25 FK-376 are bound south of Little San Salvador and Eleuthera, west of Cat Island: Two No.111 OTU aircraft patrolled in search of a reported submarine. U-boat sighted off San Salvador, diverted to search the area, signal not received by aircraft, area not searched.

22nd February 1943

B-25 FR-374 airborne for crossover patrol convoy containing 24 vessels sighted recognition signals exchanged, aircraft flew over Conception Island, dropped a stick of 12 depth charges at a sea marker 2 nm off shore, landed back at base after carrying out a descent through cloud exercise. B-25 FR-384 is airborne for cross over patrol, depth charges dropped at sea marker but overshot by 15 feet. Message received of submarine sighted on surface proceeding northeast at 5 knots from a position of 4 nm east of Hatchet Bay, Eleuthera, diversion signal sent subsequently directing aircraft to submarine position square search commenced nothing sighted search abandoned; aircraft carried out gunnery exercise before landing signal sent to Marine Craft Section ASR launch HMS *P-89* dispatched to search for U-boat, fruitless signal for return sent, came back and docked. B-25 FR-378 is airborne for navigation exercise reconnaissance was made of Georgetown, Exuma, Hatchet Bay, Eleuthera, and Harbour Island. Air Sea Rescue Launch sighted 3 nm north of Hatchet Bay, Eleuthera, creping line ahead, search for sighted submarine commenced off Little Harbour, Great Abaco, aircraft landed back at base later. SS *Flora McDonald* and USS *PC-565* are bound south in Straits towards St. Augustine. SS *Terkoele* is bound north in Straits passing Canaveral. SS *E. S. K. Bank* is bound north in Straits towards Canaveral. SS *Joseph H. Teal* is bound north in Straits near Palm Beach. SS *Empire Knight*, USS *USAT-Cuba*, and oil barge USS *YO-143* are bound south in Straits near Port Everglades. Tug USS *Montcalm* and nine USS *APC*'s are bound south in Straits near Port Everglades. SS *Gavealoide*, USS *CL-10*, and USS *SC-683* are bound north in Straits near Bimini. Tanker *Esso Augusta* north in Straits towards Port Everglades. Tanker *Esso Rochester* is bound south in Straits. Tanker *Gulfmaid* and SS *Walter Gresham* are bound north in

Straits. B-25 FR-378, between James Point [possibly James Cistern] and Harbour Island, northeast Eleuthera: One No.111 OTU aircraft U-boat reported and observed going northwest from east of Hatchet bay, 5 nm east off coast at 5 knots. No.113: An advance party arrived in Nassau from No.45 HQ in Dorval to open the Lodger Unit. Five nm east of Hatchet Bay, Eleuthera, U-boat sighted, patrol to investigate. NAS Great Exuma: *USS YAG-19* tied up then departed. Sikorsky landed and departed. One unidentified plane and B-25 sighted. Three schooners sighted.

23rd February 1943

Pupils of No.111 OTU performed night circuits and landings. B-25 FR-381 DC airborne for navigation Exercise Number 2, sailing vessel sighted 75 nm east of Harbour Island, landed back at base after gunnery exercise after completion of navigation exercise. B-25 FR-383 south for reconnaissance of Georgetown, Exuma, Hatchet Bay, and Harbour Island landed back at base after completion. B-25 FR-379 DA airborne on navigation Exercise Number 3 practice submarine attack exercise off Samaná Cay encountered heavy storm altered course landed back at base. USS *Bull Wheel* (YO-48) is bound south in Straits towards Daytona. **Convoy NK-524** is bound south in Straits towards Canaveral with US Navy plane escorting. SS *Caleb Strong*, SS *John Dickenson*, SS *G. B. Sheldon*, tanker *Stanvac-Manila*, and SS *Fern Plant* are bound north in Straits passing Palm Beach. Tanker *Esso Manhattan* and SS *Flora McDonald* are bound south in Straits passing Port Everglades. Tug *Peter Moran* is bound south in Straits. SS *Leland Stanford* is bound north in Straits. *U-68* enters region west to Mona Passage. NAS Great Exuma: One Army bomber and Two B-25's sighted. Army JRF-2 Amphibian arrived and departed for Guantanamo.

24th February 1943

Night flying exercise carried out together with general flying instructions. B-25 FR-381 O airborne for a cross over patrol navigation exercise engine unserviceable, returned to base. B-25 FR-393 airborne for visual reconnaissance of Georgetown, Exuma, and Harbour Island photographs taken bombing range circled no towed target, returned to base. B-25 FR-378 DD airborne dropped bombs on sea marker creeping line ahead search made abandoned search after a while, flare dropped at 10 nm east of Eleuthera, gunnery practice held, returned to base. SS *Leland Stanford* is bound north in Straits towards St. Augustine. USS *YP-432* south in Straits passing Daytona. USS *SC-1027* is bound south in Straits towards Canaveral. USS *YMS-69* is bound south in Straits passing Canaveral. SS *Robert F. Stockson* is bound north in Straits passing Canaveral. SS *Liebre* is bound south in Straits towards Port Everglades. **Convoy NK-524** is bound south in Straits passing Port Everglades. **Convoy Motor Torpedo Boat Squadron-15** is bound south in Straits towards Miami. SS *Fred Remington* is bound north in Straits with US Navy blimp near Bimini. Tanker *Delaware Sun* and USS *YP-426* are bound north in Straits. SS *George Washington* is bound south in Straits. SS *Alcoa Pioneer* and tanker *Petrofuel* are bound northeast in Straits. *U-183* enters area south to Windward Passage and Caicos. *U-68* is bound northeast of Mayaguana and Turks and Caicos Islands, repairing its METOX Anti-Submarine detection device. NAS Great Exuma: One Army bomber sighted. One plane landed in the bay at Great Exuma due to engine trouble.

25th February 1943

B-25 FR-383 DD airborne for navigation exercise two cruisers sighted circled for 3 nm challenged replied visually and by pyrotechnics gunner practice landed back. B-25 FR-393 airborne cross over patrol exercise set course for Rum Cay exercise completed landed back. Report received of submarine sighting at 5 nm south of Pittstown Point, Crooked Island, 2 nm northwest of Long Cay, Crooked Island. B-25 FR-379 airborne antisubmarine sweep, square search nothing sighted, returned to base. USS *PC-559* is bound north towards Daytona. USS *SC-674* and USS *SC-711* are bound northeast in Straits. **Convoy KN-224** is bound north in Straits with US Navy blimp near Matanilla Shoal, northwest Bahamas. USS *Redwood* (YN-25), SS *Arizona*, SS *William Boyes Thompson*, SS *Roger Williams,* and SS *Britamsea* are bound north in Straits near Port Everglades. Tug USS *Icarus* (YAG-16) is bound north in Straits. Tug *Peter Moran* is bound southeast in Straits. Tanker *Goldheels,* SS *King S. Woolsey* and SS *Swivel* are bound northeast in Straits. *U-183* is bound east of Bahamas in transit south to Windward Passage. Northwest of Castle Island Light, Acklins, No.111 OTU B-25 F3-379 investigates a U-boat sighting.

26th February 1943

B-25 FR-376 CK is airborne gunnery practice and navigational exercise, returned to base after completion. SS *King S. Woolsey* and USS *YMS-226* are bound north in Straits passing Daytona. British Royal Navy motor torpedo boats HMS *MTB-333*, HMS *MTB-343*, and HMS *MTB-84* and SS *Chilbar* are bound north in Straits near Canaveral. USS *SC-1472,* USS *SC-1475,* and destroyer USS *Barney* are bound south in Straits towards Palm Beach. SS *Toteco,* SS *Polonaise* and tanker *Esso Hartford* are bound north in Straits near Palm Beach. SS *Conastoga* and tanker *Texas Sun* are bound south in Straits near Palm Beach. USS *Arapaho* (AT) is bound north in Straits passing Palm Beach. SS *Pat Harrison*, SS *Alcoa Prospector*, and SS *Caribbean* are bound north in Straits. Tug *Peter Moran* is bound south in Straits. Submarine USS *Scorpion* and USS *SC-674* and USS *SC-711* are bound west in Straits. Report received that aircraft had crashed at Delaware, FK-176 had been on night flying training, crew uninjured. Occupants of house, it had crashed into, all killed. NAS Great Exuma: A Sikorsky landed in Great Exuma's bay and departed.

27th February 1943

B-25 FR-374 is airborne for navigation Exercise Number 3, circled Hatchet Bay, 26-ship convoy sighted, visual signals exchanged with escort ship, gunnery practice at Samaná Cay landed back at base. B-25 FR-376 airborne first night navigational exercise, lost in the Atlantic unfavourable conditions, radio contact difficult, landed at Fort Pierce, Florida. SS *Pat Harrison* is bound north in Straits towards St. Augustine. SS *Toteco* is bound north in Straits towards Canaveral. SS *Dione*, USS *Icarus* (YAG-16), and SS *American Trader* are bound south in Straits passing Canaveral. Tanker *Gulfport* is bound north in Straits near Canaveral. Ferry *Jean Brillant* is bound south near Eleuthera. USS *YP-451* is bound north in Straits towards Miami. USS *YAG-19* south in Straits. USS *SC-703* is bound northwest from Cuba near Nicholas Channel. Army tug USS *SC-47* is bound southeast towards Nicholas Channel. *U-155* enters area west to Windward Passage. *U-183* is bound north of Turks and Caicos. *U-86* is northeast of Bahamas, where it encounters a Consolidated flying boat, sighted from clouds, that then circled the U-boat.

28th February 1943

B-25 FR-374 airborne, reconnaissance of Eleuthera, Exuma, and Harbour Island, landed back after exercise completion. B-25 FR-364 CC and B-25 FR-375 CJ airborne for navigation exercise around Samaná Bay, landed back after exercise completion. **Convoy Charleston-59** is bound south in Straits towards Palm Beach. USS *SC-638,* USS *SC-676*, and USS *YMS-34* are bound north in Straits towards Palm Beach. SS *American Trader* and USS *PC-1193* are bound south in Straits. SS *Dunlemia* and tanker *Esso Wilmington* are bound northeast in Straits towards Bimini. Oil barge USS *YO-51* is bound southwest in Straits. SS *A. S. Hansen* is bound northeast out of Straits. USS *SC-703* and USS *YDG-1* are bound southwest in Straits passing Nicholas Channel. *U-183* is bound east of Inagua, bound south to Windward Passage. *U-155* is bound northeast of Bahamas, west to Windward Passage. No.113: Making necessary contacts with No.111 OTU, Pleasantville contractors of the US, and post engineer for the official opening. NAS Great Exuma: Passenger steamer *Jean Brillant* went out on a southeast course and anchored off the Blue Hole. A plane secured alongside a gasoline barge. A station pilot boarded *Jean Brillant.*

1st March 1943

Training exercise through the day and night. Navigation exercise carried out. Aircraft PBY-2 FP-532 is airborne for Miami. Hudson FK-657 lands from Morrison Field. B-25 FR-31 DC is airborne for navigation Exercise Number 3, astrograph navigation exercise between Samaná Cay and Rum Cay, air firing practice, landed back at base after carrying out a descent through cloud exercise. B-25 FR-380 DB airborne, reconnaissance of Harbour Island, landed back. B-25 FR-376 CK on night navigation exercise, set course and sent meteorological report, tired descent through clouds, could not see base continued for 17 minutes base still not visible radio compass useless from static interference, turns towards Miami and decides to head towards mainland, flare dropped to see if there was land, set to safe. Was north of Miami after passing coast, however Miami not sighted was likely over Florida Keys, saw lighted field with flashing beacon, landed at Fort Pierce, refuelled, airborne for Morrison Field, picked up remaining crew, landed back at base. SS *A. S. Hansen* and SS *Dunlemia* are bound north in Straits passing Canaveral. SS *Virginia* is bound north in Straits. **Convoy Charleston-50** is bound south in Straits towards Palm Beach. Oil barge USS *YO-142* and USS YP-*408* are bound east in Straits near Port Everglades. Tanker *Gulfland* is bound south in Straits towards Port Everglades. Tanker *L. L. Abshire* is bound north in Straits bound towards Port Everglades. USS *YAG-19* is bound north in Straits. Mine layers USS *ML-1171* and USS *ML-1172* are bound west in Straits. Sail Rock, Exumas, north of Ship Channel Cay and the Exuma Cays, and roughly midway between Cape Eleuthera and southeast New Providence (Nassau) is designated as an airplane bombing target practise range [for the duration of the war, into 1946]. No.113: Permanent station staff moved to Nassau, worked on building program, signals arrangements on the Southern Route [South Atlantic transport, Brazil *et al* to Africa]. NAS Great Exuma: Ferry *Jean Brillant* departed, escorted by USS *SC-1063.* A plane sighted a 2-masted schooner with greenish gray hull and eight fuel drums on deck. A plane went to cover **Convoy GN-44.**

2nd March 1943

Training exercise through the day, local circuits and landings made at night. B-25 FR-383 DE airborne for navigation exercise for five hours over sea, landed back. B-25 FR-364 CC set course for Man Island Light, north of Harbour Island, radius of action exercise for five hours, landed back. Tanker *L. L. Abshire* is bound north in Straits towards St. Augustine. *USCG-83379* is bound south in Straits towards Canaveral. USS *SC-977* is bound north in Straits towards Canaveral. SS *Montgomery City* is bound north in Straits passing Palm Beach. USS *YAG-19* and USS *Q-065* are bound east near Walker Cay, north of Abaco. Tanker *Esso Rochester* and SS *Sparrows Point* are bound north in Straits near Matanilla Shoal, northwest Bahamas. Ferry *Jean Brillant* and USS *SC-1053* are bound southwest in Straits towards Miami. USS *YAG-21* and USS *YP-574* are bound west in Straits. Troop transport USS *USAT-Henry Gibbons* is bound north in Straits. **Convoy KN-225** is bound east in Straits with a navy blimp. SS *Malang* is bound southeast in Straits with US Navy plane escorting, passing Nicholas Channel. **Convoy KG-625** is bound southeast in Straits passing Nicholas Channel. *U-171* is bound north of Abaco, and a destroyer comes towards U-boat, which attempts to get free by turning, but is forced off from the formation. The shadows were a warship formation, consisting of two large units that are protected on their port side by four destroyers, still recognized from the Morse traffic of the destroyers, which turned away. *U-171* pursues the formation, with no result. Then, off Jacksonville, at periscope depth two depth charge detonations in vicinity of the U-boat, which was believed to have been detected by the airship and therefore went to depth. NAS Great Exuma: A plane went to cover **Convoy GN-44** and practiced night landings. A plane went on a search mission.

3rd March 1943

Training exercise through the day local circuits and landings made at night. B-23 FR-364 CC airborne, set course for Man Island Light, north of Harbour Island landed back at base after exercise. B-25 FR-380 DB airborne for Exercise Number 7 small motor vessel sighted photographs taken of Georgetown and Exuma, reconnaissance made of Hatchet Bay, Eleuthera, and Harbour Island, photographs taken, conducted gunnery practice, landed back at base. B-25 FR-383 DX airborne for Navigation Exercise 4b, creeping in a line ahead, search with visibility of 10 nm commenced, landed back at base after exercise completion. USS *PC-591* and USS *PC-561* are bound south in Straits toward Daytona. SS *Montgomery City* and USS *SC-639* are bound north in Straits passing Daytona. **Convoy KN-225** is bound northeast in Straits with US Navy plane escorting. **Convoy SC-684** is bound north in Straits past Palm Beach. USS *Surprise* (PG-63) and tanker *Esso Manhattan* are bound north in Straits passing Port Everglades. Oil barge USS *YO-142* is bound southwest in Straits towards Miami. USS *YP-432* is bound southeast in Straits near Miami. SS *Exchange* and SS *Mormacmoon* are bound north in Straits. No.113: First delivery of RAF Transport Command aircraft accepted at Nassau. Temporary accommodation obtained from No.111 OTU in two huts next to Satellite Field [Windsor Field] HQ. One was allotted to equipment officer for temporary stores, the other allotted to the commanding officer, the adjutant the accountant officer and navigation officer. The original control building was allotted to the engineering officer, radio maintenance, aircraft checkers, operations, despatchers and test department. *U-455* is bound north of Jacksonville, nights back and forth there, yet nothing except one aircraft, from which avoided detection. *U-155* is bound north of Puerto Rico; when surprised by aircraft in poor visibility, crash dived. NAS Great Exuma: A US destroyer and light cruiser arrive and depart, a plane covered them. A plane practiced anti - submarine activities.

4th March 1943

General flying from Main Field at day, night exercises also carried out. B-25 FR-383 DE airborne to escort southbound convoy, set course for the point of interception of convoy could not sight convoy, flew up convoy, tract square search box, patrol flown around convoy after it was met left it, no relieving aircraft fixes were obtained ascertained convoy was 35 nm back on its estimated position, landed back with nothing to report. B-25 FR-378 CL airborne for convoy escort message from base regarding where convoy was intercepted aircraft could not find convoy visibility was 3 nm returned to base. USCG *CG-83381* and destroyer USS *Leary* are bound southeast in Straits towards Palm Beach. SS *Seatrain New Orleans* and tanker *Spencer Kellogg* are bound north in Straits. SS *Ariel* is bound west in Straits passing Nicholas Channel with US Navy plane escorting. *U-185* is bound northeast of Bahamas, south to Windward Passage. *U-155* is off Turks and Caicos, bound south to Caicos Passage. NAS Great Exuma: A plane went to cover **Convoy NG-346**. Another plane sighted a small freighter. HMS *Attacker* passed through the area. Destroyer USS *Dyson* (DD-572) also passed through the incident area.

5th March 1943

B-25 FR-381 DC airborne to escort southbound convoy escort vessels sighted 15 nm either side of convoy, convoy met and patrolled, left convoy at coded position ZVQC and at 11:04 am landed back at base. B-25 FR-384 CM airborne, set course for Devils Point, Cat Island, then set course for Nuevitas Cuba passed Exuma and set course for Bimini, set course for southwest Point, Abaco, gunnery practice before landing at base after descent through cloud practice. Tanker *William G. Warden*, tanker *W. H. Ferguson*, SS *Edward L. Doheny*, SS *Garnet Hulings* are bound south in Straits passing Daytona. **Convoy NK-526** is bound south in Straits passing Daytona, with US Navy plane escorting. Irish SS *Irish Oak* is bound south in Straits near Bimini. USS *YAG-16* and four US Coast Guard vessels are bound southeast in Straits passing Port Everglades. USS *YAG-21* and USS *YP-574* are bound north from Nassau in Northeast Providence Channel. No.113: Two Senior NCO's [non-commissioned officers] and 49 airmen arrived from HQ, RAF FC Dorval for ground duties. *U-155* is bound north of Bahamas; crash dived for aircraft flying west, unrecognized type; U-boat was not noticed. NAS Great Exuma: **Convoy SS *Pollox*** crossed the area. A plane covered **Convoy NG -346** and sighted four navy cargo ships. A plane went on a search mission. Another plane practiced Anti-Submarine exercises.

6th March 1943

B-25 FR-376 CK airborne patrol from New Providence over Eleuthera to Cat Island to Nuevitas to Bimini landed back at base. B-25 FR-384 CM airborne for Exercise Number 5 approached Devils Point, Cat Island passed Cuba north of Camaguey,

Cuba, returned to base descent through cloud exercise, crew had nothing to report after exercise. USS *YT-347* and SS *Furnifold Simmone* are bound south in Straits past Canaveral. USS *SC-1288* is bound south in Straits passing Palm Beach. Tanker *Gulf Caribbean* is bound southeast from Matanilla Shoal, northwest Bahamas. USS *YAG-21* and USS *YP-574* are bound west in Northwest Providence Channel. Ferry *Jean Brillant* and USS *Q-075* are bound southwest in Straits, past Northwest Providence Channel towards Miami. **Convoy NK-526** is bound south in Straits with US Navy blimp near Miami. Destroyer USS *Leary* is bound south in Straits passing Miami. Tanker *Esso Pittsburgh* is bound southwest in Straits bound past Bimini. SS *Alpha* and tanker *Gulftide* are bound southwest in Straits near Bimini. Tanker *Edward L. Doheny* and SS *W. H. Ferguson* are bound southwest in Straits. Twenty-six USS *LCI*'s are bound northeast in Straits. SS *Colorado* and SS *Irish Oak* are bound west in Straits passing Cay Sal Bank. Oil barge USS *Gulfspray* (YO-142), USS *Q-080*, USS *YAG-28*, and four US Coast Guard vessels are bound south in Straits. No.113: First dispatches of aircraft from Nassau to southern Atlantic route made. NAS Great Exuma: **Task Force-39** and **Convoy GN-45** arrived in area. A plane went to cover **Task Force-39**. A plane practiced aerial gunnery.

7th March 1943

General flying throughout the day, pupils of Number 1 course finished conversion on B-25 aircraft moved to satellite for training on Liberator aircraft. B-25 FR-379 DA airborne for exercise over Abaco, Long Island to Cuba set course for Bimini varied course to northwest to avoid storm radio reception and transmission poor because of electrical interference returned to base due deteriorating weather conditions. NAS Great Exuma: A task force unit departed. A plane went to cover **Convoy GN-45** which departed.

8th March 1943

USS *SC-1061* is bound northeast in Straits. **Convoy KN** is bound north with US Navy plane escorting. **Task Group TG.3011** is bound northeast in Straits passing Fort Everglades. USS *PC-1123* south in Straits from near Miami. USS *YO-142* is bound northwest in Straits. USS *YAG-28* is bound northwest in Straits. **Convoy GK-726** is bound northwest in Straits passing Nicholas Channel with US Navy plane escorting. *U-185* is bound south in Crooked Island Passage for Windward Passage. *U-185*, east of Bermuda: crash dive! Catalina, then another crash dive for, this one lower. Then east of Bermuda, surfaced and observed up to seven strong searchlight cones over Bermuda, various intervals. NAS Great Exuma: Two planes practiced aerial gunnery. A plane practiced Anti-Submarine bombing.

9th March 1943

Local flying throughout the day. Report received that escort vessel had not arrived with ferry *Jean Brillant* at Nassau, last position was near Gear Stirrup Cay Light, Berry Islands, FR-378 CL diverted whilst flying to last reported position to search for missing vessel, B-25 FR-383 DE airborne for search between New Providence and Great Stirrup Cay Light, Berry Islands, message received that vessel had docked at Nassau, two aircraft recalled landed back at base. US Navy and USAAF aircraft attack *U-185* in Windward Passage off northeast corner of Cuba. The U-boat counter-attacks, is forced to abort its second attack on **Convoy KG 123**. Report received that escort vessel B-25 FR-383 had not arrived with *Jean Brillant* at

Nassau, last position was near Great Stirrup Cay Light, Berry Islands, FR-378 CL diverted to search and, B-25 FR-383 DE airborne for search between New Providence and Great Stirrup Cay, Berry Islands, however vessel docked at Nassau, so aircraft recalled. *U-185* is off Bermuda on a reconnaissance mission, observes auto traffic and lights at Fort St. Catherine, behind St. George's. NAS Great Exuma: A plane practices instrument flying; another plane arrived. A plane went to cover **Convoy NG-347,** which arrived in area.

10th March 1943

General flying only on B-25 from Main Field, with flights on Liberators from Satellite under instructions. SS *Virginia Sinclair*, in **Convoy KG-123**, is hit by *U-185* off Cape Maysí, Cuba. Four crew lost, eight officers, 23 crew, four armed guards, two naval signalmen abandon ship in three lifeboats and one raft. Survivors picked up by USS *SC-742*, landed in Guantanamo. *U-185* attacks **Convoy KG-123** for second time south east of Guantanamo, torpedoes SS *James Sprunt*. All eight officers, 36 crew, and 25 armed guards survive. Tanker *Gulfstate* is sunk by *U-155*, 60 nm west southwest of Orange Cay, Bimini, and west of Andros. Eight officers, 34 crew, and 19 armed guards are discovered by a USN blimp, which drops two rubber life rafts. A US Coast Guard seaplane picks up three, takes them to Miami. Other 15 are picked up by USS *YP-351*. Eight officers, 26 crew, nine armed guards are lost. *U-155* is at Caicos Passage. Ferry *Jean Brillant* and USS *SC-752* are bound southwest from Northwest Providence Channel, bound towards Port Everglades. *U-155* is bound north of Bahamas, where it crash dives for single-engine bi-plane flying south: U-boat was not noticed. *U-185* is in the Old Bahama Channel off Cuba; a low shadow proved to be a naval patrol vessel, PC-boat, that moved off northerly on the surface. NAS Great Exuma: A plane went to cover **Convoy NG-347**, then another plane departed. A plane practiced aerial gunnery. A plane sighted a small British merchant vessel, two US Navy destroyers and a patrol craft boat.

11th March 1943

General flying from Main Field and Satellite Field. USS *USAT-Dupage* (APA-41) is bound north near Matanilla Shoal, northwest Bahamas. SS *Tampico* is bound south in Straits towards Port Everglades. USS *SC-692*, USS *YMS-62* and USS *YMS-69* are bound north in Straits, towards Port Everglades. **Convoy NK-527** is bound south in Straits passing Bimini with US Navy plane escorting. USS *USAT-Yarmouth* is bound north in Straits with US Navy blimp escorting. NAS Great Exuma: Two planes went to cover **Convoy GN-46**. Two planes practiced Anti-Submarine bombing. A plane sighted a US Navy destroyer.

12th March 1943

Training pupils on Liberators from Satellite continued. USS *SC-526* and USS *SC-535* are bound north in Straits towards Canaveral. SS *Conastoga* and **Convoy Motor Torpedo Boat Squadron-15** are bound north in Straits, passing Palm Beach. Tugs USS *Humrick*, *Ontario*, and USS *Caloria* are bound south in Straits past Port Everglades. Army tug USS *LT-113* is bound north in Straits passing Miami. Army tug USS *LT-331* is bound north out of Straits. USCG *GC-Micawber* is bound southwest in Straits. Tanker *Gulfland* and USS *Absecon* are bound northeast in Straits. NAS Great Exuma: Two planes went to cover **Convoy GN-46.** A plane reported dropping of depth charges on a submarine.

13th March 1943

Convoy Charleston-65 is bound south in Straits to Canaveral with US Navy plane escorting. USS *PYC-28* and Army tug USS *Lt. Thornton* are bound south in Straits passing Canaveral. **Convoy KN-227** is bound north in Straits with US Navy plane escorting. **Convoy Port Everglades-8** is bound south in Straits with US Navy plane escorting. USS *YAG-26* is bound southeast in Northwest Providence Channel. **Convoy Motor Torpedo Boat Squadron-15** is bound north in Straits. Fleet Exercise (Flex) Number 6 was carried out at Culebra, Puerto Rico, practices include rubber boat landing, and moving ship to shore and heavy equipment landing for amphibious attack. No.113: Work on Lodger Unit buildings, parking area and nose hangars. Royal Victoria Hotel provided for transient aircrews. Lack of motor transport, maintenance equipment. NAS Great Exuma: Two planes went to cover a **Convoy GK-65**. A plane sighted a 100' diesel-powered ship flying the Cuban flag, escorting 19 merchant vessels.

14th March 1943

Local air exercise and navigation carried out, general local training of B-25 carried out from Main Field, Liberator exercises from Satellite Field. Tug USS *Sampson* and tow and **Convoy Charleston-66** are bound south in Straits towards Daytona. **Convoy Motor Torpedo Boat Squadron-15** is bound north towards Canaveral. **Convoy Charleston-65** is bound south in Straits passing Port Everglades. SS *Gaspe County* and USS *PYC-28* are bound north in Straits, passing Port Everglades. USS *YO-51* and USS YP-*332* are bound south in Straits passing Miami. Army tug USS *Lt. Thornton* is bound west in Straits. **Convoy Havana North Coast Cuba-2** is bound northwest in Straits passing Nicholas Channel. **Convoy GK-65** is bound west in Old Bahama Channel passing Cayo Francés. USS *SC-503*, USS *SC-532*, and USS *SC-534* are bound north in Straits. NAS Great Exuma: A plane went to cover **Convoy GK-65**. Two planes went to cover **Convoy NG-348**. USS *YAG 20* arrived. A plane sighted an empty wooden raft.

15th March 1943

B-25 FR-375 CJ airborne for navigation Exercise Number 1 set course for Sail Rock, Exumas dropped 500-pound bomb, returned to base nothing to report. B-25 FR-380 DB airborne for navigation Exercise Number 1, set course for Sail Rock, Exumas, dropped 500-pound bomb, gunnery practice and returned to base. B-25 FR-378 CL airborne for navigation Exercise Number 1a, set course for Sail Rock, Exumas, did not complete bombing practice, bomb hung up, two attempts made but unsuccessful, gunnery exercise carried out against a sea marker off Sail Rock, Exumas, set course for Rum Cay, meteorological report sent to base, on return attempts at bombing again made unsuccessful returned to base. B-25 FR-384 and B-25 FR-381 carried out short air exercises. B-25 FR-393 DF is airborne, carried out photographic visual reconnaissance of Exuma, Harbour Island, and Eleuthera descent through cloud exercise, returned to base. New B-25 delivered to this unit. *U-183* leaves Bahamas south via Windward Passage. Tug USS *Sampson* is bound south in Straits towards Palm Beach. USS *YT-160* and USS *YT-185* are bound north in Straits passing Palm Beach. USS *SC-503,* USS *SC-532*, and USS *SC-534* are bound north in Straits. SS *Hawaiian Merchant* is bound west passing Walker Cay, north of Abaco. Tanker *Gulf Caribbean* is bound north out of Straits. SS *Gaspe County* is bound south in Straits. **Convoy Havana North Coast Cuba-2** is bound west in Straits, passing Nicholas Channel with Cuban plane escorting. NAS Great Exuma: A plane covered **Convoy NG-348**. USS *YAG 20* departed. A plane practiced aerial gunnery.

16th March 1943

B-25 FR-378 CL airborne for navigation Exercise Number 1, dropped 500-pound bomb on Sail Rock, Exumas set course for base when wireless became unserviceable. B-25 FR-375 CJ airborne for navigation Exercise Number 1a, dropped 500-pound bomb on Sail Rock, Exumas photographic records taken, meteorological report sent to base returned to base. B-25 FR-371 CF airborne for navigation Exercise Number 1, bomb dropped at Sail Rock, Exumas, dark patch sighted, descended to check it was floating weed, gunnery practice against sea marker, guns unserviceable after first round, exercise abandoned and landed back at base. B-25 FR-381 DC airborne reconnaissance exercise of Harbour Island, Eleuthera, and Exumas. Photographs taken of Georgetown, Hatchet Bay, cloud exercise during descent, landed back at base. SS *Kennebeck* is bound southwest near Walker Cay, north of Abaco with US Navy plane escorting. **Convoy NK-528** is bound south in Straits towards Palm Beach with US Navy blimp escorting. USS *YO-51* is bound east in Straits near Port Everglades. SS *Cornell* is bound north in Straits. NAS Great Exuma: A plane practiced instrument flying and Anti-Submarine warfare. A plane went to cover **Convoy GN-47**.

17th March 1943

B-25 FR-379 DK is airborne for photographic reconnaissance exercise of Hatchet Bay, Harbour Island, Georgetown, Exuma, then gunnery practice, landed back at base. B-25 FR-378 CL is airborne for navigation Exercise Number 1a, bomb dropped at Sail Rock, Exumas target. Merchant vessel with single black funnel amidships with little superstructure sighted high in water, seemed unladen, aircraft completed patrol, set course for base, descent through cloud exercise, landed back at base. B-25 FR-371 CF from Main Field for navigation Exercise Number 1 set course for Sail Rock, Exumas photographs taken, bomb dropped on target at Sail Rock, Exumas, found craft, had an oil leak, returned to base, pilot took off in B-25 FR-381 DC to complete exercise through cloud descent landed back at base crew, had nothing to report. B-25 FR-375 CJ airborne for cross over navigation exercise, stick of four depth charges dropped, bombs dropped in a salvo instead of 36-foot stick, landed back at base, nothing to report. USS *YP-438* is bound south in Straits passing St. Augustine. **Convoy KN-228** is bound north in Straits near Miami with US Navy blimp. USS *YAG-20* is bound west in Straits near Andros. **Convoy NK-528** is bound southwest in Straits with US Navy plane escorting. NAS Great Exuma: A plane went to cover **Convoy GN-47**. A plane practiced Anti-Submarine bombing. A plane sighted a Norwegian tanker in the area.

18th March 1943

General flying from Main Field and Satellite Field, local circuits made at night. B-25 FR-375 CJ airborne for navigation Exercise Number 1, practice bomb dropped at Sail Rock, Exumas, stick undershot, set course back to base. Convoy sighted, 26 merchant ships and three escorts circled convoy at 5,000 yards, resumes course for base, passed Eleuthera, landed back at base. B-25 FR-371 CF airborne for navigation Exercise Number 1, practice bomb dropped at Sail Rock, Exumas, undershot, exercise completed, sea marker dropped for air firing exercise. Convoy sighted, 26 merchant ships and four escorts at distance of 5 nm starboard, descent through cloud exercise, landed back at base. B-25 FR-376 CK airborne for navigation Exercise Number 1, dropped 500-pound bomb on Sail Rock, Exumas, target direct hit registered, exercise completed, sea marker dropped for air firing. Check on the dead reckoning (DR) position found to be incorrect Eleuthera pinpointed descent though cloud exercise, landed back at base. B-25 FR-380 DB airborne for navigation Exercise Number 7, reconnaissance made of Harbour Island, Exuma, Eleuthera, photographs taken of Georgetown, passed San Salvador, photographs taken of Hatchet Bay and Harbour Island, landed back at base. A Baltimore expected to arrive from Wilmington by US ATC [Atlantic Transport Command], request made to search for aircraft from Nassau on the probable route of aircraft, Morrison Field instituted search from Florida coast. B-25 FL-199 BD airborne for search, visual reconnaissance and search made from Nassau to Andros and then Williams Island [northwest Andros], creeping line ahead search made visibility of 3 nm, message received that Morrison Field had located craft, which landed at Jacksonville. Aircraft called back to base, landed back having sighted nothing. Baltimore FA-354 ferried by RAF Ferry Command preparing to land at Satellite Field, overshot runway, pilot endeavoured to bank aircraft to another circuit, failed to respond to controls, crashed a mile away from Field, both occupants badly burnt; one died at hospital other made partial recovery, aircraft caught fire and was totally destroyed: had been *en route* for Trinidad. Tanker *Gulfmaid* and SS *Murena* are bound north in Straits passing Daytona. USS *YAG-19* and USS *Q-065* are bound south in Straits from Port Everglades. **Convoy KW-130** is bound northwest in Straits towards Port Everglades. No.113: On Satellite [Windsor] Airfield A-30 No.FA354, P/O. D.T. Hollowell, Pilot F/Sgt. Atwell, A.F. WAG, whilst doing circuits and landings, bounced on landing. The pilot endeavoured to take off again, but stalled the aircraft in so doing. The aircraft crashed and burst into flames, being totally destroyed. Hollowell succumbed to his injuries on 9.3.43. Atwell was injured less seriously. Baltimore bomber reported overdue from Wilmington, NC, landed in Jacksonville. B-25 FL-199 searched Andros and Williams Island, westernmost of north Andros. NAS Great Exuma: A US Navy transport and two US Navy destroyers passed. **Convoy GN-47** departed from the area. A plane practiced instrument flying, and another plane went to cover **Convoy KW-129.**

19th March 1943

Night circuits carried out from Main Field. B-25 FR-375 CJ airborne on navigation Exercise Number 2, course set for Rum Cay convoy sighted 4 nm to starboard, photographic reconnaissance made of convoy with 26 merchant ships and four escorts in eight lines, with two stragglers a mile behind. Convoy again sighted later 10 nm to port same course, speed, composition, grouping. Catalina is observed on escort duties. B-25 FR-376 CK airborne on navigation Exercise Number 1, a 500-pound bomb dropped at target on Sail Rock, Exumas, direct hit claimed, set course for Rum Cay, completed exercise, returned to base, carried out descent through cloud exercise, crew had nothing to report. B-25 FR-378 CL airborne for navigation Exercise Number 1, convoy sighted 8 nm to 10 nm on starboard after exercise completion, photographs taken of seaplane base [NAS] at Georgetown, Exuma. Visual reconnaissance and photographs taken of Hatchet Bay, small grey painted motorboat is sighted two circuits made of Harbour Island and photographs taken, gunnery exercise carried out before landing. B-25 FR-383 DX airborne on navigation Exercise Number 7, reconnaissance exercise, photographs taken of Exuma Air Base, cloud cover used for instruction in photography in enemy territory, Hatchet Bay and Harbour Island photographed, exercise in descent through cloud cover, landed back at base. B-25 FL-697 crashed Soldier Road, had taken off for local flying had made one circuit, one engine caught fire, aircraft side-slipped into the scrub, burnt out, both occupants killed. USCG *CG-Balsam* and USCG *CG-83399* are bound south in Straits towards Canaveral. USS *YP-438* is bound south in Straits towards Palm Beach. SS *Colonel James A. Moss* is bound south towards Palm Beach. **Task Group TG-3016** is bound north in Straits towards Palm Beach. USS *YT-349* and SS *Runswick* are bound south entering Straits. USS *YAG-16* is bound south in Straits. **Convoy GK-727** is bound northwest in Straits passing three US Navy blimps near Nicholas Channel. No.113: Funeral of D. T. Hollowell in the RAF Cemetery, RAF Station, Nassau. NAS Great Exuma: Three planes went to cover **Convoy NG-349**, and another plane went to cover **Convoy KW-129**.

20th March 1943

B-25 FR-384 CM airborne on-air navigation Exercise Number 7, contacted launch towing target in position, released all bombs photographs taken of Georgetown and Exuma, gunnery practice visual reconnaissance of Hatchet Bay and Harbour Island, returned to base. B-25 FR-375 CJ airborne on navigation Exercise Number 2 carried out full patrol bombing and gunnery exercise against sea marker, landed back at base. B-25 FR-378 CL airborne on navigation Exercise Number 2a, set course for Rum Cay, bombing and gunnery exercise against sea marker, descent through cloud exercise, returned to base. B-25 FR-380 DB airborne on navigation Exercise Number 2a, set course for San Salvador after Rum Cay, descent through cloud exercise, returned to base. B-25 FR-376 airborne on navigation Exercise Number 2, patrol completed satisfactorily, four depth charges dropped against sea marker over shot target, descent through cloud exercise, returned to base. **Convoy NK-529** is bound south in Straits towards Canaveral, US Navy plane escorting. Tug USS *Humrick* is bound south in Straits passing Port Everglades. SS *Bennington* is bound south in Straits near Bimini. Four US Navy USS *C-Crash Boats* are bound south in Straits. NAS Great Exuma: Two planes covered **Convoy NG-349**, which departed, and a plane went on a search mission.

21st March 1943

B-25 FL-199 BD airborne for air Exercise Number 7, towed target for bombing practice, visual reconnaissance made and photographs taken of Georgetown, Exuma, flew over San Salvador, photographic reconnaissance made of Hatchet Bay, gunnery practice, descent through cloud exercise returned to base. B-25 FR-380 DB airborne from Main Field for navigation Exercise Number 2 course set for Rum Cay, depth charge dropped from 35 feet, descent through cloud exercise returned to base. B-25 FR-384 CM airborne on navigation Exercise Number 2, depth charge dropped from 50 feet, tug towing barge sighted 2 nm to starboard, descent through cloud exercise returned to base. USS *YT-328* is bound south in Straits passing St. Augustine. USS *YP-427* is bound north in Straits towards Daytona. USCG *CG-86* and USCG *CG-87* are bound north in Straits towards Daytona. SS *Cache* is bound southwest in Straits towards Canaveral navy plane escorting. Tanker *Gulfmaid* and two US Coast Guard cutters are bound south in Straits passing Palm Beach. **Convoy NK-529** is bound south in Straits with US Navy blimp escorting near Palm Beach. Army tug USS *LT-331* towing USS *BC-653* and USS *BC-655* are bound north in Straits passing Miami. USS *YO-51* and USS *YP-433* are bound south in Straits, passing Miami. SS *Tarazed* and USS *Palmer* are bound west in Straits with US Navy blimp near Nicholas Channel. The RAF completes its own Station Hospital between Windsor and Oakes fields, on Prospect Ridge. On August 5, 1943, S/L C. R. McLaughlin assumed command of Station Hospital from Vice Wing Commander R. H. Pratt, who fell ill. Nursing sisters (roughly six at a time) from the PM RAF NS (Princess Mary's Royal Air Force Nursing Service) were allocated. NAS Great Exuma: A plane went on a search mission, then two planes went to cover **Convoy GN-48.**

22nd March 1943

B-25 FR-376 CK airborne on navigation Exercise Number 3, had to find a given island, completed, bombs dropped, landed back at base. B-25 FR-383 DX airborne on navigation Exercise Number 2a, completed sea marker, bombs, and depth charges dropped overshot target with unserviceable radio, no signals received from aircraft, gunnery practice, landed back at base. B-25 FR-380 DB airborne on navigation Exercise Number 3 located Samaná Cay bomb exercises 10 nm to the west of cay, landed back at base. B-25 FL-199 BD airborne on-air Exercise Number 7, photographic reconnaissance taken of Georgetown, Exuma, Hatchet Bay, Harbour Island, landed back at base. B-25 FR-382 DD airborne on navigation Exercise Number 3, completed, landed back at base. Court of Inquiry assembled at this unit to inquire into the circumstances of an accident that occurred on 19th March, resulting in the loss of aircraft B-25 FL-967. New station hospital opened. USS *YT-328* is bound south in Straits passing Canaveral. USS *YP-427* is bound north in Straits towards Canaveral. **Convoy KW-133** is bound north towards Palm Beach, with US Navy plane escorting. **Convoy NK-529** is bound southwest with US Navy blimp escorting in Straits. NAS Great Exuma: Two planes went to cover **Convoy GN-48** and departed.

23rd March 1943

B-25 FR-371 CF airborne on navigation Exercise Number 2, patrol completed, four practice bombs and four depth charges dropped, Eleuthera pinpointed, descent through cloud exercise back to base. B-25 FR-375 CJ airborne on navigation Exercise Number 3, sea marker dropped, 1 nm east of Hatchet Bay practice bombs dropped on southern tip of Long Island, pin located, visual and photographic reconnaissance taken of Clarence Town Long Island, gunnery commenced, descent through cloud exercise, then back to base. B-25 FR-376 CK airborne on navigation Exercise Number 2a exercise adapted to cover area for convoy sighting square search, failed to sight convoy, practice bombs dropped, gunnery practice, descent through cloud exercise, back to base. B-25 FR-383 airborne on navigation Exercise Number 7 reconnaissance of Harbour Island, Eleuthera, and Exumas. **Convoy KW-133** is bound north in Straits passing Daytona. SS *Winooski* and SS *Chicopee* are bound southwest in Straits towards Palm Beach. SS *Tarazed* is bound north in Straits passing Palm Beach. **Convoy KN-229** is bound north with US Navy blimp near Bimini. Tug USS *Hercules* is bound south in Straits passing Port Everglades. **Task Unit TU-0316** is bound southwest in Straits towards Nicholas Channel with US Navy plane escorting. Twenty one merchant ships escorted by USS *SC-1029* are bound southwest in Straits towards Nicholas Channel, with a US Navy blimp escorting. SS *Colonel James A. Moss* is bound southwest in Straits passing Nicholas Channel. NAS Great Exuma: Two planes covered **Convoy GK-71.**

24th March 1943

B-25 FR-380 DB airborne on navigation Exercise Number 3, completed, bombing and gunnery practice off Samaná Cay, landed back at base. B-25 FR-382 DD airborne on navigation Exercise Number 2, convoy sighted, consisting of 10 merchant ships and three escorts in double line abreast. A Catalina was observed escorting convoy, sea marker dropped, bombing and gunnery practice carried out, landed back at base. B-25 FR-381 DC airborne on navigation Exercise Number 2a, set course for Rum Cay, convoy sighted, 19 merchant ships and four escorts. Catalina and B-25 observed in vicinity, depth charges dropped on wave crests, gunnery practice carried out, landed back at base. B-25 FR-371 CF airborne on navigation Exercise Number 4 practice bombs dropped against sea marker landed back at base. B-25 FR-375 CJ airborne on navigation Exercise Number 4 exercise carried out satisfactorily near Samaná Cay landed back at base. B-25 FR-181 delivered from Morrison Field. USS *AT-138* is bound south in Straits towards Canaveral. USS *Kennebeck* (AO-36) is bound northeast in Straits off Bimini. **Convoy NG-350** is bound south in Straits. NAS Great Exuma: A plane went to cover **Convoy GK-71**, and two planes went to cover **Convoy NG-350.** USS *Harrison* (DD-573) escorted ships from **Convoy KG-627.**

25th March 1943

General flying during day, and local night flying from both Main Field and Satellite Field. B-25 FR-394 CN is airborne on navigation Exercise Number 7, practice bombs dropped with clouds at 10 am at 4,000 feet, photographs taken over Georgetown, Exuma, unidentified motor vessel sighted half mile to starboard, crew held up board with *Whipple* painted in black. On aircraft's closer approach, ascertained that the ship had a grey hull, two small sails, humped bow, narrow amidships blue and white ensign flown from stern, which resembled Salvador mercantile stern. Photographs taken of Hatchet Bay and Harbour Island, then landed back at base. B-25 FR-393 DX airborne on navigation Exercise Number 3, gunnery and bombing practice held, set course for Samaná Cay, landed back at base after completion of the exercise. B-25 FR-381 DC airborne on navigation Exercise Number 3 practice bombs dropped, set course for Samaná Cay landed back at base after descent through cloud exercise. B-25 FR-384 CM airborne on navigation Exercise Number 3 submarine attacks made and bombs dropped for exercise, gunnery practice held 10 nm from coast of Samaná Cay, landed back at base after descent through cloud exercise. B-25 FR-393 DF airborne on navigation Exercise Number 3, bombs dropped for exercise, set course for Samaná Cay, gunnery practice held, landed back at base after descent through cloud exercise. Baltimore FA-340 airborne for air test from Satellite Field, but on landing bounced at runway, caught fire. The sole occupant was thrown from the aircraft and killed [funeral held next day, on 26th]. USS *SC-1062* is bound south in Straits, passing St. Augustine. USS *YAG-14* is bound south in Straits passing Daytona. SS *W. W. Mills* and USS *SC-1267* are bound south in Straits towards Canaveral. Ferry *Jean Brillant* and USS *SC-980* are bound west from Abaco in Northwest Providence Channel. SS *Cornell* is bound south in Straits. USS *YAG-28* and USS *YP-218* are bound south in Straits from Port Everglades. Two US Navy tugs and USS *YP-426* are bound northeast in Straits. **Convoy GK-71** is bound northwest in Straits near Nicholas Channel with US Navy blimp. Tug *Edmund P. Moran* towing barge is bound west in Straits past Nicholas Channel. No.113: Satellite Field; A-30 FA-340, under Pilot Sgt. Glen. I. S., on circuits and landings, bounced on landing. The pilot stalled the aircraft, which crashed and burst into flames, being totally destroyed. The pilot was killed instantly. NAS Great Exuma: Two planes went to cover **Convoy NG-350**, then **Convoy GN-49** arrived, after a plane went to cover it. One plane sighted a Nicaraguan merchant vessel, and another reported a Radar contact.

26th March 1943

B-25 FR-383 DX airborne on navigation Exercise Number 3, gunnery and firing practice held exercise completed aircraft over Samaná Bay, landed back at base, nothing to report. B-25 FR-381 DC airborne on navigation Exercise Number 3, practice bombs dropped, dummy submarine attacks carried out against a sea marker, aircraft circled over Samaná Cay and Rum Cay, landed back at base. B-25 FR-383 DD airborne on navigation Exercise Number 3, bombing and air firing over Samaná Cay, landed back at base after descent though cloud exercise. B-25 FR-384 CM airborne on navigation Exercise Number 4, gunnery and bombing exercise carried out against sea marker, creeping line ahead search in accordance with detailed instructions landed back at base. Funeral is held for sergeant killed in accident to aircraft of RAF Ferry Command yesterday. He is buried in Western [RAF] Cemetery. HMS *Kenkora* is bound south in Straits passing Daytona. **Convoy Charleston-74** is bound south in Straits passing Canaveral navy plane escorting. USS *SC-1029* and USS *SC-508* are bound north from Matanilla Shoal, northwest Bahamas. SS *W. W. Mills* escorted by USS *SC-1267* is bound south towards Port Everglades. USS *Onika* (YO-143) and USS *YP-408* are bound south in Straits near Miami. **Convoy KN-230** is bound north in Straits near Miami. USS *YAG-14* is bound southwest with US Navy blimp near Bimini. Tug USS *Humrick,* tug *Ontario* and tug USS *YP-426* are bound north in Straits towards Miami. USS *Pioneer* (AM-105) is bound east in Straits. USCG *CG-83351* is bound west in Straits passing Nicholas Channel. Tug *M. Moran* towing SS *Alex* is bound west in Straits passing Nicholas Channel. USS *YAG-28* and USCG *CG-83397* are bound southeast towards Nicholas Channel. Tanker *Gulf Caribbean* is bound south near Cay Sal Bank. **Convoy KG-628** is bound southeast with US Navy blimp towards Nicholas Channel. No.113: Funeral of Sgt. I. S. Glen took place at RAF Cemetery. NAS Great Exuma: Two planes went to cover **Convoy GN-49** and departed.

27th March 1943

Local flying by pupils from Main Field under instructions revisionary gunnery exercise carried out by aircraft of B Squadron, local night flying arranged from Satellite Field cancelled due to unfavourable weather conditions. B-25 FR-371 CF airborne on navigation Exercise Number 3, practice bombs dropped on wave crests, poor visibility, exercise abandoned, landed back at base after descent through cloud exercise. **Convoy NK-530** is bound south with US Navy blimp near Palm Beach. **Convoy Charleston-74** is bound north in Straits passing Port Everglades. Tanker *Atlantic States* is bound south in Straits passing Port Everglades. USS *PC-625* is bound north in Straits. Tanker *J. H. Tuttle* is bound north in Straits. **Convoy Havana North Coast Cuba-34** is bound north in Straits. NAS Great Exuma: A plane went to cover **Convoy KG-628**. Submarine USS *R-10* was escorted through area.

28th March 1943

B-25 FR-384 CM airborne on navigation Exercise Number 4 sea marker, dropped gunnery and bombing exercise completed carried out satisfactorily flare released and searchlight avoiding action, practice was carried out, landed back at base. B-25 FR-381 DC airborne on navigation Exercise Number 4, exercise carried out satisfactorily, landed back at base. B-25 FR-364 CC airborne on navigation Exercise Number 4, sea marker dropped, gunnery and bombing exercise completed, exercise carried out satisfactorily, landed back at base. B-25 FR-376 CK airborne on navigation Exercise Number 4, sea marker dropped, gunnery and bombing exercise completed, flame float dropped and practice searchlight avoiding action was taken, landed back at base. B-25 FR-394 CN is airborne on navigation Exercise Number 7, bombing practice against moving object, photographic reconnaissance of Georgetown, Exuma, craft passed Little San Salvador, reconnaissance of Hatchet Bay and Harbour Island, homed successfully from Harbour Island on the Radio Compass descent through cloud exercise, landed back at base. B-25 FR-364 CC airborne on navigation Exercise Number 4, dummy U-boat attacks made against sea marker, destroyer sighted 3 nm away, vessel on course north, exercise completed, landed back at base. B-25 FR-393 DF airborne on navigation Exercise Number 4, destroyer sighted 5 nm away investigated ship, took photographs, continued patrol, exercise completed, landed back at base. B-25 FR-381 DC airborne on navigation Exercise Number 4, gunnery and bombing exercise carried out, escort vessel sighted in southerly course, craft diverted from exercise to investigate, took photographs, landed back at base. Baltimore FA-427 airborne for Trinidad then caught fire, with two crew killed, one seriously burnt, who was taken to RAF hospital. [The two deceased were buried on 29th March]. USS *PYC-9* north towards Daytona. USS *YP-427* is bound north towards Canaveral. USS *PC-625* is bound north in Straits passing Palm Beach. Tanker *J. H. Tuttle* northeast towards Matanilla Shoal, northwest Bahamas. SS *Tampico* is bound northeast in Straits near Miami. SS *Fort Lee* is bound south in Straits towards Miami. Tug USS *Meteor* and tug USS *Humrick* are bound south in Straits passing Miami. **Convoy NK-530** is bound south in Straits with US Navy blimp. USCG *CG-Balsam* is bound north in Straits. SS *Veracruz* is bound north in Straits. Passenger liner SS *Lady Nelson* is bound north in Straits. **Convoy NG-351** is bound south in Straits towards Guantanamo. No.113: Satellite Field. A-30 FA-427 under Pilot Capt. E. P. Walmsley, with Australian Sgt. P. L. Taylor, Navigator,

Mr. W.F. Morris, Radio Operator, whilst taking off on a delivery flight, swung on take-off onto rough ground. The aircraft burst into flames, being totally destroyed. Walmsley and Morris were killed, and P. L. Taylor was injured, being placed on the D. I. List [Dangerously Ill]. NAS Great Exuma: A plane went to cover **Convoy KG-628**, then **Convoy NG-351** arrived. A plane went to cover **Convoy NG-351**, then **Convoy GK-726** and **Convoy GK-728** passed through the area.

29th March 1943

B-25 FR-364 CC airborne on navigation Exercise Number 4, carried creeping line search ahead off Eleuthera, completed patrol, landed back at base. B-25 FR-384 CM airborne on navigation Exercise Number 4, escort vessel sighted, landed back at base after exercise completion. B-25 FR-383 DX airborne on navigation Exercise Number 5, landed back at base after detailed exercise completion. B-25 FR-376 CK airborne on navigation Exercise Number 4, landed back at base after detailed exercise completion. SS *Solitaire* is bound south towards Palm Beach. **Task Unit TU-0296** is bound south in Straits towards Palm Beach. Tanker *Bennington* is bound north in Straits passing Palm Beach. USS *SC-691* is bound north in Straits near Palm Beach. USS *Rockaway* (AVP-29) is bound north in Straits near Palm Beach. USS *YAG-20* and barge are bound northeast towards Matanilla Shoal, northwest Bahamas. SS *David L. Swain* is bound south in Straits towards Miami. SS *Clanranald* is bound north in Straits passing Miami. HMS *Tracker* is bound northeast near Cay Sal Bank, with a US Navy plane escorting. **Convoy GK-728** is bound northwest nearing Nicholas Channel with US Navy plane escorting. No.113: Airmen's Quarters, consisting of three Barrack Huts (No.817, No.819 and No.820) and one Ablution Hut (No.818), completed and taken over, or occupied. NAS Great Exuma: USS *Stevens* reported a sub-sea Sonic contact with a submarine. A plane went to cover **Convoy NG-351**. A plane practiced aerial gunnery, while another plane sighted a US Navy light cruiser, and a plane. A plane went on a search mission. Oceangoing US Navy tug USS *Montcalm* (AT-39) arrived to tow gasoline to Guantanamo. A plane arrived.

30th March 1943

Liberator FL-970 LH is airborne from Satellite Field, navigation Exercise Number 9, bombs dropped precisely on Sail Rock, Exumas, flew over Riding Rock, San Salvador, homing air to air exercises were carried out, aircraft LE homed over FL-970 LH, and then circled FL-970 LH. Then aircraft LE resumes respective courses navigation abandoned, aircraft homed on radio beacon, landed back at Satellite Field. Liberator FL-992 LE is airborne from Satellite Field, bombs dropped precisely on Sail Rock, Exumas detailed exercise carried out, FL-970 LH was circled, signals exchanged, landed back at base after exercise completion. SS *Clanranald* is bound north in Straits passing Canaveral. Tanker *E. R. Kemp* is bound south in Straits towards Palm Beach. **Convoy KW-139** is bound north in Straits. SS *Solitaire* is bound southwest in Straits. USS *YAG-28* and USCG *CG-63397* is bound southeast in Straits, passing Nicholas Channel.

31st March 1943

B-25's FR-378 CL, FR-382 DD, FR-383 DX, and FR-393 DF are airborne carried out navigation Exercise Number 5, covered area bounded by Eleuthera, Long Island exercise competed satisfactorily landed back to Main Field, nothing to report. Tanker *Bostonian*, SS *Jamestown*, and SS

Diploma are bound south in Straits towards Palm Beach. SS *Chicopee* is bound north in Straits passing Port Everglades with US Navy plane escorting. USS *SC-1063* is bound south in Straits. *U-68* clears Anegada Passage bound north, exiting region for France. Crown Lands Officer William N. Aranha declares a right-of-way for a new road "*....from a point on the western Boundary of Satellite Field to the corner of the Brady Estate at Old Fort, to a point on west Bay Street ...in the Western District. He secured "a tract of 500 acres known as Wilderness, granted unto Josiah Tatnell in 1789, and a tract of 1,000 acres known as Old Fort or Charlottesville, granted unto the Heirs of Rush Tucker in 1790 .... possession to be taken in behalf of His Majesty the King*." NAS Great Exuma: Submarine USS *S-42* was sighted. A plane covered **Convoy GN-50**; another plane searched the area of Crooked Island for survivors of a downed US Army Air Force bomber.

1st April 1943

Liberators FL-942 LC and FL-970 LH airborne for navigation Exercise Number 9, exercise competed satisfactorily, LC carried out homing exercise over FL-970 LH failed to meet difference in height. Continued respective patrols, landed back at base. B-25's FR-376 CK, FR-378 CL, FR-383 DK, and FR-393 DF are airborne, carried out navigation Exercise Number 5, exercises checked by radio fixes in Miami and Nassau. DX completed descent through cloud exercise, abandoned due to instruction from base, weather unfavourable, especially near Cuba, where there was heavy rainfall. **Task Group TG-20-14** (USS *LST*s, Landing Ships Tank) north from US Navy blimp. **Convoy NK-531** is bound southwest. Tanker *Esso Nashville* is bound northeast in Straits from Palm Beach. **Convoy KN-231** is bound north in Straits towards Bimini with US Navy plane escorting. USCG *CG-Blaze* is bound south in Straits. Tanker *Gulfmaid* is bound northeast in Straits. SS *Jamestown* is bound south in Straits. Tug USS *Meteor* is bound north in Straits. **Convoy Guantanamo-78** is bound east from Nicholas Channel, a US Navy plane escorting. Ferry *Jean Brillant* is bound southeast from Nassau towards Cat Island. Submarine USS *S-42* is bound north, to the east of Cat Island, in the Atlantic. NAS Great Exuma: A plane made a test flight, while another plane went to cover **Convoy GN-50**. Then **Convoy GN-50** departed, and **Convoy NG-352** arrived. A plane went to cover **Convoy NG-352**.

2nd April 1943

Liberators FL-976 LJ on navigation Exercise Number 9 practice bombs dropped on Sail Rock, Exumas, exercise took craft to San Salvador, where a convoy is sighted, consisting of 28 merchant ships and four escorts. An escort vessel sweeping to east of convoy, with signal exchanged. Then homed on radio beacon, and landed back at base. SS *John Appleby* is bound north in Straits near Daytona. Troop transport USS *USAT-Alamo* is bound south in Straits bound towards Palm Beach. SS *Elmera Sperry* and SS *George H. Williams* are bound north in Straits from Palm Beach. British Royal Navy Motor Launches **HMML Flotilla-73** is bound north in Straits towards Palm Beach. USS *YAG-20* and USS *YP-574* are bound northeast in Straits from Port Everglades. USS *YP-142* and USS *YP-574* are bound east in Straits from Port Everglades. USS *PC-1251* is bound west in Northwest Providence Channel towards Palm Beach. **Convoy NK-531** is bound south in Straits near Miami with US Navy blimp escorting. Tug USS

Meteor and barge and SS *San Pablo* are bound north in Straits towards Port Everglades. Tanker *Pan Virginia* is bound north in Straits near Bimini. USS *YAG-14* is bound southwest in Straits. USS *PC-1227* is bound northeast in Straits. USS *YAG-16* and USS *YP-218* are bound south from Cay Sal Bank. SS *Colony Trader* is bound north from Nicholas Channel. **Convoy KW-143** is bound south in Nicholas Channel, with US Navy plane escorting. *U-68* is attacked by an American Mariner aircraft (USN Squadron VP-204, P-9) with three bombs and suffers light damage off Puerto Rico. *U-155* re-enters area off Key West, sinking SS *Lysefjord*, then is bound to Cay Sal past Key West. NAS Great Exuma: Two planes went to cover **Convoy NG-352**. Ferry *Jean Brillant* arrived with freight. Tug USS *Montcalm* (AT-39) departed, towing gasoline barge USS *YAG-2*, escorted by a NAS Great Exuma plane.

3rd April 1943

B-25's FR-383 DX, FR-376 CK, and FR-393 DF airborne on navigation Exercise Number 5, DX returned to base artificial horizon, unserviceable. Captain took off again in FR-371 CF all aircraft completed exercise, landed back at base scattered showers during exercise heavy rain and lightning in south of Exuma. Report received that U-boat was sighted, request received from US Navy for cooperation in search Northeast Providence Channel and Northwest Providence Channel. Decided to send aircraft at dawn to commence patrol of area. Fifteen *Gulfstate* survivors (five wounded) picked up by USS *YP-351*, all were landed at Key West. USS *PC-509* is bound south in Straits towards Canaveral. SS *San Pablo* is bound north in Straits near Canaveral. Spanish SS *Mar Rojo* is bound north near Matanilla Shoal, northwest Bahamas. SS *Gargoyle* is bound north in Straits. Tug USS *Meteor* and tow USS *YP-452* are bound south in Straits from Port Everglades. Tug USS *Humrick* and tow are bound north in Straits. SS *Swivel* and HMS *Kenkora*, Troop transport USS *USAT-Alamo* with USS *SC-1276* are bound southwest in Straits. Tug USS *Partridge* and tow USCG *CG-83381* are bound southwest towards Old Bahama Channel, near Cuba. *U-68* is well east of Turks and Caicos and north of Anegada, bound east for France. B-25 FL-976 and B-25 FR-375 are bound north of Nassau. Two No.111 OTU aircraft investigate a U-boat reported transiting northeast to Northwest Providence Channel at dawn [next day]. Planes spent 10.5 hours searching from 100-fathom-line east to west, with nothing sighted. Northeast Providence Channel to Northwest Providence Channel, a U-boat is still reported in transit, B-25 FL-976 and B-25 FR-375 performed search. NAS Great Exuma: Ferry *Jean Brillant* is escorted by plane. Planes practiced flying in two teams.

4th April 1943

Liberator FL-962 LG airborne from Satellite Field on navigation Exercise Number 9, set course for Sail Rock, Exumas, two practice bombs released ,set course for base when weather turned unfavourable and aircraft experienced electrical storm, propellers were illuminated with three to four inches of electrical discharge. Liberators FL-976 LJ airborne from Satellite Field patrol, which was to be a creeping line ahead search instead over Northeast Providence Channel and Northwest Providence Channel, visibility of 5 nm when patrol commenced a single-masted, dark-grey fishing vessel was observed at anchor approximately 50 feet long. A search was carried out as far as 100 fathom lines on each side of the channels, landed back at

Satellite Field. Single-engine exercises were used throughout, nothing sighted. B-25 FR-375 CJ airborne on Anti-Submarine sweep in Northeast Providence Channel and Northwest Providence Channels area covered by a creeping line ahead search, exercise completed, nothing sighted, landed back at base. USS *SC-1023* is bound south in Straits from near Canaveral towards Palm Beach. SS *Moysalen* is bound north in Straits towards Port Everglades. **Convoy Guantanamo [GK]** is bound north in Straits from Nicholas Channel. SS *Marjory* is bound west from Old Bahama Channel in Straits, US Navy plane escorting. *U-185* re-enters area bound north, northwest of Haiti from Windward Passage, where it had been patrolling three days. NAS Great Exuma: Two planes practiced landings and Anti-Submarine bombing by dropping water-filled bombs.

5th April 1943

B-25 FR-380 DB airborne on Anti-Submarine sweep to cover both Northeast Providence Channel and Northwest Providence Channels, completed patrol, merchant ship sighted flying hoist signal BJYN. Attempts made to exchange signals unsuccessful, landed back at base, nothing to report. USS *YP-408* is bound south in Straits towards Canaveral. **Convoy Motor Torpedo Boat Squadron-15, Group 6** is bound north in Straits from Palm Beach. USCG *CG-83348* is bound west north of Walker Cay, north of Abaco. USS *YAG-20* is bound west of Walker Cay, bound to Palm Beach. SS *J. H. Tuttle* is bound south in Straits towards Port Everglades. Tanker *Atlantic States* is bound south in Straits from Bimini. USS *YP-444* is bound north in Straits. *U-68* exits the region east after attacks from Allied planes. USS *YP-453*, ex-*Pleiades* was stripped and sank off a cay south of Bimini, USCG *CGR-7007* assisted. NAS Great Exuma: A plane went to cover **Convoy GN-51**.

6th April 1943

Pupils of course Number 3 flying FR-380 DB, FR-379 DA, FR-378 CL, FR-371 CE, and FR-394 CN carried out air and navigation exercises. US-flagged *John Sevier* is torpedoed by *U-185* 20 nm southwest of Great Inagua while it was the lead ship in four ship **Convoy Guantanamo-83**. All 57 hands rescued by USS *Bennett* (DD-473) taken to Guantanamo next day. USS *YP-427* is bound south in Straits towards Canaveral. **Convoy Motor Torpedo Boat Squadron-14** is bound south in Straits towards Canaveral. **Convoy NK-532** is bound south in Straits from Canaveral with US Navy plane escorting. USS *SC-1016* is bound south in Straits from Canaveral. SS *Fort Lee* is bound north in Straits, west of Bimini. USS *Elmore* (APA-42) is bound west of Bimini in Straits bound north. **Convoy KN-232** is bound north in Straits past Bimini with US Navy plane escorting. **Convoy *Nassau*** is in Northwest Providence Channel bound west to Straits. Two tugs and USS *YP-426* are bound southwest in Straits. USS *YAG-18* and USS *YP-437* are bound southeast in Nicholas Channel. USS *YAG-16* and USS *YP-216* are bound north in Straits. *U-155* is north, to the northwest of Grand Bahama, then it will turn east for France. NAS Great Exuma: Two planes went to cover **Convoy GN-51**, then a plane went to cover **Convoy KG-629.**

7th April 1943

Pupils of Number 3 course flew FR-364 CC, FR-384 CM, FR-375 CJ, FR-376 CK, and FR-382 DD carried out air and navigation exercises. Tug USS *Dandy* and tow with USCG *CG-83401* and USCG *CG-83403* are bound south in Straits from Canaveral with US Navy plane escorting. USS *Threat* (AM-

124) is bound south in Straits from Canaveral with US Navy plane escorting. USS *YAG-13* is bound south in Straits from Canaveral. SS *Nickeliner* is bound south in Straits towards Palm Beach. SS *Virginia Lee* and USS *SC-1289* are bound south in Straits from Palm Beach. **Convoy NK-532** is bound south in Straits from Port Everglades. The house boat USS *YHB-14* and USS *YP-612* are bound west in Straits with US Navy blimp near Bimini. Oil barge USS *YO-51* and USS *YP-218* are bound north in Straits towards Port Everglades. USCG *CG-Spruce*, USS *SC-682* and USS *SC-684* are bound south in Straits towards Cuba. **Convoy Port Everglades-96** is bound west in Straits. USS *YMS-37* is bound northwest in Old Bahama Channel from Cayo Francés towards Cay Sal Bank. *U-155* is 50 nm north of Hope Town, Abaco, bound east for France. NAS Great Exuma: Two planes covered **Convoy KG-629**. A plane searched Crooked Island and Acklins Island for a US Army bomber which may have crashed. A plane went to cover **Convoy NG-353**, then another plane was sent on a search and rescue mission.

8th April 1943

B-25's FR-380 DB, FR-378 CC, FR-375 CJ, FR-376 CK, FR-384 CM, FR-371 CF, and FR-383 DX are airborne carried out air and local navigation exercises. Message received from Comgulfseafron that a B-26 Marauder *en route* from Morrison Field to Borinquen reported to have crashed near Castle Island and Acklins Island with four occupants: it is now believed the crew parachuted, per Controller of Operations, USAF. Morrison Field requested RAF to search aircraft, endeavoured for information, but had been unsuccessful as no radio communication with Acklins Island to augment search. Aircraft were dispatched from Morrison Field and another from Guantanamo, a USN Beechcraft airborne from Nassau on survey work participates in search. B-25 FR-371 CF is airborne and set course for Castle Island. Messages dropped to islanders for information; message received from Guantanamo that the B-26 Marauder was located, damaged on Acklins Island, one survivor found. Catalina was sent to pick up crew, CF continued search for missing occupants, encountered convoy of 23 merchant ships and three escorts sighted. Aircraft challenged ship of SNO, had no information regarding survivors, messages exchanged by Aldis lamp creeping line ahead, search started, message received from CF that B2-6 had crashed on sandy flat, and was badly damaged, no propellers visible, islanders had no news regarding crew, aircraft circled Castle Island and set off for Jamaica Bay. Catalina sighted in water, crew of missing aircraft waved from shore and a dinghy. With crew found, message sent to abandon search, aircraft landed back at base, message received from Catalina that all crew of B-26 picked up with only slight injuries. SS *Buena Vista* is bound south in Straits towards Canaveral. USS *Rockaway* (AVP-29) is bound south in Straits from Palm Beach. Ferry *Jean Brillant* and USS *IX-75* are bound east in Straits from Miami to Bimini. USS *YO-51* and USS *YO-142* are bound north in Straits from Miami. Tanker *Nickeliner* is bound southeast in Straits from Miami. Tanker *Gulfland* is bound north into Straits, with US Navy plane escorting. USS *YAG-16* and USS *YP-218* are bound north in Straits from Cay Sal Bank towards Miami. USCG *CG-Spruce* is bound south in Straits towards Cuba. *U-155* is bound east of Abaco, east for France. *U-185* is bound north of Dominican Republic, bound east for France. NAS Great Exuma:

Two planes went to cover **Convoy NG-353**. A plane departed for Acklins Island. A plane landed in Pompey Bay, Acklins and rescued the pilot of a downed B-26 plane from there. A plane conducted a search and rescue mission and retrieved balance of the crew of the B-26.

9th April 1943

B-25's FR-393 DF, FR-380 DB, and FR-371 CF carried out air and local navigation exercises. B-25 FR-379 DA carried out navigation Exercise Number 5, landed back at base, took off again for the range nothing heard of aircraft on radio or radio, overdue action taken search organized for four aircraft was to cover area between cardinal points by individual patrols. B-25's FL-180 BA, FR-371 CF, 75 CJ, and FR-383 DX are airborne for air sea rescue search. Area searched, nothing found, landed back at base message received from FK-174 AL, patch of oil was sighted at 215° to 220°, 20 nm distance, captain of aircraft decided that the sighting was sea-weed returned to base was decided to search for the aircraft on the morning of 10th April. Tanker *Gulfland* is bound north in Straits towards Jacksonville with US Navy plane escorting. USS *YO-51* is bound east in Northwest Providence Channel towards Abaco. SS *Manuel Rionda* is bound east in Straits from Miami. **Convoy Miami-110** is bound south in Straits. SS *Buena Vista* is bound southwest in Straits. **Convoy Motor Torpedo Boat Squadron-14,** USS *SC-1001*, USS *SC-1059*, and USCG *CG-Spruce* are bound south in Straits towards Cuba. No.113: The first aircraft for delivery by the South Atlantic Wing were received. A high subsequent volume of aircraft throughput was caused by the release of a large portion of the back-log, and the US Army Air Force finding it easier, of course, to deliver to Nassau than to Accra. Thus No.113 was able to deliver a large number of aircraft with comparatively small number of crews. NAS Great Exuma: A plane conducted a scouting search for a convoy route.

10th April 1943

FR-398 CF, FR-382 DD, and FR-375 CJ airborne at different times for air exercise and air sea rescue search, carried out practice bombing over the range commenced creeping line ahead search visibility was 3 nm, made a reconnaissance of all the small islands during its return back to base landed back at base having sighted nothing. Tug USS *Sampson* and USCG Cutter *CG-Crawford* are bound south in Straits towards Canaveral. SS *Moysalen* and USS *PC-551* are bound north in Straits towards Canaveral. USS *SC-506* is bound north in Straits from Port Everglades. USS *PE-100* is bound east in Straits from Port Everglades towards Bimini. **Convoy KN-233** is bound north in Straits towards Bimini, with US Navy plane escorting. SS *Manuel Rionda*, USS *SC-1019*, and USS *PC-1227* are bound south in Straits. NAS Great Exuma: Two planes went to cover **Convoy GN-52.** A plane went to cover **Convoy GK-730.**

11th April 1943

B-25's FR-383 DX, FR-371 CF, FR-375 CJ, FR-381 DC, and FR-393 DF airborne for local and air navigation exercise, FR-379 DA carried out air sea rescue. Message received from civilian Pan American aircraft that small pieces of wreckage are sighted afloat south of New Providence. Large yellow-green dyed area and oil flecks also observed, could not be ascertained if men were swimming near the wreckage yellow green dyed area was thought to be a dinghy. FR-375 CJ airborne for local air exercise, diverted to last reported position, ASR launch HMS *P-190* is dispatched from

Lyford Cay for search as well later FL-199 BD airborne for search. ASR launch reported dinghy was empty, CJ recalled and landed back at base nothing to report, ASR launch also recalled dinghy, brought it to Main Field. It was found to belong to FR-379 DA, an aircraft had been missing two days, since 9th April. The dinghy had been inflated, but there was no further evidence of the fate of the lost crew. **Convoy NK-533** is bound south in Straits towards Canaveral. SS *A. S. Hansen* and USS *YMS-64* are bound northwest in Straits, off Canaveral. Tug USS *Sampson* and USCG Cutter *CG-Crawford* are bound south in Straits towards Palm Beach. USS *YMS-36* and USS *YMS-37* are bound north in Straits near Port everglades. Tug USS *Dandy* and tow with two USCG vessels are bound south in Straits towards Miami. USS *SC-651*, USS *SC-655* and USS *SC-666* are bound north in Straits. Oil barges USS *YO-51* and USS *YO-142* with USS *YP-513* and USCG *CG-83397* are bound west in Northwest Providence Channel to the Straits and Port Everglades. USS *YO-143* and USS *YP-603* are bound southeast in Straits from Miami. USS *YAG-28* and USS *YP-408* are bound northeast towards Bimini. USS *YAG-16*, USS *YP-218*, and USS *SC-500* are bound south in Straits from Miami. Tug USS *Meteor* and tow, USS *YP-426*, and USS *YAG-13* are bound southwest in Straits. NAS Great Exuma: Two planes went to cover **Convoy GN-52**, which then departed.

12th April 1943

General flying instructions given to pupil during day. B-25 FR-373 CJ and B-25 FR-381 DC airborne for local and air navigation exercise. B-25 FR-76 CK is airborne to escort convoy sailing from Nassau to Miami, met convoy, box patrol carried out as air coverage was 20 nm wide and 30 nm long, FR-81 DC relieved duty of CK, then CK landed back at base. DC provided cover until convoy was met by US Navy blimp, DC left convoy and landed back at base. Tanker *Gulfland* is bound south in Straits towards Palm Beach with US Navy plane escorting. **Convoy NK-533** is bound south in Straits towards Port Everglades, US Navy blimp escorting. USS *SC-522* is bound north in Straits with US Navy blimp escorting. USS *YAG-16* is bound south towards Cuba. SS *Manuel Rionda* is bound north from Nassau. Tug *Nassau* is bound north towards Abaco with US Navy plane escorting. **Convoy GK-730** is bound northwest with US Navy blimp escorting. NAS Great Exuma: A plane went to cover **Convoy NG-354**, which entered the area seven hours ahead of schedule.

13th April 1943

B-25 FR-394 CN airborne on air navigation Exercise Number 7, photographs taken of Hatchet Bay and Georgetown, bombing and gunnery exercise carried out, landed back at base. B-25 VT-381 DC is airborne air navigation Exercise Number 7, bombing and gunnery exercise carried out commencement of air navigation exercise descent though cloud exercise, landed back at base. FR-376 CK, FR-381 DC, and FR-378 CL airborne for local and air navigation exercise. SS *W. C. Atwater* is bound south in Straits towards Palm Beach. Tug USS *YAG-20* is bound north in Straits from Port Everglades. Tug USS *Sampson* is bound south in Straits from Port Everglades. USS *YAG-16* is bound south in Straits towards Cuba. SS *Trondanger* is bound east in Straits, with US Navy blimp escorting. *U-155* exits area, bound east for France. NAS Great Exuma: A plane went to cover **Convoy NG-354**, then a plane sighted a tug with two barges and another plane practiced instrument flying and Anti-Submarine warfare (ASW).

14th April 1943

B-25's FR-394 CN and FL-199 BD are airborne at different times, air navigation Exercise Number 7, photographs taken of Harbour Island and Hatchet Bay and Georgetown, landed back at base. FR-378 CL, FR-376 CK, and FR-381 DC airborne for local for air navigation exercise. Tug USS *San Pedro* and USS *SC-1277* are bound south in Straits towards Canaveral. SS *Merida* and USS *YP-501* are bound south in Straits passing Canaveral. USS *SC-657* and tanker *A. S. Hansen* are bound south in Straits from Palm Beach. USS *YO-143* and USS *YP-603* are bound northwest in Straits towards Port Everglades. USS *PE-10*5 is bound southwest in Straits with US Navy plane escorting. USS *YO-51*, USS *YO-142* and USS *YP-395* are bound southwest in Straits. SS *J. H. Tuttle* is bound north in Straits with US Navy plane escorting. USCG *CG-83390* is bound southeast from Nicholas Channel. *U-183* re-enters area via Mona Passage, bound north, to the east to France. NAS Great Exuma: A plane practiced landing and anti-submarine bombing.

15th April 1943

B-25's FR-376 CK and FL-199 BD are airborne for air navigation exercise. B-25 FR-373 CG, FR-378 CK, FR-394 CN, and FR-378 CL are airborne at different times photographic and visual reconnaissance Exercise Number 7, photographs taken of Georgetown, gunnery exercise, descent through cloud exercise, then landed back at base. B-25 FL-199 BD is airborne for navigation Exercise Number 1, bombing practice against Sail Rock, Exumas target complete exercise, descent through cloud exercise landed back at base. No.111 OTU has expanded rapidly, now including 84 officers and 1,300 other ranks inhabited it with some 400 trainees. *U-183* is bound north from Mona Passage, bound east to France. SS *Andree* is bound south in Straits towards Canaveral. Tanker *Churubusco* is bound south in Straits passing Canaveral. Tug *Ontario*, tug USS *Humrick* and SS *Merida* are bound south in Straits towards Miami. Tug USS *San Pedro* is bound south in Straits passing Miami. USS *YAG-28* is bound east in Straits from Miami to Bimini. USS *PC-1194* is bound southwest in Straits. USS *YN-25* is bound southeast from Miami towards Cuba. Tanker *A. S. Hansen* is bound southeast in Straits. NAS Great Exuma: A plane went to cover **Convoy GN-53**, however night coverage was cancelled, due to weather conditions.

16th April 1943

Local night flying carried out by two aircraft. B-25's FR-376 CK and FL-199 BD carried out air and local navigation exercise. B-25's FR-372 CG and FR-383 DX are airborne for photographic and visual reconnaissance over Georgetown and Hatchet Bay. Gunnery exercise, descent through cloud exercise, then landed back at base. B-25 FR-381 DC is airborne for navigation Exercise Number 1, bombing practice against Sail Rock, Exumas target complete exercise, descent through cloud exercise, landed back at base. B-25 FR-372 CG is airborne for navigation Exercise Number 5, exercise was over Abaco, gunnery practice exercise completed, landed back at base. Tug USS *Hopi* (AT-71) is bound south in Straits towards Canaveral with CAP (Civil Air Patrol) plane escorting. Tanker *Birch Coulee* is bound north in Straits off Canaveral. **Convoy NK-534** is bound south in Straits towards Palm Beach with US Navy plane escorting. SS *Lake Ormoc* is bound south in Straits towards Palm Beach. **Convoy KN-234** is bound north in Straits passing Palm Beach with US Navy plane escorting. USS *AM-124* and USS *SC-1018* are bound north in Straits passing Port Everglades. Tanker *Esso Portland* and SS *Andree* are bound south in Straits towards Port Everglades. Tanker *Churubusco* is bound south in Straits passing Port Everglades. USCG *CG-83348* and USS *YAG-20* are bound east from Walker Cay, north of Abaco. Tug *Ontario* and tug USS *Hercules* are bound southwest in Straits. USS *YAG-28* is bound southeast from Nassau towards Exuma. SS *Tristan* and USS *YAG-16* are bound north in Old Bahama Channel from Cayo Francés towards Cay Sal Bank. *U-129* enters area south from Hatteras bound to northeast of Bahamas. NAS Great Exuma: A plane went to cover **Convoy KG-630**. A plane practiced landing and anti-submarine warfare (ASW), and USCG *CG-Spruce* and tug USS *YAG-28* arrived

17th April 1943

B-25 FR-372 CG and FR-393 DF are airborne at different times, navigation Exercise Number 1, bombing practice against Sail Rock, Exumas target complete exercise, descent through cloud exercise, landed back at base. B-25 FR-384 CM is airborne for navigation Exercise Number 2, bombing practice complete exercise, descent through cloud exercise landed back at base. B-25's FR-383 DX and FL-199 DX airborne carried out detailed navigation exercise landed back at base. Trawler *E. M. Humphries* is bound south in Straits, towards Palm Beach. Two tugs and two barges are bound south in Straits from Port Everglades. SS *Sogoland* is bound north in Straits. USS *YN-25* is bound south in Straits from Miami. Tug USS *Hopi* (AT-71) is bound southwest in Straits. Tanker *Gulfmaid* is bound northeast in Straits. USS *YAG-16* is bound north in Straits. Tug *Thomas E. Moran* is bound northeast from Nicholas Channel. No.113 base's name is changed from RAF Ferry Command to RAF Transport Command. NAS Great Exuma: Two planes covered **Convoy KG-630**. USS *YAG-28* departed, then a 3-masted sailing schooner was sighted and determined to be on course. **Convoy NG-355** entered the area.

18th April 1943

B-25 FR-372 CG, FR-393 DF, FR-376 CK, and FR-384 CM are airborne at different times. Navigation Exercise Number 1, bombing practice against Sail Rock, Exumas target, complete exercise, descent through cloud exercise landed back at base. FR-372 CG also homed on radio compass. B-25 FR-379 CJ airborne for navigation Exercise Number 7, photographs taken of Harbour Island, Hatchet Bay, and Georgetown, completed exercise descent through cloud exercise landed back at base. USS *YAG-28* is bound north in Straits towards Northwest Providence Channel. Oil barge USS *YO-143* is bound south in Straits from Miami. USS *YT-379* is bound north in Straits. Crash boat USS *C-9488* is bound south in Straits. **Convoy Key West-162** is bound southwest in Straits, past Cay Sal Bank. USS *YT-328* is bound west in Nicholas Channel. USS *YAG-16* is bound north in Straits past Cay Sal Bank. *U-183* exits area, bound east to France. Message received that a US Navy PBT aircraft had force-landed on Whale Cay, so B-25 FR-381 DC and ASR launch HMS *PK-180* marine craft diverted from local flying area and were sent to search for area message received that the PBT was safe, so aircraft were recalled. NAS Great Exuma: Three planes went to cover **Convoy NG-355**. A plane sighted a US Navy yard patrol craft, boat designated USS *YP-*.

19th April 1943

FR-384 CM is airborne on Anti-Submarine patrol Northwest Providence Channel area covered by creeping line ahead search nothing sighted, landed back at base. FR-375 CJ airborne for navigation Exercise Number 2, smoke float dropped and bombing exercise carried out against sea marker, descent through cloud exercise, landed back at base. FR-376 CK airborne for navigation Exercise Number 2, bombing exercise carried out, descent through cloud exercise landed back at base nothing to report. FR-382 DD and FR-381 DC airborne at different times navigation Exercise Number 2, exercise carried out, descent through cloud exercise landed back at base nothing to report. Liberator FL-970 airborne from Satellite navigation Exercise Number 5 completed, landed back at base. Troop transport USS *USAT-Brooklyn Heights* is bound south in Straits, towards Canaveral. USS *SC-1278* and tanker *E. R. Kemp* are bound south in Straits towards Palm Beach. Tug USS *YT-359* is bound north in Straits from Port Everglades to Palm Beach. Ferry *Jean Brillant* is bound west in Northwest Providence Channel towards Port Everglades. USS *SC-1027* is bound south in Straits off Miami. Tanker SS *L. L. Abshire* is bound north in Straits. SS *Seatrain New Orleans* is bound east from Key West, in Straits. USS *YO-143* is bound south in Straits towards Old Bahama Channel. *U-129* is 200 nm northeast of Abaco, zig-zagging. NAS Great Exuma: A plane covered **Convoy NG-355**, then a convoy departed from the area. USS *Fury* reported a sound contact on a submarine.

20th April 1943

B-25 FR-371 CF is airborne for Exercise Number 2 cross over patrol, exercise completed, descent through cloud exercise landed back at base. B-25 FR-382 DD is airborne for Exercise Number 3, bombs dropped on sea marker, exercise completed, descent through cloud exercise, landed back at base. B-25 FR-374 CJ airborne for Exercise Number 2, bombing and gunnery exercise, exercise completed, landed back at base. B-25 FR-381 DC airborne for Exercise Number 2, bombing and gunnery exercise, completed, landed back at base. B-25 FR-393 DF airborne for Exercise Number 2 bombing and gunnery exercise. Exercise completed, landed back at base. *U-129* is bound northeast of Abaco, zig-zagging. NAS Great Exuma: USCG *CG-Spruce* departed, escorted by USS *YP-441*. A small yacht was sighted, then a plane went to cover **Convoy GN-54**, and another plane went to cover **Convoy GK-731**.

21st April 1943

FR-378 CL carried out air exercise and local navigation flight, local night flying instructions given to pupils. B-25 FR-393 DF is airborne for Exercise Number 2a, practice south of Abaco from 50 feet, exercise completed, landed back at base. B-25 FR-381 DC is airborne for Exercise Number 2a, navigation exercise, completed bombing practice, descent through cloud exercise, landed back at base. B-25 FR-382 DD is airborne for Exercise Number 3, bombs dropped, exercise completed, descent through cloud exercise, landed back at base. B-25 FL-184 BB was on local flying, when no signal received from aircraft for landing. So overdue action was taken, with aircraft KK-180 AS, FK-177 AO, FK-171 AK, and FK-181 ST airborne to search waters around the island, and ASR marine craft standing by contacted AT, directed to empty dinghy that was sighted 10 nm east of New Providence, message on radio from AT to AH intercepted and broadcast to other aircraft 'dinghy found, aircraft and floating objects circling and directing marine craft.'

Message from marine craft was that empty dinghy found 8 nm southeast of Booby Island, dinghy bore number of missing aircraft, wreckage sighted, no survivors located, aircraft returned to base. Aircraft is possibly submerged in 20 feet of water, thorough search again made, no further wreckage came to surface. British Motor Minesweepers HMS *BYMS-43*, HMS *BYMS-44*, HMS *BYMS-45*, and HMS *BYMS-46* are bound south in Straits passing Daytona to Canaveral. USS *SC-1278*, USS *SC-736*, and SS *Cockerel* are bound south in Straits towards Palm Beach. **Convoy NK-535** is bound south in Straits, towards Palm Beach, with US Navy plane escorting. **Convoy KN-235** and HMS *Chaser* are bound north in Straits with US Navy plane escorting. British Royal Navy Motor Launch HMS *HMML ST-64* is bound northeast in Straits. USS *YAG-28* is bound northeast in Straits from Port Everglades. Tug *Ontario*, tug USS *Hercules*, and SS *Solitaire* are bound southwest in Straits from Port Everglades. USS *YP-441* is bound west in Northwest Providence Channel. HMS *Fencer* is bound north from Straits. USS *YO-143* is bound north in Old Bahama Channel from Cayo Francés. **Convoy GK-731** is bound north from Old Bahama Channel towards Cay Sal Bank. NAS Great Exuma: Two planes covered **Convoy GN-54**, then the convoy departed from area.

22nd April 1943

B-25 FR-381 DC airborne for Exercise Number 2, cross over patrol bombing over sea marker, exercise completed landed back at base. B-25's FR-393 DF, FR-32 DD, and FR-371 CF are airborne at different times. Exercise Number 3, exercise completed, descent through cloud exercise, landed back at base. B-25 FR-384 CM is airborne for Exercise Number 3 bombing exercise over range arrived at Great Sale Cay landed back at base. B-25's FR-383 DX, FR-384 CM, and FR-381 DC, airborne for Exercise Number 4. USS *PC-575* is bound south in Straits towards Canaveral. SS *Mambi* is bound north in Straits towards Canaveral. USS *PC-492* is bound north in Straits from Port Everglades. Tanker *J. H. Tuttle* is bound south in Straits towards Cay Sal Bank with US Navy plane escorting. SS *Cockerel* is bound south in Straits with US Navy plane escorting. Tanker *Gertrude Kellogg*, SS *Samuel Heintzelman*, SS *J. G. Bennett,* and SS *Joseph Gale* are bound northeast in Straits. **Convoy Key West-165** is bound southeast from Cay Sal Bank towards Old Bahama Channel. **Convoy Havana North Coast Cuba-9** is bound northwest from Nicholas Channel, with Cuban plane escorting. B-25 FL-942 FC Liberator, USS *Swanson* are in Straits of Florida. One No.111 OTU aircraft and USS *Swanson*, which reported a submarine sighted searched the following day, with nothing further sighted. Bahamas, in an encoded position, B-25 FL-942 FC Liberator, US Navy sent position to search. *U-185* is in the Windward Passage, where it fired single shot from tube VI on a destroyer which is stopped alongside a lifeboat. Shortly after the shot, the destroyer gets underway; the shot was a miss, then the U-boat saw a Consolidated flying boat. NAS Great Exuma: A plane made a training flight, then two planes covered **Convoy NG-356**.

23rd April 1943

Signal received from US Navy destroyer which had attacked a submarine; sweep requested at daybreak. Liberator FL-942 LC is airborne from Satellite Field, carried out creeping line ahead search, aircraft patrolled the area, destroyer was sighted, was USS *Swanson*, aircraft flashed letter of the day, and destroyer replied, aircraft landed at Satellite Field having sighted nothing, S/E [sensing device] was used through-out. Liberator BZ-722 LK was airborne from Satellite Field to relieve aircraft FL-942 LC on the anti-submarine patrol, creeping line ahead patrol aircraft returned to Satellite Field, sighted nothing. S/B used throughout patrol. B-25 FR-393 DP airborne on navigation Exercise Number 3, bombing and gunnery exercises carried out against sea marker off Great Sale Cay, exercise completed; aircraft returned to base. B-25 FR-383 DX airborne on navigation Exercise Number 3, merchant vessel sighted 2 nm distant on port bow, vessel was a small steam ship of over 1,000 tons, painted grey and of British ownership, had one funnel amidships and two masts, was lightly laden, aircraft completed exercise and landed, after carrying out bombing exercises and descent through cloud exercise. B-25 FR-371 CF airborne, navigation exercise, set course for Harbour Island, bombing and gunnery exercise carried out, landed back at base. B-25 FR-332 DD airborne exercise. Number 3, merchant ship sighted 3 nm to 4 nm distant on port beam, vessel was coloured and over 1,000-ton. Aircraft completed exercise, carried out bombing and gunnery, landed back at base. Tanker *Mobiloil* is bound north in Straits. USS *SC-1028* is bound south in Straits. Tanker *18 De Marzo* is bound south towards Palm Beach, US Navy plane escorting. SS *Lyman Hall* is bound south in Straits from Palm Beach to Miami. USS *YAG-28* is bound south past Matanilla Shoal, northwest Bahamas towards Miami. SS *Britamer* is bound north in Straits off Bimini. Tug USS *Meteor* is bound south in Straits off Miami. Tanker *Atlantic States* is bound south in Straits towards Cay Sal Bank. SS *Dartmouth* is bound northeast in Straits off Cay Sal Bank. USS *PC-575* is bound west in Straits towards Key West. **Convoy Havana North Coast Cuba-10** is bound east towards Nicholas Channel, with a Cuban navy plane escorting. Court of Inquiry into loss of B-25 FL-184 and five lives is carried out by a Wing Commander, Chief Technical Officer, Chief Flying Officer, and S/L Vare. NAS Great Exuma: Two planes went to cover **Convoy NG-356**. A plane sighted four tankers and three destroyers. Tug USS *Montcalm* (AT-39) was escorted by British minesweepers of the Royal Navy. Night coverage of **Convoy NG-356** was cancelled, due to weather conditions.

24th April 1943

B-25 FR-38I DC airborne on navigation Exercise Number 3, completed exercise, landed back at base. B-25 FR-383 DX on navigation Exercise Number 4, bombing and gunnery exercises, creeping line ahead patrol, completed exercise, landed back at base. B-25 FR-371 CF airborne on navigation Exercise Number 4, bombing and gunnery exercises carried out off Palmetto Point, exercise completed, landed back at base. B-25 FR-382 DD airborne on navigation exercises Number 4, bombing and gunnery exercises off Harbour Bay, exercise completed descent through cloud exercise, landed back at base. B-25 FR-393 DF airborne on navigation exercises Number 4, exercise completed landed back at base. SS *Santa Catalina* is torpedoed by *U-129* 400 nm northeast of Abaco. Ten officers, 47 crew, 28 armed guards, and 10 passengers abandon ship. SS *Venezia* rescues survivors northeast of Abaco and takes them to San Juan. USS *YP-375* and USS *Turquoise* (PY-18) are bound south in Straits towards Canaveral. USS *SC-1279* is in Straits off Canaveral. USS *YP-429*, British

Motor Torpedo Boats HMS *MTB-88* and HMS *MTB-91* are bound north in Straits from Canaveral. SS *R. C. Stoner* is bound south in Straits from Port Everglades with US Navy plane escorting. British Motor Minesweepers HMS *BYMS-43*, HMS *BYMS-44*, HMS *BYMS-45*, and HMS *BYMS-46* are bound south in Straits from Port Everglades. **Convoy Havana North Coast Cuba-10** is bound southeast in Nicholas Channel. SS *Merida* is bound south in Nicholas Channel. USS *Portent* (AM-106) is bound northeast in Straits. Tanker *18 De Marzo* is bound southwest in Straits towards Key West with US Navy plane escorting. *U-185* is bound northeast of Bahamas, reporting that a US Navy ship has either located or seen us, and does not know if we are his starboard escort. The destroyer rushes forward, stops straight course, does not turn after. Stern fan ready to shoot at the destroyer. NAS Great Exuma: A plane practiced Anti-Submarine warfare.

25th April 1943

USCG *CG-83402* and USS *YP-375* are bound south in Straits towards Palm Beach. USS *Hopi* (AT-71) is bound north in Straits. **Convoy NK-536** is bound south in Straits towards Palm Beach with US Navy plane escorting. **Convoy KN-236** is bound north towards Bimini with US Navy plane escorting. USS *SC-1066* and USS *SC-1267* are bound southwest in Straits. **Convoy KG-631** is bound southwest in Nicholas Channel with US Navy blimp escorting. *U-129* is bound east northeast of Bahamas. NAS Great Exuma: A plane went to cover **Convoy GN-55.** A plane sighted a heavy transport and a US Navy destroyer, and another plane practiced Anti-Submarine warfare (ASW).

26th April 1943

Local flying through the day. USCG *CG-Juniper* and USS *YP-603* are bound northwest in Straits passing Canaveral. **Convoy Port Everglades-113** is bound south in Straits from Port Everglades. USS *PC-1227* is bound southwest in Straits. *U-129* is bound northeast of Bahamas, where it reports an escorting destroyer astern, then two vessels, the nearest has a cruiser silhouette with two smokestacks, and escorting destroyer. Morse calls with dimmed searchlight from the cruiser. No BOLD device ejected, because the boat went deep too quickly. First two explosives, with locating, interval approximately 2 seconds, then depth charges. Without speeding up to throwing speed, three depth charges apparently set deeper, heavy concussion, no damage. Propeller noises astern at regular intervals, apparently driving search curves. Aircraft detection searching. Since apparently there has been all sorts of disturbance, dived, proceeded submerged. NAS Great Exuma: A plane went to cover **Convoy GN-55.** A plane sighted two tankers. A plane arrived to escort tug USS *Montcalm* (AT-39) to Guantanamo.

27th April 1943

B-25 FR-375 CJ airborne on navigation exercises Number 5, exercise completed descent through cloud exercise, landed back at base. B-25 FR-383 DD airborne navigation exercises Number 5, motor was giving trouble when over San Salvador sent signal to base that it was returning due to motor trouble unable to gain height on one engine. *In extremis*, depth charges jettisoned, landed back at base, captain airborne again on FR-380 DD, completed exercise landed back at base. B-25 FR-378 CL airborne, engine developed trouble shortly after take-off returned to base. B-25 FR-383 DX airborne on navigation exercises Number 5, exercise completed descent through cloud exercise, landed back at base. USS *IX-75* and USS *PC-594* are bound northwest in Straits, passing Canaveral. USS *YO-143* and USS YP-*612* are bound south in Straits from Port Everglades. USS *SC-743*, USS *SC-1062*, and USS *SC-1045* are bound southwest in Straits. NAS Great Exuma: Two planes covered **Convoy NG-357**.

28th April 1943

Local flying through the day, night flying training also conducted. B-25 FR-375 CJ airborne for navigation exercises Number 4, part exercise carried out, returned back to base without transmitter unserviceable. B-25 FR-384 CM airborne for navigation exercises Number 4, exercise completed, landed back at base. B-25's FR-380 DB, FR-381 DC, FR-384 CM, and FR-375 CJ are airborne for navigation exercises Number 5, exercise completed, landed back at base. USS *PC-594* is bound south in Straits. British Royal Navy Motor Launches HMS *HMML-1173* and HMS *HMML-1174* are bound south in Straits from Canaveral. Tanker *Pan Massachusetts* is bound south in Straits towards Palm Beach. USCG *CG-Juniper* is bound south in Straits from Palm Beach. USS *YT-338*, USS *YF-311* and USS *YP-408* are bound east towards Walker Cay, north of Abaco. Irish SS *Irish Oak* is bound north in Straits from Miami. Ferry *Jean Brillant* and USCG *CG-83405* are bound west in Straits from Bimini to Miami. USS *SC-1276* and USS *SC-1059*, USS *YAG-20* and USS *YP-395* are bound southwest in Straits. *U-129* is bound northeast of Bahamas, where it crash dives for aircraft. NAS Great Exuma: A plane covered **Convoy NG-357**, which then departed.

29th April 1943

B-25's FR-375 CJ, FR-384 CM, and FR-380 DB airborne for navigation exercises Number 5. Liberator FL-974 LC airborne from Satellite Field, in creeping line ahead exercise. B-25 FR-375 CJ airborne for navigation exercises Number 4. All landed back at base after completing exercise. USS *Surprise* (PG-63) is bound south in Straits. SS *Cerro Gordo* is bound south in Straits with US Navy plane escorting. USS *YP-513* is bound north in Straits towards Canaveral. USS *SC-512* is bound south in Straits towards Palm Beach. British Royal Navy Motor Launches HMS *HMML-1173* and HMS *HMML-1174* are bound south in Straits towards Palm Beach. SS *W. S. USS Young* is bound north in Straits towards Palm Beach. SS *Colony Trader*, USCG *CG-83406*, and USS *SC-1001* are bound south in Straits from Palm Beach to Port Everglades. USS *YMS-74* is bound north in Straits. Ferry *Jean Brillant* and USCG *CG-83405* are bound northeast from Miami to Northwest Providence Channel. Tanker *Gulf Caribbean* is bound north in Straits with US Army plane escorting. SS *J. H. Tuttle* is bound north in Straits. USS *Samaná* (YAG-28) is bound west in Straits from Bimini to Miami. **Convoy Port Everglades-116** is bound southwest in Straits. **Convoy Havana North Coast Cuba-61** is bound northeast in Straits. *U-176* enters region for a patrol of Old Bahama Channel. NAS Great Exuma: A plane practiced Anti-Submarine bombing.

30th April 1943

Liberator FL-970 LH is airborne for navigation Exercise Number 9 completed patrol dropped bomb on Sail Rock, Exumas, landed back at Satellite Field. Liberator BZ-722 LK in creeping line ahead exercise from Satellite Field, dropped bomb on Sail Rock, Exumas landed back at Satellite Field, nothing to report. Liberator FL-942 LC is airborne for navigation Exercise Number 9, dropped bomb on Sail

Rock, Exumas descent through cloud exercise, landed back at Satellite Field. B-25's FR-384 CM and FR-380 DB airborne at different times navigation Exercise Number 5, exercise completed descent through cloud exercise, landed back at base. B-25 FR-380 DB airborne, navigation Exercise Number 2 depth charges dropped, exercise completed descent through cloud exercise, landed back at base. SS *Horace Williams* is bound south in Straits towards Canaveral with US Army plane escorting. USCG *CG-83396* is bound west towards Matanilla Shoal, northwest Bahamas. USS *YP-42* is bound northwest from Abaco towards Walker Cay, north of Abaco. USS *SC-1046* is bound south in Straits towards Miami. SS *Atlantic States* is bound north in Straits with army plane escorting. USS *YO-143* and USS YP-612 are bound northwest in Straits towards Miami. USS *Surprise* (PG-63) and USCG *CG-Juniper* are bound south in Straits. **Convoy KN-237** is bound northeast from Straits. SS *James G. Birney* is in Straits northwest from Nicholas Channel, with US Navy plane escorting. First seven months of No.111 OTU are characterized by "serious shortages of equipment and persistent shortages of staff." Countered by improvisation and long hours of work. Decision to double size of base was not feasible without US engineering division's Pleasantville Construction. Three officers and crew of Hudson bomber FK-365 [which crashed day before], were buried at RAF Cemetery with full military honours. NAS Great Exuma: Two planes went to cover **Convoy GN-56**. A plane sighted USS *Stevens*. The ferry *Jean Brillant* is escorted by US Coast Guard cutter.

1st May 1943

Liberator FL-970 LH is airborne from Satellite Field, a creeping line ahead exercise, contact made suggesting that convoy was in the vicinity in three lines. A merchant ship was sighted, stationary, 10 nm to the starboard, lit from bow to stern with series of white lights. The sides of the ship appeared to be outlined by lights, outline of hull clearly visible when passing the bow, aircraft circled twice passing immediately over the ship, received no challenge nor other signal from ship. Then completed exercise and landed back at Satellite Field. It was later ascertained ship was an Irish vessel. Liberator BZ-722 LK airborne from Satellite Field for navigation Exercise Number 9, pilot's compass discovered to be inaccurate, practice bombs dropped on Sail Rock, Exumas target, found that the wireless transmitter was unserviceable, carried out the exercise, homed on aircraft LL, signal strength was weak, contact was impossible, radio was weak. Captain decided to abandon the exercise, returned to base and landed at Satellite Field. Liberator BZ-744 LL airborne from Satellite Field, navigation Exercise Number 9, practice bombs dropped on Sail Rock, Exumas, homing practice commenced abandoned aircraft, completed the patrol and landed at Satellite Field. US Marine special Small Operations Group (SOG) establishes Navy Seal type training base on Blue Lagoon Island off Nassau until March 1944 and moved from San Diego, California, Camp Pendleton. **Convoy NK-537** is bound south in Straits towards Canaveral with US Navy plane escorting. **Convoy KN-237** is bound northeast in Straits with US Navy plane escorting. USS *PC-1227* is bound northeast in Straits from Palm Beach. SS *R. C. Stoner* is bound north in Straits near Port Everglades. USCG *CG-Spruce* and USS *YP-441* are bound west towards Northwest Providence Channel.

USS *YO-143* and USS *YP-612* are bound west in Straits towards port Everglades. Tanker *Gulfmaid* is bound north in Straits towards Port Everglades with US Navy blimp. SS *State of Delaware* and USS *PC-1252*, tug USS *Humrick* with barge, and **Convoy Havana North Coast Cuba-63** are bound south in Straits. SS *Horace Williams* is bound southwest in Straits, with US Navy plane escorting. **Convoy GK-732** is bound northwest in Old Bahama Channel passing Cayo Francés with US Navy plane escorting. *U-129* exits area north then east for France. NAS Great Exuma: A plane went to cover **Convoy GN-56.** The ferry *Jean Brillant* and a ship weighed anchor. A plane departed for 3 hours of training. Tug USS *Montcalm* (AT-39) and an escort lifted anchor. USS *Antens* and two destroyers entered area.

2nd May 1943

No flying training carried out. Tanker *Gulfmaid* is bound north in Straits passing Canaveral with US Navy plane escorting. **Convoy NK-537** is bound south in Straits passing Palm Beach with US Navy blimp escorting. Ferry *Jean Brillant* is bound west in Straits from Bimini towards Miami. USS *SC-1019* is bound south in Straits. USCG *CG-83404* is bound southwest in Straits. SS *Gabriel Duval* is bound north in Straits. East of Bermuda, eight USAAF B-24's of the 25th Air Wing go on ASW patrol to interrupt U-supply boats, *Milch-Kuh*, or Milk Cows. NAS Great Exuma: A plane went on a search, and then was instructed to return to Guantanamo. Two planes went to cover **Convoys NG-358**. A plane returned to base. USCG *CG-Spruce* and another escort ship stood in from sea. A plane sighted a troop ship and a destroyer, then a plane sighted another plane.

3rd May 1943

B-25's FR-371 CF, FR-376 CK, FR-384 CM, and FR-393 DF are airborne for local exercises. The Windsors leave Nassau on their third US tour aboard their motor yacht *Gemini*. Tug USS *Valley Forge* and USS *YO-59* are bound south in Straits towards Canaveral. SS *Gabriel Duval* is bound north in Straits, passing Canaveral with US Navy plane escorting. Tanker *Mobiloil* is bound south in Straits towards Bimini with US Navy plane escorting. Tug USS *Humrick* is bound north in Straits towards Miami. USS *SC-512* is bound south in Straits. **Convoy NK-537** is bound north in Straits with US Navy blimp. USS *PO-1252* is bound northwest in Straits towards Miami. Tug USS *Willet* is bound northwest from Nicholas Channel with US Navy plane escorting.

4th May 1943

B-25's FR-381 DC, FR-376 CK, FR-371 CF, and FR-394 CN are airborne for local exercises. British Royal Navy Motor Launches HMS *HMML-1175* and HMS *HMML-1178* are bound south in Straits towards Canaveral. USS *Velocity* (AM-128) is bound north in Straits passing Palm Beach. Tug USS *Valley Forge* is bound south in Straits passing Port Everglades. USCG *CG-Humphries* is bound southeast in Straits from Miami. USS *SC-1028*, USS *SC-1302*, USS *PC-623* and USS *YMS-373* are bound south in Straits from Miami. USS *SC-1034* is bound northwest in Straits towards Miami. USS *YT-338* is bound west in Straits. USS *PC-1254* is bound northeast in Straits. *Vessel 28* is bound northwest in Straits, passing Nicholas Channel. NAS Great Exuma: A plane sighted a 2-masted sailing schooner flying the ensign of Great Britain.

5th May 1943

B-25's FR-376 CK, FR-366 CK, FR-383 DX, FR-381 DC, and FR-371 CF are airborne for local and navigation exercises. Tanker

Gulfmaid is bound south in Straits towards Canaveral. SS *Mambi* and USS *PC-1195* are bound south in Straits, passing Palm Beach. Tug USS *Humrick* is bound south in Straits passing Port Everglades. USCG *CG-83401*, USCG *CG-83403*, and USCG *CG-83405* are bound west north of Nassau towards Northwest Providence Channel. USS *PC-1251*, USS *SC-1001*, and tug USS *Valley Forge* are bound south in Straits. NAS Great Exuma: A plane departed on patrol and was instructed to return to Guantanamo. A plane departed on a training flight, then a convoy entered the area. A plane went to cover **Convoy GN-57**.

6th May 1943

B-25's FR-378 CL, FR-383 DX, FR-384 CM, FR-393 DF, FR-376 CK, and FR-371 CF are airborne for local and navigation exercises. B-25 FR-381 DC is airborne, escorting ferry *Jean Brillant* which met at Harbour Island, escorted until coded position KNLP3535, then returned to base, with no relieving aircraft, nothing to report, after return to base. SS *Nickeliner* and SS *George Washington* are bound south in Straits towards Canaveral. **Convoy KN-238** is bound north passing Matanilla Shoal, northwest Bahamas, with US Navy plane escorting. **Convoy NK-538** is bound south in Straits towards Palm Beach with US Navy plane escorting. SS *Hasting* is bound north in Straits towards Canaveral. SS *Jose Bonifacio* is bound south in Straits passing Palm Beach with US Navy plane escorting. SS *Edward L. Doheny*, tanker *Esso Belgium*, SS *William H. Wilmer*, tanker *L. L. Abshire*, and USS *PC-1195* are bound south in Straits passing Palm Beach. SS *Julia Luckenbach*, SS *Christopher Gale*, and SS *Bostonian* are bound south in Straits passing Port Everglades. Oiler USS *YO-143* and tanker *Gulfmaid* are bound south in Straits passing Miami. USS *SC-1047* and tug USS *Humrick* are bound southwest in Straits. USS *YO-142* is bound north in Straits. Tanker *Comol Cuba* is bound northwest in Nicholas Channel. *U-176* enters area bound northwest up the Old Bahama Channel. NAS Great Exuma: A plane went to cover **Convoy GN-57**. A plane sighted a Canadian pilot boat and a British merchant vessel.

7th May 1943

B-25's FR-366 CX, FR-378 CL, FR-376 CK, FR-393 DF, and FR-380 DB are airborne for local and navigation exercises. USS *SC-1067* and USS *SC-512* are bound south in Straits towards Palm Beach. SS *Nickeliner* is bound south in Straits towards Port Everglades. Tanker *Pan Massachusetts* is bound south in Straits passing Port Everglades. **Convoy GN-57** is bound north in Straits with US Navy blimp near Bimini. USS *SC-1268* is bound south in Straits passing Miami. **Convoy Key West-171** is bound north in Straits. Tug *Ontario* and tug USS *Hercules* are bound north in Straits. USS *SC-984* and USS *Sumpter* (APA-52) are bound north out of Straits. **Convoy NK-538** is bound southwest in Straits with US Navy plane escorting. NAS Great Exuma: A plane held bombing practices, and a plane went to cover **Convoy NG-359.**

8th May 1943

B-25's FR-384 CM, FR-378 CL, FR-380 DB, FR-381 DC, and FR-366 CX airborne for local and navigation exercises. USS *Choctaw* (AT-70) is bound south in Straits towards Canaveral. USS *PC-622* is bound northeast in Straits. Tanker *Fort Niagara* and SS *San Antonio* are bound south in Straits towards Palm Beach. **Convoy KW-171** and USS *SC-1276* are bound north in Straits passing Palm Beach. Tanker *Atlantic States* is bound south in Straits towards Bimini. USS *YAG-28* is bound west in Straits from Bimini to Miami. Tanker *Cerro Gordo* is bound north in Straits towards Bimini. Tanker *Gulfland* is bound south in Straits passing Miami. **Convoy Miami-127** is bound southwest in Straits. **Convoy KW-174**, SS *Cartierdoc* and escorts, two tugs with two barges, USS *YP-612*, USS *YO-59* and USS *SC-1025* are bound north in Straits. USS *Fayette* (APA-43) is bound southeast in Straits towards Cay Sal Bank. Tanker *Pan Orleans* and escorts are bound west towards Nicholas Channel. Tanker *Francis R. Hart* is bound north from Cayo Francés in Old Bahama Channel. NAS Great Exuma: A plane reports Radar contact with a small coastal vessel moving to the northwest.

9th May 1943

Night flying conducted by five aircraft from Main Field. B-25's FR-378 CL and FR-381 DC airborne for local and navigation exercises. B-25 FR-378 CL airborne for night flying exercise signal received for aircraft with a fire on starboard engine. Marine craft section instructed to stand by for air sea rescue duty. FR-382 DD is dispatched to look for missing aircraft CL, started decent through cloud exercise, message received from craft that it was over large fires (proved to be Andros later), message received that aircraft was circling large airfield trying to determine position, CL signalled that it had landed at Boca Raton. USS *YP-375* is bound north in Straits passing Daytona. **Convoy Charleston-85** is bound south in Straits passing Canaveral. USS *SC-1325* is bound south in Straits towards Palm Beach with US Navy plane escorting. Submarine USS *R-12* and USS *PC-622* are bound southwest near Matanilla Shoal, northwest Bahamas with US Navy plane escorting. SS *Buffalo Wallow* is bound north in Straits towards Palm Beach. USS *YAG-21* and USCG *CG-Wanigan* are bound east in Straits from Miami. USS *YAG-28* is bound south from Miami. Ferry *Jean Brillant* and USCG *CG-83406* are bound south towards Nassau. SS *Francis R. Hart* is bound northwest from Cayo Francés and Old Bahama Channel. NAS Great Exuma: A plane departed on the search pattern codenamed Reds, and returned to Guantanamo. A plane departed to provide coverage for a cruiser and destroyer. A plane held a runway bounce drill.

10th May 1943

No flying from Main Field. B-25 FR-378 CL took off from Boca Raton, landed back at Main Field. USS *SC-1290* and USS *SC-1235* are bound south in Straits passing Canaveral. USS *PE-124* is bound north in Straits towards Canaveral. Ferry *Jean Brillant* is bound west in Northwest Providence Channel. **Convoy KN-239** is bound north in Straits with a US Navy blimp near Bimini. **Convoy Charleston 85** is bound south in Straits, passing Port Everglades with US Navy plane escorting. Submarine USS *R-12* and USS *PC-623* are bound south in Straits. *U-176* enters Bahamas bound west from north coasts of Hispaniola and Puerto Rico. NAS Great Exuma: Two planes cover **Convoy GN-58**.

11th May 1943

B-25's FR-371 CF, FR-376 CK, and FR-365 CD are airborne for local and navigation exercises. USS *PE-124* is bound north in Straits towards St. Augustine. USS *LST-17* is bound north in Straits towards Canaveral. USS *PC-1260* is bound south in Straits passing Canaveral. **Convoy NK-539** is bound south in Straits towards Palm Beach with US Navy plane escorting. Anti-Submarine Group 2 is bound north in Straits passing Port Everglades. USS *PC-1252* is bound southeast in Straits from Miami. USS *SC-1032* is bound south in Straits. NAS Great Exuma: A plane departed to cover **Convoy GN-58** and then returned to Guantanamo. Another plane covered **Convoy GN-58.** USS *YAG-21* stood in and docked.

12th May 1943

No flying from Main Field. SS *City of Dallas* is bound south in Straits passing Canaveral. Destroyer USS *Tarbell* (DD-142) is bound north near Matanilla Shoal, northwest Bahamas. **Convoy GN-58** is bound north in Straits with US Navy plane escorting. Tug USS *Hercules* and barges are bound north, out of Straits. USS *SC-1278* is bound south in Straits. **Task Unit TU-2910** is bound west in Straits. SS *Matitlan*, tanker *Cities Service Ohio*, tanker *Gulf Caribbean*, SS *Susan V. Luckenbach*, SS *Marina*, and SS *Orari* are bound southwest in Straits. USS *YO-143* and USS YP-*218* are bound north in Straits near Cay Sal Bank. Tanker *Mobiloil*, USS *SC-1471*, USS *SC-1472*, and USS *SC-1473* are bound north in Straits. NAS Great Exuma: A convoy left the area and USS *YAG-21* stood out to sea.

13th May 1943

Aircraft PK-180 AS is airborne for local and navigation exercises. *U-176* attacks **Convoy Havana North Coast Cuba-18** and sinks both the US-flagged tanker *Nickeliner* and the Cuban coastal tanker SS *Mambi*, 6 nm north of Manati, Cuba. All *Nickeliner*'s eight officers, 15 crew, seven armed guards, and one passenger abandon ship in two lifeboats. Survivors are picked up by a Cuban submarine chaser and landed at Nuevitas. On *Mambi* among 29 Cuban crew and five US Navy armed guard, only the master, nine crew and one armed guard survive. *U-176* eludes destroyer escort USS *Brennan* (DE-13), eight submarine chasers, and a US Navy blimp. **Convoy BT-203** sails from New York to the Pacific; this is the first troopship convoy on that route for over a year. **Convoy NG-360** is bound south, east of Eleuthera. USS *APC-85* and USS Y*P-513* are bound south in Straits, passing Canaveral. SS *Robert Toombs* and USS *SC-658* are bound south in Straits, towards Palm Beach. The house boat USS *YHB-23* and USS *PC-1251* are bound southeast in Straits from Miami. USS *PC-1253*, USCG *CG-42307*, SS *City of Dallas*, and USS *SC-1276* are bound southwest in Straits. USS *Samaná* (YAG-28) is bound north in Straits. Oiler USS *YO-142* and USS *YP-432* are bound east in Straits, to Nassau. NAS Great Exuma: A plane went to cover **Convoy NG-360**.

14th May 1943

B-25's FR-380 DB, FR-371 CF, FR-376 CM, FR-382 DD, and FR-365 CD are airborne on Exercise Number 6, for local and navigation exercises. B-25'S FR-371 CF, FR-376 CK, FR-365 CD, FR-380 BB and FR-382 DD are airborne for air Exercise Number 6. USS *PC-491* is bound south in Straits towards Canaveral. SS *Samoset* and USS *SC-1298* are bound south in Straits passing Canaveral. USCG *CG-83406* is bound northeast towards Matanilla Shoal, northwest Bahamas. USS *YO-142* is bound west in Northwest Providence Channel. USS *YAG-21* is bound west in Straits, west of Andros. USS *SC-1028* is bound south in Straits. USS *APC-85* is bound west in Straits. NAS Great Exuma: A plane went to cover **Convoy NG-360.** A plane covered the north coast of Cuba.

15th May 1943

B-25 FR-366 CX is airborne for navigation Exercise Number 1, bombing practice carried out off Sail Rock, Exumas, gunnery exercises were carried out against sea marker, exercise completed, landed back at base after descent through cloud exercise. B-25 FR-365 CD is airborne, navigation Exercise Number 1, bombs dropped on Sail Rock, Exumas exercise was completed, landed back at base. B-25 FR-381 DC is airborne for navigation Exercise Number 1a, bombing exercises carried out off Sail Rock, Exumas whilst aircraft was commencing its exercises another B-25 aircraft was observed to have completed the bombing exercise and resumed its course for the exercise, navigation exercise completed, and aircraft landed back at base after completing a descent through cloud exercise. B-25 FR-382 DD is airborne for navigation Exercise Number 1a, bombing practice carried out off Sail Rock, Exumas, navigation exercise completed aircraft landed back at base after descent through cloud exercise. B-25 FR-380 DB has bombing practice carried out against Sail Rock, Exumas, gunnery practice navigation exercise was completed and the aircraft landed back at base after descent though cloud exercise. A US Navy OS2U Kingfisher float plane of Navy Scouting Squadron VS-62 and Cuban Submarine Chaser *CC-13* sink *U-176* off Cuba, and south of Cay Sal Bank, Bahamas. *Cuban CS-13* under Mario Ramirez Delgado delivers a brief yet destructive depth charge attack against *U-176*. SS *Samuel Heintzelman* is bound south in Straits towards St. Augustine. USS *YT-338* is bound southwest in Straits towards Port Everglades. USS *PC-491* is bound west in Straits. **Convoy Motor Torpedo Boat Squadron-17** is bound north in Straits. **Convoy KN-240** is bound north in Straits with US Navy blimp escorting. Tanker *Atlantic States* is bound north in Straits with US Navy plane escorting. Destroyer escort USS *Brennan* (DE-13) is bound northwest in Old Bahama Channel passing Cayo Francés. B-25's FR-378 CL, ER-371 CF, and FR-303 DX, are 7 nm east of San Salvador; three RAF No.111 OTU aircraft respond to a report by US Navy at Guantanamo of a U-boat. Sighting proved to be a USCG Cutter escorting a convoy nearby. After an attack was initiated, signals were exchanged and the air attack broken off. Five nm east of San Salvador, B-25 FR-378 CL, B-25 ER-371 CF, and FR-303 DX searched south-eastern Bahamas, finding the USCG Cutter, not a U-boat. NAS Great Exuma: A convoy entered the area and two planes covered **Convoy GN-59**.

16th May 1943

B-25 FR-383 DX airborne, bombing exercises carried out off Sail Rock, Exumas, navigation exercise completed, aircraft landed back at base after descent through cloud exercise. B-25 FR-362 CA is airborne, carries out bombing exercises off Sail Rock, Exumas, navigation exercise completed, aircraft landed back at base after descent through cloud exercise. B-25 FR-366 DX is airborne for navigation Exercise Number 1, bombing exercises carried out off Sail Rock, Exumas, navigation exercise completed, aircraft landed back at base after descent through cloud exercise. B-25 FR-393 DF airborne for navigation Exercise Number 1, bombing exercises were carried out off Sail Rock, Exumas, navigation exercise completed, aircraft landed back at base after descent through cloud exercise. B-25 FR-371 CF airborne for navigation Exercise Number 1a, bombing exercises carried out off Sail exercise completed, aircraft landed back at base after descent through cloud exercise. **Convoy KN-240** is bound north in Straits passing Palm Beach with US Navy plane escorting. Oiler USS *YO-142* and USS *YP-432* are bound northeast near Matanilla Shoal, northwest Bahamas. **Convoy NK-540** is bound south in Straits, passing Palm Beach with US Navy plane escorting. USS *SC-1270* is bound south in Straits passing Palm Beach. SS *Elizabeth Kellogg* is bound south in Straits, passing Port Everglades.

USS *SC-753* and USS *SC-1280* are bound southeast in Straits, passing Miami. SS *Manuel Rionda* is bound northeast in Straits, towards Bimini with US Navy plane escorting. SS *Samuel Heintzelman*, USS *PC-594*, and SS *Contreras* are bound south in Straits. USS *PC-1193*, USS *PC-611*, and USS *PC-561* are bound northwest in Straits towards Miami. USS *SC-743* is bound southeast in Old Bahama Channel towards Cayo Francés. USS *YP-579* is bound west in Straits towards Key West. NAS Great Exuma: A plane searched the area from west and north. A plane went to cover **Convoy GN-59**.

17th May 1943

Heavy rain and thunderstorms restricted flying throughout the day. USS *SC-658* is bound north in Straits passing Daytona. SS *Fort Niagara* is bound north in Straits near Canaveral. USS *Tumult* (AM-127) and six APC [patrol craft]'s are bound south in Straits passing Palm Beach. SS *San Antonio* is bound north in Straits near Miami. **Convoy GN-59** is bound north, east of Bahamas. **Convoy NG-361** is bound south, east of Bahamas. NAS Great Exuma: An unidentified freighter is reported. A plane went to cover **Convoy NG-361**, which then entered the area along with **Convoy GK-734**.

18th May 1943

USS *SC-1298* is bound south in Straits passing Daytona. USS *Choctaw* (AT-70) is bound north in Straits near Palm Beach. Tanker *R. P. Smith* is bound north in Straits passing Port Everglades with US Navy plane escorting. USS *PC-1196* is bound south in Straits. **Convoy GK-94** is bound northwest in Straits passing Nicholas Channel with US Navy plane escorting. NAS Great Exuma: Two plane went to cover **Convoy NG-361**. A plane sighted a 2-masted schooner. A plane covered a convoy. USS *Ariel* and an escort went southward from Crooked Island.

19th May 1943

Flying restricted during the day. Tanker *R. P. Smith* is bound north in Straits, passing Daytona. USS *SC-1236* is bound south in Straits, passing Canaveral. USS *AM-107* is bound north in Straits, passing Port Everglades. Ferry *Jean Brillant* is bound west in Straits towards Miami with US Navy plane escorting. Tug USS *Humrick* and tug USS *Meteor* are bound south in Straits. NAS Great Exuma: A plane went to cover **Convoy NG-361**, then departed.

20th May 1943

B-25 FR-393 DF is airborne for navigation Exercise Number 3, photographs taken of Harbour Island, navigation exercise completed satisfactorily landed back at base. B-25 FR-382 DD, FR-380 DB, FR-375 CJ, and FR-362 CA are airborne at different times navigation Exercise Number 3, all completed exercise then landed back at base. Two Liberators, BZ-759 and BZ-543, are delivered to this unit today. SS *Chester* and Cuban *CS-21* and Cuban *CS-22* are bound south in Straits, towards Canaveral. USS *YMS-75* and USS *YMS-164* are bound north in Straits, passing Canaveral. Tanker *Goldheels*, USS *YP-498*, and USS *SC-1321* are bound south in Straits, passing Canaveral. Oiler USS *YO-142* and USS *YP-432* are bound southwest passing Matanilla Shoal, northwest Bahamas with US Navy plane escorting. USS *YAG-28* is bound west in Straits from Miami towards Bimini. **Convoy KN-241** is bound north in Straits with US Navy plane escorting. SS *Polonaise*, USCG *CG-83372* and USCG *CG-83375* are bound north in Straits. **Convoy KG-634** is bound southeast in Straits towards Nicholas Channel, with US Navy plane escorting. Tanker *Pan Orleans* and Cuban *CS-11* and Cuban *CS-13* are bound west in

Straits passing Nicholas Channel. NAS Great Exuma: A plane went to cover **Convoy GN-60**. A plane covered **Convoy GK-734**. The tug USS *Cormorant* and a destroyer passed through the area. USCG *CG-Spruce* and escort got underway for Guantanamo. **Convoy GN-60** entered area.

21st May 1943

B-25 FR-362 CA is airborne for navigation Exercise Number 2, bombing practice carried out off Rum Cay, a convoy of 31 merchant ships was sighted in nine columns with five escorts, ships were between 5,000 and 8,000 tons. Depth charges dropped in coded position JRMC1827, aircraft landed back at base after having completed the exercise. B-25 FR-393 DF is airborne for navigation Exercise Number 2, convoy reported by CA was sighted, exercises completed and aircraft landed back at base. B-25 FR-376 CK airborne for navigation Exercise Number 2, bombing exercises carried out off Rum Cay, convoy reported by the other aircraft, was sighted in coded position PXTF4012, one depth charge was dropped, aircraft landed back at base. B-25 FR-365 CD is airborne for navigation Exercise Number 2, bombing practice carried out off Rum Cay, aircraft sighted the convoy, previously reported by the other aircraft, aircraft passed over the convoy depth charge was dropped against a flame float in coded position FXVH0011, aircraft landed back at base. B-25 FR-382 DD airborne for navigation Exercise Number 2, the convoy of 31 merchant ships and five escorts was circled aircraft landed back at base after carrying out bombing practice. B-25 FR-362 CA is airborne for navigation Exercise Number 4, same convoy as was sighted by the aircraft on navigation Exercise Number 2 in the afternoon was seen, exercise completed, aircraft landed back at base. B-25 FR-381 DC airborne for navigation Exercise Number 4, exercise was completed aircraft landed back at base with nothing to report. B-25 FR-384 CM airborne for navigation Exercise Number 4, aircraft was over a convoy, reported by CA, exercise completed, aircraft landed back at base. B-25 FR-382 DD airborne for navigation Exercise Number 4, exercise completed, convoy was sighted, aircraft landed back at base with nothing further to report. British Royal Navy minesweepers HMS *HMM-101*, HMS *HMM-99*, HMS *HMM-104*, and HMS *HMM-108* are bound south in Straits towards St. Augustine. USS *SC-1309* is bound south in Straits towards Canaveral. **Convoy NK-541** is bound south in Straits towards Canaveral with US Navy plane escorting. USS *Christiana* (IX-80) is bound south in Straits passing Canaveral. SS *Polonaise* and USCG *CG-83348* are bound north in Straits towards Canaveral. SS *Comas Meadows* is bound north in Straits with US Navy plane escorting. USCG *CG-18338* and USCG *CG-18342* are bound north in Straits towards Canaveral. USS *YP-498* is bound north in Straits passing Palm Beach. SS *Goldheels* and USS *YP-248* are bound south towards Port Everglades. Tanker *Mobiloil* is bound south in Straits with US Navy plane escorting. Ferry *Jean Brillant* and USCG *CG-83405* are bound west in Northwest Providence Channel. **Convoy KW-182** is bound north in Straits. **Convoy Jacksonville-42** and USS *SC-1046* are bound south in Straits. USS *YAG-28* is bound south, passing Nassau. **Convoy GN-60** is bound north, east of Eleuthera. **Convoy GK-734** is bound west in Old Bahama Channel passing Cayo Francés with US Navy plane escorting. NAS Great Exuma: Two planes went to cover **Convoy GN-60**.

22nd May 1943

B-25 FR-365 CD is airborne for navigation Exercise Number 4, failed to return from the flight. A message was received giving position, parallel track search was laid on to be flown, B-25 FR-331 DC is airborne on search for missing aircraft, patrolled area was covered by a parallel track search, nothing seen during the patrol, another B-25 was observed during the patrol at 10 nm distant, aircraft passed over what seemed to be oil residue in the wake of convoy (which was sighted on 21st May) in several lines, somewhat distorted by wind and current when in visible distance, oil residue was 2 nm distant and 1 nm wide. Message was intercepted from CA 'Dinghy sighted,' so aircraft circled once and resumed its course, landed back at base having seen nothing. B-25 FR-382 DD airborne to carry out parallel track search, detailed search completed and nothing was seen, message was intercepted from CA 'Oil patch in coded position KVNH2044.' Aircraft circled the reported position and sighted CA circling at 3 nm on starboard, and observed that a smoke float had been dropped, DD circled until the float had burnt out, and nothing was seen, landed back at base having seen nothing. B-25 FR-384 CM is airborne on Air Sea Rescue patrol, carried out a visual reconnaissance of east and south side of San Salvador, without seeing anything, parallel track search carried out, gave no results, sighted from 3 nm range on starboard bow what appeared to be mirror flashes from sea level. There were several meaningless flashes, then pause, 30 seconds later the aircraft having turned towards the source of flashes and dropped a smoke float, a second series of flashes was seen, aircraft searched the area but was unable to ascertain the source of the flashes. An intercepted message was received from CA 'Over Oil patch.' A third series of flashes were sighted, aircraft turned to investigate flashes but it had ceased, and landed back at base with nothing further to report. B-25 FR-376 CK airborne on a parallel track search, detailed area was patrolled and nothing was sighted, message was intercepted from CA 'Oil streak 200 yards long and two pieces of floating yellow wood.' Aircraft had flown south along the east and south coast of Eleuthera and carried out reconnaissance of entire coast. B-25 FR-362 GA is airborne to carry out a parallel track search for missing aircraft FR-395 CD, and covered the detailed area and sighted nothing. On return flight, oil patch sighted, this was 200 yards [600 feet] long, message was sent to base, and this was intercepted by the other searching aircraft, investigated and circled the area base. Instructed aircraft to take closer investigation and report on investigation, it was found that there were two yellow planks 10 feet long in the oil patch, one broken, no oil slicks in the vicinity, other than 120 to 300 yards from the main patch. Search plane resumed its track to base, was instructed to investigate until PLE [plane's limit of endurance] or relief. Aircraft FR-362 GA returned to the area in which the oil was sighted and carried out a square search. This search continued until it was recalled to base, landed back. On receipt of message from CA of the sighting of an oil patch, B-25 FR-376 CJ was despatched, a square search was commenced in coded position KNVH0300, the ASR Air Sea Rescue launch was sighted. A convoy of 22 merchant ships and three escorts was also sighted after completing the search, a visual reconnaissance was made of the eastern shore of Eleuthera to Harbour Island, aircraft landed back at base with nothing seen of the oil patch or of the sea marker. SS *Polonaise* is bound north in Straits passing Canaveral. USS *YMS-163* is bound northeast in Straits. **Convoy NK-541** is bound southwest in Straits towards Palm Beach with US Navy plane escorting. Sixteen merchant ships from **Convoy NK-541** are proceeding independently south in Straits, towards Palm Beach. British Royal Navy minesweeper HMS *HMMS-101* is bound south in Straits, passing Palm Beach. Tanker *R. P. Smith* is bound southwest passed Matanilla Shoal, northwest Bahamas. SS *Husvik* is bound north in Straits passing Port Everglades. Tug *Ontario* and tug USS *Hercules* are bound northeast in Straits towards Miami. Oiler USS *YO-142* is bound west in Straits with US Navy plane escorting. Tanker *Goldheels* is bound west in Straits. USS *YMS-73* is bound west in Straits passing Nicholas Channel. Tanker *Pan Orleans* is bound west in Straits passing Nicholas Channel with US Navy plane escorting. **Convoy GN-60** is bound north, to the east of Bahamas with US Navy plane escorting. NAS Great Exuma: A convoy departed. A plane had a night bounce drill. A plane covered **Convoy NG-362**.

23rd May 1943

B-25 FR-382 DD is airborne on a parallel track search for aircraft FR-365 DC missing since the previous night on an exercise patrol carried out as detailed. Aircraft turned to starboard from its course to investigate a glint seen in the sea, no results, disturbance was observed on the water and on investigation this prove to be a shark, landed back at base having sighted nothing. B-25 FR-384 CM is airborne on an air sea rescue search, object sighted on port bow was a small motor vessel, oil streaks seen gave no results on investigation, patrol completed, and aircraft landed back at base with nothing further to report. B-25 FR-381 DC is airborne on air sea rescue search, area detailed covered by parallel track search, nothing sighted, landed back at base. B-25 FR-375 CJ is airborne on a search, completed the detailed patrol and having sighted nothing landed back at base. USS *SC-1291* is bound south in Straits towards Canaveral. SS *Husvik* is bound north in Straits passing Daytona. USS *SC-1309* is bound south in Straits passing Canaveral. USS *YP-498* is bound north in Straits passing Palm Beach. Tug *Lt. Colonel Herbert S. Kidwell* is bound north in Straits towards Palm Beach. SS *Montana* is bound south towards Port Everglades with US Navy plane escorting. The house boat USS *YHB-13* is bound south in Straits passing Miami. USS *SC-1067* and USS *SC-1279*, USS *Schuykill*, and USS *SC-984* are bound south in Straits. **Convoy NK-541 is bound west** in Straits towards Key West with US Navy plane escorting. USS *SC-1278* is bound southeast in Straits passing Nicholas Channel. **Convoy NG-362** is bound south, east of Bahamas.

24th May 1943

B-25 FR-375 CJ is airborne for navigation Exercise Number 4, a creeping line ahead exercise, exercise completed, landed back at base. B-25 FR-362 CA is airborne for navigation Exercise Number 4, detailed exercise completed, landed back at base. B-25 FR-393 DF is airborne for navigation Exercise Number 4, exercise completed, landed back at base with nothing further to report. B-25 FR-375 CJ airborne for navigation Exercise Number 4, creeping line ahead exercise completed, landed back at base with nothing further to report. B-25 FR-381 DC airborne for navigation Exercise Number 1, practice bombs dropped, Great Sale Cay located [north of Grand Bahama], depth charge dropped on flame float, exercise completed, landed back at base. B-25 FR-375 CJ airborne for air sea rescue patrol and navigation Exercise Number 3, exercise included search in vicinity of Mayaguana, where crashed aircraft was reported. Area thoroughly searched nothing located in coded position PXVHO129, white planks with smaller pieces of painted wood seen floating, circled, nothing sighted. Aircraft continued in its course, landed back at base, failed to locate missing aircraft. B-25 FR-393 DF is airborne for navigation Exercise Number 3, exercise completed, landed back at base. B-25 FR-381 DC airborne for navigation Exercise Number 4, bombing exercise against sea marker off Governor's Harbour, Eleuthera, creeping line ahead, exercise completed, landed back at base. B-24 BZ-746 LM is airborne from Satellite Field, navigation Exercise Number 9, bombing exercise off Sail Rock, Exumas, exercise completed, landed back at base. B-24 FL-967 LJ is airborne from Satellite Field navigation Exercise Number 9, exercise completed, landed back at base. Tug USS *Buttercup* with tow and USCG *CG-83348* are bound south in Straits towards Palm Beach with US Navy plane escorting. USS *SC-18343*, USS *SC-18344*, and USS *SC-18355* are bound north in Straits towards Palm Beach. British Royal Navy rescue tug HMS *HMRT Advantage* is towing USS *YW-65* north, out of Straits. Tanker *Atlantic States* is bound south in Straits towards Cay Sal Bank. USS *SC-753* is bound southwest in Straits. Tanker *Gulfmaid* is bound northeast in Straits. Ferry *Jean Brillant* and escort are bound east in Straits towards Nassau. NAS Great Exuma: A plane departed to cover **Convoy NG-362** and then moved to Guantanamo. Two planes had a bounce drill. One plane had a night drill.

25th May 1943

B-25 FR-384 CM is airborne for navigation Exercise Number 3, bombing practice off South Abaco Light [aka Hole-in-the-Wall Light], exercise completed, landed back at base. B-24 BZ-946 LM is airborne from Satellite Field, navigation Exercise Number 9, bombing exercise off Sail Rock, Exumas. Other Liberator on navigation Exercise Number 9, spotted one mile to port, exercise completed, landed back at base. B-24 FL-967 LJ airborne from Satellite Field, navigation Exercise Number 9, bombing exercise off Sail Rock, Exumas, exercise completed, landed back at base. B-25's FR-384 CM, FR-276 CX, and FR-380 DB are airborne for navigation Exercise Number 7, exercise completed, landed back at base. SS *Troubadour* is bound north in Straits passing Canaveral. SS *Gargoyle*, SS *John C. Calhoun*, SS *Fresno State*, SS *John Davenport*, SS *John Woolman*, SS *Michiel Stone*, and SS *Stephen J. Field* are bound south in Straits, passing Canaveral. Tanker *Gulfmaid* is bound north in Straits towards Canaveral. Tanker *Britamsea*, **Convoy NK-542**, and SS *Mapia* are bound south in Straits passing Palm Beach. USS *YP-440* is bound north in Straits towards Palm Beach. USS *Christiana* (IX-80) is bound west in Northwest Providence Channel. Tug USS *YT-338* is towing USS *YAG-28*, southwest in Straits from Bimini to Miami. SS *New Zealand Star* is bound south in Straits, passing Port Everglades. USS *SC-1000* and USS *SC-1280* are bound southwest in Straits. Tug USS *Buttercup* with tow and USCG *CG-83402* are bound southwest in Straits with US Navy plane escorting. NAS Great Exuma: Two planes went to cover **Convoy GN-61.**

26th May 1943

B-25's FR-380 DB, FR-375 CJ, FR-366 CX, FR-384 CM, and FR-393 DF are airborne for navigation Exercise Number 7, exercise completed, landed back at base. Liberators BZ-761 LM, BZ-722 LK, and FL-942 LC are airborne from Satellite Field for navigation Exercise Number 9, exercise completed, landed back at base. USS *PC-492* is bound south in Straits towards St. Augustine. USS *SC-1329* is bound south in Straits towards Daytona. Tanker *Alpha* is bound south in Straits towards Canaveral. SS *Polonaise* is bound south in Straits, passing Canaveral. SS *Balboa* is bound north in Straits passing Canaveral. SS *Kent* is bound southwest in Straits towards Palm Beach. **Convoy KN-242** is bound north in Straits with US Navy plane escorting. British Royal Navy HMS *BYM-155*, HMS *BYM-156*, and HMS *BYM-171* are bound northeast in Straits passing Palm Beach. Tanker *Mobilight* is bound southwest in Northwest Providence Channel towards Miami. USS *SC-1290* and USS *SC-736* are bound west in Straits. Ten vessels from **Convoy NK-542** are proceeding independently west in Straits. Tug *Peter Moran* is bound north in Straits. **Convoy GK-735** is bound northwest in Straits passing Nicholas Channel. USS *SC-609* is bound north in Old Bahama Channel passing Cayo Francés. **Convoy GN-61** is bound northwest. NAS Great Exuma: A plane went to cover **Convoy GN -61**.

27th May 1943

B-25's FR-366 CX, FR-384 CM, FR-393 DF, and FR-380 DB airborne for navigation Exercise Number 5, exercise completed, landed back at base. Liberator FL-970 LH, FL-942 LC, BZ-744 airborne from Satellite Field navigation Exercise Number 9, exercise completed, landed back at base. USS *YP-440* is bound north in Straits passing Canaveral. USS *SC-609* is bound north in Straits passing Palm Beach. USS *SC-1303* is bound south in Straits passing Palm Beach. USS *PC-492* is bound south in Straits. Tug *Peter Moran* is bound north in Straits towards Miami. SS *Kent* is bound southwest in Straits. Tanker *Gulfland* is bound north in Straits. USS *YAG-16* is bound south from Bimini. **Convoy GN-61** is bound northwest near Abaco with US Navy plane escorting. *U-66* enters area northeast of Bahamas west towards Straits. NAS Great Exuma: A plane went to cover **Convoy NG-363.**

28th May 1943

B-25's FR-381 DC, FR-380 DB, FR-393 DF, FR-366 CX airborne for navigation Exercise Number 5, exercise completed, landed back at base. Short navigation exercise carried out by FK-180 AS, FR-394 CN, FR-181 AV, and FR-182 AU; aircraft of 'A' squadron. Liberators BZ-722 LK and FL-970 LH are airborne from Satellite Field exercise completed, landed back at base. USS *SC-986* is bound south in Straits towards Daytona. Tanker *Gulfmaid* is bound south in Straits, passing Canaveral. SS *A. S. Hansen* is bound north in Straits towards Canaveral. USS *YAG-16* is bound south in Straits. NAS Great Exuma: Two planes went to cover **Convoy NG-363** and returned back to Guantanamo. A plane covered a convoy.

29th May 1943

B-25 FR-366 CX is airborne for navigation Exercise Number 5, exercise completed, landed back at base with nothing to report. USS *SC-1021* and **Convoy Motor Torpedo Boat Squadron-26** with six fast torpedo PT-boats, USS *PT*'s is bound south in Straits towards Canaveral. Tug *Peter Moran* is bound north in Straits towards Canaveral. USS *Tide* (AM-125) is bound south in Straits towards Palm Beach. USS *SC-1271* is bound south in Straits passing Palm Beach. Tanker *Mobiloil* is bound north in Straits. Tanker *Gulfmaid* is bound west in Straits towards Key West. *U-66* is northeast of

Grand Bahama, bound west. NAS Great Exuma: **Convoy NG-363** departed

30th May 1943

No flying during the day. Number 3 course completed conversion and training on Liberator aircraft and had returned to Main Filed, preparatory and proceeding to operational stations Number 4 course completed training on B-25's and moved to Satellite to train on Liberators. Number 5 course completed familiarization exercises and preparing to undertake navigation exercises. USS *YMS-399* and SS *John Owen* are bound south in Straits towards St. Augustine. USCG *CG-Belleville* is bound south in Straits towards Canaveral with US Navy plane escorting. USS *Umpqua* (AT-25) is bound north in Straits towards Canaveral. USS *SC-1271* is bound south in Straits towards Palm Beach. SS *Montana* is bound north in Straits with US Navy plane escorting. Tug USS *Syosset* is bound southeast in Straits from Miami with US Navy plane escorting. USS *SC-1068* and USS *SC-1325* are bound south in Straits. **Convoy KN-243** is bound north in Straits with a navy blimp. **Convoy KG-636** is bound southeast in Straits passing Nicholas Channel with US Navy plane escorting. SS *Libertad* is bound northwest in Straits passing Nicholas Channel. USS *YAG-16* is bound north in Old Bahama Channel from Cayo Francés. NAS Great Exuma: Two planes went to cover **Convoy *GN-62*.** A plane sighted US-flag tanker *Beta* (former US Navy tanker USS *Hisko* (ID-1953). A small US Navy boat was sighted by a plane.

31st May 1943

B-25's FR-366 CX, FR-388 DX, FR-378 CL, and FR-380 DF carried out air exercises and local flying. Local flying training commenced by pupils of Number 6, course. Minesweeper HMS *MMS-100* and USS *YP-224* are bound south in Straits towards Canaveral. Tanker *Permian,* tanker *Pan North Carolina* and SS *Cape Douglas* are bound south in the Straits passing Canaveral. **Convoy NK-543** is bound south in Straits passing Canaveral with US Navy plane escorting. USS *YMS-399*, SS *Rananga* and SS *Hoegh Silvercloud* are bound south in Straits towards Palm Beach. SS *Laura Keene* is bound north in Straits towards Palm Beach. USS *SC-1021* is bound south in Straits towards Port Everglades. SS *Molina Del Rey* and SS *Atlantic Coast* are bound south in Straits. SS *John Owen* is bound west in Straits with US Navy plane escorting. Tug USS *Syosset* and SS *Charlestown* are bound south in Straits. NAS Great Exuma: Two planes went to cover **Convoy GN-62**. A plane departed to cover the convoy.

1st June 1943

B-25's FR-384 CM, FR-388 DX, FR-381 DC, FR-376 CK, and FR-378 CL are airborne in evening navigation Exercise Number 3, landed back at base after completion same aircraft carried out navigation Exercise Number 1, during morning. *U-527* is transiting the Northeast Providence Channel and Northwest Providence Channel, bound to Straits. USS *SC-998* is bound south in Straits towards Palm Beach. Tanker *Permian* is bound south in Straits passing Port Everglades. Ferry *Jean Brillant* is bound southwest in Straits from Northwest Providence Channel towards Port Everglades with US Navy plane escorting. **Convoy NK-543** is bound southwest in Straits with US Navy plane escorting. **Convoy GK-736** is bound northwest in Straits passing Nicholas Channel with US Navy plane escorting. Tanker *Comol Cuba* is bound southwest from near Cay Sal Bank towards Nicholas Channel. **Convoy GN-62** is bound northwest, east of Eleuthera near entry to Northeast Providence Channel with US Navy plane escorting. *U-66* is 300 nm north of Matanilla Shoal, northwest Bahamas. NAS Great Exuma: A plane went to cover **Convoy GN-62**. A plane went to cover **Convoy NG-364**. A plane landed in Great Exuma. Tug USS *B. F. Moody* anchored in sea. A plane arrived from Guantanamo.

2nd June 1943

B-25's FR-378 CL, FR-376 CK, FR-384 CM, FR-381 DC, and FR-383 DX are airborne for navigation Exercise Number 2, again airborne later for navigation Exercise Number 3, and Number 4. *U-539* sinks SS *Pillory* is 30 nm north of Mona Passage. *U-67* is undertaking repairs in Straits. Tanker *A. S. Hansen* is bound south in Straits towards Palm Beach. Tanker *Esso Gettysburg* is bound south in Straits towards Palm Beach. SS *Jason Lee* is bound north in Straits towards Palm Beach. Tanker *Atlantic States* is bound north in Straits near Miami. USS *PC-1254* and USS *SC-1068* are bound south in Straits. Tug USS *Sampson* and tanker *Mobilight* are bound north in Straits. *U-527* is bound east of Bahamas via Northwest Providence Channel and Northeast Providence Channel. NAS Great Exuma: Two planes went to cover **Convoy NG-364.** A plane landed in Great Exuma after 6.8 hours in air. Lt. Cdr. H. C. Delong, USN was relieved of his command as commanding officer by Lt. D. C. Higgins. A plane went on a search mission. A destroyer was sighted.

3rd June 1943

B-25's FR-378 CL, FR-376 CK, FR-360 CX, FR-380 DB, FR-383 DX are airborne for navigation Exercise Number 2, again airborne later for navigation Exercise Number 4. SS *Cedaraloide* is bound south in Straits towards Canaveral. USS *SC-1466* is bound north in Straits passing Palm Beach. USS *YAG-20* is bound west near Matanilla Shoal, northwest Bahamas. Ferry *Jean Brillant* is bound south towards Nassau. USS *YAG-16* and USS *SC-1034* are bound south in Straits passing Miami towards Cay Sal Bank. USS *YP-332*, USS *YO-142*, and SS *Gulfspray* are bound south in Straits. NAS Great Exuma: A plane departed to search Great Exuma. Squadron VP-34 detachment and returned to Guantanamo. Compation 32 determined the feasibility of routine beaching. Beaching planes were tested.

4th June 1943

B-25's FR-366 CX, FR-381 DC, FR-376 CK, FR-378 CL, and FR-380 DB are airborne for local and navigation exercises. USCG *CG-83422*, USCG *CG-83423* and destroyer USS *Edgar G. Chase* (DE-16) are bound south in Straits towards Palm Beach. SS *Cedaraloide* is bound south in Straits past Palm Beach. SS *Delaires* is bound south in Straits near Palm Beach. USCG *CG-83348* is bound north in Straits towards Walker Cay, north of Abaco. Ferry *Jean Brillant* is bound southwest in Straits from Northwest Providence Channel towards Miami. USS *C-1125*, USS *PC-1124* and **Convoy NK-244** are bound north in Straits. NAS Great Exuma: Two planes went to cover **Convoy GN-63.** A plane departed for Guantanamo. A plane landed at Great Exuma. A plane and Lt. (jg) Trum, Senior Naval Aviator, arrived at Great Exuma.

5th June 1943

B-25's FR-378 CL, FR-380 DB, FR-362 CA, FR-383 DX, and FR-381 DC airborne for navigation Exercise Number 5, and Number 6. SS *John Greer Hibben* is bound south in Straits passing St. Augustine. USS *SC-1466* is bound south in Straits towards Canaveral with US Navy plane escorting. **Convoy KN-244** is bound north, passing Matanilla Shoal, northwest Bahamas with US Navy plane escorting. USCG *CG-160* is bound south in Straits passing Palm Beach. **Convoy NK-544** is bound south in Straits with US Navy blimp near Palm Beach. Canadian SS *Algonquin Park*, British SS *Laurentide Park*, and SS *Phillip Dodridge* are bound south in Straits towards Port Everglades. **Convoy Key West-199** is bound north in Straits towards Miami, with US Navy plane escorting. Tanker *Camas Meadows* is bound north out of Straits. NAS Great Exuma: Two planes went to cover **Convoy GN-63.** A plane landed at Great Exuma. Destroyer USS *Badger* sighted. Lt. Christensen, Senior Naval Aviator, departed for Guantanamo.

6th June 1943

B-25's FR-378 CL, FR-362 CA, FR-381 DC, and FR-383 DX are airborne for air and local navigation exercises Number 5 and Number 6. Local flying training and bombing practice carried out from Main Field and Satellite Field during the day. USS *SC-1271* is bound south in Straits towards Daytona. USS *SC-1466* is bound south in Straits passing Palm Beach. SS *John Greer Hibben* is bound south in Straits passing Port Everglades. USCG *CG-74300* is bound north towards Palm Beach. USS *SC-1269*, USS *SC-1270*, and USS *SC-1351* are bound south in Straits. USS *YAG-16* is bound northeast in Straits towards Miami. USS *SC-1466* and British SS *Laurentide Park* are bound southwest in Straits. USS *Fremont* (APA-44) and USS *PC-493* are bound north in Straits. **Convoy GK-737** is bound north in Old Bahama Channel, passing Cayo Francés. Tanker *Pan Orleans* is bound west in Straits passing Nicholas Channel. **Convoy NG-365** is bound south in Straits with US Navy plane escorting. **Convoy GN-63** is bound northwest east of Abaco, with US Navy plane escorting. NAS Great Exuma: A convoy of 14 merchant vessels and six escorts entered the area. A plane departed to cover **Convoy NG -365.**

7th June 1943

B-25's FR-378 CL, FR-381 DC, FR-362 CA, FR-371 CF, and FR-375 CJ are airborne on local air and navigation exercises. Tanker *R. P. Smith* is bound south in Straits towards Canaveral. USS *PC-493* is bound north in Straits towards Canaveral. USS *SC-1070* is bound north in Straits towards Palm Beach. Ferry *Jean Brillant* is bound southwest in Straits from Northwest Providence Channel towards Miami with US Navy plane escorting. Tanker *Gulfmaid* and USCG *CGR-531* are bound north in Straits. SS *Samuel Nelson* is bound northwest in Straits near Cay Sal Bank. *U-129* is northeast of Bahamas, where it sights an Allied submarine, of the *Narwhal* type, and dives to attack. Whether he dived or turned away, he has certainly seen us. The low rising sun had the boat illuminated, standing on the eastern horizon against blue sky and blue sea. That's what happens when you try to force it. NAS Great Exuma: Two planes went to cover **Convoy NG-365** of 13 merchant vessel and five escorts. One enemy submarine was spotted between Key West and Havana. Ship named *Metals* and minesweeper USS *Nuthatch* (AM-60) were northbound through Mayaguana Passage.

8th June 1943

B-25's FR-371 CF, and FR-362 CA are airborne bombing exercises against a moving target. USS *Velocity* (AM-128) is bound south in Straits towards Daytona. Tanker *Gulfmaid* is bound north in Straits towards Daytona. USCG *CG-Vigilant* and USS *PC-612* are bound south in Straits passing Palm Beach. Oiler USS *YO-142* is bound southeast in Northwest Providence Channel towards Nassau. USS *SC-1281* and tanker *R. P. Smith* are bound south in

Straits. Tanker *Charleston*, tanker *Atlantic Coast*, and destroyer escort USS *Jacob Jones* (DE-130) are bound north in Straits. SS *John G. Carlisle* is bound northwest near Cay Sal Bank. USS *Albermarle* (AVS) is bound north, north of Bahamas. *U-527* exits the area bound west, south of Key West. NAS Great Exuma: A plane departed and then returned to Great Exuma.

9th June 1943

FR-383 DX is airborne on a special Anti-Submarine exercise at Key West, visual contact with submarine, made dummy attacks, first attack made 30 seconds after submarine had submerged, second attack made when the conning tower was awash, the stick straddled the submarine, exercises completed, aircraft returned to base. FR-362 CA is airborne on a special Anti-Submarine training exercise at Key West, submarine was sighted, first attack made at 2,000 feet, bombs being dropped from 50 feet, submarine submerged and stick undershot by 100 yards, on the second attack the submarine did not submerge, third attack submarine submerged, the stick undershot, exercises completed aircraft returned to base. Liberators BZ-746 LM and BZ-760 LP are airborne from Satellite Field for Boca Chita returned to base after submarine exercise. FR-393 DD is airborne for navigation Exercise Number 1, bombing practice carried out off Sail Rock, Exumas, exercise completed, aircraft landed back at base. FR-371 CF and FR-382 DD are airborne at different times for navigation Exercise Number 1, bombs dropped on Sail Rock, Exumas exercise completed aircraft landed back at base. SS *Ole Bull* is bound south in Straits towards Daytona. SS *Moleno Del Rey* is bound north near Walker Cay, north of Abaco. Destroyer escort USS *Andres* (DE-45) is bound south towards Walker Cay, north of Abaco. Two tugs and USS *YP-408* are bound southeast in Straits from Port Everglades. **Convoy Miami-141** and **Convoy Miami-139** are bound south in Straits. USS *SC-1291* is bound southwest in Straits. British Royal Navy Minesweepers HMS *HMM-99*, HMS *HMM-100*, HMS *HMM-101*, HMS *HMM-104*, and HMS *HMM-108* are bound south in Straits towards Cay Sal Bank. USS *YMS-1* is bound northwest in Straits. **Convoy KN-245** is bound northeast with US Navy blimp near Cay Sal Bank. USS *Velocity* (AM-128) is bound west in Straits. NAS Great Exuma: Two planes went to cover **Convoy GN-64** of 39 merchant vessels and five escorts. USS *YAG 20* departed Miami for Great Exuma. Lt. Koontz, Senior Naval Aviator, arrived from Guantanamo.

10th June 1943

FR-366 CX is airborne for special Anti-Submarine exercise at Key West, submarine sighted, challenges exchanged, submarine submerged and the aircraft dropped four wooden bombs on the submarine, this was not the target ship, target submarine sighted, practice bombs were dropped and dummy attacks carried out, exercise completed, aircraft landed at base. FR-383 DX is airborne on special submarine training exercise at Key West, on course 259°, an unidentified aircraft was sighted 10 nm to 12 nm, distant on a northerly course, merchant ship sighted had one funnel, painted grey and appeared to be a troop ship, several barges observed in the vicinity, target submarine sighted, exercise could not commence until CX completed the exercise, submarine attack exercises completed, aircraft returned to base. Liberator BZ-757 and BZ-746 LR are airborne returned to base after submarine exercises at Boca Chica. B-25 FR-371 CP is airborne for navigation Exercise Number

1a, bombing practice was carried out off Sail Rock, Exumas, convoy was sighed, aircraft was challenged by escort vessel with roughly 30 ships in convoy, which was circled and then the aircraft continued on the exercise, aircraft landed back at base after having completed exercise. B-25's FR-381 DC and FR-366 DX are airborne at different times for navigation Exercise Number 1, bombing exercises carried out of Sail Rock, Exumas, exercise completed, aircraft landed back at base. Tanker *Esso Gettysburg* proceeding south in Straits, carrying crude, is hit by two torpedoes from *U-66* north of Grand Bahama and sunk. Of eight officers, 37 men, and 27 armed guards, only 15 men survive; they are picked up by SS *George Washington* the next day. USS *SC-1292* is bound south in Straits towards Daytona. SS *Chiwawa* is bound south near Matanilla Shoal, northwest Bahamas with US Navy plane escorting. **Convoy NK-545** is bound south in Straits with US Navy blimp near Palm Beach. USS *Umpqua* (AT-52) is bound north in Straits towards Palm Beach. Tanker *Manuel Rionda* is bound north in Straits passing Miami. SS *Ole Bull* and escort is bound south in Straits passing Miami. USS *PC-595* and USS *SC-1060* are bound south in Straits. SS *George Washington* is bound north in Straits. **Convoy Miami-139** is bound southeast in Straits towards Nicholas Channel. USS *YAG-20* is bound south passing Nassau. *U-66*, Straits: from periscope depth and inbound aircraft type Catalina, followed by another US Navy aircraft and a semi-rigid airship (blimp) are sighted. NAS Great Exuma: Three planes went to cover **Convoy GN-64**. A plane landed at Great Exuma.

11th June 1943

Liberators BZ-757 LR and FL-976 LJ are airborne, returned to base after having carried out submarine exercises off Boca Chica. B-25 FR-381 DC is airborne to convoy the tanker *Manuel Rionda*, vessel sighted, air escort provided by the aircraft patrolling the deep waters up to 20 nm ahead of the ship, aircraft left the ship after escort and landed at base. B-25 FK-166 AD is airborne to escort tanker *Manuel Rionda*, aircraft commenced a creeping line a-head search, sent signal to base "Not met," search was continued, ship sighted, signal sent to base "Met," box patrol with visibility of 5 nm was carried out, message received from base "No relief aircraft coming, when task completed return to base," aircraft left tanker when 1 nm from Hog Island, Nassau Harbour and landed back at base. B-25 FR-378 CL is airborne on a submarine co-operation exercise at Key West, aircraft sighted the target ship and contact, exercises commenced, first attack was made, the bombs undershot, beam attack was made up track and it was estimated that the attack was successful, fourth attack was from the beam, all attacks made from 50 feet, aircraft circled submarine whilst carried out its exercise and returned to base. B-25 FR-366 GX is airborne on a submarine co-operation exercise off Key West, target ship sighted, a B-25 was observed to be carrying out the exercises, exercises were commenced, returned to base after exercise completion. B-25 FR-383 DX is airborne for navigation Exercise Number 2e, convoy sighted, consisted of 13 merchant vessels and five escort vessels, bombing and gunnery exercises carried out, landed back at base after a descent through cloud exercise with nothing to report. B-25 FR-378 CL is airborne for navigation Exercise Number 2a, bombing and gunnery exercises were carried out 5 nm due north of Spanish Wells, aircraft diverted from its course to avoid a heavy squall, exercise was completed and the aircraft landed back at base. B-25 FR-383 DX airborne for navigation Exercise Number 2c, patrol commenced convoy sighted consisting of 20 merchant ships and five escorts, bombing and gunnery exercises carried out, exercise was completed and the aircraft landed back at base. USS *Umpqua* is bound northwest in Straits towards St. Augustine. HMS *Ravager* is bound north in Straits with US Navy plane escorting. Tanker *Gulfmaid* is bound south in Straits passing Canaveral. British Royal Navy minesweepers HMS *HMYMS-172* and HMS *HMYMS-157* are bound north in Straits near Palm Beach. SS *Manuel Rionda* is bound east in Northwest Providence Channel towards Nassau. **Convoy GN-64** is bound northwest in Straits with US Navy plane escorting. **Convoy NG-366**, destroyer USS *Dahlgren*, and USS *SC-1064* are bound south in Straits. Oiler USS *YO-142* is bound south in Straits near Port Everglades. SS *Eastern Guide* and SS *Port Sydney* are bound south in Straits. **Convoy GK-738** is bound northwest in Straits passing Nicholas Channel. *U-66* in the Straits sights one Catalina flying boat, then another flying boat, and two more flying boats; then the U-boat dove. NAS Great Exuma: A plane went to cover **Convoy GN-64**. A plane went to cover **Convoy GK-738**.

12th June 1943

B-25 FR-378 CL airborne for navigation Exercise Number 2, bombing and gunnery exercises carried out patrol commenced, aircraft landed back at base after completion. B-25 FR-366 CX airborne for navigation Exercise Number 2, bombing and gunnery exercises carried out, navigation exercise commenced, aircraft deviated from its course to avoid a heavy local storm, aircraft landed back at base after completion. B-25 FR-383 DX airborne for navigation Exercise Number 2f, crossover patrol completed, aircraft landed at base after completing descent through cloud exercise, bombing and gunnery exercises were carried out. B-25 FR-383 DX airborne for navigation Exercise Number 4, exercise completed satisfactorily, landed back at base. B-25 FR-366 CX airborne for navigation Exercise Number 2e, exercise was completed, landed back at base with nothing to report. B-25 FR-375 CJ is airborne for navigation Exercise Number 4, bombing and gunnery exercises carried out off Eleuthera, exercise commenced, part of the creeping line ahead patrol was curtailed due to shortage of time, landed back at base with nothing to report. British Royal Navy rescue tug HMS *HMRT Aspirant* is bound north in Straits towards Palm Beach. USS *SC-1069* is bound south in Straits. Tug USS *Humrick,* tug *Ontario,* and tug USS *Hercules* are bound north in Straits. British Royal Navy Motor Launches HMS *HMML-1176* and HMS *HMML-1177* are bound south in Straits. **Convoy Cuban Coast-24** is bound north in Old Bahama Channel nearing Nicholas Channel with US Navy plane escorting. *U-66*, Straits: two fast land-based aircraft forced a crash dive, and again later, also for two fast aircraft. NAS Great Exuma: USS *YAG-20* departed Great Exuma for Miami. Three planes went to cover **Convoy NG-366.** A plane landed at Great Exuma.

13th June 1943

B-25 FR-381 DC is airborne for navigation Exercise Number 2, aircraft completed patrol and landed back at base PBY was sighted bombing and gunnery exercises were carried out prior to landing. B-25 FR-

375 CJ is airborne for navigation Exercise Number 2, bombing and gunnery exercises carried out over the range before aircraft set course for exercise. PBY aircraft was sighted, exercise was completed, landed back at base. B-25 FR-366 CX, FR-362 CA, FR-378 CL airborne at different times navigation Exercise Number 2, bombing, and gunnery exercises carried out exercise was completed, landed back at base. B-25 FR-362 CA is airborne for navigation Exercise Number 4b, bombing and gunnery exercise, exercise was completed, landed back at base after a descent through cloud exercise. B-25 FR-378 CL airborne for navigation Exercise Number 4, bombing and gunnery exercises were carried out over the bombing range against a flame float exercise was completed, landed back at base after a descent through cloud exercises. B-25 FR-380 DB airborne for navigation Exercise Number 4a, bombing and gunnery exercises were carried out against a flame float, exercise was completed, landed back at base after a descent through cloud exercise, nothing to report. USCG *CG-83420* bound southeast in Straits passing St. Augustine. USS *Arapaho* (AT-68) bound southeast in Straits passing St. Augustine. British Royal Navy rescue tug HMS *HMRT Aspirant* with barge and USS *SC-1467* are bound north in Straits near St. Augustine. Tanker *Esso Washington* is bound south with US Navy plane escorting. USCG *CG-83348* is bound west near Matanilla Shoal, northwest Bahamas. SS *Stanley Mathews* is bound north towards Canaveral. USS *SC-717* south in Straits passing Canaveral. US Coast Guard vessels *CG-83434, CG-83432*, and *CG-83433* are bound south in Straits towards Palm Beach. USCG *CG-83435* is bound south towards Port Everglades. USS *YAG-20* is bound northwest in Straits towards Miami. USS *PC-1197* is bound south in Straits. USS *YAG-28* and USS *YP-602* are bound north near Cay Sal Bank. Submarine USS *R-10* and USS *SC-500* are bound northwest in Old Bahama Channel, passing Cayo Francés. USS *Lapwing* (AVP-1) is bound west in Straits passing Nicholas Channel. *U-66* is in the Straits and first sights a white light which was an airship, followed by seeing two other airships. Then a Martin-type flying boat which flew out of sight. Three flying boats followed, the first very low, forcing *U-66* to submerge deep. Then an airship, and another was heard, but out of sight, then one more. Next, three fighter bombers, very high. An airship on the horizon turned away and passed from sight.

14th June 1943

Submarine patrols "Find the island" exercise. B-25 FR-380 DB is airborne on Anti-Submarine Patrol Number 3, exercise was completed, landed back at base after a descent through cloud exercise. B-25 FR-381 DC is airborne on Anti-Submarine Patrol Number 3, exercise completed satisfactorily landed back at base after a descent through cloud exercise. B-25 FR-371 CF airborne on antisubmarine Patrol Number 3, exercise completed satisfactorily landed back at base after a descent through cloud exercise. B-25 FR-362 CA is airborne on Anti-Submarine Patrol Number 3, convoy of three tankers and five escort vessels sighted, the exercise was completed and the aircraft landed at base. Liberators BZ-746 LM, FL-967 LJ, BZ-759 LP, and FL-970 LH airborne on a submarine co-operation exercise at Key West. B-25 FR-362 CA airborne on Anti-Submarine Patrol Number 4, patrol was completed and aircraft with nothing to report. B-25 FR-381 DC airborne for Anti-Submarine patrol Number 4, aircraft completed patrol nothing to report. B-25 FR-378 CL airborne on Anti-Submarine Patrol Number 4, commenced patrol, but when radio became u/s [unserviceable; his radio broke], the captain decided to return to base, aircraft landed and took off again carried out a descent through cloud and flare dropping exercise then finally landed back at base. USCG patrol trawler *CG-E. M. Joe* (WYP-356) is bound south in Straits towards Daytona. **Convoy Jacksonville-45** is bound south in Straits towards Daytona with US Navy plane escorting. SS *Sweep*, USS *SC-1292*, and USS *Arapaho* (AT-68) are bound south in Straits passing Canaveral. **Convoy NK-546** is bound south in Straits towards Palm Beach with US Navy plane escorting. USS *YMS-196*, USS *YMS-197*, and USS *SC-1330* are bound south in Straits towards Palm Beach. Tanker *Mobiloil* is bound south near Matanilla Shoal, northwest Bahamas with US Navy plane escorting. USS *SC-997* and USS *SC-1303* are bound south in Straits near Miami. Two tugs, two tows, USS *YP-612*, USS *SC-997*, and USS *SC-1303* are bound southwest in Straits. **Convoy KN-236** is bound north in Straits. **Convoy KG-639** is bound southeast in Straits towards Nicholas Channel with US Navy plane escorting. NAS Great Exuma: Surface craft attacked an enemy submarine. Two planes went to cover **Convoy GN-65.** A plane landed at Great Exuma.

15th June 1943

Convoy escort B-25 FR-380 DB is airborne to escort ferry *Jean Brillant*, vessel was met and escorted, aircraft left the ship landed back at base. Anti-Submarine patrols on find the sland exercise, B-25 FR-381 DC is airborne on Anti-Submarine Patrol Number 3, port engine cut picked up almost immediately, captain decided to return to base as this was the fourth time the engine had cut, message was sent to base "slight engine trouble," finally landed back at base. B-25 FR-382 CA airborne on Anti-Submarine patrol, merchant ship was sighted, exercises completed, aircraft landed back at base. B-25 FR-378 CL airborne on Anti-Submarine patrol, exercises completed, aircraft landed back at base after descent through cloud exercise. Liberator BZ-759 airborne for Miami returned to base. Liberator BZ-761 airborne for Boca Chica, landing there then proceeded to Miami and landed there. Hemingway's biographer says *U-527* sent four torpedoes at an unverified merchant ship off Havana. SS *Bencas* is bound south in Straits towards Canaveral. USS *YMS-247* is bound south in Straits towards Palm Beach. USS *PC-595* and USCG patrol trawler *CG-E. M. Joe* (WYP-356) are bound southwest in Straits towards Port Everglades. USS *SC-1341* is bound southwest in Straits with US Navy plane escorting. USS *SC-1473* is bound north in Straits. USS *LST-72* is bound west in Straits. SS *Vigil* is bound southeast in Straits towards Nicholas Channel with US Navy plane escorting. Ferry *Jean Brillant* is bound south in Straits exiting Northwest Providence Channel towards Nassau with US Navy plane escorting. **Convoy GN-65** is bound north, to the east of Bahamas. NAS Great Exuma: Four planes went to cover **Convoy GN-65.** A plane went to cover **Convoy NG-367.** One US Navy aircraft carrier and two destroyers entered area.

16th June 1943

B-25's FR-378 CL, BZ-371 CF and FR-383 DX is airborne for submarine Patrol Number 5, message received from US Navy Guantanamo about submarine. Contact message sent to aircraft diverting them from patrol to the position of a reported submarine, message received correctly by

CF, which proceeded to the stated position, message incorrectly received by DX and XL, broadcast message sent to three aircraft instructing them to carry out individual search, first message was received by DX together with the second message operator in CL failed to pass the second message to pilot CF and DX eventually reached area, search carried out, an unidentified object was sighted believed to be surfaced U-boat by CF, vessel was identified as a US Coast Guard cutter, message exchanged, convoy consisting of 27 merchant ships and five escorts was sighted. Two ships sighted by DX, oil patch sighted, running almost due north and south, search carried out nothing sighted. FR-378 CL set course for position of reported submarine, two escort vessels sighted, broadcast message sent to the aircraft to return to base aircraft, returned to base, landed back with nothing to report. USS *SC-1473*, USCG *CG-83363* and tanker *E. R. Kemp* are bound north in Straits passing Canaveral. USS *Strenuous* (BAM-129) is bound north in Straits. Tanker *Atlantic Coast* and tanker *Charlestown* are bound south in Straits near Palm Beach. SS *Bencas* is bound south in Straits, passing Palm Beach with US Navy plane escorting. Tug USS *Meteor* is bound south in Straits towards Miami. Ferry *Jean Brillant* is bound west in Straits towards Miami with US Navy plane escorting. USS *SC-1021* is bound south in Straits. SS *W. E. Ogilvie* is bound west in Straits passing Nicholas Channel. **Convoy NG-367** is bound south, to the east of Bahamas. *U-66* is in Straits reports one US Navy airship and a flying boat detected. NAS Great Exuma: USS *Rockaway* passed. A plane went to cover **Convoy NG-65.** Two planes went to cover **Convoy NG-367**. Contact with a submarine was made. One light cruiser and two destroyers were passed through Crooked Island. Three RAF B-25 bombers were sent for search and rescue.

17th June 1943

B-25 FR-383 DX is airborne for Anti-Submarine Patrol Number 5, radio found to be unserviceable, aircraft continued on exercise, captain decided to abandon as the radio was still u/s, landed back at base, radio was u/s from the time of leaving the circuit, on inspection the set was found to be serviceable on the ground, but all coils had been tried in the air. B-25 FR-380 DB is airborne for Anti-Submarine Patrol Number 5, patrol complete, radio contact with base was impossible, message sent to base via CX landed back at base with nothing to report. B-25 FR-362 CA airborne for Anti-Submarine Patrol Number 5, radio contact with base was impossible, continued the patrol captain decided to return to base as the radio was still u/s landed back at base. B-25 FR-371 CF airborne for Anti-Submarine Patrol Number 5, exercise completed, aircraft was in radio contact with DB throughout the patrol, landed back at base. B-25 FR-381 DC airborne for Anti-Submarine Patrol Number 3, message was sent to base reporting engine trouble, landed back at base. B-25 FR-380 DB airborne for Anti-Submarine Patrol Number 3, message intercepted from DC reporting engine trouble to acknowledge the signal, action taken by DB, exercise completed, landed back at base. B-25 FR-362 CA is airborne for Anti-Submarine Patrol Number 3, message intercepted from DC, reporting engine trouble, no action was taken by CA, exercise was completed and the aircraft landed back at base. *U-67* is in Straits. USS *SC-1064* is bound south in Straits towards St. Augustine. Argentinian SS *Rio Primero* is bound south in Straits passing Canaveral with US Navy plane escorting. Tug USS *Humrick* with barge and USS *YP-612* are bound north in Straits towards Port Everglades. USS *SC-1260* is bound south in Straits. USS *SC-1326* is bound west in Straits, USS *29* is bound southwest in Straits. The Argentinian SS *Rio Dulce* is bound west in Straits past Nicholas Channel, with US Navy plane escorting. SS *W. E. Ogilvie* is bound west in Straits past Nicholas Channel. NAS Great Exuma: A plane landed in Great Exuma. Two planes went to cover **Convoy NG-367**. A plane went on a search and rescue mission. A plane returned to base due to engine failure.

18th June 1943

B-25 FR-380 DB is airborne for Anti-Submarine Patrol Number 5, radio contact with base was unobtainable, aircraft continued course, captain decided to return to base as the radio was still unserviceable [u/s], aircraft set course for base, re-established radio contact and landed back at base. B-25 FR-383 DX airborne for Anti-Submarine Patrol Number 5, exercise completed radio contact was difficult, landed back at base with nothing to report. B-25 FR-371 CF is airborne for Anti-Submarine Patrol Number 5, exercise completed, circled base, set course to escort the ferry *Jean Brillant,* ship was met and escorted, aircraft left convoy, landed back at base. USCG *CG-83419* is bound southeast in Straits passing St. Augustine. Tanker *E. R. Kemp* and USS *SC-1018* are bound south in Straits passing Canaveral. USS *SC-1299* is bound south in Straits towards Palm Beach. USS *YP-272* is bound north in Straits towards Canaveral. USS *SC-1272* is bound south in Straits towards Port Everglades. **Convoy SS *Chiwada*** is bound north in Straits towards Port Everglades with US Navy plane escorting. Tug USS *Humrick* with three tows and USS *YP-426* are bound southeast in Straits passing Fort Everglades. USS *PC-1198* is bound southeast in Straits passing Miami. USS *ASR-3* and USS *SC-1473* are bound south in Straits with US Navy plane escorting. Two tugs with three tows and USS *YP-513* are bound north in Straits. NAS Great Exuma: A plane departed to cover a **Convoy GN-66** consisting of 23 merchant vessels and five escorts. Another convoy entered the area.

19th June 1943

Training throughout the day from Satellite Field and Main Field. B-25 FK-380 DU, B-25 FR-382 DD, and FR-371 CF are airborne at different times Anti-Submarine patrol exercise completed, landed back at base. All aircraft experienced difficulty in reception of radio signal. USS *SC-1293* and USS *PC-1200* are bound south in Straits towards Daytona. USS *PC-1199* is bound south in Straits towards Canaveral. SS *Atlantic States* is bound southwest in Straits towards Palm Beach. SS *Plato* is bound northeast towards Matanilla Shoal, northwest Bahamas. USCG *CG-83419* is bound southwest in Straits towards Port Everglades. Ferry *Jean Brillant* is bound west in Straits from Bimini towards Miami with US Navy plane escorting. **Convoy Miami-147** is bound south in Straits. Tug USS *J. E. Cartaya* is bound south in Straits towards Cay Sal Bank. Tanker *E. R. Kemp* is bound west in Straits with US Navy plane escorting. SS *Pueblo* is bound east in Straits. **Convoy KN-247** is bound north in Straits with US Navy plane escorting. SS *Mobilight* is bound west in Straits. **Convoy KG-640** is bound southeast in Straits, with US Navy blimp near Nicholas Channel. The house boat USS *YHB-14* is bound southeast from Cay Sal Bank towards Old Bahama

Channel and Cayo Francés. NAS Great Exuma: Three planes went to cover **Convoy GN-66.** A plane landed at Great Exuma.

20th June 1943

USS *Kiowa* (AT-72) is bound south in Straits towards Canaveral. USS *PC-1199* south in Straits passing Canaveral. **Convoy NK-547** is bound south in Straits towards Palm Beach with US Navy plane escorting. USS *SC-659* is bound south in Straits towards Palm Beach. Tanker *R. P. Smith* is bound north in Straits towards Palm Beach. USS *SC-1321* is bound north in Straits passing Port Everglades. USS *SC-1329* is bound north in Straits passing Miami. Tanker *A. S. Hansen,* USS *SC-986* and USS *SC-754*, and tug *Ontario* are bound south in Straits. Tug USS *Meteor* and USS *SC-1002* are bound north in Straits. **Convoy Key West-215** is bound southeast in Straits passing Nicholas Channel. SS *Plato* is bound south passing Abaco towards Northeast Providence Channel. NAS Great Exuma: Two planes went to cover **Convoy GN-66.** Two planes went to cover **Convoy NG-368.**

21st June 1943

Original programme curtailed as weather became unfavourable to flying. Liberator FL-962 LG is airborne from Satellite Field for Anti-Submarine Patrol Number 9, bombing practice carried out off Sail Rock, Exumas, convoy sighted consisting of 27 merchant ships and five escorts Catalina seen to escorting convoy, exercise completed, landed back at Satellite Field, although electrical storm experienced. Liberator FL-942 LC airborne from Satellite Field for Anti-Submarine Patrol Number 9, bombing practice carried out off Sail Rock, Exumas, creeping line ahead, exercise completed, had nothing to report on landing at Satellite Field, again airborne later homing practice carried out off Sail Rock, Exuma Cays. B-25's FR-366 CX and FR-362 CA are airborne for air Exercise Number 7, area bounded by the patrol was Great Stirrup Cay and Whale Cay, Berry Islands, and Morgan's Bluff Andros. Completed patrol and landed back at base. B-25 FR-371 GF is airborne for Anti-Submarine patrol, patrol carried out as creeping line ahead, patrol completed without incident, landed back at base. B-25 FR-393 DF is airborne for Anti-Submarine Patrol Number 4, creeping line ahead, patrol carried completed without incident, landed back at base. B-25 FR-371 CF is airborne for Anti-Submarine Patrol Number 4, patrol carried and completed, landed back at base. B-25 FR-382 DD is airborne for Anti-Submarine Patrol Number 4, patrol carried out as creeping line ahead patrol, patrol completed without incident and landed back at base. B-25 FR-38I DC is airborne to escort ferry *Jean Brillant* creeping line ahead patrol commenced on the track of convoy, with visibility of 5 nm ship contacted and escorted to Nassau, air coverage provided by the aircraft carrying out a box patrol. USS *SC-718* is bound south in Straits towards Canaveral. Tanker *A. S. Hansen* and USCG *CG-83363* are bound north in Straits towards Canaveral. Patrol Craft *Greek King George II* (ex-USS *PC-622*) and USS *SC-1322*, tanker *18 De Marzo*, and USS *SC-511* are bound south in Straits towards Palm Beach. USCG *CG-83348* is bound west near Walker Cay, north of Abaco. USCG *CG-83420* is bound south in Straits towards Port Everglades. USS *SC-1282* is bound south in Straits passing Miami. **Convoy NK-547** is bound south in Straits passing Miami with US Navy plane escorting. USS *Kiowa* (AT-72) and **Motor Torpedo Boat Squadron-26, Group A** are bound south in

Straits. **Convoy GK-740** is bound north in Old Bahama Channel passing Cayo Francés with US Navy plane escorting. NAS Great Exuma: Two plane went to cover **Convoy NG-368.** USS *Plato* and a US Coast Guard escort stood in and anchored in the Exuma basin. A plane landed in Great Exuma.

22nd June 1943

Night Anti-Submarine patrols cancelled due to unfavourable conditions. B-25 FR-371 GF airborne for Anti-Submarine Patrol Number 3, exercise completed, landed back at base. Liberator BZ-757 LH airborne from Satellite Field for Anti-Submarine Patrol Number 9, bombing practice carried out off Sail Rock, Exumas exercise completed, landed back at Satellite Field, homing exercise attempted on LJ, no contact made with Liberator FL-967 LJ which was airborne from Satellite Field on Anti-Submarine Patrol Number 9, exercise completed, landed back at Satellite Field, nothing to report. B-25's FR-366 CX, FR-384 CM, FR-380 DB airborne Exercise Number 7. Patrol craft *Greek King George II* (ex-USS *PC-622*) south. USCG *CG-83348* is bound northeast in Straits from near Palm Beach. USS *PC-592* is bound north in Straits mid-channel. Tug USS *Humrick* with two tows and USS YP-426 are bound southwest in Straits. USCG *CG-83376* is bound south in Straits towards Cay Sal Bank. USS *SC-659*, tug USS *Meteor* and tow and USS *YP-606* are bound west in Straits. Destroyer escort USS *Robert E. Peary* (DE-132) is bound north in Straits. NAS Great Exuma: USS *Plato* and a US Coast Guard escort weighed anchor and stood out to sea.

23rd June 1943

Night Anti-Submarine patrols cancelled due to unfavourable conditions. B-25's FR-366 CX, FR-384 CM, and 380 DB are airborne Exercise Number 7. Patrol craft *Greek King George II* (ex-*PC-622*) is bound south in Straits towards Canaveral. USS *SC-662* is bound south in Straits passing Canaveral. USS *SC-1049* is bound southwest from Matanilla Shoal, northwest Bahamas towards Port Everglades. USS *SC-718* is bound south in Straits passing Palm Beach. USS *IX-80* is bound west in Straits from Bimini to Miami. Tug USS *Hercules*, USS *SC-998*, and USS *PC-1124* are bound southwest in Straits. SS *Charlestown* is bound north in Straits. NAS Great Exuma: An Army JRF-5 Amphibian plane with two commanders and the captain arrived from Guantanamo.

24th June 1943

Night Anti-Submarine patrols cancelled, as weather is unfavourable. Liberator FL-942 is airborne from Satellite Field for Anti-Submarine Patrol Number 9, bombing exercises carried out off Sail Rock, Exumas, compass found to be erratic, patrol abandoned owing to unreliability of the compass and impossibility of avoiding cumulonimbus clouds, aircraft homed on the radio compass and landed back to Satellite Field. Liberator BZ-757 LR is airborne from Satellite Field for Anti-Submarine Patrol Number 9, a steam ship contact at 10 nm on 50 nm scale is made, 10-mile scale was brought into operation and the aircraft homed on the blip, contact disappeared in 7 minutes. No surface vessels homing exercise was commenced, no contact established with LT which was passed at 6 nm distance, bombing exercises carried out, aircraft landed at Satellite Field, nothing to report. Steam ships have been used throughout the patrol. Liberator BZ-760 LT is airborne from Satellite Field for Anti-Submarine Patrol Number 9, bombing exercises carried out off Sail Rock, Exumas, homing exercises carried out successfully on aircraft FL-970 LH, completed the patrol and landed back to Satellite Field. Liberator FL-967 LJ airborne from Satellite Field for Anti-Submarine Patrol Number 9, bombing exercises were to have been carried out off Sail Rock, Exumas, visibility was so poor that the target was not visible, completed the patrol and landed back to Satellite Field with nothing to report. USS *SC-1327* is bound south in Straits towards Canaveral. **Convoy Jacksonville-51** is bound south in Straits passing Canaveral. USS *Admirable* (AM-136) is bound north in Straits midway Port Everglades and Bimini. **Convoy Port Everglades-148** is bound south in Straits from Port Everglades with US Navy plane escorting. USS *SC-642*, USS *SC-1057*, USS *SC-1271*, USS *SC-1228* and tug *Ontario* are bound south in Straits. **Convoy KN-248** is bound north in Straits, US Navy blimp escorting, past Cay Sal Bank. Tug USS *Humrick* is bound west in Straits. **Convoy KG-641** is bound southeast, passing Nicholas Channel with US Navy plane escorting. NAS Great Exuma: Two planes went to cover **Convoy GN-67**, 28 Merchant Vessels and five escorts. A plane landed at Great Exuma.

25th June 1943

Liberator FL-942 LC is airborne from Satellite Field for Anti-Submarine Patrol Number 9, bombing exercises carried out off Sail Rock, Exumas, creeping line ahead, patrol commenced, carried out with visibility at 10 nm southeast contact was made, further contact made, convoy sighted consisted of 23 merchant vessels and seven escorts; photographs taken, completed patrol and returned to base. Liberator BZ-757 LR from Satellite Field for Anti-Submarine Patrol Number 9, reports inter-island merchant ship sighted, bombing exercises carried out off Sail Rock, Exumas, results were unobserved southeast. Contact was made, aircraft homed on the contact convoy consisting of 28 merchant ships and five escorts sighted, there were two signals were exchanged with the Senior Naval vessel, two destroyers were sighted, aircraft passed over ships, completed the patrol, landed back at Satellite Field. Liberator BZ-760 LT is airborne from Satellite Field for Anti-Submarine Patrol Number 9, bombing exercises carried out off Sail Rock, Exumas, creeping line ahead, patrol commenced, southeast contact was made, convoy consisting of 28 merchant ships and five escorts sighted, completed the patrol, landed back at Satellite Field. Liberator BZ-967 LJ is airborne from Satellite Field for Anti-Submarine Patrol Number 9, with starboard engine burnt out, it had to return to base, airborne after changing the generator. Convoy consisting of 25 merchant ships and four escorts sighted, completed the patrol, landed back at Satellite Field. B-25 FL-200 RX airborne for Miami landed at Miami, stayed overnight as weather was unfit for return flight. Tanker *R. P. Smith* and USCG *CG-83362* are bound south in Straits towards Daytona. SS *Chippewa* is bound northwest in Straits with US Navy plane escorting. USS *SC-1283* is bound north in Straits near Canaveral. SS *Chiloil* and USS *SC-1329* are bound north in Straits passing Palm Beach with US Navy plane escorting. Ferry *Jean Brillant* and USCG *CG-83435* and USS *PC-612* are bound southwest in Straits towards Miami with US Navy plane escorting. Anti-Submarine Group-13 is bound north in Straits. Oiler USS *YO-142* and USS YP-*408* and **Convoy Jacksonville-51** are bound west in Straits. Tanker *Pan Orleans* and Cuban *CS-11*, Cuban *CS-32*, and Cuban *CS-33* are bound west in Straits passing Nicholas Channel with US Navy plane escorting. USS *Lapwing* (AVP-1) is bound southeast in Straits passing Nicholas Channel. Oiler USS *YO-143* and USS *YP-602* are bound north from Cayo Francés and Old Bahama Channel. **Convoy GN-67** is bound north, east of Bahamas. NAS Great Exuma: Two planes went to cover **Convoy GN-67.** A plane landed at Great Exuma. Two destroyers joined **Convoy GN-67.** A plane went to cover **Convoy NG-369**, of 23 merchant vessels and five escorts entered area.

26th June 1943

B-25 FL-200 BX returned from Miami, landed at the Main Field, brought the Officer Commanding back from Miami. **Convoy NK-180** is bound south in Straits passing Canaveral with US Navy plane escorting. SS *Jalapa* is bound south in Straits passing Canaveral. USS *YMS-247* is bound north in Straits near Matanilla Shoal, northwest Bahamas. USCG *CG-83348* is bound west near Walker Cay, north of Abaco. **Convoy NK-548** is bound south in Straits to Palm Beach with US Navy plane escorting. USS *Chewink* (AR-83) and USCG *CG-Marnell* are bound south in Straits, passing Palm Beach with US Navy plane escorting. Tanker *R. P. Smith* and USCG *CG-83362* are bound south in Straits to Port Everglades. Tug *Ontario* with tow and USS *YP-612* are bound southeast in Straits towards Miami. USS *YO-143* and USS *YP-602* are bound northwest in Straits to Miami. SS *Plato*, USCG *CGR-7007*, and USCG Cutter *CG-Yeaton* are bound northwest from Cayo Francés in Old Bahama Channel. **Convoy GN-67** is bound north, to the east of Bahamas, with US Navy plane escorting. SS *Charles Kurz* and USS *PC-619* are bound north, to the east of Bahamas. NAS Great Exuma: Three planes went to cover **Convoy NG-369.** Two planes landed Great Exuma.

27th June 1943

US Navy cruiser USS *Cincinnati* passes along Gulf Stream between Bahamas and US at night reports "not rough" conditions. Yacht *Gemini*, with HRH with Duke of Windsor and Duchess, under Captain Herman, leave Miami for Nassau with US Navy escort provided by Rear Admiral James L. Kauffman. **Convoy KW-226** is bound north in Straits from Port Everglades. **Convoy Norfolk-180** is bound south in Straits towards Miami. Tug USS *Meteor* is bound north in Straits near Port Everglades. USS *SC-1261* and **Convoy SS *Housatonic*** are bound south in Straits. **Convoy KM** is bound south in Straits with US Navy plane escorting. **Task Unit TU-02910** is bound south in Straits with US Navy plane escorting. Tug USS *Humrick* is bound north in Straits. Ferry *Jean Brillant* is bound east in Straits towards Nassau. NAS Great Exuma: Two planes went to cover **Convoy NG-369**, of 23 merchant vessels and five escorts. A plane landed Great Exuma.

28th June 1943

FR-375 CJ, FR-366 CX, FR-381 DC, FR-362 CA, FR-384 CM, FR-378 CL, FR-382 DD and FE-380 DB are airborne for air Exercise Number 1. FR-366 CX, FR-954 FB, FR-381 DC, FR-362 CA, FR-84 CM, FR-393 DF and FR-380 DB are airborne for air Exercise Number 3. FR-382 DD is airborne for air firing and gunnery Exercise Number 3, aircraft was overdue from the exercise went missing, aircraft was called on all frequencies, aircraft were detailed to be airborne on a search, four aircraft detailed were instructed to search an area 20 nm square around the Main Field. The aircraft searching were FR-362 CA, FR-371 CF, FV-951 FC, and FV-953 FA, signal was received from FR-371 CF that an empty dinghy had been sighted 5 nm north of Morgan's Bluff, so CA diverted to position in which dinghy was sighted on instructions received from OC"B" Squadron. All aircraft were diverted to the position of the sighted dinghy, signal sent to FR-371 CF instructing it to remain over the dinghy, and that other aircraft were proceeding to the area in which CF was searching aircraft. Then FA and FL recalled to base, signal was received from CA stating that it was circling the position to which it had been directed and that only oil patches were to be seen, signal sent to CP, requesting if there was anything further to report at roughly the same time, CA signalled that it was returning to base, signal was sent to CF recalling it to base, contact by radio and radio could not be established due to electrical storms, aircraft landed back at base on sighting the dinghy by CF, aircraft FL-208 BF was airborne to augment the search in the area of the dinghy, aircraft landed having sighted a damaged dinghy, without occupants, and a large oil patch nearby, signal transmitting station was instructed to cease calling back DD, search was suspended until dawn of 29th June. USCG *CG-83421* escorts Windsor yacht *Gemini* from Miami, collides with USS *SC-1330* 7 nm north of Great Isaac Light. Both vessels are part of escort of ferry *Jean Brillant*, which is *en route* from Miami to Nassau. Twenty-two minutes before the collision, *Jean Brillant* changed her course passing through the point designated for turning on routing instructions previously handed to all five escorts. All vessels darkened out. Twelve minutes later the USS *SC-1330* changed course. Sometime between 22 and 24 minutes before the collision USCG *CG-83421* had changed course. When sighted, USS *SC-1330* was only 35 yards distant USCG *CG-83421* then swung left, and USS *SC-1330* stopped engines and backed. Collision not avoided; it caused the stern of USCG *CG-83421* with two watertight compartments, to float away, yet remaining afloat. Crew were taken off, vessel taken in tow by USS *SC-1330* however it sinks, so the tow line is cut. No loss of life or serious injury. Commanding officer of USCG *CG-83421*, Ensign Lawrence E. Gallagher, USCGR was exonerated. US Navy blimp assigned to escort the Duke of Windsor's convoy, assisted in the rescue of 13 crew from USCG *CG-83421* by using its searchlight. USS *PC-613* is bound south in Straits, passing Canaveral. USCG *CG-83420* is bound south in Straits towards Port Everglades. Oilier USS *YO-143* is bound southeast from Port Everglades. Ferry *Jean Brillant* is bound southwest in Straits from Northwest Providence Channel towards Miami. USCG Cutter *CG-Micawber* and USS *SC-717* are bound south in Straits. **Convoy Havana North Coast Cuba-109** is bound north in Straits. Tug USS *Humrick* is bound north in Straits. *U-66* is in the Straits, observing continuous aircraft, airships, flying boats, and fighter bombers. Enemy Radar detection up to now was only by night. Flying boat flying low causes *U-66* to crash dive. One new aircraft.

29th June 1943

The OC"B" Squadron instructed Marine Craft Section to have two launches in area where the dinghy was sighted at dawn, aircraft FV-953 FA was airborne maintained radio contact with the launches throughout the patrol signal from ASR launch HMS *P-190* was received to the effect that the dinghy was recovered, number of dinghy was requested, found that it belonged to FR-382 DD. Search aircraft instructed to continue search for a further two hours, signal sent recalling the marine craft ASR launch HMS *P-190* and ASR launch HMS *P-63* to base, aircraft FA was instructed to search the north and east coasts of Andros before returning to base, signal received from 1 ASR launch HMS *P-63* that a wheel and spar had been recovered, FV-953 FA landed back with nothing further to report. B-25's FR-371 CF, FR-366 CX, FR-384 CM, FR-381 DC, FV-954 FB, FR-383 DX, and FR-380 DB are airborne for air Exercise Number 1, all completed the exercises and landed back at base. B-25's FR-366 CX, FR-371 CF, FR-383 DX, FR-384 CM, and FR-380 DB are airborne for air Exercise Number 2. USS *SC-1315* is bound south in Straits passing St. Augustine. **Convoy NK** is bound south in Straits with US Navy plane escorting. USS *SC-987* and SS *Chiloil* are bound south in Straits, passing Canaveral. USS *PC-1128* is bound north in Straits. **Convoy KN-249** is bound north in Straits with US Navy plane escorting. SS *Theodore Weld* is bound west in Straits. **Convoy KG-642** is bound southeast in Straits passing Nicholas Channel with US Navy plane escorting. USS *YO-143*, USS *YP-338*, USS *YOG-46*, USS *PC-1126*, and USS *YP-218* are bound south in Straits. *U-759* enters area south to Windward Passage. *U-66*, Straits: two Consolidated PB-2Y flying boats, and later more flying boats out of sight. NAS Great Exuma: Two planes went to cover **Convoy GN-68,** 32 merchant vessels and five escorts. USS *YP-42* stood in from sea and dropped anchor.

30th June 1943

B-25 FV-954 FB is airborne to escort ferry *Jean Brillant,* convoy met consisted of one merchant ship and three escorts signals exchanged, message sent to base regarding meeting the convoy. Convoy left, impossible to contact base, convoy again met later, signal sent regarding meeting the convoy, convoy entered the Harbour and aircraft left, set course for base with nothing to report on landing. B-25's FV-949 FB, FR-371 CF, FR-381 DC, FV-362 CA, FR-383 DX, and FV-375 CJ are airborne for air Exercise Number 2, again airborne later for air Exercise Number 4. Windsors return to Nassau from USA. USS *Kiowa* (AT-72) with tows and USS *SC-1060* are bound north in Straits passing Canaveral. USS *B-90*, USS *B-91*, USS *B-92*, and USS *B-93* are bound north in Straits passing Palm Beach. **Convoy NK-549** is bound south in Straits passing Palm Beach with US Navy plane escorting. Tanker *Fenja* and tanker SS *New Orleans* are bound south in Straits passing Palm Beach. SS *Colabee* is bound north in Straits towards Palm Beach. USS *PC-1125* is bound south in Straits. Tug *Ontario* with tows and USS *YP-612* are bound north in Straits. *U-527* re-enters area northeast in Straits past Key West. *U-66* is off Jacksonville and observes a flying boat, followed by two airships. NAS Great Exuma: Three planes went to cover **Convoy GN-68.** A plane landed at Great Exuma. A plane went to cover **Convoy NG-370**, of 33 Merchant Vessels, five escorts, then departed.

1st July 1943

B-25's FR-371 DP, FR-375 CJ, FR-381 DC, FR-352 CA, FR-393 DF, FV-946 FF, FR-383 DX, and FV-952 FX are airborne for air Exercise Number 3. B-25's FR-371 CF, FR-375 CJ, FR-381 DC, FR-362 CA, FR-393 DF, FV-946 FF, FR-383 DX and FF-952 FI are airborne for air Exercise Number 4. *U-159* enters via Anegada Passage west for the Northeast Providence Channel and then Northwest Providence Channel. Tanker *Bloody Marsh* is bound northeast out of Straits. USS *AM-137* is bound northeast in Straits. **Convoy NG-370** is bound south in Straits. SS *Atlantic Coast* is bound southwest in Straits towards Palm Beach. USS *SC-1070* and USS *PC-1241* are bound southwest in Straits. **Convoy NK-549** is bound west in Straits with US Navy plane escorting. SS *Peter V. Daniel* is bound southwest towards Key West. USS *Henrico* is bound northeast in Straits passing Cay Sal Bank. *U-759* is bound north of San Juan west for Windward Passage. *U-527* is in Straits towards Northwest Providence Channel and Northeast Providence Channel, bound east for France. *U-66* is off Jacksonville and observes three airships, the final of which motored out of sight, and a flying boat thought to be a Consolidated PBY-1, which turned towards the U-boat. NAS Great Exuma: Four planes escorted **Convoy NG-370.** A plane landed at Great Exuma. A plane had engine trouble yet finally landed.

2nd July 1943

B-25's FR-371 CF, FR-375 CJ, FV-395FC, FR-384 CM, FV-953 FA, FV-948 FF and FR-381 DC are airborne for air Exercise Number 5. Tanker *Bloody Marsh* is sunk by *U-66* 75 nm east of Savannah, well north of Matanilla Shoal, northwest Bahamas. USS *SC-1048* rescues then lands survivors at Charleston following day; 69 killed. USCG *CG-83446* is bound south in Straits passing Canaveral. Oiler USS *YO-143* and destroyer escort USS *Pillsbury* (DE-133) are bound north in Straits. Tanker *Atlantic Coast* is bound southwest in Straits near Cay Sal Bank. Tanker *Shiloh* is bound west in Straits. *U-66* is off Jacksonville, and observes an airship in sight, followed later by another airship. NAS Great Exuma: A plane escorted **Convoy NG-370** and then departed for training loop. A convoy of 19 merchant vessel, five escorts departed from area. Cargo transport vessel USS *YAG-28*, which was built as *Samaná* in 1940 for Austin Levy by Victor Cleare in Hatchet Bay, dropped anchor in Great Exuma.

3rd July 1943

B-25's FR-362 CA, FV-952 FF, FR-381 DC, and FR-384 CM are airborne for air exercises Number 5. B-25's FV-953 FA, FR-375 CK, FR-371 CF, and FR-380 DB are airborne for air Exercise Number 5. B-25 FV-954 FB is airborne for Anti-Submarine patrol, square search was commenced oil patch was observed directly beneath the aircraft circled but nothing was sighted landed back at base. SS *E. M. Northumberland* is bound south in Straits passing Daytona. Tanker *Gulfmaid* is bound north in Straits passing Canaveral with US Navy plane escorting. USS *SC-717* is bound northeast in Straits. SS *Steel Voyager* is bound northeast in Straits towards Canaveral. USCG *CG-83446* is bound south in Straits towards Palm Beach. USS *PC-612* is bound south in Straits. Oiler USS *YO-142* and USS *YP-408* are bound north in Straits. *U-527* is bound northeast of Abaco, east to France. B-25 FV-954 FH is 120 nm north-northeast of San Salvador. A No.111 OTU aircraft on Anti-Submarine patrol observed an oil patch. *U-66* is in the Straits, observing one destroyer masthead in sight, with three different sightings of airships, as well as one flying boat. South of Bermuda are two attacks, bombs jammed, as George W. Brown engages *U-134*, possibly *U-66*. Then, 150 nm east of Eleuthera, B-25 FV-954 FH saw an oil patch. NAS Great Exuma: A plane departed from training loop and landed in Great Exuma. USS *YAG-28* (ex- *Samaná*) lifted anchor and departed Great Exuma. A merchant vessel was sighted by a plane. **Convoy GN-69**, of 27 merchant vessels and five escorts entered the area.

4th July 1943

Convoy NK-550 is bound south in Straits passing Canaveral with US Navy plane escorting. Tanker *Jalapa* is bound north near Matanilla Shoal, northwest Bahamas.

USS *YAG-28* is bound west in Straits north of Andros. USS *SC-737* and USS *SC-1020* are bound southeast in Straits passing Miami. USS *YT-338* with tows is bound north in Straits near Miami. Tanker *Great Meadows* is bound south in Straits. USCG *GR-7007* is bound north in Straits. **Convoy KN-250** is bound north in Straits with US Navy plane escorting. **Convoy KH-643** is bound southeast in Straits passing Nicholas Channel with US Navy plane escorting. *U-159* enters area northeast of Bahamas, bound west. *U-759* exits area from Old Bahama Channel, bound south via Windward Passage. *U-66* is in Strait, with six flying boats observed; one from periscope depth, another that flew out of sight, and one which circled the U-boat. NAS Great Exuma: Two planes went to cover **Convoy GN-69**. A commercial airline pilot reported sighting a submarine. A plane departed to search area. A plane landed in Great Exuma.

5th July 1943

B-25 FR-381 DC is airborne to carry out an exercise in co-operation with a US Navy submarine off Key West, submarine sighted in the company of an escort vessel, aircraft circled whilst exercise was being carried out by another aircraft, DC commenced exercise, one attack was made, could not be completed, aircraft left submarine and returned to base. B-25 FR-366 CX is airborne for Anti-Submarine exercise in co-operation with US Navy off Key West, target ship was sighted, signals exchanged, attack, exercises carried out, aircraft circled and photographed the exercises carried out by FA, returned to base. B-25 FV-953 FA is airborne for submarine co-operation exercise in conjunction with a US submarine off Key West, two attacks were directed against the submarine one when submarine was at periscope depth, and the other following on a crash dive, exercise completed, returned to base. B-25 FR-362 CA is airborne for submarine co-operation exercise off Key West, aircraft set course for Sand Key, submarine sighted, aircraft had to circle and wait for completion of another aircraft's exercise, signals were exchanged, two dummy runs made, submarine appeared to be on its way to the base, and the submarine did not make any attempt to dive, exercise abandoned, aircraft set course for base, landed back. B-25's FR-393 DF, FR-376 CK, and FR-380 DB are airborne for air Exercise Number 6. *U-527* is bound northeast of Abaco, east to France. Tanker *Gulfmaid* is bound south in Straits towards Daytona. **Convoy KN-250** and tanker *Pan Georgia* are bound northeast in Straits. **Convoy SS *Housatonic*** is bound northeast in Straits, passing Port Everglades with US Navy plane escorting. **Convoy NK-550** and HMS *King George II* are bound southwest in Straits with US Navy planes escorting. USS *PC-1129* and USS *SC-1035* and **Convoy Motor Torpedo Boat Squadron, Group A** and HMS *Empire Pintail* are bound north in Straits. Anti-Submarine Group 6 is bound southeast in Straits towards Nicholas Channel. *U-732* enters area northeast of Bahamas south to Windward Passage. *U-159* is bound south to Windward Passage. Then, 310 nm southwest of Bermuda, 105-B-8, attacked *U-527*, yet had no bombs left from an earlier attack, as Pilot John R. Steetle, south of Bermuda also saw and attacked *U-159* with depth charges. NAS Great Exuma: Two planes departed to cover **Convoy GN-69** and **Convoy NG-371**. A plane departed to cover **Convoy GK-743.** A plane landed at Great Exuma.

6th July 1943

B-25 FV-952 FX is airborne for submarine co-operation exercise off Key West, target submarine sighted with escort vessel, practice attacks made against submarine in detailed positions, one stick was observed to straddle to submarine's conning tower, exercise completed, aircraft landed back at base. B-25 FV-946 FF is airborne for similar exercise contact made with submarine off Key West only one attack made, exercise terminated aircraft landed back at base. B-25 FV-953 FA is airborne for Key West submarine co-operation exercise target submarine sighted with escort vessel three attacks delivered against submarine when at periscope depth and again when crash diving exercise completed, aircraft landed back at base. B-25 FR-375 CJ is airborne for submarine co-operation exercise off Key West, contact made with submarine, exercise commenced, beam attacks, exercise terminated, returned to base, firing attacks against flame float, landed back at base. B-25's FR-375 CJ, FR-376 CX, and FV-951 FC are airborne for air Exercise Number 6. Liberator FL-970 LH is airborne for Key West, then returned at night. SS *Perote* is bound south in Straits towards Canaveral. HMS *Tourmaline* is bound north in Straits. Tanker *Gulfmaid* is bound south in Straits near Port Everglades with US Navy plane escorting. USS *PC-610* is bound north in Straits passing Port Everglades. USS *SC-9* is bound south in Straits passing Miami. USS *PC-1128* is bound south in Straits. *U-84* enters area bound south to Windward Passage. *U-159* exits area bound south via Anegada Passage. *U-66* is in Straits and observes three flying boats sighted, one of which flew in circles, another flew out of sight. NAS Great Exuma: Two planes departed to cover **Convoy NG-371**. A plane escorted **Convoy GK-743**. Two planes landed at Great Exuma. A patrol craft escorting a convoy made contact with a submarine.

7th July 1943

B-25 FV-952 FX is airborne for submarine co-operation exercise contact made with submarine was with an escort exercise completed, landed back at base. B-25 FV-953 FA is airborne for submarine co-operation exercise off Key West, contact made with w/t, [radio, wireless telegraph] was unsuccessful, signal passed by Aldis lamp, exercise commenced and completed, landed back at base. B-25 FR-375 CK is airborne for submarine co-operation exercise, had to force land with an oil leak in port engine at Boca Chica. B-25 FR-378 CL is airborne for submarine co-operation exercise, three island type vessels spotted, submarine spotted with escort, exercise commenced and completed as per detailed instructions, landed back at base. B-25 FR-381 DC is airborne for Anti-Submarine Exercise Number 1, patrol is a line patrol, gunnery exercise carried out, exercise completed, landed back at base with nothing to report. B-25 FV-949 FV is airborne for Anti-Submarine Exercise Number 1a, uneventful, landed back after completion, nothing to report. B-25 FV-946 FF airborne for Anti-Submarine Exercise Number 1, u/v [uneventful] sighted patrol, completed, landed back after gunnery exercise. Liberator BZ-759 LP is airborne, search for an enemy submarine reported to be following enemy routes, square search carried out, merchant ships and escort vessel sighted, merchant ship was a troop ship with raked funnels and two masts, grey colour escort vessel was a US Navy destroyer of the *Taylor* class, two masts. Schooner was sighted, it had been located by the aircraft homing on a SE [sound-emitting], contact.

Patrol completed and aircraft landed back at base with nothing to report. *U-134* is northeast of Bahamas where it is hit by six to eight depth charges from Bermuda-based USN Squadron VP-201, and survived. Oiler USS *YO-142*, USS *SC-1467*, and USS *SC-1292* are bound south in Straits. SS *Perote* is bound west in Straits. *U-732* is northeast of Bahamas, bound south for Long Island. *U-84* is off Mayaguana, bound south to Windward Passage. *U-527* exits area northeast of Bahamas, bound east to France. *U-759* is in and out via Windward Passage, tracking convoys. Liberator BX-759 LP is 200 nm east of Abaco; one No.111 OTU aircraft was following a convoy which enemy sub was reported to be following. Signals exchanged with convoy, nothing further. Then, 150 nm northeast of Abaco, Liberator BX-759 LP saw submarine following convoy. NAS Great Exuma: A plane escorted **Convoy NG-371**. A plane landed at Great Exuma. A plane departed on a training hop and returned. A sound contact was made on an enemy submarine. In a murder which has yet to be solved, the wealthiest and ascendant most powerful man in the Bahamas, American-Canadian gold tycoon Sir Harry Oakes was found dead, tarred and feathered in the bed of his Cable Beach Estate. [The resulting murder-mystery and who-done-it trial brought global publicity and attention to the small colony which continued for years].

8th July 1943

B-25 FV-952 FX is airborne for submarine co-operation exercise at Key West, submarine was sighted fully surfaced, attacks were made on the submarine at periscope depth and on a crash dive, aircraft landed back after the completion of exercise. B-25 FR-382 DX airborne for submarine co-operation exercise at Key West, submarine was sighted off Sand Key, radio contact could not be established with the submarine and the exercise could not be carried out, aircraft returned to base. B-25 FF-949 FB is airborne for Anti-Submarine Patrol Number 1b, patrol completed without incident, aircraft landed back at base after air firing exercises. B-25 FV-951 FC airborne for Anti-Submarine Patrol Number 1, patrol completed, landed back at base with nothing to report, air-firing exercises carried out prior to landing. B-25 FR-378 CL airborne for Anti-Submarine Patrol Number 1, landed back after completing the exercise. B-25 FV-954 FB airborne for Anti-Submarine Patrol Number 1a, patrol completed without incident, yellow orange streak was observed, aircraft investigated but no definite source could be ascertained, landed back at base. B-25 FV-946 FF is airborne for Anti-Submarine Patrol Number 1, patrol was completed, was uneventful, landed back at Satellite Field as the weather had closed down at the Main Field, airborne later for the Main Field, landed there. B-25 FR-375 CK is airborne for submarine exercise off Key West; a convoy was sighted 10 nm south of Key West consisted of nine merchant ships and three escorts, contact was made with target submarine, two attacks were made against the submarine, exercise completed, aircraft landed at Boca Chica with an oil leak in the port engine, was airborne from Boca Chica to return to base, landed at Satellite Field at 5.46 pm, airborne from Satellite Field and landed at the Main Field. Tanker *A. S. Hansen* is bound north in Straits towards Palm Beach with US Navy plane escorting. Tanker *R. P. Smith* is bound north in Straits towards Palm Beach with a US Navy blimp near Bimini. SS *Extavia* is bound north in Straits towards

Palm Beach. Tanker *Atlantic States* is bound south in Straits passing Palm Beach. USS *SC-1316* is bound south in Straits towards Port Everglades. SS *Jeannette S. Skinner* is bound north in Straits passing Port Everglades. Tanker *Manuel Rionda* and destroyer escort USS *Sturtevant* (DE-239) are bound north out of Straits. SS *Andrew Carnegie* is bound northwest in Straits past Nicholas Channel. SS *Merida* and three USS *CS*'s are bound northwest in Straits passing Nicholas Channel with US Navy plane escorting. *U-759* is in and out via Windward Passage, tracking convoys. *U-66* is in Straits and observes two aircraft followed by a third which flew out of sight. Then, 400 nm southeast of Bermuda USN PBM-3 #201-P-3 engaged *U-134*. NAS Great Exuma: A plane went on a search mission. A plane departed for Guantanamo and went on a training flight. A plane landed at Great Exuma.

9th July 1943

B-25 FW-362 CA is airborne for Anti-Submarine Patrol Number 2c, patrol uneventful aircraft landed back with nothing to report. B-25 FR-366 CX is airborne for Anti-Submarine Patrol Number 2a, patrol completed without incident, landed back, had spotted a tanker. B-25 FR-378 CL airborne for Anti-Submarine Patrol Number 2d, patrol completed without incident, landed back. B-25 FV-954 FB airborne for Anti-Submarine Patrol Number 2d, patrol completed without incident, landed back. B-25 FV-951 FC airborne for Anti-Submarine Patrol Number 2a, patrol completed without incident, landed back after air-firing exercises. B-25 FV-949 CK is airborne for Anti-Submarine Patrol Number 2e, patrol completed without incident, landed back. Liberator BZ-797 LR and FL-964 LD are airborne from Satellite Field for a submarine co-operation exercise off Key West, aircraft returned to Satellite Field. Tug USS *ATR-44* is bound north in Straits passing Palm Beach. SS *Buffalo Wallow* is bound south in Straits towards Port Everglades. SS *Manuel Rionda* is bound south towards Nassau. SS *Atlantic Coast* and SS *Fenja* are bound north in Straits. **Convoy KN-251** is bound north in Straits with US Navy plane escorting. **Convoy KG-644** is bound southeast in Straits towards Nicholas Channel with US Navy plane escorting. **Convoy Havana North Coast Cuba-32** is bound southeast in Straits passing Nicholas Channel with US Navy plane escorting. *U-615* enters area north via Anegada Passage. *U-759* is in and out via Windward Passage, tracking convoys. *U-66* is in the Straits where it reports a flying boat. NAS Great Exuma: **Convoy GN-70** and **Convoy GK-744** entered the area. A plane departed for training. A plane arrived to increase compliment of planes at Great Exuma. A plane departed to cover **Convoy KG-644**, of six merchant vessels and four escorts, which entered the area.

10th July 1943

B-25 FR-366 CX is airborne for Anti-Submarine Patrol Number 1e, visibility poor and the aircraft returned to base with radio failure. B-25 FR-376 CK is airborne for Anti-Submarine Patrol Number 2a, cross over patrol, patrol carried out as detailed, passed over a vessel: painted grey with funnel slightly aft amid ships, landed back at base with nothing to report. B-25 FR-954 FB is airborne for Anti-Submarine Patrol Number 2f, completed patrol without incident, message received from DC to pass to base a signal to the effect that the radio was non-operational (n/e) in DC, a vessel of 1,500 tons was sighted, landed back at base. B-25 FV-951 PC is airborne for Anti-Submarine Patrol Number 2. A destroyer was sighted ship was painted grey, and had a single funnel patrol was completed and the aircraft landed back to Satellite Field as the weather was unfit at the Main Field, it landed at the Main Field later. B-25 FV-949 FB is airborne for Anti-Submarine Patrol Number 2, patrol was completed and the aircraft landed back with nothing to report. B-25 FR-362 CA is airborne for Anti-Submarine Patrol Number 1b, completed the patrol landed at Satellite Field as the weather was unfit at the Main Field due to heavy rain landed at Main Field later. US Navy aircraft Lockheed Ventura missing from Guantanamo Cuba while on a search mission; five aviators were lost. Tanker *Atlantic Coast* is bound north in Straits passing Daytona. **Convoy NK-551** is bound south in Straits passing Canaveral with US Navy plane escorting. USS *SC-1311* is bound south in Straits passing Canaveral. USS *SC-1471* is bound southwest in Straits towards Palm Beach. SS *Robin Tuxford* is bound south in Straits passing Palm Beach. Oiler USS *YO-143* is bound south from Port Everglades. SS *Shiloh* is bound north in Straits passing Port Everglades. USS *SC-1341* is bound south in Straits. *U-134* enters area east of Bahamas, west. *U-615* exits area south via Anegada Passage. NAS Great Exuma: Three planes covered **Convoy KG-644**, a plane sighted on merchant vessel, two escorts, a Cuban vessel, USS *Ingot* (716) and a friendly vessel. A plane went to cover **Convoy GN-70.** A plane escorted **Convoy GK-744**. One destroyer sighted an enemy submarine estimated moving towards the Windward Passage.

11th July 1943

B-25 FV-954 FB is airborne for Anti-Submarine Patrol Number 1C, line patrol had to curtail the patrol as the radio was inoperative, so landed back at base. B-25 FV-951 FC airborne for Anti-Submarine Patrol Number 2, cross over patrol. Convoy, consisting of 27 merchant ships and five escorts was sighted ships were in nine groups of three ships, single line one stern, there were 18 tankers and steam ships, a Catalina aircraft was escorting the convoy, aircraft completed the patrol and returned to base, heavy rain experienced over the base, aircraft was instructed to circle and wait until the storm had abated, before landing, then landed back. B-25 FR-393 DF is airborne for Anti-Submarine Patrol Number 3, line patrol convoy that was sighted by aircraft with tail number FC was sighted by aircraft DF, composition of the convoy was the same as when previously reported, convoy was on the same course, aircraft returned to base, was ordered to wait in the vicinity of the island until the storm, that was over the Main Field, had cleared aircraft landed at the Main Field later. B-25 FR-383 CX is airborne for Anti-Submarine Patrol Number 2, cross over patrol convoy reported by FC, and DF was sighted, patrol was completed, and the aircraft was near base when it was instructed to remain in the air until the storm had cleared landed later. Two US Navy OS2U Kingfisher scouting aircraft attacked a U-boat, slight damage west of Port-au-Prince, Haiti in Windward Passage. USS *Bannock* (AT) and tanker *A. S. Hansen* are bound south in Straits passing Canaveral. USS *USAT-Thomas H. Barry* is bound north in Straits towards Palm Beach with US Navy plane escorting. SS *Osmond*, **Convoy NK-551**, USS *PC-1199*, USS *SC-718*, and two British Royal Navy Motor Launches, HMS *HMML-1180* and HMS *HMML-1181*, are bound south in Straits. SS *Port Chalmers* is bound north in Straits. **Convoy Havana**

North Coast Cuba-32 is bound northwest in Straits, passing Nicholas Channel. **Convoy NG-372** is bound south in Straits near Key West. *U-732* is south of Inagua, bound south to Windward Passage. *U-134* is northeast of Bahamas to transit Northeast Providence Channel and Northwest Providence Channel. *U-84* is south of Inagua and Turks and Caicos, bound south for Windward Passage.

12th July 1943

B-25's FR-366 CX, FV-953 FA, FV-951 FC, FR-383 DX and FV-952 FI are airborne Patrol Number 3, patrols were completed individually aircraft landed back at base, crews had nothing to report. B-25 FV-949 FD is airborne for Anti-Submarine Patrol Number 1 patrol completed without incident aircraft landed back at base. USS *PC-1234* is bound south in Straits passing Daytona. SS *Margrethe Bakke* is bound north in Straits passing Palm Beach. SS *Santa Cecilia* is bound north in Straits near Miami. Destroyer escort USS *Andres* (DE-45), USS *SC-1293*, USS *SC-1310*, USS *SC-1327* and USS *Bannock* (AT), tanker *A. S. Hansen*, and USS *SC-996* are bound south in Straits. SS *Elihu Root* is bound northwest in Straits passing Nicholas Channel. *U-732* exits area bound south via Windward Passage after being attacked by two US Navy OS2U Kingfisher aircraft, causing slight damage in Windward Passage. *U-84* exits area south via Windward Passage. *U-134* is bound west, east of Eleuthera for Northeast Providence Channel. *U-66* is in the Straits and reports land-based aircraft, then a flying boat turned towards the boat, which then went to depth

13th July 1943

B-25 FR-375 CJ is airborne for Anti-Submarine Patrol Number 1, patrol was completed, and the aircraft landed back at base after carrying out gunnery exercises nothing to report from the patrol. B-25 FR-393 DF airborne for Anti-Submarine Patrol Number 3, patrol was completed without incident aircraft landed back at base. B-25 FR-362 CA is airborne for Anti-Submarine Patrol Number 3, completed the patrol, carried out gunnery exercises off Great Sale Cay [Grand Bahama], aircraft landed back at base. B-25 FR-384 CM is airborne for Anti-Submarine Patrol Number 3, patrol was uneventful and the aircraft landed back at base with nothing to report. B-25 FV-949 FD is airborne for Anti-Submarine Patrol Number 3, descent through cloud exercise was completed before the aircraft set course for the patrol area, patrol was completed, aircraft landed back after air firing exercises. B-25 FR-366 CX is airborne for Anti-Submarine Patrol Number 3, exercise was completed and the aircraft landed back after unsuccessful attempt at a descent through cloud exercise. B-25 FR-383 DX airborne for Anti-Submarine Patrol Number 3, sent signal reporting position over Great Sale Cay, requested permission for descent through cloud exercise. Then aircraft failed to reply to signal to commence, exercise, was not heard again, failed to return to base, so overdue, action was taken. Four aircraft dispatched to search waters adjacent to island. B-25 FV-946 FF is airborne to search for missing aircraft 3 nm southwest of Main Field, when smoke is sighted through the trees, crew caught sight of wrecked plane in the area, circled aircraft found to be a Baltimore aircraft, signal sent to base conveying this information. Aircraft continued search, oil patch was sighted, patch was 50 yards wide and seemed to extend in an up wind direction, aircraft returned to base and the Captain gave fuller details to the Controller on the information passed. The Marine

Craft at Lyford Cay was ordered to put to sea to investigate the oil patches, aircraft set course for the position of the oil patch, aircraft passed the coast, the ASR launch from Lyford Cay was seen heading towards the oil patch, aircraft was directly over the oil patch, and a smoke float was dropped, aircraft flew to the launch and headed it towards the patch, and then dived to 100 feet over the patch, continued to search the oil patch, and the launch was observed to have reduced speed and to be proceeding to the southeast up the patch. Instructions were received to continue the search of the area, and a sailing vessel was observed 4 nm to the north of the patch, heading towards Nassau. Aircraft landed back with nothing further to report. B-25 FR-393 DF is airborne to search south of the island to a distance of 15 nm off shore for oil patch, aircraft circled to investigate, message was received from base, message was sent to base that the ASR launch was standing by and that the aircraft was over the oil patch, oil patch extended 3 miles in length in a west-to-east direction, patch appeared to extend up wind from narrow end, no sign of a dinghy, survivors or wreckage. Aircraft remained over patch for 45 minutes, a constant series of orange flashes were seen emerging from two points near the narrow point of the oil patch, flashes appeared to be on the surface and appeared frequently, but at irregular intervals, message was sent to base that the aircraft was returning. The ASR boat was still standing by. Aircraft landed back with nothing further to report whilst the aircraft was over oil patch, it was joined by two other aircraft. B-25 FV-949 FD is airborne to search for the missing aircraft on receipt of a signal from base to search in the area of the oil patch, the aircraft proceeded to the reported position and observed two other aircraft in the vicinity. The ASR launch was seen to be searching the oil patch series of yellow flashes on the surface of water appearing at the easterly end of the oil patch; these are presumed to be smoke floats, and the aircraft returned to base. **Convoy SS *Chepachet*** is bound southwest in Straits near Matanilla Shoal, northwest Bahamas with US Navy plane escorting. USS *SC-1284* is bound southwest in Straits nearing Miami. Bahamas [Kelly] ferries *Ena K.* and *Betty K.* are bound west in Straits towards Miami with US Navy plane escorting. USS *SC-1471* and USS *SC-1272* are bound south in Straits. *U-66* in the Straits reports that the U-boat is detected due to bubbles, and seen from an airplane, which tracks the U-boat, and turns away. After discovering the boat, the aircraft shoots with artillery and anti-aircraft weapons at *U-66*, which escapes.

14th July 1943

Air Sea Rescue launch dispatched for Lyford Cay to search for dinghy in the reported position of the oil patch, no dinghy or survivors found, sundry pieces of an aircraft including an observer's seat, a note book, sundry pieces of plywood recovered, launch, searched same area at dawn of the 14th nothing found. B-25 FV-949 FD and FR-393 DF are airborne at different times for Anti-Submarine Patrol Number 3, patrol commenced descent through cloud exercise, landed back at base with nothing to report. *U-66* is east of Turks and Caicos bound south, harassed by Bahamas-based aircraft. Tanker *R. P. Smith* is bound south in Straits passing Canaveral. USS *YMS-248* is bound northeast in Straits. Tanker *Perote* is bound north in Straits towards Bimini. Oiler USS *YO-142* is bound south in Straits. Tanker *Manuel Rionda* is bound southwest in Straits. **Convoy KN-252** is bound north in Straits. SS *William Windom* north in Straits with US Navy plane escorting. USS *YAG-28* is bound north in Straits. *U-134* is northeast of Eleuthera, bound west to the Northeast Providence Channel. *U-634* is bound east off Mona Passage, north of Virgin Islands and Puerto Rico. NAS Great Exuma: A plane sighted two escorts. Two planes departed on training loops. Two planes went to cover **Convoy GN-71.**

15th July 1943

FR-381 DC is airborne for Anti-Submarine Patrol Number 5, yet had to return to base, as radio was found to be unserviceable, aircraft landed with nothing to report. B-25's FR-393 DF, FR-366 CX, FR-375 CJ, and FV-949 FD are airborne for Anti-Submarine Patrol Number 5, exercise was completed, aircraft landed with nothing to report. *U-134* is bound north in Straits, forced north by Puerto Rico aircraft. **Convoy NK-552** is bound south in Straits with US Navy blimp escorting. USS *SC-1282* is bound northeast in Straits to north of Bahamas. **Convoy KN-252** is bound northeast in Straits with US Navy plane escorting. USS *YMS-166* is bound north in Straits near Port Everglades. Tanker *R. P. Smith* is bound west in Straits towards Key West. *U-634* enters region north via Mona Passage. *U-527* is bound northeast of Bahamas. *U-134* is off Nassau, bound east toward Northeast Providence Channel. NAS Great Exuma: Three planes went to cover **Convoy GN-71**. A plane went to cover **Convoy NG-373**. Lt. Craig arrived from Guantanamo to relieve Lt. McGuire as Senior Naval Aviator. Eleven merchant vessels and five escort vessels were sighted.

16th July 1943

B-25 FV-953 FV is airborne to search for a straggling tanker which was originally with a northbound convoy, aircraft carried out a creeping line ahead patrol, tanker was sighted, aircraft contacted the tanker by Aldis lamp, vessel replied by flags "met," signal was sent to base from 30 nm off the convoy. Aircraft instructed to tanker to proceed on course to make Mayport, Florida, aircraft left the tanker and landed back at base. Liberator BZ-759 LP is airborne from Satellite Field to relieve aircraft FA on escort duties to a tanker, a creeping line ahead search was commenced with visibility 10 nm, creeping line ahead search was abandoned and a square search was started, position of the tanker was calculated from the intercepted message of FA to base square search was started with visibility of 4 nm, contact was made and the aircraft altered course and homed from 2,000 feet, ship's wake was sighted 8 nm ahead, aircraft circled the tanker and challenged, the answer was made by flags "meet" signal was sent to base aircraft circled the tanker vessel was seen to stop and then to continue on its way at 3 knots, box patrol was carried out over the ship to provide air cover, signal was sent to the aircraft to land, aircraft signalled to the tanker it was leaving, landed back to Satellite Field. *U-67* is bound east, east of Bahamas, northeast of Antigua. Tanker *Atlantic States* is bound northwest in Straits towards Daytona with US Navy plane escorting. USS *SC-1341* and USS *SC-718* are bound north in Straits towards Canaveral. USCG *CG-83402* is bound west in Straits. Tanker *Esso Belgium*, tanker *Azerbaijan*, and tanker *Beta* are bound south in Straits passing Palm Beach. USS *YMS-166* is bound north in Straits near Port Everglades. USS *SC-1293* is bound north in Straits passing Miami. Tanker *Port Phillip* is bound south in Straits near Bimini. USS *SC-685* is bound west in Straits. Tanker *Calliope* is bound northwest in

Straits, with an RAF plane from Nassau escorting. *U-134* is in Northwest Providence Channel then bound south. NAS Great Exuma: **Convoy GN-71**, of 24 merchant vessels and four escorts, departed, with a plane sent to cover it. Two planes went to cover **Convoy NG-373**.

17th July 1943

B-25's FV-593 FA and FR-381 DC are airborne between air Exercise Number 5, aircraft landed back after completion. Liberators FL-970 and FL-964 LD are airborne for navigational Exercise Number 6, completed without incident, landed back at Satellite Field. B-25's FV-953 FA and FR-381 DC are airborne for air Exercise Number 5n. **Convoy Jacksonville**, USS *SC-1333*, and USS *PC-1201* are bound southeast in Straits passing Daytona. SS *Calliope* is bound northwest in Straits with US Navy plane escorting. SS *Buffalo Wallow* is bound northeast in Straits near Miami. Tug USS *Syosset* is bound south in Straits passing Port Everglades. SS *McClellan Creek* is bound northeast in Straits towards Bimini. SS *Betty King* and SS *R. Campbell* are bound northwest in Straits towards Miami. Tanker *Beta* and SS *Settler* are bound west in Straits. *U-134* is in Northwest Providence Channel, bound west, then south in Straits to Cay Sal Bank. NAS Great Exuma: Three planes went to cover **Convoy NG-373**. A plane departed and returned after training. A plane sighted two small British vessels.

18th July 1943

Night flying took place, solo flights took place from the Main Field. B-25's FR-375 CJ, FR-362 CA, FR-366 CX, FV-952 FX and FR-384 CM are airborne for Anti-Submarine Patrol Number 5n, patrols completed individually and without incident, landed back at base. B-25 FV-951 FC is airborne for Anti-Submarine Patrol Number 4d; patrol was completed and the aircraft landed back with nothing to report. B-25 FR-38I DC is airborne for Anti-Submarine Patrol Number 4d; patrol was completed and the aircraft landed back. A 300-ton motor vessel, painted grey was sighted. Liberator FL-962 airborne for Miami and landed there. Liberators FL-970 LH, BZ-759 LP, and BZ-757 LR are airborne from Satellite Field on Anti-Submarine patrols landed back at Satellite Field. *U-134* attacked and sank US Navy blimp USS *K-74* [the only US airship shot down with a fatality in the war]. It was 5 nm west of Elbow Cay, Cay Sal Bank. US Navy doctrine required blimps to stay out of range of surfaced submarines and guide aircraft or ships to attack. Blimp pilot broke protocol to prevent *U-134* from reaching a tanker and steamship ahead of the submarine. None of the 10-man blimp crew was injured at first. *U-134* pulled part of the wreckage aboard for photographs and cut pieces for later evaluation. All but one of *K-74*'s crew was rescued the following day by USS *SC-657* and destroyer USS *Dahlgren* (DD-*187*). Aviation machinist's mate second class Isadore Stassel drowns off Cay Sal minutes before rescue; only USN airship-man to die as a result of enemy action in WWII. *U-134* is then attacked by USN Ventura aircraft in Straits, with its battery damaged. Tanker *Chiloil* is bound south in Straits towards Daytona. Tanker *Atlantic Coast* is bound south near Matanilla Shoal, northwest Bahamas with US Navy plane escorting. SS *Santa Monica*, USS *SC-1299*, USS *J-856* and USS *J-975* are bound north in Straits passing Palm Beach. USS *YMS-158* is bound south in Straits towards Port Everglades. **Convoy SS *Housatonic*** and USS *SC-987* are bound south in Straits. USS *YAG-19*, USS *Y-16*, and one US Army tanker are bound north out of Straits. Tug USS *Syosset* is bound southwest in Straits. NAS Great Exuma: A plane is escorting **Convoy NG-373**. Two planes departed and returned after training, then went to Guantanamo. USS *YP-42* dropped anchor in Elizabeth Harbour, Exuma.

19th July 1943

BZ-757 LR is airborne from Satellite Field on Anti-Submarine patrol, patrol was completed and the aircraft landed back not having seen anything, southeast was used throughout the patrol. Liberator FL-970 LK is airborne, five minutes past its ETA at San Salvador, aircraft altered course in an attempt to locate San Salvador, which was not located and the aircraft signalled to base "returning to base." Liberator BZ-759 LP is on patrol was completed and the aircraft landed back with nothing to report. B-25's FR-375 CJ, FR-381 DC, FR-362 CA, FV-953 FS and FV-954 FF are airborne and carried out individual antisubmarine patrols across the northeast New Providence Channel and the Tongue of the Ocean, patrols were completed and the aircraft landed back with nothing to report. B-25's FR-375 CJ, FV-953 FA, and FR-362 CA are airborne, aircraft completed air Exercise Number 7, landed back at base. Liberators BZ-759 LP, BZ-757 LR, BZ-746 LM and FL-970 LH are airborne from Satellite Field, carried out individual Anti-Submarine patrols landed back at base with nothing to report. **Convoy NK-533** is bound south in Straits passing Canaveral with US Navy plane escorting. Tug USS *Hercules* and tug USS *Meteor* with tows, escorted by USS *YP-603* are bound southeast in Straits passing Port Everglades. **Convoy Jacksonville**, SS *Chiloil*, and USS *SC-755* are bound south in Straits. Key West contingent of **Convoy KN-253** is bound north out of Straits with US Navy plane escorting. Key West contingent of **Convoy KG-646** is bound southeast in Straits to Nicholas Channel with US Navy plane escorting. Cuban *CS-33* and Cuban *CS-41* are bound west in Straits passing Nicholas Channel. USCG *CG-83420* is bound north from Nassau for Northwest Providence Channel. USS *Samaná* (YAG-28) is bound southeast in Straits. *U-134* is in Straits when it is bombed by a US Ventura responding to the downing of *K-74*. *U-190* enters area to the north of Walker Cay off Abaco, bound to the northeast of Eleuthera. NAS Great Exuma: Two planes went to cover **Convoy GN-72**. A plane arrived from Guantanamo to replace a plane. USS *YAG 28* dropped anchor at Great Exuma.

20th July 1943

Night flying carried out from Main Field. Liberator BZ-759 LP is airborne from Satellite Field on Anti-Submarine patrol, a tanker of 5,000 to 6,000 tons was sighted, circled and photographs were taken, aircraft completed the patrol and landed at Satellite Field. Liberator FL-970 LH is airborne on Anti-Submarine patrol, a creeping line ahead patrol was commenced, visibility was 20 nm, patrol was completed without incident and the aircraft landed back. Liberator BZ-757 LR and BZ-746 LK are airborne on Anti-Submarine patrol at different times, patrol was completed without incident and the aircraft landed back. B-25's FV-954 FB, FV-952 FX, and FR-38I DC are airborne for Anti-Submarine patrol, patrol was completed without incident and the aircraft landed back with nothing to report. B-25'S FR-38I DC, FR-362 CA, FV-952 FX, and FR-384 CM are airborne for air Exercise Number 7, exercise was completed without incident and the aircraft landed back. B-25's FV-953 FA, FV-

946 FF, and FR-38I DC airborne for air Exercise Number 7, exercise was completed and the aircraft landed back. Liberators BZ-756 L3, BZ-757 LR, FL-970 LH and BZ-746 LK are airborne on Anti-Submarine patrol, aircraft landed back at Satellite Field after exercise completion. USCG *CG-83363* is bound northeast, tanker *Alar*, tanker *Polycastle*, and tanker *Gulftide* are bound northeast in Straits towards Daytona. SS *China Mail* is bound south in Straits, with US Navy plane escorting. SS *Sydney Star* is bound southwest in Straits near Matanilla Shoal, northwest Bahamas with US Navy plane escorting. Army tanker USS *Y-16* and escorts are bound west in Straits passing Northwest Providence Channel. **Convoy NK-553** is bound south in Straits, passing Port Everglades with US Navy plane escorting. USS *SC-1471* is bound south in Straits passing Port Everglades. Oiler USS *YO-142*, USS *SC-1467*, USS *PC-1229*, and USS *PC-1315* are bound south in Straits. Tug USS *Humrick* with two tows and one escort is bound north in Straits. Two tugs with tows and escorts are bound southwest in Straits. **Convoy Havana North Coast Cuba-36** is bound west , passing Nicholas Channel with US Navy plane escorting. **Convoy GN-72** is bound north, east of Eleuthera. SS *Pastoris* and escort are bound northeast in Straits, near Key West. NAS Great Exuma: A plane sighted a transport. Four planes departed to cover **Convoy GN-72**. A plane escorted **Convoy NG-374**. USS *PC-620* dropped anchor in Elizabeth Harbour, Exuma. USS *Pastoris* and USS *Pheasant* were sighted.

21st July 1943

B-25's FR-393 DF, FV-946 FF, FV-954 FB, FR-381 DC and FR-384 CM are airborne for air Exercise Number 7, exercises were completed and the aircraft landed back. Liberator BZ-757 LR is airborne from Satellite Field on Anti-Submarine patrol, patrol was completed without incident and the aircraft landed back sighted nothing. Liberator BZ-758 LS airborne on Anti-Submarine patrol, aircraft homed on LR, LR was located patrol was completed and the aircraft landed at Satellite Field, southeast was unserviceable throughout the flight, nothing to report. Liberator BZ-746 LM Is airborne from Satellite Field sighted a convoy consisting of one merchant ship and an escorts signals exchanged, Catalina aircraft was sent to be on escort duties, creeping line ahead patrol was carried out, patrol was completed without incident, landed back at base. Liberator FL-970 LH is airborne from Satellite Field merchant ship was sighted with an escorts photographs were taken, patrol was completed, and the aircraft landed back at Satellite Field. Liberator FL-970 LH is airborne for Anti-Submarine patrol, returned to base, magnetic compass unserviceable. Aircraft LR was substituted but this aircraft was found to have radio failure prior to its estimated time of take-off, patrol was then cancelled. USS *YMS-159* and SS *Husvik* are bound north in Straits passing Canaveral. **Convoy SS *Capachet*** is bound north*east* in Straits near Matanilla Shoal, northwest Bahamas, with US Navy plane escorting. SS *Bannock* is bound north in Straits passing Port Everglades. Ferry *Jean Brillant* is bound west towards Port Everglades with US Navy plane escorting. **Convoy Miami-162**, tug USS *Humrick* are bound south in Straits. SS *China Mail* is bound south in Straits with US Navy plane escorting. USS *SC-1071* and USS *SC-661* are bound north in Straits. USS *FT-21* is bound west in Straits towards Key West. USS *YAG-28* is bound west in Northwest Providence Channel. SS *Georgios P.* is bound southwest in Northeast Providence Channel. **Convoy**

GN-72 is bound northwest in Straits. *U-134* is bound north in Straits for Northwest Providence Channel, outbound with damaged batteries. *U-66* is bound south into area from north of Grand Bahama to east of Abaco. NAS Great Exuma: Three planes went to cover **Convoy NG-374**. A plane went to cover **Convoy GN-72.** Thirteen merchant vessels and five escorts departed. Three planes returned and departed for training and landed at Great Exuma.

22nd July 1943

Night flying was carried out by aircraft of Number 1 with squadron from the Main Field. B-25's FR-393 DF, FR-366 CX, and FR-371 CF are airborne, landed back having completed individual patrols over the Northwest Providence Channel and the Tongue of the Ocean, they had nothing to report on landing. Tanker *Cherry Valley* is attacked and damaged by *U-66* 250 nm north-northeast of Grand Turk. Despite two torpedo strikes with no casualties, the ship makes it to San Juan [it is later repaired and returns to service]. USS *YT-180* is bound south in Straits passing Miami. US Coast Guard *CG-Boone* is bound north in Straits. Tug USS *Humrick* is bound west past Nicholas Channel. *U-190* is bound east from northeast of Abaco. NAS Great Exuma: A plane escorted **Convoy NG-374**. A plane landed at Great Exuma. A J4F-2 Amphibian arrived with Lt. R. M. Miller from Miami to compliment planes.

23rd July 1943

Liberator FL-942 LC returned from Miami and landed at Satellite Field. USS *YMS-198,* USS *YMS-239,* and USS *SC-1072* are bound south in Straits towards Palm Beach. SS *Cross Keys* is bound south towards Palm Beach with US Navy plane escorting. Bahamas mailboat *Content S.* and USS *YP-408* are bound west in Straits from Bimini towards Miami. USS *YAG-20* and USS *YP-426* are bound southeast in Straits, passing Miami. Tug USS *Syosset* and tug USS *Hercules* with tows, escorted by USS *YP-603* are bound north out of Straits. SS *Nidaried* and Cuban *CS-21,* Cuban *CS-22,* and Cuban *CS-43* are bound east in Straits passing Nicholas Channel with US Navy plane escorting. SS *Colony Trader* and USS *YMS-158* are bound east in Straits passing Nicholas Channel. *U-66* is bound east, east of Eleuthera. *U-134* is in Northwest Providence Channel, bound east to France. *U-190* is northeast of Abaco, bound east to France. NAS Great Exuma: Three planes went on training and returned. A plane went to bombing hop.

24th July 1943

Liberator BZ-761 LN returned from Miami and landed at Satellite Field. A Station Defence exercise took place at the Main Field and involved the defence of the town. Tug USS *ATR-43* is bound south in Straits, passing Daytona. Tanker *A. S. Hansen* and escort are bound north in Straits towards Canaveral with US Navy plane escorting. SS *Georgios P.* and escort are bound south in Straits towards Port Everglades. Two tugs with tows and USCG *CG-Wanigan* are bound south in Straits, passing Port Everglades. USS *SC-646* is bound south in Straits, passing Miami. Mailboat *Richard Campbell* and escort are bound west in Straits with US Navy blimp near Bimini. USS *SC-1342* and USS *PC-1200* are bound south in Straits. **Convoy Motor Torpedo Boat Squadron-20, Group B** is bound north in Straits. **Convoy KN-254** is bound north in Straits with US Navy plane escorting. **Convoy KG-647** is bound southeast in Straits towards Nicholas Channel with US Navy plane escorting. USS *YAG-20* is bound southeast in Straits. *U-190* exits area northeast of Abaco, east for

France. RAF Station Defence Exercise held. Alarm sounded, troops placed in defensive positions. All manned. Three minutes later a Striking Force left Government House to defend Clifton Pier. The all-clear is given an hour later. NAS Great Exuma: Two planes went to cover **Convoy GN-73**. Twenty nine Merchant vessels and six escorts entered. A plane departed for Miami. A plane practiced bombing drill. A plane sighted a CV [a US Navy aircraft carrier] and three destroyers.

25th July 1943

Number 5 course having completed training on Liberator aircraft returned to the Main Field, preparatory to embarking on pasting. Number 6 Course moved to Satellite Field. The field commences its conversion to accommodate Liberator aircraft after completing the B-25 course at the Main Field. Number 7 Course having completed the familiarization, flights were transferred to Number 2 Squadron for more advanced flying training. Number 8 Course commenced its early training on B-25 aircraft with familiarisation flights. SS *King George II* is bound north in Straits towards Daytona. **Convoy NK-554** is bound southwest in Straits towards Palm Beach. USS *YMS-401* is bound southwest in Straits towards Port Everglades. USS *YAG-28* is bound northeast in Straits towards Matanilla Shoal, northwest Bahamas. SS *Exhibitor* is bound southwest in Straits towards Miami with US Navy plane escorting. USS *YP-408* is bound northwest in Straits towards Port Everglades. Bahamas ferry *Ena K.* and mailboat *Content S.* are bound west in Straits towards Port Everglades with US Navy blimp escorting. Tug USS *Humrick* is bound north in Straits. Tug USS *Hercules*, tug USS *Syosset*, tug USS *ATR-43*, and USS *SC-1036* are bound south in Straits. USS *SC-1322* is bound west in Straits. *U-66* exits region, bound east to France. NAS Great Exuma: Three planes went to cover **Convoy GN-73** and returned. J4F-2 Amphibian returned from Miami. USS *YAG 20* dropped its anchor at Great Exuma with a plane. A plane took off for a bombing drill.

26th July 1943

B-25's FR-362 CA, FR-371 CF, FR-378 CL, FR-384 CM, FV-946 and FV-953 FA are airborne for air Exercise Number 1, landed back at base after completion. Air Exercise Number 3 carried out by B-25's FR-362 CA, FR-371 CF, FR-393 DF, FR-384 CM, FV-946 FF and FV-954 FB. B-25's FR-362 CA, FR-371 CF, and FR-378 CL are airborne between air exercise, landed back at base after completion. B-25's FR-384 CM, FV-946 FF, and FV-954 FB carried out air Exercise Number 2. B-25's FR-362 CA, FR-393 DF, and FR-371 CF were airborne for air Exercise Number 3, landed back at base. Air Exercise Number 4 carried out by B-25's FR-384 CM, FV-946 FF and FV-954 FB. Tanker *Cedar Mills* is bound north in Straits passing Canaveral with US Navy plane escorting. SS *Isolator* is bound south in Straits passing Canaveral. **Convoy SS *Dobler*** is bound north in Straits towards Canaveral with US Navy plane escorting. **Convoy GN-73** is bound northwest near east coast of Abaco, with US Navy plane escorting. **Convoy NG-375** is bound south in Straits near Key West, with US Navy plane escorting. USCG *CG-Euphane* is bound south in Straits passing Palm Beach. Tanker *Atlantic States* is bound south past Palm Beach with US Navy plane escorting. **Convoy NK-554** is bound south in Straits passing Palm Beach with US Navy plane escorting. Cuban gunboat *Cuban CS-Cuba* and *Cuban CS*-41 are bound south in Straits. NAS Great Exuma: Two planes went to cover **Convoy GN-73**. A plane practiced bombing drill and then returned. A plane arrived from duty in Guantanamo.

27th July 1943

B-25's FR-371 CF, FR-384 CM, FR-362 CA, FR-366 CX, and FR-376 CK are airborne for air Exercise Number 1, landed back at base after completion. B-25's FR-371 CF, FR-384 CM, FR-362 CA, FR-J66 CX and FR-383 DF are airborne on air Exercise Number 3, landed back at base after completion. Air Exercise Number 1 completed by B-25's FV-954 FB, FR-384 CM, and FR-382 CA. Air Exercise Number 2 was carried out by B-25's FR-366 CX, FR-393 DF and FR-376 CK. B-25's FV-953 FB, FR-384 CM, and FR-362 CA are airborne for air Exercise Number 3. Air Exercise Number 4 completed by B-25's FR-371 CF, FV-946 FF, and FR-393 DF. Aircraft FV-946 FF failed to maintain U/T or B/T [communications] contact for an hour and 15 minutes, overdue action was taken on this aircraft, aircraft FV-952 FX was dispatched to search the bombing range, message was received from FX "For mating on another aircraft over bombing range," flying control contacts FV-946 FF as it passed over the airfield, FF and then FX landed. *U-732* lined up to attack a small naval convoy of three repair ships and three destroyers. The destroyers managed to keep the boat down and depth charged her. Then another boat in the same pack, *U-553* escaped around Cape Maysí, eastern Cuba after attacking a convoy in a three-U-boat wolf pack. USS *SC-1285* and **Convoy Jacksonville-55** are bound south in Straits passing Canaveral. SS *Atlantic Coast* is bound north in Straits passing Canaveral with US Navy plane escorting. Destroyer escort USS *J. R. Ward* (DE-243) is bound northeast near Walker Cay, north of Abaco. **Task Group TG-29-6**, with three USS *ARV*'s and a destroyer is bound southeast in Straits. SS *Tropical* and USS *SC-673* are bound north in Straits towards Miami. SS *Isolator* and USS *SC-685*, and USS *YP-612* escorting two tugs with tows are bound southwest in Straits. USS *PC-1230* is bound east in Straits. **Convoy SC-1292-747** is bound west in Straits passing Nicholas Channel. **Convoy GK** is bound west in Straits passing Nicholas Channel with US Navy plane escorting. *U-732* re-enters area in Windward Passage to be attacked by destroyers off Cape Maysí. NAS Great Exuma: Two planes went to cover **Convoy NG-375.** A plane went to cover three US Navy ships and two escorts. ***USS YAG 20*** lifted anchor and departed Great Exuma. A plane departed for training and returned.

28th July 1943

B-25's FR-362 CA, FR-366 CX, FR-372 CF, FR-393 DF, and FV-954 FB are airborne for air Exercise Number 2. B-25's FR-362 CA, FR-366 CX, FR-371 CF, FR-375 CE, FR-393 DF, and FV-954 FB are airborne for air Exercise Number 4, landed after completion of exercise. B-25's FR-376 CK, FR-393 DF, FV-954 FB, FV-951 FC, FV-946 FF, FV-952 FX, FR-362 CA, FR-366 CX, and FR-371 CF are airborne on air exercises. USCG *CG-Boone* is bound north in Straits towards Canaveral. Anti-Submarine Group 5 is bound south in Straits towards Palm Beach. Ferry *Jean Brillant* and escorts are bound southwest in Straits from Northwest Providence Channel with US Navy plane escorting. USS *YAG-20* and USS *YP-606* are bound northwest in Straits towards Miami. USS *YMS-169* is bound north in Straits passing Miami. **Convoy Jacksonville-55**, USS *YAG-19*, USS *PC-1127*, and USS *SC-1330* are bound south in Straits. Tanker *Bull Run* is bound north in Straits. Anti-Submarine Group 3 is bound southwest in Straits passing Cay Sal Bank. **Convoy Havana North Coast Cuba-43** is bound west in Straits passing Nicholas

Channel with US Navy plane escorting. *U-732* is off Cape Maysí, northeast Cuba. NAS Great Exuma: A plane departed to cover three Navy Ships and two escorts. A plane landed at Great Exuma. J4F-2 Amphibian departed and returned. A plane departed for Guantanamo. A US Navy destroyer was escorted. **Convoy GN-74** entered area.

29th July 1943

Air exercises and descent through cloud exercises carried out by B-25's FR-362 CA, FV-951 FC, FR-366 CX, FV-952 FX, and FV-946 FF. Then B-25's FV-951 FC, FV-946 FF, FV-952 FX, FR-362 CA, and FR-366 CX are airborne for air Exercise Number 4, then landed after completion of exercise. B-25's FR-362 CA, FR-366 CX, FR-371 CF, FR-375 CE, FR-393 DF, and FV-951 FC are airborne on air exercises later in the day. Liberator BZ-800 LY is airborne from Satellite Field for Bermuda, and landed there. B-25 FV-952 FX airborne escort tanker *Manual Rionda*, and motor vessel sighted at 4 nm, distance blimp was sighted, an airship circling a smoke float that was floating on the water. It was observed that a US Navy aircraft was floating beneath, wings submerged dinghy with survivors was sighted near the wrecked aircraft US Navy Kingfisher, float plane was to alight on the water and pick up the dinghy and the survivors, aircraft began creeping line ahead search to contact convoy. Tanker was sighted, contact was made, vessel escorted and aircraft landed back at base. SS *Cockerel* is bound south in Straits passing St. Augustine. USS *SC-1294* and USS *SC-1331* are bound south in Straits towards Canaveral. **Convoy NK-555** is bound south towards Canaveral with US Navy plane escorting. USCG *CG-83363* is bound north in Straits passing Palm Beach. USS *Noa* is bound north*east* in Straits near Matanilla Shoal, northwest Bahamas. USS *YMS-169* is bound north in Straits towards Palm Beach. SS *Manuel Rionda* is bound west towards Northwest Providence Channel with US Navy plane escorting. USS *YT-338* is bound west passing Northwest Providence Channel. Anti-Submarine Group 1 is bound south in Straits towards Port Everglades. USS *SC-1316* and oiler USS *YO-143* are bound south in Straits. Tug USS *Hercules* is bound north in Straits. **Convoy Havana North Coast Cuba-41** is bound northwest in Straits passing Nicholas Channel with US Navy plane escorting. *U-732* is south of Inagua, bound north from Windward Passage then east for France. NAS Great Exuma: Two planes went to cover **Convoy GN-74** and returned. Partridge towed USS *YO-83* and USS *SC-1281*.

30th July 1943

B-25's FR-395 CC, FR-376 CK, FR-366 CX, FV-951 fc, FR-371 CF and FR-362 CA airborne for air Exercise Number 5, completed the exercise and landed back. Air exercises were carried out by B-25's FV-952 FX, FR-376 CK, FR-393 DF, FR-366 CX, FR-371 CF and FR-362 CA. Liberator FL-962 LG returned from Miami and landed at Satellite Field. HRH Governor of the Bahamas inspected the RAF Transport Command Headquarters of Number 113 South Atlantic Wing at Satellite Field, made into United States Air Transport Command. Tanker USS *Balls Bluff* is bound south in Straits towards Canaveral with US Navy plane escorting. USS *YMS 239*, USS *YMS 198* and **Convoy KN-255** are bound north in Straits passing Palm Beach with US Navy plane escorting. SS *Cockerel*, USS *SC-988* USS *SC-1294*, USS *SC-1331*, and USS *Moonstone* (PYc-9) are bound south in Straits, passing Palm Beach. **Convoy NK-555** is bound south in Straits passing Palm Beach with US Navy plane escorting. Ferry

Jean Brillant is bound southeast to Nassau in Northwest Providence Channel. Tug USS *Hercules*, USS *SC-1359*, USS *SC-1071*, and USS *SC-1049* are bound south in Straits. Tug USS *Meteor* is bound north in Straits. **Convoy Havana North Coast Cuba-41** is bound northwest in Straits passing Nicholas Channel. *U-84* is bound east past Key West, then northeast in Straits to Northwest Providence Channel. NAS Great Exuma: Two planes went to cover **Convoy GN-74** and returned with engine failure. Two planes were towed and departed Great Exuma. A plane went to cover **Convoy GK-748**.

31st July 1943

B-25's FR-362 CA, FV-951 FC, FV-946 FF, FR-376 CK, and FV-954 PB are airborne, then landed back having completed their exercises. Similar exercises carried out by B-25's FR-375 CK, FV-954 FB, FV-946 FF, FV-951 and FR-366 CX. SS *Balls Bluff* is bound south in Straits towards Canaveral with US Navy plane escorting. SS *Bostonian* and USS *SC-1322* are bound north in Straits passing Canaveral. USS *YMS-192*, USS *YMS-183* and USS *YMS-184* are bound south in Straits towards Palm Beach. USCG *CG-83402* is bound west near Matanilla Shoal, northwest Bahamas. Ferry *Jean Brillant*, USCG *CG-83348* and USCG *CG-83346* are bound west passing Northwest Providence Channel with US Navy plane escorting. SS *Manuel Rionda* and USS *SC-1060* are bound west passing Northwest Providence Channel with US Navy plane escorting. USS *YT-338* towing USS *YF-367* escorted by USS *YP-606* are bound west in Straits towards Miami. **Convoy KW-247** is bound north in Straits passing Port Everglades. Tug USS *Humrick* and tug USS *Syosset* with three barges escorted by USS *YP-612* are bound north out of Straits. USCG *CG-83430* is bound south in Straits. USCG *CG-84001*, USS *SC-1002*, and USS *SC-1022* are bound west in Straits. Tug USS *AT-147* towing USS *YC-783* and USS *YOG-10* escorted by USS *SC-541* are bound southeast in Straits passing Nicholas Channel. NAS Great Exuma: Two planes went to cover **Convoy NG-376**. The J4F-2 Amphibian departed. A plane landed at Great Exuma.

1st August 1943

B-25 FV-949 FD airborne for air Exercise Number 5. Liberator BZ-508 LT is airborne from Bermuda and landed at Satellite Field. Two electricians were accidently killed when the carburettor of the aircraft in which they were working caught fire, airmen left the aircraft and walked into the moving propellers, and were killed. USCG *CG-Madrona*, SS *Tristan* and USCG *CG-83363* are bound south in Straits towards Canaveral. Destroyer escort USS *Stewart* (DE-238), USCG *CG-Majrona*, and USS *SC-1334* are bound south in Straits towards Palm Beach. Oiler USS *YO-142*, USS *YT-338*, and USS *YP-606* are bound southeast in Straits from Port Everglades. SS *Odivar II* and USS *YMS-401* are bound north in Straits passing Port Everglades. SS *Balls Bluff* is bound south in Straits passing Port Everglades with US Navy plane escorting. USS *YO-143*, USS *YAG-28*, and USS *YP-408* are bound northwest in Straits towards Port Everglades. Tug USS *Meteor* with tow, USS *YP-452* and USCG *CG-Citrus* are bound west in Straits. SS *Cross Keys* is bound northeast in Straits. **Convoy Havana North Coast Cuba-46** is bound east in Straits, passing Nicholas Channel with US Navy plane escorting. SS *Libertad* and Cuban *CS-31* and Cuban *CS-42* are bound east in Straits passing Nicholas Channel. USS *YAG-20* is bound southeast in Straits, west of Andros. *U-134* exits area, bound east to France. *U-84* enters Straits north of Cay Sal, then bound east for France. *U-732* is bound southwest of Inagua, north then east for France. NAS Great Exuma: Three planes went to cover **Convoy NG-376**. A plane arrived from Guantanamo and took off to return there. A plane departed for Williamstown. USS *YAG 20* dropped anchor. A convoy with 27 merchant vessels and five escorts departed.

2nd August 1943

B-25's FR-378 cl, FV-952 FX, FR-366 CX, FV-954 FB, FV-951 FC and FV-949 FD are airborne, and landed back after completing air Exercise Number 6. B-25's FV-952 FX, FV-954 FB, FV-946 FF, FR-395 CO, FV-951 FC, and FV-949 FD are airborne for air Exercise Number 6, aircraft completed the exercise and landed back. B-25 FV-952 FX airborne and landed back after completing Exercise Number 6. SS *Odivar II* is bound north in Straits passing Daytona. Tug USS *Syosset* is bound south in Straits passing Palm Beach. SS *Tristan* and **Convoy KW-250** are bound north in Straits passing Miami. Tanker *Gulfmaid* is bound south in Straits passing Miami. Tug USS *Hercules* is bound north out of Straits. USS *PC-613* is bound south in Straits. USS *YO-142* is bound west in Straits towards Key West. SS *Ogilvie* is bound northeast in Straits passing Nicholas Channel. *U-732* is southwest of Inagua, south of Crooked Island Passage, bound north then east for France. NAS Great Exuma: A plane landed at Guantanamo after night coverage of a **Convoy NG-376**. A plane departed on training loop and returned after 8.5 hours. A plane dropped anchor. Sixteen merchant vessels and five escorts departed the area.

3rd August 1943

B-25's FR-37I CF, FR-378 CL, FV-953 FA, PV-954 FB, FV-949 FD, and FV-952 FX are airborne and landed back after completing air Exercise Number 6. B-25's FV-952 FX, FR-378 CL, and FV-954 FB airborne carried out air Exercise Number 6, landed back after completion. Air Vice Marshal, Air Officer in charge Administration Royal Air Force Delegation, Washington, arrived with Air Vice Marshal to inspect the Station, met by the Officer Commanding, Group Captain, The Chief Flying Instructor, Wing Commander, and the Chief Technical Officer. **Convoy KN-250** is bound north in Straits towards Daytona. USCG *CG-83363* and USS *SC-1286* are bound north in Straits passing Canaveral. **Convoy NK-556** is bound north in Straits. Tanker *Atlantic Coast* and Anti-Submarine Group 1 are bound south in Straits passing Palm Beach. USS *SC-1058*, USS *SC-1311*, and **Convoy Miami-168** are bound south in Straits. Tanker *Georgia* is bound west in Straits with US Navy blimp escorting. **Convoy KN-256** is bound north in Straits with US Navy plane escorting. **Convoy KG-649** is bound southeast in Straits towards Nicholas Channel with US Navy plane escorting. **Convoy Havana North Coast Cuba-46** is bound west in Straits passing Nicholas Channel with US Navy plane escorting. *U-84* is in Northeast Providence Channel bound east [soon to be sunk off Bermuda]. *U-732* is bound south of Inagua, north from Windward Passage, then east for France. NAS Great Exuma: Two planes depart to cover **Convoy GN-75.** *USS YAG-20* gets underway and departed Great Exuma. **Convoy KG-649**, of 19 merchant vessels and five escorts entered.

4th August 1943

B-25's FR-371 CF, FR-395 CO, FV-953 FA, FV-954 FB, and FV-949 FD are airborne for air Exercise Number 6, landed back at base after completion. B-25's FR-371 CF, FR-378 CL, FV-953 FA FR-376 CK, FR-362 CA and FV-954 FB are airborne for on air Exercise Number 6, landed back at base after completion. US Naval Ventura aircraft landed from Boca Chica aircraft arrived to carry out convoy escort duties to **Convoy NG-377** [the following day]. USS *Marnell* (PYC-39) is bound north in Straits passing Canaveral. **Convoy GN-75** is bound north near Northeast Providence Channel. USS *YAG-20* and USS YP-*426* are bound west in Straits near Bimini. Tug USS *Hercules* with barges and USS *YP-427* are bound south in Straits passing Port Everglades. Tug USS *Meteor* with barge escorted by USS *YP-452* are bound north in Straits towards Miami. USCG *CG-83499* is bound south in Straits passing Miami. USS *YAG-21*, house boat USS *YHB-13*, and USS *YP-357* are bound north in Straits. **Convoy NK-556** is bound west in Straits with US Navy plane escorting. *U-84* is northeast of Bahamas, bound east to its fate. *U-732* is west of Inagua, south of Crooked Island, bound north and east for France. NAS Great Exuma: A plane escorted **Convoy NG-75**. Another plane went to cover **Convoy GN-75**. A plane departed to cover **Convoy KG-649**. A plane carried Lt. Rawson from Guantanamo for duty. Periscope sighted. Convoy consists of 16 merchant vessels and five escorts passed.

5th August 1943

Night flying programme was cancelled, as the weather was thought to be unsuitable. Thunderstorms were experienced during the early hour of the morning. Ventura 29896 was airborne to escort **Convoy NG-377**. B-25's FV-946 FF, FV-954 FB, FR-375 CF, FR-378 CL, FR-362 CA, and FV-953 PA are airborne for air Exercise Number 6, landed back having completed the exercise. B-25's FR-395 CO, FV-946 FF and FV-954 FB are airborne for air Exercise Number 6, landed back having completed the exercise. US Navy Ventura HSN 23896 airborne to escort a convoy from New York to Guantanamo, convoy met USN Ventura 29863, arrived from Boca Chica for escort duties, aircraft was airborne and carried out an escort of the convoy, landing back at the Main Field after duty completion. Air Vice Marshall left the Station in a Beechcraft aircraft for Jacksonville *en route* for Washington, aircraft landed at Jacksonville. Air Vice Marshal left for Miami landed there. *U-732* is attacked by Bahamas-based aircraft west of Inagua, south of Crooked Island Passage. *U-230* is bound east of Bahamas, south. USCG *E. M. Dow* (*CG-353*) is bound south in Straits passing Canaveral. USS *SC-1300* is bound south in Straits towards Palm Beach. SS *Yankee Sword* and USS *SC-1359* are bound north in Straits passing Port Everglades. Tug USS *Meteor* with barge and USS *YP-426* are bound southeast in Straits passing Port Everglades. Tug USS *Hercules* and barges and USS *YP-427* are bound southwest in Straits. **Havana North Coast Cuba-48** is bound southeast in Straits passing Nicholas Channel with US Navy plane escorting. USS *Chippewa* (AT-69) is bound southwest passing Walker Cay, north of Abaco. **Convoy NG-377** is bound south, to the east of Bahamas. Swedish SS *Scania* is stopped by *U-564* well east of Bahamas and released. NAS Great Exuma: A plane escorted **Convoy KG-649.**Three planes went to cover **Convoy NG-377**. A plane carrying Lt. Craig departed for Guantanamo.

6th August 1943

B-25 FV-949 is airborne for Anti-Submarine Patrol Number 1c, could not maintain radio contact with captain, decided to return to base, landed back. B-25 FR-371 CF is airborne for Anti-Submarine Patrol Number 1b, a naval vessel that appeared to be a converted merchant vessel was circled, aircraft resumed the patrol, landed back having completed the patrol nothing to report. B-25 FR-378 CL airborne for Anti-Submarine Patrol Number 1a, tug of 6-600 tons sighted, towing a flat-topped barge. B-25 FR-362 CA airborne for Anti-Submarine Patrol Number 1, tug and barge as reported by CL were sighted aircraft carried out air firing practice prior to landing. B-25 FR-393 DF is airborne for Anti-Submarine Patrol Number 1b, completed the patrol and landed back after air firing exercises. B-25's FV-949 FD, FR-362 CA, FR-393 OF, FR-376 CK, FR-366 CX and FV-946 FF airborne and carried out, individually, Anti-Submarine patrols Number *3*. A merchant ships and two escorts were sighted, aircraft landed back with nothing to report from the patrols. USN Ventura 29863 was airborne to escort the US cruiser *Cincinnati,* vessel met and escorted successfully, aircraft landed back. USN Ventura 23896 was airborne to continue the escort of the *Cincinnati,* aircraft contacted the cruiser, completed the escort duty and landed back. US Liberator 32023 landed from Bermuda. SS *Yankee Sword* is bound north in Straits passing Canaveral. USS *Carib* (AT-82) is bound south in Straits passing Canaveral. SS *Julian Alonzo* is bound south in Straits towards Palm Beach. USS *Keith* is bound north in Straits towards Palm Beach. **Convoy PE-175** is bound south in Straits from Port Everglades. **Convoy KW-256** is bound north in Straits. Tug USS *ATR-46* is bound north in Straits. *U-732* is north of Inagua, northeast for France. USN Ventura 29863 is escorting the cruiser USS *Cincinnati.* NAS Great Exuma: Two planes escorted **Convoy NG-377**. A plane departed for training loop and returned after training. Two navy ships and one escort entered the area.

7th August 1943

Night flying carried out from Main Field. B-25 FV-946 FF airborne for Anti-Submarine Patrol Number 1, patrol completed without incident, aircraft landed back. B-25 FR-376 CK is airborne for Anti-Submarine Patrol Number 1a, contact lost, patrol completed, aircraft landed back. B-25 FR-362 CA airborne for Anti-Submarine Patrol Number 1, patrol completed without incident, aircraft landed back. B-25 FV-949 FD airborne for Anti-Submarine Patrol Number 1a, patrol completed without incident, aircraft landed back. B-25 FR-393 DF is airborne for Anti-Submarine Patrol Number 1b patrol completed without incident aircraft landed back. B-25 FR-366 CX Anti-Submarine Patrol Number 1, patrol completed without incident aircraft, landed back. B-25's FR-393 DF, FR-362 CA, FR-376 CK, FV-949 FD airborne for Anti-Submarine Patrol Number 3, completed patrol individually aircraft landed back with nothing to report. B-25's FV-951 FC is airborne to escort the tanker USS *Y-18*, sighted, signals exchanged, signal sent to base advising the meeting of the tanker, vessel was escorted, aircraft landed back at base with nothing to report. US Navy Ventura 29863 is airborne to escort cruiser USS *Cincinnati*, cruiser was escorted, aircraft completed the patrol and landed back. Ventura 29863 and Ventura 23896 are airborne for Boca Chica, having completed their duties at this unit. A report was received that an aircraft had crashed east of the aerodrome, B-25 FL-200 BX, which was on local flying, reported that the aircraft was destroyed by fire, fire tender and crash, party were dispatched to the scene of the accident, further report was made by B-25 FL-187 DC that the aircraft had crashed 2 nm east of the airfield, crash tender and fire tender were observed to be at the scene of the accident. Flying Control were in contact with all the aircraft in the air by radio, it was ascertained that crashed aircraft was B-25, category FA FV-953, airframe type Mitchell B-25D, engine type Cyclone, airframe and engine both totally damaged after take-off, they reached a height of 1,000 feet, rose sharply, nose-dived aflame two minutes after take-off near MA. Cause of accident unknown, killed six, court of inquiry submitted. In 2008 Dr. Axel Niestlé proposed to amend the records to show that *U-84* under Horst Uphoff was lost on 7th August 1943 east-northeast of Bahamas and south-southwest of Bermuda to a Mk-24 homing torpedo dropped by the Liberator B-4 of VB-105 USN. In November 2019 the US Secretary of the Navy formally recognized USNR Lt. Thomas Rudolph Evert and his nine men for sinking *U-84*. The US Navy History Heritage Command then awarded the Distinguished Flying Cross and Air Medals to next of kin at the US Capitol Building, Washington, DC in December, 2019. USN Lockheed Ventura with five crew, reported shot down by U-boat, position not given. **Convoy KW-256** is bound north in Straits towards Daytona. USS *Cincinnati* (CL-6) west in Northwest Providence Channel towards Straits with US Navy plane escorting. USS *YPC-1263* is bound north in Straits. *U-634* re-enters area via Anegada Passage north then west. NAS Great Exuma: **Convoy NG-377** of 18 merchant vessel and five escorts departed. A plane departed for Nassau. Two US Navy ships departed.

8th August 1943

B-25 FR-393 DF is airborne to escort the ferry *Jean Brillant*, convoy was not sighted in the reported position, aircraft flew up track and sighted the convoy, vessel was challenged but no reply received, "met" signal was sent to base, no acknowledgement was received, no radio contact could be maintained with base and the aircraft set course for base, landed back. B-25 FR-376 CK was substituted and was airborne to intercept the convoy in the last position in which DF had left it, convoy contacted, patrol was commenced, tanker was sighted, message was received from base "What is your ETA," message was sent to base, ETA 6 pm, convoy was seen to enter Nassau. Harbour aircraft landed back at base. B-25's FV-948 FF and FV-952 FA airborne, landed back having completed exercises. An airman working in the Direction Finding Station at San Salvador was reported to be seriously ill. RAF Transport Command Catalina was dispatched to bring the airmen back to the RAF Hospital, Nassau. Catalina was damaged on being waterborne at San Salvador, a Walrus aircraft left the Main Field for San Salvador with the CSO [signals officer] to carry out an inspection of the Direction Finding Station, damage to the Catalina was more extensive than anticipated, signal was received from San Salvador for the dispatch of spares for the repair of the aircraft. B-25 FR-366 CX airborne with spares for San Salvador, spares were packed for dropping by parachute, aircraft observed that the packages were picked up, after they had been dropped. Walrus returned to base, B-25 landed later. USS *PC-614* is bound south in Straits passing Daytona. SS *Columbia Star* is bound south in Straits passing Canaveral with US Navy plane escorting. USS *YMS-249* is bound north in Straits passing Port Everglades. Tug USS *Meteor* is bound north in Straits toward Miami. Tug USS *Humrick* is bound southwest in Straits. Tug USS *Hercules* is bound north in Straits. **Convoy KN-257** is bound north in Straits with US Navy plane escorting. USS *PC-1234* is bound south in Straits. **Convoy KG-650** is bound southeast towards Nicholas Channel with US Navy plane escorting. USS *Y-18* is bound north from Nassau. *U-732* is 300 nm east of Eleuthera, east to France. *U-634* is bound north from north of Anegada. RAFTC Catalina dispatched to San Salvador to evacuate seriously ill airman working there. As Catalina was damaged on water landing, a Walrus was sent to investigate. B-25 FR-366 CX was sent with spares for the Catalina which were dropped by parachute. Walrus and B-25 CX returned. Funerals for J. Hadravek and J. Mares, killed the day before, Roman Catholic Chaplain officiating. NAS Great Exuma: Two planes went to escort **Convoy GN-76**. Destroyer joined two US Navy ships. **Convoy KG-650**, eight merchant vessels and six escorts, entered area.

9th August 1943

B-25 FV-946 FF is airborne on Anti-Submarine Patrol Number 1c, descent through cloud exercise was carried out before the commencement of the patrol, a 3-masted schooner was sighted, patrol was completed and the aircraft landed back after completing gunnery practice. B-25 FV-951 FS is airborne for Anti-Submarine Patrol Number 4d, creeping line ahead patrol, patrol was completed without incident landed back with nothing to report. B-25 FR-362 CA is airborne for Anti-Submarine Patrol Number 1b, completed, landed back. B-25 FR-393 DF is airborne for Anti-Submarine Patrol Number 4d, creeping line ahead patrol was commenced, crew had nothing to report from the patrol. B-25 FV-952 FX is airborne for Anti-Submarine Patrol Number 4e, patrol commenced, 3-masted schooner was sighted, aircraft completed the patrol without incident and landed back after a descent through cloud exercise. B-25 FR-376 CK is airborne for Anti-Submarine Patrol Number 4e, 3-masted schooner was sighted, aircraft returned to base with radio unserviceable, aircraft took off again later, was over convoy 28 merchant ships and five escorts three larger liners or troopers, no stragglers, no reply was received to a challenge. Aircraft flew back along the track of the convoy for 5 nm, no stragglers sighted, aircraft landed back. B-25 FR-362 CA is airborne for Anti-Submarine patrol, aircraft was over a convoy, completed exercise and landed back. B-25 FR-366 CX is airborne for Anti-Submarine patrol, convoy of 10 merchant ships and five escorts is sighted, patrol was completed, and the aircraft landed back. B-25 FV-952 FX is airborne for Anti-Submarine Patrol Number 2f, 3-masted schooner was sighted, patrol was completed and the aircraft landed back. B-25 FV-946 FF is airborne for Anti-Submarine Patrol Number 2e, no radio contact could be maintained with base, permission was granted to continue the patrol on B/T control over the convoy of 25 merchant ships and six escorts patrol was completed, and the aircraft landed back. B-25 FR-393 DF is airborne for Anti-Submarine Patrol Number 2d, patrol was completed, and the aircraft landed back. The repairs to the damaged Catalina at San Salvador were not completed, signal received requesting that supplies of food for the crews be delivered by air, B-25 FR-381 DC was airborne for San Salvador, mission was completed, food dropped and received, aircraft landed at base. Tanker *A. S. Hansen* is bound north in Straits passing Canaveral. USS *YMS-199* and USS *YMS-193*, SS *Delaware*, and SS *William C. Atwater* are bound south in Straits towards Palm Beach. **Convoy NK-557** is bound south in Straits towards Palm Beach with US Navy plane escorting. USS *AM-131* and USS *YO-143* are bound northeast in Straits passing Palm Beach. SS *Herbert C. Jones* is bound northeast from near Matanilla Shoal, northwest Bahamas. SS *Trodanger* is bound north in Straits passing Port Everglades. USS *YMS-176* and USS *YMS-160* are bound northeast in Straits passing Fort Everglades with US Navy plane escorting. Ferry *Jean Brillant* is bound northeast in Straits from Miami. USS *PC-1202* is bound south in Straits. As Catalina on San Salvador was not yet repaired, food was dropped to the aircrew by B-25 FR-381 DC. NAS Great Exuma: A plane returned after a search and rescue mission. Two planes went to cover **Convoy GN-76.** A plane departed on training loop and returned. A plane went to cover **Convoy KG-650**.

10th August 1943

B-25 FV-952 FX is airborne for Anti-Submarine Patrol Number 1c, two destroyers and three sub-chasers sighted, patrol was completed, and the aircraft landed back. B-25 FV-954 FB airborne on Anti-Submarine Patrol Number 1c, aircraft circled two destroyers and three sub-chasers, remaining part of the patrol was completed, and aircraft landed back with nothing to report. B-25 FR-366 CX airborne for Anti-Submarine Patrol Number 1a, patrol was completed satisfactorily, convoy of 28 merchant ships and five escorts sighted, aircraft landed back with nothing to report. B-25 FV-946 FF is airborne for Anti-Submarine Patrol Number 1, the patrol was completed satisfactorily, aircraft landed back with nothing to report. B-25 FR-393 DF is airborne for Anti-Submarine Patrol Number 1, convoy was sighted with 28 merchant ships, and four escorts patrol was not completed, returned to base with nothing to report. With the advent of heavy thunderstorms, flying training was curtailed during the day. Flying training was continued during the clearer periods from both Main Field and Satellite Field. Tug USS *ATR-44* is bound south in Straits passing Daytona. SS *Grand Manan* is bound south in Straits passing Canaveral. USCG cutter *CG-Calypso* is bound south in Straits towards Palm Beach. USS *PC-615* is bound southwest in Straits passing Palm Beach. USS *YO-143* is bound west near Matanilla Shoal, northwest Bahamas. USCG Cutter *CG-Stephen McKeever* is bound southwest in Straits towards Port Everglades. Tug USS *Meteor* and tug USS *Hercules* are bound south in Straits passing Miami. **Convoy NK-557** is bound south in Straits with US Navy plane escorting. SS *Cross Keys* is bound south in Straits. USCG *CG-Vernon McNeal* and SS *Corinth* are bound west in Straits. Cuban *CS-12*, Cuban *CS-13*, and Cuban *CS-41* are bound west in Straits passing Nicholas Channel. **Convoy GN-76** is bound northwest near Abaco with US Navy plane escorting. **Convoy NG-378** is bound south, to the east of Bahamas with US Navy plane escorting. NAS Great Exuma: Tug transiting Mayaguana Passage reported a US Navy destroyer and patrol craft boats in area. A plane escorted **Convoy NG-378**. A plane sighted five US Navy destroyers. **Convoy KG-650** and **Convoy GN-76** departed.

11th August 1943

B-25's FV-954 FB, FV-952 FI, FR-366 CX, FR-378 CL, FV-951 FC, and FR-381 DC are airborne, landed back having completed individually Anti-Submarine Patrol Number 3. USS *SC-1292* is bound north in Straits passing Canaveral. USCG Cutter *CG-Warren Edwards,* USS *SC-1074*, and USS *YN-15* are bound south in Straits passing Canaveral. **Task Unit TU-2931** is bound north in Straits passing Palm Beach with

US Navy plane escorting. SS *Georgia* and SS *Balls Bluff* are bound northeast towards Matanilla Shoal, northwest Bahamas. Tug USS *Humrick* is bound north in Straits towards Miami. SS *Atlantic Coast* is bound north in Straits passing Miami. USS *PYC-38* is bound south in Straits passing Miami. USS *ATR-44* is bound south in Straits. USCG cutter *CG-Calypso* is bound southeast in Straits. SS *Peter Donahue* is bound northwest in Straits passing Nicholas Channel. *U-634* and *U-732* exit the area northeast of Bahamas, east to France. NAS Great Exuma: Three planes went to cover **Convoy NG-378**. USS *Antaeus* and escort passed through the Crooked Island Passage.

12th August 1943

B-25 FV-952 FX is airborne for Anti-Submarine Patrol Number 2, aircraft completed the patrol without incident and landed back with nothing to report. B-25 FR-366 CX airborne for Anti-Submarine Patrol Number 2, aircraft landed back after completed the patrol nothing to report. B-25 FV-949 FD is airborne for Anti-Submarine Patrol Number 2, patrol was completed without incident aircraft landed back. B-25 FR-375 CL is airborne for Anti-Submarine Patrol Number 2, patrol was carried out satisfactorily, Ventura-type aircraft appeared to be force-landed, its turret and engine had been removed and the tall plane was observed to be in a lake nearly, FR-375 CL landed back at base. B-25's FR-366 CX, FR-375 CL, FV-952 FX, and FR-381 DC airborne and landed back having completed Anti-Submarine Patrol Number 3. B-25's FR-366 CX, FR-378 CL, FV-949 FD, FR-381 DC, and FV-952 FX are airborne for night navigation exercise, landed back after completion. Liberator FL-993 LB is airborne, landed at Miami *en route* for San Antonio. Aircraft of category FA FL-199, airframe type Mitchell B-25D, engine type Cyclone, airframe slightly damaged, engine not damaged, pilot made normal approach but rounded out too low, aircraft landed on all three wheels at the same time as MA [accident] occurred, because of the error of judgment on part of pilot in touching down prior to reaching runway occurred while carrying out flapless practice landings, pilot inexperienced, and no disciplinary action taken. USS *SC-1304* is bound south in Straits passing Daytona. USS *YMS-166* and USS *SC-1313* are bound south in Straits towards Canaveral. USCG *CG-83451* is bound west north of Matanilla Shoal, northwest Bahamas. Tanker *A. S. Hansen* is bound south in Straits towards Palm Beach. Tanker *Gulfland* is bound north in Straits passing Miami. USS *SC-1333* is bound south in Straits. HMS *Athene* is bound northeast out of Straits. Two aviators uninjured following botched landing of Mitchell B-25C FL-199 Cyclone, crew Vince McNabb. NAS Great Exuma: **Convoy NG-378**, with 10 merchant vessels and five escorts departed. Three planes landed and then departed.

13th August 1943

Liberator FL-970 LH airborne from Satellite Field special Anti-Submarine patrol to cover the convoy routes, altered course to home over blimp, merchant ships and escorts sighted on a southerly zig zag course, merchant ship2,000 tonnes and escorts larger than usual subchaser escorts challenged aircraft which replied, aircraft completed patrol and landed back at base. B-25 FR-376 CK is airborne for Anti-Submarine Patrol Number 2f, merchant ships and escorts sighted aircraft, challenged escorts no reply got, convoy again sighted later, patrol completed, aircraft landed back with nothing to report.

B-25 FR-381 DC airborne for Anti-Submarine Patrol Number 2e, merchant ship sighted with escort vessel, patrol completed, gunnery exercises carried out against a flame float, landed back after exercise completion. Liberator BZ-746 LM and BZ-757 LR are airborne from Satellite Field for Anti-Submarine Patrol Number 9, and homing exercise landed back at base. Liberator BZ-759 LF airborne from Satellite Field special Anti-Submarine patrol to cover the convoy routes, landed back at base. Liberators BZ-746 LM and BZ-757 LR are airborne for submarine exercise off Key West exercise completed, landed back to Satellite Field. Liberators BZ-810 and BZ-808 LY are airborne from Satellite Field for submarine recognition and attack exercises of Key West, landed back to Satellite Field after completion of exercise. Tug USS *Point Loma* with tow and USS *SC-1292* are bound south in Straits towards Canaveral. Troop transport USS *USAT-Etolin* and SS *American Packer* are bound south in Straits passing Canaveral with US Navy plane escorting. USS *PC-522* is bound north in Straits passing Palm Beach. **Convoy KW-259** is bound north in Straits passing Miami. USS *PC-1201* is bound west in Straits. **Convoy Motor Torpedo Boat Squadron, Group A**, SS *Henry Dearborn*, SS *Hiram S. Maxim*, SS *Jeff Davis*, SS *Gaspar De Portola*, and destroyer escort USS *Keith* (DE-241) are bound north in Straits. **Convoy KN-258** is bound north in Straits with US Navy plane escorting. **Convoy KG-651** is bound southeast in Straits passing Nicholas Channel with US Navy plane escorting. NAS Great Exuma: **Convoy GN-77**, 31 merchant vessels and five escorts, departed, escorted by two planes from the base. A plane sighted a 3-masted schooner, then another plane departed to cover **Convoy GN-77**.

14th August 1943

Night navigation training by B-25's FR-376 CX, FR-381 DC, and FV-946 FF. Liberator BZ-759 LF is airborne from Satellite Field on a special Anti-Submarine patrol to cover all convoy routes, creeping line ahead patrol carried out, no incident, recording was made which faded out, aircraft sighted, this might have made the recording, landed back at base with nothing to report. Liberators BZ-746 LM and BZ-757 LH are airborne from Satellite Field on Anti-Submarine patrol and homing exercise, small merchant ship spotted, recordings made on the S/E [sound receiver], contacts faded out, no visual contact, homing exercise completed satisfactorily, landed back with nothing to report. B-25 FF-946 FF is airborne for Anti-Submarine Patrol Number 2e, convoy of 30 merchant ships and four escorts sighted, patrol and gunner exercise completed, landed back at base. B-25 FR-381 DC is airborne for Anti-Submarine Patrol Number 2f, oil patch observed running east to west 1 nm long and 1 nm wide, a merchant ship and three escorts sighted, aircraft observed on escort duty, no reply received from escorts merchant ship was painted grey, well laden with deck cargo, probably timber, aircraft landed back after uneventful patrol. B-25 FR-376 CK is airborne for Anti-Submarine Patrol Number 2d, patrol completed without incident and the aircraft landed back with nothing to report. B-25 FR-383 DF is airborne for Anti-Submarine Patrol Number 3, landed back with nothing to report from this patrol. B-25 FR-383 DF airborne for Anti-Submarine Patrol Number 2, patrol was competed, and the aircraft landed back with nothing to report. B-25 FR-381 DC airborne for Anti-Submarine Patrol Number 3, merchant ships and escorts sighted, merchant ship grey, lightly laden, single funnel, three island type steam ship, convoy sighted with 30 merchant ships and four escorts two ships were 10,000 tons but average was 6,000 to 8,000 tons, landed back at base after exercise completion. B-25 FV-946 FF airborne for Anti-Submarine Patrol Number 2, patrol completed, landed back with nothing to report. B 25 FR 376 CK is airborne for Anti-Submarine Patrol Number 2f, patrol completed, landed back with nothing to report. Liberators BZ-746 LM and BZ-757 LR are airborne from Satellite Field for Anti-Submarine patrol and homing exercise, landed back at base after completion. Liberator BZ-959 LF is airborne from Satellite Field, carry out patrol covering all convoy routes, landed back at base after completion. Liberator FL-964 LD landed at Satellite Field, returned from San Antonio, there modifications to the aircraft was made. *U-598* sank several ships off Ragged Islands and Cay Santo Domingo, southern Bahamas from **Convoy TAW 12J** (Trinidad, Aruba, Key West); British SS *Michael Jebsen* and British tanker *Standella* (damaged and towed to Key West), and British SS *Empire Corporal*. Master, 35 crew, and four gunners from *Michael Jebsen* taken by US Navy patrol boat to Guantanamo. Six crew and a gunner lost. Master, 39 crew, and nine gunners from SS *Empire Corporal* picked up by USS *PT-498* transferred to destroyer USS *Fletcher* (DD-445), then to Guantanamo. **Convoy KN** bound north in Straits passing Canaveral. USCG *CG-Stephen McKeever* is bound south in Straits passing Canaveral. SS *Henry Dearborn*, SS *Hiram S. Maxim* and SS *Jeff Davis* are bound north in Straits passing Canaveral. SS *Gaspar De Portola*, USS *Carib* (AT-82), and troop transport USS *USAT-Aconcagua* are bound north in Straits passing Palm Beach. **Convoy NK-558** is bound south in Straits towards Palm Beach with US Navy plane escorting. Tanker *Faireno* and tanker *Hat Creek* are bound north in Straits passing Miami. Oiler USS *YO-142*, USS *SC-1467*, tug USS *Point Loma* with tow and USS *SC-1292* are bound south in Straits. Two tugs with five tows and USCG *CG-H. M. Messick* are bound southeast in Straits passing Nicholas Channel. **Convoy GN-77** is bound northwest east of Bahamas. NAS Great Exuma: Two planes went to cover **Convoy GN-77**. Navy ship and escort departed. Convoy of 30 merchant vessels and four escorts departed.

15th August 1943

Liberators BZ-757 LR and BZ-746 LM are airborne from Satellite Field for Anti-Submarine patrol and homing exercises, landed back at base after completion with nothing to report. Liberator BZ-759 LF is airborne from Satellite Field for Anti-Submarine patrol over convoy routes, creeping line ahead search completed, no incident landed back at base with nothing to report. B-25 FR-381 DC is airborne for special cross over patrol, convoy of 22 merchant ships and five escorts sighted, again sighted later with air escort while aircraft was returning to base, landed back at base after completion with nothing to report. B-25 FR-946 FF airborne for special cross over patrol, landed back at base after completion with nothing to report. B-25 FR-381 DC airborne for Anti-Submarine Patrol Number 2, convoy of 24 merchant ships and five escorts sighted landed back at base after completion with nothing to report. B-25 FR-381 DF airborne for Anti-Submarine Patrol Number 2, landed back at base after completion with nothing to report. B-25 FV-946 FF airborne for Anti-

Submarine Patrol Number 2, convoy of 22 merchant ships and five escorts sighted later, a single merchant ship was sighted, landed back at base after completion with nothing to report. Liberators BZ-757 LR and BZ-746 LM airborne from Satellite Field for Anti-Submarine and homing exercises, landed back after completion. Aircraft of category FA BZ-761, airframe type Liberator, engine type Pratt and Whitney, airframe slightly damaged, engine not damaged, Gunnery instructor cleared, stoppage in the starboard beam gun, subsequent testing burst, aircraft hit air pocket, gunnery instructor momentarily lost control of gun, two bullets entered starboard wing at the firing area, investigation held established that there was not negligence on part of gunnery instructor, accident due to loss of control and due to air disturbances, no disciplinary action taken. British tanker *Standella* hit by *U-598* arrives in Key West from Ragged Island. **Task Unit TU-2933** is bound south in Straits towards Port Everglades with US Navy plane escorting. **Convoy NK-558** and USCG *CG-Stephen McKeever* are bound west in Straits. **Convoy GK-751** is bound northwest in Straits near Cay Sal Bank with US Navy plane escorting. **Convoy GN-77** is bound northeast in Straits with US Navy plane escorting. **Convoy NG-379** is bound south, to the east of Bahamas with US Navy plane escorting. NAS Great Exuma: Two planes escorted **Convoy NG-379**. A plane departed for Guantanamo. A plane landed at Great Exuma.

16th August 1943

Liberators BZ-757 LR and BZ-746 LM airborne from Satellite Field, convoy sighted by both, mariner aircraft was seen to be on escort duty of the convoy, heavy thunder storm was sighted, characteristic of the weather being a heavy cumulus frontal condition which appeared to be curving north and south, heavy swell observed on the sea, visibility nil in cloud, continuous lighting, not possible to judge changing winds in darkness, homing exercise abandoned due to bad weather, landed back to Satellite Field. Night navigation exercise carried out by B-25's FR-362 CA, FR-393 DF, and FV-946 FF all three again airborne, completed Air Exercise No. 7. Liberator BZ-744 LL is airborne for meteorological flight, landed back having made a survey of the weather, at this point ascertained that the centre of the tropical disturbance was moving away from the Bahamas toward Bermuda. New chop line and Bahama patrol area established. NAS Great Exuma: Two planes went to cover **Convoy NG-379**. A plane landed at Great Exuma.

17th August 1943

Night patrols carried out by B-25's FV-949 FD, FR-384 CM, FR-378 CL, FR-362 CA FD and CL, landed earlier, rest completed and landed, FD and CL again airborne, landed back after completion of exercise. B-25's FK-173 AK and FL-208 BF airborne completed short navigation flight landed back at base. B-25's FR-362 CA, FR-378 CL, FR-384 CM, and FV-949 FD are airborne for air Exercise Number 7, landed back after completion. B-25's FR-392 DF, FR-362 CA, FV-949 FD, FR-378 CL, and FR-384 CM airborne and landed back after exercise completion. Anti-Submarine patrols are carried out by Liberator aircraft from Satellite Field. Liberators BZ-757 LR and BZ-746 LM airborne from Satellite Field for Anti-Submarine Patrol Number 9, landed back after completion. Liberators BZ-759 LF and FL-964 LD are airborne, creeping line ahead patrol and homing exercise, landed back after completion. USS *SC-1353* and USS *SC-1051* are bound south in Straits towards Daytona. USS *SC-1274* is bound south in Straits towards Canaveral. USS *SC-1352*, USS *SC-1283*, USS *SC-1284*, USS *SC-1285* and USS *AM-138* and USS *AM-139* are bound north in Straits passing Canaveral. Tanker *Gulfland* is bound south in Straits from Palm Beach. USS *YAG-28* and USS *SC-1037* are bound south in Straits from Port Everglades. USS *YMS-177* is bound northeast in Straits. Oiler USS *YO-142* escorted by USS *SC-1467* is bound north in Straits. NAS Great Exuma: Three planes departed on training loop and returned. **Convoy GN-78** of 14 merchant vessel and five escorts, entered.

18th August 1943

Liberator BZ-946 LD is airborne from Satellite Field for creeping line ahead patrol heavy rain clouds down to sea level. Visibility nil, so patrol abandoned found impossible to see a flame float that was dropped into the sea low stratus clouds moderate swell on the sea heavy rain and lightning patrol abandoned landed back at base. Liberators BZ-757 LF airborne creeping line ahead search patrol completed, landed back at base. BZ-746 LM airborne for Anti-Submarine Patrol Number 9, landed back, homing exercise commenced, no contact could be maintained on radio, exercise abandoned, aircraft continued patrol. B-25's FR-384 CM, FV-949 FD, FR-378 CL, and FR-362 CA airborne for Anti-Submarine Patrol Number 5, patrols completed, landed back. B-25's FV-949 FD, FR-378 CL, FR-393 DF, and FR-384 CM is airborne for air Exercise Number 7. Patrols completed, landed back. B-25's FV-949 FD, FR-362 CA, and FR-393 DF is airborne for air Exercise Number 7, patrols completed, landed back. Liberator BZ-840 MA is airborne from Satellite Field, carried out creeping line ahead search, landed back after completion. SS *Chiloil* is bound south towards Canaveral. USS *YMS-177* is bound north out of Straits. USCG *CG-83451* is bound east in Straits. SS *Corinth* is bound north in Straits passing Bimini. USS *SC-1312* is bound south in Straits. SS *American Sun,* destroyer escort USS *Douglas L. Howard* (DE-138) and SS *Delaware* are bound north in Straits. Tug *Mike Moran* with four tows escorted by USS *YMS-73* are bound southeast to exit west of Key West. *U-107* enters area west north of Walker Cay north of Abaco. NAS Great Exuma: Two planes went to cover **Convoy GN-78**. **Convoy KG-652**, with 10 merchant vessels and five escorts, entered the area.

19th August 1943

Night navigation flights carried out by B-25's KK-180 AS, FK-183 AU, FK-172 AJ, and FK-173 AA. Anti-Submarine Patrol Number 5, carried out by individual aircraft. B-25's FR-384 CM, FR-393 DF, FR-381 DC, FR-378 CL, FR-362 CA, and FV-952 FX are airborne and landed back after completion. Liberator BZ-840 MA is airborne from Satellite Field, carried out creeping line ahead search, landed back after completion with nothing to report. Tanker *Pan Pennsylvania* is bound south in Straits. SS *Delaware* is bound north in Straits towards Daytona. **Convoy NK-559** is bound south in Straits towards Canaveral with US Navy plane escorting. Troop transport USS *USAT-Monterey* is bound south in Straits passing Canaveral. Tanker *Appomattox* is bound north in Straits passing Canaveral. USS *SC-1274* and tanker *Gulf Belle* are bound south in Straits towards Palm Beach. USCG *CG-Juniper* and tanker *Cross Keys* are bound north in Straits towards Palm Beach. Tanker *Chiloil* is bound south in Straits passing Miami. SS *Seven Pines*, Anti-Submarine Group 1, and USS *PC-1262* are bound south in Straits.

USS *SC-1273* is bound southwest in Straits. **Convoy KN-259** is bound north in Straits with US Navy plane escorting. NAS Great Exuma: Two planes went to cover **Convoy GN-78**. A plane escorted **Convoy KG-652**, convoy of 14 merchant vessel and five escorts, was sighted.

20th August 1943

B-25's FV-952 FX, FR-362 CA, FR-381 DC, FR-384 CM, FV-949 FD, FV-951 FC, and FR-378 CL are airborne for Anti-Submarine Patrol Number 5, landed back after completion with nothing to report. B-25's FL-180 BA, FK-183 AU, FK-181 AT, and FK-172 AJ airborne on short navigation flight, landed back after completion. Aircraft of category FA FR-375, airframe type Mitchell B-25D, engine type cyclone, airframe slightly damaged, engine seriously damaged, hydraulic line burst, wheels lowered in emergency method, main heels refused to lower completely and would not lock lower wheels, attempted for 1.5 hours, landing made, main wheels collapsed, aircraft ended up on its nose at MA. Accident occurred due to failure of normal hydraulic system and emergency under-carriage lowering system, no blame to personnel. *U-107* is bound northeast of Abaco, northwest to exit area. USCG *CG-Stephen McKeever* is bound north in Straits passing Canaveral. Tug USS *Ada Elizabeth* and tanker *Atlantic Coast* are bound south in Straits passing Palm Beach. USS *YP-603* is bound north in Straits. USS *SC-686* is bound south in Straits. HMS *Engadine* north in Straits with US Navy plane escorting. USS *SC-756* is bound southeast in Straits passing Cay Sal Bank. NAS Great Exuma: **Convoy CL-62** and destroyer USS *DD-576* entered the area, and two planes were sent to cover it with destroyer USS *DD-576*. A plane departed to cover **Convoy NG-380**.

21st August 1943

Anti-Submarine Patrol Number 5, carried out by B-25's FV-946 FF and FV-951 FC, landed back after completion with nothing to report. No day flying from Main Field, new lighting installations undertaken on the runways of the Main Field. Flying training for exercise numbers 6, 7, and 8 completed. Number 6 completed, training on Liberator from Satellite Field and returned to the Main Field preparatory to posting. Number 7 course completed in B-25 and was moved to Satellite Field for a conversion course on to Liberator aircraft. Number 9 completed familiarization course and were transferred to Number 2, Squadron for Anti-Submarine patrols. Number 10 course commenced initial lectures. Report received from Meteorological Bureau of Miami in afternoon of a large tropical storm speed was 15 knots direction northwest from reports, it seemed that storm would reach New Providence, necessary precautions taken, arrangements made for diversion of aircraft to airfields in the US. Tanker *Goldheels* is bound south in Straits towards Palm Beach. Oiler USS *YO-143* and USS *YP-612* are bound east near Matanilla Shoal, northwest Bahamas. USS *YP-603* towing tank barge, or lighter USS *B-10* is bound northwest in Straits towards Miami. SS *Shiloh* and tanker *Atlantic Coast* are bound south in Straits. SS *Manuel Rionda* and USCG *CG-Boutwell* are bound north in Straits. NAS Great Exuma: Two planes went to cover **Convoy NG-380**. Two plane went to cover **Convoy NG-380**. A plane landed at Great Exuma.

22nd August 1943

Reports received from Miami that storm had changed course and was 200 nm to 300 nm away and heading in a north-westerly direction. No flying took place from Main Field during day, constructional work continued in the runways, and heavy rains delayed work. SS *Rupert E. Craft* and USS *APC-95* are bound south in Straits towards Canaveral. Destroyer escorts USS *DE-139* and USS *DE-142* are bound east near Matanilla Shoal, northwest Bahamas. Tanker *Hat Creek* and USS *Spry* (PG-64) are bound south in Straits towards Palm Beach. USS *PC-1242* is bound south in Straits. USS *YAG-28* is bound northwest in Straits towards Miami. Tanker *Manuel Rionda* is bound south towards Nassau. NAS Great Exuma: **Convoy NG-380** arrived with 23 merchant vessel and five escorts. A plane departed **Convoy NG-360.** USS *YAG-21* dropped anchor at Great Exuma. A convoy of six navy ships entered.

23rd August 1943

B-25's FR-371 CF, FR-381DC, FR-384 CM completed air Exercise Number 1, landed back at base after completion. FV-946 FF and FV-952 FX are airborne for air Exercise Number 2, landed back at base after completion. Flying cancelled after early morning detail, so the work could be continued throughout the day on the runways at Main Field. SS *William L. Yancey* and tug USS *Point Vincent* are bound south in Straits towards Daytona. SS *Port Dunedin*, SS *Idomeneus*, SS *Alar*, SS *Thompson Lykes*, and SS *Britamer* are bound south in Straits passing Canaveral. Tanker *Gulfmaid* is bound north in Straits towards Canaveral. SS *Kennebec* is bound northeast in Straits near Canaveral with US Navy plane escorting. Oiler USS *YO-143* is bound west from Walker Cay, north of Abaco. SS *Talisse* is bound south in Straits towards Palm Beach. SS *Rupert E. Craft* is bound south in Straits passing Palm Beach. USS *Cincinnati* (CL-6) is bound east in Straits passing Miami with US Navy plane escorting. USS *SC-1238*, SS *Rio Neuquen*, and USCG *CG-83450* are bound south in Straits. USS *Reuben James* is bound north in Straits. *U-107* is northwest of Grand Bahama, bound north and outward. Court martials convened for breaking out of camp, going over the hill, fraternizing with local women at the Weary Willie's Hotel. Verdict: guilty, sentence: detention for six months. NAS Great Exuma: A plane departed on administrative flight to Miami. A plane went to cover the naval convoy of six ships. Two planes departed to cover **Convoy GN-79.** A plane sighted a 70' fishing trawler. A plane sighted a 3-masted schooner. Cargo transporter USS *YAG-21* got underway and departed.

24th August 1943

B-25's FV-951 FC, FV-946 FF, and FV-952 FX are airborne for air Exercise Number 1, landed back after completion. B-25's FR-378 CL and FR-384 CM airborne for air Exercise Number 2, CL landed radio unserviceable CM landed back at base after completion. B-25 FV-949 FD completed air Exercise Number 4. B-25's FV-952 FX, FV-951 FC, and FV-946 FF airborne for air Exercise Number 3, landed back at base after completion. B-25 FR-384 CM completed air Exercise Number 4. Air Exercise Number 1, completed by B-25's FR-384 CM, FV-951 FC, FV-949 FD, FV-952 FX, FV-946 FF and FR-381 DC. Air Exercise Number 3 completed by B-25's FV-949 FD, FF-946 FF, FV-952 FX, FV-951 FC, FR-384 CM, and FR-381 DC. USS *SC-1332* is bound south in Straits towards St. Augustine. USS *Saluda* (IX-87) is bound north in Straits towards St. Augustine. Tanker *Corinth* is bound south in Straits towards Canaveral. USCG *CG-83365* is bound north in Straits passing Canaveral. **Convoy KN-260** is bound northeast in Straits with US Navy plane escorting. **Convoy NK-560** is bound south in Straits towards Palm Beach, with

US Navy plane escorting. SS *John Paul Jones* is bound north in Straits passing Palm Beach. Tug USS *Point Vincent* is bound north in Straits towards Palm Beach. Oiler USS *YO-143* is bound south in Straits towards Port Everglades with US Navy plane escorting. Tanker *Manuel Rionda* is bound west in Northwest Providence Channel. USS *PC-1230* and USCG *CG-E. M. Dow* are bound south in Straits. SS *Rupert E. Craft* and SS *William L.*, and USS *Yancey* are bound west in Straits. SS *Idomeneus* is bound west in Straits with US Navy plane escorting. USS *YAG-21* is bound north, passing Nassau. NAS Great Exuma: Two planes went to cover **Convoy GN-79**. A plane departed on training loop. A plane sighted a US Navy light cruiser. A convoy of 19 merchant vessels and five escorts were sighted.

25th August 1943

B-25's FR-384 CM and FL-187 BC airborne for air Exercise Number 1, completed the exercise and landed back. B-25's FL-157 BC, FL-208 HF, FV-952 FX, FR-381 DC, FV-949 FD and FV-946 FF are airborne for air Exercise Number 1. B-25's FV-952 FX, FV-946 FF, and FV-949 FD airborne and completed air Exercise Number 2. B-25's FR-381 DC airborne and landed, having completed air Exercise Number 3. B-25 FL-208 BF are airborne landed with radio unserviceable, unable to complete this exercise. B-25 FL-200 BX airborne landed, having completed the exercise. B-25's FV-949 FD, FV-946 FF, FR-371 CF, FL-200 BX and FR-393 DF airborne and landed after completing this exercise. USS *PC-1204* is bound south in Straits passing Daytona. Tug USS *ATR-45* is bound south in Straits towards Palm Beach. USS *YAG-21* is bound west in Northwest Providence Channel. SS *Seven Pines* is bound north in Straits passing Port Everglades. Tug USS *Point Vincent* towing USS *YFB-48* and escorted by USCG *CG-83365* is bound west in Straits. USS *SC-686* is bound northeast in Straits. Anti-Submarine Group 4 is bound northwest in Straits passing Nicholas Channel. USS *PC-1252*, SS *Polycastle*, and **Convoy GN-79** are bound northwest in Straits off Jacksonville with US Navy plane escorting. *U-107* exits area east of Savannah. NAS Great Exuma: A plane departed for **Convoy NG-381**. A plane landed at Great Exuma.

26th August 1943

B-25's FR-371 CF, FR-381 DC, FR-393 DF, FV-946 FF, FV-952 FX and FL-200 BX are airborne and landed after completing the night navigation exercise. B-25's FR-393 DF, and FV-949 FD are airborne, completed the air Exercise Number 1, landed back at base. B-25's FV-952 FX, FV-946 FF, and FR-381 DC are airborne for air Exercise Number 2, landed back after completion. B-25's FV-946 FF, FV-952 FX, FR-371 CF, FV-949 FD and FR-393 DF are airborne for air Exercise Number 4, landed back after completion. B-25's FV-952 FX, FR-381 DC, are FV-946 FF airborne for air exercise, landed back after completion. B-25's FR-393 DF, FR-381 DC, FR-371 CF, FV-946 FF and FV-952 EX airborne for air exercise, landed back after completion. Tanker *Gulfmaid* is bound south in Straits towards St. Augustine. Tanker *Pan Pennsylvania* is bound north towards Canaveral. USS *SC-686* is bound north in Straits near Canaveral. USS *AO-35* and USS *Hogan* are bound southwest in Straits towards Palm Beach. Tanker *McCleeland Creek* and USCG *CG-Dix* are bound north in Straits passing Palm Beach. SS *Cape Charles* is bound south in Straits nearing Bimini. Tanker *Cross Keys* is bound south in Straits passing Bimini. Tanker *Delaware*, tanker *Appomattox*, USS *PC-1203*, and USS *SC-1334* are bound southwest in Straits. Tanker *Zoella Lykes* is bound east in Straits. NAS Great Exuma: Three planes went to cover **Convoy NG-381**. Plane J4F-2 Amphibian departed and returned after training. A convoy of 10 US Navy vessels and eight escorts entered.

27th August 1943

Local night navigation flights cancelled, as weather became unfit heavy thunderstorms and rain. B-25's FR-375 CJ, FR-393 DF, FV-949 FD, FV-946 FF, and FR-381 with FV-952 FX airborne for air Exercise Number 5, CJ returned from the exercise with a leaking petrol feed pipe, aircraft circled airfield, preparatory to landing, found that the under carriage hydraulics were inoperative, aircraft continued to circle in an endeavour to release the under carriage, two-way contact could not be maintained on radio and M/T, aircraft could receive by radio only and transmit by E/T only, another aircraft was airborne to lead CJ on to the runway, CJ made a successful crash landing, crew were unhurt. B-2's FR-381 DC, FR-393 DF, FV-951 FC, and FV-946 FF are airborne for air Exercise Number 2, landed back after completion. USS *YMS-169* is bound south in Straits towards Canaveral. USS *YMS-166* is bound northeast in Straits from Palm Beach. Tanker *Gulfmaid* is bound south in Straits passing Palm Beach. USS *SC-1050* and USS *SC-720* are bound southwest in Straits towards Port Everglades. SS *Albert Gallatin* is bound northeast in Straits near Bimini. USS *YO-143* is bound south in Straits. USS *Otterstetter*, Cuban *CS-22*, Cuban *CS-32*, and Cuban *CS-43* are bound northeast in Straits. SS *Cape Charles* is bound southwest in Straits near Cay Sal Bank. **Convoy Motor Torpedo Boat Squadron-20** is bound southeast, entering Straits. SS *Rupert E. Craft* is bound southeast in Straits passing Nicholas Channel. *U-161* is off Providenciales, where it crash dives for aircraft, Providenciales Island, Turks and Caicos, is in sight. NAS Great Exuma: A plane departed to cover the convoy of 10 US Navy vessels and five escorts. Another plane escorted **Convoy NG-381**, of 14 merchant vessels and five escorts, which departed.

28th August 1943

Local night navigation flights carried out by B-25's FK-173 AK, FK-171 AH, FK-174 AL, FR-393 DF, FV-946 FF, and FR-381 DC, as aircraft airborne and landed back after completion of exercise. B-25's FR-393 DF, FV-951 FC and FR-384 CK are airborne for air Exercise Number 5, landed back after completion. B-25's FR-393 DF and FR-376 CK are airborne for air Exercise Number 5, landed back after completion. USS *SC-1323* is bound south in Straits passing Canaveral. SS *Point Halifax*, SS *Cape Spencer*, and tanker *Mobilgas* are bound south in Straits passing Palm Beach. USS *YMS-178* is bound north in Straits towards Palm Beach. SS *Chirripo* is bound south in Straits towards Port Everglades. USS *YMS-169* is bound south in Straits. Tanker *Little Big Horn* is bound north in Straits. **Convoy KN-261** is bound north in Straits with US Navy plane escorting. **Convoy KG-654** is bound northwest towards Key West with US Navy plane escorting. USS *PC-462* is bound northwest in Straits passing Nicholas Channel. Tanker *Paco* is bound southeast in Straits passing Nicholas Channel. NAS Great Exuma: Two planes escorted **Convoy GN-80**. A plane arrived with Lt. Cdr. Upham, from Guantanamo.

29th August 1943

Local night navigation flights carried out by B-25's FK-173 AK, FK-174 AL, FK-171 AH, FR-393 DF and FR-317 CF, which landed back after completion. Local night navigation flights were detailed to be carried out, local weather was unfit for flying at the detailed time of take-off, heavy thunderstorms and rain in the vicinity of the airfield, Flight commander carried out a weather test and considered the conditions fit for flying remaining aircraft then commenced the exercises. B-25 FR-376 CK is airborne carried out air Exercise Number 5, landed back after completion. B-25 FR-393 DF is airborne for air Exercise Number 3, landed back after completion. B-25 FR-371 CF is airborne for air Exercise Number 1, landed back after completion. B-25's FR-384 CM, FR-371 CF, and FR-366 CM airborne for air Exercise Number 4, landed back after completion. B-25's FR-371 CF, FR-366 CX, FR-393 DF, FL-187 BC, FR-376 CK, and FL-200 BX are airborne for air Exercise Number 5, then landed back after completion. USS *PC-606* is bound south in Straits passing Canaveral. SS *Mormacdale* is bound north in Straits towards Canaveral. SS *Tosari* is bound south in Straits towards Palm Beach. USCG *CG-83451* is bound east in Straits. SS *Hat Creek* is bound north near Matanilla Shoal, northwest Bahamas with US Navy plane escorting. **Convoy NK-561** is bound south in Straits passing Palm Beach with US Navy plane escorting. SS *Shiloh* is bound northwest in Straits off Bimini with US Navy plane escorting. Ferry *Jean Brillant* and USS *SC-656* are bound southwest in Straits towards Port Everglades. Free French Navy corvette FFS *Aconite* is bound south in Straits, passing Miami. USS *SC-1300* and SS *Chirripo* are bound south in Straits. **Convoy GN-80** is bound north, to the east of Bahamas. NAS Great Exuma: Two planes went to escort **Convoy GN-80**. A plane landed at Great Exuma. Convoy of 21 merchant vessels and five escorts departed.

30th August 1943

B-25's FR-366 CX, FR-371 CF, FR-376 CK, FR-384 CM, and FL-187 BC are airborne and landed back having completed air Exercise Number 6. B-25's FR-384 CM, FR-371 CF, and FL-200 BX are airborne for air Exercise Number 6 then landed back after completion. An aircraft of RAF Transport Command sighted a submarine, on the receipt of this report three aircraft were detailed to be airborne to search in the area in which the submarine was sighted and to attack if found to be a U-boat. B-25's FV-951 FC, FR-366 CX, and FR-380 DB are airborne on a sweep of the area around Long Island and Rum Cay, an Allied destroyer was sighted, landed back, nothing to report from the patrol. Tanker *E. H. Blum* is bound south in Straits towards Canaveral. USS *YAG-28* is bound north in Straits towards Bimini. SS *Tosari* is bound south in Straits. HMS *Victorious* and three USS destroyers are bound north in Straits. B-25's FV-951 FC, FR-366 CX, and FR-380 DB conduct exercises with RATC or remote air-traffic control between Long Island and Rum Cay. An No.113 aircraft of RAF Transport Command sighted a submarine, and three No.111 OTU aircraft sighted a US Navy destroyer, between Rum Cay and Long Island. NAS Great Exuma: Six planes departed and returned. An enemy submarine was reported.

31st August 1943

Local night navigation exercises were carried out by B-25's FL-200 BX, FR-384 CM, FL-187 BC, FR-366 GX, and FV-951 FC, which are airborne and landed back after completion. B-25's FR-379 CL, FV-951 FC, FR-380 DB, FR-384 CM, FL-200 BX, and FV-952 FX are airborne for air Exercise Number 6, landed back after completion. USS *YN-22* is bound southwest in Straits towards Palm Beach. USS *Bull* is bound northeast in Straits. A straggler from **Convoy NG-382** is bound southeast in Straits, while the actual **Convoy NG-382** is bound southeast, east of Bahamas with US Navy plane escorting. USS *J-858*, USS *J-961*, and USS *J-967* are bound northeast in Straits near Port Everglades. SS *E. H. Blum*, USS *SC-1331*, and USS *PC-1231* are bound south in Straits. USS *YMS-2* is bound west in Straits. USS *SC-1334* is bound east in Straits. Oiler USS *YO-143* is bound northwest in Straits towards Miami. *U-107* is northeast of Abaco, bound east to France. NAS Great Exuma: Two planes went on a search mission, and landed at Great Exuma. A plane sighted a convoy of 22 merchant vessels with four escorts. An Anti-Submarine search was carried out. A plane returned from Miami. A plane departed to cover **Convoy NG-382**. A plane came from Guantanamo to re-establish the compliment of four planes.

1st September 1943

B-25's FV-951 FC, FV-949 FD, and FV-952 FX are airborne for air Exercise Number 6, landed back after completion. B-25's FR-362 CA, FR-384 CM, FV-946 FF, FV-949 FD, FV-951 FC, and FV-952 FX are airborne for air Exercise Number 6, landed back after completion. Liberator FL-970 LH is airborne for Miami, landed there *en route* San Antonio. Liberator BZ-758 LS is airborne for Bermuda, landed there. Tug USS *Lenape* is bound south in Straits passing St. Augustine. Tanker *Appomattox* and USS *SC-1334* are bound north in Straits, passing Canaveral. Tanker *Pan Georgia* is bound south in Straits towards Palm Beach. SS *Colin* and tanker *Paco* are bound northwest in Straits passing Nicholas Channel. *U-171* is northeast of Jacksonville, reports a steamer makes large zig zags, while being protected by a US Navy blimp, or airship, which circles it. NAS Great Exuma: Two planes went to cover **Convoy NG-382**. A JRF-2 Amphibian arrived from Miami with Cdr. Sedgwick. A plane sighted one medium-sized merchant vessel. Then a convoy of 25 merchant vessels and six escorts entered the area.

2nd September 1943

B-25's FR-366 CX, FR-384 CM, FL-200 BX, FV-951 FC, FV-949 FD, and FV-952 FX are airborne for air Exercise Number 6, landed having completed the exercise. B-25's FR-380 DB, FV-946 FP, FV-952 FX, FR-362 CA, and FV-951 FC are airborne for air Exercise Number 6, then landed having completed the exercise. Liberator BZ-758 LS returned from Bermuda and landed at Satellite Field. US Navy Ventura 29871 and USN Ventura 298897 landed from Boca Chica, aircraft were again airborne to carry out an Anti-Submarine sweep to a sea-plane tender and escort sea-plane; USS *Albermarle* (AV-5) was contacted and was escorted in accordance with the detailed instructions, aircraft landed back, aircraft left the Main Field for Boca Chica. **Convoy NK-562** is bound south in Straits with US Navy plane escorting. USS *YMS-250* is bound north in Straits. SS *Housatonic* (AO-35) and escort USS *Hogan* (DMS) and destroyer escort USS *Hill* (DE-141) are bound north in Straits with US Navy plane escorting. SS *Shabonee* is bound southeast in Straits, passing Canaveral. Tug USS *Lenape* towing a dredge with escort USS *YP-606* are bound south in Straits passing Canaveral. **Convoy Norfolk M-196** is bound south in Straits towards Palm Beach with US Navy plane escorting. SS *Delaware* is bound north in Straits passing Palm Beach. USS *PC-1211* is bound south in Straits towards Port Everglades. USS *FT-12* (APC-66) is bound south in Straits passing Miami. SS *Cross Keys* is bound north in Straits near Miami. USS *SC-1355*, USS *SC-988* and tanker *Pan*

Georgia are bound southwest in Straits. Oiler USS *YO-142* and escort USS YP-218 are bound northeast in Straits. SS *Joseph E. Emory* is bound northwest in Straits towards Key West. SS *David Bushness* is bound northwest in Straits passing Nicholas Channel. NAS Great Exuma: Three planes went to cover **Convoy GN-81**. A plane landed at Great Exuma.

3rd September 1943

B-25's FV-949 TD, FR-384 CM, and FR-380 DB are airborne for Anti-Submarine Patrol Number 5, patrol was completed, and the aircraft landed back with nothing to report from the patrol. B-25 FV-952 FX is airborne for Anti-Submarine Patrol Number 1, a patrol craft and corvette were sighted. A sea plane and sea plane tender USS *Albermarle* (AV-5) were circled, convoy consisted of the tender and two destroyers as escort, patrol was completed, and the aircraft landed back. B-25 FV-946 FF is airborne on Anti-Submarine Patrol Number 1b, patrol was commenced and completed without incident, a patrol craft and corvette were sighted, aircraft landed at 3.03 pm with nothing to report from the patrol. B-25 FV-951 PC is airborne on Patrol Number 2, aircraft circled a convoy of 24 merchant ships and five escorts, then aircraft circled USS *Albermarle* (AV-5) and two US Navy destroyers, left USS *Albemarle* (AV-5), and returned to the convoy. Then aircraft set course for base, landed with nothing to report. B-25 FR-380 DB is airborne for Anti-Submarine Patrol Number 3, a destroyer and patrol craft sighted, aircraft was over Great Sale Cay, Grand Bahama, and carried out air-firing exercises, landed back at base. B-25 FV-951 FC is airborne for Anti-Submarine Patrol Number 3, patrol was completed satisfactorily, and the aircraft was over Great Sale Cay, landed at base with nothing to report. B-25 FV-946 FF is airborne for Anti-Submarine Patrol Number 3, patrol was completed, and the aircraft landed back, during exercise two small coastal vessels were sighted, the remaining part of the patrol was uneventful. B-25 FR-384 CM is airborne on Anti-Submarine Patrol Number 3, patrol completed with gunnery exercises off Spanish Wells, aircraft landed after an uneventful patrol. B-25 FR-376 CK is airborne for Anti-Submarine patrol for the sea plane tender USS *Albermarle* (AV-5) and its destroyer escorts were sighted, signals were exchanged, "Met" signal was sent to base, box patrol was flown to cover the convoy, which consists of 24 merchant ships (10 tankers) and five escorts sighted, Aircraft left USS *Albemarle* (AV-5) and set course for base, landed back. B-25 FR-362 CX is airborne on escort to USS *Albemarle* (AV-5) and as first relief aircraft to CK, a destroyer and a patrol craft were sighted, the vessels were circled and photographed, the seaplane tender USS *Albemarle* (AV-5) and its escort was sighted, box patrol (20 nm ahead, astern, and on either side) was carried out, convoy consisting of 24 merchant ships and six escorts was sighted, convoy was observed to have changed course, convoy was seen to be proceeding on a course of 346°, speed 7 knots, aircraft left USS *Albermarle* (AV-5), set course for base, landed. B-25 FR-302 CA is airborne as second relief aircraft for the escort of USS *Albermarle* (AV-5), a convoy was sighted, the seaplane tender and destroyer escort was sighted, signals were exchanged, box patrol was commenced around a convoy consisting of 16 merchant ships (four tankers) and three escorts was sighted, on course of escorting USS *Albemarle* (AV-5) a second convoy was sighted consisted of 29 merchant ships and six escorts, aircraft left USS *Albermarle*

(AV-5), landed at base. B-25 FR-376 CK was the third relieving aircraft was airborne, a square search was made for the seaplane tender and escorts, convoy was contacted, relieving, aircraft contacted the convoy and later left USS *Albemarle* (AV-5), convoy consisting of 24 merchant ships and five escorts sighted, aircraft continued on its course and returned to base. B-25 FR-366 CX is airborne to escort USS *Albemarle* (AV-5), convoy of 24 merchant ships and six escorts was sighted, the sea plane tender and two destroyers were sighted, aircraft CK was observed to be on escort duties, signals were exchanged between the aircraft and CX took over the patrol from CK aircraft, set course for base, landed. USS *SC-721* is bound south in Straits towards Canaveral. USS *SC-1076* is bound southeast in Straits passing Canaveral. Tanker *Chiloil* and USS *SC-1300* are bound northwest in Straits towards Canaveral. **Convoy KN-262** is bound north in Straits with US Navy plane escorting. USS *Albemarle* (AV-5) and two destroyers and **Convoy GN-81** are (separately) north in the Atlantic, to north of Bahamas. Tug USS *Lenape* with derrick in tow, and USS *YP-606* are bound south in Straits passing Miami. **Convoy NK-562** is bound south in Straits with US Navy plane escorting. SS *Seven Pines* is bound south in Straits. USS *Y-8,* escorting USS *YMS-169* are bound south in Straits to Key West, with US Navy plane escorting. NAS Great Exuma: A plane went to cover **Convoy GN-81**. A plane departed and returned to Great Exuma. A convoy of 25 merchant vessels and six escorts departed. A plane went to cover **Convoy NG-383.**

4th September 1943

B-25 FR-371 CF is airborne for Anti-Submarine Patrol Number 1c, patrol completed without incident, aircraft landed back. B-25 FV-952 FX airborne for Anti-Submarine Patrol Number 1, patrol completed without incident, aircraft landed back. B-25 FR-395 CO is airborne for Anti-Submarine patrol Number 1b, completed the patrol without incident. B-25 FR-362 CA is airborne for Anti-Submarine Patrol Number 1c, patrol was completed, air-firing exercises were carried out prior to the aircraft landing. B-25 FV-946 FF is airborne for Anti-Submarine Patrol Number 1c, aircraft was over a convoy consisting of 16 merchant ships, including tankers, aircraft landed with nothing to report. B-25 FR-384 CM is airborne for Anti-Submarine patrol Number 1, patrol was completed, landed with nothing to report. B-25's FR-376 CK, FV-946 FF, and FR-371 CF are airborne for Anti-Submarine Patrol Number 3, patrols completed, individually landed back. A signal was received from the Commander of GSF (Gulf Sea Frontier) requesting that air cover be provided for the cruiser USS *Marblehead.* B-25 FR-362 CA was airborne to escort, cruiser was sighted, signals were exchanged between the ship and the aircraft, ship was escorted, aircraft returned to base. B-25 FV-952 FX is airborne to contact the cruiser and to take over the escort, signal was sent to the aircraft from base "Cruiser is well ahead possibly 90 nm," signal was received corruptly and could not be deciphered. Creeping line ahead patrol was carried out, cruiser was not met, aircraft set course for base, visibility was poor due to the approaching darkness, landed. USCG *CG-83451* and USCG *CG-Juniper* are bound west near Walker Cay, north of Abaco. SS *Julesburg* is bound northeast in Straits from Port Everglades. Oiler USS *YO-143* is bound north in Straits from Miami towards Northwest Providence Channel. Tug USS

Lenape and USS *PC-614* are bound south in Straits. NAS Great Exuma: Three planes went to cover **Convoy NG-383**. A plane arrived and departed to Guantanamo with Lt. Cdr. Whistler.

5th September 1943

B-25's FB-366 CX, FB-376 CM, FR-381 DC, FR-395 CO, and FR-378 CL carried out night navigation patrol, airborne and landed back after exercise completion. B-25's FR-362 CA, FR-371 CF, FR-376 CL, FV-946 FF and FR-380 DB are airborne on air Exercise Number 4, aircraft landed having completed the exercises. B-25's FR-371 CF, and FR-366 CX are airborne and landed, having completed air Exercise Number 5. B-25's FR-376 CK and FR-362 CA are airborne for air Exercise Number 4, landed after completion. Liberator FL-993 LB returned from San Antonio and landed at Satellite Field. B-25 FR-381 DC is airborne to provide escort for US Army tanker USS *Y-20, en route* for Nassau, aircraft sighted the tanker, signals were exchanged, ship was escorted, DC was relieved by FX, set course for base, landed. B-25 FV-952 FX is airborne to relieve DC on escort duties to USS *Y-20*, aircraft contacted the tanker, signals were exchanged with DC and the aircraft covered the tanker, aircraft landed at base. Tanker *Julesburg* is bound north in Straits passing Canaveral. SS *Reciefeloide* is bound north in Straits passing Palm Beach. Tanker *Shabonee* is bound south in Straits, passing Palm Beach. Ferry *Jean Brillant* is bound west in Northwest Providence Channel. Army tanker USS *Y-20* is bound east in Northwest Providence Channel with US Navy plane escorting. Cuban gunboat *CS-Patria* is bound south in Straits. USS *SC-1294* is bound south in Straits. **Convoy Key West-269** is bound southwest in Straits passing Nicholas Channel. *U-171* is west of Bermuda and reports three U-boat crew wounded by battle with three destroyers in a line, failed to line up shot, distance was too far. NAS Great Exuma: A plane departed **Convoy NG-383**, 16 merchant vessels and five escorts. A plane landed at Great Exuma. Two planes departed and returned after training. A British boat arrived from Nassau.

6th September 1943

B-25's FR-380 DB, FR-371 CF, and FR-376 CK landed having completed air Exercise Number 6. B-25 FL-187 BC airborne for Anti-Submarine Patrol Number 1, returned with wireless failure, air firing exercises carried out landed. B-25 FR-395 CO airborne for Anti-Submarine Patrol Number 1, aircraft set course for base, radio compass and petrol gauges unserviceable, small merchant ship(2,000 to 3,000 tons) was sighted, vessel was a cargo ship with a single funnel, and had a deck cargo of timber. B-25 FR-362 CA airborne for Anti-Submarine Patrol Number 2, uneventful patrol completed, landed. B-25's FR-378 CL, FR-380 DB, FV-946 FF, FR-371 CF airborne for Anti-Submarine Patrol Number 2, completed the detailed patrols and landed back with nothing to report. B-25 FK-187 BC airborne for Anti-Submarine Patrol Number 2, returned with wireless failure landed at base. B-25 FR-381 DC airborne to escort the ferry *Jean Brillant,* aircraft met convoy ship, was escorted, aircraft landed back at base. Tanker *Chiloil* is bound south in Straits towards Daytona. USS *SC-1275* is bound south in Straits passing Canaveral. SS *Corinth* is bound north in Straits towards Canaveral. Ferry *Jean Brillant* escorted by USCG *CG-83450* and USCG *CG-83452* are bound south towards Nassau with US Navy plane escorting. SS *Rioloide* is bound north in Straits towards Miami. NAS

Great Exuma: The British boat which arrived from there departed for Nassau. A plane went to cover **Convoy GN-82**. A plane sighted HMS *BA-37* and two sails. A joint US and British defence plan was drawn up in connection with Capt. Waite.

7th September 1943

Local night navigation exercises were carried out by B-25's FL-164 AB, FL-200 BX, FK-L79 AR and FL-183 which were airborne and landed having completed the exercise. B-25 FR-376 CK, FR-378 CL, FR-381 DC, FR-371 CF, FR-362 CA and FV-946 FF are airborne for Anti-Submarine Patrol Number 5, completed the patrols individually and landed. B-25 FR-381 DC airborne for Anti-Submarine Patrol Number 2, completed the patrol and landed with nothing to report. B-25 FR-371 CF airborne for Anti-Submarine patrol, patrol was abandoned due to the lack of fuel, landed. B-25 FR-362 CA is airborne for Anti-Submarine Patrol Number 2, small merchant ship was sighted, a destroyer and a sub-chaser were sighted, no acknowledgement was given to a challenge by CA, cross over patrol was completed, two PBY-8's were seen to escorting the convoy, aircraft landed. B-25 FR-376 CK is airborne for Anti-Submarine Patrol Number 2b, aircraft tried to make W/P contact with base and returned contact was established when over base, aircraft continued the patrol, landing having completed, an uneventful patrol. B-25 FV-951 FC is airborne, returned to base with engine trouble. B-25 FV-946 FF is airborne for Anti-Submarine Patrol Number 2d, patrol was completed, and the aircraft landed. B-25 FR-391 DC airborne for Anti-Submarine Patrol Number 2f, patrol was completed, and the aircraft landed. B-25 FR-376 CK is airborne for Anti-Submarine Patrol Number 2d, radio was found to be unserviceable, the fuses were continually blowing, to repair them would have caused the generator to burn out, aircraft set course for base, flew over Bridge Point, Eleuthera bearing on the beacon to confirm observation, aircraft began to home in on the Radio Compass. It was found that the aircraft was 20 nm off Miami, set course for Miami, and landed there. B-25 FL-200 BX is airborne for leaflet dropping exercise, aircraft was detailed to drop leaflets over Governor's Harbour, Eleuthera, Port Howe, Cat Island, Clarence Town, Long Island, Georgetown, and Exuma, exercise completed, landed back. B-25 FL-187 BC is airborne for similar exercise, returned with radio failure, airborne later, located Nicholls Town, Andros, Hope Town, and Abaco then landed back at base. USS *SC-1301* is bound south in Straits passing Daytona. SS *Rioloide* is bound north in Straits passing Canaveral. USS *SC-1356* is bound south in Straits towards Palm Beach. Tanker *Chiloil* is bound south in Straits. USS *PYC-38* is bound southeast in Straits from Miami. USCG Cutter *CG-Unalga* and USCG Cutter *CG-Rush* are bound northwest in Straits towards Miami. USS *SC-1038* is bound southwest in Straits. NAS Great Exuma: Three planes went to cover **Convoy GN-82.** A plane sighted a US-flagged freighter. A plane sighted the transport USS *F-21* and its destroyer escort USS *DD-61*.

8th September 1943

B-25's FR-381 DC, FV-949 FD, and FR-371 CF are airborne for Anti-Submarine Patrol Number 5, patrol was completed, and the aircraft landed back. B-25's FL-200 BX airborne for Anti-Submarine Number 3, aircraft completed the patrol and landed back with nothing to report. B-25 FR-371 CF is airborne for Anti-Submarine Patrol Number 2, landed back with nothing to report from the completed patrol. B-25 FV-949 FD is airborne for Anti-Submarine Patrol Number 2, a small merchant ship sighted, ship had stopped but was heading on a course of 360°, patrol was completed and the aircraft landed back. B-25 FL-187 BC airborne for Anti-Submarine Patrol Number 2c, four destroyers sighted, signals were exchanged, patrol was completed, and aircraft landed back. B-25 FR-381 DC airborne for Anti-Submarine Patrol Number 3, patrol was uneventful, and aircraft landed back. B-25 FV-949 FD is airborne for Anti-Submarine Patrol Number 2, patrol was completed, and aircraft landed at back. B-25 FL-200 BX airborne for Anti-Submarine Patrol Number 3, aircraft completed the patrol and landed back. B-25 FR-371 BF airborne for leaflet dropping exercise, leaflets were dropped from 500 feet over Georgetown, Exuma and over Clarence Town, Long Island, aircraft completed the exercise and landed back. B-25 FV-949 FD is airborne on Anti-Submarine Patrol Number 3 and leaflet dropping exercise, leaflets were dropped over Hope Town, Abaco from 500 feet, aircraft located Great Sale Cay, and completed the patrol, landed back. B-25 FL-187 BC airborne for Anti-Submarine Patrol Number 3, aircraft landed back after an uneventful patrol. SS *G. W. Goethals* is bound north in Straits passing Palm Beach. **Convoy KN-263** is bound northeast in Straits with US Navy plane escorting. **Convoy NK-663** is bound north in Straits passing Palm Beach. SS *Jamestown* is bound south in Straits passing Palm Beach. SS *Monocacy* is bound northeast in Straits near Palm Beach. Tug USS *ATR-47* is bound north in Straits passing Port Everglades. USS *SC-1355* and USS *SC-1313* are bound south in Straits. *U-107* exits the area, bound east for France. NAS Great Exuma: A plane went to cover **Convoy GN-82**. An enemy submarine was sighted. A plane went on a search mission, sighting a 2-masted schooner. A plane landed at Great Exuma.

9th September 1943

B-25 FL-200 BX airborne for Anti-Submarine Patrol Number 3, found that radio contact could not be maintained, and the aircraft set course for base, exercise was not completed, air firing exercises carried out before the aircraft landed. B-25 FL-187 BC airborne for Anti-Submarine Patrol Number 3, exercise not completed, radio contact could not be maintained, landed back at base. B-25 FL-200 BX airborne for Anti-Submarine Patrol Number 3, located Great Sale Cay landed back with nothing to report. B-25 FR-381 DC is airborne for Anti-Submarine Patrol Number 1b, ferry *Jean Brillant* sighted with two escorts 5 nm from Nassau, patrol completed, landed back. B-25 FL-187 BC airborne for Anti-Submarine Patrol Number 3, located Great Sale Cay, landed back with nothing to report. B-25 FL-200 BX airborne for Anti-Submarine Patrol Number 3, located Great Sale Cay landed back with nothing to report. B-25 FV-949 FD is airborne to escort **Convoy NG**, convoy had 21 merchant ships and five escorts convoy contacted, signals exchanged, signal was received from base "No relief aircraft being dispatched," landed back at base. B-25 FV-954 FB airborne for escort duty to **Convoy NG**, carried out a search, met convoy, patrolled the track of the convoy as air coverage, aircraft FD took over from FB, which returned to base. B-25 FV-949 FD is airborne to escort **Convoy NG**, which was met, escorted composition, course and speed of the convoy were the same as when contacted, aircraft landed back. USCG cutter CG-*Unalga* and USCG cutter

CG-Rush are bound north in Straits passing Palm Beach. Tanker *Pan Pennsylvania* and tanker *Goldheels* are bound north in Straits passing Port Everglades. SS *William C. Atwater* is bound north in Straits passing Miami. **Convoy NK-563** is bound south in Straits with US Navy plane escorting. **Anti-Submarine Group 1** and tanker *Manuel Rionda* are bound southwest in Straits passing Nicholas Channel. NAS Great Exuma: A plane departed on a search mission and landed at Great Exuma. An Anti-Submarine search was secured. **Convoy NG-384**, with 21 merchant vessel and five escorts entered.

10th September 1943

B-25's FV-949 FD, FV-951 FC, FL-180 BA, FV-954 FB, FL-187 BC and FL-200 BX are airborne for Anti-Submarine Patrol Number 5, completed aircraft landed back. B-25 FV-949 FD airborne for Anti-Submarine Patrol Number 2f, patrol completed, landed back. Liberator FL-964 LD airborne for Satellite Field Exercise Kingsley in cooperation with HMS *Calder,* escorts was contacted on w/t [radio, or wireless telegraphy], a 3-island type radio sighted, a 9,000 to 10,000 ton tanker was sighted on course 190°, speed 12 knots, signals were exchanged, aircraft failed to locate the co-operating vessel and returned to base, landed at Satellite Field. Liberator BZ-869 MX is airborne from Satellite Field, for the Kingsley co-operation exercise, rendezvous with HMS *Calder,* ship not in sight, square search commenced, radio contact established, signals exchanged, ship circled and photographed ship left after visual signals were exchanged, aircraft set course for base, landed back to Satellite Field. Liberator BX-810 MA is airborne from Satellite Field, for the Kingsley co-operation exercise, radio contact established, aircraft homed onto the ship, ship sighted 6 nm distant, challenge signals exchanged, photographs taken of HMS *Calder,* aircraft left after exchanging visual signals, set course for base, landed back at Satellite Field. Liberator BZ-808 LY is airborne for the Kingsley co-operation exercise, set course for rendezvous, square search was commenced, sighted ship 5 nm distant, signals exchanged, triangular patrol commenced, tanker or steam ship of 5,000 to 6,000 tons is sighted, vessel appeared to be heavily laden, position of tanker was visually signalled to HMS *Calder,* ship left and signals exchanged, aircraft landed back at Satellite Field. B-25's FL-187 BC, FL-200 BX, and FV-965 FB are airborne for air Exercise Number 3, landed back after completion. B-25 FR-366 CX airborne for air Exercise Number 4, landed back after completion. *U-518* is bound north in Straits to Northwest Providence Channel. SS *William C. Atwater* escorted by USS *SC-1294* is bound north in Straits, passing Canaveral. **Convoy Motor Torpedo Boat Squadron-26, Group B**, USS *PC-1232*, and USS *SC-1299* are bound south in Straits towards Canaveral. USCG Cutter *CG-Blackrock* is bound south in Straits passing Canaveral. Minesweeper USS *YMS-179* is bound north in Straits passing Port Everglades. SS *Cross Keys* is bound south in Straits passing Port Everglades. USS *SC-1304* is bound west in Straits. NAS Great Exuma: Two planes went to cover **Convoy NG-384**.

11th September 1943

B-25's FV-952 FX, FR-393 DF, FR-363 CX, FL-187 BC, FL-200 BX, and FV-954 FB completed Patrol Number 5, individually landed back after completion. B-25 FV-949 FD airborne for Anti-Submarine Patrol Number 1b, set course for base as no radio contact could be maintained, contact established when over base, patrol resumed, bumpy conditions, experienced, returned to base. B-25's FL-200 BX, FL-187 BC, and FV-952 FX are airborne for air Exercise Number 7, landed back after completion. B-25 FV-951 FC is airborne for air Exercise Number 5, landed back after completion. B-25's FV-946 FF, FL-200 BX, FR-393 DF, and FR-366 CX, are airborne for air Exercise Number 7, landed back after completion. Liberators FL-964 LD, BX-869 MX, and BZ-810 MA are airborne from Satellite Field. Aircraft FL-964 LD landed at Bermuda with a fuel shortage, MX and MA landed at Satellite Field. **Convoy tanker *Hagan*** is bound southwest, entering area from north with US Navy plane escorting. USS *YMS-170* and USS *YMS-176* are south in Straits passing Canaveral. Tanker *Pine Ridge* is bound north near Matanilla Shoal, northwest Bahamas. Oiler USS *YO-143* and USS *YP-603* are bound west in Straits towards Miami. USCG Cutter *CG-Blackrock* is bound west in Straits. Destroyer escorts USS *DE-142* and USS *DE-143* are bound north in Straits. NAS Great Exuma: A plane departed for Guantanamo and detached three planes. USS *YAG-21* dropped anchor at Great Exuma.

12th September 1943

Night exercise cancelled due to unfavourable weather conditions. B-25 FV-952 FX airborne for air Exercise Number 7, completed the exercise and landed back. B-25 FV-946 FF airborne for air Exercise Number 7, completed the exercise and landed back. B-25 FR-366 CX airborne for Anti-Submarine Patrol Number 2, convoy sighted consisting_of one merchant ship and one escort signals exchanged, patrol completed, and the aircraft landed back. B-25 FV-949 FD airborne for Anti-Submarine Patrol Number 3 to Great Sale Cay, located small merchant ships and a tug towing a barge was observed, on course, landed after an uneventful patrol. B-25 FV-946 FF is airborne for Anti-Submarine Patrol Number 3, aircraft returned, heavy storm was encountered, took off again on a special patrol which was completed, a small steam ship sighted, completed the patrol and landed back. B-25 FV-954 FB airborne for Anti-Submarine Patrol Number 3, located Great Sale Cay, landed after completing the exercise. Liberator FL-964 LD left Bermuda and returned to Satellite Field. Liberator BZ-808 LY airborne from Satellite Field on the Kingsley exercise co-operating ship was contacted on W/T, aircraft had successfully homed on HMS *Calder,* signals were exchanged, at 7 nm photographs were taken, exercises was completed, aircraft left the ship and landed back. Liberator BZ-746 LM airborne from Satellite Field, radio contact established with HMS *Calder,* aircraft endeavoured to home on the ship, was unsuccessful, position signal was received from the ship via Bermuda, aircraft landed at Satellite Field. Liberator BZ-746 LM airborne from Satellite Field, a tanker and a destroyer were sighted, W/P contact was established with HMS *Calder,* homing exercise was successfully completed, ship was sighted 15 nm distance, photographs taken, exercise *was* completed, and the aircraft set course for base, landed at Satellite Field. Liberator BZ-758 LS was airborne from Satellite Field, radio contact was established, homing exercise was commenced and was completed satisfactorily, when the ship was sighted signals exchanged, dummy mast high attacks carried out, cobra exercise commenced, aircraft left ship and set course for base, landed at Satellite Field. SS

Arguello is bound south in Straits towards Canaveral. **Convoy NK-564** is bound south in Straits towards Canaveral with US Navy plane escorting. SS *Julesburg* is bound southwest in Straits. USCG *CG-Juniper* is bound south in Straits passing Palm Beach. SS *Ashtabula* is bound south in Straits. USS *SC-1299* is bound southeast in Straits passing Miami. USS *SC-1359* and USS *SC-1341* are bound southeast in Straits off Miami. USS *PC-1131*, USS *SC-1274*, and USS *PC-1132* are bound south in Straits. **Convoy KN-264** is bound north in Straits with US Navy plane escorting. SS *Comol Cuba* is bound northwest in Straits passing Nicholas Channel. **Convoy KG-657** is bound northwest in Straits passing Nicholas Channel with US Navy plane escorting. NAS Great Exuma: Two planes escorted **Convoy NG-384** at the designated position.

13th September 1943

B-25's FR-366 CX, FR-395 CO, FV-952 FX, FR-380 DB, FV-954 FB, FV-946 FF airborne for air Exercise Number 7, landed back after completion. Liberator BZ-757 LR airborne from Satellite Field for Jamaica, landed there, again airborne, detailed to carry out a convoy escort on return flight to base, endurance of aircraft was limited, signals sent to aircraft to return, landed back to Satellite Field. Tanker *Gulfland* is bound north in Straits towards Palm Beach. Tanker *Atlantic Coast* is bound south in Straits passing Palm Beach. **Convoy NK-564** is bound south in Straits with US Navy plane escorting. Tanker *Julesburg* is bound west in Straits. NAS Great Exuma: Two planes went to cover **Convoy GN-83**. Inter-island transport USS *YAG-21* got underway and departed. A plane went to cover **Convoy NG-385.**

14th September 1943

B-25's FV-952 FX, FR-380 DB, FR-366 CX, FR-378 CL, FV-946 FF airborne for night navigation exercise, landed back after completion. B-25 FV-952 FX, FR-366 CX, FR-380 DB airborne at different times for air Exercise Number 7, landed back after completion. B-25 FV-952 FX airborne for air Exercise Number 5, landed back after completion. Liberators BZ-764 LM and BZ-807 LV airborne for homing and navigation exercise, landed at Satellite Field having completed the exercise. Liberator BZ-757 LR and BZ-760 LT airborne for creeping line ahead navigation exercise, completed the exercise and landed back. USCG cutter *CG-Juniper* and USCG cutter *CG-Vigilant* are bound west towards Matanilla Shoal, northwest Bahamas. SS *William Byrd* is bound south toward Straits. Oiler USS *YO-143* and USS *YP-395* are bound southwest in Straits. USS *YAG-21* is bound northwest, passing Nassau. NAS Great Exuma: Four planes went to cover **Convoy NG-385**. A plane dropped anchor in Great Exuma's protected bay.

15th September 1943

Liberators BZ-764 LM and BZ-807 LV are airborne for homing exercise and navigation exercise, exercise was completed, aircraft failed to home on to each other, landed at Satellite Field with nothing to report from the exercise. Liberators BZ-757 LR and BZ-760 LT are airborne from Satellite Field for creeping line ahead, patrol was completed and the aircraft landed at Satellite Field. Night navigation exercises were carried out by B-25's FR-366 CX, FR-380 DB, FR-378 CL, FV-952 FX and FV-946 FF; patrol completed and the aircraft landed back. Later same aircraft were airborne on the same exercise and the crews completed the exercise, landed back. Tanker *Gulfmaid* is bound north in Straits passing Palm Beach.

SS *Edouard Jeramec* and escort are bound south in Straits towards Palm Beach with US Navy plane escorting. USS *Cowanesqe* (AO-79) and escort are bound southwest in Straits. USS *YAG-21* is bound west in Straits towards Miami. USS *SC-1051* is bound south in Straits. Cuban gunboat *CS-Patria* is bound west in Straits. SS *Courageous* is bound northwest in Straits, passing Nicholas Channel towards Key West. NAS Great Exuma: A plane departed on training loop and returned. USS *YP-42* dropped anchor at Great Exuma.

16th September 1943

Liberators BX-760 LT and BZ-810 MA are airborne for homing and navigation exercise, then returned to Satellite Field. Liberators BZ-811 MB and BZ-833 LW are airborne for creeping line ahead navigation exercise, landed at Satellite Field. Local night navigation flights were made by B-25's FK-164 AB, FK-163 AK, FK-166 AD and FL-208 BF, having completed the exercises the aircraft landed back. Further night exercises carried out by B-25's FR-366 CX, FV-951 FC, FH-384 CM, and FR-380 B, FR-376 CK and FV-946 FF. Special short night navigation exercises were carried out by aircraft of Number 1 Squadron. B-25 FR-378 CL airborne for calibration test for the Main Field, Direction Finding Station, exercise was completed, and the aircraft landed back. Liberator BZ-810 MA airborne from Satellite Field for Jamaica, landed there, returned to Satellite Field later same day. Liberators BZ-808 LY and BX-883 LW airborne from Satellite Field for homing and navigation exercise with Anti-Submarine Patrol Number 9, aircraft landed back at Satellite Field. Liberator BZ-760 LT and BZ-810 MA are airborne from Satellite Field creeping line ahead patrol, patrol was completed, but the aircraft were unable to carry out the exercises, landed at Satellite Field with nothing to report. Liberators BZ-883 LW and BZ-811 MB airborne from Satellite Field for homing and navigation exercise with Anti-Submarine patrol, aircraft completed the detailed patrol, homing exercise was unsuccessful, landed back MB, completed the exercise and set course for base, aircraft was due to reach base but failed to do so, radio compass was found to be inaccurate, homing procedure was requested from the Main Field Direction Finding Station but was not available, aircraft was unable to maintain its course and the captain decided to make for the base on various islands, returned on the southeast, captain then decided to make for the US coast. Aircraft homed in on the Miami beacon and was over Miami aircraft, then set course base landing there at Satellite Field. Tanker *Pan Pennsylvania* is bound southwest in Straits. SS *Appomattox* is bound northeast in Straits from Palm Beach. Tanker *R. P. Smith* is bound north in Straits towards Palm Beach. USS *SC-1306* is bound southwest in Straits towards Port Everglades. SS *Kenesaw Mountain* is bound northeast in Straits near Miami. SS *Monocacy* is bound southwest in Straits towards Cay Sal Bank. SS *Edouard Jeramec* is bound southeast in Straits with US Navy plane escorting. *U-518* enters area northeast of Bahamas, west to Northeast Providence Channel. NAS Great Exuma: A plane departed to cover **Convoy GN-84**, and a convoy of 25 merchant vessels, five escorts, entered area.

17th September 1943

Short night navigation exercises were carried out by B-25's FK-166 AD, FK-164 AB, FK-182 AU, and FK-170 AG of Number 1 Squadron. B-25 FV-946 FF is airborne on a calibration test, completed the exercise and landed back. Liberators BZ-808 LY and BZ-883 LW are airborne, completed the detailed navigation exercise, homing exercise unsuccessful, landed back to Satellite Field. Tug USS *Anacapa* and USS *SC-1319* are bound south in Straits towards Palm Beach. **Convoy NK-565** is bound south in Straits towards Palm Beach with US Navy plane escorting. Tanker *Shabonee* is bound north in Straits towards Canaveral. Tanker *Cross Keys* is bound north in Straits passing Palm Beach. Tanker *Gulfland* is bound south in Straits past Palm Beach. Tanker *Britamsea* is bound south in Straits passing Port Everglades. USCG *CG-Warren Edwards* is bound north in Straits towards Palm Beach. USS *SC-989* and USS *SC-1323* are bound northeast in Straits towards Northwest Providence Channel. **Convoy KN-265** is bound north in Straits with US Navy plane escorting. SS *Ignace Paderewski* is bound north in Straits. SS *Montgomery City* and Argentinian SS *Buenos Aires* are bound northwest in Straits passing Nicholas Channel. **Convoy KG-658** is bound southeast towards Nicholas Channel with US Navy plane escorting. *U-171* is northeast of Abaco, and reports aircraft in sight. NAS Great Exuma: Three planes went to cover **Convoy GN-84**. USS *YP-42* weighed anchor and departed. Two planes landed at Great Exuma.

18th September 1943

Course Number 8 completed, conversion course on B-25 Aircraft at, for conversion on to Liberator aircraft. Number 7 course completed their training at Satellite Field, and returned to Main Field. Number 9 course prepared to commence further training on B-25 aircraft having completed familiarization course. B-25 FV-951 FC is airborne to escort ferry *Jean Brillant,* escorted from Nassau Harbour, ship was left after a point, signals were exchanged, air observance provided by box patrol, visibility of 10 nm, patrol completed without incident. With the changing of courses little flying took place from either field. Tanker *Gulfmaid* is bound south in Straits passing St. Augustine. USCG Cutter *CG-Nike* is bound northeast in Straits. USS *YMS-177* is bound south in Straits passing Canaveral. USS *PC-607* is bound south in Straits passing Palm Beach. **Convoy GN-84** is bound north in Straits with US Navy plane escorting. **Convoy NK-565** is bound south in Straits with US Navy plane escorting. Oiler USS *YO-142* and USS *YP-602* are bound west in Straits. Anti-Submarine Group 3 is bound southeast in Straits near Nicholas Channel. USCG *CG-Yeaton* is bound south in Straits. NAS Great Exuma: Seven planes arrived and departed on a training loop.

19th September 1943

An informational signal was received from Guantanamo that a search was being conducted by Exuma for a U-boat, Commanding Officer decided that two Liberator aircraft be dispatched, Liberators BZ-810 MA, and BZ-757 LR are airborne from Satellite Field. Aircraft carried out a detailed creeping line ahead patrol, aircraft completed the patrol and landed back at Satellite Field. USS *YMS-251* is bound north in Straits near Palm Beach. Tanker *A. S. Hansen* is bound south in Straits towards Miami. Tug USS *Point Vincent* and USCG *CG-Colfax* are bound north in Straits with US Navy plane escorting. *U-518* is 400 nm south-southeast of Bermuda and reports the failed pursuit of a fast US Navy tanker. NAS Great Exuma: Four planes arrived and departed.

20th September 1943

Number 10 course commenced the familiarization course on B-25 aircraft with Number 1 Squadron. Number 9 course moved to Number 2 Squadron and commenced the navigation training course with B-25 aircraft. Number 8 course, which had moved to Satellite Field, commenced training on Liberator aircraft. B-25's FR-376 CK, FR-366 CX, FR-362 CA, FR-37I CF, and FL-187 BC on Exercise Number 1, aircraft completed the exercise and landed back. B-25's FR-376 CK, FL-200 BX, FR-366 CX, FR-384 CM, and FR-971 CF are airborne for Exercise Number 3, aircraft completed the exercise and landed back. B-25 FL-200 BX are airborne and landed, having completed air Exercise Number 7. B-25's FK-370 CF, FR-384 CM and FR-395 CO are airborne for Exercise Number 2, aircraft landed back, having completed the exercise. B-25's FR-382 CA, FR-366 CX, and FV-949 FD are airborne, landed back after completion of air Exercise Number 1. B-25's FR-376 CK, FL-187 BC, and FR-395 CO are airborne for air Exercise Number 4, landed back at base. B-25's FR-362 CA, FF-951 FC, FF-952 FX airborne and landed back after completing air Exercise Number 3. German Gunther Krech, commander of *U-558*, which was sunk on July 20 1943, is imprisoned near DC [he kills himself in prison]. USCG *CG-83417* is bound north in Straits towards Daytona. SS *Dudley M. Hughes* and tanker *R. P. Smith* are bound south in Straits towards Palm Beach. SS *Leon* is bound north in Straits passing Palm Beach. Tug USS *Point Vincent* is bound north in Straits passing Port Everglades. USS *PC-1204* is bound south in Straits. **Convoy KW-272** is bound southeast in Straits passing Nicholas Channel. SS *Spidoleine* is bound northwest in Straits passing Nicholas Channel. NAS Great Exuma: A plane went to cover **Convoy NG-386**. Four planes arrived and departed.

21st September 1943

US Naval Liaison Officer informed air Controller that a Baltimore for delivery to air traffic control (ATC) at Satellite Field was overdue, was ascertained that aircraft was coming from Morrison Field or Jacksonville. Duty Flying Control Officer at Satellite Field informed Controller aircraft had called on radio and that two-way communication had been established, nothing had been heard since that time, officer commanding aircraft had requested departure and ETA times be dispatched from Morrison Field. Aircraft on air Exercise Number 1 are instructed to look out for the missing aircraft, they landed without having seen anything. B-25's FF-395 CO, FF-362 CA and FV-952 FX are dispatched to search area of 30 nm radius of the airfield, landed back without sighting anything, signal received from Comgulfseafron requesting that search be made from Nassau by the No.111 OTU aircraft, signal was received from the Bahamas Government's Telecommunications department that a signal was intercepted from a ship off Inagua reporting that a crashed aircraft had been sighted on Castle Island. [On April 8th, 1943, a B-25 had crashed on this Island and had been located; it was thought the report was of this old crash, air traffic control (ATC) Operations Officer thought aircraft would not have been so far off course]. To augment the search party the Officer Commanding RAF Transport Command put two C-47 (Dakota) aircraft at the disposal of the Controller, Comgulfseafron instructed the Comfairship Group to search with three airships [US Navy blimps], on receipt of the signal reporting crashed aircraft on Castle

Island, signals were dispatched to Exuma requesting that reported crash be investigated, aircraft searched and confirmed crash was an old aircraft. B-25's FV-951 FC, FR-384 CM, FR-380 DB, and FV-949 FD are airborne to search neighbouring island and surrounding waters, returned to base having sighted nothing. Dakota FL-534 and Dakota FL-535 are airborne but returned to Satellite Field having sighted nothing, three airships airborne from Exuma, reports received via RAF TC that blimp had sighted a raft, and flying boat was proceeding to reported position. Commanding Officer instructed ASR launch HMS *P-339* be dispatched from Montagu Bay to the scene of the raft, flying boat instructed not to become waterborne as sea was choppy, RAF TC ascertained that Baltimore had been landed on beach on northwest point of Andros, dinghy had been put to sea to attract attention. ASR launch HMS *P-339* unable to approach near shore because of shallow water, launch instructed to stand-by and await high tide and then approach, signal received through US Navy that blimp had picked up two survivors and was landing on Main Field. Airship landed and survivors taken to US E. D. Hospital, signal sent to ASR boat HMS *P-339* to return to base. B-25 FL-187 BC is airborne for air Exercise Number 3, completed, landed back. B-25's FR-384 CM, FR-376 CK, FR-395 CO, FR-371 CF, and FV-952 FX are airborne for air navigation Exercise Number 1 and Number 2e, landed back after completion. B-25's FR-395 CO and FV-951 FC are airborne for air Exercise Number 3, completed, landed back. B-25's FR-376 CK, FV-952 FX, and FR-380 DB airborne for air Exercise Number 4, completed, landed back. Tug USS *Point Vincent* is bound north in Straits passing Canaveral. Destroyer escorts USS *DE-250* and USS *DE-248* are bound northeast near Matanilla Shoal, northwest Bahamas. **Convoy Motor Torpedo Boat Squadron-33** is bound southeast in Straits passing Miami. USS *SC-1353*, USS *SC-606*, and USS *YAG-19* are bound south in Straits. Tanker *R. P. Smith* and SS *Dudley M. Hughes* are bound west in Straits. Argentinian SS *Rio Dulce* is bound southeast in Straits passing Nicholas Channel. SS *Atlantic City* is bound northwest in Straits passing Nicholas Channel. **Convoy NG-386** is bound west in Straits past Key West. USS *Memphis* (CL-13) is bound northwest in Straits. USS *Marblehead* (CL-12) is bound southeast in Straits. USS *Christiana* (IX-80) dispatched by Commander, GS along with FairWing-12 from Walker Cay to join US Navy blimp USS *K-40* to patrol Northwest Providence Channel for an enemy submarine estimated to be north of Bahamas. NAS Great Exuma: Two planes went to cover **Convoy NG-386**. Two planes went on a search mission and then landed. A plane sighted a small merchant vessel.

22nd September 1943

Signal received from Comgulfseafron regarding U-boat sighting. It was decided that Liberator would be dispatched to search area of reported submarine, later augmented by B-25 aircraft. B-25 FV-954 FB is airborne on Anti-Submarine sweep creeping line ahead, patrol commenced, visibility 9 nm, patrol was to cover Northwest Providence Channel, oil slick observed, further investigation revealed that it was from a merchant ship in the vicinity, aircraft landed back with nothing to report. Liberator BZ-760 LT airborne from Satellite Field observed ferry *Jean Brillant,* creeping line ahead patrol carried out is bound east of Abaco, completed patrol and landed back to Satellite Field, nothing to report. B-25's FR-376 CK, FR-

384 CM, FV-951 FC, FV-949 FD, FR-380 DB airborne for air Exercise Number 2, landed back after completion. B-25's FR-376 CK, FR-384 CM, FR-380 DB, FV-949 ED, and FV-953 FX are airborne for air Exercise Number 4, landed back after completion. B-25's FR-362 CA, FR-373 CF, FL-187 BC, FV-949 FD, FR-384 CM, FR-380 LB airborne for air Exercise Number 1 and Number 2, landed back after completion. B-25's FR-381 DC, FR-384 CM, FV-351 FC, FR-362 CA, FR-371 CF, and FL-187 BC airborne for air Exercise Number 4, landed back after completion. Tanker *Chiloil* escorted by USS *SC-711* is bound south in Straits passing St. Augustine. USS *SC-1052* is bound south in Straits passing Canaveral. SS *Chickamauga* is bound north in Straits. SS *Indiana* and USS *SC-1335* are bound south in Straits. USS *YAG-19* is bound north towards Miami. SS *Lee S. Overman* is northwest in Straits near Nicholas Channel. NAS Great Exuma: Two planes went to cover **Convoy NG-386** and **Convoy GN-85**. Six planes arrived and departed. US-flagged merchant vessel *Mariposa* arrived and dropped anchor off Georgetown's dock.

23rd September 1943

B-25's FR-366 CK, FR-371 CF, FV-951 FC, FV-949 ED, FR-381 DC, and FR-362 CA are airborne, landed back after completion. B-25's FR-362 CA, FR-371 CF, FR-376 CK are airborne for air Exercise Number 1, landed back after completion. B-25 FR-384 CM is airborne for air Exercise Number 3, landed back after completion. B-25 FV-949 ED is airborne for air Exercise Number 2, landed back after completion. B-25's FR-362 CA and FR-371 CF are airborne for air Exercise Number 3, landed back after completion. B-25 FR-384 CM airborne for air Exercise Number 1, landed back after completion. B-25 FR-376 CK is airborne for air Exercise Number 3, landed back after completion. B-25 FV-951 FC is airborne for air Exercise Number 4, landed back after completion. B-25's FR-381 DC and FL-187 BC are airborne, landed back after completion. B-25 FV-951 FC is airborne for air Exercise Number 2, landed back after completion. B-25 FR-362 CA is airborne for air Exercise Number 2, landed back after completion. B-25's FV-949 ED, FR-381 DC, FL-187 BC, FL-208 BF, and FR-376 CK are airborne for air Exercise Number 3, and Number 4, landed back after completion. HRH the Governor approved re-naming the two airfields at New Providence. USS *SC-1357* and USCG *CG-Woodbine* are bound southeast in Straits passing Canaveral. **Convoy NK-566** is bound south in Straits towards Palm Beach, with US Navy plane escorting. USCG *CG-83340*, tanker *Chiloil* and USS *SC-711* are bound south in Straits, passing Palm Beach. Tanker *Seakay* is bound north in Straits passing Palm Beach with US Navy plane escorting. SS *Hogan,* tanker *Julesburg,* and tanker *Front Royal* are bound north in Straits passing Palm Beach. USS *YAG-19* is bound south in Straits. Tanker *Powder River* is bound northeast in Straits. Tug USS *Southwest Pass,* its tow, and USS *SC-1051* are bound southeast in Straits towards Nicholas Channel. *U-518* is bound east of Eleuthera west to Northeast Providence Channel. Windsor's yacht *Gemini* departs Nassau for Miami via Cat Cay, Bimini. NAS Great Exuma: Two planes went to cover **Convoy GN-85**. A plane departed on administrative flight to Nassau. A plane sighted a combat vessel and two US Navy destroyers. The US-flagged merchant vessel SS *Mariposa* departed.

24th September 1943

Local night navigation continued. B-25's FR-362 CA, FR-366 CK, FL-208 BF, FV-951

FC, FV-949 FD, and FR-381 DC are airborne for local night navigation exercise, and returned to base after completion. B-25's FR-376 CK, FL-187 BC, FR-362 CA, FR-371 CF, and FR-366 CK are airborne returned to base after completion. B-25 FR-384 CM is airborne for air Exercise Number 2, landed back after completion. B-25's FR-362 CA, FR-371 CF, FR-384 CM, FL-187 BC, FV-949 FD, FR-381 DC, and FR 366 CK are airborne for air Exercise Number 5, landed back after completion. The Chief Flying Instructor reported that the Officer Commanding the Army Transport Command at Windsor Field had left Lyford Cay to inspect the wreck of the Baltimore on Andros, yet the boat had not been heard off, was requested to dispatch aircraft to search for the boat between Nassau and Andros. B-25 FR-393 BF is airborne to search for the boat in the waters surrounding Nassau and Andros. It landed, but having made a thorough search, failed to locate it. Liberator airborne from Windsor Field to search for vessel, aircraft landed back without sighting anything. Air traffic Control (ATC) informed Controller that boat had been located 8 nm west of Nassau. SS *Little Big Horn* is bound southwest entering Straits from the north. **Convoy NG-387** is bound southwest, east of Inagua towards Windward Passage. **Convoy KN-266** is bound northeast in Straits near Palm Beach with US Navy plane escorting. USS *PC-1263* is bound northeast in Straits near Palm Beach. USS *SC-757* is bound south in Straits. **Convoy NK-566** is bound southwest in Straits with US Navy plane escorting. **Convoy GK-759** is bound northwest in Straits passing Nicholas Channel with US Navy plane escorting. Windsor's yacht *Gemini* spent night at Cat Cay, southern Bimini *en route* from Nassau to Miami. NAS Great Exuma: Flares were sighted by the NAS control tower. Two planes went to cover **Convoy NG-387**. A plane sighted a small vessel. A plane returned from an administrative flight to Miami.

25th September 1943

Local night navigation exercises carried out by B-25's FR-366 CK, FL-208 BF, FR-371 CF, and FV-949 FD are airborne, landed back after completion. B-25's FR-366 CK, FV-949 FD, FL-208 BF, and FR-381 DC are airborne for air Exercise Number 2, landed back after completion. B-25's FV-954 FB, and FL-187 BC airborne for air Exercise Number 5, landed back after completion. B-25's FV-949 FD, FR-366 CK, FL-208 BF are airborne for air Exercise Number 4, landed back after completion. B-25's FR-366 CK, FL-208 BF, FV-954 FB, FV-949 FD, FL-187 BC are airborne for air Exercise Number 5, landed back after completion. SS *Cross Keys* is bound south in Straits passing Canaveral. USS *PC-1236* is bound south in Straits towards Palm Beach. Tanker *Pan Pennsylvania* is bound north in Straits towards Palm Beach. USS *SC-1280* and USS *SC-1292* are bound north in Straits towards Palm Beach. SS *Monocacy* is bound north in Straits towards Palm Beach with US Navy plane escorting. USS *Moonstone* (PYC-9) is bound south in Straits passing Port Everglades. USS *SC-1468* is bound south in Straits passing Miami. USS *SC-1323*, USS *SC-989* and SS *Little Big Horn* are bound southwest in Straits. Ferry *Jean Brillant* is bound southeast from Northwest Providence Channel towards Nassau. *U-518* is in Straits, off Port Everglades. An outdated report from U-boat HQ (*B-Dienst*) has a U-boat attack halfway between Bermuda and Anegada, and could only have been *U-518* under Wissmann at that time, as U-518 was making an attack on a tanker. NAS

Great Exuma: A plane went to cover **Convoy NG-387**. A convoy of 33 merchant vessels and five escorts departed.

26th September 1943

No flying took place from Oakes Field, as runways were under repair. USS *SC-1335* is bound north in Straits passing Port Everglades. Ferry *Jean Brillant* is bound west in Straits from Bimini towards Miami with US Navy plane escorting. SS *Kenesaw Mountain* is bound southwest in Straits nearing Cay Sal Bank. *U-214* enters area south via Anegada Passage. *U-518* exits Northwest Providence Channel bound south in Straits, reporting that a US Navy patrol boat followed by a navy corvette-type destroyer caused an alarm. The patrol vessel had a small narrow and flat silhouette, with a very sharp outline, probably a US Navy PC-boat. At periscope depth, abeam to port, one tanker protected by a corvette, which moved along quickly. NAS Great Exuma: Three planes went to cover **Convoy GN-86**, and then landed at Great Exuma.

27th September 1943

B-25's FL-187 BC, FR-384 CM, FV-949 FD, FR-380 DB, FR-366 CX, FL-298 and BF are airborne for air Exercise Number 6, landed back after completion. B-25's FV-954 FB, FR-384 CM, FV-952 FX, FR-395 CO, and FV-949 FD are airborne for air Exercise Number 5, landed back after completion. SS *Lake George* is bound south into Straits. **Convoy NK-567** is bound south in Straits with US Navy plane escorting. SS *General Fleischer* is bound south in Straits, passing Canaveral. SS *Rapidan* is bound southeast in Straits, with US Navy plane escorting. SS *Jerome K. Jones* is bound south in Straits passing Port Everglades. Anti-Submarine Group 4 is bound north in Straits passing Port Everglades. USS *YAG-19* and USS *SC-1050* are bound south in Straits. Argentinian SS *Rio Dulce* is bound northwest in Straits, passing Nicholas Channel. *U-214* exits area via Anegada Passage east for France. *U-518* is in Straits off Miami, where it observes ahead to starboard the bright lights of Miami, aircraft beacons, and on both sides aircraft. U-boat is illuminated by aircraft. It surfaced, ran further to the centre of the Straits. Several times aircraft with position lights, however not at a sharp target angle to shoot at. Remained on the surface. A few aircraft, mainly transport aircraft from Miami to the Lesser Antilles. Then a destroyer, and twice they heard dull sounds and small concussions in the boat, like a soft impact, as if the boat had touched something. At periscope depth, they then find ahead to port an airship. NAS Great Exuma: Two planes went to cover **Convoy GN-86**. A plane sighted, one small merchant ship heading in a northwest direction. Plane JR4-F Amphibian departed on training. A plane departed to cover the navy tanker USS *Rapidan* (AO-18), with two escorts, and then returned back.

28th September 1943

Local night navigational training carried out by B-25's FV-952 FX, FV-954 FB, FR-395 CO, FR-378 CL, FL-187 BC, FR-380 DB, and FR-366 CX, which are airborne, then landed back after completion. B-25's FV-952 FX, FR-378 CL, FR-366 CX, FR-395 CO, and FL-187 BC are airborne for air Exercise Number 6, landed back after completion. B-25's FR-380 DB, FR-366 CX, FR-378 CL, FL-187 BC, FV-954 FB, and FR-362 CA are airborne for air Exercise Number 6, landed back after completion. USS *YMS-363* is bound southeast in Straits passing St. Augustine. **Convoy GN-86** is bound northwest north of Bahamas with US Navy plane escorting. **Convoy NK-567** is bound south in Straits passing Palm Beach with a

US Navy plane escorting. **Convoy KN-267** is bound north in Straits past Port Everglades with US Navy plane escorting. USS *YMS-180* and SS *Veracruz* are bound north in Straits passing Port Everglades. USS *PC-1211* and USS *YAG-19* are bound south in Straits. USS *SC-1305* is bound west in Straits. British Royal Navy motor minesweepers HMS *BYMS-59* and HMS *BYMS-162* with tanker *A. S. Hansen* are bound north in Straits. SS *Libertad* is bound northwest in Straits, passing Nicholas Channel. **Convoy KG-660** is bound southeast in Straits near Key West, with US Navy plane escorting. *U-518* is bound west from Nicholas Channel. NAS Great Exuma: A plane went to cover **Convoy GN-86.** Another plane departed to cover the navy tanker USS *Rapidan* (AO-18), with two escorts. Convoy of 18 merchant vessels, five escorts departed.

29th September 1943

B-25's FR-366 CX, FR-376 CK, FR-378 CL, FV-954 FC, and FR-395 CO are airborne for air Exercise Number 6, landed back after completion. B-25 FR-381 DC is airborne for air Exercise Number 6, returned without completion air engine trouble airborne later completed exercise. B-25's FV-954, FB, FV-952 FX, FR-366 CX, FR-376 CK, FR-395 CO, and FR-362 CA airborne for air Exercise Number 6, landed back after completion. USS *SC-1347* is bound south in Straits towards Canaveral. Tanker *A. S. Hansen* is bound north in Straits towards Canaveral. **Convoy Canaveral** is bound south in Straits towards Palm Beach. USS *YMS-363*, oiler USS *YO-69*, and USS *PC-1235* are bound south in Straits. SS *Indiana* is in Straits bound northeast towards Bimini. USS *YO-142* is bound east in Straits to Nassau. USS *Breckenridge* is bound north of Bahamas. **Convoy GK-760** is bound north in Straits near Nicholas Channel. SS *David Stone* is bound west in Nicholas Channel. *U-518* exits area west past Key West. NAS Great Exuma: A plane went to cover a convoy, and then arrived back. The navy tanker USS *Rapidan* (AO-18), with two escorts, entered. Plane JR4-F Amphibian departed on an administrative flight to Guantanamo.

30th September 1943

B-25's FV-954 FB, FR-376 CK, FR-371 CF, FV-952 FX, and FR-366 CX are airborne for air Exercise Number 6, landed back after completion. B-25's FV-954 FB, FR-371 CF, FR-366 CX, FV-952 FX, FR-380 DF, are FR-395 CO airborne for air Exercise Number 6, landed back after completion. B-25's FV-954 FB, FL-208 DF, FR-371 CF, FV-951 FC, FR-395 CO, and FV-952 FX airborne for night navigation exercise, area of exercise bounded by Bimini to west, Eleuthera to east, Andros to south, and Abaco to north. USCG *CG-83400* is bound south in Straits passing St. Augustine. USCG *CG-Agassiz* is bound south in Straits towards Canaveral. Tanker *Pan Georgia* and USS *SC-647* are bound north in Straits passing Canaveral with US Navy plane escorting. USS *YMS-178* is bound south in Straits towards Palm Beach. USS *YAG-19* is bound south in Straits. Tanker *Brandywine II* is bound southwest in Straits near Cay Sal Bank. SS *Charles M. Schwab*, SS *Sarazen* and SS *Oregon* are bound northwest in Straits near Nicholas Channel. Tanker *Chilbar* is bound southeast in Straits. NAS Great Exuma: Three planes went to cover **Convoy NG-388**. Plane JR4-F Amphibian departed and returned after training.

1st October 1943

B-25 FV-954 FB is airborne for Anti-Submarine Patrol Number 1, patrol was completed without incident and the aircraft landed back, latter part of the patrol was curtailed due to adverse weather conditions. B-25's FB-380 DF, FV-952 FX, FR-376 CK, FV-951 FC and FV-949 FD are airborne for Anti-Submarine patrols Number 1, patrols were completed individually, aircraft landed back. B-25's FV-952 FX, FR-376 CK, FV-951 FC, FR-395 CO, and FR-393 DF are airborne individually for Anti-Submarine Patrol Number 1, aircraft landed back after completion. Oiler USS *YO-142* and USCG *CG-83450* are bound west in Straits near Bimini. USS *YAG-21* is bound east in Straits towards Bimini. Russian RU *SC-720*, USS *SC-1344*, tanker *Front Royal* and tanker *Julesburg* are bound south in Straits. USS *YMS-170* is bound southeast in Straits passing Nicholas Channel. No.113: Windsor [Satellite] Field; B-26 Marauder FB-457 crashed on take-off due to co-pilot retracting under-carriage too soon. Aircraft burned, Category E. Crew were P/O F. R. McGill, pilot. F/Lt. J. Gellner, Co-Pilot. G. D. Boyd, WAG. R., WO2. J. T. D'Altroy, Navigator. P/O Boyd sustained slight burns; remainder of crew uninjured. NAS Great Exuma: A plane departed to cover **Convoy NG-388**. Another plane departed on an administrative flight to Nassau. A plane departed on a training loop, and then returned.

2nd October 1943

Five aircraft detailed for Anti-Submarine patrols, cancelled due to adverse weather conditions and uncertainty. B-25's FR-395 CO, FR-376CK, FR-371 CF, FV-949 FD, FB-393 DF, and FV-952 FX are airborne. Tanker *A. S. Hansen* is bound south in Straits passing Canaveral. **Convoy NK-568** is bound south in Straits passing Canaveral with US Navy plane escorting. SS *Caesar Rodney* is bound north in Straits towards Canaveral. USS *PC-1083* and USS *SC-647* are bound south in Straits passing Palm Beach. Ferry *Jean Brillant* and escorts are bound west in Straits towards Miami with US Navy plane escorting. Tanker *Cross Keys* is bound northeast in Straits. SS *Elihu Root* is bound west in Straits towards Key West. No.113: Ceremony of the dedication of the RAF Cemetery, Nassau. NAS Great Exuma: A plane went to cover **Convoy GN-87**. A convoy of 22 merchant vessels and five escorts entered the area. A plane departed on training, then departed again. Transport USS *YAG-21* dropped anchor at Great Exuma.

3rd October 1943

B-25's FR-393 DF, FV-952 FZ, FV-951 FC, FR-376 CK, and FR-381 DC are airborne for Anti-Submarine Patrol Number 1, aircraft landed back after completion. B-25 FV-949 FD is airborne for Miami and landed there, airborne again later, carried out Anti-Submarine patrol, landed back at base with nothing to report. Tanker *Occidental* is bound south in Straits near Canaveral. Tanker *Pine Ridge* and tanker *Monocacy* are bound south near Matanilla Shoal, northwest Bahamas. Tanker *Quemado Lake* is bound north in Straits near Palm Beach. USCG *CG-83474*, SS *Erastus Smith* and USCG *CG-83404* are bound south in Straits, passing Palm Beach. Tanker *Kenesaw Mountain* is bound north in Straits passing Bimini. Tanker *Pan Pennsylvania* is bound south in Straits towards Old Bahama Channel. USS *SC-1332* is bound south in Straits. **Convoy NK-568** is bound south in Straits with US Navy plane escorting. SS *Dudley M. Hughes* is bound northwest in Straits passing Nicholas Channel. NAS Great Exuma: Two planes went to cover **Convoy GN-87.** Three planes departed and three planes arrived at NAS Great Exuma.

4th October 1943

B-25 FR-376 CK is airborne and landed, having completed Anti-Submarine Patrol

Number 2b. B-25 FV-951 FC is airborne, carried out Anti-Submarine Patrol Number 2, saw HMS *Slinger* with tows. No aircraft were seen on the decks, destroyer was sighted, signals were exchanged, aircraft landed back with nothing to report. B-25 FR-393 BF is airborne and landed after completing Anti-Submarine Patrol Number 2e. B-25 FR-381 DC is airborne for Anti-Submarine Patrol Number 2d, tanker was sighted proved to be a straggler from a convoy, consisting of 22 merchant ships and five escorts that was sighted immediately after the tanker was seen. Convoy was circled and signals exchanged, radio contact had been lost and was re-established later. B-25 FL-187 BC is airborne and landed having competed Anti-Submarine Patrol Number 2. B-25 FR-381 DC is airborne for Anti-Submarine Patrol Number 2c, returned to base with radio failure. B-25 FR-374 CF is airborne for Anti-Submarine patrol 2f, landed back with nothing to report. B-25 FR-951 FC is airborne for Anti-Submarine Patrol Number 2d, during which a convoy was sighted consisting of 22 merchant ships and five escorts. Signals were exchanged with the convoy and aircraft landed back with nothing to report. B-25's FV-949 FD, and FE-378 CL were airborne for Anti-Submarine Patrol Number 2, aircraft landed back with nothing to report. B-25 FR-376 CK is airborne for Anti-Submarine Patrol Number 2, patrol is completed. A convoy of 22 ships and five escorts was sighted the ships were in seven columns of three ships. SS *Swivel* is bound south in Straits towards Daytona. Tanker *Little Big Horn* is bound north in Straits off Jacksonville. Tanker *Mary Ashley Townsend* is bound north in Straits towards Canaveral. Tanker *Winchester* is bound south in Straits towards Palm Beach. Tanker *Murfreesboro* is bound south in Straits passing Palm Beach. Destroyer USS *Howard* (DMS-7) is bound south in Straits passing Port Everglades with US Navy plane escorting. **Convoy KN-268** is bound north in Straits passing Miami with US Navy plane escorting. USS *YAG-19* is bound south in Straits. USS *SC-1314* is bound south in Straits. **Convoy KG-661** is bound southeast in Straits passing Nicholas Channel with US Navy plane escorting. **Convoy GK-761** is bound north in Straits with US Navy plane escorting. **Convoy GN-87**, troop transport USS *USAT-Frank Leisenring*, USS *SC-1280* and HMS *Trumpeter* are bound north in Straits. NAS Great Exuma: Two planes went to cover **Convoy GN-87**. A convoy of 22 merchant vessels and five escorts departed. Inter-island transport USS *YAG-21* dropped anchor and moved out. Six planes arrived and departed. One large merchant vessel was sighted.

5th October 1943

B-25's FL-187 BC, FV-951 FC, and FR-378 CL are airborne for Anti-Submarine Patrol Number 2, landed back with nothing to report. B-25 FR-366 CX is airborne for Anti-Submarine Patrol Number 2d, carried out the detailed patrol, coastal vessel sighted, aircraft landed back after an uneventful patrol. B-25 FR-381 DC is airborne for Anti-Submarine Patrol Number 2, two destroyers and a minelayer sighted, ships did not reply to aircraft's challenge. A convoy of 11 merchant ships and four escorts sighted, signals exchanged with escorts, flying beat observed to be on escort duty, landed back at base with nothing to report. B-25's FL-200 BF, FV-951 FC, and FR-381 DC are airborne for Anti-Submarine Patrol Number 2, landed with nothing to report. B-25 FR-366 CX is airborne for Anti-Submarine Patrol

Number 2f, landed with nothing to report. A B-25 was sighted dropping depth charges, and then to head towards land. Patrol completed without incident and the aircraft landed back. B-25 FR-384 CM is airborne for Anti-Submarine Patrol Number 2f; its radio contact stopped after a while. Signal was sent to CX requesting news of CM, aircraft had no news, so then aircraft were detailed for an ASR air search; they were to fly along the outward leg of the patrol as far as Rum Cay. Aircraft sent were B-25's FR-394 CN, FR-378 CL, and FL-187 BC; are airborne, signal sent to RAF detachment at San Salvador instructing the Widgeon aircraft to search in the Rum Cay area, and the track to its base for the missing aircraft. Marine ASR launch craft at Harbour Island and at the Montagu Bay, Nassau were ordered to stand by for ASR air-sea rescue, captain of CX returned from flight, reported seeing CM jettison its depth charge and head for land. Then a signal was received from captain of CM via Harbour Island that the aircraft had been ditched on Eleuthera's east shore and that all were safe, though one man badly injured. It was requested that the Walrus be dispatched, this was not possible as that aircraft was unserviceable, signal sent, missing aircraft had been located, aircraft landed back. Port engine of CM had failed and to rectify the drag, the captain had full boost to the starboard engine, which then began to show signs of seizing up. It was then decided to make a forced landing, depth charges were jettisoned, and the aircraft was ditched on a rock, 100 yards from the shore. Crew were able to leave the aircraft, rear gunner was unconscious with head injuries, and was extracted from the aircraft by the other members. Dinghy was launched and the crew paddled ashore, captain obtained assistance from a native constable at Bannerman Town, medical assistance was sent from Rock Sound, crew were taken to Rock Sound Hospital and there their injuries rare attended to. ASR launch HMS *P-89* at Harbour Island was instructed to proceed to the scene of the crash and to pick up the airmen vessel, was then to rendezvous at Ship Channel Cay, Exuma, with ASR launch HMS *P-191* from Nassau, which proceeded with a medical officer to Ship Channel Cay to render any assistance necessary. ASR launch HMS *P-89* failed to locate the wrecked aircraft, signal was received that all the crew were comfortable and in hospital at Rock Sound, so ASR launch HMS *P-89* signalled to proceed to Rock Sound to pick up crew. It was decided to send ASR boat HMS *P-191* to Rock Sound and if insufficient petrol for return, it was to await reloading. The Medical Officer arrived with ASR launch HMS *P-191* and arrangements for the crew to be brought to Nassau were made. Then HMS *P-191* left Rock Sound Hospital, one member injured docked, crew were taken to the RAF Station Hospital, one member was detained there. Some ships in **Convoy KN-268** split off bound north in Straits towards Daytona. USS *YMS-364* is bound south in Straits, passing Canaveral. USCG *CG-83398* is bound northeast in Straits from Palm Beach. HMS *Swivel* is bound south in Straits, passing Port Everglades. Destroyer USS *David Taylor* (DD-551) is bound north in Straits near Miami. Tanker *Lake George* is bound north in Straits near Bimini. USS *YAG-21* is bound northwest in Straits towards Miami. USS *PC-1133* and USS *PC-1237* are bound south in Straits. USS *YAG-19* and Anti-Submarine Group 3 are bound north in Straits. SS *Joshua W. Alexander* and SS *Jesse Applegate* are bound northwest near Nicholas Channel. NAS

Great Exuma: Three planes went to cover **Convoy NG-389**. A plane sighted two destroyers and one minelayer. A plane departed on an administrative flight to Miami. A plane landed at Great Exuma.

6th October 1943

B-25's FR-371 CF, FV-949 FD, FL-208 BP, and FR-366 CX are airborne for Anti-Submarine Patrol Number 3, aircraft landed back having completed the patrol and bombing practice. USS *SC-1064* is bound north in Straits passing Canaveral. USCG *CG-83400* is bound northwest in Straits from Walker Cay, north of Abaco. USS *PC-560* is bound south in Straits towards Palm Beach. British Royal Navy motor minesweepers HMS *BYMS-860* and HMS *BYMS-968J* are bound north in Straits, towards Palm Beach. Tanker *Crown Point* is bound south in Straits passing Palm Beach. Oiler USS *YO-142* is north of Nassau bound west towards Northwest Providence Channel. SS *Monroe* is bound north in Straits, passing Port Everglades. USS *YAG-21* is bound west in Straits towards Miami. Cuban *CS-32* and Cuban *CS-43*, USS *SC-1314* and USS *SC-1318* are bound south in Straits. SS *Santa Barbara* is bound north in Straits. NAS Great Exuma: Two planes went to cover **Convoy NG-389**. A convoy of 14 merchant vessel and five escorts departed. USS *YP-42* dropped anchor.

7th October 1943

Flying training cancelled due to lack of aviation spirit. **Convoy NK-569** is bound south in Straits towards Canaveral. SS *Monroe* is bound north in Straits passing Canaveral. Ferry *Jean Brillant* is bound south towards Nassau. Tanker *Brandywine II* is bound north out of Straits. USS *YAG-19* and USS *SC-1290* are bound south in Straits. Tanker *Atlantic States* is bound south in northern Straits. NAS Great Exuma: Three planes went to cover **Convoy GN-88** and then landed back. Three planes dropped anchor in Elizabeth Harbour, Exuma.

8th October 1943

Air cover was requested by the US Navy authorities for cruiser USS *Memphis* and its destroyer escort USS *Somers,* patrol was to be until the ships left the GSF. B-25 FV-949 FD airborne cruiser and destroyer sighted, signals were exchanged, aircraft covered the ship as detailed, vessels were left relieving, aircraft did not appear and the plane's Limit of endurance (PLE) of FD was reached; the aircraft set course for base, and landed. B-25 FR-366 CX is airborne to relieve FD, aircraft failed to meet FD, ships sighted, signals were exchanged and photographs taken, visual box patrol was conducted, aircraft left the ships and set course for base, landed back. Tanker *Spidoileine* is bound south in Straits towards Daytona. Tug USS *Point Arguello* and USS *SC-1064* are bound south in Straits passing Canaveral. **Convoy KN-269** is bound northeast near Matanilla Shoal, northwest Bahamas with US Navy plane escorting. **Convoy KN-269** breakoff is bound northwest in Straits towards Canaveral with US Navy plane escorting. SS *La Paz* and **Convoy NK-569** are bound south in Straits passing Palm Beach. Oiler USS *YO-142* and ferry *Jean Brillant* are bound west in Straits towards Miami. Tanker *Occidental* is bound south in Straits passing Miami. USS *SC-1295* and tanker *Fort Lee* are bound south in Straits. Tanker *Julesburg* and tanker *AthelRegent* are bound north in Straits. SS *Jerome K. Jones* is bound northwest in Straits passing Nicholas Channel. USS *Memphis* and USS *Somers* are bound southeast passing near Eleuthera with US Navy plane escorting. **Convoy GN-88** is bound northwest east of

Bahamas with US Navy plane escorting. SS *Edwin Christensen* is bound northwest. USS *General John Pope* (AP-100) is bound southwest with US Navy plane escorting. NAS Great Exuma: A plane went to cover **Convoy GN-88.** A convoy of 20 merchant vessels and four escorts departed. Plane J4F-2 Amphibian departed on training loop. A plane departed to cover transport. A sailboat was sighted.

9th October 1943

B-25's FR-371 CF, FL-208 BF, FR-393 DF, and FR-362 CA are airborne and landed, having completed the exercise. B-25's FV-951 FC, FR-366 CX and FR-378 CL are airborne for air Exercise Number 7, landed back. B-25's FV-946 FF, FR-371 CF and FR-394 CN are airborne, completed air Exercise Number 7, landed back. Liberators BZ-869 MX and BZ-883 LW are airborne from Windsor Field and landed back. Tanker *R. P. Smith* is bound south in Straits passing Canaveral. Tug USS *Point Arguello* and USS *PC-1233* are bound south in Straits passing Palm Beach. SS *Spidoleine* is bound south in Straits passing Port Everglades. SS *Knox* is bound north in Straits passing Miami. SS *Mary Ashley Townsend* is bound south in Straits passing Miami. USS *SC-1301* is bound south in Straits. SS *Esselen* is bound north in Straits. SS *Sam Houston II* and **Convoy GK-762** are bound northwest passing Nicolas Channel. NAS Great Exuma: A sailboat was mistakenly called in as a sighting of a submarine. A plane departed to escort a transport ship. USS *YP-42* lifted anchor and departed. Six planes departed and six planes arrived.

10th October 1943

Liberators BZ-883 LW and BZ-860 MX are airborne from Windsor Field on homing exercise and Anti-Submarine patrol, LW completed the patrol and landed, MX was unable to maintain contact and captain decided to set course for base, aircraft landed, homing exercise was not carried out. B-25's FL-208 BF, FR-362 CA, FR-393 BF, FR-371 CF, FR-394 CN, and FR-366 CX are airborne, exercise was completed and the aircraft landed. B-25 FR-381 DC is airborne for Anti-Submarine Patrol Number 3, patrol was abandoned due to bad weather, steady rain was experienced, visibility at 1,000 yards, aircraft landed. B-25's FV-951 FC, and FR-393 BF are airborne for air Exercise Number 7, aircraft completed the exercise and landed. A request was received from Comgulfseafron to provide air coverage for aircraft carrier HMS *Khedive,* vessel was out of the frontier, B-25 FV-946 FP is airborne to contact the HMS *Khedive* and to provide air cover. A schooner was sighted, aircraft tried to contact ship, but the vessel was not sighted and a search was commenced up track with visibility of 8 nm. Then the ship was sighted, signals were exchanged, track of the ship was patrolled, relieving aircraft arrived, ship was left, and the aircraft landed. B-25 FV-945 FB is airborne as a relief to FF on escort duty, signal was intercepted from FF giving the position of contact of the ship, contacted signals were exchanged, box patrol was commenced, aircraft covered the ship until recall signal was received, ship was left, aircraft landed at base. SS *Indiana* is bound south in Straits towards Canaveral. SS *Seakay* is bound north in Straits. USS *AM-341* is bound north in Straits passing Palm Beach. USS *SC-1320* is bound south in Straits passing Palm Beach. Destroyer escort USS *DE-129* is bound south in Straits. HMS *Khedive* is bound north in Straits with US Navy plane escorting. NAS Great Exuma: A plane landed after an 8.5-hour-long search. Two planes went to cover **Convoy NG-390**. A plane landed and departed from training loop. A convoy of 38 merchant vessel and five escorts departed.

11th October 1943

B-25's FR-362 CA, FR-378 CF, FR-393 DF, FR-394 CM, and FL-208 BF are airborne, completed the night navigation flight, landed. B-25's FV-951 FC, FR-362 CA, and FR-393 DF are airborne for air Exercise Number 7, aircraft recalled, weather was unfit to continue the exercise, aircraft landed. B-25 FV-951 FC is airborne, carry out meteorological check on the night navigation route between Nassau, Abaco and Bimini, weather reports were satisfactory, and aircraft landed. SS *Edward D. White* is bound north in Straits passing Canaveral. USS *AM-144*, USS *AM-143*; USS *SC-721*, USS *SC-1356*, USS *SC-1074*, and USS *SC-1287* are bound south in Straits towards Canaveral. Tanker *Monocacy* is bound north in Straits. Oiler USS *YO-126* is bound north in Straits towards Canaveral. USCG *CG-83404* is bound north near Matanilla Shoal, northwest Bahamas. USS *Y-21* and USCG *CG-83454* are bound east in Straits from Miami. Tanker *Rosebud* is bound north in Straits near Miami. USS *YO-143* and USS *YP-603* anchored due to weather while west near Northwest Providence Channel. USS *Lapwing* (AVP-1) is bound southwest in Straits past Key West. USS *YAG-19*, USS *SC-1360*, and USS *SC-1336* are bound south in Straits. SS *James Harlan* is bound north in Straits. Tanker *Pine Ridge*, destroyer escort USS *Brough* (DE-148), and USS *YAG-28* are bound north in Straits. NAS Great Exuma: A flight of two planes departed for Norfolk, Virginia.

12th October 1943

B-25's FR-366 CX, FR-394 CN, FL-208 BF, FR-362 CA, and FR-378 CL are on night navigation exercise, aircraft landed having completed the exercise. B-25's FR-378 CL, and FV-941 airborne for air Exercise Number 7, aircraft completed exercise, landed. B-25's FR-366 CX, FR-394 CN and FR-362 CA are airborne for air Exercise Number 7, exercise was completed, and the aircraft landed. B-25's FR-381 DC, FV-944 FF and FV-954 FB airborne for Anti-Submarine Patrol Number 3, aircraft completed the patrol without incident and landed. Liberators BZ-869 MX, and BZ-811 MB are airborne from Windsor Field, creeping line ahead patrol, aircraft landed. Liberators BZ-757 LR and BZ-746 LM are airborne from Windsor Field, completed the Anti-Submarine patrol, landed. SS *James Harlan* is bound north in Straits towards St. Augustine. USS *YMS-346* is bound south in Straits passing Canaveral. **Convoy NK-570** is bound south in Straits passing Canaveral with US Navy plane escorting. USS *SC-1296* is bound south in Straits towards Palm Beach. Tanker *Chiloil* is bound south in Straits passing Palm Beach. SS *Murphreesboro* is bound north in Straits near Palm Beach. Destroyer USS *Howard* (DMS-7) is bound north in Straits near Palm Beach with US Navy plane escorting. USS *YO-143* escorted by USS *YP-603* are bound southeast towards Nassau. SS *Winchester* is bound north in Straits near Miami. USS *YAG-19* and USS *PC-607* are bound south in Straits. USS *SC-1306* and USS *SC-1319* are bound southwest in Straits. **Convoy KW** and **Convoy KN** are bound southeast in Straits with US Navy planes escorting. Destroyer escort USS *Rich* (DE-695) is bound north in Straits. SS *Spencer Kellogg* is bound northwest in Straits, passing Nicholas Channel. NAS Great Exuma: A plane went to cover **Convoy GN-89**. A convoy of 34 merchant vessels and four escorts were sighted. A plane took administrative flight to Miami. A plane searched Darby Island, in the Exuma Cays, then lost altitude and provided negative results.

13th October 1943

B-25's FV-946 FF, FR-394 CN, FR-395 CO, FL-208 BF, FR-366 CX, FR-371, FR-378 CL, FR-362 CA, FV-951 FC, FR-381 DC, and FV-954 FB are airborne, completed the exercise and landed. Liberator BZ-869 MX is airborne from Windsor Field for Anti-Submarine patrol, creeping line ahead patrol was commenced, homing exercise was attempted, no radio contact could be established until aircraft was within 30 nm of base, aircraft landed back. Liberator BZ-746 LM is airborne from Windsor Field for Anti-Submarine patrol was carried out, aircraft was awaiting to commence a homing exercise on aircraft BZ-757 LR, exercise was not carried out, set course for base, landed. Liberator BZ-757 LR is airborne from Windsor Field, detailed Anti-Submarine patrol carried out, aircraft was forced to return to base. Liberator BZ-811 MB is airborne from Windsor Field, creeping line ahead patrol commenced, patrol was completed, and the aircraft landed. B-25's FV-954 FB and FV-946 FF are airborne, Patrol Number 3 was completed, and the aircraft landed. Liberators BZ-759 LP and BZ-811 MB are airborne, carried out Anti-Submarine Patrol Number 9, aircraft landed. B-25's FR-395 CO, FR-381 DC, FL-208 BF, and FR-371 CF are airborne, completed the exercise and landed. B-25's FV-951 FC, FV-952 FX and FR-362 CA airborne for air Exercise Number 7, landed having completed the exercise. Liberator BZ-808 LY airborne from Windsor Field. A convoy was sighted, signals were exchanged HMS *Essington* was contacted, exercise was completed, aircraft set course for base, and landed. Liberator BZ-812 MC is airborne for Kingsley Exercise, radio contact was established with the ship, homing exercise commenced, ship was not sighted, creeping line ahead patrol was commenced, ship was not located and the aircraft set course for base. Convoy of 33 merchant ships and five escorts is sighted, aircraft continues on its course to base when over Lyford Cay a smoke float and a patch of oil was sighted. Message was intercepted from a Liberator in the vicinity "Searching for survivors," ASR launch from Lyford Cay was observed to be approaching the oil patch, aircraft landed. A report was received from the guard at Clifton Pier, further report was received from DFCO of Windsor Field, that the missing aircraft is a B-25, belonging to RAFTC, which crashed in flames into the sea between Lyford Cay and Golding's Cay. Liberator on local flying circled the scene of the area. ASR launch was dispatched from Lyford Cay to the scene of the crash, and recovered the bodies of two of the occupants; it is thought that there were four occupants of the aircraft. Arrangements were made for a further search to be carried out at dawn. US Navy Ventura 29897 and USN Ventura 29901 landed from Boca Chica. USN Ventura 97 was airborne for escort duties to **Convoy NG-391**, which was met as escorted as detailed. Aircraft OL was airborne to relieve USN Ventura 97, which then landed having completed the patrol, USN Ventura 011 landed after completing the patrol. USN Ventura 29836 and USN Ventura 29867 landed from Boca Chica, then USN Ventura 67 was airborne to provide air escort to **Convoy GN-89**, which was located, end escorted until relieved by aircraft USN Ventura 36. Aircraft USN Ventura 67 landed after completing the escort, aircraft was relieved by aircraft USN Ventura 29901 and landed. **Convoy KN-270** is bound north in Straits passing Canaveral with US Navy plane escorting. Tanker *C. B. Watson* is bound south in Straits towards Palm Beach. USS *SC-1344* is bound north in Straits passing Palm Beach. USS *Aquarius* (AKA-16) is bound southwest in Straits towards Port Everglades. SS *Dwight W. Morrow* is bound south in Straits passing Port Everglades. **Convoy NK-570** is bound south in Straits passing Miami with US Navy plane escorting. Cuban *CS-22* and USS *YMS-346* are bound southwest in Straits. Tug USS *Cubits Cap* and USCG *CG-Vigilant* are bound north in Straits. SS *Sam Houston* is bound northwest in Straits near Nicholas Channel. **Convoy GN-89** is bound north in Straits. **Convoy NG-391** is bound south in Straits with US Navy plane escorting. No.113: Windsor Field; B-26 Marauder FB-454 dived into sea 1 nm due west of Windsor Field during training flight. Cause unknown. One engine, believed port, caught fire in the air. Crew were F/Lt. J. G. Owen, DFC, Pilot and Captain, F/O R. H. Barber, pilot under instruction, F/O D. Durward, pilot under instruction, F/O D. W. Cormack, pilot under instruction. Owen and Barber were killed, and their bodies recovered. Durward and Cormack are missing. NAS Great Exuma: A plane sighted destroyer escort USS *Buckley* (DE-51). Three planes went to cover **Convoy GN-89.** A plane sighted a destroyer. Plane J4F-2 Amphibian returned from an administrative flight to Miami. A plane sighted a B-24 bomber.

14th October 1943

Night navigation exercises were carried out by B-25's FV-954 FB, FL-208 BF, FV-946 IF, FR-393 DF, FV-952 FX and FR-375 CJ. US Navy Ventura 29901 is airborne to cover **Convoy GN-89**, aircraft contacted the convoy and completed the detailed escort, aircraft was relieved by USN Ventura 29867, then USN Ventura 29901 landed at base, USN Ventura 29867 was forced to return to base with radio, USN Ventura 29901 was airborne to complete the escort, and landed. Liberator BZ-809 LZ is airborne from Windsor Field on a Kingsley Exercise, radio contact was established with the co-operating ship, visual contact was established, homing and Anti-Submarine exercise was completed as detailed, aircraft set course for base on the completion of the exercise and landed at Windsor Field. Liberator BZ-759 LP is airborne from Windsor Field for the Kingsley Exercise. A tanker was sighted 16 nm distant, ship was of 7,000 tons, coloured grey, engines aft, with the bridge forward amidships, tanker signalled the aircraft "Can you stand by?" tanker was moving very slowly and starting to turn to starboard, tanker signalled "Can you send help?" was ascertained that the tanker was in distress and the aircraft agreed to send for assistance. Position of the tanker was passed to base with a request for instructions, patrol was carried out around the tanker, signal was sent to the aircraft to remain in position with the tanker until relieved by a US Navy Ventura aircraft, signal was received from base, aircraft returned to the tanker and signalled that it was leaving and a relief was coming, square search was carried out around the tanker, USN Ventura aircraft was observed to be approaching tanker. A US Navy Mariner aircraft was seen to circle the tanker, the Liberator left the tanker and set course for base, and landed at Windsor Field. Liberator BZ-846 LM was airborne for Kingsley Exercise, a signal was intercepted from LF to base "Stationary tanker requires help," aircraft set course for given position, tanker was sighted, B-24 was observed in area. Message was intercepted from base to remain with tanker until plane's limit of endurance; "PLE or relieved by USN Ventura." Contact was established with HMS *Essington* co-operating ship, yet it was not possible to maintain radio contact with the ship, due to disturbance. Anti-Submarine patrol exercise was carried out, but the ship was not sighted, aircraft returned to base, tanker was not sighted on the return flight. Liberator BZ-760 LT was airborne on Kingsley Exercise, oil pressure in the 4th engine dropped, and it was not possible to feather the propeller. Since the aircraft lost altitude, the aircraft set course for base and sent the SOS signal, depth charges were jettisoned, and the crew were instructed to prepare for ditching. A signal was sent to base "ditching," and they aircraft began to maintain height. Then the aircraft sent a signal to base to this effect, so the aircraft was instructed to transmit its position every 15 minutes, signal was sent to base stating that the 4th engine had failed and that the aircraft was over the northern tip of Eleuthera, then the aircraft landed. On the receipt of the SOS, signal preparations were made to dispatch two B-25 aircraft to the position given signal was sent to LP and LM on the receipt of the ditching signal from LT, giving the position of the Liberator LT and instructing the aircraft to search in that area. US Naval Air Stations at Guantanamo and Great Exuma were requested to give any assistance that might be required, and the RAF ASR marine craft sections were ordered to have all available boats standing by, B-25 FV-952 FX is airborne to search in the area in which LT was ditching, after it was ascertained that LT was still in the air and maintaining height, a signal was sent to FI to continue the exercise, aircraft landed back. Liberator BZ-759 LP is airborne from Windsor Field for creeping line ahead patrol, Liberator was sighted visually, homing exercise was not practicable, patrol was completed without incident and the aircraft landed. Liberator BZ-811 MB is airborne from Windsor field for creeping line ahead patrol. A US Mariner aircraft was seen to be escorting a convoy was impossible to ascertain the composition of the convoy aircraft returned to base without completing the patrol failing to maintain radio contact. B-25 FV-954 FB is airborne for Jamaica, landed there, and returned. Receipt of the signal from Liberator LP of the sighting of a distressed tanker, signal was dispatched to Comgulfseafron by the US Naval Liaison Officer, requesting instructions regarding the action to be taken regarding the distress tanker. Instructions were received to dispatch a US Navy Ventura aircraft to investigate the state of the tanker, and USN Ventura 29897 was airborne to proceed to the reported position and ascertain details of the tanker, signal sent to the Liberator LP that a USN Ventura was on its way as relief. Signal was received from the USN Ventura, that the tanker appeared to be alright, but that the ship was drifting, was ascertained from the crew of the Liberator LP that the tanker was stationary, and that the engines appear to have failed. USN Ventura 29901 was airborne to relieve the USN Ventura 97 on the escort to the tanker, USN Ventura 97 landed, USN Ventura 01 landed, was ascertained that the tanker was the US Navy tanker *Balls Bluff*. Liberator BZ-882 LW was airborne from Windsor Field to relieve USN Ventura 01 on escort to *Balls Bluff*, tanker was sighted stationary, the USN Ventura was observed to be escorting the tanker. A box patrol was commenced, aircraft left the tanker, no reply was received from the escort vessel on the transmission of a departure signal, and aircraft landed at Windsor Field. USCG Cutter *CG-E. M. Dow* grounded near Mayaguez, Puerto Rico, abandoned, entire crew removed by the USCG Cutter *CG-Marion. U-134* is bound east to France. SS *Crown Point* is bound north in Straits. **Convoy GN-89** is bound northwest with US Navy plane escorting. SS *Aconite* is bound north in Straits near Palm Beach. Tanker *Manuel Rionda* is bound west in Northwest Providence Channel. USS *Petit* and USS *Neunzer* are bound north in Straits passing Port Everglades. Tanker *Pan Pennsylvania* is bound south in Straits passing Bimini. USCG *CG-Lotus* is bound north in Straits. *U-518* re-enters area from Key West northeast in Straits. No.113: Windsor Field. The funeral of F/Lt. J. G Owen, DFC and F/O R. H. Barber took place in the RAF Cemetery, with full-service honours by S/Ldr. The Rev. E. J. Jay. NAS Great Exuma: Two planes went to cover **Convoy NG-391**, then returned back. A plane departed to cover the disabled US-flagged merchant tanker *Balls Bluff.*

15th October 1943

Station Commander issued a preliminary Hurricane Warning, and the decision was taken after consultation with the Meteorological Officer, tropical disturbance was moving along a line of Longitude 68 west, were indications that the storm might develop into hurricane force and turn towards the Bahamas, B-25 FR-393 DF is airborne to carry out a Meteorological flight to San Salvador and Turks Island to ascertain the state of the weather in this area. Aircraft landed, hurricane moved north, Station Commander again consulted the Met [Meteorological] Officer, decided that the hurricane had passed from the area and the Preliminary Hurricane warning was cancelled. US Navy Ventura 29867 is airborne to cover the disabled tanker, tanker was sighted with a destroyer escort, destroyer signalled that it was awaiting the arrival of tugs, aircraft landed, US Navy Ventura 29956 was airborne to carry out Anti-Submarine patrol to SS *Balls Bluff,* aircraft landed, ship had drifted 20 nm south of the position in which aircraft USN Ventura 67 had sighted it, USN Ventura 29897 was airborne and landed later, tanker had not changed its position and the destroyer was still in escort, signal was received from Comgulfseafron stating that a tug and a destroyer would tow the *Balls Bluff* to Charleston. As the range of the Ventura, was insufficient to cover the convoy and it was requested that the PTO cover the ships during daylight hours, it was agreed to use two B-25's and a Liberator. Liberator BX-722 LK was airborne for Anti-Submarine Patrol Number 9, patrol was completed, contact was established with LR, LK was alongside LR, aircraft continued the patrol and landed at Windsor Field. Liberator BZ-757 LR is airborne for Anti-Submarine Patrol Number 9, patrol was completed. Homing exercise was started on LK, aircraft contacted, LK landed at Windsor Field with nothing to report. Liberator BZ-759 LF airborne on Anti-Submarine Patrol Number 9, patrol was completed, aircraft landed after an uneventful patrol, homing exercise was not carried out, homing aircraft LD failed to transmit. Liberator FL-964 LD is airborne for Anti-Submarine Patrol Number 9, patrol was completed, homing exercise was not carried out, aircraft landed. Liberator BZ-809 LZ airborne, creeping line ahead patrol, patrol was not completed captain decided to return to base, radio contact could not be maintained, aircraft landed. Liberator BZ-869 MX is airborne for creeping line ahead patrol completed, aircraft landed back with nothing to report. B-25 FV-951 FC is airborne for Warner Robbins Field, southeast of Atlanta, and landed there. Tug USS *Point Sur* is bound south in Straits towards Daytona. USS *YMS-308* and USS *YMS-365* are bound south in Straits passing Canaveral. Army Dredge USS *Hoffman* and tanker *Occidental* are bound north in Straits, passing Port Everglades. Tanker *Fort Lee* and USS *YAG-19* are bound north in Straits. SS *Mount Evans* is bound southwest in Straits towards Key West. *U-518* re-enters area in Nicholas Channel, bound east. NAS Great Exuma: A plane departed for the US-flagged tanker *Balls Bluff*. A plane departed for an administrative flight to Nassau. A naval plane and Cdr. Goodwin arrived from Key West and departed for Nassau.

16th October 1943

US Navy Ventura BK-J01 is airborne to provide air cover for tanker *Balls Bluff*, tanker was met, was drifting, aircraft was relieved by USN Ventura BK-J97. BK-J01 landed, USN Ventura BK-J97 sighted the convoy, BK-J97 was due to be relieved by BK-J36 relieving aircraft was unable to meet the convoy, BK-J97 landed, BK-J36 was airborne, BK-J36 failed to meet the convoy and landed, B-25 FV-952 FX is airborne to cover tanker *Balls Bluff*, ship of 5,000 tons was sighted, vessel was a three island type steam ship, no reply was received from a signal to the ship requesting its name, aircraft commenced a square search for tanker *Balls Bluff*, ship was not located, aircraft returned to base. B-25 FR-381 DC is airborne to relieve FX; motor vessel sighted by FX was seen by DC; search was carried out, aircraft landed having failed to meet the tanker, nothing to report. USN Ventura aircraft returned to Boca Chica. B-25 FR-362 CA is airborne for night navigation Exercise Number 5, aircraft landed after exercise completion. US Navy Dakota 6091 landed from Washington with an Air Marshall. Tug USS *Point Sur* towing one section of *USS YFD-59* and tug USS *YMS-179* are bound south in Straits, passing Canaveral. USS *SC-1336* is bound north in Straits towards Canaveral. Destroyer escort USS *Petit* (DE-253) and destroyer escort USS *Neunzer* (DE-160) are bound northeast near Walker Cay, north Abaco. Army tanker USS *Y-24* and USS *SC-1473* are bound northeast in Straits towards Bimini. Oiler USS *YO-69* and barge with USS *SC-1467* escorting are bound southwest in Straits passing Miami. Minesweepers USS *YMS-308* and *USS YMS-365* are bound west in Straits towards Key West. *U-518* is north of Straits and Grand Bahama, bound east for France. NAS Great Exuma: A plane arrived from Guantanamo.

17th October 1943

B-25 FV-952 FX is airborne to escort the convoy. Tanker *Balls Bluff* was sighted, tanker was being towed by a tug, and escorted by a destroyer and a sub-chaser, signals exchanged between the ships, aircraft hovered the convoy until lost in a front, visibility in the cold front was zero, aircraft landed. B-25 FV-954 FB is airborne to relieve FX on escort duty, convoy was not met in the reported position, search was carried out on down track of the convoy, which was sighted, no reply was received to challenge signals, relieving Liberator was sighted, aircraft returned to base after handing over patrol. Liberator BZ-809 LZ is airborne to relieve aircraft B-25 FB, aircraft sighted the convoy, FB was observed on escort duties, escort plied to the challenge signal, box patrol, with visibility of 20 nm was carried out, convoy left when it was too dark to continue the patrol ships were proceeding on a course of 317°, speed 5 knots to 6 knots when left, aircraft landed. US Navy Dakota 6091 left for Washington. B-25 FR-372 CJ returned from Warner Robbins Field, near Atlanta. USS *SC-1307* is bound west in Straits towards St. Augustine. SS *Kota Baroe* is bound south in Straits, towards Canaveral. Tug USS *Point Sur* and tow are bound south in Straits passing Palm Beach. USS *Mindful* (BATR-48) is bound north in Straits passing Palm Beach. USS *J-861* and USS *J-981* are bound north in Straits towards Palm Beach. Oiler USS *YO-142*, USS *YP-408*, tanker *Gulfmaid*, tanker *Alpha*, and tanker *A. S. Hansen* are bound north from the Straits. **Convoy KN-271** is bound northeast in Straits with US Navy plane escorting. Destroyer escort USS *Buckley* (DE-51) is bound northeast in Straits. Barge USS *YO-69*, USS *SC-1467*, and USS *YMS-346* are bound south in Straits. **Convoy KG-664** is bound southeast in Straits passing Nicholas Channel with US Navy plane escorting. Tanker *Manuel Rionda* is bound north from Nassau for Northwest Providence Channel. NAS Great Exuma: Two planes went to cover **Convoy GN-90.** A convoy of 14 merchant vessels and five escorts sighted. A plane departed to Nassau and went back. A plane arrived and departed.

18th October 1943

B-25's FV-946 FF, FR-376 CK, FR-375 CJ, FR-371 CE, and FR-378 CL are airborne and landed, having completed air Exercise Number 1. B-25's FR-378 CL, FR-381 DC, FR-395 CO, FV-952 FX, FR-372 CG, and FR-376 CK are airborne and landed after completing air Exercise Number 3. B-25 FV-951 FC returns from Warner Robbins Field and lands at Oakes Field [aka Main Field]. Tanker *A. S. Hansen* is bound north in Straits, passing Daytona. Tug USS *Point Vincent* is bound south in Straits passing Daytona. Tanker *E. R. Kemp* is bound north in Straits passing Canaveral. **Convoy NK-571** is bound south in Straits passing Canaveral with US Navy plane escorting. USS *YMS-346* is bound north in Straits towards Canaveral. USS *SC-1348* is bound south in Straits towards Palm Beach. USCG *CG-83397* is bound northeast in Straits. Tanker *Manuel Rionda* is bound southwest in Straits towards Port Everglades. Tanker *Mary Ashley Townsend* and SS *Indiana*, and USS *PC-1232* are bound north in Straits passing Miami. USS *YAG-19* is bound south in Straits. SS *Kota Barde* is bound west in Straits near Cay Sal Bank. Army tug USS *Y-24* is bound north from Nassau. *U-518* is northeast of Bahamas, bound east for France. *U-214* re-enters area briefly via Anegada Passage, bound east for France. NAS Great Exuma: A plane went to cover **Convoy KG-664**. A plane escorted **Convoy GK-764.** Three planes went to cover **Convoy GN-90.** A plane departed to Nassau and landed at Great Exuma. A plane sighted a small island freighter and a fishing schooner.

19th October 1943

B-25's FV-946 FF, FR-371 CF, FL-208 BF, FR-394 CN, FR-378 CL, and FR-395 CO are airborne and landed having completed air Exercise Number 1. B-25's FL-187 BC, FR-378 CL, FV-946, FR-381 DC, and FR-371 CF are airborne, carried out air Exercise Number 3, on completion of the exercise they landed. US Navy Ventura 29901 landed from Boca Chica for convoy escort duties, aircraft was airborne and carried a patrol over **Convoy GN**, aircraft completed the detail and landed. US Navy Ventura 29897 is airborne to relieve USN Ventura

29901 on the escort duty to **Convoy GN**, aircraft completed the task and landed. USS *YMS-360* is bound south in Straits. Irish SS *Irish Elm* is bound south in Straits towards Palm Beach. USCG *CG-83404* is bound west from Walker Cay, north of Abaco. USS *YAG-28* is bound north in Straits, passing Port Everglades. Tug USS *Point Vincent* is bound south in Straits passing Port Everglades. USS *SC-1472* is bound north in Straits near Miami. **Task Unit TU-02910**, tanker *Cities Service Koolmotor*, SS *Thomas H. Wheeler,* and USS *PC-1238* are bound south in Straits. USS *Poole* and destroyer escort USS *Howard D. Crow* (DE-252) are bound north in Straits. SS *Leonidas Pole* is bound southwest in Straits towards Key West. SS *Joseph Storey* is bound south in Straits nearing Nicholas Channel. SS *James Fenimore Cooper* is bound northwest in Straits passing Nicholas Channel. **Convoy GN-90** is bound north and is north of Bahamas. *U-214* is bound east for France. *U-518* is northeast of Bahamas, bound east for France. NAS Great Exuma: A plane went to cover **Convoy GN-90**. Three planes landed and departed. A convoy of 19 merchant vessels and five escorts was sighted, and a plane went to cover **Convoy NG-392**.

20th October 1943

B-25's FR-378 CL, FV-946 FF, FL-200 BX, FL-187 BC, FV-952 FX and FR-371 CF are airborne for air Exercise Number 5 on completion of the exercise the aircraft landed. Tanker *A. S. Hansen*, destroyer USS *Noa* (DD-341), USS *Sumpter* (APA-52), **Convoy NG-392**, and Anti-Submarine Group 7 are bound south in Straits. SS *W. H. Ferguson* and tanker *Gulfbelle* are bound southeast in Straits near Canaveral. SS *Jose Bonefacto* is bound north in Straits. USS *SC-1360* is bound northeast in Straits. USS *YMS-347* is bound south in Straits towards Palm Beach. RPT's Group USS *B-271* and USS *B-276* are bound north in Straits towards Canaveral. USS *YAG-28* is bound west passing Walker Cay, north of Abaco. SS *Lyman Hall* is bound northeast in Straits near Port Everglades. SS *Kerry Patch* and tanker *AthelRegent* are bound south in Straits passing Miami. Free French Corvette *Aconite*, tanker *Gulfland*, SS *Leonidas Polk,* and SS *Josiah Macy* is bound north in Straits. Ferry *Jean Brillant,* USCG *CG-83455* and USCG *CG-83456* are bound south to Nassau. NAS Great Exuma: Three planes went to cover **Convoy NG-391**. A plane departed and returned. A plane landed at Great Exuma.

21st October 1943

B-25's FR-381 DC, FV-946 FF, FL-200 BX, FR-375 CL, FV-961 FC, and FR-376 CK are airborne, carried out air Exercise Number 5, landed on completion of the exercise. B-25 FR-381 DC is airborne and landed on the completion of the Exercise Number 4. B-25's FR-954 FB, FR-371 CF, FR-375 CJ, FR-378 CL and FL-200 BX are airborne for local night navigational exercise, was completed, and the aircraft landed. Tanker *Gulfbelle* and tanker *Gulfland* ram each other off Lake Worth Inlet, came to rest on wreck of SS *Republic* [burned for 52 days, wrecked near Hope Sound]. Tanker *E. R. Kemp* is bound south in Straits towards Daytona. SS *Port Freemantle* is bound north in Straits. USCG *CG-Dix* (W-136) is bound south in Straits near Canaveral. Destroyer escort USS *Rueben James* (DE-153), oiler USS *Casing Head* (YO-47) and tanker *Petrofuel* are bound south in Straits near Canaveral. SS *Josiah Macy,* and SS *Leonidas Polk* are bound north in Straits towards Canaveral. Tanker *A. S. Hansen* is bound south in Straits towards Port Everglades. SS *Joseph Jefferson* is bound north in Straits near Palm Beach. SS *Cape*

Icvak is bound north in Straits passing Port Everglades. USS *SC-687* is bound south in Straits passing Port Everglades. USS *Esselin* (AT-147), USS *SC-1357*, and USS *YAG-19* are bound south in Straits. SS *Edward D.* White, SS *El* Coston, and USCG *CG-Legare* (WSC-144) are bound north in Straits. Tanker *W. H. Ferguson* is bound west in Straits. USS *PC-1205* is bound southeast in Straits passing Nicholas Channel. NAS Great Exuma: A plane went to cover **Convoy NG-392.** A convoy of 19 merchant vessels and five escorts departed. A battle damage repair ship (ARB) departed for Darby Island, Exuma Cay to pick Air Commodore Lang, which then departed for San Salvador. A plane arrived from an administrative flight to Miami.

22nd October 1943

B-25's FR-375 CJ, FR-371 CF, FL-187 BC, FV-949 FD, FL-200 BX, and FR-362 CA are airborne, completed air Exercise Number 2, aircraft landed. B-25's FB-366 CX, FR-376 CK, FR-378 CL, FR-372 CG, FV-951 FC and FV-954 FB are airborne, carried out air Exercise Number 4, aircraft landed. B-25's FV-954 FB, FV-187 BC, FR-362 CA, FR-371 CF, FR-372 CG, and FR-378 CL are airborne for night navigation exercises, exercises were completed, and the aircrafts landed. USCG Cutter *CG-Legare* (WSC-144) is bound north in Straits passing Canaveral. SS *Asphalion* is bound south in Straits passing Canaveral. SS *El Coston* is bound north in Straits. USCG *CG-83397* is bound south in Straits towards Port Everglades. Tanker *E. R. Kemp* and SS *Sharpsburg* are bound south in Straits. SS *Sam H. Walker* is bound north in Straits with US Navy plane escorting. **Convoy KN-272** is bound north in Straits towards Bimini with US Navy plane escorting. SS *Rosebud* is bound west in Straits near Cay Sal Bank. **Convoy KG-665** is bound southeast in Straits passing Nicholas Channel with US Navy plane escorting. NAS Great Exuma: A plane went to cover **Convoy GN-91**. Three planes arrived and then departed. The battle damage repair ship (ARB) returned from San Salvador. A convoy of 22 merchant vessels and five escorts departed.

23rd October 1943

Request was received from the US Navy that air coverage be provided for **Convoy NG**. B-25 FV-954 FB is airborne, convoy was sighted, composed of 24 ships, including 12 tankers, and five escort vessels, signals were exchanged creeping line ahead. A sweep was carried out 100 nm ahead of the convoy, which is on course 320°, speed 8 knots, was sighted as the aircraft left. **Convoy GN**; this southbound convoy consisted of 21 ships, including 14 tankers, and 15 escort vessels, then the observing aircraft landed. B-25's FR-371 CF, FV-949 FD, FR-372 CG, FR-378 CL, and FV-952 are airborne for aircraft completed the exercise and landed. US Navy Ventura 2987 and USN Ventura 2830 landed from Boca Chica. Oiler USS *YO-47* is bound south in Straits towards Palm Beach. **Convoy NK-572** is bound south towards Palm Beach with US Navy plane escorting. Tanker *Occidental* is bound south in Straits passing Port Everglades. Tanker *Pan Georgia* is bound north in Straits passing Port Everglades. USS *PC-1134*, SS *Asphalion* and USS *SC-1347* are bound south in Straits. NAS Great Exuma: Four planes went to cover **Convoy GN-91**. A plane went to cover **Convoy NG-393**. A plane sighted small banana boat. A convoy of 24 merchant vessels and five escorts entered area. A plane returned to Great Exuma, due to Radar failure.

24th October 1943

US Navy Ventura 29897 airborne to provide air cover to **Convoy GN**, aircraft met the convoy, aircraft completed the patrol and handed over to the US Navy Ventura 29830, first aircraft landed, second aircraft relieved by USN Ventura 29897, USN Ventura 29830 landed, third aircraft landed on the completion of the air cover. Liberator BZ-883 LW is airborne for Morrison Field with Air Commodore, aircraft returned to base. USS *YP-448* is bound south in Straits towards Daytona. USS *SC-1473* is bound north in Straits passing Daytona. USS *SC-1337* is bound south in Straits towards Palm Beach. SS *Courageous* is bound north near Matanilla Shoal, northwest Bahamas. USS *SC-1338* is bound south in Straits passing Palm Beach. Ferry *Jean Brillant*, USS *SC-1342*, and USCG *CG-83455* are bound southwest past Northwest Providence Channel with US Navy plane escorting. SS *Thermo* and SS *Fort Niagara* are bound south in Straits towards Cay Sal Bank. SS *Fort Lee* is bound west in Straits passing Cay Sal Bank. SS *Dwight W. Morrow* is bound north in Straits. **Convoy GK-765** is bound northwest in Straits passing Nicholas Channel with US Navy plane escorting. **Convoy GN-91** is bound north, , to the north of the Bahamas. NAS Great Exuma: Three planes went to cover **Convoy NG-393**. One plane departed on a training loop. Two planes landed.

25th October 1943

B-25's FV-952 FX, FR-371 CF, FR-3V8 CL, FR-376 CK, FR-362 CA, and FR-366 CX are airborne and completed air Exercise Number 6, aircraft landed. USN Ventura 28935 and USN Ventura 2987 were airborne for Boca Chica. USCG *CG-Boutwell* is bound south in Straits towards Palm Beach. Tanker *Kittaning* is bound south in Straits passing Palm Beach. SS *Dwight W. Morrow* is bound north in Straits passing Port Everglades. USS *YP-42* is bound southeast in Straits passing Miami. USS *YAG-19* is bound south in Straits. Tug USS *Willet*, SS *Libertad*, and USS *SC-1001* are bound north in Straits. Tanker *Gertrude Kellogg* is bound northwest towards Key West. NAS Great Exuma: A plane departed to cover **Convoy NG-393**. Three planes arrived and departed from a training loop to Great Exuma.

26th October 1943

B-25's FV-997 FG, FW-104 FH, FR-371 CF, FR-366 CX and FV-949 FD airborne for aircraft completed air exercise and landed. The exercises detailed for the afternoon were cancelled due to the unfavourable weather conditions. USS *YOG-5* is bound south in Straits towards Daytona. SS *Mahana* and SS *Fort Lee* are bound north in Straits passing Palm Beach. SS *Red Canyon* is bound north in Straits near Miami. USS *PC-1236* and USS *SC-758* are bound south in Straits. SS *David Holmes* is bound north out of Straits. Tanker *Chester Valley* is bound south in in Straits past Cay Sal Bank. Tanker *Pan Massachusetts* is bound west in Straits near Cay Sal Bank.

27th October 1943

B-25's FR-378 CL, FV-954 FB, and FR-376 CK are airborne for air Exercise Number 6, aircraft landed having completed the air exercises. B-25's FR-362 CA, FR-366 CX, FR-378 CL, FR-381 DC, FV-949 FD, and FV-997 FC are airborne for Exercise Number 6, completed and the aircraft landed. USN Ventura 29872 and USN Ventura 29897 landed from Boca Chica. A defence exercise was conducted in co-operation with the civil authorities. The exercise was commenced and was completed. **Convoy NK-573** is bound south in Straits passing Canaveral with US Navy plane escorting. Tanker *David Holmes* is bound north in northern Straits. Tanker *Indiana* is bound south in Straits, passing Palm Beach. SS

Belva Lockwood is bound northeast in Straits near Port Everglades. **Convoy Port Everglades**, tanker *C. B. Watson*, and tanker *Gulfmaid* are bound north in Straits passing Miami. **Convoy KN-273** is bound north in Straits near Miami, with US Navy plane escorting. SS *Palmer* is bound south in Straits with US Navy plane escorting. SS *Cape Igvak* is bound north in Straits. **Convoy KG-666** is bound southeast in Straits, passing Nicholas Channel with US Navy plane escorting. SS *Andalien* is bound southeast in Straits. Library: F/Lt. W. W. Sheppard reported to the RAF unit as an Education Officer, and a Station Library has been created, with provision made for writing and reading rooms for the airmen. Furthermore, over six months, 20 or so classes have been organized for the airmen, with English, Mathematics, Foreign Language, and even matriculation through London Matriculation and London University. Trade training is also provided with general education. The Education Officer is supported by airmen librarians and voluntary instructors, and the base newspaper *Liberator* created and circulated. NAS Great Exuma: A plane went to cover **Convoy GN-92**, and a convoy of 17 merchant vessels and five escorts arrived.

28th October 1943

B-25's FR-362 CA, FR-378 CL, FV-997 FG, FV-952 FX, FR-366 CX, and FV-954 FB are airborne for air Exercise Number 6, aircraft completed the exercise and landed. B-25's FR-375 CJ, FV-949 FD, and FR-104 FH are airborne and landed, having completed air Exercise Number 6. USN Ventura 29897 is airborne to escort the light cruiser USS *Omaha* (CL-4), met ship and escorted until it was relieved by second USN Ventura, 29872. The first landed, second escorted and carried out detailed exercise, landed while on convoy escort, another convoy was sighted, convoy consisted of 15 merchant ships and five escorts, USN Ventura aircraft was airborne for Boca Chica. Aircraft of Number 1 Squadron B-25's FL-172 AJ, FL-183 AV airborne, completed local night navigation exercise and landed. SS *Willie Jones* is bound south in Straits towards St. Augustine. Tanker *Gulfmaid* is bound north in Straits towards St. Augustine. SS *C. B. Watson* is bound north in Straits passing Daytona. Minesweeper USS *YMS-347* is bound northwest in Straits. SS *Samuel Walker* escorted by USS *SC-1357* and USS *SC-1467* is bound north in Straits with US Navy plane escorting. Tanker *Sabine Sun* escorted by destroyer USS *Howard* (DMS-7) is bound north in Straits, with US Navy plane escorting. SS *Tongariro* is bound north passing Matanilla Shoal, northwest Bahamas. SS *Cape Blanco* is bound south, passing Matanilla Shoal, northwest Bahamas. USS *PC-1213* and tanker *Gulftide* are bound south in Straits towards Palm Beach. USS *Mango* (YN-19) is bound south in Straits passing Palm Beach. **Convoy NK-573** is bound north in Straits, passing Port Everglades. USS *YAG-19* is bound south in Straits with US Navy plane escorting. USS *Aimack* (AKA-10) escorted by destroyer escort USS *Snowden* (DE-246) and USS *SC-1052* are bound southwest in Straits. Tanker *AthelRegent* is bound northwest in Straits, passing Nicholas Channel. **Convoy GN-92** is north of Bahamas, bound north. NAS Great Exuma: Two planes depart to cover **Convoy GN-92**. A plane sighted a large merchant ship and then a medium-sized merchant vessel. The light cruiser USS *Omaha* (CL-4) was sighted.

29th October 1943

B-25's FK-175 AM, FK-166 AD, FK-171 AH, FK-183 AV, and FK-172 AK airborne for night navigation exercises, completed the exercises and landed. B-25 FR-366 CX is airborne for Anti-Submarine Patrol Number 1, patrol was completed and landed with nothing to report. B-25 FV-949 FD is airborne for Anti-Submarine Patrol Number 1a, uneventful patrol completed, and the aircraft landed, destroyer was sighted. B-25 FR-378 CL airborne, carried out Anti-Submarine Patrol Number 1b, aircraft landed after an uneventful patrol. B-25 FR. -362 CA airborne for Anti-Submarine Patrol Number 1, aircraft completed the patrol and landed, B-25 FV-952 FX airborne for Anti-Submarine Patrol Number 1b aircraft completed the patrol and landed. B-25 FR-381 DC airborne landed after completing Anti-Submarine Patrol Number 1. USS *YMS-309* is bound south in Straits towards Daytona with US Navy plane escorting. USS *Willet* (ARS-12) is bound south in Straits passing Daytona. SS *Willie Jones* and USS *YMS-361* are bound south in Straits passing Canaveral. Destroyer USS *Breckenridge* (DD-148), destroyer escort USS *Stewart* (DE-238), and tug USS *YO-74* are bound south in Straits passing Palm Beach. USCG *CG-83453* is bound northwest in Straits near Palm Beach. USS *YO-143* and USS *YP-395* are bound northwest towards Matanilla Shoal, northwest Bahamas. **Convoy GN-92** is bound northwest in Straits with US Navy plane escorting. Tanker *Sharpsburg*, tanker *John Stagg,* and tanker *Rosebud* are bound north in Straits. NAS Great Exuma: Three planes departed and returned.

30th October 1943

B-25 FR-362 CA is airborne for Anti-Submarine Patrol Number 1b, merchant ship was sighted, a convoy of 33 merchant ships and five escorts was sighted, aircraft completed the patrol and landed. B-25 FR-366 CX is airborne for Anti-Submarine Patrol Number 1c, aircraft landed after an uneventful patrol. B-25 FR-372 CG is airborne and landed after completing Anti-Submarine Patrol Number 1. B-25 FR-381 DC is airborne for Anti-Submarine Patrol Number 1b, a 3-island type merchant ship was sighted; ship was of 4,000 tons and was painted grey aircraft landed after completing the patrol. Request was received from Comgulfseafron to cover a southbound convoy, but it was found impossible to comply with the request to provide cover from dawn to dusk. It was arranged that an aircraft be provided for air coverage. B-25 FW-104 FH was airborne to provide air cover to the **Convoy NG** which was sighted, convoy consisted of 33 merchant ships and five escorts, convoy was escorted by a USN Mariner aircraft, which then landed. B-25 FV-949 FD is airborne to relieve FH on convoy duty, signal was sent to FH to return to base when relieved by USN plane, convoy was left 10 nm northeast of San Salvador, aircraft landed. B-25's FV-946 FF, FV-947 FG, FR-381 DC, FW-104 FH, FR-376 CK, FR-366 CX, and FR-372 CG are airborne, aircraft completed the exercise and landed. USS *YMS-366* and SS *Bostonian* are bound south in Straits towards Palm Beach. USS *YMS-309* and USS *PC-1181* are bound south in Straits. Royal Navy minesweepers HMS *J-375* and HMS *J-982* are bound northwest in Straits near Cay Sal Bank. **Convoy NG-394** is bound south from Inagua to Windward Passage. NAS Great Exuma: Two planes escorted **Convoy NG-394**. A plane departed. Inter-island transport USS *YAG-20*, ex- *Samaná*, dropped anchor at Great Exuma.

31st October 1943

An informational signal was received from Number 45 (Air Transport) Group to No.113 South Atlantic "King Instructing that all B-25 Aircraft are to be grounded pending investigations into defective ailerons" [Not clear if royal or admiral King]. Station Commander gave instruction that all B-25 Aircraft were to be grounded and inspections were to be carried out on the aircraft. Tanker *C. B. Watson* is bound south in Straits passing Canaveral. Tanker *Gulfmaid* is bound south in Straits passing Palm Beach. British Royal Navy minesweepers HMS *J-375*, HMS *J-982* and tanker *Occidental* are bound north in Straits. Oiler USS *YO-143* is bound west near Walker Cay, north of Abaco. SS *Knute Nelson* and tanker *Fort Lee* are bound north in Straits passing Port Everglades. British Royal Navy motor minesweepers HMS *BYMS-253* and tanker *Bostonian* are bound south in Straits. Tanker *Reginald A. Fessenden* is bound north out of Straits. SS *Joseph Storey*, tanker *Chilbar*, and SS *Unico* are bound northwest in Straits, passing Nicholas Channel. **Convoy GK-766** is bound northwest in Straits passing Nicholas Channel with US Navy plane escorting. NAS Great Exuma: Two planes went to cover **Convoy NG-394**. A plane went on a search mission. A plane sighted one merchant vessel and a destroyer escort. The B-25 bomber was not found. A plane arrived from an administrative flight. A convoy of 33 merchant vessels and five escorts departed.

1st November 1943

B-25 FR-381 DC airborne for Orlando, aircraft landed at Morrison Field, *en route* landed at Orlando. B-25's FV-951 FC, FV-946 FF, FR-376, CK, FV-997 FG, and FW-104 FH airborne and carried out air exercises Number 2 and Number 4, landed after completion of the exercise. USCG *CG-Marita* is bound south in Straits towards Daytona. SS *Tioloide* is bound south in Straits towards Daytona. SS *Fort Niagara* is bound north in Straits near Palm Beach. SS *Cerro Gordo* is bound south in Straits towards Bimini. SS *Kittaning* north in Straits passing Bimini. USS *YOG-5* escorted by USCG *CG-DIX* is bound south in Straits passing Port Everglades. USS *YO-69* and USCG *CG-83455* are bound southeast in Straits passing Northwest Providence Channel. Destroyer USS *Barney* (DD-149) is bound northwest in Straits. USS *YAG-19* and SS *C. B. Watson* are bound south toward Straits. Tanker *Mary Ashley Townsend* is bound north in Straits. **Convoy KN-274** is bound north in Straits with US Navy plane escorting. SS *Walter Hines Page* is bound north in Straits. **Convoy KG-667** is bound southeast in Straits passing Nicholas Channel with US Navy plane escorting. NAS Great Exuma: A plane went to cover **Convoy GN-93**. A plane departed for Guantanamo. Inter-island transport USS *YAG-20*, ex-*Samana*, weighed anchor and stood out. A plane departed and arrived.

2nd November 1943

B-25 FR-362 CA is airborne for Anti-Submarine Patrol Number 2d, aircraft landed at 3.13 pm, an oil slick was sighted running for 2 nm in an easterly direction, width 20 feet to 30 feet. B-25 FV-997 FG is airborne for Anti-Submarine Patrol Number 2d, aircraft completed the patrol and landed with nothing to report. B-25 FV-954 FB is airborne for Anti-Submarine Patrol Number 2d, aircraft landed after an uneventful patrol. B-25 FR-366 CX is airborne for Anti-Submarine Patrol Number 2e, convoy consisting of 13 merchant ships and five escorts was sighted, the escort vessels did not answer the aircraft's challenge, aircraft landed with nothing further to report. B-25 FV-951 FC is airborne for Anti-Submarine Patrol Number 2e, aircraft landed after completing an uneventful patrol. B-25 FW-104 FH airborne for Anti-Submarine Patrol Number 2E, convoy of 13 merchant ships and five escorts was sighted, aircraft landed back. B-25 FR-381 DC airborne from Orlando and landed at Oakes Field. The Operations Officer, air traffic control (ATC), reported that a distress signal had been intercepted from the Morrison Field, crash boat was thought that this might have originated from the life raft from the missing B-24, overdue at Borinquen, on 29th October, and it was requested that assistance might be given to search area. B-25's FV-997 FG, FV-949 DF, FR-366 CX, FV-946 FF, FV-954 FB, and FR-376 CK, are airborne, carried out thorough search without sighting anything, convoy of five ships sighted, ships were not investigated. Liberator BZ-760 LT is airborne from Windsor Field, standard creeping line ahead patrol, patrol completed without incident landed back. B-25 FR-395 CO is airborne to proceed to San Salvador and drop supplies to RAF, detachment operation was completed at aircraft landed. Tanker *E. R. Kemp* is bound north in Straits towards Daytona. **Convoy NK-574** is bound south in Straits towards Palm Beach with US Navy plane escorting. SS *Rioloide* and tanker SS *James Harlen* are bound south in Straits towards Palm Beach. USCG *CG-Marita* and USS *SC-1297* are bound south in Straits passing Palm Beach. Irish SS *Irish Elm* is bound northeast in Straits. SS *Joseph Storey* is bound north in Straits. Tanker *Cornell* is bound south in Straits passing Bimini. USS *YAG-20* is bound west in Straits passing Andros. The house boat USS *YHB-14* is bound northwest in Straits towards Port Everglades. USS *SC-1001*, USS *SC-1345*, and USS *PC-1243* are bound south in Straits. NAS Great Exuma: Two planes went to cover **Convoy *GN-93***. A plane went to cover **Convoy KG-667**. A convoy of 12 merchant vessels and five escorts departed.

3rd November 1943

Request was received from Comgulfseafron for air coverage to a northbound convoy from dawn to dusk. B-25 FW-104 FH is airborne to escort **Convoy GN-93**, convoy is sighted, consists of 13 merchant ships and five escorts, visual signals were sent to the senior naval escort, no reply was received. Box patrol was carried out around the convoy, ships were left after exchange of signals, when aircraft FD took over the escort FH landed at base. B-25 FV-949 FD is airborne as first relief for escort to **Convoy GN-93**, aircraft relieved FH, signals were exchanged with the escorts, box patrol was carried out and the ships were escorted, aircraft landed. Liberator BZ-758 LS is airborne from Windsor Field as second relief for **Convoy GN-93**, convoy was met and a "Met" signal was sent, air cover was provided by a creeping line ahead patrol and finally by a box patrol, convoy was left, creeping line ahead patrol was carried out, in the rear of the convoy, and the aircraft set course for base, aircraft landed. Comgulfseafron requested that air cover be provided for **Convoy SS *Dickerson***, which consists of two ships and two escort vessels. Liberator BZ-869 MX is airborne for Windsor Field aircraft failed to meet convoy, commenced a search for the convoy and flew up track of convoy, visibility falling rapidly from 2 nm to 1 nm, heavy rain and low clouds to sea level, returned to base, did not sight convoy. B-25's FR-376 CK, FR-395 CO, FR-381 DC, and FV-954 FB are airborne for air Exercise Number 5, landed after completion. B-25's FR-395 CO, FR-381 DC, FV-946 FF, and FR-362 CA are airborne for compass swinging exercise, landed after completion. B-25's FV-946 FF, FR-395 CO, FR-381 DC, FR-375 CJ, FR-393 DF, and FL-187 BC are airborne for night navigation exercise, landed after completion. SS *Andrew Marschal* is bound south in Straits towards Canaveral. USS *PT-552* and USS *PT-553* are bound south in Straits passing Canaveral. **Convoy GN-93** is bound northwest in Straits with US Navy plane escorting. Anti-Submarine Group 7 is bound south in Straits to Port Everglades. SS *Indiana* and SS *Charles Crocker* are bound north in Straits passing Bimini. USS *YO-69* is bound west passing Northwest Providence Channel. **Convoy NK-574** is bound south in Straits with US Navy plane escorting. SS *Rio Loide*, **Convoy NG-395**, and USS *YT-338* are bound south in Straits. SS *Quemado Lake* is bound southwest in Straits near Cay Sal Bank. USS *YAG-28* is bound east in Straits towards Andros. USS *Tomich* is bound north in Straits. NAS Great Exuma: A plane departed **Convoy KG-667**. Two planes went to cover **Convoy NG-395**. Six planes arrived and departed.

4th November 1943

Request received from Comgulfseafron to continue escort of **Convoy SS *Dickerson***. B-25 FV-997 FG is airborne, convoy consisting of one attack and one commando transport sighted with two escorts. Patrol carried out ahead of the convoy escort, continued until ships in a steady position, aircraft landed back. B-25's FV-949 FD, FV-954 FB, and FR-381 DC are airborne, landed back after completion of air Exercise Number 1. B-25's FR-362 CA and FV-946 FF are airborne for air Exercise Number 3, landed back after completion. B-25's FV-949 FD, FV-954 FB, FR-381 DC airborne, landed back after completion of air Exercise Number 2. B-25's FR-362 CA, FV-946 FF airborne, landed back after completion of air Exercise Number 4. B-25's FR-395 CO, FR-393 DF, FR-375 CJ, FV-954 FB, FV-997 FG and FV-949 FD airborne and landed after completing compass swing exercises. Liberator BZ-758 LS airborne from Window Field to carry out the homing navigation exercises, signal was sent to LS, diverting it to a reported position of a U-boat, signal was received, aircraft landed. Liberator BZ-6722 LK airborne from Windsor Field, broadcast signal regarding the reporting of a U-boat, aircraft was instructed to search in the area until dark and then to continue the detailed patrol, search was continued, nothing was seen of the reported submarine, heard LS faintly, but no contact was made, aircraft landed. Liberator BZ-746 LM airborne from Windsor Field creeping line ahead exercise was completed, homing signals were transmitted but no reply was received, exercise was completed and the aircraft landed. B-25's FR-362 CA, FV-946 FF, FR-381 DC, FR-393 DF, and FW-123 FJ are airborne for local night navigation exercises, landed having completed the exercises. SS *Umgent* is bound southwest in Straits. SS *Mahia* is bound north in Straits passing Palm Beach. SS *Fredericksburg* is bound north in Straits passing Port Everglades. SS *Palmer* is bound north in Straits towards Port Everglades. USS *PC-1138*, tanker *Andrew Marschall*, and tanker *Red Canyon* are bound south in Straits. Tanker *Pan Georgia* and tanker *Svithoid* are bound north in Straits. Ferry *Jean Brillant* is bound south towards Nassau. NAS Great Exuma: A plane departed for Guantanamo. A plane went to escort **Convoy NG-395**. Four planes arrived and departed. Admiral Whyler arrived aboard a plane.

5th November 1943

B-25 FR-393 DF is airborne for Anti-Submarine Patrol Number 2f, detailed patrol was abandoned due to low cloud, aircraft continued exercise and landed. B-25 FR-376 CK is airborne for Anti-Submarine Patrol Number 2d, detailed patrol was completed, two frigates were sighted, aircraft landed after exercise completion. B-25 FR-362 CA is airborne for Anti-Submarine Patrol Number 2k, aircraft landed. B-25 FR-375 CJ is airborne, was detailed to carry out Anti-Submarine Patrol Number 2e, aircraft landed back with engine trouble. B-25 FV-949 FD airborne for Anti-Submarine Patrol Number 2F, aircraft completed the patrol and landed with nothing to report. B-25 FW-123 FJ airborne for Anti-Submarine Patrol Number 2d, aircraft landed after an uneventful patrol. B-25's FW-123 FK, FV-952 FX, and FW-123 FJ are airborne and completed air Exercise Number 3, landed back after completion. B-25'a FR-381 DC, FR-393 DF are airborne and landed having completed air Exercise Number 1. B-25's FW-952 FX, FW-132 FK, and FV-946 FF airborne and landed between completed air exercise No.4. B-25's FR-381 DC and FB-393 DF landed after completing air exercise No.2, B-25's FW-132 FK, FR-372 CG, FW-123 FJ, FR-366 CX, FV-946 FX and FV-946 FF are airborne for local night navigation flight, landed after completion of the exercise. Liberators BZ-811 MB and BZ-746 LM are airborne from Windsor Field, competed the creeping line ahead patrol and landed with nothing to report. Liberator BZ-758 LS is airborne from Windsor Field for homing exercise, which could not be completed due to adverse weather conditions, aircraft landed back. Liberator BZ-757 LR is airborne from Windsor Field, attempt was made to contact LS, but adverse conditions prevented the completion of the exercise aircraft landed. SS *Svithoid* is bound north in Straits towards Canaveral. Tanker *E. R. Kemp* is bound south in Straits is towards Canaveral. SS *Chirripo* is bound southwest in Straits. Oiler USS *YO-69* escorted by USCG *CG-83455* is bound southeast in Straits, passing Fort Everglades. USS *SC-1320* and tug USS *Willet* (ARS-12) escorted by USCG *CG-Vigilant* are bound south in Straits. USCG *CG-83374* is bound north in Straits. Destroyer USS *Howard* (DMS-7) is bound southwest in Straits. SS *Umgeni* is bound southwest in Straits, with US Navy plane escorting. NAS Great Exuma: One plane arrived and departed from training loop. One plane arrived from Miami.

6th November 1943

B-25's FR-366 CX, FR-372 CG, FR-381 DC, FW-123 FJ, FW-132 FK, and FV-946 FF are airborne for air Exercise Number 7, landed after completion. B-25's FR-366 CX airborne for air Exercise Number 1, landed after completion. B-25's FW-132 FK and FV-952 FX are airborne for air Exercise Number 3, landed after completion. B-25's FR-376 CJ and FW-123 FJ airborne for air Exercise Number 1, landed after completion. B-25's FW-132 FK, FV-952 FX, FR-375 CJ, FR-372 CG, and FL-187 BC are airborne for air Exercise Number 2 and Number 4, landed after completion. Night navigation flights that were detailed to be carried out cancelled, weather conditions were found to be unsuitable, prospect of heavy thunderstorms in the area of the exercises. Liberator BZ-757 LR is airborne from Windsor Field, completed the creeping line ahead exercise, aircraft landed with nothing to report. Liberator B-2746 LM is airborne from Windsor Field to carry out creeping line ahead exercise, exercise was uneventful and the aircraft landed, these aircraft were recalled to base as the weather was expected to become unsuitable for flying. Liberator BZ-811 airborne from Windsor Field was recalled to base, due to deteriorating weather conditions aircraft landed. Liberator BZ-758 LS airborne from Windsor Field for homing exercise, aircraft did not receive the recall to base signal, aircraft MB relayed a signal to LS which was received as "returning to base," LS took no action on the recall signals and continued the navigation exercise, aircraft homing exercise was not carried out. Liberator FL-994 LA airborne from Windsor Field for Miami where it landed proceeded to Daytona. landing there. USS *SC-1361* and tanker *David Holmes* are bound southwest in Straits towards Palm Beach. USCG *CG-Marita* is bound south in Straits passing Port Everglades. USS *SC-1296* and tanker *E. R. Kemp* are bound west in Straits. **Convoy KN-275** is bound northeast in Straits with US Navy plane escorting. **Task Unit TU-0312** is bound northwest in Straits. SS *Chirripo* is bound southwest in Straits passing Cay Sal Bank. NAS Great Exuma: Two planes went to cover **Convoy GN-94**. A plane sighted a light cruiser, one medium-sized merchantman, 19 other merchant vessels, and a US Navy freighter. Three planes arrived and three planes departed.

7th November 1943

Liberator FL-970 LH is airborne from Windsor Field for homing exercise, homing exercise was commenced, no contact was established with BZ-883 LW, exercise was abandoned, aircraft landed. Liberator BZ-883 LW is airborne from Windsor Field for homing exercise, navigation exercise was completed, homing exercise was not carried out, aircraft landed. Liberator BZ-811 MB is airborne from Windsor Field on a creeping line ahead search exercise, convoy consisting of 17 merchant ships and five escorts sighted, escort end, aircraft exchanged signals, navigation exercise completed, aircraft landed. Liberator BZ-811 MA is airborne for creeping line ahead exercise, completed, aircraft landed with nothing to report. Liberator FL-994 LA returned from Daytona. SS *Sharon* and USS *SC-1345* are bound north in Straits passing Canaveral. USS *YMS-367* and USS *YMS-348* are bound south in Straits passing Canaveral. SS *George Washington* is bound northwest in Straits with US Navy plane escorting. Tanker *Pan Massachusetts* and SS *Hobart Baker* are bound northwest in Straits. **Convoy NK-575** are bound south in Straits passing Port Everglades with US Navy plane escorting. USS *PC-1233* is bound south in Straits. SS *Occidental* is bound southwest in Straits towards Cay Sal Bank. SS *American Press* is bound northwest in Straits nearing Nicholas Channel. NAS Great Exuma: Two planes went to cover **Convoy GN-94**. A plane arrived from an administrative flight. Another plane landed at Great Exuma.

8th November 1943

A request was received from Comgulfseafron to carry out a sweep 80 nm ahead of a southbound convoy. B-25 FV-951 FC is airborne for the Anti-Submarine sweep, convoy was sighted, composition of **Convoy NG** was 33 merchant ships and six escorts convoy escorted, patrol was abandoned due to adverse weather conditions, aircraft landed. B-25 FV-951 FC is airborne to sweep ahead of the convoy, which was sighted, signals were exchanged, convoy was escorted, aircraft landed. A request was received from Comgulfseafron to escort northbound convoy from dawn to evening. B-25 FR-366 CL is airborne to provide escort to **Convoy GN**, which was met, composition of the convoy was 17 merchant ships and five escorts, with a straggler 40 nm astern, no reply was received from the senior escort vessel to the challenge signal from CX, aircraft escorted the convoy, no reply was received from the escort vessels to the aircraft's departure signal, aircraft landed. B-25 FW-104 FH airborne, aircraft failed to meet the convoy in the estimated position at ETA thorough search was made to locate the convoy, aircraft landed having failed to locate the convoy. B-25's FL-200 BX, FW-123 FJ, and FV-949 FF are airborne for air Exercise Number 6, aircraft landed after completing the exercises. B-25's FR-372 CG, FR-393 DF, FW-132 FK and FR-375 CJ are airborne, carried out air Exercise Number 7, after completion the aircraft landed. Liberator FL-970 LH is airborne from Windsor Field carried out a creeping line ahead exercise, exercise was completed and the aircraft landed, a convoy consisting of 31 merchant ships and five escorts sighted, convoy again sighted during return flight, navigation exercise was completed and aircraft landed. Liberator BZ-869 MX is airborne from Windsor homing exercise, aircraft completed navigation exercise and landed, homing exercise was not carried out. Liberator BZ-931 MF airborne from Windsor Field for Miami and landed, aircraft left and landed at base, brought Air Marshal Sir William L. Welsh, KGB, DSC, AFC, to this unit on an official visit to this Station. B-25's FV-949 FD, FV-952 FX, and FR-362 CX are airborne and carried out night navigation exercises, aircraft landed. Liberators BZ-809 LZ, BZ-757 LR, and BZ-807 LV were airborne, aircraft landed at Boca Chica, were briefed there for the submarine exercises in co-operation with the US Navy, and returned to base. USS *PC-1280* and SS *Thomas Nelson* are bound south in Straits passing Canaveral. SS *George Berkley* is bound northwest in Straits. Tanker *Sharpsburg* is bound south in Straits, passing Bimini. Tanker *Mill Spring* is bound north in Straits passing Bimini. USS *YAG-19*, SS *Chancellorsville*, tanker *Bennington*, USCG *CG-83402*, and tanker *Fort Lee* are bound south in Straits. SS *Flomar* is bound northwest in Straits near Nicholas Channel. NAS Great Exuma: Four planes arrived from Nassau, and departed to perform a training loop.

9th November 1943

Request received from Comgulfseafron to continue air cover to southbound **Convoy NG-396**, it was requested that cover be provided from dawn. B-25 FW-104 FH is airborne, convoy consisting of 30 merchant ships and five merchant ships sighted, no reply was received to the challenge signal, aircraft continued the patrol, when the aircraft left it was on course 175°, speed 8 knots, aircraft landed. Liberator BZ-818 MA is airborne from Windsor Field, a B-25 was passed on its way to base from the convoy, which was contacted, no reply was received to the challenge signal although the correct procedure had been adopted, convoy was observed consisted of 27 merchant ships and five escorts, box patrol was commenced 20 nm astern, add 10 nm on either side, convoy was escorted, aircraft landed after completion of escort duty. Liberator BZ-807 LV, BZ-883 LW and BZ-746 LM are airborne, submarine exercises to be carried out with the co-operation of the US Navy at Key West, aircraft landed at Boca Chios and were briefed there for the exercises, aircraft returned to base after completion of exercise. B-35's FR-376 CL, FW-132 FK, FL-200 BX, FR-366 CX, and FV-

949 FD are airborne for air Exercise Number 6, landed back after completion. Lodestar aircraft left for Morrison Field with Air Vice Marshall, who had been visiting this unit. Night navigation exercises cancelled due to adverse weather conditions. SS *Fort Colville* and SS *Bostonian* are bound north in Straits passing Bimini. SS *Cerro Gordo,* destroyer escort USS *Leopold* (DE-319) and destroyer escort USS *Kirkpatrick* (DE-318) are bound north in Straits towards Bimini. SS *Charles R. McCormick* and USS *YAG-19* are bound northwest in Straits towards Miami. USS *PC-1239* is bound south in Straits. SS *Columbia Star* is bound east in Straits. **Convoy NG-396** is bound south in Straits with USN and RAF Nassau planes escorting. NAS Great Exuma: Two planes went to cover **Convoy NG-396**. Two planes departed on a training loop and returned to base.

10th November 1943

B-25's FR-366 CX, FV-949 FD, FL-200 BX, FW-104 FH, and FW-132 FK are airborne for air Exercise Number 6, landed back after completion. B-25's FK-163 AA, FK-166 AD, FR-395 CO, and FK-170 AG are airborne for local night navigation flights, landed after completion. Liberators BX-809 LZ, BZ-869 MX, BZ-811 MB, and FL-970 LH are airborne from Windsor Field, MX returned with engine trouble, did not proceed with exercise, remaining aircraft landed at Boca Chica, briefed for submarine co-operation exercise with US Navy submarine, returned to base after completion of exercise. Night navigation exercises cancelled due to adverse weather conditions. Tanker *Gulfmaid* is bound north in Straits passing Canaveral with US Navy plane escorting. USS *YMS-349*, USS *SC-687*, and USS *SC-1356* are bound south in Straits passing Canaveral. SS *William Dean Howell* is bound southwest in Straits. USS *PC-1182* is bound south in Straits passing Palm Beach. Tanker *Quemado Lake* is bound north in Straits near Palm Beach. SS *Fairfax* and tanker *Sappa Creek* are bound north in Straits passing Port Everglades. USS *PC-1172* is bound south in Straits. Tanker *Cornell* is bound northwest in Straits. USCG *CG-Crawford* and six US Coast Guard *CGR's* are bound north in Straits. No.113: Dakota FL-528, when returning from shuttle trip to Miami, taxied into a stationary truck causing considerable damage to port main plane, port outboard flap and aileron. Crew: F/O J. W. Narburgh, pilot, Sgt. J.G. Bruen, co-pilot, P/O. K. W., Shergold. (Australia), WAG. All uninjured. NAS Great Exuma: Two planes went to cover **Convoy NG-396**. A plane sighted one small freighter and a medium-sized merchant vessel. Three planes went on a training loop and returned.

11th November 1943

Liberators BZ-931 MF, BZ-758 LS, BZ-757 LR and BZ-807 LB are airborne from Windsor Field for Boca Chica, crews were briefed for an exercise in co-operation with the US Navy, exercises with the submarine were completed, aircraft returned to base. Air exercises detailed to be carried out were cancelled due to heavy seas, unsuitable for surface target towing, and weather conditions unsuitable for flying. B-25's FR-372 CG, FK-L66 AD, FK-172 AJ, FL-187 BC are airborne and landed after completing the local night navigation exercises. B-25's FW-123 FJ, FK-378 CL, FR-376 CK, FR-371 CF, FV-944 FF, FW-132 FK, FR-362 CA, FR-366 CX, and FV-997 FG are airborne and landed after completing uneventful night navigation patrols. Tanker *Petro Fuel* is bound south in Straits past Daytona. USS *YMS-379* is bound south in Straits towards Canaveral. SS *Gains Mills* and SS *Irwin Russell* is bound northeast passing Matanilla Shoal, northwest Bahamas. **Convoy KN-276** is bound north in Straits towards Bimini with US Navy plane escorting. USS *YAG-19*, USS *PC-1083* and USCG *CG-Ingham* are bound south in Straits. Tanker *A. S. Hansen* is bound north from Straits. SS *William Dean Howells* is bound southwest in Straits near Cay Sal Bank. SS *Rio Diamante* is bound west in Straits near Nicholas Channel. **Convoy KG-669** is bound southeast in Straits with US Navy plane escorting. SS *Minnesotan* and USS *YT-338* are bound north in Straits. NAS Great Exuma: Three planes departed to cover **Convoy GN-96**. A convoy of 15 merchant vessels and five escorts entered. A plane returned to Great Exuma due to oil leak. Five planes departed on a training loop and returned back to base.

12th November 1943

Request was received from Comgulfseafron for cover and a sweep 100 nm ahead of a southbound convoy. Liberator BZ-809 LZ is airborne to carry out the sweep to **Convoy GN**, which was contacted, and consists of 16 merchant ships and five escorts no reply was received from the escort vessels to the challenge signals, detailed air escort completed and aircraft landed, a 5,000 to 6,000 ton steam ship was sighted on course 185°, speed 8 knots. B-5 FV-946 FF is airborne to relieve Liberator LZ, convoy was sighted, recognition signals were exchanged, convoy was escorted, aircraft landed back after escort duty. Request was received for air escort to the northbound convoy, from Comgulfseafron 100 nm sweep ahead was needed. Liberator BZ-883 LH is airborne, convoy was sighted, consisted of 15 merchant ships and five escorts signals exchanged, patrol completed, aircraft landed back after duty, nothing to report. B-25 FV-951 EC is airborne for Anti-Submarine Patrol Number 2, aircraft landed after an uneventful patrol. B-25 FV-997 FC is airborne for Anti-Submarine Patrol Number 1e, aircraft landed with nothing to report from the patrol. B-25 FW-104 FH is airborne for Anti-Submarine Patrol Number 2f, a 4,000 to 5,000 ton merchant ship was sighted, patrol completed and the aircraft landed with nothing to report. B-25 FV-949 FD is airborne for Anti-Submarine Patrol Number 1, aircraft landed on the completion of patrol with nothing to report. Tanker *A. S. Hansen* is bound north in Straits towards St. Augustine. USS LCI Group 8 (seven Landing Craft Infantry's) and **Convoy NK-576** are bound south in Straits towards Palm Beach with US Navy plane escorting. **Convoy Key West-282** is bound north in Straits passing Palm Beach. Tanker *Petrofuel* is bound south in Straits passing Palm Beach. Tanker *Red Canyon* and tanker *Abiqua* are bound north in Straits towards Palm Beach. Submarine USS *Flasher* is bound south*west* in Straits towards Palm Beach with US Navy plane escorting. USS *Courageous* and tanker *Fredericksburg* are bound south in Straits. USS *YAG-19*, USS *YT-338*, and USS *YMS-379* are bound north in Straits towards Miami. USS *SC-1307* and USCG *CG-Juniper* are bound west in Straits. Tanker *Gertrude Kellogg* is bound southwest in Straits near Nicholas Channel. Argentinian SS *Rio Diamante* is bound west in Straits near Nicholas Channel. NAS Great Exuma: A plane departed to cover **Convoy KG-669**, 11 merchant vessels and five escorts. A plane went to cover **Convoy GN-95**. Inter-island transport USS *YAG-20* arrived at Great Exuma. A plane departed to cover **Convoy**

KG-669. A plane escorted **Convoy NG-397**. Two planes landed at Great Exuma.

13th November 1943

B-25 FR-366 CX is on Anti-Submarine Patrol Number 2e, patrol was completed, and the aircraft landed with nothing to report. B-25 FR-378 CL is airborne for Anti-Submarine Patrol Number 1, merchant ship was sighted vessel was of 4,000 tons and on course 230°, speed 10 knots, on the completion of the patrol the aircraft landed. B-25 FV-949 FD airborne for Anti-Submarine Patrol Number 1b, aircraft landed after an uneventful patrol. B-25 FV-951 FC is airborne for Anti-Submarine Patrol Number 1b, aircraft landed having completed the patrol without incident. B-25 FW-104 FH is airborne for Anti-Submarine Patrol Number 2 landed. B-25 FV-997 FG airborne for Anti-Submarine Patrol Number 1a, aircraft landed after completion, nothing to report from the patrol. Liberator FL-994 LA is airborne from Windsor Field and landed at Batista Field, Cuba. Tanker *Gulfmaid* is bound south in Straits towards Canaveral. USS *YMS-348* is bound north in Straits passing Palm Beach. USS *SC-1349*, USCG *CG-Vigilant,* and USS *YMS-350* are bound south in Straits passing Palm Beach. SS *City of Canberra* is bound south in Straits nearing Palm Beach. USCG *CG-Crawford* (W-134) is bound north in Straits passing Port Everglades. Spanish SS *Mar Cantabrico* is bound north in Straits passing Port Everglades with US Navy plane escorting. USS *YAG-28* is bound west, towards Nassau. USS *SC-1324* is bound south in Straits. **Convoy NK-576** is bound southwest, entering Straits with US Navy plane escorting. Tanker *Pan Virginia* and tanker *Petrofuel* are bound southwest entering Straits. USS *YMS-367* is bound southeast in Straits towards Nicholas Channel. USS *SC-1059* is bound northwest in Straits passing Nicholas Channel. **Convoy GK-769** is bound northwest in Old Bahama Channel with US Navy plane escorting. NAS Great Exuma: Two planes covered **Convoy NG-397**. A plane sighted a medium naval vessel. Two planes returned to Great Exuma.

14th November 1943

B-25's FK-393 DF, FR-362 CA, FV-954 FB, FR-366 CX, and F7-952 FX are airborne, completed air exercises Number 6, landed back. Request received from Comgulfseafron for coverage to the SS *Pastores* in the afternoon. B-25 FV-949 FD airborne to cover the steamship, and convoy was sighted, steamer was escorted by a destroyer, signals were exchanged between the escort vessel and the aircraft convoy was escorted, aircraft returned to base with nothing to report. USS *YMS 362* and *YMS 310* are bound south in Straits passing Canaveral. USS *PC-1085* is bound south in Straits passing Palm Beach. Tanker *Occidenta*l is bound north in Straits passing Palm Beach. SS *Empire Castle* and SS *Joseph Goldberger* are is bound north in Straits passing Bimini. Tanker *Lafcadio Hearn* is bound south in Straits passing Port Everglades. USS *SC-1348* and tanker *Gulfmaid* are bound south in Straits. SS *James Harlan* is bound north in Straits. SS *City of Canberra* is bound southwest in Straits.

15th November 1943

B-25's FV-952 FX, FR-376 CK, FR-366 CX, FV-946 FF, FW-104 FH and FR-375 CJ are airborne for air exercises Number 1, aircraft landed having completed the exercises. B-25's FV-954 FB, FR-393 DF, FR-362 CA, FV-946 FF, FL-200 BZ and FR-375 CJ airborne for air Exercise Number 2, aircraft landed after completion. B-25's FV-

123 FJ, FV-946 FF and FV-954 FB are airborne, landed having completed air exercise number 6. B-25 FV-949 FD are airborne for Anti-Submarine Patrol Number 2b, aircraft completed an uneventful patrol and landed. B-25 FV-951 FC airborne, carried out Anti-Submarine Patrol Number 2a, aircraft landed after completing the patrol with nothing to report. B-25 FR-381 DC airborne for Vernam Field, Jamaica aircraft landed there. B-25's FW-104 FH, FW-123 FJ, FV-954 FB, FR-393 DF, FR-378 CJ, and FR-376 CK are airborne and completed the night navigation exercises without incident and landed. SS *Irwin Russel* is bound south in Straits towards Canaveral. Tanker *Chiloil* and USS *YMS-368* are bound south in Straits passing Canaveral. SS *Chemung* is bound southwest in Straits with US Navy plane escorting. SS *Y. M. H. Ferguson* is bound southwest in Straits. SS *Alar* is bound northwest in Straits. SS *Fort Lee*, SS *Empire Castle*, and SS *Chancellorsville* are bound north in Straits. Tanker *A. S. Hansen* is bound south in Straits passing Palm Beach. SS *Menelaus* is bound north near Bimini. SS *Patria* and USS *YAG-19* are bound south in Straits. USS *YMS-362* and USS *YMS-310* are bound west in Straits. SS *Bennington* is bound east in Straits. SS *Jacob Thompson* is bound northeast in Straits. NAS Great Exuma: **Convoy GN-96**, of 15 merchant vessels and five escorts entered area.

16th November 1943

B-25's FV-954 FB, FR-362 CA, FR-375 CJ, FL-200 BX, and FV-946 FF are airborne for air Exercise Number 1, aircraft landed. B-25's FR-376 CK, FR-362 CA, FR-375 CJ, FL-200 BX, FV-946 FF, and FW-104 FH are airborne for air Exercise Number 2, aircraft landed. B-25's FR-375 CJ, FV-949 FD, FV-997 and FV-954 FB are airborne for Anti-Submarine Patrol Number 2, aircraft were recalled, the weather became unsuitable for flying. B-25 FR-381 DC returned from Jamaica, and landed. **Convoy NK-577** is bound south in Straits towards Daytona with US Navy plane escorting. Tanker *Jacob Thompson*, tanker *Sharpsburg* and tanker *Bennington* are bound north in Straits. Tanker *Pan Georgia* and tanker *Petrofuel* are bound south passing Matanilla Shoal, northwest Bahamas. SS *Oliver Evans* is bound north in Straits passing Bimini. Tanker *David Holmes* is bound south in Straits passing Bimini. The Miami Contingent of **Convoy KN-277** is bound southeast in Straits from Port Everglades. USS *YMS-368* is bound north in Straits passing Miami. USS *YAG-19* is bound south in Straits passing Miami. SS *H. M. Fredricksen* is bound north out of Straits. Cuban SS *Patria*, Cuban *CS-21*, Cuban *CS-22*, **Convoy SS *Chemung*** and tanker *William H. Ferguson*, and tanker *Chiloil* are bound south in Straits. Inter-island transport USS *Samaná* (YAG-28) is bound northeast in Straits. NAS Great Exuma: Three planes went to cover **Convoy GN-96**. A plane went to cover **Convoy GN-90**. A convoy of 15 merchant vessels and five escorts departed. Three planes went on training loop and returned. Inter-island transport USS *YAG-28*, formerly *Samaná*, dropped anchor.

17th November 1943

The morning flying details were cancelled due to unsuitable weather conditions. B-25's FR-362 CA, FR-381 DC, FL-200 BX, FV-949 FD, FV-946 FF, and FR-378 CL are airborne for air Exercise Number 5, aircraft were recalled as the weather at base had deteriorated, aircraft landed. Heavy rain swept the airfields and flying was cancelled for the remaining part of the day. Liberator BZ-758 LS is airborne from

Windsor Field and landed at Bermuda. **Task Group TG-20-21** (six USS LCI(L)'s) is bound south in Straits towards St. Augustine. SS *H. M. Friedrichsen* is bound north in Straits passing Daytona. USS *PC-1214* is bound south in Straits towards Daytona. Free French corvette SS *Renocule* is bound south in Straits passing Daytona. **Convoy KN-277** is bound north in Straits with US Navy plane escorting. Russian PT Group of six fast torpedo boats RU *PT-289*, RU *PT-290*, RU *PT-291*, RU *PT-292*, RU *PT-293*, and RU *PT-294* is bound northwest in Straits towards Canaveral. USS *YMS-379* and USS *YMS-349* are bound north in Straits towards Canaveral. SS *Libertad* is bound south in Straits passing Palm Beach. **Convoy NK-577** is bound south in Straits passing Palm Beach with US Navy plane escorting. USS *SC-1307*, USS *PC-1083*, USS *PC-1084*, and SS *Fairfax* are bound north in Straits passing Palm Beach. USS *YAG-20* are bound west in Straits from Andros. SS *George Davidson* is bound northwest in Straits towards Bimini. Anti-Submarine Group 6 is bound southeast in Straits towards Nicholas Channel. Tanker *Gertrude Kellogg* is bound northwest in Straits from near Nicholas Channel. *U-193* enters area west from northeast of Bahamas to Northeast Providence Channel. NAS Great Exuma: Inter-island transport USS *YAG-28*, ex-*Samaná* stood out to sea. A plane went on administrative flight. A plane went to depart **Convoy GN-96**. A plane went to cover **Convoy KG-670**. Two planes weighed anchor and stood out. A plane returned due to defective Radar.

18th November 1943

The Colonial government subsidized the struggling Bahamas Airways [BAL], ostensibly private. Request was received from Comgulfseafron to provide an aircraft to sweep an area 100 nm ahead of a southbound convoy. B-25 FR-376 CK is airborne to escort **Convoy NG**, which was met, composition was 24 merchant ships and four escorts, air cover was provided, aircraft landed with nothing to report from the patrol. B-25 FR-375 CJ was airborne for Anti-Submarine Patrol Number 2, patrol was uneventful and aircraft landed, a signal was intercepted on 500 kcs. The text of the message was indecipherable due to static, but it was thought that it contained 12 letters, aircraft was unable to obtain a fix on the signal, aircraft landed. B-25 FV-954 FB airborne for Anti-Submarine Patrol Number 2f, aircraft completed the patrol, convoy of 24 merchant ships and four escorts was sighted, aircraft landed with nothing to report. B-25's FR-946 FF, FR-378 CL, FW-123 FJ, FR-362 CA, FR-381 DC and FV-946 FF airborne for air Exercise Number 5, aircraft landed after completing the exercises. B-25's FR-362 CA, FR-386 CK, FR-371 CF and FL-200 BX were airborne for compass swinging exercises, aircraft landed on the completion of the exercise. Liberator BZ-722 LK airborne from Windsor Field for Miami *en route* for San Antonio aircraft landed at Miami. B-25 FV-997 FG airborne and landed after a flight to San Salvador. Night navigation exercises detailed for Number 2, Squadron were cancelled, weather was unsuitable for flying, local night flying training for Number 1 Squadron cancelled due to bad weather. Tanker *Delaware Sun* is bound south in Straits towards Palm Beach. SS *Windrush* is bound northwest in Straits. Tug USS *Willet* and USS *SC-1276* are bound north in Straits towards Palm Beach. Tanker *Mary Ashley Townsend* is bound south in Straits passing Miami. **Task Group TG-22.1**, of six USS *LCI(L)*'s and USS *YAG-19*, are bound south in Straits. Free French Navy corvette FFS *Renocule* is bound southwest in Straits. Destroyer escort USS *Menges* (DE-320) is bound northeast in Straits. **Convoy GK-770** is bound northwest in Straits near Cay Sal Bank with US Navy plane escorting. *U-129* enters area northeast of Bahamas, bound south towards San Salvador. NAS Great Exuma: A plane sighted the destroyer USS *Ingersoll*, with a small launch. A plane went to escort **Convoy KG-670**. A plane departed to cover **Convoy NG-398**. A plane arrived from Florida.

19th November 1943

Request was received from Comgulfseafron to escort ferry *Jean Brillant* until the ship docked at Nassau. B-25 FV-954 FB is airborne to escort found that the visibility was poor due to rain and the aircraft followed the Berry Islands, ship was sighted, and was escorted by a subchaser and a US Navy blimp, which left the convoy on the arrival of the B-25. Weather was bad and visibility low due to heavy rain, aircraft escorted the convoy and then landed back. It was decided not to send another aircraft on the escort, as the weather was unsuitable for flying. B-25's FR-362 CA, FR-123 FJ, and FR-376 CK are airborne for air Exercise Number 3, aircraft landed after being recalled due to unsuitable weather. Night flying training detail for Number 2, Squadron was cancelled due to adverse weather conditions. USS *YMS-180* is bound south in Straits passing Palm Beach. SS *Fredericksburg* is bound north in Straits near Palm Beach. SS *Dan Georgia* is bound southwest in Straits nearing Bimini. Ferry *Jean Brillant* is bound east in Straits, passing Northwest Providence Channel with US Navy plane escorting. USS *YAG-19* and tanker *Mount Evans* are bound north in Straits. Tanker *Delaware Sun* is bound west in Straits. NAS Great Exuma: Two planes departed to cover **Convoy NG-398**. Three planes went to search for an enemy submarine, and then landed. A plane departed to search the area, with negative results. A plane arrived from an administrative flight to Miami.

20th November 1943

B-25's FV-946 FF, FK-123 FJ, FR-375 CJ, FR-362 CA, FR-376 CK, FV-997 FQ, and FR-381 DC airborne for air exercises Number 3, aircraft landed. Air Exercise Number 4, completed by B-25's FV-997 FG, FV-946 FF, FR-381 DC, FR-375 CJ, FR-362 CA, and FR-376 CK. Air Exercise Number 7, completed by B-25's FV-997 FG, FL-200 BX, FR-371 CF, FR-378 CL, and FR-362 CA. Request received from Comgulfseafron to cover ferry *Jean Brillant* from Nassau to Miami, cover required from noon to dark. B-25 FV-954 FB airborne for the escort, ship escorted, aircraft completed detailed patrol and landed. US B-25 87471 and US B-25 81474 are delivered from Kansas City. Night navigation exercise for Squadron Number 2 cancelled due to adverse weather conditions. SS *H. M. Fredricksen* is bound south in Straits towards Canaveral. USS *SC-1276* is bound south in Straits passing Canaveral. USS *YMS-351* is bound south in Straits near Canaveral. SS *Cape Igvac* and tug USS *Anacapa* are bound south in Straits passing Palm Beach. Submarine USS *Angler* is bound south in Straits passing Palm Beach with US Navy plane escorting. USS *YT-338* is bound southwest in Straits. Destroyer escort USS *Ingersoll* (DE-652) is bound northeast in Straits with US Navy plane escorting. Navy tug USS *Montcalm* (AT-39) is bound north in Straits. NAS Great Exuma: A plane went to cover **Convoy NG-398**, of 24 merchant vessels and five escorts. A plane went to cover **Convoy GN-**

97, which then departed. Three planes arrived and then departed.

21st November 1943

B-25 FV-946 FF is airborne on Anti-Submarine Patrol Number 3, completed the patrol and landed. B-25 FV-997 FG is airborne for Anti-Submarine Patrol Number 3 completed the patrol and landed. B-25 FW-104 FH is airborne for Anti-Submarine Patrol Number 3, completed the patrol and landed. B-25 FR-362 CA airborne for Anti-Submarine Patrol Number 3, completed the patrol and landed. B-25 FR-381 DC is airborne for Anti-Submarine Patrol Number 3, completed the patrol and landed. B-25's R-376 CK, FW-132 FK and FW-183 FJ are airborne for air Exercise Number 5, landed after completion. The Grumman G-21 Goose was airborne for San Salvador landed airborne again later landed at base. B-25's FR-381 DC, FW-123 FJ, FV-946 FF, FW-132 FK, and FR-376 CL are airborne for night navigation exercises, aircraft FK and CL detailed cancelled, remaining completed the exercise and landed. Tug USS *St. Simon* towing tanker *Gulfbelle* escorted by USS *YP-433* is bound north in Straits passing Canaveral. SS *Florida* is bound south in Straits passing Canaveral. SS *William L. Davidson* is bound north passing Matanilla Shoal, northwest Bahamas. The Miami Contingent of **Convoy KN-278** is bound southeast in Straits passing Miami. USS *PC-1136* is bound north in Straits. USS *YMS-351* is bound west in Straits. **Convoy KN-278** is bound northeast in Straits near Cay Sal Bank with US Navy plane escorting. Troop transport USS *USAT-Bienville* is bound north in Straits passing Old Bahama Channel. Tug USS *Anacapa* towing USS *YR-43*, USS *YR-44*, and USS *YR-45* escorted by USS *SC-1322*, USS *PC-1137*, and USS *SC-1337*, plus the tanker *H. M. Fredricksen*, and USCG *CG-Vigilant* are bound south in Straits. NAS Great Exuma: Two planes went to cover **Convoy GN-97.** A plane went on submarine search. A plane departed to cover troop transport USS *Orizaba (AP-24)*.

22nd November 1943

Request received from the Comgulfseafron to sight and sweep ahead of the northbound convoy B-25 FW-104 FH airborne for a creeping line ahead sweep to **Convoy GN-97**, which was sighted, the sweep was completed, when the aircraft set course for base a two-funnel 7,000 ton merchant vessel, with one escort was sighted, no reply was received to the challenge signals, aircraft landed. Request was received from the Comgulfseafron to provide escort for **Convoy USS *Orizaba* (AP-24)**. B-25 FV-952 FX airborne, a cruiser and two destroyers were sighted, signals were exchanged, **Convoy USS *Orizaba* (AP-24)** which was sighted, convoy consisted of a troop transport and one escorts signals were exchanged, air escort was provided, aircraft landed with nothing to report. B-25 FR-378 CL is airborne to continue the escort to **Convoy USS *Orizaba* (AP-24),** which was sighted creeping line ahead patrol was carried out, aircraft landed after completion of duty. Request from Comgulfseafron to provide air Cover to the southbound convoy. Liberator BZ-757 LH is airborne from Windsor Field, convoy was contacted, composition of convoy was 24 merchant ships and four escorts, signals exchanged, air escort was provided by a creeping line ahead patrol, aircraft landed. B-25 FR-371 CF, and FL-200 BX are airborne for air exercises Number 7, aircraft landed after the completion of the escort. B-25's FW-132 FK, FW-123 FJ, and FW-104 FH airborne and landed after the completion of air exercise. Number 6. B-25's FL-200 BX, FR-371 CF, FW-123 FJ, FW-132 FK, FV-951 FC and FW-104 FH airborne for local night navigation exercise, landed after completion. US B-25 87476, US B-25 87477, and US B-25 87473 are delivered from Kansas City. Liberator BZ-758 LS lands at Windsor Field from Daytona at 8.26 pm. USS *SC-1322* is bound north in Straits passing Daytona. **Convoy KN-278** is bound northeast in Straits with US Navy plane escorting. USS *YW-87* and USS *SC-1354* are bound north in Straits towards Canaveral. USCG *CG-Bowstrings* is bound south in Straits towards Palm Beach. USCG cutter *CG-Boone*, and USS *YMS-362* are bound north in Straits passing Port Everglades. **Convoy NK-578** is bound south in Straits towards Port Everglades with US Navy plane escorting. Tanker *Dartmouth* is bound south in Straits passing Port Everglades. USS *YAG-19*, SS *Florida* and tanker *Perrysville*, tanker *Fort Lee*, and tanker *Pan Massachusetts* are bound south in Straits. USS *PC-1136* is bound north in Straits. *U-193* approaches Northeast Providence Channel, bound west towards Northwest Providence Channel and the Straits. NAS Great Exuma: Two planes went to cover **Convoy GN-97**. A plane escorted **Convoy KG-671**. Two planes landed at Great Exuma. A plane sighted a tug towing three oil barges. A plane arrived and departed to cover **Convoy NG-399**.

23rd November 1943

Request received from Comgulfseafron to provide air cover to ferry *Jean Brillant*, cover required until ship docked in Nassau, Harbour ship was escorted by three escort vessels. B-25 FK-378 CL is airborne, is sighted, composition was one merchant ship and escorts, a US Navy blimp providing air cover. The blimp left convoy on the arrival of CL, signals were exchanged, convoy escorted to Nassau Harbour and left at the bar [outside entrance], and aircraft landed. B-25's FL-200 BX and FK-375 CJ are airborne and landed on the completion of Exercise Number 6. B-25 FK-376 CK is airborne and landed after completing air Exercise Number 7. B-25's FR-371 CF, FV-951 EC, FW-204 FH, FK-375 CJ, FR-38l DC, and FW-123 FJ are airborne for air Exercise Number 6, aircraft landed on completion off the exercises. B-25's FR-371 CF, FR-375 CJ, and FW-104 FH are airborne and completed and the aircraft landed. B-25's FV-954 FB, FV-951 FC, FW-123 FJ, and FW-132 FK were airborne for night navigation exercises. Aircraft landed on the completion of the exercise. B-25 FV-952 FX is airborne for night navigation exercises, but failed to return, signal was received by radio which terminated abruptly and only the first letter was received which was "F," it was presumed to be the first letter of the aircraft's Cell Sign, signal was received from Miami to the effect that the Coast Guard had reported a plane crashing into the sea near Northwest Light, Bimini, signal was sent to the aircraft FK, on the night navigation exercise, diverting it to Northwest Light, and was instructed to search the area. B-25 FK-381 DC was airborne to carry out a search of the area in which the aircraft FX was reported to have crashed, yet nothing sighted and the aircraft landed. The search for FV-952 FX was then postponed until the following day. Liberator FL-992 LX returned from 36th Street Airport, Miami after modification in San Antonio. USS *YT-349* is bound north in Straits, passing Canaveral. SS *Fort Prudhomme* is bound northeast in Straits. USS *YMS-390* is bound south in

Straits towards Palm Beach. USCG *CG*-83481, USCG *CG*-83482, and USCG *CG*-83483 are bound south in Straits passing Port Everglades. SS *Henry L. Ellsworth* is bound north in Straits nearing Bimini. USS *YAG-19* and SS *Mill Spring* are bound north in Straits. USS *PC-1240* is bound southwest in Straits. **Convoy GK-771** is bound northwest in Straits near Cay Sal Bank with US Navy plane escorting. *U-129* is 200 nm northeast of Abaco, bound toward Northeast Providence Channel. *U-193* is bound west to transit Northeast Providence Channel and Northwest Providence Channel. NAS Great Exuma: A plane departed to cover **Convoy NG-399**. A plane sighted a medium-sized merchant vessel. A plane departed to cover **Convoy KG-671**.

24th November 1943

B-25 FW-104 FH is airborne to search for missing aircraft B-25 FV-942 FX; a thorough search was made of Bimini in daylight, no evidence found to presume existence of crew of missing aircraft, aircraft landed. B-25 FW-132 FK, FV-951 PC, and FW-123 FJ airborne on a combined air sea rescue and Exercise Number 7, aircraft searched the Bimini area landed after the completion of air exercise. B-25's FK-176 AX, FK-175 AM, FR-395 GO, and FK-372 CG are airborne and landed. B-25's FK-167 AX, FK-170 AG, and FR-375 CG are airborne and landed on the completion of the exercises. B-25's FV-997 FG, FV-951 FC, FR-371 CF, FK-381 DC, FV-954 FD, and FW-123 FK are airborne and landed on the completion of the exercises. The US B-25 87505 was delivered to the unit from Kansas City and landed at Oakes Field. SS *Fenja* is bound south in Straits passing Canaveral. Tug USS *Dry Tortugas* is bound north near Matanilla Shoal, northwest Bahamas, with US Navy plane escorting. Tanker *Axtell J. Byles* is bound south in Straits passing Palm Beach. *USS YAG-19* is bound south in Straits. USS *YMS-390* is bound west in Straits. USS *SC-1057* and USS *SC-1058*, and USS *YAG-20* are bound north in Straits. Tanker *Caribbean Gulf* is bound south in Straits towards Cay Sal Bank. SS *Seven Pines* is bound west in Straits. USS *YO-142* is bound southeast in Straits passing Nicholas Channel. *U-193* is bound south in Straits from Northwest Providence Channel. *U-129* is northeast of Abaco, west for the Straits via Bahamas. B-25 FW-104 FH is airborne to search for missing aircraft FX, thorough search made of Bimini, no evidence found to presume existence of crew of missing aircraft, aircraft landed. B-25 FW- 132 FK, FV-951 PC, and FW-123 FJ are airborne, searched the Bimini area. NAS Great Exuma: Three planes arrived and departed.

25th November 1943

B-25's FV-997 FG, FW-I04 FK, FV-949 FD, FR-378 CL, FW-132 FK, and FV-951 FG are airborne on air Exercise Number 6, aircraft landed. B-25's FR-372 CG, FK-163 AA, FK-170 AG, and FK-167 AX are airborne on local navigation exercises. B-25's FK-170 AG, FR-371 CF, FR-372 CG, and FR-395 CO are airborne landed on the completion of the exercises. **Task Group TG-202.2** of six USS *LCI(L)*'s, is bound south in Straits towards Canaveral. SS *Pheasant* is bound southwest near Matanilla Shoal, northwest Bahamas with US Navy plane escorting. USS *SC-1337* is bound northeast towards Matanilla Shoal, northwest Bahamas. SS *W. H. Ferguson* and SS *Sweetwater* are bound northeast in Straits towards Bimini. SS *Fenja* is bound southwest in Straits. USS *Tattoo* (BAM-32) is bound northeast in Straits towards Bimini. USS *YMS-367* towing USCG *CG-83373* escorted by USCG *CG-83402* is bound northwest in Straits towards Miami. NAS Great Exuma: A plane went to cover **Convoy GN-98** and departed. A plane departed for Guantanamo.

26th November 1943

Request received from Comgulfseafron to provide air cover to submarine USS *Burrfish* escorted by two escorts; coverage was requested until evening, then it will be taken over by US Navy blimp. B-25 FV-997 FG is airborne to provide air escort, convoy was sighted, composition of the convoy one submarine and two escorts, aircraft carried out the correct recognition procedure, but no reply was received, box patrol was commenced, after sighting the submarine submerged and the convoy stopped imposition, submarine surfaced and convoy got underway relieving aircraft FG, was observed to be preparing to take over the patrol, FG left the convoy after patrol landed at base. B-25 FV-949 FB was airborne as relief to FG, convoy was sighted, escort was taken over from FD, signals were exchanged with the SNO, convoy was escorted where a US Navy blimp relieved FG, aircraft landed. Request received from Comgulfseafron for an aircraft to sight and sweep 100 nm ahead of the minelayer USS *Keokuk* (AKN-4) which is accompanied by one escort vessel. Liberator BZ-760 LT is airborne at 11.33 am from Windsor Field, convoy was not contacted in the estimated position, square search was made, minelayer convoy was sighted, correct recognition procedure was carried out but no reply was received from the escort vessel, air escort was provided, aircraft landed at Windsor Field with nothing to report. Grumman G-21 Goose FP-438, which was part of the Station Flight, was airborne for 36th Street airport, Miami, and landed there. B-25's FW-104 FH, FW-132 FK, FV-951 FG, FV-954 FB, FK-378 CL, and FB-371 CF are airborne for air exercises Number 6, aircraft landed having completed the exercise. USS *YP-433* is bound south in Straits towards St. Augustine. Tanker *Esso Wilmington* is bound southwest in Straits. USS *Keokuk* (AKN-4) and USS *Palmer* (DMS-5) are bound south in Straits with US Navy plane escorting. USCG Cutter *CG-Bowstring* (CGC-365), and SS *Joseph Goldberger* are bound south in Straits towards Palm Beach. USS *SC-1338* and USS *YMS-351* are bound north in Straits passing Palm Beach. Tanker *Gulfmaid* north towards Palm Beach. Submarine USS *Burrfish,* with sub-chasers USS *SC-1063* and USS *SC-1064* are bound northwest towards Northwest Providence Channel, with US Navy plane escorting. SS *Tobias E. Stansbury*, USS *YO-142*, its tow, and USCG *CG-83348* are bound north in Straits. **Convoy KN-279** is bound north in Straits with US Navy blimp ,near Cay Sal Bank. SS *McDowell* and tug USS *Boone Island* are bound south in Straits. USS *LCI(L)-459* is bound southeast in Straits towards Nicholas Channel. USCG *CG-Vigilant* is bound northwest in Straits passing Nicholas Channel. USS *Lapwing* (AVP-1) is bound southeast in Straits passing Nicholas Channel. *U-193* exits area from the Straits, bound west past Key West. NAS Great Exuma: Three planes went to cover **Convoy GN-98**. Five planes arrived and departed.

27th November 1943

Request was received from Comgulfseafron to sight and sweep ahead of a northbound convoy. B-25 FW-949 FD airborne to sight and sweep ahead of **Convoy GN-98**, convoy was not sighted on arrival at the point of intersection, search was carried out. Then convoy was sighted, composed of 24 merchant ships and five escorts. A sweep was carried out and the aircraft set course for base, landed with nothing to report. B-25's FR-371 CF, FW-104 FH, FV-949 FD, FV-951 FC, and FR-393 DF are airborne for compass-swinging exercise, landed after completion. Liberator FL-994 LA is airborne for Miami, landed there. SS *Ezra Meeker*, tug USS *Willet*, and tug USS *Boone Island* are bound south in Straits passing Canaveral. **Convoy NK-579** is bound south in Straits passing Canaveral with US Navy plane escorting. **Convoy KN-279** is bound northeast in Straits with US Navy plane escorting. USS *SC-1292* is bound southwest in Straits towards Palm Beach. SS *William Phipps* is bound north in Straits passing Port Everglades. Submarine USS *Burrfish* is bound southwest in Straits towards Key West. USS *LCI(L) 459* is bound southeast in Straits. **Convoy KG-672** is bound southeast with US Navy plane escorting. NAS Great Exuma: A plane departed to cover **Convoy GN-98**. A plane departed on administrative flight to Miami. Three planes, one with Lt. Arle, arrived for duty.

28th November 1943

Courses changed Squadron Number 11 course moved to Number 3, Squadron at Windsor Field for training on Liberator Number 13 moved to Number 2, to commence training in navigation and air exercises Number 15 moved to Number 1 Squadron for familiarization flights on B-25 aircraft. USS *YT-349* is bound south in Straits towards Canaveral. Tug USS *Boone Island* is bound south in Straits towards Palm Beach. Tanker *Hadnot* and SS *Petersburg* are bound north in Straits towards Palm Beach. USS *PC-1207* is bound south in Straits passing Port Everglades. Tanker *18 De Marzo* is bound north in Straits passing Port Everglades. SS *William E. Pendleton* and USS *SC-759* are bound south in Straits. USS *Aegur* (AS-23) is bound northeast in Straits. USS *Lapwing* (AVP-1) is bound northwest in Straits near Nicholas Channel. NAS Great Exuma: Two planes went to cover **Convoy NG-400**. A plane returned due to defective Radar. Four planes arrived and departed on training loop.

29th November 1943

B-25's FR-371 CF, FW-104 FH, FV-949 FD, FV-946 FF, FR-375 CJ, and FV-997 FG are airborne swinging exercise, landed after completion. B-25's FR-371 CF, FW-104 FH, FV-951 FC, FV-997 FG, and FV-949 FD are airborne for air exercise, landed after completion. SS *Perrysville*, SS *Fort Lee*, and tanker *Pan Massachusetts* are bound north in Straits. USS *YMS-389* is bound south in Straits towards Palm Beach. SS *Kenyon* is bound south in Straits towards Port Everglades. USS *YAG-28* is bound southwest towards Port Everglades. USS *YAG-19*, tug USS *Boone Island* towing USS *YFD-59* and escorted by USS *YP-433* are bound south in Straits. Tanker *E. R. Kemp* is bound north in Straits. **Convoy GK-722** is bound southwest in Straits towards Key West with US Navy plane escorting. USS *PC-1230* and USS *PC-1172* are bound southeast in Straits passing Nicholas Channel. Tug USS *Montcalm* (AT-39) towing oiler USS *YOG-11* and escorted by USS *YMS-390* are bound southeast in Straits. NAS Great Exuma: A plane sighted a small sailboat and another small merchant ship. A 3-masted schooner was sighted. Two planes went to cover **Convoy NG-400.** An inter-island ship, or freighter (mailboat) was sighted by a plane. Five planes arrived and departed.

30th November 1943

Request received from Comgulfseafron for a sweep to be carried out, a D/F, Direction Finding plot indicated probable presence of a submarine. Liberator BZ-757 LR airborne from Windsor Field, carried out search, landed back with nothing to report. B-25's FV-949 FD, FW-132, FK, and FR-393 DF are airborne, and completed Anti-Submarine Patrol Number 1a and Number 1b, landed after completion. B-25 FV-997 FG is airborne for air Exercise Number 5, landed after completion. B-25 FV-946 FF is airborne for compass swinging exercise, landed after completion. B-25 FR-375 CJ, FV-951 FC, FV-997 FG, FV-946 FF, FR-393, and FW-132 FX are airborne for air Exercise Number 1, landed after completion. Night flying programme cancelled due to adverse weather condition. USCG *CG-83495* and USCG *CG-83496* are bound south in Straits towards St. Augustine. SS *Cape Henry* is bound south in Straits towards Canaveral. Tanker *E. R. Kemp* is bound north in Straits towards Canaveral. Tanker *Chapultepec* is bound southwest in Straits. USS *YMS-369* is bound south in Straits passing Palm Beach. Tanker *Gulfmaid* is bound south in Straits passing Port Everglades. SS *David O. Saylor* and destroyer escort USS *George* (DE-697) are bound north towards Port Everglades. USS *YAG-19* and USCG *CG-Cuyahoga* are bound north in Straits. USS *SC-1276* is bound west in Straits. NAS Great Exuma: A plane departed to cover **Convoy GN-99**, 20 merchant vessels and five escorts. A plane departed to cover USS *Bearns*. Three planes arrived and departed.

1st December 1943

B-25 FW-132 FK is airborne for air Exercise Number 3, aircraft landed after completion. B-25 FW-148 FL is airborne for air Exercise Number 4, landed having completed the exercise. B-25's FW-149 FN, FV-949 FD, FV-951 FC, FW-154 FR, and FW-145 FW are airborne for air Exercise Number 5, landed having completed the exercise. B-25 FV-949 FD is airborne for Anti-Submarine Patrol Number 1, landed with nothing to report. B-25 FW-149 FN is airborne for Anti-Submarine Patrol Number 1, landed with nothing to report, firing practice was completed at the termination of the exercise. B-25 FR-375 CJ is airborne for Anti-Submarine Patrol Number 1, landed with nothing to report. B-25 FV-951 FC airborne for Anti-Submarine Patrol Number 1, landed with nothing to report. Local night flying was carried out by five B-25 aircraft. USS *SC-1350* is bound south in Straits towards Daytona. **Convoy NK-580** is bound south in Straits towards Daytona with US Navy plane escorting. SS *Potter* and SS *Jacob Thompson* are bound south in Straits passing Daytona. SS *David O. Saylor* is bound north in Straits towards Canaveral. USS *SC-1297* is bound southeast in Straits near Port Everglades. Tanker *Esso Hartford* is bound southwest in Straits passing Bimini. SS *Seven Pines* is bound northeast in Straits passing Bimini. SS *Caribbean* is bound northeast in Straits towards Bimini. USS *PC-1244* is bound southwest in Straits. Anti-Submarine Group 5 is bound west in Straits. **Convoy KN-280** is bound northeast in Straits with US Navy plane escorting. **Convoy KG-673** is bound southeast in Straits passing Nicholas Channel with US Navy plane escorting. NAS Great Exuma: Three planes went to cover **Convoy GN-99**, one plane sighted two heavy destroyers and SS *Worden*. A plane went to cover two US Navy ships and three escorts. Three planes arrived and departed.

2nd December 1943

Liberator BZ-760 LT is airborne from Windsor Field. Homing exercise commenced, convoy consisting of two ships was sighted, signals were exchanged, radio communication was established aircraft was instructed to carry out Viper exercise, a successful attack with depth charges was carried out against a smoke float dropped by the SNO, flares were dropped from the aircraft for firing practice of the anti-aircraft grows, aircraft exchanged signals with the SNO, and set course for base, on the return flight an 800-ton mine-sweeper was sighted, no reply was received from the ship to the challenge of the aircraft. Liberator BZ-758 LS is airborne from Windsor Field, convoy of 21 merchant ships and three escorts sighted, convoy was being escorted by a B-25 aircraft, radio contact was established with HMS *Goodall,* convoy of 15 merchant ships and four escorts was sighted, no reply was received from the escorts to the aircraft's challenge, two frigates were sighted, signals were exchanged between the aircraft and the ships, attack was made with one depth charge on a smoke float dropped by the SNO, depth charge ricocheted and was a 20 yards overshot, instructions were received iron the SNO regarding the releasing of flares for anti-aircraft practice for the ship's crew, first flare was released and firing practice took place, instructions received by the aircraft to carry out Viper patrol, aircraft left the ships and set course for base, landed. Liberator BZ-809 LZ is airborne from Windsor Field, contact established with co-operating ships, two frigates were sighted, signals were exchanged between the aircraft and the ships aircraft was instructed to carry out Lizard patrol, a depth charge dropped against smoke float, depth charge failed to detonate, bounced and disappeared. The smoke float exercise was completed and the aircraft set course for base, landed. Request was received from Comgulfseafron for an aircraft to sight and sweep ahead of a northbound convoy, exposition of the convoy was 20 merchant ships and five escorts. B-25 FV-946 FF airborne to carry out the sweep, signal was received from base that convoy was 46 nm ahead of the estimated position, convoy was sighted, composition was 20 merchant ships and five escorts, signals were exchanged, aircraft commenced a creeping line ahead patrol, patrol was abandoned, aircraft set course for base, convoy of 22 merchant ships and five escorts was sighted, aircraft landed with nothing to report. B-25 FV-997 FG airborne, carried out a reconnaissance exercise, photographs were taken of small islands, aircraft landed with nothing to report. B-25's FR-393 BF, FR-375 CJ, and FW-123 FJ are airborne, aircraft landed after the completion of the exercises. B-25's FV-946 FF, FW-144 FA, FW-145 FV, FR-375 CJ, and FR-393 DF were airborne and landed. Compass swing exercises were carried out by B-25's FV-154 FR and FW-148 FL. Then B-25's FW-154 FR, FW-123 FJ, FW-149 FI, FV-946 FF and FR-393 DF were airborne for local night navigation exercise, landed. Liberator FL-994 LA is airborne for Miami and landed there. Liberator BZ-744 LL is airborne for Miami *en route* to San Antonio, aircraft landed at Miami. **Convoy KN-280** is bound southeast in Straits with US Navy plane escorting. **Convoy NK-580** is bound south in Straits passing Palm Beach with US Navy plane escorting. USS *YAG-19*, USS *YAG-21,* SS *Potter*, and SS *Jacob Thompson* are bound south in Straits. Anti-Submarine Group 5 is bound north in Straits passing Miami. Destroyer USS *Alden* (DD-211) is bound north in Straits near Cay Sal Bank. SS *Christopher Gale* is bound north in Straits towards Bimini. NAS Great Exuma: A plane departed to cover **Convoy GN-99**.

A plane went to cover **Convoy NG-401**. Four planes arrived and departed.

3rd December 1943

B-25's FW-154 FR, FW-123 FJ, FW-149 FW, FR-393 DF, and FR-375 CJ airborne for local night navigation exercise, landed. B-25 FW-123 FJ is airborne for Anti-Submarine Patrol Number 2d, aircraft completed the patrol and landed. B-25 FW-149 FN is airborne for Anti-Submarine Patrol Number 2, aircraft completed the patrol and landed. B-25 FV-997 FG is airborne for Anti-Submarine Patrol Number 2e, aircraft completed the patrol and landed. B-25 FR-375 CJ is airborne for Anti-Submarine Patrol Number 2c, aircraft completed the patrol and landed. Compass swinging exercises were carried out by B-25's FW-148 FL and FW-147 FO. B-25's FR-381 DC, FW-149 FN, FW-123 FJ, FR-375 CJ, and FV-997 FG are airborne and landed after completing air Exercise Number 4. B-25's FW-145 FM, FV-954 FB, FV-997 FG, FR-381 DC, FW-147 FO, and FW-123 FJ are airborne, landed after the completion of the night exercises. B-25 FR-375 CJ is airborne and landed after local night navigation exercise. Seven B-25's of Number 1 Squadron carried out night navigation exercises. B-25 FV-946 FF was airborne for Jamaica and landed there. Liberator BZ-309 LV is airborne from Windsor Field for Kingsley Exercise, W/T [radio] contact was established with the ships, ships were contacted, SNO instructed the aircraft to carry out Mamba exercise, instructions were not understood by the aircraft instructions were then issued by visual signals in plain language patrol was carried out and a depth charge was dropped against a sea marker, aircraft left the convoy, set course for base and landed. Liberator BZ-746 LK is airborne from Windsor Field for the Kingsley Exercise, radio contact was established with the convoy, two frigates were sighted, patrol instructions were given to the aircraft to carry out exercise, Mamba smoke fleets were dropped for target practice by the SNO, the attack by the aircraft with depth charges was a 10 yard [30 feet] overshoot, aircraft completed the exercise, set course for base, landed. Liberator BZ-760 LT is airborne radio contact was established with the ships, homing procedure "B" was carried out, frigates were sighted, signals were exchanged, smoke float was to have been dropped by the SNO, but the ship was recalled to base aircraft, then instructed to return to base, aircraft landed. Tanker *18 De Marzo* and **Task Group TG-201.6, Group-37**, of eight USS *LCI(L)*'s, are bound south in Straits passing Canaveral. Tanker *E. R. Kemp*, USS *SC-1350*, tug USS *Point Loma* towing oil barges USS *YOG*'s are bound south in Straits towards Palm Beach. USCG *CG-83404* is bound east towards Walker Cay, north of Abaco. USS *YP-603* is bound south in Straits from Port Everglades. Tanker *Esso Wilmington* is bound northeast in Straits. Anti-Submarine Group 3 is bound southeast in Straits passing Nicholas Channel. NAS Great Exuma: Two planes went to cover **Convoy NG-401.** Five planes arrived and departed for Guantanamo.

4th December 1943

B-25's FV-954 FB, FW-145 FM, FR-375 CJ, FW-123 FJ, FV-149 FN and FW-147 F0 air exercises Number 3, landed on the completion of the exercises. Signal was received from Comgulfseafron that an enemy submarine was reported, further signal was received reporting an attack by a Naval plane on the reported submarine observed to be on course of 270°, before submerging. B-25 FW-150 KP is airborne on a search for the reported submarine in the New Providence Channel, thorough search was made of the area aircraft sighted nothing and landed. B-25 FW-147 FO is airborne, searched area, landed with nothing to report. B-25 FW-145 FW is airborne searched area landed with nothing to report. B-25 FW-123 FJ is airborne, searched area, landed with nothing to report. Liberator BZ-746 LM is airborne from Windsor Field for homing and navigation Exercise 1 was completed and the homing exercise was commenced had to be abandoned, aircraft landed. Liberator BZ-757 LR is airborne from Windsor Field for homing and navigation exercise, exercise was abandoned due to poor reception, aircraft landed on the completion of the navigation exercise. B-25's FV-954 FV, FW-149 FN, FR-381 DC airborne, landed after completion of night navigation exercise. B-25 FV-946 FF landed from Jamaica. Escort patrol craft USS *PCE-869* and tanker *Touchet* are bound north in Straits towards Palm Beach. Tanker *18 De Marzo* is bound south in Straits. **Task Group TG-2-16, Group 37**, USS *SC-1361*, Anti-Submarine Group 1, are bound southwest in Straits. Tanker *E. R. Kemp* is bound west in Straits. USS *Nuthotch* (AM-60) is bound west in Straits near Cay Sal Bank. Tug USS *Cardenas* with tow escort is bound southeast in Straits passing Nicholas Channel. NAS Great Exuma: Three planes arrived and departed.

5th December 1943

Liberator BZ-758 LS is airborne from Windsor Field to carry out a sweep in the waters east of Abaco and search for reported submarine, creeping line ahead search was made, aircraft failed to locate the U-boat, and landed at base. B-25 FW-147 FO is airborne for sweep in the Northwest Providence Channel, creeping line ahead patrol was carried out, to cover channel, aircraft returned to base with nothing to report. B-25 FW-123 FJ is airborne on a search of the waters to the east of Abaco, creeping ahead search aircraft landed having failed to locate the submarine. B-25 FW-145 FM airborne search the waters northeast of Eleuthera, creeping line ahead patrol, aircraft landed with nothing to report. USCG *CG-Cuyahoga* is bound north in Straits passing Canaveral. SS *Lookout Mountain* and SS *Zoella Lykes* are bound south in Straits towards Palm Beach. SS *McDowell* is bound north in Straits. SS *Pheasant* is bound north in Straits with US Navy plane escorting. USS *Lansing* is bound northeast towards Matanilla Shoal, northwest Bahamas. USS *YAG-28* is bound southeast towards Nassau. Oiler USS *YO-143* is bound east in Straits from Port Everglades. SS *Enoree* is bound southwest in Straits with US Navy plane escorting. USS *PC-1213* is bound south in Straits. SS *Spotsylvania* is bound northeast in Straits passing Cay Sal Bank. NAS Great Exuma: A plane went to cover **Convoy GN-100**, 19 merchant vessels and five escorts, and then departed.

6th December 1943

B-25's FR-393 DF, FV-946 FF, FV-954 FB, FW-148 FL, FW-104 EH, and FW-150 FF are airborne and landed on the completion of the aim Exercise Number 7. B-25's FR-381 DC, FW-148 FL and FR-393 DF are airborne, then the aircraft landed on the completion of air Exercise Number 6. B-25's FH-104 FH, and FV-147 FO are airborne and landed on the completion of air Exercise Number 6a. Liberators BZ-746 LH, BZ-757 LR were airborne from Windsor Field for air submarine exercises at Boca Chica, in co-operation with the United States Navy, landed at Boca Chica and were briefed for the exercise there, on the completion of the exercise the aircraft returned to base. B-25's FR-362 CA FR-378 CA, FW-147 F0, FR-381 DC, FW-148 FL and FL-150 FP airborne for night navigation exercise and landed after completion, FP was recalled earlier. **Convoy NK-581** is bound south in Straits towards Canaveral with US Navy plane escorting. USS *YP-434*, USS *YMS-369* are bound north in Straits towards Canaveral. USS *J-862* and USS *J-1054* are bound north in Straits towards Palm Beach. USS *YAG-19* and tanker *Petersburg* are bound south in Straits. Army tanker USS *Y-21* is bound northeast out of Straits. **Convoy KN-281** is bound north in Straits with US Navy plane escorting. **Convoy KG-674** is bound southeast in Straits passing Nicholas Channel with US Navy plane escorting. USS *YAG-28* is bound north from Nassau. NAS Great Exuma: Three planes went to cover **Convoy GN-100**. A plane departed to cover USS *Florida* and an escort. Seven planes arrived and departed from training loop.

7th December 1943

Request was received from Comgulfseafron for an aircraft to sight and sweep ahead of northbound convoy, composition of the convoy was 21 merchant ships and five escorts later signal was received stating that the is estimated to be 30 nm behind earlier given position, Liberator BZ-809 LZ is airborne from Windsor Fields, convoy was sighted, composition of the convoy was 25 merchant ships and five escorts no acknowledgement received to the challenge by the aircraft, aircraft was in position which was the estimated position of the convoy, but the ships were not sighted, aircraft carried out a visual search for convoy, signal was received from base "Have you met convoy?" signal was sent to base "Have not met convoy, visibility 15 nm," northbound convoy was contacted later, signals were exchanged, captain decided to escort the ship for one hour, signal was sent to base stating that the convoy had been met and giving the position, aircraft left the convoy, after exchange of signals, landed at base with nothing to report from the patrol. Request was received from Comgulfseafron requesting that an aircraft sight and sweep ahead of a southbound convoy, composition of the convoy was 25 merchant ships and five escorts, signal was received from Comgulfseafron to the effect that Great Exuma Naval Air Station was covering the convoy and that this Unit was no longer required to provide air escort to the convoy. B-25's FR-393 DF and FV-946 FF are airborne for air Exercise Number 7, aircraft completed exercise and landed. B-25 FW-132 FK is airborne for air Exercise Number 7. B-25's FW-150 FF, FR-378 CL, FR-381 DC, FR-375 CJ, FR-362 CA and FW-132 FK were airborne for air Exercise Number 4, landed on the completion of the exercises. B-25 FR-362 CA is airborne for local night navigation exercise, landed on the completion. B-25's FR-378 CL, FR-393 DF, FR-l32 FK, FV-946 FF, FW-150 FP, FW-174 FL and FR-362 CA are airborne for night navigation exercise, landed on the completion of the exercise. Liberators BZ-883 LW, BZ-746 LM, BZ-760 LT and BZ-931 MF are airborne for air submarine exercise at Boca Chica in co-operation with the US Navy, aircraft landed at Boca Chica briefed on the exercise, exercise completed returned to base. SS *Mark Keppel* is bound south in Straits towards St. Augustine. Submarine USS *S-1276* is bound north in Straits passing Canaveral. Tug USS *St. Simon* is bound south in Straits passing Canaveral. SS *Kenyon* is bound north in

Straits passing Bimini. **Convoy NK-581** is bound south in Straits with US Navy plane escorting. SS *Dartmouth*, destroyer USS *Biddle* (DD-151), USS *PC-1182*, and SS *Kota Gede* are bound south in Straits. USS *YAG-19* is bound north in Straits. SS *Fort Leed* is bound southwest in Straits passing Cay Sal Bank. NAS Great Exuma: Two planes departed to cover **Convoy NG-402**. A plane went to cover **Convoy GN-100**. A plane escorted **Convoy KG-674**. Five planes arrived and then departed.

8th December 1943

B-25's FV-951, FR-366 CX, FR-393 DF, FR-378 CL and FV-949 FD are airborne and landed on the completion of air Exercise Number 6. B-25'an FK-163 AA, FK-166 AD, and FK-174 AL are airborne, completed the local navigation exercises allocated to Number 1 Squadron aircraft landed. B-25's FV-951 FC, FR-378 CL, FW-104 FH, FW-150 FP, and FR-362 CA are airborne for night navigation exercise, then landed. B-25's FK-166 AA, and FL-108 BA are airborne for local night navigation exercises, landed on the completion. Anti-Submarine patrols, B-25's FD, FF, FB, FG, and FC are airborne between 11:30 and 11:45 on Patrol Number 1. Aircraft completed uneventful patrol, landed between 4:25 and 4:51. Grumman Goose GA airborne at 2:57 for San Salvador and returned at 10:48. US Army USS *Aristides* is bound south in Straits towards Daytona. Tanker *R. P. Smith* is bound north in Straits passing Palm Beach with US Navy plane escorting. USS *SC-1330* is bound north in Straits passing Palm Beach. Destroyer escorts USS *DE-390*, USS *DE-323*, and USS *DE-389* are bound north in Straits towards Palm Beach. Tanker *Paco* is bound south in Straits passing Port Everglades. Tanker *Chiloil* is bound north in Straits near Miami. Tanker *Perryville* is bound south in Straits near Miami. USCG *CG-83402* is bound west passing Northwest Providence Channel. USS *YAG-28* is bound southeast in Straits passing Bimini. USS *CL-47*, USS *CVL-27*, USS *DD-675* and USS *DD-688* and tug USS *St. Simon* are bound south in Straits. USS *PC-1173* is bound north in Straits near Miami. Tanker *George W. Kendall* is bound north in Straits. NAS Great Exuma: Two planes went to cover **Convoy NG-402**. A plane went to cover **Convoy KG-674**. Four planes landed at Great Exuma.

9th December 1943

B-25's FW-154, FR, FV-951 FC, FV-949, FD, FR-362 CA, FR-366 CX, and FW-104 FH are airborne for air Exercise Number 6, landed. B-25's of Number 1 Squadron FK-L74 AL, FL-180 BA, FK-166 AD, and FK-163 AA are airborne and landed. B-25's of Number 1 Squadron FR-376 CK, FL-180 BA, FL-187 BC, FR-371 CF, and FK-174 AL are airborne and landed on the completion of local night navigation exercise. B-25's FR-378 CL, FW-132 FK, FW-104 FH, FR-393 DF, and FV-951 FC airborne, completed night navigation patrol landed. Liberator BZ-757 LR is airborne from Windsor Field and carried out Anti-Submarine Patrol Number 9a, navigation exercise was completed, and the aircraft landed, after an uneventful patrol. B-25's FV-945 FD, FV-946 FF, FW-176 FS, FW-147 FO, FW-l45 FM, and FW-154 FH are airborne, landed on the completion of night exercises. Liberator FL-967 LJ is airborne and landed at Boca Chica, briefed for the air submarine exercise in co-operation with US Navy, exercise was completed. Anti-Submarine patrols, B-25's FC, FB, FF, FG, and FD are airborne between 11:25 and 11:36 on Anti-Submarine Patrol Number 1, aircraft completed patrol, nothing reported and landed between 4:31 and 4:46. Tanker *Chiloil* is bound north in Straits passing

Daytona. Tug USS *Point Cabrillo* is bound north in Straits passing Daytona with US Navy plane escorting. Tug USS *Baldhead* is bound south in Straits passing Canaveral. Army training boat USS *Aristides* south in Straits passing Palm Beach. Canadian warship HMCS *Wetaskiwin* is bound southwest towards Matanilla Shoal, northwest Bahamas. SS *Sweetwater* is bound south in Straits passing Port Everglades. USS *YAG-19*, USS *PC-1174* south, USS *SC-1349*, and USS *YO-69* are bound southwest in Straits. USS *YP-426* and SS *William A. Jones* are bound northwest in Straits. SS *Flomar* is bound northwest in Straits near Nicholas Channel. *U-129* is bound northeast of Abaco southwest to San Salvador. NAS Great Exuma: A plane departed and arrived from an administrative flight to Nassau. Inter-island transport USS *YAG-28*, ex-*Samaná*, dropped anchor at Great Exuma.

10th December 1943

Request was received from Comgulfseafron for an aircraft to sweep area. Liberator BZ-757 LZ is airborne from Windsor Field area swept without incident, aircraft landed with nothing to report. Liberator BZ-883 LW was airborne from Windsor Field reconnaissance of Inagua, cloud base was low and the visibility poor and in consequence the photographs taken were not satisfactory, aircraft landed. Compass swinging exercises were carried out by B-25's FR-378 CL, FR-366 CX, FR-362 CA, FV-949 FD and FW-132 FX. Liberator BZ-722 returns from San Antonio and landed at Windsor Field. Liberator BZ-758 LS is airborne for Miami *en route* to San Antonio aircraft landed. No flying from No. 111 OTU. **Task Unit TU-2034**, USS *YMS-70*, and tug USS *Point Cabrillo* are bound south in Straits passing Palm Beach. USS *Humboldt* (AVP-21), USS *PC-1206* and USS *PC-1245* are bound south in Straits. SS *George E. Badger* is bound north in Straits passing Port Everglades. Tug USS *Baldhead* and Army training boat USS *Aristides* are bound west in Straits. USS *YAG-19* is bound north in Straits. HMCS *Wetaskiwin* is bound southwest in Straits. USS *YG-29* and USS *F-127* are bound southeast in Straits passing Nicholas Channel. NAS Great Exuma: A plane went to cover **Convoy GN-101**. A plane arrived and departed from Guantanamo.

11th December 1943

Request was received from Comgulfseafron for an aircraft to sweep, Liberator BZ-813 MD is airborne from Windsor Field to sweep the area, detailed, small US Coast Guard boat was sighted, patrol completed without incident, aircraft landed. B-25's FR-393 DF, FW-123 FJ, FR-366 CX, FW-104 FH, FW-154 FR, and FW-132 FK are airborne for air Exercise Number 6, landed after completion. Anti-Submarine patrols, B-25's FB, FC, FBD FF, and FG airborne between 11:23 and 12:39 on Air Exercise Number 5, landed between 2:22 and 3:49. Anti-Submarine patrols, B-25's FB, FJ, FD, FU, FG, and FK airborne between 5:29 and 5:35 on Air Exercise Number 1, landed between 9:48 and 10:22. **Task Unit TU-2034**, eight USS LCI's escorted by the high-speed transport USS *Tattnall* (APD-19) is bound south in Straits passing Canaveral. USS *SC-1346* is bound north in Straits passing Canaveral. USCG *CG-83454* and USCG *CG-83455*, USS *J-863*, USS *J-864*, and USS *PC-1182* are bound north in Straits passing Palm Beach. USS *SC-1342* is bound southeast in Straits passing Miami. Tug USS *Point Cabrillo* escorted by USS *SC-1276* is bound west in Straits. **Task Unit TU-0318** is bound northwest in Straits towards Key West.

Tanker *Paco* is bound northwest in Straits passing Cay Sal Bank. NAS Great Exuma: Four planes went to cover **Convoy GN-101**. Inter-island transport USS *YAG*-28, ex-*Samana*, of Hatchet Bay, weighed anchor and departed. Two planes landed at Great Exuma.

12th December 1943

Anti-Submarine patrols, B-25's FV, FD, FG, FH, and FK are airborne between 11:22 and 11:59, landed between 3:19 and 3:37 after exercise completion. Anti-Submarine patrols, B-25's FC, FS, FD, FU, and FV are airborne between 5:16 and 5:47 on Air Exercise Number 1, completed exercise and landed between 9:51 and 10:18. Grumman Goose GA airborne at 2:15 on a flight to Grand Bahama, returned at 5:02. B-24 LO airborne at 9:53 for Miami, returned at 11:23. Liberator LH, for Radar and navigation exercise, airborne at 6:36, landed at 8:26 after exercise completion. SS *Hobart Baker* is bound south in Straits passing St. Augustine. Tanker *R. P. Smith south* in Straits towards Canaveral. **Convoy NK-582** is bound south in Straits towards Palm Beach with US Navy plane escorting. Tanker *Chiloil*, USS *YMS-369*, and USS *PC-1215* are bound south in Straits passing Palm Beach. USS *YMS-76* is bound north in Straits passing Palm Beach. **Convoy KN-282** is bound north in Straits passing Palm Beach with US Navy plane escorting. USCG *CG-83406* is bound north in Straits then east towards Matanilla Shoal, northwest Bahamas. Inter-island transport USS *YAG-28*, ex-*Samana*, is bound west in Straits towards Miami. **Task Unit TU-2034**, USCG *CG-83454* and USCG *CG-83455* are bound south in Straits. SS *Enroee* is bound northeast in Straits with US Navy plane escorting. Tanker *Gertrude Kellogg* is bound northwest in Straits passing Nicholas Channel. NAS Great Exuma: Two planes departed to cover the lend-lease aircraft carrier HMS *Ameer* (D-01) and an escort. A plane went to cover **Convoy GN-101**. A plane sighted one merchant vessel. A plane departed to cover **Convoy NG-403**. A plane departed to cover **Convoy GK-775** and **Convoy KG-75**. A convoy of one small merchant ship and five escorts were sighted.

13th December 1943

B-25 FV-949 FD is airborne, carried out Anti-Submarine Patrol Number 1(1), captain decided to return to base as the PLE of the aircraft had been reached, aircraft landed with nothing to report. B-25 FW-97 FG is airborne for Anti-Submarine Patrol Number 1(3), convoy was sighted, composition of the convoy was 15 merchant ships and two escorts, a destroyer was sighted, patrol was completed and the aircraft landed after an uneventful patrol. B-25 FW-148 FL airborne, Anti-Submarine Patrol Number 1(2), aircraft landed after completing the patrol without incident. Request was received from Comgulfseafron for an aircraft to escort the US submarine USS *Bashaw* (SS-241) until dark, it was in company of two escort vessels, B-25 FR-381 DC is airborne to escort, submarine was sighted with two escort vessels, signals were exchanged with the escort vessels and the aircraft commenced the patrol, convoy was left, aircraft landed. B-25 FW-104 FH is airborne as second part of the escort to the submarine, convoy was met, signals were exchanged, aircraft continued the escort until 10.30 pm, when the convoy was left, aircraft landed with nothing to report. B-25's FW-132 FK, FR-362 CA, FW-132 FJ, FW-149 FN, and FW-145 FM are airborne for air Exercise Number 1, exercise was completed, and the aircraft landed. Liberator BZ-762 LO is airborne and landed at Miami. Liberator BZ-764 LM was airborne and landed at Miami *en route* for San Antonio. Anti-Submarine patrols, B-25's FC, FF, FH, FJ, FK, and FU airborne between 11:54 and 12:07 on Air Exercise Number 7, landed between 3:05 and 3:46 after exercise completion. Anti-Submarine patrols, B-25's FC, FF, FH, FY, and FU were airborne between 4:56 and 5:06 on Anti-Submarine Patrol Number 2, landed between 9:51 and 10:03 after uneventful patrol. Grumman Goose GA airborne between 6:35 and 8:17 on a flight to Rock Sound and back. SS *Francis Parkman* is bound south in Straits towards St. Augustine. SS *Caliente* is bound southwest in Straits. USS *Tattnall* (APD-19), USS *PC-1208*, and HMCS *Oakville* are bound south in Straits towards Palm Beach. USS *Chambers* is bound north in Straits passing Palm Beach. SS *Hobart Baker* is bound south in Straits passing Port Everglades. USS *YAG-19* is bound southwest in Straits. **Convoy NK-582** is bound west in Straits with US Navy plane escorting. SS *Algoma* is bound north in Straits. Submarine USS *Bashaw* (SS-241) is bound northwest passing Northeast Providence Channel. *U-129* is bound east of Abaco. NAS Great Exuma: Two planes went to cover **Convoy NG-403**. A plane went to cover **Convoy KG-75.** A plane departed to cover aircraft carrier HMS *Ameer* (D-01) and one escort. Seven planes landed Great Exuma and departed on training loop.

14th December 1943

Compass swinging carried out by B-25's FV-954 FB, FR-366 CX, FR-381 DC, FV-951 FC, FR-375 CJ, and FV-997 FG, aircraft landed. B-25's FW-104 FB, FV-149 FN, FW-145 FL, FW-154 FN, FW-132 FI, and FL-187 DC are airborne for air Exercise Number 1, aircraft completed the exercise and landed. Request was received from Comgulfseafron for an aircraft to sweep an area, Liberator BZ-760 LT airborne from Windsor Fields, commenced creeping line ahead sweep of area, patrol was completed, and the aircraft landed. A US B-25 3288 was delivered from Kansas City. Liberator BZ-809 LZ is airborne from Windsor Field on creeping line ahead patrol, patrol commenced and carried out, aircraft landed with nothing to report. Night navigation exercise by Anti-Submarine patrols, B-25's FC, F, H, U, and K airborne between 11:49/13 and 00:01. Landed between 03:38 and 04:20 after exercise completion. Anti-Submarine patrols, B-25's FV, FH, FJ, FK, and FM are airborne between 11:58 and 12:36 on Air Exercise Number 7, landed between 3:15 and 4:22 after exercise completion. Air exercise by B-25 FF carried out Air Exercise Number 1, between 11:42 and 4:38. Air Exercise Number 2, carried out by Anti-Submarine patrols, B-25's FV, FJ, FK, FY, and FC between 4:51 and 5:04, landed between 9:49 and 10:06 after uneventful patrol. Liberator LH, for Radar and navigation exercise, airborne at 1:10 and landed at 3:06. Grumman Goose GA airborne at 4:23 on flight to Andros and returned to base at 6:35. USS *YP-426* and USS *YP-435* are bound north in Straits passing Daytona. USCG *CG-83491* and USCG *CG-83492* are bound south in Straits, passing Palm Beach. Tanker *Bulklube* is bound south passing Matanilla Shoal, northwest Bahamas. SS *Francis Parkman* is bound south in Straits towards Port Everglades. USS *YP8-448* and patrol craft USS YP-603 are bound north in Straits towards Port Everglades. USS *Algoma* (AT-34) is bound north in Straits passing Port Everglades. USS *YAG-19* and USS *PC-1175* are bound north in Straits. SS *W. Honaker* is bound northwest in Straits passing

Nicholas Channel. NAS Great Exuma: A plane departed on administrative flight to Nassau. Inter-island transport USS *YAG 21* dropped anchor at Great Exuma. A plane departed to cover **Convoy TG29.13**. Amall merchant vessel sighted. A plane arrived from an administrative flight to Nassau. Inter-island transport USS *YAG-21* lifted anchor and departed.

15th December 1943

Signal was received from Comgulfseafron that a U-boat had been reported by the Direction Finding plot was requested that an aircraft be detailed to sweep this area, B-25 FK-104 FH was airborne to carry out a creeping line ahead sweep of the area, aircraft commenced patrol, patrol was completed and the aircraft landed with nothing to report. B-25 FW-149 FN was airborne to sweep the area creeping line ahead, sweep was commenced, patrol was completed, and the aircraft landed after an uneventful patrol. B-25 FR-366 CZ was airborne for Anti-Submarine Patrol Number 2(3)f, patrol uneventful and was completed, aircraft landed. B-25 FW-123 FJ airborne for Anti-Submarine Patrol Number 2(2), completed landed with nothing to report. B-25 FV-951 FC airborne, completed Anti-Submarine Patrol Number 2(1), without incident landed. B-25's FW-150 FP, FV-954 FB, FV-951 FC, FW-132 FK and FW-145 FM are airborne, landed after completing air Exercise Number 4. Liberator BZ-883 LW is airborne from Windsor Field to carry out Kingsley Exercise, homing procedure "B" was commenced, co-operating ships were contacted, recognition procedure was carried out, signals were exchanged with ships, aircraft was instructed to carry out Mamba, patrol was completed, smoke floats were dropped by HMS *Balfour,* aircraft carried out an attack with one depth charge which did not detonate, aircraft was instructed to carry out a dummy attack on HMS *Balfour,* ship signalled to the aircraft that exercise was completed, aircraft set course for base and landed. Liberator BZ-931 MF is airborne from Windsor Field on the Kingsley Exercise, radio contact was established with the SNO and homing procedure was commenced, ships were contacted, identification signals were exchanged, aircraft was instructed to carry out Cobra patrol, aircraft dropped a parachute flare for the target practice of the ship's crew, another flare was dropped for the port gunner, smoke float was dropped by HMS *Keats,* aircraft carried out an attack with one depth charge, exercise was completed and the aircraft set course for base, landed. Liberator BZ-808 LY is airborne from Windsor for the Kingsley Exercise, radio contact was established with the co-operating ships, ships were contacted, signals were exchanged between SNO and the aircraft, aircraft was instructed to carry out Mamba patrol, which was completed, aircraft was instructed to carry out Lizard, patrol was carried out and completed, a smoke float was dropped by the SNO, aircraft attacked these floats with a depth charge, was established by the ships as a direct hit, exercise was completed and the aircraft set course for base, landed. The first aircraft homed on eight bearings detected by Radar and sighted bearing 290°, aircraft carried out recognition procedure correctly, good radio communication established between aircraft and ship, aircraft remained on patrol for two hours, during which time patrol Mamba was carried out, depth charges dropped on smoke floats, second aircraft homed on nine bearings, detected by Radar sighted bearing 328°, aircraft established radio contact before carrying out recognition procedure, remained for two hours during which time patrols Cobra and Viper were carried out, aircraft also dropped two parachute flares and anti-aircraft shooting was carried out on these smoke flares, smoke floats were dropped and good depth charge attacks were carried out. Third aircraft homed on eight bearings, called on radio before rendezvous with ships, stating course which it was flying, carried out Mamba and Lizard patrols, smoke float was dropped, carried out a good depth attack, returned base. B-25's FV-945 FD, FV-946 FF, FW-176 FJ, and FW-147 FO are airborne, landed after completion of night navigation exercise. Liberator FL-967 LJ airborne, landed at Boca Chica, briefed regarding exercise in cooperation with the US Navy, exercise completed. Night navigation exercise by Anti-Submarine patrols, B-25's FV, FJ, FK, FM, and FF between 11:45 and 11:59/14. Landed between 03:46 and 04:04 after exercise completion. Air exercise by Anti-Submarine patrols, B-25's FB, FU, FK, FF, FH, and FJ between 11:57 and 12:24 on air Exercise Number 6, and landed between 3:04 and 3:47. B-25's FB, FC, FF, FH, and FJ airborne between 5:20 and 5:30 on Air Exercise Number 8, and landed between 8:45 and 9:10. Liberator LH, for Radar and navigation exercise, airborne at 6:39 and landed at 8:40. Grumman Goose GA airborne at 2:25 for Rock Sound and returned at 8:04. USS *Algoma* (AT-34) is bound north in Straits passing Daytona. USS *SC-83486* is bound south in Straits passing Palm Beach. SS *Dartmouth* is bound north in Straits passing Palm Beach. USS *YAG-28* bound southeast in Straits passing Port Everglades. USS *YAG-21* is bound west in Straits. Tanker *Perryville* is bound east in Straits. NAS Great Exuma: A plane went to cover **Convoy GN-102**. Two planes arrived and departed.

16th December 1943

Request was received from Comgulfseafron for an aircraft to sweep an area, B-25 FH-123 FJ airborne on a sweep of the area as detailed, creeping line ahead sweep was commenced and the area covered patrol was completed, aircraft landed with nothing to report. B-25 FR-366 CX airborne, carried out a creeping line ahead sweep, patrol was completed without incident, aircraft landed. B-25 FW-149 FN was airborne for Anti-Submarine Patrol Number 1(2). The captain of the aircraft decided to return to base from position where a front was encountered with cloud base at 300 feet, aircraft landed. B-25 F-132 FK was airborne for Anti-Submarine Patrol Number 1(3)4, completed the patrol although adverse conditions were encountered, aircraft landed. B-25 FV-949 FD airborne for Anti-Submarine Patrol Number 1(l). B-25 FV-954 FB is airborne for Anti-Submarine Patrol Number 1(4), patrol without incident. B-25's FV-997 FG, FW-123 FC, FW-149 FN, FV-949 FD and FV-946 FF are airborne and landed after exercise completion. Liberator BX-809 LX airborne from Windsor Field on Kingsley Exercise, contact was established and homing procedure "B" was commenced, co-operating ships were sighted, recognition signals were exchanged, aircraft was instructed to carry out Mamba patrol whilst on Mamba patrol a merchant ship of 5,000 tons with one escort was sighted, patrol was completed, smoke float was dropped by SNO and an attack was carried out by the aircraft with one depth change, exercise was completed and the aircraft set course for base, landed. Liberator BZ-807

LV is airborne on Kingsley Exercise, radio contact was established with SNO homing procedure was commenced, cooperating ships were sighted, signals exchanged, smoke float was dropped by SNO for target practice, attack carried out by aircraft with one depth charge, ricocheted before detonating, signal received from base to return to base, visibility was deteriorating when aircraft left, convoy landed. Liberator BZ-812 MC is airborne from Windsor Field Kingsley Exercise, recalled, visibility was deteriorating, landed. Night local flying training by Number 1 Squadron was cancelled. Night navigation exercises by Number 2 Squadron were cancelled. B-25 FL-167 AX is airborne for local flight, landed and damaged the nose wheel, aircraft took off again and crashed, landed, no casualties. The Kingsley exercise was completed this day, first aircraft homed on 10 bearings, detected by Radar, sighted, bearing 214°, carried out recognition procedure, but too far away, radio communication was established by both ships, aircraft remained in air for only 2 hours 30 minutes, during which time ordered to carry out Mamba patrol, did not understand this and had to be put on bearing from which he lost himself and had to be re-homed, carried out excellent depth charge attack on smoke float and returned to base. The second aircraft was homed on 11 bearings and detected by Radar 9 nm and sighted on bearing 140° was twice asked for magnetic courses, good radio communication was established, too far away for recognition procedure, remained for 1 hour 30 minutes in air, visual signalling communication was very poor, carried out fair depth charge attack but short. Neither aircraft reported radio callsign nor remained until asked. Third aircraft did not arrive being recalled due to weather. Anti-Submarine patrols, B-25's FB, FK, FF, FH, and FJ airborne between 12:20 and 12:35 on Air Exercise Number 8, landed between 3:46 and 4:09 after exercise completion. Anti-Submarine patrols, B-25's FB, FK, FC, FH, and FU airborne between 5:18 and 5:38 on Air Exercise Number 6 and returned after exercise completion between 8:22 and 9:07. Destroyer escort USS *Lowe* (DE-325) and USCG *CG-83403* are bound northeast passing Matanilla Shoal, northwest Bahamas. Ferry *Jean Brillant* and USS *SC-1292* are passing Northwest Providence Channel west then south in Straits. USS *YAG-19* is bound south in Straits. **Convoy KN-283** is bound east in Straits with US Navy blimp. **Task Unit TU-0314** is bound northwest in Straits. NAS Great Exuma: Four planes escorted **Convoy GN-102.** Two planes landed at Great Exuma.

17th December 1943

Request was received from Comgulfseafron for an aircraft to provide air escort to **Convoy GN-102** consisting of 25 merchant ships and four escorts, cover was required from dawn to noon. An aircraft was detailed to be airborne to provide escort to the convoy, signal was received reporting that the convoy was behind time, and that cover was requested from 8.00 pm, take off of the first aircraft was postponed, second aircraft was detailed to provide cover from 8.00 pm, but was cancelled due to adverse weather conditions. Request received from Comgulfseafron for an aircraft to sight and sweep ahead of SS *Chiksakia*, convoy consisted of a naval tanker and a destroyer, detailed aircraft was cancelled due to adverse weather conditions. B-25's FR-362 CA, FV-997 FG, FW-132 FJ, FH-149 FN, and FV-951 FC are airborne for air Exercise Number 3 and Number 4, aircraft landed.

The afternoon programme was cancelled due to adverse flying conditions, heavy rains. **Convoy NK-583** is bound southwest in Straits towards Palm Beach with US Navy plane escorting. Tanker *E. R. Kemp* is bound northwest near Matanilla Shoal, northwest Bahamas. **Convoy KN-283** is bound north near Matanilla Shoal, northwest Bahamas with US Navy plane escorting. **Convoy GN-102** in Straits is bound north with US Navy plane escorting. SS *Esselen* is bound northwest in Straits towards Miami. USS *SC-1308* and USS *SC-1339* are bound north in Straits. USS *YAG-19* is bound southwest in Straits. SS *George W. Barnes* is bound southeast in Straits. NAS Great Exuma: Two planes departed to cover **Convoy GN-102**. A plane landed at Great Exuma. A plane departed on a training loop and returned.

18th December 1943

Request was received from Comgulfseafron for an aircraft to sweep an area, two aircraft were detailed to take off, cancelled in view of the adverse weather conditions, two further aircraft were detailed to be airborne to carry out the sweep. B-25 FH-123 FJ was airborne and carried out a sweep of the area, bounded aircraft completed the patrol and landed. B-25 FV-949 FD airborne, completed a sweep of area, aircraft landed. B-25 FW-104 FB airborne for Anti-Submarine Patrol Number 1(2), convoy of 21 merchant ships and five escorts was sighted, patrol completed landed. B-25 FW-149 FN is airborne for Anti-Submarine Patrol Number 1(3), convoy of 21 ships and five escorts was sighted, ships did not reply to aircraft challenge, the convoy was again sighted, consisted of 19 merchant ships and four escorts, patrol was completed without incident and the aircraft landed. B-25 FV-951 FC Anti-Submarine Patrol Number 1(4), convoy of 21 merchant ships and five escorts was sighted, patrol was completed without incident, convoy was again sighted, was observed that two of the ships had left the convoy, remaining ships were continuing on a course of 322°, aircraft landed. B-25 FV-946 FF airborne for Anti-Submarine Patrol Number 1(1), convoy of 19 merchant ships and four escorts sighted, patrol was completed, and the aircraft landed. B-25's FR-366 CX, FV-997 FG, FV-954 FB, FL-200 BX, and FW-148 FL and FW-147 FO airborne, completed air exercises Number 6 and landed. SS *E. R. Kemp* escorted by USS *SC-1282* is bound north in Straits towards St. Augustine. SS *Sweetwater* is bound north in Straits. SS *Missionary Ridge* is bound southwest in Straits. **Convoy NG-404** is bound south, to the east of Bahamas with US Navy plane escorting. USS *Tattnall* (APD-19) is bound north in Straits towards Palm Beach. **Convoy NK-583** is bound southwest in Straits with US Navy plane escorting. Tug USS *Esselen* (AT-147) with two tows, assisted by USS *YT-338* escorted by USCG *CG-Marion* are bound north in Straits. SS *Potter* and **Convoy GK-776** are bound southwest towards Key West. SS *W. Grama* is bound west in Nicholas Channel. NAS Great Exuma: A plane arrived from San Juan with the Commandant's Aide.

19th December 1943

B-25's FV-951 FC, FW-123 FJ, FV-956 FB, FW-149 FN, and FV-946 FF are airborne and landed on the completion of air Exercise Number 6. B-25's FR-362 CA, FL-200 BX, FV-997 FG, FW-148 *FI,* FW-154 FK, FV-951 *FC,* and FR-366 CX are airborne and carried out air exercises Number *3* and Number 4, aircraft landed. Similar exercises were carried out by the same aircraft later. B-25'8 FV-946 FF, FV-954 FB, FV-949 FD, and FW-149 FN were airborne, completed night navigation exercise, landed. USS *YMS-311* is bound south in Straits passing Canaveral. Ferry *Jean Brillant* and USS *SC-1064* are bound south towards Nassau. USS *PC-1214* is bound south in Straits. Destroyer escorts USS *Gary* (DE-326) and USS *Merrill* (DE-392) are bound northeast in Straits. SS *Ile De Re* and SS *Vermont* are bound southwest towards Key West. NAS Great Exuma: Two planes went to cover **Convoy NG-404**. A plane departed for San Juan with the Commandant's Aide. A plane landed at Great Exuma.

20th December 1943

B-25's FV-946 FF, FW-123 FJ, FW-149 FN, FV-949 FD, FW-148 FL, and FR-366 CX are airborne for air Exercise Number 7, landed after completion. B-25's FW-145 FM, FW-128 FJ, FW-149 FN, FR-393 DF, and FR-362 CA CX are airborne for air Exercise Number 7, landed after completion. Liberator BZ-722 LK are airborne for Miami, and landed there. Liberator BZ-744 LL airborne from Windsor Field, landed at Miami. Local night flying exercise carried out from Oakes Field. SS *J. D. Boss* is bound north in Straits towards Canaveral. Tanker *E. R. Kemp* is bound south in Straits towards Palm Beach. USS *Tattnall* is bound north in Straits passing Palm Beach. SS *Wellesley* is bound southwest in Straits towards Palm Beach. Ferry *Jean Brillant* is bound southwest in Straits towards Miami with US Navy plane escorting. USS *YAG-19* is bound south in Straits. USS *YAG-52* is bound northeast in Straits. USS *YMS-311* is bound southwest in Straits. SS *Fairisle* is bound southwest in Straits. SS *George W. Barnes* is bound southwest in Straits passing Nicholas Channel. USS *PC-1245* and USS *PC-496* and USS *YOG-6* are bound southeast in Straits. NAS Great Exuma: A plane departed to cover **Convoy NG-404**. Two planes arrived and departed. Inter-island transport USS *YAG-20* dropped anchor at Great Exuma.

21st December 1943

B-25's FV-946 FF, FR-366 CX, FW-154 FR, FW-145 FM, and FR-375 CJ are airborne for air Exercise Number 7, landed after completion. B-25's FV-946 FF, FW-148 FL, FV-147 FO, FV-949 FD airborne for air Exercise Number 6, landed after completion. Liberators BZ-768 LT, BZ-808 LY, and BZ-931 MF are airborne for Boca Chica, landed, briefed regarding exercise in cooperation with US Navy on completion, aircraft returned to base. B-25's FK-163 AN, FK-173 AK, FK-177 A0, FK-166 AD, and FR-380 DB are airborne, completed the exercises and landed. B-25's FW-145 FM, FV-997 FG, FW-176 FS, FW-132 FK, FW-154 FB, FV-946 FF, and FW-148 FL are airborne for night navigation, flights landed after completion. USS *SC-1340* is bound south in Straits towards Canaveral. SS *Oklahoma* is bound southwest in Straits towards Palm Beach. USS *YMS-406* is bound north in Straits passing Port Everglades. Destroyer USS *Gary* (DE-326) and tanker *Gulfmaid* and **Convoy KN-284** are bound northeast in Straits. SS *Samoset* is bound south in Straits toward Cay Sal Bank. USCG *CG-83491* and USCG *CG-83492* are bound west in Straits. **Task Unit TU-0314** is bound northwest in Straits toward Key West. **Convoy KG-677** is bound southeast in Straits passing Nicholas Channel with US Navy plane escorting. NAS Great Exuma: Two planes went to cover **Convoy GN-103**. A plane went on administrative flight to Nassau. Inter-island transport USS *YAG-20* lifted anchor and put out to sea. Two planes arrived and departed.

22nd December 1943

Liberators FL-970 LH, BZ-883 IK, BZ-931 MF, and BZ-812 MC are airborne for Boca Chica aircraft landed at Boca Chica and were briefed for the exercises in co-operations with the US Navy on the completion of the exercises the aircraft returned to base. B-25's FV-946 FF, FW-176 FS, FW-154 FR, FN-154 FM, FW-104 FH, and FW-147 FD are airborne for compass-swinging exercise aircraft landed on the completion of the exercises. B-25's FR-393 DF, FR-380 DB, FK-173 AK, and F-180 BA are airborne for local navigation flight landed after completion. Further exercises were carried out by Number 1 Squadron. B-25 FR-376 CK is airborne for local navigation Exercise Number 3, route between base, South Bight, Andros, and Elbow Cay, Abaco; aircraft failed to return from the flight. Tanker *Gulfmaid* is bound north in Straits towards Daytona. **Convoy NK-584** is bound south in Straits passing Palm Beach with US Navy plane escorting. SS *Dartmouth* is bound south in Straits towards Bimini. SS *Bulklube* is bound north in Straits towards Bimini. Ferry *Jean Brillant* is bound south towards Nassau. USS *YP-430*, USS *YP-428*, and USS *YP-431* are bound north in Straits towards Miami. SS *Evans Creek* is bound south in Straits towards Cay Sal Bank. **Convoy GN-103** is bound north with US Navy plane escorting. *U-129* is bound north exiting area north of Bahamas. NAS Great Exuma: Two planes departed to cover **Convoy GN-103**. A plane departed to cover **Convoy NG-405.** Six planes arrived and departed from Guantanamo.

23rd December 1943

A search was conducted for the missing aircraft FR-376 CK, which failed to return from a local night navigation flight. Eleven B-25'A aircraft from Oakes Field, and two Liberators from Windsor Field took part in the search, B-25's was airborne to search an area bounded by Southwest Point, Grand Bahama, Morgan's Bluff, North Andros, High Point Cay, South Andros, and Harvey's Cay, Central Exuma Cays. Aircraft landed at base; nothing was sighted. The two Liberators were airborne to search an area surrounding that searched by the B-25 aircraft, then landed, having failed to locate the missing aircraft. Signals requesting information of any crashed aircraft were sent to the Commissioners on Bimini, Andros, Abaco and Eleuthera, yet produced no results. Eleven B-25 aircraft and two Liberators were airborne to carry out a repetition of the searches conducted in the morning. The aircraft landed between 22:04 and 01:56/24, but again nothing was sighted, appeal for any information regarding the lost aircraft was broadcast by the Government Radio Station. Liberator BZ-758 LS returned from Miami and landed at Windsor Field. B-25's FV-945 FD, FV-946 FF, FW-147 FO, FW-145 FM, FW-154 FK airborne landed after completion of night navigation exercise. USS *YP-430*, USS *YP-428*, and USS *YP-431* are bound north in Straits towards Daytona. Irish SS *Irish Plane*, USS *YAG-20* and ferry *Jean Brillant* escorted by USS *SC-1342* are bound south in Straits to Miami. USS *YAG-19* is bound south in Straits. USS *PCS-1424* is bound northeast in Straits. **Convoy NK-584** is bound west in Straits with US Navy plane escorting. SS *Colabee* escorted by USS *SC-1058* is bound northwest in Straits near Nicholas Channel. USCG *CG-83400*, USCG *CG-83401*, USCG *CG-83398*, and USCG *CG-83399* are bound southeast in Straits near Nicholas Channel. **Convoy Key West-301** is bound southeast in Straits near Nicholas Channel. USS *Custer* (APA-40), USS *Mauna Loa* (AE-58), and high-speed transport USS *Orizaba*

(AP-24), escorted by destroyers USS *DD-583*, USS *DD-584* and destroyer escort USS *DE-179* are bound south in Straits with US Navy plane escorting. NAS Great Exuma: Two planes went to cover **Convoy NG-405**. Five naval escorts arrived. A plane departed to cover **Convoy TG.29.7** and departed. Four ships were sighted.

24th December 1943

B-25 FW-147 FO is airborne for Anti-Submarine Patrol Number 1(3), a merchant ship of 6,000 tons was sighted. Aircraft completed the patrol and landed. B-25 FW-149 FM is airborne for Anti-Submarine Patrol Number 1(4), a tanker of 10,000 tons was sighted, aircraft landed on the completion of the patrol. B-25 FR-375 CJ is airborne for Anti-Submarine Patrol Number 1(1), patrol is uneventful and the aircraft landed with nothing to report. B-25 FR-393 DF airborne for Anti-Submarine Patrol Number 1(2), merchant ship of 2,000 tons was sighted, tanker of 6,000 tons to 8,000 tons was sighted, aircraft landed after completing the patrol. B-25 FW-150 FF is airborne for Anti-Submarine Patrol Number 1(6), landed after completing the patrol without incident. B-25 FW-123 FK is airborne for Anti-Submarine Patrol Number 1(f), landed, tanker of 6,000 tons to 8,000 tons was sighted. Liberator BZ-761 LN is airborne for Miami and landed there. SS *George Davidson* and SS *Vacuum* are bound south in Straits passing St. Augustine. SS *Monterey* is bound southwest in Straits towards Palm Beach. SS *Opequon* is bound southwest in Straits towards Palm Beach. USCG *CG-83431* is bound northeast in Straits. USS *YAG-19* is bound north in Straits towards Miami. USS *YO-142*, USS *YP-579*, USS *YOG-6* escorted by USCG *CG-83454*, and HMS *Kilmartin* (ex-USS *PC-838*) are bound north in Straits. Irish SS *Irish Plane* is bound south in Straits. NAS Great Exuma: A plane went to cover **Convoy NG-405.** Two planes departed to cover **Convoy TG.29.7** with ships Gilmartin and two escorts. Three planes arrived and departed.

25th December 1943

USS *Anthedon* (AS-24) is bound north in Straits. Tanker *Gulfmaid* is bound south in Straits passing Canaveral. Destroyer USS *DD-217* and destroyer escort USS *DE-169* are bound northwest. Tanker *Vacuum* is bound south in Straits passing Port Everglades. SS *George Davidson* is bound south in Straits. Tug USS *Warbler* is bound northeast in Straits. SS *Hawaiian Shipper* is bound southwest in Straits near Cay Sal Bank. Destroyer escorts USS *DE-327* and USS *DE-790* are bound north in Straits near Miami. Troop transport USS *USAT-McClure* is bound southeast in Straits. *U-516* enters area south via Anegada Passage. No.113: Christmas; a unit-wide holiday, except for personnel on essential duty. HRH Duke of Windsor visited Windsor Field and the Airmen's Mess during the serving of the airmen's Christmas dinner. NAS Great Exuma: A plane went to cover **Convoy GN-104**. Two planes arrived and departed from an administrative flight to Nassau.

26th December 1943

Tug USS *Warbler* is bound north in Straits passing Canaveral. USCG *CG-83348* is bound north in Straits towards Canaveral. SS *Fairland* is bound southwest in Straits. SS *Samuel Seabury* and SS *Howell Lykes* are bound south in Straits towards Palm Beach. Destroyer escort USS *Buckley* (DE-51) is bound north in Straits towards Palm Beach. Tanker *Pan Massachusetts* is bound south in Straits passing Palm Beach. Ferry *Jean Brillant* is bound south towards Nassau. USS *YMS-60* is bound northeast in Straits from Miami. USS *YP-579* is bound south from Miami. Tanker *Gulfmaid* is bound south in Straits. SS *Sloterdyk* is bound southwest in Straits. Tanker *Hadnot* is bound northwest in Straits passing Nicholas Channel. *U-516* exits area north via Anegada Passage, east for France. A tank found by a civilian on the shore near Dick's Point, Montagu Bay, Nassau. Tank was found to be from the missing aircraft CK which failed to return from a flight on 22nd December 1943. NAS Great Exuma: Two planes went to cover **Convoy GN-104**. A plane arrived from an administrative flight to Guantanamo and landed.

27th December 1943

USS *YT-338* is bound north in Straits passing Canaveral. **Convoy NK-585** is bound southeast in Straits passing Canaveral. Ferry *Jean Brillant* escorted by *SC-1292* is bound southwest passing Northwest Providence Channel with US Navy plane escorting. **Convoy KN-285** is bound north in Straits passing Bimini with US Navy plane escorting. USS *YO-70* is bound south in Straits passing Palm Beach. USS *YAG-19* is bound south in Straits. Cuban *CS-11,* Cuban *CS-12,* Cuban *CS-13,* and Cuban *CS-23* are bound northeast in Straits. **Convoy KG-678** is bound southeast in Straits with US Navy plane escorting. No.113: Category A aircraft accident reports; a Baltimore Mk.V. FW-603's undercarriage collapsed during run-up of engines on the runway apron. The only occupant was leading aircraft man LAC J. F. Jackson (IIE), was uninjured. NAS Great Exuma: A plane went to cover **Convoy GN-104**, of 32 merchant vessels and five escorts. A plane landed at Great Exuma. Four planes arrived and departed.

28th December 1943

Two days previously, on 26th December, 1943 a self-sealing petrol tank was found, by a civilian on the shore near Dick's Point, Montagu Bay, Nassau. Tank was found to be from the missing aircraft FR-376 CK, which failed to return from a flight on 22nd December 1943. Today it is reported that a dinghy had been found floating in Montagu Bay. The dinghy had been taken to Athol Island, where the Commanding Officer and the MCS Officer investigated it. On interrogation of a native on the island, it was ascertained that an aircraft had crashed into the sea off Athol Island; a visual search was made for the aircraft by the ASR, launch, and many pieces of wreckage were seen, including part of the fuselage bearing the letters CK. No trace of any bodies was visible aircraft appeared to have disintegrated. B-25 FR-375 CJ is airborne for Anti-Submarine Patrol Number 2(1), completed uneventful patrol and landed. B-25 FW-176 FS airborne for Anti-submarine Patrol Number 2(3), US naval tug was sighted, and aircraft landed after patrol completion. The B-25 FW-123 FK is airborne, Anti-Submarine Patrol Number 2(4), small craft sighted, and aircraft landed after patrol completion. B-25 FW-104 FH airborne for Anti-Submarine Patrol Number 2(6), completed uneventful patrol and landed. B-25 FW-147 FO is airborne on Anti-Submarine Patrol Number 2(5), completed uneventful patrol and landed. B-25 FR-393 DF is airborne for Anti-Submarine Patrol Number 2(2). merchant ship sighted, then completed uneventful patrol and landed. USCG *CG-83487*, USCG *CG-83493*, and USS *YP-191* are bound south in Straits passing Daytona. USS *Kilmarnock* (BCE-11) is bound north in Straits passing Daytona. SS *Missionary Ridge* is bound north passing Matanilla Shoal, northwest Bahamas. Tug USS *Owl* (AT-137) is bound north in Straits. **Convoy NK-585 is bound south** in Straits passing Port Everglades with US Navy plane escorting. SS *H. G. Blasdel* is bound north in Straits towards Bimini. USS *PC-1176* is bound south in Straits. SS *Brandy Station*, destroyer escorts USS *Spangler* (DE-696) and USS *Neundrop* (DE-200) and destroyers USS *Capps* (DD-559), USS *Corry* (DD-463), USS *David W. Taylor* (DD-551), and USS *Pitch* (DD-462) are bound north in Straits. SS *Wellesley* is bound northeast in Straits near Cay Sal Bank. **Convoy GK-778** is bound northwest in Straits passing Nicholas Channel with US Navy plane escorting. NAS Great Exuma: Three planes escorted **Convoy NG-406**. Five planes landed and departed from Great Exuma.

29th December 1943

B-25 FW-176 FS is airborne for Anti-submarine Patrol Number 1(1), however aircraft was unable to continue the patrol due to adverse weather conditions, so it returned to base and landed. B-25 FW-149 FH is airborne for Anti-Submarine Patrol Number 1(5), completed the patrol without incident and landed. B-25 FW-150 FP is airborne for Anti-Submarine Patrol No.1(3): the captain decided to abandon the patrol in view of the bad visibility, and the aircraft landed. B-25 FH-147 FO airborne for Anti-Submarine Patrol Number 1(2), patrol was completed and uneventful, and the aircraft landed. B-25 FW-104 FH, FK-149. FW-132 FK, FR-375 CJ, FW-147 FO and FW-148 FL airborne landed on the completion of air Exercise Number 1, Liberator BZ-762 LO airborne for Miami landed there. SS *Harpers Ferry* is bound southwest in Straits. USS *YP-191*, USCG *CG-83487* and USCG *CG-83493* are bound south in Straits towards Palm Beach. Ferry *Jean Brillant* is bound south towards Nassau. SS *Pine Ridge* is bound southwest in Straits towards Cay Sal Bank. *U-193* re-enters area bound north from the Windward Passage to south of Turks and Caicos, then heads east for France. No.113: Wireless Operator P/O W. S. Steele, was today travelling by RAF bus from the Royal Victoria Hotel to Windsor Field when he was unfortunately injured when the bus struck a lorry. P/O Steele's arm was outside the window, and was seriously crushed, necessitating its amputation. He has since been discharged from the hospital and proceeded to Dorval, outside Montreal, for Medical Boards. NAS Great Exuma: A plane went to cover **Convoy NG-406.** Two planes arrived and departed from Guantanamo.

30th December 1943

B-25's FW-104 FH, FW-149 FN, FW-132 FK, FW-147 FO, FW-176 FS, and FR-375 CJ are airborne for air exercises Number 7, landed on the completion of the exercise. B-25's FW-104 FH, FW-150 FP, FR-393 DF, FW-147 FO, and FR-366 CX are airborne for air Exercise Number 5, aircraft landed on the completion of the exercise. B-25's FV-945 FD, FV-946 FF, FW-176 FS, FW-347 FO, FW-145 FM, and FW-154 FN are airborne, aircraft landed on the completion of the night navigation exercise. Liberator FL-967 LJ is airborne, and landed at Boca Chica. Then briefed for the exercises in co-operations with the US Navy. On the completion of the exercises, the aircraft returned to base. B-25's FV-946 FF, FW-132 FK, FR-393 DF, FV-954 FB, and FW-104 FH are airborne, aircraft landed after completion. USS *PC-1413* and USS *YMS-391* are bound south in Straits towards Canaveral. USCG *CG-83491* is bound south in Straits passing Palm Beach. USCG *CG-83434* is bound northeast in Straits. Tanker *Quemadolake* is bound south in Straits. *U-193* is southeast of Turks and Caicos, off northwest Haiti, bound east for France. NAS Great Exuma: Three planes went on training loop and then returned.

31st December 1943
B-25's FV-946 FF, FV-954 FB, FW-132 FK, FK-362 CA, FW-148 FL, and FK-393 DF are airborne for air Exercise Number 5, then aircraft landed. Liberator FL-994 LA and FL-968 LF are airborne from Windsor Field for Miami. Then the aircraft landed at Miami and ultimately returned to base. B-25 FW-104 FH is airborne for Anti-Submarine Patrol Number 1(1), convoy of 23 merchant ships and five escorts was sighted, convoy was escorted by a US Navy Mariner aircraft. Patrol was completed and the aircraft landed, nothing to report. B-25 FW-150 FP airborne for Anti-Submarine Patrol Number 1(2), convoy was sighted, composition was 23 merchant ships and five escorts, a naval escort vessel was sighted, aircraft landed. B-25 FW-132 FK airborne for Anti-Submarine Patrol Number 1(5), patrol was completed and the aircraft landed with nothing to report. B-25 FV-951 FC is airborne for Anti-Submarine Patrol Number 1(3), during which a convoy of 23 merchant ships and five escorts is sighted. The patrol is completed without incident, and aircraft landed. B-25 FV-954 FB airborne for Anti-Submarine Patrol Number 1(4), aircraft landed after completing and uneventful patrol. Tanker *Chiloil* is bound south in Straits towards Daytona. USS *SC-1353* is bound south in Straits passing Daytona with US Navy plane escorting. *USS PC-469* is bound south in Straits passing Daytona. USS *YO-142* escorted by USCG *CG-83456* is bound southwest in Straits passing Northwest Providence Channel. USS *YMS-391* is bound south in Straits. **Convoy KN-286** contingent, with one merchant ship and USS *SC-1064* escorting, is bound north in Straits. Destroyer escort USS *Haverfield* (DE-393) and destroyer escort USS *Swenning* (DE-934) are bound east in Straits. **Convoy KN-286** is bound northeast in Straits with US Navy plane escorting. **Convoy KG-679** is bound southeast in Straits towards Nicholas Channel with US Navy plane escorting. Minor accident, 1943: Liberator bomber BZ-806 possibly LU, Pratt and Whitney engines, had its airframe, but not engines, slightly damaged on landing at No.111 OTU. On approach to land, the undercarriage was lowered and checked the usual way, by green light. Hydraulic action test was made, and visual check was made for locks, however shortly after touch down, the nose wheel gear collapsed. When the aircraft came to rest, the underside of its nose structure was damaged. The aircraft was fitted with old style locking latches. The latch jumped from locked position in response to the impact of landing. A local modification was fitted to latch after the incident, and an investigation was carried out. At No.113: The 1,000th aircraft to be dispatched from the unit since its formation departed. S/Ldr. D. Jeremiah died in RAF Station Hospital, Nassau from natural causes, namely Addison's disease. NAS Great Exuma: Three planes went to cover **Convoy GN-105**. Six planes departed for, and landed at, the US Navy base NAS Great Exuma.

Part V

Stability:
January 1, 1944 to December 30, 1944

1st January 1944
B-25's FV-951 FC, FV-946 FF, FW-123 FK, FR-393 DF, and FV-954 FB are airborne and landed after completion of air exercise. B-25 FR-393 DF is airborne for air Exercise Number 3 and landed after completion of air exercise. B-25's FV-946 FF, FV-951 FC, FV-949 FD, FW-123 FK, and FV-954 FB are airborne for air Exercise Number 14 and landed after completion of air exercise. Tanker *18 De Marzo* is bound south in Straits passing Daytona. USS *SC-1280* and Russian RU *AM-145* are bound north in Straits passing Canaveral. **Convoy KN-286** is bound northeast in Straits, US Navy plane escorting. USS *PC-1176* is bound north in Straits near Palm Beach. SS *Chiloil* is bound south in Straits passing Palm Beach. **Convoy NK-586** is bound south in Straits passing Miami with US Navy plane escorting. NAS Great Exuma: Five planes returned to base, two planes departed to Banana River, one to Nassau and the other to Guantanamo.

2nd January 1944
B-25's FV-946 FF, FW-123 FK, FV-954 FB, FV-951 FC, FW-132 FK, and FW-149 FN are airborne for air Exercise Number 3 and Number 4 and landed after completion of air exercise. B-25's FW-154 FR, FW-104 FH, FV-949 FD, FB-954FB, FV-951 FC, and FW-145 FM are airborne for night navigation exercise landed after completion of air exercise. SS *Eano* is bound south in Straits towards Canaveral. USS *YMS-353* is bound south in Straits towards Palm Beach. USS *YT-338* is bound south in Straits towards Port Everglades. SS *George W. Kendall* is bound north in Straits nearing Bimini. Tanker *18 de Marzo* is bound south in Straits. SS *Malay* and USS *Lucid* (AM-259) are bound northeast in Straits. **Convoy GK-779** is bound northwest in Nicholas Channel with US Navy plane escorting. **Convoy NG-407** is bound southeast in Straits with US Navy plane escorting. NAS Great Exuma: Two planes went to cover **Convoy NG-407** at different times, and then arrived back to base.

3rd January 1944
Liberators FL-964 LD, BZ-758 LS, and BZ-744 are airborne from Windsor Field. They landed at Boca Chica and were briefed regarding exercises, then returned after completion. Liberator FL-994 LA is airborne from Windsor landed at Miami, returned from Miami, and landed at base. B-25's FV-946 FF, FW-123 FJ, FW-149 FN, FV-951 FC, and FW-123 FK are airborne for air Exercise Number 6 and landed after completion. B-25's FV-946 FF, FW-123 FJ,

FW-149 FN, FR-381 DC, and FV-949 FD are airborne for night navigation exercise, landed after completion. B-25's FV-945 FD, FV-946 FF, FW-176 FS, FW-147 FO, FW-145 FM, and FW-154 FR are airborne and landed after exercise completion. Liberator FL-967 LJ is airborne and landed at Boca Chica, briefed regarding exercise, completed exercise, then landed back at base. Tanker *Oklahoma* is bound north in Straits passing Palm Beach. Tanker *Evans Creek* is bound north passing Matanilla Shoal, northwest Bahamas. USS *PC-1212* and USS *PC-1195* are bound north in Straits. Tanker *Eano* and USS *YT-338* are bound south in Straits. Cuban gunboat *CS-Cuba* and tanker *Opequon* are bound north in Straits. USS *Knave* (AM-256) are bound west in Straits. SS *Seatrain New Orleans* is bound southwest in Straits. NAS Great Exuma: One plane went to cover **Convoy NG-407** and then returned. Three planes departed for training. Seven USN PBY aircraft arrived back from Jacksonville. Four other planes returned back to base.

4th January 1944

B-25's are airborne for air Exercise Number 7 and landed after completion. B-25's are airborne for air Exercise Number 6 and landed after completion. Liberators are on airborne navigation and homing exercise, landed after completion. Local night flying carried out. SS *Roger Sherman* is bound south in Straits passing St. Augustine. Tanker *Cedar Creek* is bound southwest in Straits. USS *PCS-1388* is bound south in Straits towards Palm Beach. Tanker *Vitruvius* is bound north in Straits past Port Everglades. Destroyer escort USS *Gustafson* (DE-182) is bound south in Straits passing Miami. USS *LCI(L)-372* is bound south in Straits. Destroyer USS *Evans* (DD-552) and tanker *Pan Massachusetts* are bound northeast in Straits. USS *PC-1085* is bound east in Straits. NAS Great Exuma: Fire drill. Seven PBY's departed for Key West. Three planes departed for training, and returned. One plane departed for Banana River.

5th January 1944

B-25's are airborne for air Exercise Number 6 and landed after completion. Aircraft of Number 6 Squadron airborne for air Exercise Number 1, Number 2, and Number 3 and landed after completion. Aircraft of Number 2 Squadron are airborne completed local night navigation exercise landed. Liberator BZ-761 LW returned from San Antonio via Miami and landed in Windsor Field. SS *Cape Alva* is bound south in Straits, passing St. Augustine. Tanker *Vitruvius* is bound north in Straits, passing Daytona. USS *PC-611* is bound north in Straits, passing Canaveral. USCG *CG-83434* is bound west near Walker Cay, north of Abaco. Destroyer escort USS *Gillette* (DE-681) and USS *PC-1139* are bound south in Straits. SS *Roger Sherman* south in Straits passing Palm Beach. SS *William B. Wilson* is bound northeast in Nicholas Channel. NAS Great Exuma: Two planes departed to cover **Convoy GN-106** and returned. One plane arrived from Guantanamo and another departed to there. One plane departed for training, and another to Banana River.

6th January 1944

B-25's of Number 1 Squadron and Number 2 Squadron are airborne at different times for local night navigation exercise, landed after completion. B-25's FV-946 FF, FW-149 FN, FW-951 FG, FW-123 FJ, FR-366 CK, and FR-378 CL are airborne for air Exercise Number 6 and landed after completion. B-25's of Number 1 Squadron are airborne for local navigation exercise, landed after completion. Night navigation exercise carried out by B-25's FW-149 FN, FW-951

FG, FV-946 FF, and FW-123 FJ. FR-381 LG is airborne for night navigation exercise, landed after completion. Liberator BZ-759 LP is airborne, *en route* to San Antonio landed at Miami. SS *George W. Kendall* is bound south in Straits towards Daytona. USS *PC-576* is bound south in Straits passing Canaveral. Tug USS *YMS-407* is bound north in Straits. **Convoy KN-287** is bound north in Straits with US Navy plane escorting. Irish SS *Irish Plane* is bound north in Straits passing Palm Beach. SS *Pardo* is bound south passing Matanilla Shoal, northwest Bahamas. USS *YAG-19* is bound south in Straits. SS *Cape Alva* is bound west in Straits. HMS *Kimelford* is bound northeast in Straits. *U-193* exits area, bound east for France. NAS Great Exuma: Two planes departed to cover **Convoy GN-106** and returned. A Bahamas Airways seaplane arrived, and then returned to Nassau. A Naval Air Transport seaplane arrived from Miami, then flew to Guantanamo.

7th January 1944

Liberators FL-970 LH, FL-967 LJ, and BZ-812 MC are airborne from Windsor Field and landed at Boca Chica. There they were briefed for the Air Submarine Exercises. Aircraft LJ landed at Windsor on the completion of the exercise, and FL-970 LH. Later BZ-812 MC returned to Boca Chica with engine defects and stayed overnight. Liberator BZ-813 MD, which landed at Bermuda on the completion of the first day's Kingsley exercise, returned to base and landed at Windsor Field. B-25's FW-179 FT, FW-176 FS, FW-145 FM, FR-366 CX, and FW-123 FJ are airborne and landed on the completion of the compass swinging exercises. Liberator BZ-806 LU is airborne from Windsor Field for the Kingsley Exercise, radio contact was established, convoy bearings were obtained and homing procedure "B" was carried out, bearings were received, aircraft did not meet the convoy, "Not met" signal was sent to base, aircraft set course for base and landed. Liberator BZ-869 MX was airborne, radio contact established, signals were exchanged homing procedure was commenced two ships were sighted visual signals were exchanged, aircraft was instructed to carry out Mamba patrol which was completed Cobra and Lizard patrols were carried out under the instructions of the SNO. Before the aircraft left the ships, two depth charges were dropped against a smoke float aircraft landed at Windsor Float. Liberator BZ-746 LM airborne from Windsor Field, radio contact established, homing exercise commenced, bearings given to the aircraft which was located visual exchange with sighted ships, aircraft instructed to carry out Cobra patrol, smoke float dropped by the SNO, attack made by aircraft with one depth charge, Lizard patrol was carried out on instructions of the SNO. Aircraft requested to drop flare for A-A practice, convoy left, and aircraft landed at Windsor Field. Liberator BZ-931 MF is airborne, radio contact established, homing procedure "S" was commenced, aircraft homed on bearings given, convoy located, visual signals exchanged, aircraft instructed to carry out Cobra patrol, attack with depth charges carried out against smoke float dropped by SNO, left convoy and landed. Destroyer escort USS *Buckley* (DE-51) and USS *Lance* (AM-257) are bound south in Straits. Tug USS *Scotch Cape*, **Task Unit-0293**, tanker *Baldbutte*, tanker *Britamsea*, tanker *Point Pelee Park*, SS *Libre*, SS *Colin*, SS *New York*, and SS *Llangollen* are bound south in Straits towards Palm Beach. USCG *CG-Ironwood* and USCG *CG-Mesquite* are bound south in

Straits towards Port Everglades. Tanker *George W. Kendall* and tug USS *Raddy* (ATR-97) are bound southwest in Straits. SS *George M. Cohan* is bound north in Straits, passing Miami. **Convoy GK-780** is bound northwest in Straits with US Navy plane escorting. **Convoy NG-408** is bound south in Straits navy plane escorting. NAS Great Exuma: Battle drill. Inter-island transport USS *YAG-20* arrived with supplies, one plane went to cover **Convoy NG-408**, two planes arrived from Guantanamo, and one went to Florida.

8th January 1944

B-25's FW-145 FM, FV-951 FC, FV-954 FB, FV-997 FG, FW-154 FR, and FW-104 FH are airborne and landed on the completion of the compass swinging exercise. Liberator BZ-812 MC, which has been at Boca Chica with engine defects since yesterday, returned to base today and landed. USS *YP-603* and USS *YMS-392* are bound south in Straits. USS *John M. Howard* (IX-75) is bound south in Straits, passing St. Augustine. Destroyer escort USS *Willis* (DE-395) and destroyer escort USS *Kretchmer* (DE-329) are bound northeast in Straits. USS *Apollo* (AS-25) is bound north in Straits. SS *Quemado Lake* is bound north near Matanilla Shoal, northwest Bahamas. Submarine USS *Bluegill* (SS-242) escorted by USS *SC-1060* and USS *SC-1064* are bound northwest in Straits, passing Northwest Providence Channel, with a US Navy plane escorting. Tug USS *Scotch Cape* towing the auxiliary floating dock USS *AFD-10* and **Task Group TG-23**, escorted by USS *SC-1282*, are bound southeast in Straits with a US Navy plane escorting. SS *Fishers Hill* is bound southwest in Straits. SS *New York*, SS *Liebre*, SS *Colin*, and SS *Baldbutte* are bound west in Straits. SS *Llangollen* is bound east in Straits. NAS Great Exuma: Two planes went to escort **Convoy NG-408**. One plane sighted a small tanker and a US Navy aircraft carrier.

9th January 1944

FL-994 LA is airborne from Windsor Field, landed at Miami again, then was airborne and landed at Windsor Field. SS *John Owen* is bound south in Straits, passing Canaveral. Tanker *E. R. Kemp,* SS *David O. Saylor*, and USS *John M. Howard* (IX-75) are bound south in Straits towards Palm Beach. Destroyer escort USS *Edsall* (DE-129) is bound north in Straits, passing Palm Beach. HMS *Kilmartin* (BEC-12) and USS *PC-1208* are bound southwest in Straits. Submarine USS *Bluegill* (SS-242) with escorts USS *SC-1060* and USS *SC-1064* are bound west in Straits, with a Us Navy plane escorting. USS *PC-1141* is bound northeast in Straits. SS *Manassas* is bound southwest in Straits. NAS Great Exuma: Inter-island transport USS *YAG-20* departed for Miami. USS *Flounder* is moving near the base, and one plane departed to cover **Convoy GN-107**.

10th January 1944

Message was received from the Officer Commanding air traffic control (ATC) requesting news of an overdue B-24 aircraft, request was received for aircraft to carry out a search for the missing aircraft in an area southeast of New Providence. Six aircraft should cover area, B-25's FV-997 FG, FV-949 FD, FW-147 FO, FW-145 FM, FR-381 DC, and FV-954 FB are airborne, and cover the area with a distance of 2 nm between the tracks of the patrol. Then aircraft landed, having failed to locate anything. A Liberator and six US Navy Venturas searched from Boca Raton, Florida, and landed, having failed to locate anything, the aircraft left for Boca Raton. Three USN Hudsons and a USN Ventura landed after carrying out a search of the waters between Boca Raton and Nassau, aircraft returned to Boca Raton. B-25's FW-132 FK, FW-154 FR, FW-104 FH, FW-147 F0, and FV-954 FB are airborne and completed air Exercise Number 1, then landed. B-25's FW-145 FM, FW-132 FK, FV-997 FG, FR-393 DF, FW-154 FR airborne and landed on the completion of the exercises. USCG *CG-Mesquite* is bound south in Straits towards Canaveral. USS *RPC-1*, USS *RPC-2*, USS *RPC-3*, and USS *RPC-4* are bound north in Straits passing Canaveral. USS *YAG-19* is bound southwest in Straits. **Convoy NY-52** is bound southwest in Straits, with a US Navy plane escorting. SS *John Owen* is bound west in Straits. The Bahamas Air Service Squadron was formed in Nassau. NAS Great Exuma: Three planes departed to cover **Convoy GN-107** and returned. Submarine USS *Flounder* (SS-251), is moving near the base.

11th January 1944

B-25's FW-145 FW, FW-132 FK, FV-997 FG, FR-393 DF, FW-154 FR, and FR-381 DC are airborne, and air Exercise Number 1 was completed and the aircraft landed. Request received from Officer Commanding Air Traffic Control (ATC) for an aircraft to search Andros for Liberator missing from two days previously. Liberator BZ-744 LL is airborne, carried out local flying and a search of Andros for the missing aircraft, and landed having sighted nothing. B-25's FV-954 FB, FV-104 FH, FV-997 FO, FR-393 DF, FR-381 DC, and FW-132 FK are airborne, navigation flight was completed and the aircraft landed. **Convoy KN-288** is bound north in Straits ,passing Palm Beach with US Navy plane escorting. USS *YMS-392* is bound south in Straits, passing Port Everglades. USS *Patroclus* (ATR-91) is bound north in Straits passing Port Everglades. USS *PC-576* is bound south, to the east of Eleuthera from Northeast Providence Channel. Tanker *Five Forks* is bound southwest in Straits. USCG cutter *CG-Mesquite* is bound west in Straits. **Convoy Key West-310** is bound southeast in Straits, passing Nicholas Channel. NAS Great Exuma: One plane went to cover **Convoy GN-107** and another to cover **Convoy KG-681**. A plane arrived from Guantanamo. One plane departed to Florida and other to Guantanamo.

12th January 1944

B-25's FV-104 FH, FW-132 FK, FR-393 DF, FW-145 FM, and FV-997 FG are airborne for air Exercise Number 5, completed the exercise and landed. The night flying navigation exercises were cancelled. The master of mailboat *Noel Roberts*, which arrived at Rawson Dock in Nassau in the morning, picked up an American-type rubber dinghy off Great Stirrup Cay, Berry Islands, on the previous day. On inspection by the RAF, it is ascertained that the dinghy belongs to the missing Liberator B-24 aircraft. Tanker *AthelRegent* is bound southwest in Straits. Destroyer escort USS *Wilhoite* (DE-397) is bound northeast in Straits. SS *Tydolgas* is bound south in Straits, passing Canaveral. SS *Cities Service Ohio*, USCG cutter *CG-Thetis* and USCG cutter *CG-Argo* are bound south in Straits passing Palm Beach. USCG *CG-Boone* is bound north in Straits passing Port Everglades. USS *YAG-28* is bound southeast in Straits passing Miami. USS *PCS-1424S* is bound southwest in Straits. SS *Nishmaha* and SS *Folmar* are bound northwest in Straits. **Convoy GK-781** is bound northwest in Straits with US Navy plane escorting. Destroyer USS *Herbert* (APD-22) is bound south in Straits. No.113: Funeral of S/Ldr. D. Jeremiah; a short service was held in the RAF Chapel, and the body was interred in the RAF, or Western Cemetery. The Rev. Burns officiated and S/Ldr. Blanford was Officer in Charge of the

Parade. NAS Great Exuma: Two planes went to cover **Convoy NG-409**, and one to cover **Convoy KG-681**. One plane sighted an unidentified US Navy destroyer. Two planes arrived from Guantanamo, and one went to Florida.

13th January 1944

Request was received from Comgulfseafron for aircraft to sweep an area 15 nm either side of the line 25° north latitude. B-25 FW-132 FK is airborne covered the area landed with nothing to report. Liberator US 32746 is airborne from Windsor Field, carried out a creeping line ahead patrol and covered the area detailed aircraft landed with nothing to report from the sweep. B-25's FW-104 FH, FW-154 FR, FV-997 FG, and FR-393 DF are airborne completed the Anti-Submarine Patrol Number 1, then landed with nothing to report. Liberator BZ-810 LA is airborne, swept the area swoop was completed without incident and the aircraft landed at Windsor Field. B-25's FV-954 FF, FR-361 FC, FW-145 FR, FW-150 FP, FV-949 FD, FW-147 FO are airborne for air Exercise Number 5, then landed after completion. B-25's FR-381 DC, FV-997 FG, FW-104 FH, FR-393 DF and FV-946 FF carried out local night navigation exercises. SS *John M. Clayton* is bound south in Straits, passing St. Augustine. USS *YP-448* is bound south in Straits towards Daytona. Tanker *A. S. Hansen* and escort are bound north in Straits near Palm Beach. Ferry *Jean Brillant* and escort are bound south in Northwest Providence Channel. USS *SC-1353*, USS *SC-1290*, and USS *YAG-19* are bound south in Straits. Tanker *Tydolgas* is bound west in Straits. USS *AM-256* and USS *AM-257* are bound southeast in Straits. NAS Great Exuma: Two planes took off on a search mission, and a plane went to cover **Convoy NG-409**. Two planes departed to Banana River, and two arrived from Guantanamo. A Naval Air Transport seaplane arrived from Miami and went to Guantanamo. The Naval Air Transport seaplane arrived back from Guantanamo and went to Miami.

14th January 1944

At 7.20 am a request was received from Comgulfseafron for an aircraft to sweep ahead of a convoy consisting of a merchant ship and an escorts ship. Liberator BZ-746 LM is airborne from Windsor Field, the convoy sighted, signals were exchanged, area was swept, aircraft landed with nothing to report. Request was received from Comgulfseafron for an aircraft to sweep area, Liberator BZ-812 MC is airborne from Windsor Field for area, swept creeping line ahead patrol aircraft, landed with nothing to report from the patrol. One merchant ship and one escort were sighted, no reply was received to the challenge signals. B-25's FW-154 FR, FW-123 FJ, FV-997 FG, FW-148 FL, and FR-393 CA are airborne for air Exercise Number 3, aircraft completed the exercise and landed. B-25's FR-395 DF, FV-946 FF, FW-150 DF, and FW-104 FR are airborne for air Exercise Number 2, and landed after completion. B-25's FW-123 FJ, FW-132 FK, FW-154 FR, FV-946 FF, and FW-104 FH are airborne for air Exercise Number 4 and landed after completion. USS *Alabaster* (PYC-21) is bound south in Straits towards Canaveral. USS *YAG-20* and USS *YT-338* are bound northwest in Straits towards Canaveral. Ferry *Jean Brillant* is bound southwest in Straits passing Bimini with US Navy plane escorting. SS *John W. Clayton* is bound south in Straits passing Port Everglades. USCG *CG-Woodbine* is bound northwest towards Miami. Destroyer escort USS *DE-396* and destroyer escort USS *DE-328* are bound northeast in Straits. NAS Great Exuma:

Two planes arrived from Guantanamo and departed for Banana River. One plane arrived from Cayo Francés Cuba, and then took off and flew back.

15th January 1944

Request was received from Comgulfseafron for an aircraft to sweep area, and B-25 FW-176 FS is airborne and commenced a creeping line ahead sweep, sweep was abandoned due to bad weather. Report received at noon, air traffic control (ATC) stated that an aircraft was overdue on a flight from Morrison Field to Borinquen, Puerto Rico. ATC stated that 20 aircraft were searching the area from Boca Baton to Nassau although it was requested that RAF aircraft assist in the search, it was impracticable due to the scanty information regarding the available aircraft. US Navy Ventura aircraft landed back, sighted nothing. B-25's FV-945 FD, FV-946 FF, FW-176 FS, FW-I47 FO, FW-145 FM, and FW-154 FR are airborne and landed on the completion of the night navigation exercise. Liberator FL-967 LJ is airborne and landed at Boca Chica, where the crew were briefed regarding exercise in cooperation with US Navy, landed back at base after exercise completion. Liberator FL-992 LX is airborne, aircraft sighted nothing to report, and landed at Windsor Field. Ferry *Jean Brillant* takes RAF personnel and other passengers from Miami to Nassau. Motor yacht *Gemini* is sold in Miami and left with crew in US, as Windsors plan to leave the Bahamas. Destroyer escort USS *Cates* (DE-763) is bound northeast in Straits. USCG *CG-74328* is bound south in Straits towards Palm Beach. USCG Cutter *CG-Woodbine* is bound north from Miami. Tanker *Cedar Creek* is bound north in Straits passing Miami. Tanker *Bulklube* southwest in Straits. Tanker *Gertrude Kellogg* and SS *Tampa* are bound northwest in Straits passing Nicholas Channel. NAS Great Exuma: Two planes went to cover **Convoy GN-108** and one plane arrived from Guantanamo and departed for Banana River.

15th January 1944 – roughly

RAF prepared to hand over ground crew duties to 250 Bahamians, called the Bahamas Air Service Detachment in 1944 [later called Bahamas Air Service Squadron, from 14 September 1945].

16th January 1944

B-25 FW-l04 FH that left Nassau for Jamaica, returned and landed at base. US Navy Ventura aircraft which landed after searching the waters between Nassau and Florida coast for missing ATO aircraft, returned to Boca Raton. USCG cutter *CG-Legare* (WSC-144) is bound south in Straits towards Palm Beach. **Convoy KN-289** is bound south in Straits, passing Palm Beach with US Navy plane escorting. **Convoy NK-589** is bound north in Straits, passing Palm Beach. USS *Notable* (AM-267) is bound south in Straits, passing Palm Beach with US Navy plane escorting. SS *Cornell* is bound south in Straits, passing Palm Beach. Troop transport USS *USAT-Mexico* is bound south in Straits. USS *YAG-28* northwest in Straits. NAS Great Exuma: Two planes went to cover **Convoy GN-108** and one went to cover **Convoy KG-682**. A plane went to search survivors of a crashed US Army plane.

17th January 1944

Request was received from Morrison Field for an aircraft to search an area. Three B-2's, FW-123 FJ, FW-148 FL, and FW-145 FM are airborne aircraft covered the area, and landed having sighted nothing. B-25 FW-150 FP and FW-104 FH are airborne and completed Anti-Submarine Patrol Number 2. Landed with nothing to report. Liberator BZ-722 LX is airborne from

Windsor Field for Miami. Tanker *A. S. Hansen south* in Straits passing Canaveral. USS *YO-107* south in Straits, passing Palm Beach. USS *Alabaster* (PYC-21) is bound south in Straits. USS *AO-68* and destroyer escort USS *DE-387* are bound north near Bimini, a US Navy plane escorting. **Convoy NK-589** is bound west in Straits. Destroyer escort USS *Cockrill* (DE-398) is bound northeast in Straits. USS *SC-1301* is bound northwest in Straits passing Nicholas Channel. **Convoy GK-782** is bound northwest in Strait, passing Nicholas Channel with US Navy plane escorting. NAS Great Exuma: Two planes went to cover **Convoy NG-410**. Three went to Guantanamo and two arrived from Guantanamo.

18th January 1944

B-25's FR-378 CL, FS-132 FK, FW-149 FH, FV-946 FF, FW-123 FJ, and FR-393 DF are airborne, then landed on the completion of air Exercise Number 7. B-25's FW-154 FA, FR-372 CG, FK-148 FL, FW-179 FT, FV-954 FB and FV-951 FC are airborne and landed on the completion of the air Exercise Number 6. Liberator BZ-746 LM is airborne from Windsor Field for Bermuda, and returns the same day. Navigation exercise for No. 1 and Squadron detailed for the afternoon cancelled due to adverse weather conditions. USCG *CG-83502* is bound south in Straits towards Palm Beach. SS *Manassas* is bound north in Straits. Ferry *Jean Brillant* is bound west in Straits, towards Port Everglades with US Navy pane escorting. SS *Skull Bar* is bound north out of Straits. Tanker *A. S. Hansen* is bound southwest in Straits towards Key West. NAS Great Exuma: One plane went to cover **Convoy KG-314** and three arrived from Guantanamo.

19th January 1944

B-25's FW-150 FP, FW-179 FT, FW-123 FJ, FW-149 FN, and FV-946 FF are airborne on air Exercise Number 7, it was completed and the aircraft landed. B-25's FR-393 DF, FV-951 FC, FW-123 FJ, FW-148 FL, FW-149 FN, and FW-147 FO are airborne, aircraft completed air Exercise Number 7, and landed. Aircraft of Number 1 Squadron B-25's FL-172 AJ, FL-180 BA, and FR-362 CA carried out local navigation exercise. B-25's FR-381 DC, FW-148 FL, FV-951 FC, FV-946 FF, FW-147 FO, and FW-123 FJ are airborne, then aircraft landed on the completion of the exercises. USS *LST-283* and USS *PC-1168* are bound south in Straits towards Daytona. SS *Exford* and SS *Reinhold Richter* are bound south in Straits, towards Palm Beach. USS *LST-49* is bound north in Straits, passing Palm Beach. SS *Paranaloide* is bound south in Straits, towards Port Everglades. Destroyer escort USS *Haines* (DE-792) is bound northeast passing Matanilla Shoal, northwest Bahamas. Tanker *Cities Service Kansas* is bound north in Straits, passing Port Everglades. Landing craft USS *LST-50* is bound north in Straits, near Bimini. **Task Unit TU-2934** is bound north in Straits near Miami. Tanker *Five Forks* is bound north in Straits, near Miami with US Navy pane escorting. USS *LST-133* is bound northeast in Straits. USS *SC-1340* is bound south in Straits. USS *LST-283* is bound east in Straits. NAS Great Exuma: Two planes departed for Banana River and one departed for Morrison Field.

20th January 1944

B-25's FV-954 FB, FW-132 FK, and FV-951 FC and airborne and landed on the completion of the Air Exercise Number 7. B-25's FW-149 FN, FV-946 FF, FR-147 FO, and FR-393 DF airborne on the completion of the air Exercise Number 6, the aircraft landed. Number 1 Squadron B-25 FK-371

CF, FL-180 BA, and FR-372 CG carried out local navigation exercises. *U-154* enters Crooked Island Passage south to exit southbound via Windward Passage. SS *Joe C. S. Blackburn* and USS *Menkar* (AK-123) are bound south in Straits, towards St. Augustine. SS *Paranaloide* is bound north in Straits, towards St. Augustine. **Convoy NK-590** is bound south in Straits with a US Navy pane escorting. USS *LST-133* is bound north in Straits, towards Daytona. USS *LST-283* is bound north in Straits. USCG Cutter *CG-Vigilant* is bound north in Straits, passing Palm Beach. USS *LST-49* is bound north in Straits passing Matanilla Shoal, northwest Bahamas. SS *Cities Service Kansas* is bound south in Straits, passing Palm Beach. **Convoy KN-290** is bound north in Straits, passing Port Everglades with US Navy plane escorting. Tug USS *YAG-19* is bound southwest in Straits. USS *AT-147* towing USS *YCV-6* escorted by USCG Cutter *CG-Legare* are bound southeast in Straits, passing Miami. SS *Empire Rapier* is bound east in Straits. **Convoy KG-683** is bound southeast in Straits, passing Nicholas Channel with US Navy plane escorting. Ferry *Jean Brillant* escorted by USS *SC-1342* is bound south towards Nassau. NAS Great Exuma: Two planes went to cover **Convoy GN-109**. Submarine USS *Pampanito* is positioned near the base, and a Naval Air Transport seaplane arrived from Miami and went to Guantanamo. USCG *CG-PH-2* and USCG *CG-PR-1* arrived from Nassau with USS *Trippe* (DD-403). Seven US Navy USS *PBY*'s arrived from Jacksonville.

21st January 1944

Number 1 Squadron conducted night flying training. B-25's FV-951 FC, FW-149 FN, FW-147 FO, FW-946 FF, FR-393 DF, and FW-179 FT are airborne, completed air Exercise Number 6 and landed. B-25's FV-949 FD, FR-381 DC, FW-I46 FK, FV-951 FC, FV-954 FB, and FW-132 FK airborne for compass swinging exercise completed the exercise and landed. B-25'a FV-945 FD, FV-946 FF, FW-176 FS, FW-147 FO, FW-145 FM, and FW-154 F airborne aircraft landed on the completion of the night navigation. B-25's FL-187 BC, FL-180 BA, and FL-172 BJ carried out two circuits of the local night navigation exercise details for Number 1 Squadron. Liberator BZ-809 LZ was airborne from Windsor Field radio contact established, with the SEO, aircraft located one ship, signals were exchanged, ascertained that this was HMS *Dakins*, which was not the SEO, aircraft was informed by the ship of this error after exchanging signals, aircraft left and continued the search for the SEO HMS *Dakins* was located, signals were exchanged and radio contact established, SEO inquired of the position of HMS *Dakins*, aircraft was instructed to carry out Viper patrol and search for HMS *Dakins*, Viper patrol was completed and at 3 pm the ship dropped a smoke float which was attacked from 1,500 feet depth charge ricocheted, aircraft located the other vessel and signalled its position to the co-operating ship, flare was dropped for A-A practice prior to the aircraft leaving the convoy, which was left on course 230° with speed of 12 knots, aircraft landed. Liberator BZ-758 LS is airborne from Windsor Field on the Kingsley Exercise, convoy of eight merchant ships was sighted, no investigation was carried out, contact was established and on homing on the bearings given by the ship, the convoy was located, signals were exchanged, radio communication was established, ship requested news of the other vessel, aircraft had not sighted this escort vessel, aircraft completed the Serpent patrol under the instructions of the SEO, smoke float was dropped by the ship and an attack was delivered with one depth charge by the aircraft, aircraft had to curtail the exercise with the convoy as the PLE [endurance] had been reached, aircraft left the convoy, aircraft landed. Liberator BZ-813 MD was airborne from Windsor Field on the Kingsley Exercise found that the radio was unusable, attempts had been made to contact the co-operating ships for over an hour, but had been unsuccessful, convoy was located, signals were exchanged and radio communication was established, Serpent patrol commenced and second ship sighted, challenge signals were exchanged and it was observed later that both ships had met, aircraft was instructed to carry out Mamba patrol, patrol was commenced, aircraft re-joined the convoy when the ship dropped a smoke float, attack was carried out with one DC, attacks with nose and turret guns were carried out at the request of the SEO against the smoke float flare was released by the aircraft for A-A practice by the ships gunners, aircraft left the convoy on the completion of the exercise and landed at Windsor Field. Destroyer escort USS *Esdall* (DE-129) is bound south in Straits. USCG *CG-83500* is bound south in Straits, passing Palm Beach. **Convoy NK-590** is bound south in Straits, passing Palm Beach with US Navy plane escorting. Tanker USS *Evangeline* is bound south in Straits towards Bimini. Ferry *Jean Brillant* and USS *SC-1342* is bound west in Straits passing Northwest Providence Channel. USS *LST-284* is bound north in Straits towards Port Everglades. SS *Joe C. S. Blackburn*, SS *Lautaro* and tanker *Pan Massachusetts* are bound southwest in Straits. Destroyer escorts USS *DE-330* and USS *DE-331* are bound northeast in Straits. USS *AT-147* towing USS *YCV-6*, and *USCGC-Legare* (WSC-144) are bound southeast in Straits. Three US Navy USS *AM*'s, three US Navy USS *SC*'s, and USS *SC-1295* are bound east in Straits. **Convoy GN-109** is bound north in Straits with US Navy plane escorting. NAS Great Exuma: One plane went to cover **Convoy KG-683**, one plane went to cover **Convoy NG-411**, seven PBY's went back to Key West. Submarine USS *Pampanito* is positioned near the base, Naval Air Transport seaplane arrived from Guantanamo and went to Miami, *USCG-PH-2* and *USCG-RD-1* departed for Cayo Francés, Cuba with destroyer USS *Trippe* (DD-403). Four planes arrived from Guantanamo and one from Jacksonville.

22nd January 1944

B-25'S FR-176 FS, FV-147 FO, FR-378 CL, FV-946 FF, FV-149 FN, and FW-154 FR are airborne for compass swinging exercise, aircraft landed on the completion of the exercises. Liberator BZ-806 LU is airborne from Windsor Field for Kingsley Exercise, aircraft returned to Bermuda and landed there returned to base and landed later. Liberator BZ-869 MX is airborne from Windsor Field on the Kingsley Exercise convoy of 17 merchant ships and three escorts was sighted, signals were not acknowledged, homing procedure was carried out, two frigates were contacted, signals were exchanged, aircraft dropped a flare for the AA gunners, instructions were given to the aircraft to carry out Alligator patrol which was completed, smoke float was dropped by the ship and an attack was carried out with one DC, exercises were completed and the aircraft left the merchant ships, aircraft landed. Liberator BZ-746 LM is airborne from Windsor Field on the Kingsley Exercise, homing exercise was commenced, ships were contacted after obtaining bearings from the ships, signals were exchanged, SEO dropped a smoke float for an attack by the aircraft with one depth charge, aircraft was instructed to carry out Serpent patrol, which was completed, aircraft carried out Adder patrol, under the instructions of the SEO exercise was completed and the aircraft left the convoy, aircraft landed. Liberator BZ-746 LM is airborne from Windsor Field on the Kingsley Exercise, homing exercise was commenced, ships were contacted in position after obtaining bearings from the ships signals were exchanged, SEO dropped a smoke float for an attack by the aircraft, aircraft was instructed to carry out Serpent patrol, which was completed aircraft commended patrol Adder under the instructions of the SEO, patrol was completed and the aircraft left the convoy, consisting of 16 merchant ships and three escorts was sighted aircraft landed. USS *PC-1215* is bound south in Straits. USS *Halsey Powell* is bound south in Straits. Anti-Submarine Group 6 is bound west in Straits. **Convoy NK-590** is bound southwest in Straits, US Navy plane escorting. Anti-Submarine Group 1 is bound southeast in Straits passing Nicholas Channel. **Convoy GK-783** is bound northwest in Straits with US Navy plane escorting. NAS Great Exuma: Two planes went to cover **Convoy NG-411** and two planes arrived from Guantanamo and then went to Banana River.

23rd January 1944

Liberator BZ-886 LU returned from Bermuda and landed at Windsor Field. SS *Francis E. Warren* is bound south in Straits. Tanker *R. P. Smith* is bound south in Straits passing Canaveral. Tanker *Bulklube* is bound north in Straits. USS *LST-134* is bound north in Straits passing Palm Beach. USS *PC-1183* is bound southwest in Straits. USS *PC-469* is bound south in Straits. NAS Great Exuma: Inter-island transport USS *YAG-21* arrived from Miami with supplies. Bahamas Airways' *RD-4* arrived from Nassau and then returned.

24th January 1944

B-25's FV-951 FC, FV-949 FD, FW-147 FO, and FV-104 FH are airborne for air Exercise Number 1 and landed after completion. Widgeon is airborne waterborne at San Salvador took off again and landed at base. B-25's FV-945 FD, FV-946 FF, FW-176 FS, FW-147 FO, FW-145 FM, and FW-154 FR are airborne for night navigation exercise, landed after completion. Liberator BZ-760 LT is airborne for San Antonio, and landed at Miami *en route*. SS *Harpers Ferry* is bound north in Straits. Tanker *Gulfmaid* is bound south in Straits, passing Palm Beach. USCG *CG-83401* is bound northeast towards Matanilla Shoal, northwest Bahamas. USS *SC-1340* is bound north in Straits towards Palm Beach. SS *Francis E. Warren* is bound south in Straits. Tanker *R. P. Smith* is bound west in Straits. Tanker *Paco* is bound northwest in Straits passing Nicholas Channel. NAS Great Exuma: A plane flew to Guantanamo and returned back to Exuma.

25th January 1944

B-25's FV-951 FC, FW-149 FN, FV-949 FD, FW-145 FM, and FV-954 FB are airborne completed Anti-Submarine Patrol Number 1, aircraft landed with nothing to report. B-25's FW-123 FJ, FV-997 FG, FW-104 FH, FW-154 FR, FW-149 FN and FV-951 FC are airborne aircraft landed on the completion of the Air Exercise Number 1. USCG cutter *CG-Buttonwood* is bound south in Straits. Tanker *18 De Marzo* is bound south in Straits towards Canaveral. **Convoy KN-291** is bound north passing Matanilla Shoal, northwest Bahamas with US Navy plane escorting. USS *YMS-275* is bound north towards Matanilla Shoal, northwest Bahamas. USS *PC-1178* and USS *PC-576* are bound north in Straits. **Convoy KG-684** is bound southeast in Straits with US Navy plane escorting. NAS Great Exuma: the inter-island transport USS *YAG-21* went back to Miami. Two planes went to cover **Convoy GN-110**. One plane arrived from Guantanamo and then went to Banana River.

26th January 1944

Request was received from Comgulfseafron for aircraft to sight a convoy and sweep ahead. B-25 FV-954 FB is airborne, visibility was poor, convoy was not located, aircraft flew up track of the convoy for 30 nm, returned to first position, commenced a CLA, sweep was completed, aircraft landed, failed to locate convoy but completed sweep without incident. B-25 FW-148 FL airborne, CLA was commenced with the visibility of 6 nm, sweep commenced, aircraft landed with nothing to report. B-25's FW-150 FF, FW-l04 FH, and FV-997 FG are airborne for air Exercise Number 7 completed exercise and landed. B-25's FR-378 CL, FR-381 DC, FW-149 FN, FV-951 FC, FW-154 FR, and FW-123 FJ are airborne for air Exercise Number 5 completed exercise and landed. B-25's FW-123 FK, FV-951 FC, FW-154 FR, FW-150 FP, and FW-123 FK are airborne for night navigation exercises were completed, aircraft landed. HMS *Kilmartin* is bound north in Straits. USS *YAG-28* is bound southeast towards Walker Cay, north of Abaco. SS *Cyrus Field* is bound south in Straits towards Palm Beach. SS *Fort Fetterman* is bound northeast in Straits. Tanker *18 de Marzo* and USCG Cutter *CG-Buttonwood* are bound west in Straits. NAS Great Exuma: One plane went to cover **Convoy GN-110**, one went to cover **Convoy GK-794.** Submarine USS *Parche* (SSN-384) dived near the base, and two planes departed for Banana River.

27th January 1944

B-25's FW-176 FS, FW-150 FP, FV-951 FC, FW-123 FJ, FW-154 FR, and FV-997 FG are airborne completed air Exercise Number 5 aircraft landed. **Convoy NK-591** is bound south towards Palm Beach navy plane escorting. Tanker *Fort Fetterman* and tanker *Esso Hartford* are bound north in Straits. Submarine USS *Parche* (SS-384) is bound northwest passing Northwest Providence Channel. Destroyers USS *Runels* and USS *Stockdale* are bound north in Straits towards Bimini. USS *YAG-19* is bound south in Straits. Tanker *Fishers Hill* is bound southwest in Straits. **Convoy GK-784** is bound northwest in Straits passing Nicholas Channel with US Navy plane escorting. NAS Great Exuma: Two planes arrived from Guantanamo and departed to Banana River, one arrived from Cayo Francés, one arrived from Key West, and one plane departed to Nassau. A Naval Air Transport seaplane arrived from Miami and went to Guantanamo, and a Naval Air Transport seaplane departed to Miami.

28th January 1944

B-25's FW-104 FR, FV-997 FG, FW-147 FO, FW-145 FM, FW-167 FS, and FV-951 FC are airborne completed the-local night navigation exercise and landed. B-25's FV-949 FD, FW-148 FL, and FW-132 FK are airborne for Anti-Submarine Patrol Number 2 patrol completed aircraft landed. B-25's FW-l04 FH, and FW-154 FK are airborne for air Exercise Number 7, aircraft completed the exercise and landed. B-25's FV-954 FB, FV-997 FG, FW-147 F0, FW-150 FP, FW-176 FS and FW-104 FH are airborne for air Exercise Number 3, aircraft landed on the completion of the exercises. B-25's FW-176 FS, FV-954 FB, FV-997 FG, FW-147 FO, FR-378 CL, end FW-104 FH airborne for air Exercise Number 4, aircraft landed on the completion of the exercise. B-25's FV-945 FD, FV-946 FF, FW-176 FS, FW-147 FO, FW-145 FM, and FW-154 FH are airborne and landed after completion of night navigation exercise. Request was received from Comgulfseafron to provide air cover for **Convoy SS *Olay***, consisting of two merchant ships and three escorts, cover was required from dawn to dusk. Liberator BZ-810 MA is airborne from Windsor Field for the escort of **Convoy USS *Olay***, which was sighted, signals were exchanged, box patrol was commenced, aircraft homed on a SE [sound emission], blip destroyer was sighted on course, signals were exchanged, aircraft returned to the convoy and endeavoured to pass the position of the destroyer, no visual contact could be established, aircraft left the convoy, landed with nothing to report from the patrol. Liberator BZ-813 MD is airborne for relief to BZ-810 MA, destroyer was sighted, convoy was sighted, composition was same, aircraft landed. Liberators BZ-808 LY and BZ-811 MB are airborne completed homing and navigation exercise, LY landed, MB was diverted to investigate a ship off Clifton Pier, then landed back, having sighted nothing. Destroyer USS *MacLeish* (DD-220) is bound south towards Canaveral. Tug USS *Fire Island* towing USS *YFD-3(H)* and USS *YFD-3 (M)* and escorted by USS *YMS-275* is bound south in Straits passing Canaveral with US Navy plane escorting. USS *YMS-361*, tug USS *Sand Point* towing USS *YR-46*, USS *YR-47*, and oil barge USS *YOG-20* escorted by USS *SC-1357* are bound south in Straits passing Canaveral. Landing ship USS *LST-513* is bound north in Straits towards Palm Beach. British rescue tug HMS *HMRT Athlete* is bound north in Straits passing Port Everglades. Tanker *Pan Massachusetts* is bound north in Straits passing Port Everglades. USS *Logic* (AM-258) is bound south in Straits passing Miami. USS *YF-443* is bound north in Straits towards Miami. SS *William M. Burton*, tanker *Vacuum*, SS *Vistula*, and tanker *Bohemian Club* are bound west in Straits. Landing ship USS *LST-492* is bound northeast passing Cay Sal Bank. NAS Great Exuma: One plane went to cover **Convoy NG-410**, two departed to Guantanamo, two arrived from Guantanamo and then departed to Banana River, one departed to Nassau and seven US Navy PBY's arrived from Jacksonville.

29th January 1944

B-25's FW-148 FL, FR-381 DC, and FV-9A9 FD are airborne for Anti-Submarine Patrol Number 2, aircraft landed with nothing to report from the completed patrols. B-25's FW-l45 FM, FV-954 FB, FW-179 FT airborne for air Exercise Number 7 and landed after completion. B-25's FR-378 CL, FW-104 FH, FW-145 FM, FV-997 FG, FW-176 FS, FR-393 DF, FW-150 FP airborne for air Exercise Number 3, aircraft landed on the completion of the exercise, B-25's FW-104 FH, FW-145 FM, FR-378 CL, FV-997 FG, FW-176 FS, FW-147 FO are airborne completed the Exercise Number 4 landed. USCG *CG-74323* is bound north in Straits, passing Daytona. USS *YMS-355* is bound south in Straits, passing Daytona. **Convoy NY-61** is bound southwest in Straits with US Navy plane escorting. USS *LST-492* is bound north in Straits passing Palm Beach. USCG *CG-83398* is bound northeast in Straits near Palm Beach. USS *SC-681* is bound north in Straits towards Palm Beach. Oiler USS *YO-69* is bound southwest in Straits towards Miami. USS *PC-1251*, USS *PC-1172*, and USS *PC-1183* are bound north in Straits passing Miami. SS *Five Forks* is bound south in Straits passing Bimini. Tanker *Alpha* is bound northeast in Straits. Tug USS *Sands Point* with tows and escort and USS *PCS-1414* are bound south in Straits. SS *Empire Anvil* is bound northeast in Straits. No.113: RAF held a ceremony for the dedication of the RAF Ensign in Nassau Cathedral. NAS Great Exuma: Power House Number 2 caught fire, then the building was secured with slight damage. Seven US Navy PBY's departed to Guantanamo, seven PBY's arrived from Jacksonville, and two planes arrived from Guantanamo then departed to Banana River.

30th January 1944

Liberator BZ-813 MD airborne from Windsor Filed homing and Anti-Submarine Patrol aircraft returned to base with radio unserviceable, aircraft landed. Liberator BZ-806 LH is airborne from Windsor Field on a homing and Anti-Submarine Patrol, patrol was completed, homing exercise was cancelled on instructions from base aircraft landed. Liberator BZ-810 MA is airborne from Windsor Field for homing and Anti-Submarine Patrol small merchant ship was sighted, message was from base cancelled the homing exercise, aircraft completed the Anti-Submarine Patrol and landed. SS *Minatitlan* and SS *Dona Aniceta* are bound south in Straits towards St. Augustine. USS *YP-603* is bound north in Straits towards Daytona. **Task Unit TU-2036** is bound south in Straits passing Daytona. Anti-Submarine Group 4 is bound north in Straits passing Palm Beach. USCG *CG-83398*, USCG *CG-83401*, and USS *YMS-76* are bound south in Straits passing Palm Beach. USS *SC-1357* is bound north passing Matanilla Shoal, northwest Bahamas. **Convoy KN-292** is bound north in Straits towards Palm Beach with US Navy plane escorting. USS *YP-441*, USS *YP-437*, USCG *CGR-1509* and USCG patrol trawler *CG-E. M. Joe* (WYP-356) are bound north in Straits towards Palm Beach. USS *PCS-1413* and USS *YMS-355* are bound southwest in Straits. **Convoy KG-685** is bound southeast in Straits with US Navy plane escorting. NAS Great Exuma: One plane went to cover **Convoy GN-111**, seven PBY's departed for Guantanamo, one plane picked up a Radar contact and blip, however it disappeared. One plane arrived from Jacksonville.

31st January 1944

Request received from Comgulfseafron for an aircraft to provide air cover to **Convoy GN-111** of 17 merchant ships and five escorts, Liberator BZ-813 MD is airborne from Windsor Field on air escort of **Convoy GN-111**, did not meet convoy at estimated position, "not met" signal sent to base, convoy met later, composition was same, landed at base after air escort was completed. Troop transport USS *USAT-Apache* is bound south in Straits towards Canaveral. **Convoy NK-592** is bound south in Straits towards Canaveral with US Navy plane escorting. SS *Minatitlan* is bound south in Straits passing Palm Beach. SS *Manuel Rionda* is bound north in Straits passing Port Everglades with US Navy plane escorting. Tanker *Cities Service Ohio* is bound north in Straits passing Port Everglades. Anti-Submarine Group 7 is bound northwest in Straits towards Miami. USS *YP-441* and USS *YP-437* are bound northwest in Straits. USS *LST 51* is bound northeast in Straits. SS *Dona Aniceta* is bound southwest in Straits. Bahamas (position encoded), B-25 FW-154 FR saw feather behind U-boat periscope. B-25's FR-381 DC and FR-393 DF, and FV-997 FG patrolled, no sighting, RAF No.111 OTU. NAS Great Exuma: One plane went to cover **Convoy KG-685** and another to cover **Convoy GK-785**. Two planes departed for Guantanamo.

1st February 1944

In view of the negative results from the sweep by the three B-25 aircraft for the suspected submarine sighted by FW-154 FR yesterday, whilst on Anti-Submarine Patrol Number 1 it was decided to detail two B-24 aircraft to carry out Radar sweep. Liberator BZ 806 LU is airborne from Windsor Field to sweep an area, nothing to report from patrol. Liberator B-274 LL was airborne from Windsor Field carry out an Anti-Submarine sweep. Then aircraft, landed on the completion of the sweep, nothing to report from the patrol. B-25's FW-179 FT, FW-145 FM, FW-150 FF, FV-997 FG, and FW-154 FR are airborne for air Exercise Number 6 on the completion of the exercise the aircraft landed. B-25's FV-945 FD, FV-946 FF, FW-176 FS, FW-147 FW-145 FM, and FW-154 FR airborne, then landed after completion of night exercise. Liberator FL-967 LJ is airborne and landed at Boca Chica and was briefed for the air submarine exercise in co-operation with US Navy, then landed at base after completion of exercise. USS *LST-51* is bound northwest in Straits, northwest of Little Bahama Bank. Patrol vessel HMS *Kimelford* (Z-13) escorted by the escort patrol craft USS *PEC-839*, and HMS *Kilmington* (BEC-14) are bound north in Straits, west of Little Bahama Bank. USS *YAG-21* is bound north in Straits northwest of Bimini. **Convoy NK-592** is bound south in Straits off Port Everglades. **Convoy GK-785** is bound northwest in Nicholas Channel. USS *PC-1254* is bound southeast in Old Bahama Channel. NAS Great Exuma: The stores ship USS *Yukon* (AF-9), minesweeper USS *Chickadee* (AM-59), and Flower-Class corvette USS *Tenacity* (PG-71), passed by the base. A plane went to cover **Convoy NG-413**, and two planes arrived from Guantanamo and departed to Banana River.

2nd February 1944

Liberator BZ-807 LO is airborne from Windsor Field, landed at Boca Chica and was briefed for the air submarine exercise in co-operation with US Navy, then landed at base after completion of exercise. B-25 FR-393 DF is airborne for Anti-Submarine Patrol Number 1, aircraft landed with nothing to report from the patrol. B-25's FV-997 FG, FW-179 FT, FW-150 FP, FW-147 PO, FV-946 FF and FW-154 FB are airborne for air Exercise Number 6, aircraft completed the exercise and landed. Aircraft of Number 1 Squadron B-25's FR-380 DB, FL-187 BC, and FL-702 BJ are airborne completed the exercises and landed, second circuit was completed later in the day. B-25's FV-946 FF, FW-147 F0, FW-123, FR-381 DC, FW-150 FP and FW-179 FT are airborne aircraft completed the exercise and landed. **Convoy 11** is bound east in Straits, bound north of Little Bahama Bank. Tanker *E. R. Kemp* is bound north in Straits, west of Little Bahama Bank. Oiler USS *YO-69* is bound north in Straits west of Bimini. Destroyer escort USS *Strickland* (DE-333) is bound north in Straits. USS *YP-445* and USS *YP-443* are bound northwest in Old Bahama Channel. USCG *CG-83455* is bound southeast in Old Bahama Channel. No.113: Liberator BZ-718 was taxiing behind another Liberator when, two thirds of the way down runway, the brakes failed, and the pilot swung off the runway to avoid the other Liberator. After crossing some rough coral, the airplane ground the nose wheel. The accident is due to failure of brakes owing to hydraulic booster pump not being switched on, placing it in damage category AC. Crew: Capt. O. Rector, civilian pilot. Capt. J. W. Ross, civilian first officer, H. W. McKercher, civilian radio operator, and R. S. Fisher, civilian flight clerk. All were uninjured. NAS Great Exuma: Eight US Navy PSP's arrived from Jacksonville, two planes arrived from Guantanamo, one plane departed to Guantanamo and one flew to Banana River.

3rd February 1944

Request was received from air traffic control (ATC) at Morrison Field for aircraft to search for the crew of a B-18 aircraft which has been found floating 150 nm off Jacksonville. The aircraft was *en route* from Havana to Miami. Liberator FL-994 LA was airborne from Windsor Field and searched the area detailed, then aircraft landed with nothing sighted. Liberator FL-757 LR was airborne and landed at Boca Chica briefed for the Air Submarine exercises, aircraft completed the exercises and landed at base. B-25's FV-997 FG, FW-149 FT, FR-381 DC, FW-123 FJ, FW-150 FP and FR-393 DF are airborne and completed air Exercise Number 6, then aircraft landed. Aircraft of Number 1 Squadron B-25's FR-366 CX, FL-173 AK, and FL-702 BJ are airborne and then landed after completion. B-25's FW-104 FK, FV-997 FG, FW-123 FJ, FV-946 FF, FH-147 FI, and FR-393 DP are airborne, then aircraft landed on the completion of the exercise. B-25's FR-362 BA, FW-163 AN, FW-183 AV, FL-702 BJ and FR-366 CX are airborne, then aircraft landed on the completion of the exercise. USS *YMS-312* is bound south in Straits, west of Little Bahama Bank. USS *LST-54* is bound north in Straits, west of Little Bahama Bank. USS *YAG-19* is bound north in Straits. USS *YMS-76* is bound southeast in Old Bahama Channel. NAS Great Exuma: Eight US Navy PSP's departed to Guantanamo, two planes arrived from Guantanamo, one plane departed to Nassau, and one plane departed to Banana River.

4th February 1944

Liberator FL-767 LJ is airborne from Windsor Field landed at Boca Chica briefed for the Air Submarine exercises, aircraft completed the exercises and landed at base. B-25's FH-148 FL, FR-393 DF, FV-946 FF, FR-381 DC, FW-179 FT and FW-123 FJ are airborne, completed compass-swinging exercise, landed after completion. **Convoy NK-593** is bound south in Straits off Canaveral with US Navy plane escorting. SS *Fishers Hill* is bound north in Straits west of Little Bahama Bank. **Convoy KN-293** is bound north in Straits off Miami with US Navy plane escorting. SS *Montana* is bound south in Straits north of Bimini. Oiler USS *YO-69* is bound south in Straits from Northwest Providence Channel. USS *LST-502* and USS *LST-493* are bound north in Straits west of Bimini. SS *Usahs Algonquin* is bound north in Straits. Anti-Submarine Group 2 is bound southeast in Nicholas Channel. USS *Sturdy* (PYc-50) is bound northwest in Nicholas Channel. **Convoy KG-686** is bound southeast in Old Bahama Channel, with US Navy plane escorting. SS *Cape Romano* is bound north, to the north of Eleuthera. NAS Great Exuma: Two planes went cover **Convoy GN-112**, one plane departed for Banana River and two planes went for training, returning safely.

5th February 1944

B-25 FW-132 FK airborne for Anti-Submarine Patrol Number 1 a merchant ship was sighted being escorted by a subchaser, patrol was completed and aircraft landed with nothing to report. B-25's FV-951 FC, FW-150 FP, FW-149 FR-393 DF, FW-145 FM and FR-381 DC are airborne for compass-swinging exercise, landed after completion. USS *Astute* (AM-148) is bound north in Straits northwest of Little Bahama Bank. Tug USS *YMS-355* is bound north in Straits northwest of Little

Bahama Bank. USCG *CG-Vigilant* is bound south in Straits west of Little Bahama Bank. **Convoy NK-593** is bound south in Straits off Port Everglades with US Navy plane escorting. SS *Cape Romano* is bound south in Straits. USS *YOG-6* escorted by USS *YP-351* is bound north in Straits north of Santaren Channel. USCG *CG-83419* is bound southeast in Nicholas Channel. USS *Logic* (AM-258) is bound southeast in Old Bahama Channel. NAS Great Exuma: One plane went to cover **Convoy KG-686** and another went to cover **Convoy NG-414**. One plane arrived from Guantanamo, one plane departed to Miami, another plane departed to Banana River, and seven US Navy PSP's arrived from Jacksonville.

6th February 1944

Grumman Goose landed from Miami; aircraft was delivered from the manufacturer. USS *Flaunt* (ATR-94) is bound north in Straits northwest of Little Bahama Bank. USS *Nucleus* (AM-268) is bound north in Straits west of Little Bahama Bank. Tanker *Paulsboro* is bound southwest in Straits from Northwest Providence Channel to Miami. SS *Salamonie* and SS *Cape Romano* are bound north in Straits west of Bimini. SS *Maria Sanford* is bound northwest in Nicholas Channel. SS *Notos* and USS *PCS-1388* are bound southeast in Old Bahama Channel. **Convoy GK-786** is bound northwest in Old Bahama Channel with US Navy plane escorting. USCG *CG-83419* is bound southeast in Old Bahama Channel. USS *Fitch* (DD-462) is bound south, to the east of Eleuthera. NAS Great Exuma: Two planes went to cover **Convoy NG-414**, seven US Navy PSP's departed for Guantanamo, and one plane went for training, returning safely.

7th February 1944

B-25's FW-123 FK, FW-149 FN, FV-946 FF, FW-179 FT, FW-148 FL and FV-951 FC are airborne for Anti-Submarine Patrol Number 1 patrol was completed without incident and the aircraft landed. B-25's FW-145 FM, FV-951 FC, FW-150 FP, FW-149 FN, and FR-393 DF are airborne aircraft completed air Exercise Number 1 and landed. Liberator FL-994 LA is airborne from Windsor Field for Miami, then aircraft landed there. USCG *CG-83451* is bound south in Straits north of Little Bahama Bank. SS *Fort Fetterman* is bound south in Straits north of Bimini. USS *YMS-409* is bound north in Straits. NAS Great Exuma: Two Naval Air Transport seaplanes moved between Guantanamo and Miami. A US Navy submarine was positioned near the base. Two planes arrived from Guantanamo, one plane departed for Jacksonville, one plane departed for Guantanamo, and one plane took off for a training flight.

8th February 1944

Request was received from Comgulfseafron for an aircraft to sight and sweep ahead the submarine USS *Gabilan* (SS-252). B-25 FV-946 FF is airborne to sight and sweep ahead of the submarine USS *Gabilan* (SS-252) submarine was sighted, signals were exchanged, sweep was carried out, aircraft landed after an uneventful sweep. B-25 FW-179 FT airborne to carry out a sweep ahead of the track of the US submarine, where a small tanker of 2,000 tons was sighted, sweep was carried out, aircraft landed on the completion of the sweep. B-25's FW-149 FN, FW-147 F0, and FR-393 BF are airborne, aircraft completed air Exercise Number 6 and landed. B-25's FV-951 FC, FW-147 F0, FW-132 FK, FW-123 FJ, FR-149 FB and FW-145 FM are airborne for air Exercise Number 1, aircraft landed. B-25

FW-148 FL is airborne for Anti-Submarine Patrol Number 1, patrol was completed and the aircraft landed with nothing to report. Liberator BZ-760 LT is airborne from Miami and landed at Windsor Field on return from San Antonio. Liberator FL-994 LA is airborne from Miami, and landed at Windsor Field. USS *Raven* (AM-55) is bound north in Straits northwest of Little Bahama Bank, exiting area. USS *LST-503* is bound north in Straits, northwest of Little Bahama Bank. Tanker *Gulfmaid* is bound south in Straits. USS *Courage* (PG-70) is bound north, to the north of Eleuthera. NAS Great Exuma: A plane went to cover **Convoy GN-113**, a plane went to cover **Convoy GN-114**, two planes arrived from Guantanamo, one plane departed for Banana River, and one plane departed for Nassau.

9th February 1944

B-25's FW-148 FL, FR-378 CL, and FR-393 DF are airborne for Anti-Submarine Patrol Number 1, aircraft completed the patrol and landed. B-25's FW-147 FO, FW-123 FJ, FW-132 FK, BW-149 FN, and FV-951 FC are airborne for air Exercise Number 5, then aircraft landed. Request was received for an aircraft to provide air cover for the US submarine USS *Gabilan* (SS-252) and B-25 FV-946 FF is airborne on the escort to the submarine, submarine is sighted, air cover was provided, FF was relieved by FD, then FF landed. B-25 FV-949 FD is airborne as first relief air escort for submarine two escorts located aircraft escorted submarine landed after duty was over. B-25 FW-179 FT is airborne as second relief convoy sighted, signals exchanged landed after duty was over. **Convoy KN-294** is bound north in Straits northwest of Little Bahama Bank with US Navy plane escorting. USS *PC-611* is bound south in Straits west of Little Bahama Bank. Submarine USS *Gabilan* (SS-252) escorted by USS *SC-1295* and USS *SC-1064* are bound west in Northwest Providence Channel. USS *AG-28* is bound north in Straits, north of Cay Sal Bank. USCG *CG-E. M. Vernon McNeal* and USCG cutter *CG-Northumberland* are bound north in Santaren Channel. **Convoy KG-687** is bound southeast in Old Bahama Channel with US Navy plane escorting. Tanker *Paco* and SS *Narbo* are bound southeast in Old Bahama Channel. NAS Great Exuma: Submarine USS *Gabilan* (SS-252) is positioned near the base, two planes went to cover **Convoy GN-113**, eight PBY's arrived from Jacksonville, a plane departed to Miami and one plane arrived from Guantanamo.

10th February 1944

Request received from Comgulfseafron for aircraft to provide air escort to **Convoy GN-113**, B-25 FR-378 CL is airborne, returned to base with engine trouble, landed. B-25 FW-149 FN is airborne for Anti-Submarine Patrol Number 2 on the return of CL with engine trouble. FN was diverted to contact **Convoy GN-113** and to provide air escort until plane's limit of endurance (PLE). Convoy is located composition was 21 merchant ships and five escorts, "met" signal was transmitted to base air cover was provided, aircraft landed after duty was over. Request was received from Comgulfseafron for an aircraft to sight the **Convoy NG-415** and to sweep ahead, convoy consisted of 14 merchant ships and five escorts, B-25 FW-104 FH is airborne, convoy located, composition was same, landed with nothing to report. B-25's FW-148 FL, FR-393 DF, and FW-949 FD are airborne for Anti-Submarine Patrol Number 2, aircraft landed on the completion of the patrol, nothing to report. B-25's FV-951 FC, FW-

132 FK, FW-176 FS, FW-123 FJ and FV-954 FB are airborne for routine Anti-Submarine Patrol, and landed after completion. B-25's FV-951 FC, FW-132 FK, FW-176 FS, FW-104 FH, FW-123 FJ, and FV-954 FB are airborne for air Exercise Number 5 and landed after completion. B-25's FW-132 FK, FW-176 FS, FW-104 FH, FW-149 FN, FS-123 FJ and FV-954 KB are airborne, aircraft landed on the completion of night navigation exercise. Liberator FL-994 LA is airborne from Windsor Field for Jamaica, aircraft landed there, returned to base later. **Convoy NK-594** is bound south in Straits at northwest of Little Bahama Bank with US Navy plane escorting. SS *Shiloh* is bound south in Straits west of Little Bahama Bank. Tug USS *Farallon*, its tow, and USS *YMS-276* are bound north in Straits west of Little Bahama Bank. USS *YP-443* and USS *YP-445* is bound north in Straits west of Bimini. USS *YAG-19* is bound south in Straits. Destroyer escort USS *Holder* (DE-401) is bound north in Straits. Tanker *Pan Virginia* is bound south in Straits. SS *K. I. Luckenbach* is bound south in Straits north of Cay Sal Bank. Oiler USS *YO-107* and USCG cutter *CG-Crawford* are bound south in Straits entering Old Bahama Channel. USS *YMS-312* is bound southeast in Old Bahama Channel. USCG cutter *CG-Spencer* is bound north, to the east of Eleuthera. NAS Great Exuma: Merchant ships SS *Frustem*, SS *Hanging River*, SS *Lucullus*, and SS *Britanon* are positioned near the base. Eight PBY's departed to Guantanamo, two planes arrived from Guantanamo, one plane departed to Banana River, and one plane departed to Guantanamo.

11th February 1944

B-25's FV-954 FB, FR-388 CL, FW-104 FH, FV-951 FC, and FW-176 FS airborne for air exercises. Number 3, aircraft completed the exercise and landed. B-25's FV-954 FB, FW-148 FC, FW-104 FH, FW-149 FN and FW-123 FJ are airborne for air Exercise Number 3 and Number 4 on the completion of the exercise the aircraft landed. B-25's FV-954 FB, FV-951 FC, FW-123 FJ, FW-176 FS, FW-104 FH and FV-949 FD are airborne on the completion of the exercises the aircraft landed. B-25 FN-123 FK is airborne and completed the night navigation exercise and landed. Liberator FL-968 LF is airborne for Boca Chica, landed there returned to Windsor Field later. SS *Cayuse* is bound north in Straits northwest of Little Bahama Bank. SS *Charles A. Wickliffe* is bound north in Straits. **Convoy NK-594** is bound south in Straits with US Navy plane escorting. SS *Montana* is bound north in Straits north of Cay Sal Bank. Tanker *Paco* and USS *YP-442* are bound north in Nicholas Channel. **Convoy GK-787** is bound north in Old Bahama Channel with US Navy plane escorting. Tug USS *Willet* and SS *Ensley City* are bound southeast in Old Bahama Channel. **Convoy NG-415** is bound south, to the east of Eleuthera with US Navy plane escorting. SS *Chilbar*, SS *Sunnetta*, SS *Durango*, SS *Tennessee*, and tanker *J. A. Mowinckel* are bound south, to the east of Eleuthera. NAS Great Exuma: Merchant ships SS *Durango* and tanker *Gulftide* are positioned near the base, three planes went to cover **Convoy NG-415**, a tug was sighted, then three large tankers were sighted, and a plane arrived from Guantanamo.

12th February 1944

B-25's FW-150 FP, FV-951 FC, FW-949 FD, FW-104 FH and FW-150 are airborne for air Exercise Number 3 completed the exercise and landed. B-25's FV-949 FD, FR-378 CL, FV-951 FC, FW-104 FH, and FW-150 FP are airborne for air Exercise

Number 3, exercise was completed and the aircraft landed. B-25's FV-949 FD, FR-378 CL, FW-954 FC, FW-104 FH, FW-150 FP and FV-997 FG are airborne on the completion of air exercises Number 3 aircraft. A Widgeon was airborne and landed at base, having landed on the lake at San Salvador. USS *LST-52* is bound north in Straits northwest of Little Bahama Bank. Tanker *Chiloil* is bound south in Straits west of Little Bahama Bank. USCG *CG-83452* is bound north in Straits west of Little Bahama Bank. USCG cutter *CG-Vernon* and USCG cutter *CG-McNeal* are bound north in Straits northwest of Bimini. Tanker *Bohemian Club* is bound northeast in Straits. NAS Great Exuma: Seven USN PBY's arrived from Jacksonville for training, and aircraft PSP *J-2 P-20* is missing. Two planes went to cover **Convoy NG-415**, two planes arrived from Key West, and a large tanker [with callsign] *MRQL E3L* was sighted. One plane arrived from Guantanamo, one plane departed to Guantanamo, and one plane went on training flight, then returned.

13th February 1944

Liberator FL-968 FL returned from Boca Chica and landed at Windsor Field. Tanker *Pan Massachusetts* and tug USS *ATR-98* are bound north in Straits, northwest of Little Bahama Bank. SS *Great Meadows* is bound south in Straits west of Bimini. USCG cutter *CG-E. M. Humphries* and USCG cutter *CG-E. M. Messick* are bound south in Straits north of Cay Sal Bank. USS *PC-1231* is bound northwest in Old Bahama Channel. NAS Great Exuma: Six USN PBY's went to find the missing PSP plane *J-2P-20*. Two planes went to cover **Convoy AM-239**, two planes went to Key West, a small motor vessel was sighted, and two planes went on training flight, then returned.

14th February 1944

Liberator EV-877 NX is airborne from Windsor Field for La Guardia Air Field, New York. B-25's FW-176 FS, FV-951 FC, FW-104 FH, FV-949 FD, FV-997 FC, and FV-954 FB are airborne aircraft completed air Exercise Number 7 and landed. B-25's FR-381 DC, FW-123 FJ, FV-951 FB, FW-101 FB, FV-9A9 FD and FV-952 XC are airborne for air Exercise Number 6, aircraft completed the exercise and landed. B-25 FL-180 BA is airborne and landed on the completion of the local navigation exercise. Tanker *Spencer Kellogg* is bound north, west of Little Bahama Bank. Destroyers USS *DD-334*, USS *DD-335*, USS *DD-794*, and USS *DD-700* are bound north in Straits, west of Little Bahama Bank. **Convoy KN-295** is bound north in Straits with US Navy plane escorting. **Convoy KG-688** is bound southeast in Nicholas Channel with US Navy plane escorting. Tug USS *Willett* (ARS-12) and USS *YMS-220* are bound northwest in Old Bahama Channel. NAS Great Exuma: Two Naval Air Transport seaplanes moved between Guantanamo and Miami via this base, one plane arrived from Key West, one plane departed to Jacksonville, two planes went to cover **Convoy GN-114** and two US Navy ships positioned near the base, along with an Allied Free French Allied submarine FFS *Amazone*.

15th February 1944

B-25's FW-150 FP, FW-123 FJ, FW-176 PS, FV-997 FG, FV-949 FD and FV-954 FB are airborne for air Exercise Number 7, aircraft landed on the completion of exercise. B-25's FW-104 FH, FW-145 FM, FR-381 DC, FR-378 CL, and FV-949 FD are airborne for air Exercise Number 6, aircraft landed after completion. B-25 FL-174 AL and FL-199 DB are airborne for local navigation exercise, landed after completion. Grumman Goose airborne, and a Harbour Island aircraft returned to Oakes Field and landed. B-25's FW-104 FH, FV-997 FG, FW-154 FR, FV-954 FU, FW-150 FP and FW-176 FS are airborne for night navigation exercise, aircraft completed the exercise and landed. Liberator BZ-806 LU is airborne from Windsor Field on Kingsley Exercise, radio contact was established, bearings were obtained and the convoy consisting of two frigates was sighted, signals were exchanged, aircraft was instructed to carry out Cobra patrol, aircraft contacted SEO on radio and stated that the southeast was not working and that contact with the SEO had not been maintained, visibility was 2 nm to 3 nm, aircraft was instructed to return to base, aircraft landed at Windsor Filed. Liberator BZ-808 LX is airborne from Windsor Field for the Kingsley Exercise, radio contact was established with the convoy homing procedure "B" commenced, convoy located, signals exchanged, aircraft instructed to carry out Platter patrol, smoke float was dropped by the SEO. and the aircraft attacked with one DC was requested to drop a flare for AA [anti-aircraft], practice by the ship's gunners exercise was completed and the aircraft left the convoy, aircraft landed. **Convoy KN-295** is bound north in Straits northwest of Little Bahama Bank, with US Navy plane escorting, exiting the area. USS *LST-287* is bound north in Straits northwest of Little Bahama Bank. USCG Cutter *GC-Vigilant*, tug USS *Point Judith*, and SS *Agwil Prince* are bound south in Straits west of Little Bahama Bank. SS *Lillemore* is bound southeast in Old Bahama Channel. **Convoy KG-688** is bound southeast in Old Bahama Channel with US Navy plane escorting. USS *PCS-1450* is bound north in Straits. Tanker *Norstank* is bound south, east of Eleuthera. USS *Santee* is bound south, to the east of Eleuthera with US Navy plane escorting. NAS Great Exuma: Two US Navy ships position near the base, with Allied Free French submarine *Amazone* to cover **Convoy USS *Santee* Group**. One plane went to cover **Convoy USS *Santee Group***, one plane departed to Key West, two planes arrived from Key West, one plane arrived from Guantanamo, and the merchant tanker *Norstank* is positioned near base.

16th February 1944

Liberator FL-93I MF is airborne from Windsor Field on the Kingsley Exercise, radio contact established, after receiving sundry bearings the aircraft located the convoy, consisting of two frigates, aircraft was instructed to carry out Adder patrol, on the completion of the Adder patrol Lizard and Viper patrols were completed, SEO dropped smoke float, aircraft attacked with one DC whilst the aircraft was carrying out a dummy attack on the ship on request of the SNO, relieving, aircraft arrived after completion, landed at Windsor Field. Liberator EV-879 is north airborne from Windsor Field as first relief for the Kingsley Exercise, radio contact was established with the SEO, aircraft honed and located the two frigates, aircraft exchanged signals with the SEO and was instructed to carry out Cobra patrol, SEO dropped a smoke float and the aircraft carried out an attack with one DC on the completion of the exercise the aircraft left the convoy and landed. B-25 FH-177 A0 is airborne on a drogue towing flight for the practice of the ship's gunners, homing procedure "B" was carried out and convoy was sighted, instructions were received from SEO to commence drogue towing, winches were found to be in-operative and the aircraft returned to base. Liberator BZ-

896 HA was airborne on Kingsley exercise, radio contact established, and homing procedure commenced, two frigates sighted, signals were exchanged, aircraft instructed to carry out Lizard patrol, tanker of 10,000 tons sighted, sighting report passed to SEO, smoke float was dropped by the SEO, attack carried out by the aircraft with one DC, exercises completed and the aircraft set course for base. Liberator BZ-810 MA is airborne from Windsor Field, an aircraft carrier and two escorts were sighted, radio contact was established with the convoy and homing procedure "B" was commenced, signals were exchanged, two frigates were sighted, aircraft was instructed to carry out Cobra patrol, patrol Cobra was completed and the aircraft commenced patrol Mamba, SEO dropped a smoke float and an attack was carried out by the aircraft with one DC, exercise was completed and the aircraft set course for base and landed. Liberator BZ-806 LU is airborne from Windsor Field for fourth relief of Kingsley Exercise, radio contact was established homing procedure "B" commenced, two frigates were contacted, signals were exchanged, aircraft was instructed to carry out patrol Salamander and followed by patrol Frog, smoke fleet was dropped by the SEO and the aircraft attacked with one DC, exercise was completed and the aircraft landed. B-25's FV-946 FD, FW-104 FH, FR-381 DC, FV-954 FG, FW-145, FW-154 FR, FR-380 DB, and FR-395 CO are airborne for air Exercise Number 6, aircraft landed. B-25's FR-362 CA, FR-380 DB, and FR-395 CO are airborne for local navigation exercise, aircraft completed patrol and landed. SS *Kernstown* is bound south in Straits northwest of Little Bahama Bank. SS *West Grama* and tanker *Mount Bruce Park* are bound north in Straits northwest of Little Bahama Bank. USS *Roper* (APD-20) is bound north in Straits off Palm Beach [she later gained fame against U-boat off Hatteras; her propeller is outside MIT]. SS *Wilcox* is bound north in Straits northwest of Bimini. USS *LST-494* and tanker *Great Meadows* are bound north in Straits west of Bimini. Allied Free French submarine FFS *Amazone* is bound northwest in Northwest Providence Channel with US Navy plane escorting. **Convoy NK-595** is bound south in Straits with US Navy plane escorting. Tug USS *Point Judith* is bound south in Straits. **Convoy GK-788** is bound northwest in Old Bahama Channel with US Navy plane escorting. USS *YOG-33* is bound southeast in Old Bahama Channel. SS *Narbo* is bound northwest in Old Bahama Channel. NAS Great Exuma: One plane departed to Nassau, one plane went to cover **Convoy USS *Santee Group***, one plane arrived from Morrison Field, one plane departed to Key West, one plane arrived from Guantanamo, and a plane departed to Banana River.

17th February 1944

Liberator BZ-931 MF is airborne from Windsor Field for the Kingsley Exercise, radio contact established, frigates were sighted, signals were exchanged, aircraft instructed to curry out Viper patrol, Lizard and Alligator patrols were completed under the instruction of the SEO, flare dropped and A-A practice by the ship's gunners, aircraft attacked with one BC, a smoke float that was laid by the SEO, aircraft left convoy after end of duty then landed. Liberator BZ-896 NA airborne from Windsor Field as first relief for the Kingsley, radio contact was established and homing procedure "B" was commenced. Two frigates were intercepted, signals were exchanged, Viper patrol commenced and completed, Reptile patrol was commenced and was completed, DC [depth charge] was dropped on a smoke float laid by the SEO, exercises were completed and the aircraft left the convoy then landed. Liberator BZ-879 is airborne from Windsor Field as second relief for the Kingsley Exercise, radio contact was established and homing procedure "B" was commenced, aircraft sighted two frigates, signals were exchanged, aircraft was instructed to carry out Viper patrol, patrol was completed, aircraft dropped a flare for the gunnery practice by the ship's crews, Alligator patrol was commenced and completed smoke float laid by SEO attacks were carried out by the aircraft with one BC mid-upper and beam gunners, aircraft was instructed to carry out Platter patrol, aircraft landed after exercise completion. Liberator BZ-810 MA is airborne from Windsor Field as fourth relief for the Kingsley Exercise the two frigates were located, signals were exchanged, the aircraft was instructed to carry out Viper patrol, Platter and Lizard patrols were also completed, a DC was dropped on a smoke float laid by the SEO after completion aircraft landed. An American Liberator VB-113 was airborne from Windsor Field as fifth relief for the Kingsley Exercise, aircraft carrier was sighted, was not possible to estimate the speed due to the darkness, homing procedure "B" was commenced frigates were sighted, signals were exchanged, aircraft was unable to carry out patrol exercises due to the approaching darkness. B-25 FV-997 FG is airborne from Oakes Field for San Juan, landed. B-25's FV-954 FB, FW-145 FW, FW-145 FR, FV-949 FD, and FW-150 FP were airborne for air Exercise Number 6, aircraft completed the exercise and landed. B-25's FR-371 CF, FK-174 AL, FR-362 CA, and FK-395 CO were airborne for local navigation exercise, landed. B-25's FR-366 CX, FR-362 CA, FK-174 AL, and FR-371 CF were airborne for local night navigation exercise, landed after completion. SS *Four Lakes* is bound south in Straits west of Little Bahama Bank. USS *LST-54* is bound north in Straits west of Little Bahama Bank. SS *Cornell* is bound south in Straits west of Bimini. Tanker *Bulkfuel* is bound south in Straits towards Miami. USS *YO-142* escorted by USCG cutter *CG-Wanigan* and SS *Shiloh* are bound north in Straits west of Bimini. **Convoy SS *Winooski*** is bound south in Straits north of Cay Sal Bank with US Navy plane escorting. SS *Fort Fetterman* is bound north in Straits north of Cay Sal Bank. SS *F. H. Bedford*, SS *Osthav* and SS *Dean Emery* are bound south, to the east of Eleuthera. NAS Great Exuma: Merchant ships the tanker *Tustem,* SS *Comos Cuba,* SS *Peter Hull* and SS *E. H. Bedford* are positioned near the base. Two planes went to cover **Convoy NG-416**, a small motor vessel was sighted, the inter-island transport USS *YAG-28*, ex-*Samana*, arrived, and a US Army crash boat departed to Mayaguana.

18th February 1944

B-25's FW-145 FW, FW-154 FR, FV-954 FB, HD-308 FU, FV-949 FD, and FW-147 FO are airborne for compass swinging exercise, landed after completion. B-25's FW-154 FR, FV-949 FD, FV-954 FB, FW-145 FW airborne for night navigation exercise, landed after completion. Grumman Goose is airborne for Miami landed there. USS *Tattnall* (APD-19) is bound south in Straits north of Little Bahama Bank. Landing ship USS *LST-288* is bound northwest in Straits northwest of Little Bahama Bank. USS *LST-511*, USS *LST-505* and USCG *CG-83376* are bound north in Straits west of Little Bahama Bank. USCG cutter *CG-E. M. Messick* is bound north in Straits northwest of Bimini. SS *Bahram* is bound north in Straits west of Bimini. Tanker *Petrofuel* is bound north in Straits. Tanker *Julesburg* and tanker *Comol Cuba* are bound south, to the east of Eleuthera. Tanker *Tustem* is bound north, to the east of Eleuthera. NAS Great Exuma: The inter-island transport USS *YAG-21* departed for Miami, one plane took off on a training flight, a plane arrived from Guantanamo and departed for Banana River, and eight PSP's arrived from Jacksonville.

19th February 1944

Grumman Goose is airborne and landed at base from Miami. USS *PC-619* is bound south in Straits north of Little Bahama Bank. USS *LST-505* is bound north in Straits, west of Little Bahama Bank. Tanker *Potrero del Llano* and USS *LST-53* are bound north in Straits, northwest of Bimini. **Convoy KN-296** is bound north in Straits with US Navy plane escorting. **Convoy KG-689** is bound southeast in Nicholas Channel with US Navy plane escorting. SS *Thorshov* is bound north south of Abaco. Kindley Field, Bermuda's USAAF sends its last ASW mission from Bermuda. NAS Great Exuma: Two planes went to cover **Convoy GN-115**, eight USN PSP's departed to Guantanamo, a plane departed to Nassau then returned, and one plane arrived from Guantanamo and departed to Banana River. The merchant ships SS *Baxil* and SS *Biaston* are positioned near the base.

20th February 1944

Grumman Goose is airborne for Miami landed and returned to Oakes Field. **Convoy NK-596** is bound south in Straits northwest of Little Bahama Bank with US Navy plane escorting. USCG *CG-83486* and landing ship USS *LST-55* are bound north in Straits west of Little Bahama Bank. Tanker *Ensley City* is bound northwest in Straits, entering Nicholas Channel. SS *W. Honaker* is bound northwest in Old Bahama Channel. Tanker *F. Q. Barstow* is bound southeast, to the east of Eleuthera. **Convoy GN-115** is bound north, to the east of Eleuthera. NAS Great Exuma: One plane departed for Guantanamo, two USN PSP's departed for Jacksonville, one plane went to cover **Convoy NG-417**, and the merchant tanker *J. A. Mowinckel* is positioned near the base. A Naval Air Transport seaplane arrived from Miami and went to Guantanamo, and then arrived from Guantanamo and departed to Miami.

21st February 1944

B-25's FR-378 CL, FV-954 FB, FW-176 FS, FW-308 FU, and FW-148 FL are airborne for compass swinging exercises, aircraft landed on the completion. B-25's FR-378 CL, HO-308 FU, FV-951 FC, FW-145 FM, FR-393 DF and FW-147 are airborne for air Exercise Number 1, aircraft completed the exercise and landed. USS *Dynamic* (AM-91) and USS *LST-289* are bound north in Straits northwest of Little Bahama Bank. The Royal Norwegian Navy escort ship HNoMS *King Haakon VII* (G3-F6) is bound south in Straits northwest of Little Bahama Bank. USS *SC-678* is bound south in Straits off Miami. **Convoy NK-596** is bound south in Straits off Port Everglades with US Navy plane escorting. *SS Cayuse* is bound south in Straits north of Cay Sal Bank. USS *LST*-497 is bound north in Straits north of Cay Sal Bank. USS *LST*-436 northwest in Santaren Channel. **Convoy GK-789** is bound northwest in Nicholas Channel with US Navy plane escorting. USCG cutter *CG-Edwards* is bound northwest in Old Bahama Channel. **Convoy NG-417** and tanker *Esso Aruba* are bound south, to the east of Eleuthera. NAS Great Exuma: The yard oiler, or fuel barge USS *YO-127* arrived, one plane departed for Guantanamo, two PSP's departed for Jacksonville, one plane took off on a local training flight, eight PSP's arrived from Jacksonville, and tanker *J. A. Mowinckel* positioned near base. A Naval Air Transport seaplane arrived from Miami and went to Guantanamo, and then arrived from Guantanamo and departed to Miami.

22nd February 1944

B-25's FW-150 FP, FV-951 FC, FW-154 FR, FW-123 FJ and FW-148 FJ are airborne for Anti-Submarine Patrol Number 1, aircraft completed the patrol without incident and landed destroyer was sighted. B-25's FV-954 FB, FW-123 FK, FW-l76 FS, FW-145 FM, HO-308 FU and FK-393 DF are airborne for air Exercise Number 1, aircraft completed the patrol and landed. B-25 FV-997 FQ returned from San Juan, Puerto Rico and landed. Landing ships USS *LST-504* and USS *LST-292* are bound north in Straits Off Canaveral. Bahamas ferry *Betty K.* and escort USCG *CG-83402* are bound west in Straits towards Miami. Tanker *Pan Massachusetts* is bound south in Straits north of Cay Sal Bank. USS *Siren* (PY-13) is northwest in Nicholas Channel. Tanker *Esso Bolivar* and tanker *J. S. Mowinckel* are bound north, to the east of Eleuthera. NAS Great Exuma: The oil barge USS *YO-127* departed for Guantanamo, two planes went to cover **Convoy NG-417**, eight PSP's departed for Guantanamo, one plane arrived from Guantanamo, one plane set out on a training flight, and a plane departed for Nassau. The merchant tankers *Esso Aruba* and *Esso Bolivar* are positioned near the base.

23rd February 1944

B-25's FW-147 FO, FW-123 FJ, FW-l50 FP, FW-148 FL and FV-951 FC are airborne for Anti-Submarine Patrol Number 1, aircraft landed on the completion of exercise, a tanker of 6,000 tons was sighted. B-25's FR-999 DF, FW-145 FW, FV-997 FG, FW-149 FW, FW-176 FS, and HD-308 FU airborne for air Exercise Number 5, aircraft completed the exercise and landed. B-25's FW-176 FS, FW-l48 FL, FW-145 FW, FV-997 FG, FW-149 FN, and FW-179 FT are airborne for night navigation exercises, aircraft landed on the completion of the exercises. SS south *Mountain* and tanker *Bulkfuel* are bound south in Straits, north of Little Bahama Bank. USS *LST-507* is bound north in Straits northwest of Little Bahama Bank. USS *YMS-60* is bound south in Straits northwest of Little Bahama Bank. Tanker *Chiloil* is bound north in Straits west of Little Bahama Bank. USCG *CG-83510* is bound south in Straits northwest of Bimini. USS *YAG-21* and high-speed transport USS *Tattnall* (APD-19) are bound north in Straits, west of Bimini. USS *Coffman* is bound south in Straits, south of Bimini. **Convoy Motor Torpedo Boat Squadron-22, Group B** is bound north in Straits. NAS Great Exuma: Submarine USS *Perch* (SS-313), is positioned near the base, two planes went to cover **Convoy GN-116**, one plane departed **Convoy NG-417**, a large merchant vessels sighted, one plane went to Nassau then returned, one plane went to Key West, one plane went on training flight returned safely, one plane arrived from Key West, one plane arrived from Guantanamo and departed for Nassau and one plane arrived from Nassau.

24th February 1944

Liberators BZ-813 MD and BZ-807 LV are airborne from Windsor Fields, aircraft landed at Boca Chica and there briefed for the submarine exercises in co-operation with the US Navy on the completion of the exercises, the aircraft returned to base and landed. B-25's FV-997 FG, FW-145 FN, HD-308 FU, FW-149 FN, FW-179 FT, and FW-

123 FK are airborne for air Exercise Number 5 and landed after completion. B-25's FV-997 FG, FW-145 FN, HD-308 FU, FW-149 FN, FW-179 FT, and FW-123 FK are airborne for night navigation exercise, and landed after completion. B-25 FR-371 CF airborne for Miami, landed there, airborne, later landed back at base. USS *LST-499*, USS *LST-512*, and USS *LST-518* are bound north in Straits, northwest of Little Bahama Bank. **Convoy KN-297** is bound north in Straits, northwest of Little Bahama Bank, with US Navy plane escorting. Tanker *Athelprince* is bound south in Straits west of Little Bahama Bank. SS *Shabonee* is bound south in Straits northwest of Bimini. SS *Loreto* is bound north in Straits off Miami. Ferry *Ena K.* is bound north in Straits, west of Bimini with US Navy plane escorting. USCG cutter *CG-Papaw* is bound south in Straits. Canadian warship HMCS *Sackville* is bound south in Straits. USS *LST-508* is bound north in Straits. **Convoy KG-690** is with US Navy plane escorting, bound southeast in Old Bahama Channel. NAS Great Exuma: Submarine USS *Perch* (SS-313) is positioned near the base *en route* to Key West, two planes went to cover **Convoy GN-116**. Capt. Brockman arrived, a plane returned from Nassau, one plane went to Key West then returned, eight PSP's arrived from Jacksonville. One plane arrived from Guantanamo and went to Banana River, one plane departed for Cayo Francés, Cuba, a plane arrived from Key West and the tanker *Esso Augusta* is bound northbound toward Mayaguana Passage.

25th February 1944

Request received from Comgulfseafron for aircraft provide air cover for submarine USS *Perch* (SS-313) is escorted by USS *SC-1063* and USS *SC-1064*. B-25 FW-123 FJ is airborne convoy consisting of sub-chasers and escorts sighted, signals exchanged, aircraft provided air cover until late afternoon, left convoy, landed back at base. B-25 FW-150 FP is airborne, submarine sighted, escorted, landed back at base after completion of duty. B-25 FW-154 FR is airborne for Anti-Submarine Patrol Number 2, tanker of 5,000 tons sighted, and a 7,000 ton merchant ship also sighted, then aircraft landed at base after completion of exercise. Liberator BZ-807 LV is airborne from Windsor Field, aircraft landed at Boca Chica and there briefed for the submarine exercises in co-operation with the US Navy on the completion of the exercises the aircraft returned to base and landed. B-25's HD-308 FU and FW-179 FT are airborne for air Exercise Number 7 and landed after exercise completion. B-25's FV-951 FC, FW-149 FN, FR-393 DF, FR-378 CL, FW-132 FR, and FR-381 DC are airborne for air Exercise Number 3 and landed after exercise completion. USS *LST-509* is bound north in Straits northwest of Little Bahama Bank. USS *YAG-21* is bound east north of Walker Cay, north of Abaco. SS *Britamsea* is bound north, from a position north of Abaco. **Convoy NK-597** is bound south in Straits off Port Everglades, with US Navy plane escorting. Submarine USS *Perch* (SS-313) escorted by USS *SC-1063* and USS *SC-1064* are bound west in Northwest Providence Channel. USS *LST-291* is bound north in Straits west of Bimini. SS *Churubusco* is bound south in Straits south of Bimini. SS *Manuel Rionda* is bound north in Straits. NAS Great Exuma: Merchant ships SS *Durango*, SS *Peter Hull*, one USS *AF*, one USS *U1*, one USS *AVD*, one destroyer escort, USS *DE-219* and three destroyers, or USS *DD*s, are bound northbound through Mayaguana Passage. Also the ship SS *Fred W. Waller* is bound southbound through Crooked Island

Passage. Nine PSP's departed for Guantanamo, one plane arrived from Guantanamo and departed for Banana River, three planes took off on a training flight, and two planes departed for Key West.

26th February 1944

B-25's FV-951 FC, FW-149 FN, FW-132 FK, HD-308 FU, FW-179 FT, and FW-123 FJ are airborne for local night navigation exercise, landed after completion. B-25's FW-154 FR and FW-147 FO are airborne for Anti-Submarine Patrol Number 2 and landed after completion, nothing to report. B-25's FR-381 DC, FW-123 FK, FW-179 FT, and FW-123 FJ are airborne for air Exercise Number 7, and landed after exercise completion. B-25's FW-149 FW, FR-378 CL, FW-150 FP, FW-123 FJ, FR-393 DF, and FV-951 FO are airborne for air Exercise Number 4, and landed after exercise completion. Tanker *E. R. Kemp*, tanker *Fourlakes*, and USCG *CG-83377* are bound north in Straits west of Little Bahama Bank. SS *Julesburg* is bound north in Straits northwest of Bimini. Tanker *Pan North Carolina* is bound south in Straits. SS *Graffias* (AF-29) and destroyer escort USS *Gunarson* (DE-795) are bound north in Straits. USS *YAG-21* is bound south, to the north of Nassau. **Convoy GK-790** is bound northwest in Nicholas Channel with US Navy plane escorting. USS *YMS-60* is bound southeast in Old Bahama Channel. Tanker *Empire Chief*, USS *AG-67*, and destroyers USS *DD-210* and USS *DD-221* are bound south, to the east of Eleuthera. NAS Great Exuma: Two planes set out to cover **Convoy *GN-418***, two planes set out on a training flight, a plane departed for Miami, one plane departed for Morrison Field and one plane, sent on an emergency mission to Mayaguana, returned safely.

27th February 1944

A Beechcraft aircraft landed from Miami and returned back later. Tanker *Gulftide* is bound south in Straits northwest of Little Bahama Bank. British Royal Navy rescue tug HMS *HMRT-Cheerly* is bound north in Straits west of Little Bahama Bank. USS *SC-1059* is bound south in Straits off Port Everglades. SS *Wolf Mountain* is bound north in Straits northwest of Bimini. USCG cutter *CG-Belleville* is bound north in Straits. USS *Saluda* and USCG cutter *CG-Monomoy* are bound south in Straits. **Task Unit TU-0315** is bound north in Straits, north of Cay Sal Bank. NAS Great Exuma: One plane on training flight, returned safely.

28th February 1944

Liberator FL-931 MF is airborne from Windsor Filed for Kingsley Exercise, radio contact established, homing procedure "B" commenced, convoy located, signals exchanged, aircraft instructed to carry out patrol, Platter and Mamba exercise after this, followed after completion, smoke float dropped by SEO, aircraft attacked with one DC [depth charge], flare dropped by ship's gunners for AA [anti-aircraft] practice, first parachute failed to open, next was dropped, aircraft left convoy after it was relieved by the second landed at base. An American Liberator was airborne from Windsor Field as first relief for Kingsley Exercise convoy, radio contact established homing procedure "B" commenced, ships located signals exchanged aircraft instructed to carry out patrol Adder smoke float dropped by SEO, aircraft attacked with machine gun fires, aircraft completed Lizard patrol before leaving convoy, landed at base. Liberator BZ-807 LV is airborne from Windsor Field as second relief, radio contact established on bearing received during homing procedure "B," frigates located, signals exchanged, aircraft instructed to carry out patrol Cobra after completion patrol, Alligator commenced smoke float dropped by SEO which was attacked by aircraft, parachute flare dropped by aircraft for A-A practice by ship's gunners, first parachute to operate properly so second was dropped, aircraft left convoy after exercise was completed landed at base. Liberator BZ-809 LZ is airborne from Windsor Field as third relief, ship of 5,000 tons sighted, first bearing during homing was received, one frigate located, aircraft completed Serpent patrol under instructions of SEO, second frigate sighted attack with one DC carried out against smoke float dropped by SEO, flare dropped by aircraft for A-A practice by ship's gunners frigates left. B-25's FW-123 FJ, FV-949 FD, FV-951 FO, FW-149 FN, FW-179 FV, and FV-946 FF are airborne for air Exercise Number 6 and landed after completion. B-25's FW-132 FK, FV-946 FF, FV-954 FB, FW-123 FJ, FW-149 FN, and FW-176 FS are airborne for night navigation exercise, landed after completion. SS *Wilbur Wright* is bound north in Straits northwest of Little Bahama Bank. SS *Herman F. Whiton* and tanker *Gerald Don* are bound south in Straits off Port Everglades. USS *LST-57* and tanker *Palo Alto* are bound north in Straits northwest of Bimini. Tanker *Walter Jennings* is bound south in Straits. SS *David O. Saylor* and USCG *CG-83402* are bound north in Straits. USS *YOG-6* is bound north in Santaren Channel. Dutch minelayer KMNS *Jan Van Brakel* is bound southeast in Old Bahama Channel. Tanker *F. H. Bedford Jr.* is bound northwest, north of Abaco. Tanker *F. Q. Barstow* is bound northwest, east of Northeast Providence Channel. Tanker *Tustem* and tanker *Stanvac Capetown* are bound south, to the east of Eleuthera. **Task Group TG-29.8** is bound south, to the east of Eleuthera with US Navy plane escorting. *U-518* enters area north via Mona Passage. NAS Great Exuma: Merchant ships SS *Stanuas*, SS *Cacotorrin*, SS *Tustem*, and SS *Mount Bure Park* are bound southbound through Crooked Island Passage, with other ships SS *Clie* and tanker *Esso Aruba*, northbound through Mayaguana Passage. Two planes are on local training flight, one plane arrived from Miami, inter-island transport USS *YAG-21* arrived from Miami. Eight PSP's arrived from Jacksonville, one plane set out to cover **Convoy GN-117**, another plane to cover **Convoy Task Group TG-21.8**. A Naval Air Transport seaplane arrived from Miami and went to Guantanamo and then returned to Exuma from Guantanamo and departed to Miami.

29th February 1944

Liberator BZ-810 MA is airborne for Kingsley Exercise homing procedure "B" commenced frigates sighted, signals exchanged, aircraft instructed to carry out Cobra patrol, attack carried out against a smoke float dropped by SEO, aircraft carried out Mamba patrol under instruction of SEO, aircraft left convoy after exercise completion, a steam ship of 6,000 tons is sighted, and aircraft landed. Liberator BZ-757 LR first relief airborne for Kingsley Exercise, homing procedure "B" commenced, frigates sighted, signals exchanged, aircraft carried out Alligator patrol, patrol Python commenced and aircraft re-joined convoy, merchant ship of 7,000 tons is sighted, details of ship's position passed to SEO, flare dropped for the firing practice by SEO's gunners, second flare dropped for other escort vessel's gunners, smoke float dropped by SEO, aircraft attacked with one DC, aircraft left convoy after exercise completion, landed at base. Liberator BA-931 MF is airborne from Windsor Field as second relief for Kingsley Exercise, ships sighted, radio contact established, with SEO, homing procedure "B" commenced, signals exchanged, aircraft carried out Frog patrol, flare dropped for A-A practice by gunners or SEO, smoke float dropped by SEO, aircraft attacked with one DC after exercise completion, aircraft left convoy, landed. Liberator BZ-809 LZ airborne from Windsor Field as third relief for Kingsley Exercise, radio contact established, with SEO, homing procedure "B" commenced, aircraft carried out Platter and then Adder patrol, smoke float dropped by SEO, aircraft attacked with one DC, flare dropped for A-A practice by gunners or SEO, second flare dropped later, aircraft landed back at base after exercise completion. B-25's FW-150 FP and FW-154 FR are airborne for Anti-Submarine Patrol Number 2, a tanker was sighted, landed after completion. B-25's FV-946 FF, FW-176 FB, and FW-123 FJ are airborne for air Exercise Number 7 and landed after completion. B-25's FW-149 FN, FV-949 FD, FW-132 FJ, FV-954 FB, FW-150 FP, and FR-393 DF are airborne for air Exercise Number 7, and landed after completion. SS *Fenja* and tanker *Pan Massachusetts* are bound north in Straits northwest of Little Bahama Bank. Destroyer escort USS *Gandy* (DE-764) is bound north in Straits north of Little Bahama Bank. **Convoy KN-298** is bound north in Straits, northwest of Bimini with US Navy plane escorting. Oiler USS *YOG-6* escorted by USS *YP-332* is bound south, to the north of Nassau. SS *Kernstown* is bound north in Straits west of Bimini. USS *LST-500* is bound north in Straits off Miami. **Convoy KG-691** is bound southeast in Nicholas Channel with US Navy plane escorting. SS *Herman F. Whiton* is bound southeast in Old Bahama Channel. SS *Clio* is bound north, to the north of Eleuthera. *U-154* enters area northeast of Bahamas, bound west. NAS Great Exuma: Two planes set out to cover **Convoy GN-117**, with one plane allocated to cover **Convoy CTG-21.8**. Inter-island transport USS *YAG-21* departed for Miami, one plane departed for Guantanamo had to return due to mechanical issues. One plane departed for Guantanamo, one plane arrived from Guantanamo and went to Banana River, two PSP's arrived from Key West, two destroyers were southbound through Crooked Island Passage, and the merchant ships tanker *Gulf Breeze,* tanker *Esso Bolivia,* tanker *Pan Amoco,* SS *Natiana,* tanker *Gulfwave,* and SS *Osthau* are bound northbound through Mayaguana Passage. HMS *Trollope* and HMS *Thornborough* are on manoeuvres with RAF aircraft from Nassau.

1st March 1944

B-25's FV-954 FB, and FV-946 FF are airborne for Anti-Submarine patrol, a tanker is sighted, US Navy destroyer and escort sighted, landed after completion. Request received from Comgulfseafron for aircraft to sight and sweep ahead of **Convoy NG-419**, so B-25 FW-154 FR is airborne to provide air escort. Convoy sighted, aircraft swept ahead of the track of convoy, landed after completion, sweep completed without incident. B-25 FW-176 FS is airborne, and sweeps ahead of convoy detailed patrol completed aircraft landed, with nothing to report. B-25's FW-149 FN, and FW-123 FJ are airborne for air Exercise Number 7 and landed after completion of patrol. B-25's FR-366 CX, FK-180 BA, and FR-396 CN are airborne for local navigation exercise, landed after completion. B-25's FW-132 FX, FW-123 FJ, FW-148 FK, and FW-147 FO, are airborne on air Exercise Number 6, aircraft landed after completion. B-25's FV-951 FC, FW-132 FI, FW-176 FT, FW-145 FK, and FW-148 FV are airborne for night navigation exercises were completed aircraft landed. B-25's FK-166 AD, FR-395 CO, and FK-173 AK are airborne for local night navigation exercises, aircraft completed the exercises and landed. Liberator BZ-810 MA was airborne for Kingsley exercise, homing procedure "B" commenced, convoy was located signals were exchanged, aircraft instructed to carry out Cobra patrol, Platter and Lizard were also completed, aircraft attacked with one DC [depth charge], smoke float dropped by SEO, aircraft carried out Alligator patrol, aircraft left convoy after exercise completion, landed back to Windsor Field. Liberator of USN is airborne from Windsor Field as first relief for Kingsley Exercise, radio contact established, aircraft homed on convoy patrol, Adder commenced followed by Platter after termination of patrols smoke float dropped by SEO which aircraft dropped with one DC [depth charge], patrol Lizard is completed before aircraft left convoy, landed and returned to present unit at Boca Chica. Liberator BZ-757 LR is airborne as second relief for Kingsley exercise, homing procedure "B" commenced, convoy was located, signals were exchanged, aircraft instructed to carry out Reptile patrol, smoke float dropped by SEO, aircraft instructed to carry out Viper patrol, aircraft left convoy after patrol completion, landed back at base. BZ-760 LT is airborne as third relief for Kingsley exercise, first bearings of homing received, convoy located, signals exchanged, aircraft instructed to carry out Frog and then Alligator patrol smoke float dropped by SEO attacked by aircraft, aircraft carried out Lizard and then Viper patrol, left convoy after patrol completion, landed. District patrol craft USS *YP-453* sinks after grounding south of Bimini, 22 survivors rescued by a US Navy boat, the former yacht *Pleiades* was stripped and scuttled. **Convoy NK-598** is bound south in Straits west of Little Bahama Bank with US Navy plane escorting. SS *Susan Luckenbach* is bound south in Straits west of Little Bahama Bank. Ferry *Jean Brillant* and escort USS *SC-1295* are bound north in Straits northwest of Bimini. USS *LST 290* is bound west towards Miami with US Navy plane escorting. USS *YAG-21* is bound west towards Bimini. USS *YAG-28* and tanker *Bostonian* are bound north in Straits. Tanker *AthelPrince* and destroyer escort USS *McCann* (DE-179) are bound southeast in Old Bahama Channel. SS *Osthav* is bound north, to the east of Eleuthera. Tanker *Reginolite* is bound south, to the east of Eleuthera. NAS Great Exuma: Three planes are allocated to cover **Convoy GN-117**. The submarine USS *Dolphin* (SS-169) reported in, two planes arrived from Guantanamo, and one departed for Banana River. Another plane departed to Miami, returned, and one plane departed for Key West. HMS *Trollope* and HMS *Thornborough* reported in.

2nd March 1944

B-25's CN, CX, and BA are airborne for local Number 1 Squadron, navigation exercise landed after completion again airborne later for a similar exercise, landed after completion. B-25's FC, FU, FL, FO, and FM are airborne and completed the air exercises Number 6, landed. B-25's CN, BA, CJ, and CG are airborne for local night navigation exercises, landed on the completion of the exercise. B-25's FC, FO, FU, FD, FU and FL are airborne for night navigation exercise, aircraft landed on the completion of the exercises. Liberator LM returned from Boca Chica and landed at

Windsor Field. SS *Bulkfuel* is bound north in Straits northwest of Little Bahama Bank. Landing ship USS *LST-532* is bound north in Straits west of Little Bahama Bank. Tanker *Gulfbreeze* is bound north, northeast of Abaco. USS *Roper* (APD-20) is bound north in Straits off Port Everglades. USS *YMS-370* and oil barge USS *YOG-6* are bound west in Straits west of Bimini. USS *YP 612* is bound west in Straits south of Bimini with US Navy plane escorting. USS *LST-519* and USS *LST-58* are bound north in Straits. **Convoy GK-791** northwest in Nicholas Channel. Tanker *Artigas* is bound northwest in Old Bahama Channel with US Navy plane escorting. Tanker *Paco* is northwest in Old Bahama Channel. **Convoy NG-419** is bound south, to the east of Eleuthera with US Navy plane escorting. NAS Great Exuma: A plane departed for Guantanamo, a plane arrived from Guantanamo and departed for Banana River, a plane departed for Key West, a plane arrived from Key West and two planes went to cover **Convoy NG-419**. The merchant ships SS *Abshire* and tanker *Norkstank* are reported to be near base. A plane went to conduct search on the plan Able, and returned next day.

3rd March 1944

Signal received from Comgulfseafron of a possible U-boat entering the area. On Commanding Officer's instructions, B-25 FF was detailed to carry out search, aircraft airborne, searched the area, creeping line ahead search patrol completed without incident, landed back at base. B-25's FU, FC, FS, FV, FO, and FM airborne for compass swinging exercises, aircraft landed on the completion of exercise. B-25's FS, FM, FD, FU, FO, and FV airborne for night navigation exercise, and landed after completion. B-25 CN is airborne for Miami landed there. Destroyer escort USS *Rueben James* (DE-153) is bound south in Straits, northwest of Little Bahama Bank. British Royal Navy rescue tug HMS *HMRT-Emphatic* (BATR-96) is bound north in Straits northwest of Little Bahama Bank. USS *PC-1248* is bound north in Straits northwest of Little Bahama Bank. Destroyer escort USS *Buckley* (DE-51) is bound north in Straits off Port Everglades. Destroyer escort USS *Major* (DE-796) is bound north in Straits west of Bimini. USS *YAG-28* is bound south in Straits. Tanker *George W. Kendall* is bound north in Straits. Tanker *Gertrude Kellogg* is bound southeast in Old Bahama Channel. Tanker *Esso Bolivar* and tanker *Norkstank* are bound north in Straits, north of Abaco. Free French Navy submarine, the Allied FFS *Centaure* is west in Northeast Providence Channel with US Navy plane escorting. Then, 100 nm east northeast of Harbour Island, B-25 FF Liberator searched for a U-boat with nil result, from RAF No.111 OTU. NAS Great Exuma: Two plane went to Key West, one returned. A Free French submarine (FFS *Centaure*) and the minesweeper USS *Despite* (AM-89) are moving near the base. Two planes went to cover **Convoy NG-419**, a plane went to Key West, a plane arrived from Morrison Field with Capt. Brookman, then departed for Caicos Island with Captain Brookman. A plane went on a search, and returned next day, and a plane departed for Banana River.

4th March 1944

B-25's DC, FL, F0, FU, and FV airborne for compass-swinging exercises, aircraft landed on the completion of the exercise. B-25 CN returned from Miami and landed. USCG *CG-Papaw* is north in Straits northwest of Little Bahama Bank. USS *LST-526* north in Straits northwest of Little Bahama Bank. USS *YP-442* is north in

Straits northwest of Little Bahama Bank. USS *LST-539* is bound north in Straits, west of Little Bahama Bank, USS *LST-141* is bound north in Straits west of Little Bahama Bank. Tanker *Petrofuel* and SS *South Mountain* are bound north in Straits west of Little Bahama Bank. SS *Liebre*, USS *LST-525*, and USS *LST-140* are bound north in Straits off Miami. USS *LST-538* is bound north in Straits. Anti-Submarine Group 6 is bound south in Nicholas Channel. Tanker *AthelPrince* is bound northwest in Nicholas Channel. *U-518* exits the area south, bound through Mona Passage to the Caribbean. NAS Great Exuma: A plane departed for Miami, eight PSP's arrived from Jacksonville, a plane arrived from Key West, a plane went to cover **Convoy GN-118**, a plane went on a search and returned next day.

5th March 1944

Liberator LK is airborne from Windsor Field for Vernan Field, Jamaica landed there returned later to land at Windsor Field. Tug USS *Wood Island* with tows, and escorts are bound south in Straits, northwest of Little Bahama Bank. **Convoy KN-299**, with US Navy plane escorting, is bound north in Straits west of Little Bahama Bank. USS *LST-540* is bound north in Straits, west of Little Bahama Bank. USS *LST-515* is bound north in Straits off Palm Beach. USS *Roper* (APD-20) is bound north in Straits off Palm Beach. USS *Aquamarine* (PYc-7) is bound north in Straits, off Palm Beach. British Royal Navy motor minesweeper HMS *BYMS-258* and USCG *CG-Vigilant* are bound north in Straits northwest of Bimini. Tug *Nassau* and USCG *CG-83454* are west towards Miami with US Navy plane escorting. USS *LST-137* is bound north in Straits. Anti-Submarine Group 5 is bound north in Straits. **Convoy KG-692** is bound southeast in Old Bahama Channel with US Navy plane escorting. NAS Great Exuma: USS *TQM*, HMCS *AO-18* and USS *AM-255* are positioned near base, a plane went to cover three troopships then departed for Key West, a plane went to Guantanamo to take an injured man, two planes went to cover **Convoy GN-118**, seven PSP's departed for Guantanamo, and two planes arrived from Key West.

6th March 1944

B-25's FC, FB, FT, and FL airborne for Anti-Submarine Patrol Number 1 and landed after completion, no incident. B-25's FO, FT, FS, FF, FW, and CL are airborne for air Exercise Number 1 and landed after completion, no incident. Request was received from Comgulfseafron for an aircraft to sight and sweep given position for a convoy, composition of convoy was 25 merchant ships and five escorts B-25 FU airborne carried out sweep of area, landed after completion of uneventful patrol. B-25 FM airborne, located convoy, aircraft landed after exercise was over. Tanker *Bohemian Club* is bound south in Straits west of Little Bahama Bank. **Convoy NK-599** is bound south in Straits off Port Everglades with US Navy plane escorting. The Royal Norwegian Navy escort ship HNoMS *King Haakon VII* and tug USS *Wood Island* are bound south in Straits. USS *LST-293* is bound north in Straits. Navy plane is bound west in Straits. SS *Gertrude Kellogg* and SS *Herman Whiton* are bound northwest in Nicholas Channel. NAS Great Exuma: A plane on mission but returned due to mechanical failure then other plane went for mission, a plane departed **Convoy GN-118**, two planes on local flight and a plane returned from Miami. A Naval Transport seaplane arrived from Miami and departed to Guantanamo then arrived from there and departed to Miami.

7th March 1944

B-25 FC is airborne on Anti-Submarine Patrol Number 1, aircraft landed after patrol completion nothing to report tanker of 8,000 tons is sighted. B-25's FL, FB, FW, and FF are airborne, completed Anti-Submarine Patrol Number 1, aircraft landed after patrol completion, nothing to report. Request received from Comgulfseafron to continue sweep ahead of **Convoy NG-420**, B-25 DC is airborne, located convoy, composition was 25 merchant ships and five escorts sweep carried out over given area, aircraft landed after completion nothing to report. Navy plane south in Straits off Canaveral. SS *Boundbrook* is bound south, to the north of Walker Cay, off Abaco. USCG *CG-83405* is bound east in Straits north of Little Bahama Bank. Tanker *Edward L. Shea* is bound north in Straits west of Little Bahama Bank. Tanker *Gulftide* is bound north off Port Everglades. Ferry *Betty K.* is bound northwest in Straits towards Miami with US Navy plane escorting. SS *Menelaus* is bound north in Straits. **Convoy GK-792** is bound northwest in Nicholas Channel with US Navy plane escorting. **Convoy NG-420** is bound south, to the east of Eleuthera. NAS Great Exuma: Two planes went to cover **Convoy NG-420**, a plane departed for Key West, two planes on local flight, two planes arrived from Guantanamo and a plane arrived from Key West, then returned.

8th March 1944

Request received from Comgulfseafrom for aircraft to swoop ahead of the tender USS *Piedmont*, an aircraft was detailed to be airborne was cancelled due to adverse weather. Five aircraft detailed to complete Anti-Submarine Patrol Number 2 was cancelled due to adverse weather. B-25's FK, FL, FU, FS, FP, and FR are airborne for air Exercise Number 5 and landed after completion. B-25 CN is airborne for Miami landed there. Tanker *Pan Georgia* is bound south in Straits northwest of Little Bahama Bank. USS *Opponent* (AM-269) is bound north in Straits northwest of Little Bahama Bank. Destroyer escort USS *Osmus* (DE-701) and destroyer escort USS *Peterson* (DE-337) are bound east in Straits, north of Little Bahama Bank. USS *Despite* (AM-89) is bound north in Straits, north of Little Bahama Bank. USS *LST-510* and SS *Santiago* are bound north in Straits, west of Little Bahama Bank. SS *Cape Clear* is bound south in Straits northwest of Bimini. Tanker *Gulfgem* is bound north in Straits off Miami. USS *YAG-28* is bound northwest in Straits towards Miami. Landing ship USS *LST-533* is bound north in Straits. USS *YMS-370* is bound west in Straits north of Cay Sal Bank. SS *Kentuckian* is bound northwest in Old Bahama Channel. NAS Great Exuma: Submarine USS *Pintado* is positioned near the base, two planes went to cover **Convoy NG-420**, one of them later went to cover USS *AS-21* and sighted USS *AD-17* and destroyer USS *DD-796*. Two planes are on local training, a plane went to Banana River then returned and five PSP's arrived from Jacksonville.

9th March 1944

Request was received from Comgulfseafron for an escort to the submarine USS *Pintado,* (SS-387) and B-25 FG was airborne, escort duty submarine was contacted, air escort was provided by a box patrol with a creeping line ahead patrol, aircraft set course for base, Ferry *Jean Brillant* with one merchant ship was sighted, aircraft landed, B-25 FM was airborne as first relief for FG, submarine USS *Pintado* was sighted, submarine was accompanied by two escorts, air escort was provided, convoy was left, aircraft landed. B-25's FW, FF, FV, and DC were airborne for Anti-Submarine Patrol Number 2, aircraft landed on the completion of the uneventful patrol. B-25 CN returned from Miami and landed. B-25's FP, FR, FN, FY, FS, and FL were airborne for air Exercise Number 5, aircraft completed exercise and landed. SS *Tillie Lykes* joined **Convoy KG 693** from Puerto Plata Dominican Republic, at Key West. SS *Dorothy Luckenbach* is bound south in Straits north of Little Bahama Bank. SS *John Colter* is north in Straits, northwest of Little Bahama Bank. SS *Baldbutte* and USS *SC-681* are bound north in Straits, west of Little Bahama Bank. Submarine USS *Pintado* (SS-387) and escorts, are in Northwest Providence Channel with US Navy plane escorting. Ferry *Jean Brillant* and USS *YMS-361* are bound south towards Nassau. Tanker *Petrofuel* is bound south in Straits. SS *K. I. Luckenbach* and USS *LST-59* are bound north in Straits. NAS Great Exuma: Submarine USS Pintado (SS-387) is positioned near the base, a plane went to cover USS *AS-21* then departed to Guantanamo, a plane on local training, two planes departed for Banana River, six PSP's departed for Guantanamo, a plane arrived from Guantanamo, a plane arrived from Key West then returned then arrived over **Convoy GN-119**.

10th March 1944

B-25's FD, FF, FG, FW, FV and FP are airborne for Anti-Submarine Patrol Number 2, aircraft landed after exercise completion. B-24 Liberators LP and LV are airborne from Windsor Field for Boca Chica, Key West for the briefing prior to the participation in the submarine co-operation exercises with the US Navy, aircraft completed detailed exercise and returned to Windsor Field. B-25's FV, FY, FP, FH, FL and FH are airborne for air Exercise Number 3 and landed after completion airborne again later for air Exercise Number 4 and landed after completion. B-25 BX is airborne for Miami landed there at night. Local night navigation for six aircraft cancelled due to adverse weather. SS *Margaret Lykes* joined **Convoy KG-693** at Key West *en route* from Puerto Plata via Guantanamo. USS *LST-534* is bound north in Straits west of Little Bahama Bank. Ferry *Jean Brillant* is bound west in Northwest Providence Channel with US Navy plane escorting. **Convoy KN-300**, with US Navy plane escorting, is bound north in Straits, west of Bimini. Landing ship USS *LST-294* and SS *Fort Bridge* are bound north in Straits. SS *Dorothy Luckenbach* is bound west in Straits north of Cay Sal Bank. **Convoy KG-693** is bound southeast in Nicholas Channel with US Navy plane escorting. Tanker *AthelPrince* is bound southeast in Old Bahama Channel. B-25's FM and FH of No.111 OTU are off Abaco guarding the Northeast Providence Channel, responding to Command, Gulf Sea Frontier's sighting of U-boat approaching Northeast Providence Channel. After no initial sightings, further sweeps also produced no results. NAS Great Exuma: Three planes went to cover **Convoy GN-119**, a plane was on mission, a plane from Guantanamo and then departed for Banana River, and a plane arrived from Jacksonville.

11th March 1944

B-25's FL, FM, FD, FH, and FT are airborne, completed the air Exercise Number 7, aircraft landed. B-25's FC, FD, FY, FP, FN and FT are airborne for air Exercise Number 3, exercise was completed and the aircraft landed, same aircraft completed air Exercise Number 4 later. B-25 BX returned from Miami and landed. SS *Papanui* is bound north in Straits west of Little Bahama Bank. Tanker *Bostonian* is bound south in Straits. Tanker *Bulkfuel* is bound south in Straits north of Cay Sal Bank. USS *SC-1301* is bound north in Straits east of Berry Islands. Tug USS *Esselen* (AT-47) towing USS *YP-204*, is bound northwest in Old Bahama Channel. Request made by Colonial Secretary for aircraft to search for an overdue boat *en route* to Nassau. B-25 AN and B-25 BA are airborne, and the vessel was located off Darby Island, Exuma. NAS Great Exuma: A plane went to cover **Convoy GN-119**, two planes on local flight, a plane arrived from Guantanamo the departed for Banana River, and a plane departed for Jacksonville.

12th March 1944

B-24 Liberators LZ, HA, and LM were airborne from Windsor Field on navigation Exercise Number 14, aircraft completed the navigation exercise, all the aircraft encountered severe frontal conditions, LM and MA returned to base without completing the exercise, LZ completed patrol and landed. SS *Olney* in north in Straits northwest of Little Bahama Bank. **Convoy NK-600** is bound south in Straits off Miami, with US Navy plane escorting. SS *Swiftscout* is bound south in Straits towards Miami from Northwest Providence Channel. Destroyer escort USS *Rueben James* (DE-153) and SS *Wyoming Valley* are bound north in Straits west of Bimini. **Convoy GK-793** is bound west in Nicholas Channel with US Navy plane escorting. Anti-Submarine Group 4 is bound south in Santaren Channel towards Cuba. USCG *CG-Blackrock* is bound north in Santaren Channel. **Convoy NG-421** is bound south, to the east of Eleuthera. NAS Great Exuma: Two planes on local flight, a plane arrived from Guantanamo and a plane went to cover **Convoy NG-421**.

13th March 1944

B-25's FT, FR, FC, FD, FN and FP were airborne for air Exercise Number 6, aircraft landed after completion of exercise. Liberator LJ was airborne from Windsor Field and landed at Boca Chica crew were briefed for the exercise in co-operations with the USN submarines aircraft returned after detailed exercise was completed. B-25's FD, FH, FN, FS, FB, and FN are airborne for night navigation exercise, aircraft landed after completion. Liberators LU, LL, and MA are airborne on navigation exercises by Number 3 Squadron, Exercise Number 14 and landed after completion. SS *Corfu* is bound south in Straits, north of Little Bahama Bank. Tanker *Paulsboro* is bound north in Straits northwest of Little Bahama Bank. USCG *CG-83403* is bound northeast in Straits, northwest of Little Bahama Bank. SS *John A. Sutter* is bound north in Straits west of Little Bahama Bank. SS *Contreras* is south in Straits northwest of Bimini. Oiler USS *YOG-6* and USS *YP-351* are bound north in Straits, northwest of Bimini. Tug USS *ATR-99* is bound north in Straits. Tanker *AthelPrince* is bound west in Nicholas Channel. NAS Great Exuma: Two planes went to cover **Convoy NG-421**. Naval Transport seaplane arrived from Miami and departed to Guantanamo, then arrived from there and departed to Miami. Two planes went on local flight.

14th March 1944

B-25's FT, FH, FS, FD, FP, and FN are airborne for air Exercise Number 6 and landed after completion. B-25's FT, FG, DF, FC, FD, and FH are airborne for air Exercise Number 7 and landed after completion. Liberator LP is airborne from Windsor Field, air submarine exercise in cooperation with US Navy at Key West, returned after completion. Liberators LY, LS, and MA are airborne from Windsor

Field, exercise carried out by Number 3 Squadron landed after completion. B-25's CL, FN, FR, FP, FH, FG, DF, and FB are airborne for local night navigation exercise, landed after completion. B-25 FT airborne for Miami landed there and returned to base later. **Convoy Motor Torpedo Boat Squadron-29, Group A** is bound north in Straits northwest of Little Bahama Bank. SS *Liebre* and USS *Roper* (APD-20) are bound south in Straits west of Little Bahama Bank. SS *Williamsburg* is bound north in Straits north of Cay Sal Bank. Navy plane is bound west in Straits, north of Cay Sal Bank. **Convoy Motor Torpedo Boat Squadron-33, Group A** and USS *Sturdy* are bound southeast in Old Bahama Channel. NAS Great Exuma: Two planes on local flight, a plane arrived from Guantanamo, then went to Banana River, a plane departed for Miami, and PSP A3-270 with Capt. Brookman arrived.

15th March 1944

A Walrus was airborne for Norfolk via Miami, aircraft landed at Miami. B-25 FB is airborne for air Exercise Number 7 and landed after completion. B-25's CL, FN, FR, FT, DF, and FG, and airborne for air Exercise Number 6 and landed after completion. Aircraft of Number 1 Squadron AD, AV, and AL are airborne for local navigation exercise, landed after completion, again airborne later for the exercise. Liberator LK was airborne from Windsor Field and landed at Boca Chica, crew were briefed for the exercise in co-operation with the US Navy submarines, aircraft returned after detailed exercise was completed. B-25's FK, FP, FB, FV, and FG are airborne for night navigation exercise, landed after completion. Aircraft of Number 1 Squadron airborne for night navigation exercise, landed after completion. Liberator LZ is airborne for air Exercise Number 14 and landed after completion. Liberator LY is airborne for air Exercise Number 14, then a troop ship is sighted, signal exchanged, landed after completion. Liberator MB is airborne for air Exercise Number 14, SE [radio wave] contacts made; identified as resulting from a convoy, nothing could be made out in darkness landed after completion. B-25 FM airborne for Anti-Submarine Patrol Number 1 and landed after completion. Tanker *Pan Virginia,* **Convoy Motor Torpedo Boat Squadron-29, Group B**, and SS *Swiftlight* are bound north in Straits, northwest of Little Bahama Bank. Escort carrier **USS *Tulagi* (CVE-72) Group** is bound north in Straits northwest of Little Bahama Bank with US Navy plane escorting. Oiler USS *YO-142* is bound north in Straits north of Bimini. **Convoy KN-301** is bound north in Straits northwest of Bimini with US Navy plane escorting. USCG *CG-83457* and USS *YAG-19* are bound north in Straits. USS *PC-466* is bound northwest in Nicholas Channel. **Convoy KG-694** is bound southeast in Old Bahama Channel with US Navy plane escorting. NAS Great Exuma: A plane on mission, a plane went to Guantanamo then returned, a plane went to Morrison Field with Capt. Brookman. A plane arrived then departed, a plane arrived from Guantanamo, a plane arrived from Miami, a plane departed for Banana River, and a plane arrived from Key West.

16th March 1944

B-25's FN, FT, and DF are airborne for air Exercise Number 6 and landed after completion. B-25's FY, FV, and FB are airborne for air Exercise Number 6 and landed after completion. Liberator LM was airborne from Windsor Field and landed at Boca Chica, crew were briefed for the exercise in co-operation with the US Navy submarines, aircraft returned after detailed exercise was completed. Aircraft of Number 1 Squadron B-25's CG, AV, AU, and AD are airborne for local triangular navigation flight aircraft landed. B-25's BD, CG, CN, and AV are airborne for local night navigation exercise, landed after completion. Widgeon airborne for San Salvador, waterborne at San Salvador, took off and landed at base later. Liberators MB and LY are airborne from Windsor Field, navigation Exercise Number 14 and landed after completion, no incident to report. Tanker *Delaware Sun* is bound north out of Straits. **Convoy Motor Torpedo Boat Squadron-22, Group A** is bound north in Straits northwest of Little Bahama Bank. Navy plane is bound north from its location north of Grand Bahama. Tanker *R. P. Smith*, SS *Boundbrook*, **Convoy SS *Chemung***, and US Navy plane are bound north in Straits, northwest of Bimini. Oil barge USS *YOG-6* escorted by USS *YP-351* is bound south in Straits from Northwest Providence Channel. SS *Manuel Rionda*, tug USS *Sankaty Head*, towing four barges escorted by USS *SC-1280*, and USS *YP-408*, are bound north in Straits. Anti-Submarine Group 6, tanker *Spencer Kellogg*, and USS *PCE-842* are bound northwest in Old Bahama Channel. NAS Great Exuma: A plane went to cover **Convoy GN-120**, two planes went to cover troop transport and destroyer escort USS *DE-203*, two planes on local flight, a plane on submarine search, a plane on a mission to the Gulf, a plane arrived from Guantanamo and departed for Banana River.

17th March 1944

B-25's FB, FV, DF, FY, FU, and FG are airborne for compass swinging exercise landed on the completion of the exercises. Liberator LY is airborne from Windsor Field for NAS Boca Chica, completed the submarine recognition in cooperation with US Navy, returned to base after completion of exercise. B-25's FH and CG are airborne for Miami and landed there. USS *LST-517* is bound north, to the north of Walker Cay, off Abaco. Tug USS *Sankaty Head*, its four barge tows, and USS *SC-1280* are bound north in Straits, northwest of Little Bahama Bank. **Convoy NK-601** is bound south in Straits off Port Everglades with US Navy plane escorting. Oiler USS *YO-142* and USS *YMS-224* are bound west in Northwest Providence Channel. SS *Manuel Rionda* is bound east in Northwest Providence Channel. Mailboat *Richard Campbell* and USS *YP-218* are bound west in Straits, towards Miami with US Navy plane escorting. Tanker *Spencer Kellogg* is bound north in Straits. Anti-Submarine Group 4 is bound north in Straits, east of Cay Sal Bank. **Convoy GK-795** is bound northwest in Old Bahama Channel with US Navy plane escorting. USS *Teaberry* (AN-34) is bound northwest in Old Bahama Channel. *U-170* enters area to the northeast, bound west towards Abaco. NAS Great Exuma: Two planes went to cover **Convoy GN-120**. Two planes arrived from Guantanamo, one went to Banana River. Eight PSP's arrived from Jacksonville.

18th March 1944

B-25's FF, FG, FK, FL, FR, and FT are airborne for compass swinging exercise, landed after completion. SS *Duquesne* is bound south in Straits north of Little Bahama Bank. Tanker *Bulkfuel* is bound north in Straits northwest of Little Bahama Bank. USCG *CG-83493* and USCG *CG-83494* are bound north in Straits west of Little Bahama Bank. USS *AN-34* is bound north in Straits north of Cay Sal Bank. Landing ship USS *LST-138* is bound north in Straits. *U-170* is east of Bahamas bound west towards northeast Abaco and Eleuthera. NAS Great Exuma: Two planes went to cover **Convoy NG-422**. Eight PSP's departed for Guantanamo, a plane went to Miami, a plane arrived from Nassau, a plane went to Guantanamo, a plane arrived from Guantanamo, then went to Banana River.

19th March 1944

B-25's FH and CG returned from Miami landed at base. SS *William Cullen Bryant,* torpedoed off Cay Sal Bank in 1942, is returned to service. USS *LST-138* and USS *LST-60* are bound north in Straits northwest of Little Bahama Bank. USS *LT-231* is bound north in Straits. Anti-Submarine Group 8 is bound south in Santaren Channel. Tanker *Paco* is bound southeast in Nicholas Channel. Tanker *AthelPrince* is bound southeast in Old Bahama Channel. Troop transport USS *USAT-State of Maryland* is bound northwest in Old Bahama Channel with US Navy plane escorting. NAS Great Exuma: A plane went to Caicos Island, a plane arrived from Guantanamo then returned, eight PSP's arrived from Jacksonville, two planes arrived from Key West.

20th March 1944

B-25's FH, FK, FL, FU, FV and DC were airborne for air Exercise Number 1 and landed after completion. B-25's FW, FF, FU, FR, FO and FT airborne for Anti-Submarine Patrol Number 2 patrol was completed without incident and the aircraft landed. USS *LST-295* and USS *LST-536* are bound northwest in Straits from northwest of Little Bahama Bank. **Convoy KN-302** and USS *LST-535* are bound north in Straits, west of Little Bahama Bank with US Navy plane escorting. Ferry *Jean Brillant* escorted by USCG *CG-83348* is bound west in Northwest Providence Channel. Dutch navy ship KMNS *Van Kinsbergen* is bound west in Northeast Providence Channel. USS *LST-524* is bound north in Straits, west of Bimini. USS *YP-218* is bound west in Straits towards Miami with US Navy plane escorting. **Convoy KG-695** is bound southeast in Old Bahama Channel with US Navy plane escorting. USS *Leon* (APA-48) is bound south, from east of Eleuthera. NAS Great Exuma: Two planes went to cover **Convoy GN-121,** a plane went to cover USS *Lyon* and destroyer escort USS *DE-219.* Six PSP's departed for Guantanamo, a plane departed for Miami, a plane *en route* to Jacksonville landed due to fuel leakage, a plane arrived from Jacksonville then returned, a plane arrived from Key West, two planes departed for Guantanamo. A Naval Transport seaplane arrived from Miami and departed to Guantanamo then arrived from there and departed to Miami.

21st March 1944

B-25's FF, FU, FV, FL, FK, and FC were airborne for Anti-Submarine Patrol Number 1, aircraft landed from the uneventful patrol. B-25's FM, DC, FP, FH, and FT are airborne for air Exercise Number 1, aircraft completed the exercise and landed. B-25's DC, FF, FU, FU, FH, and FV are airborne for night navigation exercise, aircraft completed exercises and landed. Destroyer escort USS *Richard S. Bull* (DE-402) is bound north, to the north of Walker Cay, off Abaco. **Convoy NK-602** is bound south in Straits northwest of Little Bahama Bank with US Navy plane escorting. USS *Saluda* (IX-87) and USS *SC-1291* are bound east in Northeast Providence Channel. Tanker *Pan Georgia* is bound north in Straits. Tanker *R. P. Smith* is bound south in Straits. SS *Ensley City* north in Nicholas Channel. USS *PC-1231* is bound southeast in Old Bahama Channel. **Convoy GK-795** is bound northwest in Old Bahama Channel with US Navy plane escorting. NAS Great Exuma: A plane departed for Guantanamo then returned, two planes on mission, a plane arrived from Guantanamo then went to Banana River, a plane departed for Jacksonville, a plane returned from Miami, and a plane arrived from Key West.

22nd March 1944

B-25's DC, FF, FU, FV, FP, and FM are airborne for air Exercise Number 3, and landed after completion. B-25's FN, FU, FR, FT, FV, and FM are airborne for night navigation exercises, aircraft completed the exercise and landed. Tanker *Pan Georgia* and *USS Mazama* are bound north in Straits, northwest of Little Bahama Bank. Tanker *Pan North Carolina* is bound south in Straits, west of Little Bahama Bank. USS *SC-1292* and USS *SC-1277* are bound south in Straits near Miami. USCG *CG-83490* and USCG *CG-83375* are bound north in Straits, northwest of Bimini. Ferry *Jean Brillant* is bound south, from the north of Nassau. USS *LST-516* is bound north in Straits. USS *LST-543* is bound east in Straits from north of Cay Sal Bank. Tanker *Paco* is bound northwest in Old Bahama Channel. Oiler USS *YO-69* is bound southeast in Old Bahama Channel. **Convoy NG-423** is bound south, from east of Eleuthera with a US Navy plane escorting. NAS Great Exuma: A plane went to cover **Convoy NG-423**, a plane arrived from Guantanamo then went to Banana River, a plane went on mission, and a plane arrived then departed.

23rd March 1944

B-25's FF, FR, DC, FU, FS, and FO were airborne for air Exercise Number 5, aircraft completed the exercise and landed. B-25 FT airborne for Miami landed there. SS *Contreras* is bound north in Straits west of Little Bahama Bank. Tanker *Paulsboro* is bound south in Straits off Palm Beach. Ferry *Jean Brillant* is bound west in Straits towards Miami from Northwest Providence Channel. Destroyers USS *Varian* and USS *Martin H. Ray* are bound north in Straits, west of Bimini. USCG *CG-83433* is bound north in Santaren Channel. *U-170* is bound south to Eleuthera from northeast of Abaco. NAS Great Exuma: A plane went to cover **Convoy NG-423**, a plane arrived from Guantanamo then went to Banana River, two planes on local training flight, one plane went to Nassau, and a plane arrived then departed.

24th March 1944

B-25's FR, FC, and FK are airborne for Anti-Submarine Patrol Number 2, aircraft landed. B-25's FV, FU, and FS are airborne for air Exercise Number 7 and landed after completion. SS *Prometheus* is bound north in Straits northwest of Little Bahama Bank. SS *Fort Schuyler* is bound south in Straits northwest of Bimini. Tanker *H. M. Fredricksen* is bound south in Straits off Port Everglades. Free French Navy ship FFS *Pique* is bound north in Straits off Port Everglades. SS *Liebre* is bound north in Straits off Miami. SS *Chirripo* is bound north in Straits. USCG *CG-83405* is bound north in Straits, north of Cay Sal Bank. *USS PC-1390* is bound southeast in Old Bahama Channel. USS *YO-69* escorted by USS *YP-352* are bound northwest in Old Bahama Channel. Tanker *Esso Patterson* is bound south, from the east of Eleuthera. NAS Great Exuma: A plane arrived from Guantanamo then went to Banana River, two planes on local training flight.

25th March 1944

B-25's CL, FM, and FS are airborne for air Exercise Number 7 and landed after completion. A Grumman Goose is airborne and landed at San Salvador. Tanker *Axtell J. Byles* and SS *W. W. Mills* are bound north in Straits northwest of Little Bahama Bank. SS *Cornell* is bound south in Straits northwest of Little Bahama Bank. Tanker *Delaware Sun* and tanker *Esso Augusta* are bound south in Straits west of Little Bahama Bank. **Convoy KN-303** is bound north in Straits west of Little Bahama Bank with US Navy plane escorting. USS *Monitor* (AP-160) is bound north in Straits west of Little Bahama Bank. SS *Juan Casiano* is bound south in Straits off Miami. Oiler USS *YO-69* and USS *YP-352* are bound northwest in Nicholas Channel. **Convoy KG-696** is bound southeast in Nicholas Channel with US Navy plane escorting. USS *YMS-341* and tanker *Mexican* are bound northwest in Old Bahama Channel. NAS Great Exuma: A plane went to cover **Convoy GN-122**. Two planes departed for Guantanamo, three planes arrived from there, one of them departed to Banana River. Inter-island transport USS *YAG-21* arrived from Miami, two planes on local flight, three planes arrived from Key West.

26th March 1944

SS *Julia L. Luckenbach* is bound south in Straits, north of Little Bahama Bank. Tanker *18 de Marzo* is bound north in Straits, northwest of Little Bahama Bank. SS *Saluda* and USS *SC-1291* are bound south in Straits, west of Little Bahama Bank. **Convoy NK-603** is bound south in Straits off Port Everglades with US Navy plane escorting. Tanker *Esso Augusta* is bound south in Straits northwest of Bimini. Ferry *Jean Brillant* is bound west in Straits to Miami with US Navy plane escorting. SS *Swiftlight* is bound south in Straits. No.113: Ventura PVI JT-829; the tail leg of this aircraft collapsed whilst taxiing after landing. This resulted in damage to fuselage near turret gunner's position. Crew were Capt. Lee, civilian pilot, and P/O Fawcett, WAG. Both uninjured. USN Ventura JT-869; the after under carriage had been retracted on take-off, and the aircraft dropped, so that the propeller blades and underside of aircraft touched the ground. Aircraft continued on to Trinidad. Tips of all propeller blades bent, and underside of fuselage damaged. Crew were F/O E. H. Yates, pilot, F/Sgt. H. A. Morris, and Nav. Sgt. D. J. Dartmouth, WAG. All were uninjured. NAS Great Exuma: A plane went to cover **Convoy KG-696** then went to cover **Convoy GK-796**, three planes went on a mission to the Gulf, then returned, one departed for Key West, one went to cover USS *Antaeus* and two US Navy destroyers. USS *YAG-21* departed for Miami, two planes arrived from Key West, then went on a mission into the Gulf of Mexico.

27th March 1944

Request received from Comgulfseafron for aircraft to sight and sweep US Submarine *Pipefish*. B-25 FC is airborne, and a convoy of 20 merchant ships and three escorts are sighted, submarine sighted later, signals exchanged, submarine escorted, aircraft landed after duty completion, nothing to report. B-25 FT airborne, swept given area, landed back after patrol, patrol was uneventful. B-25 FW airborne for Anti-Submarine Patrol Number 2, completed patrol, landed back with nothing to report. B-25's FS, CL, and FL are airborne for air Exercise Number 7 and landed back after completion. B-25's FJ, FF, FU, FM, DF, and FC are airborne for air Exercise Number 3, and landed back after completion same aircraft airborne later for air Exercise Number 4 and landed back after completion. B-25 CX is airborne to Jacksonville and landed there. Grumman Goose is airborne, returned to base after it had been waterborne and landed at San Salvador. Liberator LZ is airborne for Kingsley Exercise, radio contact established, homing procedure "B" commenced, convoy consisting of two frigates sighted, radio contact established, aircraft instructed to carry out patrol Platter, after completion patrol Adder commenced, completed as detailed, smoke float dropped by SEO, attacked by one depth charge, aircraft instructed to carry out patrol Viper, aircraft left convoy after arrival of first relief then landed back at base. Liberator LW airborne from Windsor Field as first relief for Kingsley Exercise, unable to maintain radio contact Radar, returned, radio contact was established, homing procedure "B" commenced, convoy sighted, signals exchanged, radio contact established, aircraft instructed to carry out patrol Frog, smoke float dropped by SEO, attacked by one depth charge, exercise completed and aircraft landed at Windsor Field. Liberator NO is airborne from Windsor Field as second relief for Kingsley Exercise, radio contact established, homing procedure "B" commenced, convoy sighted, signals exchanged, radio contact established, aircraft instructed to carry out patrol Mamba, Alligator, and Lizard, smoke float dropped by SEO, attacked by one depth charge, exercise completed and aircraft landed at Windsor Field. Liberator of USN is airborne, radio contact established, homing procedure "B" commenced, convoy sighted, signals exchanged radio contact established aircraft dropped a flare for A-A practice by ship's gunners smoke float dropped by SEO attacked aircraft, aircraft instructed to carry out patrol Frog was completed landed back at base. Liberator NA airborne from Windsor Field fourth relief for Kingsley Exercise, radio contact established, homing procedure "B" commenced, convoy sighted, signals exchanged radio contact established, aircraft instructed to carry out patrol Adder and Mamba, smoke float dropped by SEO, attacked by one depth charge, exercise completed and aircraft landed at Windsor Field. SS *Elokomin* is bound south in Straits northwest of Little Bahama Bank. *Swiftscout* north in Straits northwest of Little Bahama Bank. USS *YMS-65* is bound north in Straits off Palm Beach. SS *Chickamauga* is bound south in Straits west of Bimini. Tanker *Pan Virginia* is bound south in Straits. USS *YAG-21* is bound north in Straits west of Bimini. SS *Jamestown* is bound north in Straits. **Convoy GK-796** is bound northwest in Nicholas Channel with US Navy plane escorting. **Convoy NG-424** is bound south, to the east of Eleuthera with US Navy plane escorting. Search for overdue aircraft from Bermuda abandoned after 31 aircraft were deployed from OTU 111, two USN airplanes, ASR launches, frigate HMS *Dominica*, and B-25 FS searched a box, after two days of ASR. NAS Great Exuma: A plane on a mission to the Gulf, then departed for Key West, two planes departed for Key West, a plane went to Miami, and a plane on emergency flight to Caicos Island then to Guantanamo. And one plane went to cover **Convoy NG-424**.

28th March 1944

B-25's FK, FL, DF, FM, FJ, and FS are airborne for local night navigation exercises, aircraft completed the exercises and landed. A request was received from Comgulfseafron for an aircraft to escort the US submarine *Pipefish* from dawn to dusk whilst passing through the Northwest Providence Channel and Northeast Providence Channel, B-25 FL is airborne to provide air escort to submarine USS *Pipefish* (SS-388), submarine was contacted, provided escort, aircraft landed with nothing to report from the patrol. B-25 FS was airborne as first relief, submarine contacted and escorted, landed back at base after escort duty. B-25 FO airborne, Patrol Number 2 carried out without incident and landed. B-25 FJ, DF, and FK are airborne for air Exercise Number 7, aircraft landed after exercise completion. B-25's CL, FW, FM, FO, FY and CL were airborne for air Exercise Number 6, aircraft landed after exercise completion. B-25's FJ, FL, FO, FW, FN, and FK are airborne for night navigation exercises, aircraft landed after exercise completion. B-25 CX returned from Jacksonville. Liberator LZ is airborne from Windsor Field for Kingsley Exercise, radio contact established, homing procedure "B" commenced, convoy sighted, signals exchanged radio contact established patrols Cobra, Mamba, Lizard completed in accordance with SEO's instructions attacks carried out against a smoke float dropped by SEO, aircraft landed back after completion of exercise. Liberator LT airborne from Windsor Field as first relief for Kingsley Exercise, radio contact established, homing procedure "B" commenced, convoy sighted, signals exchanged, radio contact established, visual signals exchanged patrol Adder and Python completed, attack carried out with one depth charge against smoke float dropped by SEO, aircraft landed back after completion of exercise. Liberator of USN is airborne from Windsor Field as second relief for Kingsley Exercise, radio contact established, homing procedure "B" commenced, convoy sighted, signals exchanged, radio contact established, visual signals exchanged, attack carried out with one depth charge against smoke float dropped by SEO, landed at base after completion of exercises. B-25 DB is airborne for drogue towing for the AA gunners of the frigates, exercise commenced after convoy was contacted, firing exercises by ships terminated, attack carried out with one depth charge against smoke float dropped by SEO, aircraft landed back after completion of exercise. Liberator LR is airborne from Windsor Field as third relief for Kingsley Exercise, radio contact established, homing procedure "B" commenced, convoy sighted, signals exchanged attack carried out with one depth charge against smoke float dropped by SEO, patrol python completed, aircraft landed back after completion of exercise. Liberator LU airborne from Windsor Field as fourth relief for Kingsley Exercise, radio contact established, homing procedure commenced, convoy sighted, signals exchanged attack carried out against smoke float dropped by SEO, patrols instructed by SEO completed, aircraft landed back after completion of exercise. Liberator of USN airborne from Windsor Field as third relief for Kingsley Exercise, radio contact established, homing procedure commenced, convoy sighted, signals exchanged, radio contact established, two flares dropped, firing practice by ship's gunners attack carried out against smoke float dropped by SEO with one depth charge, patrol Lizard completed, aircraft finished escort duty and landed. US Navy blimp is bound west in Straits off Canaveral. Minesweeper USS *Palisade* (AM-270) is bound north in Straits west of Little Bahama Bank. SS *Julia L. Luckenbach*, escorted by USS *YMS-343* is bound south in Straits northwest of Bimini. SS *Walter Jennings*, USS *YT-338*, and SS *Jamestown* are bound north in Straits northwest of Bimini. SS *Fort Niagara* is bound south in Straits west of Bimini. USS *YAG-28* is bound west in Straits from Northwest Providence Channel to Miami. Submarine USS *Pipefish* (SS-388) escorted by USS *SC-1295* and USS *SC-1064* are bound west in Northeast Providence Channel. *U-170* exits area bound north. *U-154* re-enters area briefly bound east via Anegada Passage. NAS Great Exuma: Three planes arrive from Guantanamo, a plane arrived from Miami then went to Guantanamo. A plane arrived from Guantanamo then went to Banana River, six PSP's arrived on training flight.

29th March 1944

B-25's DF, FY, FK, FM, FL, and FO are airborne for air Exercise Number 6 and landed back after exercise completion. B-25's AU, CO, AN airborne for local navigation exercise, landed after completion. B-25's CO, BJ, and AN are airborne for local night navigation exercise, landed after completion. Liberator LT is airborne for Miami landed there. Liberator MB is airborne from Windsor Field for Kingsley Exercise, radio contact established, first bearing for homing procedure commenced, convoy consisting of two frigates sighted, signals exchanged radio contact established sundry patrols carried out as instructed by SEO attack carried out with one depth charge against smoke float dropped by SEO, aircraft landed back at base after completion of exercise. Liberator NA airborne met convoy after completion of exercise, landed back at Bermuda. Liberator NA is airborne from Windsor Field as second relief for Kingsley Exercise, radio contact established, homing procedure commenced, convoy sighted, signals exchanged radio contact established visual signals exchanged attack carried out with one depth charge against smoke float dropped by SEO patrols Python and Serpent carried out landed at base after completion of exercises. SS *Elizabeth Blackwell* is bound south in Straits northwest of Little Bahama Bank. Tanker *Bostonian* is bound north in Straits northwest of Little Bahama Bank. Tug USS *YT-338* is bound north in Straits off Port Everglades. USS *SC-1342* is bound south in Straits, off Miami. Tanker *AthelDuke* is bound northwest in Nicholas Channel. SS *George W. Barnes* is bound northwest in Old Bahama Channel. *U-518* is bound south to exit are via Mona Passage. NAS Great Exuma: A plane arrived from Guantanamo then returned there with six PSP's. Then eight PSP's arrived from Jacksonville.

30th March 1944

B-25's FH, FJ, FO, FF, FP, and FW are airborne for night navigation exercise completed exercise and landed back. B-25's FP, FW, FY, DF, FO, and FJ are airborne for air Exercise Number 6 completed exercise and landed. Liberator NC is airborne from Windsor Field, submarine recognition exercise carried out in co-operation with US Navy, landed at Boca Chica briefed for the flight landed back at base after completion. B-25's BJ, AV, and AD are airborne for local navigation exercise, landed after exercise completion. B-25's BJ and AV are airborne for local navigation exercise, and landed after exercise completion. B-25's BJ, BX, and AN are airborne for local night navigation exercise, and landed after completion. Liberator NA returned from Bermuda, landed at Windsor Field. Liberator NA airborne from Windsor Field as second relief for Kingsley Exercise, radio contact established, homing procedure commenced ship unable to give bearings due to weak signal strength from aircraft trailing aerial unwound and another attempt made for homing exercise unsuccessful. Radio contact could not be established, so aircraft began to home on Radar returns, radio communication established, aircraft attempted to pass messages to ship, yet signals too weak for ship to pick up rectangular search carried out using Radar ship enquired if aircraft had received any bearings from the ship, aircraft replied in the negative, aircraft increased the height of the patrol, abandoned search, commenced homing on bearings, bearings became weak due to relief aircraft being homed, ship arranged to send alternate bearings to concerned aircraft, two frigates were sighted. SEO dropped smoke float as aircraft was making dummy run bomb, doors opened, four depth charges fell out, second attack was made with one depth charge, during second attempt two more depth charges fell out, one depth charge detonated, aircraft left convoy and set course for Bermuda, and landed there. Civilian motor boat *Autosee* is bound south in Straits, west of Little Bahama Bank. Tanker *18 de Marzo* is bound south in Straits off Palm Beach. USS *Seven Seas* is bound north in Straits off Miami. **Convoy KN-304** is bound north in Straits west of Bimini with US Navy plane escorting. USS *PCE-844* is bound south in Straits. **Convoy KG-697** is bound southeast in Nicholas Channel with US Navy plane escorting. *U-218* briefly enters southeast corner of area, off Anegada. *U-154* exits via Anegada, is bound east for France. NAS Great Exuma: Two planes went to cover **Convoy GN-123**, a plane went to Miami, eight PSP's departed to Guantanamo, a plane arrived from Key West, a plane went on a local flight, and a plane arrived from Key West.

31st March 1944

B-25's FH, FJ, FO, FF, FP, and FW are airborne for night navigation flights, and landed after completion. B-25's AN, BJ, and BX are airborne for local navigation exercise, and landed after completion. Six aircraft detailed to be airborne for air Exercise Number 6, cancelled due to adverse weather conditions. **Convoy NK-604** is bound south in Straits northwest of Little Bahama Bank with US Navy plane escorting. USS *YMS-371* is bound north in Straits, west of Little Bahama Bank. USS *SC-1301*, USS *SC-1282*, and destroyer escort USS *Earl Y. Johnson* (DE-702) are bound north in Straits northwest of Bimini. USCG cutter *CG-Vigilant* and former yacht USS *Saluda* (IX-87) are bound west in Northeast Providence Channel. Destroyer escort USS *O'Toole* (DE-257) is bound south in Straits. **Convoy GK-797** is bound northwest in Nicholas Channel with US Navy plane escorting. SS *Gertrude Kellogg* and USS *Esselen* (AT-147) are bound southeast in Old Bahama Channel. SS *Sage Harbor* is bound south, to the east of Eleuthera. SS *Cathlamet* is bound north, to the east of Eleuthera. His Majesty the King of Britain awarded a DFC to Acting Squadron Leader George Crawford, who was a flying instructor at this unit, and flight commander in No. 53 Squadron. He was captain of an aircraft which sighted and made two runs against a U-boat, straddling it on the second run, and showing skill, courage and resolution of the high order. NAS Great Exuma: Two planes went to cover **Convoy GN-123**. Three planes arrived from Guantanamo, two went to Banana River. A plane arrived from Key West, then went on mission. Inter-island transport USS *YAG-19* arrived from Miami.

1st April 1944

Aircraft returning from Miami sighted a PBY Catalina, and two members of its crew were seen to be pulling dinghy towards land. It was not possible to ascertain if the aircraft was in distress or not. Grumman Goose was airborne to locate aircraft and give any assistance that might be required, aircraft landed after having searched for the Catalina and found nothing. B-25's FO and FS responded to Comgulfseafron request to investigate a direction-finding fix on a U-Boat 150 nm east northeast of Hope Town. The two No.111 OTU aircraft swept the area until they found an Allied convoy. B-25's FH, FN, FP, FY, CL, and DF are airborne for Anti-Submarine Patrol Number 3 and landed after completion. USS *YT-338* towing USS *YF-646* and USS *YF-647* are bound north in Straits off Canaveral. SS *Seven Pines* is bound south in Straits, west of Little Bahama Bank. SS *Dean Emery* and tanker *Edward L. Shea* are bound north in Straits, west of Little Bahama Bank. **Convoy NK-604** is bound south in Straits off Miami with US Navy plane escorting. USCG *CG-83433* is bound north in Straits. USS *PCS-1418* is bound south in Old Bahama Channel. **Convoy NG-425** is bound south, to the east of Eleuthera with US Navy plane escorting. NAS Great Exuma: USS *YAG-19* departed for Miami, one plane went to cover **Convoy GN-425**. Two planes departed for Key West, and one arrived from Guantanamo, which departed to Banana River.

2nd April 1944

B-25's FB, FH, DF, FV, FW, and FR are airborne for air Exercise Number 6 and landed after completion. Grumman Goose is airborne for Harbour Island landed at base later. USS *SC-1276* and USS *SC-1353* are bound north in Straits off Canaveral. SS *Liebre* is bound south in Straits off Canaveral. USS *LST-528* is bound north in Straits northwest of Little Bahama Bank. SS *Julesburg* is bound south in Straits west of Little Bahama Bank. Canadian HMCS *Oakville* is bound north in Straits west of Little Bahama Bank. SS *Santiago* is bound south in Straits northwest of Bimini. Destroyer escort USS *Weeden* (DE-797), USS *LST-529*, USS *LST-600* and USS *LST-654* are bound north in Straits, northwest of Bimini. USS *LST-527*, USS *LST-682*, and SS *Cornell* are bound north in Straits. Anti-Submarine Group 1 is bound south from Santaren Channel towards Cuba. USS *YAG-19* is bound west north of Abaco. NAS Great Exuma: One plane went to cover USS *AF-21*, and one went to cover **Convoy NG-425**.

3rd April 1944

B-25's FF FV, FJ, FB, FW, and FN are airborne for Anti-Submarine Patrol Number 2 and landed after completion. B-25's FB, FR, DC, CD, and FB are airborne for air Exercise Number 1 and landed after completion. Grumman Goose is airborne for Cat Cay, Bimini, and returned to base in evening. **Convoy KN-78** is bound north in Straits, northwest of Little Bahama Bank. Tanker *Pan Georgia* is bound south in Straits northwest of Little Bahama Bank. SS *Andrew Marschall* is bound south in Straits off Port Everglades. USS *YP-448* is bound north in Straits west of Little Bahama Bank. USS *LST-521* is bound north in Straits west of Bimini. USS *LST-520* is bound north in Straits. *U-170* re-enters area bound southwards from northeast of Abaco. B-25's FB and FW are 110 nm northeast of San Salvador, responding to a Comgulfseafron direction-finding fix on a U-Boat, to which the US Navy sent sub-chasers and destroyers, with no result. NAS Great Exuma: Naval Air Transport seaplane arrived from Miami and went to Guantanamo. Another Naval Air Transport seaplane arrived from Guantanamo and also went to Miami. One plane departed to Nassau, one plane returned from Miami.

4th April 1944

B-25's FW, FB, FN, FS, FR, and FV are airborne for compass swinging exercise, landed after completion. B-25's FN, FW,

FH, FJ, FM, and FU are airborne for air Exercise Number 1, recalled shortly after take-off because of unsuitable weather, heavy and continuous rain. B-25's FH and DC airborne after storm air Exercise Number 1 and landed after completion. **Convoy NK-605** is bound south in Straits northwest of Little Bahama Bank with US Navy plane escorting. **Convoy SS *Elokomin*** is bound north in Straits, northwest of Little Bahama Bank with US Navy plane escorting. USS *Carnelian* (PY-19) is bound north in Straits west of Little Bahama Bank. Tanker *Esso Little Ro*ck is bound south in Straits west of Bimini. **Convoy KN-305** is bound north in Straits northwest of Bimini with US Navy plane escorting. USS *YMS-360* is bound north, to the east of Abaco. Tanker *Petrofuel* and tanker *British Resolution* are bound north in Straits. **Convoy KG-698** is bound southeast in Old Bahama Channel with US Navy plane escorting. SS *Baldbutte* is bound northwest in Nicholas Channel. *U-518* exits area north from the Mona Passage, bound east to France. NAS Great Exuma: Two planes departed to cover **Convoy GN-124**, one plane went to Guantanamo and returned, and one plane arrived from Guantanamo and then went to Banana River. Two planes took off on a local flight.

5th April 1944

B-25's FR, FU, FN, FV, and FW are airborne for air Exercise Number 5 and landed after completion. Grumman Goose is airborne, waterborne at San Salvador, then returned to base later at night. **Task Unit TU-20152** (five landing craft, USS LCI(L)'s) is bound south in Straits off Palm Beach. USS *Chicopee* (AO-34) is bound south in Straits, west of Bimini. Oiler USS *YO-69* escorted by USS *YP-352* is bound east in Straits, off Port Everglades. USS *Augbury* (AM-149) is bound north in Straits, west of Bimini. **Convoy NK-605** is bound north in Straits with US Navy plane escorting. USCG *CG-83457* is bound south from Santaren Channel towards Cuba. USS *PCE-844* is bound southeast in Old Bahama Channel. *U-218* exits area. NAS Great Exuma: A plane went to cover **Convoy KG-698,** one went to cover **Convoy GN-124**, one went to cover **Convoy GK-798**. Two planes arrived from Key West then went on a mission to the Gulf, a plane arrived from Guantanamo then returned, six PSP's arrived from Jacksonville, one plane departed for Banana river, and a plane took off for a local flight.

6th April 1944

B-25's FN, FR, FT, FU, FV, and FY are airborne for night navigation exercise, landed after completion. B-25's FN, FP, FJ, FS, FH, and FF are airborne for Anti-Submarine patrol failed to sight anything landed after completion. B-25's FF, CD, FV, DF, FB, and FU are airborne for air Exercise Number 1, and landed after completion. SS *Vermont* is bound south in Straits north of Little Bahama Bank. Tanker *Delaware Sun* and USS *Burleigh* (APA-95) are bound north in Straits northwest of Little Bahama Bank. Tanker *Pan Virginia*, USS *LST-601* and USS *LST-545* are bound north in Straits, west of Little Bahama Bank. SS *Swiftlight* and destroyer escort USS *Durik* (DE-666) are bound north in Straits. USS *YAG-28* is bound north in Santaren Channel. **Convoy GK-798** is bound northwest, entering Nicholas Channel with US Navy plane escorting. USS *PC-575* is bound northwest, entering Santaren Channel. Mr. Douglas Deane, an Australian of wide academic experience, arrived at RAF No.111 OTU to organize the Recreational Training Center, supported by Major Beaton, Senior Secretary of the

National War Service Committee of the YMCA in Canada. HRH the Governor and the Duchess of Windsor received Mr. Deane, and showed great interest in the formation of a Lady's Auxiliary, and the showing of films in the centre. Arrangements regarding the supply of soft drinks were to be made with the NAAFI. NAS Great Exuma: Six PSP's departed for Guantanamo, a plane went to Miami, two planes went to cover **Convoy NG-426**, one went to cover **Convoy NG-425**, a plane departed for Key West, a plane arrived from Guantanamo and departed to Banana River, and another plane arrived from Banana River.

7th April 1944

B-25's FN, FJ, and FM, are airborne for Anti-Submarine Patrol Number 1, and after an uneventful patrol, landed after completion. B-25's CL, DF, DC, FU, and FV are airborne for local night navigation exercise, landed after completion. B-25's FU, CL, and FV are airborne for air Exercise Number 7 and landed after completion. B-25's FP, FT, FR, and DC are airborne for air Exercise Number 3 and landed after completion. B-25 FB is airborne for air Exercise Number 4, and landed after completion. B-25's FB, FR, FS, DC, and FP are airborne for air Exercise Number 3 and Number 4, and landed after completion. Grumman Goose is airborne for Great Exuma *en route* for Mayaguana landed at Great Exuma, remained for the night, as the weather was adverse. USS *LCI(L)-66* is bound south in Straits off Canaveral. SS *Chippewa* and USS *PC-1234* are bound south in Straits northwest of Little Bahama Bank. USS *PCS-1380* is bound south in Straits west of Little Bahama Bank. USS *LST-544* is bound north in Straits west of Little Bahama Bank. USS *YAG-19* east in Northwest Providence Channel. USS *Saluda* (IX-87) and USCG cutter *CG-Vigilant* are bound northwest in Northwest Providence Channel. SS *Walter Jennings* is bound north in Straits west of Bimini. SS *Charles R. McCormick* is bound south in Straits off Miami. SS *Britamsea* is bound south in Straits. USS *PC-575* is bound north in Straits. Tanker *Paco* is bound southeast in Old Bahama Channel. USS *PC-593* is bound northwest in the Old Bahama Channel. NAS Great Exuma: Two planes went to cover **Convoy NG-426,** one went to cover **Convoy NG-430,** six PSP's arrived from Jacksonville, a plane arrived from Nassau, and a plane returned from Miami.

8th April 1944

B-25's FN, FR, FB, FT, FO, and FV are airborne for night navigation exercise, landed after completion. B-25's FN, FP, FR, FY, and CL are airborne for air Exercise Number 3 and Number 4 and landed after completion same exercise later completed by B-25's FN, FY, FP, FR, and FV. USS *LT-6* towing oiler USS *Y0-40* is bound south in Straits off Canaveral. SS *Walter Jennings*, and fast torpedo boats USS *PT-450*, USS *PT-451*, and USS *PT-452* are bound north in Straits northwest of Little Bahama Bank. SS *Carthage* is bound south in Straits, west of Little Bahama Bank. SS *St. James* is bound north in Straits west of Little Bahama Bank. USS *YP-563* is bound south in Straits. Tanker *Alpha* is bound north in Straits. USS *PC-593* is bound north in Santaren Channel. USS *YAG-19* is bound north, to the north of Nassau. NAS Great Exuma: Six US Navy PSP's departed for Guantanamo, a plane departed for Nassau, a plane arrived from Guantanamo and departed to Banana River, one plane on local flight. Three planes arrived from Key West, one went on a mission to the Gulf on the same day, and another went the next day.

9th April 1944

Liberator ND is airborne from Windsor Field for Boca Chica, landed there, briefed regarding the exercise submarine recognition exercise in co-operation with the US Navy returned after exercise completion. Seaplane tender USS *Lapwing* (AVP-1) is bound north in Straits northwest of Little Bahama Bank, exiting the area. Converted yacht USS *John M. Howard* (IX-75) is bound north in Straits off Palm Beach. SS *Maravi*, escorted by USS *YMS-361* is bound north in Straits northwest of Bimini. **Convoy KN-306** is bound north in Straits, northwest of Bimini with US Navy plane escorting. Ferry *Jean Brillant* escorted by Royal Navy minesweeper HMS *MS-65* is bound east in Straits, west of Bimini. SS *Santiago* is bound north in Straits. **Convoy KG-699** and USS *PCS-1429* are bound southeast in Old Bahama Channel with US Navy plane escorting. NAS Great Exuma: A plane went to cover **Convoy GN-125,** a plane departed for Key West, and two planes arrived from Key West.

10th April 1944

A signal was received from Comgulfseafron stating that two enemy submarines may be approaching the Northeast Providence Channel, and a request was made for aircraft to sweep the area. B-25 FM is airborne and landed with nothing to report. B-25 FH is airborne for Anti-Submarine sweep, aircraft also landed with nothing to report. B-25 FD was airborne and completed the Anti-Submarine Patrol Number 2 without incident and landed. Liberator NX is airborne from Windsor Field on the Kinsley Exercise, radio contact was homing procedure "B," a convoy was sighted, patrol Alligator was completed, smoke float was dropped by SEO, and an attack was made with one DC [depth charge]. Exercise was completed, and the aircraft set course for base, where aircraft landed. Liberator NC was airborne from Windsor Field as first relief for Kingsley Exercise, radio contact was established, frigate was located radio contact was established and signals exchanged, patrol Crocodile was completed aircraft attacked, a smoke float dropped by the SEO, exercise was completed, and the aircraft landed at base. Liberator MA airborne as second relief from Windsor Field for Kingsley Exercise, radio contact established, homing procedure "B" completed, one frigate sighted, signals exchanged, patrol Mamba completed, smoke float dropped by SEO, exercise completed, aircraft landed. A convoy of 32 merchant ships and two escorts was sighted by both Liberators NC and MA. USN Liberator airborne as third relief from Windsor Field for Kingsley Exercise, radio contact established, homing procedure "B" commenced, one frigate sighted, signals exchanged, detailed patrols completed, smoke float dropped by SEO, attacked with three depth charges, exercise completed, aircraft landed. Liberator MC fourth relief from Windsor Field for Kingsley Exercise, radio contact established, homing procedure "B" commenced, one frigate sighted, signals exchanged, detailed patrols completed, exercise completed, aircraft landed. B-25's FU, FV, FY, DC, FG, and FS airborne for air Exercise Number 6 and landed after completion. Grumman Goose returned from Mayaguana. USS *LST-546* and USS *LST-655*, destroyer escort USS *DE-404*, destroyer escort USS *DE-405*, and USS *PF-52* are bound north in Straits northwest of Little Bahama Bank. SS *Copiapo* is bound south in Straits, west of Little Bahama Bank. **Convoy NK-606** is bound south in Straits, west of Little Bahama Bank with US

Navy plane escorting. SS *Robert Luckenbach* is bound south in Straits northwest of Bimini. SS *Minatitlan* and USS *Aquamarine* (PYC-7) are bound north in Straits northwest of Bimini. Ferry *Jean Brillant* and escort USS *YMS-65* are bound west in Northwest Providence Channel. USS *YAG-28* is bound east in Straits off Miami. Tanker *Manuel Rionda*, USS *YMS-343*, SS *White River* are bound north in Straits. A US Navy plane is bound west in Straits. SS *Maturines* and Cuban *CS-23* are bound southeast in Nicholas Channel. NAS Great Exuma: Two planes went to cover **Convoy GN-125,** one arrived from Key West. Naval Transport seaplane arrived from Miami and departed to Guantanamo, then arrived from there and departed to Miami. Two planes left on a mission to the Gulf, and a plane arrived from Nassau, then returned.

11th April 1944

B-25's FM and FD are airborne for Anti-Submarine Patrol Number 1 and landed with nothing to report. B-25 FH airborne for Anti-Submarine Patrol Number 1, five sub-chasers sighted, signals exchanged. Liberator LW airborne from Windsor Field for Kingsley Exercise, radio contact established, one frigate sighted, signals exchanged, patrols Lizard and Mamba completed, smoke float dropped by SEO, attacked with a depth charge, exercise completed, aircraft landed. Liberator LM airborne as first relief from Windsor Field for Kingsley Exercise, radio contact established, one frigate sighted, signals exchanged, patrols completed, smoke float dropped by SEO, attacked with a depth charge, exercise completed, aircraft landed. USN Liberator is airborne as second relief for Kingsley Exercise, radio not established, ship located after a search, signals exchanged, patrol Alligator completed, smoke float dropped by SEO, attacked with a depth charge, exercise completed, aircraft landed because engine was giving trouble. Liberator ND is airborne as third relief for Kingsley Exercise, convoy of 33 merchant ships and five escorts sighted, radio contact established, one frigate sighted, signals exchanged, smoke float dropped by SEO, attacked with a depth charge, exercise completed, aircraft landed. B-25 AG airborne for drogue towing exercise with Kingsley, frigate cable snapped as drogue was streamed exercise not completed. Liberator NA is airborne fourth relief for Kingsley Exercise, radio contact established, one frigate sighted, signals exchanged, patrols Crocodile and Python completed smoke float dropped by SEO, attacked with a depth charge, exercise completed, aircraft landed. B-25's FU, FS, and FY are airborne for air Exercise Number 7 and landed after completion. B-25's FG, FT, FY, FO, and DF are airborne for air Exercise Number 6 and landed after completion. B-25 DB is airborne, made three successful runs over the ship on third run, then drogue was shot away by ship's gunners, aircraft landed. SS *Nevada* is bound north in Straits northwest of Little Bahama Bank. USS *PCS-1392* is bound south in Straits west of Little Bahama Bank. USS *LST-683* and USS *LST-730* are bound north in Straits northwest of Bimini. Tanker *Manuel Rionda* is bound north in Straits, west of Bimini. **Convoy NK-606** is bound south in Straits. USS *SC-1291* and tanker *Mary Ashley Townsend* are bound north in Straits. **Convoy GK-799** is bound northwest in Nicholas Channel, with US Navy plane escorting. SS *Eduardo Jeremac* is bound northwest in Old Bahama Channel. **Convoy NG-427** is bound north, in position east of Eleuthera, with US Navy plane escorting. NAS Great Exuma: A plane on local training, a plane departed for Mayaguana, and another departed for Key West. Then a plane arrived from Guantanamo and departed to Banana River, a plane went to cover **Convoy NG-427**, and USS *YAG-28* arrived from Miami.

12th April 1944

B-25's DC, DF, FB, FC, FT, and FY are airborne for night navigation exercise, landed after completion. B-25's FH, FL, and FM are airborne for Anti-Submarine Patrol Number 1, uneventful patrol, then landed after completion. B-25's FB, FT, and FW are airborne for air Exercise Number 7, and landed after completion. B-25's DF, FR, DC, FG, FW, and FJ are airborne for air Exercise Number 6, and landed after completion. B-25's AM, CA, and CN are airborne for short navigation exercise, landed after completion, again airborne later for similar exercise, landed after completion. Liberator MA is airborne from Windsor Field for Kingsley Exercise, radio contact established, homing procedure commenced, frigate sighted, signals exchanged, patrols Lizard and Viper completed, smoke float dropped by SEO, attacked exercise completed, aircraft landed. Liberator NC is airborne as first relief from Windsor Field for Kingsley Exercise, radio contact established, homing procedure commenced, frigate sighted, signals exchanged, patrol Adder completed, smoke float dropped by SEO, attacked aircraft, dropped parachute flare for ship's AA gunners, exercise completed, aircraft landed. US Navy PBM is airborne at Montagu Bay, second relief to Windsor Field for Kingsley Exercise, radio contact established, homing procedure commenced, frigate sighted, signals exchanged, patrol Mamba completed, smoke float dropped by SEO, attacked with six depth charge, left convoy after exercise completion, returned waterborne at Montagu Bay. Liberator LZ is airborne as third relief from Windsor Field for Kingsley Exercise, radio contact established, frigate sighted, signals exchanged, patrol Crocodile completed, smoke float dropped by SEO, attacked aircraft, dropped flare for ship's AA gunners, exercise completed, aircraft landed. Liberator of US Navy is airborne fourth relief from Windsor Field for Kingsley Exercise, radio contact established, frigate sighted, signals exchanged, patrol Viper completed smoke float dropped by SEO, attacked aircraft, dropped flare for ship's AA gunners, exercise completed, aircraft landed. SS *W. F. Burdell* is bound south in Straits northwest of Little Bahama Bank. SS *A. S. Hansen* south in Straits northwest of Little Bahama Bank. Tanker *Juan Casiano* is bound north in Straits northwest of Little Bahama Bank. USS *LST-683* north in Straits northwest of Little Bahama Bank. Tanker *Chiloil* is bound north, to the west of Little Bahama Bank. SS *Cornell* is bound south in Straits off Port Everglades. SS *Samoset* is bound south in Straits, north of Bimini. Oiler USS *YO-69* is bound north in Straits north of Bimini. Tanker *Esso Little Rock* is bound north in Straits, northwest of Bimini. USS *YAG-28* is bound north in Straits west of Exuma Cays. USS *YT-338* is bound north in Straits. Tanker *Paco* is west in Nicholas Channel. SS *Eduardo Jeremac* is bound north in Santaren Channel. Tanker *Esso Aruba* is bound north, to the east of Eleuthera. NAS Great Exuma: A plane went to cover **Convoy NG-427**, and another plane arrived from Guantanamo and departed to Banana River. USS *YAG-28* and a plane departed for Miami.

13th April 1944

B-25's BA, BJ, and BC are airborne for local night navigation exercise, landed after completion. B-25's FG, FP, FJ, FR, and FT are airborne for night navigation exercise, landed after completion. Liberator NC is airborne from Windsor Field, landed at Boca Chica, crew briefed regarding air submarine recognition exercise, returned to base, and landed. Six aircraft of Number 1 Squadron detailed to carry out local navigation exercise, flight cancelled due to adverse weather. B-25 CN is airborne, weather test, if found fit was to continue with exercise, captain decided to continue, completed the exercise. SS *Chicopee* is bound north in Straits, northwest of Little Bahama Bank, exiting area. US Navy plane is bound south in Straits northwest of Little Bahama Bank. **USS *LCI(L)* Flotilla-6** is bound north in Straits, northwest of Little Bahama Bank. Tanker *18 de Marzo* is bound north in Straits northwest of Little Bahama Bank. US Navy plane is bound south in Straits northwest of Little Bahama Bank. USS *YMS-343* is bound north in Straits west of Little Bahama Bank. USS *LST-547* is bound north in Straits northwest of Bimini. USS *YAG-28* is bound north in Straits, west of Bimini. SS *A. S. Hansen* and SS *W. F. Burdell* are bound south in Straits. USS *YT-231* towing USS *YF-331* is bound north in Straits. Tanker *British Glory* is bound northwest in Old Bahama Channel. SS *Empire Flint* is bound north, to the east of Eleuthera. NAS Great Exuma: Two planes went on a mission to the Gulf, a plane on a search mission, a plane arrived from Key West then returned, and another plane arrived from Guantanamo.

14th April 1944

Liberator LT is airborne from Windsor Field, landed at Boca Chica crew took part in submarine recognition exercise in co-operation with the US Navy, returned to base and landed. B-25's FB, FR, FT, FJ, FM, and FW are airborne for air Exercise Number 5, compass-swinging exercise, landed after completion. B-25's CA, BJ, and BC are airborne, local navigation exercise, completed exercise and landed. Grumman Goose is airborne returned to San Salvador landed there. SS *Salamonie* is bound south, north of Walker Cay, north of Abaco with US Navy plane escorting. USCG *CG-83515* is bound north in Straits north of Little Bahama Bank. SS *William H. Ferguson* is bound southwest in Straits, northwest of Little Bahama Bank. SS *Edouard Jeremac* and escort are bound north in Straits northwest of Little Bahama Bank. **Convoy KN-307** is bound north in Straits west of Little Bahama Bank with US Navy plane escorting. SS *John W. Griffiths* and escorts are bound north in Straits northwest of Bimini. USS *YAG-28* is bound west in Straits south of Bimini. SS *Manuel Rionda* west in Northwest Providence Channel. Tanker *Gulfhawk* is bound north in Straits east of Abaco. USS *LST-602* is bound north in Straits. USS *PCS-1379* is bound north in Straits. **Convoy KG-700** is bound southeast in Old Bahama Channel with US Navy plane escorting. USS *PCE-845* and USS *PCS-1431* are bound southeast in Old Bahama Channel. NAS Great Exuma: Two planes went to cover **Convoy GN-126,** a plane arrived from Guantanamo and departed to Banana River, and four PSP's arrived from Jacksonville.

15th April 1944

B-25's BJ, BA, and CA are airborne for local navigation exercise, completed exercise and landed. B-25's FY, FG, FW, FM, and FU are airborne for air Exercise Number 6, completed exercise and landed. B-25's FY, FF, FL, FD, FU, and FG are airborne for compass-swinging exercise, landed after completion. Liberator LT is airborne from Windsor Field submarine recognition exercise co-operation with US Navy, landed after exercise completion. Tanker *Gulf of Venezuela* is bound south, to the north of Walker Cay, off Abaco. USS *Sierra* (AD-18) and USS *LST-653* are bound north in Straits northwest of Little Bahama Bank with US Navy plane escorting. USS *SC-673* is bound south in Straits. SS *William H. Ferguson* is bound southeast in Santaren Channel. NAS Great Exuma: Two planes went to cover **Convoy GN-126,** a plane went to cover **Convoy GK-800**, a plane departed for Miami, and four PSP's departed for Guantanamo. Then a plane with Capt. Brookman arrived from Nassau then went to Caicos Island, a plane arrived from Guantanamo, and then returned.

16th April 1944

Convoy Motor Torpedo Boat Squadron-35, Group A, and tug USS *YT-338* are bound south in Straits northwest of Little Bahama Bank. **Convoy NK-607** is bound south in Straits west of Little Bahama Bank with US Navy plane escorting. SS *Chippewa* is bound north in Straits west of Little Bahama Bank. SS *Abiqua* is bound south in Straits, west of Little Bahama Bank. Landing ships USS *LST-684*, USS *LST-548*, and USS *LST-731* are bound north in Straits, off Port Everglades. Tanker *Gulf of Venezuela* is bound south in Straits, south of Bimini. Tanker *Delaware Sun,* SS *Swiftlight,* and SS *Arkansas* are bound south in Straits, entering the Santaren Channel. SS *Robert Lucas* and SS *David B. Henderson* are bound northwest in Santaren Channel. SS *La Paz* is bound northwest in Old Bahama Channel entering Santaren Channel. **Convoy GK-800** is bound northwest in Old Bahama Channel entering Nicholas Channel with US Navy plane escorting. SS *Moose Mountain Park* and destroyer escort USS *Brister* (DE-327) are bound northwest in Old Bahama Channel. SS *William H. Ferguson* is bound southeast in Old Bahama Channel. Destroyer escort USS *Major* (DE-796) is bound south, to the east of Eleuthera. NAS Great Exuma: A plane arrived from Guantanamo then returned, and six PSP's arrived from Jacksonville.

17th April 1944

B-25's FM, FW, FP, FJ, and FY are airborne for Anti-Submarine Patrol Number 1 and landed after patrol completion, destroyer was sighted, blimp was sighted, patrol uneventful. Liberator FL-970 LH is airborne for Radar exercise, aircraft fitted as a flying classroom for instruction of aerial operation of Radar detectors aircraft, completed and landed. B-25's FB, FO, FT, FU, FG, and FD are airborne for air Exercise Number 1 and landed after completion. A Widgeon is airborne, lands waterborne at Harbour Island. Tanker *Vacuum* is bound south in Straits off Canaveral. USS *John M. Howard* (IX-75) is bound south in Straits off Palm Beach. SS *Britamsea* and SS *Moose Mountain Park* are bound north in Straits west of Little Bahama Bank. SS *La Paz* is bound north in Straits west of Bimini. SS *Argon*, USS *YMS-224*, and SS *Wacosta* are bound north in Straits. Free French Navy ship FFS *Duc Daumale* escorted by USS *PC-1169*, and a US Navy plane and tanker *Comol Cuba* are bound northwest in Nicholas Channel. Tanker *Delaware Sun* is bound southeast in Old Bahama Channel. *U-170* is bound northeast of Eleuthera exiting and bound east. NAS Great Exuma: Two planes went to cover **Convoy NG-428,** six PSP's departed for Guantanamo, a Naval Transport seaplane arrived from Miami and departed to Guantanamo, then arrived from there and departed to Miami. A plane arrived from Key West, then went on mission to the Gulf.

18th April 1944

B-25's FL, FF, FT, FO, FW, and FY are airborne for Anti-Submarine Patrol Number 1, aircraft landed after completion. Liberator FL-970 LH, the flying classroom, is airborne completed Radar exercise, aircraft landed after completion. B-25's FB, FY, FU, FM, FV, and FD are airborne for air Exercise Number 1, and landed after completion. SS *Moose Mountain Park* and USS *SC-1292* are bound north in Straits northwest of Little Bahama Bank. Tanker *Chiloil* and destroyer escort USS *Tweedy* (DE-532) are bound south in Straits Off Canaveral. USS *PCS-1426* is bound south in Straits northwest of Little Bahama Bank. USS *LST-603* and USS *Machias* (PF-53) are bound north in Straits west of Bimini. SS *George W. Barnes* is bound southeast in Old Bahama Channel. NAS Great Exuma: A plane went to cover **Convoy NG-428,** a plane arrived from Miami, a plane twice arrived from Guantanamo, and departed to Banana River. A plane arrived from a mission to the Gulf, then departed for Key West.

19th April 1944

B-25's FC, FY, FM, FK, FU, and FY are airborne for night navigation exercise, landed after completion. B-25's FB, FC, DC, FK, and FJ airborne for air Exercise Number 1 and landed after completion. Liberator FL-970 LH is airborne, Radar and navigation flight landed after completion. Request received from Comgulfseafron aircraft sight and sweep sea plane tender escorted by two destroyers. B-25 FT is airborne, a convoy contacted, and creeping line ahead patrol completed sweep carried out without incident. Tanker *Edward L. Doheny* is bound south in Straits northwest of Little Bahama Bank. **Convoy KN-308** is bound north in Straits, northwest of Little Bahama Bank with US Navy plane escorting. SS *Robert Luckenbach*, USS *LST-656* and USS *LST-549* are bound north in Straits west of Little Bahama Bank. USS *PCE-846* is bound south in Straits. SS *Robin Adair* is bound northwest in Nicholas Channel. **Convoy KG-701** is bound southeast in Old Bahama Channel. Ferry *Jean Brillant* is bound south in Straits north of Nassau. NAS Great Exuma: Three planes went to cover **Convoy GN-127,** three planes arrived form Key West, two planes arrived from Guantanamo and departed to Banana River. A plane with Capt. Brookman arrived from Caicos Island, then went to Nassau.

20th April 1944

B-25's FB, FW, DC, FH, FD, and FV are airborne for night navigation exercise, landed after completion. B-25's DC, FY, FH, FC, FD, and FW are airborne for air Exercise Number 5 and landed after completion. Liberator FL-970 LH is airborne for exercise in Radar and navigation aircraft had Radar pupils and others landed after completion. SS *Kittanning* is bound south in Straits north of Little Bahama Bank. **Convoy NK-608** is bound south in Straits northwest of Little Bahama Bank with US Navy plane escorting. SS *William S. Clark* is bound south in Straits northwest of Little Bahama Bank. USS *Penetrate* (AM-271) is bound north in Straits northwest of Little Bahama Bank. USS *LST-685* is bound north in Straits west of Little Bahama Bank. USCG *CG-92001* is bound south in Straits. USS *PCS-1379* is bound southeast in Old Bahama Channel. USCG *CG-Papaw* is bound south, to the east of Eleuthera. NAS Great Exuma: A plane went to cover **Convoy KG-701,** then went to cover **Convoy GK-801**. A plane went to cover a US Navy ship,

Convoy TG-28.8 and two US Navy destroyers. Two planes on a mission to the Gulf, a plane went to Key West, a plane arrived from Guantanamo and departed to Banana River.

21st April 1944

B-25's FD and FC are airborne for air Exercise Number 7 and landed after completion. B-25's FR, FH, FJ, FV, CL, FW, and FC are airborne for air Exercise Number 3 and Number 4 and landed after completion. A Widgeon is airborne for San Salvador, landed waterborne, returned after completion landed at base. Liberator FL-970 LH is airborne for exercise in Radar and navigation, landed after completion. Request received from Comgulfseafron for aircraft to cover Royal Navy escort carrier HMS *Queen* (D-19). B-25 FL is airborne, convoy with escort located, exchanged signals. Patrol Viper carried out, left after patrol completion, landed back at base. B-25 FO is airborne as first relief of escort duty convoy with escort located, exchanged signals left after patrol completion, landed back at base. USS *John C. Butler* is bound north, to the north of Walker Cay, north of Abaco. **Task Unit TU-20156** is bound south in Straits northwest of Little Bahama Bank. SS *Cornell* is bound north in Straits northwest of Little Bahama Bank. USS *LST-657* is bound north in Straits west of Little Bahama Bank. British Royal Navy motor minesweepers HMS *BYMS-257* and HMS *BYMS-258* are bound south in Straits, off Port Everglades. Landing craft USS *LST-604* is bound north in Straits off Miami. USS *YMS-360* is bound west in Straits from Bimini towards Miami. **Convoy NK-608** is bound south in Straits with US Navy plane escorting. USS *YMS-372* is bound north in Straits. **Convoy GK-801** is bound northwest in Nicholas Channel. Troop transport USS *USAT-George Washington* is bound northwest in Santaren Channel. Anti-Submarine Group 5 is bound south from Santaren Channel towards Cuba. USS *PC-466* is bound southeast in Old Bahama Channel. Tug USS *YT-338* is bound east towards Nassau. NAS Great Exuma: A plane went to cover British HMS *SV-E* and USS *DL-246*, with two planes departed for Key West, a plane on local flight, a plane went to cover **Convoy NG-429**, a plane arrived from Guantanamo and then departed to Banana River.

22nd April 1944

Request received from Comgulfseafron for aircraft to cover submarine USS *Pomfret* (SS-391). B-25 FF is airborne, submarine located, signals exchanged, aircraft swept area ahead of submarine, sweep covered area of 25 nm on either side of submarine's track, aircraft landed after completion. B-25 FW is airborne for Miami, aircraft returned and landed at base. Liberators LP and LL are airborne for submarine, in co-operation with US Navy Liberator LL, returned to base with engine trouble, and LP landed at Boca Chica, completed exercise and landed at base. Liberator LT is airborne and landed at Boca Chica, completed exercise and landed back at base. B-25's FO and FL are airborne for Anti-Submarine Patrol Number 2, and landed after completion, nothing to report from patrol. B-5's FC, FD, and FJ are airborne for air Exercise Number 7, aircraft landed after completion. B-25's FC, FD, FV, FH, CL, and DC are airborne for air Exercise Number 3, aircraft landed after completion. Liberator FL-970 LH is airborne for exercise in Radar and navigation landed after completion. B-25 FY airborne calibration tests with San Salvador Direction Finding Station aircraft landed after test completion. Tanker *E. R. Kemp* is bound south in Straits, off Palm

Beach. SS *Torvanger* is bound north in Straits west of Bimini. Tug USS *Dry Tortugas* is bound south in Straits. USS *YMS-178* is bound north in Straits. *U-541* enters area north of Walker Cay, north of Abaco. NAS Great Exuma: A plane went to cover **Convoy NG-429**, a plane went to Miami, a plane went to Nassau, two planes arrived from Key West, and a plane arrived from Guantanamo.

23rd April 1944

USS *LST-732* is bound north in Straits west of Little Bahama Bank. Ferry *Jean Brillant* is bound southwest from Northwest Providence Channel in Straits with US Navy plane escorting. USS *Pomfret* (SS-391) is bound west in Northeast Providence Channel with US Navy plane escorting. Allied Free French Navy submarine FFS *Antiope* is bound west north of Eleuthera with US Navy plane escorting. *U-170* is bound east to France and out of area. NAS Great Exuma: Two planes on a mission to the Gulf, three planes on local flight, and a plane arrived from Key West.

24th April 1944

Request received from Comgulfseafron for aircraft to cover Allied Free French submarine FFS *Antiope.* B-25 FF is airborne, submarine not located in estimated position at estimated time, creeping line ahead search carried out, escort sighted, signals exchanged, aircraft remained with convoy for 15 minutes, landed at base after duty completion. B-2 FO is airborne as first relief to FF, yet failed to locate convoy in estimated position, tanker of 4,500 tons sighted 2 nm astern of submarine, two USN Vought Kingfisher aircraft observed in the area, signals exchanged, aircraft set course for base, left convoy, landed at base. B-25 FL airborne, completed Anti-Submarine Patrol Number 2 without incident, landed at base. B-25's FJ, FR, and FV are airborne for air Exercise Number 7, landed. B-25's FC, FV, FJ, FR, FF, and FL are airborne for air Exercise Number 6, completed exercise and landed. Request received from Comgulfseafron for aircraft to investigate wreck on a beach on west end of Andros. B-25 AG is airborne, failed to locate wreck, aircraft landed. Liberator LA airborne for Miami, landed there, returned to base later at night. B-25 FY airborne for calibration tests with San Salvador's Direction Finding Station, tests completed, aircraft landed. Liberator LM airborne from Windsor Field for Kingsley Exercise, radio contact established, homing exercise commenced, signals exchanged, patrols Mamba and Lizard completed, smoke float dropped by SEO, attack carried out by aircraft with one depth charge, landed after exercise completion. Liberator LT is airborne as first relief from Windsor Field for Kingsley Exercise, radio contact established, homing exercise commenced, signals exchanged with frigate, patrols Lizard, Cockerel and Cobra completed, smoke float dropped by SEO, attack carried out by aircraft, landed after exercise completion. A PB Martin Mariner of US Navy is airborne from Montagu Bay as third relief to Windsor Field for Kingsley Exercise, radio contact established, homing exercise commenced, signals exchanged, merchant ship spotted, patrol Cobra completed, landed after exercise completion. Liberator NC airborne as third relief from Windsor Field for Kingsley Exercise, radio contact established, homing exercise commenced, frigate located, aircraft instructed to search area 10 nm around last known position of Liberator LP, landed after exercise completion. Liberator LP airborne as second relief from Windsor

Field for Kingsley Exercise, radio contact established aircraft instructed by frigate to carry out patrol Python. Then contact with aircraft lost, SEO instructed to divert fifth aircraft on Kingsley to search for missing aircraft, enquiries made of Kingsley Field, Bermuda for LP, no news received of missing aircraft, arranged that frigate HMS *Dominica* should remain in vicinity of last known position of aircraft, and to carry out search pending the dispatch of Liberators to carry out an air search. The RAF B-24 Liberator LP was lost on flight from Nassau on training to Key West. USS *PCE-846* is bound north in Straits northwest of Little Bahama Bank. USS *Edmonds*, USS *Shelton*, and USS *Scroggins* are bound north, to the north of Walker Cay, off Abaco. USS *LST-550* and USS *YMS-65* are bound north in Straits west of Little Bahama Bank. **Convoy KN-309** is bound north in Straits west of Little Bahama Bank with US Navy plane escorting. Allied Free French submarine FFS *Antiope* and a US Navy plane are bound west in Northwest Providence Channel. SS *Escolante* is bound south in Straits northwest of Bimini. **Convoy KG-702** is bound southeast in Nicholas Channel with US Navy plane escorting. NAS Great Exuma: A plane went to cover **Convoy GN-128**, a plane on a local flight, a plane returned from Miami. Naval Transport seaplane arrived from Miami and departed to Guantanamo, and then arrived from there and departed to Miami.

25th April 1944

B-25 FV is airborne, on parallel track, search carried out convoy of 39 vessels sighted, aircraft landed with nothing to report. B-25 AB is airborne, fruitless search, carried out parallel track search, convoy sighted, aircraft landed. B-25 FJ is airborne, frigate HMS *Dominica* is located, aircraft landed with nothing to report from patrol. During time, aircraft were searching area frigate HMS *Dominica* was detailed to standby in vicinity of last known position of aircraft B-24 Liberator LP. An HMS ASR launch contacted the frigate, and together a search was made for the waters. ASR Launch HMS *P-712* was dispatched to Harbour Island to replace ASR Launch HMS *P-339*, and returned to Harbour Island on 27th April, then ASR Launch HMS *P-712* returned to Montagu Bay. It was decided that no result was obtained from the day's search, Kingsley Exercise was cancelled, and all available craft on air and sea were detailed to continue the searches. B-25's FC, FO, and FR are airborne for air Exercise Number 6, and landed back after exercise completion. A Widgeon is airborne for Harbour Island, and returned to base late in the day. USS *YMS-360* is bound north in Straits off Canaveral. Tug USS *YMS-65,* towing a barge, is bound west in Straits, northwest of Little Bahama Bank. **Convoy NK-609** is bound south in Straits, west of Little Bahama Bank with a US Navy plane escorting. SS *Liebre* is bound north in Straits, west of Little Bahama Bank. British Royal Navy motor minesweepers HMS *BYMS-258* and HMS *BYMS-257* are bound southeast, entering Santaren Channel. USS *Shellbark* (AN-67) is bound north in Straits. USS *PCS-1426* is bound south in Straits. Tanker *Hadnot* is bound northwest in Nicholas Channel. *U-541* is bound northeast, to the northeast of Grand Bahama and Abaco and out of area. NAS Great Exuma: Two planes went to cover **Convoy NG-430**, a plane left on a mission to the Gulf, three planes on local flight, seven PSP's departed for Guantanamo, a plane arrived from Guantanamo and departed to Banana River, and two planes arrived from Key West.

26th April 1944
ASR Launch HMS *P-339*, stationed at Harbour Island, was instructed to put to sea and to rendezvous with the frigate HMS *Dominica* to search for a dinghy of B-24 Liberator LP lost 24 April, 1944, two days previously. A Kingsley exercise for second day was cancelled, all aircraft of Number 2 and Number 3 Squadron were diverted to search for missing Liberator. Liberator LM is airborne from Windsor Field, and a steady light was sighted 3 nm to 5 nm distant, light resembled that from a flash lamp; as aircraft flew over it seemed to change to a flashlight. An attempt was made to illuminate area with landing light, unsuccessful, because of heavy haze, flare dropped nearby the supposed dinghy, navigator of aircraft believed he saw a small dark circular patch with a light to one side. A signal was flashed to the light, requested a series of dots be flashed if there were occupants in the dinghy, two series of dots observed by several members of crew, aircraft dropped parachute flare in attempt to illuminate sea position of supposed dinghy, passed to HMS *Dominica.* Transmitted facts to base, aircraft began square search, aircraft dropped two smoke floats as sea markers, aircraft contacted frigate, USN Mariner aircraft was observed to set course for the dinghy position, aircraft landed on receipt of dinghy- sighting report. ASR Launch HMS *P-339* was instructed to proceed to that position. Liberator LT is airborne from Windsor Field, HMS frigate contacted, creeping line ahead patrol carried out, convoy of 35 merchant ships and five escorts sighted, aircraft landed. Liberator LR is airborne from Windsor Field, frigate HMS *Dominica* contacted, creeping line ahead patrol carried out, signal reporting sighting of dinghy received, aircraft completed patrol in that area, square search carried out, aircraft sighted southbound convoy, convoy of 35 merchant ships and five escorts, aircraft landed with nothing to report. USN Liberator is airborne from Windsor Field, frigate is contacted, ship passed latest detail regarding sighting of dinghy, aircraft carried out square search, extensive search carried out, then landed. USN Mariner airborne at Montagu Bay, developed engine trouble and was forced to return, landed waterborne back at Montagu Bay to carry out search of detailed area. Liberator NA is airborne from Windsor Field, square search commenced, aircraft dropped flares and carried out sundry patrols under instructions of the frigate, HMS *Dominica* was sighted, aircraft landed, did not see any survivors. Comgulfseafron requested to instruct the south and other ships under his command to keep a look out for missing aircraft or any survivors. Royal Navy motor torpedo boat HMS *MTB-339* was unable to remain at sea due to oil leak, which had resulted in large oil loss. Five B-25's are detailed to carry out search on either side of drift track of reported dinghy. B-25 FV is airborne, swept area, landed back, nothing to report. B-25 FL is airborne, swept area, landed back, nothing to report. B-25 FY is airborne, small object sighted, aircraft encircled to investigate, yet lost sight of the object. A convoy was sighted of 35 merchant ships and five escorts HMS *Dominica* was sighted, the aircraft searched, detailed findings, landed with nothing to report. B-25 FR is airborne, searched area, landed with nothing to report. B-25 DC is airborne, HMS *Dominica* is sighted, area searched, aircraft homed on contact registered on radio compass, nothing seen from this contact, nor from patrol, then landed with nothing to report.

B-25 FS is airborne, a creeping line ahead patrol carried out, frigate contacted, ASR launch sighted, landed with nothing to report. Kingsley Exercise cancelled, Anti-Submarine patrols, deferred all available aircraft of Number 2 and Number 3 Squadron detailed for air sea rescue searches. B-25's and B-24's and Mariner of the USN co-operated with HMS *Dominica* and the ASR launches in an attempt to locate the missing Liberator LP, or any of the crew, however all efforts proved fruitless. B-25 FC is airborne, dinghy is sighted, sea marker seen to be 3 nm from dinghy, marker was thought to have been dropped by another aircraft, position of dinghy signalled to base, aircraft lost sight of dinghy after two minutes, aircraft landed, no evidence of the missing aircraft found. B-25 FR is airborne, searched, landed back with nothing to report. B-25 FT airborne, searched, landed back with nothing to report. B-25 FO airborne, patrolled the given area, and messages received from base regarding finding of an empty dinghy, nothing was seen, aircraft landed back. B-25 FU is airborne searched given area, contact made with frigate HMS *Dominica*, aircraft instructed to proceed to dinghy position, aircraft instructed that the object seen between two sea markers was unidentifiable and landed back, with nothing to report. Liberator LR is airborne from Windsor Field, searched given area, frigate sighted, position of reported dinghy passed to aircraft proceeded to given position, sighted nothing, landed back nothing to report. Liberator LM airborne from Windsor Field, frigate sighted, searched given area. Liberator FL-970 LH airborne, frigate located, sighted nothing, landed back nothing to report. Liberator NC airborne from Windsor Field, frigate sighted, a pole 5 feet long painted black and white was sighted, area searched, water logged rowing boat approximately 16 feet long was sighted, landed back, sighted nothing. Liberator NA, Liberator LS, B-25's FU, FY, FR, FH, FP, and FS are airborne at different times searched given area, found nothing, landed back, nothing to report. Contact was maintained with NOB Bermuda and HMS *Dominica* throughout the search, all signals were received, search carried out for two days, during daylight hours, 31 aircraft of Unit 1 and Unit 2, and American aircraft took part in the searches together with ASR launches and frigate HMS *Dominica.* The sighting of the possible dinghy was made in the hours of darkness on 25th and an empty dinghy in the sorties on 26th, so search was abandoned on April 27th. B-25's CO, CN, and AV are airborne for local navigation exercise, then landed back after exercise completion. Oiler USS *YO-69* and USS *YP-352* are bound west, north of Walker Cay, north of Abaco. USS *YMS-65* and USS *SC-1496* are bound south in Straits off Palm Beach. USS *PCE-847* is bound west in Straits. **Convoy Motor Torpedo Boat Squadron-30**, **Group B** is bound north in Straits. SS *William F. Empey* is bound north in Santaren Channel. **Convoy GK-802** is bound northwest in Nicholas Channel with US Navy plane escorting. British Royal Navy motor minesweepers HMS *BYMS-258* and HMS *BYMS-257* are bound southeast in Old Bahama Channel. **Convoy NG-430** is bound south, to the east of Eleuthera. *U-541* is bound northeast, north of Abaco and out of area. NAS Great Exuma: Three planes left on missions. Two planes arrived from Guantanamo, one returned, and another went to Banana River. Four PSP's arrived from Jacksonville.

27th April 1944

B-25's FW, FH, FR, FT, and FC are airborne for night navigation exercise, landed after completion. B-25's BC, CN, and CO are airborne for local navigation exercises landed back after completion. Liberators LM and LR airborne, air submarine exercise at Boca Chica, returned to base after completion. B-25's FB, FC, FR, FU, and FS are airborne for air Exercise Number 6, and landed after completion. SS *William F. Empey* is bound north in Straits, northwest of Little Bahama Bank. SS *Woodlake* is bound north in Straits west of Little Bahama Bank. Oiler USS *YO-69* and USS *YP-352* are bound south in Straits off Port Everglades. Tanker *George W. Kendall* is bound north in Straits west of Bimini. USS *PCS-1419* is bound south in Straits. USS *YAG-19* is bound northwest in Old Bahama Channel. NAS Great Exuma: Four PSP's departed for Guantanamo, a plane arrived from Guantanamo and departed to Banana River, a plane departed to Miami, a plane on mission, two planes on local flight.

28th April 1944

B-25's CG, CN, BC, CX, and CA are airborne, completed night navigation exercise, landed. B-25's FP, FR, FS, FT, FW, and FH are airborne, completed local night navigation exercise, and landed. Liberators MX and LS are airborne from Windsor Field, landed at Boca Chica for submarine recognition exercise in co-operation with US Navy, returned to base and landed after exercise completion. B-25's FB, FH, FS, FP, and DF are airborne for air Exercise Number 6, aircraft was recalled to base as heavy thunderstorms and showers experienced at base, which was expected to last for rest of day, aircraft landed back. Liberator LB is airborne from Windsor Field navigation Exercise Number 14, aircraft completed detailed exercise and landed back nothing to report. USS *Aucilla* (A0-56) and destroyer USS *DD-251* are bound south in Straits north of Little Bahama Bank. SS *Wilson's Creek* and tanker *Kittanning* are bound north in Straits west of Little Bahama Bank. Destroyer escort USS *Edgar G. Chase* (DE-16) is bound south in Straits west of Little Bahama Bank. USCG *CG-83457* is bound north in Straits off Port Everglades. Destroyer escort USS *DE-340* and destroyer escort USS *DE-408* are bound north in Straits. USS *YAG-19* is bound north in Straits, north of Cay Sal Bank. Spanish SS *Monte Altube* is bound northwest in Nicholas Channel. NAS Great Exuma: a plane arrived from Guantanamo and departed to Banana River, two planes on local flight, a plane arrived from Key West, then went on a mission to the Gulf, and a plane arrived from Miami.

29th April 1944

Convoy KN-310 is bound north in Straits, west of Little Bahama Bank with US Navy plane escorting. SS *George W. Kendall* is bound south in Straits off Miami. **Convoy KG-703** is bound southeast entering Old Bahama Channel with US Navy plane escorting. NAS Great Exuma: Two planes went to cover **Convoy GN-129**, two planes arrived from Key West.

30th April 1944

B-25's FH, DF, FP, FU, FT, CL, and FW are airborne for compass swinging exercise landed back after exercise completion. B-24 EV-935 is delivered from Elizabeth City [near Newark] and landed from there. USS *ATR-100* is bound north in Straits northwest of Little Bahama Bank, exiting the area. **Convoy NK-610** is bound south in Straits, west of Little Bahama Bank with US Navy plane escorting. **Convoy GK-803** is bound northwest, entering Old Bahama Channel. **Convoy NG-431** is bound south, to the east of Eleuthera. **Convoy GN-129** is bound north, to the east of Eleuthera. In

April, the Bahamas Air Service Squadron grew to 248 persons, of whom 66 completed preliminary training, and were allocated to Maintenance, Training and SHQ wings. Seventeen new recruits were enlisted. Programs for the squadron include parade and arms drill, guard duties, ceremonial parades, and trade training in the many sections of the unit, led by F/O A. F. Hawkes, Squadron Adjutant, and Vice Adjutant F/O T. E. Wynne-Evans. NAS Great Exuma: A plane went to cover **Convoy GN-129**, then went to Guantanamo, then a plane went to cover **Convoy KG-703**, a plane on a mission to the Gulf, and two planes arrived from Key West. Then two planes departed to Key West, a plane with Cdr. Armstrong arrived from Nassau, then went to Guantanamo, and a plane arrived from Guantanamo and departed to Banana River.

1st May 1944

Report received from Comgulfseafron of a DF [direction finding] fix of an enemy submarine within 50 nm. B-25 FS is airborne and covered the given area. A convoy was sighted, and the aircraft landed back. B-25 FO is airborne, swept given area, landed back after uneventful patrol. B-25's FL, FF, FB, and FW are airborne, aircraft completed Anti-Submarine Patrol Number 1 without incident, aircraft landed back. B-25's FH, FP, FT, DF, FB, and FS are airborne for air Exercise Number 1, landed after completion of exercise. USCG *CG-83303* and USCG *CG-83304* are bound south in Straits, passing Canaveral. **Convoy KH-521** is bound west in Straits with a US Navy plane escorting. **Convoy GK-803** is bound northwest in Nicholas Channel with US Navy plane escorting. Tug USS *USAT-Belle Isle* escorted by USS *PC-1245*, USS *SC-512*, and SS *William S. Clark* are bound northwest in Nicholas Channel. SS *Castana* is bound northwest entering Nicholas Channel. USS *PC-464* is bound southeast in Old Bahama Channel. SS *Colobre* is bound northwest in Old Bahama Channel.

2nd May 1944

B-25's FB, FH, FP, FT, FF, and FB are airborne for night navigation exercise, landed after completion of exercise. B-25's CL, FP, FS, FH, and FT are airborne for air Exercise Number 1, aircraft completed exercise and landed. USS *PC-490* rescues all 52 survivors of tanker *Mobiloil,* delivers them to San Juan. Naval Auxiliary Air Facility, Great Exuma is established [this NAAF will eventually replace NAS]. USS *SC-1280* is bound north in Straits, east of Canaveral. USCG *CG-80303* and USCG *CG-80304* are bound south in Straits off Palm Beach. Navy plane is bound south in Straits, west of Little Bahama Bank. USS *Escolante* (A0-70) escorted by USS *PCE-847,* is bound north in Straits, west of Little Bahama Bank. Ferry *Jean Brillant* escorted by USS *YMS-65* is bound west in Northwest Providence Channel with a US Navy plane escorting. SS *Cornell* is bound south in Straits west of Bimini. Tanker *Gulfmaid,* tanker *18 de Marzo*, and USS *Charlottesville* (PF-25) are bound north in Straits. *SS Colobre* north in Santaren Channel. SS *Aldebaren* northwest in Nicholas Channel.

3rd May 1944

A direction finding [DF] fix was made on enemy submarine, US Navy dispatched sub-chasers and destroyers to search area where it was seen, request received from Comgulfseafron to search area. B-25 FB is airborne, nothing was sighted during sweep, landed back. B-25 FW and FU airborne at different times, swept given area, landed back after uneventful patrol. B-25's CL, DF, and FS are airborne for air Exercise Number 7, landed after exercise completion. B-25's FT, FJ, FF, FG, FK, and FP are airborne for air Exercise Number 5, landed after exercise completion. Grumman Goose is airborne for San Salvador, lands waterborne later. B-25 FY airborne for calibration test with San Salvador, aircraft landed after test completion. Liberator NC is airborne for navigation and Radar exercise, aircraft returned to base and landed after exercise completion. Nassau has by now become a centre for aircrew training as well as a base for Anti-Submarine patrols and transport ferry command. USS *SC-1474* is bound south in Straits northwest of Little Bahama Bank. SS *Colobre* north in Straits northwest of Little Bahama Bank. Tanker *A. S. Hansen* is bound north in Straits, off Miami.

4th May 1944

Request received from Comgulfseafron for further sweep to be made of the given area. B-25 FW is airborne, aircraft landed, uneventful patrol. B-25 FS is airborne, carried out sweep, aircraft sighted convoy of two merchant ships and two escorts patrol completed without incident, aircraft landed back. B-25 FU is airborne, carried out Anti-Submarine sweep, landed back, nothing to report. B-25's DF and FF are airborne for Anti-Submarine Patrol Number 3, and landed. B-25's CL, FS, FK, FG, and DF are airborne for air Exercise Number 5, aircraft completed exercise and landed. Liberator LR is airborne, carried out Radar exercise, landed on completion. The Grumman Goose returned from San Salvador, landed, airborne later for Harbour Island, landed waterborne there, returned to base. Destroyer escort USS *Earle K. Olsen* (DE-765) is bound north in Straits, north of Little Bahama Bank. USS *Dade* (APA-99) is bound north in Straits, northwest of Little Bahama Bank. USS *SC-660* is bound southeast in Old Bahama Channel. SS *Fairfax* escorted by USS *PC-573* is bound northwest in Old Bahama Channel.

5th May 1944

B-25's DF, FJ, FK, FN, FU, and FG are airborne, completed night navigation exercise, landed. B-25's FN, DF, FU, FJ, FV, and FY are airborne for air Exercise Number 3, landed after completion. B-25's FN, DF, FU, CL, FV, and FJ are airborne for air Exercise Number 3, landed after completion. Liberator FL-970 LH is airborne, landed back with Radar equipment unserviceable exercise cancelled. **Convoy NK-611** is bound south in Straits northwest of Little Bahama Bank with US Navy plane escorting. **Convoy KN-311** is bound north in Straits, northwest of Little Bahama Bank with US Navy plane escorting. USS *YO-115* is bound south in Straits northwest of Little Bahama Bank. USS *YO-77* is bound south in Straits. USS *YAG-19* is bound south in Straits, west of Exuma Cays. Tanker *Manuel Rionda* is bound southeast in Nicholas Channel. **Convoy KG-432** is bound southeast in Old Bahama Channel, with US Navy plane escorting. **Convoy NG-432** is bound south, to the east of Eleuthera, with US Navy plane escorting.

6th May 1944

B-25's FG, FC, FV, FW, and FY are airborne, completed local night navigation exercise, airborne for second sortie later, landed after completion. US Navy C-47 force-landed in the unit from Miami, had been *en route* to San Juan. B-25 FT is airborne, Anti-Submarine Patrol Number 2 completed, diverted from route to drop supplies to San Salvador's Direction Finding station, landed back after task completion, nothing to report from patrol. B-25 FF is airborne for Anti-Submarine Patrol Number 2, landed back after patrol completion, no incident. B-25's FE, FY, and FV are airborne for compass swinging exercise, landed back after completion. B-25's DF, FC, FV, FW, and FY are airborne for air Exercise Number 3 and Number 4, landed back after completion, again airborne and completed same exercises later as well, landed after completion. The Widgeon is airborne for San Salvador, aircraft landed waterborne, again airborne later for base landed back. USS *PC-452* and tug USS *Humrick* are bound south in Straits off Canaveral. USCG *CG-83391* is bound west in Straits north of Little Bahama Bank. **Convoy NK-611** is bound south in Straits west of Bimini with US Navy plane escorting. USCG *CG-83348* is bound south in Straits. **Convoy GK-804** is bound west in Nicholas Channel with US Navy plane escorting. USS *SC-1064* is bound northwest in Nicholas Channel. Tanker *Manuel Rionda* and USS *LST-531* are bound southeast in Old Bahama Channel.

7th May 1944

Task Unit TU-201510 (10 USS LCI(L)'s) is bound south in Straits off Canaveral. Tanker *18 De Marzo* and tug USS *Warbler* (ARS-11) are bound south in Straits northwest of Little Bahama Bank. Submarine USS *Sterlet* (SS-392) escorted by USS *SC-499* are bound south in Straits west of Little Bahama Bank. Tanker *Gulfmaid* is bound south in Straits, northwest of Bimini. USS *YP-351*, tug USS *YT-338* towing USS *YF-483*, USS *YC-471*, and USS *YAG-28* are bound north in Straits. SS *Sea Perch* is bound north in Straits, north of Cay Sal Bank.

8th May 1944

B-25's DF, FC, FK, FW, and FY are airborne for air Exercise Number 7, landed after completion. Grumman Goose is airborne, waterborne at San Salvador, transported members of Corporation Staff which was controlling Queensley Exercise. The Widgeon is airborne for San Salvador, aircraft landed waterborne, again airborne later for base, landed back. A B-26 delivered from Wilmington, landed for RAF TC, then the aircraft was flown over to Windsor Field. Tanker *Beaver Dam* is bound north in Straits, northwest of Little Bahama Bank. USS *Sapelo* (A0-11) and escort patrol craft USS *PCE-846* are bound south in Straits west of Little Bahama Bank. Destroyer escort USS *Raymond* (DE-341) is bound north in Straits northwest of Bimini. SS *Wellesley* is bound south in Straits northwest of Bimini. USS *YAG-19* west in south of Bimini. Spanish SS *Monte Arnabal*, SS *Fairfax* and USS *PC-573* are bound southeast in Old Bahama Channel. Oiler USS *YOG-6* and USS YP-*579* are bound north in Straits. *U-541* is bound north to exit the area to the north of Walker Cay, above Grand Bahama and Abaco.

9th May 1944

B-25's FL, FC, and FF are airborne for Anti-Submarine Patrol Number 2, completed patrol without incident landed back. B-25's FW, FN, and CL are airborne for air Exercise Number 7, landed after completion. B-25's FT, FK, DC, FC, FY, and FF are airborne for air Exercise Number 6, and landed after completion. Queensley Exercise was laid on in place of Kingsley, it was a co-operative exercise with Royal Naval units in Bermuda, units moved from Bermuda and pending further arrangements, exercise was instituted with San Salvador's Direction Finding Station. The station is to perform the function of ships, Signals Officer was to be stationed at San Salvador and act as SOE, aircraft are routed via turning points to provide navigation exercise. Liberator NA is airborne from Windsor Field, radio contact established, island of San Salvador contacted, recognition procedure carried out, patrols completed under direction of SOE, dummy submarine attack carried out against smoke float, aircraft landed, exercise well carried out, radio procedure was good. Liberator MX is airborne as first relief for Queensley Exercise, radio contact maintained, carried out visual recognition procedure with San Salvador, patrol Lizard carried out, attacked smoke float, dropped by itself, landed back after exercise completion. Liberator LY is airborne from Windsor Field, radio contact established, aircraft was unable to maintain contact for homing procedure "B," passed over San Salvador due to failure to obtain bearings and the haze, identified Cat Island, located San Salvador by DR [dead reckoning]. Navigation patrols completed under direction of SOE, attacks made against smoke float, landed back after exercise was over. **Convoy KN-312** is bound north in Straits passing Miami with US Navy plane escorting. **Convoy KG-705**, tanker *AthelRegent,* and SS *Olney* are bound west in Nicholas Channel, with US Navy blimp escorting. SS *Massachusetts* is bound east in Nicholas Channel. SS *George W. Barnes* is bound southeast in Old Bahama Channel. Tug USS *Cubits Gap* and tug USS *LT-531* are bound northwest in Old Bahama Channel. US Navy plane is bound west towards Santaren Channel.

10th May 1944

B-25's FC, FT, FV, FK, and FN are airborne, completed night navigation exercise, landed back. B-25's DC, FL, FV, FY, and FC are airborne, completed air Exercise Number 6, landed back. B-25's BX, AL, BJ, CG, and AJ are airborne for local navigation exercise, landed after completion. Liberator LS is airborne from Windsor Field for Queensley Exercises, radio contact established. Homing procedure "B" commenced aircraft was over San Salvador recognition procedure carried out, parcel dropped near the hotel, patrols completed under instruction of SEO aircraft landed back. Liberator LL is airborne as first relief Windsor Field for Queensley Exercises, radio contact established, San Salvador located, signals exchanged, patrols completed under instruction of SEO, aircraft landed back. Liberator MA airborne, radio contact established, San Salvador located, signals exchanged Lizard patrol completed under instruction of SEO, direct hit was scored on smoke float, dropped by aircraft, landed after exercise completion, exercise was well performed, good radio contacts were established. RAF Liberator aircraft B-24 reported faulty hydraulic booster pump on landing. However, the next crew took off before any action was taken on landing, and it ploughed into several parked aircraft due to there being insufficient hydraulic pressure to operate aircraft's brakes. The Royal Norwegian Navy escort ship HNoMS *King Haakon VII* is bound north in Straits, northwest of Little Bahama Bank, exiting the area. USS *Elokomin* (A0-55) is bound south in Straits northwest of Little Bahama Bank. **Convoy NK-612** is bound south in Straits, west of Little Bahama Bank. Destroyer USS *Stafford* (DD) and destroyer USS SS *Lapiade* (DD) are bound north in Straits northwest of Little Bahama Bank. Tanker *Cornell* is bound north in Straits northwest of Little Bahama Bank. The tanker *Liebre* is bound north, to the west of Bimini. USS *Peril* (AM-272) is bound north in Straits. USS *YMS-220* is bound northwest in Nicholas Channel. Tug USS *LT-531* and tug USS *Cubits Gap* and tow are bound northwest in Nicholas Channel. Tanker *Manuel Rionda* is bound northwest in Old Bahama Channel.

11th May 1944

B-25's BX, BD, and CJ are airborne, completed night navigation exercise, landed. B-25's FC, FL, FF, FV, FN, and FK are airborne for night navigation exercise, landed after completion. B-25's FY, FL, DC, CL, FK, and FC are airborne for air Exercise Number 6, landed after completion. B-25's AJ, AL, and AN are airborne for local navigation exercise, landed after completion, again airborne later for same exercise, landed after completion. Liberator FL-970 LH is airborne, completed navigation and Radar exercise, landed back. The Grumman Goose returned from San Salvador and landed. Liberator LS is airborne from Windsor Field, US Naval sloop is sighted, electrical failure detected, aircraft returned to base, and failed to carry out exercise. Liberator LY is airborne, radio contact not established, homing procedure "B" could not be carried out, a convoy of 35 merchant ships and five escorts is spotted, San Salvador located, signals exchanged, Lizard patrol completed, attack with one depth charge carried out against a smoke float, aircraft landed after completing exercise. Liberator LR is airborne, radio contact established, radio contact made when San Salvador was located, patrols Lizard, Python and Viper completed, dummy Anti-Submarine attacks carried out against smoke float dropped by aircraft, then LR aircraft landed after completion. HMS *Sackville* is bound north in Straits, northwest of Little Bahama Bank. USS *PC-1264* is bound north in Straits off Palm Beach. Patrol craft escort USS *PCE(R)-849* and USS *SC-1333* are bound south in Straits. **Convoy GK-805** and tanker *Manuel Rionda* are bound northwest in Nicholas Channel. SS *Howard Stansbury* is bound north in Santaren Channel. SS *W. Kyska* is bound south, from east of Eleuthera. **Convoy NG-433** is bound south, from east of Eleuthera with a US Navy plane escorting.

12th May 1944

B-25's DC, FJ, FC, FL, and FY are airborne for compass swinging navigation, exercise completed, exercise and landed. B-24 Liberator FL-970 LH is airborne for Radar navigation exercise, completed exercise and landed. The Grumman Goose is airborne for San Salvador, landed waterborne there, completed exercise and landed at base. Tanker *Howard Stansbury,* tanker *R. P. Smith*, and **Convoy Motor Torpedo Boat Squadron-35, Group A,** are bound north in Straits, northwest of Little Bahama Bank. SS *Medina* is bound south in Straits, west of Little Bahama Bank. Oil barge USS *YOG-6* and USS *YP-579* are bound northeast in Northwest Providence Channel. Tug USS *Warbler* is bound north in Straits off Port Everglades. SS *Napier* is bound south, to the east of Eleuthera. USS *Rehoboth* (AVP-50) is bound north, to the east of Eleuthera.

13th May 1944

B-25's DC, FG, FL, FM, FR, and FJ are airborne for compass swinging exercise, landed after completion. SS *Taganrog* is bound south in Straits west of Little Bahama Bank. SS *Housatonic* is bound south in Straits, north of Cay Sal Bank. A US Navy plane is bound north in Straits to the north of Cay Sal Bank. SS *George W. Barnes* is bound west in Nicholas Channel. **Task Unit TU-03113** is bound northwest in Nicholas Channel. USS *YT-338* is bound south in Straits, north of Nassau.

14th May 1944

Tanker *Bulkcrude* is bound north in Straits northwest of Little Bahama Bank, exiting the area. Tanker *Bohemian Club* is bound north in Straits northwest of Little Bahama Bank. USS *PC-1209* is bound south in

Straits west of Little Bahama Bank. Tanker *Petrofuel* is bound north in Straits west of Little Bahama Bank. **Convoy KN-313** is bound north in Straits west of Little Bahama Bank with US Navy plane escorting. Tanker *Mary Ashley Townsend* is bound north in Straits, northwest of Bimini. USS *YAG-28* is bound east in Straits west of Bimini. Oil barge USS *YOG-6* is bound north in Straits, then north of Nassau. Tanker *Paulsboro* is bound south in Straits. **Convoy KG-706** is bound southeast in Nicholas Channel with US Navy plane escorting. USS *PC-575* is bound southeast in Old Bahama Channel. Tanker *Gulfwave,* is bound north, to the east of Eleuthera. *U-541* re-enters area bound south from north of Walker Cay.

15th May 1944

B-25's FC, FO, FL, FG, FU, and FF are airborne for Anti-Submarine Patrol Number 2, one aircraft sighted merchant ship of 1,500 tons, submarine USS *Barbel* (SS-316) is sighted, landed after exercise completion. B-25's FC, DC, CL, FJ, and FV are airborne for air Exercise Number 1, landed after exercise completion. Liberator LH is airborne on Radar and navigation exercise, landed after exercise completion. Grumman Goose is airborne to Hatchet Bay, then returned to base and landed. Liberator LA is airborne and landed at Miami. RAF Base No.111 OTU has expanded rapidly includes over 2,750 pupils including pilots of the Czech, Polish and Dutch air forces, as well as Canadian, Australian, Bahamian, and US fliers. USS *PC-1209* and SS *Vistula* are bound south in Straits off Canaveral. **Convoy NK-613** is bound south in Straits northwest of Little Bahama Bank. USS *SC-681* and *SS Bric V. Hauser* are bound south in Straits off Palm Beach. Ferry *Jean Brillant* escorted by USS *YMS-370* is bound west in Straits, towards Miami. SS *Florida* is bound north in Straits, west of Bimini. Submarine USS *Barbel* (SS-316) escorted by USS *SC-1063* is bound west in Northeast Providence Channel. SS *Liebre* is bound west in Straits. SS *Britamsea* and SS *W. F. Burdell* are bound north, to the east of Eleuthera. Tanker *Delaware Sun*, SS *Luculus,* and tanker *Gulf of Venezuela* are bound south, to the east of Eleuthera. *U-541* is bound northeast of Abaco, bound east to France and out of area.

16th May 1944

B-25's FF, FG, FL, FN, and FU are airborne for Anti-Submarine Patrol Number 1, and landed after completion, with no incident. A tanker of 6,000 tons is sighted by one of the aircraft, another aircraft also sighted tanker in the corresponding position relative to the time of sighting. B-25's FV, FS, FL, FR, and FM are airborne for air Exercise Number 1, landed after exercise completion. Liberator LH is airborne for Radar and navigation exercise, landed after completion. Grumman Goose is airborne for Harbour Island, landed waterborne there, airborne later returned to base. SS *Wellesley* and SS *Bayou Chico* are north in Straits, northwest of Little Bahama Bank. Tanker *Spencer Kellogg* is bound north in Straits, west of Little Bahama Bank. Tug USS *YTB-338* towing USS *YF-311* is bound south in Straits from Northwest Providence Channel. Destroyer escort USS *Richard W. Suesens* (DE-342) and tanker *A. S. Hansen* are bound north in Straits, west of Bimini. **Convoy NK-613** is bound south in Straits. **Convoy GK-806** is bound northwest in Nicholas Channel with US Navy plane escorting. Tanker *Britamer,* escort aircraft carrier HMS *Arbiter* (D-31) and a US Navy plane are bound north, to the east of Eleuthera. A life raft was spotted, and an aircraft carrier shortly thereafter. The life raft was empty, and the position encoded.

17th May 1944

B-25's FV, FS, FR, FJ, FM, and FH are airborne for night navigation exercise, landed after completion. B-25's DC, FJ, FY, FG, and FM are airborne for air Exercise Number 5, landed after exercise completion. Liberator LH is airborne for Radar and navigation exercise, landed after exercise completion. Liberator LA is airborne from Miami landed at base. USS *SC-1340* is bound south in Straits northwest of Little Bahama Bank. Ferry *Jean Brillant* is bound south in Straits. Navy plane is bound south in Straits. USS *Elokomin* and SS *Colabee* are bound north in Straits. Tanker *Axtell J. Byles* and SS *Fort Erie* are bound north, to the north of Eleuthera.

18th May 1944

B-25's FG, DC, CL, FR, and FS are airborne for air Exercise Number 5, landed after exercise completion. B-25's FH, FO, FU, CL, and FJ are airborne for air Exercise Number 6, landed after exercise completion. Liberator LH is airborne for Radar and navigation exercise, landed after exercise completion. USS *YAG-13* and USS *YTB-338* are bound north in Straits off Port Everglades. Ferry *Jean Brillant* is bound south in Straits off Miami. USS *PCE(R)-850* and USS *YMS-373* are bound north in Straits. Destroyers USS *Cronin* and USS *Samuel B. R. Roberts* are bound north in Straits. USS *PC-493* is bound northwest in Old Bahama Channel. Tanker *Mount Bruce Park* is bound south, to the east of Eleuthera. Tanker *Magnolia* and SS *Christopher L. Sholes* are bound north, to the east of Eleuthera.

19th May 1944

B-25's FP, DC, FM, FW, FO, and FS are airborne for local night navigation exercise, landed after completion. B-25's FO and DC are airborne for air Exercise Number 7, landed after exercise completion. B-25's DC, FM, FP, FW, and FY are airborne for air Exercise Number 3, landed after exercise completion. Then the same aircraft are airborne later for air Exercise Number 4, and landed after exercise completion. B-25's FF, FN, FU, and FJ are airborne for Anti-Submarine Patrol Number 2, during which sundry small vessels were sighted, and landed after exercise completion. Liberator LH is airborne for Radar and navigation exercise, then landed after exercise completion. USS *Sapelo* (AO-11) escorted by USS *PCE-846,* is bound north in Straits, to the north of Little Bahama Bank. USCG *CG-80303* is bound north in Straits, northwest of Little Bahama Bank. **Convoy KN-314** is bound north in Straits, west of Little Bahama Bank. Oil barge USS *YOG-6* and USS *YP-579* are bound south in Straits. USS *SE-1340* and USS *SE-713* are bound south in Straits. SS *George W. Kendall* is bound north in Straits. **Convoy KG-707** is bound southeast in the Old Bahama Channel. Oiler USS *YO-69* and USCG *CG-83489* are bound northwest in Old Bahama Channel. SS *R. G. Stewart* is bound south, to the east of Eleuthera.

20th May 1944

B-25's FN, FS, and FU are airborne for Anti-Submarine Patrol Number 2, with sundry small vessels sighted. A convoy of 30 merchant ships and five escorts are also sighted, then they landed after exercise completion. B-25's CL, FY, and FW are airborne for air Exercise Number 7, and landed after completion. B-25's FN, FO, CL, FM, and FK are airborne for air Exercise Number 3, and landed after completion. B-25's FM, FR, FC, CL, and FN are airborne for air Exercise Number 4, and landed after completion. Liberator LH is airborne for Radar and navigation exercise, then landed after exercise completion. Grumman Goose is airborne for San Salvador, landed waterborne there, returned to base, landed, then is airborne from Oakes Field later, and landed waterborne at Hatchet Bay, finally it returned to base and landed. USCG *CG-80304* is bound north in Straits, off Canaveral. USS *YT-338* is bound south in Straits, also off Canaveral. USS *Housatonic* (AO-35) and USS *PC-1181* are bound north in Straits, northwest of Little Bahama Bank. **Convoy NK-614** is bound south in Straits, northwest of Little Bahama Bank. Ferry *Jean Brillant* and USCG *CG-Vigilant* are bound south, to the north of Nassau. SS *Brentwood Bay Park* is bound north, to the north of Eleuthera. USCG *CG-83524* is bound south in Straits, off Port Everglades. USS *YO-129* is bound north in Straits, off Miami. SS *George Washington*, USS *PC-1198* and a US Navy blimp are bound north in Straits, east of Cay Sal Bank. USS *Y0-69* and USCG *CG-83489* are bound northwest in Nicholas Channel. Tanker *Esso Providence* is bound south, to the east of Eleuthera. Tanker *Gulfwing* is bound north, to the east of Eleuthera. *U-541* exits northeast of Bahamas, bound east for France and out of area. F/O Paul Remby Rackham took No.1 Course at No.111 OTU, then went to 120 Squadron where he earned a DFC for a successful airborne attack under F/Lt. Kerrigan, DFC, who was awarded the DSO. Rackham was wounded in the legs during the attack, yet continued his navigational duties.

21st May 1944

USS *LST-551* is bound north in Straits northwest of Little Bahama Bank. Tanker *Pan North Carolina* is bound south in Straits west of Little Bahama Bank. Ferry *Jean Brillant* is bound west in Northwest Providence Channel with US Navy plane escorting. **Convoy NK-614** and SS *Walter Jennings* are bound south in Straits. SS *H. M. Fredricksen* is bound north in Straits. SS *William H. Gray* and **Convoy GK-807** are bound northwest in Nicholas Channel. Navy plane is bound north above Nicholas Channel. Tanker *Clio*, SS *Mokihana*, tanker *Fenja* and tanker *Ontariolite* are bound north, to the east of Eleuthera.

22nd May 1944

Tanker *Bohemian Club* is bound south in Straits, northwest of Little Bahama Bank. USS *LSM-1* and USS *LSM-2* are bound north in Straits, northwest of Little Bahama Bank. USS *FC-577* is bound south in Straits off Palm Beach. Tug USS *ATR-7* is bound south in Straits off Port Everglades. Oiler USS *YO-129* is bound east in Northwest Providence Channel. SS *Tustem*, SS *Richard J. Cleveland*, and SS *John Rutledge* are bound north, to the east of Eleuthera. Tanker *Gulfbreeze* is bound south, to the east of Eleuthera.

23rd May 1944

SS *James Island* is bound south in Straits, north of Little Bahama Bank. Tanker *Vacuum* is bound south in Straits off Palm Beach. SS *George Washington* escorted by USS *PC-1198* is bound south in Straits. Tanker *Fred W. Weller* and tanker *Gulftide* are bound south, to the east of Eleuthera. Tanker *Gulfwax* is bound north, to the east of Eleuthera.

24th May 1944

USCG *CG-83523* is bound west north of Walker Cay, north of Abaco. USS *LST-659* and **Convoy KN-315** are bound north in Straits northwest of Little Bahama Bank. SS *Exchange* is bound south in Straits. Tanker *Paulsboro* is bound north in Straits. **Convoy KG-708** is bound southeast entering Old Bahama Channel with US Navy plane escorting. The house boat USS

YHB-13 is bound southeast in Old Bahama Channel. Tanker *Fred W. Weller* and tanker *Vacuum* are bound south, to the east of Eleuthera. USS *PCE-847* and tanker *Esso Bolivar* are bound north, to the east of Eleuthera. An Empire Day parade of the civil and military forces stationed on the island was organized. Troops and the Bahamas Police Force assembled at the eastern Parade for inspection. HRH the Governor and the Duchess of Windsor arrived at the saluting base erected by the Public Buildings in Bay Street. The military column was headed by the RAF Military Band, followed by the Officer Commanding. Pictou Highlanders were under the command of Captain Lane. Three flights of the RAF were under the command of F/Lt. Patrie. The Bahamas Volunteer Guard were represented by a platoon under the command of Lt. George Higgs. The rear guard was a detachment of the Bahamas Police Force, under the command of Lt. J. C. Douglas. Laying of Corner Stone at the RAF Cemetery: Following the march past, the HRH Governor went to the RAF Cemetery to lay the corner stone of the memorial gates. The detachment of the RAF was paraded for inspection. The ceremony was opened by prayers by the Station Chaplain, Rev. Edwards. A special trowel made from salvaged aircraft metal was used in the laying of the stone by the Governor. No.113: Ceremony of laying of Foundation Stone of the RAF Cemetery by HRH the Duke of Windsor.

25th May 1944

Liberator NC is airborne for Queensley exercise, two merchant ship sighted, radio contact established with San Salvador, patrols Lizard and Mamba completed, depth charge attack carried out against a smoke float, aircraft landed back after exercise completion. Liberator NA is airborne for Queensley exercise, homing procedure "B" commenced, San Salvador located, patrols Lizard and Python completed, depth charge attack carried out against a smoke float, aircraft landed back after exercise completion. **Convoy NK-615** is bound south in Straits west of Little Bahama Bank with US Navy plane escorting. SS *Maridal* is bound south in Straits off Palm Beach. SS *Mary Ashley Townsend* is bound south in Straits off Port Everglades. SS *H. M. Fredricksen* is bound south in Straits. SS *Gertrude Kellogg* is bound northwest in Nicholas Channel. SS *Edouard Jeremac* is bound east in Northeast Providence Channel. Tanker *British Tenacity* is bound south, to the east of Eleuthera. Tanker *Esso Aruba* is bound north, to the east of Eleuthera.

26th May 1944

B-25's FU, FS, FM, FK, FY, and FT are airborne for night navigation exercise, landed after completion. B-25's BD, AN, AM, AU, and AL are airborne for local navigation exercise, landed after completion. B-25's FB, FS, FT, FV, and FY are airborne for Anti-Submarine Patrol Number 3, landed after exercise completion. Grumman Goose GB is airborne for San Salvador, landed waterborne there, was again airborne later and returned to base. B-25 FR is airborne for calibration tests with Main Field Direction Finding, station aircraft landed back. Destroyer escort USS *Kretchmer* (DE-329) is bound north in Straits northwest of Little Bahama Bank. Tanker *Petrofuel* is bound south in Straits. **Convoy NK-615** is bound west in Straits. **Convoy GK-808** and tanker *Olney* are bound northwest in Nicholas Channel. Tanker *Alar* is bound north, to the east of Eleuthera. Tanker *Andrew Marschall*, tanker *Britamsea*, tanker *Esso Balboa*, and tanker *Axtell J. Byles* are bound south, to the east of Eleuthera.

27th May 1944

B-25's FP, FS, FT, FV, FR, and FC are airborne for night navigation exercise, landed after completion. B-25's DC, FK, FY, FP, FR, and FG are airborne for Anti-Submarine Patrol Number 3, landed after exercise completion. Destroyer escort USS *Slater* (DE-766) is bound north, to the north of Walker Cay, off Abaco. USS *Mendorino* (APA-100) is bound north in Straits, west of Little Bahama Bank. USS *YTB-338* is bound south in Straits from Northwest Providence Channel. Submarine USS *Queenfish* (SS-393) escorted by USS *SC-1064* is bound west in Straits north of Berry Islands. USS *Sapelo* (A0-11) escorted by USS *PCE-846* is bound west in Northeast Providence Channel. Escort US Navy patrol craft plus USS *PCE(R)-851* and USS *PC-1143* are bound north in Straits. SS *Vancolite* is bound northwest in Old Bahama Channel. **Task Unit TU-2969**, with repair ships USS *Oahu* and USS *ARG-5* and USS *Cebu* (ARG-6) (both on maiden voyages to the Pacific), destroyers USS *DD-228* and USS *DD-217* is bound south, to the east of Eleuthera.

28th May 1944

Grumman Goose GA returned from Palisadoes airfield, Kingston Jamaica and landed. USS *Barrier* (AM-150) is bound north in Straits, west of Little Bahama Bank. USCG *CG-83525*, USCG *CG-83526*, and USCG *CG-83527* are bound south in Straits, west of Little Bahama Bank. Tug USS *YTL-298* towing USS *YC-893*, USS *YC-993*, and tanker *Chiloil* are bound north in Straits. Troop transport USS *USAT-George Washington*, destroyer escort USS *DE-682* and SS *Vancolite* are bound north in Santaren Channel. USS *PC-494* and house boat USS *YHB-13* are bound northwest in Old Bahama Channel.

29th May 1944

B-25's FP, FK, FN, FY, FC, and FF are airborne for Anti-Submarine Patrol Number 1, landed after exercise completion, aircraft moved to home. B-25 DC is unable to take part, and this exercise was cancelled. B-25's DC, FR, FG, FF, and FT are airborne for air Exercise Number 1, landed after completion. Free French Navy submarine FFS *Le Centaure* is bound north in Straits, off Canaveral. A US Navy plane is bound north in Straits towards Canaveral. **Convoy NK-616** and SS *Wallowa* are bound south in Straits northwest of Little Bahama Bank. Tanker *A. S. Hansen* is bound north in Straits northwest of Little Bahama Bank. Navy plane is bound south in Straits north of Canaveral. USS *PC-1265* is bound south in Straits, west of Little Bahama Bank. Tanker *Olney* and SS *Touraine* are bound north in Straits, west of Little Bahama Bank. Tanker *Edward L. Dohney*, SS *Vancolite*, SS *San Melito*, tanker *Pan Virginia*, and **Convoy KN-316** are bound north in Straits northwest of Bimini. SS *Empire Regent* is bound south in Straits west of Bimini. The house boat USS *YHB-13* is bound north in Straits north of Cay Sal Bank. SS *Caldas* is bound north in Santaren Channel. **Convoy KG-709** is bound southeast in Old Bahama Channel. SS *Christopher L. Shoales* and SS *W. W. Mills* are bound south, to the east of Eleuthera. SS *Tide Penn* and SS *Swiftscout* are bound north, to the east of Eleuthera.

30th May 1944

B-25's FB, FW, FH, FM, FR, FS, and FT are airborne for night navigation exercise, landed after completion. B-25's FR, FM, FN, and DC are airborne for air Exercise Number 1, landed after completion. Tanker *Vancolite* is bound north in Straits, northwest of Little Bahama Bank. Tanker *Cities Service Fuel* is bound south in Straits, northwest of Little Bahama Bank. Tanker *San Juan Hill* is bound north in Straits, northwest of Bimini. USS *SC-1001* is bound northeast in Northwest Providence Channel. USS *Mattole* (A0-17) is bound north, to the east of Eleuthera.

31st May 1944

B-25's FK, FM, FT, FW, and DC airborne for air Exercise Number 5, landed after completion. Liberator NC airborne from Windsor Field, completed navigation exercise, detailed for Number 3 Squadron, landed after exercise completion. USS *Housatonic* is bound south in Straits, north of Little Bahama Bank. USS *YAG-21* and SS *James Island* are bound north in Straits, northwest of Little Bahama Bank. Tanker *Pan North Carolina* is bound north in Straits, west of Little Bahama Bank. Tanker *Cities Service Fuel* is bound south in Straits off Miami. Destroyer escorts USS *Oberrender* (DE-334), USS *Leroy Wilson* (DE-414) and USS *Frybarger* (DE-705) are bound north in Straits west of Bimini. SS *Juan Casiano* is bound north in Straits. **Convoy GK-809** is bound northwest in Nicholas Channel. SS *Calliope*, SS *Garnet Hulings*, tanker *J. A. Mowinckel*, and **Convoy NG-437** are bound south, to the east of Eleuthera.

1st June 1944

B-25's FH, FP, FJ, and FK are airborne for air Exercise Number 5, landed after completion. Liberator LH is airborne for Radar and navigation exercise, landed after completion. B-25's FG, FN, FP, FV, FB, FC, and FK are airborne for Anti-Submarine Patrol Number 1, landed after completion. Tanker *E. R. Kemp* is bound south in Straits towards Daytona. Tanker *A. S. Hansen* is south in Straits passing Canaveral. USCG *CG-Juniper* is bound northeast near Matanilla Shoal, northwest Bahamas. Tanker *Comol Cuba* and tanker *Massachusetts* are bound north in Straits passing Palm Beach. USS *SC-678* is bound west passing Northwest Providence Channel. USS *YAG-21* is bound south, passing Nassau. Tanker *Durango* and SS *Wilbur Wright* are bound south in Straits. *U-539* enters area from the east, bound south, towards Puerto Rico and Mona Passage.

2nd June 1944

B-25's FP, FJ, FU, FV, and FH are airborne for local night navigation exercise, landed after completion. B-25's FP, FJ, FU, and FV are airborne for local night navigation exercise, landed after completion. B-25's FN, FC, and FT are airborne for Anti-Submarine Patrol Number 2, landed after completion, a whale factory ship of 20,000 tons sighted. B-25's FJ, FV, and DC are airborne for air Exercise Number 7, landed after completion. B-25's FU, FG, FY, FV, and FB are airborne for air Exercise Number 3, landed after completion, airborne again later for air Exercise Number 14, landed after completion. Liberator LH is airborne for Radar and navigation exercise, landed after completion. SS *W. W. Bruce* and USS *Lawrence C. Taylor* are bound north in Straits. Tanker *Edward L. Doheny* is bound south in Straits passing Canaveral. Tanker *R. P. Smith* and SS *Touraine* are bound south in Straits towards Palm Beach. USS *LSM-3* is bound north in Straits near Palm Beach. SS *Wood Lake* is bound south in Straits near Palm Beach. SS *Empire Norse* and tanker *E. R. Kemp* are bound south in Straits. SS *Nevada* bound southeast in Straits passing Nicholas Channel. USS *PC-495* is bound southeast in Straits. SS *Edouard Jeremac* is bound west towards Northeast Providence Channel. Tug USS *YAG-19* is bound northwest in Old Bahama

Channel towards Cay Sal Bank. SS *Caida* is bound southeast in Old Bahama Channel. *U-539* is bound east of Bahamas, south to Mona Passage. *U-539* is north of Mayaguana, and reports from HQ, or *B-Dienst* show that a U-boat in *U-539*'s position was attacked this day or previous c.5 pm by a 2-engine US Navy PBM Mariner-type flying boat, and that U-boat's anti-aircraft weapons were manned, and the aircraft was turned away and flew back onto its old course. The boat crash dived, and later heard an unexplained detonation, which did not sound like a bomb from an aircraft.

3rd June 1944

B-25's FN, FT, FG, and FS are airborne for Anti-Submarine Patrol Number 2, landed after completion three merchant ships and two naval crafts sighted one aircraft sighted the tanker with patrol craft photographs taken aircraft landed after exercise completion. B-25's FP, FM, and DC are airborne for air Exercise Number 7, landed after completion. B-25's FC, FJ, FH, and DC are airborne for air Exercise Number 3 and later Number 4, landed after completion. Liberator LH is airborne for Radar and navigation exercise, landed after completion. Grumman Goose GB is airborne for Bahamas Island to pick up airman who was absent without leave returned to base later. **Task Unit TU-20156** is bound south in Straits passing Canaveral. **Convoy NK-617** is bound south in Straits towards Palm Beach. USS *PC-452*, **Convoy KN-317**, and tanker *Eano* are bound north in Straits towards Palm Beach. Destroyer escort USS *Lawrence C. Taylor* (DE-415) is bound northeast passing Matanilla Shoal, northwest Bahamas. Hospital ship USS *John L. Clem* (USAHS) is bound north in Straits, passing Miami. Tanker *Bostonian* is bound north in Straits. USS *SC-1474* and tanker *Daylight* are bound southwest in Straits. Tanker *R. P. Smith* is bound west in Straits. SS *T. J. Jackson* and **Convoy KG-710** are bound southeast in Straits passing Nicholas Channel. USS *YAG-21* is bound northwest in Straits passing Nassau. Tanker *Willowdale Park* and SS *Peter Cartwright* are bound south in Straits.

4th June 1944

Tanker *Pan Georgia* is bound south in Straits towards Canaveral. SS *Chiloil* and tanker *Gulfwax* are bound south in Straits passing Canaveral. Free French Navy ship FFS *Coutelas* (ex-USS *PC-1560*) is bound north in Straits passing Palm Beach. Tanker *Gulfgem*, HMCS *Midland*, and Free French Ship FFS *Lobella* are bound north in Straits towards Palm Beach. Tanker *Mary Ashley Townsend* and SS *Bahram* are bound north in Straits passing Port Everglades. SS *Bostonian* and Cuban transport *Columbia* escorted by Cuban *CS-14* are bound north in Straits. Tanker *L. L. Abshire* is bound north in Straits. USS *SC-1424*, Russian RU *SC-1496*, and SS *Cornwallis* are bound southwest in Straits. SS *Wildwood Park*, SS *Arlington Wood Park*, and tanker *Gulf of Venezuela* are bound south in Straits. SS *Desoto* is bound southeast in Old Bahama Channel.

5th June 1944

Four B-25's detailed to take off, air Exercise Number 6 are cancelled due to adverse weather condition. Liberator LS is airborne from Windsor Field for Kingsley Exercise. A convoy of 15 vessels is sighted, signals exchanged with ships, radio contact established with co-operating frigates, two frigates sighted from bearings received, ships were in line abreast, signals exchanged. Patrol Crocodile completed, SEO dropped smoke float, attacked with one depth charge, aircraft requested to drop flares for A-A practice by ship's gunners could not be carried out, flares jammed in the chute, exercise completed, aircraft left convoy, landed at base. Liberator ND is airborne from Windsor Field as first relief for Kingsley Exercise, tanker of 4,000 tons with merchant ship of 7,000 tons sighted, radio contact established with co-operating frigates, convoy contacted from bearings received, signals exchanged, patrol Alligator completed, SEO dropped smoke float, attacked with one depth charge, aircraft dropped flares for A-A practice by ship's gunners exercise completed, aircraft left convoy, landed at base. Liberator NO is airborne from Windsor Field as second relief of Kingsley Exercise, merchant ship sighted, radio contact established with convoy, signals exchanged, patrol Python completed, SEO dropped smoke float, attacked with one depth charge, aircraft dropped flares for A-A practice by ship's gunners exercise completed, aircraft left convoy, landed at base. Liberator MA is airborne from Windsor Field as third relief of Kingsley Exercise, tanker of 10,000 tons and merchant ship of 7,000 tons sighted, radio contact established with convoy, two frigates sighted from bearings received, signals exchanged, patrol Cobra completed, SEO dropped smoke float, attacked with one depth charge, exercise completed, aircraft left convoy, landed at base. Tanker *Pillory* departed San Juan and met USCG *CG-83310*. She was then torpedoed by *U-539* off Puerto Rico. Twenty one survivors picked up by a US Coast Guard cutter, one later died. USCG cutter *CG-Crawford* (WSC-134) took two more survivors to Mayaguez. Master, 20 crew, and four armed guards were lost. Tanker *Alar* is bound south in Straits. Destroyer escort USS *Melvin R. Newman* (DE-416) is bound north in Straits passing Port Everglades. USS *LSM-4* is bound north in Straits, passing Miami. USS *PC-1175* and USS *PC-1185* are bound southwest in Straits. Tanker *L. V. Stanford* is bound northeast in Straits. Tanker *Nevada* and SS *Narbo* are bound northwest in Straits passing Nicholas Channel. **Convoy GK-810** is bound northwest towards Nicholas Channel with US Navy plane escorting. Destroyer escort USS *Brister* (DE-327) is bound southeast towards Old Bahama Channel. *U-539*, Mona Passage: a flying boat approached from astern, was taken under fire. Aircraft flew out of the moon in the dark sector and turned on a brilliant bluish searchlight when approaching, followed by an inaccurate bomb throw: three bombs missed astern. Aircraft circled the boat, whose 3.7 cm-gun failed due to cartridge burner, with result that cartridge could not be extracted. After pre-flooding main ballast tank 8, and firing continuously U-boat had to crash dive. No bombs. At periscope depth a Mariner seeing flying in large circles. Some activity carried into early morning following day.

6th June 1944

B-25's FH, FJ, FN, FP, FM, and FV are airborne for night navigation exercise, landed after completion. B-25's FH, FN, and FP are airborne for air Exercise Number 7, landed after completion. B-25's FJ, FP, FM, and DC are airborne for air Exercise Number 6, landed after completion. Liberator LH is airborne for Radar and navigation exercise, landed after completion. Grumman Goose GB is airborne from Governor's Harbour, returned to base, later had been dispatched to Rock Sound to pick up injured crew from B-25 FR which ditched during Anti-Submarine Patrol Number 2. Liberator LL is airborne from Windsor

Field for Kingsley Exercise, radio contact established with convoy, homing procedure "B" commenced, two frigates located, recognition procedure carried out, patrol Crocodile completed, no attack on smoke float possible due to poor visibility, left convoy, dropped smoke float later, was attacked with one depth charge, landed back at base. Liberator LL is airborne from Windsor Field for Kingsley Exercise as first relief, instruction from base to carry out homing exercise until ship sighted, frigate sighted, convoy lost due to poor visibility, convoy searchlight located and convoy sighted again, recognition signals exchanged, attack carried out against a smoke float, signal received from SEO to call back convoy, visibility had reduced to nil, left convoy, landed back at base. Liberator NA is airborne from Windsor Field for Kingsley Exercise as third relief, recalled, landed back, didn't take part in Kingsley Exercise. Liberator LS is airborne from Windsor Field for Kingsley Exercise as second relief, recalled from base, dummy submarine attack carried out, signal received to continue exercise and navigation patrol, 7,000 ton merchant ships and 8,000 ton tanker sighted, landed back at base after exercise completion. B-25 FX is airborne for Anti-Submarine Patrol Number 2, merchant vessel sighted, aircraft landed back. B-25 FG is airborne for Anti-Submarine Patrol Number 2, merchant vessel and tanker sighted, aircraft landed back. B-25 FR is airborne for Anti-Submarine Patrol Number 2, radio contact made with aircraft, long dash was received, signal received from pilot via government channels that aircraft was ditched in sea, 2 nm north of Tarpum Bay, Eleuthera, aircraft had depth charges aboard but these were not safe, crew uninjured except the radio operator who had face injuries, ASR launch HMS *P-712* is dispatched from Harbour Island and arrived at Rock Sound then Tarpum Bay. Grumman Goose GB is airborne with Chief Ground Instructor, Medical Officer, and Staff for scene of ditching, aircraft returned to base and landed with pilot of ditched aircraft, remainder crew brought to Nassau in ASR launch HMS *P-712*, which berthed at Prince George dock. SS *Fort Cumberland* is bound southwest in Straits. Tanker *Petrofuel* is bound north in Straits passing Canaveral. **Task Unit TU-20.154** is bound south in Straits towards Palm Beach. USS *Phantom* (AM-273) is bound north in Straits near Palm Beach. USS *Sapelo* (AO-11) is bound north near Matanilla Shoal, northwest Bahamas being joined by navy aircraft from northeast for escort to the southwest. USS *YMS-65* is bound northwest in Straits towards Miami. USS *SC-1480* and USS *SC-1488* are bound southwest in Straits. SS *Cahawba* is bound northeast in Straits. USS *PC-470* is bound southeast in Straits passing Nicholas Channel. SS *William H. Ashley* is bound south in Straits. Cuban gunboat *CS-Cuba* is bound northwest in Old Bahama Channel. *U-539* exits area south via Windward Passage.

7th June 1944

B-25's of Number 1 Squadron carried out local night flying training. Three B-25 aircraft are detailed to be airborne for Anti-Submarine Patrol Number 2, detail cancelled due to adverse weather condition. B-25's FN, FM, and FU are airborne for air Exercise Number 7, aircraft landed after exercise completion. B-25's FW, FK, DC, and FJ are airborne for air Exercise Number 6, aircraft landed after exercise completion. B-25's CG, AK, and AU are airborne for local navigation exercise, aircraft landed after exercise completion, same aircraft again airborne later for same exercise, aircraft landed after exercise completion. Grumman Goose GA is airborne for San Salvador to transport the Control Staff for the Queensley Exercise, aircraft returned to base after termination of exercise. Liberator LM is airborne from Windsor Field, radio contact established, San Salvador located from bearings, signals exchanged, aircraft completed attack on smoke float, completed patrol Python and Alligator, aircraft landed after exercise completion. Liberator NO is airborne from Windsor Field as first relief for Queensley Exercise, radio contact established, San Salvador located correct recognition procedure carried out completed patrol Python and Lizard aircraft completed attack on smoke float, aircraft landed after exercise completion. Liberator MD is airborne from Windsor Field as second relief for Queensley Exercise, radio contact established, San Salvador located, correct recognition procedure carried out, completed patrol Lizard, aircraft completed attack on smoke float, aircraft landed after exercise completion. Liberator NA is airborne from Windsor Field as third relief for Queensley Exercise, radio contact established, San Salvador located, correct recognition procedure carried out, completed patrol Python and Lizard, aircraft completed bombing exercise, aircraft landed after exercise completion. USS *SC-1476* and USS *SC-1507* are bound south in Straits passing Palm Beach. USS *YMS-5* and USS *YMS-6* are bound north in Straits passing Palm Beach. Tanker *Fort Cumberland* is bound southwest in Straits. USS *SC-1480* and Russian RU *SC-1488* are bound southwest in Straits. USS *PC-1174* is bound southeast in Old Bahama Channel.

8th June 1944

B-25's AU, BA, and CX are airborne for local navigation exercise aircraft landed after exercise completion same aircraft airborne for same exercise again later landed after completion. B-25's DC, FS, FN, FJ, FW, and CL are airborne for air Exercise Number 6, landed after completion. **Task Unit TU-20152** (three USS LCI(L)'s) is bound south in Straits towards Daytona. SS *Bear Paw* and **Convoy SS *Housatonic*** are bound north in Straits. **Convoy KN-318** is bound north in Straits, passing Palm Beach. SS *Abiqua* is bound south in Straits towards Bimini. USS *Kishwaukee* (AOG-9) and USS *Genesee* (AOG-8) are bound north in Straits passing Port Everglades. SS *American Express* escorted by USS *SC-1474* is bound north in Straits passing Miami. **Convoy NK-618**, USS *YAG-21*, and USS *YMS-84* are bound south in Straits. USS *PC-489* is bound north in Straits towards Miami. SS *Moors Field* is bound southwest in Straits near Cay Sal Bank. USS *LST-690* is bound northeast in Straits. Destroyer escort USS *W. T. Powell* (DE-213) is bound southeast in Straits. **Convoy KG-711** is bound southeast in Old Bahama Channel. Troop transport USS *USAT-George Washington* escorted by destroyer escort USS *Gunarson* (DE-795) is bound northwest in Straits, US Navy plane escorting.

9th June 1944

B-25 FS is airborne for night navigation exercise, landed after completion. B-25's AU, CO, AD, AK, FA, and CX are airborne for local night navigation exercise, landed after completion. B-25's FM, FW, FG, and CL are airborne for compass swinging exercise, landed after completion. B-25's FY, FB, and FU are airborne for Anti-Submarine Patrol Number 2, landed after exercise completion, convoy of 4,000 ton barges and tugs is sighted. SS *Thomas J. Lyons* is bound south in Straits towards St.

Augustine. SS *American Express*, escorts, and tanker *Tampico* are bound north in Straits passing Daytona. Tanker *L. V. Stanford* and escort are bound south in Straits passing Daytona. USS *LST-690* is bound north in Straits passing Canaveral. Destroyer escort USS *Robert Brazier* (DE-345) and USS *Silverleaf* (AN-68) are bound north in Straits near Palm Beach. **Task Unit TU-20152**, USS *PC-1231*, and **Convoy NK-618** are bound south in Straits. USS *PC-1185* is bound north in Straits near Miami. USS *John M. Howard* (IX-75) is bound west near Nassau.

10th June 1944

B-25's FT, FN, FW, DF, and FK are airborne for compass swinging exercise, landed after completion. Grumman Goose GA is airborne Oakes Field for Harbour Island landed there. SS *William H. Dall* is bound south in Straits passing Daytona. USS *LSM-7* is bound north in Straits near Palm Beach. Tanker *L. V. Stanford* and tanker *Wood Lake* are bound south in Straits. **Convoy Motor Torpedo Boat Squadron-31, Group B** is bound northwest in Straits towards Miami. USS *YMS-220* and USS *PC-577* are bound northeast in Straits. **Convoy GK-811** is bound northwest in Straits passing Nicholas Channel with US Navy plane escorting.

11th June 1944

Grumman Goose GA is airborne from Harbour Island returned after exercise completion. USS *SC-1510* is bound south in Straits passing Daytona. USS *SC-658* is bound north in Straits passing Canaveral. **Task Unit TU-20151** is bound south in Straits passing Canaveral. Tanker *Matinicock* is bound south in Straits passing Palm Beach. SS *George Bellows* is bound south in Straits towards Port Everglades. Oiler USS *YOG-6* is bound northeast in Straits from Port Everglades. USS *SC-1001* and SS *William H. Dale* are bound south in Straits. Submarine USS *Barbero* (SS-317) is bound northwest in Northeast Providence Channel.

12th June 1944

B-25's FU, FS, FG, FM, and FB are airborne for Anti-Submarine Patrol Number 1, completed patrol without incident, landed back, a 3-island steam ship type vessel was sighted. B-25's FV, FS, DF, FC, and FB are airborne for air Exercise Number 1, completed exercise and landed. Liberator LH airborne for Radar and navigation exercise, landed after exercise completion. USS *LST-692* is bound northwest in Straits near Canaveral. Tanker *Esso Camden* is bound southwest in Straits. **Task Unit TU-2964**, of USS *PC-1598*, USS *PC-1599*, USS *PC-1600*, USS *PC-1601*, and USS *PC-1602*, are bound south in Straits passing Palm Beach. USCG tender *CG-Blackthorn* is bound south in Straits.

13th June 1944

B-25's FG, FK, FM, FU, and FV are airborne for Anti-Submarine Patrol Number 1, completed patrol without incident, landed back at base. B-25's FB, DF, CL, and FW are airborne for air Exercise Number 1, landed after exercise completion. Grumman Goose GA airborne Harbour Island returned to base. USS *PC-1087* and tanker *Argon* are bound south in Straits towards Daytona. **Convoy NK-619** is bound south in Straits. **Task Unit TU-20157** is bound south in Straits passing Canaveral. USCG *CG-83452* is bound west past Walker Cay, north of Abaco. **Convoy KN-319** is bound north in Straits passing Palm Beach. Oil barge USS *YOG-16* is bound west in Straits passing Northwest Providence Channel. USS *PC-1145* is bound north in Straits. USCG *CG-83406* is bound northwest in Straits towards Miami. Cuban *CS-31*, Cuban *CS-32*, and Cuban *CS-34*, and SS *Colobre* are bound northwest in Straits passing Nicholas Channel. **Convoy KG-712** is bound southeast in Straits passing Nicholas Channel with US Navy plane escorting. Cuban gunboat *CS-Cuba* and USS *PC-1204* are bound northwest in Old Bahama Channel with US Navy plane escorting.

14th June 1944

B-25's FW, FY, FT, CL, and FC are airborne for night navigation exercise, landed after exercise completion. B-25's CL, DF, FG, FJ, and FD are airborne for air Exercise Number 5, landed after exercise completion. Liberator LH is airborne for Radar and navigation exercise, landed after completion of exercise. SS *Gerard Dou* is bound south in Straits passing St. Augustine. SS *Spidoleine* and USS *PC-1218* are bound south in Straits towards Canaveral. USS *YO-116* is bound south in Straits passing Palm Beach. **Convoy NK-619** is bound south in Straits towards Port Everglades. USS *Bombard* (AM-151) is bound north in Straits passing Port Everglades. SS *Kewanee* is bound southeast in Straits from Port Everglades with US Navy plane escorting. USS *PC-1217* is bound southwest in Straits. SS *Colabee* is bound northwest in Straits towards Key West. SS *George Read* is bound northwest in Straits passing Nicholas Channel.

15th June 1944

B-25's CL, FS, FJ, FF, and FU are airborne for night navigation exercise, landed after completion. B-25's FU, DF, FS, and FJ are airborne for air Exercise Number 5, landed after completion. Liberator LH airborne for Radar and navigation exercise, *en route* to Boca Chica, landed there, returned to base later. Grumman Goose GA is airborne for San Salvador returned to base later. B-25 FC is airborne for Miami, landed there, returned to base later. USS *Montour* (APA-101) is bound north in Straits. USS *LST-665* and USS *LST-691* are bound north in Straits passing Palm Beach. USS *SC-1058* is bound south in Straits passing Palm Beach. Tanker *Moors Field* is bound north in Straits passing Palm Beach. USS *John M. Howard* (IX-75) is bound north from Nassau. SS *Gerard Dou*, tanker *Spidoleine* and tanker *Cities Service Kansas* are bound southwest in Straits. USS *PC-822* is bound north in Straits. **Convoy GK-812** is bound northwest in Old Bahama Channel.

16th June 1944

B-25's FV, FM, and FK are airborne for Anti-Submarine Patrol Number 2, life raft was sighted was empty aircraft carrier was sighted signals were exchanged completed patrol without incident landed. B-25's FJ and FT are airborne for air Exercise Number 7, landed after exercise completion. B-25's FF, DF, CL, FS, and FD are airborne for air Exercise Number 6, landed after exercise completion. Liberator LH is airborne for Radar and navigation exercise, returned to base after completion. USS *SC-1485* is bound south in Straits. USS *SC-708* is bound south in Straits towards Canaveral. USS YMS-65 is bound north in Straits towards Canaveral. USS *LST-561* and USS *LST-664* are bound north in Straits near Palm Beach. USS *PC-1169* is bound north near Matanilla Shoal, northwest Bahamas. USS *PC-1593* is bound south in Straits passing Palm Beach. SS *Solitaire* escorted by USCG *CG-83523* is bound south in Straits. Tanker *Cities Service Ohio* is bound north in Straits. USS *SC-1484*, USS *SC-1480*, USS *SC-1488*, and Russian RU *SC-1496* are bound east in Straits. **Havana North Coast Cuba-50** is bound southeast in Straits passing Nicholas Channel.

17th June 1944

Liberator MA airborne for Miami returned to base later. B-25's FM, FT, and FV are airborne for Anti-Submarine Patrol Number 2, returned after exercise completion. B-25's FS, FL, DF, FC, and FD are airborne for air Exercise Number 3, landed after exercise completion, same aircraft airborne later for air Exercise Number 4, landed after completion. USS *YTB-385* is bound south in Straits towards St. Augustine. Tanker *R. P. Smith* and USS *YMS-65* are bound south in Straits towards Daytona. Russian RU *SC-1484*, RU *SC-1480*, RU *SC-1488*, and RU *SC-1496* are bound north in Straits passing Canaveral. Destroyer USS *Dahlgren* (DD-187) and USS *PC-1217* are bound north in Straits passing Palm Beach. Tanker *Spencer Kellogg* is bound northwest in Straits passing Nicholas Channel.

18th June 1944

SS *Richmond P. Hobson* is bound south in Straits passing Daytona. USS *PC-1140* is bound north in Straits towards Canaveral. **Convoy KN-320** is bound north in Straits towards Canaveral with US Navy plane escorting. **Task Unit TU-20156** is bound south in Straits towards Palm Beach. SS *Frederick Lykes* is bound south passing Matanilla Shoal, northwest Bahamas. SS *Calmar* is bound north in Straits passing Palm Beach. USS *YTB-385* is bound south in Straits towards Port Everglades. Tanker *R. P. Smith* is bound south in Straits passing Port Everglades. USS *YAH-19* is bound southeast towards Nassau. Tanker *Cities Service Ohio* is bound southwest in Straits. USS *PC-1209* is bound west in Straits. USS *SC-660* and USS *SC-499* are bound northwest in Straits passing Nicholas Channel.

19th June 1944

Two B-25's detailed to carry out Anti-Submarine patrols and three B-25's detailed to carry out air Exercise Number 7 cancelled due to adverse weather. B-25's FT, FF, FC, DF, and FY are airborne for air Exercise Number 6, landed having returned to base early due to adverse weather. Grumman Goose GB returned from San Salvador, landed. Grumman Goose GA airborne waterborne at San Salvador remained there with Control Party for Queensley Exercise that was to take place on 20th and 21st of the month. Grumman Goose GB is airborne for San Salvador returned to base. SS *Calmar* is bound north in Straits passing Daytona. USS *SC-1504* and USS *SC-681* are bound south in Straits towards Palm Beach. **Convoy NK-620** is bound south in Straits towards Palm Beach with US Navy plane escorting. USS *SC-1064* and USS *PC-1210* are bound southwest in Straits. SS *Richard P. Hobson* is bound south in Straits towards Cay Sal Bank. Destroyer USS *E. C. Winslow* (DD-359) is bound northwest passing Nicolas Channel. **Convoy GN-139** is bound north in Straits.

20th June 1944

B-25's FB and FM are airborne for Anti-Submarine Patrol Number 2, completed patrol without incident, aircraft landed. B-25's FT, DC, and FC are airborne for air Exercise Number 7, landed after completion. B-25's FW, CL, FL, DF, and FF are airborne for air Exercise Number 6, landed after completion. Liberator NA airborne from Windsor Field for Queensley Exercise, submarine USS *Scabbardfish* (SS-397) sighted, radio contact established, San Salvador located, signals exchanged, patrols Mamba and Cobra completed, attack carried out against smoke float, aircraft landed after exercise completion. Liberator NC airborne from Windsor Field for what are called Queensley Exercise (similar to Kingsley Exercise) as first relief, San Salvador located after homing exercise "B" ended, patrols Lizard and Cobra completed, signal procedure smoothly carried out, attack carried out against smoke float, aircraft landed after exercise completion. Liberator NY is airborne from Windsor Field for Queensley Exercise as second relief, San Salvador located, signals exchanged, patrol Mamba completed, aircraft landed after exercise completion. B-25 CX is airborne Morrison Field landed back at base later. USS *SC-1502* is bound south in Straits towards Canaveral. Tug USS *Anacapa* towing USS *YRD-(M)4*, and USS *YRD-(H)4* escorted by USS *YMS-220* is bound south in Straits towards Palm Beach. USS *PC-1168* is bound north in Straits near Palm Beach. USS *PC-1593* is bound north in Straits passing Port Everglades. Oil barge USS *YOG-6* is bound west in Straits from Port Everglades. Mailboat *Richard Campbell* escorted by USCG *CG-83450* is bound west in Straits towards Miami with US Navy plane escorting. SS *Sharpsburg* is bound south in Straits near Miami. USS *Mizpah* is bound southwest in Straits. **Convoy NK-620** is bound southwest in Straits. Tanker *Athelviking* west in Straits passing Nicholas Channel. **Convoy GK-813** is bound west in Straits passing Nicholas Channel with US Navy plane escorting. Submarine USS *Scabbarfish* (SS-397) escorted by USS *SC-710* is bound west passing Northeast Providence Channel.

21st June 1944

B-25's FT, FF, FW, CL, and DF are airborne for air Exercise Number 6, landed after completion. B-25's FH, FC, and DC are airborne for air Exercise Number 7, landed after completion. B-25's FB, FT, and FV airborne for Anti-Submarine Patrol Number 2, uneventful patrol, landed after completion. Liberator MA airborne from Windsor Field for Queensley Exercise, radio contact established, homing procedure "B" commenced, San Salvador located, aircraft carried out correct recognition procedure, patrols carried out under direction of SOE, attack carried out against smoke float, landed back after completion. Liberator LK is airborne from Windsor Field for Queensley Exercise as first relief, homing procedure commenced, from bearings received located San Salvador, aircraft carried out correct recognition procedure, patrols carried out under direction of SOE, dummy attack made with one depth charge, aircraft completed exercise and landed. Liberator NA airborne from Windsor Field for Queensley Exercise as second relief, homing procedure commenced, San Salvador located, aircraft carried out correct recognition procedure, patrols carried out under direction of SOE, dummy attack carried out on submarine, aircraft landed back after completion of exercise. Liberator ND airborne from Windsor Field for Queensley Exercise as third relief, homing procedure commenced, San Salvador located, signals exchanged, depth charge attack carried out against a smoke float, patrols carried out under direction of SOE, landed after exercise completion. USCG *CG-83474* and USS *PC-1209* are bound north in Straits passing Palm Beach. Inter-island transport USS *YAG-19* is bound west in Straits towards Miami. Troop transport USS *USAT-George Washington* and escorts are bound northwest in Straits with US Navy plane escorting.

22nd June 1944

B-25's FC, FY, FG, FT, and FW are airborne for night navigation exercise, landed after completion. B-25's BX, CO, and CG are airborne for local navigation exercise, landed after exercise completion same aircraft completed similar exercises again later. B-25's FU, FV, FF, DC, DF, and FK are airborne for air Exercise Number 6, landed after exercise completion. Grumman Goose GB returned from San Salvador. Liberator NX is airborne from Windsor Field for Queensley Exercise, homing procedure commenced, San Salvador located, signals exchanged, dummy submarine attack carried out against a smoke float, aircraft landed after exercise completion. Liberator NC airborne from Windsor Field for Queensley Exercise as first relief radio contact established San Salvador located, signals exchanged patrols Alligator and Cobra completed a depth charge attack carried out against a smoke float, aircraft landed after exercise completion. Liberator LV airborne from Windsor Field for Queensley Exercise as second relief, homing procedure "B" commenced, San Salvador located, Alligator and Cobra completed, a depth charge attack carried out against a smoke float, aircraft landed after exercise completion. **Task Unit TU-20154** is bound south in Straits towards Canaveral. USS *PC-1603* is bound south in Straits passing Canaveral. SS *Lake Erie* is bound south near Matanilla Shoal, northwest Bahamas. **Convoy Jacksonville-71** is bound south in Straits towards Port Everglades. HMS *Prince William* (CVE-31) is bound north in Straits near Bimini. SS *Walter Jennings* is bound south in Straits passing Miami. Cuban gunboat *CS-Cuba* is bound north in Straits.

23rd June 1944

B-25's FC, FV, FF, FK, and FL are airborne for night navigation exercise, landed after exercise completion. B-25's BD, CG, BX, and BJ are airborne for local night navigation exercise, aircraft landed after exercise completion. B-25's FV, DF, FF, FO, and FC are airborne, landed after exercise completion, again airborne later for second half of air Exercise Number 6, landed after exercise completion. Grumman Goose GB is airborne for San Salvador, returned to base landed after exercise completion. B-25 CX returned from Morrison Field. **Convoy NK-621** is bound south in Straits. SS *John Alden* is bound south in Straits passing Canaveral. **Convoy KN-321** is bound north in Straits near Canaveral with US Navy plane escorting. USS *SC-681* is bound north in Straits passing Palm Beach. USS *YOG-6* is bound west in Straits towards Miami. USS *YMS-374* is north in Straits passing Miami. Bahamas mailboat *Richard Campbell* and mailboat *Lady Cordeaux* are bound northwest in Straits, towards Miami. USS *YMS-374* and tanker *Lake Erie* are bound southwest in Straits. Destroyer escort USS *Holt* (DE-706) is bound northeast in Straits.

24th June 1944

B-25's DC, FD, FL, FO, and FV are airborne for local night navigation exercise, aircraft landed after exercise completion. B-25's FB, FF, DC, FH, FL, and FY are airborne for compass swinging exercise, aircraft landed after exercise completion. USS *PC-1210* and USS *Yosemite* (AD-19) are bound north in Straits passing Palm Beach. SS *Tampico*, **Convoy NK-621**, SS *Massachusetts*, and SS *Daylite* are bound south in Straits passing Port Everglades. Tanker *Atheldukе* is bound south in Straits passing Miami. SS *W. M. Burton*, SS *Blue Jacket*, SS *Lafacadio Hearn,* and SS *Morton Prince* are bound south in Straits. Destroyer escort USS *Earl K. Olsen* (DE-765) is bound west in Straits. Dutch navy ship KMNS *Van Kinsbergen* is bound southeast in Straits. **Convoy GN-140** is bound north in Straits.

25th June 1944

B-25's FD, FJ, FU, FK, FV, and FB are airborne for compass swinging exercise, aircraft landed after exercise completion. USS *SC-1477* is bound south in Straits, towards St. Augustine. USS *PC-1186* is bound south in Straits towards Daytona. USS *SC-1489* is bound northwest in Straits near Canaveral. SS *Sea Nymph* is bound southwest passing Matanilla Shoal, northwest Bahamas. USCG *CG-83497* is bound southeast in Straits passing Fort Everglades. Mailboat *Noel Roberts* and USCG *CG-83505* are bound north in Straits passing Port Everglades. USS *YP-448* is bound west in Straits towards Miami with US Navy plane escorting. Tug USS *ATR-29* towing oil barge USS *YOG-46* is bound north in Straits towards Bimini. **Convoy GK-814** is bound northwest in Straits. Tanker *Glarona* is bound north in Straits.

26th June 1944

B-25's FH, FB, FF, FG, FU, and DC are airborne for Anti-Submarine Patrol Number 1, aircraft landed after exercise completion, FF returned to base earlier with radio failure due to heavy thunderstorm during patrol. B-25's FD, FU, FF, and FV are airborne for air Exercise Number 1, aircraft landed after exercise completion. B-25 FO is airborne for Washington via Elizabeth City landed in Washington. Grumman Goose GA is airborne for Palisadoes, Jamaica via Guantanamo landed in Jamaica, then is again airborne for a waterborne landing at Grand Turks Island. USS *SC-1477* is bound south in Straits towards Daytona. SS *Jacob Sloat Fassett* is bound south in Straits passing Daytona. USS *YP-448* is bound north in Straits towards Daytona. Destroyer escorts USS *DE-326* and USS *DE-328* are bound north in Straits. Tanker *Edward L. Doheny* is bound south in Straits past Palm Beach. USS *John Hood* and SS *Platte Bridge* are bound north in Straits near Miami. SS *Dague* is bound south in Straits. Tug USS *ATR-29* is bound east to Nassau.

27th June 1944

B-25's FB, FH, FK, FD, and FJ are airborne for Anti-Submarine Patrol Number 1, aircraft landed after exercise completion. B-25's FF, FY, DC, and FL are airborne for air Exercise Number 1, aircraft landed after exercise completion. Liberator LH is airborne for Radar and navigation exercise. Detailed tracks were completed, aircraft landed after exercise completion. Tanker *Chilbar* and tanker *Empire Norse* are bound south in Straits towards Canaveral. USS *LSM-16* is bound north in Straits towards Canaveral. Tanker *Bohemian Club* is bound south in Straits towards Palm Beach. USS *YO-129* and tanker *Chiloil* are bound north in Straits passing Palm Beach. Tanker *Oscar F. Barrett* and tanker *Permian* are bound south in Straits passing Port Everglades. USS *SC-1014* is bound south in Straits. Bahamas mailboat *Richard Campbell* and ferry *Betty K.* are bound northwest in Straits towards, Miami with US Navy plane escorting. USS *PC-1265* is bound southwest in Straits. USS *Cotinca* (AMC-43) is bound south in Straits towards Cay Sal Bank. SS *Jacob Sloat Fassett* is bound south in Straits.

28th June 1944

B-25's FD, FK, and FH are airborne for Anti-Submarine Patrol Number 2, aircraft landed after exercise completion. B-25's CL, DC, and FM are airborne for air Exercise Number 7, aircraft landed after exercise completion. Grumman Goose GA is airborne from Palisadoes, Jamaica and landed back at base. USCG *CG-83389* south towards Canaveral. **Convoy KN-322** and tanker *R. P. Smith* are bound north in Straits. SS *Niobe* and SS *La Paz* are bound south in Straits towards Palm Beach. Tanker *Gulfgem* is bound northeast in Straits. SS *Christopher L. Sholes* is bound north in Straits passing Port Everglades. USS *Sapelo* (AO-11) escorted by USS *PCE-846* is bound north in Straits passing Port Everglades. SS *Chilbar* and SS *Empire Norse* are bound southwest in Straits. USS *SC-1492* is bound northeast in Straits.

29th June 1944

B-25's FD, FB, FJ, FK, FM, and FY are airborne for night navigation exercise, aircraft landed after exercise completion. B-25's FD, FY, FP, and FM are airborne for air Exercise Number 5, aircraft landed after exercise completion. Liberator LH is airborne for Radar and navigation exercise, aircraft landed after exercise completion. Liberators NC and NA airborne from Windsor Field navigation exercises aircraft landed after exercise completion. Grumman Goose GB is airborne for flight to Great Exuma, aircraft landed after exercise completion. USS *SC-1492* is bound north in Straits passing Canaveral. Tanker *Pan North Carolina* is bound north in Straits. Tanker *Cities Service Fuel* is bound south in Straits passing Canaveral. SS *Maumee* is bound northeast in Straits. SS *Sage Harbour* is bound southwest in Straits towards Palm Beach. **Convoy NK-622** is bound south in Straits passing Palm Beach. *USS LSM-17* and tanker *L. L. Abshire* is bound north in Straits passing Palm Beach. SS *Edward Livingstone* is bound south in Straits passing Port Everglades. USS *Mahogany* (AN-23) is bound south in Straits passing Port Everglades. Tanker *Hadnot* and tanker *Gulfgem* are bound north in Straits passing Miami. SS *Sharpsburg* is bound north towards Bimini. USS *YO-129* is bound southeast towards Abacos. Submarine USS *Razorback* (SS-394) is bound northwest passing Northeast Providence Channel. SS *Ernest Hinds* is bound northwest in Straits. **Convoy GN-141** is bound northeast, to the east of Bahamas. SS *John W. Searle* and SS *Walker Taylor* are bound northeast. SS *Niobe* is bound southwest in Straits. SS *Gilbert Stuart* is bound northwest with US Navy plane escorting.

30th June 1944

B-25's FJ, FM, FL, FP, FH, and FW are airborne for night navigation exercise, aircraft landed after exercise completion. B-25's DC, CL, FH, and FL are airborne for air Exercise Number 3, then are airborne later to complete air Exercise Number 4, aircraft landed after exercise completion. Liberator LH is airborne for Radar and navigation exercise, aircraft landed after exercise completion. Liberator MD is airborne from Windsor Field night navigation exercise, aircraft landed after exercise completion. **Task Unit TU-201513** is bound south in Straits passing Canaveral. SS *Lookout Mountain* is bound southwest in Straits. USS *YMS-370* is bound south in Straits towards Palm Beach. SS *Swiftscout* is bound north in Straits near Palm Beach. Free French navy ship FMS *Dague* is bound north near Matanilla Shoal, northwest Bahamas. SS *Bessemer* is bound north in Straits towards Palm Beach. USS *YO-129* is bound southeast towards Eleuthera. SS *William W. Campbell* is bound south in Straits. USS *PC-1221*, USS *SC-1475*, USS *SC-1490*, **Convoy NK-622**, and tanker *Cities Service Fuel* are bound west in Straits. Tanker *Lake Erie* and SS *El Oriente* are bound north in Straits. SS *Ravenfjell* is bound northwest in Straits passing Nicholas Channel. **Convoy NG-443** is bound south in Straits with US Navy plane escorting. SS *Polycastle* is bound south in Straits.

1st July 1944

B-25's FG, FL, and FW are airborne for local night navigation exercise, aircraft landed after exercise completion. B-25's FC, FP, and FM are airborne for Anti-Submarine Patrol Number 2, aircraft landed after exercise completion, a merchant ship sighted, armed merchant vessel sighted. B-25's FW, FH, and FL are airborne for air Exercise Number 7, aircraft landed after exercise completion. B-25's CL, DF, FG, and FV are airborne for air Exercise Number 3, same aircraft completed air Exercise Number 4, aircraft landed after exercise completion. Grumman Goose GB are airborne for Great Ragged Island, aircraft landed after exercise completion. B-25 FG returned from Washington. Tug USS *Acushnet* (AT-63), tanker *Pan Georgia*, and tanker *Chiloil* are bound south in Straits, passing Canaveral. Destroyer escort USS *Robert Braziere* (DE-345), SS *Robert G. Ingersoll*, SS *Bahram*, and USS *PC-1265* are bound north in Straits passing Palm Beach. SS *Bostonian* and SS *H. H. Raymond* are bound north in Straits, near Miami. USS *PC-1603* is bound north in Straits. Tanker *Athelprince* and SS *T. J. Jackson* are bound north in Old Bahama Channel. Oiler USS *YO-129* is bound south passing Northeast Providence Channel. SS *General A. E. Anderson* is bound south in Straits, towards Key West, US Navy plane escorting.

2nd July 1944

Task Unit TU-201514 is bound south in Straits passing Canaveral. SS *Shirabank* is bound south in Straits towards Palm Beach. Tanker *R. P. Smith* is bound south in Straits passing Palm Beach. Troop transport USS *USAT-Sea Owl* is bound north passing Matanilla Shoal, northwest Bahamas. USS *YO-129* is bound west passing Northeast Providence Channel. SS *James A. Wilder*, tanker *Pan Georgia, and* USS YAG-21 are bound south in Straits. Tug USS *Acushnet* (AT-63), USS *PC-1144*, **Task Unit TU-03111**, and SS *Chiloil* are bound southwest in Straits. SS *Lafayette* is bound southeast in Straits passing Nicholas Channel. SS *William W. Campbell* and SS *H. H. Raymond* are bound southeast in Old Bahama Channel.

3rd July 1944

B-25's DF, FV, FC, and FU are airborne for air Exercise Number 6, aircraft landed after exercise completion. B-25 FJ is airborne for Boca Chica, returned to base later. Liberator MD airborne from Windsor Field for Kingsley Exercise, radio contact established, homing procedure "B" commenced, visual contact made with convoy, attack with one depth charge carried out against smoke float dropped by SOE, carried out patrols as instructed by SOE, aircraft landed after exercise completion. Liberator ND airborne from Windsor Field for Kingsley Exercise as first relief, small frigate of 800 tons sighted, merchant ship sighted, radio contact established with convoy, homing procedure "B" commenced, convoy located, aircraft carried out patrols as detailed, dummy attack carried out against smoke float dropped by SOE, aircraft landed after exercise completion. Liberator MF is airborne from Windsor Field for Kingsley Exercise as second relief, Liberty ship sighted, 8,000 ton tanker sighted, radio contact established with convoy, homing procedure commenced, convoy located, aircraft carried out patrols as detailed by SEO, dummy submarine attack carried out against sea marker, aircraft landed after exercise completion. Liberator MC is airborne from Windsor Field for Kingsley Exercise as third relief, radio contact established with convoy, frigates sighted visually, dummy submarine attack carried out, aircraft landed after exercise completion. USCG *CG-83528*, SS *Adoniram Judson*, and tanker *Edward L. Shea* are bound south in Straits towards Daytona. **Convoy NK-623** is bound south in Straits, passing Canaveral. SS *Charles W. Wooster* is bound north in Straits in Straits near Canaveral. Tanker *Ardmore* is bound north in Straits. Tanker *Malabar* is bound south in Straits towards Palm Beach. Tanker *Point Pleasant* is bound southwest from Matanilla Shoal, northwest Bahamas. **Convoy KN-323** is bound north in Straits passing Port Everglades. Tanker *Washita* is bound south in Straits passing Bimini. USS *YO-129* is bound southwest in Straits towards Miami. USS *PC-1264* is bound southwest in Straits. USS *Density* (AM-218) is bound northeast in Straits. Tanker *Britamsea*, tanker *San Vulfrano*, and tanker *Mount Bruce Park* are bound northeast in Straits.

4th July 1944

B-25's FV, FW, FO, CL, FK, and FM airborne for night navigation exercise, aircraft landed after exercise completion. B-25's CL, DF, FV, and FO are airborne for air Exercise Number 6, aircraft landed after exercise completion. B-25's FJ, FW, and FK are airborne for air Exercise Number 7, aircraft landed after exercise completion. B-25's FU and FM are airborne for Anti-Submarine Patrol Number 2, aircraft landed after exercise completion. Liberator MA airborne for Kingsley Exercise, radio contact established, frigates located, recognition procedure carried out, carried out procedure as directed by SOE, dummy submarine attack against smoke float flare dropped by aircraft for AA for practice by ship's gunners, aircraft landed after exercise completion. Liberator NC is airborne from Windsor Field for Kingsley Exercise as first relief, homing procedure "B" commenced, convoy located, signals exchanged patrols carried out as directed by SOE, dummy submarine attack carried against smoke float, aircraft landed after exercise completion. Liberator ND is airborne from Windsor Field for Kingsley Exercise as second relief, radio contact established, convoy located, patrols carried out as directed by SOE, dummy submarine attack against sea marker, aircraft landed after exercise completion. Liberator LK is airborne from Windsor Field for Kingsley Exercise as third relief, radio contact established frigates located from bearings received, exercises completed after aircraft carried out dummy submarine attacks, aircraft landed after exercise completion. SS *Vistula* and SS *Beacon* are bound north in Straits. USCG *CG-83528* is bound south in Straits, passing Palm Beach. Tanker *A. S. Hansen* is bound north in Straits passing Palm Beach. **Convoy NK-623** is bound southwest in Straits with US Navy plane escorting. SS *C. A. Canfield* and SS *Adoniram Judson* are bound southwest in Straits. SS *Massachusetts* is bound northeast in Straits. Tanker *Edward L. Shea* is bound west in Straits. SS *Lafcadio Hearn* is bound northeast passing Cay Sal Bank.

5th July 1944

B-25's FV, FC, FP, FO, and FJ are airborne for night navigation exercise, aircraft landed after exercise completion. B-25 FM airborne for Anti-Submarine Patrol Number 2, aircraft landed after exercise completion. B-25's FJ and FV are airborne for air Exercise Number 7, aircraft landed after exercise completion. B-25's DF, FO, FC, FF, and FW are airborne for air Exercise Number 6, aircraft landed after exercise completion. B-25 FV is airborne for air Exercise Number 7, aircraft landed after exercise completion. B-25's AL, AM, and AD are airborne for local navigation exercise, aircraft landed after exercise completion, similar exercises carried out by same aircraft again later. Not possible to arrange third day's Kingsley Exercise, so in lieu it was arranged that a homing and navigation exercise should take place, five aircraft would fly to a datum line 30 nm east of Nassau, aircraft were to set course for points 30 nm apart on this line, aircraft were to then fly on easterly course, one of the aircraft was to home, the others sending a practice sighting an attack submarine report, aircraft to return to base after exercise completion via turning points. Liberator LU is airborne from Windsor Field, Liberator NA is heard to be sending practice sighting and attack report. Contact made with aircraft on radio, no visual contact could be made, abandoned search and dropped sea marker, carried out dummy submarine attack, aircraft landed after exercise completion. Liberator NC airborne from Windsor Field, contact made with NA on radio, no visual contact could be made, dummy submarine attack made against smoke float, aircraft landed after exercise completion. Liberator NA airborne, merchant ship sighted, message sent to base with practice sighting and attack report, changed frequency and tried to home the other aircraft, it was not successful, carried out dummy submarine attack against smoke float, six depth charges fell out, aircraft then landed after exercise completion. Liberator LL is airborne, signal from NA to base was intercepted, aircraft changed frequency, nothing further was heard on this frequency, tanker sighted, transporter was sighted, attack carried out against smoke float, aircraft landed after exercise completion. Liberator MA airborne from Windsor Field, message from NA intercepted, no contact made with the aircraft, square search carried out, NA sighted, aircraft carried out attack against sea marker dropped by NA, aircraft landed after exercise completion. Request received from Comgulfseafron for aircraft to escort troopship with a destroyer, escort estimated position of convoy conveyed. B-25 FG is airborne, convoy located, no acknowledgement received to recognition signals, aircraft left ship and landed back at base. B-25 FU airborne sighted convoy, signals exchanged, air escort provided, aircraft landed after exercise completion. SS *Christopher L. Shoals* is bound south in Straits. SS *Calliope*, SS *Bostonian*, USS *Palm* (AN-28) are bound south in Straits passing Canaveral. Tanker *Atheldulke* is bound north in Straits. Destroyer escort USS *Oswald* (DE-767) is bound northeast in Straits. Tanker *Athelprince* is bound south in Straits towards Palm Beach. SS *Lafcadio Hearn*, SS *Maridal*, SS *Daylight*, and SS *Sea Nymph* are bound north in Straits passing Palm Beach. Tug USS *YTL* is towing USS *YF-367* and USS *YF-492* are bound southeast in Straits from Port Everglades. SS *Richard Upjohn* is bound southwest in Straits. USS *SC-1475* and USS *SC-1490* are bound northeast in Straits. **Convoy New Orleans-41** is bound northeast in Straits. Destroyer escort USS *Oliver Mitchell* (DE-417) is bound northeast in Straits. USS *Thrush* (AVP-3) and SS *Andoniram Judson* are bound southeast in Straits. **Convoy HK-816** is bound northwest in Straits.

6th July 1944

B-25's FD and FF are airborne for air Exercise Number 3, aircraft landed after exercise completion. FV didn't complete patrol, as just after take-off two depth charges fell from bomb doors that opened up, due to a faulty mechanism. One depth charge exploded in bush land west of airfield, the other didn't explode, aircraft carried out air firing and low flying exercises, landed back at base. Unexploded depth charge was detonated during the course of the day, with no casualties. B-25's DC, FF, FK, FO, and FW are airborne for air Exercise Number 6, aircraft landed after exercise completion. B-25's AS, BJ, and AD are airborne for local navigation exercise, aircraft landed after exercise completion. USS *SC-1475* and Russian RU *SC-1490* are bound north in Straits towards Daytona. SS *Chilbar* and tanker *E. R. Kemp* are bound north in Straits. SS *Tekoa* is bound south in Straits. SS *Medina* is bound south near Matanilla Shoal, northwest Bahamas. USCG *CG-83524* is bound east passing Walker Cay, north of Abaco. Destroyer escort USS *Gentry* (DE-349) is bound north near Matanilla Shoal, northwest Bahamas. SS *W. M. Burton* is bound north in Straits passing Port Everglades. SS *Sag Harbour* is bound north in Straits passing Bimini. SS *Calliope*, tug USS *YTB-338* towing USS *YP-367*, USS *YP-492*, and tanker *Esso Memphis* are bound southwest in Straits. Tanker *18 de Marzo* is bound north in Straits.

7th July 1944

B-25's CG, BJ, AD, CF, CJ, and CX are airborne for local night navigation exercise, aircraft landed after exercise completion. B-25's DC, CL, FW, and FC are airborne for compass swinging exercise, aircraft landed after exercise completion. Grumman Goose GA is airborne to Harbour Island, returned to base, aircraft landed after exercise completion. Tanker *18 De Marzo* is bound north in Straits towards Daytona. Tanker *Edward L. Doheny*, USS *Porate* (AM-275), HMS *Reaper* (CVE), and HMS *BAM-25* are bound north in Straits. SS *Edward Livingston* is bound north near Matanilla Shoal, northwest Bahamas. Tanker *Spidoliene* is bound south in Straits towards Palm Beach. Destroyer escort USS *Andres* (DE-45) and tanker *Liebre* are bound south in Straits towards Port Everglades. USS *PCER-851* and USS *PCER-852* are bound north in Straits near Miami. USS *Palm* (AN-28) is bound southeast in Straits. Tanker *Esso Augusta* is bound southwest in Straits. Tanker *AthelRegent* is bound northwest in Straits, passing Nicholas Channel.

8th July 1944

B-25's FC, FB, FG, FU, and FW are airborne for night navigation exercise, aircraft landed after exercise completion. B-25's FB, FK, FF, FC, and FL are airborne for compass swinging exercise, aircraft landed after exercise completion. SS *Prometheus*, SS *A. S. Hansen* are bound south in Straits passing Canaveral. USS *YTB-338* is bound northwest in Straits past Canaveral. **Convoy KN-324**, USCG cutter *CG-Calypso*, and SS *Empire Norse* are bound north in Straits. SS *Juan Casiano* is bound south in Straits towards Port Everglades. Tanker *Gulfdisc* is bound north in Straits near Miami. USS *Robert F. Keller* is bound northeast in Straits. USS *State of Maryland* is bound southeast in Straits passing Nicholas Channel with US Navy plane escorting. SS *W. W. Bruce* and tanker *Gulfgem* are bound south in Straits.

9th July 1944

Tug USS *ATR-21* is bound south in Straits towards Daytona. Tanker *E. R. Kemp* is bound south in Straits towards Canaveral. Tug USS *Point Sur* and **Task Unit TU-201515** are bound south in Straits. Tanker *Horace. H. Harvey* and tanker *Lookout Mountain* are bound north in Straits. Tanker *AthelRegent* and tanker *Tule* are bound north in Straits passing Palm Beach. **Convoy NK-624** is bound south in Straits passing Palm Beach. Submarine USS *Seacat* (SS-399) is bound northwest towards Northwest Providence Channel with US Navy plane escorting. USS *YO-142* is bound northeast in Straits near Miami. USS *PC-*

1563 is bound north in Straits. Tanker *Docket Canyon* is bound northeast in Straits. Tanker *Josiah C. Holland* and USCG cutter *CG-Cobb* are bound southwest in Straits. Tanker *Motrix* is bound northwest in Straits near Nicholas Channel. SS *City of Savannah*, tanker *James Cook* and tanker *Alar* are bound south in Straits. **Convoy GN-143** is bound north in Straits.

10th July 1944

B-25's FG, FD, FY, FF, and FK are airborne for Anti-Submarine Patrol Number 1, aircraft landed after exercise completion, a hospital ship was sighted, the rest of patrol uneventful. B-25's FN, DF, CL, FW, and FH are airborne for air Exercise Number 1, aircraft landed after exercise completion. Liberator LH airborne for Radar and navigation exercise, aircraft landed after exercise completion. B-25 FS is airborne from Morrison Field, *en route* to Albuquerque, New Mexico, then Dallas, aircraft landed after exercise completion. Tug USS *YTB-338* towing USS *YSR-10* is bound south in Straits towards Canaveral. Tug USS *ATR-21* towing two sled targets is bound south in Straits towards Palm Beach. Destroyer escorts USS *DE-347*, USS *DE-350*, and USS *DE-420* are bound northeast in Straits. Submarine USS *Icefish* (SS-367) is bound south in Straits towards Port Everglades. Tug USS *ATR-7* towing tug USS *ATR-29* and escorted by USS *SC-453* are bound southeast in Straits from Miami. SS *George Uhler* and SS *Point Pleasant* are bound north in Straits, passing Miami. **Task Unit TU-201515**, **Convoy NK-624**, and tanker *E. R. Kemp* are bound southwest in Straits. SS *Mount Royal Park* is bound south in Straits. **Convoy GK-817** is bound northwest in Straits passing Nicholas Channel. SS *Cape Gaspe* is bound southeast in Straits passing Nicholas Channel.

11th July 1944

B-25's FD, FH, FF, and FG are airborne for Anti-Submarine Patrol Number 1, aircraft landed after exercise completion. B-25's CL, DF, DC, FD, FW airborne for air Exercise Number 1, aircraft landed after exercise completion. Tanker *Bahram* is bound south in Straits. Tanker *18 De Marzo* and USS *LSM-257* are bound south in Straits towards Canaveral. Tanker *Edward L. Doheny* is bound south in Straits passing Canaveral. Tanker *Bidwell* is bound northwest in Straits towards Canaveral. USCG *CG-80303* and tanker are bound southeast towards Matanilla Shoal, northwest Bahamas. Tanker *Cities Service Fuel* and SS *M. Michael Edelstein* are bound north in Straits. USS *Hitchiti* (ATF-103) with tow and escort is bound south in Straits towards Palm Beach. USS *YTB-338* towing USS *YSR-10* is bound south in Straits passing Palm Beach. Tug USS *Point Sur,* tow and escort are bound west in Straits. Tanker *Washita* is bound northeast in Straits. SS *John R. R. Hannay* is bound northwest in Straits.

12th July 1944

B-25's FF and FY airborne for Anti-Submarine Patrol Number 2, aircraft landed after exercise completion. B-25's CL, FP, and FW are airborne for air Exercise Number 7, aircraft landed after exercise completion. B-25's FL, DC, FM, FP, and FT are airborne for air Exercise Number 5, aircraft landed after exercise completion. Liberator LH is airborne for Radar and navigation exercise, aircraft landed after exercise completion. B-25 FS proceeded to Morrison Field on 10th July, and landed at Albuquerque. Request received from Comgulfseafron for close air escort of ferry *Jean Brillant* from time of departure from Nassau in a convoy of one merchant ship and two escorts. B-25 FM is airborne and met convoy leaving Nassau Harbour, close escort carried out, aircraft landed after exercise completion. B-25 FF is airborne, sighted convoy at Nassau Harbour mouth, carried out sweep ahead, aircraft landed after exercise completion. SS *John B. Hamilton* and SS *San Valerio* are bound north in Straits. USS *YO-142* is bound west, passing Northwest Providence Channel. Tanker *Bahram* and tanker *18 De Marzo* and USS *Hitchiti* (ATF-103) are bound southwest in Straits. USCG *CG-83523* are bound east in Straits. USS *Pegasus* (AK-48) is bound southeast in Straits passing Nicholas Channel. SS *Floridan* and SS *Salator* are bound south in Straits.

13th July 1944

B-25's FP and FC are airborne for night navigation exercise, then FP completed exercise and landed. FC returned to base earlier, but was again airborne, completed exercise and then landed back. Liberator MA is airborne from Windsor Field to continue air escort of ferry *Jean Brillant,* convoy sighted, escort completed, the convoy left, and aircraft landed after exercise completion. B-25's FT, FD, FP, DC, and CL are airborne for air Exercise Number 5, aircraft landed after exercise completion. Liberator LH is airborne for Radar and navigation exercise, aircraft landed after exercise completion. **Convoy Jacksonville** is bound north in Straits. Tug USS *ATR-6* and tows and tanker *Pan North Carolina* are bound south in Straits, towards Palm Beach. **Convoy KN-325** is bound north in Straits towards Palm Beach. Tanker *R. P. Smith* is bound south in Straits passing Miami. Tanker *C. A. Canfield* is bound north in Straits passing Miami. Tanker *Beacon* and USCG *CG-Vigilant* are bound southwest in Straits. Tanker *Bulkoil* is bound northeast in Straits. SS *James A. Wilder* is bound northwest in Straits passing Nicholas Channel. USS *YAG-21* is bound southeast passing Nassau. Tanker *Jean Baptiste Lemoyne*, SS *Tungsha*, tanker *Esso Aruba*, tanker *Esso Bolivar*, tanker *Clearwater Park*, and tanker *Nevada* are bound south, to the east of San Salvador.

14th July 1944

B-25's FN, FY, and FF are airborne for Anti-Submarine Patrol Number 3, aircraft landed after exercise completion. B-25's FP and FD airborne for air Exercise Number 7, aircraft landed after exercise completion. B-25's FN, DC, FC, FL, and FT are airborne for air Exercise Number 5, aircraft landed after exercise completion. B-25's FN, FT, FC, FL, and FM are airborne for air Exercise Number 4, aircraft landed after exercise completion. Grumman Goose GA is airborne for Governors harbour returned to base later in the day. Four P-40 aircraft and one B-25 bomber landed from Morrison Field, then participated in fighter affiliation exercises with aircraft of Number 3 Squadron. Liberator LH is airborne for Radar and navigation exercise, aircraft landed after exercise completion. SS *W. S. Halstead* is bound south towards Canaveral. **Task Unit TU-201518**, tug USS *Arkab* (AK-130) towing USS AFD-20 and escorted by destroyer escort USS *DE-342* are bound south in Straits, passing Canaveral. Landing ship USS *LSM-134* and **Convoy NK-625** are bound south in Straits towards Palm Beach. USS *SC-1476*, Russian RU *SC-1507* and tanker *Bulkoil* are bound north in Straits, near Palm Beach. SS *Britamsea* is bound south in Straits passing Port Everglades. SS *Tampico*, USS *SC-1498* and Russian RU *SC-1481* are bound north in Straits passing Port Everglades. SS *Richard Upjohn* and tanker *Petrofuel* are bound north in Straits towards Miami. USS *ATR-6* towing USS *YO-101* and USS *YO-102* escorted by USCG *CG-83529* are bound south in Straits. SS *Christopher L. Shoals* is bound north in Straits. SS *Minnesotan* is bound southwest in Straits. **Convoy GK-818** is bound northeast in Old Bahama Channel.

15th July 1944

B-25's FH, FT, FL, FK, and FC are airborne for local night navigation exercise, aircraft landed after exercise completion. B-25's FY and FJ are airborne for Anti-Submarine Patrol Number 2, aircraft landed after exercise completion. B-25's FO, FG, FB, DC, and FC are airborne for air Exercise Number 3, aircraft landed after exercise completion, same aircraft completed air Exercise Number 4 later in the day. Signal received of a tropical storm cantered 200 nm to 250 nm east northeast of Nassau apparently moving in a north-easterly direction, advisory stated precaution should be exercised in extreme north-eastern Bahamas region, Station Commander instructed Hurricane warning White be put into operation, Station continues normal routine but is prepared for further developments of a hurricane. Liberator MA is airborne from Windsor Field, carry out reconnaissance of given area, weather condition survey made, aircraft landed after exercise completion. Aircraft report and a report of the movement of the tropical storm away from the Bahamas, Commanding Officer gave instructions that Station was to revert to normal. SS *Morton M. McCarver* is bound south in Straits passing Canaveral. Tanker *Walter Jennings* is bound north in Straits passing Canaveral. Tanker *Christy Payne*, USS *ATR-60*, and USS *PC-593* are bound south in Straits. USS *Arkab* (AK-130) is bound south in Straits passing Port Everglades. Tanker *L. V. Stanford* is bound north in Straits. Tanker *Chatham* north out of Straits.

16th July 1944

Tanker *Vacuum* is bound south in Straits near Canaveral. SS *Castana*, USS *PC-1145*, and USS *PC-1218* are bound south in Straits. USS *YO-129* is bound west in Straits towards Port Everglades. USS *YP-580* is bound northwest in Straits towards Port Everglades. **Convoy Motor Torpedo Boat Squadron-32, Group A**, SS *Liebre* and SS *Malabar* are bound north in Straits. USS *PC-1087* and USS *PC-1218* are bound west in Straits.

17th July 1944

B-25's FF, FJ, and FL are airborne for Anti-Submarine Patrol Number 2, aircraft landed after exercise completion. B-25's FK and FC are airborne for air Exercise Number 7, aircraft landed after exercise completion. B-25's FW, FT, FF, DC, and FC are airborne for air Exercise Number 6, aircraft landed after exercise completion. Liberator NA is airborne for Kingsley Exercise from Windsor Field. Contact established with SOE of convoy, two frigates sighted, signals exchanged, patrols completed under SOE. Direction depth charge attack made against smoke float laid by SOE, aircraft landed after exercise completion. Liberator NC is airborne for Kingsley Exercise from Windsor Field as first relief, merchant ship of 1,000 tons sighted, radio contact established with convoy, frigates sighted, signals exchanged patrols completed under SOE, direction depth charge attack made against smoke float laid by SOE, aircraft landed after exercise completion. Liberator ND is airborne for Kingsley Exercise from Windsor Field as second relief, earlier aircraft was still homing so this aircraft could not carry out homing procedure to locate convoy, square search carried out, frigates sighted, Crocodile patrol completed under SOE, direction depth charge attack made against smoke float laid by SOE, aircraft landed after exercise completion. Liberator NX is airborne for Kingsley Exercise from Windsor Field as third relief, homing procedure "B" commenced, two frigates located but they were stragglers to the original Kingsley Exercise convoy, main part of convoy located, patrols completed under SOE, direction depth charge attack made against smoke float laid by SOE, aircraft landed after exercise completion. Liberator NO is airborne for Kingsley Exercise from Windsor Field fourth relief homing commenced after receiving bearings convoy located patrols completed under SOE direction depth charge attack made against smoke float laid by SOE aircraft landed after exercise completion. Tanker *Pan Georgia* is bound north out of Straits. USS *Design* (AM-219) is bound north passing Matanilla Shoal, northwest Bahamas. USS *UAG-21* is bound northwest in Straits passing Northwest Providence Channel. Tanker *Solsten* and USS *Javelot* (ex-USS *PC-1562*) are bound south in Straits. SS *John A. Rawlins* and tanker *Britamsea* are bound northwest in Straits passing Nicholas Channel. *U-539* re-enters north via Mona Passage, to northeast Dominican Republic.

18th July 1944

B-25's FK, FT, FC, FN, and FL are airborne, aircraft landed after exercise completion. Liberator MA is airborne for Kingsley Exercise from Windsor Field, homing commenced, patrols completed under SOE, direction depth charge attack made against smoke float laid by SOE, aircraft released flares for A-A practice by ship's gunners, aircraft landed after exercise completion. Liberator LL is airborne for Kingsley Exercise from Windsor Field as first relief homing procedure "B" commenced, patrols completed under SOE, direction depth charge attack made against smoke float laid by SOE, aircraft released flares for A-A practice by ship's gunners, aircraft landed after exercise completion. Liberator NA is airborne for Kingsley Exercise from Windsor Field, as second relief, radio contact established, convoy of three frigates sighted, patrols completed under SOE, direction Liberty ship of 7,000 tons sighted. Depth charge attack made against smoke float laid by SOE, aircraft released flares for A-A practice by ship's gunners, aircraft landed after exercise completion. Liberator NC is airborne for Kingsley Exercise from Windsor Field as third relief, radio contact established, convoy located from bearings received patrols completed under SOE, direction depth charge attack made against smoke float laid by SOE, aircraft released flares for A-A practice by ship's gunners, aircraft landed after exercise completion. Liberator NX is airborne for Kingsley Exercise from Windsor Field as fourth relief, radio contact established convoy located from bearings received, patrols completed under SOE, direction depth charge attack made against smoke float laid by SOE, aircraft landed after exercise completion. **Convoy NK-626** is bound south in Straits towards Canaveral. **Convoy KN-326** is bound north in Straits towards Palm Beach. SS *Ben Holladay* is bound north in Straits passing Port Everglades. USS *PC-1546*, USCG *CG-83528*, USS *PCS-1376*, and tug USS *ATR-21* are bound south in Straits. SS *Cedar Mountain* is bound southwest in Straits nearing Cay Sal Bank. **Convoy KG-719** is bound southeast in Old Bahama Channel to exit via Windward Passage. No.113: His Excellency the Acting Governor, accompanied by Group Captain R. M. Field, Commanding Officer of No. 111

OTU, inspected RAF Transport Command Unit in Nassau.

19th July 1944

B-25's FT, FH, FK, FN, FL, and FO are airbornc for night navigation exercise, adverse weather and poor visibility, aircraft completed southerly part of exercise only, aircraft landed after exercise completion. B-25's FC, FW, FK, FO, FL, and FT are airborne for air Exercise Number 6, aircraft landed after exercise completion. B-25's CF, AS, DF, and CF are airborne, aircraft landed after short navigation exercise completion. Heavy rain and poor visibility restricted flying programme during afternoon and evening flight by six B-25's detailed to take place cancelled. SS *A. S. Hansen* is bound north out of Straits. SS *W. F. Burdell*, SS *John F. Crittenden*, and SS *Express Hill Park* are bound south in Straits. USS *LST-571* is bound northwest in Straits past Canaveral. SS *Niobe* is bound south in Straits passing Canaveral. USS *YAG-28* is bound east passing Matanilla Shoal, northwest Bahamas. SS *Jamestown* is bound south in Straits passing Palm Beach. **Convoy NK-626** is bound south in Straits. SS *Amasa Delano* and SS *Oscar F. Barrett* are bound north in Straits. USS *SC-660* and **Task Unit TU-0313** are bound southeast in Straits. USS *SC-1292* is bound east in Northeast Providence Channel. SS *Escolar* is east of the Bahamas, bound west, towards Northeast Providence Channel.

20th July 1944

B-25's AK, DF, CJ, and AS are airborne short navigation exercise, aircraft landed after exercise completion. B-25's FK, FW, FH, FL, FG, and FO are airborne for air Exercise Number 6, aircraft landed after exercise completion. B-25 FS returned from Albuquerque via Miami landed. SS *Vistula* is bound south in Straits. USS *LST-702*, USS *LST-674*, SS *Nick Stoner*, SS *Juan Casiano*, USS *PC-1218* and USS *PC-1087* are bound north in Straits. USS *YO-117* is bound south in Straits passing Canaveral. Tanker *L. V. Stanford* and SS *Niobe* are bound south in Straits passing Palm Beach. Destroyer escort USS *Maurice J. Manuel* (DE-351) is bound north in Straits near Miami. SS *John J. Crittenden* and tanker *Esso Roanoke* are bound southwest in Straits. **Convoy GK-819** is bound northwest in Straits, passing Nicholas Channel. **Convoy NG-447** is bound south in Straits.

21st July 1944

B-25's CO, AS, CJ, BJ, and CG are airborne for local night navigation exercise, aircraft landed after exercise completion. B-25's FK, FN, FH, FL, FF, and FO are airborne for night navigation exercise, aircraft landed after exercise completion. B-25's FF, FO, and FH are airborne for compass swinging exercise, aircraft landed after exercise completion similar exercise completed by B-25's FY, FN, FW, and FL. **Task Unit TU-201519** and SS *Chilbar* are bound south in Straits passing Canaveral. USS *YAG-28* are bound southwest passing Matanilla Shoal, northwest Bahamas. Landing craft USS *LCI (L)-570* and USCG *CG-Sagebrush* are bound south in Straits towards Palm Beach. USS *LST-572*, USS *LST-621*, SS *Rufus W. Peckham,* Free French Navy ship FFS *Javelot* are bound north in Straits, passing Palm Beach. Tug USS *YTB-338* and tow are bound east in Straits from Port Everglades. Oiler USS *YO-129* is bound south in Straits passing Miami. SS *George W. Kendall* is bound north in Straits passing Miami. SS *Duc D'Aumale* and USS *PC-822* are bound southwest in Straits. Tanker *Chiloil* is bound north in Straits. SS *Samtapa* is bound northwest in Straits.

22nd July 1944

B-25's FC, FF, FM, FP, FG, and FW are airborne for night navigation exercise, aircraft landed after exercise completion. B-25's FM, FP, FB, FG, FC, and FT airborne for compass swinging exercise, aircraft landed after exercise completion. USS *LSM-135* and *LMS-136* and USS *PC-1219* are bound south in Straits, passing Canaveral. SS *Prometheus*, tanker *Bidwell*, and SS *Eano* are bound north in Straits. SS *John D. Whidden* and USS *YO-167* are bound south in Straits, towards Palm Beach. SS *A. S. Hansen* is bound south in Straits passing Port Everglades. Tanker *Edward L. Dohney* and tanker *R. P. Smith* are bound northeast in Straits.

23rd July 1944

Grumman Goose GB is airborne for Governor Harbour, aircraft landed after exercise completion, again airborne for Miami, and aircraft landed after exercise completion. SS *Bostonian* and **Task Unit TU-2052** are bound south in Straits, passing Canaveral. SS *Beacon* is bound north in Straits. USS *YTB-338* and tow are bound west passing Walker Cay, north of Abaco. **Convoy KN-327** is bound north in Straits passing Palm Beach with US Navy plane joining from the north. **Task Unit TU-02910** is bound south in Straits passing, Port Everglades. USS *PC-1186* is bound southwest in Straits. SS *Duc D'Aumale* is bound southeast in Straits, passing Nicholas Channel.

24th July 1944

B-25's FN, FF, FG, FJ, FO, FT, and FU are airborne for Anti-Submarine Patrol Number 2. Convoy of two merchant ship sighted, uneventful patrol, aircraft landed after exercise completion. Grumman Goose GB is airborne for Harbour Island, aircraft landed after exercise completion. Tug USS *Humrick* and three barges are bound south in Straits towards St. Augustine. SS *C. A. Canfield* is bound southeast in Straits. Tanker *Esso Philadelphia* is bound southwest in Straits. USS *SC-1505* is bound south in Straits towards Palm Beach. SS *Western Queen* is bound north near Matanilla Shoal, northwest Bahamas. USS *YTB-338* and tow are bound south in Straits towards Port Everglades. USS *LCI(L)-1012* is bound north in Straits passing Bimini. SS *Britamsea*, SS *Maridal* and Tanker *Paco* are bound north in Straits. **Task Unit TU-20152** and SS *Bostonian* are bound west in Straits.

25th July 1944

B-25's FF, FJ, FH, FO, FT, and FY are airborne for Anti-Submarine Patrol Number 2, uneventful patrol, aircraft landed after exercise completion, convoy of 24 merchant ship sighted by FY. B-25's FB, FP, FL, and FW are airborne for air Exercise Number 1, aircraft landed after exercise completion. Liberator LH is airborne for Radar and navigation exercise, aircraft landed after exercise completion. Grumman Goose GB is airborne for Great Inagua, carrying the Acting Governor of the Colony on government affair. It landed at Great Inagua, and Hidden the following afternoon proceeded to Great Exuma and landed there. USS *John M. Howard* (IX-75) is bound north in Straits, passing Daytona. Tanker *Pan Georgia* is bound south in Straits passing Canaveral. Tug USS *Humrick* towing three barges is bound south in Straits towards Palm Beach. SS *Trimountain* is bound north in Straits passing Palm Beach. SS *Cedar Mountain* north in Straits. USCG cutter *CG-Sagebrush* is bound southeast in Straits. **Convoy GK-820** is bound northwest in Straits passing Nicholas Channel. **Convoy NG-448**, SS *C. A. Canfield*, SS *Point Saint John*, and **Task Unit TU-2961** are bound south in Straits.

26th July 1944

B-25's FG, FH, FO, FT, FY, and FL are airborne for Anti-Submarine Patrol Number 1, aircraft landed after exercise completion. B-25's FW, FD, FM, and FP are airborne for air Exercise Number 1, aircraft landed after exercise completion. Liberator LH is airborne for Radar and navigation exercise, aircraft landed after exercise completion. Grumman Goose GB returned from Great Exuma landed. Tanker *Edward L. Doheny* is bound southeast in northern Straits. SS *Cyrus W. Field* and SS *Thomas Guardia* are bound north in Straits. **Task Unit TU-20157** is bound south in Straits, passing Canaveral. USS *Device* is bound northeast near Matanilla Shoal, northwest Bahamas. SS *William H. McGuffey* is bound south in Straits passing Palm Beach. USS *SC-1058* is bound southwest in Straits. SS *George W. Barnes* is bound northeast in Straits. Tanker *Pan Georgia* is bound west in Straits. USS *YO-69* is bound southeast to Windward Passage. *U-539* exits from northeast of Bahamas towards the east for France.

27th July 1944

B-25's FY, FT, FL, FG, FJ, and FH are airborne for Anti-Submarine Patrol Number 1, aircraft landed after exercise completion. B-25's FY, FW, FB, and FD are airborne for air Exercise Number 5, aircraft landed after exercise completion. Liberator LH is airborne for Radar and navigation exercise, aircraft landed after exercise completion. SS *Herbert L. Pratt* is bound south in Straits, passing St. Augustine. SS *Liebre* and tanker *Petrofuel* are bound south in Straits, passing Canaveral. SS *Josiah Macy* is bound north near Matanilla Shoal, northwest Bahamas. SS *Tampico* is bound south in Straits, towards Palm Beach. SS *Chilore* and tanker *Edward L. Doheny* are bound south in Straits, passing Port Everglades. Tanker *Pan Virginia* is bound north in Straits, towards Port Everglades. Tanker *Esso Little Rock* is bound southwest in Straits, towards Cay Sal Bank. SS *Jacob Luckenbach* is bound south in Straits.

28th July 1944

B-25's FP, FH, FO, FV, FW, and FN are airborne for night navigation exercises aircraft landed after exercise completion. B-25's FW, FS, FC, are FM are airborne for air Exercise Number 5, aircraft landed after exercise completion. Liberator LH is airborne for Radar and navigation exercise, aircraft landed after exercise completion. Tanker *Esso Roanoke* is bound north in Straits with US Navy aircraft joining from N. **Task Unit TU-20159** is bound south in Straits towards Palm Beach. SS *Maridal* is bound south in Straits passing Miami with US Navy aircraft joining from **Convoy KN-328** and tanker *Vacuum* are bound north in Straits passing Port Everglades. USS *PC-1171* is bound north in Straits. USS *YMS-245*, USS *SC-1485*, USS *SC-1510*, SS *Tampico* and tanker *Petrofuel* are bound west in Straits. SS *Delmar* is bound northwest in Straits towards Key West. USS *YAG-21* is bound north in Straits. USS *YAG-19* is bound south from Northwest Providence Channel to Nassau. SS *Samtapa* is bound southeast in Straits. SS *Aquila* is bound south in Straits. Tanker *L. L. Abshire* is bound north, out of Straits. *U-518* enters area west, north of Walker Cay near the destroyer USS *Warrington* convoy.

29th July 1944

B-25's FW, FL, FS, FN, FH, and FM are airborne for night navigation exercise, aircraft landed after exercise completion. B-25's FC, FU, FS, and FL are airborne for air Exercise Number 3, aircraft landed after exercise completion same aircraft airborne later for air Exercise Number 4, aircraft landed after exercise completion. Liberator LH is airborne for Radar and navigation exercise, aircraft landed after exercise completion. SS *Carolinian* and tanker *Trontolite* are bound south in Straits. Tanker *Wahoo Swamp* and tanker *Castana* are bound north in Straits. **Convoy NK-628** is bound south in Straits, passing Palm Beach. USS *SC-1276* is bound northeast passing Northeast Providence Channel. Oiler USS *YO-69* is bound south in Straits. The film *The Battle For Russia* was shown to airmen at the Nassau YMCA, with other light entertainments arranged during evenings.

30th July 1944

Grumman Goose GB is airborne for Bimini, aircraft landed back after exercise completion. SS *Frank J. Sprague* is bound south in Straits, passing St. Augustine. **Task Unit TU-20155** and tanker *Trontolite* are bound south in Straits passing Canaveral. Destroyer escort USS *Edgar G. Chase* (DE-16) and **Convoy Motor Torpedo Boat Squadron-32, Group B** are bound north in Straits. SS *Toteco* and USS *YO-117* are bound south in Straits. SS *Carolinian* is bound south in Straits passing Palm Beach. Tanker *Bohemian Club* is bound north in Straits, passing Miami. SS *Tampico* and **Convoy NK-628** are bound west in Straits, towards Key West. **Convoy GK-821** is bound northwest in Straits, passing Nicholas Channel.

31st July 1944

B-25's FP, FF, FO, FW, FS, and FU airborne for air Exercise Number 7, aircraft landed after exercise completion. B-25 FO is found unserviceable and replaced by FH. B-25's FS, FU, FN, and FM are airborne for air Exercise Number 6, aircraft landed after exercise completion. B-25 FV is airborne for Miami, aircraft landed after exercise completion. Advisory warning regarding a tropical disturbance received by Meteorological Station, so Liberator LS is detailed to be airborne to investigate area where storm was reported. Aircraft LS landed after exercise completion reporting that the storm was observed to be northeast of Bahamas and travelling in a north-westerly direction. Eight B-25 aircraft detailed to be airborne for local night navigation exercise had their flights cancelled due to adverse weather. Four B-24 Liberator aircraft's navigation exercise by Number 3 Squadron substituted for Kingsley Exercise, due to adverse weather were cancelled, and they were detailed to fly on 3rd August. USS *SC-1486* is bound south in Straits towards St. Augustine. USS *SC-1497*, USS *PC-1222*, and tanker *Marathon* are bound south in Straits towards Daytona. Tanker *Walter Jennings* is bound south in Straits towards Canaveral. Tanker *Juan Casiano* is bound south in Straits, passing Canaveral. Tanker *Niobe* is bound north in Straits. USS *YAG-21* is bound east passing Matanilla Shoal, northwest Bahamas. Tanker *Trontolite* is bound south in Straits passing Palm Beach. Submarine USS *Besugo* (SS-321) and USS *SC-1276* are bound west towards Northwest Providence Channel. USCG *CG-Cobb* is bound northeast in Straits from Miami. SS *Frank J. Sprague* southwest in Straits. Tanker *Spidoleine* is bound north in Straits. **Task Unit TU-20155** is bound west in Straits towards Key West. USS *YO-69* is bound west in Straits passing Nicholas Channel. RAF Bahamas Air Service Squadron: Recruiting for this Detachment continued through-out the month. On an average, 12 men pass through the preliminary training each month and are posted to the Wings for trade, training, and other employment as suitable. An establishment of 40 coloured batmen ["soldier-servant"] has been authorized by the RAF Delegation Washington. At the end of the month 12 vacancies had been filled. [The name "native detachment" was dropped, as was perceived to be disincentive]. Defence Training Program: Sten and Browning guns, revolver firing, training and assault courses were provided by officers and [non-commissioned officers, or] NCO's.

1st August 1944

B-25's FF, FK, FH, FT, FP, and FM are airborne for air Exercise Number 7, aircraft landed after exercise completion. B-25's FN, FB, FK, FJ, FU, and FS are airborne for air Exercise Number 6, aircraft landed after exercise completion. Kingsley exercise could not be arranged at completion of Course Number 27, navigation exercise was substituted with homing exercise with Main Field Direction Finding station. Liberator MA, NX, NA, and LV are airborne at different times from Windsor Field, completed navigation and homing exercise as detailed, LV completed depth charge attack against smoke float, aircraft landed after exercise completion. USS *SC-1491* is bound south in Straits. USCG cutter *CG-Cobb*, destroyer USS *Phelps* (DD-360), and SS *Walker D. Hines* are bound north in Straits. SS *Marathon* is bound south in Straits passing Canaveral. **Task Unit TU-20156** is bound south in Straits towards Palm Beach. SS *Glorieta* is bound south near Matanilla Shoal, northwest Bahamas. USS *Pivot* (AM-276) and SS *William L. Halstead* are bound north in Straits near Palm Beach. USCG *CG-Sweetgum* is bound south near Matanilla Shoal, northwest Bahamas. Tug USS *YAG-19* is bound west in Straits west of Andros. Tanker *Esso Philadelphia* is bound north in Straits near Miami. SS *Juan Casiano* and SS *Walter Jennings* are bound southwest in Straits. Submarine USS *Besugo* (SS-321) and USS *SC-647* are bound west in Straits. USS *Woonsocket* (PF-32) is bound north in Straits near Cay Sal Bank. SS *George C. Childress* is bound north in Straits.

2nd August 1944

B-25's FM, FN, FO, FS, FH, and FF are airborne for night navigation exercise, aircraft landed after exercise completion. B-25's FN, FB, FF, FH, FU airborne for air Exercise Number 6, aircraft landed after exercise completion. Grumman Goose GB is airborne for Bimini, aircraft landed after exercise completion. B-25 FV returned from Miami, landed at Oakes Field. Liberator NO, NX, NA, MA airborne at different times for Windsor Field, completed navigation and homing exercise as detailed, aircraft landed after exercise completion. Ship of 3,000 tons sighted, 3-island type vessel of 3,000 tons sighted. Tanker *R. P. Smith* is bound south in Straits. Landing craft flotilla USS *LCI(L)-352*, USS *LCI(L)-640*, and USS *LCI(L)-641* are bound south in Straits, passing Daytona. USS *SC-1491* is bound south in Straits, passing Palm Beach. SS *John H. Drummond* and USS *Diploma* (AM-221) are bound north in Straits near Palm Beach. **Convoy KN-329** is bound north in Straits, near Port Everglades. Cuban gunboat *CS-Patria* is bound south in Straits from Miami. Tanker *Walter Jennings* and tanker *Marathon* are bound southwest in Straits. Tanker *Jamestown*, SS *Cape Frio*, tanker *Mount Royal Park* and destroyer escort USS *Doyle C. Barnes* (DE-353) are bound north in Straits. **Convoy KG-722** is bound southeast in Old Bahama Channel with US Navy plane escorting. SS *George W. Ather* is bound southeast in Old Bahama Channel. USS *Menkar* (AK-123) and tanker *Paco* are bound south in Straits.

3rd August 1944

B-25's FB, FO, FH, and FF are airborne on a night navigation exercise, aircraft landed after exercise completion, then same aircraft did similar exercise again later. B-25's FF and FE were substituted for B-25's FB and FF. B-25's FH, FM, FO, and FK are airborne for air Exercise Number 3, aircraft landed after exercise completion. B-25's FB, FF, FP, FO, FY, and FU are airborne for air Exercise Number 6, aircraft landed after exercise completion. B-25's CL, AU, and AD are airborne for local navigation exercise, aircraft landed after exercise completion B-25's CL, DF, and AD are airborne for similar exercise later, aircraft landed after exercise completion. Liberator LK, LR, NO, and LY are airborne at different times for Windsor Field, completed navigation and homing exercise as detailed. Then LR sighted a convoy of 20 merchant ships and five escorts, with no signals exchanged, LY sighted, convoy of 21 merchant ships and five escorts, and aircraft landed after exercise completion. Tanker *James Cook* and tanker *Arlington Berch Park* are bound south in Straits. Tanker *Spidoleine* is bound south in Straits, passing Canaveral. USS *YAG-21* is bound west in Straits. **Convoy NK-629** is bound south in Straits, passing Palm Beach. Tanker *R. P. Smith* is bound south in Straits passing Port Everglades. Tanker *C. A. Canfield* is bound north in Straits, towards Miami. Landing craft USS *LCI(L)-352*, USS *LCI(L)-640*, and USS *LCI(L)-641* are bound west in Straits. USS *PC-594* is bound southeast in Straits. USS *YAG-28* is bound south, towards Nassau. Tanker *Swiftscout* is bound south in Straits. **Convoy GN-148** is bound north in Straits.

4th August 1944

B-25's CF, CL, BD, DF, CO, and BJ are airborne for local night navigation exercise, aircraft landed after exercise completion. Liberator LA is airborne from Windsor Field for New York with a stretcher patient, and landed in New York. Grumman Goose GB is airborne for Governor's Harbour, and the aircraft landed after exercise completion. A Beechcraft aircraft landed in Washington with Group Captain Spaight and Group Captain Lotte. B-25's FY, FL, FF, FO, and FK are airborne for compass swinging exercise, aircraft landed after exercise completion. Liberator LA is airborne for New York, and landed there. Two B-25's are delivered from manufacturers in the US. Tanker *Virginia*, USS *LSM-258*, and tanker *Tustem* are bound south in Straits, passing Canaveral. Tanker *Esso Little Rock*, destroyer escort USS *Douglas A. Munro* (DE-422), and SS *Henry Wells* are bound north in Straits. USS *SC-1485* and USS *SC-1510* are bound north in Straits, towards Palm Beach. USS *YAG-32* is bound southwest in Straits towards Port Everglades. USS *YSD-66* is bound north from Straits. Tanker *James Cook* is bound southwest in Straits. Russian RU *SC-1517* and tanker *Speidoline* are bound southwest in Straits. Tanker *A. S. Hansen* is bound west in Straits. **Convoy GK-822** is bound northwest in Straits towards Key West. **Convoy NG-450** is bound south in Straits.

5th August 1944

B-25's FG, FY, FJ, FO, FK, and FS are airborne for night navigation exercise, aircraft landed after exercise completion. B-25's FY, FF, FO, are FJ airborne for air Exercise Number 14, aircraft landed after exercise completion. B-25's FS, FD, FK, FG, FP, and FJ are airborne for compass swinging exercise, aircraft landed after exercise completion. A Beechcraft airborne for Morrison Field, *en route* to Washington, is returning with Group Captains Spaight and Lotte. Tanker *Pan Virginia* is bound south in Straits, passing Daytona. Tug USS *YTB-386* is bound south in Straits, passing Canaveral. USS *LSM-144* is bound south in Straits, towards Palm Beach. Destroyer escort USS *Eert* (DE-768) is bound northeast, passing Matanilla Shoal, northwest Bahamas. Tanker *Bahram* is bound north in Straits, near Palm Beach. SS *Tustem* and USS *PC-1187* are bound south in Straits, passing Port Everglades. Tanker *Gulfmaid*, tanker *Cities Service Ohio*, tug USS *YTB-338* and tows are bound southwest in Straits. SS *White Bird Canyon* is bound north in Straits, with US Navy aircraft joining it. USS *YAG-28* is bound northeast in Straits, towards Bimini. USS *YF-443*, SS *Mount Bruce Park*, tanker *Olney*, and tanker *L. V. Stanford* are bound north in Straits. SS *Virginia* is bound west in Straits, towards Key West. Tanker *Britamsea* is bound south in Straits.

6th August 1944

Grumman Goose GA is airborne for Bimini, then aircraft landed after exercise completion. A US-made B-25 is delivered from Morrison Field. SS *Trimountain* and SS *Cities Service Koolmotor* are bound south in Straits. SS *Morton Prince* is bound south in Straits, towards Canaveral. **Task Unit TU-201511** (two USS LCI(L)'s) is bound south in Straits, towards Canaveral. SS *Swiftlight* is bound south in Straits, passing Palm Beach. SS *Kettle Creek* and SS *Manvilla* are bound south, passing Matanilla Shoal, northwest Bahamas. USS *YMS-365* is bound north from Nassau to Northwest Providence Channel. SS *Maridal* is bound north in Straits. Tanker *Pan Virginia* is bound southwest in Straits.

7th August 1944

B-25's FT, FF, FM, FW, FP, and FG are airborne for Anti-Submarine Patrol Number 2, aircraft landed after exercise completion due to faulty navigation equipment, so B-25 FF returned to base later than its given ETA. B-25's FO, FG, FB, FT, FK, and FL are airborne for Exercise Number 1, aircraft landed after exercise completion. Liberator LH is airborne for Radar and navigation exercise, aircraft landed after exercise completion. Request received from Comgulfseafron for aircraft to sweep given area. B-25 FT is airborne, landed back at base after uneventful sweep, sundry units of shipping sighted after patrol. Grumman Goose GB is airborne for Cat Cay, aircraft landed after exercise completion. SS *Argon* is bound south in Straits, towards Canaveral. **Task Unit TU-20153** (two USS LSM's), SS *Paul H. Harwood*, and SS *Trimountain* are bound south in Straits passing Canaveral. Submarine USS *Seacat* (SS-399) and USS *SC-653* are bound northwest in Straits near Canaveral with a US Navy plane escorting. Tanker *Pan North Carolina*, SS *Cities Service Koolmotor*, and USS *PC-577* are bound south in Straits, passing Palm Beach. SS *Solsten* is bound north in Straits, passing Palm Beach. **Task Unit TU-201514**; USS *Rushmore* (LSD-14), destroyer escort USS *Gillion* (DE-508), and SS *Morton Prince* are bound south in Straits. **Convoy KN-330**, SS *Sam Tampa*, and USS *Acushnet* (ATO-63) are bound north in Straits near Miami. USS *YAG-21* is bound northwest in Straits towards Miami. Oiler USS *YO-142* is bound northeast in Straits from Miami. **Task Unit TU-201511** (two USS LCI(L)'s) is bound west, in Straits.

8th August 1944

Liberator LA is airborne from New York, and landed. B-25's FF, FP, FT, FJ, and FY are airborne for Anti-Submarine Patrol Number 1, aircraft landed after exercise completion. B-25's FC, FL, FD, FG, and FV are airborne for air Exercise Number 1, aircraft landed after exercise completion. Liberator LH is airborne for Radar and navigation exercise, aircraft landed after exercise completion. USS *LSM-137* and **Task Unit TU-201516** are bound south in Straits passing Canaveral. Tanker *Hadnot* and SS *Josiah Macy* are bound south in Straits, passing Palm Beach. SS *Joshua Hendy* is bound north in Straits, near Palm Beach. SS *Luuluro* is north in Straits, passing Palm Beach. Tanker *Petrofuel* is bound north in Straits, passing Port Everglades. SS *John J. Crittendon* is bound north in Straits, passing Port Everglades. Army dredge USS *San Pablo* and tanker *Cities Service Koolmotor* are bound south in Straits, passing Miami. USS *ATA-174* and USS *YF-443* are bound north in Straits, passing Miami. USS *YAG-28* is bound northwest towards Northwest Providence Channel. Tanker *Trimountain* and tanker *Paul H. Hardwood* are bound southwest in Straits. Tanker *Argon* and tanker *Pan North Carolina* are bound west in Straits.

9th August 1944

B-25's FU, FD, FV, FY, FG, and FC are airborne for air Exercise Number 3, aircraft landed after exercise completion same aircraft airborne later for air Exercise Number 4, aircraft landed after exercise completion. Tanker *Edward L. Shea* is bound north in Straits near Palm Beach. **Convoy NK-630** and **Task Unit TU-03112** are bound south in Straits. SS *Zona Gale* and SS *O. B. Sorensen* are bound south in Straits. USS *LSM-137*, SS *Josiah Macy*, and Tanker *Hadnot* are bound west in Straits. **Convoy GK-823** is bound northwest in Straits passing Nicholas Channel with a US Navy plane escorting. Destroyer escorts USS *DE-423*, USS *DE-354*, and the SS *Angus McDonald* are bound north in Straits.

10th August 1944

B-25's FF, FH, FY, and FU are airborne for night navigation exercise, aircraft landed after exercise completion. B-25's FF, FJ, FU, FM, FF, and FD are airborne for air Exercise Number 3, aircraft landed after exercise completion. B-25's FF, FK, FU, FM, and FD are airborne for air Exercise Number 4, aircraft landed after exercise completion. Liberator LH is airborne for Radar and navigation exercise, aircraft landed after exercise completion. Tanker *L. V. Stanford* is bound south in Straits passing Daytona. USS *PC-1549*, USS *LSM-139* and tanker *Clio* are bound south in Straits, passing Canaveral. USCG *CG-83505* is bound east in Straits, from Palm Beach. Tanker *Vacuum* is bound south in Straits towards Palm Beach. Tanker *Clam* is bound south in Straits, passing Palm Beach. Tanker *Bohemian Club* is bound south in Straits, passing Port Everglades. Oiler USS *YO-142* is bound west passing Northwest Providence Channel. Tanker *Axtell J. Byles* and tanker *Castana* are bound south in Straits. USS *PC-823* is bound north in Straits, to Miami. USS *SC-1517*, Russian RU *SC-1504*, and tanker *El Caney* are bound north in Straits. SS *Cape Nun* is bound southeast in Straits passing Nicholas Channel.

11th August 1944

B-25's FM, FB, FF, FP, FT, and FY are airborne for Anti-Submarine Patrol Number 2, aircraft landed after exercise completion. Six B-25 aircraft detailed to be airborne for air Exercise Number 5, cancelled due to adverse weather. USS *SC-1517* and Russian RU *SC-1504* are bound north in Straits, towards Daytona. SS *Henry L. Ellsworth* is bound south in Straits passing Canaveral. Tug *Margot Moran* with tows escorted by USCG *CG-83418* is bound south in Straits towards Palm Beach. SS *Petter II* is bound south in Straits passing

Palm Beach. USCG *CG-Cobb* and tanker *L. V. Stanford* are bound south in Straits. SS *Clio*, USS *SC-1502*, and Russian RU *SC-1477* are bound west in Straits. Destroyer escort USS *Parle* (DE-708) is bound north in Straits. Destroyer USS *Amesbury* (DE-66) is bound west passing Northeast Providence Channel. *The Liberator,* the RAF station newspaper, has been organized by a number of enthusiastic journalist airmen of this unit, under the direction of the Education Officer. The First Edition was printed and published on August 11th. The printing is carried out by members of the editorial staff on the presses of the local daily [the *Nassau Guardian*]. The paper is sold for a penny, the additional cost of printing being borne by the advertisers. The editors invite contributions from members of the services, these have included articles, poems, observations and editorials on world affairs and sport [Sir Arthur A. Foulkes, who had recently arrived from Inagua, worked on the paper with Mrs. Hicks, an RAF officer. He later became a Governor General].

12th August 1944

Request received from Comgulfseafron to escort submarine USS *Atul* (SS-403) through Northwest Providence Channel, eastbound. B-25 FY is airborne, submarine located, carried out air escort, aircraft landed after exercise completion. B-25 FJ is airborne, relief sighted submarine, took over escort, aircraft landed after exercise completion. B-25 FU is airborne for Anti-Submarine Patrol Number 2, steamship sighted, aircraft landed after exercise completion. B-25 FS is airborne for Anti-Submarine Patrol Number 2, two tankers sighted, aircraft landed after exercise completion. Grumman Goose GB airborne for Harbour Island, aircraft landed after exercise completion. B-25 FY, FK, FT, FB, and FF are airborne for air Exercise Number 5, aircraft landed after exercise completion. Tug USS *ATR-61*, **Task Unit TU-20152**, USS *LSM-260*, tanker *Bostonian*, and whale oil tanker *Suderoy* are bound south in Straits passing Canaveral. Tanker *Herbert L. Pratt* is bound north in Straits near Canaveral. USS *SC-1508* is bound south in Straits passing Palm Beach. **Convoy KN-331** is bound north in Straits near Palm Beach. Tanker *Oscar F. Barrett* is bound north in Straits passing Port Everglades. USS *YF-443* is bound southwest in Straits towards Port Everglades. USS *YAG-21* is bound southeast in Straits from Miami. Tug *Margot Moran* is bound south in Straits. USS *PC-489* is bound southeast in Straits. Tanker *Esso Springfield* is bound west in Straits. SS *W. E. Ogilvie* and escorts are bound southeast in Straits passing Nicholas Channel. USS *YTB-383* is bound southeast towards Old Bahama Channel. Oil barge USS *YOG-6* is bound northwest in Old Bahama Channel. **Convoy KG-724** is bound southeast in Old Bahama Channel, with US Navy plane escorting. Submarine USS *Atule* (SS-403) and USS *PC-1144* escorting are bound east, passing through Northeast Providence Channel. USS *SC-1063* is bound east passing Northeast Providence Channel, with US Navy plane escorting.

13th August 1944

B-25's FF, FO, FN, FP, and FH are airborne for night navigation exercise, aircraft landed after exercise completion, airborne again later for completion of same exercise. Liberator LV is airborne from Windsor Field for Kingsley Exercise, steamship sighted, contact established with convoy, homing exercise "B" commenced, convoy of 21 ships sighted, didn't acknowledge signals from aircraft recognition procedure with Kingsley

Exercise convoy, patrol Cobra completed, aircraft attacked with a depth charge, aircraft landed after exercise completion. Liberator ND is airborne from Windsor Field for Kingsley Exercise as first relief, radio contact established, homing procedure "B" commenced, from bearing received two frigates sighted, patrol Alligator and Mamba completed, aircraft attacked with a depth charge aircraft landed after exercise completion. Liberator NX is airborne from Windsor Field for Kingsley Exercise as second relief, radio contact established, homing procedure "B" commenced, from bearing received convoy located, signals exchanged, patrols Viper and Mamba completed, aircraft attacked with a depth charge from smoke float dropped by SOE, merchant ship sighted, factory ship sighted, **Convoy GN** of 13 ships sighted, aircraft landed after exercise completion. Liberator MD is airborne from Windsor Field for Kingsley Exercise as third relief, homing procedure "B" commenced, convoy located, signals exchanged, carried out dummy submarine attacks against smoke float dropped by SOE, detailed exercises carried out, **Convoy GN** of 13 ships sighted, aircraft landed after exercise completion. Liberator LR airborne from Windsor Field for Kingsley Exercise as fourth relief, radio contact established, convoy located, signals exchanged, patrol Alligator and Crocodile completed, aircraft attacked with a depth charge, aircraft landed after exercise completion. B-25's FO, FF, FB, FJ, FN, and FY are airborne for air Exercise Number 5, aircraft landed after exercise completion. SS *C. A. Canfield* is bound south in Straits near Daytona. SS *Cathlamet* and USS *YP-448* are bound south in Straits towards Canaveral. **Convoy NK-631** is bound south in Straits towards Port Everglades with US Navy plane escorting. Tanker *L. L. Abshire* is bound south in Straits towards Port Everglades. SS *Marathon* is bound north in Straits towards Palm Beach. SS *Mauvilla* and SS *James Cook* are bound north in Straits near Bimini. SS *Jamestown* is bound south in Straits passing Miami. Destroyer escort USS *Earl K. Olsen* (DE-765) is bound east in Straits. SS *Mount Royal Park*, SS *Moose Mountain Park* and SS *Rhea* are bound south in Straits. Tanker *R. P. Smith*, SS *Walter Jennings*, SS *Pueblo* Number 2, and **Convoy GN-150** are bound north in Straits.

14th August 1944

B-25's FH, FO, FU, and FT are airborne for air Exercise Number 7, aircraft landed after exercise completion. Liberator NA airborne from Windsor Field for Kingsley Exercise, recognition signals exchanged, completed Alligator patrol under direction of SOE, aircraft landed after exercise completion. Liberator NO is airborne from Windsor Field for Kingsley Exercise as first relief, radio contact established, homing procedure "B" commenced, frigates located, patrols Python and Frog completed under direction of SOE, merchant ship sighted, aircraft landed after exercise completion. Liberator LK is airborne from Windsor Field for Kingsley Exercise as second relief, homing procedure "B" commenced, frigates located, signals exchanged, carried out dummy submarine attack, aircraft landed after exercise completion. Liberator LV is airborne from Windsor Field for Kingsley Exercise as third relief, homing procedure "B" commenced, convoy located, patrols Cobra completed under direction of SOE, two flares dropped for AA for practice by ship's gunners, smoke float attacked with one depth charge, aircraft landed after exercise completion. Liberator LR airborne from Windsor Field for Kingsley Exercise fourth relief, merchant ship of Liberty type sighted, homing procedure "B" commenced, convoy located, patrols Cobra and Alligator completed under direction of SOE, smoke float attacked with one depth charge tanker sighted, aircraft landed after exercise completion. SS *Ferdinand R. Hassler* and SS *Kettle Creek* are bound north in Straits passing Palm Beach. SS *Indiana* is bound south in Straits near Miami. SS *Cathlamet* and SS *C. A. Canfield* are bound southwest in Straits. SS *Rhea,* tug USS *ATR-61*, and **Convoy NK-631** are bound west in Straits. **Convoy GK-824** is bound northwest in Straits towards Key West. SS *24 De Mayo* is bound west in Straits passing Nicholas Channel. SS *San Gaspar* and SS *W. W. Mills* are bound south in Straits. Tanker *E. Bidwell* is bound north in Straits.

15th August 1944

B-25's FC, FB, FG, FH, FK, and FU are airborne for air Exercise Number 7, aircraft landed after exercise completion. B-25's FO, FH, FY, FS, FV, and FB are airborne for air Exercise Number 6, aircraft landed after exercise completion. Liberator LH is airborne for Radar and navigation exercise, aircraft landed after exercise completion. SS *Eano* is bound south in Straits towards Canaveral. Tanker *Beaconoil*, SS *San Nelito*, and SS *Christopher* are bound south in Straits, passing Canaveral. **Task Unit TU-2053** is bound south, towards Palm Beach. Tanker *A. S. Hansen* is bound north in Straits passing Palm Beach. USS *PC-1564* is bound southwest in Straits. SS *Medina* is bound south in Straits. Troop transport USS *USAT-George Washington*, destroyer escort USS *Jaccard* (DE-355), and USS *PC-1145* are bound north in Straits. SS *Tampico* is bound north in Straits passing Port Everglades. SS *Bostonian* is bound west in Straits.

16th August 1944

B-25's FV, FH, FN, and FS are airborne for night navigation exercise, aircraft landed after exercise completion. B-25's BJ, AD, and CJ are airborne for local navigation exercise, aircraft landed after exercise completion, same aircraft completed similar exercises again later. B-25's FB, FC, FJ, FO, and FP are airborne for air Exercise Number 6, aircraft landed after exercise completion. Liberator LH is airborne for Radar and navigation exercise, aircraft landed after exercise completion. Tanker *Empire Chief* is bound south in Straits. Tanker *Gulf of Venezuela*, tanker *R. P. Smith*, tanker *Liebre*, military cargo ship USS *Syrma* (AK-134) and USCG cutter *CG-Icarus* are bound south in Straits passing Canaveral. SS *George L. Curry* is bound north in Straits towards Palm Beach. Tanker *Beaconoil* and tanker *San Melito* are bound west in Straits. Tanker *Morton* Prince, USS *YAG-21*, and USS *PC-1564* are bound north in Straits. SS *Edwin A. Stevens* is bound northwest in Straits passing Nicholas Channel. Cuban *CS-33* and Cuban *CS-34* are bound west in Straits passing Nicholas Channel. Tanker *Gertrude Kellogg* is bound southeast in Straits towards Old Bahama Channel. SS *William H. Seward* and SS *Washington*, tanker *W. G. Warden*, USS *PGM-9*, tanker *Washita*, tanker *Eano*, tanker *Quemado Lake*, and SS *Wilwood Park* are bound southwest in Straits.

17th August 1944

B-25's FB, FH, FK, FN, FP, FO, and FV are airborne for local night navigation, aircraft landed after exercise completion. B-25's BJ, AS, and AM are airborne for local navigation exercise, aircraft landed after exercise completion landed at Oakes Field.

B-25's FH, FT, FN, FC, FV, FK, and FS are airborne for air Exercise Number 6, aircraft landed after exercise completion. USS *PC-1220* and SS *Fort George* are bound southeast in Straits near Daytona. USS *SC-1502* and USS *SC-1477* are bound northwest in Straits passing Canaveral. Tanker *Pan Virginia*, SS *Tustem* and **Convoy KN-332** are bound north in Straits. **Task Unit TU-20156** is bound south in Straits towards Palm Beach. USS *Pledge* (AM-277) and cargo ship USS *Syrma* (AK-134) are bound north near Matanilla Shoal, northwest Bahamas. Tanker *Alpha* is bound north in Straits near Port Everglades. Oil barge USS *YOG-6* is bound east in Straits from Port Everglades. SS *Empire Chief* is bound southwest in Straits. Tug *Margot Moran* and tanker *R. P. Smith* are bound west in Straits. SS *Kobkirk's Hill* and troop transport USS *USAT-Copiapo* are bound southwest in Straits passing Cay Sal Bank. SS *Tide Penn*, SS *Solsten*, and USS *Diphada* (AKA-59) are bound south in Straits.

18th August 1944

B-25's FT, FV, FH, FO, FL, and FK are airborne for night navigation exercise, aircraft landed after exercise completion. B-25's FB, FC, FH, FN, FT, and FV are airborne for compass swinging exercise, aircraft landed after exercise completion. Tanker *Pan Georgia* is bound south in Straits. USS *SC-685*, SS *Bahram*, and SS *Phoebus* are bound south in Straits towards Palm Beach. Tanker *Pan North Carolina* and SS *Swiftlight* are bound north in Straits. Tanker *Edward L. Shea* and SS *Fort George* are bound south in Straits passing Port Everglades. USS *PCE-871* is bound northwest in Straits.

19th August 1944

B-25's BJ, AD, and BC are airborne for local night navigation exercise, aircraft landed after exercise completion. B-25's AS, CG, and CO are airborne for similar exercises, aircraft landed after exercise completion. B-25's FS, FF, FM, FN, FK, and FG are airborne for compass swinging exercise, aircraft landed after exercise completion. Grumman Goose GA airborne transit flight to San Salvador to take Station Medical Officer for inspection to the island, aircraft landed after exercise completion. Tanker *Esso Springfield*, tug USS *ATR-7*, and SS *Argon* are bound north in Straits. **Task Unit TU-20181** is bound south in Straits towards Palm Beach. Tanker *A. S. Hansen* south in Straits passing Palm Beach. Tanker *E. R. Kemp* north in Straits passing Palm Beach. Tanker *Pan Georgia* is bound south Port Everglades. Tanker *L. V. Stanford* is bound southwest in Straits. Tanker *Vacuum* is bound northeast in Straits. Tanker *Bahram* is bound west in Straits. **Convoy GK-825** is bound northwest in Straits passing Nicholas Channel.

20th August 1944

SS *Oldham* is bound south in Straits passing St. Augustine. USCG *CG-Pandora* is bound south in Straits towards Canaveral. **Task Unit TU-20191** is bound south in Straits passing Canaveral. Tanker *Gulfhawk* is bound south in Straits, near Canaveral. Tanker *Hadnot* is bound north in Straits near Miami. Oil barge USS *YOG-6* is bound west in Straits. Tanker *Chiloil* and tanker *Cities Service Koolmotor* are bound north in Straits. SS *Svealand* is bound south in Straits.

21st August 1944

B-25's FH, FC, FP, FT, FK, and FO are airborne for Anti-Submarine Patrol Number 1, aircraft landed after exercise completion. B-2's FW, FM, FV, and FS are airborne for air Exercise Number 1, aircraft landed after exercise completion.

Liberator LH is airborne for Radar and navigation exercise, aircraft landed after exercise completion. SS *Tobias E. Stansbury* is bound south in Straits. Tanker *Chiloil*, SS *Indiana*, tanker *Cities Service Koolmotor*, and SS *Vistula* are bound north in Straits. USS *YMS-360* and USS *YMS-353* are bound south in Straits, passing Daytona. SS *Walter Jennings* is bound south in Straits near Canaveral. Tanker *L. V. Stanford* is bound north in Straits near Palm Beach. SS *Oldham* is bound south in Straits passing Palm Beach. SS *Oscar F. Barrett* is bound south in Straits passing Port Everglades. USS *YDG-1* is bound north and tanker *Clio* are bound north in Straits passing Port Everglades. Tanker *Petrofuel* is bound south in Straits passing Miami. SS *Henry L. Ellsworth* is bound north in Straits towards Miami. **Task Unit TU-0532** is bound northwest in Straits towards Miami. SS *Mexican* and USCG cutter *CG-Pandora* are bound southwest in Straits.

22nd August 1944

B-25's FF, FH, FK, FT, FS, and FU are airborne for Anti-Submarine Patrol Number 1, aircraft landed after exercise completion. Grumman Goose GA is airborne for Harbour Island, aircraft landed after exercise completion. B-25's FP, FM, FC, FW, and FK are airborne for air Exercise Number 1, aircraft landed after exercise completion. Liberator LH airborne for Radar and navigation exercise, aircraft landed after exercise completion. SS *Clam* is bound north in Straits passing Palm Beach. SS *Josiah Macy* is bound north in Straits towards Palm Beach. Oil barge USS *YOG-6* is bound northeast in Straits towards Northwest Providence Channel. SS *Bennington*, SS *Hanging Rock* are bound south in Straits passing Bimini. Destroyer escort USS *Haas* (DE-424) is bound north in Straits near Bimini. SS *William Crompton*, SS *Fishers Hill*, and USS *YF-443* are bound south in Straits passing Port Everglades. **Task Unit TU-0291** is bound south in Straits. SS *Tobias E. Stansbury* and SS *Sweetwater* are bound southwest in Straits. USS *Van Valkenburg* is bound north in Straits. **Convoy KG-726** is bound southeast in Straits.

23rd August 1944

B-25's CH, DT, SN, and M are airborne for Anti-Submarine Patrol Number 2, aircraft landed after exercise completion, sundry small ships sighted in patrol vicinity, two tankers and two merchant ship sighted, one ship was a 3-island steam ship. B-25's FU, FL, FW, and FJ are airborne for air Exercise Number 5, aircraft landed after exercise completion. Liberator LH is airborne for Radar and navigation exercise, aircraft landed after exercise completion. SS *Cavelier* is bound south in Straits. SS *Empire Viscount* is bound south in Straits towards Canaveral. **Task Unit TU-20192** is bound south in Straits towards Palm Beach. **Task Unit TU-0532**, **Convoy KN-333** and SS *Juan Casiano* are bound northeast in Straits. SS *The Cabins* is bound northwest in Straits near Bimini. Whale oil tanker *Suderoy* is bound north in Straits near Bimini. USS *SC-1493*, Russian RU *SC-1482* and SS *Walter Jennings* are bound southwest in Straits.

24th August 1944

B-25's FN, FO, FT, and FH are airborne for night navigation exercise, aircraft landed after exercise completion. Signal received from Comgulfseafron that a Mariner aircraft was on the sea, position given, crew safe. Assistance was requested from Harbour Island. B-25 FC is airborne for night navigation exercise, diverted to reported position, signal sent to ASR launch at Harbour Island, instructing it to put to sea to locate Mariner aircraft. B-25 located Mariner, B-25 FC returned to base and landed, US Navy depot ship stationed at Royal Island, near Spanish Wells, Eleuthera, took over from ASR launch HMS *P-339* as its tow-rope was unable to stand the strain, HMS *P-339* returned to base. B-25's FF, FC, FT, FV, and FH are airborne for air Exercise Number 5, aircraft landed after exercise completion. Liberator LH airborne for Radar and navigation exercise, aircraft landed after exercise completion. Grumman Goose GA is airborne for San Salvador, aircraft landed after exercise completion. **Task Unit TU-0532** is bound north in Straits passing Canaveral. USS *LSM-151* and tanker *Morton Prince* are bound south in Straits, towards Palm Beach. Tanker *The Cabins*, tanker *Washita*, tanker *Rhea*, tanker *Christopher L. Sholes*, and tanker *Quemado Lake* are bound north in Straits. USS *PC-1147* is bound north in Straits, passing Palm Beach. USS *YMS-353* and USS *YMS-360* are bound southeast in Straits. USS *YAG-28* is bound south in Straits, near Andros. USS *NY-116* is bound south in Straits with US Navy plane escorting. Tanker *Thomas H. Wheeler* and SS *Chomedy* are bound south in Straits.

25th August 1944

B-25's FF, FN, FL, and FO are airborne for air Exercise Number 3, aircraft landed after exercise completion. Liberator LH is airborne for Radar and navigation exercise, aircraft landed after exercise completion. B-25's FJ, FH, FO, and FL are airborne for air Exercise Number 4, aircraft landed after exercise completion. USS *LSM-148* and SS *Prometheus* are bound south in Straits towards Canaveral. SS *Hobkirk's Hill* is bound north in Straits. SS *Bidwell* and USS *PCS-1383* are bound south in Straits towards Palm Beach. SS *Tustem* is bound south in Straits passing Palm Beach. SS *Jamestown* is bound north in Straits passing Palm Beach. Tanker *Gulfcrest* is bound north in Straits.

26th August 1944

B-25's FK, FL, FD, FN, FB, and FV airborne for local night navigation exercise, aircraft landed after exercise completion same aircraft these exercises again later. B-25's FC and FG are airborne for Anti-Submarine Patrol Number 2, aircraft landed after exercise completion. B-25's FD and FN are airborne for air Exercise Number 7, aircraft landed after exercise completion. B-25's FU, FL, FM, FV, and FW are airborne for air Exercise Number 3, aircraft landed after exercise completion. B-25's FF, FL, FM, FV, and FW are airborne for air Exercise Number 4, aircraft landed after exercise completion. Liberator LH is airborne for Radar and navigation exercise, aircraft landed after exercise completion. Request received from Comgulfseafron for aircraft to escort submarine USS *Caiman* (SS-323) through the Northwest Providence Channel. B-25 FG is airborne for Anti-submarine sweep, submarine located, escorted, nothing to report from exercise, aircraft landed after exercise completion. SS *Lillian Wald* is bound south in Straits towards Daytona. Tanker *Pan North Carolina* is bound south in Straits towards Canaveral. **Task Unit TU-20193** is bound south in Straits passing Canaveral. SS *Touraine* and SS *Durango* are bound south in Straits passing Palm Beach. SS *Empire Chief* is bound north in Straits passing Palm Beach. SS *Prometheus* is bound south in Straits towards Port Everglades. SS *Nevada* is bound south in Straits passing Port Everglades. SS *Bidwell* is bound west in Straits. USS *PC-1147* and SS *Carlos J. Finlay* are bound north in Straits.

27th August 1944

Grumman Goose GB is airborne for Palisadoes, Jamaica was to refuel at Guantanamo. SS *New York* is bound south in Straits. SS *Swiftscout*, SS *Cities Service Oklahoma*, SS *Gabriel Franchere*, SS *W. H. Libby* are bound south in Straits towards Canaveral. SS *W. F. Burdell* is bound northwest in Straits towards Canaveral. USS *YO-109*, USS *LSM-264*, **Convoy NK-634**, and SS *Watertown* are bound south in Straits towards Palm Beach. Free French Navy submarine FFS *Amazone* escorted by USS *SC-1059* is bound north in Straits passing Palm Beach. Tanker *Eastern Sun* is bound south in Straits passing Palm Beach. Tanker *Pan North Carolina* is bound south in Straits towards Miami. USS *YF-994* is bound southeast in Straits from Miami. USS *YP-580* is bound west in Straits towards Miami. USS *Caiman* (SS-323) escorted by USS *SC-647* is bound west towards Northwest Providence Channel. SS *Lillian Wald* is bound southwest in Straits. USS *PC-1547* is bound west in Straits. Tug USS *YTB-281* and tow are bound northeast in Straits.

28th August 1944

B-25's FJ and FL are airborne for Anti-Submarine Patrol Number 2, aircraft landed after exercise completion. B-25's FB, FF, and FM is airborne for air Exercise Number 7, aircraft landed after exercise completion. B-25's FS, FL, FF, FV, and FN are airborne for air Exercise Number 6, aircraft landed after exercise completion. Liberators MC, MD, MF, and LR are airborne from Windsor Field for navigation exercise which was the substitute of Kingsley exercise, carried out homing exercise "B" prior to landing, all dropped depth charge against smoke float on return, leg to base navigation and homing exercise successfully carried out. USS *ATR-16* with tows and escorts are bound south in Straits passing St. Augustine. USS *SC-1059* is bound south in Straits passing Daytona. USS *SC-1493* is bound north in Straits passing Daytona. USS *YTB-281* towing USS *YF-652* is bound north in Straits passing Canaveral. **Task Unit TU-20194** is bound south in Straits towards Palm Beach. **Convoy KN-334** is bound north in Straits near Port Everglades. Tanker *Cities Service Oklahoma* and tanker *Pan* Virginia, tanker *Swiftscout*, tanker *New* York and SS *Tidewater* are bound south in Straits. Tanker *Blackwater* is bound southwest in Straits. USS *ATR-82* is bound southeast in Straits near Miami. SS *Empire Gannet* is bound north in Straits. Oiler USS *YO-69* is bound south in Nicholas Channel. USS *YF-294* is bound southeast in Old Bahama Channel. **Convoy GK-827** is bound northwest in Old Bahama Channel.

29th August 1944

B-25's FD, FG, FF, FV, and FW are airborne for air Exercise Number 7, aircraft landed after exercise completion. B-25's FK, FD, FG, FL, and FJ are airborne for air Exercise Number 6, aircraft landed after exercise completion. B-25's FF and FN are airborne for air Exercise Number 7, aircraft landed after exercise completion. B-24's ND, LU, MD, and MF of Number 3, Squadron are airborne from Windsor Field for navigation exercise, carried out homing exercise "B," aircraft landed after exercise completion. Oiler USS *YO-118* is bound south in Straits passing St. Augustine. USS *PYC-4* is bound north in Straits passing Canaveral. SS *Phoebus* and SS *Empire Gannet* are bound north in Straits. Tug USS *ATR-16* and tows and **Task Unit TU-20195** are bound south in Straits towards Palm Beach. USCG *CG-83434* is bound southwest in Straits. SS *Tidewater* is bound west in Straits near Cay Sal Bank. SS *Indiana* is bound south*west* in Straits towards Cay Sal Bank.

30th August 1944

B-25's FD, FL, FO, FF, FN, and FJ are airborne for air Exercise Number 6, aircraft landed after exercise completion. B-25's AU, AR, AS, and CF are airborne for local navigation exercise, aircraft landed after exercise completion. B-24's LR, ND, and LU are airborne from Windsor Field navigation exercise carried out homing exercise "B," aircraft landed after exercise completion. USS *SC-1499* and USS *SC-1506* are bound south in Straits towards Canaveral. SS *Sweetwater* is bound north in Straits. USS *YO-118* is bound south in Straits passing Palm Beach. Tug USS *Humrick* and tows are bound north in Straits. USS *PC-1142* is bound west in Straits.

31st August 1944

B-25's FD, FF, FG, FH, and FN are airborne for night navigation exercise, aircraft landed after exercise completion. Liberator LA is airborne for Bermuda, and landed there. Grumman Goose GB is airborne from Jamaica, landed at Oakes Field. B-25's FD, FL, FH, FO, and FF are airborne for air Exercise Number 6, aircraft landed after exercise completion. B-25's AS, AM, and AK are airborne for local navigation exercise, aircraft landed after exercise completion same aircraft did same exercises again later. SS *William M. Stewart* and USS *LSM-150* are bound south in Straits passing St. Augustine. **Task Unit TU-20182** is bound south in Straits passing Canaveral. Tug USS *Humrick* and tows, SS *Rich Mountain*, and destroyer escort USS *George E. Davis* (DE-357) are bound north in Straits. Tug USS *Warbler* (ARS-11) is bound south in Straits passing Palm Beach. Tanker *Cities Service Ohio* is bound northwest in Straits towards Port Everglades. **Convoy Motor Torpedo Boat Squadron-32, Group C** is bound south in Straits.

1st September 1944

B-25's BD, CX, CF, BA, BC, and AS are airborne for local night navigation exercise, aircraft landed after exercise completion. B-25's FF, FJ, FO, FS, FP, and FU are airborne for compass swinging exercise, aircraft landed after exercise completion. Liberator NO is airborne from Windsor Field for navigation exercise 14, aircraft landed after exercise completion, nothing to report. Tanker *R. P. Smith* is bound south in Straits northwest of Little Bahama Bank. USS *Aucilla* (AO-56) is bound east in Straits north of Little Bahama Bank. **Convoy KN-335** is bound north in Straits, northwest of Little Bahama Bank. Landing ship USS *LSM-150* and SS *William M. Stewart* are bound south in Straits. Tug USS *Warbler* and **Convoy KG-728** are bound southeast in Old Bahama Channel. Aircraft carrier USS *Ticonderoga* (CV-14) escorted by destroyers USS *O'Brien* (DD-751), USS *Maddox* (DD-721) and USS *Walker* (DD-723) are bound south, to the east of Eleuthera. NAAF Great Exuma: A plane arrived from Caicos Island and departed to Nassau. Two planes arrived from Fort Lauderdale (Port Everglades) and then returned back. A plane arrived from Guantanamo on a training flight. Forty-nine planes from the aircraft carrier USS *Ticonderoga* (CV-14) carried out practice attack, a plane made [an actual] crash landing, and a pilot was rescued. A plane departed to Banana River.

2nd September 1944

B-2's FF, FP, FO, and FB, and FW are airborne for compass swinging exercise, aircraft landed after exercise completion. Liberator LA returned from Bermuda, landed at Windsor Field. SS *Nordahl Grieg* is bound north in Straits northwest of Little Bahama Bank. **Convoy NK-635** is bound south in Straits northwest of Little Bahama Bank. SS *C. A. Canfield* is bound north in Straits northwest of Little Bahama Bank. **Task Unit TU-20196** is bound south in Straits off Palm Beach. Tanker *R. P. Smith* is bound south in Straits. Argentinian SS *Quilmec* is bound northwest in Nicholas Channel. **Convoy GK-828** and USS *SC-663* are bound northwest in Old Bahama Channel. Tanker *Comol Cuba* is bound south in Old Bahama Channel. **Convoy NG-456** is bound south, to the east of Eleuthera. NAAF Great Exuma: A plane arrived from Guantanamo and departed to Banana River.

3rd September 1944

USS *Device* (AM-220) and USS *LSM-205* are bound south in Straits west of Little Bahama Bank. Troop transport USS *USAT-George Washington* is bound north in Straits north of Little Bahama Bank. **Convoy NK-635** and USS *YAG-28* are bound south in Straits. USS *PC-1142* and Navy plane are bound north in Straits. SS *Antinous* is bound west in Nicholas Channel. Tug USS *Warbler* (ARS-11), towing SS *Gertrude Kellogg* and tug USS *Willett* (ARS-12) are bound northwest in Old Bahama Channel. NAAF Great Exuma: A plane departed to patrol, and another plane arrived from Guantanamo and departed to Banana River on a training flight.

4th September 1944

B-25's FF, FL, FB, FO, and FJ are airborne for Anti-Submarine Patrol Number 1, aircraft landed after exercise completion, a US submarine was sighted, rest of patrol uneventful. B-25's FF, FN, FS, FH, and FO are airborne for air Exercise Number 1, aircraft landed after exercise completion. Liberator LH is airborne for Radar and navigation exercise, aircraft landed after exercise completion. SS *Tampico* is bound south in Straits north of Little Bahama Bank. SS *Indiana* is bound north in Straits west of Little Bahama Bank. Submarine USS *Blenny* (SS-324) and USS *SC-711* are bound west in Northwest Providence Channel with US Navy plane escorting. Tug USS *Moctobi* (ATF-105) is bound south in Straits off Miami. SS *Groveton* is bound west in Straits north of Cay Sal Bank. Argentinian SS *Rio Diamante* is bound northwest in Nicholas Channel. USS *YAG-28* is bound southeast in Old Bahama Channel. SS *Andalien* northwest in Old Bahama Channel. NAAF Great Exuma: A plane arrived from Miami. A plane departed to local flight. A plane departed to Guantanamo. The plane on local flight returned to base. Army received supply from boat. A plane arrived from Guantanamo. A plane departed to Miami. A plane arrived from Guantanamo and departed to Banana River.

5th September 1944

B-25's FK and FW are airborne for air Exercise Number 1, aircraft landed after exercise completion. B-25's FC, FV, and FT are airborne for air Exercise Number 1, aircraft landed after exercise completion. B-25's FG, FD, FJ, FL, and FM are airborne on Anti-Submarine Patrol Number 1, aircraft landed after exercise completion, patrol uneventful except a 10,000-ton tanker was sighted. Liberator LH is airborne for Radar and navigation exercise, aircraft landed after exercise completion. SS *Blackwater* is bound north in Straits northwest of Little Bahama Bank. SS *Mauma* is bound south in Straits off Palm Beach. Destroyer escort USS *Mack* (DE-358) is bound north in Straits west of Bimini. USS *Ability* (PYc-28) is bound north in Straits. USS *PC-469* is bound southeast in

Old Bahama Channel. NAAF Great Exuma: A plane arrived from Guantanamo and departed for Banana River on a training flight.

6th September 1944

B-25's FC, FD, FH, FL, and FW are airborne for night navigation exercise, aircraft landed after exercise completion. B-25's FB, FH, FK, FU, and FT are airborne for air Exercise Number 5, aircraft landed after exercise completion. Liberator LH is airborne for Radar and navigation exercise, aircraft landed after exercise completion. USS *Abilitz* (PYC-28) is bound north in Straits off Canaveral. **Task Unit TU-20197**, USS *SC-704* and USS *SC-528* are bound south in Straits northwest of Little Bahama Bank. **Convoy KN-336** is bound north in Straits west of Little Bahama Bank. SS *Josiah G. Holland* is bound north in Straits off Port Everglades. **Convoy KG-729** is bound southeast in Nicholas Channel. Tanker *AthelRegent* is bound northwest in Nicholas Channel. NAAF Great Exuma: Gasoline barge USS *YOG-5* and USS *SC-1313* arrived from Guantanamo and departed for Nassau, an Army PBY arrived from Morrison Field and went to Mayaguana, a plane departed to Guantanamo, a plane arrived from Guantanamo and departed to Banana River.

7th September 1944

B-25's FH, FM, FN, FB, FS, and FU are airborne for night navigation exercise, aircraft landed after exercise completion. B-25's FB, FP, FH, FK, and FN are airborne for air Exercise Number 5, aircraft landed after exercise completion. B-25 FW is airborne for air Exercise Number 7, aircraft landed after exercise completion. B-24 LH is airborne for Radar and navigation exercise, aircraft landed after exercise completion. **Convoy NK-636** and **Task Unit TU-20193** are bound south in Straits northwest of Little Bahama Bank. USS *PC-1198* and USS *PC-1213* are bound south, to the north of Walker Cay, off Abaco. USS *LCI(L)-570* is bound north in Straits northwest of Bimini. Tanker *Esso Washington* is bound north in Straits, west of Bimini. **Convoy GK-829** is bound northwest in Old Bahama Channel. **Convoy GN-155** is bound north, to the east of Eleuthera. NAAF Great Exuma: A plane departed and returned to base, a plane arrived from Guantanamo and then departed for Banana River, an Army PBY aircraft arrived from Caicos Island, and departed for Morrison Field.

8th September 1944

B-25's FG, FJ, and FM are airborne for Anti-Submarine Patrol Number 1, aircraft landed after exercise completion, steam ship of 5,500 tons sighted. B-25's FH and FC are airborne for air Exercise Number 7, aircraft landed after exercise completion. B-25's FN, FY, FW, FV, and FP are airborne for air Exercise Number 3, aircraft landed after exercise completion, again airborne later for air Exercise Number 4, aircraft landed after exercise completion. Liberator LH is airborne for Radar and navigation exercise, aircraft landed after exercise completion. USS *SC-1511*, USS *SC-1482*, and Russian RU *SC-1505* are bound north in Straits, northwest of Little Bahama Bank, exiting the area. USS *YO-168* is bound south in Straits off Port Everglades. **Convoy NK-636** is bound south in Straits. NAAF Great Exuma: A plane arrived from Guantanamo and departed for Banana River.

9th September 1944

B-25's FM, FG, FP, and FH are airborne for local night navigation exercise, aircraft landed after exercise completion. B-25's FJ, FB, and FU are airborne for Anti-

Submarine Patrol Number 2, aircraft landed after exercise completion. B-25's FG, FM, and FH are airborne for air Exercise Number 3 and later for air Exercise Number 4, aircraft landed after exercise completion. Liberator LH is airborne for Radar and navigation exercise, aircraft landed after exercise completion. Liberator MD is airborne from Windsor Field for Kingsley Exercise, radio contact established, frigates located from bearing received, patrols completed under direction of SOE, depth charge attack made against a smoke float laid by SOE, aircraft landed after exercise completion. Liberator MF is airborne from Windsor Field for Kingsley Exercise as first relief, homing procedure "B" commenced, frigates located, patrols completed under direction of SOE, machine gun and depth charge attack made against a smoke float laid by SOE, aircraft landed after exercise completion. Liberator MC is airborne from Windsor Field for Kingsley Exercise as second relief, radio contact established, frigates located from bearing received, patrols completed under direction of SOE, dummy submarine attack made against a smoke float laid by SOE, aircraft landed after exercise completion. Liberator LC is airborne from Windsor Field for Kingsley Exercise as third relief, frigates located, signals exchanged, patrols completed under direction of SOE, dummy submarine attack made against a smoke float laid by SOE, aircraft landed after exercise completion. SS *Barren Hill* is bound north in Straits northwest of Little Bahama Bank. USS *YAG-32* is bound east in Northwest Providence Channel. USS *Y0-129* is bound north in Straits off Port Everglades. Tanker *Gulfmaid* is bound north in Straits off Miami. Submarines USS *Seacat* and USS *Dragonet* are bound south in Straits with US Navy plane escorting. USS *PC-1148* is bound north in Straits. NAAF Great Exuma: Two planes arrived from Guantanamo one departed for Banana River. A plane departed for Guantanamo, and the Army supply boat USS *WT-76* arrived from Caicos Island.

10th September 1944

Liberator MT is airborne from Windsor Field for Kingsley Exercise, frigates located, recognition signals exchanged, patrols completed under direction of SOE, depth charge attack made against a smoke float laid by SOE, aircraft landed after exercise completion. Liberator LK is airborne from Windsor Field for Kingsley Exercise as first relief, homing exercise "B" commenced, frigates located, patrols completed under direction of SOE, Anti-Submarine attack made against a smoke float laid by SOE, aircraft landed after exercise completion. Liberator NX is airborne from Windsor Field for Kingsley Exercise as second relief, homing exercise "B" commenced, radio contact established, frigates located, patrols completed under direction of SOE, depth charge attack made against a smoke float laid by SOE, aircraft landed after exercise completion. Liberator MD is airborne from Windsor Field for Kingsley Exercise as third relief, homing exercise "B" commenced, frigates located, patrols completed under direction of SOE, depth charge attack made against a smoke float laid by SOE, aircraft landed after exercise completion. Destroyer USS *Warrington* (DD-383) departs Norfolk for Trinidad with USS *Hyades* [she is sunk in a hurricane in three days, 200 nm northeast of Abaco. Seventy-three men were later rescued by the US Navy]. USS *SC-1503* is bound south in Straits off Miami. **Task Unit TU-20198** is bound south in Straits off Miami. SS *C. A. Canfield* is bound south in

Straits off Port Everglades. USS *SC-1497* and USS *SC-1486* are bound north in Straits west of Bimini. USS *Project* (AM-278), SS *George Ade*, tanker *Pan Georgia*, tanker *W. H. Libby*, tanker *Atheldulke*, tanker *Bidwell*, and tanker *AthelRegent* are bound north in Straits. SS *Chagres* is bound northwest in Old Bahama Channel. USS *PC-1253* is bound southeast in Old Bahama Channel. No.113: Hurricane Blue warning received. Aircrew instructed to pack four days' kit and stand by. CTO [Chief Technical Officer] is instructed to prepare all aircraft for departure. Maintenance crews recalled from Nassau; aircraft servicing commenced. NAAF Great Exuma: Army supply boat USS *WT-76* departed. A plane departed and returned to base.

11th September 1944

Weak disturbance reported in region of Guadeloupe, Leeward Islands in morning of 9th September, moving north-westerly, increased in intensity, Nassau was threatened, Commanding Officer of Unit in consultation with Meteorological Officer instructed Station, Hurricane warning Blue to be issued. Arrangements were made to make aircraft serviceable and evacuate airfields, airmen of Maintenance Wing recalled to camp and aircraft servicing work began, orders issues to cease all flying until further notice, arrangements made to evacuate all aircraft to refuge airfields, decided to delay take off until later during the 12.th ASR launches are reported to be at the Hurricane anchorage. With uncertain prevailing condition, Meteorological officer requested aircraft be detailed to investigate tropical disturbance located to east of Bahamas. Liberator LA is airborne, sent half hourly weather reports, landed back with further developments of disturbances, another aircraft was dispatched to make reports of conditions to east and southeast of Nassau, aircraft landed after exercise completion. Due to adverse weather, Chief Instructor cancelled Kingsley Exercise, signal received from HMCS *LaSalle* that ship HMCS *Levis* had developed condenser failure, was able to proceed at low speed. SOE decided to make for Nassau and then dock there during storm return to Bermuda after repair ships docked at Nassau exercise cancelled in view of diverting of co-operating ships left Nassau and arrived at Miami repairs were effected. A US Navy *Dauntless* aircraft is missing east of Naval Air Station (NAS) Miami. Tanker *Fenja* is bound south in Straits off Canaveral. SS *William Crompton*, SS *William Ford Nicholas*, tanker *Eastern Sun*, SS *Watertown*, SS *New York*, and tanker *Pan Virginia* are bound north in Straits northwest of Little Bahama Bank. SS *Swiftscout* and **Convoy KN-337** are bound north in Straits west of Little Bahama Bank. USS *PC-824*, USS *Stagbrush* (AN-69), and SS *Oscar F. Barrett* are bound north in Straits. USS *YF-443* is bound south in Straits. USS *PC-1142* is bound southeast in Nicholas Channel. **Convoy KG-730** is bound southeast entering Old Bahama Channel with US Navy plane escorting. USS *PC-609* is bound southeast in Old Bahama Channel. USS *PC-1191* is bound northwest in Old Bahama Channel. No.113: Hurricane level Blue warning still in force. CTO reported 28 aircraft serviceable and ready for departure, two unserviceable. Crew assignments reported 28 Crews, including test crews available and allotted to aircraft. Crews ordered to remain stand by. Forty Ground Crew personnel detailed to remain on standby to proceed with aircraft if evacuated. All other personnel ordered to be in camp. Arrangements are made to evacuate all aircraft on the 12th. Refuge airfields, ranging from Cuba to the US, were studied. NAAF Great Exuma: A plane arrived from Guantanamo and departed for Nassau, arrived from Nassau and returned there to Guantanamo. Naval Air Transport seaplane arrived from Miami and went to Guantanamo. A Naval Air Transport seaplane arrived from Guantanamo and went to Miami. A plane arrived from Guantanamo and departed for Banana River.

12th September 1944

No immediate menace from Hurricane, but Commanding Officer decided that all crew be briefed pending further developments for evacuation of all aircraft to refuge fields. The crew of B-24's are briefed at Oakes and Windsor Fields, crews of B-25's briefed at Oakes Field. Thirty-one B-24's, two Grumman Goose aircraft, and the Oxford are airborne for Miami, and landed there. Forty-two B-25's airborne for Batista, Cuba, landed there. Later at night, Hurricane warning Red issued, all necessary action under this warning carried out except evacuation of ground personnel, general order given to evacuate fields, personnel of Maintenance and Windsor Field accommodated at British Colonial Hotel, Training Wing accommodated at Bahamian Club, SEO was in Fort Montagu Hotel, MT [mechanized transport] section transported all personnel to respective quarters and then vehicles not earmarked for special duties were parked at Curry Hill Pit, all fields are manned by skeleton staff in the Signals, Operations Room, Hospital and other essential buildings. Neutral Spanish SS *Marques De Comillas* from Veracruz to Cadiz is observed by Hemingway halting off Havana. His subsequent vigilante accusations filed to FBI and US Ambassador in Havana of SS *Marques De Comillas* supplying U-boats and aiding Germans remain unsubstantiated. USS *Warrington* abandoned *in extremis* by escort, drifting keeled over in a hurricane, over 300 sailors on board, remains afloat for 24 hours. Searches by provisions vessel *USS Hyades*, destroyer escorts USS *Frost* (DE-144), USS *Nuse* (DE-145), USS *Inch* (DE-146), USS *Snowden* (DE-246), USS *Swasey* (DE-248), USS *Woodson* (DE-359), and USS *Johnnie Hutchins* (DE-360), plus tug USS *ATR-9*, and tug USS *ATR-62* rescued five officers and 68 men of the destroyer's 20 officers and 301 men. USS *LSM-267* is bound south in Straits off Canaveral. USS *Stagbrush* (AN-69) is bound north in Straits northwest of Little Bahama Bank. **Convoy NK-637** is bound north in Straits west of Little Bahama Bank. SS *Baldbutte* is bound south in Straits northwest of Bimini. Navy plane is bound east in Straits northwest of Bimini. SS *Empire Norse* and SS *San Melito* are bound south in Straits west of Bimini. SS *John S. Sargent*, tanker *Gulf of Venezuela* and tanker *Gulfhawk* are bound south in Straits. Tanker *R. P. Smith* is bound north in Straits. USS *PC-1549* is bound west in Straits. Tug USS *Scotch Cap* and tows are bound southeast in Old Bahama Channel. SS *Earlscourt Park*, tanker *Edward L. Shea*, SS *Tobias E. Stansbury*, USS *Acubens* (AKS-5), SS *George Taylor*, SS *Cleodora*, SS *Moose Mountain Park*, and SS *Erastus Smith* are bound south, to the east of Eleuthera. SS *Permian* and **Convoy GN-156** are east of Eleuthera, bound north. No.113: Controller reported Hurricane is now definitely threatening Nassau. Crews ordered to report to Windsor Field, and stand by for briefing. Air Field to evacuate aircraft to Morrison Field, West Palm Beach. Last aircraft away, however two aircraft developed last-minute faults which could not be remedied in time, making four unserviceable aircraft left on the field. These were picketed down. Preliminary battening down and stowing away of loose equipment. Hurricane Red warning received. Secret and Confidential files and publications placed in C&C Strong Room. Battening down completed. Ground personnel instructed to pack kit and report to RAF TC HQ with small kit and blanket . Roll call of 176 Ground Personnel to be evacuated. First transport left for British Colonial Hotel. Evacuation completed, CO and WAdO made final inspection of buildings and barrack huts. Order given for all aircraft to avoid hurricane by evacuating to refuge airfields. Only skeleton crew remained at No.111 OTU Nassau. Buildings battened, ground personnel evacuated. Remaining aircraft picked down. General order to evacuate the airfield was given. Maintenance went to British Colonial Hotel and Training wing to the Bahamian Club, and SHQ to the Fort Montagu Hotel. NAAF Great Exuma: One plane arrived from Guantanamo to relieve another standby plane, however then both planes returned Guantanamo because of danger of Hurricane activity in the Bahamas.

13th September 1944

Advisory from Miami received to the effect that the hurricane is turning north, hurricane alert cancelled on Florida coast, Georgia and Carolina, standing by. Arrangements made to recall all aircraft from refuge airfields and to bring ground staff to their respective airfields. Commanding Officer consulted Meteorological Officer, cancelled hurricane warnings Blue and Red, officers in hurricane quarters instructed to arrange for return of troops to airfields, signals dispatched to senior RAF officer at Miami and Batista Fields recalling all to Nassau, ground personnel brought back to their respective actions station reverted to normal during day. B-24's returned from Miami, Grumman Goose and Oxford aircraft returned during later part of day. Forty-two B-25's returned from Batista. Tanker *Olney* and SS *Indiana* are bound south in Straits northwest of Little Bahama Bank. USS *Thrush* (AVP-3) is bound north in Straits northwest of Little Bahama Bank. SS *Alar* is bound north in Straits north of Abaco. **Task Unit TU-20199** is bound south in Straits off Palm Beach. Tanker *L. V. Stanford* is bound north in Straits west of Little Bahama Bank. Submarine USS *R-13* and USS *PC-1145* are bound south in Straits off Miami with US Navy plane escorting. SS *Turkey Island* is bound north in Straits. Tanker *Gulfwax* and SS *Silver Star Park* are bound north, to the east of Eleuthera. No.113: Adjutant informed by officer in charge at British Colonial Hotel of the withdrawal of the Red warning as the storm had veered north and west during the night and was missing the Bahamas. Transport Commend personnel left for Windsor Field. Signal sent ordering all aircraft to return. All buildings un-battened, Secret and Confidential documents and files returned to Sections and normal work commenced. Last aircraft returned from Morrison Field. Hurricane heading north to Carolina and Georgia, so aircraft were recalled, and cancellation of warnings Blue and Red issued. Station reverted to normal, with aircraft returned.

14th September 1944

B-25's AD, AH, CO, CF, BA, and AR are airborne for local navigation flights, aircraft landed after exercise completion. B-25's FT, FK, FB, FV, FJ, and FO are airborne for air Exercise Number 6, aircraft landed after exercise completion. Request received from Comgulfseafron to search for destroyer and tug which were in distress. Liberators NC, LK, NX, and MF are airborne for detailed search of give area, 6,000 ton tanker sighted, 3,000 ton steam ship was sighted, 7,000 ton Liberty ship sighted, heavy tug sighted, it was ascertained that commander of tug had not sighted destroyer USS *Warrington.* B-24's landed after exercise completion. Tanker *Beacon Oil* is bound south in Straits Off Canaveral. *SS Empire Success* is bound south in Straits northwest of Little Bahama Bank. SS *Annibal* is bound north in Straits northwest of Little Bahama Bank. Tug USS *ATR-7* is bound northeast in Northeast Providence Channel. USCG *CG-83487* is bound north in Straits off Palm Beach. Tanker *Alpha*, destroyer escorts USS *DE-359*, and USS *DE-360* are bound north in Straits, west of Bimini. SS *New Orleans* is bound southeast in Nicholas Channel. SS *Evagoras* is bound northwest in Old Bahama Channel. Tanker *Cities Service Kansas* and tanker *Hagood* are north in Straits, bound east of Cat Island. Request received from Comgulfseafron to search for destroyer [USS *Warrington*] and tug which were in distress due to hurricane. Liberators NC, LK, NX, and MF are airborne for detailed search of give area. Sighted vessels included a 6,000-ton tanker, a 3,000-ton steamer, and a 7,000-ton Liberty ship, and a heavy tug. It was ascertained that commander of tug had not sighted destroyer. In the aftermath of hurricane: the Acting Governor was stranded with his yacht whilst making an official tour of the Out Islands, and brought back to Nassau in a Grumman Goose aircraft. Nassau Radio had called the yacht to warn him of the imminent danger, and the Governor took shelter in the island of Andros. Also, as to the search for personnel of USS *Warrington*, destroyer lost in hurricane; there has been no contact between Bahamas and survivors. NAAF Great Exuma: A plane arrived from Guantanamo as standby.

15th September 1944

B-25's FM, FP, FV, and FO are airborne for night navigation exercise, aircraft landed after exercise completion. Further request received from Comgulfseafron for search to be made for USS *Warrington,* area for search conveyed, agreed that four aircraft detailed for standard routine Anti-Submarine Patrol Number 2 be diverted to cover area. B-25's FU, FJ, FC, and FT are airborne, sundry ships and tankers sighted thorough search made for missing ship not located. A signal received from Comgulfseafron that survivors had been picked up by an aircraft carrier, which was searching area. B-25's FM and FP are airborne for air Exercise Number 7, aircraft landed after exercise completion. B-25's BM, JY, and W are airborne for air Exercise Number 6, aircraft landed after exercise completion. B-25's AU, CO, and CF are airborne for local navigation exercise, aircraft landed after exercise completion, again airborne later for continuation of same exercise. B-24 LA is airborne for and landed at Elizabeth City, then landed at Washington. USS *SC-1323* is bound south in Straits off Canaveral. SS *Esturia* and USS *YMS-198* are bound south in Straits, northwest of Little Bahama Bank. USCG *CG-Redbud* is bound north, to the north of Walker Cay, off Abaco. Tug USS *ATR-7* towing the small military freighter USS *FS-315* is bound south, to the north of Walker Cay, off Abaco. Tanker *Haakon Hauan* is bound south in Straits west of Little Bahama Bank. Destroyer escort USS *Slater* (DE-766) is bound south in Straits west of Little Bahama Bank. Tanker *Clio* is bound north in Straits west of Little Bahama Bank.

SS *Virginia* and tanker *Beacon Oil* are bound south in Straits. Tanker *Cities Service Kansas* and tanker *Hagood* are bound north in Straits north of Abaco. SS *Mount Bruce Park*, USS *Alacrity* (PG-87), USS *PC-1210* and SS *Elisha Mitchell* are bound south, to the east of Eleuthera. Battle of Britain Parade: members of unit paraded to the Station Square Parade ground, where colours were unfurled. Due to incessant rain, it was impossible to carry through with three days of events, but in spite of the rain the spectators gathered along the route taken. The acting governor took the salute in the march past Marlborough Street, downtown Nassau. Parade embussed to the Nassau Cathedral for the service, and returned to the camp by transport. NAAF Great Exuma: Two planes arrived and returned there for Fort Lauderdale.

16th September 1944

B-25's FY, FC, FK, FP, FO, and FW are airborne for local night navigation exercise, aircraft landed after exercise completion. B-25's FC, FK, and FW are airborne for air Exercise Number 7, aircraft landed after exercise completion. B-25's FV, FT, FY, and FO are airborne for air Exercise Number 6, aircraft landed after exercise completion. Signal received that HMCS *LaSalle* and HMCS *Levis* which were co-operating ships with aircraft of this Unit on Kingsley Exercise had proceeded to Miami for repairs for protection from the hurricane, were departing to Bermuda. It was decided that B-25 aircraft detailed for Anti-Submarine Patrol be diverted to provide for cover for and to carry out exercises in co-operation with these ships as they passed through Northeast Providence Channel. B-25 FU airborne, radio communication established with ships, ships located, signals exchanged, aircraft carried out detailed patrol under instruction of SOE, dummy submarine attack with one depth charge against smoke float dropped by SOE, aircraft landed after exercise completion. B-25 FT is airborne, met Kingsley Exercise convoy, aircraft carried out detailed patrol under instruction of SOE, dummy submarine attack with one depth charge against smoke float dropped by SOE, aircraft landed after exercise completion. Seaplane tender USS *Sandpiper* (AVP-9) is bound south in Straits northwest of Little Bahama Bank. SS *Evagoras* is bound north in Straits northwest of Little Bahama Bank. USS *ATR-7* and tow and Navy plane are bound south in Straits west of Little Bahama Bank. Destroyer escort USS *Coats* (DE-685) and **Convoy KN-338** are bound south in Straits off Port Everglades. Tanker *Olney* is bound south in Straits west of Bimini. SS *George W. Barnes* is bound southeast in Old Bahama Channel. SS *Alacrity* and USS *Hyades* (AF-28) are bound south, to the east of Eleuthera. NAAF Great Exuma: A plane arrived and returned there for Guantanamo.

17th September 1944

B-25's BC, NJ, CX, CO, and AD are airborne for night navigation exercise, aircraft landed after exercise completion. B-25's FO, FC, FW, FB, FM, and FG are airborne for night navigation exercise, aircraft landed after exercise completion. No further flying took place, detailed programmes cancelled due to adverse weather. USS *ATA-174* and tows, SS *Jamestown* are bound south in Straits northwest of Little Bahama Bank, destroyer escort USS *Jordon* (DE-204) are bound south in Straits northwest of Little Bahama Bank. SS *Lake Superior* is bound northwest in Old Bahama Channel. Tug USS *ATR-62* is bound south, to the east of Eleuthera. No.113: Battle of Britain

Commemoration Service, RAF parade held at the Cathedral, Nassau. Salute after the service was taken by H-E the Acting Governor. Senior RAF leadership. NAAF Great Exuma: US Naval Air Transport seaplane Flight 622 landed due to engine trouble. A plane arrived from Guantanamo and another returned there. A plane arrived from Guantanamo and departed for Banana River.

18th September 1944

B-25's FY, FO, FL, FV, FW, and FB are airborne for Anti-Submarine Patrol Number 3, aircraft landed after exercise completion. B-25's FY, FG, FT, FK, and FO are airborne for air Exercise Number 1, aircraft landed after exercise completion. Liberator LH airborne for Radar and navigation exercise, aircraft landed after exercise completion. SS *Cilicia* is bound north in Straits west of Little Bahama Bank. USS *YO-129* is bound south in Straits northwest of Bimini. USS *ATA-174* towing three USS YC's escorted by USS *Design* (AM-219) is bound south in Straits. *SS Lake Sapor* is bound west in Nicholas Channel. **Convoy GK-831** is bound northwest in Nicholas Channel. NAAF Great Exuma: US Naval Air Transport seaplane Flight 622 departed to Miami. A Naval Air Transport seaplane arrived from Miami and went to Guantanamo. A Naval Air Transport seaplane arrived from Guantanamo and went to Miami. A plane arrived from Guantanamo then returned there after a training flight. A plane arrived from Guantanamo and departed for Banana River. Two US Army SRBs arrived from Morrison Field.

19th September 1944

B-25's FT, FL, FC, FP, FU, and FW area airborne for Anti-Submarine Patrol Number 3, aircraft landed after exercise completion. B-25's FB, FO, FY, and JA are airborne for air Exercise Number 1, aircraft landed after exercise completion. Liberator LH is airborne for Radar and navigation exercise, aircraft landed after exercise completion. Liberator LA returned from Washington landed. SS *Marathon* is bound north in Straits northwest of Little Bahama Bank. **Task Unit TU-20184** is bound south in Straits northwest of Little Bahama Bank. Tanker *Hadnot* is bound north in Straits west of Little Bahama Bank. Tanker *R. P. Smith* is bound south in Straits. Tug *Margot Moran*, tow and escort are bound north in Straits. No.113: B-26 Marauder HD-638's undercarriage was accidentally raised by co-pilot after the pilot had made a normal landing, with no injuries to crew. Damage to aircraft included buckling of lower portion of frames, and port air-screw blades bent. Pilot was F/O C. A. Grant, and co-pilot was Sgt. F. S. Pyan. NAAF Great Exuma: A plane arrived from Guantanamo and departed for Banana River.

20th September 1944

B-25's FJ, FL, FV, FK, and JA are airborne for air Exercise Number 5, aircraft landed after exercise completion. Liberator LH is airborne for Radar and navigation exercise, aircraft landed after exercise completion. Tug *Margot Moran*, two USS YF's, and USS *SC-685* are bound north in Straits northwest of Little Bahama Bank. SS *Diamond Island* is bound south, to the north of Walker Cay, off Abaco. SS *Indiana* is bound north in Straits west of Bimini. USS *PC-610* is bound northwest in Old Bahama Channel. **Convoy NG-459** is bound south, east of Eleuthera. NAAF Great Exuma: Two planes arrived from Guantanamo and departed for Banana River, then returned back.

21st September 1944

B-25's FT, FK, FO, FP, FG, and FL are airborne for night navigation exercise, aircraft landed after exercise completion. B-25's FO, FL, JA, and JB are airborne for air Exercise Number 5, aircraft landed after exercise completion. B-25's FK, FY, FT, FV, FC, and FP are airborne for Anti-Submarine Patrol Number 1, aircraft landed after exercise completion, uneventful patrol. Liberator LH is airborne for Radar and navigation exercise, aircraft landed after exercise completion. *U-518* transited east of Bahamas in same hurricane as USS *Warrington*. SS *Medina* is bound south in Straits north of Little Bahama Bank. **Convoy KN-339** is bound north in Straits northwest of Little Bahama Bank. **Convoy NK-638** and USCG cutter *CG-Active* are bound south in Straits northwest of Little Bahama Bank. USS *Fulmar* (AMG-46) is bound south in Straits off Palm Beach. USS *Zaniah* (AG-70) is bound south in Straits off Port Everglades with US Navy plane escorting. SS *H. M. Fredricksen* is bound north in Straits northwest of Bimini. USS *PC-1220*, USS *PC-1222*, and a US Navy plane are bound south in Straits. SS *Erastus Barnes* is bound west in Nicholas Channel. SS *George W. Barnes* is bound northwest in Old Bahama Channel. **Convoy KG-732** is bound southeast in Old Bahama Channel. NAAF Great Exuma: USS *YAG-19* arrived from Miami, an Army PBY arrived from Morrison Field and departed for Caicos Island. The Army supply boat USS *WT-76* arrived, a plane *en route* from Guantanamo to Jacksonville landed due to mechanical trouble, and a plane went to patrol the route of **Convoy *NG***.

22nd September 1944

B-25's FT, FV, and FP are airborne for Anti-Submarine Patrol Number 2, aircraft landed after exercise completion, B-25 FP had difficulty in releasing undercarriage and stayed in air in a successful attempt to release mechanism, nothing to report from patrol. B-25's FU, FC, and FL are airborne for Anti-Submarine Patrol Number 1, aircraft landed after exercise completion. B-25's JA, FM, FY, FB, and FG are airborne for air Exercise Number 3, aircraft landed after exercise completion. B-25's JB, FY, FM, FW, and FG are airborne for air Exercise Number 4, aircraft landed after exercise completion. SS *Laguna* is bound south in Straits northwest of Little Bahama Bank. **Convoy NK-638** is bound south in Straits. **Convoy Motor Torpedo Boat Squadron-32, Group D** is bound north in Straits. **Convoy GN-460** is bound north, to the east of Eleuthera. NAAF Great Exuma: The Army supply boat USS *WT-76* departed for Mayaguana, USS *YAG-19* departed for Miami, a plane departed for Jacksonville, a plane arrived from Guantanamo and departed for Banana River. Another plane arrived and returned there to Guantanamo after a training flight.

23rd September 1944

B-25's FF, FL, FO, FG, FY, FV, and FK are airborne for night navigation exercise, aircraft landed after exercise completion. B-25's JB, FO, JA, and JC are airborne for air Exercise Number 3, aircraft landed after exercise completion. B-25's FO, JA, JB, and JC are airborne for air Exercise Number 3 and later for air Exercise Number 4, aircraft landed after exercise completion. B-25's CL, KV, and WY are airborne for Anti-Submarine Patrol Number 2, aircraft landed after exercise completion. Grumman Goose GA is airborne for Harbour Island, aircraft landed after exercise completion. Destroyer USS *Warrington*'s name struck from the Navy list as US Navy concedes she is a total loss. USS *LSM-268* is bound south in Straits, northwest of Little Bahama Bank. Tanker

Gulfport is bound north in Straits, west of Bimini. USS *YAG-19* is bound west in Straits north of Andros. Tug USS *ATR-29* towing oil barge YOG-45 is bound north in Santaren Channel. **Convoy GK-832** is bound northwest in Old Bahama Channel. USS *PC-461* is bound southeast in Old Bahama Channel. Tug USS *ATO-147*, towing USS *YF-493*, USS *W-1*, and USS *YC-895* are bound northwest in Old Bahama Channel. Submarine USS *Blower* (SS-325) is bound west in Northeast Providence Channel with US Navy plane escorting. USS *PC-1210* is bound west in Northeast Providence Channel. NAAF Great Exuma: An Army PBY departed for Morrison Field, a plane arrived from Guantanamo with Cdr, Carlson to make arrangements to change this base to Auxiliary Air Facility, then returned there. Two planes arrived from Fort Lauderdale and then returned, a plane arrived from Guantanamo, and a plane went on a training flight.

24th September 1944

Destroyer escort USS *Muir* (DE-770) is bound north, to the north of Walker Cay, off Abaco. Tanker *Hadnot* is bound south in Straits Off Canaveral. USCG *CG-83486* is bound north in Straits northwest of Little Bahama Bank. **Task Unit TU-20 1912** is bound south in Straits northwest of Little Bahama Bank. USS *SC-685* is bound south in Straits off Palm Beach. Tug USS *ATR-29* and tow south in Straits northwest of Bimini. SS *Malabar* is bound north in Straits northwest of Bimini. Tug USS *Montauk Point* and tows are bound north in Straits west of Bimini. USS *SC-1508* and USS *SC-1483* are bound south in Straits. USS *LST-592* is bound north in Straits. SS *Cape Sebastian* is bound southeast in Nicholas Channel. USS *PC-491* is bound southeast in Old Bahama Channel. NAAF Great Exuma: A plane departed to patrol the routes of **Convoy GN** and **Convoy NG**, one US Army battle-damage repair ship [ARB], arrived from Mayaguana, and a plane arrived from Guantanamo and departed for Banana River

25th September 1944

B-25's FV, FO, FL, FW, FM, and FC are airborne for air Exercise Number 6, aircraft landed after exercise completion. B-25's FW, FB, FY, and FT are airborne for air Exercise Number 6, aircraft landed after exercise completion. Tug USS *Montauk Point* and tanker *Gulfport* are bound north in Straits, northwest of Little Bahama Bank. SS *H. M. Fredricksen* are bound south in Straits off Port Everglades. USS *Poughkeepsie* (PF-26) and USS *ATA-176* are bound north in Straits. NAAF Great Exuma: The status of base changed from Air Facility to Auxiliary Air Facility, and Lt. Cdr. H. J. Bowker was relieved by Lt. J. O. Guthrie. A plane arrived from Guantanamo and departed for Banana River.

26th September 1944

Six B-25's are detailed to carry out night navigation exercise, cancelled due to adverse weather. B-25's FB, FO, FT, FU, and FW are airborne for air Exercise Number 6, aircraft landed after exercise completion. B-25's FY, FV, FL, FP, FB, and FM are airborne for air Exercise Number 7, aircraft landed after exercise completion. Grumman Goose GA is airborne for Harbour Island returned. USS *ATO-147* and tow, **Convoy KN-340**, and USS *PC-1220* are bound north in Straits northwest of Little Bahama Bank. USS *PCS-1384* is bound south in Straits off Port Everglades. SS *Josiah G. Holland* is bound north in Straits. SS *Elwood* bound southeast in Nicholas Channel. **Convoy KG-733** is bound southeast in Old Bahama Channel. SS *New Zealand Star* is bound south, to the east of Eleuthera. NAAF Great Exuma: Two Army repair ships, or ARBs departed for Morrison Field, and an Army PBY arrived from Morrison Field and went to Nassau. A plane arrived from Guantanamo and departed for Banana River.

27th September 1944

B-25's BL, BO, and BW are airborne for night navigation exercise, aircraft landed after exercise completion, again airborne later for same exercise. B-25's BL, BO, BW, BT, and BP are airborne for air Exercise Number 7, aircraft landed after exercise completion. B-25 FK is airborne for air Exercise Number 7, aircraft landed after exercise completion. B-25's FO, FW, JD, and JC are airborne for air Exercise Number 6, aircraft landed after exercise completion. B-25's BD, AH, and BC are airborne twice at different times for local night navigation exercise, aircraft landed after exercise completion. Grumman Goose GA is airborne for Harbour Island aircraft landed after exercise completion. A Beechcraft landed from Miami with a Major General who was making a visit to unit in connection with flying training. Liberator LT is airborne from Windsor Field for Kingsley Exercise, made radio contact with co-operating ships, frigates located, signals exchanged, patrols Adder and Cobra carried out under SOE instructions, depth charge attack carried out against smoke float dropped by SOE, aircraft landed after exercise completion. Liberator LK is airborne from Windsor Field for Kingsley Exercise as first relief, made radio contact with co-operating ships, frigates located, signals exchanged, patrols carried out under SOE instructions, depth charge attack carried out against smoke float dropped by SOE, aircraft landed after exercise completion. Liberator LR airborne from Windsor Field for Kingsley Exercise as second relief, made radio contact with co-operating ships, frigates located, signals exchanged, exercises carried out under SOE instructions, depth charge attack carried out against smoke float dropped by SOE, aircraft landed after exercise completion. Liberator LV is airborne from Windsor Field for Kingsley Exercise as third relief, made radio contact with co-operating ships, frigates located, signals exchanged, exercises carried out under SOE instructions, depth charge attack carried out against smoke float dropped by SOE, aircraft landed after exercise completion. Hurricane causes Cuban schooner *Rubicon* to break its moorings and drift into the Gulf Stream, to be found there roughly a month later by a US Navy blimp. Tug USS *YTB-338* and tug USS *ATR-9* are bound south in Straits west of Little Bahama Bank. SS *James Island* and SS *Bemes Heights* are bound north in Straits, west of Little Bahama Bank. A US Navy plane is bound west over Straits, towards Palm Beach. **Convoy NK-640** is bound south in Straits, off Port Everglades. Tug USS *ATR-29* towing gasoline barge USS *YOG-46* is bound west in Northwest Providence Channel. SS *Diamond Island* is bound north in Straits northwest of Bimini. **Convoy GK-833** is bound northwest in Old Bahama Channel. A US Navy plane is bound southwest in Old Bahama Channel. NAAF Great Exuma: A plane arrived from Guantanamo. An Army supply boat arrived.

28th September 1944

B-25's FG, FL, FB, FV, FU, and FT are airborne for night navigation exercise, aircraft landed after exercise completion. Grumman Goose GA airborne for Harbour Island, aircraft landed after exercise completion. B-25's JA, FB, FV, FO, FT, and FW are airborne for air Exercise Number 6, aircraft landed after exercise completion. B-25's CO, AD, and AS are airborne for local navigation exercise, aircraft landed after exercise completion, again airborne later for same exercise. Liberator MC is airborne from Windsor Field for Kingsley Exercise, made radio contact co-operating ships located from bearing received, signals exchanged, exercises carried out under SOE instructions, depth charge attack carried out against smoke float dropped by SOE, aircraft landed after exercise completion. Liberator NC is airborne from Windsor Field for Kingsley Exercise as first relief, homing procedure "B" commenced, made radio contact frigates located signals exchanged exercises carried out under SOE instructions, depth charge attack carried out against smoke float dropped by SOE, aircraft landed after exercise completion. Liberator LT airborne from Windsor Field for Kingsley Exercise as second relief, homing procedure "B" commenced, made radio contact frigates located signals exchanged exercises carried out under SOE instructions, depth charge attack carried out against smoke float dropped by SOE, aircraft landed after exercise completion. Liberator LK airborne from Windsor Field for Kingsley Exercise as third relief, homing procedure "B" commenced, made radio contact frigates located, signals exchanged, exercises carried out under SOE, instructions flares dropped for A-A practice by ship's gunners, depth charge attack carried out against smoke float dropped by SOE, aircraft landed after exercise completion. USS *LSM-269* is bound south in Straits off Canaveral. SS *Eli Whitney*, SS *Robert Morris*, SS *Michael Pupin*, and SS *David Lubin* are bound south in Straits, north of Little Bahama Bank. USS *Bath* (PF-55) is bound north in Straits, north of Little Bahama Bank. USS *Carib* (ATF-82) is bound south in Straits, northwest of Little Bahama Bank. Destroyer escort USS *Tweedy* (DE-532) and SS *Malabar* are bound south in Straits west of Little Bahama Bank. NAAF Great Exuma: A plane departed on patrol the route of **Convoy NG,** and then departed to Guantanamo. Another plane arrived from Guantanamo, and departed for Banana River.

29th September 1944

B-25's DC, BC, AD, BD, AH, and BA are airborne for local navigation flights, aircraft landed after exercise completion. Liberator LR is airborne from Windsor Field for Kingsley Exercise, made radio contact frigates located signals exchanged exercises carried out under SOE instructions, depth charge attack carried out against smoke float dropped by SOE, aircraft landed after exercise completion. Liberator LV airborne from Windsor Field for Kingsley Exercise as first relief, homing procedure "B" commenced, made radio contact, frigates located, signals exchanged, exercises carried out under SOE instructions, dummy submarine attack carried out against smoke float dropped by SOE, aircraft landed after exercise completion. Liberator MC is airborne from Windsor Field for Kingsley Exercise as second relief, homing procedure "B" commenced, made radio contact made, frigates located, signals exchanged, aircraft signalled ship that instruments were not opening satisfactorily, depth charge jettisoned, captain set course for base, failed to complete exercise, landed back at base. Liberator NC is airborne from Windsor Field for Kingsley Exercise as third relief, made radio contact, frigates located recognition procedure carried out, exercises carried out under SOE instructions, depth charge attack carried out against smoke float dropped by SOE, aircraft landed after exercise completion. USS *PC-1146* is bound north in Straits off Canaveral, exiting the area. USS *YMS-112* is bound north in Straits northwest of Little Bahama Bank. USS *LSM-265* is bound south in Straits northwest of Little Bahama Bank. Navy plane is bound southwest in Straits northwest of Little Bahama Bank. Destroyer escort USS *Rolf* (DE-362) is bound north, to the north of Walker Cay, north of Abaco. SS *Briar Creek* is bound north in Straits west of Little Bahama Bank. SS *Andrew Briscoe* is bound south in Straits west of Little Bahama Bank. USS *YO-142* is bound north in Straits west of Bimini. USCG *CG-Madrona* is bound south in Straits. Tanker *A. S. Hansen* is bound north in Straits. SS *David Luben*, tanker *Haakon Hauan*, SS *Robert Morris*, and SS *Michael Pupin* are bound south in Straits towards Cay Sal Bank. SS *Dry Tortugas* is bound south, to the east of Northeast Providence Channels. NAAF Great Exuma: A plane arrived from Guantanamo and departed for Banana River. A plane went to cover **Convoy TG-27.3**.

30th September 1944

B-25's GK, LY, and JB are airborne for compass-swinging exercise, aircraft landed after exercise completion. Exercises carried out between a B-24 and three P-40's from Number 3 Fighter Command HQ, USAAF. Tanker *A. S. Hansen* is bound north in Straits northwest of Little Bahama Bank. USS *LSM-127*, tug USS *Fire Island* and tows, escorted by USS *PC-1146* are bound south in Straits west of Little Bahama Bank. USS *PC-492* is bound south in Straits. Tug USS *Dry Tortugas* towing USS *ARD-2*, escorted by USS *SC-712* and USS *SC-1474* is bound south, to the east of Eleuthera. NAAF Great Exuma: Two planes arrived from Guantanamo and departed for Banana River

1st October 1944

USS *SC-512* is bound north in Straits towards Daytona. Tug USS *ATR-56* with tows and escorts, **Task Unit TU-20186** is bound south in Straits towards Palm Beach. SS *San Juan Hill* is bound southwest in Straits near Matanilla Shoal, northwest Bahamas. **Convoy KN-341** is bound north in Straits passing Palm Beach. USS *YO-142* is bound west passing Northwest Providence Channel. USS *Prime* (AM-279) is bound north in Straits. Tug USS *Fire Island* with tows and escorts is bound southwest in Straits. Troop transport USS *USAT-State of Maryland* and **Convoy KG-734** is bound southeast in Straits passing Nicholas Channel. NAAF Great Exuma: A plane arrived from patrol to relieve another plane, which departed for Guantanamo.

2nd October 1944

B-25's FC, FL, FJ, FY, FM, and FU are airborne for Anti-Submarine Patrol Number 1, aircraft landed after exercise completion. B-25's FK, FB, FC, FJ, and FD are airborne for air Exercise Number 1, aircraft landed after exercise completion. Liberator LH is airborne for Radar and navigation exercise, aircraft landed after exercise completion. **Task Unit TU-20195** is south towards Canaveral. **Convoy NK-641**, SS *Ardmore*, and SS *John H. McIntosh* are bound south in Straits passing Canaveral. SS *W. F. Burdell* is bound north in Straits. SS *Scandia* is bound south in Straits towards Palm Beach. Tanker *Clam*, tanker *Clio*, SS *Vav*, tanker *Petrofuel*, SS *Herbert L. Pratt*, tanker *Baldbutte*, SS *Petter*, and SS *William Crompton* are bound south in Straits, towards Palm Beach. SS *Solsten* and tanker *Tydolgas* are bound north in Straits passing Palm Beach. Tug USS *Yaquina Head* with tow and escorts are bound south in Straits passing Port

Everglades. SS *Christopher L. Sholes*, SS *Oscar F. Barrett*, and destroyer escort USS *Silverstein* (DE-534) are bound southwest in Straits. The house boat USS *YHB-16* is bound southeast in Straits near Miami. SS *Mardial* and tanker *Beaconoil* are bound northeast in Straits. USS *Gloucester* (FF-22) is bound northeast in Straits near Bimini. **Convoy GN-160** is bound north in Straits. NAAF Great Exuma: A plane departed on patrol of the route of **Convoy NG** and returned to base. A Naval Air Transport seaplane arrived from Miami and went to Guantanamo, then again arrived from Guantanamo and went to Miami. Another plane arrived from Guantanamo and departed for Banana River.

3rd October 1944

B-25's FJ, FT, FM, FC, and FU are airborne for compass-swinging exercise, aircraft landed after exercise completion. Radar exercise cancelled due aircraft unserviceable. Grumman Goose GA is airborne for San Salvador, aircraft landed after exercise completion. B-25's FW, FG, FN, FK, and FY are airborne for air Exercise Number 1, aircraft landed after exercise completion. B-25's FG, FJ, FY, FM, and FU are airborne for night navigation exercise, aircraft landed after exercise completion. Oiler USS *YO-169* is bound south in Straits passing St. Augustine. Tanker *A. S. Hansen* is bound south in Straits passing Daytona. USCG cutter *CG-Active* is bound north in Straits passing Canaveral. Tanker *Beaconoil*, tanker *Malay*, and tanker *Vincenzia* are bound north in Straits. **Task Unit TU-20123**, **Convoy NK-641**, and tanker *Cities Service Koolmotor,* are bound south in Straits. Tanker *Empire Norse* is bound northeast in Straits. Allied Italian submarine R. Smg. *Marea* and tug USS *ATF-82* are bound west in Straits with US Navy plane escorting. **Task Unit TU-201915** is bound west in Straits. USS *YTB-338* towing USS *YF-311* are bound north in Straits. SS *Fort Livingston* is bound northwest in Straits passing Nicholas Channel. NAAF Great Exuma: A plane departed to patrol the route of **Convoy NG**. Another plane arrived from Guantanamo patrolled the route of **Convoy NG** and relieved standby plane which returned there. A plane arrived from Guantanamo and departed for Banana River.

4th October 1944

B-25's FJ, FB, FN, JB, and JC are airborne for air Exercise Number 5, aircraft landed after exercise completion. B-24 LH is airborne for Radar and navigation exercise, aircraft landed after exercise completion. B-25's FB, FG, FO, FM, FN, and FJ airborne for night navigation exercise, aircraft landed after exercise completion. Tanker *British Prestige*, tanker *Gulfpoint*, tug USS *Point Vincent* with tows and escort, tanker *Annibal,* and tug USS *Point Loma* with tows and escort are bound south in Straits, passing Canaveral. Tanker *Empire Norse* is bound north in Straits. SS *Kina II* is bound southwest in Straits. Tanker *Tobias E. Stansbury* and tanker *Gulfdisc* are bound south in Straits towards Palm Beach. USS *YO-169* is bound south in Straits passing Palm Beach. Tanker *A. S. Hansen* and tanker *Gulfking* are bound south in Straits. NAAF Great Exuma: An Army PBY arrived from Caicos Island and departed for Nassau. A plane arrived from Guantanamo and departed for Banana River.

5th October 1944

B-25's FB, FT, FJ, FW, JB, and JD are airborne for air Exercise Number 5, aircraft landed after exercise completion. B-25's FB, FG, FO, FM, and FW are airborne on Anti-Submarine Patrol Number 1, aircraft landed after exercise completion. B-25 FY airborne for Boca Chica, aircraft landed after exercise completion. SS *Juan Casiano* and SS *Daylight* are bound south in Straits passing Canaveral. Tanker *Hagood*, SS *Tidewater*, and SS *Meline* are bound south in Straits towards Palm Beach. SS *Durango* and USCG cutter *CG-Madrona* are bound north in Straits. SS *Liebre* SS *W. W. Mills*, tanker *Edward L. Doheny*, and SS *Fort Churchill* are bound south in Straits, passing Palm Beach. USS *Newport* (PF-27) is bound north in Straits, towards Palm Beach. Tanker *British Honour* and SS *Petter II* are bound south, towards Port Everglades. Tug USS *Point Vincent* towing USS *ARCD-3* and USS *YN-118* are bound south in Straits towards Port Everglades. Tug USS *Point Loma* towing ARDC-5 and escorted by *PC-1145* are bound south in Straits, towards Miami. Tanker *British Prestige* and SS *Hoegh Hood* are bound southwest in Straits. Tanker *Gulfpoint* and SS *Annibal* are bound west in Straits. SS *James Island* is bound southwest near Cay Sal Bank. The house boat USS *YHB-13* is bound northwest in Straits. SS *F. J. Wolfe*, tanker *L. V. Stanford*, SS *Richard Coulter*, SS *Washington*, and SS *Tidewater Associated* are bound south in Straits. NAAF Great Exuma: An Army PBY arrived from Nassau. A plane arrived from Guantanamo, then departed for Banana River.

6th October 1944

B-25's BG, OM and NO are airborne for Anti-Submarine Patrol Number 2, aircraft landed after exercise completion. A Liberty ship sighted by the aircraft at different times in relative positions. B-25's FC, FJ, FY, JA, and JB and airborne for air Exercise Number 3, aircraft landed after exercise completion. B-25's FB, FV, FM, FN, and FU are airborne for local night navigation exercise, aircraft landed after exercise completion. SS *Benjamin J. Grieson* and USS *SC-1487* are bound south in Straits, passing St. Augustine. SS *San Valerio* is bound south in Straits past Key West. USS *SC-1293* is bound south in Straits, passing Daytona. **Convoy NK-642** is bound south in Straits near Canaveral. **Task Unit TU-291915** is bound south in Straits, passing Canaveral. **Task Unit TU-20187** and tanker *L. L. Abshire* are bound south in Straits, towards Palm Beach. SS *Andrew Stevenson* and **Convoy KN-342** are bound north in Straits, passing Palm Beach. Tanker *Gulfmaid* is bound north in Straits, passing Port Everglades. SS *Western Park* and SS *Richard Coutler* are bound south in Straits. SS *Argon* is bound north in Straits. SS *Daylight* and tanker *Hagood* are bound west in Straits. SS *David F. Houston* is bound northwest near Nicholas Channel. USS *YAG-21* is bound east towards Eleuthera. SS *Bethore*, SS *Britamer*, and SS *Hiram S. Maxim* are bound south in Straits. NAAF Great Exuma: Army supply boat USS *FS-138* arrived. A plane arrived from Guantanamo and departed for Banana River.

7th October 1944

B-25's FY, FJ, FM, FV, and FN are airborne for local night navigation exercise, aircraft landed after exercise completion. B-25's FM, FG, FK, FO airborne for Anti-Submarine Patrol Number 2, aircraft landed after exercise completion. B-25 FM airborne for air Exercise Number 7, aircraft landed after exercise completion. B-25's FB, FV, FU, JA, JB airborne for air Exercise Number 3 and later for air Exercise Number 4, aircraft landed after exercise completion. USS *PY-12* and USS *YMS-80* are bound south in Straits passing St. Augustine. USCG *CG-83347* is bound south in Straits towards Canaveral. Tanker *Henry L. Ellsworth* is bound south in Straits towards Palm Beach. Tanker *Rhea* and tanker *Gulfhawk* are bound south in Straits passing Palm Beach. Tanker *San Valerio* is bound south in Straits passing Port Everglades. **Convoy NK-642**, tanker *F. Q. Barstow* and SS *Benjamin J. Grieson* are bound south in Straits. Tanker *E. T. Bedford* is bound west in Straits. Tanker *L. V. Stanford,* tanker *Northern Sun,* and tanker *Josiah G. Holland* are bound north in Straits. NAAF Great Exuma: The Army supply boat USS *FS-138* departed for Mayaguana. A plane arrived from Guantanamo and departed for Banana River. Gasoline barge USS *YOG-5* arrived from Guantanamo with a plane.

8th October 1944

Found impractical to organize Kingsley Exercise in co-operation with units of Royal Canadian Navy stationed at Bermuda, arrangement made for similar exercise to be carried out in co-operation with units of USN operating out of Key West, seaplane tender seaplane tender USS *Sandpiper* (AVP-9) left Key West and co-operated with aircraft of this Unit and sailed in Bahamas region. Liberator ND airborne from Windsor Field exercise with USS *Sandpiper,* radio and radio contact established with ship, homing commenced by Radar, aircraft found it had homed on wrong ship, carried out square search and located convoy ship met too late to carry out exercise, attack made against towed target, landed at Boca Chica to refuel, aircraft landed back at base. Liberator LK airborne from Windsor Field as first relief, radio maintained, aircraft carried out intensive search for convoy and co-operating ship, was late on patrol, could not complete whole exercise, carried out sundry patrols as directed by SEO, made dummy attack against towed target, aircraft landed after exercise completion. Liberator NX airborne from Windsor Field as second relief, convoy located from bearing received, recognition signals exchanged, aircraft completed detailed exercise, landed at Boca Chica to refuel on outward track, radio contact established, unable to carry out homing due to technical failures, convoy located by DR [dead reckoning], carried out exercises made dummy attack against towed target, aircraft landed after exercise completion sundry units of merchant ships observed. SS *Josiah G. Holland* is bound northwest in Straits. SS *Edward L. Shea* is bound south in Straits passing Canaveral. Tanker *L. V. Stanford* is bound north in Straits near Canaveral. SS *Russell H. Crittenden* is bound southwest in Straits. Tanker *Eastern Sun* is bound south in Straits towards Palm Beach. SS *David F. Houston* is bound north passing Matanilla Shoal, northwest Bahamas. SS *Herbert L. Pratt* is bound south in Straits towards Port Everglades. SS *W. H. Libby* is bound north in Straits passing Port Everglades. Submarine USS *Blueback* (SS-326) with escort, and SS *Arlington Beach Park* are bound north in Straits near Miami. SS *O. N. Bernuth* is bound north in Straits near Miami, US Navy plane joining from east for escort. USS *YAG-21* is bound west in Straits. SS *William L. Garrison* is bound south in Straits with US Navy plane escorting. **Convoy NG-463**, SS *El Aleto*, and SS *Matinicock* are bound south in Straits. Tanker *Cities Service Kansas*, and SS *Benjamin J. Coston* are bound north in Straits. NAAF Great Exuma: A plane went to cover **Convoy NG-463**, a plane arrived from Guantanamo and departed for Banana River.

9th October 1944

B-25's FG, FM, FO, and FK are airborne for air Exercise Number 7, aircraft landed after exercise completion. B-25 FN is airborne for Anti-Submarine Patrol Number 2, aircraft landed after exercise completion, uneventful patrol except a tanker of 15,000 tons was sighted. Liberator NA is airborne from Windsor Field, radio contact established, ship located from bearing received, exercised with ship and carried out depth charge attack against towed target, exercise completed, set course for Boca Chica for refuelling *en route*, developed engine trouble, depth charges are jettisoned, and shortage of petrol experienced, signals sent to base, aircraft limped to Boca Chica. They the crew refuelled aircraft, took off, and then landed back at base. Liberator LY is airborne from Windsor Field as first relief escort duty, radio contact established, ship located by DR [dead reckoning], signals exchanged, flare dropped for ship's gunners completed exercises as directed by SOE, landed at Boca Chica, refuelled aircraft landed back at base. Liberator MX airborne from Windsor Field as second relief escort duty, emergency signals intercepted from NA, aircraft instructed to proceed with exercises after requesting instructions, commenced homing procedure, convoy located, completed exercises as directed by SOE, landed at Boca Chica, refuelled, aircraft landed back at base. Liberator MX is airborne from Windsor Field as third relief, escort duty landed at Boca Chica for refuel before proceeding with exercise, attempt made to carry out homing procedure "B" with ship, wasn't possible, ship located by DR [dead reckoning], completed exercises as directed by SOE, dummy attack carried out against towed target, aircraft landed after exercise completion. SS *O. N. Bernuth* is bound north towards Daytona. SS *Christy Payne* and SS *Vistula* are bound south in Straits passing Daytona. SS *Benjamin J. Coston* is bound north in Straits. SS *William L. Garrison* is bound south in Straits. SS *El Aleto*, tanker *Bohemian Club*, tanker *Gulfmaid*, and tanker *Gulfcrest* are bound south in Straits passing Canaveral. Tanker *Hadnot* is bound north in Straits near Palm Beach. SS *Algonquin* is bound south in Straits, passing Matanilla Shoal, northwest Bahamas. SS *Hover Lea*, SS *Francis Lewis*, SS *James B. Eads*, and SS *John Dorkweller* are bound south in Straits passing Bimini. SS *Elias Boudinot* is bound northeast in Straits. Tanker *Edward L. Shea* is bound southwest in Straits. SS *Matinicock* and SS *Russell H. Crittenden* are bound west in Straits. SS *Chatterton Hill* is bound east in Straits. NAAF Great Exuma: One plane arrived from Guantanamo and returned there with Commander Carlson. Another plane arrived from Guantanamo and departed for Banana River.

10th October 1944

B-25's JA, JB, JC, JD, and FB are airborne for air Exercise Number 7, aircraft landed after exercise completion. B-25's UV, HY, and JB are airborne for air Exercise Number 4, aircraft landed after exercise completion. B-25's UV, WN, and SP is airborne for night navigation exercise, aircraft landed after exercise completion. Liberator LU is airborne from Windsor Field for escort duty, homing procedure "B" commenced, convoy located, completed exercises as directed by SOE, attack made against dummy submarine, aircraft landed after exercise completion. Liberator LK is airborne from Windsor Field for escort duty as first relief, ships located from bearing obtained, signals exchanged, completed exercises as directed by SOE, aircraft landed after exercise completion. Liberator NX airborne from Windsor Field for escort duty as second relief, radio contact established, ships located from bearing obtained, completed exercises as directed by SOE, aircraft landed after exercise completion. Liberator LH is airborne for Radar and navigation exercises, aircraft landed after exercise completion. SS *Phoebus* is bound south in Straits passing Canaveral. **Task Unit TU-20188**, SS *Virginia*, USS *Dont* (AG-80), and SS *Marigold* are bound south in Straits towards Palm Beach. SS *Trimountain* is bound northeast in Straits near Palm Beach. SS *Gylfee* and USS *SC-675* are bound south in Straits passing Port Everglades. SS *John S. Sargent* is bound north in Straits passing Port Everglades. SS *Vistula* and USS *Knoxville* (PF-64) are bound north in Straits. USS *Shreveport* (PF-23) and tanker *Gulfmaid* are bound west in Straits. Destroyer escorts USS *Pratt* (DE-363) and USS *Romback* (DE-364) are bound northeast in Straits. SS *James Harvson* and tanker *Esso Bolivar* are bound south in Straits. NAAF Great Exuma: Two planes arrived and departed on a training flight. An Army PBY departed for Caicos Island and returned to base because of engine trouble, then later was repaired and able to return there. A plane arrived from Guantanamo and then departed for Banana River.

11th October 1944

B-25's DC, BC, and BJ are airborne for local navigation exercise, aircraft landed after exercise completion. B-25's DC and BC are airborne, and later for similar exercise, aircraft landed after exercise completion. B-25 FN is airborne Anti-Submarine Patrol Number 2, aircraft landed after exercise completion. B-25's JA, JC, FK, FB, FP, and FU are airborne for air Exercise Number 6, aircraft landed after exercise completion. Six B-25's are detailed for night navigation exercise, adverse weather cancelled flights, so no flying took place from either field, due to severe thunderstorms in the airfields. Tanker *L. V. Stanford*, tanker *Cities Service Oklahoma*, SS *Geronimo,* SS *Phoebus*, USS *YTM-330* and SS *Lake George* are bound south in Straits. SS *Jamestown* and **Convoy KN-343** are bound north in Straits. SS *Bahram* and SS *Olney* are bound south in Straits passing Canaveral. SS *San Tirso* is bound south in Straits, past Canaveral. **Task Unit TU-201917**, USS *LSM-130*, and SS *Ponca City* are bound south in Straits towards Palm Beach. USCG cutter *CG-Boone* and tanker *Axtell J. Byles* are bound south in Straits passing Palm Beach. SS *Samskern* is bound north in Straits passing Palm Beach. SS *Olney* is bound north in Straits passing Port Everglades. SS *Virginia* is bound southwest in Straits. SS *Alice Fortier* is bound northeast in Straits. NAAF Great Exuma: A plane arrived from Guantanamo to relieve a plane which departed to cover **Convoy GN-162**, an Army PBY arrived from Caicos Island, and a plane landed because of weather conditions over Guantanamo.

12th October 1944

B-25's AD, AR, and AH are airborne for local navigation exercise, aircraft landed after exercise completion, again airborne later for same exercise. B-25's JC, JD, FB, FN, FU, and FL are airborne for air Exercise Number 6, aircraft landed after exercise completion. SS *William H. Prescott* and SS *Marigold* are bound south in Straits. SS *Petter*, USS *Blower*, SS *Herbert L. Pratt*, SS *James Island*, and SS *Albert C. Titchie* are bound north in Straits. SS *Swiftlight*, SS *Newata Park*, SS *Ben Robertson*, and USS *ATR-83* are bound south in Straits passing Canaveral. USS *LSM-166* is bound south in Straits, passing Canaveral with US Navy plane escorting. SS *Josiah C. Holland*, **Convoy NK-643**, and SS *Solsten* are bound south in Straits towards Palm Beach. Tanker *L. V. Stanford*, and USS *ATR-63* are bound south in Straits towards Port

Everglades. Tanker *Cities Service Oklahoma* is bound south in Straits, passing Port Everglades. SS *Baldbutte* is bound north in Straits passing Miami. SS *California* and tanker *Cities Service Ohio* are bound northeast in Straits near Miami. SS *Cacalilao* is bound northwest in Straits towards Miami. USS *YTM-330*, SS *Ban Tirso*, SS *Olney*, and SS *Bahram* are bound southwest in Straits. SS *Kaikoura* and SS *Steelore* are bound south in Straits. NAAF Great Exuma: A plane arrived from Guantanamo and departed for Banana River, a plane departed for Guantanamo, and an Army PBY arrived from Nassau.

13th October 1944

Six B-25's are detailed for Anti-Submarine Patrol Number 3, however adverse weather delayed fight until later in the day. B-25's FG, FK, FC, FV, FU, and FS are airborne for Anti-Submarine Patrol Number 3, aircraft landed after exercise completion. A warning of a tropical storm, practically stationary, that could move in direction of Bahamas, poses no immediate danger. To be sure, the Station Commander put station under hurricane warning, While sundry consultations held, situation changed a little. USCG cutter *CG-Valor* (CG-3080), SS *W. W. Bruce*, USS *PC-574*, and SS *Beacon* are bound south in Straits. SS *Caclailao*, tanker *British Honour*, and SS *Valverde* are bound north in Straits. SS *Prometheus* and SS *Rhode Island* are bound south in Straits passing Canaveral. SS *Malay* and tanker *Gulfdisc* are bound south in Straits towards Palm Beach. USS *LST-644* and SS *Edward G. Janeway* are bound northeast in Straits near Palm Beach. SS *Horace H. Harvey* and tanker *Hadnot* are bound south in Straits towards Port Everglades. SS *Scandia* is bound northwest in Straits towards Port Everglades. Tanker *Paulsboro*, SS *Swiftlift*, SS *Ben Robertson*, and SS *W. F. Burdell* are bound southwest in Straits. SS *James Ives* is northwest in Straits. NAAF Great Exuma: Two Army PBY's arrived from Nassau.

14th October 1944

B-25's FC, FU, FS, FK, FP, and FT are airborne for Anti-Submarine Patrol Number 3, aircraft landed after exercise completion. Hurricane situation reviewed, Station Commander decided flying training should cease, crew should stand-by. SS *Wilson P. Hunt* is bound south in Straits. SS *Hoegh Hood* and SS *Medina* are bound north in Straits. SS *Arizona* and SS *Tustem*, **Task Unit TU-201918**, and USS *LSM 223* are bound south in Straits passing Canaveral. USCG cutter *CG-Valor* (CG-3080) is bound southeast in Straits towards the Northwest Providence Channel. Tanker *Gulfpoint*, SS *Esturia*, SS *W. W. Mills*, SS *Prometheus*, and SS *Herbert D. Croly* are bound north in Straits, passing Port Everglades. SS *W. W. Bruce*, tanker *British Glory*, SS *Rhode Island* and SS *Malay* are bound south in Straits. SS *Walter Jennings* is bound northeast in Straits. SS *Bennington* is bound north in Straits. NAAF Great Exuma: An Army PBY departed for Caicos Island and then departed for Nassau after returning to Exuma. A plane departed for Guantanamo.

15th October 1944

No flying training took place from the unit, as the hurricane situation remained static. The Station Commander decided to leave station under warning White. Cuban cargo schooner *Rubicon* is discovered by USN blimp, adrift in Gulf Stream with dog, but no persons. It drifted from Havana following a hurricane. Tanker *Gulfwing* is bound south in Straits passing Canaveral. **Task Unit TU-20189** is bound south in Straits towards Palm Beach. SS *W. M. Burton* and SS *Petter II* are bound north in

Straits towards Miami. USS *Diploma* (AM-221), USS *SC-1063* and USS *PC-1188* are bound southwest in Straits. Tanker *L. L. Abshire*, SS *Daylight*, and destroyer escort USS *McGinty* (DE-365) are bound northeast in Straits. SS *Tustem* is bound south in Straits towards Cay Sal Bank. NAAF Great Exuma: Two planes arrived from Caicos Island and departed for Nassau.

16th October 1944

B-25's FU, FY, FH, FO, FN, and FM are airborne for Anti-Submarine Patrol Number 2, aircraft landed after exercise completion submarine USS *Devilfish* (SS-292) escorted by submarine chaser and a Mariner aircraft are sighted. B-25's FK, FP, FG, FL, and FC are airborne for air Exercise Number 1, aircraft landed after exercise completion. B-25's FC, FL, FP, FK, FB, and FU are airborne for night navigation exercise, aircraft landed after exercise completion. Liberator LH is airborne for Radar and navigation exercise, aircraft landed after exercise completion. SS *Daylight*, tanker *British Prestige*, and **Convoy KN-344** are bound north in Straits. Tanker *Motocarline* and **Task Unit TU-201810** are bound south in Straits passing Canaveral. Tanker *Cities Service Fuel* is bound south in Straits towards Canaveral. SS *Marita* and USS *AM-280* are bound northwest in Straits near Palm Beach. SS *Empire Success* is bound south in Straits passing Palm Beach. SS *Elizah White*, tanker *Gulfwing*, SS *George Gershwin*, and SS *Edward* are bound south in Straits passing Port Everglades. SS *Bidwell* is bound north in Straits near Miami. USS *Charlotte* (PF-60) and tanker *Hagood* are bound southwest in Straits. Submarine USS *Devilfish* (SS-292) is bound west towards Northwest Providence Channel with US Navy plane escorting. SS *Erina*, Free French steam ship FFS *Sieur De La Salle* and **Convoy GN-163** are bound north in Straits. NAAF Great Exuma: Naval Air Transport seaplane arrived from Miami and went to Guantanamo then arrived from Guantanamo and went to Miami. An Army PBY arrived from Nassau.

17th October 1944

B-25's BC, KG, and V are airborne for night navigation exercise, aircraft landed after exercise completion. Liberator LH is airborne for Radar and navigation exercise, aircraft landed after exercise completion. Hurricane warning cantered off Great Cayman Island, and Station Commander decided not to alter state of hurricane warning until late evening of the 19th [in two days]. SS *Massachusetts* and tanker *Trontolite,* SS *Tide Penn* and SS *San Valerio* are bound south in Straits. Tanker *Hagood* and tanker *Eastern Sun* are bound north in Straits. Tanker *R. P. Smith* is bound northwest in Straits, passing Canaveral. **Task Unit TU-20 1811** is bound south in Straits passing Canaveral. SS *Arthur Riggs* is bound south in Straits towards Palm Beach. Tanker *Bidwell* and tanker *E. R. Kemp* are bound north in Straits, passing Port Everglades. Tanker *Motocarline* is bound southwest in Straits. No.113 reports that a B-26 Marauder HD-664, crashed into sea 400 yards off shore, shortly after take-off. Aircraft sank immediately, with all crew missing and believed killed. Crew were F/O. J. W. Wood, RCAF, pilot, F/O M. F. O'Neill, RCAF, co-pilot. Also, Hurricane status Blue warning issued, due to hurricane raging over Cuba showing signs of veering to the northwest. Tally: 28 aircraft on field, of which two are unserviceable, 26 crews, including eight US Army Air Corps, available and ordered to report for briefing 7 am 18.10.44

[tomorrow]. NAAF Great Exuma: An Army PBY departed for Morrison Field.

18th October 1944

B-26 Marauder of RAF Transport Command crashed into sea 500 yards off north shore of New Providence, to the west of north Cay [Balmoral Island], yesterday. Aircraft entirely submerged, no survivors recovered from aircraft [the two crew KIA were memorialized in a wreath-laying ceremony on the spot on 23 November, 1944 by the Commanding Officers, a Padre, and six other officers on ASR launches]. B-25's FO, FY, and FM airborne for Anti-Submarine Patrol Number 1, aircraft landed after exercise completion. B-25's FK and FU are airborne for Anti-Submarine Patrol Number 2, aircraft landed after exercise completion. B-25's FC, FB, FL, and FG are airborne for air Exercise Number 1, aircraft landed after exercise completion. After a discussion with Meteorological Officer, the Station Commander decided to issue hurricane warning of Blue, all airmen recalled to camp, and Maintenance Wing commenced servicing aircraft for evacuation later in the day. The crews of B-24's and B-25's briefed for take-off to Smyrna Field and Turner Field [both in Georgia], crew awaited instruction of Station Commander. Sundry consultations with Meteorological Officers conditions studied, decided to leave evacuation of aircraft, hurricane observed to be moving away from airfield, warning Blue left in force. SS *Chrysler Field,* SS *Cacalilao*, SS *Samuel Johnson*, SS *Norbris*, SS *City of Savannah*, and SS *Empire Lord* are bound south in Straits. Tanker *Edward L. Shea* is bound north in Straits, near Palm Beach. SS *Carleton Ellis,* SS *Argon,* SS *Vermont,* and SS *Fisher Ames* are bound south in Straits passing Palm Beach. SS *Juan Casiano* is bound north in Straits, passing Miami. SS *Massachusetts* and SS *Oklahoma* are bound south in Straits passing Bimini. No.113: Hurricane status Blue in force. Crews on standby. Probable that hurricane would not unduly affect Bahamas. All aircraft picketed down, in case of high winds at night.

19th October 1944

After consultations with Meteorological Officer the Station Commander decided to cancel hurricane warning Blue, and revert to normal flying, so that training resumed. B-25's OW and TP are airborne for air Exercise Number 5, aircraft landed after exercise completion. B-25's BL, NC, GM airborne for Anti-Submarine Patrol Number 1, aircraft landed after exercise completion. Night flying detail of Number 2 Squadron and Number 1 Squadron cancelled adverse. SS *Joseph Jefferson* and SS *Amos George Throop* are bound south in Straits. SS *Nevada* is bound north in Straits. SS *Trimountain* is bound south in Straits, passing Canaveral. SS *C. A. Canfield* is bound north in Straits towards Canaveral. SS *Oscar F. Barrett* and SS *Vav* are bound northeast in Straits, passing Palm Beach. SS *Virginia* is bound northwest in Straits passing Palm Beach. Tanker *Cities Service Kansas* south in Straits passing Palm Beach. Tanker *Gulfmaid* is bound north in Straits passing Port Everglades. No.113: The All Clear from hurricane signal is issued.

20th October 1944

B-25's FH, FV, FU, JA, and JC airborne aircraft landed after exercise completion. B-25's FH, FC, FG, FL, FM, and FO are airborne for Anti-Submarine Patrol Number 2, aircraft landed after exercise completion, sundry small ships sighted. B-25's FG, FW, FT, FD, and FP are airborne for local navigation exercises, aircraft landed after exercise completion. Liberator LH is airborne for Radar and navigation exercises, aircraft landed after exercise completion. SS *Malabar,* tanker *Edward L. Shea* and SS *Virginia* are bound north in Straits. USS *Appanoose* (AK-226), tanker *Esso Bolivar,* and tanker *Esso Aruba* are bound south in Straits. SS *Christy Payne* and SS *Lake George* are bound north in Straits near Palm Beach. SS *Vav* is bound north in Straits passing Port Everglades. SS *Fort Stevens,* tanker *Axtell J. Byles,* SS *Gylfee,* and tanker *Pan North Carolina* are bound north in Straits passing Miami. Tanker *Gulftide* and SS *Argon* are bound south in Straits. SS *Rhea* is bound northeast in Straits.

21st October 1944

B-25's FG, FL, FM, and FP are airborne for local navigation exercise, aircraft landed after exercise completion. B-25's FG, FP, FT, FH, FL, and FW are airborne for compass-swinging exercise, aircraft landed after exercise completion. B-25's JA, JB, JD, and JC are airborne for air Exercise Number 3, aircraft landed after exercise completion. B-25's FP, FT, FD, and JC are airborne for air Exercise Number 4, aircraft landed after exercise completion. Tanker *Pan North Carolina* is bound north in Straits towards Daytona. Tanker *Gulfleet* is bound north in Straits passing Canaveral. Submarine USS *Blower* (SS-325) is bound northeast in Straits with a US Navy plane escorting. SS *Herbert L. Pratt* is bound southwest in Straits. USS *Brunswick* (PF-68) is bound northeast in Straits. Tanker *Gulfpoint* and SS *Edward Canby* are bound south in Straits towards Palm Beach. SS *Algonquin* is bound north in Straits passing Palm Beach. SS *G. B. Watson* is bound south in Straits passing Palm Beach. **Task Unit TU-201811** is bound southwest in Straits towards Miami. USS *SC-1499,* SS *William Wirt,* and USS *SC-1506* are bound south in Straits. SS *Bahram,* SS *Lake George,* and SS *Rhea* are bound northeast in Straits. Tanker *Cities Service Kansas* is bound southwest in Straits. NAAF Great Exuma: A plane arrived from Guantanamo, a plane arrived from San Francisco de Macoris, Dominican Republic, an Army JRF-5 Amphibian plane arrived from Morrison Field, departed for Caicos Island, and then arrived back.

22nd October 1944

Liberator NA is airborne from Windsor Field for Kingsley Exercise, radio contact made, homing procedure "B" commenced, convoy located, patrols carried out under instruction of SOE, attack made against smoke float exercise, U-boat attack signals sent to base, aircraft and ship should have completed homing "C," but was unsuccessful, and aircraft landed after exercise completion. Liberator LK is airborne from Windsor Field for Kingsley Exercise as first relief, bearings received, frigates located, signals exchanged, smoke float dropped and attack made with a depth charge and machine gun fire, exercise U-boat attack signals sent base, co-operating ship homed by procedure "C," after exercise completion aircraft dropped flare for A-A practice by ship's gunners, aircraft landed after exercise completion. Liberator MC is airborne from Windsor Field for Kingsley Exercise as second relief, homing procedure "B" commenced, convoy located, signals exchanged, smoke float attacked with front and mid-upper guns, first depth charge hung up on first attack, third attack delivered and depth charge dropped, 10 yards overshoot estimated, aircraft sent relevant attack signals to base for exercise purpose carried out exercises as detailed by SOE, aircraft dropped flare for A-A practice by ship's gunners, aircraft landed after exercise completion. Liberator MX is airborne from Windsor

Field for Kingsley Exercise as third relief, radio contact made, homing procedure "B" commenced, convoy located, attack made against smoke float, first depth charge failed to detonate, second attack made, failed, third attack made, exercise U-boat attack signals sent to base, completed exercises, aircraft dropped flare for A-A practice by ship's gunners, aircraft landed after exercise completion. Grumman Goose GD is airborne for Grand Bahama, aircraft landed after exercise completion. B-25 FM is airborne for Miami, landed there and stayed overnight. Tanker *E. R. Kemp* is bound north in Straits. SS *Jamestown* is bound south in Straits. USCG *CG-Pandora* is bound south in Straits towards Palm Beach. SS *San Tirso* northeast in Straits near Palm Beach. SS *George Gipp* and SS *Hoegh Hood* are bound south in Straits passing Palm Beach. USS *PC-1243* south in Straits passing Port Everglades. SS *Tustem* is bound northeast in Straits near Port Everglades. USCG *CG-Valor* (CGR-3080) is bound west passing Northwest Providence Channel. Tanker *Bohemian Club* is bound northeast in Straits. Oiler USS *YO-169* is bound south in Straits towards Cay Sal Bank. SS *Chagres* is bound northwest passing Cay Sal Bank. Tanker *Gulfgem*, tanker *Vacuum*, SS *Henry M. Roberts*, tanker USS *Roxane* (AKA-37), SS *Herbert L. Pratt*, and tanker *J. A. Mowinckel* are bound south in Straits. NAAF Great Exuma: An Army JRF-5 Amphibian plane departed for Nassau, a plane departed for San Francisco de Macoris, a plane arrived from Guantanamo and a plane departed there.

23rd October 1944

Liberator NX is airborne from Windsor Field, homing exercise "B" commenced, ships located, depth charge attack carried out against a smoke float, practice U-boat reports transmitted to base, aircraft homed co-operating ships to the scene of attack, completed exercises as detailed by SOE, aircraft landed after exercise completion. Liberator NA is airborne from Windsor Field for Kingsley Exercise convoy as first relief, radio contact established, ships located after receiving bearings, attack carried out against a smoke float, practice U-boat reports transmitted to base, aircraft landed after exercise completion. Liberator LU airborne from Windsor Field for Kingsley Exercise convoy as second relief, radio contact established homing exercise "B" commenced ships located after receiving bearings Anti-Submarine attack carried out against a smoke float practice U-boat reports transmitted to base flare dropped for A-A practice by ship's gunners, aircraft landed after exercise completion. Liberator LK is airborne from Windsor Field for Kingsley Exercise convoy, third relief homing exercise "B" commenced convoy located signals exchanged attack carried out against a smoke float completed exercises as detailed by SOE, aircraft landed after exercise completion. B-25's JC, JB, FP, FF, FK, and FV are airborne for air Exercise Number 7, aircraft landed after exercise completion. B-25's FU, FT, FL, and JC are airborne for air Exercise Number 6, aircraft landed after exercise completion. Liberator LH is airborne for Radar and navigation exercise, aircraft landed after exercise completion. B-25's CX, BD, and DC are airborne for local navigation exercise, aircraft landed after exercise completion same aircraft carried out, same exercise again later. B-25 FM returned from Miami, where it had taken officers. Tanker *British Glory*, SS *Jamestown*, SS *Simon Newcomb*, and SS *Allegan* are bound south in Straits. SS *Horace H. Harvey* and **Convoy SS *Dry***

Tortugas are bound north in Straits. SS *General George S. Simonds* is bound south in Straits with a US Navy plane escorting. Tanker *Gulfdisc* and USS *YTB-338* are bound south in Straits passing Canaveral. Tanker *R. P. Smith* and SS *Solsten* are bound north in Straits towards Canaveral. **Task Unit TU-2019 19** is bound south in Straits passing Palm Beach. Tanker *Paulsboro,* USCG cutter *CG-Dione,* USS *YO-142*, USS *YO-129*, and tanker *Hadnot* are bound north in Straits, passing Port Everglades. SS *Rhode Island*, USS *Uniontown* (PF-65), and tanker *Cities Service Oklahoma*, are bound northeast in Straits. **Convoy GK-838** is bound northeast in Straits passing Nicholas Channel. NAAF Great Exuma: A Naval Air Transport seaplane arrived from Miami and went to Guantanamo then arrived from Guantanamo and went to Miami.

24th October 1944

Liberator MC is airborne from Windsor Field for Kingsley Exercise convoy-homing exercise "B" commenced, requested ships located from bearings received, attack carried out against a smoke float, practice U-boat reports transmitted to base, flare dropped for A-A practice by ship's gunners, exercises completed as detailed by SOE, aircraft landed after exercise completion. Liberator MA is airborne from Windsor Field for Kingsley exercise as first relief, homing exercise "B" commenced, attack carried out against a smoke float, practice U-boat reports transmitted to base, exercises completed as detailed by SOE, aircraft landed after exercise completion. Liberator NX is airborne from Windsor Field for Kingsley exercise as second relief, homing exercise "B" commenced, ships located, signals recognition procedure carried out, attack carried out against a smoke float, aircraft completed exercises as detailed by SOE, flare dropped for A-A practice by ship's gunners, aircraft landed after exercise completion. Liberator NA is airborne from Windsor Field for Kingsley exercise as third relief, ships located from bearings received, signals recognition procedure carried out, depth charge attack carried out against a smoke float exercise, signals transmitted, aircraft completed exercises as detailed by SOE, aircraft landed after exercise completion. B-25's FG, FP, FV, FD, and JB are airborne for air Exercise Number 3 and then for Number 4, aircraft landed after exercise completion. B-25's FL, FK, FS, FM, FF, and FY are airborne for compass-swinging exercise, aircraft landed after exercise completion. B-25's BA, AS, and CX are airborne was airborne twice for same exercise but with different crew for local navigation exercise, aircraft landed after exercise completion. Liberator LH is airborne for Radar and navigation exercise, aircraft landed after exercise completion. Grumman Goose GB is airborne Grand Bahama Island, aircraft landed after exercise completion. B-25 JA is airborne for Miami *en route* to Batista Field, landed at Miami, stayed overnight. Grumman Goose GB is airborne from Harbour Island, aircraft landed after exercise completion. SS *C. A. Canfield* is bound south in Straits, passing Daytona. **Task Unit TU-20191** and SS *Gilbert Costin* are bound south in Straits, passing Canaveral. SS *Good Gulf*, and SS *Prairie Grove* are bound south in Straits. Tanker *Beaconoil*, SS *Laguna*, and SS *Fredrick E. Ives* are bound north in Straits. SS *Abel Sterns* and SS *W. H. Libby* are bound south in Straits passing Palm Beach. USS *ATA-181* is bound north in Straits passing Miami. Tanker *British Glory* and tanker *British Honour* are bound west in Straits. SS *Malay* is bound northeast in Straits. NAAF Great

Exuma: A plane arrived from Guantanamo and returned there. A plane arrived from Guantanamo and then departed for Banana River.

25th October 1944

B-25 FB airborne for Anti-Submarine Patrol Number 2, aircraft landed after exercise completion. B-25's FW, FP, FT, FK, and FU are airborne for air Exercise Number 6, aircraft landed after exercise completion. Liberator LH is airborne for Radar and navigation exercise, aircraft landed after exercise completion. B-25 JA is airborne for Miami, aircraft landed back at base after exercise completion. B-25's BX, BD, BC, AR, and CX are airborne for local night navigation exercise, aircraft landed after exercise completion. B-25's FG, FB, FM, FV, FT, and FL are airborne for night navigation exercise, aircraft landed after exercise completion. USS *YO-511*, USCG cutter *CG-Argo, SS Tullahoma, SS El Ocena*, SS *Narbo*, SS *Tennessee*, and **Convoy NK-110** are bound south in Straits. SS *Malay*, SS *Swiftlight* and **Convoy KN-9** are bound north in Straits. SS *Setter II*, USS *LSM-224*, SS *James Cook*, and SS *Salvator* are bound south in Straits passing Canaveral. SS *James Bayard* and tanker *Gulfmaid* are bound south in Straits towards Palm Beach. SS *Walter Williman* is bound north passing Matanilla Shoal, northwest Bahamas. SS *Petter*, SS *Megna*, and tanker *Edward L. Doheny* are bound south in Straits passing Port Everglades. SS *Ponca City* is bound north in Straits passing Port Everglades. SS *C. A. Canfield* and USS *PC-1563* are bound southwest in Straits. NAAF Great Exuma: An Army JRF-5 Amphibian plane arrived from Nassau.

26th October 1944

B-25's HB, JD, FT, FN, FG, and FD are airborne for air Exercise Number 7, aircraft landed after exercise completion. B-25's FL, FM, FV, JB, and JD are airborne for air Exercise Number 6, aircraft landed after exercise completion. Grumman Goose GB is airborne for San Salvador, aircraft landed after exercise completion. B-25's FL, FC, FT, FV, FW, and FV are airborne for night navigation exercise, aircraft landed after exercise completion. SS *Dripa*, SS *Tidewater*, and **Convoy NK-646** are bound south in Straits. SS *Imperial Transport*, **Task Unit TU-201812**, and **Convoy KN-346** are bound south in Straits towards Palm Beach. SS *Petter II* is bound north in Straits, passing Palm Beach. USS *YO-170* is bound south in Straits towards Port Everglades. SS *Ansel F. Haines* and SS *Olney* are bound north in Straits passing Port Everglades. Tanker *Gulfwing* is bound north in Straits towards Miami. SS *Josiah G. Holland* and USS *Chippewa* (AO-68) are bound north in Straits. SS *Imperial Transport* is bound south in Straits. SS *Petter II* is bound southwest in Straits. SS *Elwood* is bound northwest in Straits passing Nicholas Channel. NAAF Great Exuma: A plane arrived from Guantanamo to relieve the standby plane, which then departed for Guantanamo.

27th October 1944

B-25's of Number 2 Squadron fighter affiliation exercises with B-24's of Number 3 Squadron. B-25's FJ, FU, FK, FB, and FN are airborne for compass-swinging exercise, aircraft landed after exercise completion. Grumman Goose GB is airborne for Harbour Island, aircraft landed after exercise completion. SS *Point Breeze* is bound north in Straits. Tanker *Hadnot* is bound south in Straits, passing Canaveral. SS *Empire Norse*, tanker *Gulfking*, SS *Reginald A. Fessenden*, and destroyer escort USS *Andres* (DE-45) are bound south in Straits, towards Palm Beach. SS *Opequon* is bound south in Straits passing Matanilla Shoal, northwest Bahamas. Oiler USS *YO-142* is bound southwest towards Port Everglades. USS *PCS-1385* is bound south in Straits passing Port Everglades. SS *Powder River* is bound south in Straits, passing Bimini. SS *Daylight*, SS *Mississippi*, USCG cutter *CG-Sweetbrier* (WLB-405), tanker *Pennsylvania Sun*, SS *Oscar F. Barrett*, SS *City of Montgomery*, SS *Dripa* and **Convoy Motor Torpedo Boat Squadron-32 Group C and Group D** are bound south in Straits. SS *Elwood* is bound north near Cay Sal Bank. NAAF Great Exuma: USS *YAG-19* arrived from Miami with supplies, then returned there, a plane went to cover **Convoy GN-164**. An Army JRF-5 Amphibian plane arrived from Morrison Field and then departed for Port au Prince, Haiti. An Army JRF-5 Amphibian plane departed for Caicos Island then arrived back.

28th October 1944

Aircraft of Number 2 Squadron carried out formation flying. B-25's FO, FM, FD, FH, FP, FW, and FL are airborne for compass-swinging exercise. Then FL failed to return for exercise, and was later reported crashed near bombing range. The rest of aircraft landed after exercise completion. B-25 FL was airborne for compass-swinging exercise, detailed to commence bombing on target at Fox Hill Beach, radio contact maintained, report received from civilian that aircraft had been observed to have crashed to the south of Fox Hill Road, and smoke was observed in that direction. B-25's FH and FP, which had re-joined circuit on completion of exercise diverted to accident scene, reported that object identifiable as aircraft was on fire in swampy bush, disintegrated, difficult to identify aircraft type, ambulance and fire tender dispatched to scene. Flying Instructors followed them, it was found to be a B-25 on close inspection of aircraft's wreck, and six crew killed, no recognizable distinguishing marks on aircraft to identify, it was established that all aircraft had landed back from day's flying except FL, it was presumed to be the crashed aircraft, later this was confirmed. Later reports said that aircraft had commenced bombing practice then banked steeply, crashed into bush, caught fire. Observers said that craft was not on fire while on flight. SS *James Island* is bound northeast in Straits. SS *Cacalilao* is bound northwest in Straits. SS *Malabar*, SS *A. O. Bedford*, SS *Hanleu*, and SS *Jacques Carter* are bound south in Straits. SS *Liebre* and tanker *Gulfgem* are bound south in Straits towards Palm Beach. SS *Chrysler Field* is bound north passing Matanilla Shoal, northwest Bahamas. SS *Meline* is bound south in Straits passing Port Everglades. SS *Daylight* and *Vessel 9* are bound south in Straits passing Miami. *Vessel 126* and tanker *L. L. Abshire* are bound northeast in Straits. USS *SC-1292* is bound southwest in Straits. USS *YAG-19* is bound north in Straits near Andros. NAAF Great Exuma: A plane arrived from Guantanamo with Capt. Weld and Cdr. Carlson then returned there with them. A plane arrived from Guantanamo to relieve a plane which went to cover the routes of **Convoy NG** and **Convoy GN**.

29th October 1944

SS Silver Star Park is bound south in Straits, near Key West. SS *R. G. Stewart* is bound northeast in Straits. SS *George Rogers Clark*, SS *Vermont*, tanker *Gulfwave*, and SS *Thomas McDonough* are bound south in Straits. SS *Race Point* and tanker *R. P. Smith* are bound north in Straits towards Canaveral. Tanker *Pan Virginia* and USS *YP-448* are bound south in Straits, towards

Palm Beach. USS *Sheboygan* (PF-57) is bound northeast in Straits, near Port Everglades. Tanker *Gulftide* is bound north in Straits, towards Port Everglades. USS *YAG-19* is bound southwest in Straits, past Northwest Providence Channel. SS *Jacob Thompson* is bound southwest in Straits. USS *YO-69* is bound southeast in Straits passing Nicholas Channel. NAAF Great Exuma: A plane arrived from Guantanamo to relieve a plane which went to cover the routes of **Convoy NG** and **Convoy GN**. Two planes arrived and departed on a training flight. An Army JRF-5 Amphibian plane departed for Caicos Island, and then arrived back.

30th October 1944

B-25's FB, FN, FS, FU, and FW are airborne for Anti-Submarine Patrol Number 1, aircraft landed after exercise completion. B-25's FH, FK, FU, FV, and JD are airborne for air Exercise Number 1, aircraft landed after exercise completion. Tanker *Motocarline* is bound north in Straits, northwest of Little Bahama Bank. USS *ATA-182* and tanker *Axtell J. Byles* are bound north in Straits west of Little Bahama Bank. Tanker *Delaware Sun* is bound south in Straits off Port Everglades. SS *J. D. Yeager* is bound north in Straits west of Bimini. SS *George E. Badger*, SS *John B. McCuaig* and SS *David B. Henderson* are bound south, to the east of Eleuthera. NAAF Great Exuma: A Naval Air Transport seaplane arrived from Miami and went to Guantanamo, then arrived from Guantanamo and went to Miami. An Army JRF-5 Amphibian plane departed for Nassau.

31st October 1944

B-25's BN, WK, and M are airborne for Anti-Submarine Patrol Number 1, aircraft landed after exercise completion, a tanker of 10,000 tons, a light cruiser of Omaha class, and a submarine in convoy were sighted. B-25's FW, FH, FB, FS, and JD are airborne for air Exercise Number 1, aircraft landed after exercise completion. Tanker *Motocarline* and USS *ATA-182* are bound north in Straits west of Little Bahama Bank. SS *Vermont* is bound south in Straits, off Palm Beach. SS *Axtel J. Byles* is bound north in Straits, off Palm Beach. Tanker *Delaware Sun* is bound south in Straits, off Port Everglades. SS *J. D. Yeager* is bound north in Straits west of Bimini. SS *George E. Badger*, SS *John B. McCuaig*, and SS *David B Henderson* are bound south, to the east of Eleuthera. NAAF Great Exuma: An Army JRF-5 Amphibian plane arrived from Morrison Field, two planes arrived from Guantanamo, and one departed for Banana River, then a plane departed for Guantanamo.

1st November 1944

B-25'S FB, FN, FT, FV, and FK are airborne for air Exercise Number 5, aircraft landed after exercise completion. B-25's JD, FP, FW, FH, and FV are airborne for air Exercise Number 6, aircraft landed after exercise completion. Liberator LH is airborne for Radar and navigation exercise, aircraft landed after exercise completion. Tanker *Beaconoil* is bound south in Straits, off Canaveral. SS *North Point* is bound north in Straits, northwest of Little Bahama Bank, exiting the area. SS *Prairie Glory* is north in Straits, northwest of Little Bahama Bank. Tanker *Bidwell* and SS *Horace H. Harvey* are bound south in Straits, off Palm Beach. SS *Hoegh Hood* and USS *Covington* (PF-56) are bound north in Straits, northwest of Bimini. SS *Thorshammer* is bound south in Straits, off Miami. Tanker *R. P. Smith* and SS *Christopher L. Sholes* are bound south in Straits. USS *SC-685* and *Fred J. Evans* are bound north in Straits. USS *OC-595* and **Convoy GK-840** are bound northwest in

Old Bahama Channel. NAAF Great Exuma: Army JRF-2 Amphibian departed, two planes arrived from Guantanamo, and one departed for Banana River. A Navy PBN departed for Guantanamo.

2nd November 1944

B-25's FN, FT, FB, JC, and JB are airborne for air Exercise Number 5, aircraft landed after exercise completion. B-25's FK, FU, FY, FW, and FP are airborne for Anti-Submarine Patrol Number 2, aircraft landed after exercise completion, merchant ship of between 3,000 tons and 4,000 tons was sighted. Liberator LH is airborne for Radar and navigation exercise, aircraft landed after exercise completion. Tanker *L. L. Abshire*, USS *LSM-208*, and tanker *Edward L. Shea* are bound south in Straits northwest of Little Bahama Bank. USS *YTB-8282*, tanker *Gulfdisc*, and SS *Tullahoma* are bound north in Straits northwest of Little Bahama Bank. SS *William Crompton* is bound north in Straits west of Little Bahama Bank. **Task Unit TU-20193** and tanker *Bohemian Club* are bound south in Straits, off Palm Beach. SS *Richard J. Hopkins* and tanker *British Glory* are bound north in Straits, northwest of Bimini. Tanker *Beaconoil* is bound south in Straits. USS *YMS-418* is bound north in Straits. **Convoy GK-840** is bound northwest in Nicholas Channel. SS *Francis M. Smith* is bound south, to the east of Eleuthera. NAAF Great Exuma: A plane arrived from Guantanamo and departed for Banana River.

3rd November 1944

B-25's FD, FM, FY, and FT are airborne Anti-Submarine Patrol Number 2, aircraft landed after exercise completion, convoy of a merchant ship of 7,000 tons, a tanker of 4,000 tons, and a steam ship of 5,000 tons is sighted. B-25's FK, FB, JB, and JC are airborne for air Exercise Number 3, and later Number 4, then aircraft landed after exercise completion. Liberator LH is airborne for Radar and navigation exercise, aircraft landed after exercise completion. Demonstration of modern warfare arranged on 10th November at Clifford Park with co-operation of members of this unit and Bahamas Defence Battalion, bombing rehearsal, took place. For this, B-25 FW is airborne and then dropped six empty practice bombs on Silver Cay, charges of 50 pounds of explosives detonated on Silver Cay by the Armament Officer and times to detonate as each bomb struck the target, aircraft landed after exercise completion. B-25's FB, FK, FM, FT, and FY are airborne for local navigation flights, aircraft landed after exercise completion. SS *Sands Point* and USS *YMS-418* are bound north in Straits, to the northwest of Little Bahama Bank. SS *El Aleto* is bound north in Straits west of Little Bahama Bank. SS *Cacalilao* is bound south in Straits off Palm Beach. SS *Nevada* is bound south in Straits, off Port Everglades. SS *Petter* and SS *W. F. Burdell* are bound north in Straits west of Bimini. SS *Five Forks* is bound south in Straits, west of Bimini. SS *W. F. MacLennan* is bound north in Straits. Troop transport USS *USAT-Gatun* is bound north in Santaren Channel. SS *Cape Nun* is bound south in Old Bahama Channel. SS *Furniford M. Simmons* is bound south, to the east of Eleuthera. Tanker *Mobilgas* is bound north, to the east of Eleuthera. NAAF Great Exuma: One plane arrived from Guantanamo and departed for Banana River, and four OS2U's arrived.

4th November 1944

B-25's FK, FM, FH, and FY are airborne for local navigation flights, aircraft landed after exercise completion. B-25's FN, FW, FU, JA, and JB are airborne for air Exercise Number 3 and later Number 4, aircraft landed after exercise completion. Liberator LH is airborne for Radar and navigation exercise, aircraft landed after exercise completion. Exercise in co-operation with two ships of Royal Canadian Navy organized HMCS *St. Stephen* and HMCS *Royal Mount*, ships sent on a pre-determined track at a distance of 15 nm apart, one ship detailed as Escort Commander for each day, non-conducting ship was to drop smoke float for anti-U-boat practice. Liberator MX is airborne from Windsor Field for Kingsley Exercise, radio contact established, homing "B" started after bearing received, convoy located, ship not located in calculated position on ETA, aircraft sent 'not met' signal, signals exchanged on location, non-conducting ship sighted was 9 nm away from Escort Commander, patrol Python completed, conducting ship was homed onto a smoke float by procedure "C," float had been laid by non-conducting ship, aircraft attacked with a depth charge, treated float as a U-boat, sent relative attack signal to base, exercise completed with transmission of 'drill cancelled' signal, exercise completed as detailed, aircraft left convoy, aircraft landed after exercise completion. Liberator MB airborne from Windsor Field for Kingsley Exercise as first relief, radio contact established, homing "C" started after bearing received, convoy sighted, aircraft exercised by Escort Commander completed patrol Cobra, dummy submarine attack carried out, practice signals sent to base, aircraft landed after exercise completion. Liberator NX is airborne from Windsor Field for Kingsley Exercise as second relief, homing procedure "C" requested, not completed, located convoy by DR [dead reckoning], recognition procedure completed, smoke float laid by non-conducting ship, dummy submarine attack made by aircraft, relative attack signals sent to base, exercise completed as detailed by Escort Commander, aircraft left convoy, aircraft landed after exercise completion. Liberator MC is airborne from Windsor Field for Kingsley Exercise as third relief, co-operating ship located, aircraft experienced difficulty in locating convoy, sent 'not met' signal to base during course of flight of MC, to the effect that HMCS *St. Stephens* left convoy and set sail for Nassau with an emergency appendicitis case aboard. Then ship arrived, and the patient was taken to the unit's hospital. Royal Mount was signalled to aircraft that one ship had set course for Nassau and that smoke float would be dropped for the U-boat attack, flare dropped for ship's AA gunners practice, U-boat attack signals were sent by aircraft to base during attack smoke float, aircraft landed after exercise completion. HMCS *St. Stephens* arrived at Nassau found that bearings in the engine had been burnt out due to excessive speed, repairs have to be affected before the ship could be put to sea again, ship was repaired, then sailed. SS *Pine Bluff* is bound south in Straits north of Little Bahama Bank. Tanker *Paulus Hook* is bound north in Straits northwest of Little Bahama Bank. SS *George E. Waldo* and SS *Markay* are bound south in Straits northwest of Little Bahama Bank. SS *W. H. Libby* and SS *Petter II* are bound north in Straits west of Little Bahama Bank. Tanker *Cities Service Oklahoma* is bound south in Straits off Palm Beach. SS *James Island* is bound south in Straits northwest of Bimini. SS *Oklahoma* is bound south in Straits south of Bimini. Tanker *Gulfmaid* is bound north in Straits. SS *Gylfe* and SS *New York* are bound south in Straits. HMS *HMSS Heatherton* is bound southeast in Old Bahama Channel.

SS *Elias Boudinot*, SS *Jeremath*, SS *S. Black*, SS *Ralph A. Cram*, SS *Scandia*, and SS *Permian* are bound south, to the east of Eleuthera. NAAF Great Exuma: An Army JRF-5 Amphibian plane arrived and departed, a plane arrived from Guantanamo and departed for Banana River.

5th November 1944

Liberator LY is airborne from Windsor Field for Kingsley Exercise homing "B" commenced frigate sighted, recognition signals exchanged, patrols Cobra and Python completed, smoke float was laid, aircraft carried out dummy U-boat attacks and transmitted practice attack signals to base, aircraft landed after exercise completion. Liberator LU is airborne from Windsor Field for Kingsley Exercise as first relief, homing "B" procedure requested, frigate sighted, aircraft carried out dummy U-boat attacks against a smoke float dropped by ship, practice signal relative to an attack were sent to base, patrol Cobra was completed and the exercise was terminated, aircraft landed after exercise completion. Liberator LM is airborne from Windsor Field for Kingsley Exercise as second relief, homing "B" procedure requested, frigate sighted, signals exchanged, aircraft carried out dummy U-boat attacks against a smoke float dropped by ship, practice attack signals sent to base, patrol Viper completed, aircraft landed after exercise completion. Liberator MX is airborne from Windsor Field for Kingsley Exercise as third relief, radio contact established, homing "B" procedure requested, was unsuccessful, ship located by DR [dead reckoning], aircraft late to arrive at the convoy, unable to carry out exercises, aircraft landed after exercise completion. A Beechcraft aircraft landed from Morrison Field came with Group Captain and others for conference on training policy. SS *American Merchant Mariner* is bound north in Straits northwest of Little Bahama Bank, exiting the area. **Convoy KN-348**, SS *San Tirso*, and SS *Jamestown* are bound north in Straits, northwest of Little Bahama Bank. USS *Grand Rapids* (PF-31) is bound north in Straits north of Little Bahama Bank. SS *James Hawson* is bound south in Straits, west of Little Bahama Bank. SS *Powder River* and SS *Opequon* are bound north in Straits, west of Little Bahama Bank. USS *LSM-174* is bound south in Straits off Palm Beach. SS *Francis Lewis* and SS *Massachusetts* are bound north in Straits west of Bimini. SS *Vermont* is bound south in Straits. SS *Castana* is bound north in Santaren Channel. Tanker *Esso Aruba*, SS *William J. Bryan*, SS *Jeremiah O'Brian*, and SS *Clark Howell* are bound south, to the east of Eleuthera. SS *Castana* is bound north, to the east of Eleuthera. Armistice Day Parade; a parade to Nassau Cathedral was held, under the command of W/C J. R. Bland, and was composed of four flights, including one from the Bahamas Air Service Squadron. No.113: Armistice Day Service held in Nassau Cathedral attended by RAF leadership. NAAF Great Exuma: A plane arrived at Guantanamo to relieve the plane on standby. Another plane arrived from Guantanamo and departed for Banana River.

6th November 1944

B-25's JA, JB, JD, FW, and FP are airborne for air Exercise Number 7, aircraft landed after exercise completion. B-25's FF. FD, FV, FH, FT, and FU are airborne for air Exercise Number 6, aircraft landed after exercise completion. B-25's FH, FM, FN, FP, FV, and FW are airborne for night navigation exercise, aircraft landed after exercise completion. B-24 LA airborne for

Boca Chica, landed there, aircraft landed after exercise completion. Liberator MB is airborne from Windsor Field for Kingsley Exercise, frigate sighted, signals exchanged, dummy attacks against a smoke float dropped by frigate carried out, ship's gunners given firing practice against flare dropped by aircraft, aircraft landed after exercise completion. Liberator LK is airborne from Windsor Field for Kingsley Exercise as first relief, homing "B" procedure started, frigates sighted, carried out details from the Escort Commander, dummy attacks against a smoke float carried out, practice U-boat attack signals sent to base aircraft landed after exercise completion. Liberator LY is airborne from Windsor Field for Kingsley Exercise as second relief, homing "B" procedure started, ship located, recognition signals exchanged, dummy submarine attack carried out against a smoke float, practice attack signals sent to base, flare dropped for A-A practice by ship's gunners, patrols Lizard and Crocodile completed, aircraft landed after exercise completion. Windsors return from New York via Miami, possibly aboard ferry *Jean Brillant*. Their biographer Bloch notes "there was a brief scare as German U-Boats, thought to have been long since driven from local waters, suddenly began a new long-range offensive." SS *Vav* is bound south in Straits off Canaveral. SS *Reginald A. Fessenden* and SS *Henry L. Ellsworth*, SS *Drupa* are bound north in Straits northwest of Little Bahama Bank. SS *Malay* is bound south in Straits northwest of Little Bahama Bank. Tanker *British Hope* is bound south in Straits west of Little Bahama Bank. USS *Phenakite* (PYC-25) and tanker *Gulfking* are bound north in Straits west of Little Bahama Bank. SS *Fort Churchill* is bound north in Straits off Port Everglades. SS *Bessemer* is bound north in Straits. Tanker *18 de Marzo* and USS *YO-170* are bound southeast towards Santaren Channel. Tanker *British Honour* and USS *SC-1002* is bound north in Straits north of Cay Sal Bank. **Convoy GK-841** are bound northwest in Old Bahama Channel. SS *William Cullen Bryant* is bound south, to the east of Eleuthera. Tanker *Gulfwing* and **Convoy CORT Div-67** are bound south, to the east of Eleuthera. **Convoy GN-167** and SS *Robert Stuart* are bound north, to the east of Eleuthera. NAAF Great Exuma: A Naval Air Transport seaplane arrived from Miami and went to Guantanamo, then arrived from Guantanamo and went to Miami. A plane arrived from Guantanamo, and departed for Banana River. A plane went to cover the routes of **Convoy NG** and **Convoy GN**, a US Navy PBN arrived from Guantanamo and then returned there with two more planes at different times.

7th November 1944

B-25's JB, JD, FV, FP, and FS are airborne for air Exercise Number 7, aircraft landed after exercise completion. B-25's JB, JD, FT, FV, FP, and FB are airborne for air Exercise Number 6, aircraft landed after exercise completion. B-25's FH, FN, FM, FV, and FU are airborne for night navigation exercise, aircraft landed after exercise completion. SS *Daylight* is bound north in Straits northwest of Little Bahama Bank. USS *LSM-176* and USS *LSM-226* are bound south in Straits off Palm Beach. SS *Josiah G. Holland* and tanker *Olney* are bound south in Straits off Port Everglades. SS *Salvator* is bound north in Straits northwest of Bimini. SS *Jacob Thompson*, tanker *Edward L. Doheny*, SS *Baldbutte* and SS *Sharon* are bound north in Straits west of Bimini. SS *Benjamin Silliman* is bound north in Straits. USS *YO-170* is bound southeast in Old Bahama Channel. SS *Phoebus* is bound north, to the east of Eleuthera. NAAF Great Exuma: Two planes arrived from Guantanamo, one returned back, and the other departed for Banana River. An Army PBY departed.

8th November 1944

B-25's JD, JA, PN, and TV are airborne for air Exercise Number 6, aircraft landed after exercise completion. Grumman Goose GB is airborne for Governor's Harbour returned to base later. A Beechcraft with Group Captain is airborne for Morrison Field and landed there. Tanker *Gulf of Venezuela* is bound north in Straits, northwest of Little Bahama Bank, exiting the area. Tanker *Gulftide* is bound south in Straits off Palm Beach. USS *Abiline* (PF-58) and SS *Edgehill* are bound north in Straits west of Little Bahama Bank. Tanker *Axtell J. Byles* is bound south in Straits northwest of Bimini. SS *Arkansas* is bound south in Straits off Port Everglades. SS *David Thompson* is bound south in Straits. SS *Hughli* and SS *Liebre* are bound north in Straits. SS *Cape Texas* is bound southeast in Nicholas Channel. SS *Jacob Luckenbach* is bound south, to the east of Eleuthera. SS *Steelore* is bound north, to the east of Eleuthera. NAAF Great Exuma: A plane arrived from Guantanamo and departed to Banana River.

9th November 1944

B-25's JD, FT, FO, FS, FU, and FV are airborne for air Exercise Number 6, aircraft landed after exercise completion. B-25's FT, FO, FS, FU, FV, and FD are airborne for night navigation exercise, aircraft landed after exercise completion. USS *LSM-179*, USS *ATR-75* and tows are bound south in Straits northwest of Little Bahama Bank. SS *Tustem* is bound north in Straits northwest of Little Bahama Bank. **Task Unit TU-20196** is bound south in Straits off Palm Beach. SS *Esturia* and SS *Garnet Huelings* are bound south in Straits northwest of Bimini. USCG cutter *CG-Active* is bound north in Straits west of Bimini. SS *Paul Bunyan*, SS *Matinicock*, tanker *Gulfhawk*, SS *Watertown* are bound south in Straits. **Convoy KG-742** is bound southeast in Nicholas Channel. **Task Unit TU-29611**, SS *Jose J. Acosta*, SS *Norbris*, and SS *William G. Warden* are bound south, to the east of Eleuthera. Air Chief Marshall Sir Frederick Bowhill, GBE, KCB, CMO, DSO, with Squadron Officer Lady Bowhill, WAAF, accompanied by Air Vice Marshall Marix, CP, DSO, and Air Commodore Jordan, with Group Captain McDougall, visited the unit whilst the Air Officer Commanding RAF Transportation Command was making a staff visit to No.113 South Atlantic Wing, situated at Windsor Field. NAAF Great Exuma: A plane arrived from Guantanamo and departed for Banana River.

10th November 1944

B-25 FS is airborne for Anti-Submarine Patrol Number 1, aircraft landed after exercise completion. Liberator LA is airborne on a flight to Miami, aircraft landed there. B-25 FT took part in a tattoo, held in co-operation with Pictou Highlanders, the Royal Bahamas Police Force, and Bahamas Defence Battalion at Clifford Park. The aircraft was airborne and landed. The duty of aircraft was to carry out bombing attacks, charges were detonated from ground as the dummy bombs dropped from aircraft struck target. Tanker *Eastern Sun* and **Convoy KN-349** are bound north in Straits, northwest of Little Bahama Bank. Tanker *Hagood* is bound north in Straits west of Little Bahama Bank. USS *LSM- 275*, USS *LSM-175*, **Task Unit TU-20197** are bound south in Straits off Palm Beach. SS *Glenaffaric* is bound south in Straits off Port Everglades. SS *Horace H. Wheeler* is bound north in Straits. SS *Chomedy* and tanker *Esso Bolivar* are bound south, to the east of Eleuthera. SS *Buffalo Park* is bound north, to the east of Eleuthera. NAAF Great Exuma: A plane arrived from Guantanamo and departed for Banana River, a PBY arrived and then departed, a plane took flight to test compass.

11th November 1944

B-25's FM, FS, FT, FU, FY, and FV are airborne for compass swinging exercise, aircraft landed after exercise completion. Liberator LA returned from Miami and landed. Grumman Goose GB is airborne for San Salvador and returned. Tanker *Edward L. Dohney* is bound south in Straits off Canaveral. SS *Fort Augustus*, **Convoy NK-649**, and SS *James Island* are bound north in Straits, northwest of Little Bahama Bank. Tanker *Gulfpoint* south in Straits northwest of Bimini. SS *Steel Engineer* and USS *LSM-178* are bound south in Straits. SS *Herbert. L. Pratt* and tanker *Beaconoil* are bound north in Straits. **Convoy GK-842** is bound northwest in Old Bahama Channel. USS *PCE-871* is bound southeast in Old Bahama Channel. Tanker *J. A. Mowinckel*, SS *Niobe* and SS *Georgian* are bound south, and are east of Eleuthera. NAAF Great Exuma: Two planes arrived from Guantanamo, and one departed for Banana River.

12th November 1944

SS *Morton Prince* is bound south in Straits northwest of Little Bahama Bank. SS *Markay* is north in Straits northwest of Little Bahama Bank. SS *Robert S. Bean* is bound north in Straits and northwest of Little Bahama Bank. **Task Unit TU-20198** and tanker *A. S. Hansen* are bound south in Straits, west of Little Bahama Bank. SS *Pine Bluff* is bound north in Straits, west of Little Bahama Bank. SS *William Crompton* is bound south in Straits off Port Everglades. SS *Cacalilao* and SS *Nevada* are bound north in Straits. SS *Walter Jennings* and SS *Moose Mountain Park* are bound south, to the east of Eleuthera. NAAF Great Exuma: A plane went to cover the routes of **Convoy NG** and **Convoy GN**, a plane arrived from Guantanamo, and then departed for Banana River. Two US Navy OS2U's arrived, and then departed.

13th November 1944

B-25's FD, FM, FS, FU, FV, and FY are airborne for Anti-Submarine Patrol Number 1, aircraft landed after exercise completion. B-25's FD, FU, FS, FV, and FY are airborne for air Exercise Number 1, aircraft landed after exercise completion. Grumman Goose GB is airborne for San Salvador returned later. Tanker *British Merchant* is south in Straits, northwest of Little Bahama Bank. SS *James Howson* north in Straits northwest of Little Bahama Bank. **Task Unit TU-20199** is bound south in Straits west of Little Bahama Bank. USS *YTB-283* is bound south in Straits off Palm Beach. Oiler USS *YO-142* is bound south in Straits off Miami. SS *Tidewater* and SS *Gylfe* are bound north in Straits, west of Bimini. SS *New York* and SS *Arthur Riggs* are bound north in Straits. Troop transport USS *USAT-Monterey* is bound northwest in Nicholas Channel. SS *W. F. Burdell* is bound south, to the east of Eleuthera. NAAF Great Exuma: Naval Air Transport seaplane arrived from Miami and went to Guantanamo then arrived from Guantanamo and went to Miami. A plane arrived from Guantanamo, then departed for Banana River.

14th November 1944

B-25's CL, FS, FV, FW, and FD are airborne for compass-swinging exercise, aircraft landed after exercise completion. B-25's FV, FD, FP, FS, and FW are airborne for air Exercise Number 1, aircraft landed after exercise completion. Tanker *Cities Service Kansas* is bound north in Straits, northwest of Little Bahama Bank exiting the area. SS *Tennessee* is bound south in Straits, northwest of Little Bahama Bank. Tanker *Pan North Carolina* and destroyer escort USS *Parker* (DE-369) are bound north in Straits, northwest of Little Bahama Bank. SS *Solarium* is bound south in Straits, off Palm Beach. SS *Anson Mills* and SS *Vermont* are bound north in Straits, west of Bimini. SS *Robert Neighbors* and SS *Empire Success* is bound north in Straits. SS *Toteco* northwest in Old Bahama Channel. SS *Castana* bound southeast in Old Bahama Channel. NAAF Great Exuma: Two planes arrived form Guantanamo, two departed for Banana River. An Army JRF-2 Amphibian arrived, however, its landing gear collapsed while it was taxiing up the ramp from the sea.

15th November 1944

B-25's CL, FN, FT, FP, FW, and FC are airborne for night navigation exercise, aircraft landed after exercise completion. B-25's NP and TV are airborne for air Exercise Number 3, aircraft landed after exercise completion. B-25's FW and FC are airborne twice at different times for air Exercise Number 4, aircraft landed after exercise completion. SS *Massachusetts* is bound south in Straits off Canaveral. SS *Chesapeake* is bound north in Straits, northwest of Little Bahama Bank, exiting the area. SS *William Penn* is bound south in Straits, northwest of Little Bahama Bank. **Convoy NK-650**, tanker *Beaconoil* and SS *Gabasco* are bound south in Straits, west of Little Bahama Bank. **Convoy KN-350** is bound north in Straits, west of Little Bahama Bank. USS *LSM-131* is bound south in Straits, off Palm Beach. SS *Fort Binger* is bound south in Straits off Port Everglades. tanker USS *Roxane* (AKA-37) is bound south in Straits off Miami. Tanker *Cities Service Oklahoma* and SS *Ponca City* are bound south in Straits. SS *George W. Barnes* and **Convoy KG-743** are bound southeast in Old Bahama Channel. Tanker *Beta* is bound south, to the east of Eleuthera. Capt. Warner and Capt. Adams, accredited War Artist for the Australian Government, visited the unit to gather material for pictures from the activities of the Australian troops in training at this unit. NAAF Great Exuma: An Army JRF-5 Amphibian plane departed. A plane arrived from Guantanamo and departed for Banana River.

16th November 1944

B-25's CL, FM, FP, FN, and FC are airborne for night navigation exercise, aircraft landed after exercise completion. B-25's FN, FP, FT, and FV are airborne for air Exercise Number 4, aircraft landed after exercise completion. B-25's FW and FC are airborne for air Exercise Number 3, aircraft landed after exercise completion. B-25's FC, FD, FY, FN, and FP are airborne for Anti-Submarine Patrol Number 1, aircraft landed after exercise completion. B-25 FN developed engine trouble, then set course for base. B-25 FP followed aircraft back, neither completed exercise, others completed and landed back. SS *Taishan* is bound south in Straits, north of Little Bahama Bank. SS *Frederick W. Galbraith*, SS *W. W. Bruce*, and SS *Bidwell* are bound south in Straits, northwest of Little Bahama Bank. USS *SC-512* and SS *Malay* are bound north in Straits west of Little Bahama Bank. Tanker *L. L. Abshire* is bound north in Straits, off Port Everglades. Oiler USS *YO-129* is southwest in Straits, northwest of Bimini. SS *Knute Nielsen* is bound north in Straits, west of Bimini. Tanker *Tydolgas*, SS *Massachusetts* and SS *Saint Onge* are bound south in Straits. Tanker *L. J. Vicat* is bound northwest in Nicholas Channel. SS *Fort Binger* is bound southeast in Old Bahama Channel. SS *Kent*, SS *White Rock Park*, and **Convoy GN-169** are bound north, to the east of Eleuthera. No.113: An Annual Transport Command Dance for airmen was held at the Jungle Club. The evening proved to be a huge success. Base commander attended with a party of friends. NAAF Great Exuma: A plane went to cover **Convoy NG** and **Convoy GN** routes. An Army supply boat departed, an Army PBY arrived and departed, the tug USS *ATR-21* with two lighters or barges in tow arrived from Guantanamo, a plane arrived from Guantanamo, and then departed for Banana River.

17th November 1944

B-25's FM, FC, FD, FT, and FU are airborne for Anti-Submarine Patrol Number 2, aircraft landed after exercise completion, sundry units of ships sighted. B-25's FP, FF, FT, FM, FW, and FU are airborne for air Exercise Number 2, aircraft landed after exercise completion. SS *Elisha Walker* is bound south in Straits off Port Everglades. SS *San Tirso* is bound north in Straits northwest of Little Bahama Bank. Tanker *British Honour* is bound south in Straits west of Little Bahama Bank. **Convoy KS-182** is bound south in Straits, off Palm Beach. SS *Jamestown* is bound south in Straits northwest of Bimini. Tanker *Cities Service Koolmotor* and SS *Petter* are bound south in Straits northwest of Bimini. SS *Antietam* is bound south in Straits. Tanker *Pan Virginia* is bound north in Straits. USCG *CG-Cuyahoga* is bound southeast in Old Bahama Channel. Tanker *Gulf of Venezuela* and SS *Eimindo* are bound north, to the east of Eleuthera. Flight Lt. D. Lester Brown of Nassau is awarded the Distinguished Flying Cross, or DFC. NAAF Great Exuma: Three planes arrived from Guantanamo, two planes departed there, and a plane departed for Banana River.

18th November 1944

Liberator NA is airborne from Windsor Field for Kingsley Exercise, radio contact established convoy located from bearing received, signals exchanged, radio contact established, patrol Python and other exercises carried out under direction of Escort Commander, smoke float dropped by non-conducting ship, dummy submarine attacks made by aircraft, practice U-boat signals sent to base, conducting ship homed by procedure "C" to the scene of smoke float, aircraft landed after exercise completion. Liberator LR is airborne from Windsor Field for Kingsley Exercise as first relief, homing procedure "B" used to locate ships, recognition procedure completed, patrol Python and Lizard carried out, smoke float dropped by non-conducting ship, dummy submarine attacks made by aircraft, practice U-boat signals sent to base, conducting ship homed by procedure "C," aircraft landed after exercise completion. Liberator LZ is airborne from Windsor Field for Kingsley Exercise as second relief, homing procedure "B" commenced, ships located, recognition procedure completed, patrol Python and Adder carried out under direction of Escort Commander, smoke float dropped by non-conducting ship, dummy submarine attacks made by aircraft practice, U-boat signals sent to base, conducting ship homed by procedure "C," aircraft landed after exercise completion. Liberator LY is airborne from Windsor Field for Kingsley Exercise as third relief, convoy located from bearing received, signals exchanged, recognition procedure completed, patrol Python and other exercises carried out under direction of Escort Commander, smoke float dropped by non-conducting ship, dummy attacks made by aircraft with depth charge and machine guns, practice U-boat signals sent to base, flares dropped by aircraft for A-A practice by ship's gunners, aircraft landed after exercise completion. B-25's FF, FM, FP, FV, and FW are airborne for air Exercise Number 2, aircraft landed after exercise completion. Liberator LY is airborne from Windsor Field for Kingsley Exercise, convoy located, homing procedure "B" commenced, recognition procedure completed, patrol Python and Adder carried out under direction of Escort Commander, smoke float dropped by non-conducting ship, dummy attacks made by aircraft with depth charge and machine guns, practice U-boat signals sent to base, conducting ship homed to scene of attack, aircraft landed after exercise completion. SS *Red Bank*, SS *Spondilus*, and SS *Vav* are bound north in Straits northwest of Little Bahama Bank. SS *William N. Pendleton*, SS *Phoebus*, and SS *Richard M. Pierson* are bound south in Straits northwest of Little Bahama Bank. SS *John L. Williamson* is bound north in Straits west of Little Bahama Bank. SS *Race Point* is bound south in Straits west of Little Bahama Bank. SS *Carleto Nellis* is bound south in Straits off Palm Beach. SS *Phoenix* is bound north in Straits west of Bimini. SS *Trimountain*, SS *Matinicock*, and *Swiftlight* are bound north in Straits. Tanker *Comol Cuba*, SS *Joseph Hamilton*, SS *Polycastle* and SS *Stephen J. Field* are bound south, to the east of Eleuthera. NAAF Great Exuma: Two US Navy OS2U's arrived and departed, a plane arrived from Guantanamo and departed for Banana River.

19th November 1944

Liberator LU is airborne from Windsor Field for Kingsley Exercise, convoy located, homing procedure "B" commenced, recognition procedure completed, patrol Python and Adder carried out under direction of Escort Commander, smoke float dropped by non-conducting ship, dummy attacks made by aircraft, practice U-boat signals sent to base, conducting ship homed to scene of attack, aircraft landed after exercise completion. Liberator NA is airborne from Windsor Field for Kingsley Exercise as first relief, convoy located, homing procedure "B" commenced, recognition procedure completed, patrol Python and Crocodile carried out under direction of Escort Commander, smoke float dropped by non-conducting ship, dummy attacks made by aircraft with depth charge and machine guns practice, U-boat signals sent to base, conducting ship homed to scene of attack, aircraft landed after exercise completion. Liberator NC airborne from Windsor Field for Kingsley Exercise as second relief, convoy located, homing procedure "B" commenced, recognition procedure completed, patrol Python and Viper carried out under direction of Escort Commander, smoke float dropped by non-conducting ship, dummy attacks made by aircraft with depth charge and machine guns practice, U-boat signals sent to base, conducting ship homed to scene of attack using procedure "C," aircraft landed after exercise completion. SS *Daylight* is bound south in Straits west of Little Bahama Bank. SS *Tide Penn* is bound north in Straits west of Little Bahama Bank. Tanker *Axtell J. Byles* and SS *Phoebus* are bound south in Straits northwest of Bimini. SS *Clam* is bound north in Straits northwest of Bimini. SS *Tobias E. Stansbury*, USS *YAG-19* and SS *Race Point* are bound south in Straits. SS *Elwood* is bound southeast in Nicholas Channel. SS *W. W. Mills*, SS *Wahoo Swamp*, SS *Tustem*, SS *William MacClay* and SS

Touraine are bound south, to the east of Eleuthera. NAAF Great Exuma: An Army plane arrived then departed, two planes arrived from Guantanamo, one departed there and a plane departed for Banana River.

20th November 1944

Liberator NA is airborne from Windsor Field for Kingsley Exercise, convoy located, homing procedure "B" commenced, recognition procedure completed, patrol Python carried out under direction of Escort Commander, smoke float dropped by non-conducting ship, dummy attacks made by aircraft, practice U-boat signals are sent to base, conducting ship homed to scene of attack using procedure "C," aircraft landed after exercise completion. Liberator LY is airborne from Windsor Field for Kingsley Exercise as first relief, convoy located, homing procedure "B" commenced, patrol Lizard and Alligator carried out under direction of Escort Commander, escort changed, ships slightly delayed, aircraft carried out recognition procedure with other ship, smoke float dropped dummy attacks made by aircraft with depth charge, practice U-boat signals sent to base, conducting ship homed to scene of attack, aircraft landed after exercise completion. Liberator LZ airborne from Windsor Field for Kingsley Exercise as second relief, homing commenced, ships located, smoke float dropped, dummy attacks made by aircraft with depth charge, practice U-boat signals sent to base, homed conducting ship to scene of float, escort commander directed aircraft to return to base due to adverse weather, aircraft landed after exercise completion. B-25's FU, FW, FY, FC, and FN are airborne for Anti-Submarine Patrol Number 2, aircraft landed after exercise completion, a ship of 6,000 tons is sighted. B-25's FN, FF, FD, FO, and FY are airborne for air Exercise Number 7, aircraft landed after exercise completion. Tanker *Gulfmaid* and **Convoy KN-351** are bound north in Straits, northwest of Little Bahama Bank. SS *Bostonian* is bound south in Straits, west of Little Bahama Bank. USS *YTM-460* is bound north in Straits, west of Little Bahama Bank. SS *Charles William Post* and SS *Tobias E. Stansbury* are bound south in Straits off Port Everglades. Tanker *Bohemian Club* and tanker *Gulfcrest* are bound north in Straits, west of Bimini. **Convoy Key West** is bound south in Straits. USS *Kochab* (AKS-6) and SS *Fenja* are bound north in Straits. **Convoy KG-744** is bound southeast exiting Nicholas Channel. SS *Empire Voice* and SS *Philips Wouwerman* are bound north, to the east of Eleuthera.

21st November 1944

B-25's FF, FY, FP, FM, FW, and FN are airborne for air Exercise Number 7, aircraft landed after exercise completion. B-25's FN, FC, FG, FD, FO, and FV are airborne for air Exercise Number 6, aircraft landed after exercise completion. SS *Salvator* is bound south in Straits northwest of Little Bahama Bank. SS *Yellow Tavern* and SS *William Crompton* are bound north in Straits northwest of Little Bahama Bank. Tug USS *ATA-214* and **Convoy NK-651** is bound south in Straits northwest of Little Bahama Bank. SS *Prometheus* and SS *Milledgville* are bound south in Straits off Palm Beach. SS *Point Breeze* and tanker *Gulfwax* are bound north in Straits west of Little Bahama Bank. Tanker *Gulfpoint* and SS *Rhode Island* are bound north in Straits. SS *George W. Barnes* is bound northwest from Old Bahama Channel, entering Santaren Channel. **Convoy GN-170** is bound north, to the east of Eleuthera. NAAF Great Exuma: A plane arrived from Guantanamo to relieve second plane.

22nd November 1944

B-25's FN, FD, FP, FU, and FT are airborne for air Exercise Number 7, aircraft landed after exercise completion. B-25's FN, FC, FP, FD, and FT are airborne for air Exercise Number 6, aircraft landed after exercise completion. USCG cutter *CG-Argo*, a navy plane, SS *Niobe*, SS *Tidewater*, and USCG *CG-83505* are bound south in Straits northwest of Little Bahama Bank. **Task Unit TU-201913** is bound south in Straits off Palm Beach. SS *Oldham* is bound north in Straits west of Little Bahama Bank. Tanker *Hagood* and tanker *Eastern Sun* are bound south in Straits. SS *Francis B. Ogden* is bound north in Straits. SS *Fort Binger* and **Convoy GK-844** and northwest in Old Bahama Channel. NAAF Great Exuma: A plane went to cover the **Convoy NG** and **Convoy GN** routes, then departed to Guantanamo. A plane arrived from Guantanamo, and departed for Banana River.

23rd November 1944

Liberator LF is airborne for Miami and landed there. Grumman Goose GA is airborne for Guantanamo *en route* for Palisadoes, Jamaica, then landed there. B-25's FN, FC, FD, FF, FW, and FU are airborne for air Exercise Number 7, aircraft landed after exercise completion. USS *YTM-460* is bound north in Straits, northwest of Little Bahama Bank. **Task Unit TU-201914** is bound south in Straits, west of Little Bahama Bank. SS *Phoebus* is bound north in Straits west of Little Bahama Bank. SS *San Melito* and USS *ATA-180* are bound south in Straits, west of Little Bahama Bank. Bahamas ferry *Ena K.* is bound north in Straits west of Bimini. SS *Toteco* is bound north in Straits. Troop transport USS *USAT-John R. R. Hannay* is bound southeast in Old Bahama Channel. SS *Josiah Macy* and tanker *Flomar* are bound south, to the east of Eleuthera. No.113: A memorial service was held just west of North Cay [Balmoral Island], on one of the RAF Air Sea Rescue launches over the spot where B-26 Marauder RD-664 crashed into the sea on 17 October, 1944 [killing F/O. J. W. Wood, RCAF, pilot, and F/O M. F. O'Neill, RCAF, co-pilot]. The service was taken by Padre Flint of No.111 OTU, and the Commanding Officer, W/Cdr. Baskett, DFC and six officers were on board. A wreath was dropped overboard and photographs of the ceremony were taken. NAAF Great Exuma: A plane arrived from Guantanamo and departed for Banana River.

24th November 1944

B-25's FG, FF, FC, FT, and FN airborne for Anti-Submarine Patrol Number 1, aircraft landed after exercise completion. B-25's FF, FC, FD, FG, and FT are airborne for local navigation exercise, aircraft landed after exercise completion. B-25's FW, FP, FY airborne for formation flying, Duty Flying Control Officer reported that radio conversation had been intercepted, appears that an aircraft had ditched. The captain of B-25 FY confirmed that B-25 FP had ditched, and B-25 FW had gone straight to sea, so B-25 FS circled scene of crash air, the ASR air sea rescue launch was dispatched from Montagu Bay, picked up two survivors of FW, no trace of FP, and no survivors were found. Grumman Goose GB is airborne to co-operation search, B-25 FF airborne to relieve B-25 FS ,which was circling position of crash. SS *Herbert L. Pratt* and tanker *Gulfmaid* are bound south in Straits northwest of Little Bahama Bank. Tanker *Pan North Carolina,* SS *Toteco* and tanker *A. S. Hansen* are bound north in Straits northwest of Little Bahama Bank. SS *Grand River* is bound north in Straits west of Little Bahama Bank. SS *Nevada* is bound south in Straits. SS *Tennessee* is bound north in Straits. USS *YAG-21* is bound south, to the east of Exuma Cays. Tanker *Esso Aruba* is bound south, to the east of Eleuthera. SS *John Bidwell* is bound north, to the east of Eleuthera. NAAF Great Exuma: Two OS2U's arrived and departed. One plane arrived from Guantanamo, and departed for Banana River. USS *YAG-21* arrived from Miami with supplies. Tug USS *ATR-21* departed to Guantanamo with two barges.

25th November 1944

B-25's FC, FN, FT, FG, and FD are airborne for local navigation flights, aircraft landed after exercise completion. B-25's FY, FC, FF, and FT are airborne for Anti-Submarine Patrol Number 2, aircraft landed after exercise completion. Tanker *Beaconoil* is bound north in Straits northwest of Little Bahama Bank, exiting the area. SS *Josiah G. Holland* and SS *Reginald A. Fessenden* are bound north in Straits northwest of Little Bahama Bank. **Convoy KN-352** and tanker *Petrofuel* are bound north in Straits west of Little Bahama Bank. Tanker *Trontolite* is bound south in Straits west of Little Bahama Bank. SS *Massachusetts* is bound north in Straits, west of Bimini. Tanker *British Merchant* and destroyer escort USS *Pesley* (DE-371) are bound north in Straits. **Convoy KG-745** is bound southeast in Old Bahama Channel. SS *Rondeau* is bound south, to the east of Eleuthera. NAAF Great Exuma: USS *YAG-21* departed for Miami, two planes arrived from Guantanamo, a plane departed there, and a plane departed for Banana River.

26th November 1944

Grumman Goose GA is airborne for Grand Turk Island, then landed at Nassau. SS *Solsten* is bound south in Straits off Canaveral. SS *Jean Baptiste Lemoyne*, SS *Edward R. Squibb*, and tanker *Cities Service Kansas* are bound south in Straits, northwest of Little Bahama Bank. **Convoy NK-652** and USS *LSM-129* are bound south in Straits west of Little Bahama Bank. SS *Permian* is bound south in Straits, northwest of Little Bahama Bank. USS *YAG-21* is bound west in Northwest Providence Channel. SS *Tydolgas* is bound north in Straits, northwest of Bimini. SS *New York* is bound south in Straits, northwest of Bimini. SS *Ponca City* is bound north in Straits, west of Bimini. SS *William Penn*, SS *Petter*, and SS *Esturia* are bound north in Straits. Destroyer USS *Broome* (DD-214) is bound north in Santaren Channel. SS *Thorsholm* and SS *Clearwater Park* are bound south, to the east of Eleuthera. **Convoy GN-171** and tanker *Comol Cuba* are bound north, to the east of Eleuthera. USCG cutter *CG-Ingham* is bound south, to the east of Eleuthera. NAAF Great Exuma: A plane arrived from Guantanamo, a plane departed there, and an Army PBY arrived.

27th November 1944

B-25's FC, FD, FT, FY, FO, and FS are airborne for air Exercise Number 5, aircraft landed after exercise completion. B-25's FT, FO, FU, FS, and FF are airborne for air Exercise Number 1, aircraft landed after exercise completion. SS *Bostonian* is bound north in Straits northwest of Little Bahama Bank, exiting the area. Tanker *Hagood* is bound north in Straits northwest of Little Bahama Bank. Tanker *A. S. Hansen* is bound south in Straits northwest of Little Bahama Bank. SS *Bidwell*, SS *William B. Travis*, and SS *Swiftscout* are bound south in Straits, west of Little Bahama Bank. SS *John H. Hammond* and SS *Vermont* are bound north in Straits, northwest of Bimini. SS *Solsten* is bound north in Straits. **Convoy GK-845** is bound northwest in Nicholas Channel. NAAF Great Exuma: A US Naval Air Transport seaplane arrived from

Miami and went to Guantanamo, then arrived from Guantanamo and went to Miami. Two OS2U's arrived then departed, a plane arrived from Guantanamo and departed for Banana River, an Army PBY departed, and another plane departed on a mission.

28th November 1944

B-25's DC, FT, FO, FV, and FG are airborne for air Exercise Number 5, aircraft landed after exercise completion. B-25's FT, FO, FV, FG, and FF are airborne for air Exercise Number 1, aircraft landed after exercise completion. USS *LST-773* is bound north in Straits, northwest of Little Bahama Bank, exiting the area. SS *Empire Chief* is bound south in Straits, northwest of Little Bahama Bank. SS *E. R. Kemp* and tanker *Cities Service Koolmotor* are bound north in Straits, west of Little Bahama Bank. Tanker *Axtell J. Byles* and destroyer escort USS *Andres* (DE-45) are bound north in Straits northwest of Bimini. Tanker *Edward L. Doheny* is bound south in Straits. USS *PC-1265* is bound west in Straits. USS *PC-593*, USS *PC-493*, and USS *PC-494* are bound northwest in Santaren Channel. Tug USS *ATA-182* is bound south, to the east of Eleuthera. NAAF Great Exuma: One plane arrived from San Juan and departed for Banana River. A plane arrived from San Juan, and departed for Banana River.

29th November 1944

B-25's FT, FO, and FU are airborne for air Exercise Number 3, then aircraft landed after exercise completion. B-25's FG, and FD are airborne for air Exercise Number 6, aircraft landed after exercise completion. B-25's FT, FO, and FU are airborne for air Exercise Number 4, aircraft landed after exercise completion. B-25's FG, and FT are airborne for air Exercise Number 3, aircraft landed after exercise completion. Liberator NJ are airborne for Boca Chica, aircraft landed after exercise completion. B-25's BO, SV, and JG are airborne for night navigation exercise, aircraft landed after exercise completion. Tanker *Gulftide* is bound north in Straits, northwest of Little Bahama Bank, exiting the area. SS *Cottonwood Creek* is bound north in Straits northwest of Little Bahama Bank. **Task Unit TU-201818** is bound south in Straits off Palm Beach. Submarine USS *R-5* is bound north in Straits west of Little Bahama Bank. US Navy plane is bound north in Straits, west of Little Bahama Bank. **Task Unit TU-201116** is bound south in Straits off Port Everglades. SS *Empire Chief* is bound south in Straits. SS *Cape Pembroke* bound southeast in Old Bahama Channel. SS *Knute Rockne* and SS *Swiftlight* are bound south, to the east of Eleuthera. SS *W. F. Burdell* is bound north, to the east of Eleuthera. NAAF Great Exuma: An Army JRF Amphibian arrived then departed, and another plane arrived from Guantanamo and departed for Banana River.

30th November 1944

B-25's FB, FO, FU, FG, and FN are airborne for air Exercise Number 3 and Number 4, aircraft landed after exercise completion. B-25's FB, FT, FS, FO, and FU are airborne for night navigation exercise, aircraft landed after exercise completion. **Convoy NK-653** and tanker *Pan Virginia* are bound south in Straits, northwest of Little Bahama Bank. SS *Stephen Austin* is bound south in Straits, west of Little Bahama Bank. **Convoy KN-353** is bound north in Straits, west of Little Bahama Bank. SS *Tide Penn* is bound south in Straits off Port Everglades. Tanker *Edward L. Shea* is bound north in Straits west of Bimini. **Convoy KG-746** is bound southeast in Old Bahama Channel. SS *Pleiades* is bound south, to the east of Eleuthera. Tanker

Gulfwave is bound north, to the east of Eleuthera. NAAF Great Exuma: A plane arrived from Guantanamo and departed for Banana River.

1st December 1944

B-25's BV, OU, and DJ are airborne for Anti-Submarine Patrol Number 2, then the aircraft landed after exercise completion convoy of three ships sighted. B-25's FU, FS, FO, FV, and FY are airborne for air Exercise Number 2, aircraft landed after exercise completion. B-25 BX is airborne and landed at Miami. Tanker *Edward L. Shea* is bound north in Straits, northwest of Little Bahama Bank, exiting the area. SS *Tidewater* and tanker *Gulfcrest* are bound north in Straits, northwest of Little Bahama Bank. SS *Clam* and SS *Murfreesboro* are bound south in Straits, northwest of Little Bahama Bank. **Task Unit TU-201917** is bound south in Straits, off Palm Beach. SS *Daylight* and tanker *Bostonian* are bound north in Straits, west of Little Bahama Bank. Tanker *E. R. Kemp* and SS *Garnet Hulings* are bound south in Straits northwest of Bimini. Tanker *Paco* is bound south in Straits. SS *Glenaffarice* is bound north in Straits. USS *Chub* (SS-329) is inbound, west in the Northeast Providence Channel. SS *F. H. Bedford* Jr., tanker *Reginolite* and SS *Gulf of Mexico* are bound south, to the east of Eleuthera. NAAF Great Exuma: One plane arrived from Guantanamo and returned there. Another plane arrived from San Juan and departed for Banana River.

2nd December 1944

B-25's FO, FU, FS, FC, and FT are airborne for Anti-Submarine Patrol Number 2, however FC and FT returned early, due to adverse weather. Sundry units of single ships were sighted, others landed after exercise completion. B-25's FU, FD, FT, and FB are airborne for air Exercise Number 2, aircraft landed after exercise completion. B-25's FT and FO are airborne for air Exercise Number 7, aircraft landed after exercise completion. B-25 FX returned from Miami and landed back at base. An exercise in co-operation with seaplane tender USS *Sandpiper* (AVP-9)'s cooperation is organized so that four aircraft took part for three consecutive days. Liberator NA is airborne from Windsor Field for Esco exercise, radio and homing commenced, ships located, signals and recognition procedures completed, aircraft completed exercise as directed by ship. However the sea is too rough for lowering of target from co-operating ship, sea marker dropped, attack made by aircraft with depth charge, practice U-boat attack carried out, signals sent to base, aircraft landed after exercise completion. Liberator NO is airborne from Windsor Field for Esco exercise as first relief, convoy sighted homing commenced, patrol Adder and other exercises completed, aircraft unable to carry out dummy U-boat attacks, smoke float failed to ignite when lowered from ship, left convoy, landed at Boca Chica, refuelled, aircraft landed after exercise completion. Liberator NC airborne from Windsor Field for Esco exercise as second relief, developed engine trouble, jettisoned the depth charges, captain decided to land at Boca Chica, repaired, retuned to base. Liberator LR airborne from Windsor Field for Esco exercise as third relief, ship located, aircraft unable to carry out dummy U-boat attacks, signals transmitted to base, patrol Lizard and other exercises completed, aircraft landed after exercise completion. SS *Mobile City* is bound north in Straits west of Little Bahama Bank. SS *Trimountain* and USS *Pirate* (AM-275) are bound south in Straits off Port Everglades.

Task Unit TU-201817 is bound south in Straits. Tanker *Eastern Sun*, SS *Niobe*, and SS *Cacalilao* are bound north in Straits. USS *PC-492* is bound northwest in Santaren Channel. USS *PCE-842* northwest in Old Bahama Channel. NAAF Great Exuma: A plane arrived from Guantanamo and departed for Banana River.

3rd December 1944

Liberator LU is airborne from Windsor Field for Esco exercise, homing procedure "B" commenced, convoy located, recognition signal sent, carried out patrol, Adder and Alligator carried out under direction of co-operating ship, practice U-boat attack carried out against smoke float dropped by ship practice signals sent to base aircraft landed after exercise completion. Liberator NO is airborne from Windsor Field for Esco exercise as first relief, homing procedure "B" commenced, convoy located, recognition signal carried out, practice U-boat attack carried out against smoke float dropped by ship, practice signals sent to base, aircraft landed after exercise completion. Liberator MD is airborne from Windsor Field for Esco exercise as second relief, ship located, recognition signal carried out, patrol Alligator completed, practice U-boat attack carried out against smoke float dropped by ship, practice signals sent to base, landed at Boca Chica to refuel aircraft, landed after exercise completion. Liberator LT is airborne from Windsor Field for Esco exercise as third relief, homing procedure "B" commenced, dummy U-boat attack carried out against smoke float dropped by ship, practice signals sent to base, exercises carried out under direction of Escort Commander, aircraft landed after exercise completion. Tanker *Eastern Sun* is bound north in Straits northwest of Little Bahama Bank, exiting the area. SS *San Tirso* is bound south in Straits north of Little Bahama Bank. Tanker *Hagood* is bound north in Straits north of Grand Bahama. SS *Gold Mouth*, USS *LCI(L)-398*, destroyer escort USS *Brennan* (DE-13), and tanker USS *Roxane* (AKA-37) are bound south in Straits west of Little Bahama Bank. USS *ATA-200* and USS *YO-129* are bound north in Straits northwest of Bimini. Tanker *Goldwyne* is bound north in Straits. Tug USS *Iuka* (ATO-37) towing USS *YC-727* is bound northwest in Santaren Channel. SS *Michigan* and tanker *Gulf Breeze* are bound south, to the east of Eleuthera. NAAF Great Exuma: Two planes arrived from Guantanamo, a plane departed there, and a plane departed for Banana River.

4th December 1944

B-25's FB, FG, FT, FO, and FD are airborne for air Exercise Number 2, aircraft landed after exercise completion. B-25's FG, FO, FS, FT, and FV are airborne for air Exercise Number 7, aircraft landed after exercise completion. Liberator LR airborne from Windsor Field, convoy located, signals exchanged, smoke float laid by ship, dummy U-boat attack carried out against smoke float dropped by ship, practice signals sent to base, aircraft landed after exercise completion. Liberator MX is airborne from Windsor Field as first relief, homing procedure "B" commenced, ship located, patrols and exercises completed under direction of ship's commander, smoke float dropped by ship, practice signals sent to base, aircraft landed after exercise completion. Liberator NA is airborne from Windsor Field as second relief, homing procedure "B" commenced, ship located, smoke float dropped by ship, practice signals sent to base, aircraft landed after exercise completion. Liberator MB airborne from Windsor Field as third relief, contact made with ship, smoke float dropped by ship, practice signals sent to base, aircraft landed after exercise completion. Tanker *Pan North Carolina* is bound north in Straits northwest of Little Bahama Bank. SS *William Crompton* is bound south in Straits west of Little Bahama Bank. SS *Telfair Stockton* is bound south in Straits off Palm Beach. SS *Durango* north in Straits northwest of Bimini. Tanker *L. V. Stanford* is bound south in Straits off Miami. Tug USS *Irka* (AT0-37) is towing USS *YG-55* northwest in Straits. SS *Little Butte* is bound north in Straits. Tanker *Gulfpride* is bound south, to the east of Eleuthera. NAAF Great Exuma: Naval Air Transport seaplane arrived from Miami and went to Guantanamo then arrived from Guantanamo and went to Miami.

5th December 1944

B-25's DC, FB, FO, FS, FJ, and FD are airborne for air Exercise Number 8, aircraft landed after exercise completion. B-25's FT, FV, FB, FO, and FN are airborne for air Exercise Number 7, aircraft landed after exercise completion. SS *Durango* is bound north in Straits northwest of Little Bahama Bank, exiting the area. SS *Waxhaws* is bound south in Straits northwest of Little Bahama Bank. **Convoy SS *Nola*-58** is bound south in Straits northwest of Grand Bahama. SS *Massachusetts*, **Convoy KN-354**, and USS *PC-1276* are bound south in Straits west of Little Bahama Bank. SS *Vav* and SS *Ponca City* are bound south in Straits. SS *Watertown* and tanker *W. F. Burdell* are bound north in Straits. SS *Josiah Macy* is bound northwest in Old Bahama Channel. **Convoy KG-747** is bound southeast in Old Bahama Channel. NAAF Great Exuma: A plane arrived from Guantanamo and departed for Banana River.

6th December 1944

B-25'S FN, FB, FJ, FD, and FF are airborne for air Exercise Number 6, aircraft landed after exercise completion. B-25's FN, FB, FV, FC, and FU airborne for local night navigation flight, aircraft landed after exercise completion. SS *Josiah G. Holland* is bound south in Straits west of Little Bahama Bank. USS *YO-171* is bound north in Straits off Palm Beach. SS *Nevada*, USS *ATA-216*, SS *Watertown*, USS *YMS-419* and SS *Josiah Macy* are bound north in Straits west of Little Bahama Bank. SS *Josiah G. Holland* is bound south, towards Bimini. Tug USS *ATR-76* and tow are bound south in Straits off Port Everglades. Tanker *Trontolite* is bound north in Straits off Miami. **Convoy GK-847** and USS *PC-489* are bound northwest in Old Bahama Channel. USS *Esclante* (A0-70) is bound north, to the east of Eleuthera. Tanker *Beta* and ferry *Ena K.* are bound south, to the east of Eleuthera. USS *Arizona* is bound north, to the east of Eleuthera. NAAF Great Exuma: An OS2U arrived from Fort Lauderdale with report that two OS2U's were down. An OS2U, an Army JRF Amphibian, a plane and two rescue boats departed on search for those two OS2U's, both sighted by an Army JRF Amphibian, one of which was at Great Ragged Island.

7th December 1944

B-25's FN, FB, FV, FC, and FU airborne for local night navigation flight, then aircraft landed after exercise completion. B-25's FC and FN airborne for Anti-Submarine Patrol Number 2, aircraft landed after exercise completion. B-25's FN, FB, FD, FC, and FF are airborne for air Exercise Number 6, and aircraft landed after exercise completion. B-25 FH is airborne to provide radio for guard for four Royal Navy Harvards, flying to Trinidad as far as Gamaquey, then aircraft landed after exercise completion. B-24 NJ is airborne for Boca Chica, aircraft landed after exercise completion. SS *Fort Augustus* and tanker *Bidwell* are bound north in Straits northwest of Little Bahama Bank. Tanker *Axtell J. Byles*, and SS *Antietam* are bound south in Straits west of Little Bahama Bank. Tanker *Alpha* and SS *Permian* are bound north in Straits, west of Little Bahama Bank. Tanker *Gulfking* is bound south in Straits, northwest of Bimini. Oiler USS *YO-129* is bound south, to the north of Nassau. Tanker *Beaconoil* is bound south in Straits, off Miami. SS *Olney* and SS *Narbo* are bound north in Straits. SS *Nemiskam Park* is bound northwest in Old Bahama Channel. SS *Aquarius*, tanker *Gulf of Venezuela*, SS *Edward Livingstone*, **Convoy NG-475** and SS *Moose Mountain Park* are bound south, to the east of Eleuthera. No.113: Four Royal Navy Harvard aircraft landed at this field en route Trinidad. NAAF Great Exuma: An Army rescue boat departed for Great Ragged Island to refuel an OS2U which then arrived at base, two planes arrived from Guantanamo, one was damaged on the seaplane ramp, and the other departed for Banana River. Gasoline barge USS *YOG-5* and submarine chaser USS *SC* escort arrived from Guantanamo.

8th December 1944

B-25's DF, BG, and C are airborne for Anti-Submarine Patrol Number 1, aircraft landed after exercise completion. Grumman Goose GA born for San Salvador returned to base later. SS *Petter* is bound south in Straits west of Little Bahama Bank. SS *Fort Rideley* and SS *Fort Augustus* are bound north in Straits northwest of Bimini. SS *Nemiskam Park* and SS *Swiftscout* are bound north in Straits. SS *William Penn* and SS *H. M. Fredricksen* are bound south in Straits. USS *PC-495*, USS *PC-609*, USS *PC-594*, and USS *PCE-844* are bound northwest in Old Bahama Channel. SS *Pocahontas* is bound south, to the east of Eleuthera. NAAF Great Exuma: Three OS2U's departed, two planes arrived from Guantanamo, a plane departed there, and a plane departed for Banana River.

9th December 1944

B-25's FC, FB, FF. FG, and FD are airborne for Anti-Submarine Patrol Number 1, aircraft landed after exercise completion. USS *YMS-419* and SS *Empire Chief* are bound north in Straits west of Little Bahama Bank. USCG *CG-83456*, USCG *CG-83504*, and USCG *CG-83454* are bound south in Straits off Port Everglades. USS *New Bedford* (PF-71), and SS *James Howson* are bound north in Straits northwest of Bimini. SS *New Durn Park* and SS *George E. Badger* are bound north, to the east of Eleuthera. **Task Group TG-22.5**, with SS *Cape Boyer* and SS *John Drake Sloat* are bound south, to the east of Eleuthera. **Convoy KG-748** is bound west in Straits, west of Cay Sal Bank. NAAF Great Exuma: A plane arrived from Guantanamo and departed for Banana River.

10th December 1944

SS *George Kenny*, SS *James G. Maquine*, SS *James Bowie*, SS *John E. Ward*, SS *John G. Whittier*, and SS *William Moultrie* are bound south in Straits north of Little Bahama Bank. **Convoy KN-355** and USCG cutter *CG-Argo* are bound north in Straits northwest of Little Bahama Bank. SS *Rhode Island* is bound south in Straits west of Little Bahama Bank. **Task Unit TU-20184** is bound south in Straits off Palm Beach. SS *Murfreesboro* is bound north in Straits northwest of Bimini. Oiler USS *YO-129* is bound west in Northwest Providence Channel. USS *Reading* (PF-66) is bound north in Straits west of Bimini. SS *James Island* is bound south in Straits west of Bimini. SS *Charles L. McNary* is bound south in Straits. Tanker *Gulfdisc* is bound north in Straits. **Convoy KG-748** is bound southeast in the Old Bahama Channel. SS *W. W. Millis* and SS *Fort LeHavre* are bound south, to the east of Eleuthera. NAAF Great Exuma: Two planes arrived from Guantanamo, two planes departed there, and a plane departed for Banana River.

11th December 1944

B 25's FB, FC, FD FF, and FG are airborne for air Exercise Number 5, aircraft landed after exercise completion. B-25's FB, FJ, FD, FU, FG, and FK are airborne for air Exercise Number 1, aircraft landed after exercise completion. USS *SC-1059* sinks after running aground between Nassau and Harbour Island, at Fleeming Channel. Over weeks was salvaged, towed to Miami and repaired. SS *Tide Penn* is bound north in Straits northwest of Little Bahama Bank. SS *Tustem*, tug USS *ATR-10* towing three USS *YCF*'s, USS *Nimble* (AM-266), SS *Baldbutte*, and **Convoy NK-655** are bound south in Straits northwest of Little Bahama Bank. Tanker *E. R. Kemp* is bound north in Straits northwest of Little Bahama Bank. Free French Navy submarine *Amazone,* USS *YO-171*, SS *Malabar* and SS *Cacalilao* are bound south in Straits west of Little Bahama Bank. SS *Edward L. Shea* is bound south in Straits off Palm Beach. SS *Medina* south in Straits northwest of Bimini. **Task Unit TU-0318** is bound east in Northwest Providence Channel. SS *W. M. Burton* is bound south in Straits off Port Everglades. SS *Matinicock* is bound south in Straits. USS *PCE-845* and troop transport USS *USAT-Florida* are bound northwest in Nicholas Channel. No.113: Nassau Harbour; a Coronado JX-494 carried away from anchorage, close to shore when chain from buoy to the buoy anchor snapped during sudden squall. Starboard elevator was damaged, with no casualties. NAAF Great Exuma: A Naval Air Transport seaplane arrived from Miami and went to Guantanamo. A plane arrived from Guantanamo and the tug USS *ATR-21* departed there.

12th December 1944

B-25'S FV, FD, FG, FH, and FK area airborne for air Exercise Number 5, aircraft landed after exercise completion. B-25's FC, FU, FD, FS, and FV are airborne for air Exercise Number 1, aircraft landed after exercise completion. Grumman Goose GA is airborne on a flight to Grand Bahama returned later. B-24 LO is airborne for Miami returned to base later. Liberator LH is airborne for Radar and navigation exercise, aircraft landed after exercise completion. SS *Clam* is north in Straits northwest of Little Bahama Bank, exiting the area. SS *Mississippi* is bound north in Straits northwest of Little Bahama Bank. Tanker *Bostonian*, tanker *Gulfcrest*, and tug USS *ATR-10* are bound south in Straits northwest of Little Bahama Bank. SS *Wauhatchie* is bound north in Straits northwest of Little Bahama Bank. USS *PGM-21* is bound south in Straits off Palm Beach. SS *Hoegh Hood* is bound north in Straits west of Little Bahama Bank. USS *ATR-29* is bound east in Northwest Providence Channel. USS *YTB-338*, SS *Herbert L. Pratt*, and SS *C. B. Watson* are bound north in Straits west of Bimini. SS *Garnet Hulings* is bound north in Straits. SS *William Brewster* is bound southeast in Nicholas Channel. SS *Oldham* and SS *George W. Barnes* are bound south in Straits, east of Eleuthera. After running aground in formation at night in the Fleeming Channel, near Current Island, north Eleuthera, USS *SC-1059* under Lt. jg Charles A. Tobin was salvaged by USS *ATR-29*. A Bahamian government power craft, believed to one of the ASR boats out of Montagu or Harbour Island, stood by offering assistance. No.113: HRH the Duke and Duchess of Windsor, paid an informal visit to the unit to inspect a British de Havilland DH.98 Mosquito aircraft. The crew, Capt. Gragg and R/N. J. S. Tegart, were presented to HRH. NAAF Great Exuma: Two OS2U's arrived and departed. A plane departed for Banana River and other for Guantanamo.

13th December 1944

B-25's FC, FF, FH, FJ, FK, and FU are airborne for air Exercise Number 7, aircraft landed after exercise completion. B-25's FC, FF, FH, FY, and FU are airborne for Anti-Submarine Patrol Number 2, aircraft landed after exercise completion. Grumman Goose GA is airborne flight to Rock Sound returned later. SS *Washington* and SS *Empire Norse* are bound north in Straits northwest of Little Bahama Bank, exiting the area. SS *Waxhaws* and SS *C. A. Canfield* are bound north in Straits off Palm Beach. Tug USS *YTB-404*, SS *Tidewater*, and SS *Edward H. Allen* are bound south in Straits off Palm Beach. SS *George W. Barnes* is bound southeast in Old Bahama Channel. SS *Castana*, tanker *Cities Service Koolmotor*, USS *Aquila* (AK-47), SS *City of Montgomery* and SS *Colabee* are bound south, to the east of Eleuthera. NAAF Great Exuma: A plane arrived from San Juan and departed for Banana River, a plane arrived from Guantanamo, gasoline barge USS *YOG-5* and submarine chaser USS *SC* escort departed for Guantanamo.

14th December 1944

B-25's FC, FF, FH, FU, and FK are airborne for night navigation exercise, aircraft landed after exercise completion. B-25's FV, FH, FJ, FK, and FM are airborne for air Exercise Number 7, aircraft landed after exercise completion. B-25's FV, FJ, FK, FY, and FC are airborne for Anti-Submarine Patrol Number 2, aircraft landed after exercise completion. Liberator LH is airborne for Radar and navigation exercise, aircraft landed after exercise completion. Grumman Goose GA is airborne flight to Andros, and returned to base later. SS *Empire Chief* is bound south in Straits, northwest of Little Bahama Bank. SS *W. F. Burdell* is bound north in Straits, off Palm Beach. SS *Tidewater* and SS *El Mundo* are bound south in Straits, northwest of Bimini. SS *Rum River* and tanker *Haakon Hauan* are bound north in Straits west of Bimini. Tanker *Eastern Sun* and SS *Josiah Macy* are bound south in Straits. **Convoy KG-749** is bound southeast entering Nicholas Channel. SS *Elwood* is bound northwest in Old Bahama Channel. Tanker *Paco* is bound southeast in Old Bahama Channel. Discipline, crime and punishment; various base Courts Martial were convened over several months, with crimes ranging from flying an aircraft below 1,000 feet; even as low as 50 feet, to striking an officer in the face in broad daylight on Bay Street (the officer, too, was reprimanded for not reacting with adequate discipline), to a home invasion in the dwelling of a Swedish hotel manager, theft of bicycles and military trucks, consorting with local women 'over the hill,' going AWOL off base, signing a colleague "present" on base when in fact they were not, and stealing from other aviators or civilians. Sentences were often "Not Guilty" on most counts, however those who were found guilty were docked pay, sentenced to 90 days in confinement (often reduced to 60 or 30 days), or demoted. For losses of aircraft, a special inquiry might be instigated with senior investigators being flown in from the likes of Bermuda, the US or UK. Though personnel did resign, and many were posted throughout the empire and promoted, there are almost no references to forced demotions, dismissals, or banishment from bases. (One exception occurred when a vehicle damage an aircraft tail on the tarmac at night a party of senior brass en route to Canada was delayed, but the culprit was never caught). Matters seemed to have been adjudicated locally, judges flown in for significant loss of life and only prisoners from off-island, such as the German from Jamaica, imprisoned and flown off-island. NAAF Great Exuma: An Army PBY arrived and departed. A plane arrived from Guantanamo and departed for Banana River.

15th December 1944

B-25's FV, FJ, FK, FM, and FF are airborne for night navigation exercise, aircraft landed after exercise completion. B-25's FB, FU, FK, FF, FH, and FJ are airborne for air Exercise Number 6, aircraft landed after exercise completion. B-25's FB, FC, FF, FH, and FJ are airborne for air Exercise Number 8, aircraft landed after exercise completion. Liberator LH is airborne for Radar and navigation exercise, aircraft landed after exercise completion. Grumman Goose GA is airborne flight to Rock Sound, returned to base later. SS *Spondilus* is bound north, tanker *Goldmouth* and SS *Scandia* are bound north in Straits northwest of Little Bahama Bank. SS *Mayrants* is bound west in Straits north of Little Bahama Bank. USS destroyer escorts USS *DE-135*, USS *DE-134*, and destroyer USS *DD-136* are bound south in Straits, west of Little Bahama Bank. SS *Vav*, SS *La Paz*, USS *LST-852*, and **Convoy KN-356** are bound north in Straits west of Little Bahama Bank. Tanker *Hadnot* and SS *Fenja* are bound north in Straits northwest of Bimini. SS *Massachusetts* is bound north in Straits. SS *Niobe* is bound south, to the east of Eleuthera. No.113: A Mosquito KB-589 swung on take-off, crashed, and burst into flames, the aircraft becoming a total loss. Crew were pilot, civilian Capt. Lee, slightly injured, and Radio/Navigator, F/Lt. A. H. Manwaring, seriously injured. Dedication of the Lych-gate at the RAF Cemetery: The Lord Bishop of Nassau officiated at the dedication of the Lych-gate at the entrance of the unit's cemetery. Members of the local clergy attended with representative senior officers of this unit. A Guard of Honour, under the command of F/Lt. Page, Station Adjutant, was inspected by HRH the Governor. NAAF Great Exuma: Two Army PBY's arrived, one departed. A plane arrived from San Juan and departed for Banana River, and a US Navy PBN arrived from Jacksonville.

16th December 1944

B-25's FB, FK, FF, FH, and FJ are airborne for air Exercise Number 8, aircraft landed after exercise completion. B-25's FB, FK, FC, FH, and FU are airborne for air Exercise Number 6, aircraft landed after exercise completion. Liberator LH is airborne for Radar and navigation exercise, aircraft landed after exercise completion. SS *Niarchos Kundouritious*, SS *Art Young*, and USS *LCI(L)-596* are bound north in Straits, west of Little Bahama Bank, exiting the area. Tanker *Paulsboro* is bound north, to the east of Abaco. SS *Josiah G. Holland*, USS *LSM-103,* and Spanish SS *Mar Negro*, tug USS *ATR-10* and tow are bound north in Straits. Tanker *Esso Bolivar*, SS *Megara*, SS *Christopher L. Sholes* and SS *Medrs City* are bound south, east of Eleuthera. **Convoy GN-175** and SS *William MacClay* are bound north, to the east of Eleuthera. NAAF Great Exuma: A Navy PBN departed for San Juan, an Army PBY departed other one arrived, a plane arrived from Guantanamo and departed for Banana River.

17th December 1944

Tanker *Pan Virginia* is bound north in Straits, northwest of Little Bahama Bank. USS *LCI(L)-425* and USS *LCI(L)-423* are bound south in Straits off Palm Beach. USS *YAG-21* is bound east in Northwest Providence Channel. **Task Unit TU-20197** is bound north in Straits northwest of Bimini. USS *LCI(L)-423* is bound south in Straits off Miami. Tanker *Axtell J. Byles* is bound north in Straits. SS *Touraine* and SS *Hubbardton* are bound south in Straits. SS *George W. Barnes* is bound northwest entering Santaren Channel. SS *Colborne*, tanker *Hagood*, tanker *Pan Gulf*, **Task Group TG-27-3**, and tanker *Britamer* are bound south, to the east of Eleuthera. NAAF Great Exuma: An Army PBY departed, two planes arrived from Guantanamo, a plane departed there, and a plane departed for Banana River, an Army PBY departed, inter-island transport USS *YAG-19* arrived with supplies.

18th December 1944

B-25's BS, FK, and HG are airborne for air Exercise Number 3 and Number 4, aircraft landed after exercise completion. It was not possible to arrange Kingsley exercise with HMCS *Somers Isles* or Esco exercises in co-operation with US Naval Units at Key West, so navigation exercise were carried out with dummy submarine attacks, homing exercise and radio exercise also carried out. B-24 Liberators NA, NC, and ND are airborne from Windsor Field, aircraft landed after exercise completion, B-24 NA returned earlier, due to engine failure. Tanker *British Hope* is bound south in Straits northwest of Little Bahama Bank. USS *PCE-875*, USS *PCE-874*, and SS *James Island* are bound north in Straits northwest of Little Bahama Bank. SS *Swiftarrow* is bound south in Straits, west of Little Bahama Bank. SS *Eano* is bound north in Straits west of Little Bahama Bank. SS *Durango* is bound south in Straits northwest of Bimini. *USS LSM-104* is bound north in Straits off Port Everglades. Spanish SS *Urumea* is bound south in Straits west of Bimini. USS *Nevada*, SS *Christy Payne*, and SS *Ponca City* are bound south in Straits. SS *Comedian* is bound south in Straits east of Abaco. The two-thousandth aircraft dispatched by No.113 took off from Windsor Field. NAAF Great Exuma: USS *YAG-19* departed for Miami, a plane departed on mission then returned, a plane arrived from Guantanamo and then departed for Banana River.

19th December 1944

Liberator LH is airborne for Radar and navigation exercise, aircraft landed after exercise completion. B-25's FB, FS, FF, FG, FU, and FK are airborne for air Exercise Number 3 and Number 4 later, aircraft landed after exercise completion. B-25's NX, MB, and NA are airborne from Windsor Field for navigation exercises, aircraft landed after exercise completion. B-25 MB returned earlier, due to generator failure, failed to complete exercise, which consisted of homing exercise "B," dummy U-boat attack against smoke float, and transmission practice signals. SS *Eleftheria* is bound north in Straits northwest of Little Bahama Bank. **Task Unit TU-201919**, SS *Spring Hill*, USS *LST-1020*, USS *LST-197*, USS *LST-926*, USS *LST-995*, USS *LST-994* and USS *LST-913* are bound south in Straits northwest of Little Bahama Bank. Tanker *A. S. Hansen* and tanker *Charles H. Marshall* are bound north in Straits west of Little Bahama Bank. Tug USS *ATR-29* is bound south in Straits west of Bimini. USS *YAG-19* is bound west in Northwest Providence Channel. SS *Petter* is bound north in Straits. USS *YF-443* is bound north, exiting Santaren Channel. Spanish SS *Urumea* is bound southeast in Santaren Channel.

Tanker *Paco* is bound northwest in Nicholas Channel. SS *Castana* is bound northwest exiting Old Bahama Channel into the Straits. Tanker *British Hope* is bound west in Straits. Tanker *Gulfdisc*, SS *James Howson*, SS *Hollywood* and USS *Arizona* are bound south, to the east of Eleuthera. NAAF Great Exuma: A plane arrived from Guantanamo for refuelling and then departed for Banana River.

20th December 1944

B-25's FB, FF, FK, FV, FJ, and FS are airborne for air Exercise Number 5, aircraft landed after exercise completion. B-25's FB, FD, FK, FV, FS, and FU are airborne for air Exercise Number 2, aircraft landed after exercise completion. Liberator LH is airborne for Radar and navigation exercise, aircraft landed after exercise completion. Tanker *Pan North Carolina* is bound south in Straits off Canaveral. USS *LCI(L)-398* and **Convoy KN-357** are bound north in Straits northwest of Little Bahama Bank. USS *LST-914* is bound south in Straits west of Little Bahama Bank. SS *Rouge River* is bound north in Straits west of Little Bahama Bank. SS *Malay* and SS *Dynastic* are bound south in Straits, west of Little Bahama Bank. USS *LSM-110* is bound north in Straits off Port Everglades. **Task Unit TU-20188** is bound south in Straits. USS *PCS-1387* is bound north in Straits. **Convoy KG-750** is bound southeast in Old Bahama Channel. Tug USS *ATR-77*, SS *Thomas McDonough*, and SS *City of St. Louis* are bound south, to the east of Eleuthera. SS *Voncenzia* is bound north, to the east of Eleuthera. NAAF Great Exuma: An Army JRF Amphibian arrived from Caicos Island.

21st December 1944

B-24's LT, ND, NC, and MD are airborne from Windsor Field for navigation exercises, aircraft landed after exercise completion, B-24 LT returned earlier due to radio failure. Exercises consisted of homing exercise "B," smoke float with depth charge, transmission practice signals. B-25's FB, FV, FH, FG, FD, and FU are airborne for air Exercise Number 5 and Number 2, later aircraft landed after exercise completion. Tanker *Comol Cuba* is bound north in Straits north of Little Bahama Bank. SS *Cuculllao* and tanker *Bostonian* are bound north in Straits northwest of Little Bahama Bank. **Task Unit TU-20191** and SS *Bidwell* are bound south in Straits, northwest of Little Bahama Bank. USS *LSM-111* is bound north in Straits northwest of Bimini. SS *Kawartha Park*, SS *Joseph Mckenna*, and SS *John Bidwell* are bound south, to the east of Eleuthera. NAAF Great Exuma: An Army PBY arrived from Morrison Field then departed for Caicos Island, two planes arrived from Guantanamo and one departed for Banana River.

22nd December 1944

B-25's FM, FY, FB, FD, FG, and FC are airborne for Anti-Submarine Patrol Number 1, aircraft landed after exercise completion. Grumman Goose GB is airborne for San Salvador returned to base later. SS *Kaimata* is bound north in Straits northwest of Little Bahama Bank, exiting the area. Tanker *Edward L. Shea* is bound south in Straits northwest of Little Bahama Bank. SS *Matiniclock* and SS *Swiftscout* are bound north in Straits northwest of Little Bahama Bank. SS *Castana* and SS *W. M. Burton* are bound north in Straits west of Little Bahama Bank. SS *Oldham* is bound north in Straits northwest of Bimini. Tanker *A. S. Hansen* is bound south in Straits off Port Everglades. **Task Unit TU-0377** is bound north in Straits west of Bimini. Tanker *Paulsboro* and SS *Hoegh Hood* are bound south in Straits. **Convoy**

GK is bound northwest in Nicholas Channel. USS *YF-294* is bound northwest entering Old Bahama Channel. SS *Corinth*, SS *Comet*, and USS *Aquila* (AK-47) are bound south, to the east of Eleuthera. NAAF Great Exuma: An Army JRF Amphibian departed for Morrison Field, a plane arrived from San Juan and departed for Banana River.

23rd December 1944

B-25's FB, FM, FC, FG, FH, FS, and FK are airborne for Anti-Submarine Patrol Number 2, aircraft landed after exercise completion. SS *Baldbutte*, tug USS *ATA-181* towing *Cape Henry* (YC-113) are bound north in Straits west of Little Bahama Bank. USS *LSM-107* is bound south in Straits off Palm Beach. SS *Mizra* is bound south in Straits west of Bimini. Submarine USS *Hackleback* (SS-295) is bound west in Northwest Providence Channel off Berry Islands Bahamas. Tug USS *YFB-338* towing *Cuban CS-021*, and tanker *Athelduke* are bound north in Straits. SS *Vitorialoide*, SS *Barnett* and SS *Cantuatia* are bound south, to the east of Eleuthera. NAAF Great Exuma: A plane arrived from Guantanamo on a training flight and departed for Banana River, and US Coast Guard tug USCG-*Sagebrush* arrived in Elizabeth Harbour, Exuma.

24th December 1944

Tanker *Beaconoil* and SS *Hubbardston* are bound north in Straits northwest of Little Bahama Bank, exiting the area. SS *Tide Penn* and Tanker *Hadnot* are bound south in Straits northwest of Little Bahama Bank. **Task Unit TU-20193** is bound south in Straits off Palm Beach. SS *Mississippi* and SS *Morton Prince* are bound south in Straits northwest of Bimini. SS *Tidewater* is bound north in Straits west of Bimini. SS *Winston* is bound south in Straits west of Bimini. SS *Jacob Chandler Harper*, SS *Trimountain*, and SS *Hakan Havan* are bound north in Straits. Tug USS *Kalmia* (AT0-23) and tow are bound south in Straits. **Convoy KN-358** is bound north in Straits west of Cay Sal Bank. **Convoy KG-751** is bound south in Straits west of Cay Sal Bank. USS *C-56193*, USS *C-56194*, USS *C-56195*, USS *C-56196* are bound southeast in Old Bahama Channel. SS *Fenja* is bound south, to the east of Eleuthera. NAAF Great Exuma: A plane arrived from Guantanamo. Another plane then departed to Guantanamo for a mission.

25th December 1944

Convoy NK-658 is bound south in Straits northwest of Little Bahama Bank. SS *John Hammond*, SS *William Crompton*, SS *Josiah Macy*, **Convoy KN-358** and tanker *Gulfcrest* are bound north in Straits northwest of Little Bahama Bank. USS *Foss* is bound south in Straits northwest of Little Bahama Bank. The tug *Ontario* and tows are bound south in Straits off Palm Beach. Tanker *L. V. Stanford*, SS *Vermont* and SS *Telfair Stockton* are bound south in Straits, west of Little Bahama Bank. USS *Tench* is bound west in Northeast Providence Channel, with US Navy plane escorting. SS *Massachusetts* is bound south in Straits off Port Everglades. SS *Scandia* is bound south in Straits, west of Bimini. **Task Unit TU-20111** is bound north in Straits, west of Bimini. Tanker *Hadnot* is bound south in Straits. Tanker *Eastern Sun* and SS *Malabar* are bound north in Straits. SS *George K. Fitch* is bound southeast in Nicholas Channel. SS *White Clover* is bound southeast entering Old Bahama Channel. **Convoy KG-751** is bound southeast in Old Bahama Channel with US Navy plane escorting. SS *Willow Dale Park*, SS *Cypress Hills Parks*, and SS *Samuel Moody* are bound south, to the east of Eleuthera. Tanker *Esso Bolivar* is bound north, to the east of Eleuthera. Permanent Staff Officers of No.113 Wing, played football against the Airmen. Later, Christmas dinner was served to the Airmen by Officers and Senior NCOs. Christmas activities: arrangements were made for sections to cease work on 24th until 26th December, inclusive. A special menu was prepared in the various messes of the unit. An entertainment was given on the night of the 25th. NAAF Great Exuma: A plane arrived from Banana River and then departed to Guantanamo on a training flight.

26th December 1944

Task Unit TU-20194 is bound south in Straits off Canaveral. Tanker *Gulfking* is bound south in Straits northwest of Little Bahama Bank. Tanker *L. L. Abshire* and SS *Barren Hill* are bound south in Straits west of Little Bahama Bank. SS *C. A. Canfield,* SS *William B. Otwell*, USS *Moberly* (PF-63) are bound north in Straits northwest of Bimini. SS *Medina* is bound north in Straits. **Convoy NK-658** is bound south in Straits. SS *Cahlamet* is bound south, to the east of Eleuthera. SS *Charles F. Meyer* is bound north, to the east of Eleuthera. NAAF Great Exuma: A plane arrived from Guantanamo another one departed there. The US Coast Guard tug USCG-*Sagebrush* departed from Elizabeth Harbour, Exuma, for Miami.

27th December 1944

B-25's YC, MG, and FB are airborne for Anti-Submarine Patrol Number 2, aircraft landed after exercise completion, then FY and FF returned earlier, radio trouble. B-25's FS, FD, FH, FJ, and FK are airborne for air Exercise Number 1, aircraft landed after exercise completion. **Task Unit TU-20112** is bound north in Straits northwest of Little Bahama Bank. SS *New York* and tanker *Pan Virginia* are bound south in Straits west of Little Bahama Bank. SS *AthelViking* and SS *Spondilus* are bound south in Straits northwest of Bimini. USS *SC-453* is bound north in Straits northwest of Bimini. SS *Atlantic* and SS *Durango* are bound north in Straits. SS *Cape Texas* and SS *John D. Archibald* are bound south, to the east of Eleuthera. HRH the Governor is pleased to appoint F/Lt. D. Lester Brown, DFC, as his Honorary RAF *Aide-de-Camp* (ADC), signed Vice S/L David Carver. NAAF Great Exuma: A plane arrived from Guantanamo then departed to Banana River, an Army PBY arrived from Morrison Field.

28th December 1944

B-25's FM, FF, FK, FG, FU, and FC are airborne for Anti-Submarine Patrol Number 2, aircraft landed after exercise completion. B-25's FB, FJ, FS, FU, FC, and FF are airborne for air Exercise Number 1, aircraft landed after exercise completion. Liberator NJ is airborne for Radar and navigation exercise, aircraft landed after exercise completion. SS *Imperial Transport* is bound south in Straits off Canaveral. SS *Nevada* is bound north in Straits northwest of Little Bahama Bank. Tanker *Gulfbreeze* and SS *Orvilie Harden* are bound south in Straits northwest of Little Bahama Bank. SS *George K. Fitch* is bound north in Straits west of Little Bahama Bank. SS *Medina* is bound north in Straits northwest of Bimini. Tanker *British Merchant* is bound north in Straits off Port Everglades. SS *T. E. Mitchell* is bound north in Straits. **Task Group TG-22.5** is bound north, to the east of Eleuthera. SS *W. F. Burdell*, SS *Reginolite*, SS *Fort Willington*, SS *Willie Jones*, and SS *Washington* are bound north, to the east of Eleuthera. NAAF Great Exuma: A plane arrived from Guantanamo, Morrison Field, and Caicos Island each, a plane departed for Guantanamo, Morrison Field, and Caicos Island each.

29th December 1944

B-25's FO, FF, FC, FB, FH, and FS are airborne for air Exercise Number 4, aircraft landed after exercise completion. B-25's FO, FF, FU, FB, FH, and FJ are airborne for air Exercise Number 8, aircraft landed after exercise completion. Free French Navy submarine FFS *Antiope* is bound north in Straits ,west of Little Bahama Bank. USS *Y0-129* is bound south in Straits northwest of Bimini. USCG *CG-83345*, USCG *CG-83374*, USCG *CG-83376*, and USCG *CG-83390* are bound north in Straits, northwest of Bimini. Tanker *Beta* is bound south in Straits. SS *Brand Whitlock* is bound north in Straits. USS *Evansville* (PF-70) is bound east in Straits, north of Cay Sal Bank. SS *C. B. Watson* is bound south in Nicholas Channel. SS *Clearwater Park* and SS *W. Kyska* are bound south, to the east of Eleuthera. NAAF Great Exuma: A transport aircraft with BuNo-6479 (flight 630) *en route* from San Juan to Miami arrived on a forced landing with engine faults. A plane arrived from Morrison field and then returned.

30th December 1944

B-25's FY, FO, FC, FJ, FF, and FG are airborne for air Exercise Number 4, aircraft landed after exercise completion. B-25's FJ, FO, FF, FT, FY, and FH are airborne for air Exercise Number 8, aircraft landed after exercise completion. B-25's FB, FM, FT, FY, FO, and FC are airborne for local night navigation exercises, aircraft landed after exercise completion. SS *Touraine* is bound north in Straits northwest of Little Bahama Bank, exiting the area. **Convoy KN-359** is bound north in Straits west of Little Bahama Bank. Tanker *Cities Service Ohio*, SS *Michael J. Stone*, and SS *B. F. Shaw* are bound south in Straits, off Port Everglades. USS *YTB-338* is bound north in Straits. SS *W. F. Burdell* is bound northwest in Nicholas Channel. Spanish SS *Urumea* is bound northwest entering Nicholas Channel. SS *Chilbar* and **Convoy KG-752** are bound southeast entering Old Bahama Channel. SS *Daulton Mann* is bound south, to the east of Eleuthera. SS *Arizona* is bound north, to the east of Eleuthera. NAAF Great Exuma: A transport aircraft with builder's number BuNo-6473 arrived from Miami and then returned, a plane arrived from Guantanamo and departed for Banana River.

31st December 1944

USS *PGM-17* and SS *Antietam* are bound north in Straits northwest of Little Bahama Bank, exiting the area. SS *Swift Arrow* is bound north in Straits northwest of Little Bahama Bank. SS *Matinicock* is bound south in Straits northwest of Little Bahama Bank. SS *W. W. Mills*, USS *LT-874*, and **Convoy NK-659** are bound south in Straits west of Little Bahama Bank. **Task Unit TU-201819** is bound south in Straits, off Port Everglades. SS *Point Breeze* is bound north in Straits, off Port Everglades. SS *Petter* is bound north in Straits. SS *Abner Doubleday* is bound northwest in Nicholas Channel. Tanker *Comol Cuba* is bound south, to the east of Eleuthera. NAAF Great Exuma: Two planes arrived from Guantanamo, a plane departed there, and a plane departed for Banana River.

Part VI

Resurgence: January 1, 1945 to May 15, 1945

1st January 1945
B-25's FO, FT, FY, FG, FH, and FF are airborne for air Exercise Number 7, aircraft landed after exercise completion. USS *PGM-20* is bound north in Straits via Canaveral exiting the area. SS *William Penn* and SS *Tidewater Associated* are bound south in Straits, passing Canaveral. Troop transport USS *USAT-Cuba* is bound north in Straits passing Canaveral. SS *W. M. Burton* is bound south in Straits towards Palm Beach. SS *Bidwell* and SS *Edward L. Shea* are bound north towards Canaveral. USS *LST-839* is bound south in Straits passing Palm Beach. SS *Tustem* and SS *Caclilao* are bound south in Straits towards Port Everglades. **Task Unit TU-20198** is bound south in Straits. USS *LST-874* and tow are bound southwest in Straits. Tanker *Pan North Carolina* is bound northeast in Straits. SS *W. W. Mills* and SS *Matt Nicock* are bound southwest in Straits exiting the area. SS *Mather B. Brady* is bound north in Santaren Channel. SS *Quetice Park* is bound south, to the east towards San Salvador. NAAF Great Exuma: A plane arrived from Banana River and departed for Guantanamo, a plane arrived from Guantanamo.

2nd January 1945
B-25's FO, FT, FC, FY, FD, and FF are airborne for air Exercise Number 7, aircraft landed after exercise completion. B-25's FO, FM, FT, FC, FG, and FF are airborne for night navigation patrols, aircraft landed after exercise completion. Tanker *Pan North Carolina* is bound north past Canaveral, exiting the area. SS *Felix Reisenberg* is bound south in Straits near Canaveral. SS *Tidewater* is bound north in Straits, from Palm Beach to Canaveral. **Task Unit TU-201910** and SS *Malay* are bound south in Straits towards Palm Beach. SS *Empire Bombardier* is bound south towards Walker Cay, north of Abaco. Tanker *Paulsboro* is bound north in Straits off Palm Beach. **Task Unit TU-20199**, SS *Baldbutte*, and USS *YAG-19* are bound southwest in Straits. SS *William Penn* and SS *Tidewater Associated* are bound north in Straits. SS *Granet Hulings* is bound north in Northwest Providence Channel. SS *Tobiatic Park* is bound north, to the east of Bahamas passing San Salvador. NAAF Great Exuma: A plane arrived from Guantanamo, and two planes departed for Banana River.

3rd January 1945
Adverse weather restricted flying, local flying. B-24 Liberator NL is airborne and then landed at Bermuda. SS *Swiftscout* is

bound north in Straits northwest of Grand Bahama. Tanker *Pan Virginia* is bound north in Straits west of Grand Bahama. SS *Minatitlan* is bound north in Straits north of Cay Sal Bank. SS *Massachusetts* is bound northeast in Straits northwest of Cay Sal Bank. SS *Imediato Joao Silva* is bound south, to the northeast of Cat Island. SS *Narbo* is bound south, to the east of San Salvador. Squadron 3 welcomed four P-40's, and one C-47, arrived at Windsor Field from Drew Field, Tampa, for exercises. The RAF Central Band under the direction of W/C O'Donnell visited this unit and gave two musical performances at Oakes Field. NAAF Great Exuma: A plane arrived from Guantanamo, a plane arrived from Banana River then departed for Guantanamo, a plane arrived from San Juan then departed for Banana River, and a plane departed for Guantanamo.

4th January 1945

B-25's FO, FT, FK, FC, FU, and FF are airborne for air Exercise Number 6, aircraft landed after exercise completion. B-25's FM, FU, FO, FK, FG, and FT are airborne for night navigation exercise, then aircraft landed after exercise completion. Liberator FL returned from Bermuda, landed at base. USS *YTB-395* is bound south, to the north of Grand Bahama. SS *William Crompton* and SS *Crown Point* are bound south in Straits north of Grand Bahama. **Convoy KN-360** and Free French Navy submarine FFS *Bearn* are bound north in Straits northwest of Grand Bahama. **Task Unit TU-0318** is bound south in Northwest Providence Channel. SS *South Mountain* is bound south in Northwest Providence Channel. USS *YO-129* is bound north in Northwest Providence Channel. SS *Duquesne* is bound south in Straits west of Bimini. SS *Clyde Austin Dunning* is bound north in Straits west of Bimini. SS *Rebecca Boone* is bound northwest in Santaren Channel. **Convoy KG-532** is bound southeast in Santaren Channel. SS *El Oceano* is bound south in Northeast Providence Channel. Aircraft carrier HMS *Tracher* is bound south, SS *George W. Alther*, and SS *Elax* are bound south, to the east of San Salvador. NAAF Great Exuma: A plane arrived from Morrison Field then returned there. A plane arrived from Guantanamo and departed for Banana River.

5th January 1945

B-25's FO, FT, FC, FF, FK, and FG are airborne for air Exercise Number 6, aircraft landed after exercise completion. B-25's FG, FO, FK, FF, FC, and FD airborne for air Exercise Number 3 and Number 4, aircraft landed after exercise completion. SS *Bushy Run* is bound south in Northwest Providence Channel. **Task Unit TU-0312** is bound north in Straits, north of Bimini. SS *BulkOre* is bound north in Straits south of Bimini. SS *Chesapeake Capes* is bound north in Straits north of Cay Sal Bank. Argentinian SS *Rio Corrientes* is bound east in Nicholas Channel. SS *Western Queen* and USS *YTB-404* are bound east is in Old Bahama Channel. USS *PC-1143* is bound southeast in Northeast Providence Channel. Submarine USS *Tigrone* (SS-419) is bound west in Northeast Providence Channel. SS *Henry S. Groove* and SS *Atlantic* are bound south, to the east of San Salvador. No.113: A Dakota C-47 KK-204 had its starboard wing tip damaged in a taxiing accident, when the wing tip came in contact with stationary Mosquito, which was undamaged. There were no casualties; the pilot was F/Lt. W. M. Howes. NAAF Great Exuma: A plane departed and returned from a mission, a plane arrived from Mayaguana, a plane arrived from

Guantanamo then departed for Banana River.

6th January 1945

B-25's FC, FO, FK, FY, FD, and FJ are airborne twice for air Exercise Number 2, aircraft landed after exercise completion. SS *Swiftlight* is bound south in Straits off Palm Beach. SS *Kirkenes*, SS *Measure* and SS *Benedick* are bound south in Straits off Port Everglades. SS *Tidewater* and SS *Samuel Johnson* are bound north in Straits west of Bimini. SS *Samnid* is bound north in Straits. Submarine USS *Tigrone* (SS-419) is bound southwest, to the north of Cay Sal Bank. USCG *CG-Sagebrush* and the Royal Norwegian Navy escort ship HNoMS *King Haakon VII* are bound east in Old Bahama Channel. Tanker *Norsktank* and **Convoy NG-481** are bound south, to the east of Eleuthera. NAAF Great Exuma: A plane arrived from Guantanamo then departed for Banana River, then a plane departed for Morrison Field.

7th January 1945

B-25's FO, FN, FT, FY, FK, and FJ are airborne twice for Anti-Submarine Patrol Number 1, uneventful, aircraft landed after exercise completion. Liberators from Windsor Field carried out Radar and navigation exercises. SS *George Kenny* is bound north in Straits west of Bimini. SS *Elwood* is bound east in Nicholas Channel. SS *Miobe* is bound south, to the east of Cat Island. NAAF Great Exuma: A plane departed and returned from a mission. A plane arrived from Guantanamo then departed for Banana River, a plane arrived from South Caicos Island.

8th January 1945

B-25's FO, FD, FK, FJ, FF, and FN are airborne for air Exercise Number 5, aircraft landed after exercise completion. B-25's FH, FN, FK, FO, and FJ are airborne for air Exercise Number 1, aircraft landed after exercise completion. SS *Spoundilus* is bound north in Straits off Canaveral. SS *Martin Behrman* is bound north in Straits off Palm Beach. USS *YTB-395* and **Convoy KG-754** are bound south in Old Bahama Channel. SS *Thorshammer*, SS *Margaret* Brent, and SS *Washington* are bound south, to the east of Eleuthera. NAAF Great Exuma: A plane arrived from Guantanamo, another plane departed on a mission, and then flew to Guantanamo.

9th January 1945

Notice received from air traffic control (ATC) that a B-29 bomber is missing from a flight to Morrison Field from Borinquen, Puerto Rico. The last contact with aircraft was when it was 100 nm from Morrison Field on 8th (yesterday). Request received to search area between Nassau and Exuma. B-24 Liberator LZ was to sweep area, aircraft is airborne, searched, yet nothing is located. B-25 FM is airborne, searched, located nothing. B-25's FF, FH, FK, FJ, FO, and FN are then airborne for air Exercise Number 5, these aircraft landed after exercise completion. B-25's FF, FK, FJ, FT, FV, and FN are airborne for air Exercise Number 1, aircraft landed after exercise completion. Liberators from Windsor Field completed bombing exercise. B-24 Liberator NL returned from Miami and landed. **Convoy KN-361** is bound north in Straits, off Canaveral. SS *Josiah Macy* is bound north in Straits, off Port Everglades. SS *Charles S. Cramp* is bound south, to the east of Cat Island, in the Atlantic. NAAF Great Exuma: A plane departed to search for a missing B-29 Super Fortress bomber then returned, a plane arrived from Banana River then departed for Guantanamo, a plane arrived from Guantanamo and then went back.

10th January 1945

Request received from RAF Transport Command to continue search for the US B-29 Super Fortress bomber. B-24 Liberator LH is airborne to search area between new Providence and Eleuthera, landed back, sighted nothing, after search again airborne later, landed back after an uneventful search. B-25's FV, FH, FY, FM, and FK are airborne for air Exercise Number 3 and Number 4, aircraft landed after exercise completion. B-25's VH, YN, and J are airborne for air Exercise Number 3 and Number 4, then aircraft landed after exercise completion. Liberators LU and NY are airborne, landed after completing detailed exercises, bombing attacks carried out against smoke float, practice U-boat signals transmitted to base, aircraft landed after exercise completion, airborne again later for similar exercises, landed back at base after completion. B-25's FM, FD, FJ, FG, FV, and FT are airborne for local night navigation exercise, aircraft landed after exercise completion. **Convoy NK-661** is bound south in Straits north of Canaveral. USCG cutter *CG-Active* is bound north in Straits off Canaveral. Tanker USS *Roxane* (AKA-37), and SS *Chads Ford* are bound north in Straits off Miami. Troop transport USS *USAT-Monterey* is bound north in Old Bahama Channel. SS *Charles H. Cramp* is bound south, east of San Salvador. Parade to mark the first anniversary of the formation of the Bahamas Air Service Squadron was held at Oakes Field, then the Squadron marched past HRH The Governor at Government House. B-24 Liberator LH continued the search for the US B-29 Super Fortress, overdue on approach to Morrison Field, by searching from New Providence to north Eleuthera, then to south Eleuthera. NAAF Great Exuma: A plane departed and came back from a mission. Two planes arrived from Guantanamo and departed for Banana River.

11th January 1945

B-25's FM, FD, FJ, FG, FV, and FT are airborne for local night navigation exercise, aircraft landed after exercise completion. B-25's FM, FG, FV. FT, FJ, and FD are airborne for air Exercise Number 3 and Number 4, aircraft landed after exercise completion. B-25's FH, FG, FV, FT, FJ, and FD airborne for air Exercise Number 3 and Number 4, aircraft landed after exercise completion. Request received from air traffic control (ATC) to continue search for the US B-29 in the same area as 10th January (yesterday). B-24 Liberator LH is airborne, searched, landed back, sighted nothing. B-25's FM, FN, FH, FG, FD, and FV are airborne for night navigation exercise, then all aircraft landed after exercise completion. Liberators MX and MC are airborne from Windsor Field, completed navigation Exercise Number 48, carried out attacks with depth charges against smoke floats, U-boat practice attack signals transmitted to base. Liberators NH and MC are airborne later for similar exercise. SS *Edwin L. Godkin,* SS *George Gale,* SS *Meyar Lissner,* and SS *David L. Swain* are bound south, from the north of Walker Cay and north of Abaco. SS *William Tyler* is bound south, from the north of Little Bahama Bank. USS *Yokes* (APD-69) is bound north in Straits north of Little Bahama Bank. SS *Sebastian Vizcairo* is bound southeast in Old Bahama Channel. SS *Tungsha* and SS *William MacClay* are bound south, to the east of San Salvador. NAAF Great Exuma: A plane arrived from Guantanamo and departed for Banana River. A plane arrived from Morrison Field and then took off in search of the missing B-29 Super Fortress bomber with another plane; one plane returned to Exuma, the other went to Guantanamo.

12th January 1945

B-25's FM, FH, FG, FC, FN, and FT are airborne for Anti-Submarine Patrol Number 2, uneventful, aircraft landed after exercise completion. B-25's FC, FG, FH, FN, and FV are airborne twice at different times for air Exercise Number 3 and Number 4, aircraft landed after exercise completion. A C-47 arrived from Morrison Field, bringing a Major General for Staff Visit to the Allied forces on island. B-24 Liberators NC and MC are airborne for air Exercise Number 48, exercise similar to 11th January (yesterday), aircraft landed after exercise completion. Liberators NC and LU are airborne similar exercise as former aircraft landed after exercise completion. SS *Tidewater Associated* and SS *Edward L. Shea* are bound north in Straits northwest of Little Bahama Bank. SS *Clam* and tanker *Bohemian Club* are bound north in Straits west of Grand Bahama. Oiler USS *YO-129* and **Task Unit TU-0377** is bound south in Straits north of Bimini. SS *Matinicock* is bound north in Straits south of Bimini. SS *James Eason Layne* and SS *Tidewater* are bound north in Straits, west of Bimini. A US Navy plane is bound northeast in Straits, towards Bimini. SS *Paco City* and Argentinian SS *Rio Corrientes* are bound northwest in Old Bahama Channel. USS *YF-294* is bound southeast in Old Bahama Channel. SS *Richard Upjohn* is bound southwest in Northeast Providence Channel. **Task Unit TU-2965** and **Task Unit TU-20183** are bound south, to the east of Eleuthera. NAAF Great Exuma: A plane arrived from Guantanamo and departed for Banana River. A plane went to search for the missing US B-29 bomber, then returned.

13th January 1945

B-25'S FH, FC, FG, FT, FJ, and FO are airborne for Anti-Submarine Patrol Number 2, uneventful, aircraft landed after exercise completion, ship thought to be aircraft carrier sighted, sundry small units of ships observed. Grumman Goose GB is airborne for Dinner Key, off Miami, then returned to base later at night. B-25's FO, FC, FH, FG, CL, and FN are airborne for air Exercise Number 7, aircraft landed after exercise completion. Liberators of Number 3 Squadron carried out Radar and navigation exercise from Windsor Field during the day. USS *AM-256* is bound south, to the north of Walker Cay, off Abaco. Submarine USS *C-2* (SS-13) is bound north in Straits off Canaveral with a US Navy plane escorting. SS *Baldbutte* and SS *William Compton* are bound north in Straits, off Canaveral. SS *Bushy Run* is bound north in Straits west of Little Bahama Bank. SS *William Penn* is bound north in Straits off Palm Beach. **Convoy KG-755** is bound southeast in Nicholas Channel. SS *George W. Alther* is bound north in Straits east of Cay Sal Bank. SS *Mormacmar* is bound northwest in Old Bahama Channel. NAAF Great Exuma: A plane arrived from Guantanamo, and departed for Banana River, and a plane departed for Morrison Field.

14th January 1945

Destroyer escort USS *Sutton* (DE-771) and SS *South Mountain* are bound northwest of Walker Cay, north of Abaco. SS *Medina* is bound south, to the north of Little Bahama Bank. **Convoy KN-362**, SS *Irwin Russell*, SS *George Walther*, SS *Durango*, and SS *W. M. Burton* are bound north in Straits northwest of Little Bahama Bank. SS *Penja* and *San Pasqual* are bound north in Straits west of Little Bahama Bank. SS *Bostonian* is bound north in Straits northwest of Bimini. Submarine USS *Bullhead* (SS-332) and submarine USS *Lionfish* (SS-298) are bound west in Northwest Providence Channel with US Navy plane escorting. Two anonymous ships are reported east in Old Bahama Channel. No.113: One German prisoner under armed escort passed through, in transit from Jamaica to Canada. He was accommodated at Fort Charlotte Barracks, and proceeded to Dorval next day [research indicates the individual was not likely the member of a U-boat crew]. NAAF Great Exuma: A plane arrived from Guantanamo and then departed for Banana River.

15th January 1945

B-25's FO, FD, FJ, FC, FH, and FG are airborne for air Exercise Number 8, aircraft landed after exercise completion. B-25's FJ, FD, FH, FO, FC, and FB are airborne for air Exercise Number 7, aircraft landed after exercise completion. B-24 Liberators carried out Radar and navigation exercise from Windsor Field during the day. Tanker *Bulklube* is bound south in Straits south of Bimini. USS *PC-1210* is bound northwest in Santaren Channel. SS *Eliza Ann Nicholson* is bound south in Straits north of Cay Sal Bank. Tanker *E. R. Kemp* is bound north in Straits north of Cay Sal Bank escorted by a navy plane. **Convoy GN-181** is bound north, to the east of Eleuthera with a US Navy plane escorting. SS *Beta* and SS *Runnymede Park* are bound south, to the east of Eleuthera. SS *Meagro* is bound south, east of Cat Island. NAAF Great Exuma: Two planes arrived from Guantanamo, and then departed for Banana River.

16th January 1945

B-25's DC, FO, FC, FH, FN, and FG are airborne for air Exercise Number 8, aircraft landed after exercise completion. B-25's DC, FO, FC, and FJ are airborne for air Exercise Number 6, aircraft landed after exercise completion. Grumman Goose GB are airborne for San Salvador, retuned to base later at night. Liberator NX is airborne from Windsor Field for Kingsley Exercise, shortly after take-off aerial snapped, it was unserviceable throughout exercise, homing "B" commenced, three frigates sighted, recognition procedure completed, completed patrol Python, depth charge attack carried out against smoke float laid by non-conducting ship, practice U-boat signals sent to base, homing "C" procedure unsuccessful, and aircraft landed after exercise completion. Liberator ND is airborne from Windsor Field for Kingsley Exercise, yet had to return to base due to engine failure, and failed to complete exercise. Liberator LR is airborne from Windsor Field for Kingsley Exercise, homing "B" requested, no bearing received, convoy sighted after Radar location, recall to base signal received, aircraft left convoy without carrying out exercise, returned to base. Liberator NF is airborne from Windsor Field for Kingsley Exercise, was not permitted to set course for convoy, jettisoned the depth charges, and landed back at base. Adverse weather, convoy area and exercise not successful. All aircraft recalled to the base, only one having carried out exercise. SS *Richard K. Call* is bound north in Straits northwest of Little Bahama Bank. USS *PC-1547* is bound north in Straits north of Cay Sal Bank. SS *Sebastian Vizcaino* is bound west in Nicholas Channel. Tanker *Gulfbreeze* is bound south, east of Eleuthera. NAAF Great Exuma: A plane arrived from Guantanamo and then departed for Banana River.

17th January 1945

B-25's FO, FC, FF, FH, FJ, and DC are airborne for air Exercise Number 6, aircraft landed after exercise completion. B-25's FO, FC, FF, FN, and FG are airborne for local night navigation exercise, then aircraft landed after exercise completion. Liberator LY is airborne from Windsor Field for Kingsley Exercise, radio contact established, convoy located, Cobra patrol carried out, homing procedure "B" commenced under direction of Escort Commander, smoke float laid by no-conducting ship attacked with depth charges, U-boat practice signals sent to base, escort commander homed to scene of attack by aircraft, aircraft landed after exercise completion. Liberator NX is airborne from Windsor Field for Kingsley Exercise as first relief, after six bearing from homing procedure "B" convoy located, Python patrol carried out, smoke float laid by non-conducting ship, attacked with depth charges, U-boat practice signals sent to base, escort commander homed to scene of attack by aircraft, aircraft landed after exercise completion. Liberator LZ is airborne from Windsor Field for Kingsley Exercise as second relief, from homing procedure "B" convoy located, Python, Lizard, and Alligator patrol carried out, smoke float laid by no-conducting ship, attacked with depth charges, U-boat practice signals sent to base, escort commander homed to scene of attack by aircraft, aircraft landed after exercise completion. Liberator NG is airborne from Windsor Field for Kingsley Exercises third relief, homing procedure "B" commenced, smoke float laid by non-conducting ship, attacked with depth charges, U-boat practice signals sent to base, aircraft landed after exercise completion. Liberator LR is airborne from Windsor Field for Kingsley Exercise as fourth relief, radio contact established, convoy located, Python and Lizard patrols carried out, smoke float laid by non-conducting ship, attacked with depth charges, U-boat practice signals sent to base, aircraft landed after exercise completion. SS *Phoenix* is bound south in Straits west of Little Bahama Bank. SS *Mill Spring* is bound south in Straits west of Bimini. SS *Benedick* is bound north in Straits west of Bimini. SS *Peorta* is bound northeast in Straits off Miami. SS *Richard V. Oulahan* is bound north in Straits north off Cay Sal Bank. The house boat USS *YHB-16* is bound northwest in Northeast Providence Channel. **Task Group TG-21.4** is bound south, to the east of San Salvador.

18th January 1945

B-25's FD, FG, FY, FF, FO, and FN are airborne for night navigation exercise, aircraft landed after exercise completion. B-25's FH, FO, FF, FM, and FY are airborne for air Exercise Number 7, aircraft landed after exercise completion. B-25's FY, FO, FH, FC, FF, and FN are airborne for night navigation exercise, aircraft landed after exercise completion B-25's FY and FH returned earlier due to engine failure. B-24 Liberator NF is airborne from Windsor Field for Kingsley Exercise, radio contact established, convoy located, signals exchanged, patrols completed under direction of Escort Commander, smoke float attacked with depth charge, U-boat practice signals sent to base, aircraft landed after exercise completion. Liberator LY is airborne from Windsor Field for Kingsley Exercise as first relief, homing "B" commenced after six bearings, convoy located, smoke float attacked with depth charge, U-boat practice signals sent to base, aircraft landed after exercise completion. Liberator MX is airborne from Windsor Field for Kingsley Exercise as second relief, radio contact established, convoy located, smoke float attacked with depth charge, U-boat practice signals sent to base, patrol Cobra completed, aircraft landed after exercise completion. Liberator LZ airborne from Windsor Field for Kingsley Exercise as third relief, radio contact established, convoy located, signals exchanged, patrols completed under direction of Escort Commander, smoke float attacked with depth charge, U-boat practice signals sent to base, aircraft landed after exercise completion. Liberator LR is airborne from Windsor Field for Kingsley Exercise as fourth relief, convoy located, aircraft returned to base earlier, generator unserviceable, exercise not completed with ships smoke float attacked with depth charge, U-boat practice signals sent to base, aircraft landed. SS *Pavlic* is bound north in Straits north of Little Bahama Bank. A group of landing craft, **Convoy USS *LCI(L)* Group** is bound south in Straits, west of Little Bahama Bank. USS *APA-110* is bound north in Straits, west of Little Bahama Bank. **Task Group TG-22.1** is bound north in Straits, north of Abaco. SS *George T. Angell* is bound west in Nicholas Channel. SS *William J. Palmer* is bound west, in Old Bahama Channel. SS *City of Savannah*, SS *John D. Archibald*, SS *Kanawha,* and SS *William G. Fargo* are bound north, to the east of Cat Island. NAAF Great Exuma: A plane arrived from Guantanamo and then departed to Banana River on a training flight.

19th January 1945

B-25's F, FN, FC, FF, FG, and FU are airborne for Anti-Submarine Patrol Number 1, aircraft landed after exercise completion. B-25's FO and FC are then airborne for night navigation exercises, aircraft landed after exercise completion. SS *Elbridge Gerry*, tanker *Gulfcrest*, and SS *Esturia* are bound north in Straits, northwest of Little Bahama Bank, exiting the area. **Convoy KN-363** is bound north in Straits, west of Little Bahama Bank. SS *Herbert L. Pratt* is bound north in Straits, west of Little Bahama Bank. Tanker *L. V. Stanford* is bound south in Straits, off Port Everglades. Tug USS *ATR-5* is bound north in the Santaren Channel. **Convoy KG-756** is bound southeast in Old Bahama Channel. SS *George C. Childress* is bound south, to the east of San Salvador. NAAF Great Exuma: A plane arrived from San Juan and then departed for Banana River on a training flight.

20th January 1945

B-25's FN, FM, FO, FF, and FC are airborne for Anti-Submarine Patrol Number 1, aircraft landed after exercise completion. Grumman Goose GD is airborne for Rock Sound retuned to base later in the evening. SS *San Triso*, tanker *British Merchant,* SS *F. J. Wolfe*, and SS *Bidwell* are bound north in Straits. **Convoy GK-856** and USS *PC-1265* are bound northwest in Old Bahama Channel. Tanker *Gulf of Venezuela* and SS *Chagres* are bound south, to the east of San Salvador. **Convoy GN-182** is bound north, to the east of San Salvador.

21st January 1945

Grumman Goose GB is airborne for Long Island, aircraft landed back later in the day. SS *Christy Payne* and SS *Bulkero* are bound north in Straits northwest of Little Bahama Bank, exiting the area. **Task Group TG-22.7** is bound south, to the north of Abaco. SS *William Crompton* is bound south in Straits off Palm Beach. SS *C. B. Watson* is bound north in Straits off Palm Beach. SS *Ameriki* is bound north in Straits. SS *Donald W. Baine*, SS *Oldham*, SS *Roger Sherman*, and **Convoy NG-484** are bound south, east of San Salvador. NAAF Great Exuma: USS *YAG-19* arrived from Miami, then departed to Nassau, a plane arrived from Caicos Island then departed for Morrison Field, a plane arrived from Guantanamo and departed for Banana River.

22nd January 1945

B-25's FN, FO, FC, FF, and DC are airborne for air Exercise Number 5, aircraft landed after exercise completion. B-25's FO, FC, FF, FD, FT, and DC are airborne for air Exercise Number 1, aircraft landed after exercise completion. Radar exercises carried out by Liberators from Windsor Field during the day. Tanker *Good Gulf*, and USS *Arizona* are bound north in Straits, northwest of Little Bahama Bank, exiting the area. Submarine USS *Sturgeon* (SS-187) is bound north in Straits west of Little Bahama Bank, with a US Navy plane escorting, exiting the area. SS *Cacalilao* is bound south in Straits. NAAF Great Exuma: A plane arrived from Banana River and departed for Guantanamo. A plane arrived from Guantanamo and then departed for Banana River.

23rd January 1945

B-25's FC, FV, FF, FY, FK, and FM are airborne for air Exercise Number 5, aircraft landed after exercise completion. B-25's F, FV, FF, FU, FM, and FJ are airborne for air Exercise Number 1, aircraft landed after exercise completion. SS *Nevada* is bound north in Straits northwest of Little Bahama Bank, exiting the area. SS *Fenja* is bound north in Straits northwest of Little Bahama Bank, exiting the area. Destroyer USS *Kearney* (DD-432) is bound south in Straits northwest of Little Bahama Bank. SS *James Hawson* is bound north in Straits off Palm Beach. SS *Perryville* is bound north in Straits off Port Everglades. SS *Mobile Bay* is bound south in Straits west of Orange Cay, Andros. **Convoy KG-757** is bound north in Straits south of Cay Sal Bank. USS *Pegasus* (AK-48) is bound northwest in Old Bahama Channel. Tanker *Gulfbird* and SS *Dorcasia* is bound south, to the east of San Salvador. NAAF Great Exuma: A plane arrived from Guantanamo and then departed for Banana River.

24th January 1945

Navigation exercise was organized to provide a co-ordinating exercise for final training of earlier course for this Unit, DR [dead reckoning], homing navigation, depth charge and gun attacks against smoke float carried out, transmission of practice signals to base. Liberators LY and NJ are airborne twice from Windsor Field, aircraft landed after exercise completion. B-25's JT, FM, and SC are airborne for air Exercise Number 3 and Number 4, aircraft landed after exercise completion. B-25's KS, YT, and J are airborne for night navigation exercise, aircraft landed after exercise completion. Radar exercises carried out by Liberators from Windsor Field. SS *Phoenix* is bound north in Straits northwest of Little Bahama Bank, exiting the area. SS *Swiftscout* is bound north in Straits, off Palm Beach. USS *PC-1245* is bound north in Straits, southeast of Palm Beach. USS *PCE-875* and USS *PCE-874* are bound south in Straits, east of Northeast Providence Channel. NAAF Great Exuma: Tug USS *ATR-21* arrived from Guantanamo, yet it hit a reef while entering the channel leading to the base, and its rudder sheared off. A station tug and a US Navy damage repair ship, or ARB-type, were sent out to its aid. A plane arrived from Guantanamo, then departed for Banana River.

25th January 1945

Radar and local flying. B-25's FK, FY, FU, FJ, FT, and FN are airborne for air Exercise Number 3 and Number 4, aircraft landed after exercise completion. B-25's FY, FS, FH, and FM are airborne for night navigation exercise, then aircraft landed after exercise completion. Liberators MD and NC are airborne for air Exercise

Number 48, exercise same as previous day, aircraft landed after exercise completion. Liberators NK and MD are airborne, B-24 MD returned with engine trouble, landed back, other aircraft completed exercise and then landed. Tanker *Petrofuel* is bound north in Straits, northwest of Little Bahama Bank, exiting the area. SS *Nachman Syrkin*, SS *Henry L. Ellsworth*, and SS *Garnet Hulings* are bound north in Straits west of Little Bahama Bank. SS *Bohemian Club* and destroyer escort USS *Gustafson* (DE-182) are bound south in Straits, west of Little Bahama Bank. SS *Touraine* is bound west out of Northwest Providence Channel. **Convoy GK-857** is bound northwest in the Old Bahama Channel. SS *Western Queen* is bound north, to the east of San Salvador. G/C P. D. Gracroft, AFC, officially opened the Oakes Field YMCA in the presence of HRH the Governor and the Duchess of Windsor, who toured the building, which has been furnished and equipped by the Canadian YMCA. NAAF Great Exuma: A plane arrived from Guantanamo and then departed for Banana River.

26th January 1945

B-25's FY, FS, FT, FJ, and FD are airborne for Anti-Submarine Patrol Number 2, aircraft landed after exercise completion. B-25's FB, FS, FD, FT, FJ, and FG are airborne for air Exercise Number 2, aircraft landed after exercise completion. Liberators NA, MC, and LY are airborne from Windsor Field, and completed Exercise Number 48, then aircraft landed after exercise completion. SS *Reginald A. Fessenden* is bound north in Straits, northwest of Little Bahama Bank, exiting the area. USS *LSM-255* and SS *Finnmark* are bound south in Straits northwest of Little Bahama Bank. **Convoy Narragansett Group** is bound south in Straits west of Northwest Providence Channel. Tanker *Cities Service Koolmotor* is bound north in Straits west of Northwest Providence Channel. Tanker *A. S. Hansen* is bound north in Straits off Miami. Tanker *Cities Service Kansas* and tanker *Pan Virginia* are bound north in Straits. SS *Carnifex Ferry* is bound south in Straits, north of Cay Sal Bank. SS *J. Willard Gibbs,* **Convoy NG-485**, and SS *Robert Rogers* are bound south, to the east of San Salvador. NAAF Great Exuma: A plane arrived from San Juan.

27th January 1945

B-25's FT, FS, FB, FJ, FF, and FU are airborne for Anti-Submarine Patrol Number 2, aircraft landed after exercise completion. B-25's FS, FT, FJ, FN, FM, and FU are airborne for air Exercise Number 2, aircraft landed after exercise completion. Liberator LR is airborne for navigation Exercise Number 48, then aircraft landed after exercise completion. SS *Gulf of Mexico* is bound north in Straits, northwest of Little Bahama Bank, exiting the area. SS *Thomas H. Wheeler* is bound south in Straits, west of Little Bahama Bank. USS *LST-115* and SS *Jean Baptiste Lemoyne* are bound north in Straits, west of Little Bahama Bank. SS *B. F. Shaw* is bound north in Straits, west of Northwest Providence Channel. The mailboat *Joyce Roberts* is bound south in Straits, to the south of Bimini. USS *Williamsburg* (PG-56) is bound northwest in Santaren Channel. SS *Hoke Smith* and SS *Fred W. Weller* are bound south, to the east of San Salvador. SS *Lot Whitcomb* is bound south, to the east of San Salvador, exiting the area. No.113: Mosquito Conversion Training commenced at Oakes Field. NAAF Great Exuma: A plane arrived from Banana River and departed for Guantanamo. An Army PBY arrived from Mayaguana and Morrison Field, another Army PBY departed for Morrison Field and Mayaguana, and a plane arrived from Guantanamo and departed for Banana River.

28th January 1945

Mosquitoes on Conversion Course for Remote Air Traffic Control, or RATC. Pilots were airborne from Oakes Field for local flying throughout the day. Liberators MC, MB, NK, and LW are airborne from Windsor Field for Exercise Number 23A and Exercise Number 23B, aircraft landed after exercise completion. Tanker *Olney* is bound north in Straits northwest of Little Bahama Bank. SS *Gulwing* and SS *Carleton* are bound north in Straits west of Little Bahama Bank. SS *Imperial Transport* is bound north in Straits off Palm Beach. Liberty ship EC2 SS *Abel Parker Upshur* is bound north in Straits, north of Cay Sal Bank. SS *Daniel Carroll* is bound northwest in Nicholas Channel. Tug USS *YTB-405* is bound southeast in Old Bahama Channel. SS *Trinidad H. D.* is bound south, to the east of San Salvador. SS *Rampart* is bound south, to the east of San Salvador, exiting the area. NAAF Great Exuma: A plane arrived from Guantanamo and then departed for Banana River.

29th January 1945

B-25's FM, FS, FG, FF, FT, and FN are airborne for air Exercise Number 8 and later Number 7, aircraft landed after exercise completion. Liberator LH is airborne for Radar Exercise Number 2. A C-47 with G/C Lott and party is airborne for Washington DC via Morrison Field. B-25's from Oakes Field and B-24's from Windsor Field are airborne throughout the day for local flying and training. SS *William Asa Carter*, SS *Ardmore*, and SS *Tydolgas* are bound north in Straits northwest of Little Bahama Bank. **Convoy KN-365** is bound north in Straits, northwest of Little Bahama Bank. SS *Scandia* is bound north in Straits, west of Little Bahama Bank. **Task Unit TU-2821** and destroyer USS *Lea* (DD-118) are bound south in Straits, north of Bimini. USS *YO-110* is bound southeast in Old Bahama Channel. **Convoy KG-758** is bound southeast in Old Bahama Channel. SS *Silver Star Park* is bound north, to the east of Cat Island. NAAF Great Exuma: A plane arrived from Banana River and departed for Guantanamo. Two OS2U's arrived from the seaplane tender USS *Absecon* (AVP-23), then returned.

30th January 1945

B-25's TU, NO, RT, and H are airborne for air Exercise Number 8, aircraft landed after exercise completion. B-25's GT, SN, and B are airborne for air Exercise Number 7, aircraft landed after exercise completion. Liberator NA is airborne for Kingsley Exercise Number 22, homing "B" commenced, convoy sighted, recognition procedure carried out, patrol Python carried out, frigate laid smoke float, depth charge attack made, practice U-boat signals transmitted to base, homing "C" unsuccessful, aircraft landed after exercise completion. Liberator MC is airborne for Kingsley Exercise as first relief, homing "B" commenced, convoy sighted, recognition procedure carried out, patrol Python carried out, frigate laid smoke float, depth charge attack made, practice U-boat signals transmitted to base, homing "C" successful, aircraft landed after exercise completion. Liberator NX is airborne for Kingsley Exercise as second relief, homing "B" commenced, convoy sighted, recognition procedure carried out, patrol Python carried out, frigate laid smoke float, depth charge attack made, practice U-boat signals transmitted to base, homing "C" successful, aircraft landed after exercise completion. Liberator NJ is airborne for

Kingsley Exercise as third relief, homing "B" commenced, convoy sighted, recognition procedure carried out, patrol Python carried out, frigate laid smoke float, depth charge attack made, practice U-boat signals transmitted to base, homing "C" successful, aircraft landed after exercise completion. B-24 LH is airborne for two hours for detailed Radar exercise. Local flying was carried out by Mosquitoes and B-25's from Oakes Field and B-24's from Windsor Field throughout the day; one Mosquito swung on landing, damaged its undercarriage, crew were uninjured. B-25's FT, FH, FN, FM, FV, and FF are airborne for air Exercise Number 6, aircraft landed after exercise completion. Grumman Goose GD is airborne to Harbour Island, landed back at Oakes Field later in the evening. Local flying was carried out by Mosquitoes and B-25's from Oakes Field, then B-24's from Windsor Field. **Convoy NK-665** is bound south in Straits, off Canaveral. SS *Empire Chief* and SS *Watertown* are bound north, to the northwest of Little Bahama Bank, exiting the area. SS *Black Water* and SS *Bulkore* are bound south in Straits, north of Little Bahama Bank. SS *Bulkoil* and SS *William Crompton* are bound south in Straits west of Little Bahama Bank. Tanker *Cities Service Ohio,* SS *President Tyler*, and SS *Fort Robinson* are bound north in Straits, west of Little Bahama Bank. SS *Wellesley* is bound south in Straits west of Grand Bahama. Tanker *Esso Belgium* is bound north in Straits off Port Everglades. SS *Loreto* is bound west in Nicholas Channel. SS *George Washington* is bound south, to the east of Cat Island. **Convoy GN-184** is bound north, to the east of Cat Island. No.113: Training Mosquito KB-540 ground-looped on landing, owing to its right brake locking. Pilot was F/Lt. J. F. Ochalsky of Poland; there were no casualties. NAAF Great Exuma: A plane arrived from Guantanamo and departed for Banana River. Tug USS *ATR-21* and USCG cutter *CG-Legare* (WSC-144) departed for Guantanamo.

31st January 1945

Liberator LR is airborne from Windsor Field, radio contact established, homing "B" unsuccessful, incorrect bearings passed by conducting ship, ships not sighted for some time, patrol Python carried out, smoke float laid by ship, attacked with depth charge, U-boat signal transmitted to base, homing "C" procedure unsuccessful for same reason as before, aircraft landed back. Liberator NC is airborne from Windsor Field, radio contact established, conducting ship sighted, patrol Python and Adder carried out, smoke float laid by ship, attacked with depth charge, U-boat signal transmitted to base, homing "B" successful. Then patrols Cobra and Alligator were carried out, and aircraft landed after exercise completion. Liberator NA is airborne from Windsor Field, radio contact established, homing "B" commenced, patrol Python carried out, smoke float laid by ship, attacked with depth charge, U-boat signal transmitted to base, homing "C" unsuccessful, radio failure, patrol Cobra commenced, aircraft landed after exercise completion. SS *Richard Upjohn* is bound north in Straits northwest of Little Bahama Bank, exiting the area. SS *Phoenix* is bound south in Straits north of Little Bahama Bank. SS *Orville Harden* is bound north in Straits west of Little Bahama Bank. Tanker *Cities Service Ohio* is bound north in Straits off Palm Beach. SS *Bulklube* is bound south in Straits northwest of Bimini. SS *Medina* is bound north in Straits north of Cay Sal Bank. SS *Madaket* is bound southeast in Nicholas Channel. SS *Robert Watchorn* is bound north in Santaren Channel. **Convoy GK-858** is bound northwest in Old Bahama Channel. SS *H. J. Raymond* is bound south, to the east of San Salvador. NAAF Great Exuma: A plane arrived from Guantanamo and then departed for Banana River on a training flight.

1st February 1945

Training is carried out on Mosquitoes and B-25 from Oakes Field and B 24 from Windsor Field. B-25's FH, FS, FB, FC, FF, and FD are airborne for air Exercise Number 5, aircraft landed after exercise completion. B-25's FF, FS, FC, FH, FB, and FM are airborne for local navigation exercise, aircraft landed after exercise completion. C-47 9256 is airborne for Miami. Liberator NA is airborne from Windsor Field, homing "B" commenced, two frigates sighted, recognition procedure carried out, patrol Python and Cobra carried out, smoke float laid by ship, attacked with depth charge practice, U-boat signal transmitted to base, aircraft landed after exercise completion. Liberator MC is airborne from Windsor Field, homing "B" commenced, recognition procedure carried out, patrol Python and Cobra carried out, smoke float laid by ship, attacked with depth charge and guns, photographs taken, practice U-boat signal transmitted to base, aircraft landed after exercise completion. Liberator NX is airborne from Windsor Field, homing "B" commenced, recognition procedure carried out, smoke float laid by ship, attacked with depth charge and guns, photographs taken, practice U-boat signal transmitted to base, patrol Python and Cobra carried out, aircraft landed after exercise completion. USS *Odum* (APD-7) is bound north in Straits north of Cay Sal, exiting the area. Submarine USS *Bumper* (SS-333) is bound north in Straits northwest of Little Bahama Bank, with US Navy plane escorting, exiting the area. SS *Robert Watchorn* is bound north in Straits, northwest of Little Bahama Bank, exiting the area. SS *E. R. Kemp* is bound north in Straits, west of Little Bahama Bank. SS *W. M. Tyler* and SS *Empire Gulf* are bound north in Straits, off Palm Beach. USS *Williamsburg* (PG-56) is bound north in Straits off Port Everglades. USS *Kawca* (AO 15) is bound west in Old Bahama Channel. Base population, trainees: 283, officers and other ranks 465, Czechoslovakian: 10 officers and 27 men, RNZAF, New Zealand, four officers and two men, Permanent Staff: 169 officers and 1,819 other ranks, nurses: nine officers, Australian RAAF: 22 officers and 20 men, RAF: 334 officers and 2,182 men, RASC, EFI: One officer, 13 other ranks, RCAF, Canadian: 82 officers and 67 men, 250 ASR Unit: 14 other ranks. NAAF Great Exuma: A plane arrived from Guantanamo and then departed for Banana River.

2nd February 1945

Training on Mosquitoes and B-25 from Oakes Field and B-24 from Windsor Field, one mosquito damaged on undercarriage during take-off, no crew injury. B-25's FF, FS, FD, FN, FB, and FM airborne for local navigation Exercise Number 1, aircraft landed after exercise completion. B-25's FS, FV, FU, FD, FB, and FM airborne for Anti-Submarine Patrol Number 1, aircraft landed after exercise completion. SS *Carl Zachary Webb*, tanker *Pan Europe*, SS *Daylight*, SS *Point Breeze,* and tanker *Gulfmaid* are bound north in Straits northwest of Little Bahama Bank, exiting the area. Tanker *Mary Ashley Townsend*, SS *T. E. Mitchell* and SS *Castana* are bound north in Straits off Port Everglades. SS *Piek* is bound south in Straits off Miami. Mailboat *Joyce Roberts* and mailboat

Richard Campbell are bound north in Straits, off Miami. SS *Paco City* is bound southeast entering Old Bahama Channel. **Task Unit TU-2965** is bound south, to the east of Eleuthera. No.113: Training Mosquito KB-479 at Oakes Field, swung on take-off ground looped, undercarriage collapsed. Both undercarriages were torn off, port main plane near spar split, plywood torn, and fuselage extensively damaged from cockpit forward. Both engine's reduction gears were broken, and both propeller blades bent, damage is in Category E. No casualties Pilot F/L J. F. Olchalski of Poland. NAAF Great Exuma: A plane arrived from San Juan and then departed for Banana River.

3rd February 1945

Training on Mosquitoes and B-25's from Oakes Field and B-24's from Windsor Field. B-25's FB, FS, FC, FH, FU, and FV are airborne for Anti-Submarine Patrol Number 1, aircraft landed after exercise completion. Four P-40's and one C-47 arrived at Windsor Field from Draw Field for Tampa exercises with Number 3 Squadron. SS *Baltic* is bound north in Straits northwest of Little Bahama Bank, exiting the area. USS *PC-1083* is bound south in Straits off Palm Beach. SS *Empire Chief*, SS *Peter II*, SS *Toteco* and USS *PGM-23* are bound north in Straits off Port Everglades. SS *Etbedford* is bound south, to the east of Eleuthera. NAAF Great Exuma: A plane arrived from Guantanamo and then departed for Banana River.

4th February 1945

Training on Mosquitoes and B-25's from Oakes Field and B-24's from Windsor Field. B-24's NC, MD, and MB are airborne from Windsor Field for Radar exercise, aircraft landed after exercise completion. B-24's ND, NO, and MF are airborne twice from Windsor Field for Radar exercise, aircraft landed after exercise completion. SS *Spondilius* is bound north in Straits northwest of Little Bahama Bank. SS *Allatoona* is bound south in Straits, west of Little Bahama Bank. SS *W. M. Burton* is bound north in Straits passing Palm Beach. **Convoy GK-859** is bound west in Northwest Providence Channel. SS *Lloyd S. Carson* and SS *James Hanson* are bound south in Straits off Port Everglades. SS *Eliza Jane Nicholson* is bound north in Straits off Port Everglades. SS *Mexico Victory* and SS *Finnmark* are bound north in Straits off Miami. SS *Port Republic* and SS *Hellas* are bound south in Straits south of Cay Sal Bank. SS *R. G. Stewart* and tanker *J. A. Mowinckel, Jr.* are bound south, to the east of Eleuthera. NAAF Great Exuma: A plane arrived from Guantanamo and departed for Banana River. An Army PBY arrived from South Caicos Island.

5th February 1945

B-25's from Oakes Field and B-24's from Windsor Field carried out local flying training. B-25's FC, FV, FG, FB, FN, and FU are airborne for air Exercise Number 5, aircraft landed after exercise completion bombing, air firing, DTC [descent through cloud] and signal practice. B-25's FC, FH, FG, FN, FD, and FM are airborne for air Exercise Number 1, aircraft landed after exercise completion. B-24's LT, LZ, and ND are airborne for Exercise Number 42 from Windsor Field, aircraft landed after exercise completion. B-24's LZ, ND, and MB are airborne for Exercise Number 42 from Windsor Field, aircraft landed after exercise completion. B-24's NO, MD, ad MB are airborne for Exercise Number 42 from Windsor Field, aircraft landed after exercise completion. B-24 LH is airborne for Radar Exercise Number 1, aircraft landed after exercise completion. SS *Clam* is bound north in Straits northwest of Little

Bahama Bank, exiting the area. SS *C. A. Canfield* and SS *Touraine* are bound north in Straits northwest of Little Bahama Bank. Tanker *R. P. Smith* and SS *St. Dominque* are bound north in Straits west of Little Bahama Bank. SS *Irwin Russell* and SS *James W. Reese* are bound north in Straits northwest of Bimini. Tanker *Alpha* and tanker *Eastern Sun* are bound north in Straits west of Bimini. USS *YOG-6* is bound west in Northwest Providence Channel. Tanker *Pan North Carolina* is bound north in Straits north of Cay Sal Bank. SS *James Hanson* is bound south in Straits, north of Cay Sal Bank. USS *YF-336* is bound southeast in Old Bahama Channel. **Convoy NG-487** is bound southeast of Eleuthera. NAAF Great Exuma: A plane arrived from Banana River and departed for Guantanamo. Then another plane arrived from Guantanamo, and departed for Banana River. An Army PBY departed for South Caicos Island.

6th February 1945

B-25's from Oakes Field and B-24's from Windsor Field carried out local flying training. B-25's FC, FH, FG, FK, FB, and FN are airborne for air Exercise Number 5, aircraft landed after exercise completion. B-25's FD, FB, FM, FC, FO, and FH are airborne for air Exercise Number 1, aircraft landed after exercise completion. B-24 NO is airborne for Exercise Number 42 from Windsor Field, aircraft landed after exercise completion. Grumman Goose GB is airborne for Harbour Island, then landed back at Oakes Field later. SS *William H. Gray*, SS *Daniel L. Johnston* are bound north in Straits northwest of Little Bahama Bank. SS *Fort Ridely* is bound north in Straits west of Bimini. SS *Bulkoil*, SS *Bessemer*, and SS *William Penn* are bound north in Straits west of Bimini. SS *Paco City* is bound west in Nicholas Channel. Tanker *Southern Sun* and USS *Tronkaha* are bound south, to the east of Eleuthera. NAAF Great Exuma: An Army PBY arrived from Morrison Field.

7th February 1945

B-25's from Oakes Field and B-24's from Windsor Field carried out local flying training. B-25's FC, FV, FO, FB, FK, and FD are airborne for air Exercise Number 3 and Number 4, aircraft landed after exercise completion. B-25's FV, FB, FK, FO, FD, and FC are airborne for Night Patrol Number 2, aircraft landed after exercise completion. B-24 LH is airborne for Radar Exercise Number 1, aircraft landed after exercise completion. Grumman Goose GB is airborne for Harbour Island, then landed back at Oakes Field later. B-24's MB and NJ are airborne for Exercise Number 48 of homing "B," attack on float by depth charge and guns, Radar and practice U-boat signals in rehearsal for Kingsley or Esco exercises. They then landed at Windsor filed after completion, and same aircraft took off on another briefing of same exercises again, later aircraft landed after exercise completion. Ferry *Jean Brillant* transported RAF airman Tom Kayes to Nassau from Miami. SS *Bulkero* is bound north in Straits northwest of Little Bahama Bank, exiting the area. USS *New York* is bound north in Straits northwest of Little Bahama Bank. USS *Grundy* (APA-11) is bound south, to the north of Walker Cay, north of Abaco. SS *George Gale*, SS *Wellesley*, and USS *LST-1116*, SS *John G. Whittier* and SS *Hellas* are bound north in Straits, west of Little Bahama Bank. **Convoy KG-760** is bound east in Nicholas Channel. SS *Kennebec* and SS *Stephen B. Elkins* are bound south, to the east of Eleuthera. NAAF Great Exuma: A plane arrived from Guantanamo and departed for Banana River, a plane came from San

Juan then departed for Banana River, and an Army PBY departed for Mayaguana.

8th February 1945

B-25's from Oakes Field and B-24's from Windsor Field carried out local flying training. B-25's FV, FB, FF, FC, FK, FO, and FG are airborne for air Exercise Number 3 and Number 4, aircraft landed after exercise completion. B-24 LH is airborne for Radar Exercise Number 1, aircraft landed after exercise completion. B-24's MC and LY are airborne from Windsor Field for Exercise Number 48, carried out second day's briefings of the three-day rehearsal of Kingsley exercises, homing "B," attack on float by depth charge and guns, photography and practice U-boat signals, aircraft landed after exercise completion. B-24's NJ and MC took off for same exercise later, aircraft landed after exercise completion. SS *James Sullivan* is bound north in Straits northwest of Little Bahama Bank, exiting the area. SS *Pala Duro* and tanker *Esso Baytown* are bound north in Straits west of Little Bahama Bank. Argentinian SS *Argentina* is bound north in Straits west of Little Bahama Bank. **Convoy KN-367** is bound north in Straits, west of Bimini. SS *Kittanning* is bound north, to the north of Cay Sal Bank. SS *Crow Wing* is bound east, to the north of Cay Sal Bank. SS *Mirza*, tanker *Gulfbreeze*, **Task Unit TU-20185**, and USS *LSM-254* are bound south, to the east of Eleuthera.

9th February 1945

B-25's from Oakes Field and B-24's from Windsor Field carried out local flying training. B-25's FD, FB, FC, FF, FH, and FK are airborne for air Exercise Number 7, aircraft landed after exercise completion. B-25's FD, FF, FB, FK, and FJ are airborne for air Exercise Number 8, aircraft landed after exercise completion. B-24 LH is airborne for Radar Exercise Number 1, aircraft landed after exercise completion. B-24's MB and MF are airborne from Windsor Field for Exercise Number 48, carried out third day's briefings of the three-day rehearsal of Kingsley exercises Radar, depth charge attack, photography, signals and navigations, aircraft landed after exercise completion. B-24's MF and LZ did the same exercise later, aircraft landed after exercise completion. SS *California* and USS *LST-1117* are bound north in Straits, northwest of Little Bahama Bank, exiting the area. Argentinian SS *Argentina*, SS *Ardmore*, SS *Thoebus*, and USS *Solomons Group* are bound south in Straits, west of Little Bahama Bank. Tanker *Paco* is bound north in Straits, west of Little Bahama Bank. SS *Elise* and SS *Tampico* are bound south in Straits, passing Port Everglades. Tanker *A. S. Hansen* is bound north, to the northeast of Cat Island. SS *Maturin* is bound west in Nicholas Channel. **Convoy GN-186** is bound north, east of Eleuthera. SS *Pierre S. duPont* is bound south, to the east of Eleuthera.

10th February 1945

Mosquitoes and B-25's from Oakes Field and B-24's from Windsor Field carried out local flying training. B-25's FD, FC, FG, FH, FJ, and FO are airborne for air Exercise Number 7, aircraft landed after exercise completion. B-25's FC, FD, FF, FH, FJ, and FO are airborne for air Exercise Number 8, aircraft landed after exercise completion. B-24 LH is airborne for Radar Exercise Number 1, aircraft landed after exercise completion. SS *Jean L. D.* is bound north in Straits northwest of Little Bahama Bank, exiting the area. SS *Malabar* is bound north in Straits northwest of Little Bahama Bank. Tanker *Delaware Sun*, SS *Niobe*, and SS *Charlestown* are bound south in Straits, west of Little Bahama Bank. SS *Jotunfjell*, USS *LST-1118*, and USS *PCE-867* are bound north in Straits, west of Little Bahama Bank. Tanker *Cities Service Koolmotor* is bound north in Straits, off Miami. SS *Bulkcrude* is bound south in Straits, south of Bimini. SS *George R. Poole* and SS *Port Republic* are bound north in Straits, north of Cay Sal Bank. Argentinian SS *Argentina* and SS *Ardmore* are bound south in Straits, to the north of Cay Sal Bank. SS *Touraine* is bound southeast in Old Bahama Channel. **Convoy NG-488** is bound north, to the east of Eleuthera. Tanker *Gulfwax* is bound south, to the east of Eleuthera. NAAF Great Exuma: An Army PBY arrived from Mayaguana.

11th February 1945

Grumman Goose GB is airborne for harbour Island returned to Oakes Field later in the evening. B-24's MF, NK, LY, NH, and MC are airborne more than once from Windsor Field for Radar exercise, aircraft landed after exercise completion. Tanker *Gulfcrest*, SS *Esturia,* and SS *Allatoona* are bound north in Straits northwest of Little Bahama Bank, exiting the area. SS *Eskbank* and SS *Mississippi* are bound south in Straits west of Little Bahama Bank. USS *LCI-570*, USS *SC-1277*, USS *SC-1276*, USS *SC-1057*, SS *Salmon Falls*, USS *Eugene* (PF-40), USS *El Paso* (PF-41), and SS *Ponca City* are bound north in Straits, south of Bimini. Tug USS *Willet* is bound northwest in Nicholas Channel. SS *El Mundo* is bound east in Old Bahama Channel. SS *Edwin Abbey*, SS *Joseph Brown*, and SS *F. Q. Barstow* are bound south, to the east of Eleuthera.

12th February 1945

B-25's FD, FC, FF, FH, FJ, and FO are airborne for Anti-Submarine Patrol Number 2, aircraft landed after exercise completion. B-25's FM, FS, FH, FF, FK, and FJ are then airborne for air Exercise Number 2 of bombing and depth charge, aircraft landed after exercise completion. B-24 LH is airborne for Radar Exercise Number 1, and aircraft landed after exercise completion. USS *SC-1277*, USS *SC-1276*, USS *SC-1057*, and USS *SC-1058* are bound north in Straits, northwest of Little Bahama Bank, exiting the area. SS *Daylight* is bound south in Straits, northwest of Little Bahama Bank. SS *Spidoleine* and SS *Piek* are bound north in Straits west of Little Bahama Bank. SS *Russell R. Jones* and SS *Swift Arrow* are bound north in Straits, west of Northwest Providence Channel. SS *Empire Chief* is bound north in Straits, north of Cay Sal Bank.

13th February 1945

B-25's from Oakes Field and B-24's from Windsor Field carried out local flying training. B-25's FG, FF, FK, FM, and FO are airborne for Anti-Submarine Patrol Number 2, aircraft landed after exercise completion. B-24 LH is airborne twice for Radar Exercise Number 2, aircraft landed after exercise completion. B-24 MF is airborne from Windsor Field for Esco Exercise Number 3, homing "B" commenced, vessel sighted, recognition signals carried out, Viper patrol carried out, smoke float laid by ship, attacked with depth charge and gun, photograph taken, practice signals sent to base, patrol Python commenced, aircraft landed after exercise completion. B-24 MX is airborne from Windsor Field as first relief, homing "B" commenced, vessel sighted, recognition signals carried out, Alligator and Adder patrol carried out, smoke float laid by ship, attacked with depth charge, photograph taken, practice signals and weather report sent to base, aircraft landed after exercise completion. B-24 MC is airborne from Windsor Field as second relief, homing "B" commenced, vessel sighted, recognition signals carried out, Adder patrol carried out, smoke float laid by ship, attacked with depth charge, photograph taken, practice signals and weather report sent to base, aircraft landed after exercise completion. B-24 NK is airborne from Windsor Field as final relief, homing "B" commenced, recognition signals carried out, Mamba and Cobra patrol carried out, smoke float laid by ship, attacked with depth charge, photograph taken, practice signals and weather report sent to base, aircraft landed after exercise completion. SS *Good Gulf* is bound north in Straits, northwest of Little Bahama Bank, exiting the area. **Convoy KN-368**, SS *Frank E. Spencer*, and SS *Reginald J. Fessenden* are bound north in Straits west of Little Bahama Bank. Tanker *L. V. Stanford* is bound south in Straits passing Port Everglades. **Convoy KG-761** is bound southeast in Old Bahama Channel. Tug USS *ATR-5* is bound south, to the east of Eleuthera. USS *Extricate* (ARS-16) and SS *S. M. Shoemaker* are bound south, to the east of Eleuthera. Tanker *Gulfport* and SS *City of Savannah* are bound north, to the east of Eleuthera. No.113: Training Mosquito KB-340 bounced several tires on landing at Oakes Field , and its undercarriage collapsed. Severe damage to airframe and both engines. Category E. No casualties; pilot was Lt. Sodal. NAAF Great Exuma: A plane arrived from Guantanamo on an administrative flight, then returned to Guantanamo

14th February 1945

B-25's from Oakes Field and B-24's from Windsor Field carried out local flying training. B-25's FD, FS, FC, FF, FK, and FH are airborne for Anti-Submarine Patrol Number 1, aircraft landed after exercise completion. B-25's FS, FD, FG, FO, FK, and FF are airborne for night patrol, aircraft landed after exercise completion. B-24 LH is airborne from Windsor Field for Radar Exercise Number 2, aircraft landed after exercise completion. B-24 NX is airborne from Windsor Field, homing "B" commenced, recognition signals carried out, Alligator patrol carried out, smoke float laid by ship, attacked with depth charge, photograph taken, practice signals and weather report sent to base, Mamba patrol carried out, aircraft landed after exercise completion. B-24 LZ airborne from Windsor Field, homing "B" commenced, recognition signals carried out, Mamba patrol carried out, smoke float laid by ship, attacked with depth charge, photograph taken, practice signals and weather report sent to base, aircraft landed after exercise completion. B-24 NG airborne from Windsor Field, homing "B" commenced, recognition signals carried out, Mamba and Cobra patrol carried out, smoke float laid by ship, attacked with depth charge, photograph taken, practice signals and weather report sent to base, aircraft landed after exercise completion. B-24 MF is airborne from Windsor Field final relief, homing "B" commenced, recognition signals carried out, Mamba and Adder patrol carried out, smoke float laid by ship, attacked with depth charge, photograph taken, amplifying report sent to base, aircraft landed after exercise completion. SS *Vermont* is bound north in Straits northwest of Little Bahama Bank, exiting the area. SS *Durango*, SS *Empire Lord*, tanker *Cities Service Foil*, SS *Andrew A. Humphreys*, SS *Baltic*, tanker *Bohemian Club*, SS *Tide Water*, and tanker *Gulftide* are bound south in Straits, west of Little Bahama Bank. Tanker *Haakon Hauan* is bound north in Straits, off Miami. SS *Tullahoma* is bound south, entering Santaren Channel. SS *Wiley Post* is bound south, to the east of Eleuthera, exiting the area. SS *Castana* is bound south, to the east of Eleuthera. Navy tanker, or oiler, USS

Kaweah (A0-15) is bound north, to the east of Eleuthera. NAAF Great Exuma: Three OS2U-3 Kingfishers arrived from and departed back to Fort Lauderdale.

15th February 1945

B-25's from Oakes Field and B-24's from Windsor Field carried out local flying training. B-25's FD, FO, FS, FM, FN, FG, and FK are airborne for Anti-Submarine Patrol Number 1, aircraft landed after exercise completion. B-25's DC, FT, CL, FG, FN, and FM are airborne for local navigation Number 1, aircraft landed after exercise completion. B-24 LH airborne from Windsor Field for Radar Exercise Number 2, aircraft landed after exercise completion. B-24 NK is airborne from Windsor Field homing "B" commenced, recognition signals carried out, Adder patrol carried out, smoke float laid by ship, attacked with depth charge, photograph taken, amplifying report sent to base, aircraft landed after exercise completion. B-24 LY is airborne from Windsor Field, homing "B" commenced, recognition signals carried out, Cobra and Adder patrol carried out, smoke float laid by ship, attacked with depth charge, photograph taken, amplifying report sent to base, aircraft landed after exercise completion. B-24 MC is airborne from Windsor Field, homing "B" commenced, recognition signals carried out, Cobra patrol started but ship's co-operation poor, usual depth charge and gun attack not made, photograph taken, amplifying report sent to base, Mamba patrol carried out, aircraft landed after exercise completion. B-24 LZ is airborne from Windsor Field as final relief, homing "B" commenced, recognition signals carried out Cobra and Alligator patrol carried out, smoke float laid by ship, attacked with depth charge, photograph taken, amplifying report sent to base, aircraft landed after exercise completion. Liberator NL airborne from Oakes Field for Bermuda *en route* to UK landed. SS *Tide Penn* and tanker *Paulsboro* are bound north in Straits northwest of Little Bahama Bank, exiting the area. SS *Medina* is bound south in Straits west of Little Bahama Bank. SS *Elbridge Geary* and SS *Irwin Russell* are bound north in Straits off Miami. SS *Palo Duro* is bound south in Straits south of Bimini. SS *Bulklube* and SS *Phoenix* are bound south in Straits north of Cay Sal Bank. **Convoy GK-861** is bound north in Straits west of Cay Sal Bank. **Convoy KW-379** is bound northwest in Nicholas Channel. SS *City of Omaha* is bound south, to the east of Eleuthera. Tanker *Comol Cuba* is bound north, to the east of Eleuthera.

16th February 1945

B-25's from Oakes Field and B-24's from Windsor Field carried out local flying training. B-25's DC, FY, FO, FN, FS, and FV are airborne for air Exercise Number 6, aircraft landed after exercise completion. Mitchell NG is airborne for Boca Chica, landed back to Windsor on 17th. SS *Philip Doddridge* is bound north in Straits, northwest of Little Bahama Bank, exiting the area. SS *Touraine*, SS *Gulf of Mexico*, and SS *Wallace M. Tyler* are bound north in Straits, northwest of Little Bahama Bank. Tanker *L. L. Abshire*, SS *Seahawk*, and SS *Spidoeine* are bound north in Straits, west of Bimini. SS *Irwin Russell* is bound south in Straits, off Miami. SS *El Mundo* is bound north in Straits, north of Cay Sal Bank. Tanker *Stanvac Calcutta* is bound south in Straits, north of Cay Sal Bank. USS *Dauphin* (APA-97), USS *Arosa*, and USS *Omaha* (CL-4) are bound south, to the east of Eleuthera. SS *R. G. Stewart* is bound north, to the east of Eleuthera. NAAF Great Exuma: A Bahamian Airways [BAL] plane arrived from Nassau, then returned and a US Army JRF Amphibian arrived from Nassau with the American Consul [possibly John Dye], and returned to Nassau.

17th February 1945

Mosquitoes and B-25's from Oakes Field and B-24's from Windsor Field carried out local flying training. B-25's FV, FS, FY, FH, FO, and FB are airborne for air Exercise Number 6, then the aircraft landed after exercise completion. SS *W. W. Mills* is bound south in Straits, northwest of Little Bahama Bank. Tanker *Pan Virginia* and SS *Alpha* are bound north in Straits off Port Everglades. SS *Empire Saturn*, tanker *18 de Marzo* and submarine USS *S-11* are bound north in Straits, north of Cay Sal Bank. SS *Spondilus* is bound south in Straits, north of Cay Sal Bank. **Convoy KG-762** and SS *Gallant Fox* are bound southeast in Nicholas Channel. Tanker *Gulfbird* is bound south, to the east of Eleuthera. SS *Petter Hull* is bound north, to the east of Eleuthera. NAAF Great Exuma: An Army JRF Amphibian departed to Nassau, and then returned.

18th February 1945

B-24's MD, MX, and LU are airborne twice from Windsor Field for Radar Exercise Number 41, then the aircraft landed after exercise completion. B-24's NC, MB, and MC are airborne, and NC and MC are airborne twice, from Windsor Field for Radar Exercise Number 41, then aircraft landed after exercise completion. SS *Edward L. Shea* and SS *Olney* are bound north in Straits, northwest of Little Bahama Bank, exiting the area. Tanker *18 de Marzo*, tanker *Beacon Street*, and submarine USS *Narwhal* (SS-167) and tows are bound north in Straits, northwest of Little Bahama Bank. SS *Rich Mountain* is bound south in Straits northwest of Little Bahama Bank. **Convoy KN-369** is bound north in Straits west of Little Bahama Bank. USS *YAG-21* and SS *Black Jack* are bound south in Straits, north of Bimini. Tanker *Bulkoil* is bound north in Straits, north of Bimini. SS *Imperial Transport* is bound north in Straits, off Port Everglades. USS *PC-1200* and tanker *Gustafson* are bound southeast in Old Bahama Channel. Tanker *British Merchant*, SS *Cypress Hills Park*, USS *Barry* (APD-29), and SS *Artigas* are bound south, to the east of Eleuthera. NAAF Great Exuma: An Army PBY arrived from Mayaguana, and an Army JRF Amphibian departed for Morrison Field.

19th February 1945

B-25's from Oakes Field and B-24's from Windsor Field carried out local flying training. B-25's FV, FS, FC, FY, FH, and FN are airborne for air Exercise Number 5, aircraft landed after exercise completion. B-25's FV, FS, FC, FY, FN, and FO are airborne for air Exercise Number 1, aircraft landed after exercise completion. B-25's FB and FG replaced B-25's FC and FO, which were unserviceable, during the exercise. B-24's MB, NC, ND, and LH are airborne from Windsor Field for Radar and Leigh Light Exercise Number 42, aircraft landed after exercise completion. SS *Tobias E. Stansbury* is bound north in Straits northwest of Little Bahama Bank exiting the area. Tanker *Axtell J. Byles* is bound north, to the north of Walker Cay, off Abaco. Tanker *Gulfwave* is bound south in Straits, west of Little Bahama Bank. Destroyer escort USS *Gantner* (DE-60) and SS *William W. Johnson* are bound north in Straits, west of Little Bahama Bank. **Convoy GK-862** is bound northwest in Straits northwest of Great Bahama Bank. SS *Empire Marriott* is bound southeast in Old Bahama Channel. USS *SC-1000* and submarine USS *R-15* are bound southwest in Straits then the Northeast Providence

Channel. SS *Magara*, SS *Atlantic*, and SS *Beacon Hill* are bound south, to the east of Eleuthera. NAAF Great Exuma: An Army PBY departed for Mayaguana.

20th February 1945

Mosquitoes and B-25's from Oakes Field and B-24's from Windsor Field carried out local flying training. B-25's FT, FV, FS, FY, FB, and FN are airborne for air Exercise Number 5, FK replaced FS during exercise. B-25's FO, FF, FS, FV, FK, and FY are airborne for air Exercise Number 1, aircraft landed after exercise completion. B-24 LH is airborne from Windsor Field for Radar Exercise Number 1, aircraft landed after exercise completion. SS *Argon* is bound south in Straits, northwest of Little Bahama Bank, exiting the area. SS *Elsie* and SS *Point Breeze* are bound north in Straits, west of Little Bahama Bank. SS *Nach Bulk* and SS *H. M. Flagler* are bound south in Straits west of Little Bahama Bank. USS *PC-1147* and SS *Trevilian* are bound south in Straits, north of Bimini. SS *Eano* and tanker *Empire Gulf* are bound north in Straits north of Cay Sal Bank. Submarine USS *R-15* is bound south in Straits, north of Cay Sal Bank. **Convoy San Juan-354,** escorted by USS *PC-1213* and USS *Rockambean* (AP-63) are bound north, to the east of Eleuthera. **Convoy NG-490** and SS *Temple Inn* are bound south, to the east of Eleuthera. NAAF Great Exuma: A plane arrived from Guantanamo and departed for the same place. A plane arrived from Banana River and then returned there. A plane departed from Guantanamo and departed for Banana River. An Army JRF Amphibian arrived from Morrison Field and departed for Mayaguana.

21st February 1945

Mosquitoes and B-25's from Oakes Field and B-24's from Windsor Field carried out local flying training. B-25's FV, FF, FY, FO, FT, and FS are airborne for air Exercise Number 3 and Number 4, aircraft landed after exercise completion. B-25's FV, FS, FG, FY, FB, FC, and FT are airborne for night Patrol Number 2, then FT replaced FV halfway, aircraft landed after exercise completion. B-24 LH is airborne from Windsor Field for Radar Exercise Number 1, aircraft landed after exercise completion. B-24 MB is airborne from Windsor Field navigation Exercise Number 48 of rehearsal for Kingsley or Esco Exercise, depth charge and gun attack on smoke float, signal practice, photography and homing "B," aircraft landed after exercise completion. SS *Niobe* and *USS SC-499* are bound north in Straits, northwest of Little Bahama Bank, exiting the area. SS *New York* and SS *Christyn Payne* are bound south in Straits northwest of Little Bahama Bank. SS *Eano* is bound north in Straits west of Little Bahama Bank. SS *Fort Republic* is bound south in Straits, north of Bimini. SS *Joyunfjell* and SS *Cacalilao* are bound north in Straits, west of Bimini. Argentinian SS *Argentina* is bound east in Nicolas Channel. Tanker *Comol Cuba* west enters into Nicholas Channel. *SS George T. Anfell* is bound south, to the east of Eleuthera. NAAF Great Exuma: Two planes arrived from Banana River and returned there. An Army JRF Amphibian arrived from Mayaguana, and then departed for Morrison Field.

22nd February 1945

Mosquitoes and B-25's from Oakes Field and B-24's from Windsor Field carried out local flying training. B-25's FV, FS, FT, FY, FN, and FB are airborne for air Exercise Number 3 and Number 4, run consecutively with interval for re-bombing, aircraft landed after exercise completion. B-25's FC, FN, FV, FT, FJ, and FS are airborne for Night Patrol Number 2, aircraft landed after exercise completion. B-24 LH airborne from Windsor Field, carried out two briefings for Radar Exercise Number 1, aircraft landed after exercise completion. B-24's MD, MF, and NF are airborne from Windsor Field, carried out 2nd day's briefings for Exercise Number 48 of homing "B," depth charge and gun practice, signalling and photography, and navigation. B-25 FB is airborne to investigate flashing light reported east of New Providence, reported by US Navy. A square search was carried out with no result, conclusion was that it was a marine marker which had been dropped during Exercise Number 48, in the area, then aircraft landed after exercise completion. USS *Active* (W-125) and SS *Vanja* are bound north in Straits, northwest of Little Bahama Bank, exiting the area. SS *Tampico* and tanker *Mary Ashley Townsend* are bound north in Straits northwest of Little Bahama Bank. SS *Swift Arrow* is bound south in Straits northwest of Little Bahama Bank. Tanker *AthelPrince* is bound south, to the north of Walker Cay, off Abaco. SS *Pawnee Rocks* and SS *Vincenja* are bound north in Straits, west of Little Bahama Bank. SS *Phoebus* is bound south in Straits west of Little Bahama Bank. Tanker *Bohemian Club* is bound south in Straits west of Bimini. SS *Bialystok* is bound north in Straits west of Bimini. USS *Dix* (W-136) is bound north in Santaren Channel. SS *Artigas* is bound northwest in Old Bahama Channel. Argentinian SS *Argentina* is bound southeast in Old Bahama Channel. SS *Brunswick* is bound south, to the east of Eleuthera. NAAF Great Exuma: An Army aircraft rescue [ASR] boat arrived from Mayaguana.

23rd February 1945

Mosquitoes and B-25's from Oakes Field and B-24's from Windsor Field carried out local flying training. B-25's FV, FS, FT, FK, and FJ are airborne for air Exercise Number 7, aircraft landed after exercise completion. B-25's CL, FY, FV, FC, FJ, and FG are airborne for air Exercise Number 8, aircraft landed after exercise completion. B-24 LH is airborne from Windsor Field, carried out two briefings for Radar Exercise Number 1, aircraft landed after exercise completion. B-24's MX, MC, and MF are airborne from Windsor Field, carried out 2nd day's briefings Exercise Number 48 of homing "B," depth charge and gun practice, signalling and photography, and navigation. Grumman Goose GB took off for Harbour Island and returned to Oakes Field. B-24's LM and LA are airborne from Windsor Field for local Exercise Number 4 (dual), yet collided in mid-air south of the aerodrome. B-25 LA was damaged and managed to make a belly landing, with one crew injured. B-25 LM crashed on land, and all nine crew were killed. Tanker *Bulklube* is bound north in Straits northwest of Little Bahama Bank, exiting the area. USS *Dix* (W-136) is bound north in Straits west of Little Bahama Bank. SS *Bulkore* is bound north in Straits west of Little Bahama Bank. **Convoy KN-370** is bound north in Straits west of Little Bahama Bank. USS *Curl* (ARS-21) and tow is bound south in Straits, west of Little Bahama Bank. SS *Philip Doddridge* is bound north in Straits west of Grand Bahama. SS *W. M. Burton* is bound north in Straits west of Little Bahama Bank. SS *San SS Pasqual* and SS *Daylight* are bound north in Straits, west of Bimini. SS *Tidewater* and SS *Artigas* are bound west in Straits south of at Cay Sal Bank. **Task Unit TU-0315** is bound northwest in Old Bahama Channel. **Convoy KG-763** is bound southeast in Old Bahama Channel. SS *Scandia* and SS *Point Pelee Park* are bound south, to the east of Eleuthera.

24th February 1945

B-25's from Oakes Field and B-24's from Windsor Field carried out local flying training. B-25's FV, FY, FC, FJ, FT, and FK are airborne for air Exercise Number 7, aircraft landed after exercise completion. B-25's FO, FT, FH, FC, FK, and FY are airborne for air Exercise Number 8, aircraft landed after exercise completion. B-24's MF and LY are airborne for Radar Exercise Number 23, on Route A, and B-24's MC and MD are then airborne for Radar Exercise Number 23 on Route B, aircraft landed after exercise completion. B-24 LH is airborne for Radar Exercise Number 2 with two briefings. SS *George Kendall* is bound north in Straits, west of Little Bahama Bank. SS *Peter II* is bound north in Straits west of Little Bahama Bank. USS *LST-1* is bound north in Straits west of Little Bahama Bank. Tanker *Haakon Hauan* is bound north in Straits, west of Little Bahama Bank. SS *W. F. Burdell* is bound south in Straits west of Little Bahama Bank. SS *General Fleischer* is bound north in Straits north of Abaco. SS *Jacob Thompson* is bound south, to the north of Walker Cay, off Abaco. Tanker *Cities Service Koolmotor*, SS *Durango*, and tanker *Hagood* are bound north in Straits west of Bimini. USS *Pochard* (AM-375), and USS *Shellbark* (AM-67) are bound south, to the east of Eleuthera. Tanker *Esso Nashville,* SS *William MacClay,* and SS *Cape Gaspe* are bound north, to the east of Eleuthera, exiting the area. NAAF Great Exuma: A PBY arrived from Mayaguana.

25th February 1945

Mosquitoes and B-25's from Oakes Field and B-24's from Windsor Field carried out local flying training. **Convoy NK-670** at northwest of Little Bahama Bank, exiting the area. SS *Quemado Lake,* SS *Tidepenn,* SS *Vermont,* and SS *New Market* are bound south in Straits north of Little Bahama Bank. Tanker *Cities Service Ohio* is bound south in Straits west of Little Bahama Bank. SS *American Builder* is bound north in Straits west of Little Bahama Bank. USS *LST-377* is south in between Bimini and Port Everglades. SS *Ardmore* is bound north in Straits, between Bimini and Port Everglades. SS *Chads Ford* is bound north in Straits north of Cay Sal Bank. SS *Chantilly* is bound south in Straits north of Cay Sal Bank. USS *PC-1184* south in Old Bahama Channel. SS *Great Meadows* is bound south, to the east of Eleuthera. SS *New Orleans* is bound south, to the east of Eleuthera. Tanker *Gulfking* is bound south, to the east of Eleuthera. Tanker *Esso Aruba* is bound south, to the east of Eleuthera. NAAF Great Exuma: A PBY-2 departed to Mayaguana.

26th February 1945

Mosquitoes and B-25's from Oakes Field and B-24's from Windsor Field carried out local flying training. B-25's FV, FY, FC, FU, FO, and FJ are airborne on Anti-Submarine Patrol Number 2, aircraft landed after exercise completion. B-25's FK, FT, FH, FG, FO, and FC are airborne for air Exercise Number 2, landed after interval for re-bombing, B-25's FK and FT did not complete the exercise, due to bad weather. B-24's MF and MC are airborne from Windsor Field, and carried out two briefings of Radar Exercise Number 23 on Route A. B-24's LY and MX are airborne from Windsor Field, and carried out two briefings of Radar Exercise Number 23 on Route B. B-24 LH is airborne from Windsor Field, carried out two briefings of Radar Exercise Number 2. SS *Baltic* is bound north in Straits west of Little Bahama Bank. Tanker *Cities Service Oklahoma* and SS *Gulf of Mexico* are bound south in Straits west of Little Bahama Bank. Tanker *Cities Service Ohio* is bound south in Straits off Miami.

Tanker *Stanvac Calcutta* is bound north in Straits west of Bimini. SS *Leonardo Romero* is bound north in Straits north of Cay Sal Bank. SS *Seatrain Nola* is bound south in Straits. **Convoy NG-491** is bound south, to the east of Eleuthera, exiting the area.

27th February 1945

Mosquitoes and B-25's from Oakes Field and B-24's from Windsor Field carried out local flying training. B-25's FF, FV, FB, FC, FH, and FG are airborne for Anti-Submarine Patrol Number 2, aircraft landed after exercise completion, gave shipping reports. B-25's FH, FY, FC, FB, and FV are airborne for air Exercise Number 2 of navigation, bombing, and DTC [descent through cloud], aircraft landed after exercise completion. B-24's NA, MC, and MF are airborne from Windsor Field for Exercise Number 50 of navigation, depth charge and gun attacks, smoke float, photography, and homing "B," aircraft landed after exercise completion. B-24 LH is airborne for Radar Exercise Number 2, aircraft landed after exercise completion. USS *LST-1129* (ARL-29) is bound north in Straits northwest of Little Bahama Bank. SS *Edward L. Shea* and tanker *British Hope* are bound south in Straits west of Little Bahama Bank. SS *Touraine* is bound south in Straits off Port Everglades. SS *San Tirso*, USS *SC-659*, USS *SC-540*, and USS *SC-635* are bound north in Straits off Miami. SS *Niso* is bound south in Straits southwest of Bimini. Tanker *L. L. Abshire* and SS *Spondilus* are bound north in Straits north of Cay Sal Bank. Submarine USS *R-15* is bound southeast in Old Bahama Channel. SS *Papanui* and USS *Shelby* (APA-105) are bound south, to the east of Eleuthera. NAAF Great Exuma: Two plane arrived from Caicos Island and Morrison Field, and then two planes departed for Caicos Island and Morrison Field.

28th February 1945

B-25's from Oakes Field and B-24's from Windsor Field carried out local flying training. B-25's YC, UH, and MJ are airborne for Anti-Submarine Patrol Number 1, aircraft landed after exercise completion. B-24's NA, NK, and MC are airborne from Windsor Field for Exercise Number 50 of navigation, depth charge and gun attacks, smoke float, photography, and homing "B," aircraft landed after exercise completion. In conjugation with HMCS *Sussexvale* and tanker *Fort Erie* provided by HMCS *Somers Isles*, Kingsley Exercise Number 22 carried out, 11 crew, 39 courses, and four B-24's Liberators took part on 1st and 2nd day and three on 3rd. SS *San Tirso* is north in Straits, northwest of Little Bahama Bank, exiting the area. Tanker *L. L. Abshire* and SS *Egbert Benson* are bound north in Straits, northwest of Little Bahama Bank, exiting the area. SS *Alexander Wilson* is bound north in Straits northwest of Little Bahama Bank. USS *LST-1125*, SS *Minatitlan*, SS *Charles Nordoff*, and SS *Bidwell* are bound south in Straits, northwest of Little Bahama Bank. **Convoy KN-371** is bound north in Straits, west of Little Bahama Bank. SS *Wayne MacVeagh* is bound south in Straits west of Little Bahama Bank. SS *Chagres* is bound north in Straits west of Bimini. SS *Trevilian* is bound north in Straits north of Bimini. Tanker *Gulfwave* is bound north in Straits north of Cay Sal Bank. Tanker *Spencer Kellogg* is bound east entering Old Bahama Channel. **Convoy KG-764** and SS *George Washington* are bound southeast in Old Bahama Channel. SS *Quetico Park* and SS *John Bidwell* are bound south, to the east of Eleuthera.

1st March 1945

Mosquitoes and B-25's from Oakes Field and B-24's from Windsor Field carried out local flying training. B-25's FM, FC, FT, FG, and FJ are airborne for Anti-Submarine Patrol Number 1, aircraft landed after exercise completion. B-24's NJ, MF, and MD are airborne from Windsor Field for Exercise Number 50 of navigation, depth charge and gun attacks, smoke float, photography, and homing "B," aircraft landed after exercise completion. B-25's NT, FM, and DC airborne from Oakes Field for air Exercise Number 1, aircraft landed after exercise completion. B-25's FT and FD landed early, due to unserviceability. Grumman Goose GB is airborne for Harbour Island, then on return landed at Oakes Field. B-25 FC is airborne for a direction-finding Calibration Flight with San Salvador Station, aircraft landed after exercise completion. SS *Samuel H. Walker* is bound north in Straits, northwest of Little Bahama Bank, exiting the area. SS *Tustem* and *Thomas H. Wheeler* are bound south in Straits, northwest of Little Bahama Bank. SS *Jean Baptiste Lemoyne*, Tanker *Hadnot* and SS *Port Republic* are bound north in Straits west of Little Bahama Bank. USS *YAG-19* is bound south-southeast of Nassau. Tanker *Spencer Kellogg* bound southeast in Old Bahama Channel. SS *Mirza*, SS *Esturia*, tanker *Gulfbreeze*, SS *Casper S.*, and USS *Yost* are bound south, to the east of Eleuthera. SS *Madaket* is bound north, to the east of Eleuthera. NAAF Great Exuma: A plane arrived from Banana River and returned there. An Army PBY arrived from Caicos Island, and then returned there.

2nd March 1945

Mosquitoes and B-25's from Oakes Field and B-24's from Windsor Field carried out local flying training. B-25's FN, FJ, FH, FM, FB, and FD are airborne for air Exercise Number 6, aircraft landed after exercise completion. B-25's FH, FJ, FF, FN, and FM are airborne for local navigation Exercise Number 1, aircraft landed after exercise completion. SS *Piqua* is bound north in Straits northwest of Little Bahama Bank, exiting the area. Tanker *Pan Scandia*, tanker *E. R. Kemp*, SS *Baldbutte*, tanker *Delaware Sun*, and tanker *Axtell J. Byles* are bound north in Straits west of Little Bahama Bank. SS *Tecumseh Park* is bound south in Straits north of Bimini. SS *Poucou* and SS *Katy* are bound north in Straits north of Cay Sal Bank. **Convoy GK-864** is bound northwest in Nicholas Channel. USS *ATA-205* is bound north in Straits off Port Everglades. USS *YAG-19* is bound south entering Exuma Sound. **Task Unit TU-2922**, USS *Achelous* (ARL-12), USS *Duchess* (APA-98), USS *Yofes* (APA-69), SS *Nomsuco*, and **Convoy NG-492** are bound south, to the east of Eleuthera. SS *Charles Robinson* is bound south, to the east of Eleuthera exiting the area. SS *Cape Spencer* is bound north, to the east of Eleuthera. NAAF Great Exuma: A plane arrived from Guantanamo, then returned. Then two planes arrived from Banana River and returned, and USS *YAG-19* arrived from Miami, with supplies.

3rd March 1945

B-25's from Oakes Field and B-24's from Windsor Field carried out local flying training. B-25's FT, FM, FB, FF, and FN are airborne for air Exercise Number 6, aircraft landed after exercise completion. Grumman Goose GB is airborne for Harbour Island, thereafter aircraft landed after exercise completion. B-25 FC is airborne for direction finding Calibration Flight with San Salvador Station, aircraft landed after exercise completion. Submarine USS *Barbero* (SS-311) and SS *Minatitlan* are bound north in Straits, northwest of Little Bahama Bank, exiting the area. SS *H. M. Flagler*, SS *Cape Lookout*, and SS *Washington* are bound north in

Straits, west of Little Bahama Bank. SS *Vanja* is bound south in Straits west of Little Bahama Bank. Tanker *AthelDuke* is bound south in Straits. USS *Roselle* (AM-379) is bound north in Straits. SS *Castana* is bound southeast in Nicholas Channel. USS *Brock* (APD-93), SS *Samford*, tanker *Gulfwax*, and SS *Nemaha* are bound south, to the east of Eleuthera. NAAF Great Exuma: A plane arrived from Banana River and returned there. USS *YAG-19* departed for Miami.

4th March 1945

B-24's MF and NX are north airborne with three briefings of Radar Exercise Number 41, carried out two-hour patrol with Radar exercise, aircraft landed after exercise completion. No flying from Oakes Field, Sunday. SS *Cephee* and SS *William Crompton* are bound north in Straits northwest of Little Bahama Bank, exiting the area. SS *Cacalilao* is bound south in Straits, northwest of Little Bahama Bank. SS *Bessemer* is bound north in Straits, west of Little Bahama Bank. USS *Absecon* (AVP-23) is bound south in Straits west of Little Bahama Bank. Tanker *Eastern Sun* and SS *Empire Lord* are bound north in Straits west of Bimini. USS *YAG-19* is bound south in Straits west of Bimini. SS *Mary A. Townsend* is bound south in Straits off Port Everglades. SS *Tekoa*, tanker *Beaconoil* and SS *W. W. Burton* are bound south, to the east of Eleuthera. NAAF Great Exuma: A plane with an oil leak *en route* from Guantanamo to Banana River arrived then departed for Banana River.

5th March 1945

B-25's from Oakes Field and B-24's from Windsor Field carried out local flying training. B-25's FV, FF, FC, FK, and FB are airborne for air Exercise Number 5, aircraft landed after exercise completion. B-25's FU, FM, FK, FF, and FV are airborne for air Exercise Number 1, aircraft landed after exercise completion. B-24's MF, NO, NA, LW, and ND are airborne with three briefings of Radar Exercise Number 42, aircraft landed after exercise completion. B-24 LH is airborne for Radar Exercise Number 1, aircraft landed after exercise completion. SS *Prometheus* and SS *Chantilly* are bound north in Straits, northwest of Little Bahama Bank, exiting the area. SS *New Market*, SS *John L. McCarley*, tanker *Esso Wilmington,* SS *New York*, and tanker *Petrofuel* are bound north in Straits northwest of Little Bahama Bank. **Convoy NK-672** is bound south in Straits west of Little Bahama Bank. SS *Joseph P. Bradley* and SS *Empire Noble* are bound north in Straits west of Little Bahama Bank. SS *Ponca City* are bound north in Straits northwest of Bimini. SS *Arizona* is bound north in Straits west of Bimini. SS *W. F. Burdell* is bound north in Straits. USS *YF-336* is bound north in Straits north of Cay Sal Bank. SS *Gertrude Kellogg* is bound southeast in Old Bahama Channel. **Convoy KG-765** is bound south in Old Bahama Channel. Tanker *Olney* and SS *Tidewater Associated* are bound south, to the east of Eleuthera. SS *Scandia* is bound north, to the east of Eleuthera. NAAF Great Exuma: A plane arrived from Banana River and returned there. An Army JRF Amphibian arrived, and then departed.

6th March 1945

B-25's from Oakes Field and B-24's from Windsor Field carried out local flying training. B-25's FV, FS, FU, FK, FM, and FB are airborne for air Exercise Number 3 of bombing and DTC [descent through clouds], aircraft landed after exercise completion. B-25's FM, FU, FN, FV, FB, and FD are airborne for air Exercise Number 1 of general flying, aircraft landed after exercise completion. A US Navy *Dauntless* is missing 40 nm off Port Everglades in the Straits. SS *Eliza Jane Nicholas* is bound north in Straits, northwest of Little Bahama Bank, exiting the area. SS *Frank B. Kellogg*, SS *Oliver Wolcott*, SS *Harry Percy*, and SS *Woodrow Wilson* are bound south in Straits, northwest of Little Bahama Bank. Tanker *AthelPrince* is bound north in Straits northwest of Little Bahama Bank. SS *Molino* is bound south in Straits north of Little Bahama Bank. SS *Darrel M. Ritter* is bound north in Straits west of Little Bahama Bank. SS *James McHenry*, SS *Somme*, tanker *Hagood*, and SS *Benjamin Franklin* are bound south in Straits west of Little Bahama Bank. SS *Swift Arrow*, tanker *Gulfmaid*, and SS *Morton Prince* are bound north in Straits, to the north of Cay Sal Bank. SS *Empire Viscount* is bound south, to the east of Eleuthera. SS *Megara* is bound north, to the east of Eleuthera, exiting the area. **Convoy GN-191**, SS *Alexander E. Brown*, and destroyer escort USS *Cross* (DE-448) are bound north, to the east of Eleuthera. NAAF Great Exuma: A plane arrived from Banana River, and then returned there.

7th March 1945

Mosquitoes and B-25's from Oakes Field and B-24's from Windsor Field carried out local flying training. B-25's FS, FV, FN, FK, and FO are airborne for air Exercise Number 3 and Number 4, carried out bombing on fixed and moving targets, aircraft landed after exercise completion. B-24's MD and NF are airborne from Windsor Field for Exercise Number 48 of rehearsal for Kingsley and Esco exercise, depth charge and gun attack, photography, homing procedure, signals, and navigation, aircraft landed after exercise completion. B-25's FD, FK, FB, FV, FH, and FU are airborne for Night Patrol Number 2, with six details of signals and navigation, aircraft landed after exercise completion. Liberator NL, from Bermuda, is *en route* from UK, and landed at Oakes Field. B-24 LH is airborne from Windsor Field with two briefings of Radar Exercise Number 1, aircraft landed after exercise completion. SS *Vermont* and tanker *Spencer Kellogg* are bound north in Straits northwest of Little Bahama Bank. SS *New Hope* is bound south in Straits west of Bimini. Submarine USS *R-1* is bound north in Straits, off Port Everglades. SS *Samuel L. Jeffery* is bound north in Straits. SS *Gulf of Mexico* is bound north in Straits north of Cay Sal Bank. USS *Arizona* is bound southeast entering, Old Bahama Channel. SS *City of Savannah* is bound south, to the east of Eleuthera. No.113: Training Mosquito KB-381 ground-looped on landing, causing damage to the port wing and engine. Pilot was F/O R. Cripps, and there were no casualties.

8th March 1945

Mosquitoes and B-25's from Oakes Field and B-24's from Windsor Field carried out local flying training. B-25's FV, FS, FK, FY, and FM are airborne for air Exercise Number 3 and Number 4, bombing on fixed and moving targets, aircraft landed after exercise completion. B-24's MC, NC, and MJ are airborne from Windsor Field, carried out 2nd day's briefings Exercise Number 48, aircraft landed after exercise completion. B-25's FM, FJ, FK, FH, and FV are airborne for night Patrol Number 2 with five details of navigation and signals, aircraft landed after exercise completion. B-24 LH airborne from Windsor Field with two briefings of Radar Exercise Number 1, aircraft landed after exercise completion. SS *O. B. Martin*, **Convoy SS *Gulf of Mexico*** is bound north in Straits, northwest of Little Bahama Bank. Spanish SS *Urola* is bound south in Straits west of Little Bahama Bank. SS *Tide Penn* is bound north in Straits, west of Little Bahama Bank. SS *Matinicock* is bound north in Straits northwest of Bimini. Tanker *Esso Bayonne* is bound south in Straits, west of Bimini. Tanker *Cities Service Koolmotor* and SS *Corfu* are bound north in Straits, west of Bimini. SS *Sideling Hill* is bound north in Straits north of Cay Sal Bank. **Convoy SS *Bodie Island* Group** is bound south in Straits, off Palm Beach. SS *Washington*, SS *Clearwater Park*, and SS *Jacob Luckenbach* are bound south, to the east of Eleuthera. SS *R. G. Stewart* is bound north, to the east of Eleuthera. NAAF Great Exuma: A training plane arrived from Banana River, and then returned there. Tug USS *ATR-22* and its escort, USCG *CG-Colfax* arrived from Guantanamo towing yard the freighter USS *YF-309*.

9th March 1945

Censorship was eased, as a result of which, civilian telephone connections we re-opened with US and Canada. B-25's from Oakes Field and B-24's from Windsor Field carried out local flying training. B-25's FY, FH, FO, and FK are airborne for air Exercise Number 7 of compass wing, air firing, and photography, aircraft landed after exercise completion. B-25's FV, FY, FJ, FH, and FG are airborne for air Exercise Number 8 of bombing, decent through cloud (DTC), and photography, aircraft landed after exercise completion. B-24's NC and NO are airborne from Windsor Field on 3rd day's briefings of Exercise Number 48, aircraft landed after exercise completion. B-24 LH is airborne from Windsor Field with two briefings of Radar Exercise Number 2, aircraft landed after exercise completion. Grumman Goose GA is airborne to San Salvador, landed back to Oakes Field later at night. SS *Empire Chief*, USS *LST-1127*, and SS *Chancellorsville* are bound north in Straits, west of Little Bahama Bank. SS *Isaac M. Wise* and SS *Swiftlight* are bound south in Straits, west of Little Bahama Bank. Tanker *L. V. Stanford* is bound north in Straits west of Bimini. SS *Touraine* and SS *Willard R. Johnson* are bound north in Straits. Tanker *Gertrude Kellogg* is bound west in Nicholas Channel. USS *YO-129* is bound north in Straits west of Northwest Providence Channel. SS *W. H. Libby*, tanker *Pan Europe*, SS *Moose Mountain Park*, and USS *Augusta* (CA-31) are bound south, to the east of Eleuthera. NAAF Great Exuma: A plane arrived and departed for Banana River.

10th March 1945

B-25's from Oakes Field and B-24's from Windsor Field carried out local flying training. B-25's FV, FB, FH, FG, and FO are airborne for air Exercise Number 7 of bombing, decent through cloud (DTC), compass swing aircraft landed after exercise completion. B-25's FV, FG, FN, FJ, FY, and FH are airborne for air Exercise Number 8 of bombing, decent through cloud (DTC), and photography, aircraft landed after exercise completion. B-24 NO is airborne from Windsor Field, carried out one briefing of Exercise Number 48 that could not be completed on 7th March [three days prior], pilot's original and standby aircraft had become unserviceable. B-24 LH is airborne from Windsor Field with two briefings of Radar Exercise Number 2, aircraft landed after exercise completion. Grumman Goose GA is airborne for Marsh Harbour Abaco, then to Moore's Island, northwest of Sandy Point, and landed back at Oakes Field later in the evening. SS *Tourain*, SS *Tustim* and tanker *Edward L. Shea* are bound north in Straits northwest of Little Bahama Bank, exiting the area. SS *Moose Mountain Park* and tow are bound south in Straits northwest of Little Bahama Bank. SS *Piqua* is bound south in Straits west of Little Bahama Bank. *SS Andrew A. Humphrey's* is bound north in Straits west of Little Bahama Bank. Tanker *Gulfbird* is bound south in Northwest Providence Channel. SS *Jean Baptiste* and SS *Mill Spring* are bound north in Straits northwest of Bimini. SS *Pivot* is bound north in Straits west of Bimini. SS *Arthur R. Lewis* is bound south in Straits. SS *Bidwell* is bound north in Straits north of Cay Sal Bank. **Convoy KG-766** is bound southeast in Old Bahama Channel. SS *William Mulholland* is bound northwest in Old Bahama Channel. Argentinian SS *Rio Parana* is bound southeast Tustin of Cay Sal Bank. Tanker *Gulfdisc* is bound east in Nicholas Channel. Ferry *Ena* K. is bound south, to the east of Eleuthera. SS *Robert Watchorn* is bound south, to the east of Eleuthera, exiting the area. NAAF Great Exuma: A plane arrived from Guantanamo and departed for Banana River. Tug USS *ATR-22*, towing open lighter, and escort USCG *CG-Colfax* departed for Guantanamo.

11th March 1945

Grumman Goose GA is airborne for San Salvador in connection with Queensley Exercise, then landed back at base later. B-24's MB and MC are airborne from Windsor Field to search for US Navy aircraft PB4Y-2 Privateer, reported missing near Paw Paw Rocks north of Grand Bahama Island, MB carried out square search in northern part and MC in southern part, aircraft landed after exercise completion nothing to report. Tanker *British Hope* is bound north in Straits northwest of Little Bahama Bank, exiting the area. SS *Rio Pico* and SS *Ephrain Baughman* are bound south in Straits, northwest of Little Bahama Bank. SS *Petter* and SS *Pawnee Rock* are bound north in Straits west of Little Bahama Bank. SS *John F. Meyers* is bound south in Straits south of Bimini. USS *Logic* (AM-258) is bound west in Old Bahama Channel. SS *James Barbour* and **Convoy GK-866** are bound northwest in Old Bahama Channel. SS *John T. Holt* is bound south, to the east of Eleuthera. SS *Beacon Hill* is bound north, to the east of Eleuthera.

12th March 1945

B-25's from Oakes Field and B-24's from Windsor Field carried out local flying training. B-25's FV, FJ, FG, FY, FK, and FN are airborne for Anti-Submarine Patrol Number 1 with six detail of navigation, signals, and photography, usual tracks amended to search area north of Grand Bahama Island to comply with request of USN in search for missing PB4Y-2 Privateer, then aircraft landed after exercise completion, nothing to report. B-25's FB, FD, FY, FT, and FG are airborne for air Exercise Number 2 of navigation, signals, and photography, aircraft landed after exercise completion, B-24 LY is airborne and took briefing on Exercise Number 48 in Windsor Field, was to be done on 9th [three days previously] but was not possible then, because original and stand-by aircraft were unserviceable. Then the aircraft landed, after exercise completion. B-24's NM, LU, and NX are airborne for Radar Exercise Number 23, aircraft landed after exercise completion. B-24 LH completed two briefings of Radar Exercise Number 2; aircraft landed after exercise completion. Grumman Goose GA airborne for San Salvador, aircraft landed after exercise completion. Tanker *Cities Service Kansas* and SS *El Mirlo* are bound north in Straits, northwest of Little Bahama Bank, exiting the area. SS *Vanja,* SS *James Barbour*, and tanker *Paulsboro* are bound north in Straits northwest of Little Bahama Bank. USS *YTB-338* is bound south in Straits north of Little Bahama Bank. SS

William Mulholland and USS *YOG-6* are bound north in Straits west of Bimini. USS *Basset* (APD-73) is bound east in Straits north of Cay Sal Bank. SS *Empire Viscount* and SS *Emma Willard* are bound northwest in Old Bahama Channel. SS *Pocone*, SS *Louis Hennepin*, and SS *Albar* are bound south, to the east of Eleuthera. Tanker *Gulfwax* and SS *James B. Aswell* are bound north, to the east of Eleuthera. B-25's FV, FJ, FG, FY, FK, and FN are airborne on amended training tracks to search north of Grand Bahama for missing PB4Y-2 Privateer. NAAF Great Exuma: A plane arrived and departed for Banana River on a training flight.

13th March 1945

B-25's from Oakes Field and B-24's from Windsor Field carried out local flying training. B-25's VG, TD, and NO are airborne for Anti-Submarine Patrol Number 1 with five details of navigation, descent through cloud (DTC) signals, and photography, usual tracks amended, area covered was north of Grand Bahama Island. Included conducting search for missing PB4Y-2 Privateer. B-25's FV, FJ, FF, FY, and FN are airborne on air Exercise Number 2 of navigation, bombing, decent through cloud (DTC), and photography. B-25's from Oakes Field and B-24's from Windsor Field, carried out local flying training. B-24's NX, MD, and NO are airborne for Radar Exercise Number 23, aircraft landed after exercise completion. Queensley Exercise Number 5 replaced. Original flight carried out by C. Flight, 3 Squadron, of two ships could not be provided by HMCS *Somers Isles*, Bermuda. MF/DF direction finding Station at San Salvador replaced escort-vessel, visual signalling from San Salvador. B-24 NX is airborne from Windsor Field, homing "B" procedure commenced, sighted San Salvador, carried out recognition, carried out patrol Crocodile on instruction by officer in charge, flame float dropped, attacked by depth charge and guns, attack photographed, practice signals sent back to base, aircraft landed after exercise completion. B-24 LU is airborne from Windsor Field as first relief, some shipping sighted, homing "B" procedure commenced, sighted San Salvador recognition, carried out patrol Lizard and Mamba on instruction by officer in charge, flame float dropped, attacked by depth charge and guns, attack photographed, practice signals sent back to base, aircraft landed after exercise completion. B-24 LR is airborne from Windsor Field as second relief, homing "B" procedure commenced, sighted San Salvador recognition, carried out patrol Mamba on instruction by officer in charge, flame float dropped, attacked by depth charge and guns, attack photographed, practice signals sent back to base, aircraft landed after exercise completion. B-24 NA is airborne from Windsor Field as final relief, homing "B" procedure commenced, sighted San Salvador recognition, carried out patrol Mamba on instruction by officer in charge, flame float dropped, attacked by depth charge and guns, attack photographed, practice signals sent back to base, some shipping sighted, aircraft landed after exercise completion. Tanker *Gulfcrest*, SS *Clifford E. Ashby*, tanker *Cities Service Oklahoma*, tanker *Esso Belgium*, USS *Woodbury* (W-155), tanker *Bohemian Club*, and SS *Emma A. Willard* are bound north in Straits, northwest of Little Bahama Bank, exiting the area. SS *Chrysler Field*, SS *Madaket*, and tanker *Mary Ashley Townsend* are bound north in Straits west of Bimini. SS *El Aleto* is bound north in Straits. USS *YO-129* is bound south in Northwest Providence Channel. USS

Agassiz (W-126) is bound south in Straits, west of Little Bahama Bank. Submarine USS *Quillback* (SS-424) is bound south in Straits west of Bimini. USS *YBH-16* is bound east in Straits, near Cay Sal Bank. Landing ship USS *LST-42* is bound south, east of Eleuthera. B-25's VG, TD, and NN are airborne over Grand Bahama, conducting search for the missing PB4Y-2 Privateer. NAAF Great Exuma: A plane arrived from Banana River, then returned.

14th March 1945

Mosquitoes and B-25's from Oakes Field and B-24's from Windsor Field carried out local flying training. B-25's FV, FY, FF, FG, FH, and FN are on Anti-Submarine Patrol Number 1 with six details of navigation, decent through cloud (DTC), and signalling, aircraft landed after exercise completion. B-24 LH is airborne from Windsor Field for Radar Exercise Number 2, aircraft landed after exercise completion. B-24 NC airborne for Radar Exercise Number 23 on Route A, aircraft landed after exercise completion. B-24 MB airborne from Windsor Field, second day of Queensley Exercise Number 5, homing "B" procedure commenced, sighted San Salvador recognition, carried out patrol Mamba on instruction by officer in charge, flame float dropped, attacked by depth charge and guns, attack photographed, practice signals sent back to base, some shipping sighted, aircraft landed after exercise completion. B-24 NX is airborne from Windsor Field as first relief, second day of Queensley Exercise Number 5, homing "B" procedure commenced, sighted San Salvador recognition, carried out patrol Python on instruction by officer in charge, flame float dropped, attacked by depth charge and guns, attack photographed, practice signals sent back to base, some shipping sighted, aircraft landed after exercise completion. B-24 NJ is airborne from Windsor Field as second relief, second day of Queensley Exercise Number 5, homing "B" procedure commenced, sighted San Salvador recognition, carried out patrol Alligator on instruction by officer in charge, flame float dropped, attacked by depth charge and guns, attack photographed, practice signals sent back to base, some shipping sighted, aircraft landed after exercise completion. B-24 MF is airborne from Windsor Field as final relief, second day of Queensley Exercise Number 5, homing "B" procedure commenced, sighted San Salvador full recognition, carried out patrol Mamba on instruction by officer in charge, flame float dropped, attacked by depth charge and guns, attack photographed, practice signals sent back to base, some shipping sighted, aircraft landed after exercise completion. SS *El Aleto* and SS *Somme* are bound north in Straits northwest of Little Bahama Bank, exiting the area. SS *Christy Payne* is bound south in Straits off Palm Beach. Tanker *Cities Service Ohio* is bound north in Straits off Palm Beach. SS *Point Breeze*, tanker *Gulfcoast*, and naval tanker, or oiler USS *Cacalil* (A0), and Tanker *Hadnot* are bound north in Straits west of Bimini. SS *Costa Rica Victory* is bound south in Straits north of Santaren Channel. **Convoy KG-767** is bound southeast in Nicolas channel. USS *PC-608* is bound east in Old Bahama Channel. NAAF Great Exuma: A plane arrived from Banana River, and then it returned.

15th March 1945

Mosquitoes and B-25's from Oakes Field and B-24's from Windsor Field carried out local flying training. B-25's FN, FY, FG, FT, and FK are airborne for Anti-Submarine Patrol Number 1 with five details of navigation, signals practice, and decent through cloud (DTC), aircraft landed after exercise completion. B-25's FU, FY, FF FG, FO, and FT are airborne twice for local night navigation Exercise Number 1, aircraft landed after exercise completion. B-24 LH is airborne from Windsor Field for Radar Exercise Number 2, aircraft landed after exercise completion. B-24 NM is airborne from Windsor Field for Radar Exercise Number 1, aircraft landed after exercise completion. B-24-MB is airborne from Windsor Field, final day Queensley Exercise Number 5, homing "B" procedure commenced, sighted San Salvador full recognition, carried out patrol Mamba on instruction by officer in charge, flame float dropped, attacked by depth charge and guns, attack photographed, practice signals sent back to base, some shipping sighted, aircraft landed after exercise completion. B-24 LY is airborne from Windsor Field as first relief, final day Queensley Exercise Number 5, photographed passing ships, homing "B" procedure commenced, sighted San Salvador full recognition, carried out patrols completed on instruction by officer in charge, flame float dropped, attacked by depth charge and guns, attack photographed, practice signals sent back to base, some shipping sighted aircraft landed after exercise completion. B-24 NA is airborne from Windsor Field as second relief, final day Queensley Exercise Number 5, homing "B" procedure recommenced, sighted San Salvador full recognition, carried out patrol Python on instruction by officer in charge, flame float dropped, attacked by depth charge and guns, attack photographed, practice signals sent back to base, some shipping sighted, aircraft landed after exercise completion. B-24 MF is airborne from Windsor Field as final relief, final day Queensley Exercise Number 5, homing "B" procedure commenced, sighted San Salvador, full recognition carried out, patrols completed on instruction by officer in charge flame float dropped, attacked by depth charge and guns, attack photographed, practice signals sent back to base, some shipping sighted, aircraft landed after exercise completion. Tanker *Cities Service Ohio* is bound north in Straits, northwest of Little Bahama Bank. SS *William G. Warden* is bound south in Straits west of Little Bahama Bank. SS *New Hope* and **Convoy KN-374** are bound north in Straits, west of Little Bahama Bank. SS *Vincenzia* is bound north in Straits west of Bimini. Bahamas civilian boat *Autosee* is bound south in Straits. Tanker *AthelDuke* and SS *Santiago Iglesias* are bound north in Straits. SS *Billy Mitchell* is bound west entering Nicholas Channel. Destroyer USS *Cole* (DD-155) is bound northwest in Old Bahama Channel. Tanker *Gulf Gem*, SS *Irwin Russell*, SS *Salvator* and SS *Scandia* are bound south, to the east of Eleuthera. SS *Empire Pickwick* is bound north, to the east of Eleuthera. NAAF Great Exuma: A plane *en route* from Banana River to San Juan arrived in an emergency, with a damaged engine. An Army PBY arrived from Nassau then departed for Mayaguana, a plane arrived from Guantanamo and departed for San Juan. Two planes arrived from Banana River, one returned on same day and other on next day.

16th March 1945

The senior official in the Bahamas, HRH the Duke of Windsor and the Duchess, Wallis Simpson, quit their roles as Governor and head of the Bahamas Red Cross. [Ultimately leave the colony before war's end, sneaking out on a boat at dawn before VE-Day]. B-25's from Oakes Field and B-

24's from Windsor Field carry out local flying training. B-25's FH, FF, FY, FG, FT, and FC are airborne for air Exercise Number 6 of homing practice, bombing, and photography, aircraft landed after exercise completion. Grumman Goose GA is airborne and proceeds to Harbour Island and San Salvador, then the aircraft landed after exercise completion. SS *Artigas* is bound north in Straits, northwest of Little Bahama Bank, exiting the area. SS *Susan Colby* is bound south in Straits, northwest of Little Bahama Bank. SS *Meuse Argonne* is bound south, to the north of Walker Cay, north of Abaco. SS *W. F. Burdell* is bound south in Straits west of Little Bahama Bank. Tanker *Clio* and SS *Meline* are bound north in Straits west of Little Bahama Bank. SS *Belgium Airman* and SS *Trimountain* are bound south in Straits northwest of Bimini. SS *Delmar* is bound north in Santaren Channel. **Convoy GK-867** is bound east in Santaren Channel. SS *Elwood* and SS *Francis R. Hart* are bound northwest in Old Bahama Channel. SS *Edward Bruce*, SS *Henry C. Payne*, and SS *Swift Arrow* are bound south, to the east of Eleuthera. SS *San Valerio* is bound north, to the east of Eleuthera.

17th March 1945

B-25's from Oakes Field and B-24's from Windsor Field carried out local flying training. B-25's FN, FY, FF, FC, and FS are airborne for air Exercise Number 6 of homing practice, bombing, and photography, aircraft landed after exercise completion. SS *Irwin Russell*, tanker *Hagood* and SS *North Star* are bound north in Straits northwest of Little Bahama Bank. SS *Herbert H. Bancroft*, SS *Vermont*, and SS *Benjamin Siliman* are bound south in Straits west of Little Bahama Bank. SS *Malabar* and SS *George Whitehead* are bound south in Straits, northwest of Bimini. SS *C. A. Canfield* is bound north in Straits, northwest of Bimini. Tanker *Gulfhawk*, SS *Henry Watterson*, and SS *Edwin W. Scripts* are bound north in Straits. SS *Mill Springs* is bound north in Straits north of Cay Sal Bank. SS *Pittston Victory* is bound south in Straits north of Cay Sal Bank. USCG cutter *CG-Agassiz* (WSC-126) bound southeast in Old Bahama Channel. SS *Richmond Park* is bound south, to the east of Eleuthera. NAAF Great Exuma: A plane arrived from Banana River, then returned.

18th March 1945

B-24's MX, MF, ND, and LW are airborne from Windsor Field for Radar Exercise Number 41, aircraft landed after exercise completion. SS *Thomas H. Wheeler* and SS *Ardmore* are bound north in Straits northwest of Little Bahama Bank, exiting the area. SS *Tampico* is bound north in Straits northwest of Little Bahama Bank. Tanker *Petrofuel* is bound south in Straits west of Little Bahama Bank. SS *Mill Springs*, tanker *A. S. Hansen* and SS *AthelDuke* are bound north in Straits west of Little Bahama Bank. SS *Swiftscout* and tanker *Esso Bayonne* are bound north in Straits west of Bimini. SS *DelRio* and SS *Buffalo Park* are bound south, to the east of Eleuthera. SS *Zebulon Pike* is bound north, to the east of Eleuthera. NAAF Great Exuma: A plane arrived from Banana River then returned.

19th March 1945

Mosquitoes and B-25's from Oakes Field and B-24's from Windsor Field carried out local flying training. B-25's FT, FU, FS, FG, and FF are airborne for air Exercise Number 5 of bombing, photography, and decent through cloud (DTC), aircraft landed after exercise completion. B-24's FF, FC, FB, FS, FO, and FG are airborne for air Exercise Number 1 of general flying check and bombing on moving target, aircraft landed after exercise completion. B-24 LH is airborne from Windsor Field for Radar Exercise Number 2, aircraft landed after exercise completion. SS *Piqua* is bound north in Straits northwest of Little Bahama Bank. USS *New York* is bound south in Straits west of Little Bahama Bank. SS *L. L. Abshire*, SS *Niso*, and SS *Belgium Equality* are bound north in Straits west of Little Bahama Bank. SS *American Packer* is bound north in Straits, west of Bimini. SS *Swiftlight*, SS *Malay* and SS *Cyprus Field* are bound north in Straits. SS *Skandinavia* and SS *Henry L. Ellsworth* are bound north in Straits, north of Cay Sal Bank. USS *Opponent* (AM-269), SS *G. Harrison*, and SS *Thomas Macdonough* are bound south, to the east of Eleuthera. SS *Costa Rica Victory* is bound north, to the east of Eleuthera. NAAF Great Exuma: Two planes arrived from Banana River, and returned there.

20th March 1945

B-25's from Oakes Field and B-24's from Windsor Field carried out local flying training. B-25's FD, FC, FU, FM, FB, and FO are airborne for air Exercise Number 5 of bombing, decent through cloud (DTC), and photography, aircraft landed after exercise completion. B-25's FN, FC, FU, FF, FK, and FY are airborne for air Exercise Number 1 of general flying and bombing on moving targets, aircraft landed after exercise completion. Grumman Goose GA airborne for Rock Sound, aircraft landed back to Oakes Field after exercise completion. B-24 LH is airborne for Radar Exercise Number 1, aircraft landed after exercise completion. SS *Daylight*, SS *Zona Gale*, and tug USS *ATF-82* are bound north in Straits, northwest of Little Bahama Bank, exiting the area. Tanker *Spencer Kellogg* is bound south in Straits northwest of Little Bahama Bank. Tanker *Bidwell* and SS *Iris Bank* are bound south in Straits west of Little Bahama Bank. **Convoy KN-375** is bound north in Straits, west of Little Bahama Bank. USS *YMS-463*, USS *YMS-403*, USS *YMS-376*, USS *YMS-75*, USS *YMS-372*, USS *YMS-42*, and USS *YMS-461* are bound south in Straits, west of Little Bahama Bank. SS *William Seaton* is bound south in Straits off Port Everglades. SS *Alcee Fortier* and SS *Jean Baptiste Lemoune* are bound north in Straits west of Bimini. SS *Idaho Falls* is bound north in Straits north of Cay Sal Bank. SS *Samuel Ashe* is bound southeast in Nicholas Channel. **Convoy KG-768** is bound southeast in Old Bahama Channel. SS *William Window* is bound south, to the east of Eleuthera, exiting the area. SS *Cornwall* is bound north, to the east of Eleuthera. SS *Gulf of Mexico* is bound south, to the east of Eleuthera. NAAF Great Exuma: A plane arrived from Banana River then returned. A plane arrived from Guantanamo, and departed for Banana River.

21st March 1945

Mosquitoes and B-25's from Oakes Field and B-24's from Windsor Field carried out local flying training. B-25's FU, FS, FY, FK, FF, and FH are airborne for air Exercise Number 3 and Number 4 of bombing on fixed and moving targets and photography, aircraft landed after exercise completion. B-24's MX, NH, LH, and NH are airborne from Windsor Field, first day briefing of Exercise Number 42 of navigation, Radar, homing, depth charge and gun practice, photography and signals, aircraft landed after exercise completion. B-24 LH is airborne for Radar Exercise Number 1, aircraft landed after exercise completion. Grumman Goose GA airborne for Harbour Island landed back at base later. Tanker *Gulf Point* is bound south, to the north of Walker Cay, off Abaco. Tanker *Gulfbird* and

SS *Blue Licks* are bound north in Straits, northwest of Little Bahama Bank. SS *Touraine* and **Convoy NK-675** are bound south in Straits west of Little Bahama Bank. Tanker *Petrofuel* is bound north in Straits northwest of Bimini. SS *William Lloyd Garrison* and SS *Hector Garbo* are bound north in Straits. SS *Bidwell* is bound south in Straits north of Cay Sal Bank. USS *Houston* (CL 81) and USS *Reno* (CL-96) are bound north in Straits north of Cay Sal Bank. **HDML Flotilla** of British Royal Navy HMS Harbor Defence Motor Launches, and **Convoy GK-868** are bound northwest in Old Bahama Channel. SS *North Star*, SS *Henry Wright Hurley*, SS *William MacClay*, SS *Garnet Hulings*, SS *Mission Park View*, and tanker *Esso Bolivar* are bound south, to the east of Eleuthera. SS *Drefn* is bound north, to the east of Eleuthera. NAAF Great Exuma: Training plane arrived from Banana River then returned. An Army PBY arrived from Caicos Island, departed for Nassau, and a plane *en route* San Juan to Banana River made an emergency landing using only one engine.

22nd March 1945

B-25's from Oakes Field and B-24's from Windsor Field carried out local flying training. B-25's FN, FS, FU, FK, FY, and FM are airborne for air Exercise Number 3 and Number 4 of bombing on fixed and moving targets and photography, in second half FD and FO replaced FM and FN, and the aircraft landed after exercise completion. B-24's NH, MF, and NA are airborne from Windsor Field. The second day briefing of Exercise Number 48, due to bad weather, resulted in take-off being delayed, rearranged for afternoon, with only three of four exercises completed ultimately. The fourth detail is postponed to 24th, two days in the future. B-25's FU, FK, FM, FY, and FD are airborne for Anti-Submarine Patrol Number 1 with five details of navigation and astrological details, aircraft landed after exercise completion. B-24 LH is airborne for Radar Exercise Number 1 and later for Exercise Number 2 for Leigh Light practice, aircraft landed after exercise completion. SS *Clam* is bound north in Straits northwest of Little Bahama Bank, exiting the area. SS *Medina* is bound south In Straits northwest of Little Bahama Bank. SS *Egbert Benson* and SS *Harry Percy* are bound north in Straits, west of Little Bahama Bank. **Task Unit TU-20183** (USS *LSM- 165*, USS *LSM-273*, USS *LSM-391*, and USS *LSM-392*) is bound south in Straits west of Little Bahama Bank. SS *Samuel Ashe* is bound north in Straits, northwest of Bimini. SS *Empire Chief* is bound north in Straits off Miami. **Convoy NK-675** is bound south in Straits, off Miami. Tanker *Gulfpoint* is bound north in Straits, west of Bimini. SS *Ocala Victory*, tanker *E. R. Kemp*, and SS *Hagerstown* are bound north in Straits. Bahamas civilian craft *Autosee* is bound south in Straits. SS *Asa Grey* is bound northwest in Santaren Channel. USS *YAG-19* is bound south, to the west of Great Exuma. Tanker *Gulfbreeze* is bound south, to the east of Eleuthera. NAAF Great Exuma: A plane arrived from Banana River and returned there, and the USS *YAG-19* arrived from Miami with supplies and provisions.

23rd March 1945

B-25's from Oakes Field and B-24's from Windsor Field carried out local flying training. B-25's FU, FB, FH, FM, FK, and FF are airborne for air Exercise Number 7 of compass swing, air firing, and photography, aircraft landed after exercise completion. B-25's FF, FS, FM, FC, and FH are airborne for air Exercise Number 8 of bombing and decent through cloud (DTC), aircraft landed after exercise completion.

B-24's LT, MF, and LY airborne from Windsor Field, final day briefing of Exercise Number 48, aircraft landed after exercise completion. B-24 LH airborne for Radar Exercise Number 1 and later for Exercise Number 2 of Leigh Light practice, aircraft landed after exercise completion. USS *Semmes* (AG-24), SS *Asa Gray*, and SS *Imperial Transport* are bound north in Straits northwest of Little Bahama Bank, exiting the area. SS *Stella Lykes* is bound south in Straits northwest of Little Bahama Bank. USS *PC-1145* is bound west in Northwest Providence Channel. Bahamas mailboats *Content S.*, mailboat *Three Bays,* and mailboat *Ann Bonny* are bound north in Straits west of Bimini. Tug USS *YTB* towing *SS Maria Joseph Bellow* and both are bound north in Straits, to the south of Bimini. Tanker *Bohemian Club* is bound south in Straits, west of Little Bahama Bank. SS *Morton Prince*, SS *Trimountain*, and SS *Christy Payne* are bound north in Straits west of Little Bahama Bank. SS *Nevada* is bound north in Straits, west of Bimini. SS *Ponce De Leon* is bound south in Straits north of Cay Sal Bank. SS *Bardstown Victory* is bound north in Straits north of Cay Sal Bank. SS *Wayne McVeigh* and USS *Hampton* (APA-115) are bound north in Straits. SS *Robert C. Tuttle*, SS *James E. Howard*, and SS *Atlantic Coast* are bound south, to the east of Eleuthera. NAAF Great Exuma: A plane arrived from Banana River, then returned back. USS *YAG-19* departed for Miami.

24th March 1945

B-25's from Oakes Field and B-24's from Windsor Field carried out local flying training. B-25's FU, FO, FH, FK, FJ, and FG are airborne for air Exercise Number 7 of compass swing, photography, bombing, and decent through cloud (DTC), aircraft landed after exercise completion. B-25's FU, FD, FS, FK, FG, and FJ are airborne for air Exercise Number 8 of bombing, air firing, and decent through cloud (DTC), aircraft landed after exercise completion. B-24 LH is airborne from Windsor Field for Radar Exercise Number 2, aircraft landed after exercise completion. SS *Ponce De Leon* is bound north in Straits northwest of Little Bahama Bank, exiting the area. SS *St. Mihiel* is bound south in Straits north of Little Bahama Bank. SS *Salvador Braun,* SS *Thomas Kearns,* SS *William D. Hoxie,* tanker *Cities Service Oklahoma,* SS *Ponca City*, and SS *Joseph Pilitzer* are bound south in Straits west of Little Bahama Bank. Tanker *Mary A. Townsend* is bound south in Straits, west of Port Everglades. SS *Paul Buck* is bound south in Straits west of Bimini. SS *Edwin W. Moore*, tanker *British Prestige*, and tanker *Alpha* are bound north in Straits, west of Bimini. SS *Pachaug Victory* is bound south in Straits north of Cay Sal Bank. **Convoy KG-769** is bound southeast in Nicholas Channel. USS *Ptargian* (AM-376) and USS *Poseidon* (ARL-12) are bound south, to the east of Eleuthera. NAAF Great Exuma: A plane arrived from Banana River and two departed there.

25th March 1945

B-24's LU, LH, LY, and MF are airborne from Windsor Field for Radar Exercise Number 23, aircraft landed after exercise completion. Grumman Goose GA is airborne for San Salvador and Rock Sound, returned to base later at night. SS *Stephen F. Austin* is bound north in Straits west of Little Bahama Bank. SS *Bessemer* is bound north in Straits northwest of Bimini. SS *Tobias E. Stansbury* and SS *Elsie* are bound south in Straits, west of Bimini. SS *Meuse Argonne* is bound north in Straits south of Bimini. SS *F. Q. Barstow* is bound north in Straits. Tanker *Gulfwave* is bound south in Straits. Mailboat *Lady Cordeaux* and mailboat *Joyce Roberts* are bound north in Straits. SS *Edward Bruce* is bound northwest in Santaren Channel. **RN HMS HDML Flotilla** comprised of Royal Navy Harbor Defence Motor Launches HMS *HDML-1360*, HMS *HDML-1361*, HMS *HDML-1362* and HMS *HDML-1363* are bound north in Straits, north of Cay Sal Bank. Submarine USS *S-15* (SS-120) is bound west in Old Bahama Channel. SS *Wiley Post* is bound southeast of Eleuthera. NAAF Great Exuma: A plane arrived from Banana River and then returned there.

26th March 1945

B-25's from Oakes Field and B-24's from Windsor Field carried out local flying training. B-25's FS, FU, FC, FJ, FB, and FO are airborne for Anti-Submarine Patrol Number 2 with six details of navigation, signal practice, and decent through cloud (DTC), aircraft landed after exercise completion. B-25's FS, FH, FU, FG, FM, and FD are airborne for air Exercise Number 2 of navigation, bombing, and decent through cloud (DTC), aircraft landed after exercise completion. Queensley Exercise Number 6 replaced with usual Kingsley, Esco exercise, ships not available, MF/DF Direction Finding Station at San Salvador used as convoy, Operations Officer on Duty is to act as Escort Commander, and co-operate with aircraft crew of C. Flight of 3 Squadron. This is the final passing out exercise for the course. B-24 LR airborne from Windsor Field for Queensley Exercise Number 6, homing "B" procedure commenced, sighted San Salvador full recognition, carried out patrol Python on instruction by officer in charge, flame float dropped, attacked by depth charge and guns, attack photographed, practice signals sent back to base, carried out Lizard patrol, aircraft landed after exercise completion. B-24 LU is airborne from Windsor Field as first relief for Queensley Exercise Number 6, no homing due to faulty Radar and bad weather, incorrect ETA sent to Queensley, finally recognition carried out, patrol Python on instruction by officer in charge, flame float dropped, attacked by depth charge and guns, attack photographed, practice signals sent back to base, aircraft landed after exercise completion. B-24 MD is airborne from Windsor Field as second relief for Queensley Exercise Number 6, homing "B" procedure commenced, sighted San Salvador full recognition, carried out patrol Python on instruction by officer in charge, flame float dropped, attacked by depth charge and guns, attack photographed, practice signals sent back to base, carried out Cobra patrol, aircraft landed after exercise completion. B-24 NH is airborne from Windsor Field for Queensley Exercise Number 6 as third relief, homing "B" procedure commenced, sighted San Salvador full recognition, carried out patrol Mamba on instruction by officer in charge, flame float dropped, attacked by depth charge and guns, attack photographed, practice signals sent back to base, aircraft landed after exercise completion. B-24's NC, NJ, NO, NR, and NM are airborne from Windsor Field for Radar Exercise Number 23 on Route A and B, aircraft landed after exercise completion. B-24 LH is airborne for Radar Exercise Number 2 from Windsor Field, aircraft landed after exercise completion. SS *Donald S. Wright* is bound north in Straits northwest of Little Bahama Bank, exiting the area. SS *John Fairfield*, SS *Edward Bruce*, and SS *Woodrow Wilson* are bound north in Straits northwest of Little Bahama Bank. SS *Peter II*, SS *Moses Brown* is bound south in Straits west of Little Bahama Bank. Submarine USS *Carbonero* (SS-337) is bound south in Straits west of Great Harbour Cay. SS *Clio*, SS *Ephrian W. Baughman*, and tanker *Esso Caroline* are bound north in Straits northwest of Bimini. SS *Tide Penn* is bound south in Straits south of Bimini. Submarine USS *S-15* (SS-120) is bound north in Straits. SS *Samuel F. Miller* is bound south, to the east of Eleuthera. NAAF Great Exuma: A plane arrived from Banana River then returned.

27th March 1945

B-25's from Oakes Field and B-24's from Windsor Field carried out local flying training. B-25's FM, FN, FS, FJ, and FD are airborne for Anti-Submarine Patrol Number 2 with five details of navigation signals practice and decent through cloud (DTC), aircraft landed after exercise completion. B-25's FH, FC, FG, FS, FN, and FU are airborne for air Exercise Number 2 of navigation bombing and decent through cloud (DTC), aircraft landed after exercise completion. B-24 LH airborne from Windsor Field for Radar Exercise Number 2, aircraft landed after exercise completion. B-24 MX is airborne from Windsor Field for Queensley Exercise Number 6, no radio, no recognition, carried out patrol Mamba on instruction by officer in charge, flame float dropped, attacked by depth charge and guns, attack photographed, difficulty in sending practice signals back to base, aircraft landed after exercise. B-24 ND is airborne from Windsor Field for Queensley Exercise Number 6, homing procedure commenced, recognition, carried out patrol Lizard on instruction by officer in charge, flame float dropped, attacked by depth charge and guns, attack photographed, practice signals sent back to base, aircraft landed after exercise completion. B-24 MD is airborne from Windsor Field for Queensley Exercise Number 6 as third relief, homing "B" procedure commenced, sighted San Salvador full recognition, carried out patrol Alligator and Crocodile on instruction by officer in charge, flame float dropped, attacked by depth charge and guns, attack photographed, practice signals sent back to base, carried vertical speed patrol practice impossible due to Aldis lamp failure, aircraft landed after exercise completion. B-24 NH is airborne from Windsor Field for Queensley Exercise Number 6 of homing "B" procedure commenced, sighted San Salvador recognition, carried out patrol Mamba and Python on instruction by officer in charge, flame float dropped, attacked by depth charge and guns, attack photographed, practice signals sent back to base, shipping noted, weather report sent, aircraft landed after exercise completion. Grumman Goose GB airborne for Palisadoes via Guantanamo, landed there. USS *LST-1123* is bound north in Straits northwest of Little Bahama Bank. SS *Hillsboro Inlet* is bound south in Straits northwest of Little Bahama Bank. SS *Skandinavia* is bound south in Straits west of Little Bahama Bank. SS *Beacon Hill*, SS *W. F. Burdell*, and tanker *British Hope* are bound north in Straits west of Little Bahama Bank. SS *Arthur R. Lewis*, SS *Francis E. Slitz*, and SS *Malabar* are bound north in Straits west of Bimini. USS *SC-716* is bound south in Straits off Palm Beach. SS *Tolovana* is bound south in Straits, north of Bimini. SS *Silver Star Park*, **Convoy NG-497**, USS *General McMeigs* (AP-116), gasoline barge USS *Wakulla* (AOG-44), and USS *Toucan* (AM-387) are bound south, to the east of Eleuthera. NAAF Great Exuma: A plane arrived from Banana River, then returned. A plane arrived from Guantanamo, and then returned.

28th March 1945

Mosquitoes and B-25's from Oakes Field and B-24's from Windsor Field carried out local flying training. B-25's FG, FJ, FU, FT, FF, and FC are airborne for Anti-Submarine Patrol Number 1 with six details of navigation and photography, aircraft landed after exercise completion. B-25's FF, FM, FS, FJ, FT, and FU are airborne for Anti-Submarine Patrol Number 1, aircraft landed after exercise completion. Grumman Goose GB airborne from Palisadoes, proceeded via Guantanamo, went to Grand Turks, landed there. B-24 ND is airborne from Windsor Field for Queensley Exercise Number 6, homing procedure "B" commenced, recognition carried out flame float dropped, attacked by depth charge and guns, attack photographed, practice signals sent back to base, patrol Adder completed on instruction by officer in charge, aircraft landed after exercise completion. B-24 MX is airborne from Windsor Field for Queensley Exercise Number 6, homing procedure "B" commenced, recognition carried out flame float dropped, attacked by depth charge and guns, attack photographed, practice signals sent back to base, aircraft landed after exercise completion. B-24 NH is airborne from Windsor Field for Queensley Exercise Number 6, homing procedure "B" commenced, recognition carried out, a flame float was dropped, then it was attacked by depth charge and guns, attack photographed, practice signals sent back to base. Then patrol Python completed on instruction by officer in charge, aircraft landed after exercise completion. USS *Chateau Thierry* and SS *Vermont* are bound north in Straits northwest of Little Bahama Bank. SS *Washington* south in Straits northwest of Little Bahama Bank. Submarine USS *S-12* (SS-117) is bound north in Straits northwest of Little Bahama Bank. SS *Carlton Ellis* and SS *Malabar* are bound north in Straits west of Little Bahama Bank. Tanker *Beta* is bound north in Straits off Port Everglades. SS *George E. Rickett* is bound south in Straits off Miami. SS *Malden Victory* is bound south in Straits north of Cay Sal Bank. USS *YMS-418*, USS *YMS-178*, USS *YMS-371*, USS *SC-1294*, tug USS *Tybee* and tow are all bound south in Straits, off Port Everglades. Bahamas mailboats *Noel Roberts*, ferry *Betty K.* and mailboat *Ann Bonny*, escorted by USCG *CG-83488,* are bound north in Straits. USCG *CG-Hillsboro Inlet* is bound south in Straits north of Cay Sal Bank. Gasoline barge USS *YOG-129* is bound east in Straits south of Cay Sal Bank. SS *George E. Hale*, SS *Chagres*, SS *Henry L. Ellsworth*, SS *City of Montgomery*, tanker *Cancolite,* SS *Permian*, SS *Norwegian*, and SS *Reginald A. Fessenden* are bound south, to the east of Eleuthera. NAAF Great Exuma: A Naval Air Transport Service seaplane arrived from Miami for installation of new engine, then returned. A plane arrived from Banana River, then returned. A plane arrived from operational mission and then departed for Guantanamo.

29th March 1945

Mosquitoes and B-25's from Oakes Field and B-24's from Windsor Field carried out local flying training. B-25's FS, FC, FF, FJ, FT, and FH are airborne for Anti-Submarine Patrol Number 1 with six details of navigation, signals, and photography, aircraft landed after exercise completion. B-25's FB, FC, FY, FF, FT, and FO are airborne for local night navigation exercise, aircraft landed after exercise completion. Grumman Goose GB returned to Oakes Field from Grand Turk Island via Guantanamo and Palisadoes, Jamaica. Tanker *Gulfwax* and SS *Argon* are bound north in Straits northwest of Little Bahama Bank, exiting the area. SS *Jotunfjell*, SS *City of Omaha*, and SS *Ruth Lykes* are bound south in Straits northwest of Little Bahama Bank. SS *Bostonian* and SS *Coaldale Victory* are bound south in Straits northwest of Bimini. SS *Point Breeze* is bound north in Straits northwest of Bimini. SS *Swiftlight* is bound south in Straits. SS *Ardmore* is bound north in Straits. **Convoy KG-770** is bound southeast in Nicholas Channel. SS *W. L. Worspar* is bound north in Santaren Channel. USCC tug *CG-Hillsboro Inlet* and tow, USCG *CG-Colfax* (W-133), USS *LV-91*, USS *YMS-463*, and USS *YMS-461* are bound southeast in Old Bahama Channel. USS *Brunswick* is bound north, east of Eleuthera. NAAF Great Exuma: A plane arrived from Banana River and then returned.

30th March 1945

B-25's from Oakes Field and B-24's from Windsor Field carried out local flying training. B-25's FD, FC, FF, FY, FT, and FO are airborne for local night navigation Exercise Number 1 of navigation and signals, aircraft landed after exercise completion. B-25's FF, FC, FY, FM, FO, and FN are airborne for air Exercise Number 6 of homing procedure, photography, and signals, aircraft landed after exercise completion. Grumman Goose GA airborne to Harbour Island and San Salvador, aircraft landed after exercise completion. Tanker *Hadnot* and SS *W. L. Worspar* are bound north in Straits northwest of Little Bahama Bank, exiting the area. SS *Chapel Hill Victory* is bound south in Straits north of Little Bahama Bank. Tanker *Delaware Sun* is bound south in Straits, northeast of Palm Beach. SS *Oneida Victory* is bound south in Straits, northwest of Bimini. SS *R. G. Stewart* is bound north in Straits, northwest of Bimini. SS *William Crompton* and SS *Samuel Parker* are bound north in Straits west of Bimini. SS *Jacob Thompson* is bound south in Straits west of Orange Cay, Andros. Mailboat *Richard Campbell* is bound north in Straits south of Bimini. SS *Castana* is bound northwest in Nicholas Channel. SS *Clement Clay* is bound southeast in Nicholas Channel. Tanker *Beta* is bound south in Old Bahama Channel. USS *ATR-22* and USS *PC-1211* are bound west in Old Bahama Channel. SS *Gertrude Kellogg* is bound east in Old Bahama Channel. SS *Baltic* is bound south in Straits passing Key West. SS *Juanflaco Brown* is bound south, to the east of Eleuthera. SS *Trocas* is bound north, to the east of Eleuthera. NAAF Great Exuma: A plane is *en route* to Banana River to Guantanamo arrived with a defective radio, then it departed for its original destination. A plane arrived from Banana River, then returned, an Army PBY arrived from Caicos Island, and a plane departed for Banana River.

31st March 1945

Grumman Goose GA is airborne for Marsh Harbour, Abaco and landed back on 1st April [next day]. It was a general holiday for both fields, with no training. SS *John T. Holt*, tanker *Edward L. Shea*, and SS *W. S. Jennings* are bound north in Straits northwest of Little Bahama Bank. SS *Hendrik W. Van Loon* is bound south in Straits, northwest of Little Bahama Bank. SS *Benjamin Franklin*, submarine USS *R-1*, and SS *Oliver Evans* are bound north in Straits, west of Little Bahama Bank. SS *Melville Jacoby*, SS *John A. Dix*, and SS *Whitebird Canyon* are bound south in Straits west of Little Bahama Bank. USS *YMS-443* is bound south in Straits, off Port Everglades. USS *LST-1131* is bound north in Straits off Port Everglades. SS *C. A. Canfield* is bound north in Straits west of

Bimini. USS *Mississippi* is bound north in Straits north Cay Sal Bank. **Convoy Motor Torpedo Boat Squadron-39, Group A** is bound south in Straits and north of Cay Sal Bank. **Convoy GK-870** is bound north in Old Bahama Channel. SS *Cape Mohican*, SS *Thomas Hartely*, SS *Atlantic*, and SS *Temple Yard* are bound south, to the east of Eleuthera. SS *Samuel D. Champlain* is bound north, to the east of Eleuthera. NAAF Great Exuma: An Army plane departed for Caicos Island, and returned to base with defective engine. A plane arrived from Banana River and then returned.

1st April 1945

SS William Black Yates, SS *Thomas H. Wheeler*, and destroyer USS *Upshur* (DD-144) are bound north in Straits northwest of Little Bahama Bank. SS *Clyde L. Seavey*, tanker *Paulsboro*, and SS *Saint Mihiel* are bound north in Straits west of Little Bahama Bank. SS *Mahoney City Victory* is bound south in Straits west of Little Bahama Bank. SS *Antietam*, SS *Edward Payne*, SS *Arizonian*, SS *Samuel Gorton*, SS *Clam*, and tanker *Britamsea* are bound south, to the east of Eleuthera. SS *Chilibar* is bound north, to the east of Eleuthera. NAAF Great Exuma: A plane arrived from Banana River and returned. An Army PBY arrived from Morrison field and returned there.

2nd April 1945

B-25's from Oakes Field and B-24's from Windsor Field carried out local flying training. B-25's FC, FY, FJ, FB, FH, and FK are airborne for Anti-Submarine Patrol Number 2 of navigation, signals, and photography, aircraft landed after exercise completion. B-25's FH, FC, FJ, FF, and FY are airborne for air Exercise Number 1 of familiarization with descent through cloud (DTC) and bombing on moving targets, aircraft landed after exercise completion. B-24's airborne, MC, MX, MF, MD, LZ, and LU are airborne from Windsor Field for Radar Exercise Number 41, aircraft landed after exercise completion. Grumman Goose GB is airborne for San Salvador, arrived at Oakes Field at night. SS *Mary Ashley* and SS *Mardial* are bound north in Straits northwest of Little Bahama Bank. SS *Clarenville* is bound west in Old Bahama Channel. SS *San Tirso* and SS *Touraine* are bound north in Straits, west of Bimini. SS *Henry S. Foote*, SS *Petter*, SS *Antietam* and SS *Trimountain* are bound north in Straits. SS *Augustus P. Loring* is bound north in Straits north of Cay Sal Bank. Oiler USS *Patola* (A0-9) and six US Coast Guard vessels are bound northwest in Santaren Channel. USCG cutter *CG-Legare* (WSC-144) is bound southeast in Straits entering Old Bahama Channel. Destroyer USS *McCormick* (DD-223), SS *John Gallup*, tanker *Gulftide*, SS *W. H. Libby*, SS *Joseph Holt*, SS *Drafn*, SS *Sul Ross*, and SS *Anniston City* are bound south, to the east of Eleuthera. NAAF Great Exuma: A plane arrived from Banana River and returned. An Army PBY came from Morrison field and departed for South Caicos Island. A plane *en route* from Guantanamo to Banana River arrived with defective radio.

3rd April 1945

B-25's from Oakes Field and B-24's from Windsor Field carried out local flying training. B-25's FC, FY, FD, FB, FM, and FK are airborne for Anti-Submarine Patrol Number 2 with details of navigation, signals, photography, aircraft landed after exercise completion. B-25's FH, FY, FM, FN, FT, and FJ are airborne for air Exercise Number 1 of familiarization with descent through cloud (DTC) and bombing on moving targets, aircraft landed after exercise completion. B-24 LH is airborne for Radar Exercise Number 1, aircraft landed after exercise completion. SS *George Whitefield*, SS *Herbert L. Pratt*, and tanker *Gulfpoint* are bound north in Straits northwest of Little Bahama Bank. SS *Amiens* is bound south in Straits north of Little Bahama Bank. Tanker *Gulfwave* and USS *Naubue* (AN-84) are bound north in Straits west of Little Bahama Bank. SS *Fort Henry* is bound south in Straits west of Little Bahama Bank. USS *Sitka* (APA-113) is bound north in Straits northwest of Bimini. SS *Fort Fetterman* is bound south in Straits northwest of Bimini. Tanker *Gulwing* and SS *Henry W. Van Loon* are bound north in Straits west of Bimini. SS *Kokomo Victory* is bound south in Straits west of Bimini. USCG *CG-Vigilant* (W-154), SS *Clarenville*, SS *Benedick*, tanker *L. V. Stanford* and SS *W. M. Burton* are bound north in Straits. A steam ship is bound northwest in Santaren Channel. SS *Paco City* and SS *William Lloyd Garrison* are bound southeast entering Old Bahama Channel. SS *Clearwater Park*, SS *Massachusetts* and SS *William Penn* are bound south, to the east of Eleuthera. NAAF Great Exuma: An Army PBY plane came from Morrison Field and went to Mayaguana. A training plane came from Banana River then returned with another plane.

4th April 1945

B-25's from Oakes Field and B-24's from Windsor Field carried out local flying training. B-25's FO, FH, FY, FJ, and FK are airborne for air Exercise Number 7 of decent through cloud (DTC), compass swing, and photography, aircraft landed after exercise completion. B-25's FO, FB, FD, FM, FY, and FT are airborne for night Patrol Number 1 with six details of navigation with astrological information, and signals practice, aircraft landed after exercise completion. B-24's MF and NH are airborne twice from Windsor Field for Exercise Number 48 of homing procedure, signals, navigation, photography, depth charge and gun attacks, aircraft landed after exercise completion. B-24 LH is airborne for Radar Exercise Number 1, aircraft landed after exercise completion. SS *Benedick* is bound north in Straits, northwest of Little Bahama Bank, exiting the area. SS *Castana* is bound north in Straits northwest of Little Bahama Bank. SS *El Aleto* is bound south in Straits northwest of Little Bahama Bank. SS *W. M. Burton*, tanker *Esso Bayonne*, tanker *Haakon Hauan*, tanker *L. V. Stanford*, and **Convoy KN-378** is bound north in Straits west of Little Bahama Bank. SS *Fort Lennox* is bound south in Straits west of Little Bahama Bank. Tanker *R. P. Smith* north in Straits northwest of Bimini. SS *Fort Alexander* is bound north in Straits west of Bimini. SS *Ponca City*, SS *Vanja*, and tanker *Hagood* is bound north in Straits. **Convoy KG-771** is bound southeast in Nicholas Channel. SS *Baldwin Park*, and tanker *Comol Cuba* are bound south, to the east of Eleuthera. SS *Pueblo II* is bound north, to the east of Eleuthera. NAAF Great Exuma: A plane came in from Banana River and returned there. An Army PBY came from South Caicos Island and then flew to Morrison Field.

5th April 1945

B-25's from Oakes Field and B-24's from Windsor Field carried out local flying training. B-25's FG, FN, FH, FY, FJ, and FK are airborne for air Exercise Number 7 of decent through cloud (DTC), compass swing, photography, aircraft landed after exercise completion. B-24's MC, MF, and NX are airborne from Windsor Field for Exercise Number 48, fourth crew in programme could not take off, aircraft and reserve craft unserviceable, aircraft landed after exercise completion. B-25's FB, FJ, FS,

FG, FT, and FD are airborne for night Patrol Number 1 with six details of navigation with astrological and signals practice, aircraft landed after exercise completion. B-24 LH is airborne for Radar Exercise Number 1, aircraft landed after exercise completion. SS *Skandinavia* is bound north in Straits northwest of Little Bahama Bank, exiting the area. SS *Vanja*, SS *Petter II*, SS *Cape Beale*, and SS *Sidney Sherman* are bound north in Straits northwest of Little Bahama Bank. Tanker *British Prestige* and SS *Winchester Victory* are bound south in Straits west of Little Bahama Bank. Tanker *Cities Service Ohio* is bound north in Straits. *SS Coal Dale Victory* and USS *Scoter* (AM-381) are bound north in Straits north of Cay Sal Bank. SS *Moose Mountain Park* is bound south, to the east of Eleuthera. **Convoy GN-197**, USS *Impulse* (PG-68) and SS *Dean Emery* are bound north, to the east of Eleuthera. NAAF Great Exuma: A plane arrived from Banana River and returned there.

6th April 1945

B-25's from Oakes Field and B-24's from Windsor Field carried out local flying training. B-25's FB, FG, FS, FH, and FK are airborne for air Exercise Number 6 of homing procedure and photography, aircraft landed after exercise completion. B-25's FG, FC, FJ, FS, FB, and FN are airborne for Exercise Number 8 of bombing, decent through cloud (DTC), and photography, aircraft landed after exercise completion. B-24's MC, LR, MF, and NA are airborne from Windsor Field for Exercise Number 48, aircraft landed after exercise completion. B-24 LK is airborne for familiarization flight with flight captain and crew of six, no instructor, crashed on land south of Windsor Field, damaged, minor injuries to crew. SS *Elsie* and SS *Andrew A. Humphries* are bound north in Straits northwest of Little Bahama Bank, exiting the area. SS *Edwin G. Weed* and tanker *Clio* are bound north in Straits northwest of Little Bahama Bank. SS *Fred C. Stebbins* is bound north in Straits west of Little Bahama Bank. SS *Tidepenn* is bound south in Straits west of Little Bahama Bank. The tanker *Alpha* is bound north in Straits, west of Little Bahama Bank. SS *Edward D. White* is bound north in Straits, north of Cay Sal Bank. Tanker *Comol Cuba* is bound southeast in Nicholas Channel. SS *Beacon Hill*, SS *W. C. Yeager* are bound south, to the east of Eleuthera. SS *Pond Castle* is bound north, to the east of Eleuthera. NAAF Great Exuma: Two planes came from Banana River and Guantanamo then returned back.

7th April 1945

B-25's from Oakes Field and B-24's from Windsor Field carried out local flying training. B-25's FS, FJ, FC, FG, FB, and FH are airborne for air Exercise Number 6 of homing and photography, aircraft landed after exercise completion. B-25's FS, FV, FC, FN, FF, and FH are airborne for air Exercise Number 8 of homing decent through cloud (DTC), and photography, aircraft landed after exercise completion. B-24 NH is airborne from Windsor Field for Exercise Number 48, this should have been flown on 5th but postponed because crafts were unserviceable, aircraft landed after exercise completion. Grumman Goose GB is airborne for Marsh Harbour, Abaco, after which the aircraft landed after exercise completion. B-24's LH and LU are airborne from Windsor Field for Radar Exercise Number 2, aircraft landed after exercise completion. B-24 LR is airborne from Windsor Field for Radar Exercise Number 23 on Route B, aircraft landed after exercise completion. SS *Irwin Russell*, SS *Empire Chief*, and SS *Bostonian* are bound north in Straits northwest of Little Bahama

Bank. Tanker *L. L. Abshire*, SS *S. M. Shoemaker*, and SS *Cantigny* are bound north in Straits west of Little Bahama Bank. SS *Hagerstown Victory* is bound north in Straits northwest of Bimini. Tanker *British Hope* is bound north in Straits west of Bimini. SS *Bessemer* and USS *LST-980* is bound north in Straits. SS *Salator*, SS *Charles McCormick*, and SS *Robert R. Randall* are bound south, to the east of Eleuthera. NAAF Great Exuma: A plane arrived from Banana River and then returned there.

8th April 1945

B-25's FS, FT, FB, FM, FH, and FG are airborne for air Exercise Number 5, consisting of bombing, then decent through cloud (DTC), and photography, aircraft landed after exercise completion. Only B-25's FT and FF briefed out of five aircraft regarding air Exercise Number 2 of navigation, bombing, and photography, B-25 FT landed early, due to bad weather. B-25 FF crashed into the sea south of New Providence, with the loss of its crew of six and an instructor. B-24's LX, LZ, MC, and MF are airborne from Windsor Field for Radar Exercise Number 23, aircraft landed after exercise completion. B-24 LH is airborne from Windsor Field for Radar Exercise Number 1, aircraft landed after exercise completion. USS *Knave* (AM-256) is bound north in Straits, northwest of Little Bahama Bank, exiting the area. SS *American Ranger* is bound south in Straits north of Little Bahama Bank. SS *Annibal* south in Straits northwest of Little Bahama Bank. SS *William Crompton* is bound south in Straits west of Little Bahama Bank. SS *Deborah Gannett* and SS *LaSalle Seam* are bound north in Straits west of Little Bahama Bank. SS *Albion Victory* is bound south in Straits, north of Bimini. SS *Cotton Mather* is bound north in Straits off Port Everglades. Tanker *Pan North Carolina* is bound north in Straits, off Miami. SS *Swiftlight* and SS *Jotunfjell* are bound north in Straits. **Convoy KG-772** is bound southeast in Nicholas Channel. SS *Orville Harden*, tanker *Gulfpride*, SS *Atlantic Coast*, SS *Brunswick*, SS *Willow Dale Park*, and SS *R. G. Stewart* are bound south, to the east of Eleuthera. NAAF Great Exuma: A plane arrived from Banana River and then returned back.

9th April 1945

B-25's from Oakes Field and B-24's from Windsor Field carried out local flying training. B-25 FY is airborne, carried out search in area south of New Providence where B-25 FF had crashed. Oil streaks observed in sea, ASR marine craft, the launch HMS *P-2779* was directed to the area, with nil result. Only two parachutes that had been assigned to crew were located. SS *Aldebaran* and SS *Ardmore* are bound north in Straits, northwest of Little Bahama Bank. SS *Cyrus Orne Jewett*, the high-speed transport USS *Enright* (APD-66), destroyer escorts USS *Pillsbury* (DE-133) and USS *Chatelain Neunzer* (DE-150) are bound south in Straits, northwest of Little Bahama Bank. SS *White Bird Canyon* is bound north in Straits west of Little Bahama Bank. SS *Fort Clatsop* and SS *Henry H. Richardson* are bound south in Straits off Palm Beach. SS *Jean Ribaut* and SS *Edward N. Hinton* are bound north in Straits northwest of Bimini. *SS Hubert H. Bancroft* and tug USS *ATR-12* are bound south in Straits northwest of Bimini. SS *Jean Baptiste Lemoyne* and SS *War Bonnet* are bound north in Straits. SS *William W. Johnson* is bound south in Straits north of Little Bahama Bank. SS *William Lloyd Garrison* is bound northwest in Old Bahama Channel. SS *George W. Barnes* is bound southeast in Old Bahama Channel.

SS *Mississippi* and SS *John D. Archbold* are bound south, east of Eleuthera. NAAF Great Exuma: A plane *en route* from Banana River to San Juan arrived with defective radio then departed to destination. A plane arrived from Banana River and returned there. An Army PBY arrived from Morrison Field.

10th April 1945

B-25's FV, FY, FB, FH, and FN are airborne for air Exercise Number 5 of bombing and decent through cloud (DTC), aircraft landed after exercise completion. B-25's FH, FV, FC, FN, and FK are airborne for air Exercise Number 2 of navigation, bombing, and photography, aircraft landed after exercise completion. B-25 FT is airborne twice, in affiliation with B-24's at Windsor Field. B-24's are airborne from Windsor Field, B-24 LH is airborne for Radar Exercise Number 2 and B-24 LW is airborne for Radar Exercise Number 23a, aircraft landed after exercise completion. Grumman Goose GA is airborne for San Salvador, and returned to base later in evening. B-24 LF is airborne with captain for air Exercise Number 8 (local), landed with nose wheel up, some damage to aircraft, no personal injury. SS *George W. Kendall* is bound north in Straits northwest of Little Bahama Bank, exiting the area. Tanker *British Gratitude* is bound south, to the north of Walker Cay, off Abaco. SS *Edwin W. Moore* and SS *William A. Dobson* are bound north in Straits west of Little Bahama Bank. **Convoy NK-679** is bound south in Straits off Palm Beach. SS *Medina* north, SS *Minatitlan*, and SS *Big Foot Wallace* are bound north in Straits, to the northwest of Bimini. USS *LST-1133* is bound north in Straits, off Miami. Tanker *Gulfcrest* is bound north in Straits. SS *Sedan* is bound south entering Santaren Channel. SS *Queenston Heights* is bound south in Straits north of Cay Sal Bank. SS *Nelson Dingley* is bound east in Straits south of Cay Sal Bank. USS *Montcalm* (ATO-34) is bound northwest in Old Bahama Channel. Tanker *Esso Bolivar* and SS *Arlington Beach Park* are bound southeast of Eleuthera. NAAF Great Exuma: A plane arrived from Banana River then returned.

11th April 1945

B-25's from Oakes Field and B-24's from Windsor Field carried out local flying training. B-25's FM, FV, FC, FK, and FG are airborne for night local navigation Exercise Number 1, aircraft landed after exercise completion. Usual ships not available, MF/DF Direction Finding station at San Salvador used for Queensley Exercise Number 7. B-24 LY is airborne from Windsor Field, homing procedure "B" commenced, recognition performed, carried out patrol Python, flame float dropped by aircraft, attacked with depth charge and guns, attack photographed, reported to base, re-joined "ship" for further vertical speed practice, set course for base, aircraft landed after exercise completion. B-24 MF is airborne from Windsor Field, homing procedure "B" commenced, recognition performed, carried out patrol Mamba, flame float dropped by aircraft, attacked with depth charge and guns, attack photographed, reported to base re-joined "ship" for further vertical speed practice, set course for base, aircraft landed after exercise completion. B-24 LR was on third relief, recalled due to bad weather. B-24 NH was on fourth relief, yet returned due to engine trouble, and later flew on 12th [the following day]. B-24's are airborne from Windsor Field B-24 LW is airborne for Exercise Number 23a and B-24 LH is airborne for Exercise Number 2, aircraft landed after exercise completion. SS

Tustem, SS *Morton Prince*, and USS *YMS-420* are bound north in Straits northwest of Little Bahama Bank. Tanker *Gulfcoast* and SS *Wendall L. Wilkie* are bound south in Straits west of Little Bahama Bank. SS *Sea Falcon* and USS *Knave* (AM-256) are bound south in Straits west of Little Bahama Bank. SS *Nelson Dingley* and SS *Argon* are bound north in Straits. USS *Montcalm* (ATO-34) is bound north in Santaren Channel. SS *Algernon Cheves*, SS *George Abernathy*, SS *William Cullen Bryant*, SS *Quetico Park*, SS *Tara Kan*, SS *Magnolia*, and **Convoy NG-500** are bound south, to the east of Eleuthera. Colonial Secretary requests that aircraft are to search for overdue boat *en route* to Exuma from Nassau. NAAF Great Exuma: A plane arrived from Banana River then returned.

12th April 1945

B-25's from Oakes Field and B-24's from Windsor Field carried out local flying training. B-25's FC, FG, FT, FK, and FY are airborne for local night navigation Exercise Number 1, aircraft landed after exercise completion. B-25's DC, FG, FK, FT, FY, and FM are airborne for air Exercise Number 6, consisting of homing procedure and photography, aircraft landed after exercise completion. B-25's DC, FY, FC, FG, and FK are airborne for air Exercise Number 3 and Number 4 of descent through cloud (DTC) and bombing on fixed and moving targets, landed for an interval for re-bombing and finally landed later. B-24 LR is airborne from Windsor Field for Queensley Exercise Number 7, homing procedure "B" commenced, recognition performed, carried out patrol Python, flame float dropped by aircraft, attacked with depth charge and guns, attack photographed, reported to base, re-joined "ship" for further vertical speed practice, set course for base, aircraft landed after exercise completion. B-24 MD is airborne from Windsor Field for Queensley Exercise Number 7, homing procedure "B" commenced, recognition performed, flame float dropped by aircraft, attacked with depth charge and guns, attack photographed, reported to base re-joined "ship" for further vertical speed practice, set course for base, aircraft landed after exercise completion. B-24 NH is airborne from Windsor Field for Queensley Exercise Number 7, homing procedure "B" commenced, recognition performed, carried out patrol Cobra, flame float dropped by aircraft, attacked with depth charge and guns, attack photographed, reported to base re-joined "ship" for further vertical speed practice, set course for base, aircraft landed after exercise completion. B-24 LY is airborne from Windsor Field for Queensley Exercise Number 7, homing procedure "B" commenced, recognition performed, carried out patrol Mamba, flame float dropped by aircraft, attacked with depth charge and guns, attack photographed, reported to base, vertical speed practice not possible due to bad weather, set course for base, aircraft landed after exercise completion. B-24 MC is airborne from Windsor Field for Queensley Exercise Number 7, homing procedure "B" commenced, recognition performed, flame float dropped by aircraft, attacked with depth charge and guns, attack photographed, reported to base, vertical speed practice not possible due to bad weather, set course for base, aircraft landed after exercise completion. B-24's airborne from Windsor Field include NR is airborne for Exercise Number 23a and LH, which is airborne for Exercise Number 2, aircraft landed after exercise completion. SS *Trimountain* and SS *Tampico* are bound north in Straits northwest of Little Bahama Bank, exiting the area. SS *Stephen W. Kearny* , SS *Belgian Airman*, and SS *Montcalm* are bound north in Straits west of Little Bahama Bank. SS *Harold A. Jordon* is bound north in Straits northwest of Bimini. SS *Belgian Captain* is bound south in Straits south of Bimini. SS *Mirza* is bound south in Straits. SS *Fort Henry* is bound north in Straits. USS *Enright* (APD-66) and SS *Towanda Victory* are bound south in Straits north of Cay Sal Bank. USS *LST-979* is bound south, to the east of Eleuthera. NAAF Great Exuma: A plane arrived from Banana River and then returned.

13th April 1945

B-25's from Oakes Field and B-24's from Windsor Field carried out local flying training. B-25's DC, FK, FN, FV, FX and B are airborne for air Exercise Number 2 of navigation and photography, aircraft landed after exercise completion and re-bombing. B-24 MF is airborne from Windsor Field for Queensley Exercise Number 7 and final day escort duty, took off late due to turbo trouble, homing procedure "B" commenced, recognition performed, flame float dropped by aircraft, attacked with depth charge and guns, attack photographed, reported to base for vertical speed practice, set course for base, aircraft landed after exercise completion. B-24 MD is airborne from Windsor Field for Queensley Exercise Number 7, final day escort duty, homing procedure "B" commenced, recognition performed, carried out patrol Mamba, flame float dropped by aircraft, attacked with depth charge and guns, attack photographed, reported to base for vertical speed practice, set course for base, aircraft landed after exercise completion. B-24 LR is airborne from Windsor Field for Queensley Exercise Number 7, final day escort duty, homing procedure "B" commenced, recognition performed, carried out patrol Lizard, flame float dropped by aircraft, attacked with depth charge and guns, attack photographed, reported to base for vertical speed practice, set course for base, aircraft landed after exercise completion. B-24 LY is airborne from Windsor Field for Queensley Exercise Number 7, final day escort duty, homing procedure "B" commenced, recognition performed, flame float dropped by aircraft, attacked with depth charge and guns, attack photographed, reported to base for vertical speed practice, set course for base, aircraft landed after exercise completion. B-24 LH is airborne for Radar Exercise Number 2, aircraft landed after exercise completion. SS *Thomas Paine* is bound south, to the north of Walker Cay, off Abaco. SS *William W. McKee* is bound south in Straits northwest of Little Bahama Bank. SS *Hay M. Solomon* and SS *John S. Casement* are bound north in Straits west of Little Bahama Bank. USCG *CG- 83376*, USCG *CG-83406*, and USCG *CG-83405* are bound south in Straits north of Bimini. SS *Perryville* is bound south in Straits southwest of Bimini. USS *YO-129* is bound south in Straits off Miami. SS *Nicholas Herkimer* and SS *Minatitlan* are bound north in Straits. SS *Empire Glenn*, tanker *Esso Balboa*, SS *Garnet Hulings*, SS *Skandinavia*, and SS *California* are bound south, to the east of Eleuthera.

14th April 1945

B-25's from Oakes Field and B-24's from Windsor Field carried out local flying training. B-25's FB, FC, FY, FV, and FN are airborne for Anti-Submarine Patrol Number 1, aircraft landed after exercise completion. B-24 NX airborne for Corpus Christi via Morrison Field and Pensacola, landed at Corpus Christi. Grumman Goose GA airborne for San Salvador via Harbour Island and Rock Sound, landed back at base at night. Tanker *Pan Georgia* is bound north, northwest of Little Bahama Bank, exiting the area. **Convoy KN-380** is bound north in Straits northwest of Little Bahama Bank. SS *Oneida Victory* and SS *Michigan* are bound north in Straits west of Little Bahama Bank. SS *Eano* is bound north, to the west of Bimini. Tanker *E. R. Kemp* is bound north in Straits. SS *Orman E. Mack* and USS *PC-1181* are bound south in Straits. Tanker *Gulfpoint* is bound south, to the east of Cat Island. Tanker *Gulfwing* and SS *Glenpool* are bound south, to the east of Eleuthera. No.113: A Memorial service for the late President Roosevelt was conducted by the Dean of Nassau at the Nassau Cathedral. NAAF Great Exuma: A plane arrived from Banana River then returned. An Army PBY flew in from Nassau.

15th April 1945

B-24's LR, LY, MD, ND, LW, and MF are airborne from Windsor Field for Radar Exercise Number 41, aircraft landed after exercise completion. On request of USN, an ASR search was organized for a B-29, overdue while on round robin-flight from Borinquen to Charleston. B-24's LW, MD, NO, and NS are airborne, after which all returned with report that craft had landed at Borinquen. SS *Pachaug Victory* is bound north in Straits, north of Abaco. **Convoy NK-680** is bound south in Straits, northwest of Little Bahama Bank. SS *Stanford Victory* is bound south in Straits west of Little Bahama Bank. SS *Baldbutte* north in Straits northwest of Bimini. SS *Attleboro Victory* is bound south in Straits west of Bimini. SS *Kingston Victory* is bound north in Straits west of Bimini. SS *George Bellows* is bound southeast in Nicholas Channel. SS *Oliver Wendall Holmes* is bound southeast entering Old Bahama Channel. **Convoy GK-873** is bound northwest in Old Bahama Channel. SS *Theodore Sedgewick* and SS *Mark Hopkins* are bound southeast of Eleuthera. SS *Peaper* is bound north, to the east of Eleuthera. NAAF Great Exuma: A plane arrived from Banana River, then returned. An Army PBY departed to Nassau.

16th April 1945

B-25's from Oakes Field and B-24's from Windsor Field carried out local flying training. B-25's FC, FD, FG, FJ, and FT are airborne for air Exercise Number 5 of bombing and decent through cloud (DTC), aircraft landed after exercise completion. B-25's FJ, FN, FV, FC, FO, and DC are airborne for air Exercise Number 1, aircraft landed after exercise completion. B-24 LH is airborne for Radar Exercise Number 1, aircraft landed after exercise completion. Grumman Goose GA is airborne for Rock Sound and Harbour Island returned to base later in evening. SS *William W. Seaton* is bound north in Straits northwest of Little Bahama Bank, exiting the area. SS *Tide Penn* is bound north in Straits northwest of Little Bahama Bank. Tanker *Sinclair H. C.* is bound north in Straits northwest of Bimini. SS *Pittson Victory* is bound north in Straits west of Bimini. **Task Unit TU-03110** is bound west in Northeast Providence Channel. Tanker *Hagood* is bound south in Straits. Tanker *British Prestige*, SS *New York*, and SS *Kewanee* are bound north in Straits. SS *George W. Barnes*, **Task Unit TU-0316**, and USS *Grainger* (AK 184) are bound southeast in Old Bahama Channel. SS *Clan Colquhoun*, SS *John P. Mitchell*, SS *Salando*, SS *Empire Lancaster*, and **Task Unit TU-2963** are bound south, to the east of Eleuthera. NAAF Great Exuma: An Army

PBY arrived from Nassau, then departed to Morrison Field. A plane arrived from Banana River and returned there.

17th April 1945

B-25's from Oakes Field and B-24's from Windsor Field carried out local flying training. B-25's FO, FK, FN, FD, and FT are airborne for Anti-Submarine Patrol Number 1 with five details of navigation, signals, and photography, aircraft landed after exercise completion. B-25's FJ, FC, FO, FH, FN, and FU are airborne for air Exercise Number 1 of general flying, decent through cloud (DTC), and bombing, aircraft landed after exercise completion. B-25's FV, FK, and FT then replaced three aircraft that were unserviceable. B-24 LH is airborne for Radar Exercise Number 1, aircraft landed after exercise completion. Tanker *British Prestige* is bound north in Straits, northwest of Little Bahama Bank. SS *Fred Herrling* is bound south in Straits northwest of Little Bahama Bank. USS *Warbler* (ARS-11) is bound north, to the north of Abaco, exiting the area. SS *Carleton Ellis* is bound north in Straits west of Little Bahama Bank. SS *Logan's Fort* is bound south in Straits north of Bimini. SS *John Davey*, SS *Vermont*, and USS *ATR-29* are bound north in Straits. SS *Western Queen* is bound south in Straits. SS *George Westinghouse*, SS *James B. Atwell*, and SS *Del Rio* are bound south, to the east of Eleuthera. SS *G. Harrison Smith* is bound north, to the east of Eleuthera. NAAF Great Exuma: Two Army PBY's arrived from Morrison Field, and one departed. A training plane arrived from Banana River and returned there.

18th April 1945

B-25's from Oakes Field and B-24's from Windsor Field carried out local flying training. B-25's FH, FC, FJ, FT, and FV are airborne for air Exercise Number 3 and Number 4 of bombing on fixed and moving targets with decent through cloud (DTC), aircraft landed after exercise completion. B-24's LX, NH, and MC are airborne from Windsor Field for Exercise Number 48 of navigation signals, homing procedure, photography, depth charge and gun practice. B-25's FJ, FD, FC, FH, and FG are airborne for Night Patrol Number 2 with five details of navigation with astrological signals practice, aircraft landed after exercise completion. B-24 LH is airborne for Radar Exercise Number 1, aircraft landed after exercise completion. B-25 JX is airborne for Orlando, Florida and landed there. SS *George Bellows* is bound north in Straits northwest of Little Bahama Bank, exiting the area. Tanker *Cities Service Fuel* is bound north in Straits northwest of Little Bahama Bank. SS *Montgomery City Victory*, and SS *Scandia* are bound north in Straits west of Little Bahama Bank. Tanker *Pan Scandia* and SS *Sedan* are bound north in Straits west of Little Bahama Bank. SS *Melville Jacoby* and SS *George E. Pickett* are bound north in Straits northwest of Bimini. SS *Champion Clark*, SS *Wallace M. Tyler*, and USS *LST-1136* are bound north in Straits. SS *Cape Sebastian* is bound west in Straits south of Cay Sal Bank. SS *Lloyd S. Carlson* is bound south in Straits south of Cay Sal Bank. *SS Canturaria*, SS *Thorshammer*, and USS *Inflict* (AM-251) are bound south, east of Eleuthera. SS *Orville Harden* is bound north, to the east of Eleuthera. NAAF Great Exuma: A plane *en route* San Juan to Banana River made emergency landing then departed to destination, an Army PBY departed to South Caicos Island.

19th April 1945

B-25's from Oakes Field and B-24's from Windsor Field carried out local flying training. B-25's FC, FH, FG, FJ, and FD are airborne for air Exercise Number 3 and

Number 4 of descent through cloud (DTC) and bombing on fixed and moving targets, took interval for re-bombing, aircraft landed after exercise completion. B-24's LZ, LY, NH, and MD are airborne from Windsor Field for Exercise Number 48, B-24 MD then returned to base with engine trouble, crew flew on 20th [following day], B-24 LH is then airborne for Radar Exercise Number 2, aircraft landed after exercise completion. B-24 MD is airborne for Corpus Christi via Morrison Field and Pensacola, and is confirmed having landed at Corpus Christi, Texas. SS *San Tirso* is bound north in Straits northwest of Little Bahama Bank. SS *Queenston Heights* is bound north in Straits north of Little Bahama Bank. SS *George W. Lively* is bound south in Straits west of the Little Bahama Bank. SS *Henry Clay* is bound north in Straits west of Little Bahama Bank. USCG *CG-Faunce* (W-138) and tug USS *YTB-392* are bound west in Northwest Providence Channel. SS *Henry S. Grove* is bound north in Straits. **Convoy KN-381** is bound north in Straits north of Cay Sal Bank. SS *John R. R. Hannay* and **Convoy KG-774** are bound southeast in Old Bahama Channel. SS *H. M. Fredricksen* is bound north, to the east of Eleuthera. NAAF Great Exuma: A plane and an Army PBY arrived from search for crash survivors, then plane went to Guantanamo, an Army PBY arrived from San Juan, departed to Morrison Field, a plane arrived from Banana River then returned, a plane departed to Banana River.

20th April 1945

B-25's from Oakes Field and B-24's from Windsor Field carried out local flying training. B-25's FC, FH, FK, FG, FD, and FM are airborne for air Exercise Number 7 of bombing, decent through cloud (DTC), and compass swing, aircraft landed after exercise completion. B-25's FC, FJ, FH, FG, and FM air Exercise Number 8 of bombing and decent through cloud (DTC), aircraft landed after exercise completion. B-24 LY is airborne twice, and B-24 MC is airborne because it could not complete on 19th [previous day] from Windsor Field for Exercise Number 48, aircraft landed after exercise completion. B-24 LH is airborne for Radar Exercise Number 2, aircraft landed after exercise completion. B-25 JX is airborne on its return journey from Orlando, then landed back at base in the evening. USS *SC-512* is bound north in Straits northwest of Little Bahama Bank. Tanker *New Hope* and tanker *Front Royal* are bound south, to the north of Walker Cay, off Abaco. Tanker *El Aleto* is bound north in Straits, northwest of Little Bahama Bank. SS *Morton Prince* and tanker *Tustem* are bound south in Straits northwest of Little Bahama Bank. Tanker *Pan North Carolina* is bound south in Straits west of Little Bahama Bank. SS *Malabar*, SS *Hendrix William Van Loon*, and tanker *Thomas H. Wheeler* are bound north in Straits. Tanker *Chickamauga* and tanker *Spencer Kellogg* are bound south in Straits, north of Cay Sal Bank. SS *Lucius Fairchild* and SS *John A. Holland* are bound southeast in Nicholas Channel. USCG cutter *CG-Marion* is bound northwest in Old Bahama Channel. **Convoy Motor Torpedo Boat Squadron-38, Group B**, (four fast torpedo PT-boats) USS *PT*'s, is bound southeast in Old Bahama Channel. Tanker *Polycastle* and USCG *CG-Colfax* are bound southeast of Eleuthera. NAAF Great Exuma: A plane arrived from Banana River then returned. An Army PBY went to search for crash survivors then returned.

21st April 1945

B-25's from Oakes Field and B-24's from Windsor Field carried out local flying training. B-25's FC, FH, FJ, FV, FG, and FU are airborne on air Exercise Number 7 of bombing, decent through cloud (DTC), and compass swing, aircraft landed after exercise completion. B-25 FU suddenly crashed into sea with no survivors out of a captain and six crew. The plane crashed into sea north of New Providence, and an ASR marine craft launch HMS *P-170* and ASR launch HMS *P-191* searched the area. They observed oil patch, and picked up a dinghy with the plane's identification. B-25's FB, FG, FS, FH, and FC are airborne for air Exercise Number 8 of bombing and decent through cloud (DTC), aircraft landed after exercise completion. B-24 LH is airborne for Radar Exercise Number 2, aircraft landed after exercise completion. Tanker *Empire Gulf* and tanker *Cities Service Koolmotor* are bound north in Straits northwest of Little Bahama Bank, exiting the area. Tanker *Gulfcoast* is bound north in Straits northwest of Little Bahama Bank. SS *Rich Mountain* is bound south in Straits west of Little Bahama Bank. Tanker *Edward L. Shea*, SS *W. F. Burdell*, and SS *Honningsvaag* are bound north in Straits west of Little Bahama Bank. SS *Argon* is bound south in Straits, northeast of Palm Beach. SS *Kokomo Victory* is bound north in Straits, northwest of Bimini. USS *YAG-19* is bound west in Straits. SS *Winchester Victory*, SS *Antietam*, and SS *William Crompton* are bound north in Straits. Tanker *Mary A. Townsend* is bound south in Straits. **Convoy GK-874** is bound north in Straits west of Cay Sal Bank. USS *Redwood* (AM-30) and USCG cutter *CG-Dix* are bound southeast in Old Bahama Channel. Tanker *Beta*, tanker *Gulfdisc*, SS *George Durant*, tanker *Gulfcrest*, SS *Baltic*, tanker USS *Roxane* (AKA-37), SS *Schmitt* (APD-76), SS *James Howson*, and **Convoy NG-502** are bound south, to the east of Eleuthera. SS *Westview Park* is bound north, to the east of Eleuthera. NAAF Great Exuma: An Army PBY arrived from Morrison Field departed to Mayaguana, a plane arrived from Guantanamo then returned. A plane arrived from Banana River then returned.

22nd April 1945

B-25's from Oakes Field and B-24's from Windsor Field carried out local flying training. B-24's MC and NJ are airborne for Route A and B-24 LU is airborne on Route B, for Radar Exercise Number 23, aircraft landed after exercise completion. B-25 FM is airborne for Boca Chica, Key West, and landed back at Oakes Field later this evening. SS *Elusive* (AM-225) is bound north in Straits northwest of Little Bahama Bank, exiting the area. Tanker *Cities Service Ohio*, SS *Meline*, and SS *Paul Buck* are bound north in Straits northwest of Little Bahama Bank. SS *Winter Hill* is bound south in Straits northwest of Little Bahama Bank. SS *New Bern Victory* is bound north in Straits north of Little Bahama Bank. SS *City of Montgomery*, SS *Lucius Fairchild*, tanker *L. L. Abshire*, and SS *Tobias E. Stansbury* are bound north in Straits west of Little Bahama Bank. SS *Oliver H. Perry* is bound north in Straits off Palm Beach. SS *Harold Bassett* is bound south in Straits, off Port Everglades. SS *Sam Strule* is bound west in Northwest Providence Channel. SS *Mirza* is bound north in Straits. SS *Lawrence D. Tyson* is bound south in Straits north of Cay Sal Bank. USS *APC-87* is bound south, southeast of Cay Sal Bank. **Convoy Kingston-164** is bound north, southeast of Cay Sal Bank. SS *Gallant Fox* is bound southeast in Nicholas Channel. Tanker *Comol Cuba* is bound southeast in Old Bahama Channel. USS *Arizona* is bound south, to the east of Eleuthera. NAAF Great Exuma: An Army PBY arrived from South Caicos Island, then departed to Morrison field.

23rd April 1945

B-25's from Oakes Field and B-24's from Windsor Field carried out local flying training. B-25's FC, FJ, FM, FS, and FD are airborne for Anti-Submarine Patrol Number 2 with five details of navigation, signals, and photography, aircraft landed after exercise completion. B-25's FV, FM, FS, FN, and DC are airborne for air Exercise Number 2 of navigation with bombing, photography, and descent through cloud (DTC), aircraft landed after exercise completion. B-24's airborne from Windsor Field, LU and NK are on Route A and B-24 LY is on Route B, for Radar Exercise Number 23, aircraft landed after exercise completion. B-24 LH is airborne for Radar Exercise Number 2, aircraft landed after exercise completion. B-24 NO is airborne for Corpus Christi via Morrison Field and has landed there. B-24 NX is airborne from Corpus Christi and landed back at base. SS *Albion Victory* is bound north, to the north of Walker Cay, off Abaco. SS *Empire Chief* and SS *John P. Holland* are bound north in Straits ,west of Little Bahama Bank. Tanker *British Hope* and SS *Richard K. Call* are bound south in Straits, off Palm Beach. SS *William Larbey* north in Straits, off Port Everglades. SS *William J. Palmer* and SS *Perryville* are bound north in Straits. SS *Lafcadio Hearn* is bound south in Straits. SS *Charles Scribner* is bound northwest in Santaren Channel. **Convoy KG-775** is bound southeast in Nicholas Channel. NAAF Great Exuma: A plane arrived from Banana River, and then returned.

24th April 1945

B-25's from Oakes Field and B-24's from Windsor Field carried out local flying training. B-25's FC, FG, FN, FB, and FD are airborne for Anti-Submarine Patrol Number 2 with five details of navigation, photography, and signals, aircraft landed after exercise completion. B-25's FM, FD, FN, FB, FS, and FT are airborne for air Exercise Number 2 of photography, bombing, and decent through cloud (DTC), aircraft landed after exercise completion. Esco Exercise Number 4 arranged by US Navy at Boca Chica, seaplane tender USS *Sandpiper* (AVP-9) co-operated with aircraft of C. Flight Number 3 Squadron, used this exercise for passing out of Course exercise. Homing procedure, recognition, signals, patrols, navigation, photography, depth charge and gun attacks practiced by crew in three-day exercise. B-24 LR is airborne from Windsor Field, homing procedure commenced, seaplane tender USS *Sandpiper* (AVP-9) reached, recognition performed, carried out patrol Cobra under instruction of ship's commander, re-joined ship to attack *Ploughshare*, was towed by ship, attacked with one depth charge and guns, attack photographed, reported to base for Mamba patrol and vertical speed practice, set course for base, aircraft landed after exercise completion, other two aircraft on this exercise were withdrawn to take part in Operation Anti-Submarine patrol as requested by B-24 MH, which is airborne from Windsor Field, homing procedure commenced, seaplane tender USS *Sandpiper* (AVP-9) reached, recognition performed, carried out patrol Mamba and Cobra under instruction of ship's commander, re-joined ship to attack *Ploughshare*, was towed by ship, attacked with one depth charge and guns, attack photographed, reported to base, set course for base, aircraft landed after exercise completion. B-24 LH is airborne for Radar Exercise Number 2, aircraft landed after exercise completion. Aircraft from Royal Island had disappearing Radar contact and positive Sono Buoy, USN requested search, barrier patrol organized in the Northeast Providence Channel area between South Point [Hole-in-the-Wall], Abaco, and North Point [Spanish Wells] Eleuthera. B-24 MF found nothing, B-24 NC had one Radar contact which it lost, B-24 NX had nil report. Tanker *Gulfwave* and SS *W. M. Burton* are bound north in Straits, northwest of Little Bahama Bank, exiting the area. Tanker *Hadnot*, SS *Fort Lennox*, and tanker *Paulsboro* are bound north in Straits northwest of Little Bahama Bank. SS *Tonto* is bound south in Straits west of Little Bahama Bank. Tanker *Gulfmaid* and SS *Five Forks* are bound south in Straits off Palm Beach. Tanker *Stanvac Calcutta* is bound south in Straits. Fuel barge USS *YOG-6* is bound north in Straits. SS *Richard K. Call* is bound southeast in Old Bahama Channel. Destroyer USS *Beatty* (DD-756) is bound south, to the east of Eleuthera. NAAF Great Exuma: A plane came from Guantanamo and returned there. A plane came from Banana River and returned there.

25th April 1945

B-25's from Oakes Field and B-24's from Windsor Field carried out local flying training. B-25's FB, FS, FD, FM, and FN are airborne for Anti-Submarine Patrol Number 1 with five details of navigation and signals, aircraft landed after exercise completion. B-25's DC, FK, FV, FM, and FS are airborne for air Exercise Number 5 of bombing and decent through cloud (DTC), FT and FD replaced FM and FV, which became unserviceable, and the aircraft landed after exercise completion. B-24 LU is airborne for Esco Exercise Number 4, completed navigation exercise and patrols, could not complete attack, armament gear unserviceable, returned to base, special briefing of Exercise Number 48 arranged for 27th to enable crew to practice, descent through cloud (DTC) omitted from their Esco exercise. B-24 NC is airborne from Windsor Field, homing procedure commenced, seaplane tender USS *Sandpiper* (AVP-9) reached, recognition performed, carried out patrol Viper under instruction of ship's commander, re-joined ship to attack *Ploughshare*, was towed by ship, attacked with one depth charge and guns, attack photographed, reported to base for Cobra patrol and vertical speed practice, set course for base, aircraft landed after exercise completion. B-24 NF is airborne from Windsor Field, homing procedure commenced, not a great success, transmitter unserviceable, seaplane tender USS *Sandpiper* (AVP-9) reached, recognition performed, carried out patrol Viper under instruction of ship's commander, re-joined ship to attack *Ploughshare*, was towed by ship, attacked with one depth charge and guns, attack photographed, reported to base for vertical speed practice, starboard outer engine unserviceable, set course for base, aircraft landed earlier than schedule. Aircraft from Royal Island had disappearing Radar contact and positive Sono Buoy, US Navy requested search, continued barrier patrol in Northeast Providence Channel area between South Point [Hole-in-the-Wall], Abaco, and North Point [Spanish Wells] Eleuthera. B-25 MD found object which seemed like aluminium sea marker, object photographed, decided it wasn't useful, B-25 LR spotted a disappearing Radar, nil report. B-25 ND is airborne for return journey from Corpus Christi via Boca Chica, and landed at base. SS *Langston B. Evans* is bound south, to the north of Walker Cay, off Abaco. Tanker *Minatitlan* is bound south in Straits northwest of Little Bahama Bank. Submarine USS *Stingray* (SS-186) escorted by USS *PC-1142* is bound north in Straits west of Little Bahama Bank. **Convoy GK-875**, one merchant ship escorted by USS *PCS-1425* and USS *SC-678*, and SS *Cape Texas* are bound northwest in Nicholas channel. Tanker *Gertrude Kellogg* is bound northwest in Old Bahama Channel. SS *Paco City* is bound southeast in Old Bahama Channel. Tanker *Gulfwax* is bound south, to the east of Eleuthera. NAAF Great Exuma: A plane arrived from Banana River then returned, three planes in from a mission, one departed to Guantanamo, an Army PBY in from Mayaguana.

26th April 1945

B-25's from Oakes Field and B-24's from Windsor Field carried out local flying training. B-25's FV, FM, FB, FK, FJ, and FT are airborne for Anti-Submarine Patrol Number 1, with six details of navigation and signals with photography, aircraft landed after exercise completion. B-24 NH is airborne from Windsor Field for Esco Exercise Number 4, homing procedure commenced, seaplane tender USS *Sandpiper* (AVP-9) was reached, recognition performed, carried out patrol Cobra under instruction of ship's commander, re-joined ship to attack *Ploughshare*, was towed by ship, attacked with one depth charge and guns, attack photographed, reported to base Viper and Adder patrol and vertical speed practice, set course for base, aircraft landed after exercise completion. B-24 LR is airborne from Windsor Field for Esco Exercise Number 4, homing procedure commenced, point codenamed *Sandpiper* reached, recognition performed, carried out patrol Viper under instruction of ship's commander, re-joined ship to attack *Ploughshare*, was towed by ship, attacked with one depth charge and guns, attack photographed, reported to base Cobra and Adder patrol and vertical speed practice, set course for base, aircraft landed after exercise completion. B-24 LU is airborne from Windsor Field for Esco Exercise Number 4. homing procedure commenced, seaplane tender USS *Sandpiper* (AVP-9) reached, recognition performed, Smoke float dropped, attacked with one depth charge and guns, attack photographed, reported to base Viper patrol and vertical speed practice, set course for base, aircraft landed after exercise completion. B-24 NX is airborne for barrier patrol, nothing to report, landed back at base. B-24 MB is airborne for Corpus Christi via Boca Chica, landed there. B-24 NG is airborne for Boca Chica, returned back to base same day. Aircraft carrier USS *Prince William* (CVE-31) Group and **Task Unit TU-2821** are bound north in Straits, northwest of Little Bahama Bank, exiting the area. SS *Washington Irving*, SS *Tennessee*, tanker *Pan Georgia* and SS *Inverbank* are bound south in Straits northwest of Little Bahama Bank. SS *Malay* and SS *Baldbutte* are bound south in Straits west of Little Bahama Bank. Submarines USS *R-2* and USS *R-14,* escorted by USS *SC-512* and USS *SC-1000,* are bound north in Straits west of Little Bahama Bank. Destroyer escort USS *Jordan* (DE-204) is bound north in Straits west of Little Bahama Bank. **Convoy KN-382** is bound north in Straits, off Port Everglades. SS *Tide Penn* is bound south, in Straits. USS *LSM-394* is bound south in Straits. SS *Castana* is bound north in Straits south of Cay Sal Bank. Tanker *Comol Cuba* is bound north in Old Bahama Channel. USS *Bassett* (APD-23), USS *Burke* (APD-65), and tanker *Gulf Hawk* are bound north, to the east of Eleuthera. **Convoy NG-503** is bound south, to the east of Eleuthera. NAAF Great Exuma: A plane arrived from Banana River then returned, two planes out to Guantanamo, two planes on local flight.

27th April 1945

B-25's from Oakes Field and B-24's from Windsor Field carried out local flying training. B-25's FT, FS, FB, FK, and FV are airborne for night Patrol Number 1 with five details of navigation with astrological and signals, aircraft landed after exercise completion. B-25's FR, FS, FT, FB, and DC are airborne for air Exercise Number 6 of homing procedure and photography, aircraft landed after exercise completion. B-24 LY is airborne from Windsor Field for one sortie of Exercise Number 48, organized for crew of Esco Exercise Number 4 on 25th [two days earlier] in B-24 LU, which had been unsuccessful due to faulty armament, carried out full briefing of depth charge and gun attacks, signals and navigation, aircraft landed after exercise completion. B-24 MD is airborne on a barrier patrol, a square search in given area, found metal bound chest, partly submerged, Intelligence found no interest in it, landed back at base. B-24 NJ is airborne for Boca Chica, landed there, returned to base same day, landed at Windsor Field. Tanker *British Justice* is bound south in Straits west of Little Bahama Bank. Tanker *Alpha* is bound north in Straits west of Little Bahama Bank. Tanker *Empire Gulf* is bound northeast in Santaren Channel. NAAF Great Exuma: A plane out to Guantanamo. A plane arrived from Guantanamo on an administrative flight and then departed for Miami. A plane arrived from Banana River and returned there.

28th April 1945

B-25's from Oakes Field and B-24's from Windsor Field carried out local flying training. B-25's FM, FK, FB. FH, FJ, and FT are airborne for local night navigation Exercise Number 1 with two briefings, aircraft landed after exercise completion. B-25's DC, FF, FB, FT, FJ, and FM are airborne on air Exercise Number 6 of homing procedure and photography, aircraft landed after exercise completion. B-24 NK is airborne for Boca Chica landed back at base same day. USS *YMS-421* is bound north in Straits northwest of Little Bahama Bank. SS *Kathio* is bound south in Straits northwest of Little Bahama Bank. SS *C. B. Watson* is bound north in Straits, west of Little Bahama Bank. USS *YO-129* is bound south in Straits north of Bimini. Destroyer escort USS *Coates* (DE-685) is bound north in Straits northwest of Bimini. SS *East Point Victory* is bound north in Straits, south of Bimini. SS *Ponca City* is bound north in Straits, southwest of Bimini. SS *Andrew Jackson* is bound north in Straits north of Cay Sal Bank. NAAF Great Exuma: Two Army PBY's came from Morrison Field, departed for bounce drill then arrived back. A plane arrived from Banana River then returned.

29th April 1945

B-24's MC, LR, LY, MD, and LR are airborne from Windsor Field for Radar Exercise Number 41, aircraft landed after exercise completion. SS *Winter Hill* is bound north in Straits west of Little Bahama Bank. SS *Markay* is bound north in Straits north of Bimini. SS *Front Royal, SS New Hope,* **Convoy KN-383** is bound north in Straits northwest of Bimini. SS *George Washington* is bound north in Straits. SS *Paco City* is bound west in Nicholas Channel. SS *Sea Dolphin* and **Convoy KG-776** are bound southeast in Old Bahama Channel. SS *Pedro Menendez* northwest in Old Bahama Channel. NAAF Great Exuma: A plane arrived from a mission. A plane arrived from Banana River and returned there.

30th April 1945

B-25's continued training from Oakes Field. B-25's FH, FG, FT, FJ, and FK are airborne for air Exercise Number 5 of bombing aircraft landed after exercise completion. B-24 MD is airborne for Corpus Christi via Boca Chica, landed at destination at night. B-24 NO is airborne and then landed back at base from Corpus Christi. B-24 LH is airborne from Windsor Field for Radar Exercise Number 1, aircraft landed after exercise completion. **Convoy NK-683**, USS *Abnaki* (ATF-86) towing dredge *Henry Bacon*, USS *ST-841*, USS *SC-687*, and USS *SC-989* are all bound south in Straits, northwest of Little Bahama Bank. SS *Rich Mountain* is bound east in Straits, north of Walker Cay, off Abaco. SS *Woodbridge Victory* is bound south in Straits west of Little Bahama Bank. **Convoy GK-876** and tanker *Gertrude Kellogg* are bound northwest in Nicholas Channel. USS *Woodford* (AKA-86) and USS *Runel* (APD-85) are bound south, to the east of Eleuthera. NAAF Great Exuma: A plane out to San Juan, two planes in from a mission, a plane went to Guantanamo, and another plane went to make contact with it and then returned, a plane arrived from Banana River then returned.

1st May 1945

B-25's from Oakes Field and B-24's from Windsor Field carried out local flying training. B-25's FV, FH, FT, FJ, FG, and FK are airborne for air Exercise Number 5 of bombing and decent through cloud (DTC), FM and FS replaced FV and FH, because those two had become unserviceable, and all aircraft landed after exercise completion. B-25's FT, FJ, FG, FH, FK, and FV are airborne for air Exercise Number 1 of familiarization, decent through cloud (DTC), and bombing on moving target, aircraft landed after exercise completion. B-24 LH is airborne from Windsor Field for Radar Exercise Number 1, aircraft landed after exercise completion. B-25 FB is airborne, carried out search in area, Joulter Cays, north of Andros, for Bahamas mailboat *William Sayle,* overdue Nassau to Miami, yet nil report; vessel arrived later to destination. Tug USS *Abenaki* (ATF 86) and tows are bound south in Straits off Port Everglades. SS *Stanvac Calcutta* is bound north in Straits north of Walker Cay. **Convoy Miami SS *Joseph Henry*** (cable ship) escorted by USCG cutter *CG-Faunce* (W-138) is bound south in Straits north of Nassau. USS *YF-443* bound southeast in Santaren Channel. *SS Jason L. Breck* is bound northwest in Nicholas Channel. SS *Jesse Applegate* and SS *Finley P. Dunne* are bound northwest entering Nicholas Channel. SS *David L. Swain* is bound east entering Old Bahama Channel. SS *Benjamin H. Brewster* and SS *Binger Herman* are bound northwest in Old Bahama Channel. NAAF Great Exuma: An Army PBY went to Morrison Field then returned, a plane arrived from Banana River then returned, two planes out to Guantanamo, a plane in from a mission.

2nd May 1945

B-25's from Oakes Field and B-24's from Windsor Field carried out local flying training. B-25's FR, FG, FJ, FN, and FO are airborne for air Exercise Number 1 of familiarization, decent through cloud (DTC), and bombing on moving target, aircraft landed after exercise completion. B-24 LH is airborne from Windsor Field for Radar Exercise Number 1, aircraft landed after exercise completion. B-24 NT is airborne from Windsor Field to escort ferry *Jean Brillant* with HRH, the Duke and Duchess of Windsor journey to Miami, landed back after duty completion on 3rd day. Two Liberators collided mid-air, probable death toll 30. SS *Ruth Lykes* and

SS *Attleboro Victory* are bound north in Straits. Fleet ocean tug USS *Abenaki* (ATF-86) towing dredge *Henry Beacon* and destroyer USS *Noa*, DD-841 are bound south in Straits. NAAF Great Exuma: A plane arrived from Banana River then returned, an Army PBY set out to South Caicos Island, and another plane took off for Guantanamo.

3rd May 1945

B-25's from Oakes Field and B-24's from Windsor Field carried out local flying training. B-25's FJ, FY, FK, FG, FT, and FM are airborne for air Exercise Number 3 and Exercise Number 4 of descent through cloud (DTC) and bombing on fixed and moving targets with interval of re-bombing, aircraft landed after exercise completion. B-24 LH is airborne from Windsor Field for Radar Exercise Number 1, aircraft landed after exercise completion. B-24 NT is airborne, 2nd day from Windsor Field to escort the ferry *Jean Brillant* with HRH, the Duke and Duchess of Windsor on their voyage to Miami and retirement. B-24 MB returned from Corpus Christi landed at Windsor Field at night. Windsors; "slipped away from Nassau at dawn" to visit friends in Palm Beach via Miami, their tenure wound down in the ferry *Jean Brillant.* USS *Deft* (AM-126) is bound north in Straits northwest of Little Bahama Bank. SS *Five Forks* is bound north in Straits, west of Little Bahama Bank. NAAF Great Exuma: A plane arrived from Banana River then returned, an Army PBY in from Mayaguana and out to Nassau.

4th May 1945

B-25's from Oakes Field and B-24's from Windsor Field carried out local flying training. B-25's FT, FK, FY, FG, and FM are airborne for night Patrol Number 2 with five details of navigation with astrological and signals practice, aircraft landed after exercise completion. B-25's FY, FG, FH, FT, FB, and FS are airborne for air Exercise Number 6 of homing procedure and photography, aircraft landed after exercise completion. B-25's FY, FH, FG, FM, and FT are airborne for air Exercise Number 8 of bombing, decent through cloud (DTC), and photography, aircraft landed after exercise completion. B-24 LH is airborne from Windsor Field for Radar Exercise Number 1, aircraft landed after exercise completion. B-24 NA is airborne from Windsor Field for Corpus Christi via Morrison Field, landed at destination at night. SS *Spirit Lake* is bound north in Straits northwest of Little Bahama Bank, exiting the area. SS *Stanford Victory* is bound north in Straits north Little Bahama Bank. **Convoy KN-384** is bound north in Straits west of Bimini. Oil barge USS *YO-69* is bound north in Straits. USS *YAG-19* is bound south in Old Bahama Channel. USS *PC-1134* and USS *SC-1001* are bound southeast in Old Bahama Channel. NAAF Great Exuma: A plane on a mission, USS *YAG-19* sailed in from Miami, a plane arrived from Banana River then returned.

5th May 1945

B-25's from Oakes Field and B-24's from Windsor Field carried out local flying training. B-25's FY, FH, FG, FS, FB, and FT are airborne for air Exercise Number 7 of decent through cloud (DTC), compass swing, photography, aircraft landed after exercise completion. B-25's FV, FD, FT, FO, FS, and FK are airborne for air Exercise Number 8 of bombing, decent through cloud (DTC), and photography, aircraft landed after exercise completion. B-24 LH is airborne from Windsor Field for Radar Exercise Number 1, aircraft landed after exercise completion. **Convoy NK-684**, consisting of 15 Merchant ships and four escorts, is bound south in Straits Off

Canaveral. SS *Spirit Lake* is bound north in Straits, northwest of Little Bahama Bank. SS *Kittanning* and SS *Cody Victory* are bound south in Straits north of Little Bahama Bank. Oiler USS *YO-129* is bound west in Northwest Providence Channel. USS *YTB-227* is bound southeast in Northwest Providence Channel. USS *LST-512* is bound south in Straits. SS *Sebastian Vizcaino* and SS *Laura Keene* are bound west entering Nicholas Channel. **Convoy GK-877** is bound northwest in Old Bahama Channel. SS *John A. Sutter* and SS *Paul Hamilton Hayne* are bound north in Old Bahama Channel north of Cay Lobos. NAAF Great Exuma: A plane in from a mission, a plane out to Guantanamo, a plane arrived from Banana River then returned.

6th May 1945

B-25's from Oakes Field and B-24's from Windsor Field carried out local flying training. B-25's FN, FG, FM, FO, and FK are airborne for Anti-Submarine Patrol Number 2 with five details of navigation and signals practice, B-25 FN returned with engine trouble, and the rest of aircraft landed after exercise completion. B-25 FY then took off for Boca Chica, and landed back at base late evening. B-24 LZ is airborne from Windsor Field and landed at Bermuda. B-24 MD is airborne for its return journey from Corpus Christi, and landed at base at night. Tanker *Drewrys Bluff* is bound south in Straits northwest of Little Bahama Bank. USS *Saluda* (IX-28) is bound north in Straits northwest of Bimini. Tanker *Thomas H. Wheeler* is bound north in Straits. Tanker *Nordahl Crieg* and SS *Benjamin Lundy* are bound northwest in Nicholas channel. Tanker *George W. Barnes* is bound southeast in the Old Bahama Channel. NAAF Great Exuma: Inter-island transport USS *YAG-19* sailed for Miami, and a plane took off for Guantanamo.

7th May 1945

B-24 LH is airborne from Windsor Field for Radar Exercise Number 1, aircraft landed after exercise completion. SS *Chilbar* is bound north in Straits west of Little Bahama Bank. Destroyer escort USS *Jordon* (DE-204) is bound north in Straits northwest of Bimini. Oiler USS *YO-69* is bound south in Straits northwest of Bimini. SS *Spidoleine* is bound north in Straits. SS *Rio Teugo* is bound southeast in Old Bahama Channel. NAAF Great Exuma: A plane in from a mission, a plane out to Guantanamo, a plane arrived from Banana River then returned.

8th May 1945

V-E Day, for Victory in Europe: Germany surrenders to Allies. B-25's from Oakes Field and B-24's from Windsor Field carried out local flying training. B-24 is airborne for Radar Exercise Number 23, B-25's NT and NK are on Route A, B-25's NS and NR are on Route B, aircraft landed after exercise completion. B-24 LZ is airborne on return journey from Bermuda, landed at Windsor Field at night. Tanker *Gulf of Maracaibo* is bound north in Straits west of Little Bahama Bank. Tanker *Kaposia* is bound north in Straits west of Bimini. **Convoy KG-778**, consisting of three merchant ships and one escort, are bound southeast in Nicholas Channel. SS *Fred L. Dau* is bound southeast in Old Bahama Channel. No. 113: A general holiday was declared for V-E Day, and all personnel except those required for essential duty. Work ceased at 11.00 am and the Commanding Officer and Staff Officers provided a picnic for ground personnel at Lyford Cay [HRH the Duke of Windsor, the senior British official, had already left the colony]. NAAF Great Exuma: An Army PBY in from Morrison Field, then out to Mayaguana, a plane flew out to Guantanamo, and a plane arrived from Banana River then returned. U-Boats: At least one U-Boat was proceeding from Europe to the Bahamas region at the time of Admiral Dönitz's order for all German U-Boats to cease hostilities and surrender to the nearest Allied vessel or in the nearest allied or neutral port; none made it as far as Bahamas before then.

9th May 1945

Grumman Goose GB is airborne for Harbour Island landed back at base late evening. **Convoy KN-385** Is bound north in Straits off Port Everglades. *USS PCE(R)-857* is bound north in Straits. SS *Thomas U. Walter* is bound northwest in Nicholas Channel. SS *John McShofield* is bound northwest in Old Bahama Channel. SS *Thomas H. Wheeler* is bound southeast in Old Bahama Channel. SS *Registan*, SS *Herbert D. Croly*, and SS *Atlantic* are bound northwest in Old Bahama Channel. USS *PC-822* is bound southeast in Old Bahama Channel. No.113: A general holiday for all personnel except those on essential duty. NAAF Great Exuma: A plane *en route* from Banana River to Guantanamo arrived with mechanical trouble. A plane arrived from Banana River then returned.

10th May 1945

Local flying continued through the day from Windsor and Oakes Field. B-25's FS, FY, FV, FM, and FK are airborne for Anti-Submarine Patrol Number 2 with five details of navigation and signals practice, aircraft landed after exercise completion. B-25's FY, FV, FB, FD, FK, and FS are airborne for air Exercise Number 7 of decent through cloud (DTC), compass swing, and photography, aircraft landed after exercise completion. B-24 LH is airborne from Windsor Field for Radar Exercise Number 1, aircraft landed after exercise completion. B-25 JC took off for Boca Chica, landed, returned to base later at night. **Convoy NK-685**, 31 Merchant ships and four escorts, is bound south in Straits northwest of Little Bahama Bank. USS *Siren* (PX-13) is bound north in Straits west of Little Bahama Bank. USS *SC-665* is bound north in Straits off Miami. Tanker *Atlantic* is bound north in Santaren Channel. SS *Will R. Wood* is bound northwest in Nicholas Channel. No.113. Senor Pena, Director of Civil Aviation in Uruguay, passed through Nassau *en route* from Belem to the UK. NAAF Great Exuma: A plane arrived from Banana River then returned.

11th May 1945

B-25's from Oakes Field and B-24's from Windsor Field carried out local flying training. B-25's FY, FG, FK, FV, FB, and FS are airborne for night Patrol Number 1 with six details of navigation with astrological and signals practice, aircraft landed after exercise completion. B-25's FY, FV, FK, FG, FS, and FH are airborne for Anti-Submarine Patrol Number 1 with six details of navigation and signals practice, aircraft landed after exercise completion. B-24's airborne from Windsor Field, NT and NP are on Exercise Number 23 on Route A. B-24's NS and MD are on Exercise Number 23 on Route B, B-24 LH is airborne for Radar on Exercise Number 2, aircraft landed after exercise completion. B-24 NA is airborne on its return journey from Corpus Christi, and landed at Windsor Field in afternoon. SS *Towanda Victory* is bound north in Straits west of Little Bahama Bank. Tug USS *Dry Tortugas* and tow, escorted by USS *SC-678* are bound south in Straits. SS *George W. Barnes* is bound north in Nicholas Channel. Submarine USS *R-9* escorted by USS *SC-1000* is bound north in Old Bahama Channel. **Convoy NG-506** and **TU-0296**,

55 Merchant ships escorted by USS *PC-1084*, USS *PC-1087*, USS *PC-1546* and USS *PC-1549* are bound south, to the east of Eleuthera. NAAF Great Exuma: A plane landed in from a mission, another plane flew out to Guantanamo, and a plane arrived from Banana River then returned. An Army PBY flew in from Mayaguana.

12th May 1945

B-25's FB, FD, FG, FH, FV, and FY are airborne for local night navigation Exercise Number 1 with two briefings, aircraft landed after exercise completion. B-25's FY, FV, FD, FS, FB, and FK are airborne for air Exercise Number 3 and Number 4 of descent through cloud (DTC) and bombing on fixed and moving targets with interval of re-bombing, aircraft landed after exercise completion. Esco Exercise Number 5 is arranged in conjunction with US Navy at Boca Chica, and the seaplane tender USS *Pelican* (AVP-6) cooperated with aircraft of C. Flight 3 Squadron. B-24 LR is airborne from Windsor Field, homing "B" impossible, Radar unserviceable, shipping observed was logged, carried out recognition procedure, set for patrol Cobra, visibility poor, first smoke float laid by ship went out, second smoke float attacked with one depth charge, was an overshoot, photographs sent back to base, set course for base, aircraft landed after exercise completion. B-24 LU is airborne from Windsor Field, homing procedure commenced, seaplane tender USS *Pelican* (AVP-6) was reached, recognition performed, smoke-float was dropped, attacked with one depth charge and guns, attack photographed, reported to base. Then Adder and Viper patrol commenced, carried out vertical speed practice, set course for base, aircraft landed after exercise completion. B-24 MB is airborne from Windsor Field, homing procedure commenced, position code named *Pelican* [seaplane tender USS *Pelican* (AVP-6)] was airborne reached, recognition performed, attacked position codenamed *Ploughshare* with one depth, overshoot, attack photographed, reported to base, carried vertical speed practice, set course for base sent weather report to base, aircraft landed after exercise completion. B-24 NF is airborne from Windsor Field, homing procedure commenced, seaplane tender USS *Pelican* (AVP-6) was reached, recognition performed, attacked *Ploughshare* with one depth charge, overshoot, attack photographed, reported to base, carried vertical speed practice, set course for base sent weather report to base, aircraft landed after exercise completion. B-24 LZ is airborne final escort for the duty, due to petrol leak returned to base. Tanker *A. S. Hansen* and SS *Nonpareil* are bound north in Straits west of Little Bahama Bank. USS *Owl* (ATO-137) and tow escorted by USCG cutter *CG-Faunce* are bound south in Straits off Miami. Tug USS *ATR-29* is bound southeast in Santaren Channel. SS *Backhuysen* is bound northwest in Nicholas Channel. SS *George Washington* escorted by USS *PC-1240* is bound northeast of Eleuthera. NAAF Great Exuma: An Army PBY flew out to Mayaguana, a plane took off for Guantanamo, and a plane arrived from Banana River, then returned.

13th May 1945

B-25's from Oakes Field and B-24's from Windsor Field carried out local flying training. B-24 MD is airborne from Windsor Field, 2nd day of Esco Exercise Number 5, homing procedure commenced, seaplane tender USS *Pelican* (AVP-6) was reached, recognition performed, patrols carried out as directed, attacked *Ploughshare* with one depth charge and guns, attack photographed, reported to base, carried vertical speed practice, set course for base, aircraft landed after exercise completion. B-24 NA is airborne from Windsor Field, 2nd day of Esco Exercise Number 5, homing procedure commenced, seaplane tender USS *Pelican* (AVP-6) was reached, recognition performed, Cobra patrol carried out as directed attacked, *Ploughshare* with one depth charge did not detonate, attack photographed, reported to base, carried vertical speed practice carried out patrol Viper and Frog set course for base, aircraft landed after exercise completion. B-24 NH is airborne from Windsor Field 2nd day of Esco Exercise Number 5, homing procedure commenced, Seaplane tender USS Pelican (AVP-6) was reached, recognition performed, attacked *Ploughshare* with one depth charge direct hit attack photographed, reported to base, carried vertical speed practice, patrols Alligator and Mamba carried out set course for base, aircraft landed after exercise completion. B-24 NK is airborne from Windsor Field, 2nd day of Esco Exercise Number 5, no bearing received, homing procedure commenced, seaplane tender USS *Pelican* (AVP-6) was reached, recognition performed, attacked *Ploughshare* with one depth charge, undershot, attack photographed, reported to base, carried vertical speed practice, patrols Frog and Adder carried out, set course for base, aircraft landed after exercise completion. B-24 LU airborne from Windsor Field, 2nd day of Esco Exercise Number 5, homing procedure commenced, seaplane tender USS *Pelican* (AVP-6) was reached, recognition performed, attacked with depth charge, both failed, attack photographed, gave amplifying report to base, Alligator and Lizard patrol carried out, set course for base, aircraft landed after exercise completion. B-24 NO is airborne for Radar Exercise Number 23 on Route B aircraft landed after exercise completion. Tanker *Gylfe* is bound north in Straits off Miami. Tanker *Royal Oak* is bound north in Straits. USS *Cormorant I* (ATO-133) escorted by USS *PC-414* towing USS *YR-234* and USS *YN-64* are bound south in Straits. **Convoy KG-779**, one merchant ship escorted by *PCS-1450*, and USS *SC-989* are bound southeast in Nicholas channel.

14th May 1945

B-25's from Oakes Field and B-24's from Windsor Field carried out local flying training. B-25's JB, JC, FK, FV, FS, and FG are airborne for air Exercise Number 5 of bombing and decent through cloud (DTC), aircraft landed after exercise completion. B-25's FJ, FG, FM, FV, FY, and FT are airborne for air Exercise Number 1 of familiarization and bombing on moving target, aircraft landed after exercise completion. B-24's MD, LW, MB, LU, and LZ are airborne from Windsor Field for Radar Exercise Number 41, aircraft landed after exercise completion. B-24 LH is airborne from Windsor Field for Radar Exercise Number 1, aircraft landed after exercise completion. B-25 JX airborne for Washington DC via Jacksonville. Tug USS *Chipawa* (AO-68) and destroyer *John D. Ford* (DD-228) are bound south in Straits north of Little Bahama Bank. SS *Fort Pitt* is bound south in Straits west of Little Bahama Bank. SS *Camp Union* and **Convoy KN-386** are bound north in Straits west of Little Bahama Bank. SS *William Paca* and SS *John Laurence* are bound northwest in Old Bahama Channel. **Convoy KG-779** is bound southeast in Old Bahama Channel. Tug USS *YTB-399* is bound south in Straits. NAAF Great Exuma: An Army PBY arrived then went to Port au Prince, a plane in from

Guantanamo then departed for Nassau, a plane arrived from Banana River then returned.

15th May 1945

B-25's from Oakes Field and B-24's from Windsor Field carried out local flying training. B-25's FT, FM, FJ, FH, FY, and CX are airborne for air Exercise Number 5 of bombing and decent through cloud (DTC), aircraft landed after exercise completion. B-25's FV, FM, FT, FY, FJ, are FH airborne for air Exercise Number 1, then air check exercise, decent through cloud (DTC), bombing on moving targets, aircraft landed after exercise completion. B-24 MB is then airborne from Windsor Field for Radar Exercise Number 41, aircraft landed after exercise completion. B-25 JD is airborne for Boca Chica, landed at base later at night. B-24 LH is airborne from Windsor Field for Radar Exercise Number 1, aircraft landed after exercise completion. An RAF B-24 bomber is believed lost, due to the pilot following the Leigh Light into the sea, with eight crew lost. NAAF Great Exuma: A plane arrived from Banana River then returned.

NOTE: *On this day ship escorts from the air ended.*

Part VII

Postscript:
May 16, 1945 to May 31, 1946

16th May 1945
B-25's CX, FH, FJ, FT, FM, and FY are airborne for air Exercise Number 3 and Number 4. decent through cloud (DTC), bombing on fixed and moving targets with some interval of re-bombing, aircraft landed after exercise completion. B-25's FC, FB, F , FT, and DC are airborne for Anti-Submarine Patrol Number 1, including six details of navigation and signals, aircraft landed after exercise completion. NAAF Great Exuma: An Army PBY flew in from South Caicos Island, then returned, and a plane arrived from Banana River, then returned.

17th May 1945
B-25's from Oakes Field and B-24's from Windsor Field carried out local flying training. B-25's FH, FB, FT, FV, and FC are airborne for air Exercise Number 3 and Exercise Number 4, bombing and descent through cloud (DTC) on fixed and moving targets carried out, aircraft landed after exercise completion. B-24 LH airborne from Windsor Field for Radar Exercise Number 1, aircraft landed after exercise completion. NAAF Great Exuma: Army PBY in from Morrison Field, a plane arrived from Banana River, then returned.

18th May 1945
B-25's from Oakes Field and B-24's from Windsor Field carried out local flying training. B-25's FV, FJ, FH, FB, FO, and FS are airborne for night Patrol Number 1 including six details of navigation with astrological and signals, aircraft landed after exercise completion. B-25's FC, FJ, FB, FD, FV, and FS are airborne for air Exercise Number 7 of compass swinging and photography; aircraft landed after exercise completion. B-25's HJ, BD, and VC are airborne for air Exercise Number 8 of bombing and descent through cloud (DTC). Then B-25 FB landed early, with engine trouble. The other aircraft landed after exercise completion. B-24 LU is airborne for Radar Exercise Number 23, then aircraft landed after exercise completion. NAAF Great Exuma: An Army PBY flew out to Borinquen, another plane took off for Guantanamo, and a plane arrived from Banana River ,then returned.

19th May 1945
B-25's FH, FJ, FD, FO, and FV are airborne for air Exercise Number 6 of homing procedure and photography, aircraft landed after exercise completion. B-25's FH, FJ, FV, FC, FT, and FS are airborne for air Exercise Number 8, consisting of bombing and decent through cloud (DTC), then the aircraft landed after exercise completion. B-24 LH is airborne for Radar Exercise Number 2, aircraft landed after exercise completion. B-24's airborne for

air Exercise Number 23, were NO and NJ, which are on Route A. B-24's MF and NH are on Route B. B-25 JX is airborne from Washington, DC via Jacksonville, and landed back at base. NAAF Great Exuma: A plane flew in from Guantanamo, then returned, after that a plane arrived from Nassau and then returned, and a plane arrived from Banana River then returned.

20th May 1945 – roughly

NAAF Great Exuma: A plane in from Guantanamo then departed to Nassau, a plane arrived from Banana River then returned. RAF serviceman Harold Woodman married Bahamian LaGloria Maria Pyfrom in Nassau; one of several such matrimonial unions.

21st May 1945

B-25's from Oakes Field and B-24's from Windsor Field carried out local flying training. B-25's FF, FV, FG, FC, and DC are airborne for Anti-Submarine Patrol Number 2, with six details of navigation and signals, aircraft landed after exercise completion. B-25's FT, FH, FJ, FG, FD, and DC are airborne for air Exercise Number 2 of navigation with astrological, photography, descent through cloud (DTC) flown in two sections, aircraft landed after exercise completion. B-24's are airborne and flew sorties from Windsor Field for Exercise Number 23. B-24's NH and NO are on Route A and B-24 NJ is on Route B, following which the aircraft landed on exercise completion. NAAF Great Exuma: A plane arrived from Banana River, then returned.

22nd May 1945

B-25's from Oakes Field and B-24's from Windsor Field carried out local flying training. B-25 FH, FG, FJ, FD, CX, and DC are airborne, and carried out Anti-Submarine Patrol Number 2 with six details of navigation and signals, aircraft landed after exercise completion. B-25's FM, FS, FT, FH, FJ, and FD also became airborne for air Exercise Number 2, including navigation, photography, decent through cloud (DTC), aircraft landed after exercise completion. B-24 LH is airborne for Radar Exercise Number 2, and aircraft landed after exercise completion. Grumman Goose GB is airborne for San Salvador, where the aircraft landed and returned after exercise completion. NAAF Great Exuma: A plane arrived from Banana River, then returned.

23rd May 1945

B-25's from Oakes Field and B-24's from Windsor Field carried out local flying training. B-25's FC, FT, FG, FK, FS, and FJ are airborne for Anti-Submarine Patrol Number 1, aircraft landed after exercise completion. B-25's FC, FT, FS, CX, and DC are airborne for air Exercise Number 6, including homing procedure and photography, aircraft landed after exercise completion. B-24 LH is airborne for Radar Exercise Number 2, aircraft landed after exercise completion. Queensley Exercise Number 8, constituting a passing out of [or graduation from] course exercise, was carried out. Ships not available, MF/DF Direction Finding station at San Salvador was used. B-24 LY is airborne from Windsor Field, homing procedure commenced, station sighted, recognition performed, attacked with one depth charge carried out, attack photographed, reported to base, carried out patrols Cobra and Python and vertical speed practice [VSP], set course for base, aircraft landed after exercise completion. B-24 MF is airborne from Windsor Field, homing procedure commenced, station sighted, recognition performed, patrol Adder performed, attacked with one depth charge carried out, attack photographed, reported to base, carried out patrols Cobra and vertical speed practice, set course for base aircraft, landed after exercise completion. B-24 NO is airborne from Windsor Field, homing procedure commenced, station sighted, recognition performed, patrol Mamba performed, attacked with one depth charge carried out, undershot, attack photographed, reported to base, carried out vertical speed practice, set course for base, aircraft landed after exercise completion. B-24 NH is airborne from Windsor Field, homing procedure commenced, station sighted, recognition performed, patrol Adder performed, attacked with one depth charge carried out, attack photographed, reported to base after vertical speed practice, set course for base, aircraft landed after exercise completion. NAAF Great Exuma: A plane arrived from Banana River then returned.

24th May 1945

W/C Eric H. Coleman, AFC became the colony's first Director of Civil Aviation; he had been Officer, Commanding RAF No.111 OTU for over a year. B-25's from Oakes Field and B-24's from Windsor Field carried out local flying training. B-25's FJ, FD, FS, FC, FD, and FK are airborne for Anti-Submarine Patrol Number 1, with six details of navigation and signals, aircraft landed after exercise completion. B-24 LH is airborne for Radar Exercise Number 2, aircraft landed after exercise completion. B-24 MD is airborne from Windsor Field for Queensley Exercise Number 8, homing procedure commenced, station sighted, recognition performed, patrol Adder performed, attacked with one depth charge, undershot, attack photographed, reported to base, carried vertical speed practice, set course for base, aircraft landed after exercise completion. B-24 LY is airborne from Windsor Field for Queensley Exercise Number 8, homing procedure commenced, station sighted, recognition performed, patrol Python performed, attacked with one depth charge, overshot, attack photographed, reported to base, carried vertical speed practice, set course for base, aircraft landed after exercise completion. B-24 NC is airborne from Windsor Field for Queensley Exercise Number 8, homing procedure commenced, station sighted, recognition performed, patrol Adder performed, attacked with one depth charge, overshot, attack photographed, reported to base, carried vertical speed practice, set course for base, aircraft landed after exercise completion. B-24 NF is airborne from Windsor Field for Queensley Exercise Number 8, homing procedure commenced, station sighted, recognition performed, attacked with one depth charge, bomb door of one jammed and other undershot, attack photographed, reported to base, carried vertical speed practice, set course for base weather report sent to base, aircraft landed after exercise completion. NAAF Great Exuma: A plane arrived from Banana River then returned.

25th May 1945

B-25's from Oakes Field and B-24's from Windsor Field carried out local flying training. B-25's FH, FM, FD, FJ, FG, and FC are airborne for night Patrol Number 2 with six details of navigation with astrological and signals practice, aircraft landed after exercise completion. B-25's CX, FN, FO, FD, FG, and FX are airborne for air Exercise Number 7 of compass swing and photography, aircraft landed after exercise completion. B-24 NH is airborne from Windsor Field for Queensley Exercise Number 8, homing procedure commenced, station sighted, recognition performed, patrol Adder performed, attacked with one depth charge, overshot, attack photographed, reported to base, carried vertical speed practice, set course for base, aircraft landed after exercise completion. B-24 NO is airborne from Windsor Field for Queensley Exercise Number 8, homing procedure commenced, station sighted, recognition performed, patrol Mamba performed, attacked with one depth charge, direct hit, attack photographed, reported to base, carried vertical speed practice, set course for base, aircraft landed after exercise completion. B-24 MF is airborne from Windsor Field for Queensley Exercise Number 8, homing procedure commenced, station sighted, recognition performed, patrol Adder performed, attacked with one depth charge, direct hit, attack photographed, reported to base, carried vertical speed practice, set course for base, aircraft landed after exercise completion. B-24 MD is airborne from Windsor Field for Queensley Exercise Number 8, homing procedure commenced, station sighted, recognition performed, patrol Python performed, attacked with one depth charge, overshot, attack photographed, reported to base, carried vertical speed practice, set course for base, aircraft landed after exercise completion. NAAF Great Exuma: A plane arrived from Banana River then returned. A plane arrived from Guantanamo then returned. F/L Wilfred E. "Tubby" Welch, who had led the Guard of Honour at the RAF Cemetery Nassau while posted there, joined BAL Bahamas Airways in 1951.

26th May 1945

B-25's from Oakes Field and B-24's from Windsor Field carried out local flying training. B-25's CX, FM, FV, FD, FC, and FG are airborne for local navigation Exercise Number 1 with two briefings, navigation and signals, aircraft landed after exercise completion. B-25's DC, FC, FD, FG, FV, and FT are airborne for air Exercise Number 6, did homing procedure and photography, aircraft landed after exercise completion. Grumman Goose GB is airborne for San Salvador, landed back at base later at night. NAAF Great Exuma: A plane arrived from Banana River then returned, and the US Navy gasoline barge USS *YOG-5* arrived from Guantanamo.

27th May 1945

B-25's from Oakes Field and B-24's from Windsor Field carried out local flying training. B-24's ND, NR, LU, LY, and MX are airborne from Windsor Field for Radar Exercise Number 41, aircraft landed after exercise completion. NAAF Great Exuma: A plane arrived from Banana River then returned, and the gasoline barge USS *YOG-5* sailed out to Guantanamo.

28th May 1945

B-25's from Oakes Field and B-24's from Windsor Field carried out local flying training. B-25's DC, FB, FC, FD, FS, and FV are airborne for air Exercise Number 5 of bombing and decent through cloud (DTC), aircraft landed after exercise completion. B-24 NF is airborne for Bermuda, carried out Exercise Number 48 of homing procedure, navigation, depth charge and gun attack, photography, signals landed there in the afternoon. NAAF Great Exuma: A plane arrived from Banana River then returned, and an Army PBY flies in from South Caicos Island.

29th May 1945

B-25's from Oakes Field and B-24's from Windsor Field carried out local flying training. B-25's FB, FC, FD, FS, FV, and DC are airborne for air Exercise Number 5 of bombing and decent through cloud (DTC), aircraft landed after exercise completion.

B-25 NF is airborne to return from Bermuda, went back to Bermuda due to engine trouble. NAAF Great Exuma: A plane arrived from Banana River then returned. A plane arrived from Guantanamo then returned. An Army PBY flew out to South Caicos Island.

30th May 1945

B-25's from Oakes Field and B-24's from Windsor Field carried out local flying training. B-25's TN and CX are airborne for air Exercise Number 1 of familiarization, decent through cloud (DTC), and bombing, aircraft landed after exercise completion. B-24 LH is airborne from Windsor Field for Radar Exercise Number 1, aircraft landed after exercise completion. NAAF Great Exuma: A plane arrived from Banana River, returned next day. A plane *en route* from Guantanamo to Banana River landed with a defective radio, and then departed to its original destination.

31st May 1945

B-25's from Oakes Field and B-24's from Windsor Field carried out local flying training. B-25's FB, FS, FD, and FN are airborne for night Patrol Number 1 with four details of navigation with astrological and signals, aircraft landed after exercise completion. B-25's FB, FS, FN, and FY are airborne for air Exercise Number 1 of familiarization, decent through cloud (DTC), and bombing, aircraft landed after exercise completion. B-25 JX is airborne for Guantanamo, landed there late evening. Grumman Goose GB is airborne for San Salvador, landed back for base late at night. B-24's LU, MB, and NL are airborne for Exercise Number 48 of final rehearsal for Kingsley and Esco exercises, aircraft landed after exercise completion. B-24 LU proceeded to Bermuda, then landed in Bermuda. NAAF Great Exuma: A plane arrived from Banana River then returned.

1st June 1945

Oakes Field reclaimed by civilians, including Bahamas Airways, Civil Aviation. Bahamas Airways, or BAL, was bailed out once more by the colonial government with subsidies tied to performance and scheduling deliverables. B-25's from Oakes Field and B-24's from Windsor Field carried out local flying training. B-25's FN, FT, FG, and FB are airborne for night Patrol Number 1 with four details of navigation with astrological, signals, and patrol, aircraft landed after exercise completion. B-25's FO, FN, FG, and FY are airborne for air Exercise Number 3 and Number 4 of decent through cloud (DTC), bombing on fixed and moving targets, and signals, aircraft landed after exercise completion. B-24's MD and MX are airborne from Windsor Field on 2nd days schedule of Exercise Number 48 of homing, navigation, signals, photography, and depth charge attacks of 5-hour duration, aircraft then landed after exercise completion. B-25's CX, FY, FV, and DC are airborne on Exercise Number 8, aircraft landed after exercise completion. B-24 LH is airborne from Windsor Field for Radar Exercise Number 1, aircraft landed after exercise completion. B-24's NL and NG are airborne and carried out LORAN exercise on Section X, following which the aircraft landed after exercise completion. B-25 JX returns from Guantanamo, landing at base late at night.

2nd June 1945

B-25's from Oakes Field and B-24's from Windsor Field carried out local flying training. B-25's FS, FY, FN, and FC are airborne for air Exercise Number 3 and Number 4 of descent through cloud (DTC) and bombing on fixed and moving targets, aircraft landed after exercise completion. B-25's FO, FY, FG, and FC are airborne for air Exercise Number 8 of instructional check. Then, on completion of Number 2 Squadron period, aircraft landed. B-24 LH is airborne from Windsor Field for Radar Exercise Number 1, then aircraft landed after exercise completion. B-24's NP and NG are airborne and carried out LORAN exercise on Section X, then the aircraft landed after exercise completion. B-25's KJ-796, KJ-797, KJ-798, KJ-799, KP-308, KP-309, KP-310, and KP-311, flown by US crew, are airborne to return to Orlando, Florida. B-24 NS is airborne, and then landed at Bermuda.

3rd June 1945

B-24's airborne from Windsor Field for Exercise Number 23, B-24's LZ and LU are on Route A, while B-24's NP, LR and NG are on Route B, following which the aircraft landed on exercise completion. B-24 LH is airborne from Windsor Field for Radar Exercise Number 2, aircraft landed after exercise completion. B-24 NS is airborne from Bermuda, and landed at base late at night. B-25 FS is airborne for Palisadoes, Jamaica via South Caicos Island, and landed back at base at night. B-24 MB is airborne from Windsor Field for local Radar exercise, however it crashed into the sea 10 nm northeast of New Providence. There were no survivors of B-24 MB, and all eight trainees perished. B-24's LZ and NO are airborne and carried out a search of the area 10 nm northeast of New Providence. There was no located wreckage, and the ASR marine craft, or launches from Montagu Bay were directed to the crash scene, following which aircraft B-24's LZ and NO returned to base. Since the ASR marine craft located floating wreckage yet were unable to tow the main portion of wreck toward Nassau Harbour, a US Navy escort vessel arriving on scene later requested permission to sink it.

4th June 1945

B-25's from Oakes Field and B-24's from Windsor Field carried out local flying training. B-25's FQ, FG, FN, and FV are airborne for Anti-Submarine Patrol Number 2 with four details of navigation and signals practice, aircraft landed after exercise completion. B-25's FC, FB, FN, and FV are airborne for air Exercise Number 2 of navigation and photography, aircraft landed after exercise completion. B-24 LD made two transit flights to Miami, then landed back at base later at night.

5th June 1945

B-25's from Oakes Field and B-24's from Windsor Field carried out local flying training. B-25's FB, FN, FM, and DC are airborne for Anti-Submarine Patrol Number 1 with four details of navigation and signal practice, aircraft landed after exercise completion. B-25's DC, FD, FB, and FN are airborne for air Exercise Number 2 of navigation and photography, aircraft landed after exercise completion. B-25 FC airborne for Jamaica via South Caicos, landed back at base later at night. Grumman Goose GB is airborne for San Salvador, then landed back at Oakes Field.

6th June 1945

B-25's from Oakes Field and B-24's from Windsor Field carried out local flying training. B-25's FB, FN, FM, and DC are airborne for Anti-Submarine Patrol Number 1 with four details of navigation and signal practice, aircraft landed after exercise completion. B-24 NO is airborne for LORAN Exercise Number 7, aircraft landed after exercise completion. B-25 JB is airborne for Boca Chica, Key West, and landed back at base later in evening.

7th June 1945

B-25's from Oakes Field and B-24's from Windsor Field carried out local flying training. B-25's DC, FG, FM, and FY are airborne for Anti-Submarine Patrol

Number 1 with four details of navigation and signal practice, aircraft landed after exercise completion. A B-25 is airborne for Boca Chica, and landed back at base at night. A B-24 returned from Bermuda, and landed at Windsor Field at night. B-24 LD took two transit flights to Miami, and then returned to base at night. Esco Exercise were carried out in conjunction with seaplane tender USS *Sandpiper* (AVP-9), allowing for homing exercise included recognition, vertical speed navigation, photography, depth charge and gun attack, and signals practice by all crew, with a two-day briefing provided. B-24 NJ is airborne from Windsor Field, a homing procedure commenced, station sighted, recognition performed, patrols carried out, attacked with depth charge and guns, attack photographed, reported to base, also reported of weather and shipping logging, set course for base, aircraft landed after exercise completion. B-24 LR is airborne from Windsor Field, a homing procedure commenced, station sighted, recognition performed, Mamba patrol carried out, attacked with depth charge and guns, which it undershot, attack photographed, reported to base, carried a vertical speed drill with seaplane, patrol Crocodile carried out, set course for base, aircraft landed after exercise completion. B-24 LU is airborne from Windsor Field, homing procedure commenced, station sighted, recognition performed, Mamba patrol carried out, attacked with depth charge and guns, attack photographed, reported to base, carried vertical speed with seaplane, patrol Cobra carried out set course for base, weather report sent to base, aircraft landed after exercise completion. B-24's NS and NG are airborne for LORAN exercise on Section Y, then aircraft landed on exercise completion.

8th June 1945

B-25's from Oakes Field and B-24's from Windsor Field carried out local flying training. B-25's FG, FM, FN, and FY are airborne for local navigation Exercise Number 1 of night navigation with astrological and signals, aircraft landed after exercise completion. B-25's FC, FY, FV, and FD are airborne for air Exercise Number 7 of compass swing and photography, aircraft landed after exercise completion. B-24 LR is airborne from Windsor Field, homing procedure commenced, station sighted, recognition performed, Crocodile patrol carried out, attacked coded position *Ploughshare* with depth charge and guns, two didn't detonate and third overshot, attack photographed, reported to base, carried vertical speed with seaplane, set course for base, aircraft landed after exercise completion. B-24 LX is airborne from Windsor Field, homing procedure commenced, station sighted, recognition performed, Cobra patrol carried out, attacked with depth charge and guns, plane overshot, attack photographed, reported to base, carried vertical speed with seaplane, carried out patrol Crocodile, set course for base, aircraft landed after exercise completion. B-24 NJ is airborne from Windsor Field, homing procedure were commenced, station sighted, recognition performed, Crocodile patrol carried out, attacked with depth charge and guns, attack photographed, reported to base, carried vertical speed with seaplane, set course for base, aircraft landed after exercise completion. Grumman Goose GB is airborne for San Salvador, then it landed back at base that night.

9th June 1945

B-25's from Oakes Field and B-24's from Windsor Field carried out local flying training. B-25's DC, FG, FM, and FT are airborne for air Exercise Number 7 of compass swing and photography, aircraft landed after exercise completion. B-25 JX is airborne for Boca Chica via Miami, landed at destination, returned to base later same day.

10th June 1945

B-24's LR, LY, MX, LU, NC, and MD are airborne from Windsor Field for Exercise Number 41, aircraft landed after exercise completion. B-25 FK is airborne for Guantanamo and then landed there.

11th June 1945

B-25's from Oakes Field and B-24's from Windsor Field carried out local flying training. B-25's DC, CX, FT, FO, and FV are airborne for air Exercise Number 1 of familiarization and bombing on moving target, aircraft landed after exercise completion. B-25's FV, FM, FY, and FT are airborne for air Exercise Number 5 of bombing and decent through cloud (DTC), aircraft landed after exercise completion. B-24 LH is airborne from Windsor Field for Radar Exercise Number 1, aircraft landed after exercise completion. B-24 LD embarks on two flights to Miami. Another B-24 is airborne, and landed at Corpus Christi. B-25 FK returned from Guantanamo, and landed at base in the afternoon.

12th June 1945

B-25's from Oakes Field and B-24's from Windsor Field carried out local flying training. B-25's FM, FY, FD, and DC are airborne for air Exercise Number 5 of bombing and decent through cloud (DTC), aircraft landed after exercise completion. B-25's MD, KY, and TN are airborne for air Exercise Number 1 of familiarization and bombing on moving targets, aircraft landed after exercise completion. Grumman Goose GB is airborne for San Salvador via Rock Sound, Eleuthera, and landed back at base later the same day.

13th June 1945

B-25's from Oakes Field and B-24's from Windsor Field carried out local flying training. B-25's FJ, FK, FM, FD, FY, and FT are airborne for air Exercise Number 3 and Number 4 of descent through cloud (DTC) and bombing on fixed and moving targets with interval of re-bombing, aircraft landed after exercise completion. Then the crew of C. Flight, 3N Squadron carried out Exercise Number 48. B-24 LZ is airborne for weather report and made attack on smoke float with photography and amplifying report to base, unsuccessful with homing "B," landed back at base. B-24 NH is airborne for weather report, made attack on smoke float, the plane overshoot, sent photography and amplifying report to base, unsuccessful with homing "B," landed back at base. B-24 LZ is airborne and gave an amplifying report to base, successful with homing "B," landed back at base. B-24 NL is airborne for weather report, made attack on smoke float, gave photography and amplifying report to base, landed back at base. B-24 LH is airborne from Windsor Field for Radar Exercise Number 1, aircraft landed after exercise completion. B-24 NT is airborne for LORAN Exercise on Section X, then the aircraft landed on exercise completion.

14th June 1945

B-25's from Oakes Field and B-24's from Windsor Field carried out local flying training. B-25's FM, FY, FD, and FC are airborne for night Patrol Number 1, with four details of navigation with astrological and signals, aircraft landed after exercise completion. B-25's FV, FK, FT, FH, and FJ are airborne for air Exercise Number 3 and Number 4 of descent through cloud (DTC) and bombing on fixed and moving targets, aircraft landed after exercise completion. B-24's airborne, 2nd day briefing of Exercise Number 48. B-24 MF is airborne twice, and B-24's LR and NH were airborne once, then all aircraft landed after exercise completion. B-25 JK is airborne for Guantanamo, and landed back at base later at night.

15th June 1945

B-25's from Oakes Field and B-24's from Windsor Field carried out local flying training. B-25's FV, FY, FK, FH, and FC are airborne for air Exercise Number 7, consisting of compass swing and photography, aircraft landed after exercise completion. B-25's FK, FV, FT, and FC are airborne for air Exercise Number 8 of bombing and decent through cloud (DTC), aircraft landed after exercise completion. B-24 NX returns from Corpus Christi and lands at base on 16th [early following morning]. B-24 LH is airborne from Windsor Field for Radar Exercise Number 1, aircraft landed after exercise completion. B-24 NC is airborne with Leigh Light for Exercise Number 2, aircraft landed after exercise completion. B-24 Liberators MK, GR, and KK-289 VIII of 59 Squadron and No.111 OTU are said to have crashed.

16th June 1945

B-25's from Oakes Field and B-24's from Windsor Field carried out local flying training. B-25's FY, FC, FV, FT, and FH are airborne for air Exercise Number 7 of compass swinging and photography, then the aircraft landed after exercise completion. B-25's FK, FB, FC, and FV are airborne for air Exercise Number 8 of bombing and decent through cloud (DTC), aircraft landed after exercise completion. B-24 LH is airborne from Windsor Field for Radar Exercise Number 2, aircraft landed after exercise completion. B-24's NO, NC, and MX are airborne from Windsor Field on a Leigh Light Exercise, and thereupon aircraft landed after exercise completion.

17th June 1945

B-24 LH is airborne from Windsor Field for Radar Exercise Number 2, then aircraft landed after exercise completion. Sunday, and no training flights.

18th June 1945

B-25's FB, FD, FT, and FC are airborne for Anti-Submarine Patrol Number 2 with four details of navigation and signals practice, aircraft landed after exercise completion. B-25's FN, FJ, FB, FS, and FV are airborne for air Exercise Number 2 of navigation with astrological, signals, and photography, aircraft landed after exercise completion. B-24 MF is airborne for Bermuda, and carried a briefing on Exercise Number 48 of navigation, depth charge and guns, photography, and homing B procedure, landed at Bermuda, landed back at base on 19th [following day].

19th June 1945

B-25's from Oakes Field and B-24's from Windsor Field carried out local flying training. B-25's FB, FV, FK, and FT are airborne for Anti-Submarine Patrol Number 2 with four details of navigation and signals practice, aircraft landed after exercise completion. B-25's FJ, FH, FK, FT, and FC are airborne for air Exercise Number 2 of navigation and photography, aircraft landed after exercise completion. B-24 ND is airborne for Corpus Christi, Texas, and landed back at base on 20th.

20th June 1945

B-25's from Oakes Field and B-24's from Windsor Field carried out local flying training. B-25's FC, FT, FH, FS, and FD are airborne for Anti-Submarine Patrol Number 2 with five details of navigation and signals practice, aircraft landed after exercise completion. B-24's NJ, LX, and NL are airborne for navigational Exercise Number 50, arranged for final passing out of students of C. Flight 3 Squadron, arranged in place of Kingsley, Queensley and Esco exercises. They carried out navigation with aid, depth charge and gun attack on smoke float, photography, signals procedure and homing "B," and all three crew carried out exercise successfully. B-25's FH, FS, FO, and FD are airborne for modified night navigation patrol, adverse weather, aircraft landed after exercise completion. B-24 NO is airborne for LORAN exercise on Section Y, aircraft landed after exercise completion. USN NAS Exuma, is dis-established, after three years continuous utilization.

21st June 1945

B-25's from Oakes Field and B-24's from Windsor Field carried out local flying training. B-25's FH, FS, FD, FT, and FC are airborne for Anti-Submarine Patrol Number 2 with five details of navigation and signals practice, aircraft landed after exercise completion. B-24's NJ, MF, NX, and MX are airborne for air Exercise Number 50. B-25's FK, FM, FY, FT, and FO are airborne for local navigation Exercise Number 1, aircraft landed after exercise completion. Grumman Goose GA is airborne for Washington via Morrison Field and Charleston, and landed back to same day.

22nd June 1945

B-25's from Oakes Field and B-24's from Windsor Field carried out local flying training. B-25's FK, FH, FM, FT, and FO are airborne for air Exercise Number 6 of homing procedure and photography, aircraft landed after exercise completion.

23rd June 1945

B-25's from Oakes Field and B-24's from Windsor Field carried out local flying training. B-25's FT, FS, FB, FM, and FN are airborne for air Exercise Number 6 of homing procedure, aircraft landed after exercise completion.

24th June 1945

B-24's MF, LW, LZ, MD, LR, MX, ND, and NX are airborne from Windsor Field for Exercise Number 41 aircraft landed after exercise completion. B-25 FK is airborne for Guantanamo, Cuba landed returned to base same day at night.

25th June 1945

B-25's from Oakes Field and B-24's from Windsor Field carried out local flying training. B-25's FM, FB, FV, FS, FT, and FD are airborne for air Exercise Number 5 of bombing and decent through cloud (DTC), recalled before completion due to bad weather. B-24's NX, MF, and LZ are airborne for Radar Exercise Number 41, aircraft landed after exercise completion. B-24 LH is airborne for Radar Exercise Number 1, aircraft landed after exercise completion. Grumman Goose GA is airborne from Washington DC, and landed at base.

26th June 1945

B-25's from Oakes Field and B-24's from Windsor Field carried out local flying training. B-25's FV, FH, FB, FM, and FD are airborne for night Patrol Number 1 with five details of navigation with astrological and signals, aircraft landed after exercise completion. B-25's CX, FB, FH, FM, and FT are airborne for air Exercise Number 5 of bombing and decent through cloud (DTC), aircraft landed after exercise completion. B-24 LH is airborne for Radar Exercise Number 1, aircraft landed after exercise completion. B-25 FO and Grumman Goose GB are airborne to search for a C-47 of USAAF, which was last reported in the area of New Providence and Eleuthera near Ship Channel Cay. The Harbour Island and Montagu Bay ASR marine craft launch HMS

P-191 also took part, however nil report by all, unable to verify.

27th June 1945

B-25's from Oakes Field and B-24's from Windsor Field carried out local flying training. B-25's FD, FB, FC, FS, are FM are airborne for air Exercise Number 8 of bombing and decent through cloud (DTC), aircraft landed after exercise completion. B-25's FS, FH, FY, FB, and FD are airborne for night Patrol Number 1 with five details of navigation with astrological and signals, aircraft landed after exercise completion. B-24's NH, LR, NC, and NF are airborne, 1st day briefing of Exercise Number 48, navigation exercise with practice in bombing, depth charge and guns flown by C. Flight, 3 Squadron. Then aircraft carried out rehearsal for final, depth charge, however the DC did not detonate, and after being given rest, all completed exercise, and landed after exercise completion. B-24 LH is airborne for Radar Exercise Number 1, aircraft landed after exercise completion. B-25 JX is airborne for Guantanamo, landed in destination, then landed back at base at night.

28th June 1945

B-25's from Oakes Field and B-24's from Windsor Field carried out local flying training. B-24's LR, MD, and NH are airborne, for 1st day briefing of Exercise Number 48, Radar of NH is unserviceable, so it was given rest, and all completed exercise, and landed after exercise completion. B-25's GM, BC, and JK are airborne for air Exercise Number 8 of bombing and decent through cloud (DTC), aircraft landed after exercise completion. B-24 LH is airborne for Radar Exercise Number 1, following which aircraft landed after exercise completion.

29th June 1945

B-25's from Oakes Field and B-24's from Windsor Field carried out local flying training. B-25's FC, FV, FM, FO, and FY are airborne for Anti-Submarine Patrol Number 1 with five details of navigation and signals practice, aircraft landed after exercise completion. B-24 LH is airborne for Radar Exercise Number 1, and then aircraft landed after exercise completion.

30th June 1945

B-25's from Oakes Field and B-24's from Windsor Field carried out local flying training. B-24 NH is airborne, completed exercise of Number 48, which it had not completed on 28th [two days previously]. B-25's FM, FO, FS, FV, and FY are airborne for Anti-Submarine Patrol Number 1, with five details of navigation and signals practice. Then the aircraft landed after exercise completion. B-24 LH is airborne for Radar Exercise Number 2, following which the aircraft landed after exercise completion.

2nd July 1945

B-25's from Oakes Field and B-24's from Windsor Field carried out local flying training.

3rd July 1945

B-25's from Oakes Field and B-24's from Windsor Field carried out local flying training. B-24's airborne from Windsor Field included B-24's NL and NG which were airborne on Exercise Number 23B and B-25's NH, NO and MD, which were airborne on Exercise Number 23B, then the aircraft landed after exercise completion. B-24's NF, NO, and LZ are airborne, on first day briefing of Exercise Number 50. For this exercise replaced Kingsley and Esco, ships were not available, homing procedure "B," navigation, signals, depth charge and gun attacks, photography carried out, aircraft landed after exercise completion. B-24 LH is airborne for Radar

Exercise Number 2, then aircraft landed after exercise completion. B-24 LR is airborne for Bermuda, where it landed at destination in evening.

4th July 1945

B-25's from Oakes Field and B-24's from Windsor Field carried out local flying training. B-24's NK, NX, NC, and NH are airborne, on the second day briefing of Exercise Number 50, this exercise replaced Kingsley and Esco, ships not available, homing procedure B, navigation, signals, depth charge and gun attacks, photography carried out. Then B-24 NK returned with ASV unserviceable, put on rest, the aircraft landed after exercise completion. B-24 LR is airborne from Bermuda, and landed back at base. B-24 LO is airborne to search for B-24 MD, carrying out Exercise Number 47, to the northeast of Andros. There it located wreckage however all crew of B-24 MD were lost. B-24 LO and ATC directed ASR marine craft launch HMS *P-191*, then returned to base. The ASR launch, picked up a dinghy and saw portions of a fuselage and tall plane in the water off Andros.

5th July 1945

B-25's from Oakes Field and B-24's from Windsor Field carried out local flying training. B-24 NK is airborne, carried out sortie of Exercise Number 50, however it hadn't been able to complete exercise on 4th [the day before]. The aircraft landed after exercise completion. B-25 JD is airborne for Guantanamo.

6th July 1945

B-25's from Oakes Field and B-24's from Windsor Field carried out local flying training. B-25 JD is airborne from Guantanamo, and landed back at base.

7th July 1945

B-25's from Oakes Field and B-24's from Windsor Field carried out local flying training. Grumman Goose GA is airborne for Conch Sound, Andros [south of Nicholls Town], and returned to base in the afternoon.

8th July 1945

B-24's MF, LR, LZ, LW, NC, and NO are airborne from Windsor Field for Exercise Number 41, with Leigh Light. All the aircraft landed after exercise completion.

9th July 1945

B-25's from Oakes Field and B-24's from Windsor Field carried out local flying training. B-24's NO, LZ, LR, MF, and LW are airborne from Windsor Field for Exercise Number 42, Leigh Light exercise, aircraft landed after exercise completion. B-24 NJ is airborne for Exercise Number 30 of homing, aircraft landed after exercise completion. B-24's NK, NF, and NL are airborne for Exercise Number 25 of bombing, aircraft landed after exercise completion. B-24's LB and LC are airborne for fighter affiliation exercise, aircraft landed after exercise completion. B-24 LH is airborne for Radar Exercise Number 1, aircraft landed after exercise completion. B-25 FC is airborne for Guantanamo, where it landed at destination, then returned to base at night.

10th July 1945

B-25's from Oakes Field and B-24's from Windsor Field carried out local flying training. B-24's NL, NX, NF, LW, NO, and NK are airborne from Windsor Field for Exercise Number 42, Leigh Light exercise, aircraft landed after exercise completion. B-24's NG, NJ, and LW are airborne for Exercise Number 25 of bombing, aircraft landed after exercise completion. B-24 LC is airborne for sortie of US Navy fighter-affiliation exercises, then the aircraft landed after exercise completion. B-24 LH is airborne for Radar Exercise Number 1, aircraft landed after exercise completion.

US Navy at NAS Miami requested RAF assistance to search for a PBM Mainer which is overdue while on navigational exercise routed from Banana River over the Bahamas to Great Exuma. Joint forces of aircraft searched the main area, leaving the Bahamian area to the B-25's FC, FH, JM, JC, and JG, and the Grumman Goose GA. B-24 LZ is also airborne, however with nil report from all, and the aircraft landed back at base. B-24 NJ is airborne on Exercise Number 25, B-24 MF airborne on Exercise Number 26, and B-24 MX is airborne on Exercise Number 30, bombing, aircraft landed after exercise completion. Search for the missing USN PBM continued by B-25's FN, JA, FC, JG, and JD, however still with nil reports, and they returned to base. The PBM Mariner missing on night patrol over the Bahamas had a crew of 12.

11th July 1945

B-25's from Oakes Field and B-24's from Windsor Field carried out local flying training. B-24's NF and NC are airborne for Exercise Number 48 of rehearsal for final course navigation, signals practice, homing, photography, depth charge and gunnery, aircraft landed after exercise completion. B-24's NO, LZ, LR, NK, and LW are airborne from Windsor Field for Exercise Number 44, Leigh Light exercise, aircraft landed after exercise completion. B-24 NL is airborne for fighter affiliation exercise, and the aircraft landed after exercise completion. B-24 LH is airborne for Radar Exercise Number 1, aircraft landed after exercise completion.

12th July 1945

B-25's from Oakes Field and B-24's from Windsor Field carried out local flying training. B-24's LZ, LR, NF and NC are airborne for Exercise Number 48 of rehearsal for final course, aircraft landed after exercise completion. B-24's airborne for Leigh Light exercise, B-24 LW is airborne for Exercise Number 43, B-24's MF and WN are airborne for Exercise Number 44, B-24's MF, LR, and NO are airborne for Exercise Number 45, then the aircraft landed after exercise completion. B-24 NL is airborne for Exercise Number 34, fighter affiliation exercise, aircraft landed after exercise completion. B-24 LH is airborne for Radar Exercise Number 2, aircraft landed after exercise completion. B-24's NJ and MX are airborne for Exercise Number 31, B-24 LZ is airborne for Exercise Number 25, and B-24 NL is airborne for Exercise Number 26, following which the aircraft landed after exercise completion. RAF aircraft concluded their search for the US Navy PBM Mariner. NAS Miami then sent to Nassau its own aircraft, PBY-*48424*, PBY-*02963*, and PBY-*48291*, to continue the search.

13th July 1945

B-25's from Oakes Field and B-24's from Windsor Field carried out local flying training. B-24's MF, MZ, LD, and LW are airborne for Leigh Light exercise, aircraft landed after exercise completion. B-24 LH is airborne for Radar Exercise Number 2, aircraft landed after exercise completion. B-24 NK is airborne twice for bombing exercise and Exercise Number 26 aircraft, landed after exercise completion. B-24 NL is airborne for Exercise Number 34 of fighter affiliation, then the aircraft landed after exercise completion. PBY-*48424*, PBY-*02963*, and PBY-*48291* continued their search for their 12 comrades, with nil result.

14th July 1945

B-25's from Oakes Field and B-24's from Windsor Field carried out local flying training. B-24's NO and NC are airborne with Leigh Light for Exercise Number 46, aircraft landed after exercise completion. B-24's airborne for bombing exercise, B-24's NK-29, NJ-27, ND-30, NK, NJ-26, NO, and ND-31 then landed after exercise completion. B-24 LH is airborne for Radar Exercise Number 2, aircraft landed after exercise completion. PBY-*48424*, PBY-*02963*, and PBY-*48291* continue their ASR search for colleagues, with nil result.

15th July 1945

B-25's from Oakes Field and B-24's from Windsor Field carried out local flying training. B-24's airborne, NJ, NK, and MF which are on Exercise Number 23a. B-24's LZ and MF are on exercise Number 23B of navigation exercise, following which aircraft landed on exercise completion. US Navy PBY-*48424*, PBY-*02963*, and PBY-*48291* continued searches, with nil result, and returned to Nassau. Headquarters No.113 (South Atlantic) Wing, No.7 Flying Unit and No.7 APU disbanded officially, with effect this date.

16th July 1945

B-25's from Oakes Field and B-24's from Windsor Field carried out local flying training. B-24's NC and MX are airborne with Leigh Light for Exercise Number 47, aircraft landed after exercise completion. B-24 NF is airborne for Exercise Number 50 of navigation, final course exercise in place of Kingsley, Esco, returned to base without completing exercise due to adverse weather. B-24's airborne for bombing include B-24's ND and NK, participating in Exercise Number 27 and B-24's NC and NF-32, and the aircraft landed after exercise completion. The US Navy PBY-*48424*, PBY-*02963*, PBY-*48291* concluded searches and returned to their base at Banana River. No.113: Rudder of a South Atlantic Transport Command Liberator FL-995 was damaged by an unknown ground vehicle while parked on tarmac prior to the aircraft leaving for Dorval on 17 July, 1945 with VIP's, compounding weather delays.

17th July 1945

B-24's from Windsor Field carried out local flying training. B-24's LZ and NJ are airborne for Exercise Number 50 of navigation, then completed full schedule for 2nd day, and the aircraft landed after exercise completion. B-24's on bombing exercises include ND and NX-27, NX-29, NX-32, aircraft landed after exercise completion. B-24 LH is airborne for Radar Exercise Number 2, aircraft landed after exercise completion.

18th July 1945

B-24's airborne from Windsor Field for Leigh Light exercise are NO, MX, LW, and LC, are airborne for Exercise Number 47, and MX and NO are airborne on Exercise Number 49, aircraft landed after exercise completion. B-24's MF and LZ airborne for Exercise Number 50, both did full schedule, then B-24 LZ diverted to participate in an ASR search. B-24's airborne for bombing are NX, NO, and ND, for Exercise Number 29 and B-24's NO, NX, and ND are airborne for Number 27, aircraft landed after exercise completion. B-24 LH is airborne for Radar Exercise Number 2, aircraft landed after exercise completion. Grumman Goose GA airborne for Great Exuma, landed at destination in afternoon. B-25 FC is airborne to co-operate with aircraft and ships of the US Navy in its search for a US Liberator was overdue while on exercise to the north and east of Eleuthera. B-24 LZ is airborne, then all returned with nil report.

19th July 1945

B-24's from Windsor Field carried out local flying training. B-24's MX, MF, and NJ are airborne for Exercise Number 50 of navigation, proceeded to ASR search area after completing exercises. B-24 NX is airborne for Exercise Number 49, Leigh Light exercise, aircraft landed after exercise completion. B-24's airborne for bombing include ND and B for Exercise Number 32. B-24's NX, MD, and MX are airborne for Exercise Number 29, MX Number 27, aircraft landed after exercise completion. B-25 FH and B-24 MX, while on Exercise Number 50 cooperated with USN search for US Liberator based at Nassau, negative result.

20th July 1945

B-24's from Windsor Field carried out local flying training. B-24 NL is airborne on Exercise Number 50 of navigation to complete what had not been done on 16th, aircraft landed after exercise completion. Grumman Goose GA is airborne to Great Exuma and returned to base late evening. The only USN aircraft based in Nassau carried on search for US Liberator, with negative result.

21st July 1945

B-24's from Windsor Field carried out local flying training. USN aircraft based in Nassau carried on search for US Liberator, with negative result.

22nd July 1945

B-24's MF, MX, and ND are airborne with Leigh Light for Exercise Number 41, aircraft landed after exercise completion. USN aircraft based in Nassau carried on search for US Liberator, with continued negative results.

23rd July 1945

B-24's from Windsor Field carried out local flying training. B-24's were also airborne with Leigh Light, NC, LR, and LW are airborne for Number 42, and B-24's NC, LW, MF, and LR are airborne for Number 43; then the aircraft landed after exercise completion. B-24's MF and NJ are airborne for Number 25, NJ and NL for Number 30, then the aircraft landed after exercise completion.

24th July 1945

B-24's NX and NC are airborne from Windsor Field, for the 1st day briefing on Exercise Number 48 of four-day navigation exercise as practice for final passing out homing, signal, photography, depth charge and gun attack. B-24's airborne with Leigh Light include NC, LR, and LW for Exercise Number 42, and B-24's MF, LR, MC, and LY for Exercise Number 44, then the aircraft landed after exercise completion. B-24's airborne, for bombing practice were NJ and NL on Exercise Number 30 and NF on Exercise Number 25.

25th July 1945

B-24's NX an NC are airborne from Windsor Field on 2nd day briefing on Exercise Number 48, aircraft landed after exercise completion. B-24's airborne with Leigh Light for homing include NS, and NP, on Exercise Number 45 and B-24's MC and LR on Exercise Number 44, then the aircraft landed after exercise completion. B-24's airborne for bombing were ND and NL on Exercise Number 25, and NL and NF on Exercise Number 30, then the aircraft landed after exercise completion.

26th July 1945

B-24's NX and NC were airborne from Windsor Field on the 3rd day of briefing on Exercise Number 48, aircraft landed after exercise completion. B-24's ND and MX were airborne with Leigh Light for homing Exercise Number 45, aircraft landed after exercise completion. B-24's airborne for practise bombing were NF and LU on Exercise Number 26, and B-24 NL on Exercise Number 27, with aircraft landing after the exercise completion. USS *PT-163* arrives at the fast motor torpedo boat PT Shakedown Base in Miami, from New York via Jacksonville.

27th July 1945
B-24 NX is airborne from Windsor Field for one sortie of Exercise Number 48, aircraft landed after exercise completion. B-24's airborne for bombing were B-24 NF on Exercise Number 27, B-24s NF, NL, and MX on Exercise Number 26, NL on Exercise Number 30, then the aircraft landed after exercise completion. B-25 FC is airborne to South Caicos Island, where it landed at its destination in the afternoon.

28th July 1945
B-24's airborne from Windsor Field for practice bombing included NL, on Exercise Number 27, NF on Exercise Number 29, NL and NF on Exercise Number 26, and NF on Exercise Number 30, following which aircraft landed after exercise completion. No.113: The Governor-Designate of the Bahamas, Mr. W. L. Murphy, and party arrived by air from Bermuda. Swearing-in ceremony in the Legislative Council Chamber [Murphy replaced Windsor].

29th July 1945
B-24's NC, NX, and MX are airborne from Windsor Field with Leigh Light for homing on Exercise Number 46, aircraft landed after exercise completion. B-25 FC is airborne, returned from South Caicos Island and landed at base.

30th July 1945
B-24's NC, NX, and MX are airborne from Windsor Field with Leigh Light for homing, aircraft landed after exercise completion. B-24's NL, NF, and MX are airborne for Exercise Number 23 of Route A and B navigation, then the aircraft landed after exercise completion. B-24 NR is airborne for one sortie of Exercise Number 48, then aircraft landed after exercise completion.

31st July 1945
B-24's NL, NX, and LW are airborne from Windsor Field for Exercise Number 23 of Route A and B navigation, aircraft landed after exercise completion. B-24's LL, NT, LU, and NP are airborne for Exercise Number 50 of navigation, aircraft landed after exercise completion. B-24's NL and NX are then airborne for practice bombing, and the aircraft landed after exercise completion.

1st August 1945
B-24's LY, NR, MC, and NO are airborne from Windsor Field for Exercise Number 50, aircraft landed after exercise completion. B-24's are airborne for bombing, NC and NL are airborne on Exercise Number 32 and Number 27, then B-24's NF and NL are airborne on Exercise Number 31, aircraft landed after exercise completion. USS *PT-603* is bound west in Straits off Miami for night-time training off Ocean Cay, Bahamas and Straits.

2nd August 1945
B-24 MX is airborne for Exercise Number 48 of homing and Number 49 with Leigh Light, aircraft landed after exercise completion. B-24's LG, LT, LB, and LD are airborne for radius action flights undertaken for the purpose of testing loads and instructing the crew in Transport Command procedure in preparation for forthcoming transit flight to the United Kingdom.

3rd August 1945
B-24 NC is airborne with Leigh Light for homing Exercise Number 47, and the aircraft landed after exercise completion. B-24's ND, LR, NP, and LZ are airborne for radius action flights carried out in preparation for forthcoming transit flight to the UK, following which the aircraft landed after exercise completion.

4th August 1945
B-24 LV is airborne for Turks Island in order to better observe the weather, since there has been a preliminary hurricane warning in operation since 2nd August [two

days]. Grumman Goose GA is airborne to Rock Sound, Eleuthera, then returns to base early evening.

5th August 1945

B-24's are airborne from Windsor Field for Radar exercise; B-24's LW, NC, and MX on Exercise Number 41, and NC on Exercise Number 42, after which aircraft landed, the exercise complete. At the request of meteorological officer B-24 NG is airborne to investigate weather in hurricane area, and landed at base after duty over. No.113: His Excellency the Governor of the Bahamas, Mr. W. L. Murphy visited the Commanding Officer of No.88 TSP South Atlantic, in Nassau.

6th August 1945

B-24's are airborne from Windsor Field for homing, with LW, MF, and MX airborne for Exercise Number 42, and B-24's NK, LW, and MX airborne for Exercise Number 43, then the aircraft landed after exercise completion.

7th August 1945

B-24's airborne with Leigh Light for homing, LW and MF on Exercise Number 43, and MX and LW on Exercise Number 42, then the aircraft landed after exercise completion. B-24's airborne for navigation exercise included NC and NK, which carried Exercise Number 48. These are done in rehearsal for the final course exercises including navigation, signals, homing, depth charge and gun attack, aircraft landed after exercise completion.

8th August 1945

B-24's airborne for the final return flight from the Bahamas to the UK via Bermuda, Gander and the Azores are: B-24's LL, NT, NY, LU, and NP. The aircraft landed after exercise completion. B-24's are airborne for Leigh Light exercises, including tail numbers LW and NL on Exercise Number 43, LW on Exercise Number 44, then NL and LW on Exercise Number 45, with the aircraft landing after exercise completion. B-24's MF, NX, and NK are airborne for navigation Exercise Number 48, then the aircraft landed after exercise completion.

9th August 1945

B-24's airborne from Windsor Field for UK via Bermuda, Gender and Azores. LY, NR, NS, MC, and NO aircraft landed after exercise completion. B-24's airborne for Leigh Light are NL on Exercise Number 43, and LW on Exercise Number 44, then the aircraft landed after exercise completion.

10th August 1945

B-24's are airborne from Windsor Field for the UK via Bermuda, Gender and Azores are LY, LT, and LB, and these aircraft landed after exercise completion. B-25 FH is airborne for Miami, and returned to base in afternoon.

11th August 1945

B-24's airborne for UK via Bermuda, Gender and Azores: ND, LR, NH, NG, and LZ; aircraft landed after exercise completion. B-24's airborne, Leigh Light by MX and NO on Exercise Number 44, and B-24 MX on Exercise Number 45, the aircraft landing after exercise completion. LX is airborne for Miami, and returned to base later. No.113: 45 Group Staff Party is under G/Capt. Bonham-Carter, G/Capt. W.A. Lee, G/Capt. B. D. Nicholas, W/Cdr. H. Myles, W/Cdr. E. S. McQuillan, W/Cdr. H. Peterson, F/Lt. W. McKay and G/Capt. Irving, S/O Woolston. Then Mr. Dodd and Mr. Williams, Air Ministry, establishment Committee, arrived from Dorval to discuss taking over of RAF station, Nassau, by RAF Transport Command.

12th August 1945

B-24's airborne for Leigh Light exercises are MX and NO, and then aircraft landed after exercise completion. B-24's LL, NT, NY, LU, and NP, which were airborne from

Nassau for the RAF base in Prestwick, UK on 8th, landed there today, four days later. No.113: Air Chief Marshal, Sir A. T. Harris, KCB, OBK, AKC [aka Bomber Harris] arrived from Rio de Janeiro in one of two Bomber Command Lancasters, and was entertained at Government House, Nassau.

13th August 1945

B-24's NO and MX are airborne from Windsor Field for navigation, the aircraft landing on completion.

14th August 1945

VJ-Day [Victory over Japan] is declared. B-24's LY, NS, and NO, which were airborne from Nassau to the UK on 9th arrived at Prestwick, in the UK today, five days later. B-24 LG which left Nassau to the UK via Bermuda, Gander, and the Azores arrived at Prestwick.

15th August 1945

B-24 NL is airborne from Windsor Field, and goes on one sortie of Exercise Number 50 of navigation. B-24's MX, NC, and NF are airborne on 2nd day after briefing of Exercise Number 50, aircraft landed after exercise completion. B-24's NR and MC are airborne and left Nassau for UK on 9th, then B-24's LV and LB took off and left on 10th, later arriving at Prestwick, in the UK. No.113: A general holiday was declared for all personnel except on essential duty, with effect midday. The Commanding Officer and Staff Officer of this Unit gave a picnic for all ground personnel at Lyford Cay [presumably the ASR Marine base].

16th August 1945

No.113: General holiday for all personnel except those on essential duty to celebrate the end of World War II.

19th August 1945

No.113: A V-J Day Thanksgiving service was held at Nassau Cathedral with His Excellency.

23th August 1945

No.113: The final two B-24 Liberators of No.111 OTU left Nassau, meaning that a total of 41 were sent to the UK, mostly via Bermuda and other nodes.

3rd September 1945

No.113: W/Cdr. Coleman, Commanding Officer, No.88 TSP, Transport Staging Post, attended a conference with His Excellency the Governor, Mr. Murphy and the Officer in Charge of the Pictou Highlanders, to decide future policy for Oakes Field, aka Main Field.

2nd September 1945

The signing of final surrender documents between the Allies and Japan aboard USS *Missouri* in Tokyo Bay, officially ending World War II. Overall, three percent of the entire global population, civilian and military, were killed in the conflagration.

13th September 1945

A hurricane hits Miami, **Ron-41 Division Baker**, USS *PT-605*, USS *PT-606*, USS *PT-607*, and USS *PT-608*, shelter in Miami River. At No.113 TC, Hurricane alerts escalate from White to the more severe Blue warnings. Preparations are made to evacuate the Nassau bases.

14th September 1945

Hurricane Red, highest warning level, issued. Airmen and permanent staff officers of No.113 were evacuated to the safety of the Royal Victoria Hotel downtown Nassau on receipt of Red Warning. Aircraft were dispatched to the US mainland. The Bahamas Air Service Squadron was disbanded, providing a "core of trained Bahamian aero mechanics."

15th September 1945

No.113: All Clear received, personnel returned to their unit. Slight superficial damage experienced.

19th September 1945

No.113: Group Superintending Engineer, Dorval inspects Oakes Field offices for 88 TSP.

21st September 1945

USS *PT-603* commanded by USN Lt. Roger M. Jones [later Nassau resident and consul] is bound south in Straits, to Havana.

24th September 1945

No.113: Commanding Officer and his section commanders met at Oakes Field for allocation of offices for Traffic, Equipment and Engineering in the old Operations Block.

25th September 1945

USS *PT-603* is bound north in Straits, Havana to Miami.

28th September 1945

USS *PT-603* is bound south in Straits past Key West, exiting the area. No.113: No.88 Terminal Staging Post moved from Windsor Field to Oakes Field.

5th October 1945

No.113: Six Peruvian Officers arrived, entertained at British Government Expense. Capt. Garlick, RASC HQs, Jamaica, arrived to discuss changing from Lend-Lease rations to RASC European Troops' on the North Caribbean scale.

18th October 1945

USN Privateer Aircraft missing over Bahamas while on routine patrol.

5th December 1945

Port Everglades Training Squadron Flight FT-19, from which all five avenger torpedo bombers with 14 US Airmen, go missing in northeast Bahamas *en route* to decoy bomb on the *Sapona* wreck off Great Isaac Light in Northwest Providence Channel.

11th November 1945

No.113: Ceremonial Station Parade at full strength was held at the RAF Cemetery in honour of Commemoration Day. The service by Rev. F. V. Burns, RAF Chaplain and Very Rev. Dean of Nassau. His Excellency the Governor of the Bahamas and the Commanding Officer laid wreaths on the Memorial Stone after which two minutes silence was observed; the *Last Post* and *Reveille* were sounded by the bugler, whilst the RAF Guard of Honora performed Present Arms and the Lord Bishop of Nassau gave the blessing. The Cemetery was open to the public for this Service. Brigadier Jefferson, Commanding General North Caribbean Forces, arrived Nassau. Discussed personnel and the Caribbean Shuttle.

17th November 1945

No.113: Informal inspection by His Excellency the Governor and Mrs. Murphy, accompanied by the Commanding Officer, of the Cemetery, Hospital and Camp, including Laundry and Bakery.

24th November 1945

No.113: Dinner Party was given at the Prince George Hotel in honour of six Uruguayan Officers, the British Military Attaché at Montevideo and a Government Official, who were passing through Nassau from Montevideo to the UK, as guests of the British Government.

6th December 1945

US Navy Martin Mariner flying boat Training 49 with 13 persons, is sent with Training Squadron 32 to find US Navy Squadron FT-19 is also lost off northeast of Bahamas. No.113: Air search made by Dakota KN-673 under F/L Welch for five missing US Navy aircraft in Bahamas region.

7th December 1945

No.113: No.80 Staging Post arrived with football team to challenge RAF Nassau.

8th December 1945

No.113: Football match between No.88 Staging Post and No.80 Staging Post took place at eastern Parade, Nassau, Result; 88 SP scored 5 goals to 80 SP's 1 goal.

20th December 1945
No.113: Annual Transport Command Christmas Dance held at the Jungle Club, Nassau. S/L K. C. Browne, Commanding Officer and party attended.

25th December 1945
No.113: Christmas Dinner was served in the Airmen's Mess when the usual RAF Custom of Officers and NCO's waiting on Airmen, was observed. The Airmen seemed well pleased with the arrangements, and a good time was had by all.

1st January 1946
No.113: No.88 Staging Post undertook the responsibility of the guarding of 400,000 gallons fuel tank, and the area and buildings at Clifton Pier, at the west end of New Providence Island. A medical inspection was carried out by the Medical Officer of the unit and it was certified that renovations were necessary for RAF personnel. By arrangement with the local Garrison Commander, coloured troops of the local detachment continued to guard the area for a further period until the Post Engineers cleaned and renovated the building, in readiness for RAF guards.

2nd January 1946
No.113: F/Lt. Lockhart-Mummery reported No.88 SP to take over the duties of Senior Medical Officer (SMO) from S/Ldr. Rassin.

5th January 1946
No.113: Commanding Officer's Parade and March Past held on opening of new parade recently constructed outside the Headquarters Building.

8th January 1946
No.113: By request of the Garrison Commander and the Colonial Secretary, two Air Sea Rescue [ASR] craft were employed in evacuating re-patriates and released service personnel from the *Island* [mis-written *Hyland*] *Monarch*. The boat anchored outside the bar of Nassau Harbor, and an officers' reception party proceeded to welcome the repatriates.

9th January 1946
No.113: Air Officer Commanding made a tour of the RAF Station Hospital and Clifton Pier.

11th January 1946
No.113: No.45 Group rents part of Clerihew House [formerly Belmont Bahamas] for the transfer of ASR personnel from Fort Montagu Hotel. Work commenced on renovating the accommodations. A fire broke out at 10.35 pm in Hut 201 and was quickly under control after the alarm had been sounded. Considerable damage done to building, stationery, duplicating machines.

16th January 1946
No.113: Court of Enquiry in the cause of the fire in Hut 201 commenced.

17th January 1946
No.113: Last Special Flight aircraft departed Nassau, with passengers for repatriation.

18th January 1946
No.113: Director of Administration and Finance from RAFDEL arrived to discuss policy, Oakes Field, and a number of problems connected with No.111 OTU.

19th January 1946
No.113: ASR Officers took over Clerihew House as Air Sea Rescue Headquarters for the duty crew only. The remainder were transferred to barrack blocks on the camp. The moorings for the ASR vessels remain in Montagu Bay, pending the loan of the port dredger for their transfer to sites approved by the Port Captain, opposite Clerihew House. ASR anchored.

21st January 1946

No.113: Emergency call was received from the Colonial Secretary's Office and a request was made for an aircraft to proceed with a doctor, to Pine Ridge Lumber Camp on Grand Bahama where the boiler of a railway engine burst, killing four and injuring four persons. Grumman Goose FP-472 was dispatched. Captain was F/Lt. Lester Brown, Co-Pilot, Nay, F/Lt. Miles, and S/Ldr. Lockhart-Mummery, the SMO accompanied them. The four injured persons were evacuated and transported to Oakes Field for medical treatment. Plasma was administered to one of the injured, who had lost an arm as a result of the explosion, but he died on the airfield.

22nd January 1946

No.113: The last aircraft to land at Miami on the routine shuttle departed Nassau.

1st February 1946

No.113: At the request of the Colonial Secretary, the Grumman Goose FP-472 flew to Cherokee Sound, Great Abaco, in order to transfer to hospital a serious ambulance case.

2nd February 1946

No.113: At the Colonial Secretary's request, the Grumman Goose FP-472 flew to Mangrove Cay, Andros, in order to bring back a serious ambulance case.

10th February 1946

No.113: As a result of a sailing mishap, Leading Aircraftman E. J. Beeston, Cook, was drowned.

11th February 1946

No.113: Funeral for LAC E. J. Beeston, took place at the RAF Cemetery, with full military honours.

26th February 1946

No.113: Widgeon aircraft JS-996 was flown to Miami and returned to US Authorities.

8th March 1946

No.113: US War Department visited, instructed that all Post Engineers' activities cease forthwith, and that US Corps of Engineers detachment was to return by 31.3.46 [in three weeks].

9th March 1946

No.113: Vital services, US Post Engineers Department taken on by RAF civilian payroll.

11th March 1946

No.113: A Sunderland flying boat of the Dodero Airlines, Argentina, arrived at Nassau with officials from airline and South American Government. Aircraft marshalled, refuelled by RAF.

16th March 1946

No.113: No.113: Captain Rogers, British Army Staff movements Officer, arrived from Miami on board *Jean Brillant* on liaison visit.

22nd March 1946

No.113: F/Lt. E. H. Lang, Accountant Officer departed for Harbour Island to investigate, at the request of RAF Delegation, Washington, a claim for compensation submitted by the owner of The Little Boarding House in which ASR personnel were billeted 1943 to 1945.

2nd April 1946

No.113: On authority of Air Ministry four civilian guards took over guard duties at Clifton Pier from the RAF Police.

3rd April 1946

No.113: H. E. Earl of Athlone [Governor-General of Canada and South Africa] and HRH Princess Alice arrived from Trinidad on B-24 Liberator AL-574. The Commanding Officer and Staff attended an official cocktail party on board HMS *Birmingham*. A Station dance was given in honour of Petty Officers and Ship's Company of HMS *Birmingham*. F/L A. E. Miles (GDP) delivered Grumman Goose FP-470 to US Naval Authorities, Opa-Locka, Florida, then Dorval, Canada.

4th April 1946

No.113: Staff Officers return in a Lancastrian VM-727 after their tour of No.89 and No.95 Staging Posts.

6th April 1946

No.113: Lancastrian VM-727 took His Excellency the Governor and Miss E. Murphy along with other prominent citizens of Nassau, for tour of Bahama Islands.

11th April 1946

No.113: No.45 Wing staff Officers departed Nassau for Dorval in the Lancastrian VM-727.

15th April 1946

No.113: His Excellency the Governor, Miss E. Murphy, AVM R. L. G. Marix, CB, DSO, AVM D. V. Carnegie, AFC, A/Cdr. D. U. Gordon, CBE, AFC, G/C D. W. F. Bonham-Carter, DFC, G/C D. N. Roberts, OBE, AFC, S/Ldr. T. Primrose and Mr. G. M. Eaton departed Nassau for Bermuda and Dorval in KK-191, which replaced Liberator AM-262.

16th April 1946

No.113: Liberator AM-262 departed Nassau for Dorval via Bermuda.

24th April 1946

No.113: Grumman Goose FP-472 delivered to US Naval Authorities Opa-Locka, Florida.

29th April 1946

No.113: Final Trinidad Special flight departed Nassau to evacuate remaining personnel of No.89 and No.95 Staging Posts.

30th April 1946

No.113: S/Ldr. L. C. Giles (GD'P') relinquished command of No.1318 Communication Flight on obtaining local release in Nassau. F/L W. E. Welch (GDP) assumed command of No.1318 Communication Flight.

1st May 1946

Philip Farrington, back from the RAF, joined civilian passenger airline Bahamas Airways [BAL, 1936-1970] in Nassau.

8th May 1946

No.113: The Commanding Officers' parade, to commemorate VE-Day [the year before] was held on the new parade ground outside the Headquarters Building. The Rev. Dean of Nassau Cathedral addressed the parade.

9th May 1946

No.113: Officer's Mess, Oakes Field closed down and Sergeant's Mess henceforth to be used as an Officer's and Senior Non-Commissioned Officer's (SNCO's) Club.

10th May 1946

No.113: Many pilots delivered the base's remaining Dakotas to Headquarters of No.45 Wing and remained at Dorval for disposal.

18th May 1946

No.113: A C-47 aircraft of Universal Airlines made an emergency landing at Oakes Field at 11.22 pm.

22nd May 1946

No.113: F/L D. Lester Brown, DFC (GD'P') departed this unit on a posting to Headquarters No.45 Wing *en route* to United Kingdom for Class 'A' Release [He remained an ADC to the Duke of Windsor for some time, and returned to Nassau where he flew his seaplane].

23rd May 1946

No.113: KN-673 departed for Bermuda to replace KK-198.

24th May 24, 1946

No.113: RAF Hospital, Nassau, closed down. Dr. Pearce, Chief Medical Officer, Bahamas General Hospital, agreed to undertake medical care of remaining RAF personnel.

29th May 1946

No.113: S/L H. E. Lockhart-Mummery (Med) proceeded to Dorval for disposal.

30th May 1946

No.113: G/C D. W. F. Bonham-Carter, DFC, OinC No.45 Wing, and F/L Morris, Wing Signals Officer, left Nassau for Dorval, via Bermuda.

31st May 1946

No.113: The care and maintenance of Windsor Field, aka Satellite Field, is handed over to the Bahamas Government. Personnel formerly employed by Post Engineer are transferred to Bahamas Government payroll. Signed: Wing Commander, Commanding, Royal Air Force Station, Nassau, Bahamas.

****

Photographs

Virtually all images are now in the public domain through a combination of having been taken by government entities (military photographers and artists) over 75 years ago. Every effort has been made to provide proper credit and to obtain permission for their. For more full resources on how to find the image's holder of license or rights, see Bibliography Section, Websites, and Notes on Sources Section as well

Cover Photo Sources:

Front, top (aircraft in line-up): masterbombercraig.wordpress.com/raf-volunteer-reserve-40-41/111-otu-nassau-42-3, image purchased from The Australian War Memorial, Canberra, Australia, in author's collection.

Front, centre (stamp): B-24.weebly.com.

Front, bottom (RAF & Seafarer's Club): vintagebahamas.com.

Back, bottom (funeral in RAF Cemetery): Sources: fcafa.com/2017/04/18/Czechoslovak-raf-airmen-in-the-caribbean, prints purchased from The Australian War Memorial, Canberra, Australia, in author's collection.

Back, top (U-boat men interrogating men in lifeboat): Kelshall, Gaylord T. M., *U-Boot-Krieg im Der Karibik*, Verlag E. S. Mittler & Sohn, Germany, 1999, which in turn is from Kelshall, Gaylord T. M., *U-Boat War in the Caribbean*, Naval Institute Press, US, 1988 & 1994.

Photographs & Captions in Main Text

Virtually all images are now in the public domain through a combination of having been taken by government entities (military photographers and artists) over 75 years ago. Every effort has been made to provide proper credit and to obtain permission for their. For more full resources on how to find the image's holder of license or rights, see Bibliography Section, Websites, and Notes on Sources Section as well

Photo 1: Consolidated B-24 Liberator GR Mk V's & VIII's of No.111 OTU lined up on the tarmac on the RAF No. 111 Operational Training Unit, Windsor Field and Oakes Field, Nassau, Bahamas, 1943-1945.
Source: masterbombercraig.wordpress.com/raf-volunteer-reserve-40-41/111-otu-nassau-42-3, image purchased from The Australian War Memorial, Canberra, Australia, in author's collection.

Photo 2: A B-24 Liberator bomber outside the maintenance hangar at Windsor Field, RAF No.111 O.T.U., Nassau. Source: The Imperial War Museum, London, UK.

Photo 3: Two B-24 Liberator bombers with RAF No.111 OTU on the tarmac at Windsor Field, Nassau, Bahamas. The forward aircraft, LO is B.III BZ-762 and MA is GR.V BZ-810, which is logged herein as taking at least three flights on 25 January, 1944 alone.
Sources: masterbombercraig.wordpress.com/raf-volunteer-reserve-40-41/111-otu-nassau-42-3, B-24.weebly.com, fcafa.com/2017/04/18/Czechoslovak-raf-airmen-in-the-caribbean, print purchased from The Australian War Memorial, Canberra, Australia, in author's collection.

Photo 4: A commemorative stamp from 1995 for the Bahamas, showing a Liberator presumably crewed by "Bahamians [who] served in the R.A.F.," in this case likely outside the Bahamas.
Source: B-24.weebly.com

Photo 5: The RAF and Seafarer's Club, a general serviceman's canteen in Nassau. This controversial establishment met the needs of aviators, ground crew, mariners and any woman or man in the service for an alcoholic drink. The I.O.D.E. were strict teetotallers and served lemonade, and the men were strictly forbidden, and often courts-martialled, for going "over the hill" in Nassau to slake their thirst for drink, dance and so forth. The Duchess of Windsor Wallis Simpson cut through red tape to help make this canteen possible. As the head of the Red Cross, for her to overrule the I.O.D.E. and the voices of "concerned citizens" took a bit of Moxy.
Source: vintagebahamas.com

Photo 6: A B-24 Liberator bomber escorting a large convoy, one of whose ships, at bottom, appears to be emitting a rather lot of smoke.
Source: Public Domain, Quora.com, numerous sites.

Photo 7: The funeral for instructor Jan Hadyzek and other Czech pilots who perished in a crash after take-off south of Oakes Field. Held at what is known as the RAF and the Western Cemetery, Nassau. According to the Commonwealth War Graves Commission country report, "The cemetery is located off Farrington Road in the Oakes Field district of Nassau. There are 51 casualties from the Second World War buried here."
Sources: fcafa.com/2017/04/18/Czechoslovak-raf-airmen-in-the-caribbean, prints purchased from The Australian War Memorial, Canberra, Australia, in author's collection, and Users/ericw/Downloads/Country%20Report%20BAHAMAS.pdf.

Photo 8: The Canadian ferry *Jean Brillant* which made hundreds of Nassau-Miami voyages.
Source: Courtesy of Kevin Griffin, St. Lawrence Saga: the Clarke Steamship Story, clarkesteamship.wordpress.com/tag/m-v-jean-brillant.

Photo 9: The redoubtable Bahamian passenger ferry *Ena K.,* owned by the Kelly family who had a lumber yard on Nassau Harbor, Bay Street near Market Street. Named for Ena Kelly, as the *Betty K.* and *Deborah K.* were named for Kelly women, this vessel performed well over a thousand voyages on the Miami-Nassau route, before, during and after the war, and including Sir Sydney Poitier and the Duke and Duchess of Windsor.
Source: Bahamas National Archive Exhibition: *The Boat-Building Industry of the Bahamas*, p. 5, courtesy of Rev. D. Jennings and Alton Lowe. Lawlor, Anne & Jim; *The Harbour Island Story*.

Photo 10: Survivors of the US-flagged SS *Potlatch*, who had been at see a month when the photo was taken in early August, 1942 before the Symonette-owned Rozelda Hotel in Nassau. From left to right are Michael J. Kaplan (later Stark), Sawyer, Tiliacos (or Esfakis), Marion Carstairs (her niece Kim Aranha does not believe it is her), a mate named Sorensen, possibly Terese Straton, the Duchess of Windsor, another mate named Alan Holmes-Jackson, sailor John Moe, and Michael J. Carbotti, who returned to Nassau to inquire after persons who helped them. They stayed in other hotels; the Windsor among

them, and were flown back by amphibious US Navy aircraft to Miami for further recuperation. Straton sent Carbotti's family a list of crew who where patients in Nassau, and they shared it with the author. Kaplan was given a pair of the Duke's silk pyjamas.
Source: Associated Press, also Hugo Vickers, *Behind Closed Doors*, numerous other publications, including Capt. Arthur Moore's *A Careless Word, a Needless Sinking*.

Photo 11: The Duke of Windsor, formerly King Edward VIII, became the only royal to lead the Bahamas as a civilian Governor. Prime Minister Sir Winston Churchill insisted the couple leave Lisbon Portugal and it's intrigues, so they sailed via Bermuda and waited out their sentence on the rock, as they saw it. Here his wife the American Wallis Simpson, Duchess of Windsor, and he meet officers and men of the Norwegian tanker ship *Kollskegg*, which was sunk by a German U-Boat west of Bermuda. They were subsequently rescue by the US-flagged steamer SS *Bushranger* and brought to Nassau where the Windsors greeted them.
The Duke was fond of having a drink with the sailors at their hotel, and not only buying the round, but donating items including his pyjamas to them. The Duchess, head of the Red Cross, was kept very busy with administering to up to two ships a week which arrived, with nearly 300 survivors to tend and repatriate. She complained to her US-based Aunt Bessie about this. For escape and recreation they could and did utilize large yachts owned by friends, like *Rene* and *Southern Cross*, as well as their smaller power boat the *Gemini*, to go Miami or Palm Beach. Some trips were escorted by the US Coast Guard and US Navy and RAF from Nassau.
Source: Back caption reads: "Survivors greeted by royalty, Nassau." ACME, New York Bureau, April 10, 1942, from the author's collection.

Photo 12: David Parson, an assistant cook aboard SS Potlatch. He was 62 years of age, from the Norfolk, Virginia area, with no wife or children, only his spinster sister Nell, with whom he stayed in Harlem when ashore. Parson expired of exhaustion in the arms of Captain John Joseph "Jack" Lapoint as their small lifeboat was within sight of a lighthouse beam after 30 days seeking for same. He was buried at Hard Hill, Andersons Settlement, Acklins Island, Bahamas end of July, 1942, following a torchlight procession led mostly by locals as the American sailors were still too sick and exhausted to support them. **Source:** Author's collection, ancestry.com and findagrave.com; unable to locate his sister Nell.

Photo 13: Captioned "8 Flee Devil's Island," in French Guyana. Dated August 25, 1940, the Associated Press report that the man were former prisoners in the dreadful island camp, and that France "abandoned" it at the outbreak of war. They voyaged in a 30-foot sailing boat some 2,700 nautical miles via South America and the Caribbean – many of them transiting the Bahamas – to reach Miami, where these men were interned. Their

wish was to sail to the U.K. and General Charles de Gaulle to help him to reclaim France from the Germans.

These transitional voyagers are relevant to the period between the outbreak of war in September, 1939 and the actual substantive arrival of the vanguard of 112 German and Italian attack submarines in January, 1942. During that time German ships dodged UK blockades using US neutrality, as they had in World War I in Newport and off Cape Cod. Even before the outbreak of war a British "civilian" yacht named *Tai Mo Shan* left Hong Kong under the control of British naval officers ostensibly sailing home from a posting in Asia, yet in reality spying on Japanese-held islands in the Pacific. They ran aground on Acklins Island, Bahamas, but were not permitted to use the village's only shovel, as they needed it for burials.

Finally, a French aristocrat and yachtsman named Alfred de Marigny from Madagascar defied local authorities to help these hapless, but likely ruthless men and was ostracized for it in conservative, British-run Nassau. Several of the unfortunates were simply given water and food and cast off back to sea to fend for themselves. These activities were to come into further light and play during de Marigny's trial for the sensational wartime murder of the very wealthy and powerful American and Canadian gold magnate Sir Harry Oakes, whose murder and the presence of the Duke of Windsor cast attention on the Bahamas in thousands of news articles which, along with wartime air bases, would catapult the tourism business to the front of the Bahamian economy following the war.
Source: Associated Press Wirephoto, August 25, 1940.

Photo 14: A typical interrogation of a neutral or enemy merchant ship by a German U-boat in World War II on the high seas, North Atlantic, in summer and settled conditions. The Allied ship has been torpedoed or had explosive charges put in it and detonated to sink here, and the U-Boat commander has, typically, demanded an explanation of ship's name, departure port, destination, what cargo, and whether the officers and crew have any medical or navigational needs. Usually some cigarettes, blanket, water, bread and navigational device would be dispensed and the U-Boat would motor off on the surface on a "decoy direction," and call the success into head-quarters, or BdU., in France or

Germany. In the case of U-84, lost southwest of Bermuda, and many other U-Boats, radio chatter made them softer targets, easier to track by the Allies with superior code-breaking techniques to the Germans for large parts of the war.
Source: Kelshall, Gaylord T. M., *U-Boot-Krieg in Der Karibik*, Verlag E. S. Mittler & Sohn, Germany, 1999, which in turn is from Kelshall, Gaylord T. M., *U-Boat War in the Caribbean*, Naval Institute Press, US, 1988, 1994.

THE LIBERATOR

NEED FOR LIVE PAPER

NEWS FROM THE FRONTS

STATION PERSONALITIES

BROWNED OFF?

For Rent In Town

H. G. Christie

Photo 15: The very first edition of *The Liberator*, an "RAF Bahamian Weekly" reads the subtitle here. This unique four-page, 39-cm weekly newspaper created and published just for the officers and men and women at RAF No. 111 OTU, Nassau. Published in conjunction with the *Nassau Guardian*, some copies lie with the UK National Archives, the Bahamas National Archives have them on microfiche, and the Australian War Memorial has an almost complete set in Canberra, at awm.gov.au/collection/LIB100028207.
Source: U.K.'s The National Archives at Kew, London, England. Nationalarchives.gov.uk, the RAF No. 111 OTU diaries feature the first edition only. AIR 29 section; see Source Notes and Bibliography sections.

Photo 16: Survivors of another allied merchant ship sunk near the Windward Passage in World War II. Here they have been rescued and landed by an amphibious US Navy flying boat.
Source: National Archives and Records Administration (NARA) photographic records, College Park, MD, from the author's collection.

Photo 17: Windsor Field as seen from the air by a Royal Australian Air Force aviator in training at No.111 OTU during World War II. The pilot is looking southeast, towards southern New Providence and the vicinity of Coral Harbor and what is now Stuart Cove's diving operation. Originally known as Main Field, with Oakes Field being known as the Satellite Field, Windsor Field was constructed by a US engineering firm. After the war it became known as Nassau International Airport, and now the Lyndon Pindling International Airport.
Source: The Australian War Memorial, Canberra, Australia.

Photo 18: Bahamian mailboat *Stede Bonnet*, built as the Royal Navy's minesweeper HMS *MM 194* at the Symonette Shipyard on Hog Cay (Paradise Island) in 1942. Two were built in the Bahamas and neither fought for the British.
Source: Dodge, Steve: *Abaco, The History of an out Island and its Cays*, p.88, and Saunders, et al, *Islanders in the Stream,* p.318. Also Rumelier.com by Bob Davies, Oldbahamas.com, Dave Gale, and Craig Symonette helped.

Photo 19: The historic racing thoroughbred schooner *Morning Star* which at the outbreak of war was taking paying prep-school students on an extended voyage through the Bahamas. After finally returning to Miami in late spring, 1942, many of the students enrolled in the US Navy. Captain William Pond and his wife ran the program on a variety of schooners over the years, starting at Harvard, which he attended.
Source: Purchased from the archives of *The Baltimore Sun*, from the author's collection.

Photo 20: Nassau, New Providence, capital of the colony, in 1959, showing both the commercial waterfront and, in the background at left the re-purposed land which had been Oakes Field. One can see the close proximity to inhabited areas and why the Duke of Windsor is rumoured to have complained of the effects of many take-offs and landing from the field when it was run by No.111 OTU and No.113 Transport Wing, might have had on his and the duchess' sleep.
Source: Capt. Paul. C. Aranha, who kindly highlighted certain noteworthy buildings in town.

Photo 21: Students at Belmont School, Clerihew House, Nassau, probably in 1941 as it began to grow substantially from a nucleus of 16 pupils from the UK, 3 of them sisters.
Source: Courtesy of Ronald Lightbourn Jr., who attributes them to the superb photographic collection of Stanley Toogood.

Photo 22: B-25 Mitchell's of No. 111 OTU in flight. In the background is KJ-593 (B), a Mk III, and in the foreground FV-954 (FB), a Mk II.
Source: fcafa.com/2017/04/18/Czechoslovak-raf-airmen-in-the-caribbean

Photo 23: Czechoslovak airmen of No.111 OTU being inspected by the Duke of Windsor.
Source: fcafa.com/2017/04/18/Czechoslovak-raf-airmen-in-the-caribbean

Photo 24: A mixed group of trainees from Courses 30 and 31 leaving Liberator BZ-811 (MB) after a training sortie.
Source: fcafa.com/2017/04/18/Czechoslovak-raf-airmen-in-the-caribbean

Photo 25: Czech RAF officer Thomas Lom in the Bahamas. Tomáš Lom is one of the very few surviving Czechoslovaks who served in Britain's RAF during World War II. Born Tomáš Löwenstein into a Jewish family in Prague, he signed up in London the moment he turned 18 and ended up serving as a wireless operator in the Bahamas in the latter period of the conflict.
Source: Archive of Tomáš Lom, radio.cz/en/section/czech-history/i-was-very-keen-on-teaching-the-germans-a-lesson-says-tomas-lom-one-of-few-surviving-czech-raf-vets

Photo 26: Photo of RAF ground and air crew relaxed following a day of training at Windsor Field, New Providence, in 1945. Taken by Czech-British RAF aviator Tomáš Lom, who had recently been a biology student at the University of Glasgow.
Source: Archive of Tomáš Lom, memoryofnations.eu/en/lom-tomas-1924

Photo 27: Adam Heald, the young radio operator aboard the British tanker Athelqueen which was torpedoed, rammed, and sunk by the Italian submarine R.Smg. Enrico Tazzoli off Hope Town in March of 1942. He was so traumatized, despite his friendly reception in Abaco and Nassau, by the return voyages that he could never speak of them. To anyone.
Source: Alan Heald, from author's collection.

Photo 28: Ancil Pratt, former insurance salesman in Nassau and store owner south of Clarence Town, Long Island. He and several friends witnessed a German U-boat U-732 under Claus-Peter Carlsen taking its bearings from the two large churches at Clarence Town before heading for Europe.
Source: Ancil Pratt, from author's collection.

Photo 29: Young German U-boat commanders Günther Müller-Stöckheim with his colleague Carl Emmermann during an ocean rendezvous between U-67 and U-172 in mid-June 1942. This was on war patrol in the Caribbean Sea during Operation *Neuland* in early 1942, before air cover hardly existed; after the USN and RAF made 1,000 attacks in German U-boats the mood changed, with boats like U-134 which earlier had shot up an aircraft off Bermuda, being hounded non-stop, despite sinking a USN airship off Cay Sal Bank.
Source: Kelshall, *The U-Boat War in the Caribbean*, in English, and by permission of the German publisher, Verlag, of Kelshall's *U- Boot- Krieg in der Karibik 1942-1945*, also the U-Boot Museum, *(Deutsches U-Boot-Museum,* Cuxhaven-Altenbruch, Germany.

Photo 30: Crew from the British ship *Daytonian*, sunk by the Enrico Tazzoli off Hope Town and brought to Nassau aboard the *Content S.*, enjoy time ashore before being sent to Canada and the front lines of the Battle of the Atlantic, the longest continuously fought engagement of World War II. John Holmes and Henry Mappleback are among the men shown.

Source: Henry Mappleback's private collection, courtesy of his daughters, author's collection, now in the National Archives of the Bahamas and the LJM Maritime Academy Library.

Photo 31: John Holmes (far right) and Henry Mappleback (tallest), survivors of the Daytonian sinking with a host family either in Nassau or Hope Town.
Source: Henry Mappleback's private collection, courtesy of his daughters, author's collection, now in the National Archives of the Bahamas and the LJM Maritime Academy Library.

Photo 32: Rev. Clarisse Cox, Minister of the Methodist Church of Hard Hill, northeast Acklins. During an 2012 interview she recalls witnessing, along with another youngster now named Rev. Newton Williamson, 48 men washed ashore after 30 days from the sinking by U-boat of the SS *Potlatch*. Steward David Parson was buried in Andersons Hill and the emaciated crew were rescued first by Rev. Cox' father, the Rev. Capt. Collie on his schooner Go On, and then by speed boat racing heiress Marion (Joe) Carstairs, in the Bahamian-built speedboat *Vergemere IV*, from her island at Whale Cay, Berry Islands.

They were taken to Nassau and the story has spawned books, a BBC TV documentary, a script, and considerable interest given Ms. Carstairs' extraordinary life.
Source: Mari Anderson and Fitz Damler, who live on Crooked Island much of the time, at the behest of the author, spent an entire day interviewing and filming the Reverends and others.

Photo 33: Lt. Wenzel Granger was an officer of the Bahamas Defence Company, North Caribbean Force, which was formed in October of 1942, and later known as the Bahamas Battalion. Other officers included Lt. David Smith, Lt. William Pemberton, Lt. Augustus Roberts, Capt. Reney, and commanding officer Major J'Arcy Rutherford. Over 300 Bahamians trained for the battalion in Jamaica. In May of 1944 Lt. Granger commanded a contingent that saver service in Egypt and Italy.
Source: Commonwealth of the Bahamas Archives Exhibition: *The Bahamas During the World Wars, 1914-1918 and 1939-1945*. Post Office Building, 4-23 February, 1985. Department of Archives, Ministry of Education. Printed in 2004 by the Publications Section. Page 27, photo 1.

Photo 34: Entitled "Sections of an Army Camp," this shows the more humdrum aspects of maintaining a city of over 3,000 personnel. The Bahamas Defence Company and Volunteer Defence Force would have stood sentry over military and civilian sites alike. The caption suggests that Royal Army units, as well as Canadian army units were active in ground defence in the colony. The Royal Navy Town-Class Light Cruiser HMS *Birmingham* and many other naval ships transited Bahamas as well, with HMS *Birmingham*'s officers and crew both feted ashore in 1946. Source: Commonwealth of the Bahamas Archives Exhibition: *The Bahamas During the World Wars, 1914-1918 and 1939-1945*. Post Office Building, 4-23 February, 1985. Department of Archives, Ministry of Education. Printed in 2004 by the Publications Section. Page 27, photo 1, also Jane Lloyd, *The History of the I.O.D.E. (Bahamas) 1901-2005*, Media Enterprises, Nassau, 2016), p.86.

Photo 35: The RAF put on impressive parades, often marching from Windsor Field. And the crowds appreciated it, even turning out in the pouring rain to cheer some of them on. Some occasions, in remembrance of fallen comrades, were sombre, but many, like this on 7 May, 1945 for an RAF Victory March, were boisterous and happy events; VE-Day, and VJ-Day and the birthdays or royals, for example. This was on "the announcement of Germany's capitulation," and shows RAF troops marching down Bay Street in Nassau. Parades included Bahamian troops, and a designated senior official to receive the salute. Often bussing was provided for the participants back to base. During major holidays the

base was manned only by a skeleton staff, extra busses laid on to and from the beach, and days of reverie ensued.
Source: Commonwealth of the Bahamas Archives Exhibition: *The Bahamas During the World Wars, 1914-1918 and 1939-1945.* Post Office Building, 4-23 February, 1985. Department of Archives, Ministry of Education. Printed in 2004 by the Publications Section. Page 53, photo 47.

Photo 36: Part of a series of murals on panels or screens at the Bahamas Historical Society, these small yet very detailed paintings translate better in colour. This painting bears close resemblance to a photo of HRH the Duke of Windsor inspecting "A company of the Volunteer Defence Force 1943," so can be deemed accurate. In the background many RAF aircraft are being worked on and the RAF symbol is in the upper right corner. The caption recognizes the importance of the RAF to the economy as well as the war, as do National Archives exhibits. It ends with "...the defence of the Airport was largely entrusted to local troops." The ORBs or diaries of the bases substantiate the role played by local troops in base security, and although one building burned, the integrity and effectiveness of Bahamian troops was never called into doubt in the records.
Source: Painting by Diana Pullinger, text by Jim Lawlor, photography by Ronald G. Lightbourn in The History of the Bahamas in Pictures, The Bahamas Historical Society, Media Enterprises, Ltd. Nassau, 2012, p.42; Mrs. Pullinger was born in the UK in 1908 and passed in Canada in 1985 after painting a large mural for the Nassau International Airport, with other work in the Nassau Shop and St. Philip's in Inagua. The Lyford Cay Foundation assisted in the publication of the book. The photo (not shown), is on p.32, image 2 in : Commonwealth of the Bahamas Archives Exhibition: *The Bahamas During the World Wars, 1914-1918 and 1939-1945*, 2004 National Archives Publications Section.

Photo 37: "RAF dance during the Second World War." Large banners including a revered RAF Ensign hung – and hang – from the rafters of Christ Church Cathedral, Nassau to this day. The Erickson family, who reinvigorated the salt industry in Inagua, donated a book of remembrance for fallen RAF aviators from World War II in which names, ranks, places and dates of death are caringly inscribed in calligraphy. It lies in a glass case to the right as one enters the cathedral.
Source: Jane Lloyd, *The History of the I.O.D.E. (Bahamas) 1901-2005*, Media Enterprises, Nassau, 2016), p.83.

Photo 38: The Bahamian Battalion and Volunteer Defence Force protected this vital source of energy for bases and colony alike. During rationing and with U-boats threatening to sever supplies of tankers, the oil depot at Clifton Pier was critical. It was also situated between where many RAF aircraft crashed, between South Ocean and Simms Point in Lyford Cay. The original caption reads "The Bahamas Electricity Corporation's Clifton Pier Power Station on the southwestern shore of New Providence island is the nerve centre of the island's power supply." Here, presumably in the 1950's, a tanker owned by the Kulukundis family of Kasos Island, eastern Greece, discharges petroleum into the tank farm.

Source: *Bahama Islands, Report for the Years 1966-1967*, Government Information Services, Cabinet Office; photos taken by the Bahamas News Bureau photographers, Part II: Public Utilities and Public Works, p.64.

Photo 39: The Lych Gate memorial in which, after the war, descriptive bronze plaques were placed with the names of all persons, most of them RAF, killed in wartime service in or from the Bahamas. The thriving vines around the gate would have pleased the arborists and designers who started off by hewing an elegant cemetery from raw limestone, void of plant life. In its many decades the cemetery has had royal visits as well as struggled against neglect, with many organizations and individuals, from the Red Cross and IODE and Commonwealth War Graves Commission, RAF and family members and volunteers all pitching in towards resolutions. It is still a well-kept and peaceful island of tranquillity and reflection and, suitably, honour. Behind the Lych Gate can be seen the main memorial stone for fallen and missing aviators and others from all over the world; Poland, Czechoslovakia, the UK, Canada, US, Australia, New Zealand, and beyond, some of whom took off from Windsor or Oakes fields and perished in Belem, Khartoum, Georgetown, Accra or St. Vincent.

Source: Jane Lloyd, *The History of the I.O.D.E. (Bahamas) 1901-2005*, Media Enterprises, Nassau, 2016), p.183 (unfortunately the cross atop the gate was largely cropped).

Photo 40: The primary hotels in Nassau utilized by the IODE for hosting RAF dances and concerts. From upper left clockwise, the Emerald Beach was near the Hobby Horse race track at Cable Beach, the Royal Victoria lies between the Nassau Post Office and Shirley Street (survivors of allied ships sunk by U-boats stayed in the "European Annexe," which still stands), the Prince George, astride Bay Street and Nassau Harbour, and the British Colonial, so named by Sir Harry Oakes. Only the Colonial remains active as a hotel.
Source: Appropriately on p.111 of Jane Lloyd, *The History of the I.O.D.E. (Bahamas) 1901-2005*, Media Enterprises, Nassau, 2016.

Photo 41: The Duke of Windsor visits RAF No. OTU at Windsor Field in 1943 with its founder and first commander, Wing Commander John Alexander. Overcoming in illness early in the posting, Alexander travelled widely with the RAF, including to the Middle and Far East, and befriended the author and fellow RAF pilot Roald Dahl amongst other notables. The commanding officers or No. 111 OTU in particular were well known in the community.
Source: Jane Lloyd, *The History of the I.O.D.E. (Bahamas) 1901-2005*, Media Enterprises, Nassau, 2016, p.124 (her husband Chris Lloyd's uncle is W/C J. Alexander).

Photo 42: New pilots learning to fly new planes in a new country equated too often with tragic results. Four of the ten Czechoslovakian aviators photographed here in the Bahamas perished on a single crash and fire which left their aircraft unrecognizable. Seated in the front row, the men with black X's on their chests in this photo are Karel Salz, Josef Turna, instructor Jan Hadravek, and Vitezlav Briza. They and two others were buried at the RAF Cemetery with full honours, and by mutual decree between the nations it was decided that their remains would stay there, as they have.
Source: Jane Lloyd, *The History of the I.O.D.E. (Bahamas) 1901-2005*, Media Enterprises, Nassau, 2016, p.175.

Photo 43: Portraits and names of the men killed that day. While it is possible to name 183 or more of the KIA and missing from Nassau's bases, space does not permit images of each in this book. May the book itself be a remembrance of the sacrifices that they made.
Source: Jane Lloyd, *The History of the I.O.D.E. (Bahamas) 1901-2005*, Media Enterprises, Nassau, 2016, p.173.

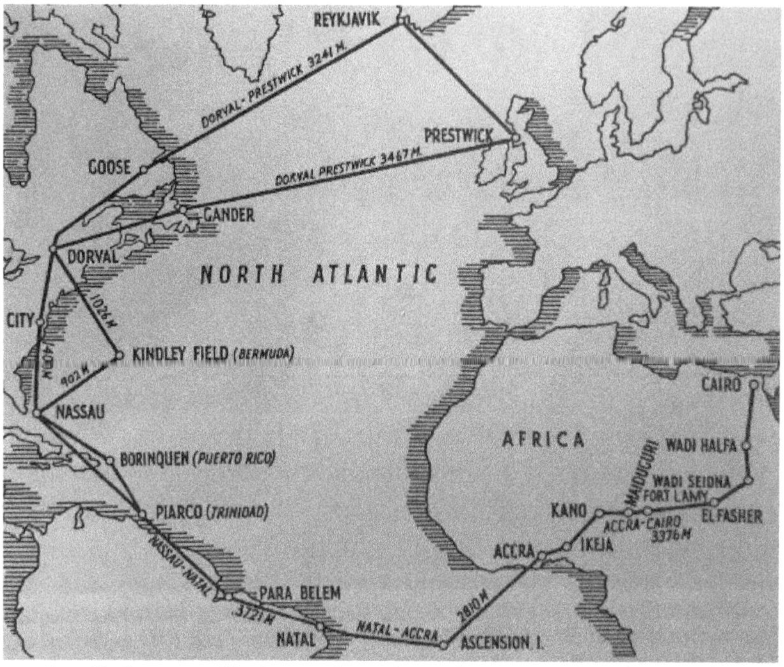

Photo 44: How the Bahamas fit into the aircraft transport (manufacture to delivery) equation in the Atlantic Basin during World War II from 1943 onwards.
Source: Arthur Pearcy, *Lend-Lease Aircraft in World War II*, Great Circle Routes, p.52.

Photo 45: "Fitters working on engines of Liberator G.R. Mk. VI EV-880 NC, No. 111 (C)OTU, the Bahamas, 1945. Note MC/n 29-73 painted aft of the Pitot tube mast. The digits were hand-painted with light colour paint inside an upper surface camouflage colour. Originally, there was a tir-cell chute next to the ASG-3 radar radome; tunnel gun scanning windows can be seen next to the open rear exit door."
Source: Caption and photo Robert Livingstone, Brisbane, Australia, whose father flew with the RAAF at No. 111 OTU, by kind permission.

Photo 46: "Airborne Liberator G.R. Mk. V BZ-746 LM of No. 111(C)OTU. The aircraft was delivered to Nassau on 25th May, 1943. On 23 February 1945 it collided mid-air with Liberator B. Mk. III FL-994 LA on approach orbit to Windsor Field. BZ-746 crashed 3 miles south of the airfield (FA Cat E). The aircraft had an antenna of ASG-1 radar below the nose. It was also equipped with an A-3C upper turret; its tail turret was a Consolidated A-6A as the plane arrived directly from North America. LM had it upper surfaces in Extra Dark Sea Grey."
Source: Robert Livingstone, Brisbane, Australia, whose father flew with the RAAF at No. 111 OTU, by kind permission.

Photo 47: "Liberators from No. 111(C)OTU, from left to right: Liberator G.R. Mk. Vs BZ-809 LY, BZ-722 LK, with an ASG-1 Radar in a chin below the nose, BZ-813, BZ-810 MA, and BZ-812 MC, Nassau, the Bahamas. Periscope tubes B-3 drift-meters can be seen at the nose sections of LY, MD, and MA."
Source: Robert Livingstone, Brisbane, Australia, whose father flew with the RAAF at No.111 OTU, by kind permission. Also the author purchased a copy from the Australian War Memorial in Canberra, author's collection.

Photo 48: "Fitters working on engine of Liberator G.R. Mk. VI of No.111 (C)OTU, the Bahamas, 1945. Liberator EV-880 NC. Note an A-6A nose turret with asymmetrically installed guns; the right gun lacks a muzzle compensator. Behind the No. 4 engine is a Leigh Light. In the background Liberator G.R. Mk. V LL, BZ-744 after an overhaul, during which the chin (housing the antenna and modulator of the ASG-1 radar) below the nose was removed. Several notable details to be seen on Liberator LL, in particular a side gun mount on the glassed nose, masts with D-1 Pitot tubes attached diagonally ahead of STA 1.0, DVW in the windscreen, blister side windows of the cockpit as well as an A-3C upper turret and propeller fitted with paddle blades."
Source: Robert Livingstone, Brisbane, Australia, whose father flew with the RAAF at No. 111, OTU, by kind permission.

Photo 49: RAF Cpl. Bryn Griffiths outside the Bahamas Club, Nassau, 1943
Source: photography by Len Baker, from p.26., *Into the Blue, From the Diary of Corporal Bryn T. Griffiths (Rigger Aero), whilst serving with the RAF in the Bahamas, 1943-1945*, by his son Geraint Griffiths, Wales, 2009

Photo 50: T.T. & Station Flight RAF Nassau, with Bermuda aircraft, 1943Source: *Into the Blue,* From the Diary of Corporal Bryn T. Griffiths (Rigger Aero), whilst serving with the RAF in the Bahamas, 1943-1945, by his son Geraint Griffiths, Wales, 2009, p.30

Photo 51: RAF Station Choir, Nassau, 1944. Source: *Into the Blue, From the Diary of Corporal Bryn T. Griffiths (Rigger Aero), whilst serving with the RAF in the Bahamas, 1943-1945*, by his son Geraint Griffiths, Wales, 2009, p.53

Photo 52: Montagu Foreshore, East Bay Street, Nassau, with Fort Montagu Beach Hotel in foreground. During World War II the RAF ASR, or Aircraft Search and Rescue base as at the dock or jetty centre-right. The Water Tower behind Nassau is at upper left, Hog Island (later Paradise Island) with its canal largely built post-war, is upper right, as is the entrance to Nassau Harbor.
Source: Bob Davis, vintagebahamas.com/hotels

Photo 53: Original caption by the Windsor's confidant and biographer, reads "The Duke pours the first bucket of cement at the Project, Oakes Field, May 8, 1942. Behind him stand Sir Harry Oakes (left) and Harold Christies (right) with Sidney Farrington between them in the background." Oakes would be murdered at his home roughly a year later. A uniformed military officers, presumably in the US Army Air Force, stands

second from right. The subsequent photo was of the Duke and Duchess with Major Hayes, US Corps of Engineers, also on-site in August.
Source: Bloch, Michael, *The Duke of Windsor's War, From Europe to the Bahamas, 1939-1945*, Coward-McAnn, Inc., New York, NY, USA, 1983 (1983 US), photo 12 following p. 270.

Photo 54: "The personnel of a United States Naval Air Station in the Bahamas (Great Exuma Island) lines up for a formal inspection on the ramp where the big flying boats are serviced. Lieut. Comdr. D.C. Higgins of Los Angeles, the commanding officer, keeps the isolated base in ship shape," October 28, 1943, see below
Source: WWII West Indies Bahamas U.S. Forces, Stock Image by Shugars for editorial use, Oct 28, 1943, found on Shutterstock.com, 4.28.2020

Photo 55: "Base crash boat runs in to see if plane photographer is aboard, or has any trouble in Great Exuma Island, Bahamas," see also below.
Source: WWII So. America Central America Bahamas U.S. Forces, BahamasStock Image by Anonymous for editorial use, Oct 4, 1943, Shutterstock.com, 4.28.2020

Photo 56: US Navy SEAL Museum caption reads "OSS Maritime Unit – Group I. With Duke of Windsor." [it concludes with "and Wally Stewart," meaning presumably the Duke's American wife, Wallis Simpson]. This particularly underwater detonation team was sent to the Pacific soon thereafter, Assigned to Navy UDT 10. US Navy Demolition squads trained at Salt Cay, aka Blue Lagoon Island, northeast of Hog Island, aka Paradise Island, in World War II. "During the Second World War, Blue Lagoon Island was used by the allies as training base for American and British underwater demolition squads. They would swim seven miles around the island every day, while regularly detonating explosives on the island. This lasted for about a year, during which time the explosions weakened a cliff that would later cause the small fort that rested on it to collapse." One of the O.S.S. trainees "...went on to Salt Cay in the Bahamas Islands where he spent weeks honing O.S.S. M.U. operations that he would later employ in combat. There they used high speed boats, motorized surfboards, explosives and limpet mines, and the "Sleeping Beauty" (or Motorized Submersible Canoe - MSC) the British built O.S.S. inventoried. This was the first one man underwater submersible vehicle for a single SCUBA wearing frogmen to perform clandestine reconnaissance or attacks against enemy vessels."
Source: Property of the US Navy SEAL Museum, catalogue #2000.0044.5. Other sources: Bahamastourcenter.com/history-of-blue-lagoon-island, McCutcheon, H. Shaw (2001). A Family Island - A Short History of Salt Cay, Bahamas. Salt Cay Publishing, maritime unit.org, specialforces.com/newsletter/2014_04

Appendices

Appendix 1: Ship Names Of All Participants In Convoys, 1942-1945 by Convoy Name

18 De Marzo - GK 856, KN 231, KN 244, KN 252, KN 265, KN 273, KN 285, KN 290, KN 291, KN 297, KN 384, NK 515, NK 555, NK 568, NK 576, NK 600
1St Lt W G Sylvester - KN 206
A C Bedford - GN 24, GN 159, KN 202, KN 386, NG 322, NG 425, NG 448, NG 455, TAW 15
A L Kent - GK 703, GN 24, GN 67, GN 84, GN 100, KG 607, NG 333, NG 374, NG 390, NG 502, WAT 13
A P Hill - GN 60, KG 613
A S Hansen - KN 212, KN 218, KN 233, KN 257, KN 260, KN 282, KN 306
Aalsum - GN 37, WAT 11
Abangarez - KN 114
Abbot L Mills - GK 800, KN 314
Abel Parker Upshur - GN 67, NG 340, WAT 14
Abiel Foster - GN 9, GN 64, NG 317
Abner Doubleday - GN 60
Abner Nash - NG 332
Abraham Baldwin - GN 26, WAT 12
Abraham Clark - GN 89, NG 351
Abraham Rosenberg - NK 685
Acasta - GN 18, NG 312
Acavus - AH 3, KN 210, KN 226, NK 508
Achilles - WAT 17
Acme - GK 765, GN 56, GN 62, GN 68, GN 77, GN 85, KN 123, KN 126, KN 201, KN 212, KN 215, KN 275, NG 355, NG 361, NG 367, NG 376, NG 384, NG 390, NG 407, NK 511, NK 514
Aconcagua - GK 732, KG 626, KG 635
Ada - GK 761, GK 796, GN 190, KG 688, KG 704, KN 341, KN 356, KN 362, NG 338, NG 488, NG 496
Adabelle Lykes - GN 6, GN 59, NG 322, NG 369
Adolph Sutro - GN 155, KN 265, NG 438, NG 462
Adoniram Judson - GN 118, GN 135, NG 425
Adula - GN 64, WAT 13
Aedanus Burke - KN 236
Aegeus - NG 301
Aelybryn - KN 129, NG 307, TAW 10
Afghanistan - GN 19, NG 345
African Prince - GN 30, NG 400
Agwidale - GK 752, GK 753, NG 379
Agwileon - GN 8
Agwimonte - GN 46, NG 353
Agwiprince - GN 55, KN 131, NG 311, NK 545, TAW 12
Agwistar - GK 754, GN 67, KN 262, KN 311, NG 307, NG 380, NG 429, NG 458, NG 468
Akaroa - GN 77, GN 137, KN 133, NG 322, NG 422, NG 455, NK 541, TAW 13
Alabaman - GN 52, WAT 11
Alamo - KG 629, NK 527
Alan-A-Dale - GN 15, GN 90, KN 256, NG 390, NK 555
Alanson B Houghton - KN 311, NG 449
Alar - AH 2, GK 769, GN 5,

Alma Ata - NG 304
Almirante Rodriguez Luis - GN 206
Alpha - KN 203, KN 207, KN 211, KN 215, KN 229, KN 238, NK 506, NK 510, NK 514, NK 526, NK 532, NK 541, NK 635, WAT 13
Alphard - NG 383
Altair(Amer) - GK 704, GN 66, GN 72, GN 78, GN 84, GN 89, GN 96, GN 203, KN 102, KN 208, KN 212, KN 215, KN 218, NG 347, NG 371, NG 377, NG 383, NG 394, NK 511, NK 514, NK 517
Am-Mer-Mar - NK 686
Amakura - WAT 15
Amastra - KN 200, KN 201
Ambrose E Burnside - GN 37
Amelia Earhart - GN 69, KG 622
America Sun - GN 3, GN 120, KN 103, NK 561
American Builder - GN 21, GN 66, KN 103, NG 329, NG 434, NK 556, WAT 2
American Fisher - GN 116, GN 121, GN 133, KG 704, KN 105, KN 129, KN 201, KN 229, KN 262, KN 268, KN 273, KN 280, KN 340, NG 386, NG 392, NG 399, NG 406, NG 415, NG 421, NG 443, NG 458, NK 511, NK 607
American Manufacturer - GN 37, KN 108, KN 260, NG 345, NG 389, NG 433
American Packer - GN 65, KN 127, NG 301, TAW 8
American Pilot - KN 211
American Press - GK 810, GN 24, KN 276, KN 318, NG 304, NG 394, NG 433
American Press Ii - GN 54, KN 112
American Robin - GN 89, GN 111, GN 139, GN 150, GN 175, KG 719, NG 376, NG 396, NG 439, NG 458, NK 625
American Sailor - GK 796
American Seafarer - KG 700
American Seaman - KN 220
American Trader - KN 122, KN 204, KN 208, KN 212, KN 227, NK 507, NK 511, NK 521
Amerika - GN 32, NG 302
Amos G Throop - KN 265
Amos Kendall - GN 62, KN 385
Ampetco - KN 124, KN 210, NG 452, NK 507, TAW 5
Amstel - NG 302
Anadyr - GN 24, NG 335, NG 427
Ancylus - GN 24, NG 321, WAT 1
Andalien - GK 763, GK 786, GN 156, KG 673, KG 693, KN 315, KN 343, KN 384, NG 442, NG 461, NG 470
Andino - GN 12, GN 64, NG 320, NG 372
Andrea F Luckenbach - GN 31, KN 112, WAT 11
Andreas - NG 306
Andrew A Humphreys - KN 280
Andrew Briscoe - KN 294
Andrew Carnegie - GK 742, KN 257
Andrew Furuseth - GN 56
Andrew Marschalk - GN 120, GN 141, KN 272, KN 277, KN 308, NG 401, NG 440, NK 531, NK 611
Andrew Moore - GN 22, KG 610
Andrew Turnbull - NK 686
Aneroid - WAT 15
Angelina - KN 128, TAW 9, WAT 2
Anghyra - KN 111, WAT 13
Angouleme - GK 755, GK 805, GK 825, GK 850, GK 862, GN 99, GN 162, GN 197, KG 664, KG 729, KG 755, NG 420, NG 467, NG 505
Anita - NG 300, NG 466
Ann Skakel - KG 642, NK 548
Anna H Branch - KN 304
Anna Howard Shaw - NG 408
Anna Knudsen - GN 18, GN 42, NG 314
Anna Marcou - GN 12
Anna Odland - GK 700, NG 328
Annibal - GN 67, KG 661, KN 242, KN 382, NG 366, NK 537, NK 635
Annik - GK 712, GN XXX, GN XXX, KN 293, KN 300, KN 308, NK 590
Anniston City - GN 24, GN 203, NG 307
Anson Jones - KN 246
Anson Mills - NK 686
Antaeus - GN 31, GN 37, NG 335
Anthony Ravalli - GN 128
Anthony Wayne - GN 16
Antietam - GN 24, GN 104, GN 111, GN 133, GN 140, KN 107, KN 119, KN 136,

WAT 13
Athelregent - GN 34, GN 47, KN 124, KN 206, KN 274, KN 293, KN 313, NG 412, NG 432, NG 444, NG 456, NK 505, NK 568
Athelsultan - KN 131, KN 134, TAW 12, WAT 12
Athelvictor - GN 67
Athelviking - KN 294, KN 321, KN 359, NG 413, NG 440
Athina Livanos - GN 1, NG 316, TAW 16
Athos - GN 43, KN 130, NG 382, TAW 7, TAW 11
Atland - KN 220
Atlanta City - GN 16, KN 266, NG 385
Atlantian - GN 30, KN 245, NG 353
Atlantic City - GN 25, GN 135, KG 707, NK 613
Atlantic Coast(Amer) - GN 202, KN 262, KN 266
Atlantic States - NK 542
Atlantic Sun - KN 126, KN 127
Atlantic Trader - GK 789, GN 59, KG 630, KG 693, KN 219, NK 523, NK 593
Atlantic(Nor) - GN 14, NG 309, NG 462, NG 498
Atlantic(Pan) - GK 790, GK 851, GK 864, GN 63, KG 695, KG 755, KG 770, KN 126, KN 298, NG 310, NG 406, NG 469, NK 601, TAW 6
Atlantida - GK 707, GK 716, GK 720, GN 43, TAW 12, WAT 3
Augustine Heard - GN 89
Augustus Saint-Gaudens - KN 308
Aun - GN 59, NG 349, NG 404
Aurora(Amer) - KN 228
Austvangen - GK 718, GK 754, KG 623, KG 662
Avila - TAW 7
Avon Coast - AH 4
Axios - NG 321
Axtell J Byles - GN 33, GN 127, KN 208, KN 212, KN 284, KN 336, KN 341, NG 330, NG 426, NG 432, NG 441, NG 446, NK 507, NK 511, NK 638, NK 643
Aymeric - GN 43
Ayuruoca - GN 51, GN 91, GN 159, GN 181, NG 363, NG 401, NG 490, WAT 17
Azerbaidjan - KG 648, NK 552
Aztec(Amer) - GK 774, KG 664, KG 668, KG 672, KN 108
B F Shaw - GN 118, KN 321, NG 438
B.P. Newton - GN 26, KN 116, KN 203, NG 323, NG 367, NK 501
Baalbeck - GK 731, GK 758, GN 19, GN 43, GN 121, GN 130, GN 138, KG 649, KG 671, KG 678, KG 685, KG 692, KN 122, KN 382, KN 385, NG 325, NG 349, NG 427, NG 436, TAW 4, WAT 17
Baarn - GN 11
Bacchus(Du) - GK 701, GK 717, GK 727, GK 734, GK 746, GK 761, GK 778, GK 791, GK 804, GK 816, GK 828, GN 204, KG 608, KG 620, KG 629, KG 637, KG 651, KG 671, KG 684, KG 696, KG 709, WAT 3, WAT 14
Bacoi - GK 700, WAT 15
Bactria - GN 30
Baharistan - GN 26
Bahram - KN 302, KN 308, KN 335, NK 599, NK 605, NK 613, NK 620, NK 643
Baialoide - GK 775, GN 43, KN 298, NG 367
Bajamar - GK 788, GN 29, GN 53, GN 131, KG 601, KG 683, KN 304, NG 337, NG 429, NG 439, NK 585, NK 628, TAW 12, WAT 3
Baldbutte - GN 19, GN 41, GN 80, GN 87, GN 103, GN 130, GN 138, GN 144, KN 111, KN 202, KN 214, KN 235, KN 242, KN 250, KN 306, KN 386, NG 311, NG 325, NG 338, NG 377, NG 385, NG 400, NG 424, NG 430, NG 436, NG 443, NG 449, NG 503, NK 528, NK 539, NK 547, NK 587, NK 635, NK 641
Baldwin Park - GN 184, GN 204, NG 479
Balfe - GN 26, NG 498
Ball'S Bluff - GN 110
Balla - KG 666, KG 711, NG 383, NG 392, NK 631
Baltic(Pan) - GN 7, GN 28, GN 59, GN 108, GN 204, NG 316, NG 333, NG 367, NG 421, NG 458
Baltimoree - GN 103
Baltyk - GN 26, NG 417
Banaderos - GN 29, GN 53, GN 118, GN 129, KN 133, NG 306, NG 336, NG 406, NG 425, NG 435, NG 442,

British Tenacity - GN 38, KN 108, NG 319
British Tradition - GN 12, GN 48, NG 309, NG 334
British Unity - KN 128, NG 316, TAW 9
British Vigilance - KN 121, KN 204, NG 328, NK 503
British Vigour - NG 352
British Zeal - KN 122, NG 312
Brockholst Livingston - GN 34
Bronnoy - GN 17, GN 189, NG 338, NG 349, NG 434, NG 485
Brooklyn Heights - GK 841
Brunswick - GN 22, GN 60, GN 89, GN 98, GN 107, GN 115, GN 123, GN 131, GN 138, GN 157, KN 108, KN 136, KN 232, KN 250, KN 260, NG 319, NG 359, NG 377, NG 388, NG 396, NG 406, NG 413, NG 421, NG 429, NG 436, NG 443, NG 456, NG 462, NG 505, NK 529, NK 547, NK 583, TAW 15, WAT 13
Brush - KN 302, KN 312, NK 606, NK 616
Buchanan - KN 127, TAW 8
Budenni - NG 306
Buena Vista - NK 537
Buenos Aires(Nor) - KN 124, NG 305, TAW 5
Bulklube - KN 295, NK 593
Bulkoil - GN 157
Bull Run - GN 82, NG 395, NK 553
Bunker Hill - KN 103, KN 118, KN 128, NG 390
Bunting - NG 322, NG 325
Bureya - NG 389
Bury(Bz) - GN 77, NG 388
C A Canfield - GN 123, KG 687, KN 106, KN 117, KN 212, KN 216, KN 321, NG 338, NK 514, NK 608
C B Watson - GK 713, KN 269, TAW 6, WAT 16
C H Spedden - WAT 3
C J Barkdull - GN 9, KN 123, NG 306
C W Pasley - KG 730
Cacalilao - GN 82, GN 88, GN 94, GN 101, GN 107, GN 113, GN 141, KN 209, KN 214, KN 217, KN 225, KN 383, NG 349, NG 386, NG 393, NG 400, NG 406, NG 412, NG 456, NG 462, NK 512, NK 519, NK 686
Cachapoal - GK 742, GK 758, GK 774, GK 792, GN 139, KG 630, KG 647, KG 679, KG 698, NG 447
Cacique - GK 703, GK 715, GK 728, GK 734, GK 744, GK 754, GK 767, GK 778, GK 789, GN 124, GN 134, GN 162, GN 204, KG 608, KG 617, KG 630, KG 637, KG 648, KG 658, KG 671, KG 684, KG 693, NG 430, NG 440, NG 449, NG 458, NG 468, NG 477, NG 502, WAT 9
Caddo - KN 118, KN 134, NG 305, TAW 14, WAT 12
Caesar Rodney - GN 64, NG 318, NG 319, NG 385
Caleb Sprague - KN 260
California(Amer) - GN 201, NG 506
Calliope - GN 30, GN 46, GN 55, GN 64, GN 71, GN 87, GN 109, GN 115, GN 121, GN 127, GN 133, GN 140, GN 158, GN 203, NG 345, NG 352, NG 360, NG 369, NG 386, NG 407, NG 414, NG 420, NG 426, NG 432, NG 457, WAT 11, WAT 12
Calmar - GN 87, GN 115, GN 124, GN 175, KN 132, KN 320, NG 313, NG 375, NG 398, NG 421, NG 434, NG 461, TAW 12
Calobre - GK 856, GN 97, KG 764, KN 320, NG 379, NG 406, NG 407, NG 437
Caloria - KG 664, KG 669, KN 205, NG 403
Calumet - GN 60, GN 200, NG 374, NG 461
Camaguey - KN 209
Camamu - GK 718, GK 813, GK 845, GN 69, GN 103, GN 197, KG 624, KG 725, KG 752, NG 380, NG 413, NG 506, WAT 15
Campfire - GK 790, KG 679, KN 305, NG 375, NG 494
Canada - GN 7, TAW 5
Canadian Star - KN 225, NG 301
Cantuaria - GN 47, GN 85, GN 117, NG 360, NG 394, NG 426, NG 456
Cap Cantin - GN 26, KN 128, NG 321, TAW 9, WAT 5
Cape Alava - GN 48, KG 604, KG 693, KN 129, NG 361, TAW 10
Cape Barrow - KG 715
Cape Beale - KN 382
Cape Bon - NK 550
Cape Breton - GN 43, GN 67, KN 204, NG 362, NK 502

KN 299, KN 339, NG 385, NG 412, NG 423, NG 430, NK 502, NK 516, NK 519, NK 532, NK 538, NK 543, NK 548, NK 578, WAT 9
Castor - GK 711, GK 722, GK 729, GK 736, GK 748, GK 759, GK 777, GK 788, GK 801, GK 812, KG 613, KG 630, KG 641, KG 652, KG 667, KG 681, KG 693, KG 705, KN 131, KN 386, NG 303, TAW 12
Cathcart - GN 18, GN 38, GN 50, GN 62, GN 73, GN 89, GN 100, GN 109, GN 118, GN 128, GN 137, GN 157, KN 136, NG 313, NG 326, NG 345, NG 357, NG 371, NG 382, NG 398, NG 407, NG 416, NG 425, NG 435, NG 443, TAW 4, TAW 15, WAT 1, WAT 11
Cathlamet - GN 93, GN 115, GN 123, GN 142, KG 725, NG 377, NG 400, NG 422, NG 436, NG 479, WAT 13
Catrine - KG 698, NK 602
Cavalcade - KN 130, KN 283, KN 292, NG 317, NG 402, NG 411, TAW 11
Cavelier - GN 15, GN 32, GN 47, GN 60, GN 73, GN 89, GN 100, GN 113, GN 128, GN 155, KG 605, NG 321, NG 342, NG 355, NG 368, NG 384, NG 398, NG 411, NG 425, NG 441, NG 461, TAW 12, WAT 9
Cearaloide - GK 856, GN 54, GN 83, GN 155, KG 638, KG 763, NG 392
Cecil N Bean - KN 309
Cefn-Y-Bryn - GN 124, KN 306, NG 419
Celeste M - NG 304
Celtic Monarch - GN 47, WAT 11
Celtic Star - GN 34
Cepalturn - WAT 2
Ceramic - KN 119
Ceres - GN 15, GN 39, NG 324, NG 347, WAT 14
Cerion - KN 245
Cerro Gordo - NG 395
Cetus - NG 327
Chagres - GK 819, GN 203, NG 443
Champ Clark - GN 47
Champoeg - GN 127
Chantilly - GN 38
Chapultepec - NG 366
Charles A Mccue - KN 386
Charles Brantley Aycock - GK 724, KG 609, KN 229
Charles C Pinckney - GN 23
Charles Carroll - GN 90, KN 136, TAW 15
Charles D Mciver - KG 640
Charles D Poston - GN 99, KN 383
Charles F Amidon - NK 640
Charles F Meyer - KG 619, KN 120, KN 206, NK 505
Charles G Coutant - GK 802, GN 140, KG 709
Charles Goodnight - GK 806, GN 125, KG 686, KN 319, NG 431
Charles Goodyear - GN 60
Charles H Cramp - GN 81, GN 103, NG 368, NG 389, NG 390, NG 426
Charles H Salter - GN 68
Charles Henderson - KN 243
Charles Hull - KG 707
Charles J Finger - KN 297, NG 466
Charles J Folger - GK 755, KN 269
Charles Kurz - GN 11, GN 22, GN 38, GN 44, GN 51, GN 57, GN 67, GN 78, GN 86, NG 316, NG 335, NG 343, NG 350, NG 356, NG 362, NG 376, NG 383, NG 404
Charles L D - GN 20
Charles Lummis - KG 711
Charles M Conrad - GN 109, GN 141, KN 250, NG 418
Charles M Hall - GN 37
Charles M Schwab Ii - KN 268, NG 386
Charles Morgan - KN 287
Charles N Mcgroarty - GN 100, NG 410
Charles Nordhoff - GN 114
Charles R Mccormick - GN 105, GN 116, GN 137, GN 143, KG 706, KN 276, NG 394, NG 402, NG 413, NG 426, NG 443
Charles Robinson - GN 106
Charles Scribner - KN 385
Charles Treadwell - GN 71
Charles Tufts - NK 686
Charles W Stiles - GN 204
Charles Willson Peale - GN 77, KN 268, NG 386, NK 683
Charlestown - NK 543
Chateau-Thierry - NK 684
Chedabucto - KN 201, KN 208
Chernyshevski - NG 367
Cherry Valley - GN 82, GN 104, KN 213, NK 512
Chester Harding - GN 85, NG 343
Chester O Swain - GN 1, GN

63, GN 69, GN 77, GN 100, GN 108, GN 114, GN 120, GN 126, GN 132, GN 137, GN 146, GN 204, KN 208, KN 212, KN 216, KN 272, KN 340, NG 338, NG 348, NG 355, NG 362, NG 368, NG 376, NG 395, NG 405, NG 413, NG 419, NG 425, NG 431, NG 445, NG 450, NG 502, NK 511, NK 515, NK 561, NK 643
City Of Adelaide - GN 31
City Of Athens - KN 114
City Of Baroda - NG 341
City Of Bombay - KN 131, NG 322, TAW 12
City Of Calcutta - GN 28
City Of Canberra - GN 31, NG 438
City Of Canton - GN 22
City Of Christchurch - KN 121, TAW 2
City Of Christiania - GN 32
City Of Dalhart - GN 60, KG 680, NG 377, WAT 14
City Of Dallas - KG 633
City Of Delhi - GN 47, KN 121, NG 307, TAW 2
City Of Dundee - GN 51
City Of Dunkirk - GN 28, GN 137
City Of Durban - GN 24
City Of Eastbourne - GN 54
City Of Elwood - GN 46, GN 80, GN 90, NG 380, NG 387, NG 402
City Of Exeter - GN 13, NG 338
City Of Flint - GN 20
City Of Florence - GN 4, GN 116, NG 315
City Of Fort Worth - NG 336
City Of Glasgow - GN 24, GN 104, NG 384, WAT 9
City Of Hankow - GN 3
City Of Hereford - GN 47
City Of Hongkong - NG 337
City Of Houston - GN 54, GN 63, KN 201, KN 213, KN 218, KN 226, NG 351, NG 361, NG 394, NK 511, NK 523, WAT 13
City Of Khios - GN 34
City Of Kimberley - GN 25, NG 415
City Of Lancaster - GN 18, GN 141, NG 303, NG 434
City Of Leicester - GK 716, KN 213
City Of Lille - GN 9, GN 153, NG 425
City Of Lyons - GN 33, NG 423
City Of Montgomery - GK 739, GK 754, GK 773, GN 130, GN 140, GN 160, KG 629, KG 643, KG 678, KG 688, KN 302, KN 382, NG 313, NG 426, NG 437, NG 447, NG 457
City Of Newcastle - GN 47
City Of Omaha - GN 24, GN 190, NG 306, NG 443
City Of Philadelphia - GN 32, GN 43, GN 54, KG 608, KN 209, NG 340, NG 350, NG 387, NK 513, TAW 9, WAT 7
City Of Savannah - GK 717, GK 735, GK 756, GK 765, GK 783, GN 118, GN 128, GN 138, GN 159, KG 620, KG 625, KG 640, KG 660, KG 669, KG 678, KG 687, NG 320, NG 424, NG 434, NG 455, NG 503
City Of St Louis - GK 719, GN 50, GN 83, GN 129, GN 203, KG 622, NG 319, NG 365, NG 390, NG 438, NG 455
City Of Yokohama - GN 29, NG 341, NG 438
Clam - GN 157, GN 205, KN 210, NG 456, NG 504, NK 641, NK 659
Clan Alpine - GN 29
Clan Cameron - GN 45
Clan Colquhoun - KN 129, KN 268, NG 443, TAW 10
Clan Farquhar - GN 14, KN 269, NG 442
Clan Macaulay - GN 102
Clan Macbean - GN 152, NG 448
Clan Macgillivray - GN 28
Clan Macilwraith - GN 37, GN SPEC2
Clan Macinnes - GN 3
Clan Maciver - GN 45
Clan Macnair - GN 24
Clan Macpherson - KN 129, TAW 10
Clan Matheson - NG 366
Clan Murdoch - GN 17
Clara Barton - GN 86
Clark Mills - GN 62
Clausina - GN 44, KN 124, KN 205, KN 261, NK 504, NK 558
Clavella - AH 1, GN 17, GN 34, GN 51, NG 314, NG 331, NG 350, NG 389, TAW 1
Clearpool - KN 211, NK 510
Clearwater Park - GN 126, GN 131, GN 204, NG 424, NG 431, NG 442, NG 448, NG 455
Clement Clay - GN 121
Cleodora - NG 458
Cleveland Abbe - GN 56

Corinto - KN 383
Cornelia - GK 709, GK 714, GK 758, GK 788, GN 31, GN 38, GN 45, GN 55, GN 63, GN 70, GN 77, GN 92, GN 100, GN 108, GN 122, GN 131, KG 611, KG 616, KN 266, KN 296, NG 336, NG 343, NG 353, NG 360, NG 368, NG 375, NG 383, NG 390, NG 399, NG 406, NG 413, NG 420, NG 428, NG 442, NG 455
Cornelia P Spencer - NK 545
Cornelius Gilliam - GN 35, WAT 15
Cornell - NK 530
Corneville - NG 326
Cornish City - GN 22, WAT 2
Cornwall - GN 68, GN 110, NG 308, NG 388
Cornwallis - GN 87, GN 105, GN 124, GN 142, GN 157, KG 652, NG 399, NG 418, NG 436, NG 451, WAT 16
Corozal - GK 716, GK 727, GK 743, GK 753, GK 761, GK 770, GK 782, GK 792, GK 803, GK 813, GK 825, GK 838, GK 850, GK 864, GK 876, KG 605, KG 622, KG 629, KG 635, KG 648, KG 656, KG 664, KG 676, KG 686, KG 696, KG 707, KG 718, KG 730, KG 743, KG 757, KG 768, TAW 11M
Coteaudoc - GK 706, KG 621, KG 648, TAW SL/1
Cottica - GK 710, GK 718, GK 723, GK 727, GK 738, GK 760, GK 769, GK 785, GK 797, GN 133, GN 143, GN 153, GN 163, GN 175, GN 196, GN 205, KG 612, KG 620, KG 629, KG 633, KG 655, KG 664, KG 674, KG 690, KG 702, NG 440, NG 450, NG 460, NG 470, NG 482, NG 492, NG 502, WAT 9
Cotton Mather - GN 75, GN 117, NG 331
Coulbeg - GN 45
Courageous - GK 756, GK 772, GN 19, GN 57, KN 265, KN 280, KN 291, NG 344, NG 383, NG 391, NG 409, WAT 17
Cowrie - GN 12, KN 120, NG 308, TAW 1
Crawford W Long - KN 244, NK 684
Crescent Park - GN 105, GN 122, GN 141, GN 154, NG 401, NG 415, NG 434, NG 448
Cressington Court - WAT 11
Cringle Reefer - NG 503
Cripple Creek - WAT 5
Cromarty - GN 1, NG 314
Cross Keys - GN 202
Crystal Park - GN 138
Cuauhtemoc - GK 750, KG 697
Cuba(Amer) - GK 748, KG 637, KG 673, KN 205, TAW 9
Cuba(Br) - GK 804, KN 312, NG 427
Cubore - GK 734, GK 745, GK 840, GN 22, GN 53, GN 83, GN 93, GN 102, GN 112, GN 123, GN 132, GN 140, GN 147, GN 154, GN 160, GN 187, GN 206, KG 609, KG 626, KG 639, KG 650, KG 747, KN 200, NG 336, NG 358, NG 390, NG 398, NG 409, NG 419, NG 429, NG 438, NG 445, NG 452, NG 459, NG 466, NG 502, NK 504, TAW S(1)
Curaca - GK 712, GK 733, GK 786, GK 804, GK 819, GK 841, GN 45, KG 674, KG 692, KG 709, KG 727, KG 747, NG 352, TAW 8
Curacao - GN 32
Cushing Eells - GN 203, KN 260
Custodian - GN 20, GN 204
Cygnet Ii - GK 833, GK 844, GK 856, GK 868, KG 737, KG 748, KG 760, NG 453
Cymbula - GN 64, KN 116, KN 120, KN 204, KN 216, NK 502
Cypress Hills Park - GN 123, GN 131, NG 430, NG 437, NG 441, NG 504
Cypria - GN 18, NG 329
Cyrus Field - KN 104, KN 217, KN 293, NK 511, NK 589
Cyrus H K Curtis - KN 384
Cyrus H Mccormick - GK 730, KN 234
Cyrus Sears - KN 250
Cyrus W Field - GK 809, KN 270, NG 397, NG 434
D L Harper - KN 130, TAW 11, WAT 7
Dageid - GN 20
Daghild - KN 123
Dagrun - GN 17, GN 131, KG 680, NG 327, NG 502, NK 582
Dahomian - NG 415
Dalcross - GN 39
Dalemoor - GN 123, NG 420

NK 510
Dilworth - KN 115, KN 202, NK 501
Dimitrios Inglessis - NG 337
Diomed - GN 15, NG 386, NG 435
Director - GK 701, KN 203, TAW 13, WAT 6
Dirkje - GK 822
Dixiano - GK 770, GK 791, GN 103, GN 111, GN 139, KG 694, KN 202, KN 268, KN 278, NG 395, NG 402, NG 410, NG 416, NG 439, NG 456, WAT 1, WAT 10
Dneprostroi - WAT 10
Dolabella - GN 68, WAT 2
Dolius - KN 105
Dolly Madison - GN 98, KG 667, NG 466
Domby - KN 206, NK 505
Domino - GK 713, GK 720, GK 727, GK 734, GK 745, GK 753, GK 776, KG 614, KG 623, KG 629, KG 638, KG 649, KG 656, TAW 13, WAT 6
Don - GK 737, GN 74, KG 644, KN 263, NG 332, NG 411, NK 560, NK 567, NK 574, NK 583, NK 598, NK 606, NK 612, NK 618
Dona Aniceta - GN 48, GN 89, KG 686, NK 537, NK 591, WAT 17
Donacilla - GN 68
Donald W Bain - KN 386
Donax - AH 2, WAT 10
Dorcasia - GN 23, GN 80, KN 111, NG 319, NG 376, NG 394
Dordrecht - GN 32
Doros - GK 732, GK 739, GK 745, GK 753, GK 763, GK 773, GK 784, GK 793, GK 802, GN 137, GN 148, GN 158, GN 167, GN 178, GN 188, GN 198, KG 629, KG 635, KG 643, KG 649, KG 658, KG 668, KG 679, KG 688, KG 697, KG 706, NG 443, NG 454, NG 464, NG 485, NG 494, NG 504
Dorothy - GN 12, GN 40, GN 89, GN 109, GN 133, GN 155, NG 322, NG 347, NG 374, NG 397, NG 419, NG 443, NG 462, NG 504, TAW 4, WAT 13
Dorothy Luckenbach - GN 51, GN 105, KN 302, NG 306, NK 541
Drafn - GN 45, GN 105, GN 122, GN 128, NG 373, NG 404, NG 414, NG 420, NG 427, NG 440, NG 505
Dramatist - GN 38, GN 71, GN 142, KN 126, KN 344, NG 320, NG 360, NG 434, NK 640, TAW 5
Drammensfjord - GN 150, NG 447, WAT 5
Dromus - KG 613, KN 136, NK 508, TAW 15, WAT 13
Drupa - GN 31, KN 110, KN 232, NG 392, NK 517
Dubrovnik - KN 123, NG 305, NG 383, NG 438, TAW 4
Duc D'Aumale - GK 831
Dudley M Hughes - KG 659, KN 269, NG 505
Dufferin Park - GN 87, GN 97, GN 110, GN 124, GN 139, GN 162, NG 385, NG 394, NG 408, NG 421, NG 437, NG 448
Duke Of Athens - NG 421
Duke Of Sparta - GN 45
Duluth - KG 632
Dumfries - NG 309
Dunboyne - GK 714, GK 728, GK 757, GK 775, GK 785, GK 816, GK 831, GN 38, GN 57, GN 71, KG 619, KG 630, KG 662, KG 670, KG 680, KG 689, KG 720, KG 736, KN 302, KN 311, KN 357, NG 345, NG 363, NG 380, NG 426, NG 437, TAW 14, WAT 5
Duncan U Fletcher - GN 89, KG 658
Dundas - KG 631
Dundee - GK 716
Dundurn Park - GN 203
Dunkeld - GN 9, NG 321
Dunster Grange - GN 22
Duquesne(Amer) - KN 304
Durango - GK 703, GN 32, GN 42, GN 49, GN 54, GN 60, GN 67, GN 73, GN 81, GN 88, GN 95, GN 101, GN 122, GN 128, GN 138, KN 204, KN 336, NG 329, NG 340, NG 348, NG 353, NG 359, NG 365, NG 372, NG 380, NG 386, NG 394, NG 400, NG 421, NG 427, NG 433, NG 443, NK 638
Dvina - NG 355
Dwight L Moody - GK 757, KG 651, KN 265, NG 421
Dynastic - GK 763, GK 791, GK 849, GN 144, KG 671, KN 307, KN 314, NG 374, NG 439, NG 468, NK 611
E A Peden - KN 254
E G Seubert - KN 120, KN 207, KN 219, NK 506
E H Blum - KN 214, KN 263, NG 387, NK 513

Eleazar Wheelock - GN 31, WAT 4
Elg - GN 141, NG 434
Elias Boudinot - KN 233
Elias G Kulukundis - GN 3, NG 318
Elihu Root - KG 643, KN 252, KN 268, NG 386
Elihu Yale - KN 240
Elijah White - GK 761, KN 273
Eliphalet Nott - KN 386
Elisabeth Dal - GN 121, NK 592
Elise - GN 68, GN 95, NG 392
Elisha Mitchell - GN 89, GN 139, NG 438, NG 459, NK 537
Elisha Walker - GN 159, WAT 15
Elisha Whitney - KG 704
Eliza Jane Nicholson - KN 284, NG 505
Elizabeth Blackwell - GK 796
Elizabeth Kellogg - GK 768, KN 206, KN 210, KN 214, KN 229, KN 237, KN 243, KN 250, KN 259, NG 383, NK 509, NK 522, NK 532, NK 540, NK 545, NK 553
Ellenor - GN 51, GN 66, GN 77, GN 87, GN 97, GN 110, GN 119, GN 129, GN 139, GN 148, GN 157, KN 382, NG 331, NG 345, NG 362, NG 372, NG 385, NG 394, NG 406, NG 417, NG 427, NG 437, NG 445, NG 454, NG 501
Elmdale - KN 131, NG 312, TAW 11
Elmer A Sperry - GK 729
Elmwood - WAT 2
Elna Ii - NG 367
Elona - GN 13, NG 310
Elqui - GK 730, GK 746, GK 763, GK 776, GK 792, KG 631, KG 650, KG 681, KG 698, KN 320, KN 335, NG 445, NG 460
Elwood(Pan) - GK 728, GN 30, GN 141, GN 153, KG 709, KN 129, NG 315, NG 345, NG 449, NK 521, TAW 10
Emilia - GN 21, GN 45, GN 75, GN 96, GN 117, NG 358, NG 383, NG 403, NG 430, NG 504, NG SPEC, TAW 7, WAT 15
Empire Addison - GN 32
Empire Advocate - GN 38
Empire Airman Ii - GN 19, KN 132, KN 225, NG 316, NK 522
Empire Almond - NG 320
Empire Arnold - WAT 6
Empire Austen - GN 28, NG 339
Empire Ballad - GN 15, GN 73, NG 323
Empire Barrie - GN 38
Empire Bede - TAW 13
Empire Blessing - GN 128, NG 423
Empire Bombardier - GN 120
Empire Boswell - GN 53, NG 347
Empire Bowman - WAT 11
Empire Bronze - KN 110, KN 204, NG 353, NK 503
Empire Bruce - KN 129, NK 507
Empire Bunting - KN 212, NK 510
Empire Cabot - KN 114
Empire Castle - GN 134, NG 414, NG 455, NK 553
Empire Caxton - GN 38, GN 64, NG 325, NG 359
Empire Chapman - GN 17, KN 130, KN 217, NG 314
Empire Charmian - NG 493
Empire Chief - GN 111, GN 119, GN 128, GN 158, KN 334, NG 410, NG 427, NG 433, NG 458
Empire Chivalry - GK 761, GN 15, GN 70, KN 269, NG 362, NG 425, WAT 13
Empire Cloud - TAW 12, TAW SL/1
Empire Coleridge - GN 13, GN 33, KN 127, NG 309
Empire Conrad - NG 300, NG 318
Empire Coral - KN 133, NK 506
Empire Corporal - TAW 12, WAT 10
Empire Cougar - GN 44, GN 121, GN 164, GN 188, KN 114, NG 338, NG 417, NG 462, NG 484
Empire Curzon - GN 129, NG 426
Empire Cutlass - GN 102
Empire Dickens - GN 11, KN 108, NG 307, NK 514
Empire Dunstan - NG 336
Empire Emerald - KN 109
Empire Faith - WAT 17
Empire Falcon - GN 17, NG 328
Empire Falstaff - NG 484
Empire Farmer - GN 183
Empire Fletcher - GN 80, GN 116, KG 670, KN 204, NG 374, NK 503, NK 574

318
Empire Splendour - NG 506
Empire Spray - GN 43
Empire Squire - GN 24
Empire Stanley - GN 44
Empire Steelhead - KN 126
Empire Stour - NG 337
Empire Strength - GN 110, NG 342
Empire Success - NG 416
Empire Sunbeam - GN 136, NG 435
Empire Swale - GN 130, NG 423, NG 498
Empire Tamar - KN 249, NK 507, NK 545
Empire Tarpon - KN 106, KN 201
Empire Tennyson - NG 300
Empire Thackeray - GN 36, NG 324
Empire Toiler - GN 49, NK 523
Empire Trader - GN 26, KN 103
Empire Traveller - GN 201
Empire Viscount - KN 311, KN 335, NG 506, NK 609
Empire Waimana - GN 37, KN 260, NG 313, NK 537
Empire Whale - KN 124, TAW 5
Empire Whimbrel - GN 24
Empire Wolfe - GN 30, NG 315
Empire Wordsworth - GN 19, NG 316
Empire Wyclif - KN 126, NK 642, TAW 5
Empire Zephyr - KN 117
Ena - GN 158
Ensis - KG 620
Ensley City - GN 62, GN 85, GN 107, GN 143, GN 167, GN 188, KN 129, KN 297, KN 303, KN 383, NG 306, NG 370, NG 392, NG 421, NG 430, NG 452, NG 475, NG 496, TAW 9
Ephraim Brevard - GN 202
Erastus Smith - KN 340, NG 458
Eric V Hauser - NK 613
Ericus - GN 67, GN 80, GN 92, GN 111, GN 128, KG 677, NG 326, NG 376, NG 387
Ernebank - NG 370
Ernest W Gibson - GN 203, NG 447
Erria - GN 37
Errington Court - GN 20
Eskbank - GK 722, KN 244
Eskdalegate - GN 38
Esso Annapolis - KN 103, KN 122, WAT 13
Esso Aruba - GN 48, GN 54, GN 60, GN 68, GN 74, GN 80, GN 85, GN 91, GN 96, GN 101, GN 107, GN 113, GN 130, GN 138, GN 153, GN 157, GN 203, KN 107, KN 119, KN 223, NG 347, NG 353, NG 359, NG 365, NG 373, NG 379, NG 384, NG 390, NG 400, NG 406, NG 412, NG 425, NG 430, NG 443, NG 448, NG 457, NG 462, TAW 15, WAT 11
Esso Augusta - GN 204, KN 109, NK 510
Esso Balboa - GN 203
Esso Baltimore - GN 7, GN 21, GN 33, GN 43, KN 130, NG 301, NG 316, NG 328, NG 341
Esso Baton Rouge - KN 214, NK 512, NK 516
Esso Bayonne - GN 12, KN 111, KN 122, KN 133, KN 276, KN 281, NG 304, NG 409, NK 573
Esso Baytown - GN 37, GN 44, GN 120, KN 104, NG 343, NG 429, TAW 12, WAT 14
Esso Bayway - GN 9, GN 42, KN 123, KN 135, NG 305, NG 399
Esso Belgium - GN 15, GN 32, GN 44, GN 84, GN 105, GN 202, KN 241, KN 256, NG 325, NG 341, NG 382, NG 393, NG 414, NG 460, NG 506, NK 530, NK 538, NK 552
Esso Bolivar - GN 4, GN 12, GN 25, GN 32, GN 38, GN 48, GN 54, GN 60, GN 66, GN 71, GN 77, GN 85, GN 91, GN 96, GN 101, GN 107, GN 113, GN 120, GN 130, GN 138, GN 153, GN 157, GN 205, KN 130, NG 300, NG 309, NG 317, NG 329, NG 336, NG 347, NG 353, NG 359, NG 365, NG 370, NG 376, NG 384, NG 390, NG 400, NG 406, NG 412, NG 429, NG 443, NG 448, NG 457, NG 462, NG 504
Esso Buffalo - GN 139, KN 248, NK 545
Esso Caracas - GK 771, GK 775, GK 798, KG 643, KG 707, WAT 14
Esso Charleston - GN 25, GN 45, KN 103, KN 113, KN 127, NG 320, NG 343, NK 686
Esso Concord - GN 58, TAW 4, TAW 6, WAT 13, WAT 14

Faireno - KG 680
Fairfax - GK 800, GN 94, GN 187, GN 195, GN 202, NG 394, NG 485, NG 493, NG 500
Fairisle - KN 244
Falcon - GK 703, GK 717, GK 725, GK 731, GK 742, GK 753, GK 763, GK 772, GK 784, GN 122, GN 132, GN 142, GN 153, GN 204, KG 609, KG 619, KG 626, KG 633, KG 647, KG 657, KG 667, KG 676, KG 689, NG 428, NG 438, NG 449, NG 459, WAT 10
Falmouth - GK 707, GK 716, GK 723, GK 733, GK 743, GK 753, GK 761, GK 770, GK 781, GK 793, GK 803, GK 811, GK 819, GK 834, GK 870, KG 609, KG 624, KG 630, KG 636, KG 646, KG 657, KG 665, KG 674, KG 687, KG 696, KG 706, KG 714, KG 727, KG 742, KG 753, KG 764, KG 774, TAW 1
Fana - GK 792, GN 128, GN 139, GN 181, KG 675, KG 686, KG 696, NG 478, NK 577, NK 613
Far - GN 66, KG 638, NG 335
Farida - KG 630, KG 648, KG 697, KG 702, NK 591
Farrandoc - GK 716, KG 630
Fawkner Park - NG 416
Federico Schwager - KN 326, NG 451
Felipi De Bastrop - KN 317
Felix Taussig - GN 18, GN 47, GN 88, GN 94, GN 104, GN 110, GN 127, GN 138, GN 148, GN 160, GN 202, KN 303, NG 329, NG 388, NG 394, NG 404, NG 410, NG 421, NG 427, NG 434, NG 444, NG 454, WAT 17
Fellipe Camarao - TAW 8
Fenja - AH 2, GK 703, GN 19, GN 29, GN 38, GN 45, GN 65, GN 77, GN 83, GN 89, GN 95, GN 106, GN 123, GN 128, GN 138, GN 144, GN 153, GN 203, HA 2, KN 204, KN 251, KN 293, KN 340, NG 302, NG 315, NG 324, NG 334, NG 343, NG 363, NG 375, NG 388, NG 394, NG 405, NG 422, NG 433, NG 443, NG 448, NK 549, NK 590, NK 637
Ferdinand A Silcox - GN 133, NG 442
Ferdinando Gorges - GN 129, NG 416, NG 437, NG 455, NG 506
Fernbank - GN 20, KN 212
Fernbrook - GN 76, NG 330
Ferncliff - KN 123, TAW 4
Ferncourt - GN 15, GN 40, KN 128, NG 312, NG 339, TAW 9, WAT 5
Fernglen - KN 134, NG 308, NG 343, NG 368, NG 393, NG 436, TAW 14
Fernhill - GK 700, KN 200, NG 315, TAW 16
Fernplant - KN 124, NG 303, NG 350, TAW 5
Fernwood(Nor) - GN 3
Filleigh - GN 42, KN 342, NG 458
Finley Peter Dunne - GK 774, KG 682
Firethorn - KN 127, NG 301, TAW 8
Firmore - GK 727, GK 751, GK 761, GK 773, GN 18, GN 62, GN 106, GN 130, GN 137, GN 144, GN 153, GN 184, KG 606, KG 631, KG 655, KG 665, KG 676, KN 129, NG 328, NG 373, NG 428, NG 435, NG 442, NG 450, NG 464, NG 472, NG 506, NK 502, TAW 10
Fisher Ames - GN 81, KN 119, NG 340
Fishpool - GN 25, NG 337
Fitzhugh Lee - KN 233
Fjallfoss - KN 327, NK 626
Flight Command - NG 321, NG 325
Flimston - GN 34, GN 138, NG 435
Flomar - GK 773, GK 781, GN 30, GN 121, GN 133, KN 276, KN 282, KN 289, KN 321, NG 304, NG 394, NG 400, NG 407, NG 418, NG 429, NG 440, NG 447, NG 461
Flora Macdonald - KG 625
Florence Crittenton - GN 102
Florida Ii - GK 728, GK 742, GK 760, GK 774, GK 798, KG 613, KG 618, KG 630, KG 647, KG 657, KG 666, KG 678
Florida(Amer) - GN 45, GN 84, KN 108, KN 119, KN 133, KN 319, NG 306, NG 354, NG 502, NK 610
Florida(Hon) - GK 730, GN 30, KG 627, KG 652, KG 672, KG 695, KN 218, KN 262, KN 271, NG 343, NG 465, NK 507, NK 567, NK

Fort Mingan - GN 131, NG 428
Fort Nakasley - KN 239
Fort Nashwaak - GN 50
Fort Nipigon - GN 141
Fort Nisqually - GN 174, GN 204
Fort Norman - GN 20
Fort Paskoyac - GN 26
Fort Pelly - GN 12, NG 343
Fort Perrot - GN 125
Fort Pine - KN 318, NK 614
Fort Pitt - GN 7
Fort Poplar - GN 28
Fort Prudhomme - KN 278
Fort Rae - GN 20
Fort Rampart - GN 48
Fort Reliance - GN 22
Fort Rupert - GN 22
Fort Saleesh - GN 108
Fort Schuyler - KN 305
Fort Simpson - GN 16
Fort Slave - GN 26
Fort Souris - GN 28
Fort St Croix - GN 110
Fort St James - NG 320
Fort St Nicolas - KN 275
Fort Stager - GN 50
Fort Steele - GN 32
Fort Stikine - GN 14
Fort Sturgeon - KN 251
Fort Thompson - GN 28, GN 201
Fort Ticonderoga - GN 127, NG 423
Fort Turtle - KN 250
Fort Venango - GN 102
Fort Vermillion - GN 22
Fort Ville Marie - NG 317
Fort Wallace - GN 124
Fort Walsh - GN 48
Fort Wedderburne - GN 26
Fort Wrigley - GN 60, GN 112
Fort Yukon - GN 82
Forthbank - GN 32
Fosna - GN 116
Framlington Court - GN 199, NG 496
Frances E Willard - KN 258
Franche-Comte - KN 112, NG 307
Francis Drake - KN 269, NG 387
Francis E Warren - GN 102, KG 685, NK 589
Francis L Lee - GN 59, WAT 17
Francis Lewis - GN 54
Francis Marion - GN 32
Francis N Blanchet - GN 202
Francis Parkman - GN 5, KG 676, NK 579
Francis R Hart - GK 716, GK 745, KG 608, KG 637, KG 671, KG 712, TAW 15
Francisco Coronado - GK 741, GN 78, KG 648
Francois Hennebique - KG 737
Frank A Munsey - GK 801, KN 315
Frank A Vanderlip - NG 442
Frank Dale - NK 685
Frank Wiggins - NG 502
Franka - GN 36, GN 142, NG 439
Franklin K Lane - NG 407
Franklin P Mall - GN 90
Franklin S Leisenring - GK 717, GK 752, GK 761, KG 627, KG 667
Frans Hals - GN 22, NG 330
Frans Van Mieris - NK 685
Franz Klasen - KN 120, KN 206, NK 505
Franz Sigel - GN 205
Fred Herrling - KN 383
Fred Morris - GN 72, GN 100
Fred W Weller - GN 11, GN 27, GN 40, GN 70, GN 78, GN 86, GN 98, GN 107, GN 122, GN 131, GN 138, GN 154, KN 135, KN 231, KN 239, NG 304, NG 323, NG 335, NG 366, NG 375, NG 383, NG 392, NG 395, NG 403, NG 412, NG 429, NG 443, NG 451, NG 459, NK 528, NK 536, TAW 14, WAT 11
Frederic A Eilers - GN 201
Frederic C Howe - GN 109, KG 678
Frederic R Kellogg - GN 28, GN 116, GN 125, KN 117, KN 135, KN 207, KN 231, NG 321, NG 332, NG 355, NG 424, NG 430, NK 502, NK 528, TAW 14, WAT 11
Frederic Remington - GN 84, NG 351
Frederick C Hicks - GN 202
Frederick C Johnson - GK 770, KG 676, NG 312
Frederick H Newell - GN 85, KG 654
Frederick L Dau - GN 131, KN 247
Frederick W Taylor - KN 383, NG 433
Fredrick Von Steuben - KN 340
Freeman - GK 723, GN 68, KG 633, NG 314, NG 315
Freja - NG 459
Fresno City Ii - GN 48, WAT 13
Fresno Star - GN 41, KN 258, NK 542

George Gershwin - GN 125, KN 270
George H Thomas - GN 59
George H Williams - GK 729
George Handley - KN 386
George Kenny - GN 107, GN 129, NG 414
George Leonard - GN 25, KG 608
George M Bibb - GK 738, KN 252
George M Cohan - NK 584
George M Pullman - NG 440
George Matthews - GN 54
George P Garrison - GN 98
George Poindexter - KN 246
George Read - KG 711, KN 270, KN 320, NK 617
George Sharswood - GN 86, NG 351
George Steers - KN 305
George Taylor - NG 412
George Thacher - KG 603, KN 136, TAW 15
George Uhler - NK 618
George Vancouver - GN 32, NG 410
George Vickers - GN 141, NG 421
George W Barnes - GN 69, KN 110, KN 128, KN 205, KN 210, KN 214, KN 227, KN 233, KN 240, KN 256, KN 264, KN 274, NG 368, NG 398, NG 450, NK 504, NK 509, NK 513, NK 537, NK 553, NK 567, NK 635
George W Campbell - GN 50
George W Kendall - GN 202, NG 502, NG 506
George W Lively - KN 274, KN 386
George W Mccrary - KN 245
George W Tucker - KG 715, KN 335, NK 640
George Washington Ii - GN 12, GN 35, GN 75, GN 81, NG 373, NG 391, TAW 13
George Westinghouse - GN 52, NK 541
George Wythe - GK 759, GN 92, KG 664, KN 121, NG 347, NK 685, WAT 14
Georgia - KN 248, NG 393, NK 544, NK 545
Georgian - GN 42, NK 545, WAT 7
Georgios P - GN 6, GN 43, GN 80, GN 108, KG 648, WAT 13
Gerard Dou - GN 123, KG 692, KG 713
Gerassimos Vergottis - NG 308
Germa - GN 138, GN 165, NG 436, NG 463
Geronimo - GN 103, GN 155, NG 441
Gertrude Kellogg - GK 705, GK 726, WAT 14
Gibara - KN 381, NK 684
Gideon Welles - GN 43
Gilbert Stuart - NG 439
Glaisdale - AH 4
Glarona - GN 32, GN 123, GN 140, KN 126, NG 328, NG 422, TAW 1, TAW 2, TAW 6
Glen Pool - AH 4
Glen White - GN 57, GN 73, GN 84, NG 338, NG 368, NG 380, NK 636
Glenaffric - GN 40, NG 425
Glenapp - GN 29, GN 114
Glenbank - GN 52, NK 613
Glenn Curtiss - GN 114
Glenpark - GN 198, NG 496
Glenpool - GN 1, GN 12, GN 26, GN 38, GN 44, GN 50, GN 56, GN 62, GN 68, GN 74, GN 80, GN 112, GN 118, KN 123, NG 306, NG 317, NG 335, NG 343, NG 349, NG 355, NG 361, NG 367, NG 373, NG 379, NG 384, NG 417, NG 427, WAT 15
Glenstrae - GN 185, NG 460
Glenwood - KN 136, NG 313, TAW 15
Glittre - GN 16, KG 600, KN 214, NK 513
Globe Ix - GN 130
Globe Viii - GN 130
Gloria - GN 30, GN 60, GN 85, GN 105, GN 129, GN 151, GN 175, NG 339, NG 369, NG 392, NG 413, NG 436, NG 457, NG 483, NG 504
Godfrey B Holt - GN 23
Gogra - GK 715, KN 215
Goiazloide - GN 30, GN 73, NG 345, NG 384, WAT 8
Gold Heels - GN 12, KN 123, KN 210, KN 231, KN 257, NG 304, NG 306, NG 406, NK 508
Gold Shell - GN 1, KN 111, KN 204, NK 503
Goldmouth - GN 56
Good Gulf - GN 25, GN 158, KN 119, KN 135, KN 384, NG 321, NG 457
Gorjistan - GN 56, KG 600, KN 121, TAW 3
Governor - GN 31, GN 61, NG 319, NG 351
Governor John Lind - GK 775, GK 785, GK 796, GK 806, GK 827, GK 841, GN 175, KG 639, KG 679, KG 691, KG 700, KG 709, KG

GN 52, GN 58, GN 64, GN 70, GN 104, GN 111, GN 118, GN 127, GN 133, GN 139, GN 157, GN 181, KG 608, KN 111, KN 126, KN 200, KN 214, KN 217, KN 259, KN 271, KN 278, NG 345, NG 351, NG 357, NG 363, NG 369, NG 375, NG 402, NG 410, NG 416, NG 424, NG 432, NG 450, NG 456, NG 462, NG 505, NK 503, NK 513, NK 519, NK 568, NK 574
Gulfprince - KN 113, KN 128, KN 202, NK 501, NK 510
Gulfstate - GN 37, KN 225, NK 523, NK 529, TAW 8, WAT 15
Gulftide - AH 2, GN 5, GN 25, GN 37, GN 89, GN 101, GN 125, GN 131, GN 204, KN 101, KN 111, KN 206, KN 229, KN 238, KN 248, KN 262, KN 276, KN 288, NG 301, NG 320, NG 333, NG 388, NG 400, NG 410, NG 412, NG 423, NG 430, NG 443, NG 449, NG 455, NG 461, NG 503, NK 505, NK 521, NK 526, NK 535, NK 545, NK 551, NK 557, NK 585, WAT 11
Gulfwave - GN 123, GN 140, GN 154, GN 158, KN 311, NG 422, NG 439, NG 444, NG 449, NG 458, NG 506, NK 608
Gulfwax - GN 204, NG 449, NG 456, NK 641
Gulfwing - AH 3, GN 122, GN 128, GN 139, GN 201, KN 115, KN 209, NG 421, NG 427, NG 433, NG 444, NG 450, NG 455, NG 506, NK 508
Gullpool - GN 26, NK 564
Gurney E Newlin - GN 9, KG 602, KN 130, NK 501
Gus W Darnell - KG 713
Gylfe - GN 57
Gypsum Empress - NG 301
Gypsum King - GK 759, GN 14, GN 122, KN 267, NG 328, NG 402, NG 454, WAT 17
H G Blasdel - GK 778
H H Rogers - KN 126
H Lechatelier - GK 858, KG 728
H M Flagler - GN 18, GN 34, KN 123, NG 308, NG 331, NG 403, TAW 2
H M Fredrichsen - KN 302, KN 320
Haakon Hauan - KN 202, KN 212, KN 341, NK 511, NK 638
Hadnot - GN 12, GN 24, GN 38, GN 45, GN 51, GN 57, GN 63, GN 69, GN 87, GN 93, GN 112, KN 112, KN 135, KN 286, KN 299, KN 314, NG 304, NG 317, NG 335, NG 343, NG 350, NG 356, NG 362, NG 368, NG 382, NG 392, NG 404, NG 412, NG 417, NG 423, NG 429, NG 446, NG 506, NK 577, NK 620, NK 636
Hagood - GK 757, GN 9, GN 47, GN 53, GN 59, GN 65, GN 71, GN 107, GN 114, GN 120, GN 126, KN 117, KN 222, KN 265, KN 271, KN 281, KN 322, KN 382, NG 304, NG 346, NG 352, NG 358, NG 364, NG 370, NG 376, NG 382, NG 405, NG 412, NG 419, NG 425, NG 445, NG 451, NG 456, NG 506, NK 568, NK 574, NK 610
Hahira - GK 703, KN 205
Hall J Kelley - GN 84
Hallanger - WAT 1
Hamlin F Mccormick - GK 701, GK 722, KG 605, KG 606, KG 624, WAT 7
Hammerfest - NG 506
Hamul - KN 217
Hanging Rock - NG 471
Hanley - GN 19, GN 73, GN 117, NG 330, NG 386
Hannibal Hamlin - KN 386, NK 683
Harberton - WAT 15
Hardanger - GK 758, GN 61, KN 266, NK 561, WAT 8
Harmala - GN 28
Harmonic - GN 32
Harold Dossett - KN 385
Harold O Wilson - KN 386
Harold T Andrews - KN 300
Harpalyce Ii - AH 4
Harpalycus - GN 195, NG 491
Harpefjell - GK 776, GK 790, GN 124, GN 135, KG 684, KG 693, KN 310, KN 322, KN 327, KN 333, KN 349, KN 355, KN 360, NG 400, NG 435, NK 608, NK 620, NK 625, NK 636, NK 646, NK 658
Harperley - GN 38
Harriet Beecher Stowe - GN 119, NK 685
Harry Lane - GN 41
Harry Luckenbach - GN 32, WAT 2

NK 613
Herman F Whiton - GK 744, GK 754, GK 766, GK 777, GK 801, GK 849, GN 30, GN 113, GN 136, GN 145, KG 630, KG 637, KG 648, KG 659, KG 670, KG 682, KG 696, KG 705, KG 757, KN 135, NG 311, NG 336, NG 418, NG 442, NG 451, NG 462, NG 471, TAW 13
Herman Frasch - GK 721, GK 728, GK 736, GK 745, GK 754, GK 764, GK 774, GK 784, GK 842, GK 865, GN 123, GN 132, GN 142, GN 151, GN 159, GN 178, GN 204, KG 630, KG 639, KG 648, KG 659, KG 667, KG 678, KG 692, KG 770, NG 429, NG 438, NG 448, NG 457, NG 465, NG 486
Herman Melville - GN 77, NG 319
Hermes(Du) - GK 710, GK 722, GK 740, GK 757, GK 772, GK 793, GK 805, KG 612, KG 632, KG 646, KG 661, KG 683, KG 697, KN 121, KN 320, NG 300, TAW 3
Hermiston - GN 38
Heron(Nor) - KN 285
Heyser - KN 300
Heywood Broun - GN 115
Hibueras - GK 701, KG 603, WAT 1, WAT 8
Hilary - NG 334
Hilton - GK 739, GK 755, GN 32, GN 54, GN 94, GN 112, GN 121, GN 129, GN 137, GN 149, GN 159, KG 627, KG 649, KG 661, KN 118, NG 327, NG 338, NG 360, NG 401, NG 410, NG 419, NG 427, NG 435, NG 443, NG 456
Hilversum - GN 26, GN 58, GN 122, KN 114, NG 316, NG 355
Hindoo - KN 297, NG 456
Hindustan - NG 420
Hinton R Helper - GN 114
Hiram - GN 22, GN 126, KN 277, KN 285, KN 291, KN 296, KN 301, KN 314, KN 320, KN 325, KN 331, KN 339, KN 346, KN 352, KN 358, KN 363, KN 369, KN 375, KN 382, NG 425, NK 574, NK 583, NK 589, NK 594, NK 599, NK 611, NK 618, NK 623, NK 628, NK 637, NK 642, NK 650, NK 655, NK 661, NK 673, NK 679
Historian - GN 43
Hjelmaren - GN 97, NG 394, WAT 17
Hm Ft 12 - KG 657
Hm Ft 17 - KG 661
Hm Ft 6 - KG 661
Hmcs Provider - KN 231
Hms Lariat - KG 603
Hms Lincoln Salvor - GN 104
Hobart Baker - KG 676, KN 275
Hoegh Hood - HA 3, KN 107, KN 339, NK 637, TAW 15
Hoegh Scout - GN 128, KN 128, NG 312
Hoegh Silvercloud - GN 14, GN 56, KN 271, NG 324, NK 543
Hoegh Silverdawn - GN 5, GN 48, NG 316, NG 357
Hoegh Silverlight - GN 14, GN 62, NG 325, NG 371
Hoegh Silverstar - GN 20, KN 250, NG 331
Hoffman(Amer) - KN 270
Hollywood - GN 107, NG 404, NG 442
Holmbury - GN 20
Hoogkerk - GN 26
Hopecrest - GK 787, GN 31, KN 295, NG 412
Hopepeak - GN 147, NG 445, WAT 10
Hoperange - NG 385
Hoperidge - GN 30, NG 300
Hopestar - WAT 11
Hopetarn - GN 29, NG 339
Horace Gray - GN 86, NG 348
Horace H Watson - KG 715
Horace Luckenbach - GN 65, KN 135, NG 308, NG 505
Horace Mann - GN 54, NG 425, NK 541
Horace Wells - KN 340
Horace Williams - GN 87, KG 632
Houston Volunteers - GN 26, KG 612
Howard E Coffin - GN 205
Howard Gray - KG 715
Howard L Gibson - KN 340
Howard Stansbury - GK 805
Howard T Ricketts - GN 125, NG 413
Hubert Howe Bancroft - GN 59
Huemul - GK 748, GK 763, GK 792, GN 138, KG 624, KG 636, KG 699, NG 446
Hugh Mcculloch - NG 439
Hugh S Legare - KN 258
Hugh Williamson - NG 437

Jade - AH 4
Jalapa - KN 250
James A Butts - NK 685
James B Aswell - GK 806, KG 711, KN 305
James B Eads - KN 262
James B Hickok - NK 636
James B Miller - GK 802, KN 317
James B Richardson - GN 64, NG 322
James B Weaver - KN 254, NG 504
James Barbour - KN 227
James Bennett Moore - KN 386
James Bowie - GN 30, KG 614
James C Cameron - GN 132, NG 431
James Caldwell - GN 103, NG 357
James Cook - GN 141, GN 146, GN 156, KN 384, NG 440, NG 456, NG 461
James De Wolf - GN 122, GN 153, NG 401, NG 431
James E Haviland - KN 248
James E Howard - KG 634
James Ellwood Jones - GN 54, GN 63, GN 74, NG 312, NG 361, NG 370
James Fenimore Cooper - GN 12, GN 45, KN 272, NG 321, NG 390
James Fergus - GN 135, GN 159, NG 419, NG 443, NG 506
James Ford Rhodes - GN 46
James G Birney - KN 238
James G Blaine - GN 64, GN 132, NG 312, NG 431
James G Maguire - GK 800, KN 312
James Gunn - GN 46, WAT 16
James Guthrie - KN 267
James H Kimball - KG 710
James Harlan - KG 668, KN 277
James Harrod - KN 268
James Hawson - GN 14, GN 203, KN 120, KN 235, NG 311, NK 532, TAW 1
James Ives - GN 80
James J Hill - KN 245
James J Maguire - KN 129
James K Polk - NG 308
James Kerney - GN 201
James L Breck - GN 108
James L Richards - GK 757, GN 30, GN 61, GN 71, KN 265, NG 321, NG 339, NG 369, NG 378
James Longstreet - GK 734
James Lykes - GN 23, GN 104, KN 250, NG 336
James M Porter - KN 271
James M Wayne - AH 4
James Madison - GK 781, GN 69, GN 133, KG 687
James Mchenry - GN 96, KG 629
James Mckay - GN 12
James Miller - NK 684
James Monroe - KN 131, TAW 12
James Oliver - NG 396
James Parker - GN 20, KG 610
James Rumsey - KN 286
James Russell Lowell - GN 50
James S Hogg - KG 635
James Schureman - GN 64, NG 313
James Shields - NG 435
James Smith - KN 258
James Swan - NK 686
James Turner - GN 51
James W Fannin - KN 245
James W Johnson - NK 686
James W Marshall - GN 60
James W Nesmith - GN 9, GN 61, NG 317
James Whitcomb Riley - GN 34
James Withycombe - KG 674
James Woodrow - NG 506
Jamestown - GK 755, GN 59, GN 65, KN 234, KN 280, KN 287, KN 297, KN 312, KN 320, KN 334, NG 358, NG 364, NG 372, NK 577, NK 584, NK 601, NK 608, NK 615, NK 637
Jan Van Goyen - GN 141, NG 437
Jane A Delano - GK 763, GN 100, KG 668
Jane G Swisshelm - KN 309
Jane Long - GN 70, KG 637
Jared Ingersoll - GN 51, GN 85, NG 308, NG 359, NK 598
Jason Lee - GK 736, GN 127, KG 706, NG 414, NK 612
Jasper Park - NG 344
Java - GN 18, KN 260
Jean - GK 791, GN 135, KN 125, KN 299, KN 304, NG 415, NG 423, NG 432, TAW 6
Jean Baptiste Le Moyne - GN 124, GN 138, KN 311, NG 423, NG 443, NK 608, NK 613
Jean L D - WAT 13
Jean Ribaut - KN 320
Jedediah S Smith - GN 99
Jeff Davis - WAT 11

6
John M Brooke - GN 132, KG 701, NG 439
John M Clayton - KG 682, KN 243
John M Parker - KN 314
John M Schofield - GN 143
John Marshall - KN 116
John Mary Odin - GN 78, GN 129, NG 414
John Mason - NK 685
John Mcdonogh - KN 253, KN 268, NG 387
John Mcloughlin - GK 711, KN 207
John Merrick - GN 186, NK 685
John Morton - GN 56
John Owen - GN 97
John P Holland - GN 56
John P Mitchell - GN 151, GN 203
John Page - KN 127, NG 436, TAW 8
John Paul Jones - KN 260
John R Park - GN 81
John R R Hannay - GK 710, GK 721, GK 738, GK 776, KG 617, KG 627, KG 633, KG 642, KG 655, WAT 9
John Rutledge - GN 56
John S Bassett - GK 791, KN 305
John S Copley - GN 71, GN 204
John S Pillsbury - GN 65
John S Sargent - NK 637
John Sedgwick - GN 85
John Sevier - KN 129, NG 305, TAW 10
John Sharp Williams - GN 78, KG 645
John Stevens - GN 37
John Swett - GK 796, KG 702
John Vining - GN 37, GN 90
John W Brown - GN 60, NG 314
John W Davis - GN 134
John W Garrett - GN 120
John W Gates - KN 288
John W Griffiths - KN 307
John W Hoyt - GN 119
John W Searles - GN 141, NG 439
John Wise - KN 249
John Worthington - GN 16, GN 41, KN 123, KN 211, NG 337, NG 349, NK 510, NK 514
Jonathan Edwards - GN 59, KN 129, NG 307, TAW 10
Jonathan Elmer - GN 62, NG 312
Jonathan Grout - GN 20, KG 608, NG 329
Jonathan P Dolliver - GN 109, GN 131, NG 417, NG 441
Jonathan Sturges - GN 29, KG 614
Jonathan Trumbull - GN 67
Jose Bonifacio - KN 319, NG 436
Jose G Benitez - GN 204, KN 301
Jose J Acosta - GN 124
Jose Marti - GN 114
Jose Navarro - KN 275
Josefina Thorden - GN 43, NG 341
Joseph Alston - NG 333
Joseph Aspdin - KN 335
Joseph E Johnston - KN 214
Joseph Francis - GK 793, KG 699
Joseph Gale - GN 54
Joseph Goldberger - KN 282
Joseph H Hollister - GN 46
Joseph Hamilton - GN 158
Joseph Hewes - GN 30
Joseph Holt - KN 265
Joseph Hooker - GN 48, KG 604, KN 136, TAW 15
Joseph M Medill - KN 247
Joseph Mckenna - KN 266
Joseph N Nicollet - KN 249, KN 383
Joseph P Bradley - GK 774, KG 681, NK 541
Joseph Priestley - GK 796, GN 131, KG 700, NG 439
Joseph Pulitzer - GN 52
Joseph R Parrott - KN 103
Joseph Reynolds - KG 711, NG 432
Joseph Rodman Drake - GK 729, KN 238
Joseph S Emery - KN 263
Joseph Smith - GK 771
Joseph Squires - NK 683
Joseph Stanton - GN 32, GN 99, KG 600, NK 541
Joseph Story - KN 274
Joseph T Robinson - KN 219
Joseph W Folk - GK 794, KN 307
Joseph Wheeler - GN 75, KG 614
Josephine Shaw Lowell - GN 139, KG 708
Joshua A Leach - KN 272, KN 321, NG 438
Joshua Hendy - GK 811, KG 707, NK 613
Joshua Seney - GN 26, KG 612
Joshua W Alexander - KN 269, NG 387
Josiah B Grinnell - GN 142
Josiah G Holland - GN 141, NG 440

NG 419, NG 425
Kerma - GN 30
Kerry Patch - KN 274, NG 400
Kevin Moran - NG 314
Kewanee - KN 216, KN 277, KN 293, KN 296, KN 319
Keyholt - GK 712, KG 655, KG 705, KG 744
Keymont - GK 712, GK 791, GK 829, KG 662, KG 707, KG 734
Kielce - KN 308
King - GK 709, GK 715, KG 611, KG 621, KG 627, KG 635, KG 642, KG 648
King Edgar - GN 182, NG 317
King Frederick - WAT 17
King James - GN 14, NG 326, NG 408
King Neptune - GN 16, NG 324
King Stephen - GN 32
King William - GN 26, NG 338
King(Amer) - GK 709, GK 715, KG 601, KG 611, KG 621, KG 627, KG 635, KG 642, KG 648, TAW 5
Kingman - GK 806, NG 429
Kingsbury - GN 40
Kingsley - GK 733, KN 240
Kingswood - GN 40, GN 69, NG 365
Kirsten B - KG 649, KG 691, NG 368
Kiruna - KN 206, KN 238, NK 505, NK 535
Kitsilano Park - GN 114, NG 432
Knute Nelson - NG 506
Knute Rockne - GN 74, GN 153, NG 444, NG 460
Kobad - KN 291
Kollbjorg - GN 12, KN 127, NG 309
Kolno - KN 308
Kolsnaren - GN 15, GN 52, GN 76, GN 100, NG 306, NG 340, NG 373, NG 397
Komiles - NG 362
Komsomolets Arctiki - NG 377
Kong Haakon Vii - GN 41, WAT 9
Kopperston - GN 73, KG 632, NG 337
Kosmos Ii - KN 104
Kota Baroe - GN 37, NK 571, WAT 14
Kota Gede - GN 89, KG 675, NG 352, WAT 12
Kota Inten - GN 18, NG 342
Kotlin - NG 362
Kotor - KG 614, NG 319, NK 510, TAW S(1)
Kowel - KN 305
Krasnoe Znamya - NG 362
Kristianiafjord - GN 1, NG 313, TAW 16
Kroman - GN 174, NG 486
Kronprinsessen - GN 54, WAT 17
Krosno - KN 309
Kumasian Ii - NG 363, NG 432
Kutno - KN 315
Kuzbass - NG 382
L H Mcnelly - KN 276
L L Abshire - GK 705, GK 767, GN 76, GN 82, GN 88, GN 120, KN 209, KN 213, KN 241, KN 250, KN 275, KN 281, KN 287, KN 296, KN 310, KN 340, NG 374, NG 381, NG 387, NG 393, NG 419, NG 445, NK 512, NK 523, NK 530, NK 538, NK 547, NK 578, NK 584, NK 590, NK 613, NK 620, NK 685
L V Stanford - GK 701, GN 32, GN 128, KN 204, KN 278, KN 287, KN 302, KN 384, NG 329, NK 575, NK 581, NK 590, NK 612, NK 635, NK 686
La Paz(Am) - GK 780, GK 800, GN 144, KG 687, KG 716, NK 623
La Playa - KG 676, KG 683, KG 689, KG 695, KG 701
La Salle - KN 129, NG 304, TAW 9
Lachinedoc - GK 776
Lackenby - KN 215, NK 512
Lacklan - GN 66
Laconikos - GN 12
Lady Denison-Pender - NG 452
Lafayette - GK 774, GN 153, KN 288, KN 310, NG 385, NG 428, NG 461, NK 620
Lafcadio Hearn - GN 119, GN 125, GN 137, KN 282, KN 312, KN 384, NG 424, NK 609, NK 621
Lagarto - GN 24, GN 84, GN 124, KN 102, NG 373, NG 410
Lago - GN 24, GN 69, NG 311, NG 367
Laguna - GK 703, GN 65, GN 129, KN 210, KN 276, NK 562, NK 640
Lake Folcroft - GN 133, GN 151, KG 635, KG 660, KG 717, KN 290, NG 439
Lake Freeland - GN 117, GN 151, KG 669, KG 701, KN 314, KN 323, KN 353, NG

GN 28, GN 47, GN 63, GN 95, GN 123, GN 135, GN 154, KG 701, KG 716, KN 108, KN 123, KN 217, NG 321, NK 607, TAW 4, WAT 2
Manitou Park - GN 144, KG 673, NK 577
Manuel - GK 705, GK 774, GK 877, GN 91, GN 134, KN 203, KN 282, NG 341, NG 400, NG 433
Mapia - GN 46, KN 115, NG 411, NK 542, WAT 17
Maplewood - GN 200
Mara - NG 504
Marathon(Nor) - AH 2, GN 21, KN 102, NG 315, NK 636, WAT 11
Maravi - GK 709, GK 728, GN 123, GN 205, KG 611, KN 274, KN 281, KN 288, KN 294, KN 299, KN 314, KN 319, NG 398, NG 406, NG 413, NG 418, NG 423, NG 433, NG 438, NG 443, NG 505, TAW 10, WAT 1
Marcus H Tracy - NK 686
Margalau - GN 22
Margaret Lykes - GK 811, GN 112, GN 120, KN 130, KN 283, KN 308, NG 410, NG 418, NG 433, WAT 5
Margot Moran - KG 603, KG 609
Maria Sanford - KG 687
Maridal - KN 320, NK 612
Marie Bakke - GN 44, NG 300
Marietta E - GN 23, NG 332
Marija Petrinovic - GK 709, GN 141, KN 207, NG 305, NG 347, NG 438
Marina - GN 48, GN 87, KN 125, NG 306, NK 539, TAW 7
Marina Raskova - GN 101
Marion Mckinley Bovard - GN 53
Mariso - NG 344
Marit Ii - AH 1, GN 82, GN 123, HA 1, NG 344, NG 422
Marjory - GK 761, GK 773, GN 7, WAT 15
Mark Hanna - KN 268
Mark Keppel - KG 675, KN 263, NK 635
Mark Twain - KN 125, NG 439, TAW 6
Markay - NK 507
Markhor - GN 29
Markland - GK 731, GK 744, GK 765, GK 774, GK 782, GK 793, GK 802, GK 812, GN 24, GN 81, KG 608, KG 626, KG 635, KG 648, KG 668, KG 678, KG 688, KG 697, KG 707, KN 134, NG 335, NG 387, NK 504, TAW 13, WAT 6
Marpesia - GK 726, GK 743, GK 773, GN 118, GN 143, KG 648, KG 680, KG 705, KN 261, KN 268, KN 275, KN 307, KN 316, KN 336, KN 350, NG 386, NG 393, NG 400, NG 441, NG 448, NG 461, NK 604, NK 653
Marpessa - GN 69
Marsa - KN 105
Marsdale - GN 28
Martaban - GN 100, GN 129, GN 153, NG 422, NG 447
Martin Bakke - KN 127, TAW 8
Marvin Lyle Thomas - NG 505
Marwarri - GN 20
Mary A Livermore - GK 802, KN 315
Mary Ashley Townsend - GN 141, KN 384, NG 440, NK 610, NK 615
Mary Austin - KN 253, KN 269, NG 387
Mary Ball - GN 106, KG 675, NG 420
Mary Kingsley - GN 28
Mary Lyon - GN 94, NG 392, NK 686
Mary M Dodge - GN 132
Marymar - GN 9, NG 301
Massachusetts - GN 60, GN 71, GN 77, GN 83, GN 89, GN 95, GN 101, GN 107, GN 113, GN 119, GN 154, KN 205, KN 384, NG 318, NG 370, NG 376, NG 382, NG 388, NG 394, NG 400, NG 406, NG 412, NG 418, NG 432, NG 446, NK 504, NK 621
Masunda - GN 22, GN 52, GN 156, NG 346, NG 453
Mataroa - GN 77
Mathew Luckenbach - KN 133
Mathilda - KN 120
Mathura - GN 48
Matinicock - GN 87, GN 94, KN 136, KN 315, KN 335, NG 384, NG 393, TAW 8
Matthew B Brady - GN 131
Matthew J O'Brien - GN 107, KG 676
Matthew Maury - GN 41, KN 130, NG 304, TAW 8
Matthew P Deady - GN 97, NG 353, NG 439
Maturines - GK 769, GK 786, GK 801, GN 26, KG

130, GN 142, GN 154, GN 159, NG 344, NG 350, NG 356, NG 362, NG 369, NG 381, NG 411, NG 423, NG 429, NG 440, NG 446, NG 451, NG 458, NK 519, NK 684, TAW 11, TAW 12, WAT 17
Mobile City - GK 846, NG 466
Mobilgas - GN 23, GN 47, GN 140, KG 692, KN 101, KN 111, KN 127, KN 200, KN 264, NG 320, NG 387
Mobilight - KN 240, NK 538, NK 547
Mobiloil - NG 367, NK 533
Moena - WAT 15
Mogy - GN 38, GN 73, GN 115, KG 603, KG 629, NG 350, NG 381, NG 423, TAW 14
Mokihana - GN 16, KG 602
Molda - KN 124, NG 502, TAW 5
Moldova - GK 713, GK 722, GK 730, GK 737, GK 747, GK 756, GK 775, GK 787, GN 123, GN 136, GN 145, GN 155, GN 164, GN 185, GN 198, KG 615, KG 625, KG 632, KG 641, KG 651, KG 669, KG 680, KG 692, KN 353, KN 367, NG 430, NG 442, NG 451, NG 461, NG 472, NG 481, NG 493, NG 505, TAW 7
Molino Del Rey - NK 543
Monkleigh - KN 218, NK 516
Monmouth - GN 81, NK 509
Monroe - GK 761, GN 4, GN 31, GN 113, GN 119, GN 126, GN 146, GN 202, KG 706, KN 315, NG 314, NG 372, NG 399, NG 419, NG 426, NG 441, NK 612, WAT 2
Mont Everest - GK 734, KG 622, KG 627, KN 242
Montana - KN 136, KN 295, NK 541
Montanan - GN 21
Montcalm - GK 712
Monterey Ii - GK 765, GK 781, KG 654, KG 668, KG 688, NG 344
Montevideo - KN 102
Montgomery City - KN 266, NG 384
Montmorency Park - GK 767, GN 110, GN 121, KN 275, NG 385, NG 403, NG 417
Montoso - GK 707, GK 715, GK 723, GK 730, GK 741, GK 757, GK 766, GK 778, GK 790, GK 799, GK 809, GK 814, GK 819, GK 832, GK 845, GK 868, KG 609, KG 617, KG 625, KG 632, KG 649, KG 661, KG 670, KG 686, KG 703, KG 714, KG 725, KG 739, KG 750, KG 776, WAT 14
Moorby - GN 63, KN 129, KN 341, NG 302, TAW 10
Moose Mountain Park - GN 120, GN 132, GN 138, GN 159, GN 200, NG 425, NG 431, NG 443, NG 458
Moreton Bay - GN 33
Morgenen - GN 31, KN 116, KN 204, NK 502
Mormacdale - NG 347, NG 388
Mormacdove - KN 243
Mormacmar - GN 50, GN 84, GN 101, GN 121, KN 311, NG 372, NG 391, NG 409, NG 430, WAT 2
Mormacrey - GN 79, GN 99, GN 119, KN 227, NG 317, NG 363, NG 387, NG 427, NG 452
Mormacrio - GK 786, GN 58, GN 84, GN 106, GN 135, KN 294, NG 368, NG 391, NG 412, NG 420, NG 442, WAT 13
Morris Sheppard - GN 122, KG 691, NG 430
Morrison R Waite - GK 811, KN 245, NG 435
Morton M Mccarver - GN 75, GN 139, NG 425, NG 445
Morton Prince - GN 137, KN 336, KN 383, NG 437, NG 446, NG 459, NK 621
Moses Austin - GN 64, KG 633
Moses Brown - GN 86, NG 385
Moses Cleaveland - GN 53, KN 386
Moses G Farmer - NG 462
Moses Rogers - GN 99, KG 627
Mosli - AH 3, TAW 3, TAW 15, WAT 3
Mossovet - NG 361
Motocarline - GN 15, GN 69, GN 111, NG 325, NG 380, NG 421
Mount Bruce Park - GN 108, GN 120, GN 127, GN 142, GN 154, GN 160, NG 425, NG 440, NG 447
Mount Douglas Park - GN 136
Mount Evans - GN 105, KN

Norden - GN 52
Nordlys - GN 26
Norefjord - KN 125
Noreg - GN 21, GN 52, KN 129, NG 319, NG 351, TAW 10, WAT 5
Norfalk - GN 129, NG 423
Norfjell - KG 602, KN 119, NK 501
Norfolk I - GK 702
Norfolk(Amer) - GK 727, KN 282, NG 423, NK 578, NK 626
Norheim - KN 102, KN 131, KN 210, NK 509
Norholm - AH 3, GN 20, NG 318
Norindies - GK 712, GK 723, GK 730, GK 742, KG 618, KG 625, KG 632, KN 126, KN 250, NG 307, NG 394
Norlom - GK 754, KN 262, NG 378
Norlys - GN 50, KN 106, NG SPEC
Norma - GN 18, KN 339, NG 332, NG 440, NG 457
Norman Star - GN 23, NG 343
Norse Lady - GN 51, NG 344, NG 345, NK 504
Norseman - NG 394
Norsk Tank - AH 3, GN 38, GN 70, GN 141, HA 3, NG 333, NG 369, NG 434
Norsol - GN 21, GN 49, KN 102, NG 318
Norteloide - GN 72, KG 666, NG 389, NG 392
North Britain - GN 24
North King - GN 97, NG 387
Northern Chieftain - KN 385
Northern Sun - KN 107, KN 128, KN 201, KN 206, KN 210, KN 215, KN 235, NK 504, NK 505, NK 509, NK 514, NK 638
Northern Sword - KG 622
Northern Traveller - KN 385
Northern Warrior - KN 384
Northland - GN 70
Northleigh - NG 311
Northumberland - GN 18, KN 238, NG 339
Nortind - GN 33, KN 205, NG 328
Norton - NG 398
Nortun - NG 318
Norvarg - GN 140, GN 157, GN 192, NG 438, NG 456, NG 489
Norvik - KN 115, WAT 16
Norvinn - AH 3, GN 203, NG 503, WAT 6, WAT 12
Norwalk - KN 214
Norwegian - KN 382
Notos - GK 738, GK 747, GK 755, GK 761, GK 777, GK 795, GN 130, GN 141, GN 150, GN 159, GN 169, GN 179, GN 190, GN 200, KG 627, KG 633, KG 642, KG 651, KG 659, KG 666, KG 699, KN 290, NG 436, NG 447, NG 456, NG 475, NG 486, NG 496, NG 506, WAT 5
Novelist - GN 68, GN 89, GN 112, NG 360, NG 406, NK 501
Nuculana - GN 69, NK 510
Nueva Granada - KN 107, KN 133, KN 208, NK 507
Nuolja - GN 175
Nurani - NG 321
Nyanza - GN 134, GN 198, KN 353, NG 308, NG 431, NG 470, NG 494
NYCO - GN 32, GN 113, GN 118, GN 140, GN 148, GN 158, GN 182, GN 192, GN 202, KN 247, KN 289, KN 304, KN 309, KN 316, NG 413, NG 418, NG 440, NG 446, NG 456, NG 464, NG 488, NG 498, NG 499, NK 542, NK 544, NK 587, NK 602, NK 607, NK 613, NK 651
Nyholm - NG 504
O B Martin - GN 155, KG 708
O B Sorensen - GN 24
O M Bernuth - GN 125, KN 106, KN 117, KN 127, KN 205, KN 233, NG 394, NG 434, NK 503, NK 530
Oakey L Alexander - GN 60, GN 70, NG 343, NG 368
Oakmount Park - NG 488, NG 506
Oakworth - GN 52, KG 677
Ob - NG 368
Ocean Angel - KN 385
Ocean Courage - KN 201
Ocean Fame - NG 439
Ocean Glory - NG 327
Ocean Gypsy - NG 326
Ocean Hope - GN 25, NG 332
Ocean Peace - GN 19, NG 328
Ocean Pride - GN 29, NG 338
Ocean Vagabond - GN 1
Ocean Vagrant - GN 52, KN 123, TAW 4
Ocean Valentine - KN 133, NG 424, TAW 13
Ocean Valley - GN 17, NG

Oscar Chappell - KN 281
Oscar F Barrett - GN 138, KN 337, NK 641
Osmond - KN 248, NK 551
Osmussar - NG 365
Osthav - GN 61, NG 431, NG 444
Otina - GN 19, KN 108
Ottern - GN 130
Owen Summers - GN 75, NG 401
Ozark - KG 635
P E Crowley - KG 631
P T Barnum - NG 434
P.L.M.17 - GN 47, NG 342, NK 504
Pacific(Pan) - GN 7, GN 9, KN 214, KN 225, KN 240, KN 246, KN 252, KN 275, NK 543, NK 549, NK 555
Pacific(Sw) - GN 199, NG 496
Paco - GK 710, GN 151, KG 603, NG 456, WAT 1
Paderewski - GK 700, NG 312, TAW 16
Paine Wingate - GN 62
Palembang - NG 425
Palo Alto - GK 777, GN 153, KN 229, NK 527
Pan Aruba - KN 106, KN 201, KN 211, NK 500, NK 510
Pan Carolina - GN 3, KN 112, KN 207, KN 211, KN 215, KN 226, KN 232, KN 238, KN 246, KN 252, KN 258, KN 264, KN 271, KN 277, KN 283, KN 294, KN 305, KN 312, KN 336, NK 506, NK 510, NK 514, NK 529, NK 535, NK 543, NK 549, NK 555, NK 561, NK 568, NK 574, NK 586, NK 608, NK 609, NK 619, NK 638
Pan Crescent - GK 720, GK 728, GK 741, GK 748, GK 793, GN 17, KG 625, KG 634, KG 690, KG 697, KG 714, NG 329, NG 459
Pan Gulf - GK 713, KN 222, NG 306
Pan Orleans - KN 103
Pan Scandia - GN 74
Pan York - GK 730, GK 742, GK 749, GK 772, GK 821, KG 626, KG 667, KG 676, KG 688, KG 696, NG 339, NK 516
Pan-Amoco - GN 63, KN 106, KN 120, KN 131, NG 302, NG 373, NG 427, NG 504
Pan-Delaware - KN 113, KN 125, KN 200, KN 211, NK 510
Pan-Florida - GN 6, GN 44, KN 128, KN 207, KN 211, KN 215, NG 343, NK 506, NK 510, NK 514, TAW 6, WAT 17
Pan-Georgia - KN 111, KN 205, KN 209, KN 214, KN 238, KN 244, KN 257, KN 280, KN 291, KN 297, KN 308, KN 315, KN 320, KN 386, NG 503, NK 504, NK 508, NK 512, NK 536, NK 541, NK 547, NK 553, NK 577, NK 583, NK 594, NK 611
Pan-Maine - KN 109, KN 136, KN 210, NK 509
Pan-Maryland - GN 32, KN 118, KN 128, KN 205, NG SPEC, NK 503
Pan-New York - KN 111, KN 204, NK 503
Pan-Rhode Island - KN 105, KN 131, KN 203, KN 224, NK 501
Pan-Virginia - GN 151, KN 111, KN 124, KN 204, KN 208, KN 216, KN 237, KN 243, KN 248, KN 255, KN 261, KN 266, KN 272, KN 282, KN 291, KN 297, KN 310, KN 316, KN 321, KN 337, NK 507, NK 511, NK 515, NK 527, NK 529, NK 534, NK 545, NK 551, NK 558, NK 563, NK 569, NK 576, NK 585, NK 594, NK 599, NK 613, NK 618, NK 653, NK 684
Panam - NG 355, NK 538
Panama - KN 135, TAW 14
Panama City - GN 97, GN 123, GN 142, KG 692, KG 706, KN 315, NG 379, NG 430, NG 440, NK 584, NK 586
Panaman - GN 46, KN 278, NG 300, NG 396, NG 466
Pandorian - GN 138, NG 434, NG 435
Para - GN 36, GN 147, KN 284, NG 400, NG 425, NG 445, NG 468, WAT 8
Paranaloide - GK 782, GN 143, GN 185, NG 386, NG 423, NG 451, NG 493
Parismina - KN 113
Parkhaven - GN 38
Patricia Skakel - NG 319
Patrick C Boyle - NK 686
Patrick Henry - NG 417
Patrick J Hurley - KN 106, KN 117
Paul H Harwood - GN 39, KN 108, KN 204

539, NK 549, NK 610, NK 637
Photinia - GK 720, GK 734, GK 746, GK 759, GK 774, GK 782, GK 790, GK 799, GK 809, GN 148, KG 623, KG 630, KG 639, KG 652, KG 667, KG 678, KG 686, KG 694, KG 704, KG 714, TAW 2, WAT 17
Phrontis - KN 102, NG 429
Pierce Butler - NG 308
Pierre Gibault - GK 742, KN 254, NG 466
Pierre Laclede - KN 237
Pierre S Dupont - GN 56, NG 442, NG 457, NG 502
Pierre Soule - KN 225
Pieter De Hoogh - GN 125, NG 422
Pilar De Larrinaga - GN 28, KN 355, NG 359
Pillory - KG 709
Pine Ridge - NG 469
Pirailoide - GK 716, KG 693
Planter - NG 500
Plato - KG 651
Platte Bridge - NG 458
Plavnik - GK 707, GK 730, GK 743, GN 30, KG 610, KG 632, KG 646, KG 707, KN 256, KN 268, KN 275, KN 282, KN 291, NG 341, NG 384, NG 433, NK 572, NK 579, NK 587, WAT 10
Pleiades - GN 89, GN 202, NG 376
Plymouth - GK 724, GK 740, GK 757, GK 768, GN 103, KG 609, KG 626, KG 649, KG 661, KG 672, NK 505
Pocone - GK 757, GN 47, GN 121, GN 157, GN 185, NG 357, NG 430
Poelau Roebiah - GN 13, NG 331
Point Arguello - NK 582
Point Breeze - GN 73, GN 147, KG 663, KN 117, KN 383, NK 508
Point Loma - KN 260
Point Pelee Park - GN 9, GN 23, GN 37, GN 46, GN 61, GN 70, GN 81, GN 116, GN 142, KN 290, NG 305, NG 319, NG 333, NG 343, NG 360, NG 369, NG 379, NG 398, NG 414, NG 440, NG 504, NK 587
Point Pleasant Park - GN 133, NG 406, NG 448
Point San Pedro - NK 534
Point Vicente - GK 766
Pol Vii - GN 130
Polarland - NK 558
Polarsol - GN 20, KN 130, NG 317, TAW 11, WAT 9
Polartank - GN 39, KN 101, KN 130, KN 209, NG 318, NG 338, TAW 11, WAT 9
Pollux(Amer) - GN 26, NG 320, NG 335
Pollux(Nor) - GK 741, GK 796, GK 805, GK 836, GN 143, KG 622, KG 635, KG 689, KG 711, KN 347, NG 334, NG 452, NK 653
Polonaise - KN 214, KN 219, KN 230, KN 235, KN 245, KN 257, KN 264, KN 270, NG 403, NK 527, NK 532, NK 538, NK 548, NK 567
Polycastle - GN 23, GN 46, GN 79, GN 108, GN 203, KN 205, KN 233, KN 238, KN 248, KN 320, NG 318, NG 344, NG 377, NG 406, NG 413, NG 455, NG 502, NK 504, NK 514, NK 530, NK 536, NK 546, NK 551
Pomona - KN 276
Ponca City - GN 12, GN 20, GN 22, GN 34, GN 39, GN 58, GN 64, GN 71, GN 83, GN 91, GN 100, GN 106, GN 113, GN 119, GN 125, GN 132, GN 137, KN 226, KN 234, KN 259, KN 384, NG 318, NG 330, NG 338, NG 357, NG 363, NG 370, NG 376, NG 382, NG 390, NG 398, NG 405, NG 412, NG 418, NG 424, NG 430, NK 514, NK 523, NK 531, NK 635
Ponce - KN 123, NG 301, NG 370, TAW 2
Ponce De Leon - NG 503
Pontfield - KN 125, KN 127, KN 207, NK 506
Port Adelaide - GN 32, GN 110, NG 385, NG 440
Port Alma - KN 131, NG 319, NG 356, TAW 12
Port Campbell - GN 3, GN 52, NG 328, NG 370
Port Caroline - GN 52, GN 116, NG 324, NG 389, NG 390, NG 491
Port Darwin - GN 32, GN 120, KN 258, NG 353, WAT 3
Port Dunedin - GN 16, GN 100, KN 243, NG 335, NG 447, NK 560, WAT 7
Port Fremantle - GN SPEC1, KG 637, NK 543
Port Halifax - GN 40, GN 104, KN 119, NK 561
Port Huon - GN 1, GN 32, GN 122, GN 160, KN 266,

Ralph Waldo Emerson - GN 88, KN 125, NG 387, TAW 6
Ramo - GN 64, NG 344, NG 379
Rancher - GK 709, GN 53, KN 208, NG 304
Randa - GN 12, GN 24, GN 34, GN 42, GN 59, GN 83, GN 135, KG 630, KN 117, NG 304, NG 319, NG 330, NG 340, NG 349, NG 379, NK 591
Randfontein - GN 22
Rangitata - GN 31
Rapana - GN 25, KN 131, NG 322, TAW 12, WAT 9
Rapel - GK 725, GK 734, GK 752, GK 786, KG 626, KG 639, KG 673, KG 691
Raranga - GN 40, GN 84, GN 164, KN 111, NG 437, NK 543
Ravnefjell - GK 805, GN 156, GN 166, GN 180, GN 191, GN 200, KG 699, KG 709, KN 114, KN 328, KN 339, KN 354, NG 417, NG 452, NG 463, NG 478, NG 487, NG 497, NG 506, NK 652
Raymond V Ingersoll - GN 204, KN 341
Reaper - GN 51, GN 114, KG 603, KG 699, KN 386, NG 363, NK 602, TAW 15, WAT 9
Reaveley - GN 40, KN 125, TAW 6
Rebecca Boone - KN 289
Recifeloide - GN 38, NG 355, NG 391
Recorder Ii - KN 121
Redgate - GN 30
Regent Lion - KN 126, KN 207, NK 506
Regent Panther - KN 110, KN 200, KN 210
Reginald A Fessenden - GN 204
Reginolite - GK 767, GN 16, GN 46, GN 56, GN 71, GN 84, GN 101, GN 108, GN 121, GN 129, HA 1, KN 135, KN 275, NG 309, NG 343, NG 355, NG 370, NG 383, NG 393, NG 399, NG 406, NG 426, NG 439, NG 457, NK 593, TAW 14, WAT 11
Reinhold Richter - GK 782, KN 290, NG 439
Reinholt - KN 131, TAW 12
Rembrandt - GN 25, NG 335
Remsen Heights - GK 703, GK 717, GK 725, GK 739, GK 746, GK 753, GK 763, GK 772, GK 781, GK 790, KG 607, KG 620, KG 627, KG 633, KG 642, KG 649, KG 657, KG 666, KG 675, KG 685, KN 302, KN 309, KN 318, KN 325, KN 332, KN 339, KN 346, KN 352, KN 359, KN 367, KN 372, KN 378, KN 385, NK 606, NK 613, NK 622, NK 635, NK 649, NK 683, WAT 2
Rena - GN 145, NG 443, NG 468
Reuben Snow - KN 254
Revolutsioner - NG 361
Reynolds - NG 300
Rhea - GN 72, GN 138, GN 158, NG 371, NG 457
Rhexenor - KN 130, NG 309, TAW 11
Rhode Island - GN 91, KN 115, KN 127, KN 201, KN 210, NG 319, NG 344, NG 399, NK 509
Ribera - GN 20
Richard Caswell - NG 340
Richard D Spaight - KG 613
Richard Henderson - GN 68
Richard Henry Dana - GN 30, GN 80, GN 117, GN 137, NG 341, NG 424, NG 446
Richard J Cleveland - GN 130, GN 141, NG 440
Richard J Oglesby - GN 131
Richard K Call - KN 386
Richard Lewis Humphrey - KG 704
Richard M Johnson - KG 649
Richard Olney - GN 47
Richard Rush - GN 73
Richard S Ewell - GN 66, GN 204, KG 632, KG 633, NG 502
Richard Stockton - GN 50
Richard V Oulahan - KG 709, KN 335
Richmond Hill - KG 600
Richmond Mumford Pearson - GN 32, KG 615
Richmond P Hobson - KN 322
Richmond Park - GN 190
Riding Mountain Park - NG 410
Rigoletto - GN 133, NG 427
Riley - GN 24, GN 161, NG 458
Rimutaka - GN 84, GN 122, GN 195, NG 439
Rio Branco - GN 63, GN 87, GN 114, GN 145, NG 345, NG 371, NG 395, NG 431, NG 454, NG 487
Rio Branco(Bz) - GK 737, GN 23, GN 97, GN 147, KG 645, NG 336, NG 407, WAT

398, NG 442
Ruben Dario - NG 461
Rufus Choate - KN 319
Rufus E Foster - GK 813, KN 340
Rufus W Peckham - GK 813, NG 437
Rugeley - GN 44, KN 125, NG 300, TAW 7
Runnymede Park - GN 205
Runswick - KG 627, NG 348
Russel Sage - KG 696
Russell A Alger - GN 81
S B Hunt - GK 712, GK 747, GN 113, GN 119, GN 125, KN 115, KN 131, KN 212, KN 215, KN 277, KN 289, NG 302, NG 341, NG 412, NG 418, NG 424, NG 430, NK 514, NK 517, TAW 12, WAT 10
Sabine Sun - KN 103, KN 113, KN 124
Sachem - NG 371
Sagittaire - GK 753, GN 151, GN 177, GN 206, KG 678, NG 432, NG 472, NG 498
Sagoland - NK 540, WAT 4
Sagua - WAT 8
Saint Bernard - AH 4
Saint Domingue - GK 859
Saintonge - GN 5, GN 48, NG 461
Sakko - NG 381
Salamaua - NG 438
Salamis - KN 109, KN 201
Salando - NG 323
Salawati - GN 17
Salland - GN 16, GN 79
Salmon P Chase - GN 67
Salt Flat - KN 294
Saltersgate - WAT 17
Saluta - GN 4, TAW 16, WAT 1
Salvador Brau - KN 297
Salvator - GN 154, KN 208, KN 216, KN 237, NG 461, NK 514, NK 519, NK 530
Salvatore - NK 685
Sam Houston Ii - KN 255, KN 271, NG 388
Samblade - GN 201
Samboston - NG 443
Sambrake - GN 187
Samcrest - NG 461
Samdonard - GN 177, NG 446
Sameden - NG 442
Samfleet - NG 506
Samindoro - GN 205, NG 502
Samluzon - NG 466
Samokla - NG 506
Samoset - GN 201, NK 540
Sampler - GN 142, NG 439
Samskern - NG 461
Samson(Amer) - KN 263
Samuel A Fabens - GN 154, KN 320, KN 381
Samuel Adams - GN 1, GN 52, NG 314
Samuel Blatchford - GN 91, NG 349
Samuel Chase - KN 268, NG 353
Samuel Colt - GN 104
Samuel De Champlain - KN 265, NG 477
Samuel F Dewing - KG 708, KN 386
Samuel F Miller - KN 245
Samuel G French - KG 704
Samuel Griffin - GN 27, KG 603
Samuel Huntington - GN 27
Samuel J Tilden - GN 65
Samuel Johnston - GN 25, WAT 11
Samuel L Fuller - GN 23, GN 30, GN 38, NG 319, NG 327, NG 334, TAW 12
Samuel Lancaster - GK 774, KG 685
Samuel Livermore - GN 30, KG 613, KN 216
Samuel Mitchell - KN 341
Samuel Moody - GK 730, KN 126, KN 240, NG 302, TAW 8
Samuel Seabury - GK 740, GK 741, KG 679, KN 249
Samuel T Darling - NG 504
Samuel V Shreve - KN 251
Samuel Very - KN 253
San Adolfo - GN 47, NG 461
San Amado - GN 3, WAT 1
San Ambrosio - GN 47, KN 112
San Andres - GN 16, GN 82, KN 123, NG 310, NG 338, NG 381, TAW 2, TAW 4
San Antonio Ii - GN 32, GN 53, GN 62, KN 205, NG 350, NG 360, NG 387, NK 509
San Antonio(Amer) - NK 539
San Benito - WAT 9
San Casto - GN 128
San Cipriano - GN 122, KN 122, KN 207, NK 506
San Cirilo - GN 3
San Cristobal - KG 668
San Eliseo - GK 706, KN 233
San Emiliano - TAW 5, WAT 8
San Ernesto - HA 3, KN 107, TAW 15
San Fabian - TAW 11, TAW 15
San Felix - GN 48, NG 443

Shelon - NG 300
Shickshinny - GN 20, NG 504, NK 589, NK 637
Shilka - NG 367
Shillong - GN 46
Shiloh - NG 374
Shirvan - KN 126, TAW 4, TAW 6
Siboney - GN 75
Sidney Lanier - NG 477, NG 506
Sidney Sherman - KN 252
Silver Star Park - GN 113, GN 123, GN 132, NG 421, NG 430, NG 441, NG 455, NG 462
Silverash - NG 435
Silverbeech - GN 26
Silverelm - GK 747, GN 199, KG 657, NG 306
Silverteak - NG 439
Simon Newcomb - KN 263
Simon Willard - KG 620, NK 567
Sinclair H-C - GN 55, GN 123, GN 203, KN 112, KN 124, NG 306, NG 431, NK 541
Sinclair Opaline - KN 113, KN 122, KN 134, KN 206, NG 439, NK 505, TAW 14, WAT 12
Sinclair Rubilene - GN 131
Sinnington Court - GN 6
Sir James Clark Ross - GN 204, KN 109, KN 223, NG 347, NG 465, NK 520
Siranger - GN 21, NG 331, NG 365, NG 388, WAT 15
Skandinavia - GN 17, KN 134, KN 216, NG 314, NG 391, TAW 14, WAT 11
Skaraas - GN 11, KN 122, NG 306
Skeldergate - KN 206, NK 505
Skjelbred - KN 225
Skotaas - KN 248
Slask - GN 34, GN 59, GN 79, GN 99, GN 117, GN 133, KN 121, NG 348, NG 368, NG 388, NG 409, NG 425, TAW 3
Slemmestad - GN 25, GN 136, GN 156, GN 205, NG 434
Smaland - KG 616
Smith Thompson - GN 54
Socony-Vacuum - GN 8, GN 21, GN 33, GN 110, KN 110, KN 121, KN 132, NG 302, NG 316, NG 328
Socrates - KG 625
Soekaboemi - GN 2
Solarium - AH 3, GN 58, KN 111, KN 209
Solfonn - GN 69, HA 3, NG 373, TAW 15
Solin - KN 251
Solitaire - KN 309
Solomon Juneau - GK 733, GK 734, KN 242
Solor - GN 30
Solstad - KN 201
Solsten - GN 154, KN 200, KN 330, NK 638
Somerville - GN 37, KN 102, NG 437
Sommerstad - KN 204, NK 502
Sonnavind - GN 37, KN 260, NG 350, WAT 9
Sonora - GK 825, KN 302, KN 324, KN 340, NK 606, NK 619
Sophocles - KN 100
Sorholt - WAT 2
Soroka - NG 367
Sorvangen - GK 723, GK 767, GK 808, GN 183, KG 629, KG 678, KG 714
Sorvard - GN 131, GN 203, NG 428, NG 501
Sourabaya - KN 122
Southern Empress - KN 128
Southern Princess - GN 8, KN 112, NG 304
Southern Sun - KN 117, KN 131, KN 202, KN 206, KN 210, NG 395, NK 505, NK 509
Southgate - GN 201, KN 206, NK 505
Spar - KN 117
Sparrows Point - NK 524
Spencer Kellogg - GK 756, GN 109, GN 127, KG 626, KG 662, KN 135, KN 203, KN 210, KN 214, KN 271, KN 278, KN 321, NG 396, NG 408, NG 421, NG 425, NK 502, NK 507, NK 612, NK 619, WAT 6
Spidoleine - GN 64, GN 80, GN 95, GN 110, GN 117, GN 130, KN 116, KN 201, KN 213, KN 217, KN 227, KN 240, KN 266, KN 321, NG 363, NG 370, NG 404, NG 416, NG 422, NG 429, NG 436, NK 519, NK 564
Spinanger - KN 129, TAW 8, TAW 10
Spokane - GK 726, KG 627, KG 630, NG 336
Spondilus - GN 115, NG 414, NG 434, NG 505, NK 635
Sreca - GN 36
St Clears - GN 18
St Cuthbert - WAT 13
St James - GN 98, KN 252,

427, NG 433, NG 439, NG 446, NG 470, NG 504
Suffolk(Amer) - GN 69, GN 87, KG 638, NG 340, NG 385
Sul Ross - KN 312
Sun - KN 207, KN 211, KN 215, NK 510, NK 514
Sun Yat-Sen - KN 260
Suncrest - GN 24
Sunetta - GN 68, GN 101, TAW 3
Sunoil - GN 13, GN 26, GN 38, KN 133, NG 308, NG 320, NG 335, TAW 13, WAT 10
Sunset Park - NG 505
Surewater - NG 347
Surville - TAW S(1)
Susan V Luckenbach - GN 43, KN 300, NK 539, NK 598
Susana - GN 1
Sutherland - GN 11, KN 320, NG 320, NG 437
Svealand - GN 19, GN 29, GN 43, GN 54, GN 61, GN 68, GN 75, GN 82, GN 89, GN 97, GN 104, GN 112, GN 120, GN 127, GN 135, GN 142, GN 155, KN 104, NG 313, NG 324, NG 339, NG 351, NG 359, NG 366, NG 373, NG 380, NG 387, NG 395, NG 402, NG 410, NG 418, NG 425, NG 433, NG 440, NG 447, NG 460
Svend Foyn - KN 125
Sveti Duje - GN 143, NG 437, NK 640, TAW S(1)
Sveti Vlaho - GN 137, NG 303, NG 435
Svithoid - GN 131, KG 682, KN 129, NG 345, NG 440, NK 585, TAW 4, TAW 9
Sweep - KG 704, NK 546
Swiftarrow - GN 125, GN 157, GN 202, KG 712, NG 423, NG 430, NG 456, NG 462, NK 686
Swiftlight - GN 129, GN 141, GN 204, NG 434, NG 439, NG 455, NK 598, NK 635
Swiftscout - GN 70, GN 80, GN 98, GN 105, GN 112, GN 128, KN 267, KN 273, KN 337, NG 379, NG 396, NG 403, NG 410, NG 427, NG 434, NK 564, NK 570, NK 600, NK 619
Swivel - KN 211, KN 213, KN 229, KN 235, KN 241, KN 255, KN 261, KN 266, NK 515, NK 518, NK 522, NK 527, NK 538, NK 550, NK 558, NK 563, TAW 12
Sydland - GN 44, GN 167, NG 465, WAT 5
Sylvester Pattie - NG 438
Sysonby - GK 712, GK 724, GK 755, GK 790, GK 820, GK 847, GK 861, KG 613, KG 626, KG 661, KG 725, KG 753, KG 766, TAW 14, WAT 6
T J Jackson - GK 805, KN 136, TAW 15, WAT 5
T W Drennen - KG 624, TAW 2
Tabasco - KG 602, KG 642
Tabian - GN 37, GN 81, KN 113, NG 348, WAT 12
Tabitha Brown - GN 49
Tabor - GN 14, NG 323
Tacoma - GN 151
Tactician - GN 44
Taganrog - KG 710
Tahchee - GN 30, GN 117, GN 159, NG 326, NG 416, NG 434
Tahsinia - GN 38
Talisman - KN 128, NG 327
Talisman(Nor) - TAW 9
Talisse - GN 14, GN 68, NG 329, NG 338, NK 560
Tamaha - GN 55, GN 110, NG 394
Tamaroa - GN 33
Tamesis - NG 312
Tampa - GN 34, GN 61, KG 678, KN 290, NG 344, NG 417, WAT 12
Tampico(Mex) - KN 126, KN 206, KN 217, KN 274, KN 282, KN 290, KN 297, KN 305, KN 312, KN 385, NK 502, NK 509, NK 520, NK 527, NK 577, NK 585, NK 600, NK 608, NK 615, NK 621
Tanafjord - GN 15
Tanamo - WAT 4
Tarakan - KN 136, KN 237, NG 311, NG 366, TAW 15
Taranaki - GN 3, GN 52, KN 269, NG 329, NG 370
Taria - GN 17, GN 35, GN 53, KN 272, NG 304, NG 352
Tarifa - KN 237, NK 554
Tarleton Brown - KG 632, KN 244
Taron - NG 323
Tasker H Bliss - KN 127, TAW 8
Tatra - GN 54, GN SPEC1, KG 637, NG 311, NK 543, TAW 16
Taubate - GK 841, GK 864, GN 57, KG 747, KG 769, KN 135, NG 318, NG 369, NG

Thomas Scott - GN 28, KG 613
Thomas Sim Lee - GN 84, NG 347
Thomas Sinnickson - GN 30, KG 613, NG 339
Thomas Sully - NG 439
Thomas Sumter - GN 35, WAT 11
Thomas T Tucker - KG 609
Thomas W Gregory - GN 98, KG 667, NK 598
Thomas W Owen - GN 138, NG 413
Thomas W Symons - GN 204
Thompson Lykes - GN 23, GN 73, GN 101, NG 337, NK 560
Thordis - GN 64
Thorfjell - GN 130
Thorgaut - GN 130
Thorhild - KN 203, KN 214, NK 513
Thoris - GN 130
Thorshammer - GK 804, KN 316
Thorshavet - NG 314
Thorshavn - KN 386
Thorsholm - GN 19, GN 29, KN 126, NG 316, NG 324, TAW 4, TAW 5
Thorshov - GN 17, GN 39, GN 108, KN 105, KN 131, NG 314, NG 337, NG 406, NG 414
Thorshovdi - GN 21, KN 128, NG 318, NK 518, TAW 9, WAT 4
Thunderer - GN 29, GN 46, KN 136, NG 311, NK 517, TAW 15
Thysville - GN 31
Tiberius - GK 700, GK 716, GK 723, GK 731, GK 741, GK 755, GK 786, GK 798, GK 811, KG 606, KG 617, KG 625, KG 633, KG 646, KG 665, KG 678, KG 691, KG 703, KG 715, KG 728, KG 766, TAW 13
Tibia - KN 119, KN 300, NG 306
Ticonderoga - KN 206
Tide Penn - GN 28, GN 38, GN 45, GN 82, GN 93, GN 100, GN 108, GN 115, GN 123, GN 154, KN 102, KN 114, KN 135, KN 203, KN 208, KN 237, KN 242, KN 248, KN 254, KN 385, NG 324, NG 334, NG 343, NG 380, NG 391, NG 398, NG 405, NG 413, NG 420, NG 432, NG 440, NG 447, NK 506, NK 534, NK 540, NK 545, NK 551, TAW 13, WAT 11
Tide Water - GN 11, GN 47, KN 122, KN 133, KN 339, NG 307, NG 455
Tide Water Associated - GN 158, NG 457
Tigre - GN 100, NG 422
Tijuca - GN 24, WAT 10
Tilsington Court - GN 45
Timothy Bloodworth - GN 64
Timothy Pickering - GN 20, KN 131, NG 308, NG 330, TAW 12
Tintagel - GN 58, GN 121, GN 131, GN 144, GN 161, GN 179, KG 629, KN 290, KN 319, NG 406, NG 419, NG 429, NG 437, NG 443, NG 451, NG 468, NG 489
Tipperary Park - NG 506
Tiradentes - NG 504
Tiradentes(Bz) - GN 32, GN 73, GN 129, GN 157, NG 344, NG 390, NG 438, NG 465
Tivives - KN 115
Tjibadak - GN 40
Tjibesar - GN 47, KG 634, KN 125, NG 309, NG 403, NK 540, TAW 7
Tjimanoek - KN 129, TAW 10
Tjisalak - GN 14
Tjitjalengka - KN 103
Tobias E Stansbury - GN 158, KN 335, NG 449, NG 458
Toledo - GN 4, NG 315
Toloa - KG 603, TAW 2, TAW 15, WAT 11
Tolten - GN 131, NG 422, WAT 6, WAT 7
Tongariro - GN 14, GN 56, GN 205, NG 339, NG 440
Toorak - KN 208, NK 507
Topa Topa - WAT 12
Topdalsfjord - GK 754, GN 9, KN 262, NG 377, WAT 17
Tordene - GN 30, GN 71
Torfinn Jarl - GN 14, WAT 15
Tornus - GN 62, KN 253, NK 547, WAT 14
Toronto - GN 8, NG 327
Torres - GN 91, GN 101, GN 112, GN 118, GN 128, GN 137, KG 683, KN 244, KN 314, KN 335, KN 341, KN 383, NG 399, NG 426, NG 433, NG 442, NG 459, NK 547
Tosari - GN 69, KN 127, NG 307, TAW 8
Toteco - GK 762, GN 50, GN

196, NG 337
Umtali - GN 25, GN 76
Umvuma - GN 44
Unaco - GK 711, GK 719, GK 727, GK 740, GK 748, GK 758, GK 766, GK 775, GK 784, GK 794, GK 804, GK 840, KG 612, KG 621, KG 636, KG 644, KG 653, KG 662, KG 671, KG 680, KG 688, KG 699, KN 315, KN 322, NG 447, NK 619, WAT 15
Unicoi - WAT 10
Uranienborg - KN 123, TAW 2
Uritski - NG 356
Us Apc 87 - NG 499
Us Atr 89 - NG 505
Uss Algorma - GK 709, KN 205
Uss Apc 19 - KG 625
Uss Aquila - GK 748, GN 24, GN 38, GN 43, GN 48, GN 53, GN 59, GN 71, GN 78, GN 88, GN 99, GN 106, KG 649, NG 321, NG 338, NG 343, NG 348, NG 353, NG 359, NG 372, NG 387, NG 394, NG 406, NK 553
Uss Bowditch - NG 344
Uss Cabildo - NG 506
Uss Castor - KN 220
Uss Chourre - NG 493
Uss Conemaugh - NG 506
Uss Enceladus - NG 403
Uss Hydrographer - NG 333
Uss Iuka - KG 677
Uss Kaweah - GN 117, KN 326, NG 416, NG 438, NK 623
Uss Klaskanine - NG 505
Uss Laertes - NG 506
Uss Laramie - GN 119, GN 156, NG 418, NG 454, NG 464, NK 643
Uss Markab - KG 609
Uss Mattole - GN 95, GN 113, GN 117, GN 119, GN 193, GN 199, NG 308, NG 368, NG 412, NG 418, NG 435, NG 440, NG 487, NG 498
Uss Maumee - GN 44, NG 343, NG 381, NG 413
Uss Montcalm - KN 208
Uss Munargo - GN 86, NG 333
Uss Nitro - GN 89
Uss Ochlockonee - NG 490
Uss Pastores - GN 17, GN 29, NG 312, NG 323, NG 352
Uss Patoka - GN 59, NG 379
Uss Pegasus - GK 718, GK 727, GK 759, GN 76, KG 621, KG 625, KG 631, KG 636, NG 325, NG 385
Uss Picket - NG 504
Uss Pinon - NG 487
Uss Pumper - GN 59
Uss Pyc 42 - KG 626
Uss Rapidan - GN 89, NG 385, NG 421
Uss Sacandaga - NG 482
Uss Salinas - GN 119, NG 410
Uss Sapelo - GN 84, GN 90, NG 383, NG 389
Uss Sarpedon - NG 504
Uss Shikellamy - KG 635
Uss Stratford - WAT 11
Uss Tallulah - NG 316
Uss Tarazed - KN 201
Uss Towaliga - NG 489
Uss Warren - KN 225
Uss Weight - NG 506
Uss Y9 - KG 661
Uss Yt 99 - KG 661
Uss Yukon - NG 337
Uss Yw 86 - KG 657
Ussuri - NG 360
Utahan - GK 788, GN 151, KG 674, KN 301, NG 369, NG 439, NG 460
Vacport - GN 92
Vacuum - GN 19, KN 116, KN 270, KN 288, KN 294, KN 306, KN 314, NG 466, NK 567, NK 583, NK 591, NK 620
Vagrant(Pan) - GK 700, GK 717, GK 742, GK 790, KG 606, KG 613, KG 623, KG 629, KG 634, KG 667, KG 674, KG 685, KG 695, KN 308, NG 433, NG 455, WAT 12
Valera - KG 677
Valldemosa - GN 63
Van De Velde - GN 200, NG 419
Van Honthorst - NG 506
Vancolite - GN 8, GN 22, GN 41, GN 50, GN 60, GN 69, GN 79, GN 93, GN 102, GN 161, KG 600, KN 123, KN 292, NG 317, NG 338, NG 349, NG 359, NG 368, NG 377, NG 392, NG 400, NG 421, NG 459, NG 467, NG 505, NK 589, TAW 2
Vancouver City - WAT 17
Vanja - GN 12, GN 154, KN 382, NG 306
Vanzetti - KG 644, NG 370, NK 551
Var - GK 751, KN 279
Vardefjell - KN 122, TAW 4
Vav - GN 17, GN 41, GN 157, KN 101, NG 312, NG 456

95, GN 102, GN 108, GN 114, GN 120, GN 141, GN 159, GN 203, KN 111, KN 132, KN 213, KN 232, NG 304, NG 355, NG 361, NG 367, NG 374, NG 400, NG 407, NG 413, NG 419, NG 425, NG 440, NG 458, NK 512, NK 529
W W Mills - GN 15, GN 25, GN 37, GN 57, GN 63, GN 71, GN 98, GN 105, GN 129, GN 158, KN 119, KN 132, KN 203, KN 228, KN 233, KN 272, KN 292, KN 383, NG 313, NG 321, NG 332, NG 356, NG 362, NG 368, NG 374, NG 396, NG 403, NG 426, NG 434, NG 441, NG 457, NK 501, NK 559, NK 589
Wade Hampton - GN 35, KG 616
Wagon Box - NK 586
Waigstill Avery - GN 98, NK 545
Waimarama - KN 109
Walker Taylor - GN 141, NG 439
Wallace E Pratt - AH 2, GN 7, GN 19, HA 2, KN 102, NG 301, NG 315
Walt Whitman - KN 125
Walter A Luckenbach - GK 762, KN 270, NG 351, NK 621, WAT 4
Walter Forward - GN 52
Walter Hines Page - NG 393
Walter Jennings - AH 3, GN 19, KN 116, KN 311, KN 320, KN 336, NG 316, NK 608, NK 640
Walter L Fleming - GN 119, KG 688
Walter Q Gresham - KN 223
Walter Raleigh - GN 202, NK 541, NK 579
Walter Reed - KN 253
Wanderer - GK 717, GN 99, KN 217, NG 390
Wanks - GK 709, TAW S(1)
Wanstead - GN 26
War Admiral - GK 763, KG 627, KG 648, KG 660, KG 675, KG 684, KG 698, NG 324, NK 566
Warbler - KN 231
Ward - GN 73, NG 304, NG 403
Warfield - GN 48
Waroonga - GN 46, KN 131, TAW 12
Warren Delano - GN 137, NG 423
Washington - GN 61, GN 67, GN 73, GN 83, GN 92, GN 104, GN 110, GN 116, GN 121, GN 127, GN 133, GN 138, GN 153, GN 158, KN 206, KN 382, NG 324, NG 366, NG 372, NG 382, NG 390, NG 397, NG 403, NG 409, NG 415, NG 420, NG 426, NG 432, NG 447, NG 457, NG 506, NK 504, TAW 12
Washington Irving - GN 11, NG 503
Watertown - GN 59, GN 65, GN 71, GN 84, GN 91, GN 97, GN 115, GN 126, GN 140, KN 205, KN 209, KN 213, KN 235, KN 286, KN 290, KN 302, KN 312, KN 384, NG 358, NG 364, NG 370, NG 382, NG 389, NG 396, NG 414, NG 420, NG 426, NG 431, NG 439, NK 508, NK 512, NK 532, NK 581, NK 634, TAW 15
Wayfarer - GN 1, GN 65, NG 324, NG 358, TAW 16
Wearpool - TAW S(1), TAW SL/1
Wellington Court - GN 16
Wendell L Willkie - KN 384
Wendell Phillips - KG 714, KN 235, NG 432
Wentworth Park - NG 506
West Celina - TAW SL/1
West Cheswald - GN 8, GN 54, GN 109, NG 317, NG 384
West Chetac - WAT 11, WAT 14
West Durfee - GN 30, GN 179, KN 132, KN 329, NG 312, NG 344, NG 443, NG 456, TAW 11
West Gotomska - GK 740, KG 652, NG 367
West Grama - GK 705, GK 730, GK 776, KG 610, KN 295, WAT 6
West Honaker - GK 775, GN 46, KN 283, KN 297, NG 401, NG 415, WAT 4
West Humhaw - KN 126, NG 303, TAW 6
West Keene - GN 45, GN 83, KG 756, KN 320, NG 382, NG 438, NG 447
West Kyska - GN 6, GN 56, GN 101, GN 125, GN 138, KN 339, NG 318, NG 384, NG 409, NG 433, NG 451, NG 465
West Maximus - GN 38, NG 321
West Nohno - GK 723
West Point - GN 79, NG 377
West Portal - KN 123, TAW

435
William Grayson - GN 66, KN 342, NG 313
William H Ashley - GK 786, KN 294
William H Aspinwall - GN 72
William H Crawford - GN 85, KG 627
William H Dall - GN 118, KG 712, NK 618
William H Gray - GN 204, KG 712, NG 433
William H Mcguffey - KG 721, KN 320, NG 438
William H Moody - GN 121, GN 159
William H Prescott - GK 731, GN 68, KG 634, KN 266, NG 384
William H Seward - GN 43, GN 120, NG 428
William H Welch - NK 538
William H Wilmer - NK 538
William Harper - GN 82, KG 627
William Hawkins - GN 27, GN 118, NG 356, NG 392
William Hume - NG 462
William J Bryan - GN 70, KG 638, NG 469
William J Duane - GN 77, NG 399
William J Worth - GK 729, GK 763, KN 235, KN 271, NG 386
William James - GN 98
William Kent - GK 753, KN 268
William King - NG 324
William L Marcy - GN 65
William L Mclean - NK 686
William L Smith - KN 222
William L Yancey - KG 654, KN 263
William Lloyd Garrison - NG 505
William M Eastland - KN 295
William M Evarts - GK 768, KG 647, KG 722
William M Meredith - KN 273, NG 424
William M Rayburn - GN 132, KN 277, NG 431
William M Stewart - KN 133, TAW 13
William Maclay - GN 83, GN 115, GN 139, NG 358, NG 393, NG 424, NG 448, NG 459, NG 505
William Moultrie - GN 81, NG 356
William N Page - GN 135, NG 433
William N Pendleton - KG 632, KN 243
William Osler - GN 93, NG 354
William P Fessenden - NG 357
William Paca - GN 53, KN 282
William Patterson - GN 7, NG 317, WAT 14
William Penn - GN 1, KN 111, KN 124, KN 204, KN 209, NK 503, NK 507, WAT 14, WAT 15
William Pepper - NK 541
William Phips - KN 279
William R Davie - GN 33, WAT 15
William R Day - GN 88, NG 350
William Rawle - GN 50, NG 312
William S Clark - KG 702, KG 708
William S Rosecrans - GN 50
William T Barry - GN 159, NG 435
William T Coleman - GK 774, KN 282, NG 400
William Terry Howell - GN 203
William Tilghman - KG 606
William W Campbell - GN 114
William W Johnson - KN 382
William W Loring - KN 304, NG 504
William W Mayo - KN 268
William W Mckee - KN 386
William Wheelwright - KN 320
William Whipple - GN 40
William Wilkins - KN 231
William Windom - KG 637
William Wirt - GN 98, GN 117, GN 139, GN 158, GN 205, NG 363, NG 407, NG 423, NG 446
William Wolfskill - NG 442
Willie Jones - GN 86, NG 352
Willis A Slater - KN 298
Willis Van Devanter - NK 686
Willowdale Park - GN 139, GN 155, NG 449, NG 462
Wilscox - GK 745, GN 24, KG 676, NG 371
Wilsons Creek - GN 204
Winamac - HA 2, WAT 12
Wind Rush - GN 96, GN 119, GN 139, KG 701, NG 379, NG 406, NG 427, NK 607
Winding Gulf - GK 741, GN 78, GN 135, KG 605, KG

Appendix 2: Convoys To Nassau

Vessel	If Esc	GRT	Flag	Dep. Port	Dep. Date	Arr. Port	Arr. Date	Notes / Escort
Black Condor	Ind.	5,358	US	Texas City	11/2/41	Nassau	15/02/41	
Black Condor	Ind	5,358	US	Nassau	18/02/41	Halifax	25/02/41	Purchased By Britain
Bruxelles	Ind	5,085	BEL	Galveston	21/06/40	Nassau	26/06/40	
Bruxelles	Ind	5,085	BEL	Nassau	28/06/40	Bermuda	2/7/42	
Cathcart	Ind	3,708	UK	Nassau	7/12/41	Halifax	13/12/41	Prev Traded E Coast Can/W Indies
Cathcart	Ind	3,708	UK	Bowden	30/12/41	Nassau		
Cathcart	Ind	3,708	UK	Nassau	2/1/42	Halifax	7/1/42	
Cathcart	Ind	3,708	UK	Bowden	28/01/42	Nassau	30/01/42	
Cathcart	Ind	3,708	UK	Nassau	31/01/42	Halifax	9/2/42	
Cathcart	Ind	3,708	UK	Hampton Roads	26/03/42	Nassau	30/03/42	
Cathcart	Ind	3,708	UK	Nassau	31/03/42	Kingston	4/4/42	
Cathcart	Ind	3,708	UK	Bowden	9/4/42	Nassau	14/04/42	
Cathcart	Ind	3,708	UK	Nassau	15/04/42	Hampton Rds	18/04/42	
Cavelier	Ind	3,663	UK	Halifax	12/11/45	Nassau	20/11/45	
Cavelier	Ind	3,663	UK	Nassau	22/11/45	Kingston	25/11/45	
Culebra	Ind	3,044	UK	Bermuda	10/10/39	Nassau	20/10/39	
Culebra	Ind	3,044	UK	Nassau	20/10/39	Kingston	25/10/39	
Culebra	Ind	3,044	UK	Bermuda	10/8/40	Nassau	14/08/40	
Culebra	Ind	3,044	UK	Nassau	15/08/40	Kingston	18/08/40	
Culebra	Ind	3,044	UK	Bermuda	21/08/41	Nassau		
Culebra	Ind	3,044	UK	Nassau	25/08/41	Kingston	28/08/41	
Culebra	Ind	5,260	PAN	Antwerp	23/01/40	Nassau	25/02/40	
Culebra	Ind	5,260	PAN	Nassau	25/02/40	Bahia Blanca	6/3/40	
Empire Confidence	Ind	5,023	UK	Bermuda	2/6/40	Nassau		
Empire Confidence	Ind	5,023	UK	Nassau	6/6/40	Kingston, Jamaica	6/6/40	
Ena K	Esc			Nassau	23/02/44	Miami	24/02/44	USS YMS 275

Fernbank	GN 020	4,333	NOR	Guantanamo	16/11/42	Pt Everglades	22/11/42	Arr Leaking, Via Nassau
Franklin S Leisenring	Esc	1,317	US	Nassau	21/12/42	Miami	22/12/42	USS PC 621
Franklin S Leisenring	Esc	1,317	US	Nassau	29/12/42	Miami	30/12/42	USS PC 488
Franklin S Leisenring	Esc	1,317	US	Nassau	2/1/43	Miami	3/1/43	USS YP 428
Franklin S Leisenring	Esc	1,317	US	Nassau	2/1/43	Miami		USS YP 428
Franklin S Leisenring	Esc	1,317	US	Nassau	9/1/43	Miami	10/1/43	USS YP 427
Franklin S Leisenring	Esc	1,317	US	Nassau	15/01/43	Miami	15/01/43	USS YP 426
Franklin S Leisenring	Esc	1,317	US	Nassau	18/01/43	Miami	19/01/43	USS YP 433
Franklin S Leisenring	Esc	1,317	US	Nassau	24/01/43	Miami	25/01/43	USS YP 428
Ganymedes	NK 561	2,682	DU	NYC	24/08/43	Nassau	30/08/43	NYC To Nassau
Ganymedes	Ind	2,682	DU	Nassau	31/08/43	Kingston	3/9/43	
Georgios P	Esc	4,052	Gk	Nassau	23/07/43	Key West	25/07/43	USS SC 661, via Miami
Gulfspray	Esc	379	US	Nassau	10/4/43	Pt Evergld s	11/4/43	USS YP 531, USCG 83397
Gulfspray	Esc	379	US	Nassau	6/6/43	Pt Evergld s	6/6/43	USS SC 1467
Heimgar	Ind	2,978	NOR	Miami	2/4/43	Nassau	3/4/43	
Heimgar	Ind	2,978	NOR	Nassau	5/4/43	Miami	6/4/43	
Hercules	Ind	2,317	DU	Bermuda	12/2/43	Nassau	16/02/43	
J. Brillant	Esc	640	UK	Nassau	24/02/43	Miami	25/02/43	USS SC 977
J. Brillant	Esc	640	UK	Nassau	5/3/43	Miami	5/3/43	Q 075
J. Brillant	Esc	640	UK	Nassau	9/3/43	Miami	10/3/43	USS SC 752
J. Brillant	Esc	640	UK	Nassau	13/03/43	Miami	14/03/43	Q 065
J. Brillant	Esc	640	UK	Nassau	21/03/43	Miami	22/03/43	USS SC 1032
J. Brillant	Esc	640	UK	Nassau	25/03/43	Miami	25/03/43	USS SC 980
J. Brillant	Esc	640	UK	Nassau	28/03/43	Miami	29/03/43	USS PC 1194
J. Brillant	Esc	640	UK	Nassau	10/4/43	Miami	10/4/43	USCG Wanigan
J. Brillant	Esc	640	UK	Nassau	16/04/43	Miami	17/04/43	USCG 83398
J. Brillant	Esc	640	UK	Nassau	23/04/43	Miami	24/04/43	USCG 83403
J. Brillant	Esc	640	UK	Nassau	28/04/43	Miami	28/04/43	USCG 83405
J. Brillant	Esc	640	UK	Nassau	6/5/43	Miami	7/5/43	USS SC 622, USCG 83401, USCG 83402,

							USCG 83403, USCG 83405
J. Brillant	Esc 640	UK	Nassau	10/5/43	Miami	10/5/43	USCG 83406
J. Brillant	Esc 640	UK	Nassau	12/5/43	Miami	13/05/43	USCG 83402
J. Brillant	Esc 640	UK	Nassau	15/05/43	Miami	16/05/43	USCG 83348, USS SC 1067
J. Brillant	Esc 640	UK	Nassau	18/05/43	Miami	19/05/43	USS SC 1473, USS SC 1471
J. Brillant	Esc 640	UK	Nassau	22/05/43	Miami	22/05/43	USCG 83405, USS PC 1195, USS SC 1290
J. Brillant	Esc 640	UK	Nassau	25/05/43	Miami	25/05/43	USCG 83401, USS SC 1034, SC training Ops.
J. Brillant	Esc 640	UK	Nassau	27/05/43	Miami	28/05/43	USCG Wanigan, USS SC 1048
J. Brillant	Esc 640	UK	Nassau	31/05/43	Miami		USCG 83402, USS SC 1060
J. Brillant	Esc 640	UK	Nassau	3/6/43	Miami	4/6/43	USCG 83402
J. Brillant	Esc 640	UK	Nassau	6/6/43	Miami	7/6/43	USS SC 1473, USS SC 997
J. Brillant	Esc 640	UK	Nassau	9/6/43	Miami	10/6/43	USCG 83402, USS SC 1321, USS SC 1326, USS SC 1298, SC's training
J. Brillant	Esc 640	UK	Nassau	12/6/43	Miami	13/06/43	USCG 83341, USS SC 1282
J. Brillant	Esc 640	UK	Nassau	15/06/43	Miami	16/06/43	USS SC 661, USCG 83431
J. Brillant	Esc 640	UK	Nassau	18/06/43	Miami	19/06/43	USS SC 661, USS SC 1271
J. Brillant	Esc 640	UK	Nassau	22/06/43	Miami	22/06/43	USS SC 661
J. Brillant	Esc 640	UK	Nassau	24/06/43	Miami	25/06/43	USCG 83435, USS PC 612
J. Brillant	Esc 640	UK	Nassau	27/06/43	Miami	28/06/43	USCG 83435, USS SC 1282
J. Brillant	Esc 640	UK	Nassau	30/06/43	Miami	07/01/43	USCG 83419, USS SC 511, USS SC 1035, USCG 83421, USCG 83421 Sank After Collision

J. Brillant	Esc 640	UK	Nassau	3/7/43	Miami	4/7/43	With USS SC 1330 USCG 83435, USS SC 718, USS SC 1327
J. Brillant	Esc 640	UK	Nassau	7/7/43	Miami	8/7/43	USS SC 685, USS SC 1282, USS SC 1299
J. Brillant	Esc 640	UK	Nassau	10/7/43	Miami	11/7/43	USS SC 661, USCG 83435
J. Brillant	Esc 640	UK	Nassau	13/07/43	Miami	14/07/43	USS SC 656, USS SC 755, USS SC 1002
J. Brillant	Esc 640	UK	Nassau	16/07/43	Miami	17/07/43	USS SC 1322, USCG 83420
J. Brillant	Esc 640	UK	Nassau	20/07/43	Miami	21/07/43	USS SC 1022, USS SC 1026, USCG 8435
J. Brillant	Esc 640	UK	Nassau	23/07/43	Miami	24/07/43	USS SC 1359, USCG 83435
J. Brillant	Esc 640	UK	Nassau	27/07/43	Miami	28/07/43	USS SC 1311, USCG 83348
J. Brillant	Esc 640	UK	Nassau	30/07/43	Miami	31/07/43	USCG 83348
J. Brillant	Esc 640	UK	Nassau	4/8/43	Miami	5/8/43	USCG 83402
J. Brillant	Esc 640	UK	Nassau	6/8/43	Miami	7/8/43	USCG 83451, USS SC 1333
J. Brillant	Esc 640	UK	Nassau	8/8/43	Miami	9/8/43	USCG 83450, USCG 83453
J. Brillant	Esc 640	UK	Nassau	28/08/43	Miami	28/08/43	USS SC 656
J. Brillant	Esc 640	UK	Nassau	1/9/43	Miami	2/9/43	USCG 83453, USS SC 1313
J. Brillant	Esc 640	UK	Nassau	4/9/43	Miami	5/9/43	USS SC 647, USS SC 1304, USS PC 1131
J. Brillant	Esc 640	UK	Nassau	6/9/43	Miami	7/9/43	USCG 83450, USCG 83451, USCG 83452
J. Brillant	Esc 640	UK	Nassau	9/9/43	Miami	10/9/43	USCG 83450, USCG 83453
J. Brillant	Esc 640	UK	Nassau	18/09/43	Miami	19/09/43	USS SC 1342, USS SC 989, USS SC 1323

J. Brillant	Esc 640	UK	Nassau	22/09/43	Miami	23/09/43	USS PC 1230, USS SC 1305, USS SC 1050
J. Brillant	Esc 640	UK	Nassau	25/09/43	Miami	26/09/43	USS PC 1230, USCG 83450
J. Brillant	Esc 640	UK	Nassau	28/09/43	Miami	29/09/43	USS SC 1342, USCG 83452
J. Brillant	Esc 640	UK	Nassau	4/10/43	Miami	5/10/43	USCG 83450
J. Brillant	Esc 640	UK	Nassau	7/10/43	Miami	8/10/43	USS PC 1205, USS PC 1238, USCG 83402
J. Brillant	Esc 640	UK	Nassau	10/10/43	Miami	11/10/43	USS SC 687, USS PC 1134, USCG 83402
J. Brillant	Esc 640	UK	Nassau	2/11/43	Miami	3/11/43	USS SC 1290
J. Brillant	Esc 640	UK	Nassau	4/11/43	Miami	5/11/43	USS SC 1290, USS SC 1348
J. Brillant	Esc 640	UK	Nassau	20/11/43	Miami	21/11/43	USS SC 1280
J. Brillant	Esc 640	UK	Nassau	23/11/43	Miami	24/11/43	USS SC 1295
J. Brillant	Esc 640	UK	Nassau	28/11/43	Miami	29/11/43	USS SC 1342
J. Brillant	Esc 640	UK	Nassau	1/12/43	Miami	2/12/43	USS SC 1295
J. Brillant	Esc 640	UK	Nassau	15/12/43	Miami	16/12/43	USS SC 1292
J. Brillant	Esc 640	UK	Nassau	19/12/43	Miami	20/12/43	USS SC 1064
J. Brillant	Esc 640	UK	Nassau	22/12/43	Miami	23/12/43	USS SC 1342
J. Brillant	Esc 640	UK	Nassau	26/12/43	Miami	27/12/43	USS SC 1292
J. Brillant	Esc 640	UK	Nassau	30/12/43	Miami	31/12/43	USS SC 1295
J. Brillant	Esc 640	UK	Nassau	13/01/44	Miami	14/01/44	USS SC 1280
J. Brillant	Esc 640	UK	Nassau	17/01/44	Miami	18/01/44	USS SC 1292
J. Brillant	Esc 640	UK	Nassau	20/01/44	Miami	21/01/44	USS SC 1342
J. Brillant	Esc 640	UK	Nassau	23/01/44	Miami	24/01/44	USS SC 1292
J. Brillant	Esc 640	UK	Nassau	26/01/44	Miami	27/01/44	USS SC 1280
J. Brillant	Esc 640	UK	Nassau	29/01/44	Miami	30/01/44	USS SC 1301
J. Brillant	Esc 640	UK	Nassau	2/2/44	Miami	3/2/44	USS SC 1292
J. Brillant	Esc 640	UK	Nassau	5/2/44	Miami	6/2/44	USS SC 1295
J. Brillant	Esc 640	UK	Nassau	8/2/44	Miami	9/2/44	
J. Brillant	Esc 640	UK	Nassau	11/2/44	Miami	12/2/44	USS SC 1280
J. Brillant	Esc 640	UK	Nassau	15/02/44	Miami	16/02/44	USS SC 1301
J. Brillant	Esc 640	UK	Nassau	18/02/44	Miami	19/02/44	USS SC 1282
J. Brillant	Esc 640	UK	Nassau	23/02/44	Miami	24/02/44	
J. Brillant	Esc 640	UK	Nassau	26/02/44	Miami	27/02/44	USS SC 1280
J. Brillant	Esc 640	UK	Nassau	29/02/44	Miami	1/3/44	USS SC 1295
J. Brillant	Esc 640	UK	Nassau	3/3/44	Miami	4/3/44	USS YMS 038
J. Brillant	Esc 640	UK	Nassau	6/3/44	Miami	7/3/44	USS SC 1292
J. Brillant	Esc 640	UK	Nassau	10/3/44	Miami		USS YMS 361

J. Brillant	Esc 640	UK	Nassau	20/03/44	Miami		USCG 83348
J. Brillant	Esc 640	UK	Nassau	22/03/44	Miami	23/03/44	USS SC 1295
J. Brillant	Esc 640	UK	Nassau	25/03/44	Miami	26/03/44	USS SC 1292
J. Brillant	Esc 640	UK	Nassau	29/03/44	Miami	29/03/44	USS YMS 345
J. Brillant	Esc 640	UK	Nassau	31/03/44	Miami	1/4/44	USS YMS 342
J. Brillant	Esc 640	UK	Nassau	3/4/44	Miami	4/4/44	USS YMS 342
J. Brillant	Esc 640	UK	Nassau	6/4/44	Miami	7/4/44	USS YMS 341
J. Brillant	Esc 640	UK	Nassau	10/4/44	Miami	11/4/44	USS YMS 065
J. Brillant	Esc 640	UK	Nassau	16/04/44	Miami	17/04/44	USS SC 1343
J. Brillant	Esc 640	UK	Nassau	19/04/44	Miami	20/04/44	USS SC 1343
J. Brillant	Esc 640	UK	Nassau	22/04/44	Miami	23/04/44	USS YMS 224
J. Brillant	Esc 640	UK	Nassau	25/04/44	Miami	26/04/44	USS YMS 224
J. Brillant	Esc 640	UK	Nassau	28/04/44	Miami	29/04/44	USS YMS 065
J. Brillant	Esc 640	UK	Nassau	1/5/44	Miami	2/5/44	USS YMS 065
J. Brillant	Esc 640	UK	Nassau	14/05/44	Miami	15/05/44	USS YMS 370
J. Brillant	Esc 640	UK	Nassau	17/05/44	Miami	18/05/44	USS YMS 370
J. Brillant	Esc 640	UK	Nassau	20/05/44	Miami	21/05/44	USCG Vigilant
J. Brillant	Esc 640	UK	Nassau	23/05/44	Miami	24/05/44	USS YMS 370
J. Brillant	Esc 640	UK	Nassau	30/05/44	Miami	31/05/44	USS YMS 370
J. Brillant	Esc 640	UK	Nassau	8/11/44	Miami	9/11/44	USS SC 710
Lady Rodney	Ind 8,194	UK	Bermuda	3/1/42	Nassau		
Lady Rodney	Ind 8,194	UK	Nassau	6/1/42	Kingston	8/1/42	
Lady Rodney	Ind 8,194	UK	Bowden	11/1/42	Nassau		
Lady Rodney	Ind 8,194	UK	Nassau	13/01/42	Bermuda	16/01/42	
Lady Rodney	Ind 8,194	UK	Bermuda	4/2/42	Nassau	8/2/42	
Lady Rodney	Ind 8,194	UK	Nassau	9/2/42	Kingston	11/2/42	
Lady Rodney	Ind 8,194	UK	Bowden	14/02/42	Nassau	16/02/42	
Lady Rodney	Ind 8,194	UK	Nassau	17/02/42	Bermuda	19/02/42	
Lady Rodney	Ind 8,194	UK	Bermuda	24/03/42	Nassau	28/03/42	
Lady Rodney	Ind 8,194	UK	Nassau	30/03/42	Kingston	2/4/42	
Lake Traverse	Ind 1,993	PAN	Nassau	27/01/42	Halifax	7/2/42	
Lancastria	Ind 16,243	UK	NYC	2/9/39	Nassau	6/9/39	
Lancastria	Ind 16,243	UK	Nassau	6/9/39	NYC	8/9/39	

Lombardy	Ind	3,379	UK	Bermuda	11/3/41	Nassau		
Lombardy	Ind	3,379	UK	Nassau	15/03/41	Kingston	18/03/41	
Lombardy	Ind	3,379	UK	Bermuda	2/7/41	Nassau	5/7/41	
Lombardy	Ind	3,379	UK	Nassau	6/7/41	Kingston	9/7/41	
Lombardy	Ind	3,379	UK	Bermuda	24/10/41	Nassau		
Lombardy	Ind	3,379	UK	Nassau	28/10/41	Kingston	31/10/41	
Madison	Esc	3,734	US	Nassau		Miami	4/9/42	USS PC 568, USS PC 483, USS SC 642
Manaqui	Ind	2,802	UK	Bermuda	21/12/41	Nassau		
Manaqui	Ind	2,802	UK	Nassau	24/12/41	Kingston	27/12/41	
Manuel Rionda	Esc	1,375	CU	Nassau	12/4/43	Miami	13/04/43	USCG 83400
Manuel Rionda	Esc	1,375	CU	Nassau	30/07/43	Miami	31/07/43	USS SC 1060
Moysalen	Esc	990	NOR	Nassau	29/06/42	Pt Evergids	30/06/42	USS YP 218
Moysalen	Esc	990	NOR	Nassau	13/07/42	Pt Evergids	14/07/42	USS YP 395
Nassau	Esc			Nassau	12/4/43	Miami	13/04/43	USS SC 1032
Orbita	Ind	15,495	UK	Havana	9/9/39	Nassau	10/9/39	
Orbita	Ind	15,495	UK	Nassau	10/10/39	Bermuda	13/09/39	
Randa	Ind	1,555	UK	NYC	26/05/42	Nassau	5/6/42	
Randa	Ind	1,555	UK	Nassau	7/6/42	Miami	8/6/42	
Sabor	Ind	5,212	UK	Bermuda	10/7/40	Nassau		
Sabor	Ind	5,212	UK	Nassau	14/07/40	Kingston Ja	17/07/40	
Sabor	Ind	5,212	UK	Bermuda	5/11/40	Nassau		
Sabor	Ind	5,212	UK	Nassau	8/11/40	Kingston Ja	11/11/40	
Samana	Esc		US	Nassau	1/12/42	Miami	2/12/42	USS YP 535
Sarthe	Ind	5,271	UK	Bermuda	6/2/40	Nassau	11/2/40	
Sarthe	Ind	5,271	UK	Nassau	11/2/40	Kingston	14/02/40	
Siris	Ind	5,242	UK	Bermuda	23/12/39	Nassau	27/12/39	
Siris	Ind	5,242	UK	Nassau	28/12/39	Kingston	31/12/39	
Siris	Ind	5,242	UK	Bermuda	8/5/40	Nassau	12/5/40	Notional Dates
Siris	Ind	5,242	UK	Nassau	12/5/40	Kingston	14/05/40	
Somme	Ind	5,265	UK	Bermuda	12/5/41	Nassau		
Somme	Ind	5,265	UK	Nassau	16/05/41	Kingston	19/05/41	
Venus	Esc	1,855	DU	Nassau	9/7/42	Key West	10/7/42	USS SC 539
Virginia	Ind	6,390	US	Pt Arthur	3/3/42	Nassau	8/3/42	
Virginia	Ind	6,390	US	Nassau	9/3/42	Para	21/03/42	Intermediate Calls Unknown
YAG 020	Ind		US	Nassau	10/1/44	Miami	11/1/44	

YAG 021	Ind	US	Nassau	27/01/44	Miami	27/01/44	
YAG 028	Ind	US	Nassau	19/02/44	Miami	20/02/44	
YAG 028	Esc	US	Nassau		Miami	13/04/43	USS YP 408
YO 051	Esc	US	Nassau	10/4/43	Pt Everglds	11/4/43	USS YP 531, USCG 83397

Appendix 3: All Convoys List

Am 239/3	Norfolk	10/2/44	Cristobal	16/02/44
Am 259/3	Norfolk	1/3/44	Guantanamo	7/3/44
Bt 202	New York	9/4/42	Cristobal	18/04/42
Bt 203	New York	15/05/43	Cristobal	21/05/43
Buttermere	Guantanamo	11/8/42	Key West	12/8/42
Charleston	Charleston Sc	3/10/42	Key West	5/10/42
Charleston 013	Charleston Sc	9/9/42	Smith Shoals	11/9/42
Charleston 015	Charleston Sc	14/09/42	Key West	16/09/42
Charleston 016	Charleston Sc	15/09/42	Key West	17/09/42
Charleston 017	Charleston Sc	19/09/42	Key West	21/09/42
Charleston 018	Charleston Sc	21/09/42	Key West	24/10/42
Charleston 019	Charleston Sc	23/09/42	Key West	26/09/42
Charleston 020	Charleston Sc	2/10/42	Key West	5/10/42
Charleston 021	Charleston Sc	3/10/42	Key West	
Charleston 022	Charleston Sc	10/10/42	Key West	14/10/42
Charleston 023	Charleston Sc	15/10/42	Fort Pierce	18/10/42
Charleston 032	Charleston Sc	18/10/42	Key West	21/10/42
Charleston 034	Charleston Sc	26/10/42	Smith Shoals	28/10/42
Charleston 035	Charleston Sc	31/10/42	Key West	3/11/42
Charleston 036	Charleston Sc	1/11/42	Guantanamo	5/11/42
Charleston 037	Charleston Sc	21/09/42	Key West	24/09/42
Charleston 040	Charleston Sc	11/11/42	Key West	14/11/42
Charleston 047	Charleston Sc	22/11/42	Key West	26/11/42
Charleston 051	Charleston Sc	28/11/42	Key West	1/12/42
Charleston 057	Charleston Sc	3/12/42	Smith Shoals	6/12/42
Charleston 058	Charleston Sc	15/01/43	Miami	18/01/43
Charleston 059	Charleston Sc	25/02/43	Key West	1/3/43
Charleston 060	Charleston Sc	28/02/43	Miami	2/3/43
Charleston 065	Charleston Sc	11/3/43	Key West	15/03/43
Charleston 066	Charleston Sc	14/03/43	Mayport	15/03/43
Charleston 074	Charleston Sc	25/03/43	Key West	28/03/43
Charleston 083	Charleston Sc	4/5/43	Key West	7/5/43
Charleston 085	Charleston Sc	7/5/43	Key West	11/5/43
Charleston 191	Charleston Sc		Cristobal	9/10/44
Charleston 222	Charleston Sc		New Orleans	3/5/45
Chicopee	Galveston	1/9/42	Boston	8/9/42
Ciudad Trujillo 032	Puerto Plata	18/01/44	Guantanamo	20/01/44

Ciudad Trujillo 037	Puerto Plata	10/3/44	Guantanamo	12/3/44
Ciudad Trujillo 069	Puerto Plata	12/8/44	Guantanamo	13/08/44
Cuban N Coast	Havana	3/6/43	Puerto Padre	
Cuban N Coast 002	Sagua	14/03/43	Havana	15/03/43
Cuban N Coast 010	Havana	23/04/43	Isabela	23/04/43
Cuban N Coast 010 A	Cayo Frances	25/04/43	Havana	26/04/43
Cuban N Coast 016	Havana	6/5/43	Cardenas	7/5/43
Cuban N Coast 018	Havana	10/5/43	Tanamo	
Cuban N Coast 019	Sagua	15/05/43	Havana	16/05/43
Cuban N Coast 021	Cayo Frances		Havana	13/06/43
Cuban N Coast 026	Havana	15/06/43	Cardenas	15/06/43
Cuban N Coast 027	Havana	19/06/43	Cardenas	19/06/43
Cuban N Coast 028	Matanzas	26/06/43	Key West	26/06/43
Cuban N Coast 030	Havana	29/06/43	Cardenas	29/06/43
Cuban N Coast 030/1	Cardenas	3/7/43	Havana	3/7/43
Cuban N Coast 031	Havana	1/7/43	Cardenas	1/7/43
Cuban N Coast 031/1	Cardenas	2/7/43	Cayo Frances	2/7/43
Cuban N Coast 031/2	Cardenas	9/7/43	Havana	9/7/43
Cuban N Coast 032	Havana	8/7/43	Cardenas	8/7/43
Cuban N Coast 032/2	Caibarien	11/7/43	Havana	12/7/43
Cuban N Coast 032/A	Cardenas	9/7/43	Caibarien	
Cuban N Coast 032/B	Caibarien	10/7/43	Havana	11/7/43
Cuban N Coast 033	Sagua		Cayo Frances	14/07/43
Cuban N Coast 033/1	Nuevitas	15/07/43	Antilla Cuba	
Cuban N Coast 035	Cardenas	17/07/43	Havana	17/07/43
Cuban N Coast 036	Havana	18/07/43	Cardenas	20/07/43
Cuban N Coast 037	Santa Lucia	21/07/43	Havana	21/07/43
Cuban N Coast 038	Antilla Cuba	23/07/43	Guantanamo	30/07/43
Cuban N Coast 038	Havana		Guantanamo	31/07/43
Cuban N Coast 040	Havana	22/07/43	Cardenas	
Cuban N Coast 041	Havana	25/07/43	Cardenas	25/07/43
Cuban N Coast 041/1	Cardenas	26/07/43	Sagua	
Cuban N Coast 041/2	Sagua	27/07/43	Cayo Frances	
Cuban N Coast 042	Havana	26/07/43	Cienfuegos	
Cuban N Coast 043	Havana	27/07/43	Antilla Cuba	
Cuban N Coast 043	Havana	28/07/43	Antilla Cuba	
Cuban N Coast 044	Cardenas	29/07/43	Havana	
Cuban N Coast 046	Havana	31/07/43	Cardenas	31/07/43
Cuban N Coast 046/1	Cardenas	1/8/43	Sagua	1/8/43
Cuban N Coast 046/2	Sagua	2/8/43	Cardenas	2/8/43
Cuban N Coast 046/3	Cardenas	3/8/43	Havana	3/8/43
Cuban N Coast 048	Havana	4/8/43	Cardenas	4/8/43
Cuban N Coast 048	Havana	5/8/43	Nuevitas	
Cuban N Coast 048/1	Cardenas	5/8/43	Cayo Frances	5/8/43

Cuban N Coast 048/2	Nuevitas	9/8/43	Havana	
Cuban N Coast 049	Cardenas	6/8/43	Havana	6/8/43
Cuban N Coast 050	Havana	15/06/44	Nuevitas	18/06/44
Cuban N Coast 051	Havana		Antilla Cuba	
GK 700	Guantanamo	1/9/42	Key West	5/9/42
GK 701	Guantanamo	6/9/42	Key West	9/9/42
GK 702	Guantanamo	9/9/42	Key West	12/9/42
GK 703	Guantanamo	14/09/42	Key West	17/09/42
GK 704	Guantanamo	17/09/42	Key West	20/09/42
GK 705	Guantanamo	21/09/42	Key West	25/09/42
GK 706	Guantanamo	24/09/42	Key West	27/09/42
GK 707	Guantanamo	28/09/42	Key West	
GK 708	Did Not Sail			
GK 709	Guantanamo	11/10/42	Key West	14/10/42
GK 710	Guantanamo	19/10/42	Key West	22/10/42
GK 711	Guantanamo	27/10/42	Key West	30/10/42
GK 712	Guantanamo	4/11/42	Key West	7/11/42
GK 713	Guantanamo	12/11/42	Key West	15/11/42
GK 714	Guantanamo	20/11/42	Key West	23/11/42
GK 715	Guantanamo	28/11/42	Key West	1/12/42
GK 716	Guantanamo	6/12/42	Key West	10/12/42
GK 717	Guantanamo	14/12/42	Key West	17/12/42
GK 718	Guantanamo	22/12/42	Key West	25/12/42
GK 719	Guantanamo	30/12/42	Key West	2/1/1943
GK 720	Guantanamo	7/1/43	Key West	10/1/43
GK 721	Guantanamo	15/01/43	Key West	18/01/43
GK 722	Guantanamo	23/01/43	Key West	26/01/43
GK 723	Guantanamo	2/2/43	Key West	8/2/43
GK 724	Guantanamo	13/02/43	Key West	16/02/43
GK 725	Guantanamo	23/02/43	Key West	26/02/43
GK 726	Guantanamo	5/3/43	Key West	8/3/43
GK 727	Guantanamo	16/03/43	Key West	19/03/43
GK 728	Guantanamo	27/03/43	Key West	31/03/43
GK 729	Guantanamo	30/03/43	Key West	2/4/43
GK 730	Guantanamo	9/4/43	Key West	12/4/43
GK 731	Guantanamo	19/04/43	Key West	22/04/43
GK 732	Guantanamo	29/04/43	Key West	4/5/43
GK 733	Guantanamo	10/5/43	Key West	13/05/43
GK 734	Guantanamo	19/05/43	Key West	22/05/43
GK 735	Guantanamo	24/05/43	Key West	27/05/43
GK 736	Guantanamo	30/05/43	Key West	1/6/43
GK 737	Guantanamo	4/6/43	Key West	7/6/43
GK 738	Guantanamo	9/6/43	Key West	12/6/43
GK 739	Guantanamo	14/06/43	Key West	17/06/43

GK 740	Guantanamo	19/06/43	Key West	22/06/43
GK 741	Guantanamo	24/06/43	Key West	27/06/43
GK 742	Guantanamo	29/06/43	Key West	2/7/43
GK 743	Guantanamo	4/7/43	Key West	7/7/43
GK 744	Guantanamo	9/7/43	Key West	12/7/43
GK 745	Guantanamo	14/07/43	Key West	17/07/43
GK 746	Guantanamo	19/07/43	Key West	22/07/43
GK 747	Guantanamo	24/07/43	Key West	27/07/43
GK 748	Guantanamo	29/07/43	Key West	1/8/43
GK 749	Guantanamo	3/8/43	Key West	6/8/43
GK 750	Guantanamo	8/8/43	Key West	11/8/43
GK 751	Guantanamo	13/08/43	Key West	16/08/43
GK 752	Guantanamo	18/08/43	Key West	21/08/43
GK 753	Guantanamo	23/08/43	Key West	26/08/43
GK 754	Guantanamo	28/08/43	Key West	31/08/43
GK 755	Guantanamo	2/9/43	Key West	5/9/43
GK 756	Guantanamo	7/9/43	Key West	10/9/43
GK 757	Guantanamo	12/9/43	Key West	15/09/43
GK 758	Guantanamo	17/09/43	Key West	20/09/43
GK 759	Guantanamo	22/09/43	Key West	25/09/43
GK 760	Guantanamo	27/09/43	Key West	30/09/43
GK 761	Guantanamo	2/10/43	Key West	5/10/43
GK 762	Guantanamo	7/10/43	Key West	10/10/43
GK 763	Guantanamo	12/10/43	Key West	15/10/43
GK 764	Guantanamo	17/10/43	Key West	20/10/43
GK 765	Guantanamo	22/10/43	Key West	25/10/43
GK 766	Guantanamo	28/10/43	Key West	31/10/43
GK 767	Guantanamo	1/11/43	Key West	4/11/43
GK 768	Guantanamo	6/11/43	Key West	9/11/43
GK 769	Guantanamo	11/11/43	Key West	14/11/43
GK 770	Guantanamo	16/11/43	Key West	19/11/43
GK 771	Guantanamo	21/11/43	Key West	24/11/43
GK 772	Guantanamo	26/11/43	Key West	29/11/43
GK 773	Guantanamo	1/12/43	Key West	4/12/43
GK 774	Guantanamo	6/12/43	Key West	9/12/43
GK 775	Guantanamo	11/12/43	Key West	14/12/43
GK 776	Guantanamo	15/12/43	Key West	18/12/43
GK 777	Guantanamo	20/12/43	Key West	23/12/43
GK 778	Guantanamo	25/12/43	Key West	28/12/43
GK 779	Guantanamo	31/12/43	Key West	2/1/44
GK 780	Guantanamo	5/1/44	Key West	8/1/44
GK 781	Guantanamo	10/1/44	Key West	13/01/44
GK 782	Guantanamo	15/01/44	Key West	18/01/44
GK 783	Guantanamo	20/01/44	Key West	23/01/44

GK 784	Guantanamo	25/01/44	Key West	28/01/44
GK 785	Guantanamo	30/01/44	Key West	1/2/44
GK 786	Guantanamo	4/2/44	Key West	7/2/44
GK 787	Guantanamo	9/2/44	Key West	12/2/44
GK 788	Guantanamo	14/02/44	Key West	17/02/44
GK 789	Guantanamo	19/02/44	Key West	22/02/44
GK 790	Guantanamo	24/02/44	Key West	27/02/44
GK 791	Guantanamo	29/02/44	Key West	3/3/44
GK 792	Guantanamo	5/3/44	Key West	8/3/44
GK 793	Guantanamo	10/3/44	Key West	13/03/44
GK 794	Guantanamo	15/03/44	Key West	18/03/44
GK 795	Guantanamo	20/03/44	Key West	23/03/44
GK 796	Guantanamo	25/03/44	Key West	28/03/44
GK 797	Guantanamo	29/03/44	Key West	1/4/44
GK 798	Guantanamo	4/4/44	Key West	7/4/44
GK 799	Guantanamo	9/4/44	Key West	11/4/44
GK 800	Guantanamo	14/04/44	Key West	17/04/44
GK 801	Guantanamo	19/04/44	Key West	21/04/44
GK 802	Guantanamo	24/04/44	Key West	27/04/44
GK 803	Guantanamo	29/04/44	Key West	1/5/44
GK 804	Guantanamo	4/5/44	Key West	6/5/44
GK 805	Guantanamo	9/5/44	Key West	12/5/44
GK 806	Guantanamo	14/05/44	Key West	17/05/44
GK 807	Guantanamo	19/05/44	Key West	22/05/44
GK 808	Guantanamo	24/05/44	Key West	27/05/44
GK 809	Guantanamo	29/05/44	Key West	1/6/44
GK 810	Guantanamo	3/6/44	Key West	6/6/44
GK 811	Guantanamo	8/6/44	Key West	11/6/44
GK 812	Guantanamo	13/06/44	Key West	16/06/44
GK 813	Guantanamo	18/06/44	Key West	21/06/44
GK 814	Guantanamo	23/06/44	Key West	26/06/44
GK 815	Did Not Sail			
GK 816	Guantanamo	3/7/44	Key West	
GK 817	Guantanamo	8/7/44	Key West	
GK 818	Guantanamo	12/7/44	Key West	
GK 819	Guantanamo	18/07/44	Key West	21/07/44
GK 820	Guantanamo	23/07/44	Key West	26/07/44
GK 821	Guantanamo	28/07/44	Key West	
GK 822	Guantanamo	1/8/44	Key West	4/8/44
GK 823	Guantanamo	7/8/44	Key West	
GK 824	Guantanamo	11/8/44	Key West	14/08/44
GK 825	Guantanamo	17/08/44	Key West	20/08/44
GK 826	Did Not Sail			
GK 827	Guantanamo	26/08/44	Key West	29/08/44

GK 828	Guantanamo	31/08/44	Key West	3/9/44
GK 829	Guantanamo	5/9/44	Key West	8/9/44
GK 830	Did Not Sail			
GK 831	Guantanamo	16/09/44	Key West	19/09/44
GK 832	Guantanamo	21/09/44	Key West	24/09/44
GK 833	Guantanamo	25/09/44	Key West	28/09/44
GK 834	Guantanamo	30/09/44	Key West	3/10/44
GK 835	Guantanamo	5/10/44	Key West	8/10/44
GK 836	Guantanamo	10/10/44	Key West	13/10/44
GK 837	Did Not Sail			
GK 838	Guantanamo	21/10/44	Key West	23/10/44
GK 839	Guantanamo	25/10/44	Key West	28/10/44
GK 840	Guantanamo	30/10/44	Key West	3/11/44
GK 841	Guantanamo	4/11/44	Key West	7/11/44
GK 842	Guantanamo	9/11/44	Key West	12/11/44
GK 843	Did Not Sail			
GK 844	Guantanamo	20/11/44	Key West	23/11/44
GK 845	Guantanamo	25/11/44	Key West	28/11/44
GK 846	Guantanamo	29/11/44	Key West	
GK 847	Guantanamo	4/12/44	Key West	7/12/44
GK 848	Did Not Sail			
GK 849	Guantanamo	14/12/44	Key West	17/12/44
GK 850	Guantanamo	20/12/44	Key West	22/12/44
GK 851	Guantanamo	24/12/44	Key West	27/12/44
GK 852	Guantanamo	30/12/44	Key West	2/1/45
GK 853	Did Not Sail			
GK 854	Did Not Sail			
GK 855	Did Not Sail			
GK 856	Guantanamo	18/01/45	Key West	21/01/45
GK 857	Guantanamo	23/01/45	Key West	
GK 858	Guantanamo	29/01/45	Key West	1/2/45
GK 859	Guantanamo	2/2/45	Key West	
GK 860	Did Not Sail			
GK 861	Guantanamo	12/2/45	Key West	15/02/45
GK 862	Guantanamo	17/02/45	Key West	20/02/45
GK 863	Did Not Sail			
GK 864	Guantanamo	28/02/45	Key West	3/3/45
GK 865	Guantanamo	4/3/45	Key West	7/3/45
GK 866	Guantanamo	9/3/45	Key West	12/3/45
GK 867	Guantanamo	14/03/45	Key West	20/03/45
GK 868	Guantanamo	19/03/45	Key West	22/03/45
GK 869	Did Not Sail			
GK 870	Guantanamo	29/03/45	Key West	1/4/45
GK 871	Did Not Sail			

GK 872	Did Not Sail			
GK 873	Guantanamo	13/04/45	Key West	16/04/45
GK 874	Guantanamo	19/04/45	Key West	21/04/45
GK 875	Guantanamo	23/04/45	Key West	26/04/45
GK 876	Guantanamo	28/04/45	Key West	1/5/45
GK 877	Guantanamo	3/5/45	Key West	6/5/45
GK 878	Records N/A			
GK 879	Did Not Sail			
GN 1	Guantanamo	5/9/42	NYC	12/9/42
GN 10	Guantanamo	6/10/42	NYC	13/10/42
GN 100	Guantanamo	5/12/43	NYC	12/12/43
GN 101	Guantanamo	10/12/43	NYC	18/12/43
GN 102	Guantanamo	15/12/43	NYC	22/12/43
GN 103	Guantanamo	20/12/43	NYC	27/12/43
GN 104	Guantanamo	25/12/43	NYC	1/1/44
GN 105	Guantanamo	30/12/43	NYC	5/1/44
GN 106	Guantanamo	4/1/44	NYC	12/1/44
GN 107	Guantanamo	9/1/44	NYC	17/01/44
GN 108	Guantanamo	14/01/44	NYC	20/01/44
GN 109	Guantanamo	19/01/44	NYC	27/01/44
GN 11	Guantanamo	10/10/42	NYC	16/10/42
GN 110	Guantanamo	24/01/44	NYC	1/2/44
GN 111	Guantanamo	29/01/44	NYC	6/2/44
GN 112	Guantanamo	3/2/44	NYC	10/2/44
GN 113	Guantanamo	8/2/44	NYC	15/02/44
GN 114	Guantanamo	13/02/44	NYC	21/02/44
GN 115	Guantanamo	18/02/44	NYC	26/02/44
GN 116	Guantanamo	23/02/44	NYC	29/02/44
GN 117	Guantanamo	28/02/44	NYC	6/3/44
GN 118	Guantanamo	4/3/44	NYC	11/3/44
GN 119	Guantanamo	9/3/44	NYC	17/03/44
GN 12	Guantanamo	15/10/42	NYC	24/10/42
GN 120	Guantanamo	14/03/44	NYC	21/03/44
GN 121	Guantanamo	19/03/44	NYC	26/03/44
GN 122	Guantanamo	24/03/44	NYC	31/03/44
GN 123	Guantanamo	29/03/44	NYC	5/4/44
GN 124	Guantanamo	3/4/44	NYC	10/4/44
GN 125	Guantanamo	8/4/44	NYC	15/04/44
GN 126	Guantanamo	13/04/44	NYC	20/04/44
GN 127	Guantanamo	18/04/44	NYC	24/04/44
GN 128	Guantanamo	23/04/44	NYC	1/5/44
GN 129	Guantanamo	28/04/44	NYC	5/5/44
GN 13	Guantanamo	18/10/42	NYC	24/10/42
GN 130	Guantanamo	3/5/44	NYC	10/5/44

GN 131	Guantanamo	8/5/44	NYC	15/05/44
GN 132	Guantanamo	13/05/44	NYC	21/05/44
GN 133	Guantanamo	18/05/44	NYC	26/05/44
GN 134	Guantanamo	23/05/44	NYC	30/05/44
GN 135	Guantanamo	28/05/44	NYC	4/6/44
GN 136	Guantanamo	2/6/44	NYC	9/6/44
GN 137	Guantanamo	7/6/44	NYC	14/06/44
GN 138	Guantanamo	12/6/44	NYC	19/06/44
GN 139	Guantanamo	17/06/44	NYC	24/06/44
GN 14	Guantanamo	23/10/42	NYC	30/10/42
GN 140	Guantanamo	22/06/44	NYC	29/06/44
GN 141	Guantanamo	27/06/44	NYC	5/7/44
GN 142	Guantanamo	2/7/44	NYC	8/7/44
GN 143	Guantanamo	7/7/44	NYC	15/07/44
GN 144	Guantanamo	12/7/44	NYC	19/07/44
GN 145	Guantanamo	17/07/44	NYC	24/07/44
GN 146	Guantanamo	22/07/44	NYC	28/07/44
GN 147	Guantanamo	27/07/44	NYC	5/8/44
GN 148	Guantanamo	1/8/44	NYC	8/8/44
GN 149	Guantanamo	6/8/44	NYC	12/8/44
GN 15	Guantanamo	26/10/42	NYC	2/11/42
GN 150	Guantanamo	11/8/44	NYC	19/08/44
GN 151	Guantanamo	16/08/44	NYC	23/08/44
GN 152	Guantanamo	21/08/44	NYC	29/08/44
GN 153	Guantanamo	26/08/44	NYC	3/9/44
GN 154	Guantanamo	31/08/44	NYC	6/9/44
GN 155	Guantanamo	5/9/44	NYC	12/9/44
GN 156	Guantanamo	10/9/44	NYC	18/09/44
GN 157	Guantanamo	15/09/44	NYC	22/09/44
GN 158	Guantanamo	20/09/44	NYC	27/09/44
GN 159	Guantanamo	25/09/44	NYC	2/10/44
GN 16	Guantanamo	31/10/42	NYC	9/11/42
GN 160	Guantanamo	30/09/44	NYC	8/10/44
GN 161	Guantanamo	5/10/44	NYC	12/10/44
GN 162	Guantanamo	10/10/44	NYC	18/10/44
GN 163	Guantanamo	14/10/44	NYC	22/10/44
GN 164	Guantanamo	20/10/44	NYC	28/10/44
GN 165	Guantanamo	25/10/44	NYC	1/11/44
GN 166	Guantanamo	30/10/44	NYC	7/11/44
GN 167	Guantanamo	4/11/44	NYC	12/11/44
GN 168	Guantanamo	9/11/44	NYC	16/11/44
GN 169	Guantanamo	14/11/44	NYC	22/11/44
GN 17	Guantanamo	3/11/42	NYC	10/11/42
GN 170	Guantanamo	19/11/44	NYC	27/11/44

GN 171	Guantanamo	24/11/44	NYC	2/12/44
GN 172	Guantanamo	29/11/44	NYC	7/12/44
GN 173	Guantanamo	4/12/44	NYC	11/12/44
GN 174	Guantanamo	9/12/44	NYC	17/12/44
GN 175	Guantanamo	14/12/44	NYC	22/12/44
GN 176	Cancelled			
GN 177	Guantanamo	23/12/44	NYC	30/12/44
GN 178	Guantanamo	30/12/44	NYC	6/1/45
GN 179	Guantanamo	3/1/45	NYC	10/1/45
GN 18	Guantanamo	8/11/42	NYC	17/11/42
GN 180	Guantanamo	8/1/45	NYC	15/01/45
GN 181	Guantanamo	13/01/45	NYC	20/01/45
GN 182	Guantanamo	18/01/45	NYC	25/01/45
GN 183	Guantanamo	23/01/45	NYC	30/01/45
GN 184	Guantanamo	28/01/45	NYC	4/2/45
GN 185	Guantanamo	2/2/45	NYC	9/2/45
GN 186	Guantanamo	7/2/45	NYC	14/02/45
GN 187	Guantanamo	12/2/45	NYC	19/02/45
GN 188	Guantanamo	17/02/45	NYC	24/02/45
GN 189	Guantanamo	22/02/45	NYC	28/02/45
GN 19	Guantanamo	11/11/42	NYC	17/11/42
GN 190	Guantanamo	27/02/45	NYC	6/3/45
GN 191	Guantanamo	4/3/45	NYC	11/3/45
GN 192	Guantanamo	9/3/45	NYC	15/03/45
GN 193	Guantanamo	13/03/45	NYC	21/03/45
GN 194	Guantanamo	19/03/45	NYC	26/03/45
GN 195	Guantanamo	24/03/45	NYC	31/03/45
GN 196	Guantanamo	29/03/45	NYC	4/4/45
GN 197	Guantanamo	3/4/45	NYC	10/4/45
GN 198	Guantanamo	8/4/45	NYC	15/04/45
GN 199	Guantanamo	13/04/45	NYC	20/04/45
GN 2	Guantanamo	8/9/42	NYC	15/09/42
GN 20	Guantanamo	16/11/42	NYC	23/11/42
GN 200	Guantanamo	18/04/45	NYC	24/04/45
GN 201	Guantanamo	23/04/45	NYC	30/04/45
GN 202	Guantanamo	28/04/45	NYC	5/5/45
GN 203	Guantanamo	3/5/45	NYC	10/5/45
GN 204	Guantanamo	8/5/45	NYC	15/05/45
GN 205	Guantanamo	13/05/45	NYC	19/05/45
GN 206	Guantanamo	18/05/45	NYC	25/05/45
GN 21	Guantanamo	19/11/42	NYC	25/11/42
GN 22	Guantanamo	24/11/42	NYC	2/12/42
GN 23	Guantanamo	27/11/42	NYC	4/12/42
GN 24	Guantanamo	1/12/42	NYC	8/12/42

GN 25	Guantanamo	4/12/42	NYC	10/12/42
GN 26	Guantanamo	9/12/42	NYC	18/12/42
GN 27	Guantanamo	12/12/42	NYC	19/12/42
GN 28	Guantanamo	16/12/42	NYC	24/12/42
GN 29	Guantanamo	20/12/42	NYC	26/12/42
GN 3	Guantanamo	12/9/42	NYC	19/09/42
GN 30	Guantanamo	24/12/42	NYC	1/1/43
GN 31	Guantanamo	28/12/42	NYC	4/1/43
GN 32	Guantanamo	1/1/43	NYC	9/1/43
GN 33	Guantanamo	5/1/43	NYC	11/1/43
GN 34	Guantanamo	9/1/43	NYC	16/01/43
GN 35	Guantanamo	13/01/43	NYC	19/01/43
GN 36	Guantanamo	17/01/43	NYC	25/01/43
GN 37	Guantanamo	22/01/43	NYC	27/01/43
GN 38	Guantanamo	29/01/43	NYC	4/2/43
GN 39	Guantanamo	3/2/43	NYC	11/2/43
GN 4	Guantanamo	15/09/42	NYC	23/09/42
GN 40	Guantanamo	8/2/43	NYC	16/02/43
GN 41	Guantanamo	13/02/43	NYC	21/02/43
GN 42	Guantanamo	18/02/43	NYC	26/02/43
GN 43	Guantanamo	23/02/43	NYC	2/3/43
GN 44	Guantanamo	28/02/43	NYC	7/3/43
GN 45	Guantanamo	5/3/43	NYC	13/03/43
GN 46	Guantanamo	10/3/43	NYC	17/03/43
GN 47	Guantanamo	15/03/43	NYC	23/03/43
GN 48	Guantanamo	20/03/43	NYC	28/03/43
GN 49	Guantanamo	24/03/43	NYC	31/03/43
GN 5	Guantanamo	19/09/42	NYC	26/09/42
GN 50	Guantanamo	30/03/43	NYC	7/4/43
GN 51	Guantanamo	4/4/43	NYC	11/4/43
GN 52	Guantanamo	9/4/43	NYC	17/04/43
GN 53	Guantanamo	14/04/43	NYC	20/04/43
GN 54	Guantanamo	19/04/43	NYC	26/04/43
GN 55	Guantanamo	24/04/43	NYC	2/5/43
GN 56	Guantanamo	29/04/43	NYC	6/5/43
GN 57	Guantanamo	4/5/43	NYC	11/5/43
GN 58	Guantanamo	9/5/43	NYC	17/05/43
GN 59	Guantanamo	14/05/43	NYC	22/05/43
GN 6	Guantanamo	22/09/42	NYC	30/09/42
GN 60	Guantanamo	19/05/43	NYC	27/05/43
GN 61	Guantanamo	24/05/43	NYC	31/05/43
GN 62	Guantanamo	29/05/43	NYC	6/6/43
GN 63	Guantanamo	3/6/43	NYC	10/6/43
GN 64	Guantanamo	8/6/43	NYC	16/06/43

GN 65	Guantanamo	13/06/43	NYC	20/06/43
GN 66	Guantanamo	18/06/43	NYC	26/06/43
GN 67	Guantanamo	23/06/43	NYC	1/7/43
GN 68	Guantanamo	28/06/43	NYC	5/7/43
GN 69	Guantanamo	3/7/43	NYC	10/7/43
GN 7	Guantanamo	26/09/42	NYC	4/10/42
GN 70	Guantanamo	8/7/43	NYC	15/07/43
GN 71	Guantanamo	13/07/43	NYC	21/07/43
GN 72	Guantanamo	18/07/43	NYC	25/07/43
GN 73	Guantanamo	23/07/43	NYC	31/07/43
GN 74	Guantanamo	28/07/43	NYC	4/8/43
GN 75	Guantanamo	2/8/43	NYC	10/8/43
GN 76	Guantanamo	7/8/43	NYC	14/08/43
GN 77	Guantanamo	11/8/43	NYC	19/08/43
GN 78	Guantanamo	17/08/43	NYC	24/08/43
GN 79	Guantanamo	22/08/43	NYC	29/08/43
GN 8	Guantanamo	29/09/42	NYC	7/10/42
GN 80	Guantanamo	27/08/43	NYC	3/9/43
GN 81	Guantanamo	1/9/43	NYC	9/9/43
GN 82	Guantanamo	6/9/43	NYC	14/09/43
GN 83	Guantanamo	11/9/43	NYC	18/09/43
GN 84	Guantanamo	15/09/43	NYC	23/09/43
GN 85	Guantanamo	21/09/43	NYC	29/09/43
GN 86	Guantanamo	25/09/43	NYC	3/10/43
GN 87	Guantanamo	1/10/43	NYC	9/10/43
GN 88	Guantanamo	6/10/43	NYC	14/10/43
GN 89	Guantanamo	11/10/43	NYC	18/10/43
GN 9	Guantanamo	3/10/42	NYC	11/10/42
GN 90	Guantanamo	16/10/43	NYC	24/10/43
GN 91	Guantanamo	21/10/43	NYC	29/10/43
GN 92	Guantanamo	26/10/43	NYC	2/11/43
GN 93	Guantanamo	31/10/43	NYC	7/11/43
GN 94	Guantanamo	5/11/43	NYC	12/11/43
GN 95	Guantanamo	10/11/43	NYC	17/11/43
GN 96	Guantanamo	15/11/43	NYC	23/11/43
GN 97	Guantanamo	20/11/43	NYC	28/11/43
GN 98	Guantanamo	25/11/43	NYC	3/12/43
GN 99	Guantanamo	30/11/43	NYC	6/12/43
GN Spec1	Guantanamo	31/08/42	NYC	6/9/42
GN Spec2	Guantanamo	24/01/43	NYC	31/01/43
Gtmo 004 Spec	Guantanamo	11/9/42	Key West	13/09/42
Gtmo 008	Guantanamo	25/09/42	San Juan Pr	27/09/42
Gtmo 009	Guantanamo	2/10/42	San Juan Pr	5/10/42
Gtmo 009 Spec	Guantanamo	28/09/42	Nuevitas	30/09/42

Gtmo 011	Guantanamo	11/10/42	San Juan Pr	13/10/42
Gtmo 013	Guantanamo	19/10/42	San Juan Pr	22/10/42
Gtmo 014	Guantanamo	22/10/42	San Juan Pr	24/10/42
Gtmo 015	Guantanamo	26/10/42	San Juan Pr	29/10/42
Gtmo 016	Guantanamo	29/10/42	San Juan Pr	31/10/42
Gtmo 017	Guantanamo	4/11/42	Key West	9/11/42
Gtmo 018	Guantanamo	6/11/42	San Juan Pr	8/11/42
Gtmo 019	Guantanamo	8/11/42	San Juan Pr	8/11/42
Gtmo 022	Guantanamo	15/11/42	San Juan Pr	18/11/42
Gtmo 023	Guantanamo	17/11/42	Key West	19/11/42
Gtmo 025	Guantanamo	23/11/42	San Juan Pr	25/11/42
Gtmo 027	Guantanamo	27/11/42	Key West	29/11/42
Gtmo 028	Guantanamo	30/11/42	Havana	
Gtmo 031	Guantanamo	6/12/42	Key West	10/12/42
Gtmo 032	Guantanamo	7/12/42	San Juan Pr	11/12/42
Gtmo 038	Guantanamo	13/12/42	San Juan Pr	16/12/42
Gtmo 042	Guantanamo	25/12/42	San Juan Pr	27/12/42
Gtmo 044	Guantanamo	2/1/43	San Juan Pr	5/1/43
Gtmo 047	Guantanamo	9/1/43	Key West	12/1/43
Gtmo 048	Guantanamo	11/1/43	San Juan Pr	14/01/43
Gtmo 050	Guantanamo	21/01/43	San Juan Pr	23/01/43
Gtmo 053	Guantanamo	31/01/43	Norfolk Va	4/2/43
Gtmo 054	Guantanamo	8/2/43	Havana	9/2/43
Gtmo 055	Guantanamo	16/02/43	Key West	20/02/43
Gtmo 056	Guantanamo	19/02/43	San Juan Pr	22/02/43
Gtmo 057	Guantanamo	19/02/43	Key West	21/02/43
Gtmo 059	Guantanamo	26/02/43	San Juan Pr	28/02/43
Gtmo 060	Guantanamo	27/02/43	Key West	2/3/43
Gtmo 063	Guantanamo	2/3/43	San Juan Pr	6/3/43
Gtmo 065	Guantanamo	13/03/43	Key West	15/03/43
Gtmo 066	Guantanamo	12/3/43	San Juan Pr	15/03/43
Gtmo 070	Guantanamo	19/03/43	San Juan Pr	22/03/43
Gtmo 071	Guantanamo	21/03/43	Key West	26/03/43
Gtmo 076	Guantanamo	26/03/43	San Juan Pr	30/03/43
Gtmo 078 A	Guantanamo	2/4/43	New Orleans	5/4/43
Gtmo 080	Guantanamo	31/03/43	Mobile	4/4/43
Gtmo 081	Guantanamo	1/4/43	San Juan Pr	5/1/43
Gtmo 083	Guantanamo	5/4/43	Key West	9/4/43
Gtmo 085	Guantanamo	15/04/43	Key West	17/04/43
Gtmo 086	Guantanamo	16/04/43	Joined Kh 468	18/04/43
Gtmo 087	Guantanamo	17/04/43	Nuevitas	19/04/43
Gtmo 088	Guantanamo	21/04/43	Miami	26/04/43
Gtmo 089	Guantanamo	20/04/43	San Juan Pr	23/04/43
Gtmo 094	Guantanamo	16/05/43	Key West	18/05/43

Gtmo 095	Guantanamo	19/05/43	San Juan Pr	26/05/43
Gtmo 096	Guantanamo	18/05/43	San Juan Pr	22/05/43
Gtmo 097	Guantanamo	23/05/43	San Juan Pr	27/05/43
Gtmo 098	Guantanamo	26/05/43	San Juan Pr	28/05/43
Gtmo 099	Guantanamo	29/05/43	San Juan Pr	2/6/43
Gtmo 102	Guantanamo	1/6/43	San Juan Pr	2/6/43
Gtmo 104	Guantanamo	7/6/43	San Juan Pr	11/6/43
Gtmo 105	Guantanamo	7/6/43	Key West	10/6/43
Gtmo 108	Guantanamo	11/6/43	San Juan Pr	14/06/43
Gtmo 112	Guantanamo	18/06/43	San Juan Pr	21/06/43
Gtmo 117	Guantanamo	22/06/43	San Juan Pr	24/06/43
Gtmo 120	Guantanamo	25/06/43	Puerto Plata Dr	27/06/43
Gtmo 125	Guantanamo	4/7/43	San Juan Pr	8/7/43
Gtmo 126	Guantanamo	9/7/43	San Juan Pr	11/7/43
Gtmo 128	Guantanamo	20/07/43	San Juan Pr	23/07/43
Gtmo 129	Guantanamo	21/07/43	San Juan Pr	24/07/43
Gtmo 133	Guantanamo	1/8/43	San Juan Pr	3/8/43
Gtmo 137	Guantanamo	8/8/43	San Juan Pr	11/8/43
Gtmo 139	Guantanamo	18/08/43	San Juan Pr	21/08/43
Gtmo 143	Guantanamo	2/9/43	San Juan Pr	5/9/43
Gtmo 148	Guantanamo	15/09/43	Chester Pa	3/10/43
Gtmo 150	Guantanamo	20/09/43	San Juan Pr	23/09/43
Gtmo 152	Guantanamo	23/09/43	San Juan Pr	26/09/43
Gtmo 153	Guantanamo	28/09/43	San Juan Pr	1/10/43
Gtmo 158	Guantanamo	23/10/43	San Juan Pr	26/10/43
Gtmo 160	Guantanamo	25/10/43	San Juan Pr	27/10/43
Gtmo 162	Guantanamo	1/11/43	San Juan Pr	4/11/43
Gtmo 164	Guantanamo	5/11/43	San Juan Pr	8/11/43
Gtmo 166	Guantanamo	11/11/43	San Juan Pr	14/11/43
Gtmo 168	Guantanamo	15/11/43	San Juan Pr	18/11/43
Gtmo 173	Guantanamo	2/12/43	San Juan Pr	4/12/43
Gtmo 174	Guantanamo	2/12/43	San Juan Pr	4/12/43
Gtmo 175	Guantanamo	5/12/43	Key West	6/12/43
Gtmo 176	Guantanamo	4/12/43	San Juan Pr	7/12/43
Gtmo 177	Guantanamo	5/12/43	Key West	7/12/43
Gtmo 179	Guantanamo	9/12/43	San Juan Pr	12/12/43
Gtmo 182	Guantanamo	15/12/43	Mobile	24/12/43
Gtmo 183	Guantanamo	15/12/43	San Juan Pr	18/12/43
Gtmo 184	Guantanamo	15/12/43	San Juan Pr	19/12/43
Gtmo 188	Guantanamo	21/12/43	San Juan Pr	24/12/43
Gtmo 193	Guantanamo	30/12/43	San Juan Pr	3/1/44
Gtmo 194	Guantanamo	5/1/44	San Juan Pr	8/1/44
Gtmo 195	Guantanamo	6/1/44	San Juan Pr	8/1/44

Gtmo 197	Guantanamo	8/1/44	San Juan Pr	11/1/44
Gtmo 200	Guantanamo	18/01/44	San Juan Pr	20/01/44
Gtmo 202	Guantanamo	23/01/44	San Juan Pr	26/01/44
Gtmo 203	Guantanamo	23/01/44	San Juan Pr	26/01/44
Gtmo 204	Guantanamo	1/2/44	San Juan Pr	2/2/44
Gtmo 205	Guantanamo	2/2/44	San Juan Pr	5/2/44
Gtmo 207	Ex Ng 415		San Juan Pr	15/02/44
Gtmo 208	Guantanamo	4/3/44	San Juan Pr	7/3/44
Gtmo 209	Guantanamo	9/3/44	San Juan Pr	12/3/44
Gtmo 213	Guantanamo	15/03/44	San Juan Pr	17/03/44
Gtmo 214	Guantanamo	19/03/44	San Juan Pr	21/03/44
Gtmo 215	Guantanamo	25/03/44	San Juan Pr	29/03/44
Gtmo 216	Guantanamo	27/03/44	San Juan Pr	30/03/44
Gtmo 217	Guantanamo	29/03/44	San Juan Pr	31/03/44
Gtmo 218	Guantanamo	3/4/44	San Juan Pr	6/4/44
Gtmo 220	Guantanamo	8/4/44	San Juan Pr	11/4/44
Gtmo 222	Guantanamo	13/04/44	San Juan Pr	16/04/44
Gtmo 223	Guantanamo	19/04/44	San Juan Pr	22/04/44
Gtmo 225	Guantanamo	23/04/44	San Juan Pr	26/04/44
Gtmo 226	Guantanamo	28/04/44	San Juan Pr	30/04/44
Gtmo 227	Guantanamo	28/04/44	Key West	1/5/44
Gtmo 228	Guantanamo	23/04/44	San Juan Pr	26/04/44
Gtmo 230	Guantanamo	14/05/44	Key West	
Gtmo 232	Ex KG 708	26/05/44	San Juan Pr	28/05/44
Gtmo 234	Guantanamo	6/6/44	San Juan Pr	8/6/44
Gtmo 237	Guantanamo	10/6/44	San Juan Pr	13/06/44
Gtmo 239	Guantanamo	13/06/44	San Juan Pr	17/06/44
Gtmo 240	Guantanamo	18/06/44	San Juan Pr	20/06/44
Gtmo 243	Guantanamo		San Juan Pr	24/06/44
Gtmo 244	Guantanamo	27/06/44	San Juan Pr	30/06/44
Gtmo 245	Guantanamo	28/06/44	Key West	30/06/44
Gtmo 246	Guantanamo		Cape Haitian	
Gtmo 251	Guantanamo	8/7/44	San Juan Pr	15/07/44
Gtmo 253	Nipe Bay		San Juan Pr	
Gtmo 254	Guantanamo	12/7/44	San Juan Pr	16/07/44
Gtmo 255	Guantanamo	18/07/44	Key West	20/07/44
Gtmo 256	Guantanamo	18/07/44	San Juan Pr	22/07/44
Gtmo 261	Guantanamo	29/07/44	San Juan Pr	1/8/44
Gtmo 262	Guantanamo	30/07/44	New Orleans	2/8/44
Gtmo 264	Guantanamo	1/8/44	Cap Haitian	2/8/44
Gtmo 266	Guantanamo	4/8/44	San Juan	6/8/44
Gtmo 267	Guantanamo	4/8/44	New Orleans	7/8/44
Gtmo 269	Guantanamo	11/8/44	San Juan Pr	13/08/44
Gtmo 270	Guantanamo	14/08/44	New York	

Gtmo 271	Guantanamo	16/08/44	San Juan Pr	
Gtmo 272	Guantanamo	16/08/44	Turks Is	
Gtmo 274	Guantanamo	28/08/44	San Juan Pr	30/08/44
Gtmo 277	Guantanamo		San Juan Pr	
Gtmo 279	Guantanamo	29/09/44	San Juan Pr	1/10/44
Gtmo 281	Guantanamo	5/10/44	San Juan Pr	7/10/44
Gtmo 282	Guantanamo	9/10/44	New Orleans	12/10/44
Gtmo 283	Guantanamo	12/10/44	New York	17/10/44
Gtmo 285	Guantanamo	27/10/44	San Juan Pr	29/10/44
Gtmo 289	Guantanamo	11/11/44	Key West	14/11/44
Gtmo 290	Guantanamo	14/11/44	Key West	17/11/44
Gtmo 291	Guantanamo	17/11/44	Key West	21/11/44
Gtmo 292	Guantanamo	23/11/44	San Juan Pr	25/11/44
Gtmo 293	Guantanamo	24/11/44	New Orleans	27/11/44
Gtmo 296	Guantanamo	8/12/44	San Juan Pr	10/12/44
Gtmo 297	Guantanamo		San Juan Pr	
Gtmo 299	Guantanamo	24/12/44	New Orleans	27/12/44
Gtmo 300	Guantanamo	27/12/44	San Juan Pr	29/12/44
Gtmo 301	Guantanamo	31/12/44	New Orleans	3/1/45
Gtmo 302	Guantanamo	9/1/45	San Juan Pr	11/1/45
Gtmo 303	Guantanamo	29/01/45	San Juan Pr	
Gtmo 305	Guantanamo	2/2/45	San Juan Pr	4/2/45
Gtmo 307	Guantanamo	9/2/45	New Orleans	12/2/45
Gtmo 308	Guantanamo	15/02/45	San Juan Pr	17/02/45
Gtmo 311	Guantanamo	11/3/45	San Juan Pr	13/03/45
Gtmo 312	Guantanamo	12/3/45	San Juan Pr	14/03/45
Gtmo 314	Guantanamo	4/4/45	San Juan Pr	6/4/45
Gtmo 316	Guantanamo	26/04/45	San Juan Pr	28/04/45
Gtmo 317	Guantanamo	25/04/45	San Juan Pr	
Gtmo Spec	Guantanamo	1/4/43	Antilla Cuba	
Gtmo Spec	Guantanamo	1/2/43	Baltimore	9/2/43
Gtmo Spec	Guantanamo	3/4/43	Baltimore	9/4/43
Gtmo Spec	Guantanamo	24/04/43	Cap-Haitien	24/04/43
Gtmo Spec	Guantanamo	7/2/43	Charleston	12/2/43
Gtmo Spec	Guantanamo	26/01/43	Galveston	1/2/43
Gtmo Spec	Guantanamo	1/4/43	Grand Turk	
Gtmo Spec	Guantanamo	8/2/43	Havana	9/2/43
Gtmo Spec	Guantanamo	17/02/43	Havana	19/02/43
Gtmo Spec	Guantanamo	21/01/43	Jacksonville	24/01/43
Gtmo Spec	Guantanamo	18/02/43	Key West	24/02/43
Gtmo Spec	Guantanamo	24/02/43	Key West	28/02/43
Gtmo Spec	Guantanamo	17/04/43	Key West	23/04/43
Gtmo Spec	Guantanamo	25/04/43	Key West	29/04/43
Gtmo Spec	Guantanamo	25/04/43	Key West	

Gtmo Spec	Guantanamo	26/04/43	Key West	4/5/43
Gtmo Spec	Guantanamo	25/02/43	Mayport	2/3/43
Gtmo Spec	Guantanamo	3/4/43	Miami	
Gtmo Spec	Guantanamo	13/04/43	Miami	
Gtmo Spec	Guantanamo	18/02/43	Mobile	23/02/43
Gtmo Spec	Guantanamo	16/04/43	Mobile	
Gtmo Spec	Guantanamo	23/04/43	Mobile	
Gtmo Spec	Guantanamo	24/04/43	Mobile	30/04/43
Gtmo Spec	Guantanamo	9/1/43	New Orleans	13/01/43
Gtmo Spec	Guantanamo	23/01/43	New Orleans	28/01/43
Gtmo Spec	Guantanamo	25/01/43	New Orleans	31/01/43
Gtmo Spec	Guantanamo	30/01/43	New Orleans	2/2/43
Gtmo Spec	Guantanamo	7/2/43	New Orleans	13/02/43
Gtmo Spec	Guantanamo	9/2/43	New Orleans	15/02/43
Gtmo Spec	Guantanamo	12/2/43	New Orleans	18/02/43
Gtmo Spec	Guantanamo	19/02/43	New Orleans	
Gtmo Spec	Guantanamo	17/04/43	New Orleans	22/04/43
Gtmo Spec	Guantanamo	29/04/43	New Orleans	
Gtmo Spec	Guantanamo	28/01/43	New York	4/3/43
Gtmo Spec	Guantanamo	5/2/43	New York	14/02/43
Gtmo Spec	Guantanamo	10/2/43	New York	
Gtmo Spec	Guantanamo	16/02/43	New York	23/02/43
Gtmo Spec	Guantanamo	20/02/43	New York	
Gtmo Spec	Guantanamo	23/02/43	New York	6/3/43
Gtmo Spec	Guantanamo	6/1/43	Norfolk Va	11/1/43
Gtmo Spec	Guantanamo	8/12/42	San Juan Pr	
Gtmo Spec	Guantanamo	13/12/42	San Juan Pr	
Gtmo Spec	Guantanamo	5/2/43	San Juan Pr	
Gtmo Spec	Guantanamo	25/09/43	San Juan Pr	1/10/43
Gtmo Spec	Guantanamo	10/10/43	San Juan Pr	12/10/43
Gtmo Spec	Guantanamo	11/10/43	San Juan Pr	13/10/43
Gtmo Spec	Guantanamo		San Juan Pr	
Gtmo Spec	Guantanamo		San Juan Pr	
Gtmo Spec	Guantanamo		San Juan Pr	
Gtmo Spec	Guantanamo	28/04/43	Santiago Cuba Port	28/04/43
Havana 014	Havana	28/10/42	Everglades Port	
Havana 022	Havana	4/12/42	Everglades	5/12/42
Havana 023	Havana	13/03/43	West Palm Beach	14/03/43
Havana 033	Havana	25/03/43	West Palm Beach	26/03/43

Havana 034	Havana	26/03/43	Port Everglades	27/03/43
Havana 042	Havana	9/4/43	Cardenas	9/4/43
Havana 042a	Cardenas	10/4/43	Key West	11/4/43
Havana 058	Havana	25/04/43	Port Everglades	26/04/43
Havana 061	Havana	28/04/43	Port Everglades	29/04/43
Havana 063	Havana	30/04/43	Port Everglades	2/5/43
Havana 109	Havana	27/06/43	Miami	27/06/43
Hogan	Baytown	20/09/43	Norfolk, Va	25/09/43
Howard	Galveston Bar	9/10/43	Norfolk, Va	15/10/43
Jacksonville 002	Jacksonville	3/9/42	Miami	4/9/42
Jacksonville 004	Jacksonville	6/9/42	Key West	9/9/42
Jacksonville 010	Jacksonville	5/10/42	Miami	7/10/42
Jacksonville 014	Mayport	21/10/42	Smith Shoals	24/10/42
Jacksonville 016	Mayport	2/11/42	West Palm Beach	4/11/42
Jacksonville 041	Jacksonville	20/05/43	Key West	23/05/43
Jacksonville 042	Mayport	20/05/43	Key West	22/05/43
Jacksonville 045	Jacksonville	14/06/43	Key West	
Jacksonville 051	Mayport	23/06/43	Miami	
Jacksonville 052	Mayport	18/07/43	Guantanamo	20/07/43
Jacksonville 055	Mayport	26/07/43	Key West	28/07/43
Jacksonville 056	Jacksonville	14/09/43	Guantanamo	18/09/43
Jacksonville 073	Jacksonville		Cristobal	22/07/44
KG 600	Key West	1/9/42	Guantanamo	4/9/42
KG 601	Key West	8/9/42	Guantanamo	12/9/42
KG 602	Key West	15/09/42	Guantanamo	19/09/42
KG 603	Key West	18/09/42	Guantanamo	21/09/42
KG 604	Key West	22/09/42	Guantanamo	25/09/42
KG 605	Key West	25/09/42	Guantanamo	
KG 606	Key West	29/09/42	Guantanamo	2/10/42
KG 607	Key West	2/10/42	Guantanamo	5/10/42
KG 608	Key West	9/10/42	Guantanamo	12/10/42
KG 609	Key West	18/10/42	Guantanamo	22/10/42
KG 610	Key West	26/10/42	Guantanamo	29/10/42
KG 611	Key West	3/11/42	Guantanamo	7/11/42
KG 612	Key West	11/11/42	Guantanamo	15/11/42
KG 613	Key West	19/11/42	Guantanamo	23/11/42
KG 614	Key West	27/11/42	Guantanamo	30/11/42
KG 615	Key West	5/12/42	Guantanamo	8/12/42

KG 616	Key West	13/12/42	Guantanamo	16/12/42
KG 617	Key West	21/12/42	Guantanamo	25/12/42
KG 618	Key West	29/12/42	Guantanamo	1/1/43
KG 619	Key West	6/1/43	Guantanamo	10/1/43
KG 620	Key West	14/01/43	Guantanamo	17/01/43
KG 621	Key West	22/01/43	Guantanamo	25/01/43
KG 622	Key West	30/01/43	Guantanamo	2/2/43
KG 623	Key West	9/2/43	Guantanamo	12/2/43
KG 624	Key West	20/02/43	Guantanamo	23/02/43
KG 625	Key West	1/3/43	Guantanamo	4/3/43
KG 626	Key West	12/3/43	Guantanamo	15/03/43
KG 627	Key West	23/03/43	Guantanamo	26/03/43
KG 628	Key West	26/03/43	Guantanamo	29/03/43
KG 629	Key West	5/4/43	Guantanamo	8/4/43
KG 630	Key West	15/04/43	Guantanamo	18/04/43
KG 631	Key West	24/04/43	Guantanamo	27/04/43
KG 632	Key West	6/5/43	Guantanamo	9/5/43
KG 633	Key West	15/05/43	Guantanamo	18/05/43
KG 634	Key West	19/05/43	Guantanamo	22/05/43
KG 635	Key West	25/05/43	Guantanamo	28/05/43
KG 636	Key West	29/05/43	Guantanamo	1/6/43
KG 637	Key West	3/6/43	Guantanamo	6/6/43
KG 638	Key West	8/6/43	Guantanamo	12/6/43
KG 639	Key West	13/06/43	Guantanamo	16/06/43
KG 640	Key West	18/06/43	Guantanamo	21/06/43
KG 641	Key West	23/06/43	Guantanamo	26/06/43
KG 642	Key West	28/06/43	Guantanamo	1/7/43
KG 643	Key West	4/7/43	Guantanamo	6/7/43
KG 644	Key West	8/7/43	Guantanamo	11/7/43
KG 645	Key West	13/07/43	Guantanamo	17/07/43
KG 646	Key West	18/07/43	Guantanamo	
KG 647	Key West	23/07/43	Guantanamo	27/07/43
KG 648	Key West	29/07/43	Guantanamo	1/8/43
KG 649	Key West	2/8/43	Guantanamo	5/8/43
KG 650	Key West	7/8/43	Guantanamo	10/8/43
KG 651	Key West	12/8/43	Guantanamo	16/08/43
KG 652	Key West	17/08/43	Guantanamo	20/08/43
KG 653	Key West	23/08/43	Guantanamo	26/08/43
KG 654	Key West	27/08/43	Guantanamo	31/08/43
KG 655	Key West	1/9/43	Guantanamo	4/9/43
KG 656	Key West	7/9/43	Guantanamo	10/9/43
KG 657	Key West	11/9/43	Guantanamo	14/09/43
KG 658	Key West	16/09/43	Guantanamo	19/09/43
KG 659	Key West	23/09/43	Guantanamo	26/09/43

KG 660	Key West	26/09/43	Guantanamo	1/10/43
KG 661	Key West	3/10/43	Guantanamo	6/10/43
KG 662	Key West	7/10/43	Guantanamo	10/10/43
KG 663	Key West	11/10/43	Guantanamo	15/10/43
KG 664	Key West	16/10/43	Guantanamo	19/10/43
KG 665	Key West	21/10/43	Guantanamo	24/10/43
KG 666	Key West	26/10/43	Guantanamo	29/10/43
KG 667	Key West	31/10/43	Guantanamo	3/11/43
KG 668	Key West	6/11/43	Guantanamo	9/11/43
KG 669	Key West	10/11/43	Guantanamo	13/11/43
KG 670	Key West	15/11/43	Guantanamo	18/11/43
KG 671	Key West	20/11/43	Guantanamo	23/11/43
KG 672	Key West	27/11/43	Guantanamo	29/11/43
KG 673	Key West	30/11/43	Guantanamo	3/12/43
KG 674	Key West	5/12/43	Guantanamo	8/12/43
KG 675	Key West	11/12/43	Guantanamo	14/12/43
KG 676	Key West	16/12/43	Guantanamo	19/12/43
KG 677	Key West	20/12/43	Guantanamo	23/12/43
KG 678	Key West	26/12/43	Guantanamo	29/12/43
KG 679	Key West	30/12/43	Guantanamo	2/1/44
KG 680	Key West	5/1/44	Guantanamo	8/1/44
KG 681	Key West	10/1/44	Guantanamo	13/01/44
KG 682	Key West	15/01/44	Guantanamo	18/01/44
KG 683	Key West	19/01/44	Guantanamo	22/01/44
KG 684	Key West	24/01/44	Guantanamo	27/01/44
KG 685	Key West	29/01/44	Guantanamo	1/2/44
KG 686	Key West	3/2/44	Guantanamo	6/2/44
KG 687	Key West	8/2/44	Guantanamo	11/2/44
KG 688	Key West	13/02/44	Guantanamo	17/02/44
KG 689	Key West	18/02/44	Guantanamo	22/02/44
KG 690	Key West	23/02/44	Guantanamo	26/02/44
KG 691	Key West	28/02/44	Guantanamo	2/3/44
KG 692	Key West	4/3/44	Guantanamo	7/3/44
KG 693	Key West	9/3/44	Guantanamo	12/3/44
KG 694	Key West	14/03/44	Guantanamo	17/03/44
KG 695	Key West	19/03/44	Guantanamo	22/03/44
KG 696	Key West	24/03/44	Guantanamo	27/03/44
KG 697	Key West	29/03/44	Guantanamo	1/4/44
KG 698	Key West	3/4/44	Guantanamo	6/4/44
KG 699	Key West	8/4/44	Guantanamo	11/4/44
KG 700	Key West	13/04/44	Guantanamo	16/04/44
KG 701	Key West	18/04/44	Guantanamo	21/04/44
KG 702	Key West	23/04/44	Guantanamo	26/04/44
KG 703	Key West	28/04/44	Guantanamo	1/5/44

KG 704	Key West	4/5/44	Guantanamo	7/5/44
KG 705	Key West	8/5/44	Guantanamo	11/5/44
KG 706	Key West	13/05/44	Guantanamo	16/05/44
KG 707	Key West	18/05/44	Guantanamo	21/05/44
KG 708	Key West	23/05/44	Guantanamo	26/05/44
KG 709	Key West	28/05/44	Guantanamo	31/05/44
KG 710	Key West	2/6/44	Guantanamo	
KG 711	Key West	7/6/44	Guantanamo	10/6/44
KG 712	Key West	12/6/44	Guantanamo	15/06/44
KG 713	Key West	17/06/44	Guantanamo	20/06/44
KG 714	Key West	22/06/44	Guantanamo	25/06/44
KG 715	Key West	27/06/44	Guantanamo	30/06/44
KG 716	Key West	2/7/44	Guantanamo	5/7/44
KG 717	Key West	7/7/44	Guantanamo	10/7/44
KG 718	Key West	12/7/44	Guantanamo	15/07/44
KG 719	Key West	17/07/44	Guantanamo	20/07/44
KG 720	Key West	22/07/44	Guantanamo	25/07/44
KG 721	Key West	27/07/44	Guantanamo	30/07/44
KG 722	Key West	1/8/44	Guantanamo	4/8/44
KG 723	Did Not Sail			
KG 724	Key West	14/08/44	Guantanamo	
KG 725	Key West	16/08/44	Guantanamo	19/08/44
KG 726	Key West	21/08/44	Guantanamo	
KG 727	Key West	26/08/44	Guantanamo	28/08/44
KG 728	Key West	31/08/44	Guantanamo	3/9/44
KG 729	Key West	5/9/44	Guantanamo	8/9/44
KG 730	Key West	10/9/44	Guantanamo	13/09/44
KG 731	Did Not Sail			
KG 732	Key West	20/09/44	Guantanamo	
KG 733	Key West	25/09/44	Guantanamo	28/09/44
KG 734	Key West	30/09/44	Guantanamo	3/10/44
KG 735	Key West	5/10/44	Guantanamo	
KG 736	Key West	10/10/44	Guantanamo	
KG 737	Key West	15/10/44	Guantanamo	18/10/44
KG 738	Did Not Sail			
KG 739	Key West	25/10/44	Guantanamo	27/10/44
KG 740	Did Not Sail			
KG 741	Did Not Sail			
KG 742	Key West	9/11/44	Guantanamo	
KG 743	Key West	14/11/44	Guantanamo	
KG 744	Key West	19/11/44	Guantanamo	
KG 745	Key West	24/11/44	Guantanamo	27/11/44
KG 746	Did Not Sail			
KG 747	Key West	4/12/44	Guantanamo	7/12/44

KG 748	Key West	9/12/44	Guantanamo	12/12/44
KG 749	Did Not Sail			
KG 750	Key West	19/12/44	Guantanamo	
KG 751	Key West	27/12/44	Guantanamo	
KG 752	Key West	29/12/44	Guantanamo	1/1/45
KG 753	Key West	3/1/45	Guantanamo	6/1/45
KG 754	Did Not Sail			
KG 755	Key West	13/01/45	Guantanamo	15/01/45
KG 756	Key West	18/01/45	Guantanamo	21/01/45
KG 757	Key West	23/01/45	Guantanamo	25/01/45
KG 758	Key West	28/01/45	Guantanamo	31/01/45
KG 759	Did Not Sail			
KN 100	Key West	15/05/42	Hampton Roads	
KN 100	Key West	15/05/42	Norfolk	20/05/42
KN 100	Key West	15/05/42	Norfolk	20/05/42
KN 100	Key West	15/05/42	Norfolk	20/05/42
KN 100	Key West	15/05/42	Norfolk	20/05/42
KN 100	Key West	15/05/42	Norfolk	20/05/42
KN 101	Key West	18/05/42	Hampton Roads	25/05/42
KN 101	Key West	18/05/42	Norfolk	25/05/42
KN 101	Key West	18/05/42	Norfolk	25/05/42
KN 101	Key West	18/05/42	Norfolk	25/05/42
KN 101	Key West	18/05/42	Norfolk	25/05/42
KN 101	Key West	18/05/42	Norfolk	25/05/42
KN 101	Key West	18/05/42	Norfolk	25/05/42
KN 101	Key West	18/05/42	Norfolk	25/05/42
KN 102	Key West	21/05/42	Hampton Roads	26/05/42
KN 102	Key West	21/05/42	Norfolk	26/05/42
KN 102	Key West	21/05/42	Norfolk	26/05/42
KN 102	Key West	21/05/42	Norfolk	26/05/42
KN 102	Key West	21/05/42	Norfolk	26/05/42
KN 102	Key West	21/05/42	Norfolk	26/05/42
KN 102	Key West	21/05/42	Norfolk	26/05/42
KN 102	Key West	21/05/42	Norfolk	26/05/42
KN 103	Key West	24/05/42	Hampton Roads	29/05/42
KN 103	Key West	24/05/42	Norfolk	29/05/42
KN 103	Key West	24/05/42	Norfolk	29/05/42
KN 103	Key West	24/05/42	Norfolk	29/05/42
KN 103	Key West	24/05/42	Norfolk	29/05/42
KN 103	Key West	24/05/42	Norfolk	29/05/42

KN 103	Key West	24/05/42	Norfolk	29/05/42
KN 103	Key West	24/05/42	Norfolk	29/05/42
KN 104	Key West	27/05/42	Hampton Roads	31/05/42
KN 104	Key West	27/05/42	Norfolk	31/05/42
KN 104	Key West	27/05/42	Norfolk	31/05/42
KN 104	Key West	27/05/42	Norfolk	31/05/42
KN 104	Key West	27/05/42	Norfolk	31/05/42
KN 104	Key West	27/05/42	Norfolk	31/05/42
KN 105	Key West	30/05/42	Hampton Roads	4/6/42
KN 105	Key West	30/05/42	Norfolk	4/6/42
KN 105	Key West	30/05/42	Norfolk	4/6/42
KN 105	Key West	30/05/42	Norfolk	4/6/42
KN 105	Key West	30/05/42	Norfolk	4/6/42
KN 105	Key West	30/05/42	Norfolk	4/6/42
KN 105	Key West	30/05/42	Norfolk	4/6/42
KN 106	Key West	2/6/42	Hampton Roads	7/6/42
KN 106	Key West	2/6/42	Norfolk	7/6/42
KN 106	Key West	2/6/42	Norfolk	7/6/42
KN 106	Key West	2/6/42	Norfolk	7/6/42
KN 106	Key West	2/6/42	Norfolk	7/6/42
KN 106	Key West	2/6/42	Norfolk	7/6/42
KN 106	Key West	2/6/42	Norfolk	7/6/42
KN 106	Key West	2/6/42	Norfolk	7/6/42
KN 107	Key West	5/6/42	Hampton Roads	10/6/42
KN 107	Key West	5/6/42	Norfolk	10/6/42
KN 107	Key West	5/6/42	Norfolk	10/6/42
KN 107	Key West	5/6/42	Norfolk	10/6/42
KN 107	Key West	5/6/42	Norfolk	10/6/42
KN 107	Key West	5/6/42	Norfolk	10/6/42
KN 107	Key West	5/6/42	Norfolk	10/6/42
KN 107	Key West	5/6/42	Norfolk	10/6/42
KN 107	Key West	5/6/42	Norfolk	10/6/42
KN 108	Key West	8/6/42	NYC	15/06/42
KN 108	Key West	8/6/42	Norfolk	15/06/42
KN 108	Key West	8/6/42	Norfolk	15/06/42
KN 108	Key West	8/6/42	Norfolk	15/06/42
KN 108	Key West	8/6/42	Norfolk	15/06/42
KN 108	Key West	8/6/42	Norfolk	15/06/42
KN 108	Key West	8/6/42	Norfolk	15/06/42
KN 108	Key West	8/6/42	Norfolk	15/06/42

KN 108	Key West	8/6/42	Norfolk	15/06/42
KN 109	Key West	11/6/42	Hampton Roads	16/06/42
KN 109	Key West	11/6/42	Norfolk	16/06/42
KN 109	Key West	11/6/42	Norfolk	16/06/42
KN 109	Key West	11/6/42	Norfolk	16/06/42
KN 109	Key West	11/6/42	Norfolk	16/06/42
KN 109	Key West	11/6/42	Norfolk	16/06/42
KN 109	Key West	11/6/42	Norfolk	16/06/42
KN 110	Key West	14/06/42	NYC	19/06/42
KN 110	Key West	14/06/42	Norfolk	19/06/42
KN 110	Key West	14/06/42	Norfolk	19/06/42
KN 110	Key West	14/06/42	Norfolk	19/06/42
KN 110	Key West	14/06/42	Norfolk	19/06/42
KN 110	Key West	14/06/42	Norfolk	19/06/42
KN 110	Key West	14/06/42	Norfolk	19/06/42
KN 110	Key West	14/06/42	Norfolk	19/06/42
KN 111	Key West	18/06/42	NYC	24/06/42
KN 111	Key West	18/06/42	Norfolk	24/06/42
KN 111	Key West	18/06/42	Norfolk	24/06/42
KN 111	Key West	18/06/42	Norfolk	24/06/42
KN 111	Key West	18/06/42	Norfolk	24/06/42
KN 111	Key West	18/06/42	Norfolk	24/06/42
KN 111	Key West	18/06/42	Norfolk	24/06/42
KN 112	Key West	21/06/42	Hampton Roads	26/06/42
KN 112	Key West	21/06/42	Norfolk	26/06/42
KN 112	Key West	21/06/42	Norfolk	26/06/42
KN 112	Key West	21/06/42	Norfolk	26/06/42
KN 112	Key West	21/06/42	Norfolk	26/06/42
KN 112	Key West	21/06/42	Norfolk	26/06/42
KN 112	Key West	21/06/42	Norfolk	26/06/42
KN 112	Key West	21/06/42	Norfolk	26/06/42
KN 112	Key West	21/06/42	Norfolk	26/06/42
KN 113	Key West	25/06/42	Hampton Roads	29/06/42
KN 113	Key West	25/06/42	Norfolk	29/06/42
KN 113	Key West	29/06/42	Norfolk	29/06/42
KN 113	Key West	25/06/42	Norfolk	29/06/42
KN 113	Key West	25/06/42	Norfolk	29/06/42
KN 113	Key West	25/06/42	Norfolk	29/06/42
KN 114	Key West	27/06/42	Hampton Roads	2/7/42
KN 114	Key West	27/06/42	Norfolk	2/7/42

KN 114	Key West	27/06/42	Norfolk	2/7/42
KN 114	Key West	27/06/42	Norfolk	2/7/42
KN 114	Key West	27/06/42	Norfolk	2/7/42
KN 114	Key West	27/06/42	Norfolk	2/7/42
KN 115	Key West	30/06/42	Hampton Roads	4/7/42
KN 115	Key West	30/06/42	Norfolk	4/7/42
KN 115	Key West	30/06/42	Norfolk	4/7/42
KN 115	Key West	30/06/42	Norfolk	4/7/42
KN 115	Key West	30/06/42	Norfolk	4/7/42
KN 115	Key West	30/06/42	Norfolk	4/7/42
KN 116	Key West	3/7/42	Hampton Roads	7/7/42
KN 116	Key West	3/7/42	Norfolk	7/7/42
KN 116	Key West	3/7/42	Norfolk	7/7/42
KN 116	Key West	3/7/42	Norfolk	7/7/42
KN 116	Key West	3/7/42	Norfolk	7/7/42
KN 116	Key West	3/7/42	Norfolk	7/7/42
KN 117	Key West	6/7/42	Hampton Roads	11/7/42
KN 117	Key West	6/7/42	Norfolk	11/7/42
KN 117	Key West	6/7/42	Norfolk	11/7/42
KN 117	Key West	6/7/42	Norfolk	11/7/42
KN 117	Key West	6/7/42	Norfolk	11/7/42
KN 117	Key West	6/7/42	Norfolk	11/7/42
KN 117	Key West	6/7/42	Norfolk	11/7/42
KN 118	Key West	9/7/42	Hampton Roads	13/07/42
KN 118	Key West	9/7/42	Hampton Roads	13/07/42
KN 118	Key West	9/7/42	Norfolk	13/07/42
KN 118	Key West	9/7/42	Norfolk	13/07/42
KN 118	Key West	9/7/42	Norfolk	13/07/42
KN 118	Key West	9/7/42	Norfolk	13/07/42
KN 118	Key West	9/7/42	Norfolk	13/07/42
KN 119	Key West	12/7/42	Hampton Roads	17/07/42
KN 119	Key West	12/7/42	Norfolk	17/07/42
KN 119	Key West	14/07/42	Norfolk	17/07/42
KN 119	Key West	12/7/42	Norfolk	17/07/42
KN 119	Key West	12/7/42	Norfolk	17/07/42
KN 119	Key West	12/7/42	Norfolk	17/07/42
KN 119	Key West	12/7/42	Norfolk	17/07/42

KN 120	Key West	15/07/42	Hampton Roads	20/07/42
KN 120	Key West	15/07/42	Norfolk	20/07/42
KN 120	Key West	15/07/42	Norfolk	20/07/42
KN 120	Key West	15/07/42	Norfolk	20/07/42
KN 120	Key West	15/07/42	Norfolk	20/07/42
KN 120	Key West	15/07/42	Norfolk	20/07/42
KN 120	Key West	15/07/42	Norfolk	20/07/42
KN 121	Key West	18/07/42	Hampton Roads	22/07/42
KN 121	Key West	18/07/42	Norfolk	22/07/42
KN 121	Key West	18/07/42	Norfolk	22/07/42
KN 121	Key West	18/07/42	Norfolk	22/07/42
KN 121	Key West	18/07/42	Norfolk	22/07/42
KN 121	Key West	18/07/42	Norfolk	22/07/42
KN 122	Key West	22/07/42	Hampton Roads	26/07/42
KN 122	Key West	22/07/42	Norfolk	23/07/42
KN 122	Key West	22/07/42	Norfolk	23/07/42
KN 122	Key West	22/07/42	Norfolk	26/07/42
KN 122	Key West	22/07/42	Norfolk	23/07/42
KN 122	Key West	22/07/42	Norfolk	23/07/42
KN 123	Key West	24/07/42	Hampton Roads	29/07/42
KN 123	Key West	25/07/42	Norfolk	26/07/42
KN 123	Key West	24/07/42	Norfolk	29/07/42
KN 123	Key West	24/07/42	Norfolk	29/07/42
KN 123	Key West	24/07/42	Norfolk	29/07/42
KN 123	Key West	24/07/42	Norfolk	29/07/42
KN 123	Key West	24/07/42	Norfolk	29/07/42
KN 124	Key West	27/07/42	Hampton Roads	31/07/42
KN 124	Key West	27/07/42	Norfolk	31/07/42
KN 124	Key West	27/07/42	Norfolk	31/07/42
KN 124	Key West	27/07/42	Norfolk	31/07/42
KN 124	Key West	27/07/42	Norfolk	31/07/42
KN 124	Key West	27/07/42	Norfolk	31/07/42
KN 124	Key West	27/07/42	Norfolk	31/07/42
KN 125	Key West	31/07/42	NYC	4/8/42
KN 125	Key West	31/07/42	Norfolk	4/8/42
KN 125	Key West	31/07/42	Norfolk	4/8/42
KN 125	Key West	31/07/42	Norfolk	4/8/42
KN 125	Key West	31/07/42	Norfolk	4/8/42
KN 125	Key West	31/07/42	Norfolk	4/8/43

KN 126	Key West	2/8/42	NYC	10/8/42
KN 126	Key West	2/8/42	Norfolk	10/8/42
KN 126	Key West	2/8/42	Norfolk	10/8/42
KN 126	Key West	2/8/42	Norfolk	10/8/42
KN 126	Key West	2/8/42	Norfolk	10/8/42
KN 126	Key West	2/8/42	Norfolk	10/8/42
KN 127	Key West	5/8/42	NYC	11/8/42
KN 127	Key West	5/8/42	Norfolk	11/8/42
KN 127	Key West	5/8/42	Norfolk	11/8/42
KN 127	Key West	5/8/42	Norfolk	11/8/42
KN 127	Key West	5/8/42	Norfolk	11/8/42
KN 127	Key West	5/8/42	Norfolk	11/8/42
KN 127	Key West	6/8/42	Norfolk	11/8/42
KN 128	Key West	9/8/42	NYC	15/08/42
KN 128	Key West	9/8/42	Norfolk	15/08/42
KN 128	Key West	9/8/42	Norfolk	15/08/42
KN 128	Key West	9/8/42	Norfolk	15/08/42
KN 128	Key West	9/8/42	Norfolk	15/08/42
KN 128	Key West	9/8/42	Norfolk	15/08/42
KN 129	Key West	11/8/42	NYC	18/08/42
KN 129	Key West	11/8/42	Norfolk	18/08/42
KN 129	Key West	11/8/42	Norfolk	18/08/42
KN 129	Key West	11/8/42	Norfolk	18/08/42
KN 129	Key West	11/8/42	Norfolk	18/08/42
KN 129	Key West	11/8/42	Norfolk	18/08/42
KN 130	Key West	14/08/42	NYC	18/08/42
KN 130	Key West	14/08/42	Norfolk	18/08/42
KN 130	Key West	14/08/42	Norfolk	18/08/42
KN 130	Key West	14/08/42	Norfolk	18/08/42
KN 130	Key West	14/08/42	Norfolk	18/08/42
KN 130	Key West	14/08/42	Norfolk	18/08/42
KN 130	Key West	14/08/42	Norfolk	18/08/42
KN 130	Key West	14/08/42	Norfolk	18/08/42
KN 131	Key West	18/08/42	Hampton Roads	23/08/42
KN 131	Key West	18/08/42	Norfolk	23/08/42
KN 131	Key West	18/08/42	Norfolk	23/08/42
KN 131	Key West	18/08/42	Norfolk	23/08/42
KN 131	Key West	18/08/42	Norfolk	23/08/42
KN 131	Key West	18/08/42	Norfolk	23/08/42
KN 132	Key West	20/08/42	Hampton Roads	25/08/42
KN 132	Key West	20/08/42	Norfolk	25/08/42
KN 132	Key West	20/08/42	Norfolk	25/08/42

KN 132	Key West	20/08/42	Norfolk	25/08/42
KN 132	Key West	24/08/42	Norfolk	25/08/42
KN 132	Key West	20/08/42	Norfolk	25/08/42
KN 132	Key West	24/08/42	Norfolk	25/08/42
KN 132	Key West	24/08/42	Norfolk	25/08/42
KN 132	Key West	20/08/42	Norfolk	25/08/42
KN 132	Key West	24/08/42	Norfolk	25/08/42
KN 132	Key West	24/08/42	Norfolk	25/08/42
KN 133	Key West	23/08/42	Hampton Roads	29/08/42
KN 133	Key West	23/08/42	Norfolk	29/08/42
KN 133	Key West	23/08/42	Norfolk	29/08/42
KN 133	Key West	23/08/42	Norfolk	29/08/42
KN 133	Key West	23/08/42	Norfolk	29/08/42
KN 133	Key West	23/08/42	Norfolk	29/08/42
KN 134	Key West	26/08/42	Hampton Roads	31/08/42
KN 134	Key West	26/08/42	Norfolk	31/08/42
KN 134	Key West	26/08/42	Norfolk	31/08/42
KN 134	Key West	26/08/42	Norfolk	31/08/42
KN 134	Key West	26/08/42	Norfolk	31/08/42
KN 134	Key West		Norfolk	31/08/42
KN 134	Key West	26/08/42	Norfolk	31/08/42
KN 134	Key West		Norfolk	30/08/42
KN 134	Key West	30/08/42	Norfolk	31/08/42
KN 135	Key West	29/08/42	Hampton Roads	4/9/42
KN 135	Key West	29/08/42	Norfolk	4/9/42
KN 135	Key West	29/08/42	Norfolk	4/9/42
KN 135	Key West	29/08/42	Norfolk	4/9/42
KN 135	Key West	29/08/42	Norfolk	4/9/42
KN 135	Key West	29/08/42	Norfolk	4/9/42
KN 135	Key West	29/08/42	Norfolk	4/9/42
KN 136	Key West	2/9/42	Norfolk	7/9/42
KN 136	Key West	2/9/42	Norfolk	7/9/42
KN 136	Key West	2/9/42	Norfolk	7/9/42
KN 136	Key West	2/9/42	Norfolk	7/9/42
KN 136	Key West	2/9/42	Norfolk	7/9/42
KN 136	Key West	2/9/42	Norfolk	7/9/42
KN 136	Key West	2/9/42	Norfolk	7/9/42
KN 200	Key West	8/9/42	NYC	13/09/42
KN 200	Key West	8/9/42	Norfolk	13/09/42
KN 200	Key West	8/9/42	Norfolk	13/09/42
KN 200	Key West	8/9/42	Norfolk	13/09/42

KN 200	Key West	8/9/42	Norfolk	13/09/42
KN 200	Key West	8/9/42	Norfolk	13/09/42
KN 200	Key West	8/9/42	Norfolk	13/09/42
KN 201	Key West	15/09/42	NYC	20/09/42
KN 201	Key West	15/09/42	Norfolk	20/09/42
KN 201	Key West	15/09/42	Norfolk	20/09/42
KN 201	Key West	15/09/42	Norfolk	20/09/42
KN 201	Key West	15/09/42	Norfolk	20/09/42
KN 201	Key West	15/09/42	Norfolk	20/09/42
KN 202	Key West	22/09/42	NYC	26/09/42
KN 202	Key West	22/09/42	Norfolk	26/09/42
KN 202	Key West	22/09/42	Norfolk	26/09/42
KN 202	Key West	22/09/42	Norfolk	26/09/29 42
KN 202	Key West	22/09/42	Norfolk	26/09/42
KN 202	Key West	22/09/42	Norfolk	26/09/42
KN 203	Key West	29/09/42	NYC	5/10/42
KN 203	Key West	29/09/42	Norfolk	5/10/42
KN 203	Key West	29/09/42	Norfolk	5/10/42
KN 203	Key West	29/09/42	Norfolk	5/10/42
KN 203	Key West	29/09/42	Norfolk	5/10/42
KN 203	Key West	29/09/42	Norfolk	5/10/42
KN 204	Key West	6/10/42	NYC	12/10/42
KN 204	Key West	6/10/42	Norfolk	12/10/42
KN 204	Key West	6/10/42	Norfolk	12/10/42
KN 204	Key West	6/10/42	Norfolk	12/10/42
KN 204	Key West	6/10/42	Norfolk	12/10/42
KN 204	Key West	6/10/42	Norfolk	12/10/42
KN 205	Key West	16/10/42	NYC	21/10/42
KN 205	Key West	16/10/42	Norfolk	21/10/42
KN 205	Key West	16/10/42	Norfolk	21/10/42
KN 205	Key West	16/10/42	Norfolk	21/10/42
KN 205	Key West	16/10/42	Norfolk	21/10/42
KN 205	Key West	16/10/42	Norfolk	21/10/42
KN 206	Key West	24/10/42	NYC	30/10/42
KN 206	Key West	24/10/42	Norfolk	30/10/42
KN 206	Key West	24/10/42	Norfolk	30/10/42
KN 206	Key West	24/10/42	Norfolk	30/10/42
KN 207	Key West	1/11/42	NYC	7/11/42
KN 207	Key West	1/11/42	Norfolk	7/11/42
KN 207	Key West	1/11/42	Norfolk	7/11/42
KN 207	Key West	1/11/42	Norfolk	7/11/42
KN 207	Key West	1/11/42	Norfolk	7/11/42
KN 207	Key West	1/11/42	Norfolk	7/11/42

KN 208	Key West	10/11/42	NYC	16/11/42
KN 208	Key West	10/11/42	Norfolk	16/11/42
KN 208	Key West	10/11/42	Norfolk	16/11/42
KN 208	Key West	10/11/42	Norfolk	16/11/42
KN 208	Key West	10/11/42	Norfolk	16/11/42
KN 208	Key West	10/11/42	Norfolk	16/11/42
KN 209	Key West	18/11/42	NYC	23/11/42
KN 209	Key West	18/11/42	Norfolk	23/11/42
KN 209	Key West	18/11/42	Norfolk	23/11/42
KN 209	Key West	18/11/42	Norfolk	23/11/42
KN 209	Key West	18/11/42	Norfolk	23/11/42
KN 209	Key West	18/11/42	Norfolk	23/11/42
KN 209	Key West	18/11/42	Norfolk	23/11/42
KN 210	Key West	25/11/42	NYC	1/12/42
KN 210	Key West	25/11/42	Norfolk	1/12/42
KN 210	Key West	25/11/42	Norfolk	1/12/42
KN 210	Key West	25/11/42	Norfolk	1/12/42
KN 210	Key West	25/11/42	Norfolk	1/12/42
KN 210	Key West	25/11/42	Norfolk	1/12/42
KN 211	Key West	3/12/42	NYC	8/12/42
KN 211	Key West	3/12/42	Norfolk	8/12/42
KN 211	Key West	3/12/42	Norfolk	8/12/42
KN 211	Key West	3/12/42	Norfolk	8/12/42
KN 211	Key West	3/12/42	Norfolk	8/12/42
KN 211	Key West	3/12/42	Norfolk	8/12/42
KN 212	Key West	9/12/42	NYC	14/12/42
KN 212	Key West	9/12/42	Norfolk	14/12/42
KN 212	Key West	9/12/42	Norfolk	14/12/42
KN 212	Key West	9/12/42	Norfolk	14/12/42
KN 212	Key West	9/12/42	Norfolk	14/12/42
KN 213	Key West	17/12/42	NYC	22/12/42
KN 213	Key West	17/12/42	Norfolk	22/12/42
KN 213	Key West	17/12/42	Norfolk	22/12/42
KN 213	Key West	17/12/42	Norfolk	22/12/42
KN 213	Key West	17/12/42	Norfolk	22/12/42
KN 213	Key West	17/12/42	Norfolk	22/12/42
KN 214	Key West	25/12/42	NYC	30/12/42
KN 214	Key West	25/12/42	Norfolk	30/12/42
KN 214	Key West	25/12/42	Norfolk	30/12/42
KN 214	Key West	25/12/42	Norfolk	30/12/42
KN 214	Key West	25/12/42	Norfolk	30/12/42
KN 214	Key West	25/12/42	Norfolk	30/12/42
KN 215	Key West	1/1/43	NYC	7/1/43
KN 215	Key West	1/1/43	Norfolk	7/1/43

KN 215	Key West	1/1/43	Norfolk	7/1/43
KN 215	Key West	1/1/43	Norfolk	7/1/43
KN 215	Key West	1/1/43	Norfolk	7/1/43
KN 216	Key West	10/1/43	NYC	15/01/43
KN 216	Key West	10/1/43	Norfolk	15/01/43
KN 216	Key West	10/1/43	Norfolk	15/01/43
KN 216	Key West	10/1/43	Norfolk	15/01/43
KN 216	Key West	10/1/43	Norfolk	15/01/43
KN 216	Key West	10/1/43	Norfolk	15/01/43
KN 217	Key West	17/01/43	NYC	23/01/43
KN 217	Key West	17/01/43	Norfolk	23/01/43
KN 217	Key West	17/01/43	Norfolk	23/01/43
KN 217	Key West	17/01/43	Norfolk	23/01/43
KN 217	Key West	17/01/43	Norfolk	23/01/43
KN 217	Key West	17/01/43	Norfolk	23/01/43
KN 218	Key West	26/01/43	NYC	1/2/43
KN 218	Key West	26/01/43	Norfolk	1/2/43
KN 218	Key West	26/01/43	Norfolk	1/2/43
KN 218	Key West	26/01/43	Norfolk	1/2/43
KN 218	Key West	26/01/43	Norfolk	1/2/43
KN 218	Key West	26/01/43	Norfolk	1/2/43
KN 219	Key West	30/01/43	NYC	4/2/43
KN 219	Key West	30/01/43	Norfolk	4/2/43
KN 219	Key West	30/01/43	Norfolk	4/2/43
KN 219	Key West	30/01/43	Norfolk	4/2/43
KN 219	Key West	30/01/43	Norfolk	4/2/43
KN 220	Key West	5/2/43	NYC	10/2/43
KN 220	Key West	5/2/43	Norfolk	10/2/43
KN 220	Key West	5/2/43	Norfolk	10/2/43
KN 220	Key West	5/2/43	Norfolk	10/2/43
KN 220	Key West	5/2/43	Norfolk	10/2/43
KN 221	Key West	9/2/43	NYC	14/02/43
KN 221	Key West	9/2/43	Norfolk	14/02/43
KN 221	Key West	9/2/43	Norfolk	14/02/43
KN 221	Key West	9/2/43	Norfolk	14/02/43
KN 221	Key West	9/2/43	Norfolk	14/02/43
KN 222	Key West	14/02/43	NYC	19/02/43
KN 222	Key West	14/02/43	Norfolk	19/02/43
KN 222	Key West	14/02/43	Norfolk	19/02/43
KN 222	Key West	14/02/43	Norfolk	19/02/43
KN 222	Key West	14/02/43	Norfolk	19/02/43
KN 223	Key West	20/02/43	NYC	25/02/43
KN 223	Key West	20/02/43	NYC	25/02/43
KN 223	Key West	20/02/43	Norfolk	25/02/43

KN 223	Key West	20/02/43	Norfolk	25/02/43
KN 223	Key West	20/02/43	Norfolk	25/02/43
KN 223	Key West	20/02/43	Norfolk	25/02/43
KN 224	Key West	24/02/43	NYC	1/3/43
KN 224	Key West	24/02/43	NYC	1/3/43
KN 224	Key West	24/02/43	Norfolk	1/3/43
KN 224	Key West	24/02/43	Norfolk	1/3/43
KN 224	Key West	24/02/43	Norfolk	1/3/43
KN 224	Key West	24/02/43	Norfolk	1/3/43
KN 225	Key West	2/3/43	NYC	7/3/43
KN 225	Key West	2/3/43	NYC	7/3/43
KN 225	Key West	2/3/43	Norfolk	7/3/43
KN 225	Key West	2/3/43	Norfolk	7/3/43
KN 225	Key West	2/3/43	Norfolk	7/3/43
KN 225	Key West	2/3/43	Norfolk	7/3/43
KN 226	Key West	7/3/43	NYC	13/03/43
KN 226	Key West	7/3/43	NYC	13/03/43
KN 226	Key West	7/3/43	Norfolk	13/03/43
KN 226	Key West	7/3/43	Norfolk	13/03/43
KN 226	Key West	7/3/43	Norfolk	13/03/43
KN 226	Key West	7/3/43	Norfolk	13/03/43
KN 227	Key West	12/3/43	NYC	17/03/43
KN 227	Key West	12/3/43	NYC	17/03/43
KN 227	Key West	12/3/43	Norfolk	17/03/43
KN 227	Key West	12/3/43	Norfolk	17/03/43
KN 227	Key West	12/3/43	Norfolk	17/03/43
KN 227	Key West	12/3/43	Norfolk	17/03/43
KN 228	Key West	16/03/43	NYC	20/03/43
KN 228	Key West	16/03/43	NYC	20/03/43
KN 228	Key West	16/03/43	Norfolk	20/03/43
KN 228	Key West	16/03/43	Norfolk	20/03/43
KN 228	Key West	16/03/43	Norfolk	20/03/43
KN 228	Key West	16/03/43	Norfolk	20/03/43
KN 229	Key West	22/03/43	NYC	28/03/43
KN 229	Key West	22/03/43	NYC	28/03/43
KN 229	Key West	22/03/43	Norfolk	28/03/43
KN 229	Key West	22/03/43	Norfolk	28/03/43
KN 229	Key West	22/03/43	Norfolk	28/03/43
KN 229	Key West	22/03/43	Norfolk	28/03/43
KN 229	Key West	22/03/43	Norfolk	28/03/43
KN 230	Key West	25/03/43	NYC	31/03/43
KN 230	Key West	25/03/43	Norfolk	31/03/43
KN 230	Key West	25/03/43	Norfolk	31/03/43
KN 230	Key West	25/03/43	Norfolk	31/03/43

KN 230	Key West	25/03/43	Norfolk	31/03/43
KN 231	Key West	31/03/43	NYC	7/4/43
KN 231	Key West	31/03/43	Norfolk	7/4/43
KN 231	Key West	31/03/43	Norfolk	7/4/43
KN 231	Key West	31/03/43	Norfolk	7/4/43
KN 231	Key West	31/03/43	Norfolk	7/4/43
KN 231	Key West	31/03/43	Norfolk	7/4/43
KN 232	Key West	5/4/43	NYC	10/4/43
KN 232	Key West	5/4/43	Norfolk	10/4/43
KN 232	Key West	5/4/43	Norfolk	10/4/43
KN 232	Key West	5/4/43	Norfolk	10/4/43
KN 232	Key West	5/4/43	Norfolk	10/4/43
KN 233	Key West	9/4/43	NYC	14/04/43
KN 233	Key West	9/4/43	Norfolk	14/04/43
KN 233	Key West	9/4/43	Norfolk	14/04/43
KN 233	Key West	9/4/43	Norfolk	14/04/43
KN 233	Key West	9/4/43	Norfolk	14/04/43
KN 233	Key West	9/4/43	Norfolk	11/4/43
KN 234	Key West	15/04/43	NYC	20/04/43
KN 234	Key West	15/04/43	Norfolk	20/04/43
KN 234	Key West	15/04/43	Norfolk	20/04/43
KN 234	Key West	16/04/43	Norfolk	20/04/43
KN 234	Key West	15/04/43	Norfolk	20/04/43
KN 234	Key West	16/04/43	Norfolk	20/04/43
KN 234	Key West	15/04/43	Norfolk	20/04/43
KN 235	Key West	20/04/43	NYC	25/04/43
KN 235	Key West	20/04/43	Norfolk	25/04/43
KN 235	Key West	20/04/43	Norfolk	25/04/43
KN 235	Key West	20/04/43	Norfolk	25/04/43
KN 235	Key West	20/04/43	Norfolk	25/04/43
KN 235	Key West	20/04/43	Norfolk	25/04/43
KN 236	Key West	24/04/43	NYC	30/04/43
KN 236	Key West	24/04/43	Norfolk	30/04/43
KN 236	Key West	24/04/43	Norfolk	30/04/43
KN 236	Key West	24/04/43	Norfolk	30/04/43
KN 236	Key West	24/04/43	Norfolk	30/04/43
KN 237	Key West	29/04/43	NYC	6/5/43
KN 237	Key West	29/04/43	Norfolk	6/5/43
KN 237	Key West	29/04/43	Norfolk	6/5/43
KN 237	Key West	29/04/43	Norfolk	6/5/43
KN 237	Key West	29/04/43	Norfolk	6/5/43
KN 238	Key West	5/5/43	NYC	10/5/43
KN 238	Key West	5/5/43	Norfolk	10/5/43
KN 238	Key West	5/5/43	Norfolk	10/5/43

KN 238	Key West	5/5/43	Norfolk	10/5/43
KN 238	Key West	5/5/43	Norfolk	10/5/43
KN 239	Key West	9/5/43	NYC	14/05/43
KN 239	Key West	9/5/43	Norfolk	14/05/43
KN 239	Key West	9/5/43	Norfolk	14/05/43
KN 239	Key West	9/5/43	Norfolk	14/05/43
KN 239	Key West	9/5/43	Norfolk	14/05/43
KN 240	Key West	15/05/43	NYC	21/05/43
KN 240	Key West	15/05/43	Norfolk	21/05/43
KN 240	Key West	15/05/43	Norfolk	21/05/43
KN 240	Key West	15/05/43	Norfolk	21/05/43
KN 240	Key West	15/05/43	Norfolk	21/05/43
KN 240	Key West	15/05/43	Norfolk	21/05/43
KN 240	Key West	15/05/43	Norfolk	21/05/43
KN 241	Key West	19/05/43	NYC	24/05/43
KN 241	Key West	19/05/43	Norfolk	24/05/43
KN 241	Key West	19/05/43	Norfolk	24/05/43
KN 241	Key West	19/05/43	Norfolk	24/05/43
KN 241	Key West	19/05/43	Norfolk	24/05/43
KN 242	Key West	25/05/43	NYC	30/05/43
KN 242	Key West	25/05/43	Norfolk	30/05/43
KN 242	Key West	25/05/43	Norfolk	30/05/43
KN 242	Key West	25/05/43	Norfolk	30/05/43
KN 242	Key West	25/05/43	Norfolk	30/05/43
KN 243	Key West	29/05/43	NYC	3/6/43
KN 243	Key West	29/05/43	Norfolk	3/6/43
KN 243	Key West	29/05/43	Norfolk	3/6/43
KN 243	Key West	29/05/43	Norfolk	3/6/43
KN 243	Key West	29/05/43	Norfolk	3/6/43
KN 244	Key West	3/6/43	NYC	9/6/43
KN 244	Key West	3/6/43	Norfolk	9/6/43
KN 244	Key West	3/6/43	Norfolk	9/6/43
KN 244	Key West	3/6/43	Norfolk	9/6/43
KN 244	Key West	3/6/43	Norfolk	9/6/43
KN 245	Key West	8/6/43	NYC	14/06/43
KN 245	Key West	8/6/43	Norfolk	14/06/43
KN 245	Key West	8/6/43	Norfolk	14/06/43
KN 245	Key West	8/6/43	Norfolk	14/06/43
KN 245	Key West	8/6/43	Norfolk	14/06/43
KN 246	Key West	13/06/43	NYC	18/06/43
KN 246	Key West	13/06/43	Norfolk	18/06/43
KN 246	Key West	13/06/43	Norfolk	18/06/43
KN 246	Key West	13/06/43	Norfolk	18/06/43
KN 246	Key West	13/06/43	Norfolk	18/06/43

KN 247	Key West	18/06/43	NYC	22/06/43
KN 247	Key West	18/06/43	Norfolk	22/06/43
KN 247	Key West	18/06/43	Norfolk	22/06/43
KN 247	Key West	18/06/43	Norfolk	22/06/43
KN 247	Key West	18/06/43	Norfolk	22/06/43
KN 248	Key West	23/06/43	NYC	29/06/43
KN 248	Key West	23/06/43	Norfolk	29/06/43
KN 248	Key West	23/06/43	Norfolk	29/06/43
KN 248	Key West	23/06/43	Norfolk	29/06/43
KN 248	Key West	23/06/43	Norfolk	29/06/43
KN 249	Key West	28/06/43	NYC	3/7/43
KN 249	Key West	28/06/43	Norfolk	3/7/43
KN 249	Key West	28/06/43	Norfolk	3/7/43
KN 249	Key West	28/06/43	Norfolk	3/7/43
KN 249	Key West	28/06/43	Norfolk	3/7/43
KN 249	Key West	29/06/43	Norfolk	3/7/43
KN 250	Key West	4/7/43	NYC	9/7/43
KN 250	Key West	4/7/43	Norfolk	9/7/43
KN 250	Key West	4/7/43	Norfolk	9/7/43
KN 250	Key West	4/7/43	Norfolk	9/7/43
KN 250	Key West	4/7/43	Norfolk	9/7/43
KN 250	Key West	4/7/43	Norfolk	9/7/43
KN 251	Key West	9/7/43	NYC	14/07/43
KN 251	Key West	9/7/43	Norfolk	14/07/43
KN 251	Key West	9/7/43	Norfolk	14/07/43
KN 251	Key West	9/7/43	Norfolk	14/07/43
KN 251	Key West	9/7/43	Norfolk	14/07/43
KN 252	Key West	13/07/43	NYC	19/07/43
KN 252	Key West	13/07/43	Norfolk	19/07/43
KN 252	Key West	13/07/43	Norfolk	19/07/43
KN 252	Key West	13/07/43	Norfolk	19/07/43
KN 252	Key West	13/07/43	Norfolk	19/07/43
KN 252	Key West	13/07/43	Norfolk	14/07/43
KN 253	Key West	18/07/43	NYC	23/07/43
KN 253	Key West	18/07/43	Norfolk	23/07/43
KN 253	Key West	18/07/43	Norfolk	23/07/43
KN 253	Key West	18/07/43	Norfolk	23/07/43
KN 253	Key West	18/07/43	Norfolk	23/07/43
KN 254	Key West	23/07/43	NYC	28/07/43
KN 254	Key West	23/07/43	Norfolk	28/07/43
KN 254	Key West	23/07/43	Norfolk	28/07/43
KN 254	Key West	23/07/43	Norfolk	28/07/43
KN 255	Key West	29/07/43	NYC	3/8/43
KN 255	Key West	29/07/43	Norfolk	3/8/43

KN 255	Key West	29/07/43	Norfolk	3/8/43
KN 255	Key West	29/07/43	Norfolk	3/8/43
KN 255	Key West	29/07/43	Norfolk	3/8/43
KN 256	Key West	2/8/43	NYC	8/8/43
KN 256	Key West	2/8/43	Norfolk	8/8/43
KN 256	Key West	2/8/43	Norfolk	8/8/43
KN 256	Key West	2/8/43	Norfolk	8/8/43
KN 256	Key West	2/8/43	Norfolk	8/8/43
KN 257	Key West	7/8/43	NYC	12/8/43
KN 257	Key West	7/8/43	Norfolk	12/8/43
KN 257	Key West	7/8/43	Norfolk	12/8/43
KN 257	Key West	7/8/43	Norfolk	12/8/43
KN 257	Key West	7/8/43	Norfolk	12/8/43
KN 258	Key West	12/8/43	NYC	17/08/43
KN 258	Key West	12/8/43	Norfolk	17/08/43
KN 258	Key West	12/8/43	Norfolk	17/08/43
KN 258	Key West	12/8/43	Norfolk	17/08/43
KN 258	Key West	12/8/43	Norfolk	17/08/43
KN 258	Key West	12/8/43	Norfolk	16/08/43
KN 258	Key West	12/8/43	Norfolk	16/08/43
KN 259	Key West	17/08/43	NYC	23/08/43
KN 259	Key West	17/08/43	Norfolk	23/08/43
KN 259	Key West	17/08/43	Norfolk	23/08/43
KN 259	Key West	17/08/43	Norfolk	23/08/43
KN 259	Key West	17/08/43	Norfolk	23/08/43
KN 259	Key West	17/08/43	Norfolk	23/08/43
KN 259	Key West	21/08/43	Norfolk	22/08/43
KN 259	Key West	21/08/43	Norfolk	22/08/43
KN 260	Key West	23/08/43	NYC	28/08/43
KN 260	Key West	23/08/43	Norfolk	28/08/43
KN 260	Key West	23/08/43	Norfolk	28/08/43
KN 260	Key West	23/08/43	Norfolk	28/08/43
KN 260	Key West	23/08/43	Norfolk	28/08/43
KN 261	Key West	27/08/43	NYC	2/9/43
KN 261	Key West	27/08/43	Norfolk	2/9/43
KN 261	Key West	27/08/43	Norfolk	2/9/43
KN 261	Key West	27/08/43	Norfolk	2/9/43
KN 261	Key West	27/08/43	Norfolk	2/9/43
KN 261	Key West	27/08/43	Norfolk	2/9/43
KN 262	Key West	1/9/43	NYC	7/9/43
KN 262	Key West	1/9/43	Norfolk	7/9/43
KN 262	Key West	1/9/43	Norfolk	7/9/43
KN 262	Key West	1/9/43	Norfolk	7/9/43
KN 262	Key West	1/9/43	Norfolk	7/9/43

KN 263	Key West	7/9/43	NYC	13/09/43
KN 263	Key West	7/9/43	Norfolk	13/09/43
KN 263	Key West	7/9/43	Norfolk	13/09/43
KN 263	Key West	7/9/43	Norfolk	13/09/43
KN 263	Key West	7/9/43	Norfolk	13/09/43
KN 264	Key West	11/9/43	NYC	17/09/43
KN 264	Key West	1/9/43	Norfolk	17/09/43
KN 264	Key West	1/9/43	Norfolk	17/09/43
KN 264	Key West	1/9/43	Norfolk	17/09/43
KN 264	Key West	1/9/43	Norfolk	17/09/43
KN 265	Key West	16/09/43	NYC	22/09/43
KN 265	Key West	16/09/43	Norfolk	22/09/43
KN 265	Key West	16/09/43	Norfolk	22/09/43
KN 265	Key West	16/09/43	Norfolk	22/09/43
KN 265	Key West	16/09/43	Norfolk	22/09/43
KN 266	Key West	23/09/43	NYC	28/09/43
KN 266	Key West	23/09/43	Norfolk	28/09/43
KN 266	Key West	23/09/43	Norfolk	28/09/43
KN 266	Key West	23/09/43	Norfolk	28/09/43
KN 266	Key West	23/09/43	Norfolk	28/09/43
KN 266	Key West	23/09/43	Norfolk	28/09/43
KN 266	Key West	23/09/43	Norfolk	24/09/43
KN 266	Key West	23/09/43	Norfolk	24/09/43
KN 267	Key West	27/09/43	NYC	3/10/43
KN 267	Key West	27/09/43	Norfolk	3/10/43
KN 267	Key West	27/09/43	Norfolk	3/10/43
KN 267	Key West	27/09/43	Norfolk	3/10/43
KN 267	Key West	27/09/43	Norfolk	3/10/43
KN 268	Key West	3/10/43	NYC	10/10/43
KN 268	Key West	3/10/43	Norfolk	10/10/43
KN 268	Key West	3/10/43	Norfolk	10/10/43
KN 268	Key West	3/10/43	Norfolk	10/10/43
KN 268	Key West	3/10/43	Norfolk	10/10/43
KN 268	Key West	3/10/43	Norfolk	10/10/43
KN 268	Key West	3/10/43	Norfolk	4/10/43
KN 269	Key West	7/10/43	NYC	13/10/43
KN 269	Key West	7/10/43	Norfolk	13/10/43
KN 269	Key West	7/10/43	Norfolk	13/10/43
KN 269	Key West	3/10/43	Norfolk	
KN 269	Key West	7/10/43	Norfolk	13/10/43
KN 269	Key West	7/10/43	Norfolk	13/10/43
KN 270	Key West	12/10/43	NYC	17/10/43
KN 270	Key West	12/10/43	Norfolk	17/10/43
KN 270	Key West	12/10/43	Norfolk	17/10/43

KN 270	Key West	12/10/43	Norfolk	17/10/43
KN 270	Key West	12/10/43	Norfolk	17/10/43
KN 271	Key West	17/10/43	NYC	22/10/43
KN 271	Key West	17/10/43	Norfolk	22/10/43
KN 271	Key West	17/10/43	Norfolk	22/10/43
KN 271	Key West	17/10/43	Norfolk	22/10/43
KN 271	Key West	17/10/43	Norfolk	22/10/43
KN 271	Key West	17/10/43	Norfolk	18/10/43
KN 272	Key West	21/10/43	NYC	27/10/43
KN 272	Key West	21/10/43	Norfolk	27/10/43
KN 272	Key West	21/10/43	Norfolk	27/10/43
KN 272	Key West	21/10/43	Norfolk	27/10/43
KN 272	Key West	22/10/43	Norfolk	27/10/43
KN 273	Key West	26/10/43	NYC	1/11/43
KN 273	Key West	26/10/43	Norfolk	1/11/43
KN 273	Key West	26/10/43	Norfolk	1/11/43
KN 273	Key West	26/10/43	Norfolk	1/11/43
KN 273	Key West	26/10/43	Norfolk	1/11/43
KN 274	Key West	31/10/43	NYC	6/11/43
KN 274	Key West	31/10/43	Norfolk	6/11/43
KN 274	Key West	31/10/43	Norfolk	6/11/43
KN 274	Key West	31/10/43	Norfolk	6/11/43
KN 274	Key West	31/10/43	Norfolk	6/11/43
KN 275	Key West	6/11/43	NYC	12/11/43
KN 275	Key West	6/11/43	Norfolk	12/11/43
KN 275	Key West	6/11/43	Norfolk	12/11/43
KN 275	Key West	6/11/43	Norfolk	12/11/43
KN 275	Key West	6/11/43	Norfolk	12/11/43
KN 275	Key West	6/11/43	Norfolk	12/11/43
KN 275	Key West	6/11/43	Norfolk	12/11/43
KN 276	Key West	10/11/43	NYC	16/11/43
KN 276	Key West	10/11/43	Norfolk	16/11/43
KN 276	Key West	10/11/43	Norfolk	16/11/43
KN 276	Key West	10/11/43	Norfolk	16/11/43
KN 276	Key West	10/11/43	Norfolk	16/11/43
KN 276	Key West	10/11/43	Norfolk	16/11/43
KN 277	Key West	16/11/43	NYC	21/11/43
KN 277	Key West	16/11/43	Norfolk	21/11/43
KN 277	Key West	16/11/43	Norfolk	21/11/43
KN 277	Key West	16/11/43	Norfolk	21/11/43
KN 277	Key West	16/11/43	Norfolk	21/11/43
KN 278	Key West	22/11/43	NYC	26/11/43
KN 278	Key West	22/11/43	Norfolk	2/11/43
KN 278	Key West	22/11/43	Norfolk	2/11/43

KN 278	Key West	22/11/43	Norfolk	2/11/43
KN 278	Key West	22/11/43	Norfolk	2/11/43
KN 278	Key West	22/11/43	Norfolk	2/11/43
KN 279	Key West	26/11/43	NYC	2/12/43
KN 279	Key West	26/11/43	Norfolk	2/12/43
KN 279	Key West	26/11/43	Norfolk	2/12/43
KN 279	Key West	26/11/43	Norfolk	2/12/43
KN 279	Key West	26/11/43	Norfolk	2/12/43
KN 280	Key West	1/12/43	NYC	6/12/43
KN 280	Key West	1/12/43	Norfolk	6/12/43
KN 280	Key West	1/12/43	Norfolk	6/12/43
KN 280	Key West	1/12/43	Norfolk	6/12/43
KN 280	Key West	1/12/43	Norfolk	6/12/43
KN 281	Key West	5/12/43	NYC	10/12/43
KN 281	Key West	5/12/43	Norfolk	10/12/43
KN 281	Key West	5/12/43	Norfolk	10/12/43
KN 281	Key West	5/12/43	Norfolk	10/12/43
KN 281	Key West	5/12/43	Norfolk	10/12/43
KN 282	Key West	11/12/43	NYC	16/12/43
KN 282	Key West	11/12/43	Norfolk	16/12/43
KN 282	Key West	11/12/43	Norfolk	16/12/43
KN 282	Key West	11/12/43	Norfolk	16/12/43
KN 282	Key West	11/12/43	Norfolk	16/12/43
KN 282	Key West	11/12/43	Norfolk	16/12/43
KN 283	Key West	16/12/43	NYC	22/12/43
KN 283	Key West	16/12/43	Norfolk	22/12/43
KN 283	Key West	16/12/43	Norfolk	22/12/43
KN 283	Key West	16/12/43	Norfolk	22/12/43
KN 283	Key West	16/12/43	Norfolk	22/12/43
KN 284	Key West	21/12/43	NYC	25/12/43
KN 284	Key West	21/12/43	Norfolk	25/12/43
KN 284	Key West	21/12/43	Norfolk	25/12/43
KN 284	Key West	21/12/43	Norfolk	25/12/43
KN 284	Key West	21/12/43	Norfolk	25/12/43
KN 285	Key West	26/12/43	NYC	1/1/44
KN 285	Key West	26/12/43	Norfolk	1/1/44
KN 285	Key West	26/12/43	Norfolk	1/1/44
KN 285	Key West	26/12/43	Norfolk	1/1/44
KN 285	Key West	26/12/43	Norfolk	1/1/44
KN 286	Key West	30/12/43	NYC	6/1/44
KN 286	Key West	30/12/43	Norfolk	6/1/44
KN 286	Key West	30/12/43	Norfolk	6/1/44
KN 286	Key West	30/12/43	Norfolk	6/1/44
KN 286	Key West	30/12/43	Norfolk	6/1/44

KN 287	Key West	5/1/44	NYC	10/1/44
KN 287	Key West	5/1/44	Norfolk	10/1/44
KN 287	Key West	5/1/44	Norfolk	10/1/44
KN 287	Key West	5/1/44	Norfolk	10/1/44
KN 287	Key West	5/1/44	Norfolk	10/1/44
KN 288	Key West	10/1/44	NYC	15/01/44
KN 288	Key West	10/1/44	Norfolk	15/01/44
KN 288	Key West	10/1/44	Norfolk	15/01/44
KN 288	Key West	10/1/44	Norfolk	15/01/44
KN 288	Key West	10/1/44	Norfolk	15/01/44
KN 289	Key West	15/01/44	NYC	20/01/44
KN 289	Key West	15/01/44	Norfolk	20/01/44
KN 289	Key West	15/01/44	Norfolk	20/01/44
KN 289	Key West	15/01/44	Norfolk	20/01/44
KN 289	Key West	15/01/44	Norfolk	20/01/44
KN 290	Key West	19/01/44	NYC	25/01/44
KN 290	Key West	19/01/44	Norfolk	25/01/44
KN 290	Key West	19/01/44	Norfolk	25/01/44
KN 290	Key West	19/01/44	Norfolk	25/01/44
KN 291	Key West	24/01/44	NYC	29/01/44
KN 291	Key West	24/01/44	Norfolk	29/01/44
KN 291	Key West	24/01/44	Norfolk	29/01/44
KN 291	Key West	24/01/44	Norfolk	29/01/44
KN 291	Key West	24/01/44	Norfolk	29/01/44
KN 292	Key West	29/01/44	NYC	4/2/44
KN 292	Key West	29/01/44	Norfolk	4/2/44
KN 292	Key West	29/01/44	Norfolk	4/2/44
KN 292	Key West	29/01/44	Norfolk	4/2/44
KN 292	Key West	29/01/44	Norfolk	4/2/44
KN 293	Key West	3/2/44	NYC	9/2/44
KN 293	Key West	3/2/44	Norfolk	9/2/44
KN 293	Key West	3/2/44	Norfolk	9/2/44
KN 293	Key West	3/2/44	Norfolk	9/2/44
KN 293	Key West	3/2/44	Norfolk	9/2/44
KN 294	Key West	8/2/44	NYC	14/02/44
KN 294	Key West	8/2/44	Norfolk	14/02/44
KN 294	Key West	8/2/44	Norfolk	14/02/44
KN 294	Key West	8/2/44	Norfolk	14/02/44
KN 294	Key West	9/2/44	Norfolk	
KN 294	Key West	8/2/44	Norfolk	14/02/44
KN 294	Key West	8/2/44	Norfolk	9/2/44
KN 295	Key West	13/02/44	NYC	19/02/44
KN 295	Key West	13/02/44	Norfolk	19/02/44
KN 295	Key West	13/02/44	Norfolk	19/02/44

KN 295	Key West	13/02/44	Norfolk	19/02/44
KN 295	Key West	13/02/44	Norfolk	19/02/44
KN 296	Key West	18/02/44	NYC	24/02/44
KN 296	Key West	18/02/44	Norfolk	24/02/44
KN 296	Key West	18/02/44	Norfolk	24/02/44
KN 296	Key West	18/02/44	Norfolk	
KN 296	Key West	18/02/44	Norfolk	19/02/44
KN 296	Key West	18/02/44	Norfolk	24/02/44
KN 297	Key West	23/02/44	NYC	28/02/44
KN 297	Key West	23/02/44	Norfolk	28/02/44
KN 297	Key West	23/02/44	Norfolk	28/02/44
KN 297	Key West	23/02/44	Norfolk	28/02/44
KN 297	Key West	23/02/44	Norfolk	28/02/44
KN 297	Key West	23/02/44	Norfolk	28/02/44
KN 298	Key West	28/02/44	NYC	5/3/44
KN 298	Key West	28/02/44	Norfolk	5/3/44
KN 298	Key West	28/02/44	Norfolk	5/3/44
KN 298	Key West	28/02/44	Norfolk	5/3/44
KN 298	Key West	28/02/44	Norfolk	5/3/44
KN 299	Key West	4/3/44	NYC	10/3/44
KN 299	Key West	4/3/44	Norfolk	10/3/44
KN 299	Key West	4/3/44	Norfolk	10/3/44
KN 299	Key West	4/3/44	Norfolk	10/3/44
KN 299	Key West	4/3/44	Norfolk	10/3/44
KN 300	Key West	9/3/44	NYC	16/03/44
KN 300	Key West	9/3/44	Norfolk	16/03/44
KN 300	Key West	9/3/44	Norfolk	16/03/44
KN 300	Key West	9/3/44	Norfolk	16/03/44
KN 300	Key West	9/3/44	Norfolk	16/03/44
KN 300	Key West	9/3/44	Norfolk	16/03/44
KN 301	Key West	14/03/44	NYC	19/03/44
KN 301	Key West	14/03/44	Norfolk	19/03/44
KN 301	Key West	14/03/44	Norfolk	19/03/44
KN 301	Key West	14/03/44	Norfolk	19/03/44
KN 301	Key West	14/03/44	Norfolk	19/03/44
KN 302	Key West	19/03/44	NYC	24/03/44
KN 302	Key West	19/03/44	Norfolk	24/03/44
KN 302	Key West	19/03/44	Norfolk	24/03/44
KN 302	Key West	19/03/44	Norfolk	24/03/44
KN 302	Key West	19/03/44	Norfolk	24/03/44
KN 302	Key West	19/03/44	Norfolk	21/03/44
KN 302	Key West	19/03/44	Norfolk	22/03/44
KN 303	Key West	24/03/44	NYC	29/03/44
KN 303	Key West	24/03/44	Norfolk	29/03/44

KN 303	Key West	24/03/44	Norfolk	29/03/44
KN 303	Key West	24/03/44	Norfolk	29/03/44
KN 303	Key West	24/03/44	Norfolk	29/03/44
KN 303	Key West	24/03/44	Norfolk	29/03/44
KN 304	Key West	29/03/44	NYC	3/4/44
KN 304	Key West	29/03/44	Norfolk	3/4/44
KN 304	Key West	29/03/44	Norfolk	3/4/44
KN 304	Key West	29/03/44	Norfolk	3/4/44
KN 304	Key West	29/03/44	Norfolk	3/4/44
KN 305	Key West	3/4/44	NYC	8/4/44
KN 305	Key West	3/4/44	Norfolk	8/4/44
KN 305	Key West	3/4/44	Norfolk	8/4/44
KN 305	Key West	3/4/44	Norfolk	8/4/44
KN 305	Key West	3/4/44	Norfolk	8/4/44
KN 306	Key West	8/4/44	NYC	13/04/44
KN 306	Key West	8/4/44	Norfolk	13/04/44
KN 306	Key West	8/4/44	Norfolk	13/04/44
KN 306	Key West	8/4/44	Norfolk	13/04/44
KN 306	Key West	8/4/44	Norfolk	13/04/44
KN 306	Key West	8/4/44	Norfolk	13/04/44
KN 307	Key West	13/04/44	NYC	18/04/44
KN 307	Key West	13/04/44	Norfolk	18/04/44
KN 307	Key West	13/04/44	Norfolk	18/04/44
KN 307	Key West	13/04/44	Norfolk	18/04/44
KN 307	Key West	13/04/44	Norfolk	18/04/44
KN 307	Key West	13/04/44	Norfolk	14/04/44
KN 308	Key West	18/04/44	NYC	23/04/44
KN 308	Key West	18/04/44	Norfolk	23/04/44
KN 308	Key West	18/04/44	Norfolk	23/04/44
KN 308	Key West	18/04/44	Norfolk	23/04/44
KN 308	Key West	18/04/44	Norfolk	23/04/44
KN 309	Key West	23/04/44	NYC	29/04/44
KN 309	Key West	23/04/44	Norfolk	29/04/44
KN 309	Key West	23/04/44	Norfolk	29/04/44
KN 309	Key West	23/04/44	Norfolk	29/04/44
KN 309	Key West	23/04/44	Norfolk	29/04/44
KN 310	Key West	28/04/44	NYC	4/5/44
KN 310	Key West	28/04/44	Norfolk	4/5/44
KN 310	Key West	28/04/44	Norfolk	4/5/44
KN 310	Key West	28/04/44	Norfolk	4/5/44
KN 310	Key West	28/04/44	Norfolk	4/5/44
KN 311	Key West	4/5/44	NYC	9/5/44
KN 311	Key West	4/5/44	Norfolk	9/5/44
KN 311	Key West	4/5/44	Norfolk	9/5/44

KN 311	Key West	4/5/44	Norfolk	9/5/44
KN 311	Key West	4/5/44	Norfolk	9/5/44
KN 311	Key West	4/5/44	Norfolk	9/5/44
KN 312	Key West	8/5/44	NYC	14/05/44
KN 312	Key West	8/5/44	Norfolk	14/05/44
KN 312	Key West	8/5/44	Norfolk	14/05/44
KN 312	Key West	8/5/44	Norfolk	14/05/44
KN 312	Key West	8/5/44	Norfolk	14/05/44
KN 313	Key West	13/05/44	NYC	18/05/44
KN 313	Key West	13/05/44	Norfolk	18/05/44
KN 313	Key West	13/05/44	Norfolk	18/05/44
KN 313	Key West	13/05/44	Norfolk	18/05/44
KN 313	Key West	13/05/44	Norfolk	18/05/44
KN 313	Key West	13/05/44	Norfolk	18/05/44
KN 314	Key West	18/05/44	NYC	23/05/44
KN 314	Key West	18/05/44	Norfolk	23/05/44
KN 314	Key West	18/05/44	Norfolk	23/05/44
KN 314	Key West	18/05/44	Norfolk	23/05/44
KN 314	Key West	22/05/44	Norfolk	22/05/44
KN 314	Key West	22/05/44	Norfolk	22/05/44
KN 315	Key West	23/05/44	NYC	29/05/44
KN 315	Key West	23/05/44	Norfolk	29/05/44
KN 315	Key West	23/05/44	Norfolk	29/05/44
KN 315	Key West	23/05/44	Norfolk	29/05/44
KN 316	Key West	28/05/44	NYC	2/6/44
KN 316	Key West	28/05/44	Norfolk	2/6/44
KN 316	Key West	28/05/44	Norfolk	2/6/44
KN 316	Key West	28/05/44	Norfolk	2/6/44
KN 317	Key West	2/6/44	NYC	7/6/44
KN 317	Key West	2/6/44	Norfolk	7/6/44
KN 317	Key West	2/6/44	Norfolk	7/6/44
KN 317	Key West	2/6/44	Norfolk	7/6/44
KN 318	Key West	7/6/44	NYC	12/6/44
KN 318	Key West	7/6/44	Norfolk	12/6/44
KN 318	Key West	7/6/44	Norfolk	12/6/44
KN 318	Key West	7/6/44	Norfolk	12/6/44
KN 319	Key West	12/6/44	NYC	17/06/44
KN 319	Key West	12/6/44	Norfolk	16/06/44
KN 319	Key West	12/6/44	Norfolk	17/06/44
KN 319	Key West	12/6/44	Norfolk	17/06/44
KN 319	Key West	12/6/44	Norfolk	17/06/44
KN 319	Key West		Norfolk	16/06/44
KN 320	Key West	17/06/44	NYC	22/06/44
KN 320	Key West	17/06/44	Norfolk	22/06/44

KN 320	Key West	17/06/44	Norfolk	22/06/44
KN 320	Key West	17/06/44	Norfolk	22/06/44
KN 320	Key West	17/06/44	Norfolk	22/06/44
KN 321	Key West	22/06/44	NYC	27/06/44
KN 321	Key West	22/06/44	Norfolk	27/06/44
KN 321	Key West	22/06/44	Norfolk	27/06/44
KN 321	Key West	22/06/44	Norfolk	27/06/44
KN 322	Key West	27/06/44	NYC	1/7/44
KN 322	Key West	27/06/44	Norfolk	1/7/44
KN 322	Key West	27/06/44	Norfolk	1/7/44
KN 322	Key West	27/06/44	Norfolk	1/7/44
KN 323	Key West	2/7/44	NYC	7/7/44
KN 323	Key West	2/7/44	Norfolk	7/7/44
KN 323	Key West	2/7/44	Norfolk	7/7/44
KN 323	Key West	2/7/44	Norfolk	7/7/44
KN 324	Key West	7/7/44	NYC	12/7/44
KN 324	Key West	7/7/44	Norfolk	12/7/44
KN 324	Key West	7/7/44	Norfolk	12/7/44
KN 325	Key West	12/7/44	NYC	17/07/44
KN 325	Key West	12/7/44	Norfolk	17/07/44
KN 325	Key West	12/7/44	Norfolk	17/07/44
KN 325	Key West	12/7/44	Norfolk	17/07/44
KN 326	Key West	17/07/44	NYC	22/07/44
KN 326	Key West	17/07/44	Norfolk	22/07/44
KN 326	Key West	17/07/44	Norfolk	22/07/44
KN 326	Key West	17/07/44	Norfolk	22/07/44
KN 327	Key West	22/07/44	NYC	28/07/44
KN 327	Key West	22/07/44	Norfolk	28/07/44
KN 327	Key West	22/07/44	Norfolk	28/07/44
KN 327	Key West	22/07/44	Norfolk	28/07/44
KN 327	Key West	22/07/44	Norfolk	28/07/44
KN 328	Key West	27/07/44	NYC	1/8/44
KN 328	Key West	27/07/44	Norfolk	1/8/44
KN 328	Key West	27/07/44	Norfolk	1/8/44
KN 328	Key West	27/07/44	Norfolk	1/8/44
KN 328	Key West	27/07/44	Norfolk	1/8/44
KN 328	Key West	27/07/44	Norfolk	1/8/44
KN 328	Key West	27/07/44	Norfolk	1/8/44
KN 329	Key West	1/8/44	NYC	6/8/44
KN 329	Key West	1/8/44	Norfolk	6/8/44
KN 329	Key West	1/8/44	Norfolk	6/8/44
KN 329	Key West	1/8/44	Norfolk	6/8/44
KN 330	Key West	7/8/44	NYC	12/8/44
KN 330	Key West	7/8/44	Norfolk	12/8/44

KN 330	Key West	7/8/44	Norfolk	12/8/44
KN 330	Key West	7/8/44	Norfolk	12/8/44
KN 331	Key West	11/8/44	NYC	16/08/44
KN 331	Key West	11/8/44	Norfolk	16/08/44
KN 331	Key West	11/8/44	Norfolk	16/08/44
KN 331	Key West	11/8/44	Norfolk	16/08/44
KN 332	Key West	16/08/44	NYC	21/08/44
KN 332	Key West	6/8/44	Norfolk	21/08/44
KN 332	Key West	6/8/44	Norfolk	21/08/44
KN 332	Key West	6/8/44	Norfolk	21/08/44
KN 333	Key West	22/08/44	NYC	27/08/44
KN 333	Key West	22/08/44	Norfolk	27/08/44
KN 333	Key West	22/08/44	Norfolk	27/08/44
KN 333	Key West	22/08/44	Norfolk	27/08/44
KN 334	Key West	27/08/44	NYC	2/9/44
KN 334	Key West	27/08/44	Norfolk	2/9/44
KN 334	Key West	27/08/44	Norfolk	2/9/44
KN 334	Key West	27/08/44	Norfolk	2/9/44
KN 335	Key West	31/08/44	NYC	6/9/44
KN 335	Key West	31/08/44	Norfolk	6/9/44
KN 335	Key West	31/08/44	Norfolk	6/9/44
KN 335	Key West	31/08/44	Norfolk	6/9/44
KN 335	Key West	31/08/44	Norfolk	6/9/44
KN 336	Key West	5/9/44	NYC	10/9/44
KN 336	Key West	5/9/44	Norfolk	10/9/44
KN 336	Key West	5/9/44	Norfolk	10/9/44
KN 336	Key West	5/9/44	Norfolk	10/9/44
KN 336	Key West	5/9/44	Norfolk	10/9/44
KN 336	Key West	5/9/44	Norfolk	10/9/44
KN 337	Key West	10/9/44	NYC	15/09/44
KN 337	Key West	10/9/44	Norfolk	15/09/44
KN 337	Key West	10/9/44	Norfolk	15/09/44
KN 337	Key West	10/9/44	Norfolk	15/09/44
KN 338	Did Not Sail			
KN 338	Key West	15/09/45	Norfolk	20/09/44
KN 338	Key West	15/09/45	Norfolk	20/09/44
KN 338	Key West	15/09/45	Norfolk	20/09/44
KN 339	Key West	20/09/44	NYC	25/09/44
KN 339	Key West	20/09/44	Norfolk	25/09/44
KN 339	Key West	20/09/44	Norfolk	25/09/44
KN 339	Key West	20/09/44	Norfolk	25/09/44
KN 340	Key West	25/09/44	NYC	30/09/44
KN 340	Key West	25/09/44	Norfolk	30/09/44
KN 340	Key West	25/09/44	Norfolk	30/09/44

KN 340	Key West	25/09/44	Norfolk	30/09/44
KN 340	Key West	26/09/44	Norfolk	30/09/44
KN 341	Key West	30/09/44	NYC	5/10/44
KN 341	Key West	30/09/44	Norfolk	5/10/44
KN 341	Key West	30/09/44	Norfolk	5/10/44
KN 341	Key West	30/09/44	Norfolk	5/10/44
KN 341	Key West	30/09/44	Norfolk	5/10/44
KN 342	Key West	5/10/44	NYC	10/10/44
KN 342	Key West	5/10/44	Norfolk	10/10/44
KN 342	Key West	5/10/44	Norfolk	10/10/44
KN 342	Key West	5/10/44	Norfolk	10/10/44
KN 342	Key West	5/10/44	Norfolk	10/10/44
KN 343	Key West	10/10/44	NYC	15/10/44
KN 343	Key West	10/10/44	Norfolk	15/10/44
KN 343	Key West	10/10/44	Norfolk	15/10/44
KN 343	Key West	10/10/44	Norfolk	15/10/44
KN 344	Key West	15/10/44	NYC	21/10/44
KN 344	Key West	15/10/44	Norfolk	21/10/44
KN 344	Key West	15/10/44	Norfolk	21/10/44
KN 344	Key West	15/10/44	Norfolk	21/10/44
KN 345	Did Not Sail			
KN 346	Key West	25/10/44	NYC	31/10/44
KN 346	Key West	26/10/44	Norfolk	31/10/44
KN 346	Key West	26/10/44	Norfolk	31/10/44
KN 346	Key West	26/10/44	Norfolk	31/10/44
KN 347	Key West	30/10/44	Charleston, Sc	2/11/44
KN 347	Key West	30/10/44	Norfolk	2/11/44
KN 347	Key West	30/10/44	Norfolk	2/11/44
KN 348	Key West	4/11/44	Boston	10/11/44
KN 348	Key West	4/11/44	Norfolk	10/11/44
KN 348	Key West	4/11/44	Norfolk	10/11/44
KN 348	Key West	4/11/44	Norfolk	10/11/44
KN 349	Key West	9/11/44	Hampton Roads	14/11/44
KN 349	Key West	9/11/44	Norfolk	14/11/44
KN 349	Key West	9/11/44	Norfolk	14/11/44
KN 349	Key West	9/11/44	Norfolk	14/11/44
KN 350	Key West	14/11/44	NYC	20/11/44
KN 350	Key West	14/11/44	Norfolk	20/11/44
KN 350	Key West	14/11/44	Norfolk	20/11/44
KN 350	Key West	14/11/44	Norfolk	20/11/44
KN 351	Key West	19/11/44	NYC	24/11/44
KN 351	Key West	19/11/44	Norfolk	24/11/44
KN 351	Key West	19/11/44	Norfolk	24/11/44

KN 351	Key West	19/11/44	Norfolk	24/11/44
KN 352	Key West	24/11/44	NYC	29/11/44
KN 352	Key West	24/11/44	Norfolk	29/11/44
KN 352	Key West	24/11/44	Norfolk	29/11/44
KN 352	Key West	24/11/44	Norfolk	29/11/44
KN 353	Key West	29/11/44	NYC	6/12/44
KN 353	Key West	29/11/44	Norfolk	6/12/44
KN 353	Key West	29/11/44	Norfolk	6/12/44
KN 353	Key West	29/11/44	Norfolk	6/12/44
KN 354	Key West	4/12/44	NYC	9/12/44
KN 354	Key West	4/12/44	Norfolk	9/12/44
KN 354	Key West	4/12/44	Norfolk	9/12/44
KN 354	Key West	4/12/44	Norfolk	9/12/44
KN 355	Key West	9/12/44	NYC	15/12/44
KN 355	Key West	9/12/44	Norfolk	15/12/44
KN 355	Key West	9/12/44	Norfolk	15/12/44
KN 355	Key West	9/12/44	Norfolk	15/12/44
KN 355	Key West	9/12/44	Norfolk	15/12/44
KN 356	Key West	14/12/44	NYC	
KN 356	Key West	14/12/44	Norfolk	19/12/44
KN 356	Key West	14/12/44	Norfolk	19/12/44
KN 356	Key West	14/12/44	Norfolk	19/12/44
KN 357	Key West	19/12/44	NYC	24/12/44
KN 357	Key West	20/12/44	Norfolk	24/12/44
KN 357	Key West	19/12/44	Norfolk	24/12/44
KN 357	Key West	20/12/44	Norfolk	24/12/44
KN 357	Key West	19/12/44	Norfolk	24/12/44
KN 358	Key West	24/12/44	NYC	28/12/44
KN 358	Key West	24/12/44	Norfolk	28/12/44
KN 358	Key West	24/12/44	Norfolk	28/12/44
KN 358	Key West	24/12/44	Norfolk	28/12/44
KN 359	Key West	29/12/44	NYC	3/1/45
KN 359	Key West	29/12/44	Norfolk	3/1/45
KN 359	Key West	29/12/44	Norfolk	3/1/45
KN 359	Key West	29/12/44	Norfolk	3/1/45
KN 360	Key West	3/1/45	NYC	8/1/45
KN 360	Key West	3/1/45	Norfolk	8/1/45
KN 360	Key West	3/1/45	Norfolk	8/1/45
KN 360	Key West	3/1/45	Norfolk	8/1/45
KN 361	Key West	8/1/45	NYC	13/01/45
KN 361	Key West	8/1/45	Norfolk	13/01/45
KN 361	Key West	8/1/45	Norfolk	13/01/45
KN 362	Key West	13/01/45	NYC	19/01/45
KN 362	Key West	13/01/45	Norfolk	19/01/45

KN 363	Key West	18/01/45	NYC	
KN 363	Key West	18/01/45	Norfolk	23/01/45
KN 364	Did Not Sail			
KN 364	Key West	23/01/45	Norfolk	27/01/45
KN 365	Key West	28/01/45	NYC	
KN 365	Key West	28/01/45	Norfolk	
KN 366	Did Not Sail			
KN 367	Key West	7/2/45	NYC	12/2/45
KN 367	Key West	27/09/43	Norfolk	3/10/43
KN 367	Key West	7/2/45	Norfolk	12/2/45
KN 368	Key West	12/2/45	NYC	
KN 368	Key West	12/2/45	Norfolk	16/02/45
KN 369	Key West	17/02/45	NYC	
KN 369	Key West	17/02/45	Norfolk	22/02/45
KN 370	Key West	22/02/45	NYC	
KN 370	Key West	22/02/45	Norfolk	
KN 371	Key West	27/02/45	NYC	4/3/45
KN 371	Key West	27/02/45	Norfolk	4/3/45
KN 372	Key West	4/3/45	NYC	9/3/45
KN 372	Key West	4/3/45	Norfolk	9/3/45
KN 373	Did Not Sail			
KN 373	Key West	9/3/45	Norfolk	15/03/45
KN 374	Key West	14/03/45	NYC	
KN 374	Key West	14/03/45	Norfolk	20/03/45
KN 375	Key West	19/03/45	NYC	
KN 375	Key West	19/03/45	Norfolk	24/03/45
KN 376	Did Not Sail			
KN 376	Key West	24/03/45	Norfolk	27/03/45
KN 377	Did Not Sail			
KN 378	Key West	3/4/45	NYC	
KN 378	Key West	3/4/45	Norfolk	
KN 379	Did Not Sail			
KN 380	Key West	13/04/45	NYC	
KN 380	Key West	13/04/45	Norfolk	
KN 381	Key West	18/04/45	NYC	
KN 381	Key West	18/04/45	Norfolk	23/04/45
KN 382	Key West	25/04/45	NYC	1/5/45
KN 382	Key West	25/04/45	Norfolk	1/5/45
KN 382	Key West	25/04/45	Norfolk	1/5/45
KN 382	Key West	25/04/45	Norfolk	1/5/45
KN 383	Key West	28/04/45	NYC	5/5/45
KN 383	Key West	28/04/45	Norfolk	5/5/45
KN 383	Key West	28/04/45	Norfolk	5/5/45
KN 383	Key West	28/04/45	Norfolk	5/5/45

KN 383	Key West	28/04/45	Norfolk	5/5/45
KN 383	Key West	28/04/45	Norfolk	5/5/45
KN 384	Key West	3/5/45	NYC	9/5/45
KN 384	Key West	3/5/45	Norfolk	9/5/45
KN 384	Key West	3/5/45	Norfolk	9/5/45
KN 384	Key West	3/5/45	Norfolk	9/5/45
KN 384	Key West	3/5/45	Norfolk	9/5/45
KN 385	Key West	8/5/45	Hampton Roads	13/05/45
KN 385	Key West	8/5/45	Norfolk	13/05/45
KN 386	Key West	13/05/45	NYC	18/05/45
KN 386	Key West	13/05/45	Norfolk	18/05/45
KN 386	Key West	13/05/45	Norfolk	18/05/45
KN 386	Key West	13/05/45	Norfolk	18/05/45
KN 386	Key West	13/05/45	Norfolk	18/05/45
Ks 23	Casablanca	4/11/39	Brest	10/11/39
Ks 24	Casablanca	8/11/39	Dakar	14/11/39
Ks 29	Casablanca	22/11/39	Dakar	27/11/39
Ks 30	Casablanca	23/11/39	Brest	30/11/39
Ks 31	Casablanca	25/11/39	Brest	30/11/39
Ks 32	Casablanca	30/11/39	Dakar	4/12/39
Ks 33	Casablanca	1/12/39	Brest	
Ks 36	Casablanca	6/12/39	Dakar	12/12/39
Ks 37	Casablanca	8/12/39	St Nazaire	
Ks 38	Casablanca	10/12/39	Brest	14/12/39
Ks 40	Casablanca	15/12/39	Brest	22/12/39
Ks 43	Casablanca	24/12/39	Brest	30/12/39
Ks 45	Casablanca	2/1/40	Brest	9/1/40
Ks 500	Hampton Roads	14/05/42	Key West	19/05/42
Ks 501	Hampton Roads	17/05/42	Key West	23/05/42
Ks 502	Hampton Roads	20/05/42	Key West	25/05/42
Ks 503	Hampton Roads	23/05/42	Key West	28/05/42
Ks 504	Hampton Roads	26/05/42	Key West	1/6/42
Ks 505	Hampton Roads	29/05/42	Key West	4/6/42
Ks 506	Hampton Roads	1/6/42	Key West	6/6/42
Ks 507	Hampton Roads	4/6/42	Key West	9/6/42
Ks 508	Hampton Roads	7/6/42	Key West	12/6/42
Ks 509	Hampton Roads	10/6/42	Key West	15/06/42
Ks 510	Hampton Roads	13/06/42	Key West	18/06/42
Ks 511	NYC	17/06/42	Key West	
Ks 512	Hampton Roads	20/06/42	Key West	26/06/42
Ks 513	Hampton Roads	23/06/42	Key West	30/06/42
Ks 514	Hampton Roads	26/06/42	Key West	2/7/42
Ks 515	Hampton Roads	29/06/42	Key West	5/7/42

Ks 516	Hampton Roads	2/7/42	Key West	7/7/42
Ks 517	Hampton Roads	5/7/42	Key West	13/07/42
Ks 518	Hampton Roads	8/7/42	Key West	14/07/42
Ks 519	Hampton Roads	11/7/42	Key West	16/07/42
Ks 52	Casablanca	14/01/40	Brest	20/01/40
Ks 520	Hampton Roads	14/07/42	Key West	21/07/42
Ks 521	Hampton Roads	17/07/42	Key West	22/07/42
Ks 522	Hampton Roads	20/07/42	Key West	25/07/42
Ks 523	Hampton Roads	23/07/42	Key West	31/07/42
Ks 524	Hampton Roads	25/07/42	Key West	31/07/42
Ks 525	Hampton Roads	29/07/42	Key West	
Ks 526	Hampton Roads	31/07/42	Key West	8/8/42
Ks 527	Hampton Roads	4/8/42	Key West	9/8/42
Ks 528	Hampton Roads	7/8/42	Key West	12/8/42
Ks 529	Hampton Roads	10/8/42	Key West	16/08/42
Ks 530	Hampton Roads	13/08/42	Key West	18/08/42
Ks 531	Hampton Roads	15/08/42	Key West	20/08/42
Ks 532	Hampton Roads	19/08/42	Key West	25/08/42
Ks 533	Hampton Roads	22/08/42	Key West	27/08/42
Ks 534	Hampton Roads	29/08/42	Key West	3/9/42
Ks 535	Hampton Roads	10/9/42	Key West	
Ks 56	Casablanca	21/01/40	Brest	28/01/40
Ks 57	Casablanca	21/01/40	La Pallice	27/01/40
Ks 59	Casablanca	27/01/40	Brest	1/2/40
Ks 61	Casablanca	2/2/40	Brest	9/2/40
Ks 63	Casablanca	8/2/40	Brest	14/02/40
Ks 65	Casablanca	14/02/40	Brest	20/02/40
Ks 68	Casablanca	20/02/40	Brest	26/02/40
Ks 70	Casablanca	26/02/40	Brest	3/3/40
Ks 72	Casablanca	2/3/40	Brest	8/3/40
Ks 74	Casablanca	9/3/40	Brest	15/03/40
Ks 76	Casablanca	16/03/40	Brest	22/03/40
Ks 8	Casablanca	17/06/40	Oran	18/06/40
Ks 80	Casablanca	24/03/40	Brest	
Ks 81	Casablanca	29/03/40	Brest	
Ks 83	Casablanca	4/4/40	Brest	11/4/40
Ks 86	Casablanca	10/4/40	Brest	16/04/40
Ks 9	Casablanca	17/06/40	Gironde	23/06/40
Ks 91	Casablanca	24/04/40	Brest	30/04/40
Ks 92	Casablanca	1/5/40	Brest	
Ks 94	Casablanca	8/5/40	Brest	14/05/40
Ks 97	Casablanca	15/05/40	Brest	21/05/40
Ks 98	Casablanca	15/05/40	Gironde	19/05/40
Ks 99	Casablanca	21/05/40	Brest	27/05/40

KW 011	Key West	26/08/42	Port Everglades	28/08/42
KW 012	Key West	28/08/42	Port Everglades	29/08/42
KW 014	Key West	30/08/42	Jacksonville	2/9/42
KW 016	Key West	4/9/42	Port Everglades	5/9/42
KW 020	Key West	10/9/42	Port Everglades	11/9/42
KW 023	Key West	16/09/42	Mayport	
KW 029	Key West	24/09/42	Guantanamo	
KW 030	Key West	25/09/42	Port Everglades	26/09/42
KW 032	Key West	2/10/42	Fernandina	5/10/42
KW 040	Key West	11/10/42	Jacksonville	14/10/42
KW 041	Key West	13/10/42	Miami	13/10/42
KW 045	Key West	20/10/42	Guantanamo	24/10/42
KW 050	Key West	25/10/42	West Palm Beach	
KW 052	Key West	31/10/42	West Palm Beach	
KW 055	Key West	2/11/42	Port Everglades	3/11/42
KW 060	Key West	10/11/42	West Palm Beach	11/11/42
KW 062	Key West	13/11/42	Miami	
KW 063	Key West	15/11/42	Port Everglades	
KW 068	Key West	18/11/42	Miami	19/11/42
KW 073	Key West	22/11/42	Guantanamo	
KW 074	Key West	22/11/42	Port Everglades	23/11/42
KW 077	Key West	26/11/42	Guantanamo	
KW 079	Key West	30/11/42	Port Everglades	1/12/42
KW 082	Key West	9/12/42	Charleston	11/12/42
KW 095	Key West	12/1/43	Guantanamo	
KW 106	Key West	4/2/43	Norfolk Va	9/2/43
KW 110	Key West	9/2/43	San Juan Pr	13/02/43
KW 120	Key West	24/02/43	Guantanamo	27/02/43
KW 122	Key West	6/3/43	Dispersed	10/3/43
KW 123	Key West	7/3/43	Guantanamo	10/3/43
KW 129	Key West	16/03/43	Guantanamo	20/03/43
KW 130	Key West	17/03/43	Miami	18/03/43

KW 133	Key West	21/03/43	Mayport	
KW 137	Key West	28/08/43	Guantanamo Port	31/03/43
KW 139	Key West	29/03/43	Everglades	30/03/43
KW 143	Key West	1/4/43	Guantanamo	5/4/43
KW 149	Key West	9/4/43	Guantanamo Port	11/4/43
KW 151	Key West	8/4/43	Everglades	9/4/43
KW 162	Key West	17/04/43	Guantanamo	20/04/43
KW 165	Key West	21/04/43	Guantanamo	
KW 171	Key West	6/5/43	Jacksonville	8/5/43
KW 174	Key West	8/5/43	West Palm Beach	
KW 178	Key West	17/05/43	Guantanamo Port	22/05/43
KW 182	Key West	20/05/43	Everglades	20/05/43
KW 186	Key West	22/05/43	Guantanamo	
KW 199	Key West	4/6/43	Jacksonville	
KW 200	Key West	7/6/43	Guantanamo	10/6/43
KW 204	Key West	14/06/43	Guantanamo	18/06/43
KW 208	Key West	14/06/43	Guantanamo	18/06/43
KW 211	Key West	15/06/43	West Palm Beach	16/06/43
KW 215	Key West	19/06/43	Guantanamo Port	23/06/43
KW 217	Key West	19/06/43	Everglades	20/06/43
KW 219	Key West		San Juan Pr	
KW 221	Key West		Nuevitas	
KW 226	Key West	26/06/43	West Palm Beach	27/06/43
KW 228	Key West		Guantanamo	
KW 231	Key West		Antilla Cuba	
KW 232	Key West		Nuevitas	
KW 235	Key West		Guantanamo	26/07/43
KW 236	Key West	23/07/43	Jacksonville	25/07/43
KW 246	Key West		Guantanamo	
KW 250	Key West	2/8/43	Savannah	
KW 254	Key West	4/8/43	Guantanamo	7/8/43
KW 256	Key West	5/8/43	Chesapeake Bay	
KW 257	Key West		Guantanamo	
KW 259	Key West	12/8/43	Savannah	15/08/43
KW 261	Key West	14/08/43	Guantanamo	17/08/43
KW 264	Key West	17/08/43	Guantanamo	20/08/43

KW 265	Key West	18/08/43	Joined KN 259	18/08/43
KW 267	Key West	26/08/43	Guantanamo	31/08/43
KW 268	Key West	2/9/43	Guantanamo	5/9/43
KW 269	Key West	4/9/43	Guantanamo	7/9/43
KW 272	Key West	19/09/43	San Juan Pr	
KW 273	Key West	22/09/43	Guantanamo	26/09/43
KW 274	Key West	29/09/43	Antilla Cuba	
KW 280	Key West	21/10/43	Guantanamo	24/10/43
KW 286	Key West	28/11/43	Guantanamo	1/12/43
KW 288	Key West	3/12/43	Guantanamo	6/12/43
KW 293	Key West	9/12/43	Guantanamo	12/12/43
KW 301	Key West	22/12/43	Guantanamo	26/12/43
KW 305	Key West	30/12/43	Guantanamo	2/1/44
KW 308	Key West		Guantanamo	
KW 310	Key West	10/1/44	Guantanamo	14/01/44
KW 313	Key West	12/1/44	Guantanamo	15/01/44
KW 314	Key West	16/01/44	Guantanamo	19/01/44
KW 322	Key West	5/2/44	Guantanamo	8/2/44
KW 337	Key West	16/03/44	San Juan Pr	20/03/44
KW 354	Key West	13/06/44	Kingston Ja	15/06/44
KW 379	Guantanamo	8/2/45	Key West	14/02/45
KW 383	Key West		Guantanamo	1/4/45
KW Spec 1	Key West	18/08/43	Joined KN 259	18/08/43
KW Spec 2	Antilla Cuba	13/12/43	Key West	14/12/43
Lw 100/1	Norfolk Va	10/6/42	Balboa	17/06/42
Lw 101	Norfolk Va	6/7/42	Cristobal	13/07/42
Lw 102	Norfolk Va	24/09/42	Guantanamo	28/09/42
M 001	Martinique	1/11/43	San Juan Pr	3/11/43
M 230	Cape Henry	16/01/44	Cristobal	22/01/44
M 239	Cape Henry	10/2/44	Cristobal	16/02/44
M 244	Norfolk	2/3/44	Cristobal	8/3/44
M 246	Cape Henry	13/03/44	Cristobal	20/03/44
M 248	Norfolk		Cristobal	16/04/44
Madison	Galveston	23/06/42	Norfolk	28/06/42
Marnell	Port Everglades	12/8/42	Miami	12/8/42
Miami 004	Miami	25/08/42	Key West	26/08/42
Miami 006	Miami		Key West	
Miami 009	Miami	5/9/42	Key West	6/9/42
Miami 012	Miami	16/09/42	Cayo Francis	
Miami 013	Miami	18/09/42	Mayport	19/09/42
Miami 024	Miami	9/10/42	Havana	
Miami 030	Miami		Great Exuma	27/10/42
Miami 047	Miami	19/06/43	Havana	

Miami 103	Miami	5/4/43	Nassau	
Miami 104	Miami	1/5/43	Great Exuma	
Miami 105	Miami	7/4/43	Joined NK 532	8/4/43
Miami 110	Miami	9/4/43	Havana	10/4/43
Miami 118	Miami	17/04/43	Key West	
Miami 127	Miami	8/5/43	Havana	9/5/43
Miami 136	Miami	30/05/43	San Juan Pr	4/6/43
Miami 139	Miami	9/6/43	Cayo Frances	11/6/43
Miami 141	Miami	9/6/43	Key West	10/6/43
Miami 147	Miami	19/06/43	Havana	20/06/43
Miami 148	Miami	19/06/43	Ciudad Trujillo	26/06/43
Miami 152	Miami	24/06/43	Havana	25/06/43
Miami 162	Miami	20/07/43	Key West	22/07/43
Miami 164	Miami	20/07/43	Key West	24/07/43
Miami 168	Miami	2/8/43	Key West	4/8/43
Miami 190	Miami	22/10/43	Kingston Ja	26/10/43
Miami 190	Miami		Kingston Ja	
Miami 190	Miami	22/10/43	Kingston Ja	26/10/43
Miami 195	Miami		Kingston Ja	
Miami 248	Miami		Kingston Ja	
Miami 253	Miami		Kingston Ja	
Miami 324	Miami		Guantanamo	
Miami 325	Miami	30/04/45	Nassau	1/5/45
Miami 325/1	Nassau	4/5/45	Miami	5/5/45
NC 001/1	Kingston Ja	1/4/42	San Juan Pr	3/4/42
NC 003/2	Kingston Ja	4/6/42	San Juan Pr	6/6/42
NC 003/3	San Juan Pr	8/6/42	Trinidad	10/6/42
NC 005/3	Trinidad	20/09/42	San Juan Pr	23/09/42
NC 005/4	San Juan Pr	25/09/42	Kingston Ja	27/09/42
NC 005/5	Kingston Ja	29/09/42	Belize	1/10/42
NG 300	NYC	27/08/42	Guantanamo	3/9/42
NG 301	NYC	30/08/42	Guantanamo	6/9/42
NG 302	NYC	3/9/42	Guantanamo	10/9/42
NG 303	NYC	6/9/42	Guantanamo	12/9/42
NG 304	NYC	10/9/42	Guantanamo	17/09/42
NG 305	NYC	13/09/42	Guantanamo	20/09/42
NG 306	NYC	17/09/42	Guantanamo	24/09/42
NG 307	NYC	20/09/42	Guantanamo	27/09/42
NG 308	NYC	24/09/42	Guantanamo	1/10/42
NG 309	NYC	27/09/42	Guantanamo	4/10/42
NG 310	NYC	1/10/42	Guantanamo	7/10/42
NG 311	NYC	4/10/42	Guantanamo	11/10/42
NG 312	NYC	8/10/42	Guantanamo	15/10/42
NG 313	NYC	11/10/42	Guantanamo	18/10/42

NG 314	NYC	15/10/42	Guantanamo	22/10/42
NG 315	NYC	19/10/42	Guantanamo	26/10/42
NG 316	NYC	24/10/42	Guantanamo	30/10/42
NG 317	NYC	27/10/42	Guantanamo	3/11/42
NG 318	NYC	31/10/42	Guantanamo	7/11/42
NG 319	NYC	4/11/42	Guantanamo	11/11/42
NG 320	NYC	8/11/42	Guantanamo	15/11/42
NG 321	NYC	12/11/42	Guantanamo	19/11/42
NG 322	NYC	16/11/42	Guantanamo	23/11/42
NG 323	NYC	20/11/42	Guantanamo	27/11/42
NG 324	NYC	24/11/42	Guantanamo	1/12/42
NG 325	NYC	28/11/42	Guantanamo	6/12/42
NG 326	NYC	2/12/42	Guantanamo	9/12/42
NG 327	NYC	6/12/42	Guantanamo	13/12/42
NG 328	NYC	10/12/42	Guantanamo	18/12/42
NG 329	NYC	14/12/42	Guantanamo	21/12/42
NG 330	NYC	18/12/42	Guantanamo	25/12/42
NG 331	NYC	22/12/42	Guantanamo	29/12/42
NG 332	NYC	26/12/42	Guantanamo	2/1/43
NG 333	NYC	30/12/42	Guantanamo	6/1/43
NG 334	NYC	3/1/43	Guantanamo	10/1/43
NG 335	NYC	7/1/43	Guantanamo	14/01/43
NG 336	NYC	11/1/43	Guantanamo	18/01/43
NG 337	NYC	15/01/43	Guantanamo	22/01/43
NG 338	NYC	19/01/43	Guantanamo	26/01/43
NG 339	NYC	23/01/43	Guantanamo	29/01/43
NG 340	NYC	28/01/43	Guantanamo	4/2/43
NG 341	NYC	2/2/43	Guantanamo	9/2/43
NG 342	NYC	7/2/43	Guantanamo	14/02/43
NG 343	NYC	12/2/43	Guantanamo	19/02/43
NG 344	NYC	17/02/43	Guantanamo	24/02/43
NG 345	NYC	22/02/43	Guantanamo	1/3/43
NG 346	NYC	27/02/43	Guantanamo	6/3/43
NG 347	NYC	4/3/43	Guantanamo	11/3/43
NG 348	NYC	9/3/43	Guantanamo	16/03/43
NG 349	NYC	14/03/43	Guantanamo	21/03/43
NG 350	NYC	19/03/43	Guantanamo	26/03/43
NG 351	NYC	23/03/43	Guantanamo	30/03/43
NG 352	NYC	28/03/43	Guantanamo	3/4/43
NG 353	NYC	2/4/43	Guantanamo	9/4/43
NG 354	NYC	7/4/43	Guantanamo	14/04/43
NG 355	NYC	12/4/43	Guantanamo	19/04/43
NG 356	NYC	17/04/43	Guantanamo	24/04/43
NG 357	NYC	22/04/43	Guantanamo	28/04/43

NG 358	NYC	27/04/43	Guantanamo	4/5/43
NG 359	NYC	2/5/43	Guantanamo	9/5/43
NG 360	NYC	7/5/43	Guantanamo	14/05/43
NG 361	NYC	12/5/43	Guantanamo	19/05/43
NG 362	NYC	17/05/43	Guantanamo	24/05/43
NG 363	NYC	22/05/43	Guantanamo	30/05/43
NG 364	NYC	27/05/43	Guantanamo	3/6/43
NG 365	NYC	1/6/43	Guantanamo	7/6/43
NG 366	NYC	6/6/43	Guantanamo	13/06/43
NG 367	NYC	11/6/43	Guantanamo	18/06/43
NG 368	NYC	16/06/43	Guantanamo	22/06/43
NG 369	NYC	21/06/43	Guantanamo	28/06/43
NG 370	NYC	26/06/43	Guantanamo	3/7/43
NG 371	NYC	1/7/43	Guantanamo	7/7/43
NG 372	NYC	6/7/43	Guantanamo	13/07/43
NG 373	NYC	12/7/43	Guantanamo	19/07/43
NG 374	NYC	16/07/43	Guantanamo	23/07/43
NG 375	NYC	21/07/43	Guantanamo	28/07/43
NG 376	NYC	26/07/43	Guantanamo	2/8/43
NG 377	NYC	31/07/43	Guantanamo	7/8/43
NG 378	NYC	5/8/43	Guantanamo	12/8/43
NG 379	NYC	10/8/43	Guantanamo	17/08/43
NG 380	NYC	15/08/43	Guantanamo	22/08/43
NG 381	NYC	20/08/43	Guantanamo	27/08/43
NG 382	NYC	26/08/43	Guantanamo	2/9/43
NG 383	NYC	30/08/43	Guantanamo	6/9/43
NG 384	NYC	5/9/43	Guantanamo	12/9/43
NG 385	NYC	9/9/43	Guantanamo	16/09/43
NG 386	NYC	16/09/43	Guantanamo	23/09/43
NG 387	NYC	19/09/43	Guantanamo	26/09/43
NG 388	NYC	25/09/43	Guantanamo	1/10/43
NG 389	NYC	29/09/43	Guantanamo	6/10/43
NG 390	NYC	5/10/43	Guantanamo	11/10/43
NG 391	NYC	9/10/43	Guantanamo	15/10/43
NG 392	NYC	15/10/43	Guantanamo	21/10/43
NG 393	NYC	19/10/43	Guantanamo	26/10/43
NG 394	NYC	25/10/43	Guantanamo	1/11/43
NG 395	NYC	29/10/43	Guantanamo	5/11/43
NG 396	NYC	4/11/43	Guantanamo	11/11/43
NG 397	NYC	8/11/43	Guantanamo	14/11/43
NG 398	NYC	14/11/43	Guantanamo	20/11/43
NG 399	NYC	18/11/43	Guantanamo	24/11/43
NG 400	NYC	24/11/43	Guantanamo	30/11/43
NG 401	NYC	28/11/43	Guantanamo	4/12/43

NG 402	NYC	3/12/43	Guantanamo	9/12/43
NG 403	NYC	8/12/43	Guantanamo	14/12/43
NG 404	NYC	13/12/43	Guantanamo	20/12/43
NG 405	NYC	18/12/43	Guantanamo	25/12/43
NG 406	NYC	23/12/43	Guantanamo	29/12/43
NG 407	NYC	28/12/43	Guantanamo	4/1/44
NG 408	NYC	2/1/44	Guantanamo	9/1/44
NG 409	NYC	7/1/44	Guantanamo	14/01/44
NG 410	NYC	12/1/44	Guantanamo	18/01/44
NG 411	NYC	17/01/44	Guantanamo	23/01/44
NG 412	NYC	22/01/44	Guantanamo	29/01/44
NG 413	NYC	27/01/44	Guantanamo	3/2/44
NG 414	NYC	1/2/44	Guantanamo	7/2/44
NG 415	NYC	6/2/44	Guantanamo	12/2/44
NG 416	NYC	12/2/44	Guantanamo	18/02/44
NG 417	NYC	16/02/44	Guantanamo	23/02/44
NG 418	NYC	21/02/44	Guantanamo	27/02/44
NG 419	NYC	26/02/44	Guantanamo	4/3/44
NG 420	NYC	2/3/44	Guantanamo	9/3/44
NG 421	NYC	7/3/44	Guantanamo	14/03/44
NG 422	NYC	12/3/44	Guantanamo	19/03/44
NG 423	NYC	17/03/44	Guantanamo	24/03/44
NG 424	NYC	22/03/44	Guantanamo	29/03/44
NG 425	NYC	27/03/44	Guantanamo	2/4/44
NG 426	NYC	1/4/44	Guantanamo	8/4/44
NG 427	NYC	6/4/44	Guantanamo	13/04/44
NG 428	NYC	11/4/44	Guantanamo	18/04/44
NG 429	NYC	16/04/44	Guantanamo	23/04/44
NG 430	NYC	21/04/44	Guantanamo	28/04/44
NG 431	NYC	26/04/44	Guantanamo	2/5/44
NG 432	NYC	1/5/44	Guantanamo	7/5/44
NG 433	NYC	6/5/44	Guantanamo	12/5/44
NG 434	NYC	11/5/44	Guantanamo	17/05/44
NG 435	NYC	16/05/44	Guantanamo	22/05/44
NG 436	NYC	21/05/44	Guantanamo	28/05/44
NG 437	NYC	26/05/44	Guantanamo	2/6/44
NG 438	NYC	31/05/44	Guantanamo	6/6/44
NG 439	NYC	5/6/44	Guantanamo	11/6/44
NG 440	NYC	10/6/44	Guantanamo	16/06/44
NG 441	NYC	15/06/44	Guantanamo	21/06/44
NG 442	NYC	20/06/44	Guantanamo	27/06/44
NG 443	NYC	25/06/44	Guantanamo	2/7/44
NG 444	NYC	30/06/44	Guantanamo	6/7/44
NG 445	NYC	5/7/44	Guantanamo	12/7/44

NG 446	NYC	10/7/44	Guantanamo	16/07/44
NG 447	NYC	15/07/44	Guantanamo	22/07/44
NG 448	NYC	20/07/44	Guantanamo	27/07/44
NG 449	NYC	25/07/44	Guantanamo	1/8/44
NG 450	NYC	30/07/44	Guantanamo	6/8/44
NG 451	NYC	4/8/44	Guantanamo	11/8/44
NG 452	NYC	9/8/44	Guantanamo	16/08/44
NG 453	NYC	14/08/44	Guantanamo	21/08/44
NG 454	NYC	19/08/44	Guantanamo	25/08/44
NG 455	NYC	24/08/44	Guantanamo	31/08/44
NG 456	NYC	29/08/44	Guantanamo	4/9/44
NG 457	NYC	3/9/44	Guantanamo	9/9/44
NG 458	NYC	8/9/44	Guantanamo	15/09/44
NG 459	NYC	15/09/44	Guantanamo	21/09/44
NG 460	NYC	18/09/44	Guantanamo	24/09/44
NG 461	NYC	23/09/44	Guantanamo	29/09/44
NG 462	NYC	28/09/44	Guantanamo	5/10/44
NG 463	NYC	3/10/44	Guantanamo	10/10/44
NG 464	NYC	8/10/44	Guantanamo	15/10/44
NG 465	NYC	13/10/44	Guantanamo	20/10/44
NG 466	NYC	18/10/44	Guantanamo	25/10/44
NG 467	NYC	23/10/44	Guantanamo	30/10/44
NG 468	NYC	28/10/44	Guantanamo	3/11/44
NG 469	NYC	2/11/44	Guantanamo	9/11/44
NG 470	NYC	7/11/44	Guantanamo	13/11/44
NG 471	NYC	12/11/44	Guantanamo	19/11/44
NG 472	NYC	17/11/44	Guantanamo	24/11/44
NG 473	NYC	22/11/44	Guantanamo	29/11/44
NG 474	Did Not Sail			
NG 475	NYC	2/12/44	Guantanamo	9/12/44
NG 476	NYC	7/12/44	Guantanamo	14/12/44
NG 477	NYC	12/12/44	Guantanamo	19/12/44
NG 478	NYC	17/12/44	Guantanamo	24/12/44
NG 479	NYC	22/12/44	Guantanamo	29/12/44
NG 480	NYC	27/12/44	Guantanamo	3/1/45
NG 481	NYC	1/1/45	Guantanamo	8/1/45
NG 482	NYC	6/1/45	Guantanamo	13/01/45
NG 483	NYC	11/1/45	Guantanamo	18/01/45
NG 484	NYC	17/01/45	Guantanamo	23/01/45
NG 485	NYC	21/01/45	Guantanamo	28/01/45
NG 486	NYC	26/01/45	Guantanamo	2/2/45
NG 487	NYC	31/01/45	Guantanamo	7/2/45
NG 488	NYC	5/2/45	Guantanamo	12/2/45
NG 489	NYC	10/2/45	Guantanamo	17/02/45

NG 490	NYC	15/02/45	Guantanamo	23/02/45
NG 491	NYC	20/02/45	Guantanamo	28/02/45
NG 492	NYC	25/02/45	Guantanamo	4/3/45
NG 493	NYC	2/3/45	Guantanamo	9/3/45
NG 494	NYC	7/3/45	Guantanamo	14/03/45
NG 495	NYC	12/3/45	Guantanamo	19/03/45
NG 496	NYC	17/03/45	Guantanamo	24/03/45
NG 497	NYC	22/03/45	Guantanamo	29/03/45
NG 498	NYC	27/03/45	Guantanamo	2/4/45
NG 499	NYC	1/4/45	Guantanamo	8/4/45
NG 500	NYC	6/4/45	Guantanamo	14/04/45
NG 501	NYC	11/4/45	Guantanamo	17/04/45
NG 502	NYC	16/04/45	Guantanamo	23/04/45
NG 503	NYC	21/04/45	Guantanamo	28/04/45
NG 504	NYC	26/04/45	Guantanamo	3/5/45
NG 505	NYC	1/5/45	Guantanamo	8/5/45
NG 506	NYC	6/5/45	Guantanamo	13/05/45
NG Spec	NYC	17/12/42	Trinidad	27/12/42
NK 500	NYC	28/08/42	Key West	3/9/42
NK 501	NYC	4/9/42	Key West	11/9/42
NK 502	NYC	11/9/42	Key West	17/09/42
NK 503	NYC	18/09/42	Key West	24/09/42
NK 504	NYC	25/09/42	Key West	2/10/42
NK 505	NYC	2/10/42	Key West	8/10/42
NK 506	NYC	13/10/42	Key West	20/10/42
NK 507	NYC	21/10/42	Key West	28/10/42
NK 508	NYC	29/10/42	Key West	5/11/42
NK 509	NYC	6/11/42	Key West	12/11/42
NK 510	NYC	15/11/42	Key West	21/11/42
NK 511	NYC	22/11/42	Key West	29/11/42
NK 512	NYC	30/11/42	Key West	
NK 513	NYC	8/12/42	Key West	
NK 514	NYC	16/12/42	Key West	21/12/42
NK 515	NYC	24/12/42	Key West	
NK 516	NYC	1/1/43	Key West	8/1/43
NK 517	NYC	9/1/43	Key West	15/01/43
NK 518	NYC	17/01/43	Key West	
NK 519	NYC	26/01/43	Key West	
NK 520	NYC	30/01/43	Key West	
NK 521	NYC	4/2/43	Key West	
NK 522	NYC	9/2/43	Key West	14/02/43
NK 523	NYC	14/02/43	Key West	
NK 524	NYC	19/02/43	Key West	
NK 525	Did Not Sail			

NK 526	NYC	1/3/43	Key West	
NK 527	NYC	6/3/43	Key West	12/3/43
NK 528	NYC	11/3/43	Key West	18/03/43
NK 529	NYC	16/03/43	Key West	22/03/43
NK 530	NYC	22/03/43	Key West	28/03/43
NK 531	NYC	27/03/43	Key West	3/4/43
NK 532	NYC	1/4/43	Key West	8/4/43
NK 533	NYC	6/4/43	Key West	13/04/43
NK 534	NYC	11/4/43	Key West	15/04/43
NK 535	NYC	16/04/43	Key West	
NK 536	NYC	21/04/43	Key West	
NK 537	NYC	26/04/43	Key West	
NK 538	NYC	1/5/43	Key West	
NK 539	NYC	6/5/43	Key West	11/5/43
NK 540	NYC	11/5/43	Key West	17/05/43
NK 541	NYC	16/05/43	Key West	21/05/43
NK 542	NYC	20/05/43	Key West	
NK 543	NYC	27/05/43	Key West	1/6/43
NK 544	NYC	31/05/43	Key West	
NK 545	NYC	5/6/43	Key West	11/6/43
NK 546	NYC	10/6/43	Key West	16/06/43
NK 547	NYC	15/06/43	Key West	22/06/43
NK 548	NYC	20/06/43	Key West	27/06/43
NK 549	NYC	25/06/43	Key West	
NK 550	NYC	30/06/43	Key West	
NK 551	NYC	5/7/43	Key West	11/7/43
NK 552	NYC	10/7/43	Key West	
NK 553	NYC	15/07/43	Key West	22/07/43
NK 554	NYC	20/07/43	Key West	27/07/43
NK 555	NYC	25/07/43	Key West	1/8/43
NK 556	NYC	30/07/43	Key West	
NK 557	NYC	4/8/43	Key West	
NK 558	NYC	9/8/43	Key West	16/08/43
NK 559	NYC	14/08/43	Key West	
NK 560	NYC	19/08/43	Key West	
NK 561	NYC	24/08/43	Key West	30/08/43
NK 562	NYC	29/08/43	Key West	
NK 563	NYC	2/9/43	Key West	
NK 564	NYC	8/9/43	Key West	
NK 565	NYC	13/09/43	Key West	
NK 566	NYC	18/09/43	Key West	
NK 567	NYC	23/09/43	Key West	
NK 568	NYC	28/09/43	Key West	6/10/43
NK 569	NYC	3/10/43	Key West	

NK 570	NYC	8/10/43	Key West	
NK 571	NYC	13/10/43	Key West	
NK 572	NYC	18/10/43	Key West	
NK 573	NYC	23/10/43	Key West	
NK 574	NYC	28/10/43	Key West	
NK 575	NYC	2/11/43	Key West	
NK 576	NYC	7/11/43	Key West	
NK 577	NYC	12/11/43	Key West	
NK 578	NYC	17/11/43	Key West	22/11/43
NK 579	NYC	22/11/43	Key West	
NK 580	NYC	27/11/43	Key West	
NK 581	NYC	2/12/43	Key West	
NK 582	NYC	7/12/43	Key West	
NK 583	NYC	12/12/43	Key West	
NK 584	NYC	17/12/43	Key West	23/12/43
NK 585	NYC	22/12/43	Key West	29/12/43
NK 586	NYC	27/12/43	Key West	2/1/44
NK 587	NYC	1/1/44	Key West	9/1/44
NK 588	NYC	6/1/44	Key West	
NK 589	NYC	11/1/44	Key West	18/01/44
NK 590	NYC	16/01/44	Key West	22/01/44
NK 591	NYC	22/01/44	Key West	
NK 592	NYC	26/01/44	Key West	
NK 593	NYC	31/01/44	Key West	
NK 594	NYC	5/2/44	Key West	11/2/44
NK 595	NYC	10/2/44	Key West	
NK 596	NYC	15/02/44	Key West	
NK 597	NYC	20/02/44	Key West	
NK 598	NYC	25/02/44	Key West	
NK 599	NYC	1/3/44	Key West	8/3/44
NK 600	NYC	6/3/44	Key West	
NK 601	NYC	11/3/44	Key West	
NK 602	NYC	16/03/44	Key West	
NK 603	NYC	21/03/44	Key West	27/03/44
NK 604	NYC	26/03/44	Key West	1/4/44
NK 605	NYC	31/03/44	Key West	
NK 606	NYC	5/4/44	Key West	
NK 607	NYC	10/4/44	Key West	17/04/44
NK 608	NYC	15/04/44	Key West	21/04/44
NK 609	NYC	20/04/44	Key West	26/04/44
NK 610	NYC	25/04/44	Key West	1/5/44
NK 611	NYC	30/04/44	Key West	
NK 612	NYC	5/5/44	Key West	11/5/44
NK 613	NYC	10/5/44	Key West	17/05/44

NK 614	NYC	15/05/44	Key West	
NK 615	NYC	20/05/44	Key West	
NK 616	NYC	25/05/44	Key West	
NK 617	NYC	30/05/44	Key West	4/6/44
NK 618	NYC	4/6/44	Key West	11/6/44
NK 619	NYC	9/6/44	Key West	
NK 620	NYC	14/06/44	Key West	20/06/44
NK 621	NYC	19/06/44	Key West	25/06/44
NK 622	NYC	22/06/44	Key West	
NK 623	NYC	29/06/44	Key West	
NK 624	NYC	4/7/44	Key West	
NK 625	NYC	9/7/44	Key West	15/07/44
NK 626	NYC	14/07/44	Key West	
NK 627	Did Not Sail			
NK 628	NYC	24/07/44	Key West	
NK 629	NYC	29/07/44	Key West	
NK 630	NYC	4/8/44	Key West	9/8/44
NK 631	NYC	9/8/44	Key West	
NK 632	Did Not Sail			
NK 633	Did Not Sail			
NK 634	NYC	23/08/44	Key West	
NK 635	NYC	28/08/44	Key West	3/9/44
NK 636	NYC	2/9/44	Key West	7/9/44
NK 637	NYC	7/9/44	Key West	13/09/44
NK 638	NYC	16/09/44	Key West	21/09/44
NK 639	Did Not Sail			
NK 640	NYC	22/09/44	Key West	
NK 641	NYC	27/09/44	Key West	
NK 642	NYC	2/10/44	Key West	
NK 643	NYC	7/10/44	Key West	14/10/44
NK 644	NYC	12/10/44	Key West	
NK 645	Did Not Sail			
NK 646	NYC	22/10/44	Key West	27/10/44
NK 647	Did Not Sail			
NK 648	NYC	1/11/44	Key West	
NK 649	NYC	6/11/44	Key West	
NK 650	NYC	11/11/44	Key West	16/11/44
NK 651	NYC	16/11/44	Key West	
NK 652	NYC	21/11/44	Key West	
NK 653	NYC	26/11/44	Key West	1/12/44
NK 654	NYC	1/12/44	Key West	
NK 655	NYC	6/12/44	Key West	
NK 656	NYC	12/12/44	Key West	
NK 657	Did Not Sail			

NK 658	NYC	21/12/44	Key West	
NK 659	NYC	26/12/44	Key West	
NK 660	Did Not Sail			
NK 661	NYC	6/1/45	Key West	
NK 662	Did Not Sail			
NK 663	Did Not Sail			
NK 664	Did Not Sail			
NK 665	NYC	26/01/45	Key West	
NK 666	Did Not Sail			
NK 667	Did Not Sail			
NK 668	Did Not Sail			
NK 669	Did Not Sail			
NK 670	NYC	19/02/45	Key West	
NK 671	Did Not Sail			
NK 672	NYC	1/3/45	Key West	
NK 673	NYC	6/3/45	Key West	
NK 674	Did Not Sail			
NK 675	NYC	16/03/45	Key West	
NK 676	Did Not Sail			
NK 677	Did Not Sail			
NK 678	Did Not Sail			
NK 679	NYC	5/4/45	Key West	
NK 680	NYC	10/4/45	Key West	
NK 681	Did Not Sail			
NK 682	Did Not Sail			
NK 683	NYC	25/04/45	Key West	
NK 684	NYC	30/04/45	Key West	5/5/45
NK 685	NYC	5/5/45	Key West	10/5/45
NK 686	NYC	10/5/45	Key West	
Nor 180	Norfolk		Miami	
Nor 255	Norfolk		Cristobal	28/06/44
Nor 258	Norfolk		Cristobal	8/7/44
Nor 306	Norfolk	13/05/45	Key West	22/05/45
NY 038	Guantanamo	5/11/43	New York	9/11/43
NY 044	New York	4/12/43	Cristobal	
NY 047	New York	20/12/43	Cristobal	27/12/43
NY 048	New York	24/12/43	Cristobal	
NY 052	New York	7/1/44	Galveston	12/1/44
NY 053	New York	9/1/44	Cristobal	16/01/44
NY 055	New York	15/01/44	Trinidad	23/01/44
NY 058	New York	20/01/44	Cristobal	26/01/44
NY 060	New York	25/01/44	Cristobal	1/2/44
NY 061 Spec	New York	26/01/44	Galveston	1/2/44
NY 069	New York	15/02/44	Galveston	22/02/44

NY 076	New York	19/03/44	Trinidad	25/03/44
NY 081	New York	11/4/44	San Juan Pr	16/04/44
NY 086	New York	2/5/44	Trinidad	7/5/44
NY 091	New York	21/05/44	St Thomas Vi	25/05/44
NY 100	New York	20/06/44	Trinidad	25/06/44
NY 106	New York	6/7/44	Cristobal	19/07/44
NY 131	New York	5/10/44	Miami	13/10/44
NY 132	New York	23/10/44	Cristobal	10/11/44
NY 133	New York	2/11/44	Cristobal	23/11/44
NY 134	New York	10/11/44	Cristobal	16/11/44
NY 137	New York	26/11/44	Trinidad	1/12/44
NY 138	New York	16/11/44	Cristobal	2/12/44
NY 142	New York	6/12/44	Cristobal	11/12/44
NY 145	New York	1/1/45	Cristobal	7/1/45
NY 150	New York	12/2/45	Cristobal	18/02/45
NY 151	New York	16/02/45	Cristobal	22/02/45
NY 155	New York	5/3/45	Cristobal	10/3/45
NY 156	New York	14/03/45	Guantanamo	18/03/45
NY 160	New York	9/4/45	Cristobal	15/04/45
O 017	San Pedro De Macoris	24/11/43	Guantanamo	26/11/43
Osprey	Guantanamo	9/7/42	Key West	12/7/42
Port Everglades 008	Port Everglades	13/03/43	Havana	15/03/43
Port Everglades 021	Port Everglades	13/09/42	Key West	14/09/42
Port Everglades 023	Port Everglades	13/09/42	Key West	14/09/42
Port Everglades 024	Port Everglades	20/09/42	Havana	22/09/42
Port Everglades 025	Port Everglades	21/09/42	Mayport	23/09/42
Port Everglades 028	Port Everglades	17/08/43	Cristobal	24/08/43
Port Everglades 035	Port Everglades	14/10/42	Key West	16/10/42
Port Everglades 040	Port Everglades	20/10/42	Jacksonville	
Port Everglades 041	Port Everglades	21/10/42	Lookout Bight	24/10/42
Port Everglades 046	Port Everglades	1/11/42	Key West	2/11/42
Port Everglades 052	Port Everglades	8/11/42	Key West	9/11/42
Port Everglades 054	Port Everglades	10/11/42	Key West	
Port Everglades 056	Port Everglades	14/11/42	Key West	15/11/42
Port Everglades 060	Port Everglades	20/11/42	Havana	22/11/42
Port Everglades 061	Port Everglades	20/11/42	Mayport	22/11/42
Port Everglades 063	Port Everglades	21/11/42	Key West	22/11/42
Port Everglades 064	Port Everglades	23/11/42	Havana	25/11/42
Port Everglades 065	Port Everglades	25/11/42	Havana	25/11/42
Port Everglades 066	Port Everglades	25/11/42	Havana	25/11/42
Port Everglades 067	Port Everglades	26/11/42	New York	
Port Everglades 070	Port Everglades	30/11/42	Nicaro	
Port Everglades 071	Port Everglades	1/12/42	Havana	3/12/42

Port Everglades 073	Port Everglades	19/12/42	Cristobal	9/1/43
Port Everglades 085	Port Everglades	19/03/43	Havana	20/03/43
Port Everglades 096	Port Everglades	7/4/43	Havana	8/4/43
Port Everglades 100	Port Everglades	10/4/43	Joined KN 233	10/4/43
Port Everglades 105	Port Everglades	13/04/43	Havana	15/04/43
Port Everglades 113	Port Everglades	26/04/43	Havana	27/04/43
Port Everglades 116	Port Everglades	28/04/43	Havana	30/04/43
Port Everglades 124	Port Everglades	9/5/43	Jacksonville	
Port Everglades 148	Port Everglades	24/06/43	Rebecca Shoal	25/06/43
Port Everglades 169	Port Everglades		Havana	
Port Everglades 175	Port Everglades		Port Arthur	10/10/43
Port Everglades 203	Port Everglades	3/12/43	Cristobal	11/12/43
Savannah 007	Savannah	30/09/42	Smith Shoals	
Savannah 076	Savannah	8/5/45	Corpus Christi	14/05/45
Savannah 077	Savannah		New Orleans	14/05/45
Smith Shoal 012	Smith Shoal	6/9/42	Jacksonville	10/9/42
Smith Shoal 025	Smith Shoal	18/10/42	Port Everglades	19/10/42
Tarbell 003	Kingston Ja	1/7/42	Key West	7/7/42
TAW 1	Trinidad	2/7/42	Key West	10/7/42
TAW 10	Trinidad	31/07/42	Key West	9/8/42
TAW 11	Trinidad	4/8/42	Key West	13/08/42
TAW 11jk	Mayaguez	6/8/42	Key West	12/8/42
TAW 12	Trinidad	8/8/42	Key West	16/08/42
TAW 13	Trinidad	12/8/42	Key West	23/08/42
TAW 14	Trinidad	16/08/42	Key West	25/08/42
TAW 15	Trinidad	23/08/42	Key West	31/08/42
TAW 16	Trinidad	25/08/42	Guantanamo	30/08/42
TAW 2	Trinidad	5/7/42	Key West	14/07/42
TAW 3	Trinidad	8/7/42	Key West	16/07/42
TAW 4	Trinidad	12/7/42	Key West	21/07/42
TAW 5	Trinidad	15/07/42	Key West	23/07/42
TAW 6	Trinidad	18/07/42	Key West	27/07/42
TAW 7	Trinidad	21/07/42	Key West	28/07/42
TAW 8	Trinidad	24/07/42	Key West	1/8/42
TAW 9	Trinidad	27/07/42	Key West	4/8/42
TAW S(1)	Guantanamo	28/08/42	Key West	3/9/42
TAW Sl/1	Trinidad	18/08/42	Guantanamo	25/08/42
TAW Sl/2	Trinidad	28/08/42	Key West	
WAT 1	Key West	1/7/42	Trinidad	9/7/42
WAT 10	Key West	28/07/42	Trinidad	6/8/42
WAT 11	Key West	1/8/42	Trinidad	10/8/42
WAT 12	Key West	5/8/42	Trinidad	14/08/42
WAT 13	Key West	9/8/42	Trinidad	19/08/42

WAT 14	Key West	13/08/42	Trinidad	23/08/42
WAT 15	Key West	18/08/42	Trinidad	27/08/42
WAT 16	Key West	21/08/42	Trinidad	31/08/42
WAT 17	Key West	26/08/42	Guantanamo	1/9/42
WAT 2	Key West	4/7/42	Trinidad	14/07/42
WAT 3	Key West	7/7/42	Trinidad	16/07/42
WAT 4	Key West	10/7/42	Trinidad	20/07/42
WAT 5	Key West	14/07/42	Trinidad	23/07/42
WAT 6	Key West	17/07/42	Trinidad	26/07/42
WAT 7	Key West	19/07/42	Trinidad	28/07/42
WAT 8	Key West	22/07/42	Trinidad	1/8/42
WAT 9	Key West	25/07/42	Trinidad	4/8/42
Z 045	Cristobal	26/01/44	Norfolk	
Z 160	Cristobal		New York	

Appendix 4: Allied Ships Sunk, Tonnage, Cargos Carried, Persons on Board, Fates

Ship Name	Tons	Ship Type	Cargos, or in Ballast	On	KIA	Surv	Ba h
Abgara	4,422	steam freighter	sugar	34	0	34	
Afoundria	5,010	freighter	7,700 tons general cargo: bombs, dynamite, food, lumber and road-building machinery	46	0	46	
Amazone	1,294	steam freighter	926 tons general cargo, mainly coffee and oil	40	14	26	
Andrew Jackson	5,990	tanker	ballast	49	3	46	
Anglo-Canadian	5,268	motor ship	ballast	50	1	49	
Astrea	3,190	steam freighter	general cargo	30	30	0	
Athelqueen	8,780	tanker	petroleum	49	3	46	46
Barbara	4,637	steam freighter	4,015 tons general cargo	85	26	59	
Bloody Marsh	10,195	tanker	102,500 barrels Navy fuel-oil	77	3	74	
Cape Of Good Hope	4,963	motor ship	7,500 tons general military cargo	37	0	37	
Cardonia	5,104	steam freighter	81 tons general cargo	38	1	37	
Carmen	84	schooner (two masted)	2000 sacks maize and 476 pieces mahogany and cedar	8	1	7	

Catahoula	5,030	tanker	molasses	45	7	38	
Charles Racine	9,957	tanker	ballast	41	0	41	
Cheerio	35	schooner	mahogany	9	0	9	
Cherry Valley	10,172	freighter	6,000 tons water ballast, 165 tons Army cargo	79	0	79	
Cities Service Empire	8,103	tanker	9,400 barrels crude oil	67	25	42	
City Of Alma	5,446	steam freighter	7,400 tons manganese ore	39	29	10	
Clan Skene	5,214	steam freighter	2,006 tons chrome ore	81	9	72	
Colabee	5,518	freighter	38,600 bags sugar	37	23	14	
Comol Rico	5,034	tanker	8,068 tons of bulk molasses	42	3	39	
Cygnet	3,628	freighter	iron ore, rubber	30	0	30	30
Darina	8,113	motor tanker	ballast	56	6	50	
Daytonian	6,434	steam ship	7,500 tons steel, lumber, cotton	58	1	57	57
Delfina	3,480	steam freighter	raw sugar	31	4	27	
Delisle	3,478	steam freighter	2,800 tons general cargo, paint in steel drums on deck	32	2	30	
Delmundo	5,032	steam ship	5,437 tons general cargo	58	8	50	
Derryheen	7,217	steam freighter	11,036 tons general cargo, nitrates, trucks	51	0	51	
Domino	3,170	steam freighter	ballast	30	0	30	
Eclipse	9,767	tanker	ballast	47	2	45	
Empire Corporal	7,000	freighter	4,532 tons motor spirit, 4,745 tons white spirit	55	6	49	

Esparta	3,365	steam ship	1,450 tons general cargo, bananas, coffee	40	1	39	
Esso Baton Rouge	7,989	tanker	ballast	68	3	65	
Esso Boston	7,699	tanker	105,400 barrels crude oil	37	0	37	
Esso Gettysburg	10,173	tanker	120,120 barrels crude oil	72	57	15	
Everalda	3,950	steam ship	general cargo	36	0	36	
Everelza	4,520	steam ship	manganese ore	37	23	14	
Fairport	6,266	freighter	8,000 tons war material and a deck cargo of tanks	123	0	123	
Fauna	1,254	freighter	general cargo	29	2	27	27
Federal	2,881	tanker	water ballast	33	5	28	
Frank B. Baird	1,748	steam ship	2,457 tons bauxite	23	0	23	
Fred W. Green	2,292	steam ship	725 tons military cargo, trucks	41	5	36	
Gertrude	16	trawler	20 tons of onions	3	0	3	
Grenanger	5,393	motor ship	general cargo, coffee	36	0	36	
Gulfamerica	8,081	tanker	101,500 barrels furnace oil	48	19	29	
Gulfstate	6,882	tanker	68,417 barrels crude oil	61	43	18	
Gunny	2,362	freighter	3,100 tons manganese ore, mahogany	42	30	12	
Hagan	6,401	tanker	22,676 barrels blackstrap molasses	44	6	38	
Halsey	7,088	tanker	40,000 barrels naphtha, 40,000 barrels heating oil	32	0	32	
Hanseat	8,241	tanker	ballast	40	0	40	

Hardwicke Grange	9,005	steam freighter	700 tons of refrigerated cargo	78	3	75	
Helen Forsey	167	schooner	180 tons molasses & rum	6	2	4	
Illinois	5,447	steam freighter	8,000 tons manganese ore	38	32	6	
James A. Moffett Jr.	9,788	freighter	water ballast	43	1	42	
James Sprunt	7,177	Liberty general cargo ship	4,000 tons explosives	69	0	69	
Java Arrow	8,327	tanker	water ballast, 1,300 drums lube oil	45	2	43	
John Sevier	7,176	Liberty general cargo ship	9,060 tons bauxite ore	57	0	57	
Kollskegg	9,858	tanker	8,000 tons crude oil, 6,300 tons fuel oil	42	4	38	30
Korsholm	2,647	steam freighter	4,593 tons phosphate	26	9	17	
L. A. Christensen	4,362	steam freighter	ballast	31	0	31	
La Paz	6,548	motor freighter	general cargo, whiskey	57	0	57	
Laertes	5,825	freighter	5,230 tons war material, 3 aircraft, 17 tanks, 20 trucks	66	18	48	
Lammot Du Pont	5,102	steam ship	6,812 tons linseed	54	17	37	
Leslie	2,609	steam freighter	3,225 tons raw sugar	32	4	28	
Letitia Porter	15	tugboat	unmanned deck cargo on SS Koenjit, sunk by U-69	0	0	0	
Lubrafol	7,138	tanker	67,000 barrels heating oil	44	13	31	

Macgregor	2,498	steam ship	ballast	31	1	30	
Maldonado	5,285	steam freighter	7,000 tons tinned meat, hides, wool, fats	49	0	49	
Mambi	1,983	coastal freighter	ballast	34	23	11	
Mamura	8,245	tanker	gasoline fully loaded	49	49	0	
Managua	2,220	steam freighter	potash	25	0	25	
Manaqui	2,802	steam ship	general cargo	41	41	0	
Manzanillo	1,025	steam freighter	general cargo	23	0	23	
Mariana	3,110	freighter	sugar	36	36	0	
Medea	1,311	steam freighter	1,636 tons of general cargo, 220 cases of dynamite	28	5	23	
Melbourne Star	12,806	motor ship	8,285 military supplies, general cargo	119	115	4	
Melpomene	7,011	tanker	petroleum	49	0	49	
Michael Jebsen	2,300	freighter	2,750 tons sugar	40	1	39	
Millinocket	3,274	freighter	4,300 tons bauxite ore	35	11	24	
Mobiloil	9,925	tanker	water ballast	52	0	52	
Montevideo	5,785	steam ship	6,000 tons wool, meats	49	14	35	
Nicarao	1,445	freighter	500 tons of fruit, bananas, coconuts, charcoal	39	8	31	
Nickeliner	2,249	chemical tanker	3,400 tons ammonia water	31	0	31	
Norlandia	2,689	freighter	ballast	30	9	21	
O. A. Knudsen	11,007	tanker	pool engine spirit, pool vaporising oil (petrol, fuel-oil)	41	2	39	39

Ocean Venus	7,174	steam freighter	9,450 tons lead, lumber, tinned herring, acetone	42	5	37	
Ohioan	6,078	freighter	6,000 tons manganese ore, wool, liquorice root	37	15	22	
Oklahoma	9,298	tanker	103,199 barrels gas, kerosene	72	50	22	
Olga	2,496	freighter	ballast	33	1	32	
Oneida	2,309	steam freighter	ballast	29	6	23	
Onondaga	2,309	freighter	magnesium ore	34	20	14	
Oregon	7,017	tanker	78,000 barrels Navy fuel oil	36	6	30	
Oscilla	6,341	motor ship	ballast	55	4	51	
Pan Massachusetts	8,202	tanker	104,000 barrels gasoline, kerosene, diesel oil	38	20	18	
Patrick J. Hurley	10,865	tanker	75,000 barrels gas, 60,000 barrels diesel	62	17	45	
Peder Bogen	9,741	tanker	petroleum	43	0	43	
Pedrinhas	3,666	steam freighter	general cargo, cotton, castor beans	48	0	48	
Peiping	6,390	motor ship	9,950 tons general cargo, wool, hides, dyes, tallow	34	3	31	
Pillory	1,507	freighter	ballast	47	25	22	
Potlatch	6,085	freighter	7,500 tons Army supplies, trucks, tanks, tires	55	8	47	47
Potrero Del Llano	4,000	tanker	6,132 tons petroleum	35	14	22	
Putney Hill	5,216	motor freighter	ballast	38	3	35	
Rapana	8,017	tanker	petroleum	35	0	35	

Raphael Semmes	6,027	freighter	7,500 tons manganese ore, tobacco, liquorice, wool, rugs	37	19	18
Republic	5,287	tanker	water ballast	29	5	24
Ruth	4,833	bulker	5,000 tons manganese ore	38	34	4
Sam Houston	7,176	steam / Liberty	10,000 tons army supplies	46	8	38
Sama	567	motor merchant	bananas	14	0	14
San Jacinto	6,069	steam passenger ship	3,200 tons cargo, 104 passengers, 32 women/childre n	183	14	169
Santa Catalina	6,507	C-2 class steam merchant	6,700 tons tanks, steel, tires, gasoline, small arms	95	0	95
Santiago De Cuba	1,685	steam freighter	general cargo	29	10	19
Sea Thrush	5,447	steam freighter	6,800 tons war material, ammunition, aircraft	66	0	66
Skane	4,258	freight steamer, passenger s	freight, passengers	36	0	36
Standella	6,197	tanker	oil, spirits	58	6	52
Stangarth	5,966	steam freighter	government stores, general cargo	46	0	46
Texan	7,005	freighter	10,915 tons general cargo	47	9	38
Thomas McKean	7,191	steam ship	9,000 tons tanks, aircraft, food	60	5	55
Tonsbergfjor d	3,156	motor ship	tea, rubber	33	1	32

Tredinnick	4,589	steam freighter	military stores	30	30	0	
Triton	2,078	steam freighter	3,100 tons bauxite, 60 tons timber	30	6	30	
Tysa	5,327	motor ship	general cargo	43	0	43	
Umtata	8,141	freighter	2,000 tons mineral ore	92	0	92	
Vineland	5,587	steam freighter	ballast	35	1	34	34
Virginia Sinclair	6,151	steam freighter	66,211 barrels aviation gasoline	44	7	37	
Vivian P. Smith	130	schooner (two masted)	260 tons salt, live donkeys	11	0	11	11
W. D. Anderson	10,227	tanker	133,360 barrels crude oil	36	35	1	
Western Head	2,599	steam freighter	3,710 tons sugar	30	24	6	
William Cullen Bryant	7,176	steam / Liberty	10,962 tons raw sugar	54	0	54	
Willimantic	4,857	steam ship	ballast	38	6	32	
Yorkmoor	4,457	freighter	6,700 tons bauxite	46	0	46	
Total Tonnage	**690,928**		**Total Persons on Board**	**5,844**	**1,238**	**4,613**	**321**

Appendix 5: U-Boat Patrol Data: Type, Base, Commander, Duration, Flotilla

Type	Sub #	Date Enters	Rank	Commander	Age	Days	Flotilla	Departed	Return /Fate
IXB	123	1/23/42	Korvettenkapitän	Reinhard Hardegen	28	1	2nd	Lorient	Lorient
IXC	156	2/4/42	Korvettenkapitän	Werner Hartenstein	33	14	2nd	Lorient	Lorient
IXC	67	2/10/42	Kapitänleutnant	Günther Müller-Stöckheim	27	3	2nd	Lorient	Lorient
IXC	129	2/11/42	Korvettenkapitän	Nico Clausen	30	4	2nd	Lorient	Lorient
IXC	128	2/14/42	Kapitänleutnant	Ulrich Heyse	35	24	2nd	Lorient	Lorient
IXC	504	2/14/42	Korvettenkapitän	Hans-Georg Friedrich "Fritz" Poske	37	32	2nd	Lorient	Lorient
IXC	126	2/28/42	Kapitänleutnant	Ernst Bauer	28	18	2nd	Lorient	Lorient
Calvi	Finzi	3/2/42	Capitano di Fregata	Ugo Giudice		11	Betasom	Bordeaux	Bordeaux
Calvi	Tazzoli	3/6/42	Capitano di Corvetta	Carlo Fecia di Cossato	33	13	Betasom	Bordeaux	Bordeaux
IXC	161	3/15/42	Kapitänleutnant	Albrecht Achilles	28	8	2nd	Lorient	Lorient
Marcello	Morosini	3/15/42	Capitano di Corvetta	Athos Fraternale	33	16	Betasom	Bordeaux	Bordeaux
IXB	124	3/20/42	Kapitänleutnant	Johann Mohr	25	1	2nd	Lorient	Lorient
Calvi	Calvi	3/25/42	Capitano di Corvetta	Emilio Olivieri		3	Betasom	Bordeaux	Sunk - Biscay
IXC	154	4/3/42	Fregattenkapitän	Walther Kölle	34	13	2nd	Lorient	Lorient
IXB	123	4/5/42	Korvettenkapitän	Reinhard Hardegen	29	10	2nd	Lorient	Lorient

IXC	130	4/14/ 42	Korvetten kapitän	Ernst Kals	3 6	16	2nd	Lorient	Lorient
VII C	201	4/21/ 42	Kapitänleu tnant	Adalbert Schnee	2 8	3	1st	Brest	Brest
VII C	654	4/24/ 42	Oberleutn ant zur See	Ludwig Forster	2 6	5	1st	Brest	Brest
VII C	402	4/25/ 42	Korvetten kapitän	Siegfried von Forstner	3 1	4	3rd	St. Nazair e	St. Nazaire
IXB	109	4/26/ 42	Kapitänleu tnant	Heinrich Bleichrodt	3 2	18	2nd	Lorient	Lorient
IXC	125	4/26/ 42	Kapitänleu tnant	Ulrich Folkers	2 7	13	2nd	Lorient	Lorient
IXC	507	4/26/ 42	Korvetten kapitän	Harro Schacht	3 4	14	2nd	Lorient	Lorient
IXB	108	4/27/ 42	Korvetten kapitän	Klaus Scholtz	3 4	22	2nd	Lorient	Lorient
VII C	98	4/28/ 42	Korvetten kapitän	Herbert Schulze	3 2	20	7th	St. Nazair e	St. Nazaire
VII C	333	4/28/ 42	Kapitänleu tnant	Peter Erich "Ali" Cremer	3 1	13	3rd	La Rochel le	La Rochell e
IXC	506	4/28/ 42	Kapitänleu tnant	Erich Würdemann	2 8	15	10th	Lorient	Lorient
VII C	582	4/28/ 42	Korvetten kapitän	Werner Schulte	2 9	1	1st	Brest	Brest
VII C	564	4/30/ 42	Kapitänleu tnant	Reinhard "Teddy" Suhren	2 6	23	1st	Brest	Brest
VII C	69	5/3/4 2	Kapitänleu tnant	Ulrich Graf	2 6	5	7th	St. Nazair e	St. Nazaire
IXB	103	5/8/4 2	Kapitänleu tnant	Werner Winter	3 0	13	2nd	Lorient	Lorient
IXB	106	5/13/ 42	Kapitänleu tnant	Hermann Rasch	2 7	16	2nd	Lorient	Lorient
IXC	502	5/13/ 42	Kapitänleu tnant	Jurgen von Rosenstiel	2 9	2	2nd	Lorient	Sunk - Biscay
VII C	751	5/14/ 42	Kapitänleu tnant	Gerhard Bigalk	3 3	8	7th	St. Nazair e	St. Nazaire
VII C	558	5/15/ 42	Kapitänleu tnant	Günther Krech	2 7	10	1st	Brest	Brest

VII C	753	5/15/ 42	Korvetten kapitän	Alfred Manhardt von Mannstein	3 4	14	3rd	La Rochel le	La Rochell e
IXB	107	5/22/ 42	Kapitänleu tnant	Harald Gelhaus	2 6	13	2nd	Lorient	Lorient
IXC	504	5/22/ 42	Korvetten kapitän	Hans-Georg Friedrich "Fritz" Poske	3 7	9	2nd	Lorient	Lorient
IXC	158	5/24/ 42	Kapitänleu tnant	Erwin Rostin	3 4	10	10th	Lorient	Sunk - Bermu da
IXC	172	6/2/4 2	Kapitänleu tnant	Carl Emmermann	2 7	8	10th	Lorient	Lorient
IXC	68	6/3/4 2	Kapitän zur See	Karl-Friedrich Merten	2 6	4	2nd	Lorient	Lorient
IXC	159	6/3/4 2	Kapitänleu tnant	Helmut Witte	2 7	6	10th	Lorient	Lorient
IXC	157	6/6/4 2	Korvetten kapitän	Wolf Henne	3 6	8	2nd	Lorient	Sunk - Key West
IXC	129	6/9/4 2	Kapitänleu tnant	Hans-Ludwig Witt	3 2	21	2nd	Lorient	Lorient
IXC	67	6/10/ 42	Kapitänleu tnant	Günther Müller-Stöckheim	2 7	15	2nd	Lorient	Lorient
VII C	584	6/15/ 42	Kapitänleu tnant	Joachim Deecke	2 9	6	1st	Brest	Brest
Cal vi	Finzi	6/20/ 42	Capitano di Fregata	Ugo Giudice		38	Betas om	Bordea ux	Bordea ux
IXC	154	6/25/ 42	Fregattenk apitän	Walther Kölle	3 4	8	2nd	Lorient	Lorient
VII C	203	6/25/ 42	Kapitänleu tnant	Rolf Mützelburg	2 9	8	1st	Brest	Lorient
VII C	332	6/27/ 42	Kapitänleu tnant	Johannes Liebe	2 8	6	3rd	La Pallice / La Rochel le	La Pallice / La Rochell e
IXC	153	6/28/ 42	Fregattenk apitän	Wilfried Reichmann	3 6	8	2nd	Lorient	Sunk - Panam a

Marcello	Morosini	6/28/42	Tenente di Vascello	Francesco D'Alessandro		30	Betasom	Bordeaux	Sunk - Biscay
IXC	505	6/30/42	Korvettenkapitän	Axel-Olaf Loewe	33	5	2nd	Lorient	Lorient
VIIC	571	6/30/42	Korvettenkapitän	Helmut Möhlmann	29	19	3rd	La Pallice / La Rochelle	La Pallice / La Rochelle
VIIC	575	6/30/42	Kapitänleutnant	Günther Heydemann	28	7	7th	St. Nazaire	St. Nazaire
VIIB	84	7/1/42	Kapitänleutnant	Horst Uphoff	25	29	1st	Brest	Brest
VIIC	134	7/1/42	Korvettenkapitän	Rudolf Schendel	28	33	3rd	La Rochelle	La Rochelle
VIIC	437	7/1/42	Korvettenkapitän	Werner-Karl Schulz	31	13	6th	St. Nazaire	St. Nazaire
IXC	166	7/7/42	Oberleutnant zur See	Hans-Günther Kuhlmann	28	9	10th	Lorient	Sunk - New Orleans
IXC	161	7/11/42	Kapitänleutnant	Albrecht Achilles	28	8	2nd	Lorient	Lorient
IXC	171	7/13/42	Kapitänleutnant	Günther Pfeffer	27	15	10th	Kiel	Sunk - Biscay
IXC	173	7/16/42	Kapitän zur See	Heinz-Ehler Beucke	38	10	2nd	Lorient	Lorient
Oceanico	Giuliani	7/16/42	Capitano di Fregata	Giovanni Bruno		10	Betasom	Bordeaux	Santander
IXC	509	7/21/42	Fregattenkapitän	Karl-Heinz Wolff	32	5	10th	Lorient	Lorient
IXC	508	7/26/42	Kapitänleutnant	Georg Staats	26	19	10th	Kiel	Lorient
VIIC	658	7/29/42	Kapitänleutnant	Hans Senkel	32	30	6th	Kiel	St. Nazaire
IXC	510	7/30/42	Korvettenkapitän	Karl Neitzel	39	7	10th	Lorient	Lorient

VII C	564	8/3/42	Kapitänleutnant	Reinhard "Teddy" Suhren	26	6	1st	Brest	Brest
VII C	98	8/5/42	Korvettenkapitän	Wilhelm Schutze	33	7	7th	St. Nazaire	St. Nazaire
VII C	654	8/6/42	Oberleutnant zur See	Ludwig Forster	26	7	1st	Lorient	Sunk - Panama
VII C	598	8/7/42	Korvettenkapitän	Gottfried Holtorf	30	21	6th	St. Nazaire	St. Nazaire
VII C	600	8/8/42	Kapitänleutnant	Bernard Zurmuhlen	33	17		Kiel	La Rochelle
IXC	511	8/11/42	Kapitänleutnant	Freidrich Steinhoff	33	13	10th	Kiel	Lorient
IXC	164	8/12/42	Korvettenkapitän	Otto Fechner	36	4	10th	Kiel	Lorient
VII C	553	8/12/42	Korvettenkapitän	Karl Thurmann	32	9	3rd	St. Nazaire	St. Nazaire
IXC	163	8/13/42	Fregattenkapitän	Kurt-Eduard Engelmann	29	12	10th	Kiel	Lorient
VII D	217	8/13/42	Kapitänleutnant	Kurt Reichenbach Klinke	25	4	9th	Kiel	Brest
VII C	558	8/19/42	Kapitänleutnant	Günther Krech	27	7	1st	Brest	Brest
VII C	94	8/22/42	Oberleutnant zur See	Otto Ites	24	6	7th	St. Nazaire	Sunk - Haiti
VII B	86	8/27/42	Kapitänleutnant	Walter Schug	31	6	1st	Brest	Brest
VII C	455	9/21/42	Korvettenkapitän	Hans-Heinrich Giessler	31	5	7th	St. Nazaire	St. Nazaire
IXC	154	11/5/42	Fregattenkapitän	Heinrich Schuch	36	3	2nd	Lorient	Lorient
VII D	214	1/19/43	Kapitänleutnant	Günther Reeder	27	4	9th	Brest	Brest
IXC	68	2/23/43	Oberleutnant zur See	Albert Lauzemis	24	13	2nd	Lorient	Lorient

IXC/40	183	2/24/43	Fregattenkapitän	Heinrich Schäfer	36	10	2nd	Lorient	Lorient
IXC	155	2/27/43	Korvettenkapitän	Adolf Cornelius Piening	32	19	10th	Lorient	Lorient
IXC/40	185	3/3/43	Kapitänleutnant	August Maus	28	18	10th	Lorient	Lorient
IXC	129	4/16/43	Kapitänleutnant	Hans-Ludwig Witt	33	15	2nd	Lorient	Lorient
IXC	176	4/29/43	Korvettenkapitän	Rainer Dierksen	35	17	10th	Lorient	Sunk - Cuba
IXC	66	5/27/43	Kapitänleutnant	Friedrich Markworth	28	60	2nd	Lorient	Lorient
IXC/40	527	5/28/43	Kapitänleutnant	Herbert Uhlig	27	16	10th	Lorient	Sunk - Azores
VIIC	759	6/29/43	Kapitänleutnant	Rudolf Friedrich	28	7	9th	Brest	Sunk - Haiti
IXC	159	7/4/43	Oberleutnant zur See	Heinz Beckman	30	3	10th	Lorient	Sunk - Haiti
VIIC	732	7/5/43	Oberleutnant zur See	Klaus-Peter Carlsen	23	24	1st	Brest	Brest
VIIB	84	7/6/43	Kapitänleutnant	Horst Uphoff	26	15	1st	Brest	Sunk - Bermuda
VIIC	615	7/9/43	Kapitänleutnant	Ralph Kapitsky	26	2	3rd	La Rochelle	Sunk - Caribbean
VIIC	134	7/10/43	Kapitänleutnant	Hans-Günther Brosin	26	21	3rd	La Rochelle	Sunk - Biscay
VIIC	634	7/13/43	Oberleutnant zur See	Eberhard Dahlhaus	22	19	9th	Brest	Sunk - Azores
IXC/40	190	7/19/43	Kapitänleutnant	Max Wintermeyer	29	6	2nd	Lorient	Lorient
IXB	107	8/18/43	Oberleutnant zur See	Volker Simmermacher	24	22	2nd	Lorient	Lorient
IXC	518	9/16/43	Kapitänleutnant	Friedrich-Wilhelm Wissmann	28	25	2nd	Lorient	Lorient
VIID	214	9/26/43	Oberleutnant zur See	Rupprecht Stock	27	4	9th	Brest	Brest

IXC /40	193	11/17 /43	Fregattenk apitän	Hans Pauckstadt	2 7	19	2nd	La Rochel le	El Ferrol
IXC	129	11/18 /43	Oberleutn ant zur See	Richard von Harpe	2 6	27	2nd	St. Nazair e	Lorient
IXC	516	12/25 /43	Kapitänleu tnant	Hans-Rutger Tillesen	2 9	2	10th	Brest	Lorient
IXC	518	2/28/ 44	Kapitänleu tnant	Friedrich-Wilhelm Wissmann	2 8	9	2nd	Lorient	Lorient
IXC	154	2/29/ 44	Oberleutn ant zur See	Gerth Gemeiner	2 5	11	2nd	Lorient	Lorient
IXC /40	170	3/17/ 44	Kapitänleu tnant	Gunther Pfeffer	2 9	33	10th	Lorient	Lorient
VII D	218	3/30/ 44	Kapitänleu tnant	Richard Becker	3 3	7	9th	Brest	Brest
VII C	541	4/6/4 4	Kapitänleu tnant	Kurt Petersen	2 5	30	10th	Lorient	Lorient
IXC /40	539	6/1/4 4	Kapitänleu tnant	Hans-Jürgen Lauterbach-Emdem	2 4	15	10th	St. Nazair e	Flensb urg
IXC	518	8/28/ 44	Oberleutn ant zur See	Hans-Werner Offermann	2 3	3	2nd	Kristia nsand	Flensb urg

Appendix 6: Commanders of U-Boats & Italian Submarines, Bio-Data, Awards

Sub #	Patrol #	Commander	Fate/DOD	Awards	Hits	Tons
68	4 of 10	Merten	5/2/93	Knights Cross with Oak Leaves, U-Boat War Badge with Diamonds	27	170,151
172	2 of 6	Emmermann	3/25/90	Knights Cross with Oak Leaves, War Merit Cross 2nd Class with Swords, U-Boat Front Clasp	26	152,080
109	5 of 9	Bleichrodt	1/9/77	Knights Cross	24	151,260
124	8 of 11	Mohr	4/2/43	Knights Cross with Oak Leaves	27	129,976
155	5 of 10	Piening	5/15/84	Knights Cross	25	126,664
159	2 of 5	Witte	10/3/05	Knights Cross	23	119,554
123	7 of 12	Hardegen	2018	Knights Cross with Oak Leaves, U-Boat War Badge with Diamonds	21	112,447
123	8 of 12	Hardegen	2018	[see above]	21	112,447
126	3 of 6	Bauer	3/12/88	Knights Cross	24	111,564
108	2 of 11	Scholtz	5/1/87	Knights Cross with Oak Leaves	24	111,546
130	3 of 6	Kals	11/2/79	Knights Cross	17	111,249
158	2 of 2	Rostin	6/30/42	Knights Cross and U-Boat War Badge 1939	17	101,321
129	5 of 9	Witt	2/13/80	Knights Cross, U-Boat Front Clasp	19	100,773
129	7 of 9	Witt	2/13/80	Knights Cross, U-Boat Front Clasp	19	100,773
156	2 of 5	Hartenstein	3/8/43	Knights Cross	20	97,504
564	5 of 9	Suhren	8/25/84	Knights Cross with Oak Leaves and	18	95,544

				Crossed Swords, War Merit Cross 2nd Class with Swords		
564	6 of 9	Suhren	8/25/84	[see above]	18	95,544
558	7 of 10	Krech	6/3/2000	Knights Cross	17	93,186
558	8 of 10	Krech	6/3/2000	[see above]	17	93,186
201	6 of 9	Schnee	11/4/82	Knights Cross with Oak Leaves	21	90,189
Tazzo li	7 of 9	Fecia di Cossato	8/27/44	Iron Cross	16	86,535
128	2 of 6	Heyse	11/19/70	Knights Cross	12	83,639
125	4 of 7	Folkers	5/6/43	Knights Cross	17	82,873
203	6 of 11	Mützelburg	9/11/42	Knights Cross with Oak Leaves	19	81,961
103	7 of 11	Winter	9/9/72	Knights Cross	15	79,302
502	4 of 4	Rosenstiel	7/5/42	U-Boat War Badge 1939	14	78,843
106	6 of 10	Rasch	6/10/74	Knights Cross	12	78,553
504	2 of 7	"Fritz" Poske	10/1/44	Knights Cross	15	78,123
504	3 of 7	Poske	10/1/44	[see above]	15	78,123
507	2 of 4	Schacht	1/13/43	Knights Cross	19	77,143
94	10 of 10	Ites	2/2/82	Knights Cross	15	76,882
129	4 of 10	Clausen	5/16/43	Knights Cross	24	74,807
508	1 of 6	Staats	11/12/43	Knights Cross	14	74,087
66	8 of 9	Markworth		Knights Cross	13	74,067
402	3 of 8	Forstner	10/13/43	Knights Cross	15	71,036
506	2 of 5	Würdemann	7/12/43	Knights Cross	14	69,893
185	2 of 3	Maus	9/28/96	Knights Cross	9	62,761
553	8 of 10	Thurmann	1/20/43	Knights Cross	12	61,390
161	2 of 6	Achilles	9/27/43	Knights Cross	12	60,107
161	3 of 6	Achilles	9/27/43	Knights Cross	12	60,107
176	3 of 3	Dierksen	5/14/43	German Cross in Gold	11	53,307
518	4 of 7	Wissmann			8	52,346
518	5 of 7	Wissmann			8	52,346
332	4 of 7	Liebe	10/18/82	Iron Cross First Class	8	46,729

582	1 of 4	Schulte	10/5/42		7	38,872
505	3 of 12	Loewe	12/18/84		7	37,832
154	4 of 8	Schuch	1/21/68		6	37,641
575	5 of 10	Heydemann	1/2/86	Knights Cross	8	36,010
Moro sini	6 of 7	Fraternale	7/30/63	Silver Medal of Military Valor "Sul Campo"	5	35,600
516	4 of 6	Tillesen	6/1/86	German Cross in Gold	7	34,632
571	5 of 10	Möhlmann	4/17/77	Knights Cross U-Boat Front Clasp	5	33,511
154	2 of 8	Kölle	1/10/92		7	31,352
154	3 of 8	Kölle	1/10/92		7	31,352
84	5 of 8	Uphoff	8/7/43	German Cross in Gold	6	29,905
84	8 of 8	Uphoff	8/7/43	[see above]	6	29,905
Finzi		Giudice			5	29,031
600	1 of 6	Zurmuhlen	11/25/43		5	28,600
68	6 of 10	Lauzemis	4/10/44	German Cross in Gold	5	27,302
615	4 of 4	Kapitsky	8/7/43		4	27,231
333	2 of 12	"Ali" Cremer	7/5/92	Knights Cross with Wounded Badge in Silver and U-Boat Front Clasp	6	26,873
753	4 of 7	Mannstein	5/13/43		3	23,117
170	3 of 4	Pfeffer	1/25/66		4	22,304
171	1 of 1	Pfeffer	4/25/66	U-Boat War Badge 1939	2	22,304
511	1 of 4	Steinhoff	5/19/45	U-Boat War Badge 1939	2	21,999
Finzi		Giudice			3	21,496
751	6 of 7	Bigalk	7/17/42	Knights Cross	5	21,412
584	6 of 10	Deecke	10/31/43	German Cross in Gold	4	18,684
214	4 of 10	Reeder	Alive	German Cross in Gold	3	18,266
654	3 of 4	Forster	8/22/42	Iron Cross 2nd Class	3	17,755
654	4 of 4	Forster	8/22/42	[see above]	3	17,755

129	9 of 10	Harpe	3/2/45	German Cross in Gold	3	17,362
67	3 of 7	Müller-Stöckheim	7/16/43	Knights Cross	13	17,138
67	4 of 7	Müller-Stöckheim	7/16/43	[see above]	13	17,138
69	9 of 11	Graf	2/17/43		6	16,627
153	2 of 2	Reichmann	7/3/42		3	16,186
Giuliani		Bruno			3	16,103
163	1 of 3	Engelmann	3/13/43	U-Boat War Badge 1939	3	15,011
510	1 of 7	Neitzel	11/13/66	Knights Cross	3	14,128
455	4 of 10	Giessler		Iron Cross 1st Class	2	13,908
759	2 of 2	Friedrich	7/15/43		2	12,764
134	6 of 9	Schendel	4/12/70		3	12,147
658	1 of 2	Senkel	10/30/42		3	12,146
217	1 of 3	Klinkc	6/5/43		3	10,651
193	2 of 3	Pauckstadt	8/14/84		1	10,172
107	6 of 13	Gelhaus	12/2/97	Knights Cross U-Boat Front Clasp	19	10,068
86	4 of 8	Schug	11/23/43		3	9,614
598	1 of 4	Holtorf	7/23/43		2	9,295
183	2 of 6	Schäfer	1/8/44		2	8,582
164	1 of 2	Fechner	1/6/43	U-Boat War Badge 1939	3	8,133
166	3 of 3	Kuhlmann	7/30/42		4	7,593
107	10 of 13	Simmermacher	8/27/99		2	7,324
634	3 of 3	Dahlhaus	8/30/43	Iron Cross 1st Class	1	7,176
190	2 of 6	Wintermeyer	Alive	U-Boat War Badge 1939	1	7,015
157	2 of 2	Henne	6/13/42		1	6,401
Morosini	7 of 7	D'Alessandro			1	5,327
527	2 of 2	Uhlig	10/20/97		1	5,242
Calvi		Olivieri			1	5,162
518	6 of 7	Offermann	4/22/45	German Cross in Gold	1	3,401

539	5 of 7	Lauterbach-Emdem		German Cross in Gold	1	1,517
541	2 of 4	Petersen		German Cross in Gold	1	214
214	7 of 10	Stock		German Cross in Gold	1	200
218	7 of 10	Becker		U-Boat Front Clasp	1	146
98	7 of 9	Schulze	Alive			
134	9 of 9	Brosin	8/24/43	U-Boat War Badge 1939		
154	7 of 8	Gemeiner	7/3/44			
159	5 of 5	Beckman	7/28/43	Iron Cross 1st Class		
173	2 of 2	Beucke	5/23/79			
437	3 of 11	Schulz	11/20/60			
509	1 of 4	Wolff	6/11/70			
732	2 of 3	Carlsen		Iron Cross 1st Class		
98	8 of 9	Schutze				
			Average Age: 30, Ship Hits: 10, Tons Sunk Per Skipper: 50,000	***Total Ships Sunk & Tonnage***	***989***	***4,996,550***

Appendix 7: Allied Ships Sunk By Date Of Attack, Submarine, Commander

#	Ship Name	Flag	Date	Submarine	Sub. Commander
1	Pan Massachusetts	US	2/19/42	U-128	Ulrich Heyse
2	Cities Service Empire	US	2/22/42	U-128	Ulrich Heyse
3	Republic	US	2/22/42	U-504	Hans-Georg Friedrich Poske
4	Mamura	Dutch	2/26/42	U-504	Hans-Georg Friedrich Poske
5	W. D. Anderson	US	2/26/42	U-504	Hans-Georg Friedrich Poske
6	Macgregor	UK	2/27/42	U-156	Werner Hartenstein
7	Oregon	US	2/28/42	U-156	Werner Hartenstein
8	Gunny	Norway	3/2/42	U-126	Ernst Bauer
9	Rapana	UK	3/3/42	Tazzoli	Carlo Fecia Di Cossato
10	Mariana	US	3/5/42	U-126	Ernst Bauer
11	O. A. Knudsen	Norway	3/5/42	U-128	Ulrich Heyse
12	Astrea	Dutch	3/6/42	Tazzoli	Carlo Fecia Di Cossato
13	Melpomene	UK	3/6/42	Finzi	Ugo Giudice
14	Skane	Sweden	3/6/42	Finzi	Ugo Giudice
15	Tonsbergsfjord	Norway	3/6/42	Tazzoli	Carlo Fecia Di Cossato
16	Barbara	US	3/7/42	U-126	Ernst Bauer
17	Cardonia	US	3/7/42	126	Ernst Bauer
18	Montevideo	Uruguay	3/8/42	Tazzoli	Carlo Fecia Di Cossato
19	Hanseat	US	3/9/42	U-126	Ernst Bauer
20	Charles Racine	Norway	3/10/42	Finzi	Ugo Giudice
21	Cygnet	Panama	3/10/42	Tazzoli	Carlo Fecia Di Cossato
22	Manaqui	UK	3/11/42	Morosini	Athos Fraternale
23	Olga	US	3/11/42	U-126	Ernst Bauer
24	Texan	US	3/11/42	U-126	Ernst Bauer
25	Colabee	US	3/13/42	U-126	Ernst Bauer
26	Daytonian	UK	3/13/42	Tazzoli	Carlo Fecia Di Cossato
27	Athelqueen	UK	3/15/42	Tazzoli	Carlo Fecia Di Cossato
28	Oscilla	Dutch	3/15/42	Morosini	Athos Fraternale

29	Stangarth	UK	3/16/42	U-504	Hans-Georg Friedrich Poske
30	Peder Bogen	UK	3/23/42	Morosini	Athos Fraternale
31	Tredinnick	UK	3/25/42	Calvi	Emilio Olivieri
32	Gulfstate	US	4/3/42	U-155	Adolf Cornelius Piening
33	Comol Rico	US	4/4/42	U-154	Walter Kölle
34	Catahoula	US	4/5/42	U-154	Walter Kölle
35	Kollskegg	Norway	4/6/42	U-754	Hans Oestermann
36	Esso Baton Rouge	US	4/8/42	U-123	Reinhard Hardegen
37	Oklahoma	US	4/8/42	U-123	Rienhard Hardegen
38	Esparta	US	4/9/42	U-123	Reinhard Hardegen
39	Grenanger	Norway	4/11/42	U-130	Ernst Kals
40	Gulfamerica	US	4/11/42	U-123	Reinhard Hardegen
41	Esso Boston	US	4/12/42	U-130	Ernst Kals
42	Korlsholm	Sweden	4/13/42	U-123	Rienhard Hardegen
43	Leslie	US	4/13/42	U-123	Rienhard Hardegen
44	Vineland	Canada	4/20/42	U-154	Walter Kölle
45	Derryheen	UK	4/22/42	U-201	Adalbert Schnee
46	San Jacinto	US	4/23/42	U-201	Adalbert Schnee
47	Mobiloil	US	4/29/42	U-108	Klaus Scholtz
48	Federal	US	4/30/42	U-507	Harro Schacht
49	La Paz	UK	5/1/42	U-109	Heinrich Bleichrodt
50	Laertes	Dutch	5/3/42	U-109	Heinrich Bleichrodt
51	Ocean Venus	UK	5/3/42	U-564	Reinhard Suhren
52	Sama	Honduras	5/3/42	U-506	Erich Würdemann
53	Eclipse	UK	5/4/42	U-564	Reinhard Suhren
54	Abgara	Latvian	5/5/42	U-108	Klaus Scholtz
55	Afoundria	US	5/5/42	U-108	Klaus Scholtz
56	Delisle	US	5/5/42	U-564	Reinhard Suhren
57	Amazone	Dutch	5/6/42	U-333	Peter-Erich Cremer
58	Halsey	US	5/6/42	U-333	Peter-Erich Cremer
59	Java Arrow	US	5/6/42	U-333	Peter-Erich Cremer
60	Ohioan	US	5/8/42	U-564	Reinhard Suhren
61	Lubrafol	Panama	5/9/42	U-564	Reinhard Suhren
62	Clane Skene	UK	5/10/42	U-333	Peter-Erich Cremer
63	Cape Of Good Hope	UK	5/11/42	U-502	Jürgen Von Rosenstiel
64	Potrero Del Llano	Mexico	5/14/42	U-564	Reinhard Suhren
65	Nicarao	US	5/16/42	U-751	Gerhard Bigalk

66	Fauna	Dutch	5/17/42	U-558	Günther Krech
67	Darina	UK	5/20/42	U-158	Erwin Rostin
68	Frank B. Baird	Canada	5/22/42	U-158	Erwin Rostin
69	Lammot Du Pont	US	5/23/42	U-125	Ulrich Folkers
70	Yorkmoor	UK	5/28/42	U-506	Erich Würdemann
71	Western Head	UK	5/29/42	U-107	Harald Gelhaus
72	Fred W. Green	UK	5/31/42	U-506	Erich Würdemann
73	Domino	US	6/2/42	U-753	Alfred Mannhard Von Manstein
74	Illinois	US	6/2/42	U-159	Helmut Witte
75	Triton	Dutch	6/2/42	U-558	Günther Krech
76	City Of Alma	US	6/3/42	U-172	Carl Emmermann
77	Delfina	US	6/5/42	U-172	Carl Emmermann
78	Letitia Porter	US	6/5/42	U-69	Ulrich Graf
79	Esso Gettysburg	US	6/10/42	U-66	Friedrich Markworth
80	L. A. Christiansen	Norway	6/10/42	U-129	Hans-Ludwig Witt
81	Hagan	US	6/11/42	U-157	Wolf Henne
82	Hardwicke Grange	UK	6/12/42	U-129	Hans-Ludwig Witt
83	Managua	Nicaragua	6/16/42	U-67	Gunther Müller Stöckheim
84	Millinocket	US	6/17/42	U-129	Hans-Ludwig Witt
85	Cheerio	US	6/19/42	U-107	Harald Gelhaus
86	Willimantic	UK	6/24/42	U-156	Werner Hartenstein
87	Anglo-Canadian	UK	6/25/42	U-153	Wilfried Reichmann
88	Pedrinhas	Brazil	6/26/42	U-203	Rolf Mützelberg
89	Putney Hill	UK	6/26/42	U-203	Rolf Mützelberg
90	Potlatch	US	6/27/42	U-153	Wilfried Reichmann
91	Raphael Semmes	US	6/28/42	U-332	Johannes Liebe
92	Sam Houston	US	6/28/42	U-203	Rolf Mützelberg
93	Sea Thrush	US	6/28/42	U-505	Axel-Olaf Loewe
94	Everalda	Latvian	6/29/42	U-158	Irwin Rostin
95	Onondaga	US	6/29/42	U-129	Hans-Ludwig Witt
96	Thomas McKean	US	6/29/42	U-505	Axel-Olaf Loewe
97	Tysa	Dutch	6/30/42	Morosini	Athos Fraternale
98	Bloody Marsh	US	7/2/42	U-66	Friedrich Markworth
99	Norlandia	US	7/4/42	U-575	Günther Heydemann
100	Umtata	US	7/7/42	U-571	Helmut Möhlmann
101	James A. Moffett Jr.	US	7/8/42	U-571	Helmut Möhlmann

102	Carmen	Dom. Rep.	7/11/42	U-166	Hans-Günther Kuhlmann
103	Onieda	US	7/11/42	U-166	Hans-Günther Kuhlmann
104	Fairport	US	7/16/42	U-161	Albrecht Achilles
105	Gertrude	US	7/16/42	U-166	Hans-Günther Kuhlmann
106	William Cullen Bryant	US	7/21/42	U-84	Horst Uphoff
107	Andrew Jackson	US	7/23/42	U-84	Horst Uphoff
108	Ruth	US	7/29/42	U-153	Wilfried Reichmann
109	Maldonado	Uruguay	8/2/42	U-510	Karl Nietzel
110	Vivian P. Smith	UK	8/10/42	U-600	Bernhard Zurmühlen
111	Manzanillo	Cuba	8/12/42	U-508	Georg Staats
112	Medea	Dutch	8/12/42	U-658	Hans Senkel
113	Santiago De Cuba	Cuba	8/12/42	U-508	Georg Staats
114	Delmundo	US	8/13/42	U-600	Bernhard Zurmühlen
115	Empire Corporal	UK	8/13/42	U-598	Gottfried Holtorf
116	Everelza	Latvian	8/13/42	U-600	Bernhard Zurmühlen
117	Michael Jebsen	UK	8/13/42	U-598	Gottfried Holtorf
118	Standella	UK	8/13/42	U-598	Gottfried Holtorf
119	Helen Forsey	UK	9/6/42	U-514	Hans-Jürgen Auffermann
120	Peiping	Sweden	9/9/42	U-66	Friedrich Markworth
121	Patrick J. Hurley	US	9/13/42	U-512	Wolfgang Schultze
122	James Sprunt	US	3/10/43	U-185	August MaUS
123	Virginia Sinclair	US	3/10/43	U-185	August MaUS
124	Melbourne Star	UK	4/2/43	U-129	Hans-Ludwig Witt
125	Santa Catalina	US	4/24/43	U-129	Hans-Ludwig Witt
126	John Sevier	US	5/6/43	U-185	August MaUS
127	Mambi	Cuba	5/13/43	U-176	Rainer Dierksen
128	Nickeliner	Honduras	5/13/43	U-176	Rainer Dierksen
129	Cherry Valley	US	7/22/43	U-66	Friedrich Markworth
130	Pillory	Panama	6/5/44	U-539	Hans-Jürgen Lauterbach-Emden

Appendix 8: Merchant Ship Losses by Number of Killed in Action (KIA) Per Ship

Melbourne Star	113
James Sprunt	69
Esso Gettysburg	57
Mamura	56
Stangarth	46
Tredinnick	46
Gulfstate	43
Mariana	36
Ruth	35
W. D. Anderson	35
Illinois	32
L. A. Christiansen	30
City Of Alma	29
Barbara	26
Pillory	25
Western Head	24
Colabee	23
Everelza	23
Mambi	23
Manzanillo	23
Onondaga	20
Pan Massachusetts	20
Gulfamerica	19
Lammot Dupont	19
Oklahoma	19
Raphael Semmes	19
Gulftrade	18
Laertes	18
Patrick J. Hurley	17
Ohioan	15
Amazone	14
Cities Service Empire	14
Gunny	14
Montevideo	14
San Jacinto	14
Lubrafol	13
Potrero Del Llano	13
Millinocket	11
Santiago De Cuba	11
Clan Skene	9
Korlsholm	9
Norlandia	9
Stanbank	9
Texan	9
Delmundo	8
Nicarao	8
Oscilla	8
Potlatch	8
Sam Houston	8
Catahoula	7
Michael Jebsen	7
Virginia Sinclair	7
Darina	6
Empire Corporal	6
Hagan	6
Oneida	6
Oregon	6
Standella	6
Triton	6
Willimantic	6
Federal	5
Fred W. Green	5
Medea	5
Ocean Venus	5
Republic	5
Thomas McKean	5
Delfina	4
Leslie	4
Santa Rita	4
Andrew Jackson	3
Athelqueen	3
Bloody Marsh	3
Comol Rico	3
Esso Baton Rouge	3
Grenanger	3
Hardwicke Grange	3
Peiping	3
Putney Hill	3
Delisle	2
Eclipse	2
Fauna	2
Helen Forsey	2
Java Arrow	2
O. A. Knudsen	2
Anglo Canadian	1
Cardonia	1
Carmen	1
Daytonian	1
Esparta	1
James A. Moffett, Jr.	1
Macgregor	1
Norwalk	1
Olga	1
Vineland	1
94 ships	***1341***

Appendix 9: RAF Servicemen Killed While Stationed in Bahamas, Named

Names	Rank	Where	Date
Aitken, Philip Raymond	Flight Sergeant	Missing	5/21/1944
Alderton, Richard	Sergeant	Trinidad	7/10/1943
Anderson, Robert Venable	First Officer (Civilian)	Nassau	4/29/1943
Atkins-Tallentine, Thomas William Vincent	Flight Sergeant	Missing	5/21/1943
Bake, Charles Edward	Flight Sergeant	Trinidad	7/10/1943
Barber, Roland Henry	Flying Officer (Pilot) RAF VR	Nassau	10/13/1943
Barker, John Langford Desmond	Flying Officer RAF VR	Nassau	3/12/1944
Bartlett, Ernest Roy	Flying Officer (Pilot) RAF VR	Nassau	10/28/1944
Beaulieu, Arthur Joseph	Sergeant	Missing	6/26/1943
Beeston, Edward James	Leading Aircraftman RAF VR	Nassau	2/10/1945
Bentick, Raymond Cyril Walter	Sergeant (W.Op./Air Gnr.) RAF VR	Nassau	10/28/1944
Birch, John Raymond	Sergeant (W.Op./Air) RAF VR	Nassau	2/23/1945
Birkett, Laurence Arthur	Flying Officer (Nav.) RAF VR	Nassau	2/23/1945
Blackburn, Stanley	Pilot Officer	Georgetown, BG	4/11/1944
Blake, John Leslie Edward	Sargeant (Nav.) RAF VR	Nassau	9/13/1943
Boynonm David Edmund	Sergeant	Missing	4/21/1945
Brdskym Petr	Sergeant	Missing	11/23/1943
Briza, Vitezslav	Sargeant RAF VR	Nassau	7/8/1943
Bullock, Douglas James	Pilot Officer	Trinidad	10/24/1943
Burchell, Robert Lloyd	Sergeant	Missing	4/9/1943
Burchfield, Lowell Luther	First Officer (Civilian)	Missing	6/22/1943
Burke, Laurie Francis	Warrant Officer	Trinidad	11/22/1943
Butler, Joseph Harrison	Aircraftman 2nd Class RAF VR	Nassau	1/4/1945

Cameron, Kenneth John	Sergeant	Missing	4/24/1944
Cantley, William Grant	Flying Officer	Khartoum, SU	6/3/1944
Carberry, John	Sargeant (W. Op./Air) RAF VR	Nassau	10/28/1944
Carter, Geoffrey John Hamilton	Sergeant	St. Vincent Isl.	4/12/1943
Chouteau, Henri	Captain (Captain) RAF Transport Command	Nassau	5/30/1943
Clarke, Philip	Sergeant	Belem, Brazil	2/18/1944
Cleary, Alastair Grant	Pilot Officer	Missing	5/21/1943
Cochrane, Ronald Thomas	Flying Officer	Missing	2/22/1944
Colls, Peter Raymond	Pilot Officer	Belem, Brazil	9/19/1943
Conroy, Charles Ewart Roger	Leading Aircraftman	Missing	11/7/1943
Cormack, Douglas Waitte	Flying Officer	Missing	10/13/1943
Craig, Charles Fairweather	Sergeant	Missing	12/22/1943
Cram, John Samuel	Pilot Officer	Accra, Ghana	9/7/1943
Cribbes, George Deryk	Flying Officer (Pilot) RAF VR	Nassau	2/23/1945
Cummings, John Kilby	Captain (Civilian)	Accra, Ghana	9/7/1943
Daiken, Alex John	Pilot Officer RCAF	Nassau	3/19/1943
Daschuk, William	Sergeant	Missing	7/13/1943
Day, Edred Frederick	Leading Aircraftman RAF VR	Nassau	1/8/1943
Deemer, Amor Meade	Captain (Civilian)	Trinidad	11/22/1943
Deleveaux, O. E.	North Caribbean Force	Private	5/5/1945
Donen, Samuel Jacob	Sergeant	Accra, Ghana	9/7/1943
Dumble, David Storar	Pilot Officer	Missing	4/9/1945
Durham, Jack Grover	Captain (Civilian)	Missing	4/9/1943
Durrant, Ernest William	Sergeant	Missing	5/21/1943
Durward, Dennis	Flying Officer	Missing	10/13/1943
Fansworth, Ronald	Sergeant	Missing	4/9/1943
Faulkner, Lennox Dane	Captain (Civilian)	St. Vincent Isl.	4/12/1943
Fenton, George	Sergeant	Missing	4/9/1943

Flood, George Andrew	Flight Sergeant	Missing	1/17/1944
Foreman, Arthur Henry	Flying Officer (Pilot) RAF	Nassau	10/28/1944
Frye, Douglas Hibbert	Flying Officer	Missing	12/22/1943
Fuchs, Pavel	Sergeant	Missing	11/23/1943
Fussell, Dennis William	Sergeant	Missing	4/24/1944
Gale, David	Sergeant	Accra, Ghana	9/7/1943
Gamble, George	Flying Officer	Missing	11/24/1944
Gardner, William Alexander	Pilot Officer	Accra, Ghana	9/7/1943
Gedelian, Meg	Sergeant	Missing	5/21/1943
Glen, Ian Simpson	Sargeant (Pilot) RAF VR	Nassau	3/25/1945
Goodman, Donald	Pilot Officer RAF VR	Nassau	9/14/1943
Goulding, Anthony Edward	Flight Sergeant RAF VR	Nassau	2/10/1945
Gowans, Henry	Sergeant RAF VR	Nassau	10/16/1944
Grant, John Dickson	Flying Officer, DFM, BEM	Missing	4/9/1943
Guinter, Howard Watson	Sergeant	Missing	5/21/1943
Hadravek, Jan	Pilot Officer RAF VR	Nassau	7/8/1943
Hallett, Dennis Peter	Pilot Officer	Missing	4/9/1945
Hamlin, Cyril Reginald	Sergeant	Missing	4/21/1945
Hanney, Robert Aubrey	Sergeant	Missing	4/21/1945
Harris, Albert	Captain (Civilian)	Nassau	4/22/1943
Harris, Albert	Captain RAF Transport Command	Nassau	4/29/1943
Haskins, William	Aircraftman 2nd Class RAF VR	Nassau	6/21/1945
Hayes, Allan Edward	Flying Officer (Flying Instr.) RCAF	Nassau	2/23/1945
Herndon, Preston Krumbhaar	Captain (Civilian)	Missing	1/17/1944
Hodgins, Norman Joseph Bernard	Pilot Officer	Khartoum, SU	6/3/1944
Hole, John Francis Seager	Corporal RAF VR	Nassau	1/8/1943
Hole, John Francis Seager	Corporal	Nassau	8/1/1943
Holland, Alan	Flight Sergeant (Flt. Engr.) RAF VR	Nassau	2/23/1945
Hollowell, Donald Terence	Flight Officer (Pilot) RCAF, RAF VR	Nassau	3/19/1943

Hutchings, Bertie Warren	Flying Officer (Pilot) RAF VR	Nassau	2/23/1945
Hutchins, John Richard	Pilot Officer	Missing	2/22/1944
Hutchinson, Sydney William	Sergeant	Missing	4/9/1945
Huxtable, Robert Gerald	Warrant Officer	Khartoum, SU	6/3/1944
Jackson, Christopher	Sergeant RAF VR	Nassau	2/23/1945
James, Thomas David	Sergeant	Missing	7/13/1943
James, Walter Thomas David	Sergeant RAAF	Nassau	7/13/1943
Jeremiah, Donald	Squadron Leader RAF	Nassau	11/1/1944
Johnson, Wilfred	Pilot Officer	Missing	7/13/1943
Jones, G.	North Caribbean Force	Private	12/16/1944
Jones, William James	Flight Lieutenant RAF VR	Nassau	6/13/1944
Keayes, James Brian	Sergeant RAAF	Nassau	6/28/1943
Koepke, John Barth	Sergeant	Missing	7/13/1943
Korns	Lieutenant, USAM Corps USA	Nassau	1/26/1944
Krupica, Robert	Sergeant	Missing	11/23/1943
Lewis, Robert Stanley	Sergeant (W.Op./Air Gnr.) RAF VR	Nassau	10/28/1944
Lounsbury, Ronald Cyrus	Flight Sergeant	Accra, Ghana	9/7/1943
Lumb, Leonard Sylvester	Captain (Civilian)	Missing	11/7/1943
Lunam, Donald	Sergeant	Missing	4/21/1945
Lynes, Joseph Paul	Test Observer (Civilian) RAF Transport Command	Nassau	5/30/1943
Lytle, John Henry	Leading Aircraftman RAF VR	Nassau	4/7/1943
MacLean, John Scott	Flying Officer	Accra, Ghana	9/7/1943
MacRae, John	Sergeant	Accra, Ghana	7/7/1943
Marek, Jaroslav	Flight Lieutenant RAF VR	Nassau	7/8/1943
McArthur, James Brewer	Pilot Officer	Missing	7/13/1943
McLean, James	Pilot Officer	Missing	4/21/1943
McLean, John Charles	Pilot Officer	Missing	12/22/1943
McLean, Robert Forrester	Sergeant	Missing	4/21/1945
Meehan, Robert Paul	Sergeant	Missing	7/13/1943
Miller, Burton Craig	Captain (Civilian)	Missing	6/22/1943

Morris, William Frederick	Radio Officer RAF Ferry Command	Nassau	3/28/1943
Moule, Thomas	Sergeant	Missing	4/9/1945
Mudie, Reginald Edward	Warrant Officer	Missing	6/22/1943
Muirhead, William Alexander	Aircraftman 1st Class RAF VR	Nassau	10/2/1943
Munroe, Charles	North Caribbean Force	Nassau	8/13/1944
Newell, Geoffrey William	Leading Aircraftman RAF VR	Nassau	12/1/1943
Ogren, Carl Eric	Sergeant	Missing	6/28/1943
O'Neill, Maurice Francis	Flying Officer	Missing	10/17/1944
Ovens, John Alan	Pilot Officer	Missing	11/7/1943
Owen, John Griffith	Flight Lieutenant RAF VR	Nassau	10/13/1943
Page, Frederick Menzies	Flight Sargeant RAF VR	Nassau	1/31/1943
Pearce, Fred Archibald	Sergeant	Missing	4/24/1944
Perkins, Denis Clement Farmer	Flying Officer	Khartoum, SU	6/3/1944
Perry, Douglas William	Sergeant	St. Vincent Isl.	4/12/1943
Petersen, Bernard Norman	Flight Sergeant	Missing	12/22/1943
Picher, Harold Alfred	Radio Operator (Civilian)	Missing	4/9/1943
Poling, William John	Sergeant	Natal, SA	4/15/1944
Powell, Kenneth Sydney	Sergeant	Missing	4/24/1944
Probert, Kenneth F.	Pilot Officer	Georgetown, BG	4/11/1944
Rafuse, Cedric Gardiner	Flight Sergeant (W.Op./Air Gnr.) RCAF	Nassau	9/13/1943
Raine, Frank George	Flying Officer	Missing	5/20/1944
Rennick, Erle Donald	Flight Sergeant	Accra, Ghana	9/7/1943
Richards, Geoffrey Edward Charles	Sargeant RAF VR	Nassau	2/23/1945
Rowe, Gene	Captain (Civilian)	Belem, Brazil	2/11/1944
Salz, Z. Karel	Sargeant RAF VR	Nassau	7/8/1943
Savage, James R.	Sergeant Royal Canadian Pay Corps	Nassau	10/7/1943
Scammell, John William Thomas	Flying Officer	Missing	11/24/1944
Schafer, Gordon Wellsley	Flying Officer	Missing	4/9/1943

Scott, Edward	Flying Officer	Missing	4/9/1945
Seaforth, Stanley Henry	Sergeant	Missing	4/24/1944
Shelly, Arthur	Leading Aircraftman RAF VR	Nassau	3/21/1944
Simandl, Josef	Flying Officer	Missing	11/23/1943
Smith, Eric Ogilvy	Pilot Officer	Accra, Ghana	9/7/1943
Smith, Kenneth Mortimer	Sergeant	Missing	4/21/1945
Smith, William Francis	Pilot Officer	Missing	4/24/1944
Snart, Charles Leslie	Sergeant	Missing	6/28/1943
Somerville, Cyril J.	Radio Operator (Civilian) RAF Transport Command	Nassau	5/30/1943
Sones, Walter George	Leading Aircraftman RAF VR	Nassau	9/5/1945
Sotola, Josef	Sargeant RAF VR	Nassau	7/8/1943
Spiller, John William James	Sergeant	Missing	4/24/1944
Stinson, William Lorne	Pilot Officer	Missing	6/28/1943
Stubner, Richard Frank	Pilot Officer	Georgetown, BG	4/11/1944
Styblik, Miloslav	Warrant Officer	Missing	11/23/1943
Swire, William	Sergeant	Missing	12/22/1943
Thomas, John Ewart	Flying Officer	Missing	5/21/1943
Thomas, Peter Henry	Flight Sergeant	Missing	4/24/1944
Thomson, George Somerville	Radio Operator (Civilian) RAF Transport Command	Nassau	4/29/1943
Tickler, John E.	Flying Officer	Missing	4/21/1943
Todd, Eric Douglas	Pilot Officer	Missing	5/21/1943
Tomek, Hanus	Sergeant	Missing	11/23/1943
Tomlinson, Arthur Ian	Flying Officer (Nav.) RAF VR	Nassau	2/23/1945
Tompkins, George William	Sergeant	Missing	4/9/1943
Toone, Reginald Seymour	Flying Officer RAF VR	Nassau	12/28/1943
Torrens, Arthur William	Sergeant	Missing	6/28/1943
Turna, Josef	Sargeant RAF VR	Nassau	7/8/1943
Uden, Ronald Jack	Sergeant	Natal, SA	4/15/1944
Vallance, John	Pilot Officer	Missing	4/21/1943
Vickery, Philip Arthur	Radio Operator (Civilian)	Missing	6/22/1943
Wallace, Thomas	Sergeant	Missing	6/28/1943
Walmsley, Elwood Palmer	Captain (Civilian)	Nassau	3/28/1943

Waton, Alvin Bernard	Pilot Officer	Missing	4/9/1943
Weiben, Howard	Captain (Civilian)	Georgetown, BG	4/11/1944
Whiffin, David William	Pilot Officer	Missing	2/22/1944
Wild, William	Flight Sergeant	Missing	4/9/1945
Wilkins, Robert Archie	Flying Officer	Missing	1/17/1944
Willcox, Arthur Jack	Pilot Officer RAAF	Nassau	12/23/1943
Wilson, Russell Graham	Flight Sergeant	Trinidad	11/22/1943
Winkler, Louis	Leading Aircraftman	Trinidad	11/22/1943
Wood, John Walter	Flying Officer	Missing	10/17/1944
Woodeson, Leslie Somerville	Flying Officer (Nav.) RAF VR	Nassau	10/28/1944
Yeldon, Willis Redmond	Sargeant RCAF	Nassau	10/26/1942
Young, Eugene Gilbert Antoine	Radio Operator (Civilian)	Missing	11/7/1943
Zdan, Benjamin Henry	Flying Officer (Pilot) RCAF	Nassau	3/19/1943

Appendix 10: US Navy Servicemen Killed in the Bahamas Area, Warships, Airships, Aircraft

Ahlers, Julian W.	Sonarman	USS *Warrington*	9/23/1944
Aldinger, Robert C.	Fireman	USS *Warrington*	9/23/1944
Allen, Lloyd F.	Fireman	USS *Warrington*	9/23/1944
Anderson, Emil E.	Seaman	USS *Warrington*	9/23/1944
Andress, Norman I.	Boilerman	USS *Warrington*	9/23/1944
Andrews, Gerard H.	Chief Machinists Mate	USS *Warrington*	9/23/1944
Andrus, Melvin E.	Fireman	USS *Warrington*	9/23/1944
Arbogast, Eugene F.	Seaman	USS *Warrington*	9/23/1944
Bagley, Donald R.	Machinists Mate	USS *Warrington*	9/23/1944
Bailey, robert L.	Seaman	USS *Warrington*	9/23/1944
Bandkau, Marvin P.	Fireman	USS *Warrington*	9/23/1944
Banke, Herberet M.	Fireman	USS *Warrington*	9/23/1944
Barker, Stephen J.	Seaman	USS *Warrington*	9/23/1944
Basuini, Berni A.	Boilerman	USS *Warrington*	9/23/1944
Baugh, Bill H.	Seaman	USS *Warrington*	9/23/1944
Bennett, Chester A.	Fireman	USS *Warrington*	9/23/1944
Bennett, Roger	Seaman	USS *Warrington*	9/23/1944
Berger, Andrew L.	Fireman	USS *Warrington*	9/23/1944
Bienapfi, James F.	Seaman	USS *Warrington*	9/23/1944
Blatchford	KIA USN in Bermuda	FAA Supermarine Walrus R6551 of 773 Squadron	6/6/41
Bond, Fred W.	Machinists Mate	USS *Warrington*	9/23/1944
Bonstrom, Harold	Fireman	USS *Warrington*	9/23/1944
Boyd, Perley	Stewards Mate	USS *Warrington*	9/23/1944
Brady, thomas R.	Signalman	USS *Warrington*	9/23/1944
Brand, James A.	Seaman	USS *Warrington*	9/23/1944
Breeze, John D.	Fireman	USS *Warrington*	9/23/1944
Brown	KIA USN in Bermuda	US Navy Vought Kingfisher BuNo. 2207 of VS-5D4	9/23/1944
Brown, Robert M.	Seaman	USS *Warrington*	9/23/1944

Brown, William E.	Fireman	USS *Warrington*	9/23/1944
Burton, John W.	Fireman	USS *Warrington*	9/23/1944
Campbell, Cimuel P.	Stewards Mate	USS *Warrington*	9/23/1944
Carlin, William S.	Radio Technician	USS *Warrington*	9/23/1944
Chapman, Billy R.	Sonarman	USS *Warrington*	9/23/1944
Coleman, Sidney W.	Stewards Mate	USS *Warrington*	9/23/1944
Collier, Henry F.	Seaman	USS *Warrington*	9/23/1944
Cooper, Joseph N.	Stewards Mate	USS *Warrington*	9/23/1944
Craig, Chester M.	Seaman	USS *Warrington*	9/23/1944
Crawford, Lige	Stewards Mate	USS *Warrington*	9/23/1944
Cross, Clyde B.	Seaman	USS *Warrington*	9/23/1944
Davidson, Joseph S.	Seaman	USS *Warrington*	9/23/1944
Davis, Donald C.	Radioman	USS *Warrington*	9/23/1944
Davis, Kenneth S.	Lt., Sonar Officer (Temp.)	USS *Warrington*	9/23/1944
de Marco, Salvatore L.	Fire Controlman	USS *Warrington*	9/23/1944
DeLee, william	Machinists Mate	USS *Warrington*	9/23/1944
Denny, John Jr.	Lt Jg. Asst. Comms. Offcr	USS *Warrington*	9/23/1944
Dicken, John R.	Lt Jg. Asst 1st Lt.	USS *Warrington*	9/23/1944
Dillon, Edward t.	Radioman	USS *Warrington*	9/23/1944
Dillong, clarence, Jr.	Seaman	USS *Warrington*	9/23/1944
DiLorenzo, Benjamin M.	Seaman	USS *Warrington*	9/23/1944
DiNardi, Joeph	Seaman	USS *Warrington*	9/23/1944
Dotson, thomas E.	Seaman	USS *Warrington*	9/23/1944
Duncan, robert A.	Machinists Mate	USS *Warrington*	9/23/1944
Duvall, Louis, Jr.	Seaman	USS *Warrington*	9/23/1944
Easton, Elmore, R.	Seaman	USS *Warrington*	9/23/1944
Edwards, Dallas f.	Seaman	USS *Warrington*	9/23/1944
England, Samuel D.	Watertender	USS *Warrington*	9/23/1944
Engleman,Jack	Electricians Mate	USS *Warrington*	9/23/1944
Erickson, Jake M.	Seaman	USS *Warrington*	9/23/1944
Ermert, donald T.	Fireman	USS *Warrington*	9/23/1944
Estes, Ernest	Seaman	USS *Warrington*	9/23/1944
Everette, Lawrence M.	Seaman	USS *Warrington*	9/23/1944
Farkas, Joseph	Fireman	USS *Warrington*	9/23/1944

Feruson, Horace g., Jr.	Seaman	USS *Warrington*	9/23/1944
Flynn	KIA USN in Bermuda	FAA Supermarine Walrus R6551 of 773 Squadron	9/23/1944
forester, Paul W.	Fireman	USS *Warrington*	9/23/1944
Gates, Wendell M.	Torpedoman	USS *Warrington*	9/23/1944
Ginnett, Ernest, Jr.	Seaman	USS *Warrington*	9/23/1944
Gschwend, Frank P.	Yoeman	USS *Warrington*	9/23/1944
Hagan, Robert H.	Machinists Mate	USS *Warrington*	9/23/1944
Hamel, Harold J.	Fireman	USS *Warrington*	9/23/1944
Hammon, Edward J., jr.	Machinists Mate	USS *Warrington*	9/23/1944
Hammond, Robert d.	Watertender	USS *Warrington*	9/23/1944
Hart, John P.	Lt Jg. Comms. Offcr	USS *Warrington*	9/23/1944
Hays, Glenn W.	Seaman	USS *Warrington*	9/23/1944
Head, Albert F.	Storekeeper	USS *Warrington*	9/23/1944
Hedrick, Carle E.	Seaman	USS *Warrington*	9/23/1944
Heglin, william C.	Machinists Mate	USS *Warrington*	9/23/1944
Hill, Lynn A.	Fireman	USS *Warrington*	9/23/1944
Hines, clayston	Seaman	USS *Warrington*	9/23/1944
Hodges, Homer s.	Fireman	USS *Warrington*	9/23/1944
Hoffman, John B.	Seaman	USS *Warrington*	9/23/1944
Hoy, Kenneth E.	Seaman	USS *Warrington*	9/23/1944
Humphrey, robert b., Jr.	Seaman	USS *Warrington*	9/23/1944
Hunter, Larry E.	Seaman	USS *Warrington*	9/23/1944
Jenkiins, Talmadge C.	Seaman	USS *Warrington*	9/23/1944
Johnson, Charles S.	Seaman	USS *Warrington*	9/23/1944
Johnston, Harvey L.	Seaman	USS *Warrington*	9/23/1944
Johson, Glenn W.	Lt. Jg Torpedo Offcr	USS *Warrington*	9/23/1944
Jones, Vaul W.	Fireman	USS *Warrington*	9/23/1944
Jones, William J.	Seaman	USS *Warrington*	9/23/1944
Kantner, Charles r.	Seaman	USS *Warrington*	9/23/1944
Kantor, Morris,	Seaman	USS *Warrington*	9/23/1944
Kaufman, Roland	Fireman	USS *Warrington*	9/23/1944
Keith, Clifford L.	Torpedoman	USS *Warrington*	9/23/1944
Kellemen, Nick	Radioman	USS *Warrington*	9/23/1944
Kelm, Harold L.	Seaman	USS *Warrington*	9/23/1944
Kenary, Arthur T.	Seaman	USS *Warrington*	9/23/1944

Kennedy, Robert M.	Lt. Jg Medical Corps	USS *Warrington*	9/23/1944
Keppel, William V.	Lt. Jg. Engnr. Offcr	USS *Warrington*	9/23/1944
Kern, robert G	Fireman	USS *Warrington*	9/23/1944
Kilmer, Reuben G.	Seaman	USS *Warrington*	9/23/1944
King, Leroy F.	Seaman	USS *Warrington*	9/23/1944
King, Rhomas J.	Seaman	USS *Warrington*	9/23/1944
Kirby, Kenneth	Signalman	USS *Warrington*	9/23/1944
Kocken, John D.	Seaman	USS *Warrington*	9/23/1944
Koelling, elmber C.	Seaman	USS *Warrington*	9/23/1944
Kohn, James W.	Seaman	USS *Warrington*	9/23/1944
Kolfcycle, John J.	Seaman	USS *Warrington*	9/23/1944
Kreiser, Lester J.	Seaman	USS *Warrington*	9/23/1944
Kroll, Louis R.	Lt. Jg./sst,. Gunnery Ofcr.	USS *Warrington*	9/23/1944
Kua, John E.	Seaman	USS *Warrington*	9/23/1944
Kulkkula, Frank A.	Seaman	USS *Warrington*	9/23/1944
Kunath, Charles V.	Seaman	USS *Warrington*	9/23/1944
LaBuda, Edward J.	Chief Storekeeper	USS *Warrington*	9/23/1944
Laird, Doyle D.	Sonarman	USS *Warrington*	9/23/1944
LaLonde, Willard J.	Seaman	USS *Warrington*	9/23/1944
Lambert, clarence F.	Seaman	USS *Warrington*	9/23/1944
Lapp, Carl C.	Seaman	USS *Warrington*	9/23/1944
LaRussa, Paul J.	Seaman	USS *Warrington*	9/23/1944
Latimer, Ross G.	Fire Controlman	USS *Warrington*	9/23/1944
Lawing, George A.	Seaman	USS *Warrington*	9/23/1944
Lease, Milton E.	Radio Technician	USS *Warrington*	9/23/1944
LeFebvre, Leonard J.	Machinists Mate	USS *Warrington*	9/23/1944
Leggett, Donald L.	Yeoman	USS *Warrington*	9/23/1944
Locke	KIA USN in Bermuda	US Navy Vought Kingfisher BuNo. 2207 of VS-5D4	3/3/1942
Long, Ernest J.	Watertender	USS *Warrington*	9/23/1944
Luerssen, Thomas S.	Ensign, Asst. Engir Offcr	USS *Warrington*	9/23/1944
Lynn, Ted	Seaman	USS *Warrington*	9/23/1944
Maewski, Ralphy P.	Seaman	USS *Warrington*	9/23/1944
Mainzer, Junior L.	Seaman	USS *Warrington*	9/23/1944
Mallon, Thomas J.	Quartermaster	USS *Warrington*	9/23/1944
Marklund, Richard P.	Watertender	USS *Warrington*	9/23/1944
Martin, John D.	Quartermaster	USS *Warrington*	9/23/1944

Matson, Ernest S.	Seaman	USS *Warrington*	9/23/1944
Matthews Charles F.	Seaman	USS *Warrington*	9/23/1944
Maydan, Stephen	Machinists Mate	USS *Warrington*	9/23/1944
McLean, Lawrence T.	Stewards Mate	USS *Warrington*	9/23/1944
Meeke, Damon L.	Seaman	USS *Warrington*	9/23/1944
Meinze, Fred K.	Gunners Mate	USS *Warrington*	9/23/1944
Meitzel, Charles F.	Storekeeper	USS *Warrington*	9/23/1944
Melson, Glenn C.	Ships Cook	USS *Warrington*	9/23/1944
Minnik William r.	Fire Controlman	USS *Warrington*	9/23/1944
Miskimmon, Roy A.	Seaman	USS *Warrington*	9/23/1944
Mitchell, Alfred B.	Seaman	USS *Warrington*	9/23/1944
Moore, Robert B.	Lt. Jg., Asst. Enginr Offcr	USS *Warrington*	9/23/1944
Morgan, Clarence C.	Electricians Mate	USS *Warrington*	9/23/1944
Munoz, John	Seaman	USS *Warrington*	9/23/1944
neaville, Elvin K.	Chief Gunners Mate	USS *Warrington*	9/23/1944
Olson, Vernon A.	Ships Cook	USS *Warrington*	9/23/1944
O'Malley, Harry J.	Boilerman	USS *Warrington*	9/23/1944
O'Neal, James T.	Stewards Mate	USS *Warrington*	9/23/1944
Owens, William F.	Machinists Mate	USS *Warrington*	9/23/1944
Pack, Coleman S	Lt. Jg. Gunnery Offcr	USS *Warrington*	9/23/1944
Pardee, William H.	Chief Yeoman	USS *Warrington*	9/23/1944
Parlsey, William E.	Seaman	USS *Warrington*	9/23/1944
Parton, Raymond	Fireman	USS *Warrington*	9/23/1944
Patterson, Bert O.	Fireman	USS *Warrington*	9/23/1944
Pennington, J. Marvin	Lt. Jg. 1st Lt. & Damage Control	USS *Warrington*	9/23/1944
Peratt, Garret H.	Ships Cook	USS *Warrington*	9/23/1944
Peristerere, Victor H.	Pharmacists Mate	USS *Warrington*	9/23/1944
Perry, Howard B.	Machinistst Mate	USS *Warrington*	9/23/1944
Perry, Norman A.	Gunners Mate	USS *Warrington*	9/23/1944
Pesek, Lawrence J.	Yeoman	USS *Warrington*	9/23/1944
Pirtle, Linder G.	Radioman	USS *Warrington*	9/23/1944
Pool, Thomas O.	Seaman	USS *Warrington*	9/23/1944
Pratt, Kenneth H.	Gunners Mate	USS *Warrington*	9/23/1944
Preuster, William J.	Electricians Mate	USS *Warrington*	9/23/1944
Proven, John C.	Seaman	USS *Warrington*	9/23/1944
Pszczola, Harry	Seaman	USS *Warrington*	9/23/1944

Rayburg, Sylvester A., Jr.	Gunners Mate	USS *Warrington*	9/23/1944
Raymer, Clair K.	Coxwain	USS *Warrington*	9/23/1944
Redhead, Benjamin E.	Watertender	USS *Warrington*	9/23/1944
Reiner, Joseph F.	Electricians Mate	USS *Warrington*	9/23/1944
Reoch, William r.	Boatswains Mate	USS *Warrington*	9/23/1944
Reveille, Robert E.	Radarman	USS *Warrington*	9/23/1944
Reynolds, Howard T.	Shipfitter	USS *Warrington*	9/23/1944
Rice, Robert H.	Watertender	USS *Warrington*	9/23/1944
Richards, Donald R.	Seaman	USS *Warrington*	9/23/1944
Ricketts, Walter W.	Fire Controlman	USS *Warrington*	9/23/1944
Ridley, Colin N.	Turret Captain	USS *Warrington*	9/23/1944
Ripppa, Victor A.	Machinists Mate	USS *Warrington*	9/23/1944
Roberge, Raymond C.	Seaman	USS *Warrington*	9/23/1944
Roberts, Charles E.	Watertender	USS *Warrington*	9/23/1944
Rockford, Joseph H.	Gunners Mate	USS *Warrington*	9/23/1944
Roellich, Stanley H.	Seaman	USS *Warrington*	9/23/1944
Rogers, William L.	Ensign, Temp.	USS *Warrington*	9/23/1944
Romeo, Paul	Boatswains Mate	USS *Warrington*	9/23/1944
Ryder, Mark P.	Seaman	USS *Warrington*	9/23/1944
Santini, Luigi G.	Fireman	USS *Warrington*	9/23/1944
Schoenfelder, Herbert P.	Seaman	USS *Warrington*	9/23/1944
Schott, Russell J.	Torpedoman	USS *Warrington*	9/23/1944
Scott, Eddie	Stewards Mate	USS *Warrington*	9/23/1944
Seeback, william J.	Seaman	USS *Warrington*	9/23/1944
Seekford, roy E.	Seaman	USS *Warrington*	9/23/1944
Selking, Gordon K.	Seaman	USS *Warrington*	9/23/1944
Senior, Clifford S.	Seaman	USS *Warrington*	9/23/1944
Shepher, George L.	Machinists Mate	USS *Warrington*	9/23/1944
Sifford, Carl W.	Chief Pharmacists Mate	USS *Warrington*	9/23/1944
Simon, Joe L.	Seaman	USS *Warrington*	9/23/1944
Simone, Onofrio A.	Seaman	USS *Warrington*	9/23/1944
Smelesky, Melvin	Seaman	USS *Warrington*	9/23/1944
Smith, Clarence H.	Seaman	USS *Warrington*	9/23/1944
Smith, Cliffort L.	Torpedoman	USS *Warrington*	9/23/1944
Smith, olan C.	Seaman	USS *Warrington*	9/23/1944

Smith, Theodore S.	Seaman	USS *Warrington*	9/23/1944
Snoddy, ames L.	Watertender	USS *Warrington*	9/23/1944
Snow, William b.	Quartermaster	USS *Warrington*	9/23/1944
Sosnak, John E.	Watertender	USS *Warrington*	9/23/1944
Spencer, Joseph C., Jr.	Radioman	USS *Warrington*	9/23/1944
Springel, Walter	Seaman	USS *Warrington*	9/23/1944
Staats, William P.	Watertender	USS *Warrington*	9/23/1944
Steinbrecker, Crosley	Seaman	USS *Warrington*	9/23/1944
Stewart, roger	Seaman	USS *Warrington*	9/23/1944
Stone, Joseph W.	Chief Machinists Mate	USS *Warrington*	9/23/1944
Stout, Everett G.	Seaman	USS *Warrington*	9/23/1944
Stout, Ira	Coxwain	USS *Warrington*	9/23/1944
Stressel, Isadore	Aviation Machnt's Mate 3c	USS K-74, USN Blimp	9/23/1944
Strong, Layton W.	Seaman	USS *Warrington*	9/23/1944
Studdards, Lonnie P.	Seaman	USS *Warrington*	9/23/1944
Sublette, Paul E.	Seaman	USS *Warrington*	9/23/1944
Sulanke, Samuel E.	Gunners Mate	USS *Warrington*	9/23/1944
Sullivan, John R.	Ensign, Training	USS *Warrington*	9/23/1944
Swartz, David H.	Seaman	USS *Warrington*	9/23/1944
Swift, Hubert H.	Fireman	USS *Warrington*	9/23/1944
Szopa, Frank S.	Carpenters Mate	USS *Warrington*	9/23/1944
Ternyak, Stephen	Fireman	USS *Warrington*	9/23/1944
Thomas, Vernon E.	Seaman	USS *Warrington*	9/23/1944
Thompson, Clifford	Seaman	USS *Warrington*	9/23/1944
Tipton, John D.	Seaman	USS *Warrington*	9/23/1944
Tolman, Arthur B.	Chief Radioman	USS *Warrington*	9/23/1944
Trousdell, George H.	Machinists Mate	USS *Warrington*	9/23/1944
Tucholka, Gerard J.	Seaman	USS *Warrington*	9/23/1944
Tucker, Earl M.	Stewards Mate	USS *Warrington*	9/23/1944
Turner, Loren D.	Watertender	USS *Warrington*	9/23/1944
Tuttle, Jack S.	Fire Controlman	USS *Warrington*	9/23/1944
Underwood, William L.	Seaman	USS *Warrington*	9/23/1944
Upton, Harold	Ships Cook	USS *Warrington*	9/23/1944
van Hoesen Frank H.	Motor Machinists Mate	USS *Warrington*	9/23/1944

Vickers, Sidney M.	Lt. Jg. Supply Corps, Supply Officer	USS *Warrington*	9/23/1944
Vipperman, Joseph H.	Seaman	USS *Warrington*	9/23/1944
Voorhies, Wasshington L.	Fireman	USS *Warrington*	9/23/1944
Wagner, Jack W.	Seaman	USS *Warrington*	9/23/1944
Waid Vernon R.	Seaman	USS *Warrington*	9/23/1944
Walker, Leroy E.	Chief Machinists Mate	USS *Warrington*	9/23/1944
Welcker, Chalres S.	Gunners Mate	USS *Warrington*	9/23/1944
Williams, Theron	Seaman	USS *Warrington*	9/23/1944
Wilson, Herbert E.	Seaman	USS *Warrington*	9/23/1944
Wintjen, Gerald C.	Seaman	USS *Warrington*	9/23/1944
Wolosin, John	Machinists Mate	USS *Warrington*	9/23/1944
Wood, Andrew E.	Carpenters Mate	USS *Warrington*	9/23/1944
Woolsey, William J.	Seaman	USS *Warrington*	9/23/1944
Wright, Stanley, Jr.	Seaman	USS *Warrington*	9/23/1944
Yarmat, Harry A. (buried Nassau)	Seaman	USS *Warrington*	9/23/1944
Zarelli, Antonio L.	Radio Technician	USS *Warrington*	9/23/1944

Appendix 11: German Navy Killed in Action Aboard U-84, U-157 and U-176: 155 Persons

Amberg, Günther	Seaman 2cl.	U-157	13.06.1942
Där, Anton	Fireman 2cl.	U-84	07.08.1943
Barthel, Christoph	Warrant Quartermaster	U-157	13.06.1942
Bartjes, Erwin	Seaman 2cl.	U-157	13.06.1942
Becks, Wilhelm	Seaman 2cl.	U-176	15.05.1943
Behne, Karl	Fireman 3cl.	U-84	07.08.1943
Berndt, Dietrich	Fireman 1cl.	U-84	07.08.1943
Beuche, Hans	Fireman 1cl.	U-176	15.05.1943
Bogenhardt, Johannes	Lt. (j.g.) Reservist	U-84	07.08.1943
Bohlen, Hermann	Fireman 1cl.	U-157	13.06.1942
Böhler, Adolf	Radioman 3cl.	U-157	13.06.1942
Böhme, Werner	Seaman 2cl.	U-176	15.05.1943
Bopp, Heinrich	Fireman 1cl.	U-176	15.05.1943
Born, Egon-Walter	Apprentice Seaman	U-157	13.06.1942
Böse, Ernst	Warrant Quartermaster	U-84	07.08.1943
Bremer, Heinz-Georg-Christ	Fireman 1cl.	U-176	15.05.1943
Brocke, Karl-Johann	Radioman 2cl.	U-84	07.08.1943
Brosch, Erwin	Fireman 1cl.	U-176	15.05.1943
Bublitz, Kurt	Seaman 1cl. (Torpedoes)	U-176	15.05.1943
Buchholz, Hans	Apprentice Seaman	U-157	13.06.1942
Budzik, Anton	Seaman 2cl.	U-176	15.05.1943
Bugert, Wilhelm	Fireman 3cl.	U-176	15.05.1943
Buhmann, Otto	Fireman 3cl.	U-176	15.05.1943
Burghardt, Heinrich	Seaman 2cl.	U-176	15.05.1943
Burgun, Hermann-Josef	Fireman 2cl.	U-84	07.08.1943
Claussen, Hermann	Seaman 2cl.	U-84	07.08.1943
Daschner, Oswald	Machinist	U-157	13.06.1942
Dierksen, Reiner	Lieutenant	U-176	15.05.1943
Diesler, Gerhard	Seaman 1cl. (Torpedoes)	U-157	13.06.1942
Döpfer, Walter	Machinist	U-84	07.08.1943
Drexler, Ludwig	Seaman 2cl.	U-176	15.05.1943
Dubiel, Albrecht	Engineering Lieutenant	U-84	07.08.1943

Dunzik, Erwin	Seaman 2cl.	U-157	13.06.1942
Erhardt, Alfred	Fireman 1cl.	U-157	13.06.1942
Erlautzki, Kurt	Apprentice Seaman (Radio)	U-176	15.05.1943
Federschmidt, Hans-Joachim	Midshipman, active duty	U-176	15.05.1943
Fehse, Paul	Fireman 3cl.	U-157	13.06.1942
Finke, Hans	Machinist	U-84	07.08.1943
Fischer, Adolf-Otto-Martin	Apprentice Seaman	U-157	13.06.1942
Fischer, Kurt	Seaman 2cl.	U-176	15.05.1943
Föde, Willy	Fireman 2cl.	U-157	13.06.1942
Franke, Heinz	Lieutenant Commander	U-84	05.04.2003
Gabel, Wilhelm	Fireman 3cl.	U-176	15.05.1943
Garves, Hermann	Seaman 2cl.	U-84	07.08.1943
Giesecke, Herbert	Coxswain	U-84	07.08.1943
Göbel, Mathias	Fireman 1cl.	U-84	07.08.1943
Gronbach, Erwin	Seaman 2cl.	U-84	07.08.1943
Grunwald, Günther	Lieutenant (j.g.)	U-157	13.06.1942
Hahn, Joachim-Günther	Lieutenant (j.g.)	U-176	15.05.1943
Haller, Josef	Fireman 3cl.	U-176	15.05.1943
Hammers, Josef	Fireman 3cl.	U-84	07.08.1943
Harney, Klaus	Lieutenant Commander	U-84	01.09.1942
Hauck, Kurt	Seaman 2cl. (Torpedoes)	U-176	15.05.1943
Hebs, Kurt-Walter-Karl	Machinist	U-84	07.08.1943
Heinzelmann, Horst	Seaman 1cl. (Radio)	U-176	15.05.1943
Henne, Wolf	Lieutenant	U-157	13.06.1942
Henseler, Hermann	Fireman 3cl.	U-84	07.08.1943
Herpers, Heinrich	Fireman 3cl.	U-157	13.06.1942
Herrle, Friedrich-Georg	Senior Lieutenant Duties	U-84	04.05.1945
Herrmann, Konrad	Seaman 2cl.	U-157	13.06.1942
Hessler, Karl	Fireman 1cl.	U-176	15.05.1943
Hey, Hans	Junior Lieutenant	U-176	15.05.1943
Hildebrandt, Erwin	Fireman 2cl.	U-176	15.05.1943
Hoelz, Hans	Machinist	U-157	13.06.1942
Höftmann, Rudi	Fireman 3cl.	U-157	13.06.1942
Holbe, Viktor	Fireman 3cl.	U-157	13.06.1942
Issinger, Wilhelm	Apprentice Seaman (Radio)	U-157	13.06.1942
Iwan, Erwin	Fireman 1cl.	U-84	07.08.1943
Jäger, Georg	Seaman 1cl. (Radio)	U-84	07.08.1943
Kallweit, Heinz	Apprentice Seaman (Radio)	U-84	07.08.1943

Karl, Helmut	Fireman 3cl.	U-157	13.06.1942
Kattwinkel, Hans-Hermann	Midshipman, active duty	U-176	15.05.1943
Keiner, Kurt	Apprentice Seaman (Radio)	U-157	13.06.1942
Kiesewalter, Friedrich	Lt. (j.g.), Engineering Duties	U-157	13.06.1942
Kind, Wilhelm	Machinist	U-176	15.05.1943
Klauer, Walter	Fireman 1cl.	U-176	15.05.1943
Kleinen, Leonhard	Fireman 3cl.	U-176	15.05.1943
Kraft, Walter Werner.	Machinist	U-84	07.08.1943
Kreymeyer, Fritz	Fireman 3cl.	U-157	13.06.1943
Kubas, Erich	Fireman 2cl.	U-157	13.06.1942
Kuhlmann, Hermann	Fireman 3cl.	U-157	13.06.1942
Kunze, Erich-Richard	Seaman 2cl.	U-157	13.06.1942
Kurth, Josef	Warrant Quartermaster	U-176	15.05.1943
Lahnstein, Christian	Seaman 1cl. (Torpedoes)	U-84	07.08.1943
Lapöhn, Werner	Fireman 3cl.	U-157	13.06.1942
Launer, Johannes	Seaman 1cl. (Torpedoes)	U-84	07.08.1943
Leistner, Friedrich	Coxswain	U-157	13.06.1942
Liebisch, Oskar	Seaman 2cl.	U-84	07.08.1943
Liptorw, Engelbert	Fireman 3cl.	U-176	15.05.1943
Lobe, Dietrich	Lieutenant (j.g.)	U-84	07.08.1943
Lotze, Hans-Heinrich	Engineering Lieutenant	U-176	15.05.1943
Machinek, Hans	Coxswain	U-157	13.06.1942
Mack, Johann	Apprentice Seaman (Radio)	U-84	07.08.1943
Madel, Gustav	Fireman 3cl.	U-157	13.06.1942
Maltzahn von Frh., Hartwig	Ensign, Engineering Duties	U-176	15.05.1943
Männer, Georg	Fireman 3cl.	U-157	13.06.1942
Marczok, Rudolf	Fireman 3cl.	U-157	13.06.1942
Mayhack, Richard	Radioman 3cl.	U-176	15.05.1943
Mazur, Norbert	Fireman 3cl.	U-176	15.05.1943
Meissner, Heinrich	Seaman 1cl. (Radio)	U-176	15.05.1943
Meyer, Wilhelm	Machinist	U-84	07.08.1943
Mobius, Oskar	Seaman 1cl.	U-157	13.06.1942
Nieswand, Karl	Junior Lieutenant	U-84	07.08.1943
Oberer, Anton	Fireman 1cl.	U-176	15.05.1943
Ollech, Ernst	Seaman 2cl.	U-84	07.08.1943
Oppermann, Horst	Apprentice Seaman (Radio)	U-157	13.06.1942
Pfann, Peter	Fireman 2cl.	U-176	15.05.1943
Pistor, Adolf	Junior Lieutenant	U-157	13.06.1942

Portz, Otto	Coxswain	U-176	15.05.1943
Possl, Leonhard	Fireman 3cl.	U-84	07.08.1943
Rah, Otto.	Fireman 2cl.	U-176	15.05.1943
Rettig, Fritz	Fireman 1cl.	U-84	07.08.1943
Richter, Alfred	Fireman 1cl.	U-176	15.05.1943
Riediger, Siegfried	Machinist	U-84	07.08.1943
Römer, Helmut	Fireman 1cl.	U-176	15.05.1943
Rühl, Herbert-Karl	Machinist	U-157	13.06.1942
Sawitzki, Bruno	Fireman 3cl.	U-157	13.06.1942
Schaller, Werner	Fireman 2cl.	U-157	13.06.1942
Schlesinger, Gottfried	Fireman 1cl.	U-84	07.08.1943
Schmidt, Otto	Lt. (j.g.) Reservist	U-176	15.05.1943
Schmitz, Peter	Fireman 3cl.	U-176	15.05.1943
Schneider, Werner	Fireman 1cl.	U-84	07.08.1943
Schnorfeil, Alois	Apprentice Seaman (Radio)	U-84	07.08.1943
Schulze, Fritz	Fireman 1cl.	U-176	15.05.1943
Seidensticker, Alfred	Seaman 2cl.	U-176	15.05.1943
Selbing, Gerhard	Fireman 3cl.	U-176	15.05.1943
Sickert, Werner	Fireman 1cl.	U-176	15.05.1943
Sikowski, Max	Fireman 3cl.	U-157	13.06.1942
Solau, Rudi	Fireman 1cl.	U-176	15.05.1943
Sommer, Herbert	Seaman 2cl.	U-176	15.05.1943
Stehrenberg, Herbert	Machinist	U-176	15.05.1943
Stolpe, Helmut	Seaman 1cl. (Radio)	U-84	07.08.1943
Stramm, Ernst	Machinist	U-157	13.06.1942
Strässer, Werner	Fireman 2cl.	U-157	13.06.1942
Stumps, Rudolf	Apprentice Seaman (Radio)	U-157	13.06.1942
Teufel, Anton	Fireman 3cl.	U-157	13.06.1942
Thilo, Georg	Midshipman, active duty	U-157	13.06.1942
Ullrich, Fridolin-Willibald	Seaman 2cl.	U-84	07.08.1943
Uphoff, Horst	Lieutenant Commander	U-84	07.08.1943
Urban, Walter	Seaman 2cl.	U-84	07.08.1943
Voigt, Heini	Fireman 3cl.	U-157	13.06.1942
Voigt, Heinz	Fireman 3cl.	U-84	07.08.1943
Vonhof, Paul	Fireman 3cl.	U-157	13.06.1942
Voss Dr., Rolf	Fireman 1cl.	U-84	07.08.1943
Wagner, Hans	Seaman 2cl.	U-84	07.08.1943
Wallbaum, Otto	Pharmacist's Mate	U-84	03.01.1945
Weimar, Franz	Seaman 2cl.	U-157	13.06.1942

Weinberg, Gerhard	Fireman 1cl.	U-176	15.05.1943
Wilitzko, Hans	Fireman 1cl.	U-84	07.08.1943
Windeis, Kurt	Seaman 1cl. (Radio)	U-157	13.06.1942
Wisiorek, Josef	Seaman 2cl.	U-176	15.05.1943
Wittig, Heinz	Fireman 2cl.	U-176	15.05.1943
Woge, Edo	Fireman 3cl.	U-84	07.08.1943
Zimmermann, Hans	Seaman 2cl. (Radio)	U-157	13.06.1942
Zimniak, Josef	Fireman 3cl.	U-84	07.08.1943

Appendix 12: Allied Merchant Mariners Killed in Action in Bahamas in World War II Memorialized

Aarbakke, Olav Theodor Madsen	Chief Officer	*Cities Service Empire*
Aasted, Christian Severin	2nd Officer	*Illinois*
Acevedo, Pedro K.	Coal Passer	*Peiping*
Adams, George James	Carpenter	*Melbourne Star*
Adamson, Emma	Passenger	*Melbourne Star*
Agentinhe, Jose	Able Seaman, AB	*Peiping*
Agero, Felipe	Fire Watcher	*Ruth*
Aguero, Leandro Huerta	2nd Cook	*Manzanillo*
Aguinaldo, Robert A.	Steward	*W. D. Anderson*
Ahlberg, Theobert Andrew	Fireman, Wiper	*Colabee*
Ahlrichs, Henry Charles August	Chief Officer	*Mariana*
Ahmed, Ali	Donkeyman	*Stangarth*
Ahquai, Frank	Purser	*Esso Gettysburg*
Aitken, George	Steward's Boy	*Melbourne Star*
Alanis, Juan Carlos	*Rank Unknown*	*Montevideo*
Albrecht, Carl W.	Oiler	*Gulfamerica*
Albrektsen, Vidar	3rd Officer	*L. A. Christiansen*
Ali, Abdulla	Fireman & Trimmer	*Stangarth*
Ali, Mohamed Saif	Fireman & Trimmer	*Stangarth*
Ali, Ranem	Donkeyman	*Stangarth*
Ali, Wazid	Fireman	*Clan Skene*
Allen, Warren	Able Seaman, AB	*Western Head*
Aller, Joseph Bright	Able Seaman, AB	*Thomas McKean*
Almond, Charles William	Chief Refrig. Engineer	*Melbourne Star*
Alpine, Craven E.	Able Seaman, AB	*Pan Massachusetts*
Altman, Ove Jan	Gunner, Helmsman	*Grenanger*
Anderson, Andrew Dalrymple	Boatswain (Bosun)	*Stangarth*
Anderson, Arnold K.	Able Seaman, AB	*City of Alma*
Anderson, Ernest B.	Fire Watcher	*Lammot DuPont*
Anderson, Thomas A.	Able Seaman, AB	*Colabee*
Andrade, Caros A.	Cook	*Patrick J. Hurley*

Annoy, Konrad	Able Seaman, AB	*L. A. Christiansen*
Anondsen, Kristoffer	2nd Engineer	*L. A. Christiansen*
Anthony, Doris.	Passenger	*Barbara*
Anthony, Everett	Able Seaman, AB	*Thomas McKean*
Aravenia, Manuel	Oiler	*Oneida*
Arentzen, Bjarne	Able Seaman, AB	*L. A. Christiansen*
Argust, Marvin	Messman	*Esso Gettysburg*
Armstron, Carl L.	Able Seaman, AB	*Virginia Sinclair*
Armstrong, Wayne Albert	Seaman 1st Class	*Gulfstate*
Arron, Ramon Lago	1st Engineer	*Mambi*
Aseltine, Wesley Raymond	Seaman 1st Class	*James Sprunt*
Atkins, Henry Joseph	Assistant Engineer	*Melbourne Star*
Aubert, John B.	Messman	*Gulfamerica*
Auchterlounie, James	3rd Radio Officer	*Tredinnick*
Austin, Milton George	Seaman 1st Class	*James Sprunt*
Ayling, Peter Anthony	Cadet	*Melbourne Star*
Ayra, Adolfe D.	Oiler	*Mariana*
Baas, Maarten	3rd Engineer Officer	*Mamura*
Bachmann, Frederick	Able Seaman, AB	*W. D. Anderson*
Baker, Herman V.	Able Seaman, AB	*Oklahoma*
Baker, James Joshua	Master (Captain)	*City of Alma*
Balfour, John Eyres	Passenger (Engineer)	*Melbourne Star*
Ballard, Joseph E.	Radio Operator	*Barbara*
Balog, William Joseph	Seaman 1st Class	*James Sprunt*
Banks, J.	Deckhand	*Putney Hill*
Banks, Robert J.	Messman	*Gulfamerica*
Barik Abdul	Fireman & Trimmer	*Tredinnick*
Barlow, Harry Talbot	3rd Engineer	*Stangarth*
Barlow, William R.	Wiper	*Gulftrade*
Barnes, George J.	2nd Radio Officer	*Lubrafol*
Barnum, Carlos	Oiler	*Ruth*
Barrett, Gordon George	Master (Captain)	*Tredinnick*
Barriel, Armando Conejero	Engineer's Helper	*Manzanillo*
Bates, William	Donkeyman & Greaser	*Melbourne Star*
Baxley, Alfred	Fire Watcher	*Cardonia*
Baxley, James Arthur	Seaman 1st Class	*James Sprunt*
Baxter, Donald R.	Workaway	*Raphael Semmes*
Baylis, Robert M.	Fire Watcher	*Millinocket*
Beacham, Day O.	Ordinary Seaman, OS	*Gulfstate*
Beaton, Francis J.	Radio Operator	*Oregon*
Bech, Jørgen Henrik Walther Nielsen	2nd Engineer	*Melbourne Star*

Bech, Jørgen Henrik Walther Nielsen	2nd Engineer	*Michael Jebsen*
Beck, *First Names Unknown*	Passenger	*Delmundo*
Beck, George Jame	3rd Officer	*Stangarth*
Beck, Kenneth	Seaman 1st Class	*James Sprunt*
Beckering, Wilbert William	Seaman 1st Class	*James Sprunt*
Beech, Horace Edward	Seaman 2nd Class	*James Sprunt*
Beech, Norman	Gunner	*Stangarth*
Beecham, Charles Albert	4th Engineer	*Tredinnick*
Bell, George Laidler	AB, DEMS Gunner	*Stanbank*
Belvin, Robert F.	Boatswain (Bosun)	*Ruth*
Ben, Isaac	Fireman & Trimmer	*Tredinnick*
Bengtsson, Olaf F.	2nd Engineer	*Onondaga*
Berchem, Charles	Oiler	*City of Alma*
Beresheim, Charles John	3rd Officer	*Illinois*
Bergh, Joseph van den	Messman	*Lubrafol*
Bernabella, Adriano	Ship's Boy	*Amazone*
Bernard, Augustine	Fireman	*Western Head*
Betts, Eric	AB, Royal Navy	*Tredinnick*
Beverly, Ernest R.	Storekeeper	*Republic*
Biagorri, Pedro	*Rank Unknown*	*Montevideo*
Birch, Richard William	Gunner, DEMS	*Mamura*
Blaisdell, Edgar H.	Able Seaman, AB	*Esso Gettysburg*
Blandford, Harry	Chief Engineer	*Melbourne Star*
Bliles, Burwell E.	3rd Engineer	*W. D. Anderson*
Blundell, George Edward	Lance Bomber, DEMS	*Mamura*
Boblit, Clarence W.	Ordinary Seaman, OS	*James Sprunt*
Boehnicke, Paul W.	Chief Engineer	*Texan*
Bolger, John Kelly	Seaman 2nd Class	*Mambi*
Bollier, Parrie B.	Carpenter	*Ruth*
Bond, Arthur	Crew	*Helen Forsey*
Bongcak, Francisco	Fire Watcher	*Barbara*
Boon, Ulysses	Carpenter	*Amazone*
Bosom, Clarence John	Apprentice Seaman	*Cities Service Empire*
Bosworth, William Frederick	1st Officer	*Tredinnick*
Boulware, John R.	Wiper	*Gulfstate*
Bowen, Orlando C.	Seaman 1st Class	*Esso Gettysburg*
Bowles, John	Chief Cook	*Melbourne Star*
Boyaji, Charles James	Seaman 1st Class	*Esso Gettysburg*
Boyd, Joseph F.	Ordinary Seaman, OS	*Oklahoma*
Boyle, Joseph B	Oiler	*City of Alma*

Boyle, Michael Christopher	Refrigerator Greaser	*Melbourne Star*
Boyle, Thomas Aloysius	Seaman 1st Class	*James Sprunt*
Bradley, Albert Joseph	Able Seaman, AB	*Stangarth*
Bradley, Edward Joseph	Messman	*Colabee*
Bradshaw, William	Assistant Steward	*Melbourne Star*
Bragg, Elmus H.	Able Seaman, AB	*Patrick J. Hurley*
Bragg, Lenton	Able Seaman, AB	*Thomas McKean*
Breeding, Allan S.	Ordinary Seaman, OS	*Korlsholm*
Brongers, Albert Eppo Anton Reinder	Radio Operator	*Mamura*
Brothers, John J.	Coxswain	*Esso Gettysburg*
Broussard, Morris Joseph	Seaman 1st Class	*James Sprunt*
Brow, Jonathan	Messman	*Lammot DuPont*
Brown, John	Seaman 1st Class	*Esso Gettysburg*
Brown, Ralph	Utilityman	*Ruth*
Brown, Walter E.	Oiler	*Onondaga*
Brownrigg, Garrett M.	1st Engineer	*Catahoula*
Brun, August	Able Seaman, AB	*Cities Service Empire*
Bryant, William J.	Fireman, Wiper	*Esso Gettysburg*
Bryson, Edward T.	Oiler	*Esso Gettysburg*
Buchanan, Willard John	Able Seaman, AB	*Western Head*
Buck, Michael	Engineer Cadet	*James Sprunt*
Buguna, Jaime Baste	3rd Officer	*Mambi*
Bulawa, Chester	Able Seaman, AB	*Esso Gettysburg*
Bullard, Joseph L.	2nd Engineer	*Barbara*
Burke, John J.	Fire Watcher	*Potlatch*
Burleson, C. J.	Passenger	*Barbara*
Burns, Edward J.	Ordinary Seaman, OS	*Pan Massachusetts*
Burston, John Clement Kennedy	Galley Boy	*Darina*
Burton, William Chalmers	Chief Engineer	*Stangarth*
Bush, Norris	Boatswain (Bosun)	*Gulftrade*
Butler, John Albert	Lance Bomber, DEMS	*Mamura*
Butts, James A.	1st Engineer	*Santa Rita*
Bynaar, Marius	Sailor	*Amazone*
Caldwell, Edward James	Fireman, Wiper	*Illinois*
Calhoun, Bill Brandon	Seaman 1st Class	*Gulfstate*
Callis, Lewis Wesley	Master (Captain)	*Millinocket*
Callis, Robert Melville	Master (Captain)	*Ruth*
Cam, Toribo	Utilityman	*Ruth*
Camara, Lionel	Fireman, Wiper	*Oneida*

Cambrigianis, John G.	1st Engineer	*Ruth*
Camilleri, William	Assistant Steward	*Melbourne Star*
Campbell, Edmond Andrew	Oiler	*Illinois*
Campbell, Hansel William	Seaman 2nd Class	*James Sprunt*
Campbell, J.	Assistant Cook	*Putney Hill*
Campo, Joseph E.	Passenger	*Barbara*
Capello, Francisco	Sailor	*Amazone*
Capo, Anthony	Fireman, Wiper	*Illinois*
Caputo, John J.	Ordinary Seaman, OS	*Gulftrade*
Caram, Alberto	*Rank Unknown*	*Montevideo*
Carbery, Maurice	Fire Watcher	*Onondaga*
Carlson, Eorge S.	Messman	*Onondaga*
Carmona, Alfred	Wiper	*Oklahoma*
Carnes, Harold J.	Seaman 1st Class	*Esso Gettysburg*
Carr, Elie Constantine	Master (Captain)	*James Sprunt*
Carrol, Alfred R.	Boatswain (Bosun)	*Tredinnick*
Carter, John McCormick	Engineer Cadet	*Esso Gettysburg*
Carter, Shell H.	1st Officer	*Ruth*
Carye, Halton R.	Ordinary Seaman, OS	*W. D. Anderson*
Casal, Jose	Cook	*Raphael Semmes*
Casal, Roger A. Lorenzo	Radio Operator	*Manzanillo*
Case, Kenneth A.	Coxswain	*Mambi*
Castillo, Amalio Pio	*Rank Unknown*	*Montevideo*
Castillo, Ramon	Fire Watcher	*San Jacinto*
Castle, Martin P.	Able Seaman, AB	*Lammot DuPont*
Castro, Apolonio	Cook	*W. D. Anderson*
Castro, Jose Corripio	Steward	*Manzanillo*
Cato, Eulalia	Passenger	*San Jacinto*
Catterall, Henry	Greaser	*Hardwicke Grange*
Cento, Joe	Ordinary Seaman, OS	*Ohioan*
Chalker, Albert R.	Cook	*Esso Gettysburg*
Chandler, John H.	3rd Asst. Engineer	*James Sprunt*
Chinese, Name Unknown	Cook	*Oscilla*
Chinese, Name Unknown	Quartermaster	*Oscilla*
Chorman, Mattias J.	Messman	*Oklahoma*
Christensen, Emil	Able Seaman, AB	*Catahoula*
Christie, Alexander Cormack	Chief Electrician	*Melbourne Star*
Christophersen, John N.	1st Officer	*Gulfstate*
Chu Chan	Fireman	*Michael Jebsen*
Chu, Chan,	Fireman	*Melbourne Star*
Chuen Law Sui	Ship's Boy	*Michael Jebsen*

Chuen, Law Sui,	Ship's Boy	*Melbourne Star*
Cintron, Saturina	Passenger	*San Jacinto*
Claggett, Owen L.	Fire Watcher	*Sam Houston*
Clapham, Lancelot William	Able Seaman, AB	*Tredinnick*
Clare, Dennis Bernard	Assistant Steward	*Laertes*
Clark, Olan S.	Passenger	*Barbara*
Clayton, Atwood Addison	2nd Asst. Engineer	*Illinois*
Clayton, George Samuel, RN	AB, DEMS Gunner	*Melbourne Star*
Clemons, Richard Jr.	Seaman 1st Class	*Gulfstate*
Clifton, Fred	Oiler	*Delmundo*
Cluff, Winifred Victoria	Passenger	*Melbourne Star*
Cochrane, Leo	Fireman	*Western Head*
Colbert, Lewis Jack	Seaman 1st Class	*James Sprunt*
Collins, Isaac	Messman	*Nicarao*
Colomar Castelló, José	Messman	*Santiago De Cuba*
Colon, Ramona Mujica Rivera	Passenger	*San Jacinto*
Colpitts, Vaughan	Able Seaman, AB	*Western Head*
Conde, Jose	*Rank Unknown*	*Montevideo*
Conkle, George H.	Ordinary Seaman, OS	*Gulftrade*
Connell, Leslie Hubert	Cook	*Tredinnick*
Connell, William	Steward	*Tredinnick*
Constancia, Tito Martir	Able Seaman, AB	*Medea*
Conwell, William Y.	3rd Officer	*Lammot DuPont*
Cook, John	Boatswain (Bosun)	*Melbourne Star*
Cook, Leslie L.	Radio Operator	*Onondaga*
Cooper, John Frank	Apprentice	*Stangarth*
Cooper, Joseph Michael	1st Assistant Engineer	*Esso Baton Rouge*
Cooper, Richard A.	2nd Officer	*Raphael Semmes*
Copple, Hurssell T.	Ordinary Seaman, OS	*Oregon*
Corbin, Jack E.	Oiler	*Esso Gettysburg*
Corbin, William H.	2nd Engineer	*City of Alma*
Corron, Reon B.	Cook	*James Sprunt*
Cortes, Carlos	*Rank Unknown*	*Manzanillo*
Cortez, Antonio	Messman	*San Jacinto*
Cotton, James T. Jr.	Radio Operator	*W. D. Anderson*
Coulthard, Alec	Greaser	*Melbourne Star*
Cowin, Richard	Lead Seaman, DEMS	*Melbourne Star*
Cowling, K. T.	4th Engineer	*Putney Hill*
Cox, Carl Vernon	Messman	*Colabee*
Cox, Jeff	Fire Watcher	*W. D. Anderson*
Cox, Kenneth G.	Able Seaman, AB	*Gulfstate*

Craen, Gommaire	Boatswain (Bosun)	*Lubrafol*
Craig, Alexander	2nd Cook	*Korlsholm*
Crawford, Joseph W., Jr.	Wiper	*W. D. Anderson*
Crespo, Francisco	Fire Watcher	*City of Alma*
Crille, Jorge De La Torre	Cook's Helper	*Manzanillo*
Crisp, L. B.	Passenger	*Barbara*
Cuervo, Caimiro Martines	Fireman	*Manzanillo*
Cuesta, Pedro De La Fuente	Cook	*Mambi*
Cullinan, Thomas	Deck Maintainer	*W. D. Anderson*
Culmsee, Emil R.	2nd Asst. Engineer	*Korlsholm*
Curiel, Johannes Josef	Messman	*Texan*
Curran, John Michael	Passenger	*Melbourne Star*
Curry, Jack B.	Fire Watcher	*Pan Massachusetts*
Custer, Charles M.	Messman	*Gulfstate*
Da Silva, Jeronimo	Able Seaman, AB	*Triton*
Dagher, Said	Cook	*Colabee*
Dahl, Alexander	Carpenter	*Stangarth*
Dahlgren, Robert Stellan	Mechanic	*Peiping*
Daniels, Adrian E.	Chief Engineer	*Onondaga*
Daniels, Bruce B.	Able Seaman, AB	*Pillory*
Daniels, Elton M.	2nd Asst. Engineer	*Oneida*
Danley, Earl	Oiler	*Colabee*
Dansereau, Philip Hercules	Messman	*Republic*
David, Geronimo	Oiler	*Millinocket*
Davies, Frederick Willington	3rd Radio Officer	*Melbourne Star*
Davies, Henry Norman	4th Officer	*Mamura*
Davies, Richard Llewellyn	Able Seaman, AB	*Melbourne Star*
Davis, Joseph	Radio Operator	*Raphael Semmes*
Davis, Sam	Cook	*Gulftrade*
Day, Arthur	Utilityman	*Gulftrade*
De Giorgio, Vincent	Assistant Cook	*Stangarth*
De Silva Lima, Antonio	Fireman	*Pillory*
Dean, John C.	Chief Engineer	*Mariana*
DeAndrade, John	Messman	*Gulftrade*
Dearborn, Stewart Morton	Messman	*Illinois*
Debomfin, Jaoa J.	Fire Watcher	*Lammot DuPont*
DeBrito, Manuel R.	Cook	*Gulfamerica*
DeCarvalha, Antonio S.	Able Seaman, AB	*Pillory*
Decker, Lester P.	Seaman 1st Class	*Esso Gettysburg*
Decker, Theodore Percival	Boatswain (Bosun)	*Illinois*
Degraaf, Joost	Sailor	*Triton*
Deischer, Paul C.	Able Seaman, AB	*Pillory*

Dejanikus, Alexis	2nd Engineer	*Ruth*
Del Campo, Enrique	Ordinary Seaman, OS	*Ruth*
Denholm, George Roland	Gunner	*Stangarth*
Denniston, Gordon	Radio Operator	*Western Head*
DeRosa, Julio P.	Cook	*James Sprunt*
Derry, Alfred	AB, DEMS Gunner	*Stanbank*
Deveaux, Wilfred John	4th Engineer	*Western Head*
Diaz, Rafael	Wiper	*Santa Rita*
Diffenderfer, Clinton M.	3rd Engineer	*Gulfstate*
Dillon, James	Able Seaman, AB	*Onondaga*
Dion, Lewis J.	Wiper	*City of Alma*
Dissel	Fireman & Trimmer	*Tredinnick*
Dobbinga, Rink	Master (Captain)	*Mamura*
Doherty, Alan J.	Ordinary Seaman, OS	*City of Alma*
Dong, Nyok Chao	Donkeyman No. 2	*Laertes*
Donnon, Alfred C.	Cook	*Mariana*
Donohue, Lawrence E.	Radio Operator	*Ruth*
Donoso, Carlos A.	Wiper	*Santa Rita*
Dooley, Richard A.	Utilityman	*Oklahoma*
Doran, Alfonso G.	Steward	*Amazone*
Dorola, Jr., Roy C.	Able Seaman, AB	*Mambi*
Doughty, Ivan S.	Messman	*Esso Gettysburg*
Doval, Enrique Gonsalez	Seaman	*Manzanillo*
Drange, Lars O.	Fire Watcher	*Leslie*
Draper, James Henry	Gunner, DEMS Army	*Melbourne Star*
Dubienski, Joseph	3rd Officer (Extra)	*Korlsholm*
Duboise, Ollie M.	Able Seaman, AB	*Mambi*
Duby, Louis Joseph	Ordinary Seaman, OS	*Cities Service Empire*
Duhamel, Donald	Galley Boy	*L. A. Christiansen*
Dunn, Daniel D.	3rd Asst. Engineer	*Esso Gettysburg*
Dunning, Clyde A.	Oiler	*Sam Houston*
Dunville, Robert Henry	Junior 3rd Engineer	*Melbourne Star*
Dyczkowski, John L.	Oiler	*W. D. Anderson*
Earnhardt, Robert L.	Ordinary Seaman, OS	*Ruth*
Earranco, Kladio Arrandondo	Fireman	*Pillory*
Eary, John R.	Gunner's Mate 3rd Class	*Esso Gettysburg*
Eaton, George B.	2nd Asst. Engineer	*James Sprunt*
Eaton, Harold G.	Master (Captain)	*Raphael Semmes*
Ebblewhite, John	1st Officer	*Fred W. Green*
Ebbs, Carter R.	3rd Engineer	*Republic*
Edgar, Arlis D.	Ordinary Seaman, OS	*Oklahoma*

Eerola, Kalle	Able Seaman, AB	*Esso Gettysburg*
Ehlers, Heinrich	2nd Officer	*Pillory*
Eichelberger, B.F.	Fireman, Wiper	*James Sprunt*
Ekendahl, Bror	Fireman	*Gunny*
Elliott	Messroom Boy	*Tredinnick*
Ellis, G.	2nd Radio Officer	*Willimantic*
Ellwakeel, Elsayed M.	Fire Watcher	*Esparta*
Elmond, Max	Sailor	*Triton*
Elsey, Frederick	Assistant Steward	*Melbourne Star*
Emery, Cecil Jeffries	Cook	*Colabee*
Emery, Howard Irvin	Seaman 1st Class	*Esso Gettysburg*
Encinas, Anthony	Seaman 1st Class	*Esso Gettysburg*
English, Louis G.	Ordinary Seaman, OS	*Lammot DuPont*
English, Stanley Willard	3rd Asst. Engineer	*Colabee*
Engstrom, Leroy Axel	Seaman 1st Class	*Esso Gettysburg*
Engstrom, Martin	Fireman, Wiper	*James Sprunt*
Erazo, Francisco	Waiter	*Barbara*
Erickson, Richard	Messman	*Illinois*
Ernst, Axel P.	1st Assistant Engineer	*Pillory*
Escarza, Andres Arias	Steward	*Manzanillo*
Espinosa, Luis E. Aguero	Fireman	*Mambi*
Espinoso, Charles	Galley Boy	*Western Head*
Esteves, Jose M. Mayer	Oiler	*Mambi*
Evans, Aaron	5th Engineer	*Empire Corporal*
Evereret, Henry A	Steward	*Onondaga*
Everhart, Roy S.	Fire Watcher	*Norlandia*
Everingham, Amos	Crew	*L. A. Christiansen*
Fallen, Clifford Willis	Seaman 1st Class	*James Sprunt*
Fat, Chan	Crew	*Standella*
Faulkner, Joseph Everett	Master (Captain)	*Western Head*
Faux, Louise	Passenger	*Melbourne Star*
Fentress, Albert R.	3rd Engineer	*Java Arrow*
Fenty, Clarence F.	Steward	*Barbara*
Fernandes, Marinno	Fire Watcher	*Lammot DuPont*
Fernandez, Jose	Ordinary Seaman, OS	*Pillory*
Fernandez, Jose	Fire Watcher	*Republic*
Ferreira, Manuel	Fireman	*Pillory*
Ferry, Bernard John	Deck Maintainer	*W. D. Anderson*
Figueroa, Jose	Ordinary Seaman, OS	*Mariana*
Firth, David W.	Greaser	*Athelqueen*
Fisher, Stanley R.	Ordinary Seaman, OS	*Potlatch*
Fisher, William Cedrick	Fireman	*Western Head*

Flanagan, Alfred	Messman	*Esso Gettysburg*
Fleming, Tenant L.	1st Asst. Engineer	*Esso Gettysburg*
Flick, Elmer M.	2nd engineer	*Gulfstate*
Florek, Theodore	Fireman, Wiper	*Illinois*
Flynn, Patrick	Oiler	*Pan Massachusetts*
Foo, Ah Tsai	Fireman	*Laertes*
Fook, Wong	Crew	*Standella*
Ford, Alphonso	Messman	*Pillory*
Foreman, John	Able Seaman, AB	*Stanbank*
Foulger, John William	Signalman 3rd Class	*Everelza*
Fowler, L.D.	Coxswain	*Esso Gettysburg*
Frady, Carlos	Able Seaman, AB	*Pan Massachusetts*
Francqui, Jules E.	Clerk	*Norlandia*
Franklin, Billie J.	Able Seaman, AB	*W. D. Anderson*
Franklin, John Maurice	Apprentice	*Stangarth*
Fredriksson, Kurt	Ordinary Seaman, OS	*Gunny*
Freeman, Thomas	Fireman & Trimmer	*Tredinnick*
Froelich, Hilmar,	Steward	*Lammot DuPont*
Fuentes, Antonio	Waiter	*Barbara*
Fuentes, Antonio A.	Ordinary Seaman, OS	*Mariana*
Fullerton, Lloyd H.	2nd Asst. Engineer	*Esso Gettysburg*
Fulton, David Hugh	Ordinary Seaman, OS	*Illinois*
Funk, Russell C.	Wiper	*Thomas McKean*
Galbraith, James	Messroom Boy	*L. A. Christiansen*
Gallagher, Jon B.	Junior Engineer	*Patrick J. Hurley*
Gamel, Wilks D.	1st Officer	*Patrick J. Hurley*
Ganduglia, Alfredo Apolinario	*Rank Unknown*	*Montevideo*
Ganly, John Richard	Able Seaman, AB	*Mamura*
Garcia, Constantino	Fire Watcher	*Mariana*
Garcia, Francisco S.	Oiler	*Lammot DuPont*
Garcia, Manuel	1st Engineer	*Delfina*
Garcia, Roberto	Passenger	*San Jacinto*
Garcia, Rogelio	*Rank Unknown*	*Montevideo*
Gården, Ole Torsteinsen	Chief Officer	*Gunny*
Gawthorpe, George Edward	Ordinary Seaman, OS	*Tredinnick*
Gay, Gaston W.	Messman	*Mariana*
Gearin, John Francis	Fireman, Wiper	*Colabee*
Geary, Joseph	Cook	*Oklahoma*
Genter, Arthur J.	Oiler	*Oklahoma*
George, Garfield Edward	Oiler	*Western Head*
Gerald, Jesse	Messman	*Gulfamerica*
Gessey, John Albert Edwin	Steward's Boy	*Melbourne Star*

Ghaleb, Mokbil	Fireman & Trimmer	*Stangarth*
Gillespie, Donald Stubbs	Seaman 1st Class	*Esso Gettysburg*
Gilloegly, Harold M.	Cook	*Pan Massachusetts*
Glacklin, John J.	3rd Officer	*W. D. Anderson*
Glenday, George	Lance Sergeant, DEMS	*Melbourne Star*
Goddard, Joseph Orlando	Fireman & Trimmer	*Tredinnick*
Goffi, Antonio	Fireman	*Amazone*
Goldsmith, Dorothea Rosemary	Passenger	*Melbourne Star*
Goldstone, Arnold	Fire Watcher	*Norlandia*
Gomez, Lopez Ignacio	2nd Machinist	*Manzanillo*
Gonu Mian	Trimmer, Engine	*Clan Skene*
Gonzalez, Hipolito	Passenger	*Barbara*
Gonzalez, Jose Atiliano	Ship's Boy	*Montevideo*
Goodwin, William	Fireman & Trimmer	*Tredinnick*
Gooijers, Wilhelmus Cornelis	2nd cook	*Lubrafol*
Gorman, William P.	3rd Engineer	*Raphael Semmes*
Gough, Edward	Master (Captain)	*Clan Skene*
Gowen, John Leonard	Seaman 2nd Class	*Cities Service Empire*
Gpshaw, Alton	Wiper	*Gulfamerica*
Grace, Thomas Walter	Assistant Cook	*Tredinnick*
Graham, Joseph	Fire Watcher	*Mariana*
Graner, James P.	Fireman, Wiper	*Cities Service Empire*
Granner, Egon Jørgen	3rd Engineer	*Melbourne Star*
Granner, Egon Jørgen	3rd Engineer Officer	*Michael Jebsen*
Grassey, Louis E.	Ordinary Seaman, OS	*Oneida*
Gray, Henry R.	Cook	*Mariana*
Gray, J.	Ordinary Seaman, OS	*Darina*
Graziano, Vincent J.	Seaman 1st Class	*Esso Gettysburg*
Greenwood, Alexander	Able Seaman, AB	*Melbourne Star*
Grieve, William Harper	3rd Officer	*Tredinnick*
Griffin, Edward Thomas	Able Seaman, AB	*Stangarth*
Groot, Bartholomeus	4th Engineer	*Laertes*
Groot, Jakob de	2nd Officer	*Mamura*
Grove, Robert C.	Wiper	*James Sprunt*
Guardiola, Antonio	Fireman	*Santiago De Cuba*
Guaudet, Ralph C.	Cook	*Patrick J. Hurley*
Gubb, William Albert	Fireman	*Eclipse*

Guiden, James	Able Seaman, AB	*Lubrafol*
Gunnink, Berend Harmen	Gunner, Dutch Army	*Laertes*
Gustaffson, K.	Chief Engineer	*Korlsholm*
Gutierrez, Pedro	Steward	*Manzanillo*
Haake, Casper	4th Engineer	*Mamura*
Hale, William J.	Fire Watcher	*Gulfstate*
Hali	Assistant Steward	*Willimantic*
Hall, Anthony George	Cadet	*Melbourne Star*
Hall, Betram D.	Utilityman	*Gulfstate*
Hall, Edward Arthur Sidney	Fireman	*Fred W. Green*
Hall, Elmer	Cook	*Ruth*
Hall, Ham Maxwell	Fire Watcher	*Sam Houston*
Hall, James Bennett	Master (Captain)	*Melbourne Star*
Hall, Willard	Oiler	*Barbara*
Halle, Erling	Mechanic	*L. A. Christiansen*
Halvorsen, Asbjorn	Radio Operator	*L. A. Christiansen*
Halvorsen, Karl	Boatswain (Bosun)	*L. A. Christiansen*
Hamilton, John Norman	Junior 3rd Engineer	*Melbourne Star*
Hamlin, William M.	Fireman, Wiper	*James Sprunt*
Hammond, William F.	Able Seaman, AB	*Onondaga*
Hampson, Eric	Assistant Steward	*Melbourne Star*
Hance, Lastie	Fireman, Wiper	*Oklahoma*
Hancock, Homer	Deck Maintainer	*James Sprunt*
Hannigan, John Richard	Seaman 1st Class	*Esso Gettysburg*
Hansen, Hans	Mechanic	*L. A. Christiansen*
Hansen, Hans H.	Ordinary Seaman, OS	*L. A. Christiansen*
Hanson, James Lawrence	Oiler	*Vineland*
Hardy, William	Fireman	*Empire Corporal*
Harelson, Henry	Chief Engineer	*Millinocket*
Harlow, Eric Hartley	3rd Radio Officer	*Stangarth*
Harnesk, Sixten E.	Motorman	*Korlsholm*
Harrell, Charles Frank	Master (Captain)	*Gulfstate*
Harrington, Joseph R.	Ordinary Seaman, OS	*Lubrafol*
Harris, Arthur L.	Messman	*Esso Gettysburg*
Harrison, George	Fire Watcher	*Mariana*
Harrison, Robert O.	Able Seaman, AB	*James Sprunt*
Harry, Walker	Fireman	*Ocean Venus*
Hart, Robret W.	Master (Captain)	*San Jacinto*
Hartley, John	Apprentice	*Tredinnick*
Hartley, K.	3rd Officer	*Willimantic*
Harvey, Leroy L.	Oiler	*James Sprunt*
Hauer, Albert	Chief Engineer	*James Sprunt*
Haugh, James	2nd Refrig. Engineer	*Melbourne Star*
Hawkins, Frank	Messman	*Gulftrade*

Hawkins, Jesse L.	Able Seaman, AB	*Norlandia*
Hawkins, Ralph	Able Seaman, AB	*Nicarao*
Hay, Daniel Watson	Boatswain (Bosun)	*Colabee*
Hay, Warren Charles	Oiler	*Western Head*
Hayes, Robert	Assistant Steward	*Melbourne Star*
Hazelden, Herbert D.	Wiper	*Ohioan*
Healey, James Gallop	Storekeeper	*Melbourne Star*
Hedley, Peter John	Midshipman Cadet	*Laertes*
Hefter, Raymond A.	3rd Officer	*Gulfstate*
Heine, Alfred	Able Seaman, AB	*Patrick J. Hurley*
Heir, Alfred W.	Wiper	*W. D. Anderson*
Henandez, Sandalio Estanislao	*Rank Unknown*	*Montevideo*
Henderson, James M.	Able Seaman, AB	*Ruth*
Henderson, Kenneth B.	Fire Watcher	*Ohioan*
Hendricks, Wardwell T.	Steward	*City of Alma*
Henry, Morris	Steward	*Gulftrade*
Henshall, Clifford	Assistant Engineer	*Melbourne Star*
Herbert, William Ewart	Master (Captain)	*Stangarth*
Hermann, Howard I.	Messman	*James Sprunt*
Hernandez, Leon	Able Seaman, AB	*Mariana*
Herrera, Juan R.	Quartermaster	*Manzanillo*
Hickman, Alvis J.	Messman	*Sam Houston*
Higbee, Henry	Boatswain (Bosun)	*City of Alma*
Hilker, Herman	Messman	*Republic*
Hillaren, Nils	OS, Lookout	*Grenanger*
Hilliard, Robert Henry	Fireman, Wiper	*Colabee*
Hillman, Louis	Cook	*Oregon*
Hirschbert, Herman	Fire Watcher	*Onondaga*
Hoben, James Robert	3rd Officer	*Western Head*
Hodges, George Dewey	Master (Captain)	*Onondaga*
Hodges, Leslie H.	1st Officer	*Oneida*
Hogan, Christopher	Greaser	*Melbourne Star*
Hogge, Narrone	Passenger	*Barbara*
Holl, Donald Dean	Seaman 2nd Class	*Lammot DuPont*
Hollger, Carl Bernard	Oiler	*Esso Baton Rouge*
Holt, Thomas L.	Chief Engineer	*Patrick J. Hurley*
Holthe, O'Leif	Able Seaman, AB	*Lubrafol*
Hood, Stuart W.	3rd Engineer	*Andrew Jackson*
Hoppen, Willard J.	Ordinary Seaman, OS	*Esso Gettysburg*
Housley	Crew	*Anglo Canadian*
Howell, William L.	2nd Asst. Engineer	*Oklahoma*
Hudgins, Genius T.	2nd Officer	*Onondaga*

Hugendubler, Roque	Fire Watcher	*Raphael Semmes*
Hugghins, Cullen Roote	3rd Engineer	*Sam Houston*
Hughes, Henry	Messroom Boy	*L. A. Christiansen*
Hughes, John R.	Watchman	*Barbara*
Hughes, William N.	1st Engineer	*Raphael Semmes*
Huisman, Roelof	3rd Officer	*Mamura*
Hurlstone, Ivan Elroy	Master (Captain)	*Mariana*
Hurst, Harvey Francis	Messman	*Colabee*
Hutchinson, Ronald Ian William	Writer	*Melbourne Star*
Hutton, William	Boatswain (Bosun)	*James Sprunt*
Ingham, Jack	Fire Watcher	*Patrick J. Hurley*
Ingram, Louis H.	Steward	*Pan Massachusetts*
Inskip, Ralph Crookshank	2nd Steward	*Melbourne Star*
Iriarte, Eugenio Jose	Able Seaman, AB	*Mamura*
Irons, Floyd R.	Machinist	*Pan Massachusetts*
Irvin, Martin D.	Fireman, Wiper	*Esso Gettysburg*
Iseminger, Alva J.	Deck Maintainer	*W. D. Anderson*
Jackson, Frank M.	Fire Watcher	*Gulftrade*
Jackson, Joseph Raymond	Ordinary Seaman, OS	*Colabee*
Jackson, Keith Harold	Passenger (Sub-Lt.)	*Melbourne Star*
Jacobs, Joseph E.	Chief Engineer	*Esso Gettysburg*
Jaeger, Julius Lucien	Chief Engineer	*Mamura*
Jahnsen, Harald	1st Engineer	*L. A. Christiansen*
James, R. S.	Able Seaman, AB	*Empire Corporal*
Jameson, Norman Allen	Able Seaman, AB	*Stanbank*
Jany, Bisari	Able Seaman, AB	*City of Alma*
Jarve, Enial	2nd Asst. Engineer	*Pillory*
Jenkins, John H.	Cook	*Ruth*
Jenkins, William R.	Fire Watcher	*Virginia Sinclair*
Jensen, Neil	Fire Watcher	*W. D. Anderson*
Jensen, Rolf	Fireman	*Pillory*
Jerman, William Faucett Jr.	Master (Captain)	*Cities Service Empire*
Johannesen, Olaus Martin Gamst	Able Seaman, AB	*O. A. Knudsen*
Johannesen, Oman W.	Fire Watcher	*Virginia Sinclair*
Johnene, Russell Giter	Ordinary Seaman, OS	*Colabee*
Johnson, Botvid R.	2nd Engineer	*Gulftrade*
Johnson, Ernest O.	Oiler	*Gulftrade*
Johnson, Frank L.	Purser	*James Sprunt*
Johnson, Fred W.	3rd Engineer	*Ruth*
Johnson, Peder A.	Master (Captain)	*Esso Gettysburg*
Johnson, Raymond	Fire Watcher	*Texan*

Johnston, Owen J.	Oiler	*James Sprunt*
Johonnet, Fred W.	Pumpman	*Patrick J. Hurley*
Jok, Wong	Crew	*Standella*
Jon, Jose Hung	2nd Cook	*Mambi*
Jones, Bert	Able Seaman, AB	*James Sprunt*
Jones, English	cook	*Nicarao*
Jones, Harold	Able Seaman, AB	*Athelqueen*
Jones, Horace W.	3rd Asst. Engineer	*Gulfamerica*
Jones, Johnny	Fireman & Trimmer	*Tredinnick*
Jones, Loring, F.	Junior Officer	*Raphael Semmes*
Jones, Thomas	Able Seaman, AB	*Daytonian*
Jones, Tom, Greaser	Greaser	*Tredinnick*
Jones, William F.	Messman	*Ohioan*
Jones, William S.	3rd Asst. Engineer	*Pan Massachuset ts*
Jordan, Harold	Oiler	*Millinocket*
Jørgensen, Birger Mikal	Chief Engineer	*Gunny*
Joseph, August J.	Radio Operator	*Mariana*
Joseph, James	Messman	*Gulfstate*
Joseph, Louis	Cook	*Gulfstate*
Joyner, Harry C.	Pumpman	*Comol Rico*
Judge, James J.	Fire Watcher	*City of Alma*
Kain, Walter Louis	Seaman 1st Class	*James Sprunt*
Kalima, William K.	Messman	*W. D. Anderson*
Kalody, Walter F.	Seaman 1st Class	*Esso Gettysburg*
Kamont, Edward F.	Ordinary Seaman, OS	*Ohioan*
Kankowski, Stanley	Utilityman	*Colabee*
Kapat, Cyril	Deck Maintainer	*W. D. Anderson*
Kappus, Paul B.	Radio Operator	*Ohioan*
Karayan, Stephen	Messman	*Mariana*
Karklin, William	Able Seaman, AB	*Mariana*
Karlsen, Arne Johannes	Donkeyman	*Gunny*
Karlsen, Arne Kolbjørn	Able Seaman, AB	*Gunny*
Karlsen, Karl	Mechanic	*L. A. Christiansen*
Kelleher, Richard O.	Fire Watcher	*Patrick J. Hurley*
Kenney, John	Able Seaman, AB	*Lammot DuPont*
Kervezee, W.	Gunner	*Fauna*
Keys, Anthony Arthur	Apprentice	*Tredinnick*
Khoma Mohammed Ahmed	Fireman & Trimmer	*Stangarth*
Kiedling, Jorgen	Able Seaman, AB	*L. A. Christiansen*
Kiehne, Herman H., Jr.	Chief Officer	*Lammot DuPont*
Kilpatrick, Hugh W.	Wiper	*Potlatch*
Kim, Lim See	Messman	*Pillory*
King, Eileen	Passenger	*Melbourne Star*
King, James William	Chief Steward	*Stangarth*
Kip, Jan	Gunner	*Amazone*
Kleinblatt, Mrs. Max	Passenger	*Delmundo*

Klemm, Charles H.	Steward	*Barbara*
Kline, Harvey	1st Officer	*W. D. Anderson*
Knell, James Sherwood	Seaman 1st Class	*James Sprunt*
Knowles, Kingsley	Oiler	*Western Head*
Kobe, Karl Gustav Max	Wiper	*Colabee*
Kokinis, Antonio	Cook	*Illinois*
Kraal, Benjamin Carlos,	Oiler	*Medea*
Kramer, Pieter	Chief Officer	*Mamura*
Kroy, Frank J.	Able Seaman, AB	*Oklahoma*
Kunst, Cornelis	3rd Engineer	*Laertes*
Kuypers, M. A. F.	Master (Captain)	*Oscilla*
Kuypers, M. A. F.	Master (Captain)	*Oscilla*
Kvale, Ludvig	3rd Engineer	*L. A. Christiansen*
Kyle, Marjorie J.	Passenger	*Melbourne Star*
Kyte, Norman James	AB, DEMS Gunner	*Melbourne Star*
Labiana, Louis	Able Seaman, AB	*Millinocket*
Laboy, David M.	Messman	*City of Alma*
LaFrance, Joseph	Oiler	*Esso Gettysburg*
Lago, Pedro Diaz	Messman	*Manzanillo*
Lamb, Frederick C.	Able Seaman, AB	*W. D. Anderson*
Lambert, Bob C.	Wiper	*Norlandia*
Lambert, George Frederick	Baker	*Melbourne Star*
Lancaster, John James	3rd Engineer	*Hardwicke Grange*
Landron, Joseph J.	Engineer Cadet	*Esso Gettysburg*
Lane, Cecil	messman	*Nicarao*
Langaas, Asbjørn	Able Seaman, AB	*Gunny*
Langmead, Cyril William	Ordinary Seaman, OS	*Laertes*
Langsdale, Jack D.	2nd Officer	*City of Alma*
Lannon, John	Greaser	*Melbourne Star*
Lassen, Thorlief	Cook	*Esso Gettysburg*
Lastrap, Lucius J.	Messman	*Gulfamerica*
Latimer, Richard	Donkeyman	*Eclipse*
Laurens, Marcus	Greaser	*Amazone*
Lawson, Thomas	Greaser	*Melbourne Star*
Leay, Eric Norman	Steward's Boy	*Melbourne Star*
Leckie, Thomas Sproat	2nd Radio Officer	*Melbourne Star*
Ledo, Juan Cal, Carlo	Fireman	*Mambi*
Leguro, Vincente	Messman	*Barbara*
Leito, Benito Filemon	Able Seaman, AB	*Medea*
Lendorf, J. P.	1st Officer	*Triton*
Leonard, Frederick Pierce	2nd Asst. Engineer	*Colabee*
Leonhard, Tom	Messman	*Lubrafol*

Leppvik, Sergei	Fireman	*Gunny*
Lester, Otis	Messman	*City of Alma*
Leunig, Norman Alfred	4th Officer, Super.	*Melbourne Star*
Lewis, Cecil	1st Officer	*Stangarth*
Lewis, Ian W.	Passenger (Engineer)	*Melbourne Star*
Lewis, John	Chief Engineer	*Western Head*
Lie, Kristoffer Gulliksen	Able Seaman, AB	*Lubrafol*
Liebermann, Frederick Edward James	2nd Engineer	*Tredinnick*
Lima, Peter S.	Steward	*Gulfstate*
Lindstrom, Elmer	Ordinary Seaman, OS	*Lubrafol*
Litch, Norris W.	Cook	*Catahoula*
Livdahl, Warren J.	Signalman 3rd Class	*Esso Gettysburg*
Loit, Charles R.	3rd Engineer	*Texan*
Lombard, Forrest	3rd Officer	*Mariana*
Loo, Tenio Chang	Fireman	*Laertes*
Lopes, Oscar Ferrer	Yeoman	*Mambi*
Lopez, Evangelista F. Ferrer	Assistant Cook	*Mambi*
Lopez, Lorenzo	Fire Watcher	*Lammot DuPont*
Lorenzo, Antonio B.	Boatswain (Bosun)	*Cities Service Empire*
Losnes, Ferdinand	Carpenter	*L. A. Christiansen*
Loyon, Thomas A.	messman	*Leslie*
Lpez, Juan	Able Seaman, AB	*Mariana*
Lucas, Leroy	1st Engineer	*Onondaga*
Ludwig, Martin T.	Fire Watcher	*City of Alma*
Luiz, Barnardo Francisco	Steward	*Ohioan*
Lumsden, Leslie	AB, DEMS Gunner	*Fred W. Green*
Lundquist, Emil	Ordinary Seaman, OS	*Gunny*
Luther, Joseph N.	Junior Engineer	*James Sprunt*
Lynck, Julius	Boatswain (Bosun)	*Catahoula*
Lynd, John Joseph	Greaser	*Hardwicke Grange*
MacDonald, John Albert	Chief Steward	*Western Head*
Macfarlane, William	Passenger	*Melbourne Star*
Macintyre, Robert Alexander Cameron	Chief Engineer	*Empire Corporal*
Mackay, Albert	Steward	*Mariana*
MacKinnon, Alexander Miller	Ordinary Seaman, OS	*Colabee*
MacKinnon, Arthur Ewen	Deck Engineer	*Colabee*
MacKinnon, Donald John	Able Seaman, AB	*Melbourne Star*
MacNeil, Donald Archibald	Able Seaman, AB	*Tredinnick*
MacPherson, Donald James	Passenger (Wing Cdr.)	*Melbourne Star*

MacPherson, Pamela Yvonne	Passenger	*Melbourne Star*
MacSherry, Dennis Louis Thomas	Surgeon	*Melbourne Star*
Madalaine, Leon	Able Seaman, AB	*W. D. Anderson*
Magana, Jose	Quartermaster	*Potrero Del Llano*
Magnusson, Henry A. M.	1st Assistant Engineer	*Korlsholm*
Mahar, Daniel	Able Seaman, AB	*Western Head*
Mahar, Douglas	3rd Engineer	*Western Head*
Mahoney, Patrick Sarsfield	Master (Captain)	*James A. Moffett*
Majba, Stanly J.	Fireman, Wiper	*Oklahoma*
Mallron, Alford	Able Seaman, AB	*Pan Massachusetts*
Malone, Theron A.	Wiper	*James Sprunt*
Manea, Vasile	Boatswain (Bosun)	*Mariana*
Manson, William	Greaser	*Empire Corporal*
Manuel, James	Messman	*Raphael Semmes*
Manuela, Alfredo	Steward	*Amazone*
Mario, Andrew	Fireman, Wiper	*Illinois*
Marland, William	Boatswain (Bosun)	*Texan*
Marsh, John Edward	AB, DEMS Gunner	*Melbourne Star*
Marsh, Walter C.	Fire Watcher	*W. D. Anderson*
Marshall, John Henry Russell	Ordinary Seaman, OS	*Stangarth*
Marshall, Lynwood V.	Oiler	*Delfina*
Marshall, Norman	Petty Officer, DEMS	*Melbourne Star*
Martin, James D'Arcy	Senior Asst. Steward	*Melbourne Star*
Martin, Robert	AB, DEMS Gunner	*Melbourne Star*
Martinez, Arturo	Fire Watcher	*Raphael Semmes*
Martinog, Alexandro	Messman	*W. D. Anderson*
Martins, Joao A.	Chief Officer	*Millinocket*
Martinsen, Agnar	Oiler	*L. A. Christiansen*
Martinson, Noel Charles	1st Radio Officer	*Stangarth*
Mason, Francis	Messman	*Esso Gettysburg*
Mathiesen, Hans	Master (Captain)	*Illinois*
Mathisen, Alf	Steward	*L. A. Christiansen*
Mattson, Alfred M.	Able Seaman, AB	*Pan Massachusetts*
Mauldin, William F.	Messman	*James Sprunt*
May, William Lawson	Junior 3rd Engineer	*Melbourne Star*
Mayer, Rolfe S.	Seaman 1st Class	*Esso Gettysburg*
Maynard, Eugene A.	Steward	*Raphael Semmes*
Mayo, Oscar K.	Fire Watcher	*City of Alma*
Mazanet, Cabino C.	Quartermaster	*Manzanillo*
Mazhar, Ali	Trimmer, Engine	*Clan Skene*

McCafferty, Gerald Columba	Able Seaman, AB	*Melbourne Star*
McCall, John Strong	Assistant Engineer	*Melbourne Star*
McCann, Adrian	Messman	*Gulfstate*
McDaniel, James	Ordinary Seaman, OS	*Raphael Semmes*
McDonald, C.K.	2nd Engineer	*Western Head*
Mcdonald, Russell L.	Ordinary Seaman, OS	*W. D. Anderson*
Mcdonough, Dewer W.	4th Officer	*Ohioan*
McFerrin, James Edwin	Gunner's Mate 3rd Class	*James Sprunt*
McGarry, John Bernard	2nd Cook	*Melbourne Star*
McGinnity, John Joseph	Able Seaman, AB	*Illinois*
McGonigle, Charles A.	Able Seaman, AB	*Gulftrade*
McGovern, Philip	Ordinary Seaman, OS	*Melbourne Star*
McGrath, Edward	Steward	*James Sprunt*
McGrath, Howard Timothy	Engineer Cadet	*James Sprunt*
McGregor, Robert M.	Wiper	*Oklahoma*
Mcilroy, Robert S.	oiler	*Norlandia*
McIntosh, Robert Wallace	3rd Officer	*Everelza*
McKenzie, Enrique	Steward	*Potlatch*
Mckern, James M.	2nd Engineer	*Pan Massachusetts*
McLeod, Bernard Maxwell	Oiler	*Western Head*
McManus, Hugh F.	Fire Watcher	*Texan*
McMillan, Alvah D.	Able Seaman, AB	*James Sprunt*
McNeil	Able Seaman, AB	*Willimantic*
McNuff, Ralph	1st Engineer	*Barbara*
McPhee, Archibald	Able Seaman, AB	*Stanbank*
McPhee, James William	Boatswain (Bosun)	*Western Head*
McQueen	Passenger	*Delmundo*
McWhae, Caroline Agnes Henty	Passenger	*Melbourne Star*
McWhae, Kenneth	Passenger	*Melbourne Star*
McWilliam, Frederick	AB, Lamp Trimmer	*Melbourne Star*
Meadows, Frank Fonzie	Cook	*Sam Houston*
Medina, Jacinto	Coal Passer	*Mariana*
Medina, Jose	Passenger	*San Jacinto*
Medina, Juanita	Passenger	*San Jacinto*
Medina, Manuel J.	Pumpman	*Gulfamerica*
Medina, Santiago	Fire Watcher	*Mariana*
Mehl, William J.	Ordinary Seaman, OS	*W. D. Anderson*
Meisner, Juilian R.	Chief Engineer	*W. D. Anderson*
Melin, Carl A.	2nd Engineer	*W. D. Anderson*
Mendalka, Joseph	Coal Passer	*Mariana*
Méndez Díaz, Ceferino	Quartermaster	*Santiago De Cuba*

Mendez, Juan	Oiler	*Gulfstate*
Menendez, Alfredo Fontana	2nd Officer	*Manzanillo*
Meyer, Arno E.	Pumpman	*Pan Massachusetts*
Meyer, Emil	Pumpman	*Pan Massachusetts*
Meyers, Samuel	Messman	*Pan Massachusetts*
Michaels, Thomas	Galley Boy	*Melbourne Star*
Michaelson, Ernesto	3rd Officer	*Montevideo*
Michelsen, Joseph	Oiler	*L. A. Christiansen*
Middleton, John	Able Seaman, AB	*Stangarth*
Miles, Alfred John Joseph	Able Seaman, DEMS	*Darina*
Miles, James William	Seaman 1st Class	*James Sprunt*
Miller, Alphonse Ignatius	Deck Cadet	*Esso Gettysburg*
Miller, Charles L.	Wiper	*Ruth*
Miller, John Martin Stinky	Ordinary Seaman, OS	*Potlatch*
Miller, William L.	Oiler	*Raphael Semmes*
Milligan, Stanley	3rd Engineer	*Ocean Venus*
Milloy, Arthur W.	Fire Watcher	*Andrew Jackson*
Miranda Digot, Julio	Cook	*Santiago De Cuba*
Miranda, Francisco	Cook	*Virginia Sinclair*
Mitchell, James B.	Fire Watcher	*Bloody Marsh*
Mix, Warren A.	2nd Engineer	*Comol Rico*
Mohammed Rasid	Fireman & Trimmer	*Stangarth*
Mohsen Mohamed	Donkeyman	*Stangarth*
Moksena, John	Fireman & Trimmer	*Tredinnick*
Molchan, Joseph P.	Able Seaman, AB	*W. D. Anderson*
Moledo, Ernesto	*Rank Unknown*	*Montevideo*
Moless, Allen	Cook	*Gulftrade*
Molina, Angel Macias	3rd Engineer	*Manzanillo*
Mombelardi, Lawrence	Machinist	*Cities Service Missouri*
Monahan, Michael J.	Machinist	*Gulfamerica*
Montes, Wilfredo	Passenger	*San Jacinto*
Montgomery, James N.	Apprentice Seaman	*Gulfamerica*
Moon, James Watson	Gunner	*Stangarth*
Moore, E.	Steward	*Amazone*
Moore, Edward R.	Fire Watcher	*Federal*
Morales, Rafael Gili	Deck Boy	*Mambi*
Morfi, Ramon	Carpenter	*Mariana*
Morgan, Lee Merchant	Master (Captain)	*Colabee*
Morris, Adam H.	Oiler	*Potlatch*
Morris, Howard D.	3rd Officer	*Ruth*
Morrison, Edward	Passenger	*Barbara*

Morrison, Roderick	Able Seaman, AB	*Stangarth*
Morton, Perry Shannon	Fireman, Wiper	*Illinois*
Mott, James E.	Fireman, Wiper	*Oklahoma*
Mow, Kouang Sing	Fireman	*Laertes*
Mubarak, Ali	Trimmer, Engine	*Clan Skene*
Mugbil Abdul	Fireman & Trimmer	*Stangarth*
Mui Lum	Crew	*Standella*
Muise, William H.	Able Seaman, AB	*Gulfstate*
Mulcahy, Ethne Mary	Passenger	*Melbourne Star*
Muley, Jose Quavado	1st Officer	*Manzanillo*
Mulvenna, J.	Able Seaman, AB	*Darina*
Mummery, J.	Able Seaman, AB	*Willimantic*
Munro, Alan C.	Purser	*Ohioan*
Munslow, Clifford V.	Able Seaman, AB	*James Sprunt*
Munson, Bernard	3rd Officer	*James Sprunt*
Murphy, Eugene F.	OIler	*Gulfstate*
Murphy, James A.	Oiler	*Patrick J. Hurley*
Murphy, Paul E.	Able Seaman, AB	*Esso Gettysburg*
Murphy, Philip	Able Seaman, AB	*Melbourne Star*
Murray, John	2nd Officer	*Tredinnick*
Murshed, Ali B.	Fire Watcher	*Onondaga*
Myers, Marvin T.	Able Seaman, AB	*Virginia Sinclair*
Nagi, Ahmed	Fireman & Trimmer	*Stangarth*
Nagi, Missula	Fireman & Trimmer	*Stangarth*
Nam Leung	Fireman	*Michael Jebsen*
Nam, Leung	Fireman	*Melbourne Star*
Name Unknown	Crew	*Carmen*
Name Unknown	*Rank Unknown*	*City of Alma*
Name Unknown	*Rank Unknown*	*City of Alma*
Name Unknown	*Rank Unknown*	*City of Alma*
Name Unknown	*Rank Unknown*	*Delmundo*
Name Unknown	*Rank Unknown*	*Delmundo*
Name Unknown	*Rank Unknown*	*Delmundo*
Name Unknown	*Rank Unknown*	*Esso Gettysburg*
Name Unknown	*Rank Unknown*	*Everelza*
Name Unknown	*Rank Unknown*	*Everelza*
Name Unknown	*Rank Unknown*	*Everelza*
Name Unknown	*Rank Unknown*	*Everelza*
Name Unknown	*Rank Unknown*	*Everelza*
Name Unknown	*Rank Unknown*	*Everelza*
Name Unknown	*Rank Unknown*	*Everelza*
Name Unknown	*Rank Unknown*	*Everelza*
Name Unknown	*Rank Unknown*	*Everelza*
Name Unknown	*Rank Unknown*	*Everelza*

Name Unknown	*Rank Unknown*	*Everelza*
Name Unknown	*Rank Unknown*	*Everelza*
Name Unknown	*Rank Unknown*	*Everelza*
Name Unknown	*Rank Unknown*	*Everelza*
Name Unknown	*Rank Unknown*	*Everelza*
Name Unknown	*Rank Unknown*	*Everelza*
Name Unknown	*Rank Unknown*	*Everelza*
Name Unknown	*Rank Unknown*	*Everelza*
Name Unknown	*Rank Unknown*	*Everelza*
Name Unknown	*Rank Unknown*	*Gulfstate*
Name Unknown	*Rank Unknown*	*Gulfstate*
Name Unknown	*Rank Unknown*	*Gulfstate*
Name Unknown	*Rank Unknown*	*Gulfstate*
Name Unknown	*Rank Unknown*	*Gulfstate*
Name Unknown	*Rank Unknown*	*Illinois*
Name Unknown	*Rank Unknown*	*James Sprunt*
Name Unknown	3rd Engineer	*Norwalk*
Name Unknown	*Rank Unknown*	*Potrero Del Llano*
Name Unknown	*Rank Unknown*	*Potrero Del Llano*
Name Unknown	*Rank Unknown*	*Potrero Del Llano*
Name Unknown	*Rank Unknown*	*Potrero Del Llano*
Name Unknown	*Rank Unknown*	*Potrero Del Llano*
Name Unknown	*Rank Unknown*	*Potrero Del Llano*
Name Unknown	*Rank Unknown*	*Potrero Del Llano*
Name Unknown	*Rank Unknown*	*Potrero Del Llano*
Name Unknown	*Rank Unknown*	*Potrero Del Llano*
Name Unknown	*Rank Unknown*	*Potrero Del Llano*
Name Unknown	*Rank Unknown*	*Potrero Del Llano*
Name Unknown	*Rank Unknown*	*Potrero Del Llano*
Name Unknown	*Rank Unknown*	*Thomas McKean*
Name Unknown	*Rank Unknown*	*Tredinnick*
Name Unknown	*Rank Unknown*	*Tredinnick*
Name Unknown	*Rank Unknown*	*Tredinnick*
Name Unknown	*Rank Unknown*	*Tredinnick*
Name Unknown	*Rank Unknown*	*Tredinnick*
Name Unknown, Chinese	Ordinary Seaman, OS	*Oscilla*
Name Unknown, Chinese	Cook	*Oscilla*
Name Unknown, Chinese	Ordinary Seaman, OS	*Oscilla*
Name Unknown, Chinese	Quartermaster	*Oscilla*
Neault, Ronald L.	Ordinary Seaman, OS	*Gulfstate*
Nedberg, Sigurd	3rd Officer	*Gunny*

Neely, Newton Davis	Seaman 1st Class	*Gulfstate*
Nehlin, Nils A.	Able Seaman, AB	*Raphael Semmes*
Ng, William	Ordinary Seaman, OS	*Gulfstate*
Nielsen, Alfred K.	Fire Watcher	*Catahoula*
Nielsen, Peter Vilh.	Able Seaman, AB	*Everelza*
Nilsen, Fridtjof	2nd Officer	*L. A. Christiansen*
Nilsen, Invald C.	Master (Captain)	*Oregon*
Nilson, Edward	Oiler	*Illinois*
Nilssen, Erling	Cook	*L. A. Christiansen*
Norris, Glen Dale	Seaman 1st Class	*James Sprunt*
Norton, George P.	2nd Engineer	*Virginia Sinclair*
Nossum, Finn	Ordinary Seaman, OS	*L. A. Christiansen*
Novo Teijeiro, Francisco	1st Assistant Engineer	*Santiago De Cuba*
Oakley, Michael J.	Chief Engineer	*Raphael Semmes*
O'Berry, Alton R.	Able Seaman, AB	*Ruth*
O'Connor, Frank	Ordinary Seaman, OS	*Melbourne Star*
O'Connor, Herbert	Chief Steward	*Melbourne Star*
O'Connor, James F.	3rd Officer	*Federal*
O'Connor, Joseph Michael	Able Seaman, AB	*Illinois*
Odegard, Hans	Stoker	*Gunny*
Oftedal, Alf	2nd Officer	*Mariana*
Øgård, Hans Ole	Fireman	*Hagan*
Oksnes, Ragnar	Mechanic	*L. A. Christiansen*
Olando, Carrol Alfanso	Ordinary Seaman, OS	*Illinois*
Olinerio, Anthony	Fireman 3rd Class	*Gulfstate*
Olivera, Lorenzo Romulo	*Rank Unknown*	*Montevideo*
Olsen, Arne Høst	Master (Captain)	*L. A. Christiansen*
Olsen, Emil T.	1st Cook	*Pan Massachusetts*
Olsen, John H.	2nd Officer	*James Sprunt*
O'Neill, Christopher	Able Seaman, AB	*Melbourne Star*
O'Neill, Hugh	Ordinary Seaman, OS	*Stangarth*
O'Neill, John	Deck Boy	*Melbourne Star*
Oram, William H.	Able Seaman, AB	*Barbara*
Orease (Orobase), Canuto O.	Messman	*Raphael Semmes*
Ouder, Lawrence G.	Chief Engineer	*Nicarao*
Owen, Eva Doreen	Passenger	*Melbourne Star*
Pabon, Juan	Cook	*Barbara*
Padilla, Emilio L.	Coal Passer	*Mariana*
Paganelli, F. L.	USCG Signalman	*Manzanillo*
Pagel, William Harry	Seaman 1st Class	*James Sprunt*
Paireire, Manuel	3rd Asst. Engineer	*Pillory*

Paleologas, Basilios	3rd Engineer	*Potlatch*
Palmer, Bartolome Moragues	Greaser	*Manzanillo*
Palmer, Eugene	Able Seaman, AB	*Western Head*
Pandolfo, Rosario Dominic	Seaman 1st Class	*James Sprunt*
Parker, Edward	Able Seaman, AB	*Illinois*
Parkes, Eric John	Able Seaman, AB	*Stangarth*
Parlatore, Francis	Wiper	*Esso Gettysburg*
Parson, David	Cook	*Potlatch*
Parsonage, Peter Matthews	4th Officer	*Melbourne Star*
Parsons, Leslie Thomas	1st Officer	*Melbourne Star*
Partridge, Eva J.	Passenger	*Melbourne Star*
Paton, George	Oiler	*L. A. Christiansen*
Paul, Guy D.	USCG Signalman	*Manzanillo*
Peace, Michsaw	Messman	*Mariana*
Pealzgraf, Harvey	Cook	*James Sprunt*
Pearson, John Amos C.	Sailor	*Laertes*
Pecot, Roderick	Oiler	*Delmundo*
Pedersen, Oivind L.	Chief Engineer	*Pan Massachusetts*
Peeters, Everardus	2nd Engineer	*Mamura*
Percival, William Henry	2nd Electrician	*Melbourne Star*
Perdiz, Arsenio L.	Cook	*Federal*
Pérez Delgado, Antonio	Oiler	*Santiago De Cuba*
Perez, German	Ordinary Seaman, OS	*City of Alma*
Perez, Jose	Oiler	*Lammot DuPont*
Perez, Roque A.	1st Engineer	*Manzanillo*
Perkins, John	Fireman, Wiper	*Colabee*
Perrault, Joseph Edward	Able Seaman, AB	*Illinois*
Perrin, Arthur Thomas	2nd Officer	*Stangarth*
Perry, Donald D.	Ensign	*James Sprunt*
Petersen, Gilbert Edward	Fireman, Wiper	*Colabee*
Petersen, Peter	Steward	*San Jacinto*
Peterson, Roy August	Chief Engineer	*Illinois*
Petrikas, Joseph A.	Ordinary Seaman, OS	*James Sprunt*
Pettersen, Leif Anton	Stoker	*Gunny*
Pettersen, Leif Anton	Fireman	*Hagan*
Phillips, Richard P.	3rd Engineer	*City of Alma*
Phillipson, George, British Army	Gunner, DEMS	*Melbourne Star*
Piaszcyzk, Victor S.	Oiler	*Onondaga*

Pickering, John A.	Purser	*Mariana*
Piedra, Charles A.	Steward	*Delmundo*
Pierce, John V.	Able Seaman, AB	*Ruth*
Pietersz, Julio Serapio Remigio	Cook	*Medea*
Pileta, Bernardo	Fireman	*Santiago De Cuba*
Pilicilda, Utique (Utika)	Utilityman	*Pan Massachusetts*
Pilkington, John Samuel	Greaser	*Melbourne Star*
Pimenta, Jose Rodrigues	Oiler	*Pillory*
Pineda, Antonio Fernandez	Quartermaster	*Mambi*
Pineiro, Eduardo	Able Seaman, AB	*Lammot DuPont*
Ping Lav	Trimmer, Engine	*Michael Jebsen*
Ping, Chan Wong	Crew	*Standella*
Ping, Lav	Trimmer, Engine	*Melbourne Star*
Pirangy, Joaquuin D.	Wiper	*Leslie*
Pitkin, James Alfred George	Able Seaman, AB	*Melbourne Star*
Pittman, Howard B.	Messman	*Gulfamerica*
Polk, Rufu	Messman	*Gulftrade*
Pomroy, Alan Browning	Passenger (Electrical)	*Melbourne Star*
Ponton, Eloisa	Passenger	*San Jacinto*
Poole, George R.	1st Engineer	*Gulfstate*
Pope, William	Cook	*Federal*
Porter, Alexander Sheridan	Passenger	*Melbourne Star*
Porter, Melford	Able Seaman, AB	*Gulfstate*
Potter, Thomas S.	Fire Watcher	*Gulfstate*
Pound, Weston C.	Radio Operator	*Esso Gettysburg*
Powell, Clifford Austin	Chief Engineer	*Tredinnick*
Prentice, Theophilus	Fireman & Trimmer	*Tredinnick*
Presfeldt, John	1st Officer	*Western Head*
Price, Frederick Leon	Gunner's Mate 3rd Class	*Esso Gettysburg*
Price, John	Oiler	*Oklahoma*
Proctor, Thomas M.	3rd Engineer	*Ohioan*
Proctor, William	3rd Engineer	*Athelqueen*
Pullin, Clyde W.	Messman	*James Sprunt*
Ramos, Guillermo	Ordinary Seaman, OS	*Comol Rico*
Ranger, Charles Frederick	Gunner, DEMS	*Mamura*
Rank, Dean O.	Wiper	*City of Alma*
Rankin, Whitney	Baker	*Barbara*
Rasmussen, Andrew Martin	Able Seaman, AB	*Cities Service Empire*
Rataal, Joseh	2nd Engineer	*Mariana*
Raudeberg, Arnfinn	AB, Helmsman	*Grenanger*

Rector, Charles Edward	Chief Cook	*Western Head*
Reeder, Hasker W.	Able Seaman, AB	*Cities Service Empire*
Reeves, James Thomas	Able Seaman, AB	*Cities Service Empire*
Reich, Douglas S.	Fire Watcher	*Lammot DuPont*
Reip, Charles J. Jr.	Ordinary Seaman, OS	*Barbara*
Respess, Mellin E.	Passenger	*Onondaga*
Rey, Antonio	Fire Watcher	*Catahoula*
Rey, Jesus Fernandez	Assistant Engineer	*Mambi*
Rey, Nicholas Penedo	3rd Engineer	*Mambi*
Reynolds, Monroe Hyres	Deck Maintainer	*Cities Service Empire*
Reynolds, Oren V.	3rd Engineer	*Barbara*
Reynolds, Thomas A.	Gunner, Naval	*Darina*
Rhodes, William Glenn	Apprentice Seaman	*Gulfamerica*
Rice, Henry	Sailor	*Melbourne Star*
Richards, William Edward	2nd Officer	*Melbourne Star*
Richardson, Frank Rea	Sailor	*Tredinnick*
Rickard, Alfred Albert	Able Seaman, DEMS	*Mamura*
Riley, James	Messman	*Oklahoma*
Rios, Carmelo	Ordinary Seaman, OS	*Pillory*
Rios, Francisco Creepo	Messman	*Cities Service Empire*
Rippon, Stanley	Senior 4th Engineer	*Melbourne Star*
Risso, Joseph E.	Able Seaman, AB	*Gulfamerica*
Riveira, Romana	Passenger	*San Jacinto*
Rivera, Joaquin	Fire Watcher	*City of Alma*
Rivers, Simeon P.	Messman	*Federal*
Rivette, Charles	Ordinary Seaman, OS	*Oklahoma*
Robbins, Raymond L.	Oiler	*W. D. Anderson*
Robertsen, Reidar	Messroom Boy	*Gunny*
Robertsen, Reidar Karl Johan	Messroom Boy	*Hagan*
Robertson, George	Cabin Boy	*Stangarth*
Robertson, John C.	Oiler	*W. D. Anderson*
Robles, Damasco	Coal Passer	*Pillory*
Robuck, Frank B.	Oiler	*Bloody Marsh*
Rodner, William L.	Oiler	*Gulftrade*
Rodrigues, Alexander	Messman	*Ohioan*
Rodríguez Lorenzo, Juan E.	Fireman	*Santiago De Cuba*
Rodriguez, Eulalio Amarante	Quartermaster	*Mambi*
Rodriguez, Santiago	Coal Passer	*Mariana*

Rogers, Anel	Able Seaman, AB	*Lammot DuPont*
Rogers, John L.	Able Seaman, AB	*Ruth*
Rogers, Leslie	Crew	*Helen Forsey*
Roja Figueroa, Diego	Boatswain (Bosun)	*Santiago De Cuba*
Rojas, Candelaria	Waiter	*Barbara*
Rojek, Romand Michael	Seaman 1st Class	*Esso Gettysburg*
Rokstad, Johan Bernhard Olsen	2nd Officer	*Colabee*
Rome, Lee J.	2nd Engineer	*Delmundo*
Romero, Manuel L.	Deck Engineer	*Lammot DuPont*
Rosario, Vigilio	Waiter	*Barbara*
Rose, James L.	Fire Watcher	*Delisle*
Rose, Orie G.	Wiper	*Esso Gettysburg*
Rosenvinge, Petter Melchior	Sailor	*Everelza*
Roshcaln, George Rudolph	1st Officer	*Illinois*
Round, John Barker	Able Seaman, AB	*Stanbank*
Rowley, James R.	Deck Cadet	*James Sprunt*
Rozeboom, Sikke	Fifth Engineer	*Mamura*
Rozwadowski, Leon Paul	Seaman 1st Class	*James Sprunt*
Ruder, Oswald	Messman	*Oklahoma*
Ruffin, Charles	Coal Passer	*Mariana*
Ruia, Pablo Seto	Purser	*Mambi*
Rumbellow, George Swan	3rd Engineer	*Tredinnick*
Runsted, Hans	Able Seaman, AB	*Ohioan*
Russell, Lawrence	Radio Operator	*Ruth*
Russell, Wilbur W.	Ordinary Seaman, OS	*Gulftrade*
Rux, Alfred L. E.	Oiler	*Pan Massachusetts*
Ryan, Stephen	Steward	*Texan*
Sabato, John J.	Able Seaman, AB	*Pan Massachusetts*
Sadlon, Michael	Ordinary Seaman, OS	*Esso Gettysburg*
Said, Hassan	Fireman & Trimmer	*Stangarth*
Salabarria, Marcelino Armada	Steward	*Mambi*
Salters, John C.	Oiler	*Ohioan*
Sampson, Arthur Gould	Master (Captain)	*Fred W. Green*
Samuelson, Carl S.	1st Engineer	*W. D. Anderson*
San Nicholas, Eulegio	Chief Engineer	*Barbara*
Sanawar, Mian	Fireman	*Clan Skene*
Sanders, Hogh H.	1st Asst. Engineer	*Gulfamerica*
Sanders, Roy	Wiper	*City of Alma*
Sanderson, David R.	Pumpman	*W. D. Anderson*
Sango, Edward	Greaser	*Tredinnick*

Sañudo Lavín, Arcadio Samudo	Steward	*Santiago De Cuba*
Saralegui, Camilo	Fireman	*Montevideo*
Schaaf, Elmer W.	Oiler	*James Sprunt*
Scheich, William John	Fireman, Watertender	*Esso Baton Rouge*
Scheie, Rolf V.	Able Seaman, AB	*L. A. Christiansen*
Schippers, Herman, Royal Dutch Army	Gunner	*Laertes*
Schoop, Lopoldo	Waiter	*San Jacinto*
Secker, Charles Archibald, RANVR	Passenger (Lieutenant)	*Melbourne Star*
Seman, Joseph	3rd Officer	*Millinocket*
Senior, Harry Ferguson	3rd Officer	*Melbourne Star*
Senn, William G.	Steward	*James Sprunt*
Serrano, Christopher S.	Chief Engineer	*Gulfstate*
Serrrano, Marcellino	Wiper	*Santa Rita*
Servant, Name Unknown ("Colored")	Servant	*Onondaga*
Sharkey, James H.	Purser	*Esso Gettysburg*
Shaw, Albion K.	Boatswain (Bosun)	*Esso Gettysburg*
Shaw, Godfrey Ambrose Samuel Wensley	1st Radio Officer	*Tredinnick*
Sheehan, Patrick	Able Seaman, AB	*Melbourne Star*
Shelford, Robert Victor Kimberley	Fireman	*Ocean Venus*
Siemon, Alfred	Deck Engineer	*Ohioan*
Signorelli, Charles	Radio Operator	*James Sprunt*
Simpson, Joseph H.	Radio Operator	*City of Alma*
Simpson, Randolph F.	Able Seaman, AB	*Raphael Semmes*
Sistrunk, Charles P.	Ordinary Seaman, OS	*Oklahoma*
Skatt, Karl E.	Carpenter	*Korlsholm*
Skevik, John M.	Carpenter	*Raphael Semmes*
Skinner, Richard	Boatswain (Bosun)	*Stanbank*
Slagtand, Harris Archibald	Fireman	*Medea*
Slidders, Elizabeth	Passenger	*Melbourne Star*
Slidders, John P.	Passenger	*Melbourne Star*
Slidders, Sandison W.	Passenger	*Melbourne Star*
Smalledge, Galen I.	Wiper	*Gulftrade*
Smith, Charles E.	Able Seaman, AB	*W. D. Anderson*
Smith, Foster C.	2nd Engineer	*Delfina*
Smith, Frederick	Asst. Cook & Butcher	*Melbourne Star*
Smith, Frederick Clarke	2nd Radio Officer	*Stanbank*
Smith, George	Deck Boy	*O. A. Knudsen*
Smith, George	Deck Boy	*Ocean Venus*

Smith, Henry Peter	Master (Captain)	*Delmundo*
Smith, Herman A.	Ordinary Seaman, OS	*James Sprunt*
Smith, Leslie	Able Seaman, AB	*Melbourne Star*
Smith, William John	Cabin Boy	*Tredinnick*
Sohid	Trimmer, Engine	*Clan Skene*
Solano, Juan	Oiler	*Ruth*
Sorhus, Ove B. Hansen	1st Officer	*L. A. Christiansen*
Soto, Frank	Able Seaman, AB	*Mariana*
Souffront, Louis	Oiler	*Ohioan*
Spangler, Ralph M.	Chief Engineer	*Ruth*
St. Leger, Wilbert P.	Utilityman	*James Sprunt*
Stampley, Percy R.	Cook	*Gulfstate*
Stanton, James	Greaser	*Ocean Venus*
Stark, William Christie	2nd Officer	*Stanbank*
Starns, Wilmer	Oiler	*Sam Houston*
Stelk, Edward	Fireman	*Amazone*
Stensrud, Guthorm	Electrician	*L. A. Christiansen*
Stephen, George	Galley Boy	*Macgregor*
Stephenson, Frank C.	Fireman	*Pillory*
Stevens, Frederick	Gunner, DEMS	*Melbourne Star*
Stine, Wilbur L.	2nd Engineer	*Delisle*
Stirling, Theodire R.	cook	*Nicarao*
Stitsel, Mahlon E.	2nd Officer	*W. D. Anderson*
Stolle, Ines	Passenger	*Barbara*
Stone, Raymond Lonnie	Coxswain	*Gulfstate*
Stoyanovich, Alex C.	Oiler	*Lubrafol*
Stranger, Hendrikus Petrus Philippus	1st Officer	*Laertes*
Stringfellow, John	Gunner, DEMS	*Melbourne Star*
Stromberg, Carl	Master (Captain)	*Patrick J. Hurley*
Stubberud, Asbjorn	Stoker	*Gunny*
Stubberud, Asbjørn	Fireman	*Hagan*
Subhan Ali,	Fireman	*Clan Skene*
Suggs, Fannie	Passenger	*San Jacinto*
Sullivan, Francis G.	Utilityman	*James Sprunt*
Sullivan, Joseph Thomas	Deck Engineer	*Illinois*
Summers, Douglas Haig	Able Seaman, AB	*Stangarth*
Tafazzul Husain,	Fireman	*Clan Skene*
Taix, Louis M.	Radio Operator	*Nicarao*
Talley, Charles R., Jr.	2nd Officer	*Gulfstate*
Tate, James	Cook	*City of Alma*
Taylor, John	Gunner, British Army	*Stangarth*
Taylor, Joseph R.	Wiper	*Ruth*
Taylor, Robert	Able Seaman, AB	*Tredinnick*

Taylor, Thomas	Ordinary Seaman, OS	*Tredinnick*
Teater, Charles L.	Able Seaman, AB	*Esso Gettysburg*
Tempel, Arie van den	Fifth Engineer	*Mamura*
Tewkesbury, Ernest Charles	Donkeyman	*Ocean Venus*
Thibodeaux, Oze	Seaman 1st Class	*James Sprunt*
Thom, Agnes W. M.	Passenger	*Melbourne Star*
Thom, Alfred Marshall Graeme	Passenger	*Melbourne Star*
Thomas, Andre Emile	Passenger	*Melbourne Star*
Thomson, Archibald	Able Seaman, AB	*Tredinnick*
Thorn, R. E.	Seaman 1st Class	*Gulfstate*
Tippet, Chester E.	Able Seaman, AB	*Norlandia*
Tiseth, Krsitian J.	Able Seaman, AB	*Gulfstate*
Torres, Gregorio S.	Oiler	*Ruth*
Toscano, Felix Ramirez	Fireman	*Manzanillo*
Tralie, Joseph	Able Seaman, AB	*Catahoula*
Treche, Richard H.	Able Seaman, AB	*Pillory*
Treliving, Harold	Senior 2nd Engineer	*Melbourne Star*
Trelles, Gerardo Fernandez	Fireman	*Mambi*
Tricomi, Salvatore	Ordinary Seaman, OS	*Gulfstate*
Trymers, Louis J.	Able Seaman, AB	*Delfina*
Tucek, John Paul	Deck Cadet	*James Sprunt*
Tuckeck, John P.	Chief Engineer	*Java Arrow*
Tucker, George H.	Baker	*Lammot DuPont*
Tucker, Norman	2nd Engineer	*Stangarth*
Turner, Jack Wiltshire	4th Engineer	*Stangarth*
Tyson, Burnett F.	Ordinary Seaman, OS	*James Sprunt*
Umbore Ullah,	Fire Tender	*Clan Skene*
Underwood, John George	Sailor	*Stangarth*
Uriarte, Gregory	3rd Engineer	*Mariana*
Vail, Leslie Gilson	Radio Operator	*Illinois*
Vaisanen, Antti	Fireman	*Pillory*
Van Baardwyk, C. C.	Trimmer, Engine	*Fauna*
Van Den Assen, Christian	Oiler	*Pillory*
Van Dessel, Pierre	1st Engineer	*Lubrafol*
Van Erkelens, Adrianus J. L.	3rd Asst. Engineer	*Korlsholm*
Van Eyck, Nickolas	Trimmer, Engine	*Triton*
Van Pelt, Richard A.	Wiper	*Gulfamerica*
Vega, Antonio Fernandez	Master (Captain)	*Manzanillo*
Veglio, Mauro	*Rank Unknown*	*Montevideo*
Velasco, Josef Von	Steward	*Illinois*
Velasquez, Pedro	Fire Watcher	*Mariana*

Verdon, William H.	2nd Engineer	*Ohioan*
Vichich, Henry C.	1st Assistant Engineer	*Pan Massachusetts*
Viglione, Albert P.	Messman	*Pan Massachusetts*
Vila, Manuel Parga	2nd Engineer	*Mambi*
Vincent, Thomas George	Watertender	*Colabee*
Viotto, Auguste D.	Fireman, Wiper	*Gulfamerica*
Vistung, Reidar	Able Seaman, AB	*L. A. Christiansen*
Wakefield, Homer C.	Wiper	*Texan*
Walker, Harold Rutledge	Coxswain	*Cities Service Empire*
Walker, James K.	Cook	*Gulfamerica*
Wallum, Elmer George	Seaman 1st Class	*James Sprunt*
Walters, Albert B.	Master (Captain)	*W. D. Anderson*
Warden, Van Leen	Gunner	*Triton*
Warner, Bertram	Wiper	*Andrew Jackson*
Warner, Charles L.	Chief Engineer	*City of Alma*
Warren, Maurice	Wiper	*Gulfstate*
Warren, Virgil Y.	Ordinary Seaman, OS	*Pan Massachusetts*
Warrington, Nathaniel L.	oiler	*Norlandia*
Wass, C.	Gunner	*Laertes*
Watler, William W.	1st Officer	*Oregon*
Watson, Basil	Fifth Engineer	*Darina*
Watson, James T.	Galleyman 3rd Class	*Pillory*
Watson, John A.	Lance Bomber, DEMS	*Empire Corporal*
Watson, Robert F.	Fire Watcher	*Hagan*
Watts, Donald B.	Fire Watcher	*Ruth*
Weaver, Claude O.	Messman	*Pan Massachusetts*
Webb, Edward John	Refrigerator Greaser	*Melbourne Star*
Weber, Harold J.	1st Assistant Engineer	*Hagan*
Weddington, Francis B.	Steward	*Oneida*
Wee Gon Kin	Cook	*Stangarth*
Weedon, Leonard Albert	Able Seaman, AB	*Illinois*
Wegbrands, Abraham	Junior Engineer	*Lubrafol*
Weide, Louis	Cook	*Illinois*
Welch, Willie R.	Able Seaman, AB	*Gulfamerica*
Wendler, Hubert G.	Oiler	*Virginia Sinclair*
Wengland, Ralph T.	1st Assistant Engineer	*James Sprunt*
West, George F.	Able Seaman, AB	*Ruth*
West, Thomas Woods	Deck Boy	*Laertes*
Wetart, Stanley J.	Fireman, Wiper	*Esso Gettysburg*

Wetzel, Erich W.	Able Seaman, AB	*Patrick J. Hurley*
Whalen, Patrick Bryan	1st Assistant Engineer	*Illinois*
Whipp, Eric	Assistant Engineer	*Melbourne Star*
Whishaw, Peter	Passenger (Lieutenant)	*Melbourne Star*
White, Selwyn William	Senior 3rd Engineer	*Melbourne Star*
Whitehead, Harold D.	Fire Watcher	*Sam Houston*
Whitehead, Ralph M.	Doctor	*Barbara*
Whiteside, Gerald M.	Passenger (Radio Off.)	*Melbourne Star*
Whitley, Julius E.	Gunner's Mate 3rd Class	*Esso Gettysburg*
Whitney, J.	Fifth Engineer	*Darina*
Whitter, John Alexander	Gunner, DEMS	*Mamura*
Whyte, George	Ordinary Seaman, OS	*City of Alma*
Whyte, Nina K.	Passenger	*Melbourne Star*
Wickberg, Idar	Master (Captain)	*Korlsholm*
Wicke, William B.	2nd Officer	*Oregon*
Wigsten, Nils Anders	Mechanic	*Pillory*
Wijers, Jean Louis	Fifth Engineer	*Mamura*
Wikitas, Marcus	Oiler	*Pillory*
Wilkinson, Clarence A.	Coxswain	*James Sprunt*
Williams	Donkeyman	*Tredinnick*
Williams, Arthur D.	messman	*Nicarao*
Williams, Ashley C.	Messman	*Ruth*
Williams, Fred E.	Oiler	*Ohioan*
Williams, James	Steward	*Ruth*
Williams, John Hugh	Fireman	*Fred W. Green*
Williams, Marshall George	Carpenter	*Illinois*
Williams, William R.	Pumpman	*Gulfstate*
Wilson, Carlton O.	1st Engineer	*Mariana*
Wilson, Carlton O.	Master (Captain)	*Melbourne Star*
Wilson, Ernest	Assistant Steward	*Melbourne Star*
Wilson, Sheila	Passenger	*Melbourne Star*
Wilson, Wardell M.	Fire Watcher	*Leslie*
Winklaar, Antonio	Sailor	*Amazone*
Winslow, Robert T.	3rd Engineer	*Bloody Marsh*
Winters, Andrew Patrick	Able Seaman, AB	*Stangarth*
Witham, Soren Albert Terkelsen	1st Radio Officer	*Melbourne Star*
Withee, Abner D. , Jr.	Ordinary Seaman, OS	*Onondaga*
Withee, Aubrey F.	Wiper	*Pan Massachusetts*
Womble, John D., Jr.	Messman	*Onondoga*
Wong, Ah Chang	Engine Fitter	*Laertes*

Wong, Sue King	Fireman Boy	*Laertes*
Wood, Francis D.	3rd Officer	*Texan*
Wood, James W.	Passenger	*Barbara*
Wood, Norman	2nd Radio Officer	*Tredinnick*
Woolf, Stanley Kay,	2nd Radio Officer	*Stangarth*
Woon Cha	Fireman No. 2	*Michael Jebsen*
Woon Cham,	Fireman No. 2	*Melbourne Star*
Wooten, Carey Lavon	Seaman 1st Class	*James Sprunt*
Wray, Harry	Able Seaman, DEMS	*Mamura*
Wright, John	Carpenter	*Tredinnick*
Wright, Joseph L.	Wiper	*Pan Massachusetts*
Wuolo, John F.	Chief Officer	*James Sprunt*
Wyman, David	Boatswain (Bosun)	*Amazone*
Xavier, Charles	Cook	*Olga*
Yardley, Albert	Able Seaman, AB	*Tredinnick*
Ybanez, Jose	Fire Watcher	*Norlandia*
Yet, Lo	Crew	*Standella*
Yong, Ah Tong	Fireman	*Laertes*
Youds, F.	Chief Radio Operator	*Willimantic*
Young, Reginald	Able Seaman, AB	*Melbourne Star*
Youngson, William Jenkins	Gunner, British Army	*Stangarth*
Zabala Petuya, Aureliano	Steward	*Santiago De Cuba*
Zambrana, Antonio J. Jimenez	Oiler	*Mambi*
Zgoda, Alfred N.	Messman	*Esso Gettysburg*
Zingg, Fritz	Salon Steward	*Amazone*
Zink, Glen A.	Galleyman	*Esso Gettysburg*
Zucowska, Joseph Charles	Able Seaman, AB	*Illinois*

Appendix 13: Allied Attacks on & U-Boat Sightings of Allied Warships & Aircraft

Date	Axis Sub-Marines	Position	Air	Vsls	Blimp	Bomb	Misc	Tally
		TOTALS	*587*	*138*	*50*	*154*	*39*	*c.1000*
2/8/42	U-128	SW of Bermuda	1			2		3
2/10/42	U-67	Anegada	4					4
2/14/42	U-128	NE of Bahamas					1	1
2/17/42	U-128	NE of Bahamas					1	1
2/19/42	U-128	Straits off Canaveral		1				1
2/20/42	U-504	Straits Jupiter Inlet	1					1
2/21/42	U-504	Straits off Canaveral		1				1
2/22/42	U-504	Straits off Canaveral		2		4		6
2/23/42	U-504	Straits off Canaveral	1					1
2/23/42	U-128	Matanilla Shoals NW Bahamas	2					2
2/24/42	U-504	N of Abaco		1		2		3
2/25/42	U-156	Mona Passage	1					1
2/26/42	U-156	Mona Passage	2					2
2/26/42	U-154	Mona	6					6
2/26/42	U-128	Straits off Canaveral	1					1
2/26/42	U-504	E of Jacksonville					1	1
2/27/42	U-156	N of Mona Passage	1					1

Date	Axis Sub-Marines	Position	Air	Vsls	Blimp	Bomb	Misc	Tally
2/27/42	U-154	Mona	1					1
2/28/42	U-128	Straits at Canaveral		1				1
3/4/42	U-504	N of Abaco						
3/5/42	U-504	N of Abaco						
3/6/42	U-504	N of Abaco					1	1
3/9/42	U-128	S of Bermuda near		1				1
3/10/42	U-504	NE of Bahamas					1	1
3/12/42	Finzi	S of Bermuda N of Anegada	1	3				4
4/9/42	U-123	off Jacksonville	1					1
4/10/42	U-123	off St. Augustine	1					1
4/11/42	U-123	off Canaveral	1					1
4/11/42	U-123	off Canaveral	10	2				12
4/12/42	U-130	NE of Anegada	1					1
4/14/42	U-123	E of Canaveral					1	1
4/14/42	U-123	Straits		1				1
4/15/42	U-130	Anegada Sombrero Passage	2					2
4/25/42	U-108	S of Bermuda close	1					1
4/26/42	U-108	Off Bermuda	1					1
4/27/42	U-109	NW of Jacksonville		1		10		11
4/28/42	U-654	Bermuda		1				1
4/29/42	U-654	Bermuda	1					1
4/30/42	U-98	E of Jacksonville					1	1
4/30/42	U-506	E of Bahamas	1					1
4/30/42	U-507	OBC	2					2
5/1/42	U-564	Straits off Jacksonville	4	1				5
5/1/42	U-507	OBC	1	2				3

Date	Axis Sub-Marines	Position	Air	Vsls	Blimp	Bomb	Misc	Tally
5/1/42	U-109	off Canaveral	1					1
5/2/42	U-506	Straits off Palm Beach	5					5
5/2/42	U-333	Off Jacksonville	1					1
5/2/42	U-507	OBC	7					7
5/3/42	U-98	E of Savannah	6					6
5/3/42	U-506	Straits, SE of Miami	10		2			12
5/3/42	U-564	Straits off Palm Beach	4					4
5/3/42	U-507	Nicholas Channel		2				2
5/4/42	U-564	Straits off Miami	6	2				8
5/4/42	U-333	Straits	5					5
5/4/42	U-98	Straits E of St Augustine		1		6		7
5/5/42	U-564	Straits S of Cat Cay Bimini	1				1	2
5/5/42	U-69	S of Bermuda midway Anegada		1				1
5/5/42	U-109	off Canaveral	1			3		4
5/5/42	U-108	Windward Passage	1					1
5/5/42	U-333	Straits		1				1
5/6/42	U-108	S of Inagua	1					1
5/6/42	U-506	Straits S of Keys	1					1
5/6/42	U-109	off Vero Beach	1			5		6
5/6/42	U-333	St. Lucie Inlet FL		2				2
5/7/42	U-108	N of Windward Passage					1	1
5/7/42	U-564	Straits off Bimini	1	1		3		5

Date	Axis Sub-Marines	Position	Air	Vsls	Blimp	Bomb	Misc	Tally
5/7/42	U-109	off Vero Beach	1			12		13
5/7/42	U-108	Windward Passage	3					3
5/7/42	U-69	N of San Juan	4					4
5/8/42	U-109	off Canaveral	1					1
5/9/42	U-103	NE of Bahamas	1					1
5/9/42	U-109	off Canaveral	7					7
5/9/42	U-564	Straits		2		10		12
5/10/42	U-98	Cape Canaveral	1	3				4
5/10/42	U-564	Straits	1		5			6
5/10/42	U-108	Windward Passage	4					4
5/11/42	U-108	Windward Passage	2					2
5/12/42	U-98	Straits N of Canaveral	1					1
5/12/42	U-98	Cape Canaveral	3	1				4
5/12/42	U-103	SW of Inagua	4					4
5/13/42	U-564	Nicholas Channel	4					4
5/13/42	U-108	E of San Salvador		1	6			7
5/14/42	U-564	Nicholas Channel	4					4
5/14/42	U-106	Straits S of Jax	4					4
5/14/42	U-108	E of Cat Island		1				1
5/15/42	U-564	Nicholas Channel	15	1				16
5/15/42	U-106	Straits off Canaveral	3	1				4
5/17/42	U-753	Straits W of Grand Bahama	5					5
5/17/42	U-106	straits off Jupiter Inlet	20	2	6			28

Date	Axis Sub-Marines	Position	Air	Vsls	Blimp	Bomb	Misc	Tally
5/18/42	U-564	Straits off Palm Beach	8	2		1		11
5/18/42	U-751	Windward Passage	1	1				2
5/18/42	U-753	Straits SE of Miami	15					15
5/19/42	U-564	Straits	1	1				2
5/19/42	U-558	Windward Passage	3					3
5/19/42	U-753	Straits SE of Miami	7	4		15		26
5/20/42	U-108	N of Puerto Rico					4	4
5/20/42	U-753	S of Key West	2					2
5/20/42	U-106	Straits off Keys		4				4
5/23/42	U-107	W of Matanillo Shoal	6					6
5/23/42	U-564	W of Bermuda N of Grand Bahama		1				1
5/24/42	U-107	Straits E of Miami	5	1				6
5/26/42	U-504	Windward Passage	4					4
5/26/42	U-107	OBC off Nuevitas		1				1
5/28/42	U-107	OBC off Cape Maysi	1					1
6/3/42	U-159	approaching Mona Passage southbound	2					2
6/4/42	U-172	NE of Puerto Rico	1					1
6/4/42	U-68	Anegada Channel, Sombrero Passage	3			4		7

Date	Axis Sub-Marines	Position	Air	Vsls	Blimp	Bomb	Misc	Tally
6/4/42	U-159	entering Mona Passage southbound	13					13
6/5/42	U-172	Mona Passage	5					5
6/6/42	U-158	Caicos Passage	5			2		7
6/6/42	U-172	Mona Passage		2				2
6/13/42	U-67	NW of Inagua	5					5
6/15/42	U-129	OBC off Cuba	1					1
6/16/42	U-67	Straits N of Matanzas Cuba	5	1			1	7
6/16/42	U-129	OBC off Nuevitas		2				2
6/17/42	U-129	OBC	5					5
6/17/42	U-67	Strait S of Keys	6					6
6/19/42	U-107	Mona Passage	3					3
6/19/42	U-129	Straits S of Key West	1					1
6/20/42	U-107	N of Puerto Rico	1					1
6/24/42	U-159	Sombrero Passage outbound	2					2
6/26/42	U-154	N of Anegada		1				1
6/30/42	U-505	NE of Anegada		1				1
7/1/42	U-332	NE of Bahamas, W of Bermuda					1	1
7/2/42	U-332	W of Bermuda					1	1
7/2/42	U-509	300 nm N of Anegada	1				1	2
7/2/42	U-84	NE of Eleuthera, SW of Bermuda	1					1
7/5/42	U-134	NE of Abaco					1	1

Date	Axis Sub-Marines	Position	Air	Vsls	Blimp	Bomb	Misc	Tally
7/5/42	U-571	W of Grand Bahama	1	1		5		7
7/5/42	U-84	NE of Grand Bahama	1					1
7/5/42	U-171	Windward Passage	1			3		4
7/6/42	U-571	Straits of Port Everglades	3	1				4
7/6/42	U-84	Straits off Jacksonville	4	3				7
7/7/42	U-571	Straits off Miami	5					5
7/7/42	U-84	Straits off Miami	10					10
7/8/42	U-571	Straits off Port Everglades	6	2		4		12
7/8/42	U-134	Straits off Canaveral	1					1
7/9/42	U-571	S of Key West	4	1				5
7/10/42	U-84	Crooked to OBC	2					2
7/10/42	U-571	S of Key West	1	1				2
7/10/42	U-134	Straits off Canaveral	5					5
7/10/42	U-166	S of Bermuda		2		6		8
7/11/42	U-161	Windward Passage	6					6
7/11/42	U-134	Straits off Palm Beach	3	1				4
7/11/42	U-84	OBC near Lobos		1				1
7/12/42	U-134	Straits off Palm Beach	4					4
7/13/42	U-134	Straits off Palm Beach	1					1
7/13/42	U-84	Nicholas Channel	2					2
7/13/42	U-161	Providenciale s	1					1

Date	Axis Sub-Marines	Position	Air	Vsls	Blimp	Bomb	Misc	Tally
7/15/42	U-161	NE of Eleuthera					1	1
7/16/42	U-161	NE of Eleuthera		5		20		25
7/16/42	U-161	NE of Eleuthera					1	1
7/16/42	U-161	312 S of Bermuda	1					1
7/16/42	*Giuliani*	190 nm ENE of Anegada, 750 S Bda.	1					1
7/17/42	U-134	NW Providence Channel	1				3	4
7/17/42	*Giuliani*	NE of PR	2					2
7/18/42	U-134	Straits off Bimini	6		1			7
7/18/42	U-161	SE of Bermuda		3				3
7/19/42	U-134	Straits off Key West	1					1
7/19/42	*Morosini*	SE of Bermuda		1				1
7/21/42	U-84	Straits off Key West	9			5		14
7/22/42	U-509	N of Puerto Rico					1	1
7/22/42	U-84	Straits off Port Everglades	1					1
7/23/42	U-84	Straits off Canaveral	1					1
7/24/42	U-509	N of Puerto Rico	1					1
7/26/42	U-508	NE of Bahamas					1	1
7/26/42	U-134	Nicholas Channel	2					2
7/27/42	U-134	OBC off Cayo Coco	6	2				8

Date	Axis Sub-Marines	Position	Air	Vsls	Blimp	Bomb	Misc	Tally
7/27/42	U-84	NE of Bahamas, SW of Bermuda		4				4
7/28/42	U-134	OBC off Nuevitas	1					1
7/28/42	U-84	55 nm S of Bermuda	1					1
7/28/42	U-84	S of Bermuda close	3	1		3		7
7/29/42	U-134	OBC off Cuba	1			4		5
7/30/42	U-134	OBC off Nicarao Cuba	2					2
8/1/42	U-134	N of Puerta Plata DR	2					2
8/1/42	U-510	315 SSE of Bermuda	1					1
8/1/42	U-508	OBC at Lobos		4				4
8/3/42	U-564	SE of Bermuda					5	5
8/3/42	U-173	Mona Passage	1					1
8/4/42	U-508	off Havana	2					2
8/4/42	U-134	S of Inagua, Windward Passage		1				1
8/5/42	U-173	NE of San Juan	1					1
8/8/42	U-463	405 SE of Bermuda	1					1
8/8/42	U-505	435 nm SE of Bermuda	1					1
8/9/42	U-508	off Key West	5					5
8/9/42	U-98	Straits off Jacksonville	15					15
8/10/42	U-98	Straits S of Jacksonville		4				4
8/11/42	U-463	SE Bermuda, NE Anegada, midway					1	1
8/11/42	U-164	SE Bermuda, NE Anegada, midway					1	1

Date	Axis Sub-Marines	Position	Air	Vsls	Blimp	Bomb	Misc	Tally
8/12/42	U-508	off Key West	4	2				6
8/12/42	U-163	345 SE of Bermuda	1			2		3
8/13/42	U-600	Windward Passage	3	3				6
8/13/42	U-508	Straits	3			5		8
8/13/42	U-163	330 SSE of Bermuda	1					1
8/13/42	U-163	300 nm SE of Bermuda (est.)	1					1
8/14/42	U-463	S of Bermuda					1	1
8/14/42	U-511	S of Bermuda					1	1
8/14/42	U-164	N of Mona Passage	1					1
8/14/42	U-600	OBC near Lobos	2					2
8/15/42	U-600	OBC near Lobos	1					1
8/16/42	U-558	ESE of Bermuda					1	1
8/16/42	U-462	ESE of Bermuda					1	1
8/16/42	U-600	OBC	1					1
8/17/42	U-553	Windward Passage	1					1
8/17/42	U-600	N coast Cuba OBC	5					5
8/18/42	U-600	N coast Cuba OBC	3	3				6
8/19/42	U-600	Straits	5					5
8/19/42	U-511	Windward Passage	1					1
8/23/42	U-558	Caicos Passage	2					2
8/23/42	U-558	Windward Passage	3	1		6		10
8/24/42	U-558	Windward Passage	2	2		18		22
8/24/42	U-558	Windward Passage		1		18		19

Date	Axis Sub-Marines	Position	Air	Vsls	Blimp	Bomb	Misc	Tally
8/26/42	U-553	N of Mona Passage, Punta Cana, DR	1					1
8/26/42	U-598	300 nm S of Bermuda	1					1
8/27/42	U-171	NE Jacksonville			1			1
8/28/42	U-86	NE of Bahamas	1					1
8/28/42	U-171	Straits off Jacksonville			1	2		3
8/30/42	U-171	N of Abaco		6				6
9/1/42	U-171	W of Bermuda					1	1
9/2/42	U-600	315 SSW of Bermuda	1					1
9/3/42	U-600	355 nm S of Bermuda	1			3		4
9/3/42	U-600	285 SE of Bermuda	2					2
9/5/42	U-171	NE of Abaco	1					1
9/20/42	U-455	W of Bermuda	1					1
9/23/42	U-455	N of Jax	1					1
9/24/42	U-455	E of Savannah	1					1
1/19/43	U-214	in Mona Passage northbound	11					11
1/20/43	U-214	NE of Mona Passage	1					1
2/27/43	U-155	N of Puerto Rico	1					1
3/1/43	U-185	ENE of Bermuda	1					1
3/2/43	U-155	N of Bahamas	1					1
3/2/43	U-185	E of Bermuda	1					1
3/2/43	U-185	E of Bermuda	1					1
3/3/43	U-155	N of Bahamas	1					1

Date	Axis Sub-Marines	Position	Air	Vsls	Blimp	Bomb	Misc	Tally
3/5/43	U-155	Windward Passage	1			30		31
3/8/43	U-185	Crooked Island Bahamas	1					1
3/8/43	U-185	Crooked Island, Castle Island	2					2
3/9/43	U-185	S of Castle Island	1	1				2
3/10/43	U-185	Windward Passage	1	3				4
3/10/43	U-185	Windward Passage	1	3				4
3/10/43	U-185	Windward Passage		1				1
4/22/43	U-185	NE of Bahamas					1	1
4/22/43	U-129	NE of Bahamas	2	2		6		10
4/24/43	U-129	NE of Bahamas	1					1
4/26/43	U-129	N of Bahamas		1				1
4/28/43	U-129	NE of Bahamas		1			1	2
6/7/43	U-66	Straits	1		1			2
6/10/43	U-66	Straits	4					4
6/11/43	U-66	Straits	4					4
6/12/43	U-66	Straits	5		3			8
6/13/43	U-66	Straits	1		1			2
6/13/43	U-66	Straits	3		3			6
6/16/43	U-66	Straits	1		7			8
6/28/43	U-66	Straits	2					2
6/29/43	U-66	Straits off Jacksonville	1		2			3
6/30/43	U-66	Straits off Jacksonville	1		3			4
7/1/43	U-66	Straits off Jacksonville			2			2
7/2/43	U-66	Straits	1	1	4			6

Date	Axis Sub-Marines	Position	Air	Vsls	Blimp	Bomb	Misc	Tally
7/3/43	U-66	Straits	6					6
7/4/43	U-66	Straits	3					3
7/5/43	U-527	300 nm S of Bermuda	1					1
7/6/43	U-66	Straits	3					3
7/8/43	U-66	Straits	1					1
7/8/43	U-134	300 nm SE of Bermuda	2			12		14
7/9/43	U-66	Straits	4					4
7/12/43	U-66	Straits	1					1
7/13/43	U-161	Providenciale s	1					1
7/18/43	U-134	1 mile west of Cay Sal Bank			1			1
7/19/43	U-134	Straits of Florida, N of Cay Sal Bank				3		3
7/20/43	U-134	Straits of Florida	2			9	1	12
7/29/43	U-134	Straits of Florida	3					3
8/14/43	U-463	S of Bermuda					1	1
8/27/43	U-171	NE Jacksonville			1			1
9/1/43	U-171	W of Bermuda	1	1				2
9/5/43	U-171	NE of Abaco	1					1
9/17/43	U-518	400 nm SSE of Bermuda		1				1
9//43	U-518	S of Bermuda	2			24		26
9/25/43	U-518	Straits		2				2
9/26/43	U-518	Straits off Miami	15	1	1			17
3/6/44	U-154	Windward Passage southbound	1					1
4/1/44	U-218	off San Juan, PR (est.)	1					1
6/2/44	U-539	N of Puerto Rico	1					1

Date	Axis Sub-Marines	Position	Air	Vsls	Blimp	Bomb	Misc	Tally
6/2/44	U-539	NE of Puerto Plata DR	1			1		2
6/5/44	U-539	Mona Passage	2	2				4
6/6/44	U-539	Mona Passage	2			5		7
5/2/45	U-506	Straits off Palm Beach	5					5

Appendix 14: Allied Air Attacks, Reported Attacks, Sweeps in Region

Date	Type	Where	What Allied	Who	What Axis
2/11/1942	Attack	E of Bermuda	merchant ship *Pan Halcyon* thanks Bermuda aircraft for deterring U-boat	USN	
3/8/1942	Attack	S of Bermuda (estimate)	2 VP-74 PBM Mariners	USN	U-128
4/7/1942	Attack	S of Bermuda (estimate)	Bermuda craft, vectored 2 others	Carlson, A. E.	U-571
4/7/1942	Attack	S of Bermuda (estimate)	Bermuda craft, 4 Mark 17 bombs	Gannon, John. W.	U-571
4/7/1942	Attack	S of Bermuda (estimate)	Bermuda craft, by Carlson, supported Soverell	Soverell, William W.	U-571
4/7/1942	Attack	S of Bermuda (estimate)	1 attack on other sub (U-571), minor damage	USN	U-571
4/7/1942	Attack	S of Bermuda (estimate)	resulted in 1 unsuccessful attack on U-754	USN	U-754
4/15/1942	Attack	S of Bermuda (estimate)	PBM-1, Bu.No. 1252, night sweep, surfaced U-boat, dived	Koch, George	
6/13/1942	Sunk	Straits of Florida E of Key West, CG Cutter *Thetis*	USCG *Thetis* WCG-115, focussed attack	McCormick, N. C.	U-157
7/8/1942	Attack	107 nm E of Bermuda	Kingfisher of VS-32 inshore patrol squadron	USN	E of Bermuda

7/16/1942	Attack	290 nm S of Bermuda		USN	U-161
7/16/1942	Attack	262 nm SE of Bermuda	VP-52 PBY-5 BuNo. 2367 (52-P-6)	USN	U-161, poss. *Giuliani*
7/16/1942	Attack	SE of Bermuda	VP-52 PBY-5 BuNo. 2367 (52-P-6)	USN	*Giuliani*, U-161
7/16/1942	Attack	NOB Bermuda Naval Operating Base	Catalina ferry crew reported sighting a U-boat	USN	
7/16/1942	Attack	185 ENE of Anegada, 750 nm S of Bermuda		USN	*Giuliani*
7/28/1942	Attack	68 nm S of Bermuda	VP-52 PBY 52-P-2	USN	U-84: unlikely: 400 nm away
7/28/1942	Attack	303 nm S of Bermuda	VP-52 PBY5s, 52-P-9 #2371	USN	U-510
8/1/1942	Attack	315 nm SSE of Bermuda	VP-52's PBY-5's P-9 (BuNo. 2371)	USN	U-510
8/8/1942	Attack	400 nm SE of Bermuda	52-P-9 2 attacks, poss. 505 and 463	USN	U-463, poss. U-505
8/8/1942	Attack	440 nm SE of Bermuda	52-P-9 2 attacks possibly on both 505 and 463	USN	U-505, poss. U-463
8/8/1942	Attack	400 nm SE of Bermuda	VP-52 52-P-9 2 attacks while looking for survivors of *Maldonado*	USN	
8/12/1942	Attack	350 nm SE of Bermuda	52-P-9 2, depth charges	USN	U-163
8/12/1942	Attack	350 nm S of Bermuda	VP-52 52-P-9, 2 depth charges	USN	U-163; U-463 & U-505 near

8/13/1942	Attack	330 nm SSE of Bermuda	52-P-10 another attack	USN	U-163
8/24/1942	Search	S of Bermuda (vague)	USAAF B-17 Kindley Bermuda to Langley VA	USN	
8/26/1942	Attack	N of Mona Passage at Punta Cana DR	USAAF B-17 on ASW mission	USAAF	U-553
8/26/1942	Attack	300 nm S of Bermuda	USAAF B-17 on ASW mission	USAAF	U-598
8/26/1942	Attack	S of Bermuda (estimate)	USAAF B-17 ASW, attack on a periscope	USAAF	U-598 or U-553
9/2/1942	Attack	310 nm S of Bermuda	USN Aircraft builder no. BuNo. 2352	USN	U-600
9/3/1942	Attack	280 nm ESE of Bermuda	2 BDA PBY-5's, U-boat dived	US	U-600
9/3/1942	Attack	360 nm SSE of Bermuda	52-P-10 attacked w/ 3 depth charges	USN	U-600
9/3/1942	Attack	700 nm SE of Bermuda	VP-52 52-P10, 3 depth charges	USN	U-505, poss. U-463
9/25/1942	Attack	S of Bermuda (estimate)	USAAF B-17 reported damaged a U-boat	USAAF	
2/21/1943	Search	Bahamas (encoded)	B25 FR-374 and B25 FK-376	RAF No. 111 OTU	
2/22/1943	Search	5 nm E of Hatchet Bay, Eleuthera	U-boat sighted 2pm, 5 nm at 5 knots, patrol to investigate	RAF No. 111 OTU	
2/25/1943	Search	NW of Castle Island Light, Acklins	investigated U-boat sighting	RAF No. 111 OTU	
4/3/1943	Search	in transit from NE to NW	B25 FL-976 and B25 FR-375, 10.5 fruitless hours looking	RAF No. 111 OTU	

		Providence channels			
4/22/1943	Search	Bahamas (encoded)	B25 FL-942 FC Liberator, USN sent position	RAF No. 111 OTU	
5/2/1943	Attack	E of Bermuda	8 X USAAF B-24's of the 25th Air Wing (ASW), after U-supply boats, or Milk Cows	RAF No. 111 OTU	
5/15/1943	Search	5 nm E of San Salvador	B25 FR-378 CL, B25 ER-371 CF, and FR-303 DX searched: was a USCG cutter, not U-boat	RAF No. 111 OTU	
5/15/1943	Sunk	Nicholas Channel NE of Cuba, S of Cay Sal	Effective depth charge attack by convoy escort	Delgado, Mario Ramirez	U-176
7/3/1943	Attack	S of Bermuda (estimate)	2 attacks, bombs jammed	Brown, George W.	U-134, poss. U-66
7/3/1943	Attack	S of Bermuda (estimate)	2 attacks, bombs jammed	Brown, George W.	U-66
7/3/1943	Search	150 nm E of Eleuthera	B25 FV-954 FH saw oil patch	RAF No. 111 OTU	
7/5/1943	Attack	310 SSW of Bermuda	105-B-8, attacked twice	USN	U-527
7/5/1943	Attack	S of Bermuda (estimate)	aw, attacked with depth charges	Steetle, John R.	U-159
7/5/1943	Attack	S of Bermuda (estimate)	saw, unable to attack due no bombs left	Steetle, John R.	U-527
7/7/1943	Search	150 nm NE of Abaco	BX-759 LP Liberator saw sub follow convoy	RAF No. 111 OTU	
7/8/1943	Attack	S of Bermuda	VP-74 Mariner/VP 201 (Hitchcock)	USN	U-134

7/8/1943	Attack	400 nm SE of Bermuda	PBM-3 #201-P-3	USN	U-134
8/7/1943	Sunk	SSW of Bermuda	PB4Y-1, 105-B-10, using straddle Mk.24 mines	Evert, Thomas R.	U-84
8/30/1943	Search	Rum Cay to Long Island	RAF Transport Command sighted sub. B25's FV951 FC, FR-366 CX, FR-380 DB patrolled, found USN destroyer	RAF No. 111 OTU	
1/31/1944	Sighting	Bahamas (encoded)	B25 FW-154 FR saw feather behind U-boat periscope. B25's FR-381 DC and FR-393 DF, and FV-997 FG patrolled, no sighting	RAF No. 111 OTU	
2/19/1944	Attack	Kindley Field, Bermuda	USAAF last ASW mission from Bermuda	USAAF	
3/3/1944	Sighting	100 nm ENE Harbour Isl.	B25 FF Liberator searched, nil	RAF No. 111 OTU	
3/10/1944	Sighting	ENE of Abaco heading for NE Providence Channel.	B25 FM & B25 FH Liberators searched	RAF No. 111 OTU	
7/28/1942	Attack	230 nm S of Bermuda	52-P-2 of VP-52 PBY		U-508; U-658, U-129 were far

Appendix 15: Allied Air Accident Details

Date	Aircraft	Who	Tally	Where	Summary
6/17/1941	FAA Supermarine Walrus R6551 of 773 Squadron	2 KIA; Flynn & Blatchford	2	at sea off Bermuda	failed to return from dummy attack mission at sea
3/9/1942	USN Vought Kingfisher BuNo. 2207 of VS-5D4	2 KIA; Brown & Locke	2	at sea off Bermuda	crashed
4/13/1942	USN Curtiss Seagull BuNo. 1175 of VCS-8 of USS *Savannah*			at sea off USS *Savannah* off Bermuda	ditched from aircraft carrier
6/3/1942	USN Martin PBM-1 of VP-74 BuNo. 1250	13 KIA; USN	13	at sea off Bermuda	lost in poor weather
9/3/1942	USN Martin PBM-1 of VP-74 BuNo. 1251 of VP-74	all saved		at sea off Bermuda	ditched
10/10/1942	USN Grumman F4F-4 BuNo. 11702 of VF-9 of USS *Ranger*			at sea off USS *Ranger* off Bermuda	collided with target sleeve, engine failed, ditched

10/13/1942	USN Curtiss SBD-3 BuNo. 06552 of VGS-26 of USS *Sagamon*			at sea off USS *Sagamon* off Bermuda	crashed at sea
12/15/1942	USN aircraft	all missing, KIA	10	off Melbourne FL, in straits	passing reference in base operational diaries
12/25/1942	RAF Catalina FP266	6 KIA, including 2 civilians	6	at sea off Bermuda	lost out of Bermuda
1/19/1943	A20 Havoc Douglas		5	at sea from Satellite, Oakes Field, Nassau	seen going down in the sea
2/6/1943	RAF Catalina FP309	6 KIA, including 3 civilians	6	at sea off Bermuda	lost out of Bermuda
2/10/1943	USAAC A20 Havoc Douglas	all survived, rescued		Andros Island, Bahamas	B25 FR-393 Capt. McBratney located them, dropped a Thornaby bag, ASR launch, then US Catalina on site
2/10/1943	USN B25 Liberator	all 8 missing, KIA	8	NE Andros	B25 FH and B24 MX while on exercises cooperated with USN in search for US Liberator based at Nassau, negative result
3/18/1943	Baltimore FA-354, RAF Ferry Command	2 KIA; 1 at hospital	1	1 mile from Satellite, Oakes Field	on flight to Trinidad overshot runway, so pilot endeavoured to bank aircraft to initiate another circuit, however aircraft failed to respond to controls, crashed 1 mile away from Windsor Field, both occupants badly burnt, aircraft caught fire totally destroyed
3/19/1943	B25 FL697	2 KIA; Zdan &	2	Soldier Road, Nassau	an engine caught fire at take-off, plane slipped into the

		Daiken, in hospital			scrub and burned out, both men made it to hospital, died there
3/25/1943	Baltimore FA340	1 KIA; Glenn	1	Main Field	took off Windsor Field, sole occupant, took a hard bounce on landing, caught fire, Glenn was thrown out and killed
3/28/1943	Baltimore FA427	2 KIA, 1 in hospital	2	Nassau	took off for Trinidad, caught fire, navigator seriously burnt, survived at RAF Hospital, Captain and Radio Operator KIA
4/8/1943	USAF B26 Marauder	all 4 rescued		Castle Island, S Acklins	between Morrison and Borinquen, PR, crashed near Castle Island, off Acklins Island, B25 FR-371 arrived Castle Island, dropped messages for islanders, who lacked radio contact, damaged Marauder was found by a plane from Guantanamo along with 1 survivor. Plane crashed on a sandy flat, badly damaged, no propellors visible, a Catalina from Cuba saw crew waving from shore and dinghy, all 4 rescued, slight injuries
4/9/1943	B25 FR379 DA	all 8 missing, KIA	8	at sea S of New Providence	after rearming aircraft not heard of again, FK-174 observed a 20-mile oil patch, FR-383 DX then investigated, determining the patch was sea-weed, Pan American aircraft sighted small pieces of wreckage afloat S of New Providence. Large yellow-green dyed area and oil flecks also observed, however could not be ascertained if men were swimming near the

					wreckage, yellow- green dyed area was thought to be a dinghy, ASR launch HMS *P-190* dispatched from Lyford Cay for search, on scene reported dinghy was empty, dinghy brought to Windsor Field, found to belong to FR-379 DA, dinghy had been inflated, however no further evidence of lost crew
4/17/1943	Baltimore FA121	pilot survived no KIA		Control Tower, Satellite, Oakes Field	crashed after take-off, developed engine trouble, hit the Control Tower, then caught fire and burnt out, pilot escaped with a fractured jaw
4/19/1943	USN PBY amphibious	all survived		Whale Cay, Berry Islands	Whale Cay, Berry Islands reported a USN PBY amphibious plane force-landed off Whale Cay, owned by Marion Carstairs. FR-381 DC and FR-380 left Nassau to support 2 waterborne craft in search, PBY was reported safe, search called off
4/21/1943	B25 FL184	all 8 KIA	8	8 nm SE of Booby Rock, island E of Nassau, NP	overdue action initiated, a marine craft found an empty dinghy 10 nm E of Nassau, B25 AT informed B25 AH that they had found the aircraft, floating debris, the dinghy, and were circling, dingy found again by ASR launch from Montagu in empty state 8 nm SE of Booby Island, bearing same number as the missing aircraft, which could be seen in just 20 feet of water

4/29/1943	Hudson bomber FK365	all 3 KIA	3	near Main, Windsor Field	crashed and caught fire after take off
5/9/1943	B25 FR378 CL	all survived		engine fire over Andros, landed at Boca Raton, FL	B25 FR-378 CL airborne for night flying exercise signalled aircraft fire on starboard engine, marine craft section Lyford Cay stood by for ASR duty. FR-382 DD dispatched, message received from craft that it was over large fires on Andros later, CL signalled that it landed at Boca Raton, Florida
5/21/1943	B25 FR365 CD	all 5 missing, KIA	5	at sea off Morgan's Bluff, Andros	After fruitless searches of Eleuthera area, B25 CF reported an empty dinghy 5 nm N of Morgan's Bluff, Andros. Other aircraft were sent to reinforce CF off Andros, where they spotted an oil slick. Electrical storms hampered radio communications. On 29 May B25 FL-208 BF was airborne at 17 minutes after midnight and found a damaged dinghy with no occupants. Two marine launches #190 and #63 were sent to arrive the next dawn. B25 FV-953 FA was airborne by 10.22 am and learned from the launch that the dinghy was recovered empty with markings from the missing plane, DD. The plane FA searched N and E Andros for 2 hours and landed at 3.24 pm. The launches retrieved a plane wheel and spares.

5/24/1943	B25 FR-375 CJ	all 8 missing, KIA	8	at sea near Mayaguana, unconfirmed	B25 FR-375 CJ searched sea around Mayaguana for reported crashed aircraft, white planks, smaller pieces of painted wood seen floating, circled, aircraft deemed failed to locate missing aircraft
7/10/1943	USN Lockheed Ventura	all 5 missing, KIA	5	Inagua area; left Guantanamo on ASR	missing from Guantanamo Cuba during a search mission for 5 other lost USN airmen
7/13/1943	B25 FR383 DX	all 8 KIA	8	3 nm S of Main, Windsor Field	search initiated, B25 EV-946 FF airborne, could see smoke through trees and wrecked plane 3 miles S of Main, Windsor Field, from a Baltimore aircraft, oil patch 5 nm from Windsor Field. Marine ASR craft at Lyford Cay were sent to site, found 1 mile long oil patch, W to E, orange flashes seen, Lyford Cay launch on scene found observer's seat, notebook, sundry pieces of plywood
7/29/1943	USN aircraft & Kingfisher float plane rescue craft	all survived, rescued		at sea off Bimini Islands	B25 FV-952 FX escorting tanker *Manual Rionda* observed blimp circling a smoke float on the water, and that a US aircraft was floating beneath it, with wings submerged, dinghy with survivors sighted near wrecked aircraft, and a Kingfisher float plane alighted on the water to pick up the dinghy and the survivors
8/7/1943	B25 FV953 FA Mitchell Cyclone	all 6 KIA	6	2 nm E of Main, Windsor	report received that aircraft crashed E of the aerodrome, B25 FL-200 BX reported that

			Field, near Soldier Road	the aircraft was destroyed by fire, fire tender and crash party were dispatched to scene of the accident, B25 FL-187 DC reported aircraft crashed 2 nm E of the airfield, crashed airframe and engine both totally damaged after take-off, reached height of 1,000 feet, rose sharply, nose-dived, aflame 2 minutes after take-off near Main Airfield, cause of accident unknown, court of inquiry submitted
8/12/1943	Mitchell B25C2 FL 199 Cyclone	injured	Windsor Field	pilot rounded out too low, and aircraft landed heavily on all 3 wheels at same time
8/12/1943	USN Ventura		Lake Killarney, New Providence	B25 FR-375 CL observed USN Ventura-type aircraft force land, then reported turret and engine lying separate from the Ventura's tail, which lay in a lake
8/15/1943	Liberator BZ 761, Pratt & Whitney	no injuries	firing area, Windsor Field	slight damage during gunnery exercise; gunnery instructor cleared stoppage in starboard beam gun, however during subsequent testing burst, aircraft hit an air pocked, instructor momentarily lost control of gun, resulting in 2 X 0.5 bullets entering starboard wing
8/20/1943	B25D FA FR-375 Mitchell	no injuries	Windsor Field	slightly damaged, engine seriously damaged, hydraulic line burst, wheels lowered in emergency method, main heels refused to lower completely and would not lock lower wheels, attempted for 1.5 hours, landing made,

				main wheels collapsed, aircraft ended up on its nose at Main Airfield, accident occurred due to failure of normal hydraulic system and emergency undercarriage lowering system, no blame to personnel
8/27/1943	B25 FR375 CJ	no injuries	Main Field	returned from exercise with leaking petrol feed pipe, circled airfield, preparatory to landing, found that the under-carriage hydraulics were inoperative, continued to circle in an endeavour to release the under carriage, 2-way contact could not be maintained, aircraft could receive; another aircraft was airborne to lead CJ on to the runway, and CJ successfully crash landed
9/4/1943	B24 FA	no injuries	Windsor Field	nose damage on hard daylight landing
9/21/1943	USN Baltimore	2 survivors rescued	NW Andros, beach	aircraft was coming from Morrison or Jacksonville, extensive searches initiated, reports received via RAF Transport Command that airship had sighted a raft, and flying boat was proceeding to position, commanding officer instructed ASR launch, HMS *P-339* from Montagu Bay to scene of the raft, flying boat instructed not to become waterborne as sea was choppy, ascertained that Baltimore had been landed on beach on NW point of Andros,

				dinghy had been put to sea to attract attention, ASR launch *P-339* unable to approach near shore because of shallow water, instructed to stand by and await high tide and then approach, signal received through USN that airship had picked up 2 survivors and was landing on Main, Windsor Field, ship landed, and survivors taken to hospital, signal sent to *P-339* to return to base
10/5/1943	B25 FR-384 CM	all crew rescued	Bannerman Town, S Eleuthera, on rock	CX returned from flight, reported seeing CM jettison its depth charge and head for land, signal was received from captain of CM that aircraft ditched E shore of Eleuthera, all were safe, one man badly injured, was requested that the Walrus be dispatched, port engine of CM had failed and to rectify the drag the captain had full boast the starboard engine, engine then began to show signs of seizing up, was then decided to make a forced landing, depth charges were jettisoned and the aircraft was ditched on a rock 100 yards from the shore, crew were able to leave the aircraft, rear gunner was unconscious with head injuries and was extracted from the aircraft by the other members, dinghy was launched and the crew paddled ashore, captain

			obtained assistance from a native constable at Bannerman Town, medical assistance was sent from Rock Sound, crew were taken to Rock Sound Hospital and there their injuries rare attended to. Marine craft launch HMS *P-89* at Harbour Island went to scene to pick up the airmen, then to rendezvous at Ship Channel Cay with HMS *P-191* from Nassau, however *P-89* failed to locate the aircraft, all the crew were in hospital at Rock Sound, *P-89* signalled to proceed to Rock Sound to pick up crew, *P-191* sent to Rock Sound and to await reloading, Medical Officer arrived with *P-191* which left Rock Sound Hospital, 1 member injured docked, crew were taken to The Station Hospital near Oakes Field, 1 member was detained
10/13/1943	B25 RAF Transport Command	Lyford Cay off Simms Point towards Goulding Cay	Liberator BZ-812 MC aircraft continued upon its course to base when over Lyford Cay, a smoke float and a patch of oil was sighted, message was intercepted from a Liberator in the vicinity "Searching for survivors," ASR launch from Lyford Cay was observed to be approaching the oil patch, report was received from the guard at Clifton Pier, further report was received from DFCO of Windsor Field, aircraft was a B25, belonging

				to RAFTC, which had crashed in flames into the sea between Lyford Cay, and Goulding's Cay. Liberator on local flying circled the scene of the area, ASR launch recovered the bodies of 2 of the occupants, thought that there were 4 occupants of the aircraft
10/14/1943	Liberator BZ-760 LT	emergency landing, no injuries	Main Field	Liberator BZ-760 LT was airborne upon Kingsley Exercise, oil pressure in the fourth 10 engine dropped and it was not possible to feather the propeller, aircraft lost altitude, aircraft set course for base and sent SOS signal, depth charges were jettisoned and the crew instructed to prepare for ditching, signal was sent to base "ditching", aircraft began to maintain height and the aircraft sent a signal to base to this effect aircraft was instructed to transmit its position every 15 minutes, signal was sent to base stating that the fourth had failed and that the aircraft was over the northern tip of Eleuthera, aircraft landed. Upon the receipt of the SOS signal preparations were made to dispatch, two B25 aircraft to the position given signal was sent to LP and LM upon the receipt of the ditching signal from LT, giving the position of the Liberator LT and instructing the aircraft to

				search in that area. US Naval Air Stations at Guantanamo and Great Exuma were requested to give any assistance that might be required, Marine craft section was ordered to have all available boats standing by, B25 FV-952 FX airborne to search in the area in which LT was ditching, after it was ascertained that LT was still in the air and maintaining height, a signal was sent to FI to continue the exercise, aircraft landed back. Liberator BZ-759 LP airborne from Windsor Field for creeping line ahead patrol, Liberator was sighted visually, homing exercise was not practicable, patrol was completed without incident and the aircraft landed.
10/24/1943	USN Grumman F6F3 BuNo. 25899, VF8, USS *Intrepid*		at sea off USS *Intrepid*, off Bermuda	crashed approaching USS *Intrepid*
11/2/1943	B24	all missing, 8 KIA	at sea, Palm Beach to Borinquen, PR	distress signal was intercepted from Morrison, might have originated from the life raft from the missing B24, overdue at Borinquen on 29th October, bombers carried out thorough search, sighting nothing
11/23/1943	B25 FV-952 FX	all missing, 8 KIA	NW Light, Bimini, Straits of Florida	failed to return at night, signal terminated abruptly, signal was received from Miami that the Coast Guard reported a

					plane crashing into the sea near NW Light, Bimini
12/16/1943	B25 FL-167 AX	no injuries		Windsor Field	landed and damaged the nose wheel, aircraft took off again and crashed, landed, no casualties
12/23/1943	B24 CK	all KIA	8	Athol Island, NE of Nassau	failed to return from a local night navigation, on 26th December 1943, a self-sealing petrol tank was found, by a civilian upon the shore near Dick's Point, Montagu Bay, Nassau. Tank was found to be from the missing aircraft CK. On 28th December 1943, a dinghy was found floating in Montagu Bay. The dinghy had been taken to Athol Island, where the Commanding Officer and the MCS Officer investigated it. Upon interrogation of a native on the island, it was ascertained that an aircraft had crashed into the sea off Athol Island, visual search was made for the aircraft by the ASR, launch, many pieces of wreckage were seen, including part of the fuselage bearing the letters CK. No trace of bodies, aircraft disintegrated.
12/29/1943	FAA Blackburn ROC L3109	all missing, KIA	8	at sea in Bahamas	ditched after engine explosion and fire
1/9/1944	B24 Liberator 744BZ	all missing, KIA	8	between Boca Raton, FL and Windsor Field	missing aircraft in an area SE of New Providence, request received for aircraft to search Andros for Liberator missing from 9th. Liberator BZ-744 LL

				carried out local flying search of Andros, sighted nothing
1/15/1944	USN aircraft, per ATC (air traffic control) in FL	all missing, 8 KIA	Morrison Air Field to Borinquen, PR	overdue from Morrison to Borinquen, RAF search impracticable due to the scanty information
4/24/1944	B25 AG		W coast Andros, beach	Comgulfseafron requested aircraft investigate wreck on beach, W end of Andros. B25 AG failed to locate wreck, which presumably was an older crash
4/24/1944	RAF B24 LP	all missing, 8 KIA	at sea between Bermuda & Windsor Field	Kingsley Field, Bermuda arranged that frigate of HMS *Dominica* should carry out search, also Liberators
5/6/1944	USN C47	all survived	forced landing at Windsor, Miami to San Juan	force-landed in the Unit from Miami, diverting en route to San Juan
6/6/1944	B25 FR	all survived, rescued	2 nm of Tarpum Bay Eleuthera	signal received from pilot via government channels that aircraft was ditched in sea 2 nm N of Tarpum Bay, Eleuthera, aircraft had depth charges aboard but these were not safe, crew uninjured except radio operator who had face injuries, ASR launch, HMS *P-712* dispatched from Harbour Island, arrived at Rock Sound, near Tarpum Bay, crew brought to Nassau in *P-712* at berthed at Prince George Dock, others rescued by Goose GB from Rock Sound via Governor's Harbour

8/7/1944	B25 FV-953 FA	all missing, 8 KIA		Bahamas	crash of B25 FV-953 FA
8/24/1944	USN Mariner	all rescued		N of Harbour Island, Eleuthera	landed on sea, crew safe, USN requested assistance from RAF ASR base Harbour Island, B25 FC sent signal to launch at Harbour Island, then located Mariner, FC returned to base, so USN depot ship stationed at Royal Island, N Eleuthera, took over from tug 339 whose its tow-line was unable to stand the strain, and 339 returned to base
9/11/1944	USN Dauntless	all missing, 6 KIA		E of Miami, in Straits of Florida	missing E of Naval Air Station (NAS) Miami
10/14/1944	B26 Marauder, RAF Transport Command	all 3 KIA	3	500 yards W of North Cay, Balmoral Isl.	crashed into sea 500 yards off N shore of New Providence to the W of North Cay [Balmoral Island], aircraft entirely submerged
10/28/1944	B25 FL	all 6 KIA	6	S of Fox Hill Road, near Oakes, Field	crashed near bombing range, on target at Fox Hill beach, civilian observed aircraft crashed to the S of Fox Hill road, smoke observed in that direction. B25 FH and FP reported an aircraft was on fire in swampy bush, disintegrated, difficult to identify aircraft type, ambulance and fire tender dispatched to scene, flying Instructors followed them, it was found to be a B25, aircraft had commenced bombing practice then banked steeply, crashed into bush, caught fire, was not on fire while in flight

11/24/1944	B25 FW	7 KIA, 2 rescued	7	E of Montagu Bay, Nassau	Duty Flying Control Officer reported aircraft ditched, captain confirmed that FP had ditched and FW had gone straight to sea, FS circled scene of crash air, ASR launch from Montagu picked up 2 survivors of FW
11/24/1944	B25 FP	all 9 KIA	9	E of Montagu Bay, Nassau	Duty Flying Control Officer reported aircraft ditched, captain confirmed that FP had ditched and FW had gone straight to sea, FS circled scene of crash air, ASR launch from Montagu found no trace of FP, no survivors
1/9/1945	USN B29 Liberator LH	all missing, KIA	8	at sea in Bahamas between FL and PR	missing on flight from Morrison to Borinquen, last contact with aircraft was when it was 100 nm from Morrison E of Florida
1/22/1945	USN VB-15 Curtiss SB2C-1 BuNo. 18235, USS *Hornet*	1 KIA; Fletcher	2	at sea off USS *Hornet* near Bermuda	crashed into the sea while approaching the carrier
2/23/1945	B24 LA	all 9 survived		Windsor Field	B24's LM and LA on local exercise Number 4 (dual), collided in mid-air S of drome; LA, damaged, made belly landing
2/23/1945	B24 LM	all 9 KIA		S of Windsor Field aerodrome	B24's LM and LA airborne for local exercise Number 4 (dual), collided in mid-air S of drome; LM crashed on land
3/6/1945	USN Dauntless	all missing, KIA	6	40 nm E of Port Everglades	missing

3/9/1945	USN PBY-5A of Hedron 9-2	all rescued		at sea between Bermuda and US	forced landing on water during flight from Bermuda to US; crew were rescued
3/11/1945	USN PBY4	all missing, KIA	10	at sea near Paw Hadron Rocks, N of Grand Bahama	PBY4 reported missing near Paw Paw Rocks, B24 MB carried out search with MC, saw nothing to report
4/6/1945	B24 Liberator LK	all 7 KIA	7	S of Windsor Field	B24 LK crashed on land, damaged, minor injuries
4/8/1945	B25 FF	all 7 KIA	7	S of New Providence	B25 FF crashed into sea, B25 FY saw oil streaks in sea, ASR craft marine craft HMS *2779* directed to area, nil result, only 2 parachutes locate that had been assigned to crew
4/21/1945	B24 MB	all 7 KIA	7	N of New Providence	FU crashed into sea, marine craft HMS *T-170* and HMS *P-191* searched, saw oil patch, picked up dinghy with identification
6/3/1945	B24 Liberator MB	all 8 KIA	8	10 nm NE of New Providence	B24 MB crashed into sea, NB located wreckage, directed ASR marine craft to scene, aircraft returned to base, unable to tow the main portion of wreck, US escort vessel requested to sink it
7/4/1945	B24 Liberator MD	all 8 KIA	8	Andros	B24 LO airborne to search for B24 MD located wreckage, all crew lost, directed ASR, launch HMS *P-191* picked up dinghy and saw portions of a fuselage and plane tail in water
7/10/1945	USN PBM Mariner	all 12 missing, KIA	12	at sea, Northern Bahamas to	USN at Miami requested search for a PBM, was overdue while on

				Great Exuma, unconfirmed	navigational exercise routed from Banana River, Florida over the Bahamas to Great Exuma; 12 July RAF aircraft concluded search
7/18/1945	USN B25 Liberator	all 8 missing, KIA	8	NE of Eleuthera	B25 FC in search for USN Liberator overdue while on exercise N and E Eleuthera, nil report, negative result
12/6/1945	USN Martin Mariner, Training 49	13 KIA	13	at sea NE of Abaco	sent with Training 32 to find USN FT-19 Squadron is also lost at sea NE of Bahamas
		TOTALS	***302***		

Appendix 16: Allied Anti-Submarine Patrols, U-Boat Searches

U-157 - 6/13/1942 - Straits of Florida E of Key West, CG Cutter Thetis

B25 FR374 - 2/21/1943 - B25 FR-374 and B25 FK-376 patrol 24.43N by 76.00W for a submarine.

2/21/1943 - Report received that U-boat had been sighted off San Salvador diverted to search the area, signal not received by aircraft area not searched

B25 FR-378 - 2/22/1943 - U-boat reported at 26.30N, 77.00W, sub observed going NW from E of Hatchet bay, 5 nm E off coast at 5 knots at 2pm.

B25 F-3379 - 2/25/1943 - U-boat reported 22.42N, 74.22W, B25 F3379 investigated

B25 FL-976 - 4/3/1943 - U-boat reported at 7.45 pm transiting Northeast to Northwest Providence channels, investigated at dawn next day, 4/4/43. 2 planes (B25 FL976 and B25 FR375) spent c.10.5 hours searching from 100-fathom-line to East to western 100-fathom line, nothing sighted.

B25 FL-942 FC - 4/22/1943 - U-boat reported by USN at 9.47 am; that destroyer USS Swanson or nearby reported a submarine sighted (position encoded). OTU111 sent B25 FL942 FC Liberator which was airborne from 6.10 pm to 4.15 am the following day, nothing sighted.

B25 FR-378 CL - 5/15/1943 - USN Guantanamo reported U-boat sighted at 5.18 pm at 24.10N, 74.20W. B25's FR378 CL, ER371 CF, FR303 DX took off at 6.31 pm. Just after 7 pm "an unidentified object was sighted, believed to be a surfaced U-Boat." Aircraft CF dove from 6,000 to 50 feet. The vessel proved to be a USCG Cutter escorting a convoy nearby and after signals were exchanged the attack broken off. Up to 10.07 pm all 3 aircraft returned to base.

U-176 - 5/15/1943 - Nicholas Channel NE of northern Cuba, S of Cay Sal

B25 FV-954 FH - 7/3/1943 - Anti-Submarine patrol from 6,000 feet B25 FV-954 FH saw oil patch at 25.40N, 73.16W

BX-759 LP Liberator - 7/7/1943 - 6.56 am BX-759 LP, started following convoy at 27.44N, 72.20W and by 12.35 pm was at 27.55N, 72.26W. Enemy submarine was reported to be following the convoy in question. Convoy sighted, signals exchanged, nothing further.

U-84 - 8/7/1943 - between Bermuda, NE Bahamas by VB-105 USN/B-4, Lt. UNSR Thomas R. Evert

RATC - 8/30/1943 - Between Long Island and Rum Cay near 23.28N, 75.02W at 2.40 pm a RAF Transport Command plane sighted a submarine. OTU111 aircraft B25's FV951 FC, FR366 CX, FR380 DB took off from 6.22 onwards between Long Island and

Rum Cay observed only a USN destroyer at 7.49 pm. They returned by 9.45 pm nothing further sighted.

B25 FM - 3/10/1944 - Comgulfseafron advised U-boat approaching Providence Channel on course 250 to 225 degrees, WSW. B25 FM airborne at 11.03 and landed at 4.15 pm without sightings. B25 FH also airborne at 11.03 am, returning 4.30 pm after a sweep with no results.

4/1/1944 - Comgulfseafron made a DF fix on a U-Boat within 50 nm of an indecipherable coded position [25N, 72W to 27N, 74W] at 1.10 am. B25 FO and B25 FS swept and found a convoy course 10 degrees NNE, speed 7 knots at 12.51 pm.

4/3/1944 - Again Comgulfseafron made a DF fix on a U-Boat within 50 nm of 25N, 72W to 27N, 74W at 1.50 am. USN sent SC's and destroyers. B25 FB and B25 FW patrolled uneventfully seeking U-boat.

Appendix 17: All RAF Personnel, Visitors, Recorded in the ORB's of No. 111 O.T.U.

Name Rank, Title/s

Abbott L.H.	F/O	
Aberdein	Wing Commander	
Abraham	J.P.	Sister
Abson F.L.	SGT	
Ackroyd	J.H.	LAC
Adams J.	S/Ldr.	
Adams A.	A/Snr/Sister	
Addiscott	F.	LAC
Addison	S.R.	F/L
Aikman	A.R.	AC1
Aimer M.C.	AC1	
Ainsworth	R.H.	SGT
Aitchison	J.M.	AC1
Aitchison	R.	LAC
Aitken P.R.	F/O	
Aitken F.	LAC	
Alcock C.	Sister	
Alcock D.C.R.	F/O	
Alcock C.R.	P/O	
Aldous D.	LAC	
Alexander	R.J.	F/Lt.
Alexander	Captain F/L	
Alexander	J.	Wing Commander
Alexander	F/L	
Allardyce	A.R.	LAC
Allen H.W.	A/F/L (F/O)	
Allen	Gunner Sgt.	
Allen E.A.	Flight Lieutenant	
Allen E.E.	F/L	
Allen		
Allies E.M.	F/Lt.	
Allingham	W.J.	P/O
Allman	J.C.	LAC
Allsop A.P.	AC2	
Allsop E.	AC1	
Aloof B.G.	CPL	
Ambrose	A.J.	SGT
Ancill H.A.	W.O.	
Anderson	D.F.	Group Captain
Anderson	W.S.	Rear Admiral
Anderson	M.O.	S/L
Anderson	E.J.	F/L
Anderson	J.	Cpl.
Andrew	K.	LAC
Andrew	J.W.	Cpl.
Andrews	A.E.	LAC
Andrews	H.	LAC (ACU)
Angell R.	LAC	
Anslemo	A.G.	P/O
Anton W.B.	CPL	
Appleton	J.H.	LAC (SGT)
Archer J.A.	AC1	
Archer F.	LAC	
Armstrong	G.E.	F/O (A/F/L)
Armstrong	J.	P/O
Armstrong	J.R.	F/O
Armstrong	C.	LAC
Armstrong	A.	Cpl.
Arnell N.	LAC	
Arnold W.C.J.	CPL	
Arnoldi	P.	P/O
Arnott T.	LAC	
Arthur J.	F/O	
Ashton	P.	SGT
Ashworth	T.A.	LAC
Aspinall	G.F.	P/O

Aston I.W. LAC
Atkins D.M. S/O
Atkinson E.B. LAC
Atwell F/Sgt
Austin J.G. CPL
Avery G.H. F/O
Avery M.J. F/Lt.
Avery M.E. F/L
Ayre A.G. S/L
Ayres W.E. CPL
Backham P.R. P/O
Backhouse S. Sgt.
Badgley J.A. P/O
Bagy W.B. F/L
Bailey J.P. P/O
Bailey J.P.V. F/O
Bailey R.G. LAC
Bain W.A. LAC
Baker A. F/O
Baker Useph Esquire
Baker R.C. F/O (A. F/L)
Baker H. LAC
Ballie F/Lt.
Balm R.L. LAC
Balsom A.E. LAC
Bancroft O.L. Esquire, Barrister-at-Law
Banks R.E. F/O
Banks W.U. P/O
Banks R.A. P/O
Banks W.U. F/O (A/F/L)
Banks W.U. F/O (A/F/L)
Banks C.W. SGT
Banner G. F/O
BarkerJ.L.D. P/O
BarkerL.A. AC1
BarlettG.C.C. S/L (A/W/C)
Barnden H.A. LAC
BarnesL.H. AC1
Barnett C.D. F/O
Barnett C.D. F/L
Barrett C.A. F/L (A/S/L)
Barrett H.E. LAC
Barrett F. LAC
Barson N. F/Lt.
Bartlett G.C.C. Wing Commander
Bartlett A.W.C. S/L
Bartlett B.C. AC1
Basker W/C
Baskett C.A. Wing Commander
Bates
Batey M. P/O
Batley B. LAC
Battersby S.H. F/O
Baxter J. F/L
Baxter E.L. AC2
Bean J.D. P/O
Beardsley L. LAC
Beasley W.E. AC2
Beaton R.A. P/O
Beaton Major
Beattie J. SGT
Beattie E. SGT
Beauchamp H.S. CPL
Beavis L.H. AC1
Beck S.E. CPL
Beck J.D. SGT
Beckingham T. Flight Lieutenant
Bee P.A. S/L (A/W/C)
Begy W.B. F/L
Beisiegel W.K. Group Captain
Bell C.A. G/C
Bell D. LAC
Bell W.W. F/S
Bell J. LAC
Bellamy H.J. AC2 (CPL)
Bennett J.T. F/O (A/F/L)
Bennett R.E. AC2
Bennett F.H. P/O
Bennett J. F/L
Bennett J.T. F/L
Bennett Cpl.
Bennett C.J. LAC
Bennett A. SGT
Bennett A.E. SGT
Bennett W.H. LAC
Bennett R.W.P. CPL
Bennett F.H. F/O

Bentley A.M. W/C
Bentley N. CPL
Berlis D.A. F/O
Berry I.F. CPL
Bertoft H. F/O
Best I.G. F/O (A. F/L)
Betts A.R. LAC
Beuter B.R. F/O
Bevis G.A. SGT
Beynon A. F/L
Bickell F.G. LAC
Bicknell R. F/L (A/W/C)
Biggs W.V. AC1 (LAC)
Billington G.C. AC2
Birch B.R. SGT
Birchall R.A. S/O
Birtwell M.E. F/O
Bishop V.C. Air Marshal
Biss E.H. LAC
Blackman J.P. S/L
Blackwood G. CPL
Blaikie J. LAC
Blair C.E. F/L
Blair J.W. LAC
Blair W.P. LAC
Bland J.R. S/L (A/W/C)
Bland J.R. Wing Commander
Blandford P.B. Squadron Leader
Bleasedale G. AC1
Blowers N.S. AC2
Blundell D.E. F/O
Blynch G. F/O
Bodecott V.E. LAC (F/S)
Bonelle G.H. F/O
Bonnar
Booth G.C. Mid
Booth A.W. F/O (A. F/L)
Bortoff H. P/O
Bosanquet A.K.J. S/O
Boulton H. LAC
Boupe C.S. W/C
Bourne Warrant Officer
Bournes F.J. LAC
Bousher A.S. AC1
Bower Group Captain
Bower E. Lt.Col.
Bowers W.E. CPL
Bowhill Sir Fredrick Officer Commanding RAF Ferry Command, Air Chief Marshall
Bowling F.F. LAC (ACU)
Bowman R. F/O
Box
Boxall P.W. LAC
Boyce D. P/O
Boyce W.H. CPL
Boyes N.W.R. P/O
Boyle D.S. F/O
Boyle G.L. P/O
Boyle D.S. F/L
Boyle D.J. F/L
Boyle P. LAC
Bracey R. P/O
Bradbrook B.O. LAC
Bradford H. SGT
Bradley Captain
Bradley J.W. F/L
Bradshaw Lt.Col.
Bradshaw P.L.C. F/O (A/F/L)
Braithwaite D.A. F/L
Bramley N. LAC
Brand J. P/O
Brand W. AC2
Bransby R. LAC
Breen F.U. P/O
Brennan J. LAC
Brent J.M. S/O
Brierley D.A.W. LAC
Briggs D.A. S/L
Briggs S.A. AC2
Briggs T.G. LAC
Brigham J. P/O
Brimble H.B. LAC
Bristow N.V. F/L
Bristow R.A. LAC
Brittem T.B. P/O
Brittenden R.T. P/O
Britton T.B. P/O

Britton	E.J.	CPL
Britton	F.A.	LAC
Broad H.G.	P/O	
Broadfoot	J.V.	P/O
Broadhurst	L.	F/O
Broadhurst	L.	F/L
Brockbank	K.N.	LAC
Brockwell	L.M.	F/L
Brodie W.	LAC	
Brook S.	LAC	
Brooker	R.D.	CPL
Brooks	R.L.	F/L
Brooks	G.H.	Reverend
Brooks	R.L.	F/L
Brooks	J.T.	AC2
Broom L.F.	AC1	
Brown J.	CPL	
Brown S.	CPL	
Brown F.K.	P/O	
Brown	Squadron Leader	
Brown G.P.	F/O	
Brown W.A.	S/L	
Brown H.E.	P/O	
Brown G.P.	F/O	
Brown H.E.	F/O	
Brown D. Lester		Flight Lieutenant
Brown J.P.	F/L	
Brown E.J.	F/O	
Brown W.E.	LAC	
Brown L.E.R.	LAC	
Brown D.C.	LAC	
Brown E.	LAC	
Browne	W.F.	F/L
Browne	W.D.	F/L (A/S/L)
Browne	D.	LAC
Browning	C.W.	LAC
Bruce D. Mc.D.		P/O
Bruce C.J.	CPL	
Bruza	Sergeant	
Bryan V.W.	CPL	
Buchanan	K.	P/O
Buchan-Hepburn	R.H.	P/O
Buckely	A.	LAC
Buckingham	T.	F/L

Buckle G.	F/L	
Buckley	A.E.	P/O
Bugden	J.	AC1
Bullis E.H.	P/O	
Bunce W.H.L.	LAC	
Burchett	J.G.	F/L
Burchett	J.D.	F/L
Burkinshaw	F.W.	LAC
Burkitt	A.S.	LAC
Burn A.P.	SGT	
Burn E.F.	LAC	
Burns T.	AC1	
Burridge	C.G.	P/O
Burrin M.H.	LAC	
Burrows	L.F.	P/O
Burrows	S.	LAC
Burt A.M.	LAC	
Burton	N.A.	F/O (A/F/L)
Burton	J.	SGT
Burton	W.	SGT
Bury B.S.	P/O	
Bush W.M.	F/O (A/F/L)	
Bush G.F.	AC1	
Butcher	Deane	W/C
Butcher		Sgt.
Butcher	H.A.	SGT
Butler I.	AC1 (SGT)	
Butler J.	LAC	
Butter H.A.	CPL	
Bywater	L.	LAC
Bywater	W.H.	LAC
Caine B.	AC1	
Cairns D.	LAC	
Caldicott	A.D.	P/O
Callaghan	A.N.	CPL
Calver J.A.	AC1	
Cambell	J.	LAC
Cameron	H.G.	F/O
Cameron	W.H.	P/O
Cameron	H.G.	F/L
Camerson	A.L.	P/O
Campbell	H.	LAC
Campbell	J.H.	CPL
Campbell	B.	LAC

Cann R.H. CPL
Cant C.H.S. F/O
Caplin A. F/S
Capon E. LAC
Capon T.A. LAC
Cappelman S. CPL
Carberry P.J. LAC
Cardle G. LAC
Carger D. F/O (A/S/L)
Carling E. P/O
Carlisle P.F. P/O
Carlow Lord Air Commodore
Carlton J.E. CPL
Carnell B.E. LAC
Carrington A.J. F/O
Carter Captain
Carter A.H. LAC
Carvar F/Lt.
Carver David Flight Lieutenant
Carver W.A. LAC
Carveth E.G. P/O
Caryer D. F/Lt.
Cassidy J. F/S
Cattell M.N. Sister
Cavalier E.T. LAC
Cawsey W.E.T. P/O
Cawthorne J. LAC
Cay-Cumming D.C. P/O
Chadwick J.E.H. F/L (A/S/L)
Chadwick R.W. F/L
Chadwick J.E. F/L (A/S/L)
Chadwick J.H. Squadron Leader
Chadwick F.E. CPL
Chalmers V. LAC
Chambers T.R. P/O
Chaplin A. LAC
Chapman D.D. Captain
Chapman D.I. Squadron Leader
Chapman T.W. P/O
Chapman F.C. P/O
Chapman T.W. P/O
Chapman A.J. CPL
Chapman F.A. CPL
Chapman T.E. LAC
Charlesworth A. CPL
Charlton M. F/L
Charman R.J. LAC
Charnock R. AC1
Charrington A.J. F/L
Chatwin F. LAC
Chave A.G. P/O
Cheese A.R. LAC
Chell E.A. AC1
Cherrill J.T. AC1 (LAC)
Cherrington G. F/Lt.
Cheshire G/C
Chester R.A. LAC
Chetwynd G.P. P/O
Chilver Sctry. to Secretary of State for Air
Chipperfield A.J. AC2
Chisholm A.R. F/L
Clafton F. P/O
Clark W.H. F/O
Clark J.F. P/O
Clark W.C. G/C
Clark J.F. F/O
Clark W.C. G/C
Clark J. P/O
Clark J.D. LAC
Clark F.A. LAC
Clark D. AC1
Clark J. LAC
Clark C.W. LAC
Clark G/Capt
Clarke W.H. P/O
Clarke T.J. P/O
Clarke W. G/C
Clarke J. LAC
Clarkson V.J. F/L
Clayton G.T. P/O
Clayton J.T. P/O
Clayton T.J. LAC
Cleall H. LAC
Cleary A.G. P/O
Clegg R. CPL
Clement S.T. F/S
Clements E.W. CPL

Clews J.G. LAC
Clifford R. CPL
Clift D.E. LAC
Coates K. F/O
Coates R. F/L
Coburn W.H. LAC
Cockburn J.H. LAC
Coe R. SGT
Coemack A.V. P/O
Cole F.W. F/Lt.
Cole W.E. AC1
Coleman E.H. Wing Commander
Coleman V. LAC
Colletts C.J. CPL
Collins K. Squadron Officer
Collins J.W. LAC
Collis A.V. F/O
Collishaw R.H. F/L
Colyer D.C. Air Marshall
Comben F/Sgt.
Compton W.N.A. CPL
Conduit H.J. F/L
Connolly J.W.A. F/O
Connor D.W. F/O
Cooil J.E. F/S
Cook R.L. F/S
Cook L.R. AC1
Cook R. AC1
Coombe P.V. LAC
Cooney V.R. P/O
Cooper C.H. LAC
Cooper S. CPL
Cooper B.H. CPL
Cooper A.V. LAC
Coote R.A. LAC
Cope A.S. LAC
Copeland W. LAC
Copley T. LAC
Coppen R. LAC
Cordery G.H. AC1
Corlett H.A. LAC
Cortnage P.W. LAC
Cottier G.G. W.O. (CPL W/Op)
Coulson J.R. P/O
Coulson W. CPL
Court J. LAC
Courtney D.A. Air Ministry
Cowan H.W. A/F/L (F/O)
Coward G.N. G/C
Cowie F.J. F/L
Cowie F.G.H. Flight Lieutenant
Cowie F.G.H. F/L
Cox P.C.E. P/O
Cox A.H. AC2
Cox H. W.O.
Coyne W.H. LAC
Cracroft P.D. Group Captain
Craig F/O
Craig A.J.L. F/Lt.
Craig K.W. F/O
Crampton W.H. AC2
Crank J.S. P/O
Cranna N.G. F/Lt.
Crano H.W.H. LAC
Crathorne C.A. F/O
Crawford W.J. P/O
Crawford George Acting Squadron Leader
Crawley E.J.M. P/O
Criswell H. F/O
Critch J.H. CPL
Critchard S.W. LAC
Crockart W. LAC
Croft J.L. F/L
Croft L.H. F/L
Croft A.E. LAC
Crombie D.M. CPL
Cromie M.J. P/O
Cross Colonel COO, Carib. Wing, US Ferry Command
Cross M.E. Sister
Cross T.W. P/O
Cross C.W. P/O
Cross T.W. F/O
Crossman W.J. AC1
Crowther J.K. F/S
Cruickshank L. SGT
Crump W.M.E. Flight Lieutenant

CrumpN. CPL
Cudworth G.R. F/O
Cullum C.R.P. F/L
Cunningham N. LAC
CurranD.L. CPL
Curtis G.T. F/O
Curtis W.C. F/O
Curtis G.C. F/O
Curtis W.C. F/O
Curtis A.F. AC1 (CPL)
Curtiss Cpl.
Cust H. A/S/L (F/Lt.)
Cust N. F/L (A/S/L)
Cust F.I. S/O
Cust N. S/L
Cuthbert G. AC1
Czuboka W. P/O
Dacey J. LAC
Dagger D. S/O
DaikenA.J. P/O
Dale S. LAC
Dales J.W. F/O
Daly Brigadier
Daniel T.A. LAC
Darby E. AC1
Dart K.C.D. F/Lt.
Davenport E.P. P/O
Davenport J.K. P/O
Davey L.B. P/O
Davidson C.D.G. P/O
Davie D.G. P/O
Davies E.V. Squadron Leader
Davies M. S/L (A/W/C)
Davies S.K. S/O
Davies E.H. CPL
Davies A.J. LAC
Davies J. AC1
Davies E.A. CPL
Davies H. CPL
Davies K.A. LAC
Davies C.S. LAC
Davies H.C. LAC
Davies G.W. SGT
Davies W. Sgt.
Davis A. P/O
Davis V.T. P/O
Davis K.G. S/O
Davison R. LAC
Dawes W. P/O
Dawson W.D. LAC
Dawson G.D. AC2
De Niverville A.V.M.
De Wolf A.W. Colonel
Dean Director Admin. and Finance
Dean J.A. LAC
Dean H. LAC
Deane D.M. Hon. F/L
Deane D.
Deane A. LAC
Deardan J. AC1
De-Larrinaga N.J. SGT
Denbigh R.H. F/O
Dent Colonel
Dernier H.C. P/O
Desilets A.A. P/O
Dicker A.H. P/O
Dickinson H.J.L. F/L
DifiliceL. AC1
Digby-Ovens A.R. F/O
Dimalore E. P/O
Dixon J. AC1
Dixon C. LAC
Dobson Sgt.
Dochert J.R.S. Mid
Dodd R.W. Squadron Leader
Dodd J.A. S/L
Dods T.J. CPL
Doey G. LAC
Dominik V.V. F/O
Donnery T.A. AC1
Donovan E. AC2
Dorey C.W. P/O
Dougall Mac F/O
Douglas J.C. Lieutenant
Dover E.J. LAC
Dowding C.A. F/L
Dowding C. F/L

Dower N.S. P/O
Dowiing C.A. F/L
Driver G.H. LAC
Drowley T.E. Air Commodore
Drummond Sir Peter Air Vice Marshal
Dubowski R. P/O
Duchess of Windsor
Ducie D.C. SGT
Duckworth W. LAC
Duggan G. AC1
Duke of Windsor HRH Governor of the Bahamas
Duncan C.A. P/O
Dunkinson W.O. Officer i/c W.O.
Dunlop I.W. P/O
Durnell W.E.G. LAC (CPL)
Dust J.L. CPL
Dutnall L. LAC
Dye
Dyer W.M. P/O
Dykes M.B. Sister
Eady G.R. CPL
Eales E.A. P/O
Eales I.A. P/O
Earicksteen L.C. P/O
Earl J.N. F/O
Earnshaw G.T. P/O
Easingwood H.E. CPL
Eastman T.H. LAC
Easton E.A. AC1
Easton A.L. A/S/L (F/L)
Easton A.W. Squadron Leader
Eatwell R.F. CPL
Eberton Captain
Echlin A.G. P/O
Eddington D.E. P/O
Edley F. AC1
Edmonson R.J. P/O
Edser W.E. F/L (A/S/L)
Edwards R.F. P/O
Edwards J.H. F/O
Edwards A.W. W/C
Edwards P.O.C. Rev. S/L
Edwards J.S. F/O
Edwards Rev. Station Chaplain
Edwards J.G. F/L
Edwards J. F/O
Edwards J.S. F/O
Edwards A.J.H. SGT
Edwards J.B. LAC
Edwards E.F. LAC
Eggington W. P/OD
Eggington D.W. F/O
Eglin J. LAC
Eickler J.O. Sergeant
Eland J.H. W/C
Elborn J. LAC
Elder R. Captain
Eley D. P/O
Ellerton G.E. F/L
Ellerton G.K. F/L
Elliott J. F/O
Elliott G.B.B. F/S
Elliott L.W. LAC
Elliott J. F/O
Ellis H.R. F/O
Ellison F. LAC
Elmes C.R.B. P/O
Elviss J.A. P/O
Emerson C.P. F/L
Emery H.D. LAC
Emms R. LAC
Enever W.B. Advocate S/L
Englert W.L. P/O
Entwistle F. G/C
Essex S. AC1
Evans T.E.W. P/O
Evans J.F. F/O
Evans D.G. P/O
Evans Vice E/O
Evans P.G. Squadron Leader
Evans T.E.W. Flying Officer
Evans P.G. Squadron Leader
Evans J.F. F/O
Evans J.A.S. F/O
Evans J.W. F/L
Evans T.K. F/O (A/F/L)

Evans P.C. Squadron Leader
Evans R.V. F/L
Evans F. F/O (A. F/L)
Evans A.G. LAC
Evans T.R. LAC
Evans I. LAC
Everall T. LAC
Everson C.M. F/L (A/S/L)
Everson W.M. F/L
Evill Air Marshall
Eyles R.E. F/Lt
Eyres G.V. AC1
Faddes
Faddes F. LAC
Fagan T.J. LAC
Fagan A.R. AC2 (FGT)
Fairweather G.W. LAC
Faragher F. A/F/L (F/O)
Farrage J. AC1
Fatherston D.P. P/O
Fawdon R.L. LAC
Fawkes J.Q. SGT
Fell J.B. F/L (A/S/L)
Fenton I.G.O. S/L
Fenwick T. LAC
Ferguson J. LAC
Ferris W.E. P/O
Fetherson D.P. P/O
Field R.C. Group Captain
Field J.M. AC2
Fielder A. LAC
Fielding H.C.R. F/O
Fielder E.C.W. P/O
Fielding H.C. F/O
Fielding R.C. F/O
Filer W.J. LAC
Finn Ryder Lt. Colonel
Finn S.T. LAC
Fisher E.V. P/O
Fisher W.J. P/O
Fisher J.J. F/L (A. S/L)
Fisher R. SGT
Fitter Cpl.
Fitzsimmons J.F. AC2
Fleming D.M. F/O
Fleming M. F/L
Fleming-Williams D.C. F/L
Fletcher J.K. P/O
Fletcher J.K. P/O
Fletcher W. A/F/S
Flett D.G. F/L
Flett D.S. F/L
Flint M.S. S/L (Revd.)
Flint R. CPL
Flower J.F. LAC
Fogarty F.S. Air Commador
Foggarty F.J. A/Cdr.
Foot R.G SGT
Ford F.E. P/O
Ford K. LAC
Ford S.J.C LAC
Ford M.F. LAC
Forster G.W. AC1
Forsyth A. P/O
Foster MacNeece Air Vice Marshall
Fouracre Cpl.
Fowler C.R. F/O
Fowler A.A. SGT
Fowler A.W. LAC
Fowler L.W. LAC
Fowler C.R. F/O
Fowles R.G. S/L
Fowlie R.W. Sgt.
Fox J.R. F/O
Fox R.W. F/O
Fox J.R. F/O
Fox H. LAC
FrancuJ. F/O
Franklin H.F. LAC
Franklin J.E. SGT
Fraser R.H. P/O
Fraser D.D. Sergeant
Fray J.E. LAC
Fray H.E. LAC
Freeman J.M. Wing Cmdr., 2nd in Cmnd. RAF Nassau
Freeman J.R. Chief Technical Officer, Wing Cmdr.

Freeman J. F/O (A. F/L)
Freeman F.A. LAC
Freer R.A. P/O
Freestone G. AC1
French L.A. LAC
Fry A.H. AC1
Fuller H.F. F/L
Gale K.F. SGT
Gallup S.A. LAC
Gamble K. P/O
Gamble G.D. F/O
Gammage J.S.D. LAC
Gardner J.W. P/O
Gardner F.J. SGT
Garland J.F. F/O
Garland F.J. F/L
Garton G.R. P/O
Garvagh W.R. LAC
Gaskell J. LAC
Gaston B.F. F/L
Gates G.G. P/O
Gates A. Honorable
Gates R.A. F/O (A/F/L)
Gauld E.N. F/O
Gauld F.N. F/O
Gay-Cuming D.C. P/O
Geddes P.T. CPL
Gelder T.V. LAC
Gibbons C. F/L
Gibbons G. F/L
Gibbons J. CPL
Gibson W.D. P/O
Gibson H.N. CPL
Gilbert K.E.W. CPL
Gilchrist D. CPL
Giles R.A. P/O
Gillham A. AC1
Gilliard L.H. LAC
Gittins P/O
Gittons C.P. P/O
Gleave-Dent H. LAC (CPL)
Glenn Sgt.
Glover R.N. Captain
Goad G.C.L. F/O
Goatley A. CPL
Goatman L.F. LAC
Goble Air Vice Marshal
Godd A.T. P/O
Goddard C. LAC
Godden W.E. LAC
Godderham R.J. P/O
Goddman B.B. P/O
Godwin W.A. LAC
Goldbrough
Golding W. LAC
Goldstone A.W. P/O
Goldsworthy S.E. P/O
Goman G.R. F/S
Good R. LAC
Goodchild G.A. LAC
Gooderham R.F. P/O
Goodman B.B. P/O
Goodman J.E. LAC
Goodrich D.H. F/L (A/S/L)
Goodrich B.H. S/L
Goodwin G.N.T. P/O
Goodwin Air Vice Marshall
Goodwin W. LAC
Gordon J.W. F/O
Gordon T.R. F/L (A/S/L)
Gordon J.W. F/Lt.
Gordon T.R. F/L (A/S/L)
Gordon-Cooper H.H. F/L
Gotham W.L. F/O
Gotham V.L. F/L
Graham E.L.M. A/Snr/Sister
Graham M.M. CPL
Graham D. LAC
Graham N. LAC
Graham W.B. LAC
Graham C.A.P. LAC
Graig A.J.L. F/O
Grant I.C. P/O
Grant D. Major General
Gray D.E. P/O
Gray P. LAC
Gray G.E. CPL
Gray G.S. LAC

Gray A.W. AC1
Green A. P/O
Green N.G. P/O
Green J.C. F/O
Green W.F. CPL
Green G.L. AC2 (AC1)
Green G. CPL
Greenacre A.L. F/O
Greenacre A.E. Flight Lieutenant
Greening G.H. LAC
Greenwood F.E. F/L (A/S/L)
Gregory J.R. CPL
Gregory D.P. LAC
Grieve W.E. G/C
Griffin J.A. P/O
Griffith T.M. F/L
Griffith T.A. F/L
Griffiths F/Lt
Grollman G.F.E. P/O
Groom D.J. SGT
Grudgings D.E. SGT
Grumnet Lt. General
Grundy K.C. F/S
Grundy F/S
Grunert Lt./General
Guest, M.B.E. C.V. S/L A/W/Cdr
Guilfoyle J.A. F/L
Guise R. LAC
Hackett W.P. SGT
Hadden A.W. CPL
Hadfield P.R. S/L
Hadfield L. P/O
Hadravek J. F/O
Hagarty Leonard Father
Hagen J.E. P/O
Haile W.S. CPL
Halbert L.C. AC1
Hales A.R. P/O
Haley T.W. AC1
Hall T.L. P/O
Hall H.M. P/O
Hall R.L. P/O
Hall A. LAC
Hall M.V. LAC
Hall D.R. AC2
Hall H.L. LAC
Haller B. F/Lt
Haller R. Flight Lieutenant
Halley R. Group Captain
Halstead R.F. LAC
Hamilton E.J. LAC
Hammond E.C. F/L
Hampton E. P/O
Hamson D.A. AC1
Hance A.J. LAC
Hannan L. LAC
Hanson C. P/O
Harber C.G.E. P/O
Harborne D.A.
Hardcastle F. LAC
Hardey D.G. LAC
Hardy G.N. F/O
Hardy J.N. CPL
Hargreaves R. LAC
Hargreaves C. LAC
Harper R.V. General
Harper R.W. General
Harper W.M. CPL
Harris C.A. P/O
Harris Air Ministry Works Directorate
Harris A.A. CPL
Harris W.T. LAC
Harris A.H. AC2
Harris V.R. SGT
Harris W.E. CPL
Harris V.E. LAC
Harrison J.B. F/O (A/F/L)
harrison R. CPL
Harrison J. LAC
Harrold P/O
Harrop M.F. LAC
Harrowell H.R. SGT
Hart R. LAC
Hart K.C. LAC
Harte-Lovelace P.L. P/O
Hartley J.R. LAC
Hartley P.J. CPL

Hartley H. CPL
Hartwell W.E. LAC
Harvey J.A. SGT
Harvey A.W.C. LAC
Harwood T.R. LAC
Hastewell D.E. F/O
Hastie R.M. P/O
Hastie R.N. Warrant Officer (ow Pilot Officer)
Hastings A.G. AC2
Hastwell D.E. F/O
Hatfield P.R. S/L
Havercroft J.W. Captain
Havis E.H. CPL
HawkeA.F. F/O
Hawkin Lt. Col.
Hayes A.E. F/O
Hayes G.D. F/O
Hazell N.C. LAC
Head H.G. A/F/L (F/O)
Head J.H. CPL
HealeyF.W. LAC
Heath H.J. AC1
Heaton Cpl.
Heckingbottom E.S. CPL
Heeker K. F/O
Heeks C.C. LAC
Heel P. Lieutenant
Hell O.H.M. P/O
Hellewell B.J. SGT
Helps Cpl.
Helyar A.R. CPL
Henderson A.A. F/O
Henderson G.C. P/O
Henderson A.A. F/O
Henderson B.K. P/O
Henderson D.K. P/O
Henderson A.A. F/L
Henderson J.M. LAC
HenleyE.P. F/L
Henry K.R. P/O
Henry J.R.
Hepplewhite F.W. LAC
Herbert W.E. LAC
Hesley I.J. P/O
Hess Alexander Colonel
Hetherington L.S.V. F/O (A/S/L)
Hewis R. P/O
Heywood J.P. F/O
Hicks E.G. CPL
Higgins P.N. P/O
Higgins H. LAC
Higgins R.H. LAC
Hignett G.W. W/C
Hill H.W. F/O
Hill R.W. AC1
Hill T. LAC
Hill W.L. AC2
Hill E. AC1
Hill T.H. AC1
Hiller J.G. F/O (A/F/L)
Hindess J.T. SGT
Hindle W.R. LAC
Hirst A.R. F/O
Hiscock F.G. F/O
Hiscock F.G. F/O
Hitchen G.C. P/O
Hobbs R.A. P/O
Hodkinson F.V. Cpl.
Hogan H.A.V. Group Captain
Hogan CPL
Hogg H.R. SGT
Holbrooks Lt. Col.
Holcombe G.E. S/O
Holden H.C.R. F/L (A/S/L)
Holdham S/Ldr.
Holland D. P/O
Holland J. LAC
Hollinghurst J.K. F/O
Hollinshead W. LAC
Hollowell P/O
Holman M.D. S/O
Holmes E.W.G. Reverend
Holmes S.P. LAC
Holtum R.G. LAC
Hood J.H. P/O
Hood Lt. Col.
Hooker K. P/O

Hopkins D.H. SGT
Hopper W.W. LAC
Horner D.K. P/O
Hornigold T.J. LAC
Horton D.C. P/O
Houghton S.J. P/O
Houghton A. LAC
Houston S.C. LAC
Howard B.J. LAC
Howard A.L. LAC
Howard C.G. LAC
Howarth
Howell A.S. AC1
Howells E. LAC
Howells P.G. LAC
Howlett E.T.G. F/L
Howorth J. LAC
Hoyle G. LAC
Hoyte G.J. SGT
Hrncir J. F/O
Hughes W.A. W/C Dr.
Hughes T.C. LAC
Hukin E. CPL
Hull C.J. AC2
Hume J.J. CPL
Hunt G.R.M. F/L
Hunt C.J. LAC
Hunter W. P/O
Hunter T.S. LAC
Hunter H.L. LAC
Huntley H. AC1
Hurley M.B. P/O
Hurll J.P. P/O
Hurst R.A. LAC
Hutchings A.F. LAC
Hutchings W.J. SGT
Hutchingson J.O. LAC
Hutchins B.W. P/O
Hutchins A.C. CPL
Hutchinson L.J. LAC
Hutchinson J.H. CPL
Hutton H. CPL
Hyde D.W. P/O
Hyndman R.H. S/L
Hyomes R. LAC
Ilines W.C. Cpl.
Inglis K.L. Sister
Inglis J. LAC
Ingr Sergej Czechoslovak Minister Ntl. Defence
Ingram J.S. P/O
Ingram A.S. F/O
Ingram H.H. F/S
Ingram R. LAC
Innes T. LAC
Instead J. Squadron Leader
Irving J.R. F/O
Irving J.D. W/C
Irwin F.H. LAC
Isaac A.C.T. S/L (A/W/C)
Ison W.H. F/O
Ison D.S. F/O (A. F/L)
Isted D.J. Squadron Leader
Izatt G. LAC
Jack P. AC1
Jackson J.L. P/O
Jackson A. P/O
Jackson G.W. P/O
Jackson L.M. Flight Officer
Jackson W.P.U. F/L
Jackson A.G.W. CPL
Jackson A.A. LAC
Jackson F.W. LAC
Jacob L.W. LAC
Jagger S.H. Flying Officer
Jakeman T. LAC
James D. SGT
James E. AC1
James D.G. CPL
James A.W. LAC
James E.J. AC1
James A.E. LAC
Jamison T.W. SGT
Janousek Air Vice Marshall
Janousek F. F/L (A. S/L)
Jaoques T.L. CPL
Jardine C.J. LAC
Jarratt L.S. P/O

Jay E.G. Reverend (S/Ldr.)
Jay W. LAC
Jebb W.E. AC1
Jefferson J. Brig.
Jeffery J.R. P/O
Jenkins A.D. F/O
Jenkins E.L.F. Sister
Jenkins J.H. CPL
Jermy J.L. Sister
Jester R.H. AC2 (CPL)
Jewers D.F.C. P/O
Jews V.C.P. F/O
Johansen E.U. F/L
Johnson W. P/O
Johnson R.H. F/O
Johnson L.V. LAC
Johnston J. P/O
Johnston R.B. AC1
Johnstone R.T. LAC
Joliffe W.J. F/L (A/S/L)
Jolliffe W.M. F/L (A/S/L)
Jones J.F. P/O
Jones W.J. A/F/O (P/O)
Jones R.F. F/Lt
Jones J.F. P/O
Jones L.R. P/O
Jones A.R. P/O
Jones K.P. F/L
Jones H. P/O
Jones F.H.I. F/L (A/S/L)
Jones R. P/O
Jones H. P/O
Jones W.D. F/O
Jones W.E.T. F/L
Jones E.G. LAC
Jones H.C. LAC
Jones W. LAC
Jones I. CPL
Jones F.H. LAC
Jones G.H. SGT
Jones C. AC1
Jones W.W. LAC
Jones L.G. SGT
Jordan W.R. CPL

Joseph J. LAC
Juden C. SGT
Judge LAC
Junor N.K. Lieutenant
Kane T. LAC
Kay M.C. P/O
Kay J.R. F/L
Kay J.R. F/L (A/S/L)
Kay F.E. LAC
Kaye H. LAC
Keenan J. P/O
Keenan J. P/O
Keenan J. P/O
Keenan J. F/O
Keenan J. P/O
Keenan J. F/O
Keill T.P. Squadron Leader
Keill T.F. F/L (A. S/L)
Keir H.P. P/O
Keleher R.J. P/O
Kelly Lowell Lt/Cdr.
Kelly S. F/O (A/F/L)
Kelly S. F/O (A/F/L)
Kelly S. F/O (A/F/L)
Kelly T.C. LAC
Kelly C.G. SGT
Kelly R.T. CPL
Kemp F.W.J. F/O
Kendal J.A. LAC
Keneally R.W. AC2
Kennaugh R.A. LAC (ACU)
Kennedy H.E. P/O
Kennedy J. LAC
Kenyon S.F. LAC
Keoppl W.E. LAC
Keppie J.L. F/O (A. F/L)
Kerr K.W.A. F/O
Kerr D.G. W/C
Kerr K.W. F/L
Key G.B.C. F/Lt.
Kibby W.A. F/S
Kidd W.P. LAC
Kidston H.R.A. Lt/Cdr.
Kie T.M. P/O

Kift E.G. A/CPL
Kilbourne Captain
Kilby R.G. LAC
Kimmins B.J. Sub-Lieutenant
King C.M. F/L
King J. F/O
King G.M. F/L
King J. F/O
King H.A. AC1
Kingdon S.C. AC1
Kipling A.C. AC1
Kirk I.G. F/S
Kirkpatrick C.W. F/O (A. F/L)
Kitchingman H. LAC
Knapman H.A.N. LAC
Knight J.T. P/O
Knight M. LAC
Knight H.E. LAC
Knights D.J. F/O
Knights F.J. LAC
Knott T.E. P/O
Knowles Lord Governor of Bermuda
Knowles Thomas Flying Officer
Korski G. P/O
Krauter F. F/L
Krauter E. P/O
Krauter F. F/L
Kyle W.A. LAC
Kyles R.K. F/L
Lacy M.J.P. F/Lt.
ladbury L. P/O
Lain Lt.
Lake H.M. SGT
Lakin R.G. SGT
Lamb C.D. P/O
Lamb P.S. P/O
Lancaster Major
Lane Captain
Lane R.G. AC1
Lane C.F. LAC
Lang Air Commodore
Lang A.D. G/C (A/A/Cdr.)
Langham P.G. P/O
Langlands H.D. LAC
Langley B.E. P/O
Large D.
LarneyN.H. P/O
Laughland A.R. F/L
Laughland G.W.K. Squadron Leader
Laughland A.R. F/L (A. S/L)
Lawrence J.T. F/L
Lawrence F.H. LAC
Lawson G.S. P/O
Lawson J. LAC
Lawson H. LAC
Lawton E. LAC
Lax C. CPL
Laycock P.S.L. P/O
Laycock W. LAC
Lea H. F/O
Leake O.C. CPL
Lee L.A. A/F/L (F/O)
Lee Wing Commander
Lee L.A. F/O (A/F/L)
Lee H. P/O
Lee P.S W/C
Lee P.A. W/C
Lee G.G. LAC
Leech W.T. AC1
Lees A.J. SGT
Legg D.C. LAC
Leggett E. LAC
Leggett D.G. LAC
Leigh J. F/O
Leishman G.E. S/L
Lennox G. F/L (A. S/L)
Lenson A.D. P/O
Lescaudron R.J. P/O
Leslie I.M. P/O
Letter R.W. CPL
Letts R. LAC
Levene J. LAC
Levick C.H. Sergeant
Levy M. F/O
Levy N. CPL
Lewcock P.F. P/O

Lewis J.H.C. F/L
Lewis A.J. F/O
Lewis H.J. F/O
Lewis H.T. CPL
Lewis J.J. LAC
Lidford Air Commodore
Lightbody A.Y. LAC
Lilley A.W. SGT
Limbrey D.H. F/L
Lindsay E.W. F/L (A/S/L)
Lindsey C.W. W/C
Ling A.R. Flight Lieutenant
Liska B.R. F/L (A/S/L) Czech
Lister D.S. P/O
Little A.E. LAC
Lizbrey D.H. F/L
Llewellin W.L. LAC
Lloyd J. P/O
Lloyd H.
Lloyd B.J. LAC
Lloyd F.A. AC1 (CPL)
Lloyd E.L. LAC
Loat J. Flight Lieutenant
Lock F.W. CPL
Lock L.W. LAC
Lockhart J. LAC
Lockley D.B. LAC
Lockyer V.C. CPL
Logan J. LAC
Logan G.D. AC1
Loneragan R.K. AC1
Lonnox G. S/Ldr.
Loomis General
Lord B.S. P/O
Lord W. LAC
Lord Bishop of Nassau CoE, Christ Church Cathedral
Lorimor G.V. P/O
Loring G.W. F/O
Lott C.G. G/C
Lough J.G. Sister
Lovell
Lovett W.E. LAC
Lowe J. LAC
Lowrey R.M. P/O
Lowther G.F. LAC
Lubin W. SGT
Luck G.W.D. F/O (A/F/L)
Luck G.W. F/O
Ludham L.A. F/O
LudlanL.A. F/O
Ludlow Hewitt Sir E. Air Chief Marshall
Lush W.R. F/L
LusherU. CPL
Lutes H.A. P/O
Lyder E.G.F. P/O
Lydford H.T. A/Cdr.
Lynch G.B. F/O
Lynch G.E. F/L
Lyon P.V. F/O
Lyons F.I. S/O
Lyons B.J. Sergeant
Lyons A.R. F/L
LysonsF.I. S/O
Lywood G. Group Captain
MacCallum D. LAC
MacDonald T. F/O (A/F/L)
MacEarlane J.F. F/O
MacFarlane K. AC1 (CPL)
MacGilp D.G. P/O
Macintosh A.W.U. P/O
Mackeen C.D. P/O
MacKInlay J.F. P/O
MacKinley J.E. F/L
Mackley C.A.R. P/O
Mackrill W.T.J. LAC
Mackworth Air Commodore
MacLaughllin C.R. Squadron Leader
MacLennan H. LAC
MacPherson D.T. S/Ldr.
Maddocks J. LAC
Malcolm M. F/O
Malcolm McG. F/O
MalkinH. W/C
Mallinson J.D. F/L
Mallott J.G. LAC
Malyon A.E. LAC

Manchester W.C. F/O
Mansbridge F.W. LAC
Manuel J.C. CPL
Map C. F/O (A/W/C)
Mares J. F/O
Marinovich I. P/O
Marix A.V.M.
Markham J. F/L
Marriot H.J. CPL
Marsdem J.O. P/O
Marshall C. P/O
Marshall R.J. SGT
Martin M.J.P. F/L (A/S/L)
Martin E.H. LAC
Martin D.T. CPL
Martin F. LAC
Marwing C.J. LAC
Mary Air Commodore
Mason J. LAC
Mason G. LAC
Mather H. LAC
Mathews H.L. F/L
Mathews M.D. Sister
Mathews M.D. T/Sister
Mathison T.A. CPL
Matthews D.J. P/O
Matthews S. LAC
Matthews L.A. SGT
Matthews L.T. CPL
Mattocks G. LAC
Maudley P.L. SGT
Mawdsley W. LAC
Maxted W/Cdr.
Maxwell P.H. Vice Wing Commander
Maxwell Lieutenant
May W.L. CPL
May C.J. CPL
Mayman R. LAC
Mc. O'Dwyer W. P/O
Mcacham H. AC2
McArthur J.B. P/O
McArthur G.E. Flying Officer
McAuley J.A. LAC
McBrainey V.H.M. President W/C, AFC Judge
McCarter J.P. LAC
McCarthy Captain
McCarthy T.G. Sgt.
McCarthy J. SGT
McCauley J.A. LAC
McCracken S. LAC
McCulloch M.J. CPL
McDermott W.J. LAC
McDiarmid P. W/C
McDill D.S. LAC
McDonald S.A. F/Lt.
McDonald W. SGT
McDonald H.J. LAC
McDougall D.O. P/O
McDowall T.J. P/O
McDowell M. S/Sister
McElligot C.R. Cpl.
McEnvoy A.J. LAC
McEwen A. F/O
McEwen A. F/L
McFarlene V. SGT
McGhee G.A. LAC
McGillivray J. LAC
McGregor H. P/O
McHutchinson J. SGT
McIntyre J. SGT
McKay A. LAC
McKean L.D. Air Vice Marshal
McKean I.J.P. SGT
McKeegan J. AC1
McKellar P/O
McKeller M.
McKenna W.A. F/O
McKeown Cpl.
Mckey Cpl.
McKie W.N. F/O
McKie T. P/O
McKinnie F.A. LAC
McKinnon N. SGT
McKinzie G. S/L
McLachlan W. LAC
McLachlan J. LAC

McLachlan J. SGT
McLaughlin C.R. S/Ldr.
McLaughlin R.L. P/O
McLaughlin C.R. Squadron Leader
McLean J.R. P/O
McLean J.S. F/O
McLellan F.I. F/O
McLellan W.T. P/O
McLennan V.A. F/Sgt.
McLeod J. F/O
McManus D. CPL
McMenemy F. LAC
McMillan D.G. P/O
McNabb H.F. P/O
McPhail C.D. P/O
McReady W.B. P/O
McTaggart C.H. P/O
McWilliam J. LAC
Meacham AC2
Mealin R.H. P/O
Mearns A.N. F/L
Mearns A.K. F/L (A. S/L)
Medland F.W. F/O (A. F/L)
Meeres C.J. P/O
Melay Col.
Melener F.E.G. F/O
Melkin H. Wing Commander
Melrose J.D. SGT
Melton H.A. P/O
Merbey C.E. F/L
Mercer A. P/O
Merrifiled R.T. F/Lt.
Merrill W.D. Colonel
Messinger H.L. Lieutenant
Methuen R.D. F/O
Methuen R. SGT
Meyer R.H. P/O
Michael C. LAC
Milborn W.F. F/O (A/F/L)
Miles J.W. SGT
Miller Captain R.M. Superintendent of Prisons
Miller F. F/O
Miller F.R. LAC
Miller H.J. LAC
Milligan W.C. P/O
Millikin C. LAC
Mills A.C.L. F/L
Mills C.H. LAC
Mills N.A. LAC (CPL)
Mills J.W. SGT
Milne K.W. P/O
Milton C.H. Squadron Leader
Milton F. LAC
Minday C.J. P/O
Mitchell V.M. S/O
Mitchell A.B. P/O
Mitchell A.D. F/O
Mitchell A.G. F/O
Mitchell V.M. S/O
Mocek J. F/O Czech
Mohring W.J. P/O
Moir J.W. P/O
Mollard L.B. P/O
Mombrun A.C. CPL
Monk G. AC1 (CPL)
Montagu-Smith A. W/C
Montgomery J.Q. F/O
Montgomery K.A. P/O
Moon M.A. S/O
Moore Captain
Moore D.J. P/O
Moore C. F/O
Moore L.W. P/O
Moore F.B. LAC
Moore F.C. LAC
Moore W. LAC
Morbey C.E. F/O (A/F/L)
Morcombe F.G. CPL
Morgan R.H. P/O
Morgan G.D. AC1
Morgan W. LAC
Morley H.K. P/O
Morley W.F.E. SGT
Morrell P.H.C. CPL
Morricks Air Vice Marshal
Morrill J.R. P/O
Morris S. AC1

Morris W.A. LAC
Morris H. LAC
Morrison J.P. F/O
Morrison T.H. LAC
Morrison M. CPL
Morrisson J.P. F/O
Mortimer D.I. P/O
Morton I.O. S/L
Moseley C.R. AC2
Motl J. P/O
Mowat G.A. CPL
Mowling L.B. F/L A/S/L
Moxham J.G. P/O
Moys J. CPL
Mudford W. LAC
Mudford K.C. AC1
Muir R.O. CPL
Muir D.G. P/O
Muirhead A.L. P/O
Muirhead J.T. LAC
Muldoon E.H. Captain
Mulheron G.A. F/Lt.
Mullally S.M. Sister
Mumford R.W. CPL
Munday P.H. LAC
Munro F.W. F/O
Murdeeh G.C. F/L (A/S/L)
Murdoch C.C. S/L
Murphy A. AC1
Mygind V.C. F/O
Nairn G.W. P/O
Nash Major
Navell H.E. G/C (AAC)
Navin T. SGT
Naxera F. P/O
Naylor F.E. CPL
Neilan F.M.D. P/O
Nelson S.C.R. W/C
Nelson G.H. LAC
Netherton L.H. P/O
Neville T.H. S/L
Newberry J.H. W/C
Newling L.B. F/L (A. S/L)
Newmarch D.A. F/S
Newton R. F/S
Newton J. LAC
Newton A. LAC
Nicholl J. P/O
Nicholls W.O.
Nicholls H.W. W.O.
Nichols G/C
Nicholson J. AC2
Nicklin Sgt.
Nielson A.R. F/L
Nightingale J.F. AC1
Nimmo C.S. P/O
Niverville De Air Vice Marshal
Nixon T.S.F. F/S
Nixon F.W.T. LAC
Norman C.A. AC1
Norman A.H. CPL
Normington S. AC2
Nowling L.B. S/Ldr.
Noyce A.E. LAC
Oakes W.A. LAC
Oakshott H.E. F/O
Oates H. LAC
O'Brien W/C
O'Dell D. F/L
O'Donnell R.P. S/L
O'Donnell W/C
O'Dwyer
Oels R.W. AC1
Oldfield E. F/O
Oldfield E. LAC
Oliver M. LAC
Olson P.A. P/O
Ord J. SGT
Orr J. S/L (A/G/C)
Osbaldstone H.G. LAC
Osborne J. P/O
O'Shea
Osmond O.M. Sister
Oustrom Brigadier General
Outram D.J. LAC
Over J. AC1
Owen H. P/O
Owen S.J. AC1

Paddison H. SGT
Padgham P.S. CPL
Page G.F.H. F/L
Page B.J.D. S/L
Page R. LAC
Paget S.H. A/F/L (P/O)
Paine S.J. P/O
Paish A.H. LAC
Palmer G.P. F/O
Palmer G. LAC (ACU)
Palmer J. LAC
Pampling E.A. LAC
Panhurst R.G. F/L
Pankhurst R.C. F/L
Paramor S.J. AC1 (CPL)
Parker S. LAC
Parker A. LAC
Parker H. LAC
Parkhouse J.E. P/O
Parkinson E. F/O (A/F/L)
Parkinson E.F. F/L
Parkinson P/O
Parmenter A.A. LAC
Parr E.D. AC1 (LAC)
Parr J. LAC
Parrish R.A. LAC
Parry H.C. LAC
Parry T.A. SGT
Parsons Sgt.
Parsons L. LAC
Parton H.D. P/O
Paruk A. P/O
Passey H.R. LAC
Patchin R. F/L (A/S/L)
Patrie Flight Lieutenant
Patterson R.E. LAC
Patterson R. SGT
Patterson N. LAC
Patton S.E. F/O
Pawley L.S. LAC
Payne H.G. LAC
Pear A. LAC
Pearce R.D. F/O
Pearn W.E. P/O
Pearson E.A. F/O
Pearson L.W. AC1 (LAC)
Pearson E.A. F/O
Peck G.R. Lt. Col.
Pedlar W.H. P/O
Peek P.F. P/O
Peel M.G. P/O
Pemberton C.W. Lieutenant
Penk V. F/L
Pennock A.C. LAC
Pepper F.M. Matron
Percival J.F. F/L
Perera K.F. P/O
Perkins H.W.R. LAC
Perrera P/O
Perret J.R. LAC
Peters M.J. P/O
Peters J.G. CPL
Peterson J.A. P/O
Peterson A.C. S/L
Petrie J.W. F/Lt.
Petrie J.W. F/O (A. F/L)
Pettitt G.E. LAC
Pettus F.A. P/O
Phillips D.E., Dr. N/Sister
Phillips Major
Phillips A.T. F/O
Phillips D.E. Sister
Phillips G.H. F/L (A/S/L)
Phillips M.N. F/L
Phillips A.J. F/L
Phillips A.T. F/O
Phillips D.G. LAC (CPL)
Phillips H.G. LAC
Phillips A. LAC
Philpott C.T. LAC
Pichard M.K. F/O
Pickard R. SGT
Pickering R.R. LAC
Pierce Major
Piggott W.T. LAC
Pilmer R.C. AC1
Pink E.W. AC1
Pitman E.F. P/O

Plant J.L. A/F/L (F/O)
Platt H. LAC (ACU)
Pollack E.R.N. W/C
Pollack G.R. W/C
Pollard A.G. LAC
Ponting R.W. P/O
Ponting R.H. F/O
Pool D.J. AC1
Poole J.A. F/O
Poole P.A. F/O
Pooley W.G. LAC
Poolo J.A. F/O A/F/L
Porter J. LAC
Porterfield W. P/O
Potter P.W. LAC
Povey G. LAC
Powell R. LAC
Powell W. AC2
Power J.M. LAC
Pratt R.H. Senior Medical Officer, W/Cdr
Pratt H.C. LAC
Prebble H.J. LAC
Price R.H.F. LAC
Price R.W.P. AC1 (CPL)
Price W.H. CPL
Prickett T.O. W/C
Prickett T.H.R. AC1
Priest J.C. W/C
Priest W/Cdr.
Prime J.C. A/S/L (F/L)
Pringle H.J. G/C
Prior R.G. LAC
Pritchard D.W. Lieutenant Col.
Procker W.R. CPL
Prockter CPL
Pulham R.B. P/O
Pulham R.P. P/O
Pulham H.W. LAC
Pulitzer H. F/O (A/F/L)
Pullen R.H. LAC
Purnell R.C. LAC
Purvis H. LAC
Pye W.H. LAC
Pyper E.S. LAC
Qhinton R.W.H. P/O
Quiddington T.W. F/O
Quinn S.H. LAC
Quinton R.W.H. P/O
Racham P.R. P/O
Radcliffe G.H. F/L
RadinaF. F/L Czech
Rake I.T. F/S
Ramsey H.J. LAC
Ramshaw W.L. LAC
Rassin M. S/L
Rattery W. F/O
Rauns H.A. CPO
Ray C.F. LAC
Raymor G. LAC
Rayner P/O
Raynor H.J. F/O (A/F/L)
Read A.C. Rear Admiral
Read W.J. F/L (A. S/L)
Read B.D. LAC
Read W.J. F/L A/S/L
Redman W.K. P/O
Redmond L.A. P/O
Reeder F.E. LAC
Reekie C.I.B. P/O
Reekie J.B. P/O
Reen L.G. P/O
Rees W/O
Rees T. CPL
Reeve C.J.M. LAC
Reilley J. SGT
Rembert J.P. Comdr.
RennieG. SGT
Rennison T.V. LAC (ACU)
Restall H.C. F/O
Reuter B.R. F/O
Reynolds R.A. AC1
Rhys R.O. W/C
Rhys Davies T. F/O (A. F/L)
Rice W.J. P/O
Richard J.W. P/O
Richards W/Cdr.
Richards J.D. F/L

Richards K.O. P/O
Richards D.J. Flight Lieutenant
Richards J.D. Flight Lieutenant
Richards E.B. Flying Officer
Richards F.M. F/O
Richards E.M. F/O
Richards E.B.
Richards E. F/O
Richards E.J. LAC
Richards W.J. LAC
Richardson T.W. LAC
Richfield E.R. F/O (A. F/L)
Richfield E.A. F/O
Richman R.J. SGT
Richmond F. F/L
Rickard J.W. P/O
Rickett
Riddleworth H.D. AC2
Ridings A.W. LAC
Rigg A.G. P/O
Rigg T.E. LAC
Rimpau Lt.
Ring R.C. Captain
Ring S.C. Captain
Ringland AC1
RitchieJ. LAC
Robbins J.W.T. F/L
Robbins K.R. LAC
Robbins J.W. F/L
Roberston E.F. P/O
Roberston A.K. P/O
Roberts D.W. P/O
Roberts I. S/L
Roberts H.W. F/L
Roberts Stanley Joseph AC1
Roberts R.H. F/O
Roberts P. LAC
Roberts J.H. LAC
Roberts T. LAC
Roberts U. LAC
Roberts G.A. CPL
Roberts W.H. LAC
Robertson A.K.N. P/O
Robertson E.F. P/O
Robertson J.P. P/O
Robertson D.E. Lieutenant
Robertson R.M. P/O
Robertson McD. LAC
Robertson R.H. SGT
Robinson A. P/O
Robinson R.M. F/L (A/S/L)
Robinson Squadron Leader
Robinson J.A. Auditor, Air
Ministry
Robinson
Robinson H. LAC
Robinson A.L. LAC
Robinson W. LAC
Roddick R.M.J. F/L
Rodecker F. S/Lt.
Rodwell D.A. S/O
Roe J.F. LAC
RogersE. LAC
RogersS. LAC
RogersJ.H.E. LAC
Roome A.J. AC1 (CPL)
Roper
Rose D.J. F/L
Rose B.J. S/O
Ross B.J. S/O
Ross G. LAC (ACU)
Rossick Flight Lieutenant
Rotheram R.G. LAC
Routledge W.A. LAC
Rowat D.A. P/O
Rowe H.G. G/C
Rowe W/C
Rowe R.J. CPL
Rowland C.A. LAC
Rowley R.B. P/O
Rowney G.R. F/O
Ruddick J.W. SGT
Rushbrooke Commander
Rusk A.M. S/L
Russsell J. AC1
Rutherford Major
Sales H.E. S/L
Salt J. F/S

Salter F.L. LAC
Sandell F.A. CPL
Sandercock P.J. F/O AFL
Sanders W. F/L (A. S/L)
Sarkies M.R. F/L
Sarson W. LAC
Saunders W. AC1
Saunders H. LAC
Saunt J.A. F/L
Savage H. CPL
Sayers R.F. F/O
Says R.G. S/L (A/W/C)
Scarr J.R. CPL
Schafer G.W. F/O
Schneider H. P/O
Schollar E.C. F/Lt.
Schor K. F/L
Schulcz P.P. F/O
Scott Lt/Col
Scott M.P. S/O
Scott J.R. P/O
Scott A.T. W/C
Scott H.V. F/O
Scott E. F/O
Scott E.W. F/O
Scott J. SGT
Scott N.W. CPL
Scott T.J. AC2 (CPL)
Scott R.G. AC1
Scott R. F/S
Scriven A.J. F/L
Scriven A.W. F/L
Sealy L.J. LAC
Seaman J. LAC
Sear J.L. F/L
Seares H.J. P/O
Sears F.Q. SGT
Sedgwick Lt. Commander
Seer J.E. F/L
Senderson J. F/L (A/S/L)
Seriven A.J. F/L
Servos R.A. P/O
Sesar J.E. F/L
Seward G.E. P/O
Seward Air Vice Marshall
Shaak H. P/O
Sharpe R.A. F/L
Sharpe F. AC1
Shaul A.W. LAC
Shaw Commander
Shaw A.
Shaw R.L. LAC
Shawyer H.R. F/S
Shepherd R.W. LAC
Shepherd A.E. LAC
Sheppard W.W. F/Lt.
Sheppard W. LAC (ACU)
Sheppherd A.E. CPL
Sherbourne H.G. LAC
Sherwood A.K. F/O
Showry R.S. P/O
Shrubsole W.H. F/O (A/F/L)
Simpson R.L. P/O
Simpson G. LAC
Simpson W.B. LAC
Simpson J.A. SGT
Sims J. LAC
Sinclair R.F. F/L
Skeggs S.P. AC1
Slack J. LAC
Slade E.M. S/O
Slade M. S/O
Sleep D.M. F/O (A/F/O)
Small M.S. P/O
Smallwood H. CPL
Smiley W.D.C. P/O
Smith M.J. LAC
Smith E.D. A/F/L (F/O)
Smith G.D. P/O
Smith J. P/O
Smith D.F. P/O
Smith W.J. P/O
Smith A. F/O
Smith G.O. LAC (ACU)
Smith H.E. LAC
Smith E. LAC
Smith S.E. CPL
Smith E.J. AC2

Smith A.T.D. AC1
Smith W.J. LAC
Smith E.P. AC1
Smith W.E. LAC
Smith E. LAC
Smith W.A. CPL
Smith S.H. AC1
Smith B.A. LAC
Smith F. LAC
Smith A.F. LAC
Smith H.W. SGT
Snell P.H. SGT
Solly Cpl.
Solly J. CPL
Somerville-Woodiwis W.G. F/L
Soper R.L. W/C
Sotola F/Sgt. Czech
Sowden Sgt.
Sowden Sgt.
Sowden Sgt.
Sowden J.L. SGT
Sowers G.T. Major
Spaight R.H.S. Group Captain
Spence E.S. CPL
Spencer F/S
Spencer T.W. F/S
Sperl A.S. AC1
Spiller E.J.J. P/O
Spink R. Wing Officer
Spragg D.H. Sergeant
Squires G.S. LAC
Stacey W. CPL
Stacherd R. LAC
Staff W.E. LAC
Stalabrass J.R. P/O
Stalker L. F/O
Stallabrass J.R. F/L
Stamp D.G. LAC
Stanhope B.B. F/S
Stanley A.D. P/O
Stanyer J. LAC
Staples P.A. P/O
Starr R.E. P/O
Stephans A.L. P/O
Stephans A.E. P/O
Stephens A.F. AC1
Stevans G.S. P/O
Stevans P.B. P/O
Stevans G.S. P/O
Stevenson R. LAC
Steventon
Steward W.J. SGT
Stewart A.J.S. F/Lt.
Stewart D.A. S/L
Stiles F.H. SGT
Stillman H.R. F/O
Stillman R.R. F/O
Stillman F/Lt.
Stinsen M.V. P/O
Stinson W.L. P/O
Stinson H.V. P/O
Stirland J. LAC
Stirling H. P/O
Stockdale F.T. AC1
Stockley E.G. LAC
Stockwell F.W. AC1 (CPL)
Stone L.G. P/O
Stoneham C.H. AC1 (CPL)
Storer A. P/O
Storey A. LAC
Storron A/Cdr, Air Ministry
Stradling F. LAC
Strange A.B. Lieutenant
Strangeway F. LAC
Stratfold J.A. P/O
Stratford J. P/O
Stratton R. AC2 (LAC)
Strauss E.W. P/O
Streatfield R.C. Very Reverend
Street Sir Arthur Under Secretary of State for Air
Strong H. LAC
Stronge J.F. LAC
Stroud B.W. F/O (A/F/L)
Strudwick E.W. LAC
Stuart A.W.G. S/Ldr., WiNG Commander
Stubbings H. AC2
Stubbs R.A. P/O

Stubbs	N.	F/O (A. F/L)
Suart	A.	LAC
Suddes	G.K.G.	LAC
Sulivan	D.J.	LAC
Sulley	A.H.	Air Vice Marshal
Sullivan	J.L.	AC1
Summerfield	C.	LAC
Sumner	R.A.C.	F/O
Sumner	R.A.	F/L
Sutcliffe	D.J.	P/O
Sutherland	H.W.	S/Ldr.
Sutherland	A.G.	SGT
Swain	A.	F/O
Swain		Lieutenant
Swann		
Sweatmore	H.	CPL
Sweetman	C.T.	LAC
Swift	H.	P/O
Swift	P.J.	AC1
Sword	G.D.	LAC
Sykes	D.	LAC
Symes	W.	LAC
Tait		Air Vice Marshall
Tait	C.S.	LAC
Tandy	C.R.	P/O
Tanner		LAC
Tanter	H.	Reverend
Targett	R.P.J.	Warrant Officer
Tarrant	G.	SGT
Taylor		AC1
Taylor	H.	S/L (A/W/C)
Taylor	A.	P/O
Taylor	A.	P/O
Taylor	T.L.	F/L (A. S/L)
Taylor	A.E.	F/O
Taylor	L.T.	F/L (A. S/L)
Taylor	M.	LAC
Taylor	E.O.	LAC
Taylor	H.	LAC
Taylor	V.F.	LAC
Taylor	J.E.	LAC
Taylor	J.W.	AC1
Taylor	W.F.	AC1 (LAC)
Taylor	S.C.	F/O
Tebbs	W.A.	LAC
Telfer	W.W.	F/O
Telmy	J.L.	Sister
Tew	C.H.	LAC
Thewles	H.J.A.	W/C
Thomas	J.E.	F/O
Thomas	H.O.G.	P/O
Thomas	R.	P/O
Thomas	G.	LAC
Thomas	L.	SGT
Thompson	J.K.	P/O
Thompson	E.N.	F/Lt.
Thompson	F.N.	F/Lt.
Thompson	K.J.	P/O
Thompson	J.K.	F/O
Thompson	F.N.	F/L
Thomson	R.	LAC
Thorington	W.J.	AC1
Thornbury	R.H.	AC1
Thorne	F.W.	AC1
Thornton	E.J.	F/O
Threlfall	A.F.	A/F/S
Threlfall	H.	LAC
Thurgood	E.C.	LAC
Thwaites	J.W.	F/L
Thynne	B.W.	P/O
Tickell	J.	LAC
Tickler	J.E.	F/O
Tickler	F.E.	P/O
Timson	G.	AC1
Todd	E.D.	P/O
Todd	J.W.	AC1
Toland	W.M.	P/O
Tollstam	A.A.	P/O
Tomilson	A.	
Tonks	A.S.	LAC
Toone	R.S.	F/O
Torbet	T.F.	SGT
Tottten	V.	Sgt.
Towell	G.	F/L
Townley	P.	SGT
Townsend	N.F	AC1
Trail	G.	Flying Officer
Trainer	W.K.	F/O

Trescowthick N.C. W/C
Trigg L.A. P/O
Trim J.A. LAC
Troops O.C. Wing Commander
Trotter Squadron Leader
Trotter A. SGT
Truscott R.W.C. P/O
Trusson S.J. Sgt.
Tucker F.D. F/Lt.
Tucker F.B. F/L
Tuckett S.G. P/O
Tuckwood H.F. F/O
Tudge W.E. LAC
Tudge C.G. LAC
Tunstall
Turna Sergeant, Czech
Turnball I.B.A. F/O
Turnball W.M. AC1
Turner R.W. F/O
Turner A.E. F/O
Turner D.P. CPL
Turpin J.E.M. 2nd Lieutenant
Turton S.G. P/O
Tuttle J. P/O
Tutton F.G. LAC
Tutty P.L. F/O
Tycho N.I. P/O
Tyers R.W. LAC
Tylee R.S. P/O
Tyler F.A. SGT
Underell J.W.J. F/O A/F/L
Unwin G.F. F/O
Utting S.J.B. LAC
Utting R.A. LAC
Vachon F. S/O
Vallance J. F/O
Van der Grupt Benedict Colonel
Van Patter G.D. P/O
Vann R.K. SGT
Vare Captain S/L
Vernon M.A. F/L (A. S/L)
Vevers E. LAC
Vince G.H. P/O
Vincent A.W. LAC
Vivian E.J.E. P/O
Vragg R.J. F/O
Wade E.D. AC1
Waite R.N. Commanding Officer, Group Captain
Waites G.G. Sergeant
Wakefield C.J.C. CPL
Waldrop C. P/O
Wales K.C. F/L
Walker D.W. LAC
Wallace T. F/O
Wallace L. LAC
Wallace T. F/L
Wallis H.C. LAC
Wallwork E. F/S
Walmsley H.C. F/O (A/F/L)
Walsh J.T. CPL
Walsh P. LAC
Waltman A.D. P/O
Wansbrough Group Captain
Warburton E.L. F/Sgt
Ward E.S. AC1
Wardell W. LAC
Wares F.W. CPL
Warne J. AC1
Warren Mr. Asst Director of Admin.
Warren D. F/O
Wasnbrought R. S/C
Wass W. LAC
Waters R. LAC
Watkins S.F. P/O
Watkins D. LAC
Watkins E.E. P/O
Waton A.B. P/O
Watson A. P/O
Watson F.V. AC2
Watson R. LAC
Watson S.D. LAC
Watts C.H. CPL
Waugh J.S. LAC
Weale A.C. LAC
Weaton T. CPL

Weavers J.L. SGT
Webb J.A. P/O
Webb R.A. F/L (A/S/L)
Webber A. W.O.
Webster W.S. LAC
Weeden P/O
Weeds J.R. F/Lt.
Weekes C.W. P/O
Weeks C.E. LAC
Weeks A.E. LAC
Weiner M. F/O
Welch H. AC2
Wells
Welsh Sir William L. Air Marshal
Werbiski N.M. P/O
Werbiski W.B. F/O
West J.H. F/L
West L. F/O
Westerby E. LAC
Westgarth-Taylor C. F/O
Westgate J.S. SGT
Westgate W.W. AC2
Westwood C. AC2
Weyl M.R.A. LAC
Whall T.E. LAC
Whalley H. F/L (A. S/L)
Wheatley J.D. AC1
Wheatley S.F. LAC (ACU)
Wheeler H.N.G. S/L (A/W/C)
Wheeler S.A.A. SGT
Wheeler E.B. SGT
Whelan J. AC1
Whitaker W.H. LAC
Whitaker J. LAC
Whitburn W. LAC
Whitcomb G.F. LAC
White A.G. P/O
White G. Pilot Officer
White A.C. P/O
White L.J. F/L
White A.G. F/O
White J.W. LAC
White E.W. LAC
White D.P. LAC (ACU)
Whitehead L.F. SGT
Whitehead M. LAC
Whitman R.W.F. Squadron Leader
Whitman J.S. S/L (A/W/C)
Whitman R.W. S/L
Whitman R.H. S/L
Whitman R.W.F. S/L
Whittaker G.C. LAC
Whittam H. LAC
Whittingham L.M. P/O
Whittingham Sir Harold Air Marshall
Wickenden S. AC2
Wickers L.C. LAC
Wickham B. P/O
Wicks M.W.G. P/O
Wicks E.W. CPL
Widdows J.S. F/L (A/W/C)
Widdows J.S. W/C
Wigg L.H. F/O
Wigg L.H. F/O (A. F/L)
Wightman R.W.F. F/L (A/S/L)
Wightman J.S. S/L (A/W/C)
Wightman R.W. S/L
Wightman R.F. F/L (A. S/L)
Wilcox C.J. LAC
Wild R.B. F/O
Wilford T.H. P/O
Wilkie
Wilkin J.D. SGT
Wilkinson Marine Officer, F/O
Wilkinson H.W. A/F/O (P/O)
Wilkinson J.S. F/L
Wilkinson J.C.C. Flight Lieutenant
Wilkinson H.W. F/O (A/F/L)
Wilkinson T. LAC
Wilkinson C. LAC (ACU)
Willcock R.P. A/V/M
Willett W. AC2
Williams N.D. P/O
Williams C.R.D. Captain
Williams R.A. Captain
Williams C.R. Captain
Williams N.D. P/O
Williams S.G. F/O

Williams N.D. F/O
Williams B. LAC
Williams T.L. LAC
Williams W.D.R. LAC
Williams C.E. LAC
Williams H. LAC
Williman F/L
Williman K.H. A/F/L (F/O)
Willis C.M. S/O
Willis H.E. LAC
Willock C.B. Air Vice Marshal
Willock A.V.M. Deputy Head of BAC RAF Delegation, Washington
Wilmot E. LAC
Wilmot-Dear A.T.C. P/O
WilsonA.B. S/O
Wilson Major General
WilsonH.S. F/O
WilsonH.L. F/O
WilsonV.D. LAC
WilsonJ.T. LAC
WilsonS. AC2
Winfield K.B. P/O
Winn A. CPL
WinterG.T. LAC
WinterG.W. F/O
Wishart J. F/L
Withers A.E. LAC
Woffinden J.H. F/L (A/S/L)
Wolstenholme F. Cpl.
Wood F.M. F/L
Wood R.J. SGT
Wood C.W. AC1
Wood J.A. AC1
Wood J. LAC
Wood F.M. F/O A/S/L

Woodcock G. SGT
Woodcock L. CPL
Woodley M. AC1
Woodman T.R. LAC
WoodsN.J. P/O
WoodsF.E. CPL
Woodward S. P/O
Woodworth C. CPL
Woolf F.G. LAC
Woolhouse C.N. F/O
Wootton K.J. F/O
Workman R.H. CPL
Wren R. P/O
Wren J.H. LAC
Wright Major
WrightB.S. Lt. Cdr.
WrightJ. F/L (A. S/L)
WrightF. S/L
WrightD.I. LAC
WrightE.S. LAC
WrightW. LAC
Wylde E.W. LAC
Wyne T.E. P/O
Yarnold S. AC2
Yeoman A.J. AC2
Young G.L. P/O
Young N. W/C
Young S.A. P/O
Young S.A. F/O
Youngblud R.W. F/O
Younger W. LAC
Youngs E.A. LAC
Zdan B.H. P/O
ZdimalV. P/O
Zesulka V. P/O
Zimmerman Captain

Appendix 18: All RAF Personnel, Visitors, Recorded in the ORB's of No. 113 Wing

Name	Rank, Title
Adams	S/L, No. 45 TC, HQ
Aitken, B.F.	F/L, Staff Navigator
Alexander, J., DFC	W/C
Allen	W/C, RCAF, HQ, Ottawa
Anderson	S/L, No. 45 TC, HQ
Anderson	W/O
Anderson	G/C
Anderson, P.S.	S/L, OinC, No. 113 Wing Nassau
Athlone, HE the Earl of	His Excellency
Aubrey, W.	F/L, Education
Ayling, Mr.	Civilian
Baillie	Capt.
Banks	Air Commodore
Bannister, G.H.	P/O, Tech(E)
Barber, R.H.	F/O
Barciszewski, Z.J.	F/L, GDP, U/T Flying Control
Barclay, D.	F/O, Operations
Barker	SMO/O, W/C
Barker, H.	Civilian
Barker, Mr.	Air Ministry Auditor
Barker, P.M.O.	W/C
Barr, J.	F/O, GDP
Barrington-Mason	W/C
Baskett, C.A., DFC	W/C
Baskett, O.C.	S/L, OinC, Saferry
Beamish, G.R., CB, CBF	Air Vice Marshall
Beaton, J.W.	Maj., YMCA, Canada
Beck, C.O.	F/L, Station Adjutant
Bell	S/L, SMO, HQ No. 45
Bennett, Mr.	Representative of Ministry of Labour
Besston, E.J.	LAC, Cook
Best, G.L.	G/C, Chief Engineering Officer, 45 Group HQ
Bevan	W/C
Biddell, DFC	W/C
Biesiegal	G/C, AF HQ, Ottawa
Bigg, W.C.	F/O, Tech(E)
Bishop, W, VC, RCAF	Air Marshall, RCAF
Bland, J.R.	W/C, Command of RAF Station, Nassau
Blanford, P.B.	S/L, OinC, No. 113 Wing Nassau
Bligh, H.E.	F/L, Operations
Bligh, S.	F/Lt., Operations
Bone	W/C
Bonham-Carter, D.W.F., DFC	G/C, Officer Commanding No. 45 Wing, Dorval
Both, T.A.V.	S/L, Wing Engineer Officer
Botting, L.E., DFC	W/C
Bowel, P.W.	WOP, AIR
Bowhill, Lady	S/O

Bowhill, Sir Frederick W., GBE, KCB, CMG, DSO	AOC in C, Air Chief Marshall
Bowyer, E.C.	Society of British Aircraft Constructors,
Boyce	A/C, G/C
Boyd, G.D.	P/O
Braithwaite, R.C.	F/O, Signals Officer
Breedon	F/O
Breedon, M.	S/L
Briggs, S.	S/L
Brochochi	F/Lt.
Brown	F/O, Flying Instructor
Brown, D.R.C.	CFI, Vice F/O
Brown, David Lester, DFC	S/L, ADC to HRH, GDP
Browne, K.G.	S/L, Senior Navigation Officer, OinC No. 113
Bruen, J.G.	Sgt.
Bryant, C.	S/L, Chief Accountant Officer
Bullen, A.G.	F/L, Provost Marshall, Nassau
Burghley, Lord	Governor of Bermuda
Burns, F.V., Rev.	Rev., RAF Chaplain
Burton	F/Lt., CTO
Byng-Hall	F/L, Operations
Calladine, G.E.	P/O
Calladine, G.K.	F/O
Carlow, Viscount	Air Commodore, Air Attache to Brazil
Carnegie, D.V., AFC	Air Vice Marshall
Charest	G/C, AF HQ, Ottawa
Chilvers, R.C.	Air Ministry Official
Clark, G.S.E.	P/O, Equipment Officer
Clayton-Parrott, W.M.	A/F/O, C&C
Clive	W/C
Cloyton-Parrott	F/O, Cypher Officer
Coleman, E.H.	W/C, OinC, No. 113 Wing
Coleman, R.A.	S/L, Fire Chief
Collie, H.M.	Vice F/Lt., Senior Navigation Officer
Colyer, D., CB, DFC	Air Marshall, AOC, Bahamas
Connelly, J.W.A.	F/O, A/F/L, Operations Officer
Connolly	Mr.
Cormack, D.W.W.	F/O
Corney	Lt. Col., RASC
Corristine, E.	S/L
Courtney, Sir Christopher L., KCB, CBE, DSO	Air Chief Marshall
Coventry	W/C
Cowie	S/L, Administrative Staff, RAFDEL
Cox, Mr.	Auditor, Headquarters No. 45 Wing
Cracroft, P.D.	G/C, AFC, OinC, RAF Station Nassau
Crankshaw, F.F. le B.	S/L
Crawford, S.S.	WOP, AG
Creaghan, G.F.	Group Superintending Engineer
Cripps, R.	F/O
Croft, J.W.	F/L, Operations Officer, 88 SP
Cross, B.C.H.	Air Advisor to Secretary of State to the Colonies
Crump, W.M.E.	F/Lt.
Cunningham	W/C, Deputy Chief Signals Officer

Cunningham-Grahame, A.B.	Rear Adm.
Currey	S/L
D'Altray, J.T.	W/O2
D'Arcy-Power	Air Vice Marshall
Darvall, L., MC	Air Commodore, SASO Transport Command
Davies	S/L, C&C Officer
Davies, E.D.H.	A/C
de Niverville	Air Vice Marshall, RCAF
de Pret-Roose, M.	S/L, Chief Instructor, Unit Commander
de Seversky, Alexander	Maj.
Dean, Mr.	Director of Administration and Finance, RAFDEL
Deane, W.W., CBE	G/C, Chief Accountant Officer, No. 45 Group
Desrosiere	Col. , Deputy Minister of National Defence of Canada
Dibble, N.	S/L, Administration
Dinwoodie	W/C, Command Armament Officer
Dipple	G/C, Sr. C&C Officer
Dodd, Mr.	Air Ministry, Establishment Committee
Dodd, R.W.	S/L, Court Martial Duties
Dodero, Señor	Owner of Dodero Argentine Airlines
Dods, Mr.	Civilian
Donaldson, T.N.	F/L, GDP
Duchess of Windsor, Wallis Simpson	Duchess, Spouse of Governor Bahamas
Duke of Windsor, HRH	His Excellency the Governor of Bahamas
Duke, DFC	W/C
Dunn	S/O, C&C
Durward	G/C, Air Ministry Meteorological Department
Durward, D.	F/O
Earle	F/O
Eaton, G.M.	Civilian
Edwards, W.H.	F/O
Entwhistle	G/C
Etteridge, P.	W/O
Evans	Capt.
Evans, Miss	Guest of Governor Murphy
Fairbrother, J.M., DFM	F/O, W/AG
Farquhar	F/Lt.
Farquhar	P/O
Fawcett	P/O, W/AG
Ferguson, G.W.	F/O, Junior Equipment Officer
Field, R.C.	G/C, OinC, RAF Station Nassau
Finlayson, P.J.	S/L, ASO/S, 45 Group HQ
Fish, J.	F/L, GDP
Fisher, R.S.	Civilian Flight Clerk
Fletcher	W/C, OinC RAFTC, Unit, Elizabeth City
Fletcher	G/C
Flint, Rev.	Padre, Minister to No. 111 OTU
Franklin	S/L
Fraser	G/C
Frosts, J.R.	F/L, Staff Officer, Tech(E)
Furner	W/C
Gains	S/L, Group Staff Equipment Officer

Garlick	Capt., RASC HQ, Jamaica
Gawler	W/C, OinC, ASR Officer
Gibbon, J.	F/O
Giles, L.C.	S/L, GDP
Gillson	S/L
Glen, I.S.	Sgt.
Gobeil	S/L, Group Investigating Officer
Goble, RAAF	Air Vice Marshall, Australian Liaison Officer
Gollner, T.	F/Lt.
Gordon, D.M., CBE, AFC	A/C, British Joint Staff Commission
Gragg	Capt.
Grant, C.A.	F/O
Green	F/Lt., Accountant Officer
Green, L.	P/O
Green, L.W.	F/L
Greigg	F/O
Grose, E.A.	A/S/S
Halley	G/C, Western Air Command
Hansell, O.C.	Brigadier General, Caribbean Wing, ATC
Hargreaves	W/C
Harris, A.M.	Works Dept. Official
Harris, Sir A.T., KCB, OBK, AKC	Air Chief Marshall
Hartford, Y.H.B.	S/L, C&C
Harvey	G/C
Haydon, J.C.	Brigadier, British Joint Staff Commission
Heale	S/L
Henderson	LAC
Henry, C.	F/Lt., Accountant Officer
Hensman	A/E, GDN
Hey	G/C
Hibbert, J.R.B.	F/O, Tech(S), 98 Staging Post
Highfield, H.	R/O
Hindley, Mr.	Civilian
Hiscock, F.	Vice F/O, Adjutant
Hopwood, K.R.	F/O, Chief Signals Officer
Howard, J.	S/L, Tech(S)
Howe, C.D.	Hon., Minister of Supply and Munitions, Ottawa
Howes, W.M.	F/Lt.
Hughes, R.	Section Officer, C&O
Hulme, H.	Vice F/Lt., Chief Signals Officer
Hutchison, J.	F/L
Hyslop, H.I.	S/L
Irving	G/C, Air Ministry, Establishment Committee
Jeff, V.V.	S/O, C&C
Jefferson,	Brigadier, Commanding General North Caribbean Forces
Jeffs	G/C
Jensen	Col., OinC, AAS Base Unit, Miami
Jeremiah, D.	S/L
Johnson, Mr.	B.A.C.
Johnston	R/O, W/AG
Jones, A.L.	F/O, CTO
Jordan	Air Commodore
Joyee-Clarke	W/C, Engineering
Judges, W.C.	P/O
Keeping	G/C
Kentfield	W/C, Group Staff Engineering Officer
L'Estrange	S/L

Lang, E.H.	F/L, Accountant Officer, Stores Accounting
Lash, A.G.	F/O, Tech(S)
Latham, Sir Gordon	Sir
Leak, J.W.	F/O, Flying Control
Lee	Capt., Civilian Pilot
Lee, W.A.	G/C, Air Ministry, Establishment Committee
LeFeurve, J.B.	F/O, Traffic, U/T Flying Control
Lemplugh, A.B.	F/O, Operations Officer
Lilley, G.B.	F/O, CTO
Ling, A.J.	S/L, Equipment Officer
Linnitt, W.	S/L
Loat	F/Lt., Court Martial Duties
Lockhart-Mummery, Dr.	S/L, SMO
Lord	Brigadier General, USAAF
Ludlow-Hewitt, Sir Edgar R. CBE, KCB, CMG, DSO, MC, ADC	Air Chief Marshall, Inspector General
Luke, Sir Harry	Sir
Luth,	Lt., Group Surgeon
Lyman, R.M.	F/O, Field Operations Officer
Lyons, A.R.	F/L, Medical
Lywood	G/C, RAF Delegation, Washington
MacKay	P/O
Maitland, J.O.	F/O, Flying Instructor
Manwaring, A.H.	Radio Nav., F/Lt.
Maple, C.S.	F/C, Safety Equipment Officer, HQ, TC
Marshall, F.L.	S/L, Technical Officer
Martin	F/Lt.
Matthews, M.D.	T/S, PM, RAF
Maxim, R.L.G., CB, DSO	Air Vice Marshall, AOC No. 45 Group
Maxwell, P.H.	W/C
Maynard	SPSO, W/C
McDonald, C.B.R.	Col.
McDonald, C.W.	S/L, Court Martial Duties
McGill, F.R.	P/O
McGill, RCAF	Air Vice Marshall
McKay, J.S.	Sgt.
McKay, W.	F/Lt., Air Ministry, Establishment Committee
McKean	Air Vice Marshall
McKercher, H.W.	1st Officer, Radio Operator, Civilian
McQuillanm, E.S.	W/C, SPSO, No. 45 Group
McShane, H.	F/L, GDN
Mehab, J.D.	P/O, Equipment Officer
Miles	F/Lt.
Miles, A.E.	F/L, Adjutant, GDP, Gumman Goose
Milton, C.H.	S/L, Court Martial Duties
Minchinton, C.F.	F/Lt., Navigator, Flying Control Tower
Moorhead, J.U.	Maj., US War Department
Morris	F/L, Wing Signals Officer
Morris, A.O.	F/O, RAAF, Staff Navigation Duties
Mossop, Rev.	S/L, Hon., Rev., Group Chaplain

Murphy, E., Miss	Daughter of Governor
Murphy, Lady	Spouse of Governor
Murray, D.B.	F/Lt., Signals
Murrell-Wright, A.S.	F/O
Mutch,	G/C
Myles, H.	W/C, Air Ministry, Establishment Committee
Narburgh, J.W.	F/O
Nay	Co-Pilot
Nay, G.D.	F/O, Staff Navigation, 88 Staging Post
Newton, W.A.	W/O, WOP/AIR
Nicholas, B.D.	G/C, Air Ministry, Establishment Committee
North, Mr.	B.A.S.C.
O'Neill, M.F.	F/O, RCAF
O'Bryan, D.E.	W/C, 45 Group Control Officer
Ochalski, J.P.	F/Lt., U/T Flying Control, GDP
Orlady	Col., OinC, USATC Dorval
Owen, J.G.	F/L
Patton	F/O, Operations
Paull	W/C, OinC, No. 95 Staging Post
Pearce, Dr.	Dr., CMO, Bahamas General Hospital
Pena, Señor	Director of Civil Aviation, Uruguay
Penfold	F/Lt., Group ASR Officer
Peters	F/L
Peterson, H.	W/C, Air Ministry, Establishment Committee

Phillips, T.	S/L, Accountant Officer, 88 TSP
Pohoski, J.	F/Lt., U/T Flying Control, GDP
Porter, L.E.	Capt., British Joint Staff Commission
Powell	Air Commodore, SASO
Price, S.	F/O, GDS
Price, W.T.	P/O
Primrose, T., Dr.	S/L, SMO
Princess Alice, HRH	Her Royal Highness
Pugh	AC2
Pyan, F.S.	Sgt.
Raleigh	Maj., OinC, USAAF, Nassau
Ralph, R.J.	W/C, ASO/S
Rassin, Dr.	S/L, SMO
Ratcliffe, F.L.	S/L, Sr. Signals Officer
Read, J.P.	S/O, C&C
Rector, O.	Capt., Civilian Pilot
Reynolds, DSO, DFC	W/C
Richards, Sir Arthur	Sir
Richardson	G/C, TC HQ, Air Routes Advisory Section
Riddell	W/C
Roberts, D.N. OBE, APC	G/C, Air Commodore, OinC, No. 113 Wing Nassau
Robinson, K.E., DFC, AFC	S/L, RAAF, Chief Flying Instructor
Rogers	Capt., British Army Staff Movements Officer
Rogers, Kelly	Capt., BOAC Representative

Ross, J.W.	Capt., Civilian Pilot
Rowney, G.	F/O, Flying Control Officer
Shaw, R.	F/O, CTO, Tech(E)
Shelley, H.S.	F/O
Shergold, K.W.	P/O, RAAF, W/AG
Shipp	Cpl.
Short, R.W.	F/O
Silveroruys, Baron	Baron, Beligan Ambassador to Canada
Smeeten, H.A.	F/O, RCAF, Adjutant
Smith, R.G.	F/Lt., Navigation Briefing Officer, HQ, 45 Group
Sodal	Lt.
Spencer, H.F.	F/L, Operations
Stacey, A.	F/O, Pilot, Flying Control Tower, GDP
Stamp, Lord	Lord
Stanley	Cpl.
Stanley, Oliver	Hon., Secretary of State for the Colonies
Steadman, F.C.	W/C, CTO, Tech(E)
Steele, W.S.	P/O
Stettinius, Edward J., Jr.	U.S. Secretary of State
Stewart	S/L
Stewart, A.W.G.	W/C, Court Martial Duties
Stewart, J.	F/L, GDN
Stonehouse, A.R.	F/Lt.
Stoney, K.T.H.	F/L
Stott, D.C.M.	S/L
Street, Sir Arthur	Sir, Permanent Under Secretary of State for Air
Stuart, W.	W/C
Sullivan, D.G.	Hon., Minister of Supply for New Zealand
Sully	Air Vice Marshall
Sultzberger, J.K.	S/L, C&C Officer
Taylor	F/Lt.
Taylor, C.V.	F/O, GD/AB
Tegart, J.S.	R/N
Telfer	Vice F/O, Accounting Officer
Thames	F/O
Thom, R.J.	F/Lt., A&SD, Adjutant, 113 South Atlantic Wing
Thomas, E.B.	F/O, A&SD
Thomson, F.	F/O, GDN
Thorn, Mr.	Air Ministry Auditor
Tolmie, D.	F/O, Operations
Trench	G/C
Trott, T.D.P.	F/O, Operations
Tuck	W/C
Turnham, S.R.	F/O
van der Helstraete, H.J.	F/L, GDN
Wahltuch, P.	Vice F/O, C&C
Waite, R.N.	Vice G/C
Waller, A.W.L.	F/O, C&C
Walton, E.G.	F/O, Link Trainer Instructor
Warburton	Air Commodore
Waszak, J.M.	F/Lt., U/T Flying Control, GDP
Watkins, D.T.	P/O
Welch	F/L
Welch, W.E.	F/L, GDP
Welsh, Sir William L., KGB, DSC, AFC	A/M, Air Officer Commanding, RAF USA, Bahamas
Whally	S/L, P1, Dorval
Whittemore, A.J.	F/O
Wilcox	G/C
Williams, Mr.	Financial Advisor, Air Ministry,

	Establishment Committee
Willock, R.P., CB	Air Vice Marshall
Wilson, J.G.	Canadian Air Ministry
Wood, J.W.	F/O, RCAF
Woolston	S/L, Air Ministry, Establishment Committee
Wright	AO2
Young, F.R.	F/L

Bibliography

American Merchant Marine Museum, *Convoy! Supplying Allied Victory WWII*, US Merchant Marine Academy, US, 2017.

Archives Department, Ministry Education, *Bahamas During World Wars 1914-18, 1939-45*, Department of Archives Exhibition, Bahamas, 1985.

Areschoug, Richard, *Dodlig Resa Swedish Ships 1939-1945, Svenskt Militarhistoriskt Bibliotek*, Sweden, 2008.

Bahamas Historical Society, *Drifting to Duchess* Potlatch *Carstairs*, Bahamas Historical Society Journal, Bahamas, 2015.

Bahamas Historical Society, *U-Boats in the Bahamas*, Bahamas Historical Society Journal, Bahamas, 2009.

Barnette, Michael C., *Encyclopaedia of Florida Shipwrecks Vol.1: Atlantic*, Association of Underwater Explorers, US, 2010.

Barratt, Peter, *Bahama Saga Epic Story of Bahama Islands*, 1stBooks, AuthorHouse, US, 2004.

Bartholomew, John C., *World Atlas, 9th Ed.*, John Bartholomew & Son Ltd., UK, 1974.

Baxter, Ian, *Images of War U-Boat War 1939-1945*, Pen & Sword Military, UK, 2008.

Bennet, G. H. & R., *Survivors British Merchant Seamen WWII*, Hambledon Press, UK, 2007.

Bercuson & David, Herwig & Holger, *Long Night of the Tankers Caribbean Oil*, University of Calgary Press, Canada, 2014.

Berg, Denise & Daniel, *Tropical Shipwrecks Bahamas & Caribbean*, Aqua Explorers Inc., US, 1989.

Berglund, David C., *Shipwrecks of Anguilla 1628-1993*, Creole Publishing Co., St. Kitts WI, 1995.

Beyer, Kenneth M., *Q-Ships Versus U-Boats America's Secret Project*, Naval Institute Press, US, 1999.

Billy, George J. & Billy, Christine M., *Merchant Mariners at War Oral History*, University Press of Florida, US, 2008.

Bishop, Chris, *Illustrated Encyclopaedia of Weapons for WWII*, Metro Books, Amber Books, US, 2014.

Blair, Clay, *Hitler's U-Boat War, the Hunted 1942-1945*, Random House, US, 1998.

Blair, Clay, *Hitler's U-Boat War, the Hunters 1939-1942*, Modern Library, Random, US, 2000.

Bloch, Michael, *Duchess of Windsor*, Phoenix Illustrated, Orion, UK, 1997.

Bloch, Michael, *Duke of Windsor's War,1939-1945*, Coward-McCann Inc., US, 1983.

Bloch, Michael, *Secret File of the Duke of Windsor 1937-1972*, Harper & Row, US, 1988.

Bourneuf, Gus Jr., *Workhorse of the Fleet Liberty Ships WWII*, American Bureau of Shipping, US, 2008.

Boyd Malone, Shelley & Campbell Roberts, Richard, *Nostalgic Nassau: Picture Postcards 1900-1940*, Nassau Nostalgia, Bahamas, 1991.

Braeuer, Luc, *Adalbert Schnee Patrols of U-201 WWII*, Schiffer Publishing, Zephyr BD, US, 2015.

Braeuer, Luc, *Rolf Mutzelburg Patrols of U-203 WWII*, Schiffer Publishing, Zephyr BD, US, 2015.

Brennecke, Jochen, *Hunters and the Hunted*, Burke Publishing, W. W. Norton, US, 1957.

Brown, Ken, *U-Boat Assault on America*, Seaforth Publishing, Pen & Sword, UK, 2017.

Buchheim, Lothar-Gunther, *U-Boat War*, Bonanza, Knopf, Crown, US, 1986.

Bunker, John, *Heroes in Dungaries WWII*, Naval Institute Press, US, 1995.

Busch & Rainer, Roll & Hans-Joachim, *German U-Boat Commanders of World War II*, Naval Institute Press, US, 1999.

Campbell, David G., *Ephemeral Islands: Natural History of the Bahamas*, MacMillan Education, UK, 1978.

Carr, J. Revell, *All Brave Sailors*, Simon & Schuster, US, 2004.

Carroll, Francis M., *Athenia Torpedoed*, Naval Institute Press, US, 2012.

Chewning, Alpheus J., *Approaching Storm U-Boats Virginia WWII*, Brandylane Publishers, US, 1994.

Clark, William Bell, *When the U-Boats Came to America*, Little Brown & Co., US, 2015.

Cleare, Angela B., History of Tourism in the Bahamas: A Global Perspective, Xlibris, US, 2007.

Coale, Griffith Baily, *North Atlantic Patrol Log of Seagoing Artist*, Farrar & Rhinehart, US, 1943.

Connelly, T. Garth, *PT-Boats in Action*, Squadron, Signal Publications, UK, 1994.

Craton, Michael, *History of the Bahamas* , San Salvador Press, Canada, 1986.

Craton, Michael, *A-Z of Bahamas Heritage*, Macmillan Education, Macmillan Caribbean, Oxford, UK, 2007

Cremer, Peter, *U-Boat Commander: Periscope Perspective*, Naval Institute Press, Jove Book, US, 1982.

Daniels, Josephus, *German Submarine Activities Atlantic Coast Vol. 1*, Forgotten Books, UK, 2015.

Dawes, Robert A. Jr., *Dragon's Breath: Hurricane at Sea*, Naval Institute Press, US, 1996.

de Quesada, Alejandro, *US Coast Guard in World War II*, Osprey Publishing, Random House, UK, 2010.

De Marigny, Alfred, *Conspiracy of Crowns*, Crown Publishers, New York, NY, US, 1990.

Dodge, Steve, *Abaco History of an Out Island and its Cays*, White Sound Press, US, 1984.

Domenech, Ligia, *Imprisoned in the Caribbean 1942 U-Boat Blockade*, iUniverse POD, US, 2014.

Duffy, James P., *Hitler's Secret Pirate Fleet*, University of Nebraska Press, US, 2005.

Durrell, Zoe C. *The Innocent Island, Abaco in the Bahamas*, Durrell Publications Inc., US, 1972

Edwards, Bernard, *Beware Raiders!*, Naval Institute Press, US, 2001.

Edwards, Bernard, *Quiet Heroes; British Merchant Seamen 1939-1945*, Pen & Sword Military, UK, 2003.

Edwards, Bernard, *War of the U-Boats; British Merchantmen Under Fire*, Pen & Sword Military, UK, 2006.

Evans, Jonathan Land, *Colony at War: Bermuda, 1939-1945*, Land Evans, Bermuda, 2013.

Evans, Jonathan Land, *Peace Prudence and Prosperity: Bermuda, 1919-1939*, Land Evans, Bermuda, 2012.

Fenby, Jonathan, *Sinking of the* Lancastria *Deadliest Naval Disaster*, Carroll & Graf, Avalon, US, 2005.

Ganeri, Anita, *Oceans Atlas Pictorial*, DK Publishing, UK, 1994.

Gannon, Michael, *Operation Drumbeat Germany's First U-Boat Attacks*, Harper Perennial, US, 1990.

Gentile, Gary, *Fuhrer's U-Boats in American Waters*, Bellerophon Bookworks, US, 2006.

German Navy, High Command, *U-Boat Commander's Handbook*, Thomas Publications, US, 1989.

Giese, Otto, & Wise, James E., *Shooting the War*, Naval Institute Press, Bluejacket Books, US, 1994.

Goebeler, Hans, *Steel Boats Iron Hearts U-Boat Life U-505*, Savas Beatie LLC, US, 2008.

Gomez Alvarez, Maximo, *U-Boats Del III Reich en Cuba*, Carmelo Segura Y. M. Eugenia Glez. Cintas, Spain, 2009.

Grant, R. G., *Battle at Sea 3,000 Years of Naval Warfare*, DK Publishing, US, 2010.

Griffiths, Geraint, *Into the Blue Diary of RAF Bahamas 1943-45*, lulu.com POD, US, 2009.

Grigore, Julius Jr., *Surcouf Conspiracy*, iUniverse, US, 2011

Hadley, Michael L., *U-Boats Against Canada, McGill-Queen's University Press*, Canada, 1990.

Hague, Arnold, *Allied Convoy System 1939-1945*, Vanwell Publishing Ltd., Canada, 2000.

Hanington, Felicity, & Kelly, Percy A., *Lady Boats Canada's West Indies Merchant Fleet*, Dalhousie University, Canada, 1980.

Hannau, Hans W., *Bahama Islands in Full Colour*, Argos Books, US, 1970.

Hannau, Hans W., *Nassau in the Bahamas*, Wilhelm Andermann Verlag, Germany, 1962.

Hardcastle-Taylor, Jean D., *Windsors I Knew: A Secretary's Memoir 1940-1944*, Saint Michael's Press Book, US, 2016.

Harding, Steve, *Gray Ghost;* RMS Queen Mary *at War*, Pictorial Histories Publishing Co., US, 2007.

Hegland, Jon Rustung, *Nortraships Flate, Vol. 1 Norwegian Fleet 1940-41*, Dreyers Forlag, Norway, 1977.

Hegland, Jon Rustung, *Nortraships Flate, Vol. 2 Norwegian Fleet 1942-1945*, Dreyers Forlag, Norway, 1977.

Hickam, Homer H. Jr., *Torpedo Junction; the U-Boat War off East Coast*, Bantam, Doubleday, Dell, US, 1989.

Hill, J. R. & Ranft, Bryan, *The Oxford Illustrated History of the Royal Navy*, Oxford University Press, UK, 1995.

Hogel, George, *U-Boat Emblems of World War II 1939-1945*, Schiffer Publications, US, 1999.

Hood (editor), Jean, *Submarine, An Anthology*, Conway, UK, 2007.

Hornfischer, James D., *Last Stand of the Tin Can Sailors*, Bantam, Dell, Random, US, 2004.

Hoyt, Edwin P., *Death of the U-Boats*, McGraw-Hill, US, 1988.

Hoyt, Edwin P., *U-Boat Wars*, Cooper Square Press, Rowman, US, 2002.

Hoyt, Edwin P., *U-Boats Offshore When Hitler Struck America*, Stein & Day, US, 1978.

Huettel, Wilfred Chuck, *War in the Gulf of Mexico U-Boat Invasion*, Hogtown Press, US, 1989.

Jackson, Robert, *Warships Inside Out*, Thunder Bay Press, Baker & Taylor, US, 2010.

Jones, Roger M. & Jones, Peggy (illustrator), *PT Boat Episodes: PT-163 in the Pacific*, Merriam Press, US, 2000.

Jordan, David, *Wolfpack: the U-Boat War 1939-1945*, Spellmount Ltd. Amber Books, UK, 2002.

Jordan, Roger W., *World's Merchant Fleets 1939*, Naval Institute Press, Chatham, US, 2006.

Kaplan, Philip & Currie, Jack, *Wolfpack U-Boat War 1939-1945*, Naval Institute Press, Aurum, US, 1997.

Karig, Walter, *Battle Report: the Atlantic War*, Farrar & Rhinehart, US, 1946.

Keegan, John, *The HarperCollins Atlas of the Second World War*, Borders Press, HarperCollins, US, 1999.

Keeney, L. Douglas, *War Against Nazi U-Boats 1942-1944*, FastPencil Inc., Premiere, US, 2012.

Kelshall, Gaylord T. M., *U-Boat War in the Caribbean*, Naval Institute Press, US, 1988, 1994.

Kelshall, Gaylord T. M., *U-Boot-Krieg in Der Karibik*, Verlag E. S. Mittler & Sohn, Germany, 1999.

Kimmett, Larry & Regis, Margaret, *US Submarines in World War II, Illustrated*, Navigator Publishing, US, 1996.

Klingel, Gilbert, *Inagua: An Island Sojourn*, Lyons & Burford, US, 1940, 1997.

Konstam, Angus & Mallman Showell, Jak P., *Seventh U-Boat Flotilla Donitz's Atlantic Wolves*, Ian Allan Publishing, UK, 2003.

Kurowski, Franz, Knight's Cross Holders of the U-Boat Service, *Schiffer Publications*, US, 1995.

Kurson, Robert, *Shadow Divers*, Random House, US, 2004.

Lawliss, Chuck, *Submarine Book Illustrated History Attack Sub*, Burford Books Inc., US, 2000.

Lawlor, Anne & Lawlor, Jim, *Harbour Island Story*, MacMillan Education, UK, 2008.

Lawson, Robert & Tillman, Barrett, *US Navy Air Combat 1939-1946*, MBI Publishing Co., US, 2000.

Lightbourn, Ronald G., *Reminiscing: Photographs of Old Nassau Vol. II*, US, 2005.

Lindbaek, Lise & Solum, Nora O., *Norway's New Saga of the Sea WWII*, Exposition Press, US, 1969.

Lloyd, Jane, *History of the I.O.D.E. in the Bahamas 1901-2005*, Media Enterprises Ltd., Bahamas, 2016.

Lonsdale, Adrian L. & Kaplan, H. R., *Guide to Sunken Ships in American Waters*, Compass Publications, US, 1964.

Lowe, Jack, *My Life: An Abaco Boy Story*, Abaco Print Shop, Bahamas, 2005.

Lundberg, Lennart, *Lejdtrafik och Kvarstad Marinlitteraturforeningen Nr 85*, Abrahamsons Tryckeri AB, Sweden, 1999.

Mallmann Showell, Jak P., *Hitler's U-Boat Bases*, Sutton Publishing Ltd., UK, 2002.

Mallmann Showell, Jak P., *U-Boats in Camera: Archive Photographs*, Sutton Publishing Ltd., UK, 1999.

Mallmann Showell, Jak P., *U-Boats of the Second World War: Longest Voyages*, Fonthill Media, UK, 2013.

Marquis, John, *Blood and Fire: the Duke of Windsor and Sir Harry Oakes*, LMH Publishing Ltd., Jamaica, 2006.

Marx, Robert F., *Shipwrecks in the Americas*, Bonanza, Crown, US, 1983.

Mason, David, *U-Boat the Secret Menace*, Ballantine Books, US, 1968.

McCoy, Samuel Duff, *Nor Death Dismay: A Record of Merchant Ships*, MacMillan Co., US, 1944.

McCutcheon, H. Shaw, *A Family Island - A Short History of Salt Cay, Bahamas*, Salt Cay Publishing, US, 2001

McKee, Fraser M., *Sink all the Shipping There*, Vanwell Publishing Ltd., Canada, 2004.

McKee, Fraser & Darlington, Robert, *Canadian Naval Chronicle 1939-1945*, Vanwell Publishing Ltd., Canada, 1996.

Melton, Herman E., *Liberty's War Engineer's Memoir Merchant Marine*, Naval Institute Press, US, 2017.

Miller, Nathan, *War at Sea: Naval History of World War II*, Lisa Drew Book, Scribner, US, 1995.

Mills, Carlton, *History of the Turks & Caicos Islands*, MacMillan Publications, UK, 2008.

Moore, Arthur R., *Careless Word.... A Needless Sinking*, American Merchant Marine Museum, US, 1990.

Morgan, Daniel & Taylor, Bruce, *U-Boat Attack Logs: Warship Sinkings*, Seaforth Publishing, US, 2011.

Morison, Samuel Eliot, *Atlantic Battle Won, May 1943-May 1945*, Little, Brown & Co., US, 1984.

Morison, Samuel Eliot, *Battle of the Atlantic, 1939-1943*, Little, Brown & Co., US, 1984.

Morison, Samuel Eliot, *Two Ocean War*, Little Brown & Co., Canada, 1963.

Morison Beck, Emily, *Sailor Historian: the Best of Samuel Eliot Morison*, Houghton Mifflin Co. Boston, US, 1977.

Mort, Terry, *Hemingway Patrols: Hunt for U-Boats*, Scribner, Simon & Schuster, US, 2009.

Moseley Moss, Valeria, with Lightbourn, Ronald, *Reminiscing: Memories of Old Nassau*, Bahamas, 1999.

Mueller, Edward A., *Savannah Line Ocean Steamship Co.*, Steamship Historical Society of America, US, 2001.

Mulligan, Timothy P., *Neither Sharks nor Wolves*, Naval Institute Press, US, 1999.

Munnings, Harold Alexander Jr., *Healthcare in the Bahamas*, Media Enterprises Ltd., Bahamas, 2014.

Munnings, Harold Alexander Jr., *Westward Walk of a Bahamian Doctor*, Xlibris, US, 2009.

Nagorski, Tom, *Miracles on the Water: Heroic Survivors U-Boat*, Hyperion, US, 2006.

Nautical Publications, *Central Bahamas Andros to Exuma Eleuthera*, Maptech Inc., US, 2003.

Nautical, Publications, *Northwest Bahamas Bimini & Berry Abaco*, Nautical Publications GmbH, Germany, 2004.

Neely, Wayne, *The Major Hurricanes to Affect the Bahamas: Personal Recollections of Some of the Greatest Storms to Affect the Bahamas*, Authorhouse, Bahamas, 2006.

Niestlé, Axel, *German U-Boat Losses During World War II*, Frontline Books, US, 2014 Noli, Jean, *Admiral's Wolf Pack*, Doubleday & Co., Fayard, US, 1974.

Offley, Ed, *Burning Shore: How Hitler's U-Boats Brought War*, Basic, Perseus Books Group, US, 2014.

Ott, Wolfgang, *Sharks and Little Fish: Novel German Sub War*, Lyons Press, Pantheon Books, US, 2003.

Owen, Weldon, *Illustrated Atlas of the Sea*, Weldon Owen Pty. Ltd., Australia, 2010.

Parker, Philip M., *Bahamas: Webster's Timeline History 1492-2007*, ICON Group International, US, 2009.

Parker, Mike, *Running the Gauntlet: An Oral History Merchant Seamen*, Nimbus Publishing Ltd., Canada, 1994.

Paterson, Lawrence, *Second U-Boat Flotilla*, Pen & Sword, Leo Cooper, UK, 2003.

Paterson, Lawrence, *U-Boat War Patrol Hidden Photographic Diary*, Greenhill, Stackpole, UK, 2004.

Pearce Jones, Guy, *Two Survived: 70 Days at Sea in Open Boat*, Lyons Press, Globe Pequot, US, 1940, 1976.

Piekalkiewicz, Janusz, *Sea War 1939-1945*, Sudwest Verlag GmbH & Co., Germany, 1980.

Platt, Owen, *Royal Governor and the Duchess in Bahamas*, iUniverse, US, 2003.

Pullinger, Diana, & Lawlor, Jim, *History of the Bahamas in Pictures*, Media Enterprises Ltd., Bahamas, 2012.

Pye, Michael, *King Over the Water: the Windsors War Years*, Holt Rinehart & Winston, US, 1981.

Quinn, William P., *Shipwrecks Along the Atlantic Coast*, Commonwealth Editions, US, 2004. Raddall, Thomas H., *Mersey Story*, Bowater Mersey Paper Co Ltd, Canada, 1979.

Robinson, Stephen, *False Flags: Disguised German Raiders WWII*, Exisle Publishing, Australia, 2016.

Rohwer, Jurgen, *Axis Submarine Successes of World War Two*, Naval Institute Press, US, 1999.

Rohwer, Jurgen, *Chronology of the War at Sea 1939-1945*, Naval Institute Press, US, 2005.

Roscoe, Theodore, *United States Destroyer Operations in World War II*, Naval Institute Press, US, 1953.

Russell, Joe, *Last Schoonerman: the Life of Captain Lou Kenedy*, Nautical Publishing, Far Horizons, US, 2006.

Satchell, Alister, *Running the Gauntlet: Liners 1942-1945*, Naval Institute Press, US, 2001.

Saunders, Gail, PhD, *Bahamas: A Family of Islands*, MacMillan Publications, UK, 2000.

Saunders, Gail & Cartwright, Donald, *Historic Nassau*, MacMillan Education, UK, 1979.

Saunders, Gail & Craton, Michael, *Islanders in the Stream Vol. 2 End of Slavery-2000*, University of Georgia Press, US, 2000.

Saunders, Gail & Huber, Linda M., *Nassau's Historic Landmarks*, MacMillan Education, Publishers, UK, 2001.

Saunders, Gail, PhD., *Bahamian Society After Emancipation, Essays in Nineteenth and early Twentieth Century Bahamian History*, Nassau, N.P., Bahamas, 1990

Savas, Theodore P., *Hunt and Kill U-505 and U-Boat War*, Savas Beati LLC, US, 2012.

Savas, Theodore P., *Silent Hunters U-Boat Commanders WWII*, Savas Publishing, US, 1997.

Sherwood, Martyn, *Voyage of the* Tai-Mo-Shan, Rupert Hart-Davis, UK, 1957.

Simpson, Wallis, *Heart Has Its Reasons*, David McKay, Van Rees Press, US, 1956.

Solly, Ray, *Athel Line Fleet History*, History Press, US, 2009.

Spector, Ronald H., *At War at Sea: Sailors and Naval Combat*, Penguin Putnam, UK, 2001.

Stafford, Edward P., *Subchaser*, Naval Institute Press, Bluejacket Books, US, 1988.

Standard Oil, Company of NJ, *Ships of the Esso Fleet in World War II*, US, 1946.

Starnes, H. Gerald, *Torpedoed for Life WWII Veterans US Merchant*, Starnes, US, 2014.

Stern, Robert C., *Battle Beneath the Waves: Arms & Armor*, Cassell Group, US, 1999.

Stern, Robert C., *U-Boats in Action*, Squadron, Signal Publications, UK, 1997.

Stevens, David, *U-Boat Far from Home: the Epic Voyage of U-862*, Allen & Unwin, Doubleday & Co., Australia, 1997.

Stilgoe, John R., *Lifeboat: Courage Cravenness Survival*, University of Virginia Press, US, 2003.

Stranack, Ian, *Andrew and the Onions: Royal Navy in Bermuda*, Bermuda Maritime Museum Association, Bermuda, 1990.

Suhren, Teddy & Brustat-Naval, Fritz, *Ace of Aces: Memoirs of U-Boat Rebel*, Chatham, Lionel Leventhal, UK, 2006.

Summerscale, Kate, *Queen of Whale Cay*, Bloomsbury, Viking, Penguin, UK, 2008.

Syrett, David, *Defeat of the German U-Boats*, University of South Carolina Press, US, 1994.

Taylor, Theodore, *Fire on the Beaches*, W. W. Norton, US, 1958.

Terraine, John, *Business in Great Waters: U-Boats 1916-45*, Pen & Sword, Leo Cooper, UK, 2009.

Thompson, R. Chester, *Long Day Wanes: Memoir of Love and War*, White Sound Press, US, 2006.

Toogood, Michael A., *Bahamas Portrait of an Archipelago*, MacMillan Caribbean, UK, 2004. Tougias, Michael J., *So Close to Home*, Pegasus Books, US, 2016.

Turnquest, Orville, *What Manner of Man is This?*, Grant's Town Press, Bahamas, 2016.

Vance, Roger L., *When the Battle of the Atlantic Turned Ugly*, *World War II Magazine*, US, 2016.

Vause, Jordan, *Wolf U-Boat Commanders in WWII*, Naval Institute Press, US, 1997.

Vickers, Hugo, *Behind Closed Doors*, Arrow Books, Random House, UK, 2011.

Villiers, Alan, *Posted Missing: Ships Lost Without a Trace*, Charles Scribner's Son, US, 1956.

War Shipping, Administration, *United States Merchant Marine at War*, American Merchant Marine Combat Veterans' Just Compensation Committee, US, 1946.

Waters, Sydney D., *Ordeal by Sea: New Zealand Shipping Company*, New Zealand Shipping Co. Ltd., New Zealand, 1949.

Werner, Herbert A., *Iron Coffins Personal Account of U-Boat WWII*, De Capo Press, US, 1969.

Westwood, David, *U-Boat War: German Submarine Service*, Casemate, Conway, US, 2005.

White, John F., *U-Boat Tankers 1941-45*, Naval Institute Press, US, 1998.

Wiberg, Eric T., *Swan Sinks*, Island Books, US, 2017.

Wiberg, Eric T., *U-Boats in the Bahamas*, Bricktower Press, US, 2016.

Wiberg, Eric T., *U-Boats off Bermuda*, Fonthill Media, UK, 2017.

Wiggins, Melanie, *Torpedoes in the Gulf*, Texas A&M University Press, US, 1995.

Wiggins, Melanie, *U-Boat Adventures: First-hand Accounts from WWII*, Naval Institute Press, US, 1999.

Williams, Andrew, *Battle of the Atlantic Versus Hitler's Grey Wolves*, Basic, Perseus Books Group, US, 2003.

Williams, Darius D., *Rail & Locomotive History of the Bahamas*, White Sound Press, US, 2007.

Williamson, Gordon, *Gray Wolf U-Boat Crewman of WWII*, Osprey Publishing, Random House, UK, 2001.

Williamson, Gordon, *U-Boat Tactics in World War II*, Osprey Publishing, Random House, UK, 2010.

Williamson, Gordon, *U-Boats vs. Destroyer Escorts Battle of Atlantic*, Osprey Publishing, Random House, US, 2007.

Williamson, Gordon & Pavlovic, Darko, *U-Boat Crews 1914-45 Fighting Elite*, Osprey, Reed Consumer Books, UK, 1995.

Wynn, Kenneth, *U-Boat Operations of the Second World War Vol. 1*, Caxton Editions, US, 1997.

Wynn, Kenneth, *U-Boat Operations of the Second World War Vol. 2*, Naval Institute Press, US, 1998.

Wynne, Lewis N., *Florida at War*, Saint Leo College Press, US, 1993.

Young, Everild, *Eleuthera Island Called Freedom*, Regency Press, London, UK, 1966.

Primary Websites

1) Aircrew-Saltire.org
2) AirForce.mil.nz
3) AFHRA.af.mil
4) Aircrewremembered.com
5) Ancestry.com
6) AWM.gov.au
7) BahamasNationalArchives.bs
8) Bahamastourcenter.com/history-of-blue-lagoon-island
9) Bermuda-online.org
10) BNL.bm
11) Circulonaval.com
12) Coastguardnews.com
13) Convoyweb.org.uk
14) Discovery.nationalarchives.gov.uk
15) DUBM.de
16) Ericwiberg.com
17) Findagrave.com
18) Fold3.com
19) Hetscheepvaartmuseum.com
20) Historianaval.org
21) History.Navy.mil
22) History.USCG.mil
23) IWM.org.uk
24) Janda.org/ships
25) Maritiemmuseum.nl
26) Masterbombercraig.wordpress.com
27) Meandthesea.com (AIDHNC, Cuba)
28) Media.Defense.gov
29) Navsource.org

30) Newspaperarchives.com
31) NMB.bm
32) Number59squadron.com
33) NYTimes.com (News Archives)
34) PBHitoryonline.org
35) RAFCommands.com
36) RAFMuseum.org.uk
37) Regiamarina.net
38) Rquirk.com/no25.htm
39) Uboat.net
40) Uboatarchives.net
41) Warsailors.com
42) Wikipedia.org
43) Wrecksite.eu

Interviews

Kim Aranha (Bahamian):

Niece of Marion Carstairs, regarding her aunt's personality, ambitions, and motivations. See *The Queen of Whale Cay*, by Kate Summerscale for further reading.

Lester Brown (British, Bahamian):

David Lester Brown, Squadron Leader in the RAF in 111 O.T.U. In the 1970's, as I child, we all greatly admired and literally looked up to Lester Brown, a neighbour by about a mile whose granddaughter was a classmate and friend of mine. Capt. Brown was larger than life and knew it; he joined the RAF 111 O.T.U. and it is evident from his several entries in the operational log that he was well liked and admired, and often promoted. Having survived at least one mishap and participated in most of the VIP events and, sadly, funerals during the war, we went to continue as a prominent citizen of the Bahamas. He kept a seaplane in a hangar on Lake Cunningham a short walk from our childhood home on Lakeview Drive, off Skyline Drive and Prospect Ridge. My older brother John was the fortunate invitee to be taken by seaplane from there to Dunmore Town, Harbour Island. So, while I did not actually interview Capt. Brown, I met him often and he was an active participant in 111 O.T.U. not only on behalf of the RAF, but of course representing local Bahamians as well.

Bahamian historians Jim and Anne Lawlor have tracked Squadron Leader Brown's achievements, and note that he "distinguished himself in The Royal Air Force." The *Nassau Tribune* of 2 October 1944 reported that Brown, was a Queens College Scholar; one of only eight to be so honoured: "Brown arrived in Nassau this week to be an instructor at the RAF Base. [He] was one of the first to volunteer in 1940. Mainly in the Middle East, he has more than a 100 raids on his score card. His family is to be congratulated on his safe return and with such an incredible record of service and rapid promotion behind him. The *Tribune* is hoping that many other young men will return to Nassau to serve their country with the same enthusiasm and courage. The fact that these men volunteered shows that they are unselfish and courageous, and that their travel and experience abroad should greatly expand their outlook on life. These are qualities - unselfishness, courage and breadth of vision – that are so badly needed in the public life of the Colony."

According to the *Nassau Tribune* of November 16, 1944, "Brown saw a year of active in the Persian Gulf and the Indian Ocean 'covering' the routes of convoys to Russia. A year as an instructor in the Middle East and six months in England where he was engaged in the Battle of France and for Germany." Brown received the Distinguished Flying Cross in December, 1944 for his leadership in his squadron of Bomber Command. The Citation reads: "The Officer has set an example of courage and resolution. He participated in a large number of sorties... Since the landing in Northern France, he was a most efficient commander, whose forceful leadership, strong sense of duty, and tenacity proved most inspiring" Brown was appointed as Honorary *Aide-de-Camp* (ADC) to His Excellency the Governor (the Duke of Windsor, a very high honour), until further notice. After the war, Brown became involved in the real estate sector.

In an RAF obituary which the Lawlors shared, it is claimed that Brown became "the most decorated World War II Bahamian of his era." Born in 1915, he volunteered with four other boys, referred to as the Gallant Five; Fane Solomon, John Maura, John Maillis and Ivor Thompson were the others. Brown's training on Tiger Moths took him to Sywell, England, then Oxford. After further training aboard Blenheim bombers and stints bombing Axis shipping in the English Channel northwards, he served in Egypt and North Africa, from Cairo then to the Sudan and West Africa.

On return to the UK his Flying Unit, Group 2 flew B-25 Mitchells to Europe. He supported the invasion of Normandy by leading wings of 54 aircraft. He returned home to his parents, Captain and Mrs. Willard Brown, after three operational tours. Starting in December, 1944, Brown continued his support of the Duke of Windsor as a trusted ADC for 23 years, until 1967. As well as a DFC, Squadron Leader Brown received the North Africa Star, Defence Medal, Burma Star, Air Crew Europe Star, and Victory Medal, with various clasps and rosettes.

Mr. Kendall Butler (Bahamian):

Mr. Butler is a pre-eminent expert on the construction, ownership, career histories and fates of hand-built wooden sailing and power vessels in the Bahamas and Turks & Caicos. His generosity sharing what he has learned is invaluable.

Nicolò Cortès (Italian):

Don Nicolò Maria Pignatelli Aragon Cortès (born 1923), 17th Prince of Noia, at 95 years. He recalled that as a young cadet aboard the Italian submarine *Atropo*, the commander came close enough to an uninhabited Bahama island *en route* to Guantanamo in late February 1945, and that he and some other young sailors joyfully ran along the sandy beach. So, soldiers did land in Bahamian sand but there were friendly allies. The interview was arranged by Capt. Paul Aranha at the Prince's home in Lyford Cay, Nassau on July 14, 2017.

Robin Grant (Scottish):

Robin Grant, from Fife, Scotland north of Edinburgh, visited the Bahamas in 2018 with his wife Sandra to honour his father Robert, who served on 111 O.T.U.. Mr. Grant (his father) was recruited in 1941 at the age of 37 by the RAF and became a Leading Aircraftman, LAC, in the ground crew in Nassau. Thereafter he returned to eastern Scotland with the Coastal Command, and survived the war. After leaving the RAF in 1945 and into the early 1960's he led the Queen's Park Football Team and promoted Alex Ferguson, who went on to coach Manchester United. His son Robin and wife Sandra were my dinner guests in New York City on their way being hosted at the Bahamas Historical Society by its President, Ms. Andrea Major and Jim and Anne Lawler, and Capt. Paul Arahna provided them with a detailed tour of the lovely RAF Cemetery.

Capt. Durward Knowles and **Senator Basil Kelly (both Bahamian)**

Provided phone interviews regarding Marion Carstairs' power yachts *Sonia, Sonia II*, and *Vergemere IV*, built on her Whale Cay fiefdom in the Berry Islands and was used to rescue survivors of *SS Potlatch* from off Landrail Point, Crooked Island, in July-August, 1942. Messrs. Kelly and Knowles either commanded some of Carstairs' several yachts or know who did.

Ancil Rudolph Pratt (Bahamian):

In February of 2013 I was able to interview Mr. Ancil Rudolph Pratt via phone in Clarence Town, Long Island, thanks to a spot-on introduction from diver Kevin James. I was told that on the morning of Saturday 7th August 1942 young Ancil would take his father's horses out to pasture in the morning. He and his friends Wellington Smith, Kipling Simms, Jeffrey Strachan and Isaac Taylor then became the only living witnesses to a German U-boat in the Bahamas, when *Oberleutnant zur See* Claus-Peter Carlsen, aged 23, was conning his submarine, U-732 back to Brest, France and threading the channels of the southern Bahamas when they passed close enough to south-eastern Long Island while lining up navigational fixes for their passage back to Europe to be seen by several boys out in the fields. They remembered it forever, down to the length between the telephone poles, and Carlsen's U-boat logbook or KTB matches Ancil's versions of events.

Craig Symonette (Bahamian):

Provided technical historical background and histories on innumerable vessels constructed at the shipyard which has father Sir Roland Symonette founded and operated for decades, including two wooden minesweepers which became mailboats.

Herbert A. Werner (German):

A wartime U-boat commander and author of the rather cynical and dark U-boat classic *Iron Coffins (1969*), Mr. Werner broke his usual silence to send me an email in 2009. In his only message to me, he methodically and categorically disavowed the possibility of Germans stepping ashore in the Bahamas in WWII. He explained that the German submariners were terrified, ill-clothed, and barred from entering shoal waters or disembarking except trained terrorists on missions, as in Florida, New York, Maine, New Brunswick, the St. Lawrence River, Greenland, and a medical emergency on Martinique.

Other

Belmont School, Nassau Generally:

At various Bahamas Historical Society presentations from 2010 onwards, I was able to interact with witnesses and participants of the war years in Nassau. Among those that attended the Belmont School with Jimmy Goldsmith and other London evacuees whose ship was attacked by U-Boats on the perilous voyage from the U.K. Those that volunteered their insight include: **Peter Christie** Belmont student, and historian lawyer **Pericles Maillis**, and RAF Cemetery volunteer and aviation historian **Capt. Paul C. Aranha**.

For the *Potlatch* and Carstairs story, I interviewed Survivor **Jake Jatho** in person at his home in Missouri, and **Estil Dempsey Ruggles** via phone from Kentucky; they reunited thanks to this research. Also I met in person with families of Capt. Lapoint and Carbotti, and received photos and letters form the families of Lybrand, Solomon, Kaplan, Jensen, and various others, including dozens of published articles and a script. In Acklins, where the men landed, Mari and Fritz Anderson very kindly interviewed Reverend and Captain Albert Samuel Collie's daughter, **Rev. Clarisse Cox, and Rev. Newton Williamson**, of the Hard Hill area, who personally witnessed the men land and Parson being buried. Through them we learned of others from the area who were children at the time and ran down to the beach to lend the sailors a hand, including Mary Cox and **Remilda Williamson**, who I interviewed via phone.

Summary of Interviews:

With nearly 150 vessels, there are too many people interviewed over more than a decade to list each one: I interviewed a survivor of SS *Daytonian*, family members of the first officer and crew members of the *Cygnet*, the family of the owners of the *Cygnet*, *Mariana*, and *Wawaloam*, and a number of other vessels. While the interviewees are not many, they do cover several perspectives.

Notes on Sources

The six primary sources for the Daily Diaries Section were:

1) Operational Record Books (ORB) for RAF No. 111 O.T.U., Windsor Field, Nassau
2) Operational Record Books (ORB) for RAF No. 113 Transport Wing, Oakes Field, Nassau
3) Gulf Sea Frontier, 7th & 8th US Navy Naval District, Commander, from Sept. 1942
4) Eastern Sea Frontier, temporary USN Defence of entire East Coast, Dec. 1941-Apr. 1942
5) U-Boat & Italian KTBs, daily war diaries, advising of movements and attacks by and upon
6) Base Diaries for all iterations of NAS Great Exuma: NAS, NAF, and NAAF, 1942-1945

1) ORBs for RAF 111 No. O.T.U., Windsor Field, Nassau

Operational Record Books (ORBs) of RAF No. 111 O.T.U., Nassau are kept in The National Archives in Kew, England, outside London. Since these thousands of pages are the bedrock of this book and the reader will naturally wish to know more. I provide the full citations given by professional UK researcher Simon Fowler, author of dozens of books:

Summary of Operations Record Books (ORBs) for RAF No. 111 O.T.U. at TNA.:

As one group: AIR 29/689, Form 540: Narrative (and unusually detailed) account of the Unit 1942-1945. AIR 29/690 Appendices: 1943 March-July, AIR 29/691: Appendices: 1943 August-December, AIR 29/692, Appendices: 1942 January-June, AIR 29/693 Appendices: 1944 July-December. AIR 29/694: Appendices: 1945 January-August. These are Form Orange; and are all summaries of sorties [missions] flown by individual aircraft. AIR 29/695: Appendices: 1943-1945. These are training syllabuses; details of students; administrative orders. These reports are rather uninformative as they mainly consist of navigational details in code.

Note: More may be learned by visiting Discovery.nationalarchives.gov.uk. A copy of the complete ORB for 111 O.T.U. has been given digitally and in paper form to the National Archives of the Bahamas for reference.

2) ORBs for RAF 113 Transport Wing, Oakes Field, Nassau

Operational Record Books (ORBs) of RAF No. 113 (Transport) Wing, Oakes Field, Nassau are kept in The National Archives in Kew, England, outside London. It is believed that they, like No. 111 O.T.U., are kept in the AIR 29 group with other RAF records, however this is unconfirmed. Professional UK researcher Simon Fowler, copied them. The reports were originally sent to regional headquarters at the air field in Dorval, Quebec, Canada, near Montreal. There were frequent flights transporting aircraft as well as personnel, between the bases. RAF No. 113 Wing was not in existence long yet they transported thousands of planes around the world on a route southeast of Bahamas to Trinidad, Brazil, Africa and beyond.

Note: More may be learned by visiting Discovery.nationalarchives.gov.uk or contacting Simon Fowler at thesimonfowler@gmail.com.

3) Daily War Diary of Gulf Sea Frontier [GSF], 7th & 8th US Navy Naval District

Located in the National Archives and Records Administration complex in College Park, Maryland, I discovered these on-site while researching *U-Boats in the Bahamas* in 2009. The first entry can describe them all and reads "Beginning this date, September 14, 1942, the *War Diary* will include daily maps showing operations a 1200 noon EWT in the Gulf Sea Frontier." On the front is a very detailed chart with colour arrows covering the entire Gulf of Mexico, Yucatan Peninsula, and, relevant to this book, the ocean from Havana to Guantanamo, to Key West to the border of Florida and Georgia. The area specifically covered in a triangle extending east from Florida extends from south of Cay Sal Bahamas, east to the southern tip of Cat Island, and all of the waters north and northeast of Grand Bahamas, Walkers Cay, Abaco, and Eleuthera, In short it covers the Nicholas Channel, the Straits of Florida, the entire east coast of Florida including the Florida Cays.

A "Legend" provides a match to the arrows which covers movements and directions of all vessels, military and civilian, as well as their escorts, air, fixed wing and blimp or airship, and watercraft. It is extremely handy, simple to read and understand, clear and concise and an invaluable tool. Every relevant word and activity has been typed into this study.

Note: These documents may be obtained on the site Fold3.com for a fee, or using a professional research firm such as Westmoreland Research, or by visiting NARA near Washington, DC, or by pursuing this link, which they sent: catalog.archives.gov/id/521028 (which refers specifically to photographs only. They ought to be with *Survivor Statements* in the series: Papers of Vice Admiral Homer N. Wallin, compiled 1941 - 1974. Record Group 38: Records of the Office of the Chief of

Naval Operations, 1875 - 2006 Entry P-13. National Archives at College Park - Textual Reference (Military) 8601 Adelphi Road, College Park, MD, USA 20740.

4) Daily War Diary of the Eastern Sea Frontier [ESF], US Navy, Dec. 1941-Apr. 1942

This was a short-lived Diary, yet it is filled with the panic, confusion, and dismay of de-facto defeat as the US forces struggle to even understand the extend, veracity and location and capabilities of an enemy they clearly barely understood in the January to April 1942 timeframe while they finally adjusted, in time for the German to go elsewhere to softer targets. As a practical matter, by far the easiest and most cost-efficient method is to visit the entire set, type-set and uploaded by Capt. Jerry Mason (USN, Ret.) and Mrs. Charla Mason at Uboatarchive.net/ESF/ ESFWarDiaryDec41.htm. Officially known as the *Eastern Sea Frontier Enemy Action and Distress Diary*, these are kept at NARA, outside DC. Department of the Navy. Office of the Chief of Naval Operations. Naval Observatory. 1942-9/18/1947 Record Group 80: General Records of the Department of the Navy, 1804 to 1983.

Note: These documents may be obtained on the site Fold3.com for a fee, or using a professional research firm such as Westmoreland Research, or by visiting NARA near Washington, DC. See also history.navy.mil, archives.gov, an overview is at RG 181 - Naval Districts and Shore Establishments, General Correspondence, Sixth Naval District at Charleston, South Carolina. They ought to be with *Survivor Statements* in the series: Papers of Vice Admiral Homer N. Wallin, compiled 1941 - 1974. Record Group 38: Records of the Office of the Chief of Naval Operations, 1875 - 2006 Entry P-13. National Archives at College Park - Textual Reference (Military) 8601 Adelphi Road, College Park, MD, USA 20740.

5) U-Boat KTBs or daily war diarics

In 2009 when this research was done, it was possible to procure single or double days of the *Kriegs Tag Buch*, or *War Day Books* (KTB) of several if not all German attack submarines which invited Bahamian, Caribbean and Bahamian waters in World War II. It was a difficult, slow, laborious process, relying on professors and friends to translate hard to read obtuse nautical language. In 2020, however, thanks to Capt. Jerry and Mrs. Charla Mason, a great number of German U-boat war patrols are known to us and translated beautifully into English as well.

In order to interpret every relevant patrol and track how many planes, planes, and blimps each U-boat saw, some 30 patrols not yet translated were summarized and simplified into English by Kai Steenbuck at the German U-Boat Museum in Cuxhaven, and Dr. Axel Niestlé, Prof. Stephen Aranha and others generously offered to help fill in the remaining gaps. Here are where the reader may obtain the KTBs or assistance in finding then, bearing in mind that:

- Uboat.net has daily positions of every U-boat in WWII, latitude and longitude figures.
- Uboatarchive.net, the Masons, share many of the KTBs in English, but not all.
- DUBM.de is the German U-Boat Museum: Kai Steenbuck or Mrs. Bredow.

It may be possible to find them on other online or German military sites, or books, and it might be worth trying US, UK, or German archives or even Fold3.com. For the half-dozen or so Italian submarines which patrolled the area, there is Regiamarina.net by Cristiano d'Adamo which gives a good broad outline of routes taken. There is a chart in this book detailing each Italian submarine in the BetaSom flotilla from Bordeaux, France. The most prominent, which inflicted the most damage on the allies in the area was R.Smg. *Enrico Tazzoli*.

Notes on Italian Submarines: Readers will find books not only dedicated to that submarine and its commander but also to the Betasom fleet, and the large Italian submarine fleet generally. One can, by digging deep enough, obtain actual original (copies of same) war diary logs from the subs, on an informal exchange network, and as I did, find a kind person fluent in Italian (historian Platon Alexiades) to translate them. But most of what you need, geographically and in terms of activities, attacks upon and by in Spring 1942 are here in the *Daily Diary Section* of this book.

6) Base Diaries for Iterations of NAS Great Exuma: NAS, NAF, and NAAF, 1942-1945

Fold3.com in under War Diaries, World War II, 1942 – 1945, and NOB Guantanamo Bay records as well, since NAS, NAF and NAAF mostly reported to NOB Guantanamo. One will be well rewarded searching keywords Great Exuma, Walkers Cay, Bimini, and Mayaguana in 1942-1944 records, as well as the Guantanamo ORBs, however Great Exuma's contributions and anti-submarine warfare ASW) were sadly not considered worthy of inclusion in the NOB Guantanamo ORBs at first.

Textual Sources Generally

In order to identify every ship sunk in the greater Bahamas area, a number of resources have been combed, page by page. This includes the *Eastern Sea Frontier's Enemy Action Diary* on www.uboatarchive.net, *Axis Submarine Success of World War Two* by Jürgen Rohwer, Samuel Eliot Morison's *History of United States Naval Operations in World War II*, and *U-Boat Operations of the Second World War*, by Wynn. Rohwer's *Chronology of the War at Sea* was also covered, as was Clay Blair's two-volume history, *Hitler's U-Boat War*.

The series *Nortaships Flate*, or Norwegian Ship Fleet, was consulted, as were Captain Arthur Moore's *A Careless Word, a Needless Sinking*. One of the most useful resources was Roger W. Jordan's *The World's Merchant Fleets, 1939*, which provides thumbnail sketches of each of the casualties. It is possible vessels which were attacked in the Bahamas area have been missed in this research, in which case either the ship was not destroyed, or it was mistakenly attributed to another region. It is hoped that this research will lead to revelations of new incidents.

Without the first-hand accounts of each loss, called *Survivors Statements*, such a book could not be written credibly. Since I relied heavily on them for most ship chapters, and often for the details in the bridge chapters about Allied losses, it will suffice to list the exact citation here, and hereinafter refer to them simply as Survivors Statements. After many years of searching in person and on line and via correspondence, it was Michael J. Constandy of Westmoreland Research (westmorealandresearch.org) who finally found them. The citation is as follows:

Survivor's Statements (1941-1942) Series: Papers of Vice Admiral Homer N. Wallin, compiled 1941 - 1974. Record Group 38: Records of the Office of the Chief of Naval Operations, 1875 - 2006 Entry P-13. National Archives at College Park - Textual Reference (Military) 8601 Adelphi Road, College Park, MD, USA 20740.

About the Author

Eric Wiberg grew up in Nassau and has published over a dozen books of maritime non-fiction. His family moved from Sweden to the Bahamas in 1955 and are on the third generation. Eric has contributed to the *Nassau Tribune* since 1989, the same year the national history journal *Concord Review* published his essay on Alexander the Great. Over 50 years Eric has come to know the Bahamas well; his family and he have voyaged to every large island in the archipelago aboard mailboats, a bulk cargo ship, and private boats. Eric has found the graves of two World War II merchant sailors in the Bahamas, and participated in TV programs there for the BBC, the Discovery Channel, and Norway's NRK2. He has lectured and published extensively in the Bahamas, including over half a dozen books about the islands or set there.

With over three million words published, Eric's work has been translated into Portuguese, Spanish, Norwegian, and French. He has participated in full-length documentaries for French and global audiences. Work on this book led to the November, 2019 correction of the official records by the US Navy for the sinking of U-84, and to the award of a Distinguished Flying Cross to Lt. Thomas Rudolph Evert (USNR) in the Senate Building in Washington, D.C. Featured on the front page, front fold on the *Providence Journal*, the *Nassau Tribune*, and on over a dozen radio programs and a podcast, Eric's work has been recognized in and by *Rolling Stone*, *Boston Globe*, and *Vanity Fair*. His last book, *U-Boats in New England* was named "Book of the Week" by the 4th-largest US daily, the *New York Post* in 2019. Eric's book *U-Boats in the Bahamas* was recently held up on Cuban national TV in Havana, and his writing was a runner up in Tulip Tree Publishing's Short Story competition, which will feature his work in an anthology.

Recently a captain in Boston Harbor, Eric has operated over 100 yachts in the Pacific and Atlantic, many of them as captain. A licensed master since 1995, he is qualified as a maritime lawyer. He commercially operated nine tankers from Singapore for three years, and worked briefly for two salvage firms in the Bahamas and US. Other jobs have included executive head-hunter, shipping news salesman, and marketer of a large tug-boat fleet. Having attended Lyford Cay International School and St. Andrew's School, he studied at six universities in three countries, including at Oxford and the New York Film Academy. A citizen of US and Sweden, he lives in Boston. His son, also an author, is Felix Dunmore.

www.ingramcontent.com/pod-product-compliance
Lightning Source LLC
LaVergne TN
LVHW081944200926
848484LV00009B/2508
* 9 7 8 0 9 9 8 3 7 5 9 3 9 *